Fundamentals of Me

Thermodynamics, Mechan

Strength of Materials and Fluid Dynamics

THIRD EDITION

G.S. SAWHNEY

Professor and Head

Department of Mechanical Engineering

Accurate Institute of Management and Technology

Greater Noida

PHI Learning Private Limited

Delhi-110092

2015

₹ 595.00

FUNDAMENTALS OF MECHANICAL ENGINEERING: Thermodynamics, Mechanics, Theory of Machines, Strength of Materials and Fluid Dynamics, Third Edition
G.S. Sawhney

ISBN-978-81-203-5133-2

The export rights of this book are vested solely with the publisher.

Fifth Printing (Third Edition) . . . . **July, 2015**

Published by Asoke K. Ghosh, PHI Learning Private Limited, Rimjhim House, 111, Patparganj Industrial Estate, Delhi-110092 and Printed by Rajkamal Electric Press, Plot No. 2, Phase IV, HSIDC, Kundli-131028, Sonepat, Haryana.

Contents

Preface *xi*

1. BASIC CONCEPTS AND ZEROTH LAW OF THERMODYNAMICS 1–43

Introduction 1
Definitions 1
Concept of Perfect Gas 7
Specific Heat 9
Energy 10
Zeroth Law of Thermodynamics 10
Temperature 10
Pressure 12
Diagrams 13
Solved Problems 16
Objective Type Questions 28
State True or False 28
Multiple Choice Questions 30
Fill in the Blanks 38
Answers 40

2. FIRST LAW OF THERMODYNAMICS 44–88

Introduction 44
First Law of Thermodynamics 44
Application of First Law of Thermodynamics 45
Flow Process 48
Enthalpy 48
Stored Energy 49
Steady Flow Energy Equation 49
Limitations of First Law of Ther5modynamics 55
Perpetual Motion Machine 55
Solved Problems 55
Objective Type Questions 77
State True and False 77
Multiple Choice Questions 79
Fill in the Blanks 83
Answers 85

3. SECOND LAW OF THERMODYNAMICS 89–147

Introduction 89
Heat Reservoir 89
Heat Engine 89
Heat Pump 91
Refrigerator 92

Statements for the Second Law of Thermodynamics 93
Carnot Cycle 94
Carnot Theorem 96
Thermodynamic Temperature Scale 97
Clausius Inequality 98
Entropy and Available Energy 100
Solved Problems 105
Objective Type Questions 127
State True or False 127
Multiple Choice Questions 131
Fill in the Blanks 137
ANSWERS 140

4. PROPERTIES OF STEAM AND THERMODYNAMICS 148–177

Introduction 148
Definition 148
Properties of Steam 151
Steam Tables and Mollier Diagram 152
Dryness Factor Measurement 153
Solved Problems 155
Objective Type Questions 166
State True or False 166
Multiple Choice Questions 168
Fill in the Blanks 172
ANSWERS 174

5. VAPOUR CYCLES 178–197

Introduction 178
Carnot Vapour Cycle 179
Rankine Cycle 180
Solved Problems 183
Objective Type Questions 192
State True or False 192
Multiple Choice Questions 193
Fill in the Blanks 194
ANSWERS 196

6. THERMODYNAMIC CYCLES 198–244

Introduction 198
Otto and Diesel Cycles 199
Engines 206
Indicated, Brake and Friction Power 211
Efficiencies 212
Solved Problems 212
Objective Type Questions 232
State True or False 232
Multiple Choice Questions 234
Fill in the Blanks 238
ANSWERS 240

7. MECHANISM AND SIMPLE MACHINES 245–302

Introduction 245

Kinematic Link or Element 246
Classification of Links 246
Types of Links 246
Kinematic Pair 247
Classification of Kinematic Pairs 247
Kinematic Chain 251
Mechanism and Machine 252
Types of Mechanisms 253
Mobility and Kutzbach Criterion 254
Equivalent Mechanisms 258
Inversion 259
The Four-bar Chain (4 Turning Pairs) 259
The Slider-Crank Chain 260
The Double Slider-Crank Chain 263
Grashof's Law 267
Pantograph 268
Classification of Machines 269
Terms Used with Lifting Machines 270
Law of a Machine 271
Maximum Mechanical Advantage 271
Variation of Mechanical Advantage 272
Variation of Efficiency 273
Loss of Effort in Friction 274
Reversibility of a Machine 278
Single Pulley 279
System of Pulleys 280
First-Order System of Pulleys 280
Second-Order System of Pulleys 283
Third-Order System of Pulleys 286
Differential Pulley Block 288
Wheel and Axle 290
Wheel and Differential Axle 291
Worm and Worm Wheel 293
Solved Problems 294
Objective Type Questions 298
State True or False 298
Multiple Choice Questions 299
Fill in the Blanks 300
ANSWERS 301

8. FORCE SYSTEM AND ANALYSIS 303–341

Introduction 303
Fundamental Laws of Mechanics 303
Force System 306
Solved Problems 311
Objective Type Questions 333
State True or False 333
Multiple Choice Questions 334
Fill in the Blanks 338
ANSWERS 340

9. FRICTION 342–379

Introduction 342
Coulomb's Law of Friction 343
The Angle of Repose and the Cone of Friction 344

Equilibrium: Block, Wedge and Ladder 345
Power Transmitted 346
Screw Jack 349
Solved Problems 351
Objective Type Questions 367
State True or False 367
Multiple Choice Questions 369
Fill in the Blanks 374
ANSWERS 376

10. ANALYSIS OF BEAMS 380–415
Introduction 380
Types of Beams 380
Types of Supports 381
Types of Loads 381
Relationship: Load Intensity, Shear Force and BM 383
Solved Problems 390
Objective Type Questions 409
State True or False 409
Multiple Choice Questions 410
Fill in the Blanks 412
ANSWERS 414

11. TRUSSES 416–450
Introduction 416
Types of Plane Truss 416
Supports 417
Analysis of Plane Truss 418
Graphical Method 419
Method of Joints 420
Method of Section 420
Other Structures 420
Solved Problems 421
Objective Type Questions 441
State True and False 441
Multiple Choice Questions 442
Fill in the Blanks 446
ANSWERS 448

12. CENTROID AND MOMENT OF INERTIA 451–527
Introduction 451
Centre of Mass 451
Centroid 452
Plane Area with an Axis of Symmetry 455
Area with Two Orthogonal Axes of Symmetry 456
Composite Areas 456
Moment of Volume 459
Composite Volumes 461
Centre of Mass 462
Moment of Inertia 464
Moment of Inertia of a Lamina 466
Parallel Axis Theorem 467
Theorem of the Perpendicular Axis 467
Radius of Gyration 468

Product of Area 468
Moment of Inertia of a Rectangular Section 469
Moment of Inertia of a Circular Section 471
Moment of Inertia of a Hollow Rectangular Section 472
Moment of Inertia of a Hollow Circular Section 473
Moment of Inertia of a Triangular Section 474
Moment of Inertia of *I*-Section 476
Centre of Mass 476
Centre of Mass of a Uniform Straight Rod 477
Centre of Mass of a Uniform Semicircle Wire 478
Centre of Mass of a Uniform Semicircular Plate 479
Mass Moment of Inertia 480
Mass Moment of Inertia of Uniform Rod 481
Mass Moment of Inertia of a Rectangular Plate 482
Mass Moment of Inertia of a Uniform Circular Ring 483
Mass Moment of Inertia of a Uniform Circular Plate 484
Mass Moment of Inertia of a Uniform Solid Cylinder 485
Mass Moment of Inertia of a Uniform Hollow Sphere (Thin Thickness Sphere) 486
Mass Moment of Inertia of a Uniform Solid Sphere 487
Mass Moment of Inertia of a Uniform Solid Cone 488
Rotation of Axes 490
Principal Axes 491
Solved Problems 497
Objective Type Questions 522
State True or False 522
Multiple Choice Questions 523
Fill in the Blanks 524
ANSWERS 526

13. KINEMATICS OF RIGID BODY 528–583

Introduction 528
Motion and Frame of Reference 528
Motions Referred to Moving Frame of Reference 529
Translation Motion 535
Rotational Motion 536
Plane Motion 537
Chasles Theorem 539
Instantaneous Centre of Rotation 540
Relative Velocity and Acceleration for Points on a Rigid Body 544
The Velocity of Piston of Reciprocating Engine 548
Acceleration of Reciprocating Piston 551
Analysis of Four-bar Mechanism 554
Solved Problems 556
Objective Type Questions 578
State True or False 578
Multiple Choice Questions 579
Fill in the Blanks 581
ANSWERS 582

14. KINETICS OF RIGID BODY 584–662

Introduction 584
Force, Mass and Acceleration 585
Rotatory Motion of a Rigid Body 585
Relation between Torque and Moment of Inertia 586
Relation between Torque and Angular Momentum 588

Work-Energy Principle 597
Conservation of Mechanical Energy 598
Work Done Against Spring Force 598
Kinetic Energy-Based on Centre of Mass 602
Work-Energy Equations for a Rigid Body 604
Applications of Impulse-Momentum Equations on Plane Motion of Rigid Body 608
Linear Impulse-Momentum Principle 612
Impulse-Momentum Equation for a System of Particles 616
Angular Impulse-Momentum 616
D'Alembert's Principle 620
Rotary Motion and D'Alembert's Principle 623
Solved Problems 629
Objective Type Questions 653
State True or False 653
Multiple Choice Questions 653
Fill in the Blanks 656
ANSWERS 658

15. STRESS AND STRAIN ANALYSIS 663–732

Introduction 663
Types of Materials 663
Types of Loads 664
Stress and Strain 665
Hooke's Law 665
Thermal Stresses 668
Deformation under Own Weight 669
Deformation under External Load 670
Shear Stress and Strain 674
Volumetric Strain, Bulk Modulus and Elastic Constants 678
Strain Energy and Resilence 681
Compound Stresses (2-D System) 683
Principal Planes and Principal Stresses 684
Mohr's Circle 685
Properties of Metal 687
Solved Problems 688
Objective Type Questions 716
State True or False 716
Multiple Choice Questions 719
Fill in the Blanks 724
ANSWERS 727

16. BENDING STRESSES IN BEAMS 733–774

Introduction 733
Theory of Bending 734
Beams of Heterogeneous Materials (Flitched Beams) 736
Flexural Rigidity and Uniform Strength 737
Eccentric Loading 738
Strain Energy in Pure Bending 739
Solved Problems 740
Objective Type Questions 759
State True or False 759
Multiple Choice Questions 760
Fill in the Blanks 765
ANSWERS 768

17. TORSION **775–818**

Introduction 775
Theory of Pure Torsion 775
Polar Modulus of Section 778
Torsional Rigidity 778
Power Transmitted by a Shaft 779
Arrangement of Shafts 779
Shafts in Series 779
Shafts in Parallel 780
Comparison between Hollow and Solid Shafts 781
Strain Energy 783
Bending and Torsion 784
Solved Problems 786
Objective Type Questions 808
State True or False 808
Multiple Choice Questions 809
Fill in the Blanks 812
ANSWERS 814

18. FLUID DYNAMICS **819–927**

Introduction 819
Compressible and Incompressible Fluids 820
Ideal Fluid 821
Viscosity and Newton's Law of Viscosity 821
Streamlines, Streakline and Pathlines 823
Laminar and Turbulent Flow 825
Steady Flow and Continuity Equation 826
Euler's Equation along a Streamline 827
Bernoulli's Principle 829
Torricelli's Theorem 834
Vena Contracta and Contraction Coefficient 836
Discharge Coefficient 836
Control Volume Analysis 838
Flow Measurement by Venturi Meter, Orifice Meter, Flow Nozzle and Pitot Tube 840
Venturi Meter 840
Orifice Meter 842
Flow Nozzle 843
Pitot-static Tube 843
The Siphon 846
Flow in Pipes and Ducts 848
Energy Grade Line and Hydraulic Grade Line 848
Friction Loss in Pipe Flow 851
Minor Head Losses 852
Pipes in Series 853
Pipes in Parallel 854
Series–Parallel Pipe Networks 854
Open Channel Flow 859
Applications of Open Channel Flow 859
Difference between Open Channel Flow and a Pipe Flow 860
Type of Channel 860
Classification of Flows 861
Laminar and Turbulent Flow 862
Geometric Properties for Analysis 863

Fundamental Equation of Flow 864
Energy and Momentum Coefficients 865
Velocity Distribution in Open Channel 865
The Chezy Equation 866
The Manning Equation 868
Froude Number 870
Specific Energy 871
Specific Energy Curve 872
Variable Discharge Condition 874
Flow over a Raised Hump 876
Weirs 878
Hydraulic Jump 879
Sluice Gate 881
Compressible Flow 888
Mach-Number 889
Isentropic Process in Compressible Flow 889
Isentropic Flow 890
Stagnation Condition 891
Equation of Motion for a Compressible Flow 893
Area Velocity Relation 894
Wave Propagation 895
Fundamental of Fluid Turbines 898
Pelton Wheel or Turbine 900
Work Done and Efficiency of Pelton Wheel 902
Francis Turbine 906
Layout of Francis Turbine Plant 906
Construction and Working of Francis Turbine 907
Velocity Diagram and Work Done by Water in Francis Turbine 909
Efficiencies 910
Comparison of Francis and Pelton Turbines 911
Kaplan Turbine 912
Work Done and Efficiencies of Kaplan Turbine 913
Comparison of Francis and Kaplan Turbines 914
Governing of Turbines 914
Governing Mechanism of Impulse (Pelton) Turbine 914
Governing of Reaction Turbine 916
Centrifugal Pumps 917
Work Done by Centrifugal Pump 918
Heads and Efficiencies 918
Reciprocating Pump 919
Construction and Working of Single Acting Reciprocating Pump 920
Indicator Diagram (Ideal) 921
Comparison of Centrifugal and Reciprocating Pumps 922
Objective Type Questions 923
State True or False 923
Multiple Choice Questions 924
Fill in the Blanks 926
ANSWERS 927

BIBLIOGRAPHY 929–930

INDEX 931–938

Preface

I am pleased to present the third edition of this book. The syllabi of several universities have been recently changed which has prompted me to include a new chapter on 'Fluid Dynamics'. I am confident that the book now covers the syllabi of most of the technical universities. Fluid dynamics is an important topic as it has a wide range of applications, including calculating pressure and moments on aircrafts, determining the mass flow rate of petroleum through pipelines and determining the water flow in the channels.

Based on my experience of teaching this subject, I have endeavoured to present a systematic explanation of the basic concepts of the subject matter. A large number of solved problems and objective type questions with explanatory answers are included in order to make the underlying principles comprehensible.

I wish to record my sincere thanks to my wife, Jasbeer, for her patience shown throughout the preparation of this book. I am also thankful to my children Jasdev, Tejmohan, Pooja and Nandini for their continuous encouragement extended to me.

I would appreciate receiving constructive suggestions and objective criticism from students and teachers alike with a view to enhance further the usefulness of the book. They may write to me on channi_sawhney@hotmail.com.

G.S. SAWHNEY

CHAPTER 1

Basic Concepts and Zeroth Law of Thermodynamics

Walk a mile in others' shoes before you say no to their request for a new pair.

INTRODUCTION

When a hot body is placed in contact with a cold body, the hot body cools down while the cold body warms up. The energy transferred from the hot body to the cold body as a result of temperature difference is called *heat energy*. The heat energy transferred is zero if both the bodies have the same temperature. This is the basis of zeroth law of Thermodynamics.

Heat is a transitory energy. It must not be confused with intrinsic (internal) energy possessed by a system. Similarly, work energy is also a transitory energy. Whenever heat or work transits a system, the state of the system changes. It is also observed that heat and work are two mutually convertible forms of energy. This is the basis of the first law of thermodynamics. It is also observed that heat never flows unaided from a hot body to a cold body. This is the basis of second law of thermodynamics.

Applied thermodynamics is the science of the relationship between heat, work and the properties of a system. It is concerned with the means necessary to convert heat energy from available sources such as fossil fuel into work energy. The application of thermodynamics is extremely wide. Its principles are used in designing of energy converting devices. These devices will be discussed in later chapters but first some fundamental definitions must be made.

DEFINITIONS

Thermodynamics: Thermodynamics is a science dealing with energy and its transformation, specially transformation of heat into other forms of energy and vice versa.

System: A system is any matter in space which is under analysis. The system may have a real or hypothetical boundary. Everything outside of system is termed surroundings: The boundary can be adiabatic which does not allow heat interaction between system and surroundings or diathermic which allows heat interaction between system and surroundings.

$$\text{Universe} = \text{System} + \text{Surroundings}$$

Systems can be of three types are shown in Figure 1.1. An *open system* allows mass and energy interaction with the surrounding. A *closed system* allows only energy interaction. An *isolated system* allows neither mass nor energy transaction. Turbines, compressors and pumps are examples of the open system. Piston-cylinder assembly is an example of the closed system. Our universe is an example of an isolated system.

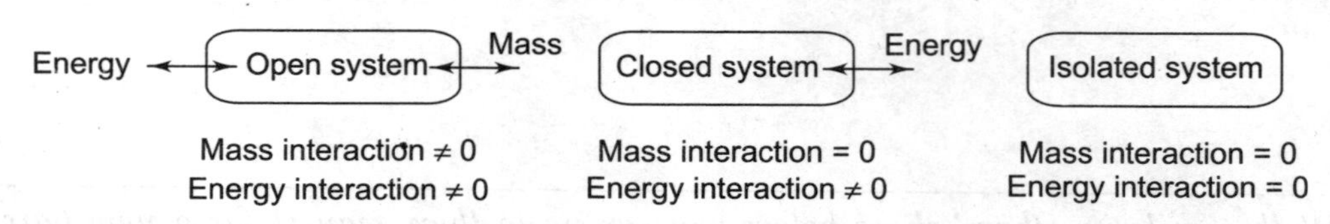

FIGURE 1.1 Types of systems.

The working fluid in a piston cylinder assembly forms a closed system as shown in Figure 1.2. The mass of the closed system remains constant. The volume of the closed system need not to remain constant. The volume changes as the piston moves up and down in the cylinder. Only energy transfer (work) between the closed system and surroundings takes place due to the movement of the boundary of the system considered.

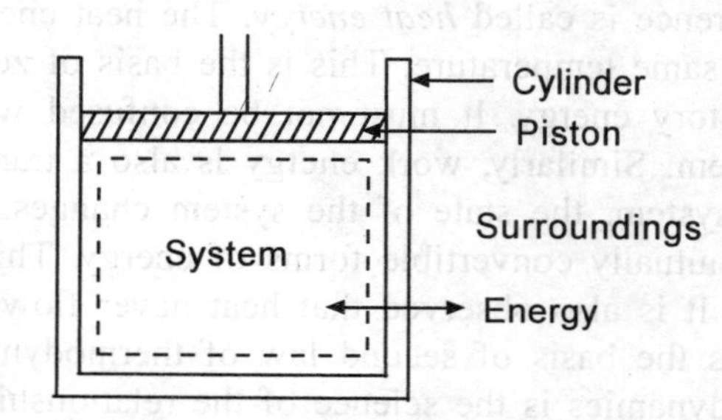

FIGURE 1.2 A closed system.

In the case of an open system, mass transfer also takes place along with energy transfer between the system and the surroundings. The boundary of the open system is known as *control volume*. The boundary of the system during the transfer of mass and energy may or may not change. The open system can be one-flow or two-flow boundary system. Air leaving a compressed air cylinder (Figure 1.3(a)) can be considered a one-flow boundary system. Here the boundary is not changed during the mass transfer (compressed air). However, if we consider air escaping from a balloon (Figure 1.3(b)) to the surroundings, the boundary shape also changes. This is an example of moving boundary one flow open

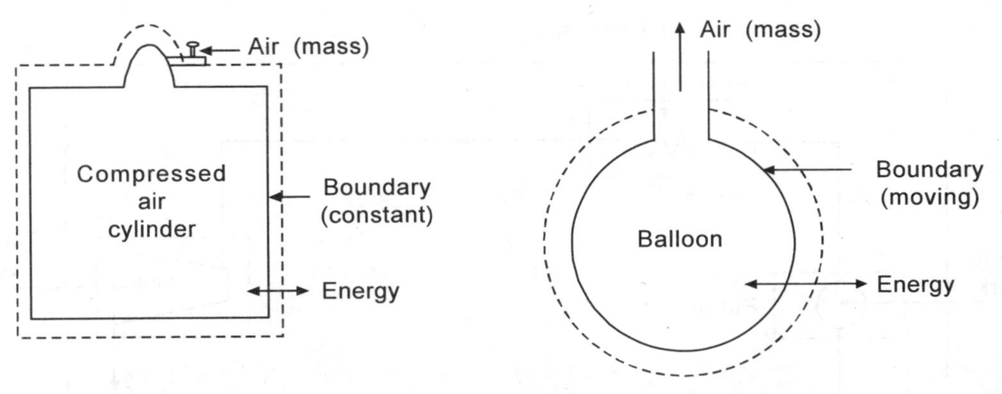

(a) Constant boundary—One flow-open system (b) Moving boundary—One-flow open system

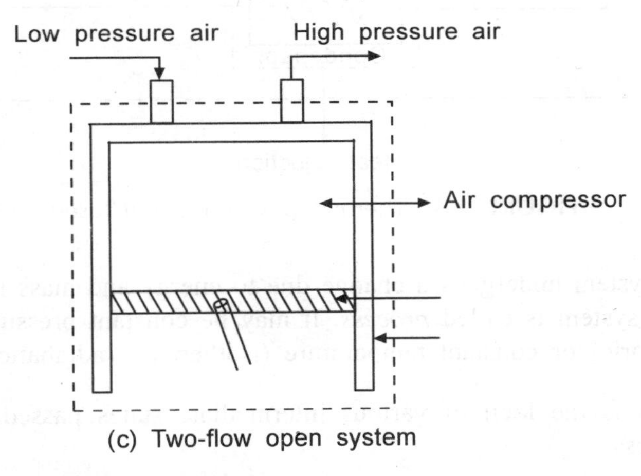

(c) Two-flow open system

FIGURE 1.3 Open system.

system. In the case of air compressor as shown in Figure 1.3(c), air is taken from the surroundings and air is sent out of the system after compression. Hence mass and energy transfer is taken place across the boundary. It is an example of two-flow open system.

A system can be considered as a closed system while its subsystems may form individually open systems. The reason is that the subsystems are inter connected. A thermal power plant consisting of a boiler, a turbine, a condensor and a pump as shown in Figure 1.4 can be considered to be a closed system. Heat transfer (through the boiler and the condensor) and work transfer (through the turbine and the pump) take place between the system (power plant) and surroundings through the boundary. No mass transfer takes place through this boundary. However, every subsystem (boiler, turbine, pump and condensor) is an open system as there is mass and energy transfer across their boundaries.

A state: A state of a system indicates the specific condition of the system. Properties like pressure, temperature and volume, etc. can define a state of system.

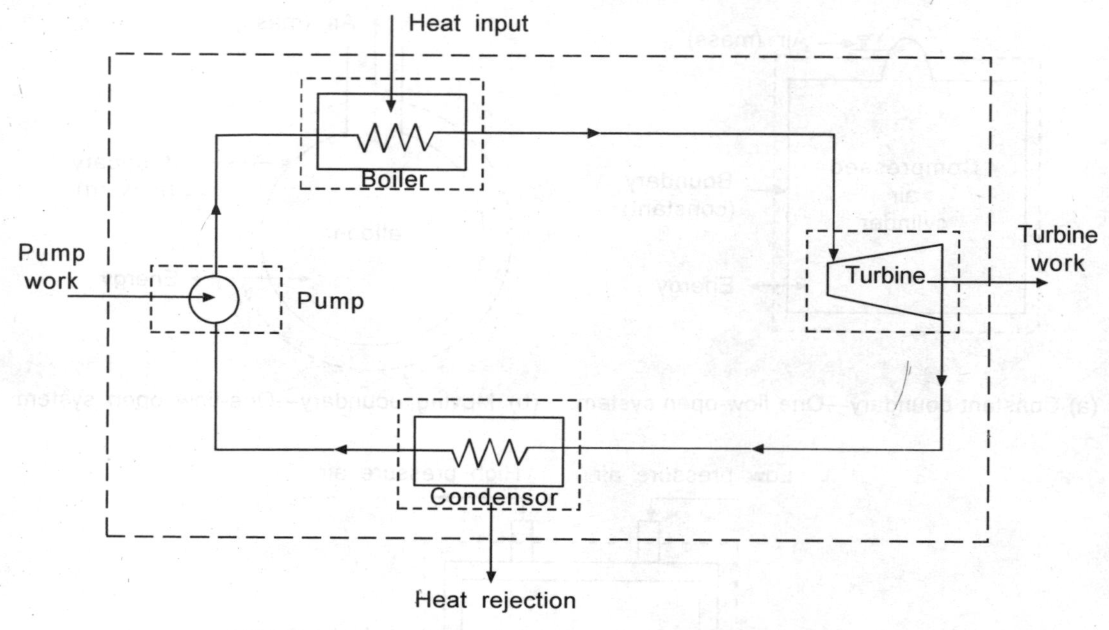

FIGURE 1.4 Thermal power plant (Closed system).

Process: A system undergoes a change due to energy and mass interaction. The mode of the change of system is called *process*. It may be constant pressure (isobaric) or constant volume (isochoric) or constant temperature (isothermal), adiabatic or isoentropic.

Path: A path is the locii of various intermediate states passed through by the system during a process.

State and path functions: The state function is independent of the path while the path function is dependent on the path. To understand this consider a person travelling from point x to point y. There can be various routes to reach point y from point x, but the travelling distances would be different for different routes. If a car is used, work done and fuel consumed are dependent on the route and the mileage of the car. However, the locations of points x and y are fixed and they are independent of the routes. Hence the positions x and y are point or state functions while fuel consumed and work done are path functions. State functions are represented by a point on a graph while path functions by area. Also $\oint$ (state function) = 0 while $\oint$ (path function) $\neq$ 0.

Mathematically a state function is an exact differential while a path function is an inexact differential. If $z = f(x, y)$, then z is an exact differential for

$$dz = \left(\frac{\partial z}{\partial x}\right) dx + \left(\frac{\partial z}{\partial y}\right) dy$$

and

$$\frac{\partial^2 z}{\partial y \partial x} = \frac{\partial^2 z}{\partial x \partial y}$$

Is $v = RT/P$ state function where, v = specific volume, T = temperature, P = pressure and R = gas constant?

Differentiating w.r.t. T, we get

$$\frac{\partial v}{\partial T} = \frac{R}{P}$$

and

$$\frac{\partial}{\partial P}\left(\frac{\partial v}{\partial T}\right) = \frac{\partial}{\partial P}\left(\frac{R}{P}\right) = -\frac{R}{P^2}$$

Similarly, differentiating w.r.t. P, we get

$$\frac{\partial v}{\partial P} = -\frac{RT}{P^2}$$

and

$$\frac{\partial}{\partial T}\left(\frac{\partial v}{\partial P}\right) = \frac{\partial}{\partial T}\left(-\frac{RT}{P^2}\right) = -\frac{R}{P^2}$$

Hence

$$\frac{\partial^2 v}{\partial P \partial T} = \frac{\partial^2 v}{\partial T \partial P}$$

and v is a state function.

Cycle: A cycle is sequence of processes undergone by the system so that initial and final states are the same. Thermodynamics properties remain unchanged on the completion of a cycle.

Thermodynamic equilibrium: It is a state of a system when its state does not change and its properties remain constant. A system is said to be in thermodynamic equilibrium if it is in state of mechanical, thermal or chemical equilibrium.

Equilibrium is a state wherein there is no tendency for a change. In other words, the rate of the process is zero where the rate is the ratio of driving force to resistance. Hence, the rate can be zero when either driving force is zero or resistance is infinitely large. Equilibrium is true when driving force is zero. It is false when resistance is infinitely large. We are interested only in true equilibrium when driving force is zero.

What is a mechanical equilibrium? In a piston and cylinder arrangement, the piston stops moving when the pressure inside the cylinder and surroundings is the same. This is called *mechanical equilibrium*. What is thermal equilibrium? When two bodies in contact attain the same temperature and heat stops transferring, it is said to be in thermal equilibrium. What is chemical equilibrium? Once chemical potential of two phases are equal, the net rate of the mass transfer between phases is zero, and thus it is called *chemical equilibrium*.

Reversible process: In a reversible process, the system can come back to the original state on removal of the factors responsible for the occurrence of the process. On reversal of the process, no trace of occurrence is left and the system follows the same path back to the initial state. It is a frictionless process.

Irreversible process: In an irreversible process, the system cannot come back to the original state on removal of the factors responsible for the occurrence of the process. Friction and dissipative effects are the causes for irreversibility.

Quasi static process: It is not possible for a system to attain equilibrium in finite time. For the sake of study and analysis, certain assumptions can make a system akin to a system in equilibrium. When a system changes its state very slowly under the influence of very small differences of temperature and pressure, the process is called *quasi static process*. Quasi means 'almost' and static means 'non-dynamic'. Hence almost non-dynamic process is a quasi static process. A reversible process is always a quasi static process.

System analysis: Macroscopic (visible to the naked eye) and microscopic (minute) approaches are the two approaches for analyzing systems. In the macroscopic approach, the structure of matter is not considered. A complete system is considered and the state is found out with measurable properties. Classical thermodynamics adopts the macroscopic approach. In the microscopic approach, the constituents and microsystem of the system are analyzed. Statistical thermodynamic adopts the microscopic approach for analysis. The result of macroscopic approach analysis is equal to the summation of microscopic approach analysis.

Concept of continuum: A substance is composed of a vast number of molecules. Most engineering systems are concerned with the macroscopic or bulk behaviour of a substance rather than the microscopic or molecular behaviour. In most cases it is convenient to think of a substance as a continuous distribution of medium or a continuum. However, there are certain instances, for example, (rocket explosion at very low pressure) in which the concept of a continuum is not valid.

In the concept of continuum, the substance is considered free from any kind of discontinuity. As the scale of analysis is large, the discontinuity of the order of intermolecular spacing or the free mean path is negligible.

State: Matter is found in three states—solid, liquid and gas. Gases have some special properties like low density, no definite volume and easy compressibility.

Properties: Properties are observable characteristics of a system like pressure, temperature and volume. Intensive properties such as pressure and temperature are independent of mass. Extensive properties such as volume, entropy and enthalpy depend upon mass.

Boyles' law: Boyles' law gives a relation between pressure and volume of a gas. For a given mass of gas at constant temperature (T), volume (V) is inversibly proportional to pressure (P):

$$V \propto \frac{1}{P}$$

Charles' law: Charles' law gives a relation between volume and temperature. For a given mass of a gas at constant pressure, volume is directly proportional to absolute temperature:

$$V \propto T(\text{K})$$

Law of pressure: The law of pressure gives a relation between pressure and temperature. For a given mass of a gas at constant volume, pressure is directly proportional to absolute temperature:

$$P \propto T(\text{K})$$

CONCEPT OF PERFECT GAS

The perfect gas equation states that volume, pressure and temperature of a given mass of a gas are interconnected as

$$P\overline{v} = \overline{R}\,T$$

where

$\overline{R}$ = universal gas constant
= 8314 joules/k mol K, here k mol = 1000 moles

$\overline{v}$ = volume per mole

The gas equation can also be written as

$$PV = mRT$$

where

m = mass ($=nM$)
V = volume ($=n\overline{v}$)
R = gas constant ($=\overline{R}/M$)
n = number of moles (mole = 6.023×10^{23} molecules)
M = molecular weight. Another form is

$$Pv = RT$$

where v is specific volume $\left(\frac{V}{m}\right)$.

As per the kinetic theory, pressure and temperature of gas are proportional to "square of root mean square (rms) velocity ($\overline{c}$)", i.e.

$$P = \frac{1}{3}\rho\overline{c}^2 = \frac{2}{3}\ \text{KE (kinetic energy)}$$

where ρ is density.

$$T = \frac{1}{3}\frac{M\overline{c}^2}{R}$$

If isothermal lines are drawn for $P\overline{v} = \overline{R}\,T$, we get curves which have maxima and minima (Figure 1.5). However, at critical point (CP) and above it, the maxima and minima on isothermal curves do not exist. A gas can be liquefied below the critical temperature (T_c). Above the critical point a gas cannot be liquefied by pressure. If the critical temperature is above room temperature, a gas is easily liquefied and stored. A domestic gas cylinder contains 14 kg LPG which is mainly butane gas. Butane can be liquefied at high pressures as its critical temperature is above room temperature. In a liquefied state, a small cylinder can hold sufficient gas. However, hydrogen can be kept in the liquefied state below critical temperature (–240°C). It is the best fuel and it is used in the liquid state in rockets. It requires either very large storage capacity at room temperature or economical cooling facility to keep it below the critical temperature before it can be used for automobiles.

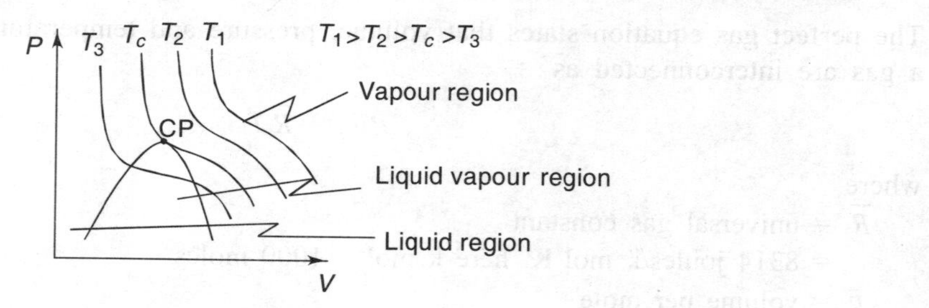

FIGURE 1.5 *PV* diagram.

A real gas does not obey the ideal gas equation $P\overline{v} = \overline{R}T$ for all pressures and temperatures. The ideal gas equation is based on two assumptions that molecules do not exert intermolecular attraction and the volume occupied by the molecules is negligible, which is not fully correct at all pressures and temperatures. At a very low pressure (pressure tending to zero) and a high temperature (temperature tending to infinity), the real gas obeys very nearly the ideal gas equation. Another deviation from ideal gases is that most of the real gases get liquefied at low temperatures.

Compressibility factor: The compressibility factor (z) is a measure of deviation of real gas from ideal gas behaviour:

$$\text{Compressibility factor } (z) = \frac{P\overline{v}}{\overline{R}T}$$

The value of z is one for ideal gas. For real gases, z can be read from P and z charts for different isothermal lines.

$$z = \frac{P\overline{v}}{RT} = \frac{\overline{v}}{RT/P} = \frac{P}{RT/P} = \frac{\text{Actual volume}}{\text{Ideal volume}} = \frac{\text{Actual pressure}}{\text{Ideal pressure}}$$

If attractive forces predominate, z is less than unity ($P_{actual} < P_{ideal}$). When repulsive forces predominate (high pressure), z is greater than unity.

As a real gas deviates from the perfect gas equation, the van der Waals equation overcomes this problem. The equation is as follows:

$$\left(P + \frac{a}{\bar{v}^2}\right)(\bar{v} - b) = \bar{R}T$$

where

a = constant to take care of attraction amongst the molecules

b = constant to take care of volume of molecules

SPECIFIC HEAT

Specific heat is the heat required to raise the temperature of a unit mass by unity. Mathematically,

$$Q\text{(heat)} = c\text{(specific heat)} \times m \text{ (mass)} \times \Delta T$$

or

$$c = \frac{Q}{m \times \Delta T}$$

If $m = 1$ and $\Delta T = 1$, then $Q = c$

If 1 kg of gas is compressed and its temperature is raised by 1°, then

$$c = \frac{0}{m \times \Delta T} = 0$$

Similarly, if a compressed gas is allowed to expand and its temperature falls, then heat (Q) is given to maintain its original temperature. Now we have

$$c = \frac{Q}{m \times 0} = \text{infinity}$$

Hence the specific heat of a gas can have value from zero to infinity. Hence we define the specific heat of gas at constant volume and at constant pressure (c_v and c_p).

c_p (specific heat at constant pressure) is greater than c_v (specific heat at constant volume). Heat at constant volume is fully used for heating gas by 1°C while heat at constant pressure is used both for heating gas and doing work ($P \times dV$) as gas expands against atmospheric pressure while heating.

We can write

$$c_p\, dT = c_v\, dT + PdV$$

As $PV = RT$, $PdV = RdT$ at constant pressure.

Therefore, the above equation becomes

$$c_p\, dT = c_v\, dT + R\, dT$$

or

$$c_p - c_v = R$$

ENERGY

Work (W) and heat (Q) are energies and equivalent to each other by the relation

$$W = JQ$$

where J is joule's constant.

$J = 1$ if W and Q have the same unit otherwise $J = 4.2$ joules/calorie when W is in joules and Q is in calories.

Work done by a system depends not only upon initial and final states but also upon the path adopted by the process. The area on a PV diagram under the process $\int_1^2 P\,dV$ is work. Similarly, $\oint P\,dV$ gives work on a PV diagram for a cyclic process.

During free expansion, work done by a gas is zero as expansion does not take place against atmospheric pressure.

The energy can be defined as the capacity to do work. Energy possessed by a system which can cross its boundary is called *transit energy*. Energy possessed by a system within its boundary is called *stored energy*. Potential energy (PE) kinetic energy (KE) and internal energy (U) are stored energies. Total stored energy is the sum of PE, KE and U.

Heat energy is transit energy as it can go out and come in the system depending on temperature. Work is also transit energy which can go out or come in the system due to difference in any intensive property other than temperature. The heat and work are not system properties but appear as a boundary phenomenon, i.e. they are observed at the boundary of the systems. Both are path functions and hence they are inexact differentials. The magnitude of heat transfer and work transfer depends upon the path followed by the systems during the process. Work is the area under the process or area enclosed in the cyclic process on a PV diagram. Similarly, heat is the area under the process or area enclosed in the cyclic process on a TS (temperature entropy) diagram.

ZEROTH LAW OF THERMODYNAMICS

If two bodies A and B are in thermal equilibrium with a third body C separately, then two bodies A and B shall also be in thermal equilibrium with each other (Figure 1.6).

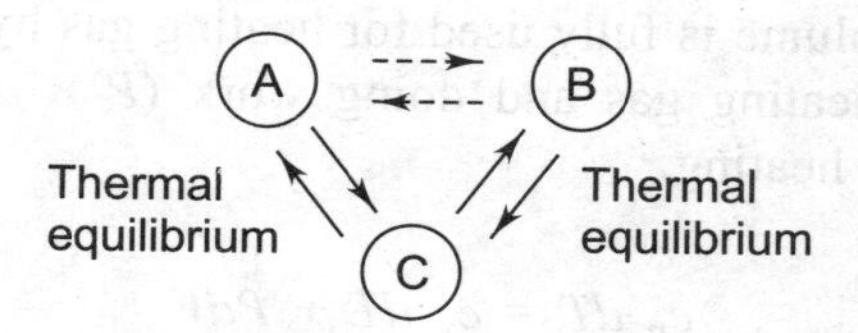

FIGURE 1.6 Zeroth law.

Temperature

Temperature is an intensive property of a system and it requires reference states for calibration. The boiling point of water and the freezing point of water are acceptable reference states.

The thermometer is a temperature measurement system which can show some change in its characteristic (termed thermometric property) due to heat interaction taking place with the body whose temperature is being measured.

Centigrade (T_c), Fahrenheit (T_f), and kelvin (T_k) temperatures are interrelated as follows:

$$\frac{T_c}{100} = \frac{T_f - 32}{180} = \frac{T_k - 273.15}{100}$$

Temperature in Rankine is equal to temperature in Fahrenheit plus 459.67.

In the mercury scale thermometer, the length of the mercury column (l) is proportional to temperature (T) and temperature is given by

$$T = \frac{100(l_T - l_0)}{l_{100} - l_0}$$

where

l_T = length at T
l_0 = length at freezing point
l_{100} = length at boiling point

Similarly, in the *constant volume thermometer*, pressure varies with temperature and temperature is given by

$$T = \frac{100(P_T - P_0)}{P_{100} - P_0}$$

where

P_T = pressure at T
P_0 = pressure at freezing point
P_{100} = pressure at boiling point

Similarly, in the *resistance thermometer*, resistance (R) varies with temperature and temperature is given by

$$T = \frac{100(R_T - R_0)}{R_{100} - R_0}$$

where

R_T = resistance at T
R_0 = resistance at freezing point
R_{100} = resistance at boiling point

In *thermocouple*, emf (electromotive force) induced is proportional to temperature difference between hot and cold junctions. Bismuth-antimony and copper-iron are a common pair of metals to form junctions. Therefore

$$\text{emf} = aT + bT^2$$

where a and b are constants.

Very high temperatures are measured by pyrometers. The pyrometer can be *total radiation pyrometer* or *disappearing filament pyrometer*. The radiated energy is proportional to the fourth power of temperature of a hot body.

Pressure

The standard atmospheric pressure is defined as the pressure produced by a column of mercury 760 mm high.

$$\begin{aligned} P_{\text{atm}} &= \rho g h \\ &= (13.6 \times 10^3) \times 9.8 \times \frac{760}{1000} \\ &= 1.01 \text{ bar (bar} = 10^5 \text{ pascal)} \end{aligned}$$

Pressure of the system can be higher or lower than atmospheric pressure. Pressure is measured by the manometer. A manometer is a U-tube containing mercury with one end opens to atmosphere and the other one is connected to a system/vessel. Refer to Figure 1.7. If pressure in a vessel (P_{abs}) is higher than P_{atm}, mercury is forced up the limb that opens to atmosphere. If pressure in a vessel (P_{abs}) is lower than atmospheric pressure, mercury is forced into the limb connected to the vessel. Higher than atmospheric pressure is known as *gauge pressure* while lower pressure than atmospheric pressure is called *vacuum pressure*.

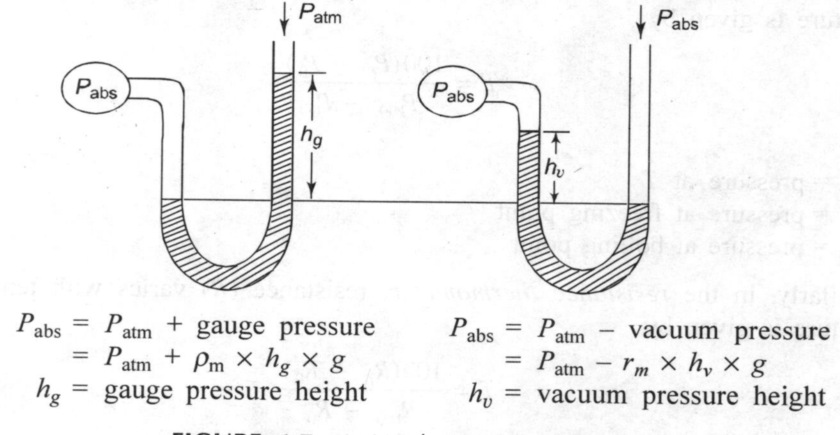

FIGURE 1.7 A U-tube manometer.

Note: NTP is normal temperature and pressure. Normal temperature is 0°C and normal pressure is 760 mm of Hg.

STP is standard temperature and pressure. Standard temperature is taken as 15°C or 25°C depending upon the geographical location and standard pressure is 760 mm of Hg.

Water is about 1300 times heavier than air. Pressure rises swiftly above the atmospheric pressure as we descend inside water. The pressure increases by one atmosphere for every 10 metres of depth inside water. On land, if we climb to a height of 150 metres the change in pressure will be slight and indiscernible. At the same depth under water, our blood vessels will collapse and our lungs will be compressed dangerously. It is impossible to go beyond a depth of about 72 metres without assistance of diving suits connected to an air pump. The average depth of oceans is about 4 km. The pressure at this depth is equivalent to about 400 atm which is sufficient to crush anything.

A diver (Figure 1.8) inside water experiences the same pressure as the surrounding water. We are made largely of water which is almost incompressible. However, it is gases inside our body, particularly inside our lungs that create problem while going down in water. The gases inside the body of the diver compress when he descends in water and the compression becomes fatal at some point. To avoid this, diving suits are used which are connected to an air pump at the surface by a long hose to develop suitable air pressure inside the body of the diver. The 'squeeze' occurs when the air pump at the surface fails which results into the loss of pressure in the suit. The air leaves the suit with such a force that the hopless diver dies instantly.

FIGURE 1.8 Diver.

We breathe air which contains 80% nitrogen. Under pressure (in deep water), nitrogen gas gets dissolved in our blood. If pressure is changed too rapidly when we ascend from deep water the nitrogen dissolved in the blood begins to liberate in the same manner of a freshly opened bottle of Coca Cola. The bubbles of nitrogen clog the blood vessels and the flow of the blood stops. The stoppage results into the deprivation of oxygen to the tissues of our body. The deprivation causes pain so excruciating that we are prone to bend double in agony. In diving, this condition is called *bends*. The bends are the occupational hazards for pearl divers and caisson workers (men working in enclosed dry chambers built on river beds).

DIAGRAMS

Processes give different curves on *PV* diagrams, *PT* diagrams, *TS* diagrams and *hS* diagrams (enthalpy–entropy). On a *PV* diagram, the isobaric process (P = constant) is shown as a straight horizontal line while the isochoric process (V = constant) is shown as a vertical line. However, the isothermal process (T = constant) is depicted as a hyperbolic curve. Refer to Figure 1.9.

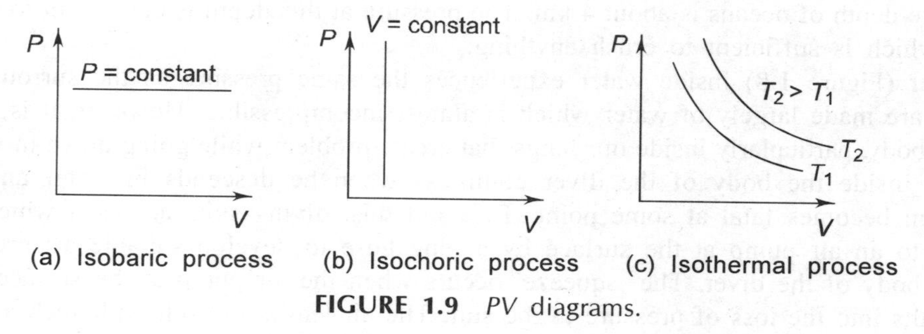

FIGURE 1.9 *PV* diagrams.

On a *VT* diagram, the isochoric process appears as a horizontal line and the isothermal process appears as a vertical line. However, the isobaric process is shown as an inclined line. Refer to Figure 1.10.

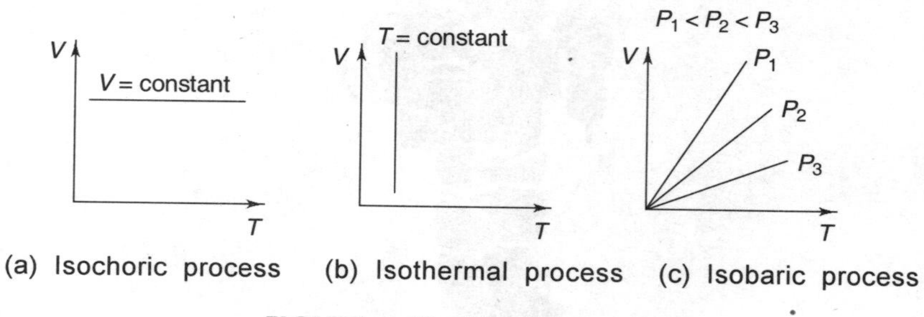

FIGURE 1.10 *VT* diagrams.

On a *PT* diagram, the isobaric process is shown as a horizontal line, the isothermal process appears as a vertical line, and the isochoric process as an inclined line (Figure 1.11).

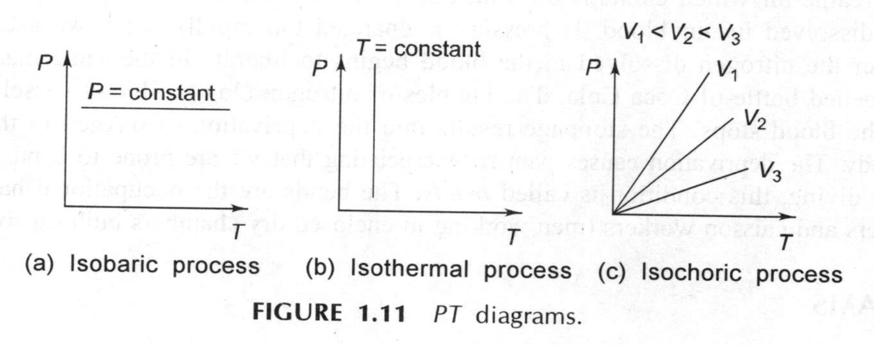

FIGURE 1.11 *PT* diagrams.

On a *TS* diagram (Figure 1.12), the isothermal process is shown as a horizontal line, and the isentropic process as a vertical line. However, the isobaric process is shown as a curve with slope = T/c_p and the isochoric process is shown as a curve with slope = T/c_v. Since $c_p > c_v$, the slope of an isochoric process is greater than slope of isobaric process.

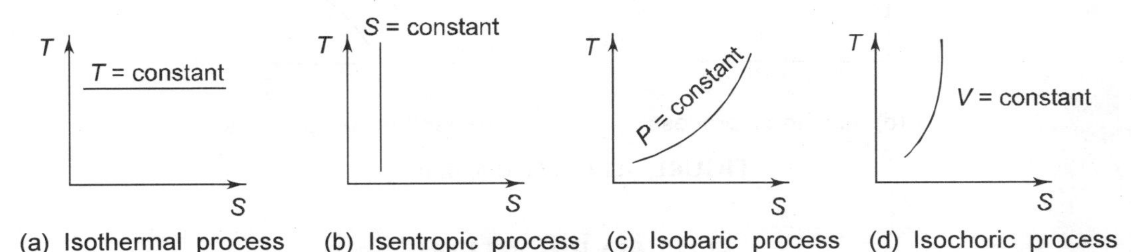

FIGURE 1.12 *TS* diagrams.

On a *PV* diagram, the isothermal process and the adiabatic process are curves as shown in Figure 1.13. The slope of an isothermal process is smaller than that of an adiabatic process.

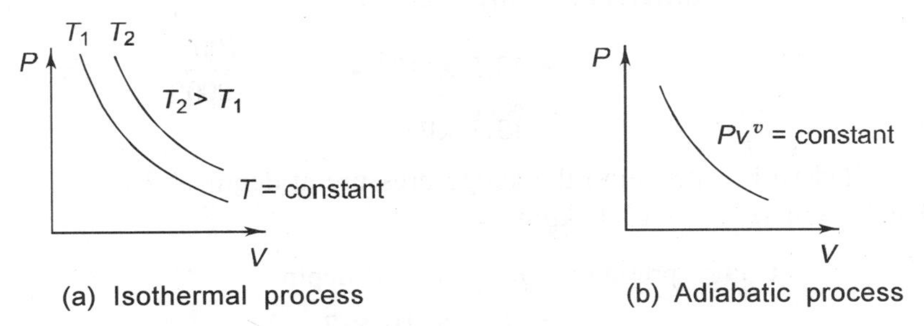

FIGURE 1.13 *PV* diagrams.

On a enthalpy–entropy (*hS*) diagram, the isentropic process appears as a vertical line while the constant enthalpy process like throttling as a horizontal line. Isobaric, isochoric and isothermal processes are also shown as in Figure 1.14. The slope of an isochoric curve is greater than that of an isobaric curve.

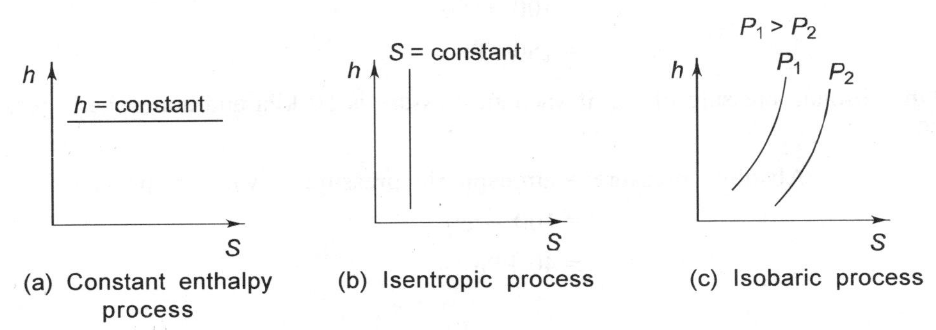

FIGURE 1.14 Contd.

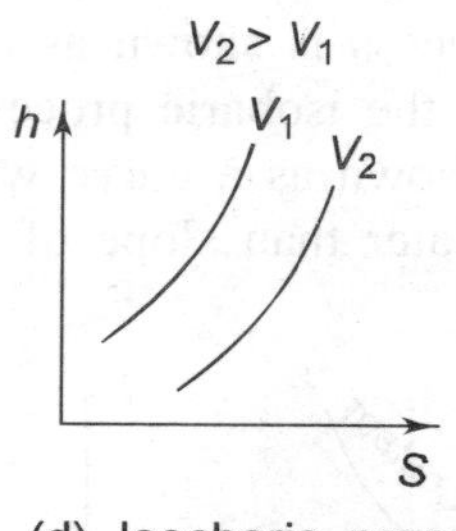

(d) Isochoric process

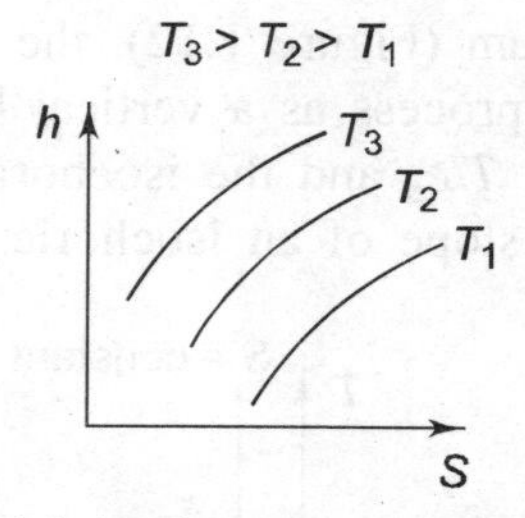

(e) Isothermal process

FIGURE 1.14 *HS* diagrams.

SOLVED PROBLEMS

1. Find the pressure difference shown by a manometer showing the difference of a mercury column of 100 mm of Hg in the limbs. Take $\rho_{Hg} = 13.5 \times 10^3$ kg/m^3 and $g = 10$ m/s^2.

$$\text{Pressure difference} = \rho_{Hg} \times g \times h$$
$$= 13.5 \times 10^3 \times 10 \times \frac{100}{1000}$$
$$= 13.5 \text{ kPa}$$

2. A tank is filled with water. Find the gauge pressure at depth of 2 m from the top. Take $g = 10$ m/s^2 and $\rho_{water} = 1000$ kg/m^3

$$\text{Gauge pressure} = \rho_{water} \times g \times \text{depth}$$
$$= 1000 \times 10 \times 2$$
$$= 20 \text{ kPa}$$

3. Find the absolute pressure of gas if a manometer reads a gauge pressure of 50 kPa and atmospheric pressure is 100 kPa.

Absolute pressure = atmospheric pressure + gauge pressure

= 100 + 50

= 150 kPa

4. Find the absolute pressure of gas if vacuum pressure is 60 kPa and atmospheric pressure is 100 kPa.

Absolute pressure = atmospheric pressure – vacuum pressure

= 100 – 60

= 40 kPa

5. A container has an absolute pressure of 40 kPa. The area of the lid is 1000 mm^2. What force is required for opening the lid?

The container has lesser pressure than atmospheric pressure. The force has to be applied to open the lid which is the pressure difference on the lid multiplied by the area of the lid.

$$\text{Force} = \text{pressure difference} \times \text{area}$$
$$= (P_{atm} - P_{container}) \times A$$
$$= (100 - 40) \times 10^3 \times \frac{1000}{1000 \times 1000}$$
$$= 60 \text{ kN}$$

6. A manometer shows a gauge pressure of 50 kPa of a gas when atmospheric pressure is 100 kPa. What will be the gauge pressure indicated by the manometer in space where residual atmospheric pressure is 50 kPa?

$$P_{abs} = P_{atm} + P_{gauge}$$
$$= 100 + 50$$
$$= 150 \text{ kPa}$$
$$\text{In space } P_{abs} = P_{res\ atm} + P_{gauge}$$
$$150 = 50 + P_{gauge}$$

or
$$P_{gauge} = 100 \text{ kPa}$$

7. A mercury manometer shows a gauge pressure of 100 mm of Hg. In case it is replaced by a water manometer, what gauge pressure in mm of water will be shown by it? Take $g = 10$ m/s^2.

$$P_{gauge} = \rho_{Hg} \times g \times hg$$
$$= 13.7 \times 10 \times \frac{100}{1000} \text{ kPa}$$
$$= 13.6 \text{ kPa}$$
$$P_{gauge} = \rho_{water} \times g \times h_{water}$$
$$= 1 \times 10 \times \frac{h_{water}}{1000}$$
$$13.6 = \frac{10}{1000} \times h_{water}$$
$$\therefore h_{water} = \frac{13.6 \times 1000}{10}$$
$$= 1360 \text{ mm}$$
$$= 1.360 \text{ metre}$$

8. Gas A and B have absolute pressures of 220 kPa and 100 kPa. What will be gauge pressure shown by the manometer that is put between gas A and B?

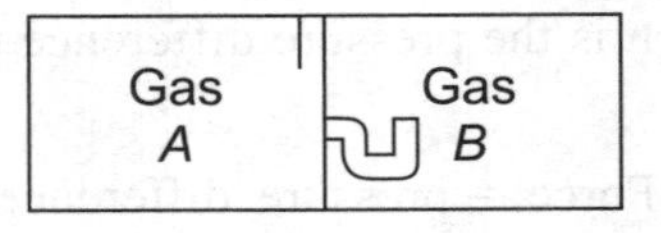

$$\text{Gauge pressure} = \text{absolute pressure of gas } A - \text{absolute pressure of gas } B$$
$$= 220 - 100$$
$$= 120 \text{ kPa}$$

9. Gas A has a vacuum pressure of 20 kPa and gas B has a gauge pressure of 20 kPa. What will be gauge or vacuum pressure shown by the manometer.

Take $P_{atm} = 100$ kPa.

$$P_1 = P_{abs} \text{ of gas } A = P_{atm} - P_{vacuum}$$
$$= 100 - 20 = 80 \text{ kPa}$$
$$P_2 = P_{abs} \text{ of gas } B = P_{atm} + P_{gauge}$$
$$= 100 + 20 = 120 \text{ kPa}$$

Gauge pressure of the manometer

$$= P_1 - P_2$$
$$= 80 - 120$$
$$= -40 \text{ kPa}$$

Therefore, the manometer will show a vacuum pressure of 40 kPa.

10. A balloon is immersed in the sea. How deep is it to be immersed so that the volume is reduced to half of its size? Take $\rho_{water} = 1000$ N/m^2, $g = 10$ m/s^2 and $P_{atm} = 100$ kPa.

The volume of the balloon is full at atmospheric pressure, i.e. 100 kPa. The volume will reduce to half in case pressure becomes double, i.e. 200 kPa. The increase of pressure by 100 kPa (200 – 100 = 100 kPa) will be given by the depth of water.

$$P = \rho_{water} \times g \times \text{depth}$$
$$100 \times 10^3 = 1000 \times 10 \times \text{depth}$$

Therefore,

$$\text{Depth} = \frac{100 \times 10^3}{1000 \times 10} = 10 \text{ m}$$

11. A balloon bursts if its volume becomes double. Find the height it climbs before it bursts if atmospheric pressure reduces by 0.1 kPa/km climb? Take P_{atm} at the earth surface = 100 kPa and temperature remains constant.

The volume of the balloon will increase with climb as atmospheric pressure will reduce with climb. The volume will be double when atmospheric pressure will be half, i.e. when atmospheric pressure is 50 kPa.

Atmospheric pressure at any height = atmospheric pressure at the earth surface–fall of atmospheric pressure for height (h)

Mathematically $$P_{res\ atm} = 100 - 0.1 \times \text{height } (h)$$

$$50 = 100 - 0.1 \times h$$

or $$h = 50 \times 10 = 500 \text{ km}$$

12. Water boils at a temperature of 100°C and a pressure of 100 kPa. What will be pressure inside a cooker in case water boils at 123°C?

$$\frac{P_1}{T_1} = \frac{P_2}{T_2}$$

Substituting the values in the above equation,

or $$\frac{100}{100 + 273} = \frac{P_2}{123 + 273}$$

or $$\frac{100}{373} = \frac{P_2}{400}$$

or $$P_2 = \frac{400 \times 100}{373}$$

$$= 107.23 \text{ kPa}$$

13. Find the work done for process a–b, c–d and e–f as indicated on the following PV diagram.

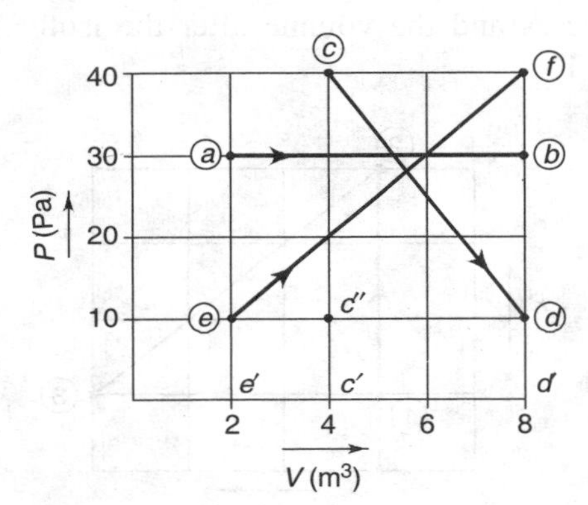

Work done (w) on the PV diagram for various processes is the area under each process. If the volume is increasing in a process, then work is done by the system. If the volume is decreasing, then work is done on the system.

Process a–b: It is a constant pressure process.

$$W_{a\text{-}b} = P \times dV$$
$$= 30 \times (8 - 2)$$
$$= 180 \text{ joules}$$

Process c–d:

$$W_{c\text{-}d} = \text{area of the triangle } cc''d + \text{area of the rectangle } c''\,c'd'\,d$$
$$= \frac{1}{2}\, cc'' \times c''d + c'c'' \times c'd'$$
$$= \frac{1}{2} \times (40 - 10) \times (8 - 4) + 10 \times (8 - 4)$$
$$= 60 + 40 = 100 \text{ joules}$$

Process e–f:

$$W_{e\text{-}f} = \text{area of the triangle } efd + \text{area of the rectangle } ed\,d'e'$$
$$= \frac{1}{2} \times ed \times df + ee' \times ed$$
$$= \frac{1}{2} \times (8 - 2) \times (40 - 10) + 10 \times (8 - 2)$$
$$= 90 + 60$$
$$= 150 \text{ joules}$$

14. An ideal gas undergoes an isochoric process from state 1 to state 2 resulting into pressure at state 2 as three times pressure at state 1. The gas is then expanded isothermally to state 3. It is then compressed isobarically to state 1. Find the temperature of the isothermal process and the volume after the isothermal process in temperature and volume of state 1.

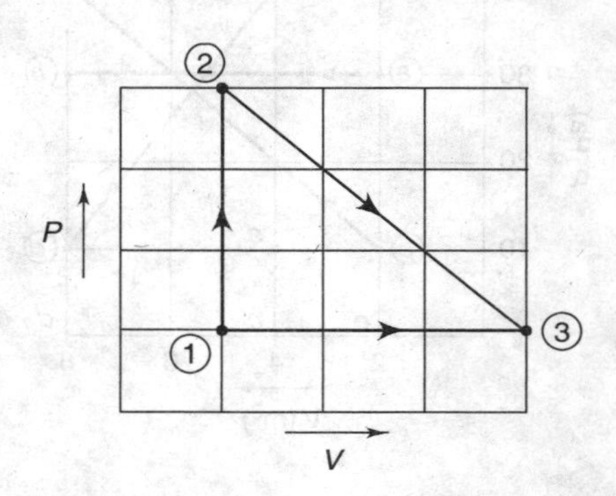

Isochoric process 1–2:

$$\frac{P_1}{T_1} = \frac{P_2}{T_2}$$

Given $P_2 = 3P_1$

Therefore, $$T_2 = \frac{T_1}{P_1} \times P_2$$

$$= \frac{T_1}{P_1} \times 3P_1$$

or $$T_2 = 3T_1$$

Isothermal process 2–3:

$$T_2 = T_3 = 3T_1$$

Isobaric process 3–1:

$$\frac{V_1}{T_1} = \frac{V_3}{T_3} = \frac{V_3}{3T_1}$$

or $$V_3 = 3V_1$$

15. A vessel of volume of 1 m^3 contains 2 moles of oxygen and 4 moles of carbon dioxide. If temperature of the mixture is 300 K, find the pressure of the mixture.

We know $$P\bar{v} = \bar{R}T$$

or $$P\frac{V}{n} = \bar{R}T$$

or $$PV = n\bar{R}T$$

For oxygen gas:

$$P_1 = 2(\text{mol}) \times 8314 \times 300$$

$$= 4.98 \text{ kPa}$$

For carbon dioxide:

$$P_2 = 4(\text{mol}) \times 8314 \times 300$$

$$= 9.96 \text{ kPa}$$

Applying Dalton's partial law

$$P_{\text{total}} = P_1 + P_2$$

$$= 4.98 \times 9.96$$

$$= 14.94 \text{ kPa}$$

16. Determine the molecular weight of a gas if $c_p = 2.286$ kJ/kg K and $c_v = 1.76$ kJ/kg K.

$$R = c_p - c_v$$
$$= 2.286 - 1.76$$
$$= 0.518 \text{ kJ/(kg K)}$$
$$\text{Molecular weight} = \frac{\overline{R}}{R} = \frac{8.3}{0.518}$$
$$= 16.05 \text{ kg}$$

17. Find the values of c_p and c_v of oxygen if $\gamma = 1.4$ and $\overline{R} = 8314$ J/(kmol K).

$$R \text{ (gas constant) of oxygen} = \frac{\overline{R}}{\text{molecular weight}}$$
$$= \frac{8314}{32 \times 10^{-3}}$$
$$\simeq 250 \text{ kJ/(kg K)}$$
$$c_p = \frac{\gamma}{\gamma - 1} \times R = \frac{1.4}{0.4} \times 250$$
$$= 875 \text{ kJ/(kg K)}$$
$$c_v = \frac{R}{\gamma - 1} = \frac{250}{0.4} = 615 \text{ kJ/(kg K)}$$

18. Find the temperature which is equal on the Fahrenheit and Centigrade scales.

Let temperature be $x°$.

$$\frac{x - 32}{180} = \frac{x}{100}$$

or $$5x - 160 = 9x$$

or $$4x = -160$$

or $$x = -40°$$

19. Find the absolute pressure (P_{abs}) of a gas if limbs of the manometer has water and mercury columns as shown in the following figure.

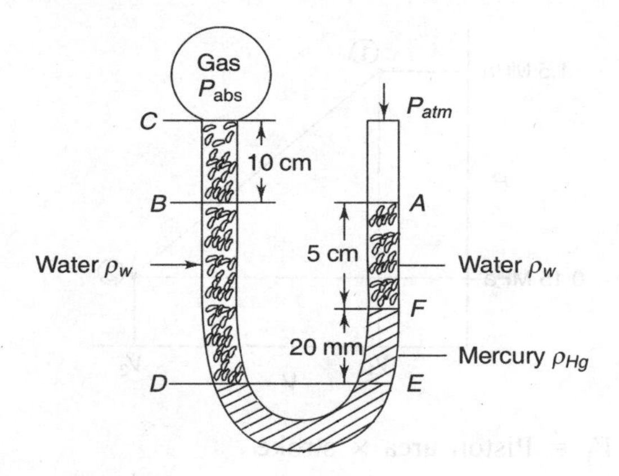

Pressure at points D and E will be the same. All pressures are calculated in mm of Hg. The water column height is divided by 13.6 (density of Hg).

$$P_{abs} + P_{CD} = P_{atm} + P_{AF} + P_{FE}$$

$$P_{abs} + \frac{170}{13.6} = 760 + \frac{50}{13.6} + 20$$

$$P_{abs} = 771.2 \text{ mm of Hg}$$

20. A diesel engine piston which has an area of 45 cm^2 moves 5 cm during part of suction stroke. 300 cm^2 of fresh air is drawn in from the atmosphere. The pressure in the cylinder during suction stroke is 0.9×10^5 N/m^2 and atmospheric pressure is 1.01325 bar. The difference between suction pressure and atmospheric pressure is due to resistance in the suction pipe and the valve. Find the net work done during the process.

(UPTU: Feb. 2001)

Work done by piston $W_1 = P_{cylinder} \times \text{area} \times \text{stroke}$

$$= 0.9 \times 10^5 \times \frac{45}{10^4} \times \frac{5}{100} \text{ J}$$

$$= 20.25 \text{ J}$$

Displacement work of free air $W_2 = P_{atm} (V_2 - V_1)$

$$= -1.01325 \times 10^5 \times \frac{300}{10^6}$$

$$= -30.475 \text{ J}$$

Net work $W = W_1 + W_2 = 20.25 - 30.475 = -10.725$ J

21. An engine cylinder has a piston area of 0.12 m^2 and contains gas at a pressure of 1.5 MPa. The gas expands according to a process which is represented by a straight line on a PV diagram. The final pressure is 0.15 MPa. Calculate the work done by the gas on the piston if the stroke is 0.3 m.

(UPTU: Dec. 2005)

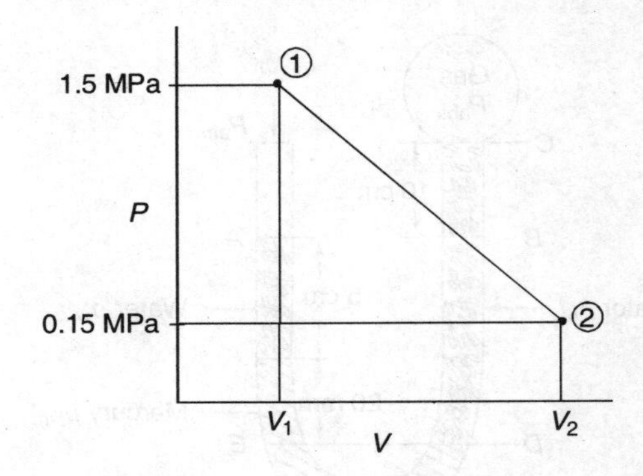

$$V_2 - V_1 = \text{Piston area} \times \text{stroke}$$

$$= 0.12 \times 0.3 = 0.036 \text{ m}^3$$

$$W_2 = \text{Area under process 1–2}$$

$$= \left[0.15\,(V_2 - V_1) + \frac{1}{2}\,(1.5 - 0.15)\,(V_2 - V_1)\right] \times 10^6 \text{ Nm}$$

$$= (0.15 + 0.675) \times 36 \text{ kJ}$$

$$= 29.7 \text{ kJ}$$

22. Determine the size of a spherical balloon filled with hydrogen at 36 °C and atmospheric pressure for lifting 400 kg pay load. Atmospheric air is at a temperature of 27°C and barometer reading is 75 cm of mercury.

(UPTU: July 2002)

$$\text{Pressure of } H_2 = \frac{75}{76} \times 1.013 \approx 1 \text{ bar}$$

$$R \text{ (gas constant) for } H_2 = \frac{8314}{2} = 4157 \text{ J/(kg K)}$$

$$\text{Mass of } H_2 = m_1 = \frac{PV}{RT}$$

$$= \frac{10^5 \times V}{4157 \times 303} = 0.0794\, V \text{ kg}$$

$$\text{Mass of air displaced } m_2 = \frac{1 \times 10^5 \times V}{287 \times 303}$$

$$= 1.16\, V \text{ kg}$$

$$\text{Pay load} = (m_2 - m_1)\, g$$

$$400 \times 9.81 = (1.16 - 0.0794)\, V \times 9.81$$

or $$V \approx 370 \text{ m}^3$$

$$\frac{4}{3}\pi r^3 = 370 \text{ m}^3$$

or $$r = 4.45 \text{ m}$$

23. A manometer measures the pressure of a tank as 250 cm of mercury for the density of mercury is 13.6×10^3 kg/m^3 and atm pressure 101 kPa. Calculate the tank pressure in MPa.

(UPTU: Dec. 2001)

$$P_{abs} = P_{atm} + P_{gauge}$$

$$= 101 \text{ kPa} + 13.6 \times 10^3 \times 9.8 \times 2.5 \times 10^{-3} \text{ kPa}$$

$$= 101 + 333.54 = 434.54 \text{ kPa} = 0.4345 \text{ MPa}$$

24. One mole of an ideal gas at 0.1 MPa and 300 K is heated at constant pressure till the volume is doubled and then it is allowed to expand at constant temperature till the volume is doubled again. Calculate the work done by the gas.

(UPTU: March 2002)

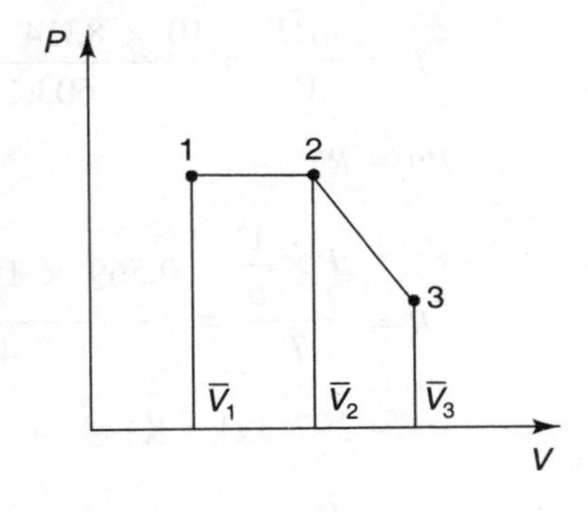

Process 1–2:

$${}_1W_2 = P_1\,(\overline{V}_2 - \overline{V}_1)$$

$$= P_1\,(2\,\overline{V}_1 - \overline{V}_1)$$

$$= P_1\overline{V}_1 = \overline{R}\,T_1$$

Also $$\frac{P_1\overline{V}_1}{T_1} = \frac{P_2\overline{V}_2}{T_2}$$

Therefore, $$\frac{P_1\overline{V}_1}{T_1} = \frac{P_1 \times 2\overline{V}_1}{T_2} \qquad (\text{as } P_1 = P_2 \text{ and } \overline{V}_2 = 2\overline{V}_1)$$

or $$T_2 = 2T_1$$

Process 2–3:

$${}_2W_3 = \int_2^3 P\,d\bar{V} = \int_2^3 \frac{\bar{R}T_2\,d\bar{V}}{\bar{V}}$$

$$= \bar{R}\;T_2\;\ln\frac{\bar{V}_3}{\bar{V}_2} = 2\,\bar{R}\,T_1\;\ln 2$$

$$\text{Now }{}_1W_3 = {}_1W_2 + {}_2W_3 = \bar{R}\,T_1 + 2\,\bar{R}\,T_1\;\ln 2$$

$$= \bar{R}\,T_1\;(1 + 2\;\ln\,2)$$

$$= 8314 \times 310\;(1 + 3.386)$$

$$= 6.15\text{ MJ}$$

25. 10 kg mol of a gas occupies a volume of 603.1 m^3 at a temperature of 140°C while its density is 0.464 kg/m^3. Find its molecular weight and gas constant and its pressure. (UPTU: 2003–2004)

$$PV = n\bar{R}\,T$$

$$P = \frac{n\bar{R}T}{V} = \frac{10 \times 8314 \times 413}{603.1} = 0.569 \times 10^5\text{ bar}$$

Also

$$Pv = RT$$

$$R = \frac{P \times \frac{1}{\rho}}{T} = \frac{0.569 \times 10^5 \times \frac{1}{0.464}}{413}$$

$$= 297\;\text{J/(kg K)}$$

We know that molecular weight $M = \dfrac{\bar{R}}{R}$

$$= \frac{8314}{297} = 28$$

26. A steel cylinder having a volume of 0.01653 m^3 contains 5.6 kg of ethylene gas (C_2H_2) whose molecular weight is 28. Calculate the temperature to which the cylinder may be heated without the pressure exceeding 200 bar, given compressibility factor z = 0.605.

$$R = \frac{\bar{R}}{M} = \frac{8314}{28} = 296.93\;\text{J/(kg K)}$$

$$v = \text{specific volume} = \frac{V}{m} = \frac{0.01653}{5.6} = 2.95 \times 10^{-3}\;\text{m}^3/\text{kg}$$

Now $$Pv = z\,RT$$

or $$T = \frac{Pv}{zR}$$

$$= \frac{200 \times 10^5 \times 2.95 \times 10^{-3}}{0.605 \times 296.93} = 328.4 \text{ K}$$

27. Determine the molecular weight of a gas if its two specific heats are: c_p = 2.86 kJ/(kg K) and c_v = 1.768 kJ/(kg K). $\bar{R}$ = 8.3143 kJ/kg mol. (UPTU: 2003)

$$R = c_p - c_v = 2.286 - 1.768 = 0.518 \text{ kJ/(kg mol K)}$$

$$M = \frac{\bar{R}}{R} = \frac{8.3143}{0.518} = 16 \text{ kg}$$

28. A tank of 0.35 m^3 capacity contains H_2S gas at 300 K. When 2.5 kg of gas is withdrawn, the temperature in the tank becomes 288 K and pressure 10.5 bar. Calculate the mass of gas initially kept in the tank and also initial pressure.

(UPTU carry over: Dec. 2005–2006)

$$R = \frac{8314}{34} = 244.5 \text{ J/(kg K)}$$

$$P_1 V = m\,RT$$

Therefore, $$P_1 \times 0.35 = m \times 244.5 \times 300$$

$$P_2 \times 0.35 = (m - 2.5) \times 244.5 \times 288 = 10.5 \times 0.35 \times 10^5$$

or $$m - 2.5 = \frac{10.5 \times 0.35 \times 10^5}{244.5 \times 288} = 5.22$$

or $$m = 7.72 \text{ kg}$$

Now $$P_1 = \frac{7.72 \times 244.5 \times 300}{0.35} = 16.2 \text{ bar}$$

29. A mass of 1.5 kg of air is compressed in a quasi-static process from 1.1 bar to 10 bar according to law $PV^{1.25}$ = constant, where v is specific volume. The initial density of air is 1.2 kg/m^2. Find the work involved in compression process.

(UPTU: Aug. 2005)

$$v_1 = \frac{1}{\rho} = \frac{1}{1.2} = 0.833 \text{ m}^3\text{/kg}$$

$$P_1 v_1^{1.25} = P_2 v_2^{1.25}$$

$$1.1 \times (0.833)^{1.25} = 10 \times v_2^{1.25}$$

or $$v_2 = \frac{1.1 \times (0.823)^{1.25}}{10} = \frac{0.784 \times 1.1}{10} = 0.086$$

We know that $$w = m\left(\frac{P_2v_2 - P_1v_1}{\gamma - 1}\right)$$

$$= 1.5 \times \frac{(10 \times 0.086 - 1.1 \times 0.833)}{1.25 - 1} \times 10^5$$

$$= \frac{1.5 \times (0.86 - 0.916) \times 10^5}{0.25}$$

$$= -\frac{1.5 \times 0.056}{0.25} \times 10^5$$

$$= -33.6 \text{ kJ}$$

OBJECTIVE TYPE QUESTIONS

Enthusiasm is the fuel of life; it helps you get where you are going.

State True or False

1. In macroscopic approach, the structure of matter is considered. (*True/False*)
2. Classical thermodynamics adopts macroscopic approach while statistical thermodynamics adopts microscopic approach. (*True/False*)
3. Thermodynamics is a science deals mainly with transformation of heat into work. (*True/False*)
4. Macroscopic approach is not equal to summation of microscopic approach analysis. (*True/False*)
5. When system like rarefied gases in rocket flight at very high attitudes are analyzed, the system cannot be considered as continuum and microscopic approach for analysis has to be used. (*True/False*)
6. Thermodynamics lets us know how much work is available from an engine at what efficiency. (*True/False*)
7. An isolated system does not permit any interaction of mass and energy. (*True/False*)
8. The state of a system can be given by any two properties like pressure, volume, and temperature. (*True/False*)
9. The state of a system cannot be given by a pair of properties like temperature–entropy or enthalpy–entropy. (*True/False*)
10. Pressure and temperature are not intensive properties. (*True/False*)
11. A quasi-static process is a reversible process. (*True/False*)
12. In a state of equilibrium, the properties of a system are uniform and only one value can be given to each of the property. (*True/False*)
13. Equilibrium means absence of any tendency for spontaneous changes of an isolated system. (*True/False*)
14. Work is energy in transit and it can interact across the boundary. (*True/False*)
15. Work done by gas during free expansion is zero. (*True/False*)
16. Heat, volume and enthalpy cannot be converted into intensive properties by dividing mass of the system. (*True/False*)
17. Kinetic energy is same for molecules of all gases at a given temperature. (*True/False*)
18. Gas constant of a gas is given by universal gas constant divided by molecular weight. (*True/False*)
19. A gas behaves as an ideal gas at a low pressure and a high temperature. (*True/False*)
20. Nitrogen will have higher root mean square velocity as compared to hydrogen at the same temperature. (*True/False*)

21. The square of root mean square velocity is proportional to temperature. (*True/False*)
22. The temperature of the sun can be measured by the pyrometer. (*True/False*)
23. The temperature of gas can be increased by keeping pressure and volume constant. (*True/False*)
24. Temperature of an isolated system cannot be kept constant. (*True/False*)
25. A room can be cooled by leaving the door of the refrigerator open. (*True/False*)
26. A system can do external work without taking heat energy but utilizing its internal energy. (*True/False*)
27. If a gas is compressed at constant temperature, its internal energy will increase. (*True/False*)
28. If two balls having masses of 10 and 20 gm collide with a target with the same velocity, the heavy ball will attain higher temperature. (*True/False*)
29. It can be found out that an iron piece has been heated by fire or by hammering. (*True/False*)
30. By rubbing hands, work can be converted into heat. (*True/False*)
31. – 40° is the same on Fahrenheit and centigrade scales. (*True/False*)
32. The total radiation pyrometer works on the principle that radiation from a hot body is proportional to the fourth power of absolute temperature. (*True/False*)
33. Absolute pressure of gas is higher than atmospheric pressure in case the manometer has higher mercury column in the limb opens to atmosphere. (*True/False*)
34. Vacuum gauge pressure is added to atmospheric pressure to find absolute pressure. (*True/False*)
35. The difference of absolute pressure and atmospheric pressure is gauge pressure. (*True/False*)
36. The difference of specific heats at constant pressure and temperature is equal to gas constant. (*True/False*)
37. Two gases with molecular weights of 28 and 12 expand at constant pressure through the same temperature range. The ratio of work done will have 12 : 28 ratio. (*True/False*)
38. The system and surrounding when put together is called universe. (*True/False*)

Multiple Choice Questions

1. The process 1–2 shown below is
   (a) isobaric (b) isochoric (c) isothermal

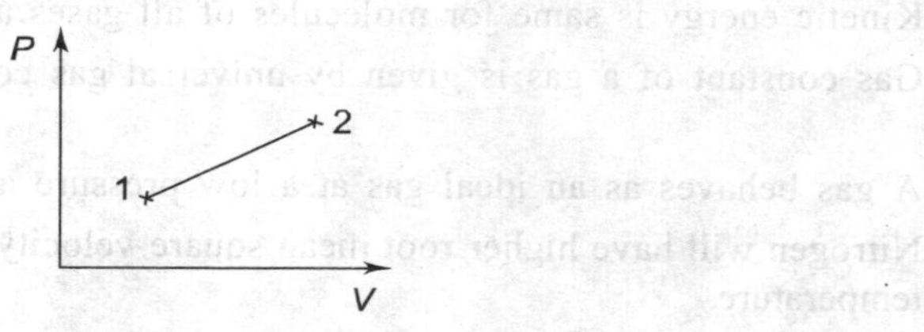

2. For two isothermals T_1 and T_2 shown below is

(a) $T_1 > T_2$ (b) $T_1 = T_2$ (c) $T_1 < T_2$

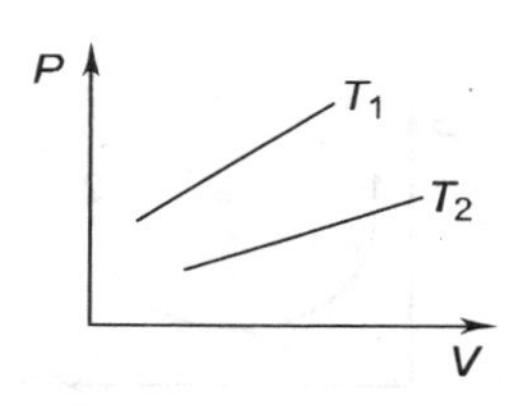

3. Identify the processes AB and BC from the diagram.

(a) AB is isothermal and AC is adiabatic.
(b) AB is adiabatic and AC is isothermal.
(c) Cannot say which is isothermal and which is adiabatic.

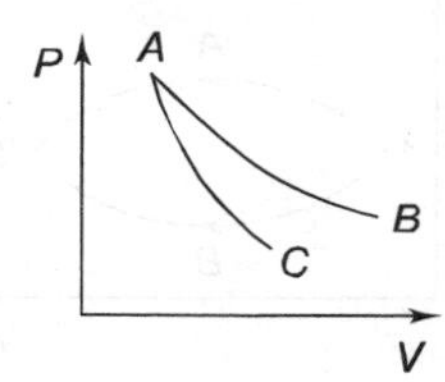

4. For two isobaric processes P_1 and P_2 shown below is

(a) $P_1 = P_2$ (b) $P_1 < P_2$ (c) $P_1 > P_2$

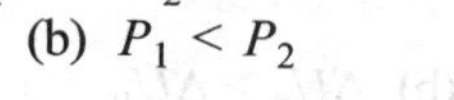

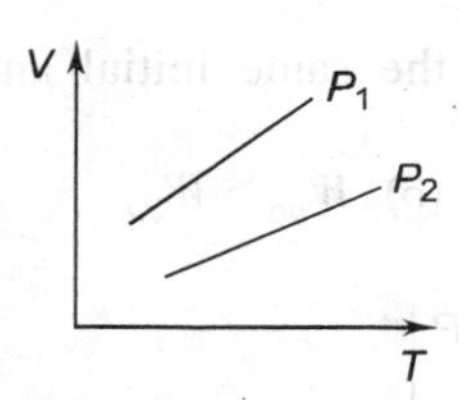

5. The work done by the system from state 1 to state 2 is

(a) positive (b) negative (c) zero

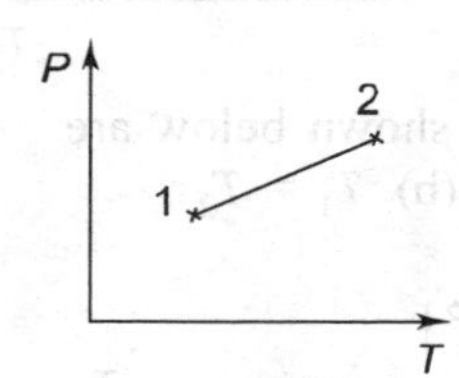

6. The work done by the system if it moves from state 1 to state 2
(a) continuously decreases (b) continuously increases
(c) first decreases then increases

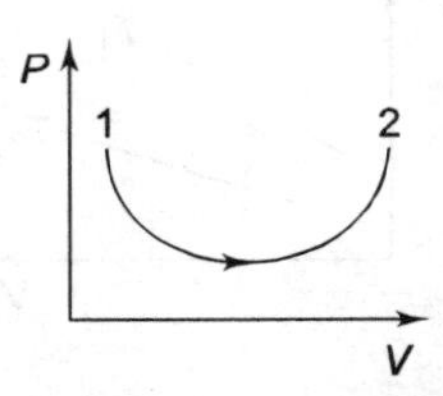

7. If the system moves from state 1 to state 2 along paths A and B and heats supplied are Q_A and Q_B respectively, then
(a) $Q_A = Q_B$ (b) $Q_A > Q_B$ (c) $Q_A < Q_B$

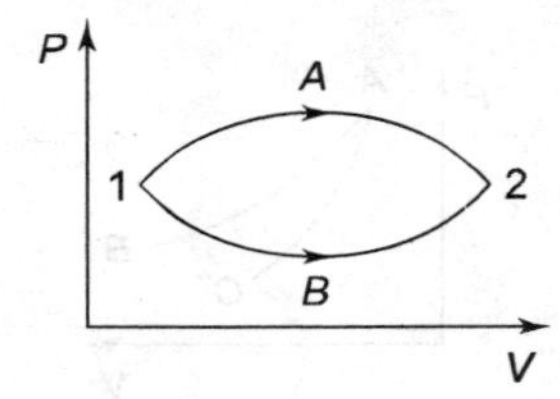

8. In the above problem, if ΔU_A and ΔU_B are changes in internal energy along paths A and B, then
(a) $\Delta U_A = \Delta U_B$ (b) $\Delta U_A > \Delta U_B$ (c) $\Delta U_A < \Delta U_B$

9. If processes ab and cd have the same initial and final volumes, then work done by processes ab and cd are
(a) $W_{ab} = W_{cd}$ (b) $W_{ab} > W_{cd}$ (c) $W_{ab} < W_{cd}$

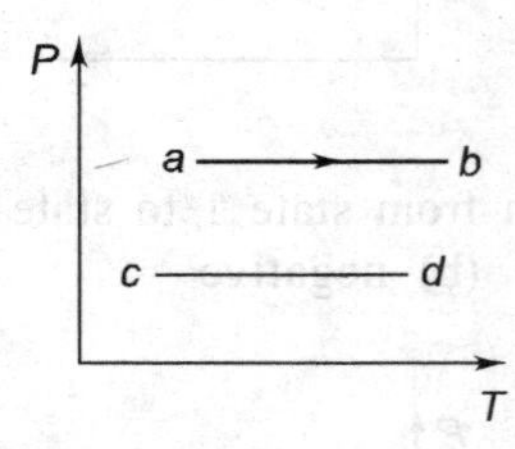

10. For two isothermals T_1 and T_2 shown below are
(a) $T_1 > T_2$ (b) $T_1 = T_2$ (c) $T_1 < T_2$

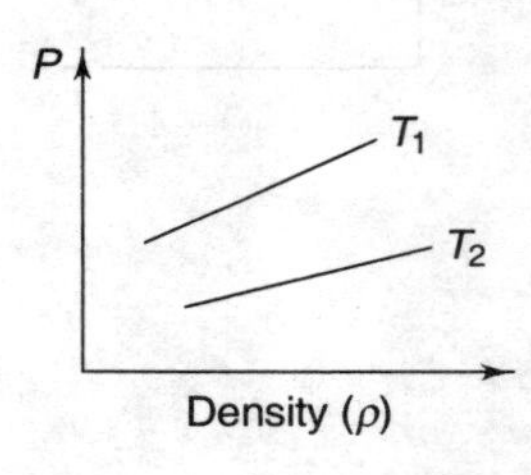

11. The rms velocity of oxygen is 400 m/s. The rms velocity of hydrogen at the same temperature is
(a) 1600 m/s (b) 2000 m/s (c) 400 m/s

12. For cyclic process 1–2–3–1, the work done is
(a) 4 joules (b) 2 joules (c) 3 joules

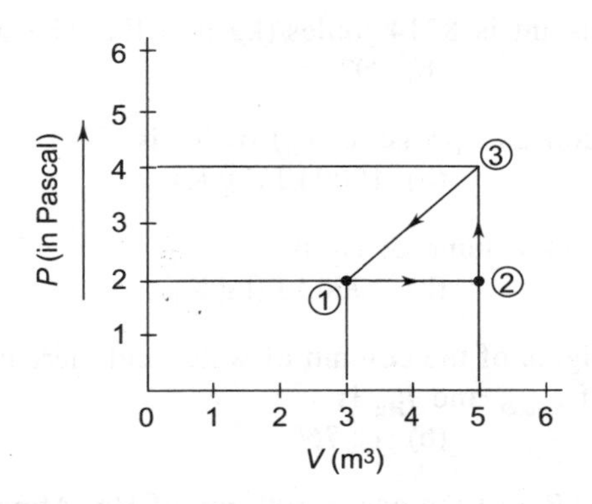

13. For cyclic process *a–b–c–d–a*, the work done by the system is
(a) 10 joules (b) 15 joules (c) 12 joules

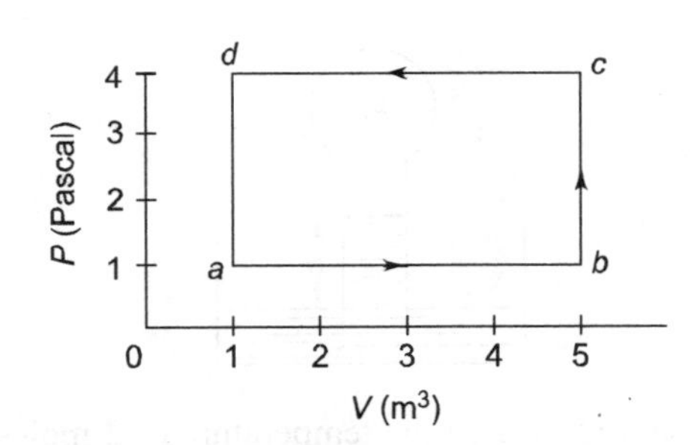

14. The temperature of the human body is 98.6°F. In centigrade it will be
(a) 38°C (b) 376°C (c) 37°C

15. If atmospheric pressure is 760 mm of Hg ($\rho = 13.6 \times 10^3$ N/m^3), then the length of liquid ($\rho = 4.53 \times 10^3$ N/m^3) which will be lifted by atmospheric pressure is
(a) 1520 mm (b) 2940 mm (c) 2280 mm

Refer to the following figure for questions 16, 17, and 18.

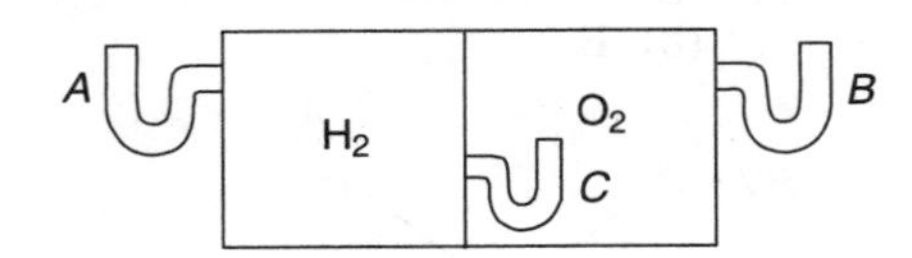

16. If atmospheric pressure is 100 kPa and pressure gauge *A* reads 140 kPa, then absolute pressure of H_2 gas is
(a) 140 kPa (b) 40 kPa (c) 240 kPa

17. If pressure gauge B reads – 60 kPa, then pressure of O_2 is
(a) 160 kPa (b) 40 kPa (c) 60 kPa

18. Pressure gauge C will read
(a) 160 kPa (b) 200 kPa (c) 240 kPa

19. The universal gas constant is 8314 joules/(kg mol K). The gas constant of N_2 is
(a) 415 (b) 297 (c) 519

20. The specific heat at constant pressure (c_p) of N_2 is
(a) 140 kJ/(kg K) (b) 1000 kJ/(kg K) (c) 743 kJ/(kg K)

21. Specific heat at constant volume of N_2 is
(a) 1040 kJ/(kg K) (b) 1000 kJ/(kg K) (c) 743 kg J/K

22. If I_{water} and l_{Hg} are heights of the column of water and mercury supported by atmospheric pressure, then ratio of l_{water} and l_{Hg} is
(a) 1 : 13.6 (b) 1 : 760 (c) 13.6 : 1

23. The absolute pressure (P_a) of the gas is 500 mm of Hg. Atmospheric pressure is 760 mm of Hg. The value of h is
(a) 1260 mm (b) 260 mm (c) 500 mm

Pa
h

24. 70 calories of heat is required to raise the temperature of 2 moles of an ideal gas at constant pressure from 30°C to 35°C. If $\gamma = 1.4$, the amount of heat required to raise the same temperature of range at constant volume is
(a) 30 calories (b) 50 calories (c) 60 calories

25. Two gases with molecular weights of 44 and 28 expand at constant pressure through the same temperature range. What is the ratio of quantity of work done by the gases?
(a) 11 : 7 (b) 22 : 14 (c) 7 : 11

26. Which line of the following diagram represents an ideal gas?
(a) A (b) B (c) C

PV
C
B
A
P

27. A cycle on a *VT* diagram is shown by *ABC*.

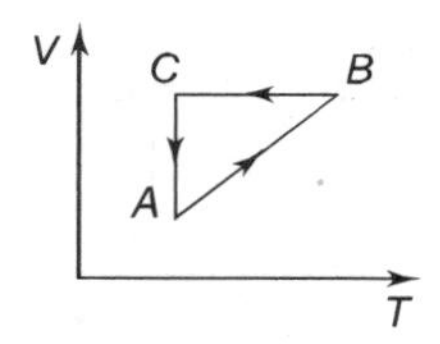

The cycle on *PV* diagrams is shown by

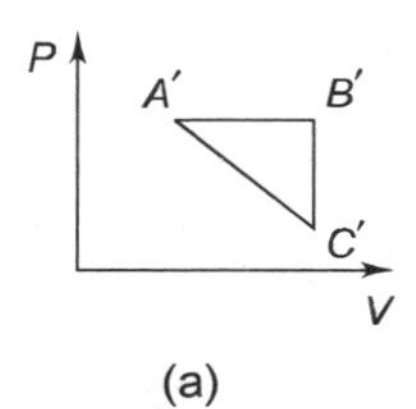

(a)

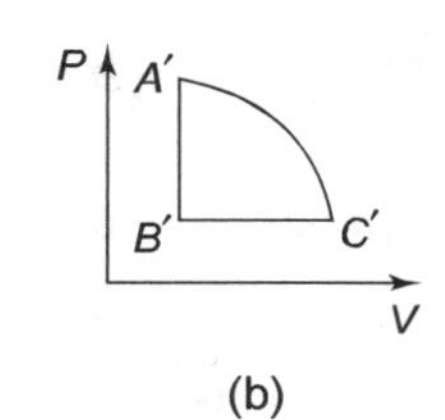

(b)

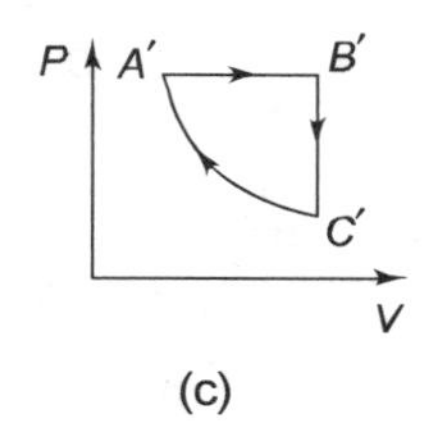

(c)

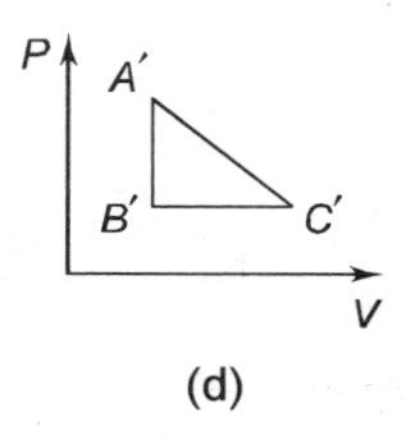

(d)

28. If two gases with the same volume (*V*), pressure (*P*) and temperature (*T*) are mixed to get the same volume (*V*), what will be the final pressure?
(a) *P* (b) 2*P* (c) 4*P*

29. Which of the following is not the property of a system?
(a) Work (b) Temperature (c) Internal energy (d) Pressure

30. Which of the following sets has all properties of a point function?
(a) Entropy, enthalpy, work
(b) Pressure, temperature, heat
(c) Heat, work, enthalpy
(d) Temperature, enthalpy, internal energy

31. Which of the following sets has all intensive properties?
(a) Pressure, volume, energy, heat, specific volume
(b) Pressure, energy, volume, heat and density
(c) Pressure, temperature, density, specific volume
(d) Pressure, volume, heat, density, specific volume

32. The difference between the pressure of fluid and the pressure of atmosphere is called
(a) barometric pressure (b) gauge pressure (c) absolute pressure

33. Time, length, mass and temperature are
(a) primary dimension (b) secondary dimension (c) none of them

34. In a constant volume gas thermometer, the thermometric property used is
(a) mass (b) length (c) volume (d) pressure

35. The thermometric property of an electrical resistance thermometer is
(a) current (b) potential difference (c) emf (d) resistance

36. A cycle on a PV diagram is shown as

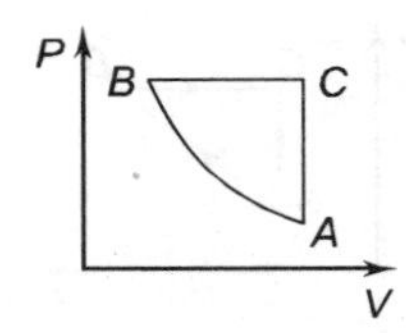

The same on a TS diagram will be shown as

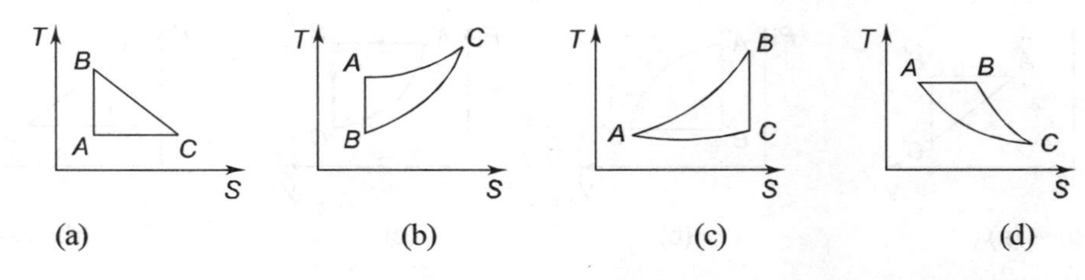

37. Which one of following PT diagrams illustrates the Otto cycle of an ideal gas?

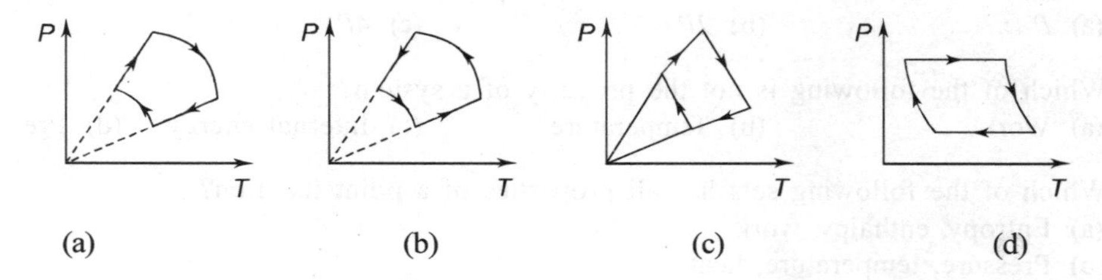

38. A cycle is shown on a VT diagram as below.

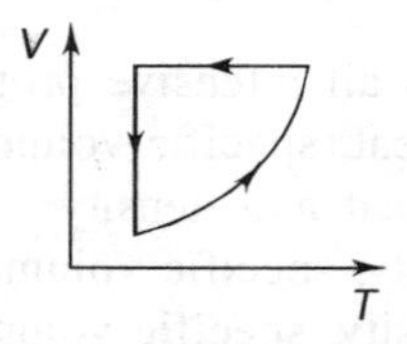

Which one of the figures on a PV diagram indicates the cycle?

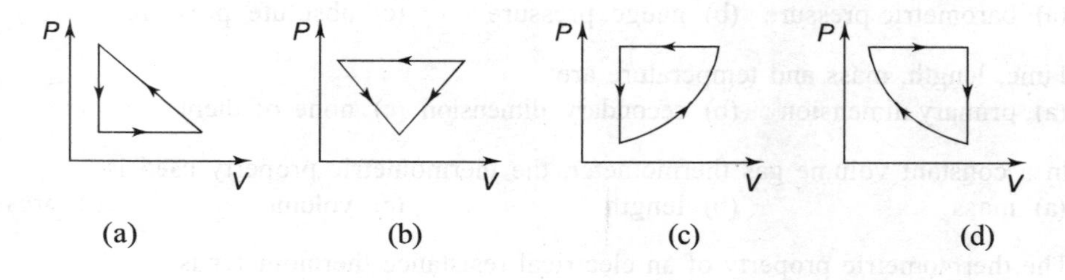

39. Match the curve on the PV diagram to the curve on the TS diagrams.

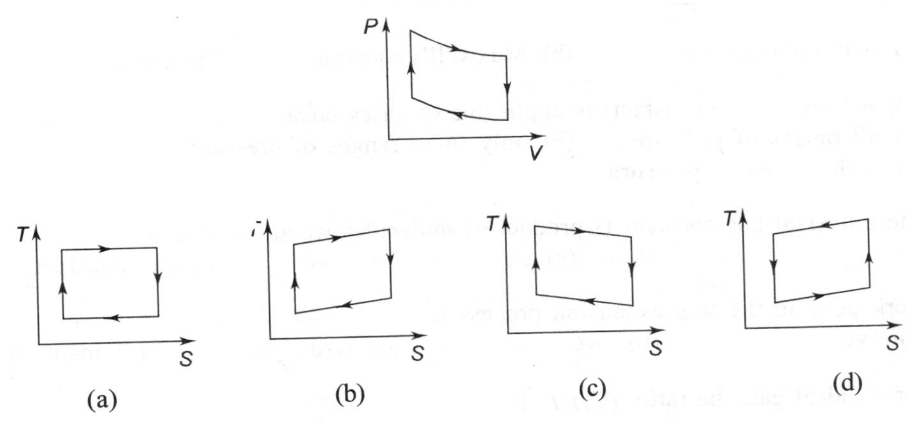

40. Match the curve on the PV diagram to the curve on TS diagrams.

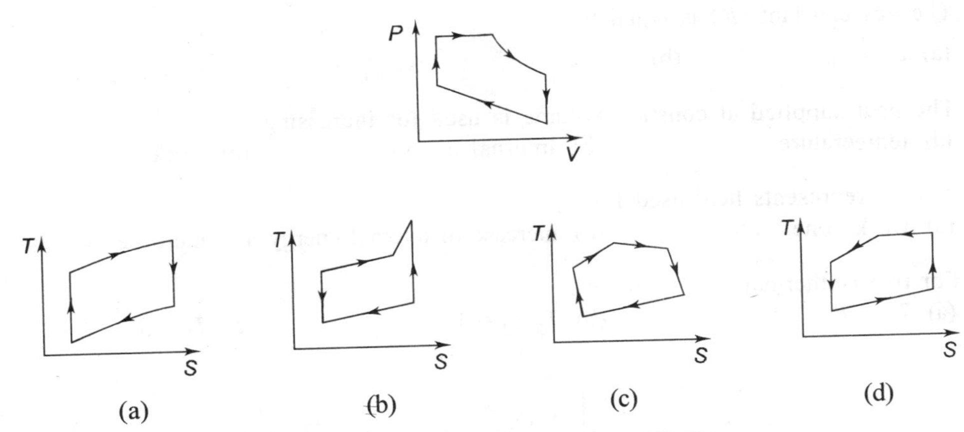

41. Match the curve on the PV diagram to the curve on TS diagrams.

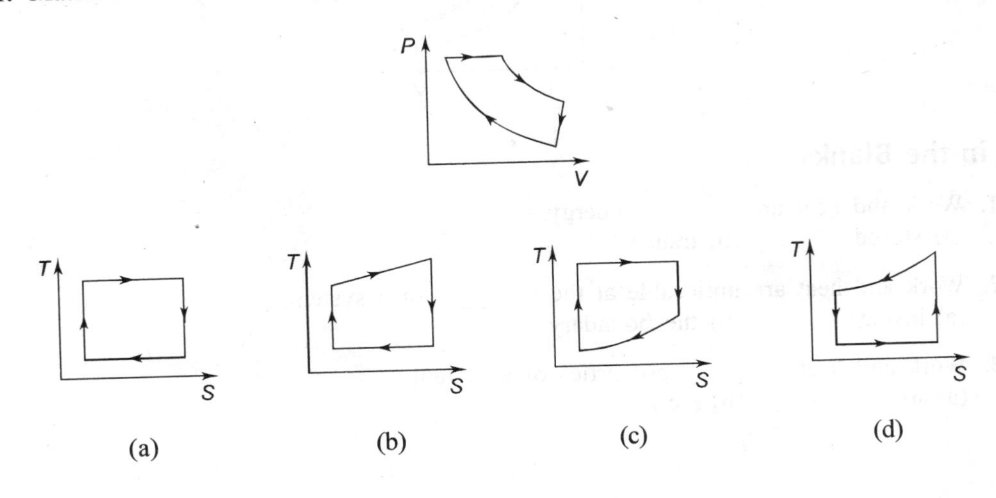

42. The equation $\left(P + \frac{a}{\overline{v}^2}\right)(\overline{v} - b) = \overline{R}T$ is known as
(a) real gas equation (b) Maxwell's equation (c) van der Wals equation

43. Boyles' law (PV = constant) is applicable to gases under
(a) all ranges of pressure (b) only small ranges of pressure
(c) high ranges of pressure

44. The universal gas constant is product of molecular weight of gas and
(a) c_p (b) c_v (c) gas constant

45. Work done in the free expansion process is
(a) +ve (b) –ve (c) zero (d) maximum

46. For an ideal gas, the ratio $P\overline{v}/\overline{R}T$ is
(a) zero (b) unity (c) tends to zero (d) tends to unity

47. The gas constant (R) is equal to
(a) $c_p + c_v$ (b) $c_p \times c_v$ (c) $c_p - c_v$ (d) $\frac{c_p}{c_v}$

48. The heat supplied at constant volume is used for increasing
(a) temperature (b) Internal energy (c) work

49. $c_p - c_v$ represents heat used for
(a) work done (b) increase of internal energy (c) increase of volume

50. For two isothermals shown below,
(a) $T_1 > T_2$ (b) $T_1\ T_2 = 1$ (c) $T_1 < T_2$

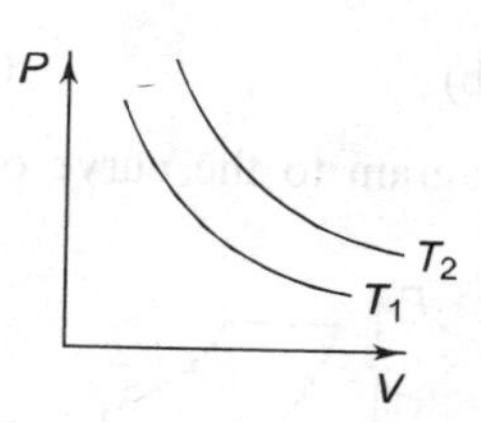

Fill in the Blanks

1. Work and heat are ________ energy.
(a) stored (b) transit

2. Work and heat are noticeable at the ________ of a system.
(a) inside (b) the boundary

3. Work and heat ________ properties of a system.
(a) are (b) are not

4. Work and heat are________ functions.
   (a) path (b) state

5. Transit of heat energy takes place due to the property of ________ difference.
   (a) pressure (b) temperature

6. Transit of work energy takes place when any other property other than ________differs.
   (a) temperature (b) volume

7. Numerically the unit of joule is ________ than calorie.
   (a) greater (b) smaller

8. In the constant volume thermometer ________ is proportional to temperature.
   (a) pressure (b) volume

9. The pyrometer is used for measuring very high temperature of body and temperature measured is proportional to ________ energy emitted by the body.
   (a) light (b) radiation

10. Constant volume gas meters are ________ thermometers.
    (a) primary (b) secondary

11. A hand water pump lifts water on account of ________ pressure.
    (a) atmospheric (b) pump

12. To control various operations on CNC (Computer Numerical Controlled) machine, ________ thermometers are used.
    (a) thermocouple (b) resistance

13. Thermometric property in the constant volume thermometer is ________, in the constant pressure gas thermometer is ________, in the electric resistance thermometer is ________, and in the mercury thermometer is ________.
    (a) resistance (b) pressure (c) length (d) volume

14. The compressibility factor (z) for ideal gas is ________.
    (a) 1 (b) 1.2

15. A real gas behaves as an ideal gas at a ________ pressure.
    (a) low (b) high

16. A real gas behaves as an ideal gas at ________ temperatures.
    (a) low (b) high

17. A gas ________ be liquefied above the critical point.
    (a) can (b) cannot

ANSWERS

The optimist fell through twelve storeys
And at each window bar
He yelled to his friends who were frightened below
Well I'm all right so far

State True or False

1. False (Structure of matter is considered in microscopic approach.)
2. True
3. True
4. False (Macroscopic approach = Σ Microscopic analysis)
5. True
6. True
7. True
8. True
9. False (State can be fixed by a pair of properties.)
10. False (Pressure and temperature are not dependent on mass)
11. True
12. True
13. True
14. True
15. True
16. False (specific heat, specific volume and specific enthalpy are based on unit mass, which is uniform in the system and hence intensive property.)
17. True
18. True
19. True
20. False (Square rms velocity is proportional to temperature for all gases)
21. True
22. True
23. False ($PV = mRT$ and T is constant if PV is constant.)
24. False (No energy can go out or come inside an isolated system.)

25. False (Room and refrigerator form a system. Energy interaction is electric energy which will increase the temperature of the system.)
26. True (Adiabatic expansion is an example.)
27. False (Internal energy depends upon temperature only.)
28. False (KE is proportional to mass and heat energy is also proportional to mass. Temperature attained will be the same.)
29. False (Heat energy does not depend upon the mode of generation. It can be felt by temperature.)
30. True
31. True
32. True
33. True (Mercury will climb towards lower pressure side)
34. False [($P_{abs} = P_{atm} - P_v$ (Vacuum gauge)]
35. True
36. True ($c_p - c_v = R$)
37. True ($P\bar{v} = RT$ or $Pd\bar{v} = \bar{R}\,dT$ or work per mole is constant or work × molecular weight is constant.)
38. True

Multiple Choice Questions

1. (c)
2. (a)
3. (a) (An isothermal curve will have a lesser slope than an adiabatic curve.)
4. (b) (For any temperature, V on $P_1 > V$ on P_2 line. Therefore $P_2 > P_1$.)
5. (c) (The volume at state 1 and 2 is the same, i.e. $\Delta V = 0$. Hence work done is zero.)
6. (b) (The volume is continuously increasing.)
7. (b) ($W_A > W_B$. Hence $Q_A > Q_B$.)
8. (a) (Internal energy is a state function. In both paths $\Delta U = U_2 - U_1$)
9. (b) (Since $P_{ab} > P_{cd}$ and ΔV is same, work is $P \times \Delta V$.)
10. (a) ($PV \propto T$ or $\frac{P}{\rho} \propto T$. Hence higher T will have a higher slope.)
11. (c) (Square of rms velocity depends upon temperature only.)
12. (b) (W = area of triangle

$$= \frac{1}{2} \times (5 - 3)\,(4 - 2)$$

$$= 2 \text{ joules})$$

13. (c) ($W = (5 - 1)(4 - 1) = 12$ joules)

14. (c) $\left[\frac{98.6 - 32}{9} = \frac{x\,°C}{5} \quad \therefore x = 37°C\right]$

15. (c) $\left[\frac{l_{Hg} \times \rho_{Hg}}{\rho_{Hg}}\right]$

16. (c) ($P_{H_2} = P_{atm} + P_{gauge} = 100 + 40$)

17. (b) ($P_{O_2} = P_{atm} - P_{vacuum} = 100 - 60$)

18. (b) ($P_{H_2} = P_{O_2} + P_{gauge}$ or $P_{gauge} = 240 - 40 = 200$)

19. (b) $\left[R_{N_2} = \frac{R}{M} = \frac{8314}{28} = 297\right]$

20. (a) $\left[c_p = \frac{\gamma}{\gamma - 1} \times R = \frac{1.4}{1.4 - 1} \times 297 = 1040\right]$

21. (c) $\left[c = \frac{R}{\gamma - 1} = \frac{297}{1.4 - 1} = 743\right]$

22. (c) ($l_{water}\,\rho_{water} = l_{Hg}\,\rho_{Hg}$)

23. (b) ($h = 760 - 500 = 260$)

24. (b) [$70 = 2 \times c_p \times 5$ or $c_p = 7$, $c_v = \frac{c_P}{\gamma}$]

$c_v = \frac{7}{1.4} = 5$. Now $Q = 2 \times c \times \Delta T = 50$ J]

25. (c) ($W_1 \times M_1 = W_2 M_2$)

26. (b)

27. (c)

28. (c) (Pressure of each gas is $2P$ for half volume. $P_{mix} = 2P + 2P = 4P$)

29. (a) (Work is a transit phenomenon at the boundary.)

30. (d) **31.** (c) **32.** (b) **33.** (a)

34. (d) **35.** (d) **36.** (b) **37.** (a)

38. (c) **39.** (b) **40.** (a) **41.** (c)

42. (c) **43.** (b) **44.** (c) ($\overline{R} = MR$) **45.** (c)

46. (b) ($P\overline{v} = \overline{R}\,T$) **47.** (c) **48.** (b) **49.** (a)

50. (c)

Fill in the Blanks

1. (b)
2. (b)
3. (b)
4. (a)
5. (b)
6. (a)
7. (a)
8. (a)
9. (b)
10. (a)
11. (a)
12. (a)
13. (b, d, a, c)
14. (a)
15. (a)
16. (b)
17. (b)

CHAPTER 2

First Law of Thermodynamics

A word of encouragement is as refreshing as a cold drink on a hot summer day.

INTRODUCTION

Thermodynamics is a science of energy transfer and its effects on the physical properties of a substance. Energy has two forms, namely transit energy and internal energy (stored energy). Internal energy is a property of a system and it is a point function. Heat and work are transiting energies. They are not the properties of a system. They are path functions.

The state of a system changes whenever heat or work transits the system. The internal energy increases whenever heat or work enters the system, and internal energy decreases whenever heat or work goes out of the system. Energy entering a system as heat may leave the system as work. Similarly, energy entering a system as work may leave the system as heat. Internal energy of the system may change as per heat or work interaction. However, energy as such is conserved. It is possible to relate heat, work and the changes in the properties for various processes in open and closed systems. Their relationship leads to the first law of thermodynamics.

FIRST LAW OF THERMODYNAMICS

If an amount of heat (Q) is given to a system, a part of it will be used in increasing the internal energy (ΔU) of the system and the rest is used in doing work (W) by the system. Thus the first law of thermodynamics is a form of the law of conservation of energy. Mathematically,

$$Q = \Delta U + W$$

Sign convention used for work and heat: Heat is positive if heat is given to the system. Heat is negative if heat is rejected by the system. Work (W) is positive if it is done by the system. Work is negative if work is done on the system.

If a system undergoes a number of processes and comes back to the initial state, then this is known as *cyclic process*. Since internal energy (U) is a state function, hence change of internal energy will be zero for a cyclic process.

$$Q = \Delta U + W$$

For a cyclic process $\Delta U = 0$. Therefore, the above equation becomes

$$Q = W$$

Hence in cyclic process, the complete heat given to a system is converted into work, i.e.

$$\oint Q = \oint W$$

In a cyclic process, the pressure and volume (PV) curve is a closed curve. Therefore, the work done in the cyclic process is equal to the area enclosed by the curve. If a closed PV curve is traced clockwise, then network is done by the system and it is positive (Figure 2.1(b)). If the curve is anticlockwise, then the network is done on the system and it is negative (Figure 2.1(a)).

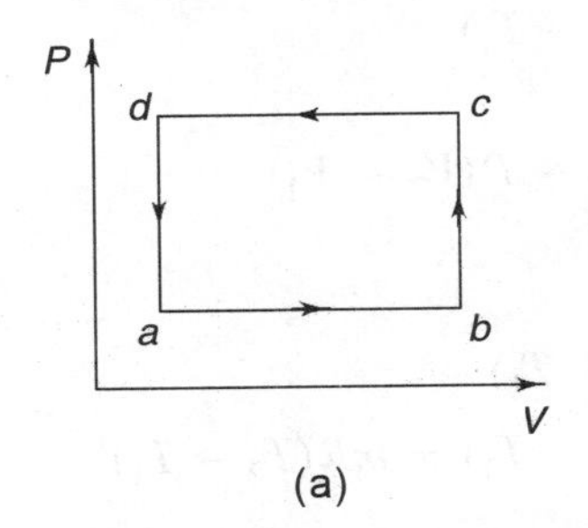

(a)

Work = area *abcd* and negative

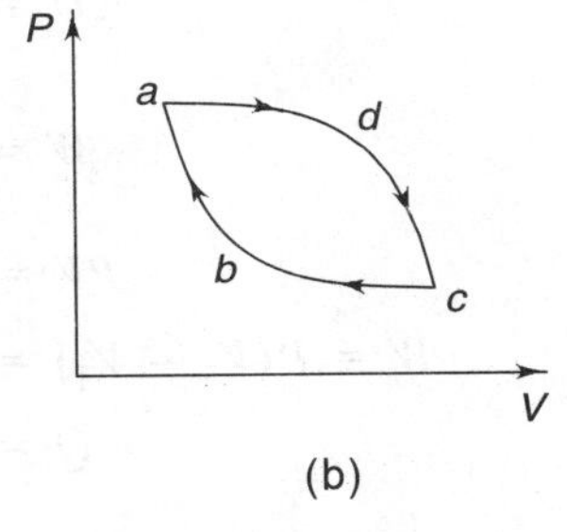

(b)

Work = area *abcd* and positive

FIGURE 2.1 Cyclic process.

If we take a system from one state to another by different processes, then the heat transferred Q and work done W are different for different paths. However, the difference $Q - W$ is the same for all processes. $Q - W$ is defined as change of internal energy (ΔU). Hence internal energy (U) of a system is a unique function and depends upon the state. Change of internal energy is zero if a system comes back to its original state.

APPLICATION OF FIRST LAW OF THERMODYNAMICS

Isolated system: An isolated system can neither take heat nor do any work ($Q = 0$, $W = 0$). Hence change of internal energy is also zero. The internal energy of an isolated system remains constant.

Consider an isolated system consisting of a vessel having two portions (Figure 2.2). The left portion has gas at pressure and the right portion has vacuum. Now the partition between the portions is removed. The gas expands to fill up the vessel. Since expansion of gas has taken place without any resistance like atmospheric pressure, gas has not performed any work in spite of expansion. This process is known as *free expansion*.

In this system:

$$W = 0,\ Q = 0,\ \Delta U = 0$$

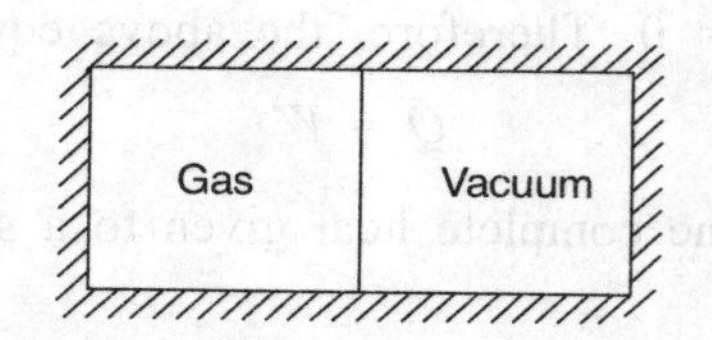

FIGURE 2.2 Isolated system.

Isobaric process: In an isobaric process, the process undergoes changes at constant pressure. Evaporation of water at the boiling point and freezing of water at the freezing point are examples of an isobaric process. In this process, heat is used both for increasing internal energy and change in work.

$$Q = \Delta U + W$$

$$\Delta U = m\ c_v(T_2 - T_1)$$

and
$$W = \int_1^2 PdV = P(V_2 - V_1)$$

Now
$$PV = mRT$$

or
$$W = P(V_2 - V_1) = mR(T_2 - T_1)$$

Therefore,
$$Q = m\ c_v(T_2 - T_1) + mR(T_2 - T_1)$$

$$= m\ \frac{R}{\gamma - 1}\ (T_2 - T_1) + mR(T_2 - T_1) \quad \left(\because\ c_v = \frac{R}{\gamma - 1}\right)$$

$$= mR\ (T_2 - T_1)\left(\frac{1}{\gamma - 1} + 1\right)$$

Isochoric process: An isochoric process undergoes charges at constant volume. Since $\Delta V = 0$,

$$W = \int PdV = 0$$

Therefore,
$$Q = \Delta U$$

Hence entire heat in an isochoric process used in increasing internal energy, i.e.

$$Q = \Delta U = m\ c_v(T_2 - T_1)$$

Isothermal process: In an isothermal process, temperature remains constant. Since internal energy depends upon temperature,

$$\Delta U = 0$$

$$Q = W = \int_1^2 PdV = \int_1^2 \frac{\text{constant}}{V} \times dV \quad (\text{Since } PV = \text{constant})$$

Therefore, $$Q = P_1 V_1 \ln \frac{V_2}{V_1} = P_1 V_1 \ln \frac{P_1}{P_2}$$

In an isothermal process, heat given to system is completely used for doing work.

Adiabatic process: In an adiabatic process, heat neither enters nor leaves the system, i.e., $Q = 0$.

Therefore, $$Q = \Delta U + W = 0$$

or $$W = -\Delta U$$

while $$W = \int_1^2 PdV = \int_1^2 \frac{\text{constant}}{V^\gamma} dV \quad (\text{as } PV^\gamma = \text{constant})$$

Therefore, $$W = \frac{P_1V_1 - P_2V_2}{\gamma - 1}$$

Internal energy: Internal energy decreases if the system does work and increases if work is done on the system. If we shake a thermos, work is done on the thermos and internal energy of the content of the thermos increases (result is increase of temperature of the content of the thermos). On the other hand if a tyre bursts, air expands adiabatically doing work against atmosphere. For doing work, air uses its internal energy resulting into the decrease of internal energy and temperature of the air.

Polytropic process: A polytropic obeys the equation

$$PV^n = \text{constant}$$

$$Q = \Delta u + W$$

Here $$\Delta U = m\ c_v(T_2 - T_1)$$

$$W = \int_1^2 PdV = \int_1^2 \frac{\text{constant}}{V^n} dV \ (\text{as } PV^n = \text{constant})$$

$$= \frac{P_1V_1 - P_2V_2}{n-1} = \frac{mR(T_1 - T_2)}{n-1}$$

Therefore, $$Q = m\, c_v(T_2 - T_1) + \frac{mR(T_1 - T_2)}{n-1} \quad (\because R = c_v(\gamma - 1))$$

or $$Q = m\, c_v(T_2 - T_1)\left(\frac{n-\gamma}{n-1}\right)$$

A polytropic process gives different curves depending upon the value of n. The prominent values of n are as follows:

$n = 0 \quad PV^0 = P = \text{constant} \rightarrow$ isobaric process

$n = 1 \quad PV^1 = PV = \text{constant} \rightarrow$ isothermal process

$n = \alpha \quad PV^\alpha = V = \text{constant} \rightarrow$ isochoric process

FLOW PROCESS

A piston having area A pushes the fluid in a pipe for a distance L against pressure P (Figure 2.3). The effort is required to flow the fluid is

$$\text{Effort} = \text{force} \times \text{distance}$$
$$= (P \times A) \times L, \quad (\because \text{volume } V = A \times L)$$
$$= P \times V$$

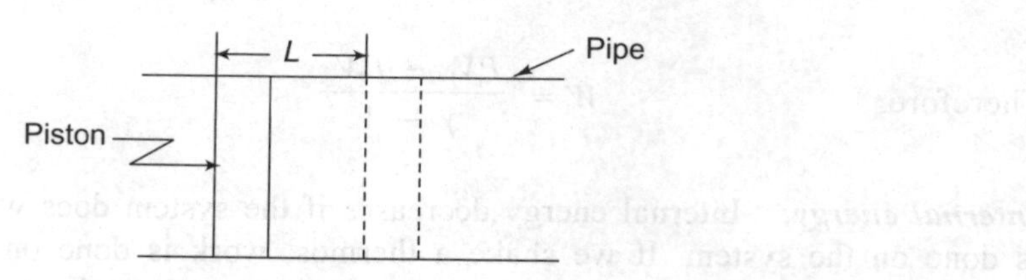

FIGURE 2.3 Flow work.

A flow process is a process in which fluid enters the system and leaves it after work interaction. As mass interaction takes place, hence each mass fraction entering or leaving the system either performs work on the system or system performs work on it. Not only that but also a portion of work is required to push the fluid into system and out of the system. This portion of work required for the flow is called *flow work*. Flow work is the product of pressure and volume, i.e. PV.

A non-flow process is a process in which no mass interaction takes place across the boundary of the system. Flow work in this case is zero.

ENTHALPY

Enthalpy of a system is quantification of energy content in it which is given by summation of internal energy and flow energy.

$$H = U + PV$$

$$h = u + Pv \quad \text{(for unit mass)}$$

On differentiating the above equation

$$dh = du + Pdv + vdP$$

At constant pressure

$$vdP = 0$$

Therefore
$$dh = du + Pdv = dQ$$

At constant pressure, change of enthalpy is heat interaction.

STORED ENERGY

At each state, system has some energy which is called *stored energy* and it is the sum of internal energy, kinetic energy and potential energy. Mathematically,

$$E = U + \frac{mc^2}{2} + mgz$$

or
$$e = u + \frac{c^2}{2} + gz \quad \text{(for unit mass)}$$

STEADY FLOW ENERGY EQUATION

Steady flow refers to flow in which its properties at any point remains constant with time (Figure 2.4). As energy is conserved, inlet total energy plus heat is equal to outlet total energy plus work. When the energy balance is applied on an open system, steady flow energy equation (SFEE) is obtained. SFEE for the open system is given by

$$Q + m\left(h_1 + \frac{C_1^2}{2} + gz_1\right) = W + m\left(h_2 + \frac{C_2^2}{2} + gz_2\right) \tag{2.1}$$

where the inlet is state 1 and outlet is state 2

h = enthalpy
C = velocity
z = potential height
Q = heat and
W = work

For unit mass the equation becomes,

$$q + \left(h_1 + \frac{C_1^2}{2} + gz_1\right) = w + \left(h_2 + \frac{C_2^2}{2} + gz_2\right)$$

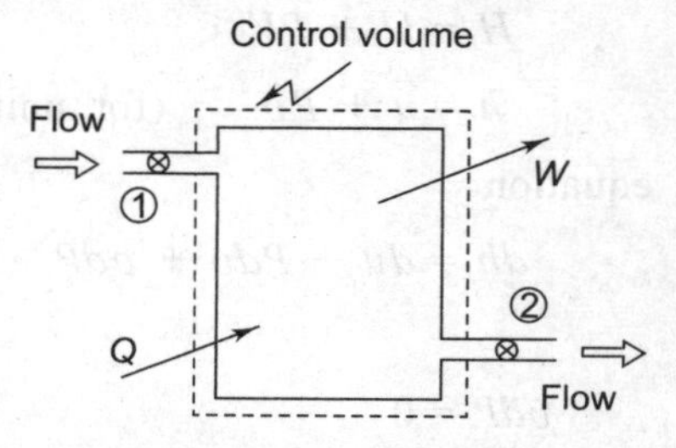

FIGURE 2.4 Steady flow.

Turbine: A turbine is a device in which fluid with high temperature and pressure is expanded to low temperature and pressure resulting into positive work at the turbine shaft (Figure 2.5). Therefore, the turbine is a work-producing device. Expansion should be adiabatic for maximum work output. The steam turbine is a turbine using steam as fluid. The turbine has a smaller cross section at the inlet and a bigger cross section at the outlet so that fluid can expand to perform work.

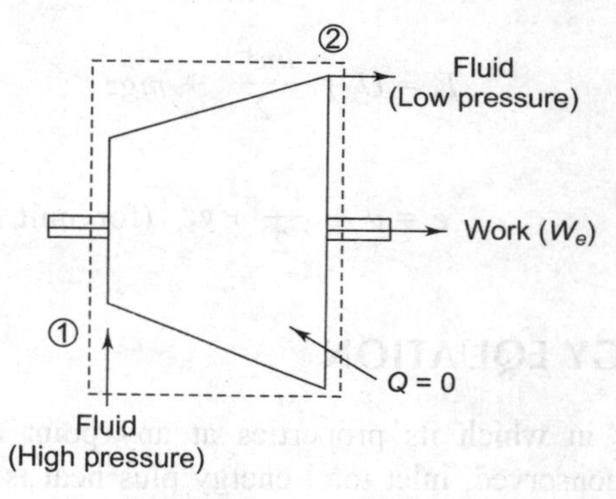

FIGURE 2.5 Turbine.

Steady flow energy equation (SFEE) for the turbine as an open system neglecting KE and PE is

$$Q + mh_1 = W_e + mh_2$$

As $Q = 0$, the above equation becomes

$$W_e = m(h_1 - h_2)$$
$$= mc_p(T_1 - T_2)$$

Compressor: A compressor (Figure 2.6) is a device for compressing or increasing the pressure of fluid by applying external work. Increase of pressure is accompanied with increase of temperature. If compression is adiabatic, then minimum work is to be applied for compression to increase pressure. The compressor has a bigger cross section at the inlet and a smaller cross section at the outlet. In a window airconditioner, the compressor is used to compress refrigerant vapour from low pressure to high pressure so that vapour at high pressure can be liquefied in the condensor at surrounding temperature.

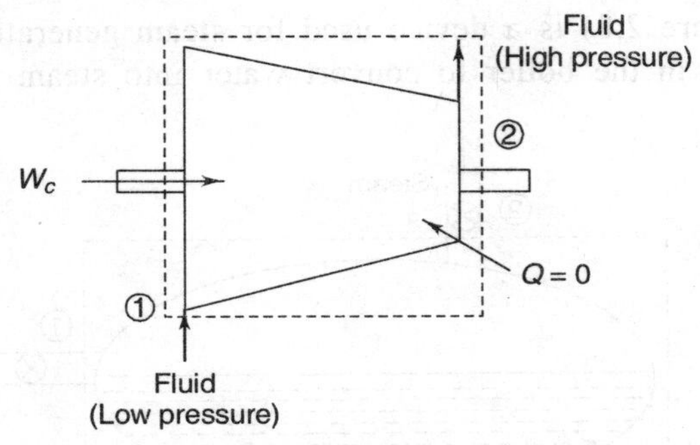

FIGURE 2.6 Compressor.

Applying a steady flow energy equation on the compressor as the system, we get

$$Q + mh_1 = -W_c + mh_2$$

As $Q = 0$,

$$W_c = m(h_2 - h_1)$$
$$= mc_p(T_2 - T_1)$$

Pump: A pump (Figure 2.7) is a device used for pumping liquid.

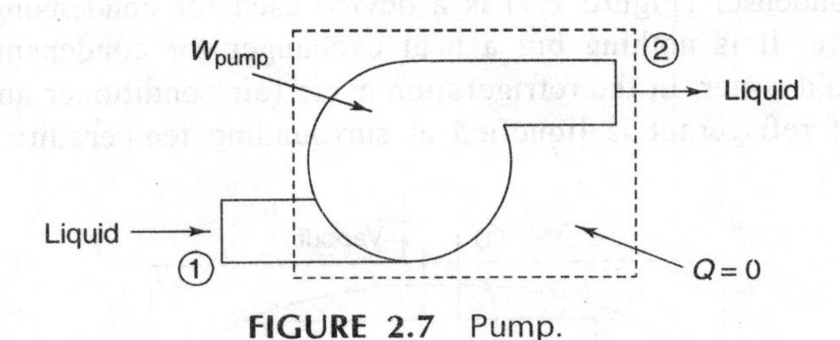

FIGURE 2.7 Pump.

Applying the SFEE on a pump, we get

$$Q + m\left(h_1 + \frac{C_1^2}{2} + gz_1\right) = -W_{\text{pump}} + m\left(h_2 + \frac{C_2^2}{2} + gz_2\right)$$

As $Q = 0$,

$$W_{\text{pump}} = m\left\{(h_2 - h_1) + \frac{c_2^2 - c_1^2}{2} + g(z_2 - z_1)\right\}$$

$$h_2 - h_1 = dh = du + d(Pv)$$
$$= du + v(P_2 - P_1) \text{ (as } v \text{ is almost constant, for liquid)}$$
$$du = 0 \text{ (if temperature is constant)}$$

Therefore, $W_{\text{pump}} = mv(P_2 - P_1)$ (if $C_1 = C_2$ and $Z_1 = Z_2$)

Boiler: A boiler (Figure 2.8) is a device used for steam generation at constant pressure. Heat is given to water in the boiler to convert water into steam at constant pressure.

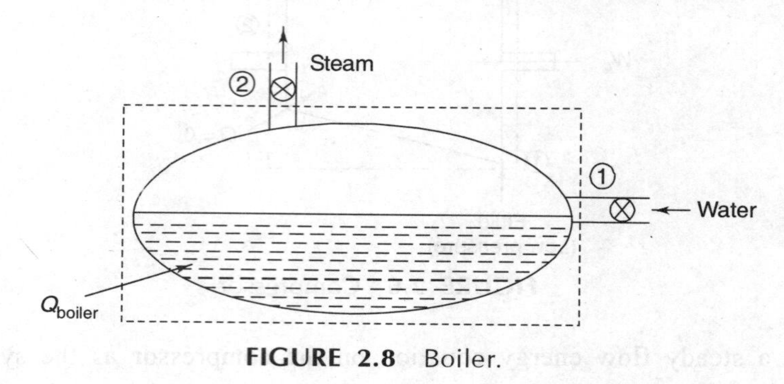

FIGURE 2.8 Boiler.

Applying SFEE on the boiler, we get

$$Q_{boiler} + mh_1 = mh_2$$

$$Q_{boiler} = m(h_2 - h_1)$$

$$= mc_p(T_2 - T_1)$$

Condenser: A condenser (Figure 2.9) is a device used for condensing vapour into liquid at constant pressure. It is nothing but a heat exchanger for condensing vapour by taking out heat through cold water. In the refrigeration cycle (air conditioner and refrigerator), high pressure vapour of refrigerant is liquefied at surrounding temperature in the condenser.

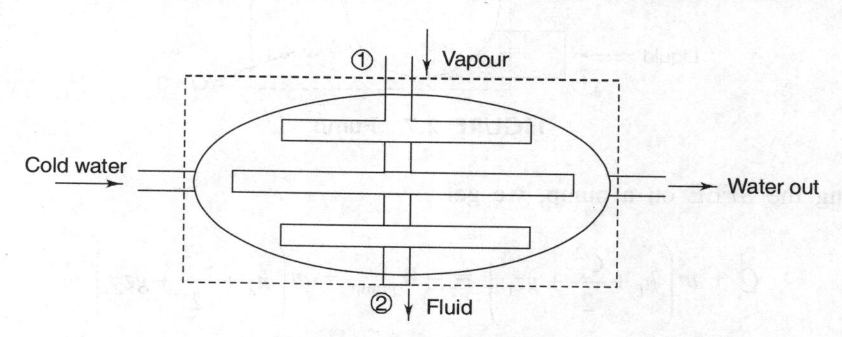

FIGURE 2.9 Condenser.

Applying SFEE on the condenser, we get

$$Q = m(h_1 - h_2) \text{ as } W = 0, \Delta\text{KE and } \Delta\text{PE} = 0$$

$$Q = \text{heat removed by cold water}$$

Nozzle and Diffuser: A nozzle (Figure 2.10) is a device for increasing the velocity of fluid at the expense of its pressure drop. The enthalpy of fluid decreases as velocity of fluid increases. The velocity of fluid increases from the inlet to the exit. For subsonic flow, the nozzle has a

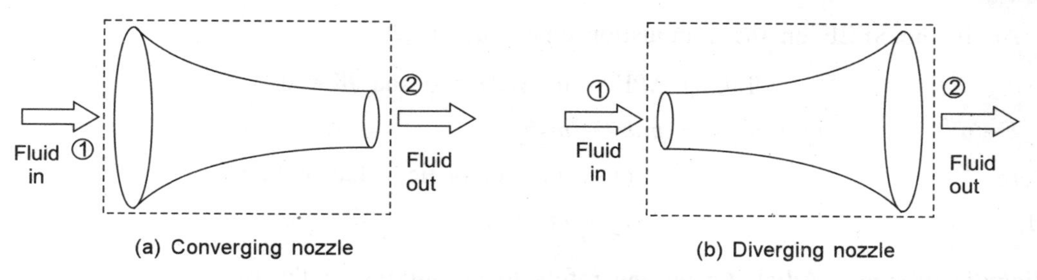

FIGURE 2.10 Nozzle.

converging cross section whereas for supersonic flow, it has a diverging cross section. The nozzle is used in a jet engine to get high thrust. Thrust is equal to the change of momentum and it depends upon higher outlet velocity. Similarly, in an impulse turbine, higher velocity of outlet steam from the nozzle gives higher output. A diffuser is a device which is required to perform the opposite of a nozzle. It has also a passage of varying cross section that serves to achieve reduction in velocity of the flowing fluid to gain pressure.

Applying the SFEE on the nozzle, we get

$$h_1 + \frac{C_1^2}{2} = h_2 + \frac{C_2^2}{2} \quad (\text{as } Q = 0,\ W = 0,\ \Delta\text{PE} = 0)$$

Generally initial velocity (C_1) is very small. Hence, $C_1 = 0$. Therefore, the above equation becomes

$$C_2 = \sqrt{2(h_1 - h_2)}$$

$$= \sqrt{2c_p(T_1 - T_2)}$$

Combustion chamber: A combustion chamber (Figure 2.11) is commonly used in gas turbine installation. Fuel is injected at high pressure into the chamber when chamber has high pressure and temperature air in it. Ignition of air and fuel mixture takes place.

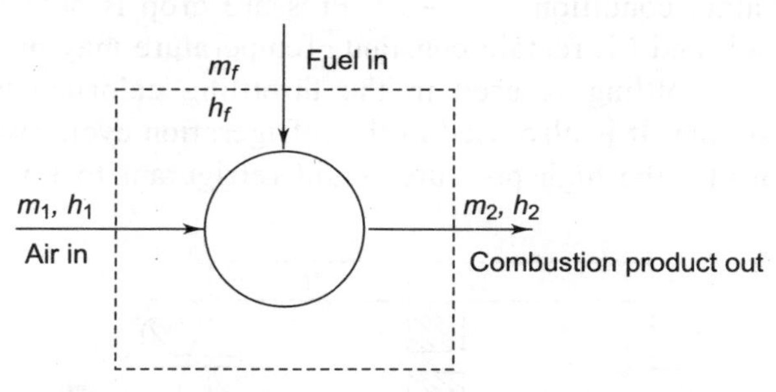

FIGURE 2.11 Combustion chamber.

Apply the SFEE on the combustion chamber, to get

Taking $\Delta PE = 0$, $\Delta KE = 0$ and $W = 0$,

$$Q + m_1 h_1 + m_f h_f = m_2 h_2$$

where $$Q = m_f \times \text{calorific value of fuel}$$

and $$m_2 = m_1 + m_f$$

Adiabatic mixing: Adiabatic mixing refers to the mixing of the two or more streams of the same or other fluids under adiabatic conditions (Figure 2.12).

Applying the SFEE on mixing, we get

$$m_1 h_1 + m_2 h_2 = m_3 h_3 \qquad [\because Q = 0 \text{ (adiabatic process)}]$$

or $$m_1 c_p T_1 + m_2 c_p T_2 = m_3 c_p T_3$$

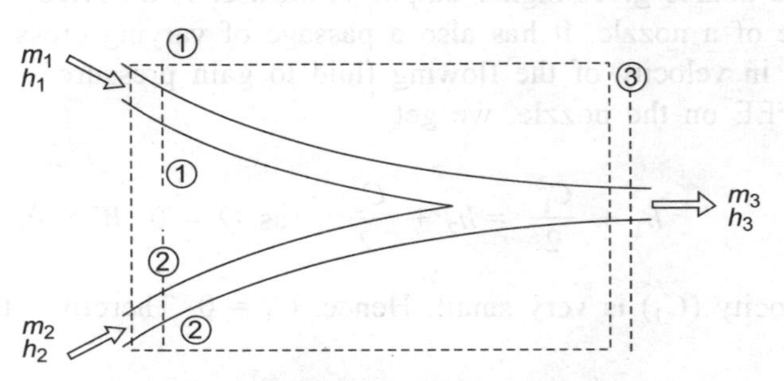

FIGURE 2.12 Adiabatic mixing.

If fluid is the same, then

$$T_3 = \frac{m_1 T_1 + m_2 T_2}{m_3}$$

Throttling: Throttling (Figure 2.13) is a process in which a fluid passes through a restricted opening under isoenthalpic condition ($\Delta h = 0$). Pressure drop is achieved without work or heat interaction. Also KE and PE remain constant. Temperature may drop or increase during the throttling process. Throttling is used in the throttling calorimeter for measuring the dryness factor of wet stream. It is also used in the refrigeration cycle (window airconditioner and refrigerator) to throttle the high pressure liquid refrigerant to low pressure evaporator to extract heat.

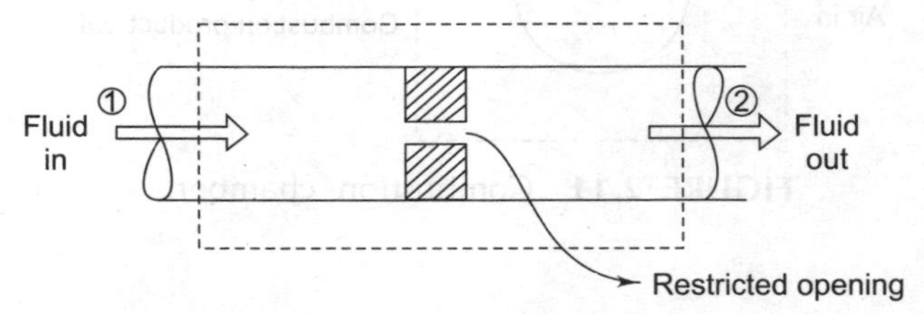

FIGURE 2.13 Throttling.

Applying the SFEE on the throttle, we get

$$h_1 = h_2$$

where $p_1 >> p_2$ and $Q = 0$, $W = 0$.

LIMITATIONS OF FIRST LAW OF THERMODYNAMICS

The first law of thermodynamics has the following limitations:

1. The first law of thermodynamics does not differentiate between work and heat. It assumes full convertibility of one with the other. Though full convertibility of work into heat is possible as work is high grade energy, but full convertibility of heat (low grade energy) into work (high grade energy) is not possible.
2. It does not explain the direction of a process. It permits (even theoretically) heat transfer from a low-temperature body to a high-temperature body which (we all know) cannot take place in general life.

PERPETUAL MOTION MACHINE

The perpetual (continuous) motion machine of the first kind (PMM–I) is a hypothetical device conceived based on violation of the first law of thermodynamics. Figure 2.14(a) shows a device which is continuously producing work ($W \neq 0$) without any supply of energy ($Q = 0$). This device is a PMM–I as it violates the first law of thermodynamic as well as the law of conservation of energy. Similarly, Figure 2.14(b) shows a device which is continuously generating heat ($Q \neq 0$) without any other form of energy ($W = 0$) supplied to it. Hence it is a PMM–I.

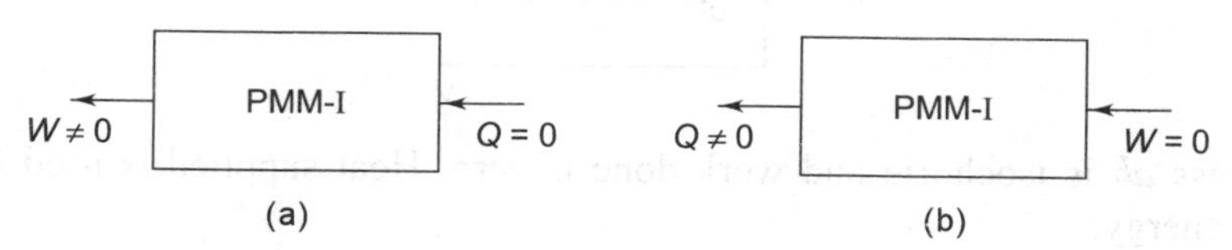

FIGURE 2.14 Perpetual motion machine.

SOLVED PROBLEMS

1. Calculate the work done by a gas as it is taken from state a to b, b to c and c to a. Process ab is a constant pressure process. Work done by the system is the area under line ab:

$$W_{ab} = 100 \times (4 - 2) = 200 \text{ kJ}$$

Process bc is a constant volume process and $W_{bc} = 0$.

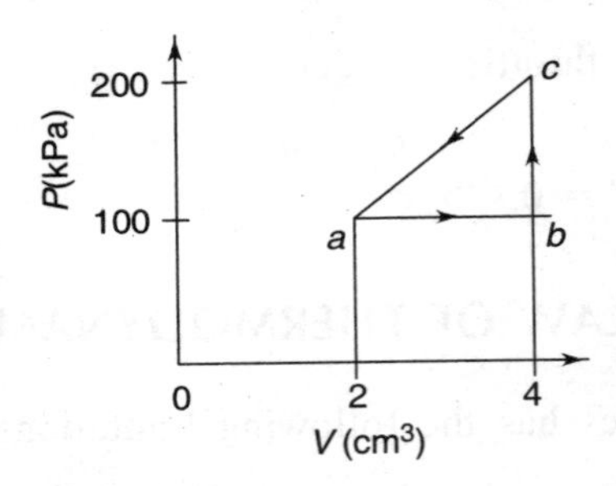

Process *ca* has volume reducing, and work done on the system is the area under line *ac*:

$$W_{ac} = -\left[\tfrac{1}{2}(200-100)\times(4-2)+100\times(4-2)\right]$$

$$= -[100 + 200] = -300 \text{ kJ}$$

Work done in the cycle = 200 – 300 = –100 kJ

The negative sign shows that work is done on the system which is also evident as the cycle is traced anticlockwise.

2. A system is taken through a cyclic process. It absorbs 100 kJ of heat in process *ab*, in process *bc* the system is compressed adiabatically, and in process *ca* the system rejects 120 kJ. If 70 kJ work is done during process *bc*, find the internal energy of the system at *b* and *c*, the internal energy at *a* is 150 kJ. Also find work done by the system during process *ca*.

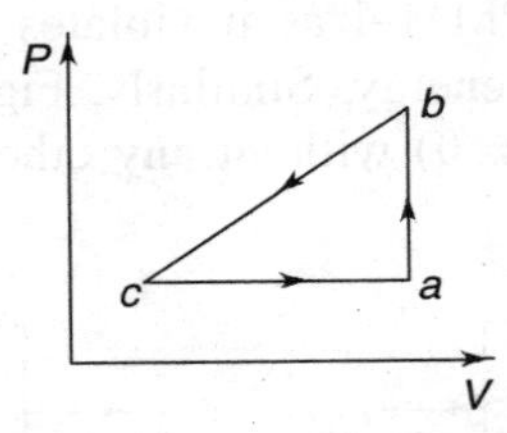

Process *ab* is isochoric and work done is zero. Heat supplied is used for increasing internal energy:

$$\Delta U_{ab} = Q_{\text{add}} = 100 \text{ kJ}$$

Therefore,

$$U_b - U_a = 100 \text{ kJ}$$

$$U_b - 150 = 100 \text{ kJ}$$

$$U_b = 250 \text{ kJ}$$

Process *cb* is adiabatic, i.e. $Q = 0$.

$$Q = \Delta U + W \text{ (volume decreasing)}$$

$$0 = \Delta U_{cb} - 70$$

Therefore,

$$u_c - u_b = 70$$

$$u_c - 250 = 70$$

$$u_c = 250 + 70 = 320 \text{ kJ}$$

Process *ca* is isobaric and volume is increasing. Therefore, work will be done by the system.

$$Q_{ca} = \Delta U_{ca} + W_{ca}$$
$$-Q_{ca} = (U_a - U_c) + W_{ca}$$
$$-120 = (150 - 320) + W_{ca}$$
$$W_{ca} = (170 - 120) \text{ kJ}$$
$$= 50 \text{ kJ}$$

3. Find out work done in processes *ab*, *bc*, *cd* and *da*. Calculate also work done in complete cycle *abcda*.

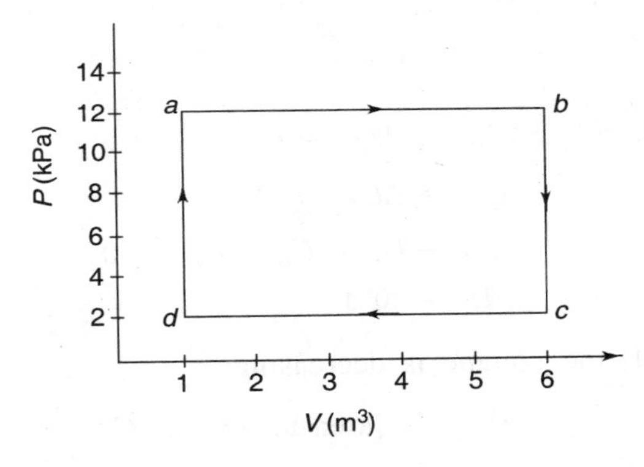

Process *ab* is an isobaric process and work done by the system is the area under line *ab*:

$$W_{ab} = P_a \times (V_b - V_a)$$
$$= 12 \times (6 - 1)$$
$$= 60 \text{ kJ}$$

Process *bc* is isochoric and $W_{bc} = 0$.

Process *cd* is isobaric and volume is reducing. Therefore, work done on the system is the area under line *cd*:

$$W_{cd} = -P_d(V_c - V_d)$$
$$= -2(6 - 1) = -10 \text{ kJ}$$

Process *da* is isochoric and $W_{da} = 0$.

$$\text{Cyclic work} = W_{ab} + W_{cd}$$
$$= 60 - 10$$
$$= 50 \text{ kJ}$$

Work is positive as evident from the cycle which is traced clockwise.

4. A system undergoes state A to states B and C and returns to A. If $U_A = 0$, $U_B = 30$ J and the heat given to the system in process BC is 50 J, then determine (i) internal energy at state C, (ii) heat given into the system in process AB, (iii) heat given into the system in process CA, and (iv) net work done in the complete cycle.

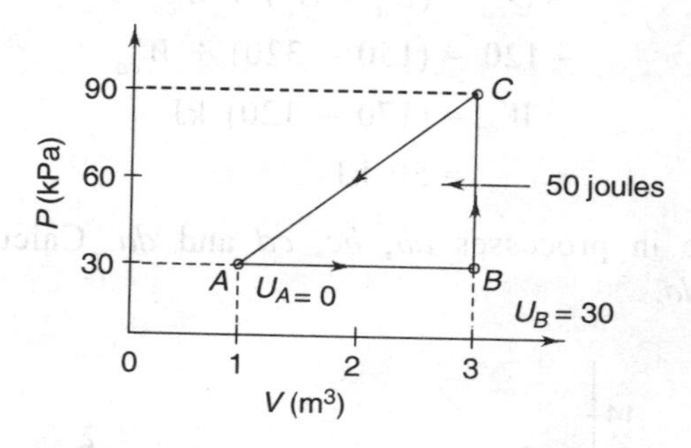

Process BC is isochoric and hence $W_{BC} = 0$.

$$Q_{BC} = \Delta U_{BC} + W_{BC}$$
$$50 = U_C - U_B = U_C - 30$$

or
$$U_C = 80 \text{ J}$$

Now for process CA, the volume is decreasing:

$$W_{CA} = -\text{ [area under line } CA]$$
$$= -\left[\frac{1}{2} \times AB \times BC + \text{area under } AB\right]$$
$$= -\left[\frac{1}{2} \times 2 \times 60 + 30 \times 2\right]$$
$$= -(60 + 60) = -120 \text{ J}$$

For process AB:

$$W_{AB} = \text{area under } AB = 60 \text{ J}$$
$$Q_{AB} = \Delta U_{AB} + W_{AB}$$
$$= (U_B - U_A) + W_{AB}$$
$$= (30 - 0) + 60 = 90 \text{ J}$$

For process CA:

$$Q_{CA} = \Delta U_{CA} + W_{CA}$$
$$= (U_A - U_C) + W_{CA}$$
$$= (0 - 80) - 120 = -200 \text{ J}$$

Q_{CA} is negative and hence heat 200 J is rejected by the system. Work done in the whole cycle is the area enclosed = 60 J. Work is positive as the cycle is traced clockwise.

5. As shown in the *PV* diagram, 100 joules of heat is given in taking a system from state *A* to state *C* along path *ADC* and 50 joules of work is done by the system. (i) If the work done by the system is 15 joules along path *ABC*, then how much heat is given in taking the system from *A* to *C*? (ii) How much heat will be absorbed or given out if the work done on the system along curved path from *C* to *A* is 15 joules, and (iii) If $U_B - U_A = 40$ joules, then how much heat will be absorbed in each of the processes *AB* and *BC*?

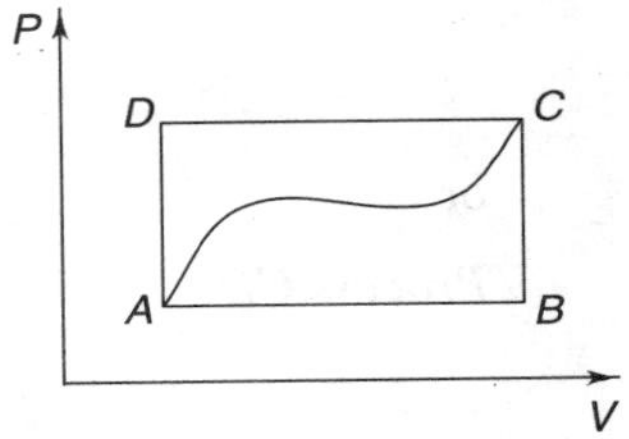

P, V, A, B, C, D

Process ADC: $Q = 100$ J, $W = 50$ J

$$Q = \Delta U_{AC} + W$$

$$\Delta U_{AC} = Q - W = 100 - 50$$

$$= 50 \text{ J}$$

Process ABC: $W = 15$ J, $\Delta u_{AC} = 50$ J

$$Q = \Delta u + W$$

$$= 50 + 15 = 65 \text{ J}$$

Process CA (curved): $W = -15$, $\Delta U_{CA} = -50$

$$Q = \Delta U_{CA} + W$$

$$= -50 - 15 = -65 \text{ J}$$

Process AB: $\Delta U_{ab} = 40$ J

$$W_{ABC} = W_{AB} \ (\because \ W_{BC} = 0)$$

$$= W_{AB} = 15$$

$$Q_{AB} = \Delta U_{AB} + W_{AB}$$

$$= (U_B - U_A) + W_{AB} = 40 + 15 = 55 \text{ J}$$

Process BC: $Q_{ABC} = 65$ and $Q_{AB} = 55$

$$Q_{BC} = Q_{ABC} - Q_{AB}$$

$$= 65 - 55$$

$$= 10 \text{ J}$$

6. The values of internal energy of a system at states *A*, *B* and *C* are 10 J, 30 J and 200 J respectively. Heat 5 J is released in process *DA*. Determine (i) internal energy of state *D*, (ii) heat released in process *CD*, (iii) heat released in process *AB*, and (iv) heat absorbed in process *BC*.

Process DA is an isochoric process. Therefore, $W = 0$.

$$Q = -5 \text{ J (Given)}$$

$$\Delta U_{DA} = Q - W$$

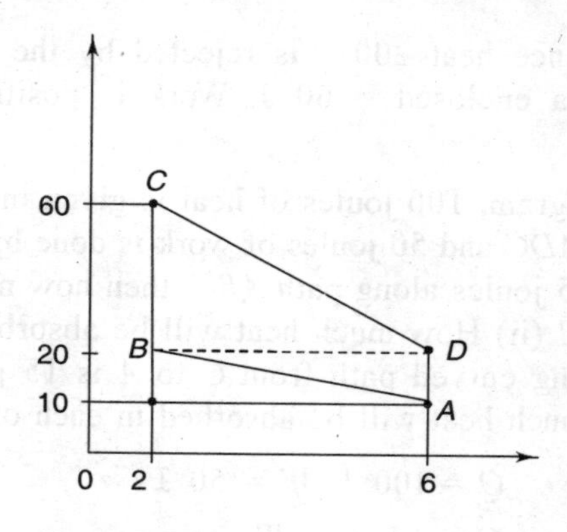

$$U_A - U_D = -5$$

or $$U_D = U_A + 5 = 10 + 5 = 15 \text{ J}$$

Process CD:

$$W_{CD} = \text{Area under line } CD$$
$$= \frac{1}{2} BD \times BC + \text{area under line } BD$$
$$= \frac{1}{2} \times (6 - 2) \times (60 - 20) + 20 \times (6 - 2)$$
$$= 80 + 80 = 160 \text{ J}$$
$$Q = \Delta U_{CD} + W_{CD}$$
$$= (U_D - U_C) + W_{CD}$$
$$= (15 - 200) + 160$$
$$= -185 + 160 = -25 \text{ J}$$

Process AB: $$\Delta U_{AB} = U_B - U_A$$
$$= 30 - 10 = 20 \text{ J}$$
$$W_{AB} = \text{Area under line } AB$$
$$= \frac{1}{2} \times (20 - 10) \times (6 - 2) + 10 \times (6 - 2)$$
$$= 20 + 40 = 60$$

Since the volume is decreasing and hence $W_{AB} = -60$.

$$Q = \Delta U_{AB} + W_{AB}$$
$$= 20 - 60 = -40 \text{ J}$$

Process BC: It is an isochoric process and hence $W = 0$.

$$\Delta U_{BC} = U_C - U_B = 200 - 30$$
$$= 170 \text{ J}$$

Therefore, $$Q = \Delta U + W$$
$$= 170 \text{ J}$$

7. Calculate the increase in internal energy of 1 kg of water at 100°C when it is converted into steam at 100°C and atmospheric pressure (100 kPa). $\rho_{water} = 1000$ kg/m^3, $\rho_{steam} = 0.6$ kg/m^3 and $h_{conversion} = 2.25 \times 10^6$ J/kg

Volume will increase when water is converted into steam.

$$V_{steam} = \frac{1}{\rho_{steam}} = \frac{1}{0.6} \text{ m}^3 = 1.67 \text{ m}^3$$

$$V_{water} = \frac{1}{\rho_{water}} = \frac{1}{1000} = 10^{-3} \text{ m}^3$$

$$V_{steam} - V_{water} = 1.67 - 0.001$$
$$\approx 1.67 \text{ m}^3$$

$$\text{Work done} = P_{atm}\,(V_{steam} - V_{water})$$
$$= 100 \times 1.67 \text{ kJ}$$
$$= 167 \text{ kJ}$$

$$Q = \text{heat given for conversion} = 2.25 \times 10^6 \text{ J}$$
$$= 2250 \text{ kJ}$$

$$Q = \Delta U + W$$

or $$\Delta U = Q - W = 2250 - 167$$
$$= 2083 \text{ kJ}$$

8. In a nozzle, air at 827°C and atmospheric pressure enters with negligible velocity (C_1) and leaves at a temperature of 27°C. Determine outlet velocity (C_2) if $\Delta Q = 0$ and $\Delta z = 0$. Take $c_p = 1$ kJ/(kg°K).

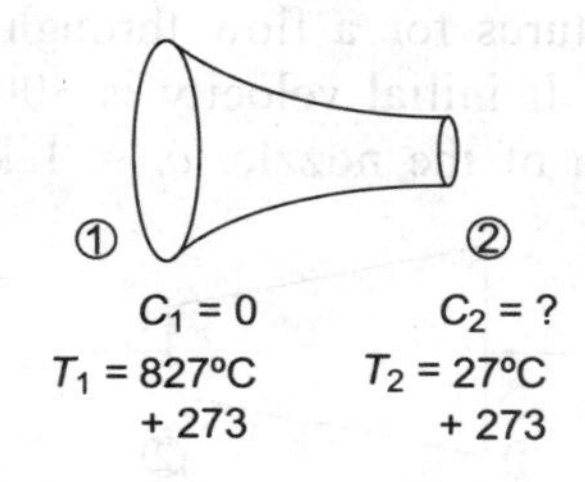

$$C_2 = \sqrt{2 c_p (T_2 - T_1)}$$

$$= \sqrt{2 \times 10^3 (1100 - 300)} = \sqrt{16 \times 10^5}$$

$$= 1265 \text{ m/s}$$

9. An air compressor takes shaft work of 200 kJ/kg and compression increases enthalpy by 100 kJ/kg of air. Cooling water picks up 90 kJ/kg of air of heat while cooling. Determine the heat transferred from the compressor to the atmosphere.

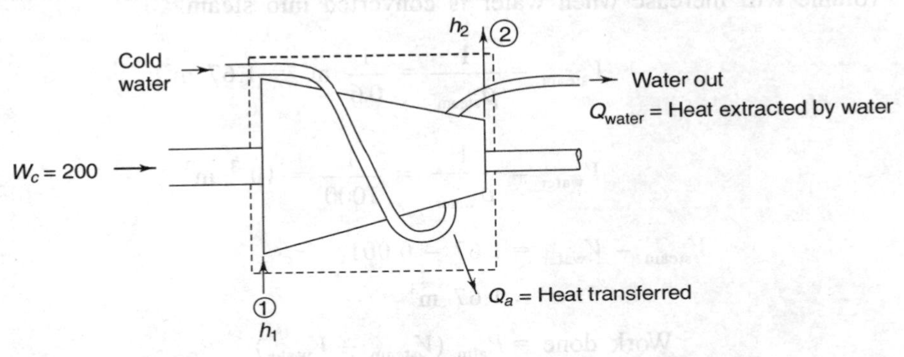

$$Q = \text{heat given out by the system} = Q_a + Q_w$$

Applying SFEE on the compressor,

$$Q + h_1 + \frac{C_1^2}{2} + gz_1 = W_c + h_2 + \frac{C_2^2}{2} + gz^2$$

$$z_1 = z_2,\ C_1 = C_2,\ W_c = -200,\ h_2 - h_1 = 100$$

$$Q = (h_2 - h_1) - 200$$

$$= 100 - 200 = -100 \text{ kJ}$$

$$Q = Q_a + Q_{\text{water}}$$

$$-100 = Q_a - 90$$

or $$Q_a = -10 \text{ kJ}$$

10. The inlet and outlet temperatures for a flow through a nozzle are 400 K and 300 K. Mass flow of air is 1.0 kg/s. If initial velocity is 300 m/s, determine exit velocity and the ratio of inlet to exit area of the nozzle. $c_p = 1$ kJ/(kg K).

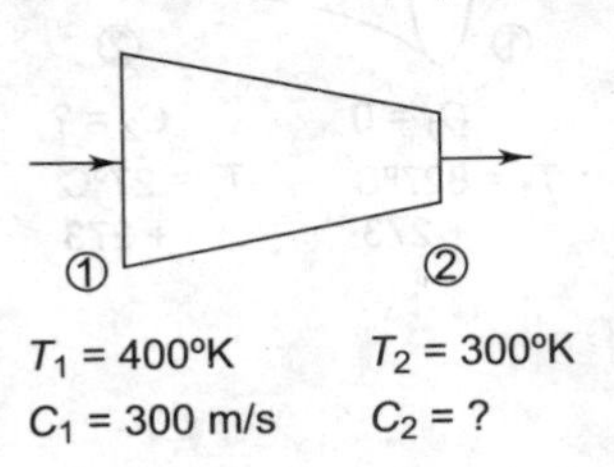

Applying the SFEE on the nozzle, we get

$$Q + m\left(h_1 + \frac{1}{2}C_1^2 + g z_1\right) = W + m\left(h_2 + \frac{C_2^2}{2} + g z_2\right)$$

$Q = 0,\ W = 0,\ z_1 = z_2,\ m_1 = m_2 = m = 1.0$

$$\frac{1}{2}(C_2^2 - C_1^2) = h_1 - h_2$$

$$= c_p(T_1 - T_2)$$

$$= 1 \times 10^3\ (400 - 300)$$

$$= 10^5$$

$$C_2^2 = 2 \times 10^5 + C_1^2 = 2 \times 10^5 + 9 \times 10^4$$

$$= 29 \times 10^4$$

or $$C_2 = 451 \text{ m/s}$$

Volume of air at the inlet = Volume of air at the outlet

Area at the inlet × velocity of the inlet = Area at the outlet × velocity at the outlet

Mathematically

$$A_1 \times C_1 = A_2\ C_2$$

or $$\frac{A_1}{A_2} = \frac{C_2}{C_1} = \frac{451}{300} \simeq \frac{1.5}{1}$$

Therefore, $$A_1 : A_2 :: 1.5 : 1$$

11. An air compressor compresses air at 0.1 MPa and 27°C by ten times the inlet pressure. During compression the heat loss to the surroundings is estimated to be 5% compression work. Air enters the compressor with a velocity of 40 m/s and leaves with 100 m/s. Inlet and outlet cross-section areas are 100 cm² and 20 cm² respectively. Estimate the temperature of air at the exit and power input to the compressor.

(UPTU: July 2002)

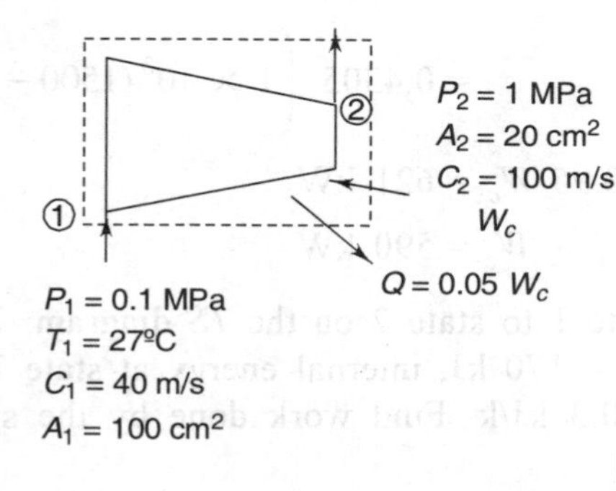

First we find out v_1 (specific volume) at the inlet so as to find mass flow (m_1).

$$P_1 v_1 = RT_1$$

$$\therefore \quad v_1 = \frac{RT_1}{P_1} = \frac{287 \times 300}{0.1 \times 10^5} = 0.861 \text{ m}^3/\text{kg}$$

$$m_1 = \frac{\text{volume/s}}{v_1} = \frac{A_1 C_1}{v_1} = \frac{(100 \times 10^{-4}) \times 40}{0.861}$$

$$= 0.4646 \text{ kg/s}$$

$$m_1 = m_2 = m = 0.4646 \text{ kg/s}$$

$$v_2 = \frac{A_2 C_2}{m} = \frac{(20 \times 10^{-4}) \times 100}{0.4646} = 0.4305 \text{ m}^3/\text{kg}$$

Now at the exit $\quad P_2 v_2 = RT_2$

or $\quad T_2 = \dfrac{P_2 v_2}{R} = \dfrac{1 \times 10^6 \times 0.4305}{287}$

$$= 1500 \text{ K or } (1227°\text{C})$$

Applying the SFEE on the compressor, we get

$$Q + m\left(h_1 + \frac{C_1^2}{2} + gz_1\right) = -W_c + m\left(h_2 + \frac{C_2^2}{2} + gz_2\right)$$

$$z_1 = z_2$$

$$Q + W_c = m(h_2 - h_1) + \frac{m}{2}(C_2^2 - C_1^2)$$

$$-0.05\, W_c + W_c = 0.4305\left(c_p (T_2 - T_1) + \frac{C_2^2 - C_1^2}{2}\right)$$

$$= 0.4305\left(1 \times 10^3 (1500 - 300) + \frac{100^2 - 40^2}{2}\right)$$

$$0.95\, W_c = 621 \text{ kW}$$

or $\quad W_c = 590 \text{ kW}$

12. A system moves from state 1 to state 2 on the *TS* diagram. $T_1 = 330$ K, $T_2 = 440$ K, internal energy at state 1 = 170 kJ, internal energy at state 2 = 190 kJ, entropy S_1 = 0.23 kJ/k, entropy S_2 = 0.3 kJ/k. Find work done by the system.

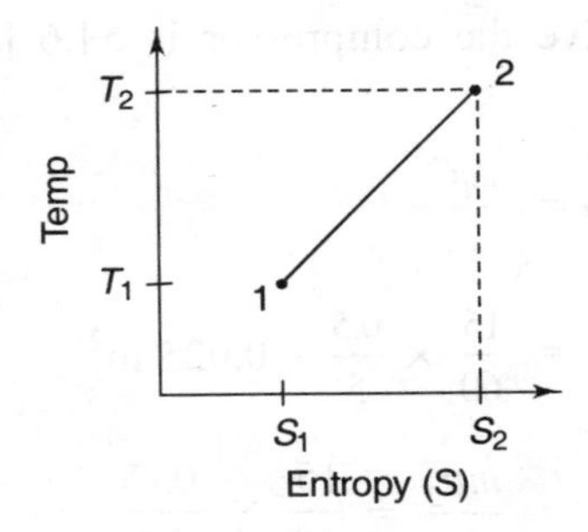

Heat given to the system, Q = area under line 1–2

$$Q = T_1(S_2 - S_1) + \frac{1}{2}(S_2 - S_1)(T_2 - T_1)$$

$$= 330(0.3 - 0.23) + \frac{1}{2}(0.3 - 0.23)(440 - 330)$$

$$= 26.95 \text{ kJ}$$

$$Q = \Delta U + W$$

$$26.95 = (190 - 170) + W$$

or $$W = 26.95 - 20 = 6.95 \text{ kJ}$$

13. A centrifugal air compressor delivers 15 kg of air per minute. The inlet and outlet conditions are as follows: At the inlet: velocity 5 m/s, enthalpy 5 kJ/kg. At the outlet: velocity = 7.5 m/s, enthalpy = 17.3 kJ/kg. Heat loss to cooling water is 756 kJ. Find (i) the power of the motor required to drive the compressor, and (ii) the ratio of the inlet pipe diameter to the outlet pipe diameter if the specific volumes at the inlet and the outlet are 0.5 and 0.15 m³/kg.

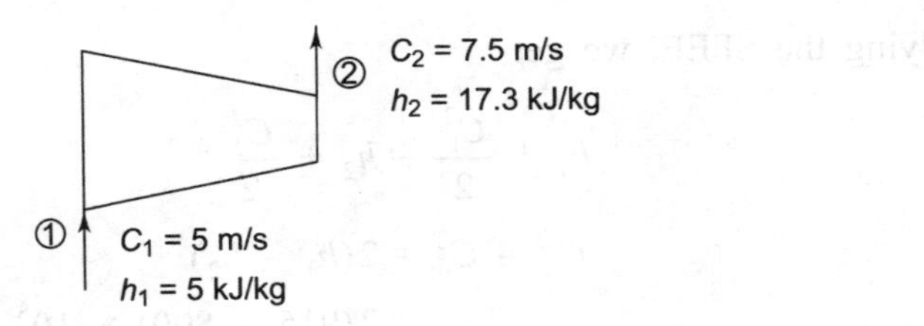

Applying the SFEE, we get

$$Q = W_c + m\left[(h_2 - h_1) + \frac{1}{2}(C_2^2 - C_1^2) + g(z_2 - z_1)\right]$$

$$-756 = W_c + \frac{15}{60}\left[(17.3 - 5) + \frac{1}{2}(7.5^2 - 5^2) \times 10^{-3}\right]$$

$\therefore$ $W_c = -54.6$ kJ/s

The power of the motor to drive the compressor is 54.6 kW. Flow rate is constant, i.e.

$$m = \frac{A_1 C_1}{v_1}$$

or $$A_1 = \frac{15}{60} \times \frac{0.5}{5} = 0.025 \text{ m}^2$$

Similarly $$A_2 = \frac{m v_2}{C_2} = \frac{15}{60} \times \frac{0.15}{7.5}$$

$$= 0.005 \text{ m}^2$$

$$\frac{A_1}{A_2} = \frac{\frac{\pi d_1^2}{4}}{\frac{\pi d_2^2}{4}} = \left(\frac{d_1}{d_2}\right)^2$$

or $$\frac{d_1}{d_2} = \sqrt{\frac{A_1}{A_2}} = \sqrt{\frac{0.025}{0.005}} = \sqrt{5}$$

$$= 2.236$$

14. Gas leaving a turbine enters a jet pipe with an enthalpy of 915 kJ/kg and leaves with an enthalpy of 800 kJ/kg. The inlet velocity is 300 m/s. Find the exit velocity.

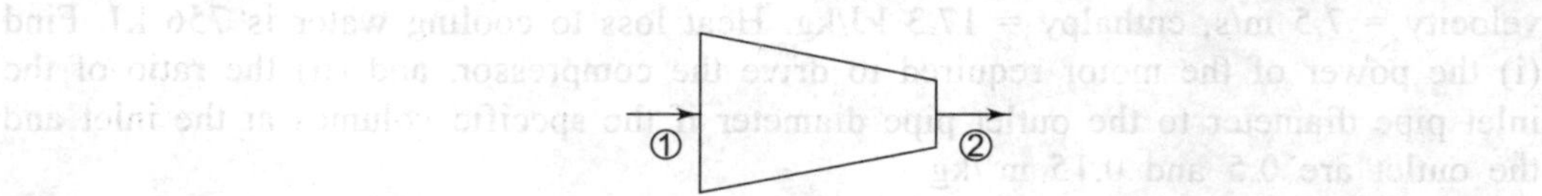

Applying the SFEE, we get

$$h_1 + \frac{C_1^2}{2} = h_2 + \frac{C_2^2}{2}$$

$$C_2^2 - C_1^2 = 2(h_1 - h_2)$$

$$= 2(915 - 800) \times 10^3$$

$$= 230 \times 10^3 = 23 \times 10^4$$

$$C_2^2 = 23 \times 10^4 + 9 \times 10^4$$

$$= 32 \times 10^4$$

$$C_2 = 565.7 \text{ m/s}$$

15. An inverter claims to invert a device which works in a cycle taking heat 50 J, 30 J and – 40 J, and giving net work of 60 J. Check the claim.

$$\Sigma Q = 50 + 30 - 40 = 40 \text{ J}$$

$$\Sigma W = 60 \text{ J}$$

Since $\Sigma W > \Sigma Q$, the device violates the law of conservation of energy. The claim of the inverter is incorrect.

16. A system undergoes a cyclic process through four states 1–2, 2–3, 3–4 and 4–1. Find the values of x_1, x_2, y_1, y_2 and y_3 in the following table.

Process	*Heat transfer (kJ/min)*	*Work tranfer (kW)*	*Change of internal energy*
1–2	800	5.0	y_1
2–3	400	x_1	600
3–4	–400	x_2	y_2
4–0	0	3	y_3

(UPTU: Dec. 2005)

Process 1–2:

$$Q = W + \Delta U$$

$$800 = 5 \times 60 + y_1$$

$$y_1 = 800 - 300 = 500 \text{ kJ/min}$$

Process 2–3:

$$Q = \Delta U + W$$

$$400 = 600 + x_1$$

$$x_1 = -200 \text{ kJ/min}$$

For a cyclic process

$$\Sigma Q = \Sigma W$$

$$800 + 400 - 400 + 0 = 300 - 200 + x_2 + 3 \times 60$$

or

$$x_2 = 800 - 280$$

$$= -520 \text{ kJ/min}$$

Process 3–4:

$$Q = \Delta U + W$$

$$-400 = y_2 + 520$$

$$y_2 = -920 \text{ kJ/min}$$

Process 4–0:

$$Q = \Delta U + W$$

$$0 = 180 + \Delta U$$

$$y_3 = -180 \text{ kJ/min}$$

17. 0.8 kg/s of air flows through a compressor in steady state conditions. The properties of air at the entry are: pressure 1 bar, velocity 10 m/s, specific volume 0.95 m^3/kg and internal energy 30 kJ/kg. The corresponding values at the exit are 8 bar, 6 m/s, 0.2 m^3/kg and 124 kJ/kg. Neglecting the change in PE, determine the power input and the pipe diameters of the entry and exit.

(UPTU: Carry over Aug. 2005–6)

$$Q + M(h_1 + \frac{C_1^2}{2} + gz_1) = W + M(h_2 + \frac{C_2^2}{2} + 9z_2)$$

$$Q = 0, z_1 = z_2$$

$$M\left(h_2 + \frac{C_2^2}{2}\right) = W + M\left(h_2 + \frac{C_2^2}{2}\right)$$

$$M(u_1 + p_1v_1 + \frac{C_1^2}{2}) = W + M(u_2 + P_2 v_2 + \frac{C_2^2}{2})$$

$$W = M[(u_1 - u_2) + (P_1v_1 - P_2v_2) + \frac{1}{2}(C_1^2 - C_2^2)]$$

$$= 0.8[(30 - 124) + (1 \times 0.95 - 8 \times 0.2) \times 10^5 \times 10^{-3} + \frac{1}{2}(10^2 - 6^2) \times 10^{-3}]$$

$$= 0.8(-90 - 65 + 32)$$

$$= 155 \text{ kJ}$$

$$M = \frac{A_1C_1}{v_1} = \frac{A_2C_2}{v_2}$$

$$A_1 = \frac{0.8 \times 0.95}{10}$$

$$A_1 = 0.076 \text{ m}^2$$

$$= 760 \text{ cm}^2$$

$$A_2 = \frac{0.8 \times 0.2}{6}$$

$$A_2 = 0.02667$$

$$A_2 = 266.7 \text{ cm}^2$$

18. A centrifugal compressor takes 16 kg of air per minute and compresses it from pressure of 1 bar to 7 bar. The flow velocity and density at the inlet are 5 m/s and 2.22 kg/m^3. The corresponding values at the exit are 8 m/s and 6.67 kg/m^3. When the air flows through the compressor, its internal energy increases by 150 kJ/kg and heat is lost to the surrounding by 800 kJ/min. Make calculations for power required to drive the compressor and the ratio of the inlet pipe diameter to the outlet pipe diameter.

(UPTU: Carry over Dec. 2005)

$$Q + M(h_1 + \frac{C_1^2}{2} + gz_1) = -W + M(h_2 + \frac{C_2^2}{2} + gz_2)$$

and

$$v_1 = \frac{1}{\rho_1}$$

$$v_2 = \frac{1}{\rho_2}$$

$$-W = \frac{-800}{60} + \frac{16}{60}[(u_1 - u_2) + (P_1v_1 - P_2v_2) + \frac{1}{2}(C_1^2 - C_2^2)]$$

$$-W = \frac{-800}{60} + 0.267\ [(1 \times \frac{1}{2.22} - 7 \times \frac{1}{6.67})] \times 10^5 \times 10^{-3} - 150 + \frac{1}{2}(5^2 - 8^2)]$$

$$-W = -13.33 + 0.267\ (-50.3 - 150 - .195)$$

$$= -13.33 - 53.48$$

$$W = +66.8 \text{ kcal}$$

$$= +15.98 \text{ KW}$$

Power required to drive compressor = 15.98 KW

$$M = \frac{A_1C_1}{v_1} = \frac{A_2C_2}{v_2}$$

$$\frac{A_1}{A_2} = \frac{v_1}{v_2} \times \frac{C_2}{C_1} = \frac{C_2}{C_1} \times \frac{\rho_2}{\rho_1}$$

$$\left(\frac{d_1}{d_2}\right)^2 = \frac{8}{5} \times \frac{6.67}{2.22} = 4.807$$

$$\frac{d_1}{d_2} = 2.19$$

19. A cylinder contains compressed helium gas which is used to inflate baloons to a volume of 0.9 m^3 at one atmosphere. Find work done by (a) the balloon, (b) the cylinder, and (c) by the system.

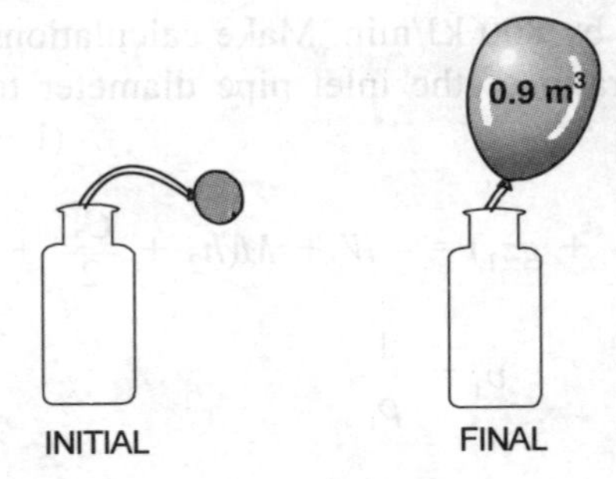

The change of volume for the cylinder is zero while for the balloon it is 0.9 m^3. The expansion of the balloon is against atmospheric pressure. Therefore,

$$W_{sys} = \int_{\text{cylinder}} P\,dV + \int_{\text{balloon}} P\,dV$$

$$= 0 + P\int dV \qquad (P = 1 \text{ bar})$$

$$= 0 + 1.01 \times 10^5 \times 0.9$$

$$= 90.9 \text{ kJ}$$

20. 2 m^3 air of a balloon enters an evacuated cylinder under one atmosphere (1 bar). Find the work done by (a) the balloon, (b) the cylinder, and (c) by the system.

$$W_{sys} = \int_{\text{cylinder}} P\,dV + \int_{\text{balloon}} P\,dV$$

$$= \int_{\text{cylinder}} P \times 0 + \int_{\text{balloon}} (1.01 \times 10^5)(-2)$$

$$= 0 - 202 \text{ kJ}$$

$$= -200 \text{ kJ}$$

The negative sign means that work is done by the atmosphere on the system.

21. A gas undergoes a reversible non-flow process according to relation $P = -3V + 5$ where V is the volume in m^3 and P is the pressure in bar. Determine the work done when volume changes from 3 to 6 m^3.

(UPTU: 2006–2007)

Work can be found out by using rotation $\int P\,dV$. P is given in terms of V which can be integrated for limit 3–6 m^3.

$$P = 3V + 5$$

$$\text{Work} = \int_3^6 P\,dV \quad (P \text{ is in bar})$$

$$= \int_3^6 (-3V + 5) \times 10^5\,dV$$

$$= \left[-\frac{3V^2}{2} + 5V\right]_3^6 \times 10^5$$

$$= [-1.5(6^2 - 3^2) + 5(6 - 3)] \times 10^5$$

$$= [-1.5(36 - 9) + 5 \times 3] \times 10^5$$

$$= [-40.5 + 15] \times 10^5$$

$$= -25.5 \times 10^5 \text{ J}$$

The negative sign indicates that the work is done on the gas.

22. A closed system whose initial volume is 50×10^4 cc undergoes a non-flow reversible process for which pressure and volume correleation is given by

$$P = 8 - 4V$$

where P is in bar and V is in m^3. If 200 kJ of work is supplied to the system, determine (a) final pressure, (b) final volume after completion of process.

(UPTU: 2006–2007)

Work supplied to system = 200 kJ

$$W = \int_{V_1}^{V_2} P\,dV$$

$$= \int_{V_1}^{V_2} (8 - 4V) \times 10^5 \times dV$$

$$-200 \times 10^3 = \int_{V_1}^{V_2} (8 - 4V) \times 10^5 \times dV$$

$$-2 = [8(V_2 - V_1) - \frac{4}{2}(V_2^2 - V_1^2)]$$

$$2 = 2[(V_2^2 - 4V_2) - (V_2^2 - 4V_1)]$$

$$V_1 = 50 \times 10^4\text{cc} = 50 \times 10^4 \times (10^{-2})^3 \text{ m}^3$$
$$= 5 \times 10^{-1} \text{ m}^3 = 0.5 \text{ m}^3$$

$\therefore$
$$2 = 2[(V_2^2 - 4V_2) - (0.5^2 - 4 \times 0.5)]$$
$$= 2[V_2^2 - 4V_2 + 1.75]$$

or
$$V_2^2 - 4V_2 + 0.75 = 0$$

$\therefore$
$$V_2 = 3.80 \text{ m}^3 \quad \text{or} \quad 0.20 \text{ m}^3$$

$\therefore$
$$P_2 = (8 - 4 \times 3.80) \text{ or } (8 - 4 \times 0.20)$$
$$= 0.80 \text{ bar or } 7.20 \text{ bar}$$

23. The internal energy of a certain substance is expresssed by the equation $u = 3.62\ pv + 86$ where P is in kPa and v is in m^3/kg. A system composed of 5 kg of this substance expands from an initial pressure of 550 kPa and volume of 6.25 m^3/kg to final pressure of 125 kPa in a process in which pressure and volume are related by $pv^{1.25}$ constant. If the expansion process is quasi-static, determine Q, Δu and w.

(UP: 2005–2006)

$$u = 3.62\ Pv + 86$$

and
$$P_1 v_1^{1.25} = P_2 v_2^{1.25}$$

or
$$v^2 = v_1 \left(\frac{P_1}{P_2}\right)^{\frac{1}{1.25}} = 6.25 \times \left(\frac{550}{125}\right)^{\frac{1}{1.25}} = 3.27 \text{ m}^3$$

$\therefore$
$$\Delta u = u_2 - u_1$$
$$= 3.62\,(P_2 v_2 - P_1 v_1)$$
$$= 3.62(125 \times 3.27 - 550 \times 6.25)$$
$$= 3.62(408.75 - 3437.5)$$
$$= -10.963 \times 10^3 \text{kJ/kg}$$

$$w = \int_{v_1}^{v_2} P\,dv \qquad \text{but } P = \frac{c}{v^{1.25}}$$

$$= \int \frac{c}{v^{1.25}} \times dv$$

$$= \frac{P_1 v_1 - P_2 v_2}{1.25 - 1} = 4(P_1 v_1 - P_2 v_2)$$

$$= -4 \times 3028.75 = -12.115 \times 10^3 \text{ kJ/kg}$$

$$Q = \Delta u + w$$
$$= -10.963 \times 10^3 - 12.115 \times 10^3$$
$$= -23.078 \times 10^3 \text{ kJ/kg}$$

For 5 kg substance, we have

$$\Delta U = -54.8 \text{ MJ}, W = -60.58 \text{ MJ and } Q = -115.4 \text{ MJ}$$

24. Calculate the work done in a piston-cylinder arrangement during expansion process 1 m^3 to 4 m^3 where process is given by $P = (V^2 + 6V)$ bar $= (V^2 + 6V) \times 10^5$ Pa

$$W = \int P\,dV = \int_1^4 (V^2 + 6V) \times 10^5 \times dV$$

$$= \left[\left(\frac{V^3}{3} + 6\frac{V^2}{2}\right) \times 10^5\right]_1^4$$

$$= \left[\frac{1}{3}(4^3 - 1^3) + 3(4^2 - 1^2)\right] \times 10^5$$

$$= \left[\frac{1}{3} \times 63 + 3 \times 15\right] \times 10^5$$

$$= (21 + 45) \times 10^5$$

$$= 6.6 \text{ MJ}$$

25. Air enters a frictionless adiabatic converging nozzle at 10 bar, 500 K with negligible velocity. The nozzle discharges to a region at 2 bar. If the exit area of nozzle is 2.5 cm^2. Find the flow rate of air through the nozzle. Assume for air $c_p = 1005$ J/kg K and $c_v = 718$ J/kg K.

(GATE: 1997)

$P_1 = 10$ bar $\qquad P_2 = 2$ bar

$$\text{v} = \frac{c_p}{c_v} = \frac{1005}{718} = 1.4 \qquad T_1 = 5000 \text{ K}$$

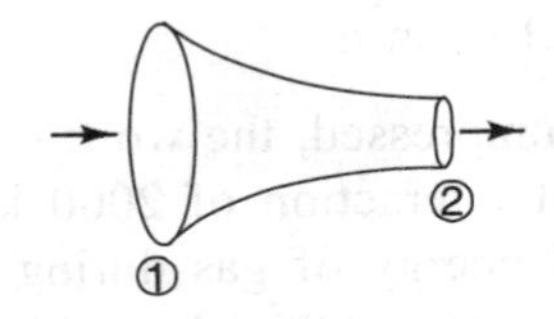

$$P_1 v_1^{\gamma} = P_2 v_2^{\gamma} \quad \text{and} \quad \frac{Pv}{T} = \text{const}$$

or $$T_1^{\gamma} p_1^{-\gamma+1} = T_2^{\gamma} p_2^{-\gamma+1} \quad \text{or} \quad v \propto \frac{T}{P}$$

$$\left(\frac{T_2}{T_1}\right)^{\gamma} = \left(\frac{P_1}{P_2}\right)^{\gamma-1}$$

or $$\frac{T_2}{T_1} = \left(\frac{P_1}{P_2}\right)^{\frac{-\gamma+1}{\gamma}}$$

or $$T_2 = T_1 \times \left(\frac{P_1}{P_2}\right)^{\frac{-\gamma+1}{\gamma}}$$

$$= 500 \times \left(\frac{10}{2}\right)^{\frac{-1.4+1}{1.4}}$$

$$= 500 \times 5^{-0.285} = 316$$

Now $$h_1 - h_2 = c_p(500 - 316) = 185 \times 10^3 \text{ J}$$

Now $$\frac{c_2^2}{2} = (h_1 - h_2) \quad [\text{where } c_2 = \text{exit velocity}]$$

or $$c_2 = \sqrt{2(h_1 - h_2)}$$

$$= \sqrt{2 \times 185 \times 10^3}$$

$$= 608.3 \text{ m/s}$$

$$\text{Flow rate} = \text{Area} \times c_2$$

$$= 2.5 \times 10^{-4} \times 608.3$$

$$= 0.152 \text{ m}^3/\text{s}$$

26. A gas contained in a cylinder is compressed, the work required for compression being 5000 kV. During the process, heat interaction of 2000 kJ causes the surroundings to be heated. The change in internal energy of gas during the process is
(a) –7000 kJ (b) –3000 kJ (c) +3000 kJ (d) 7000 kJ (GATE: 2004)

$$W = -5000 \text{ kJ (compression)}$$

$$Q = -2000 \text{ kJ (heat is rejected)}$$

$$Q = \Delta U + W \text{ by first law of thermodynamics}$$

$$-2000 = \Delta U - 5000$$

or $$\Delta U = 3000 \text{ kJ}$$

Option (c) is correct.

27. A small steam whistle (perfectly insulated and doing no shaft work) causes a drop of 0.8 kJ/kg in the enthalpy of steam from entry to exit. If the kinetic energy of the steam at entry is negligible, the velocity of the steam at exist is

(a) 4 m/s (b) 40 m/s (c) 80 m/s (d) 120 m/s

(GATE: 2001)

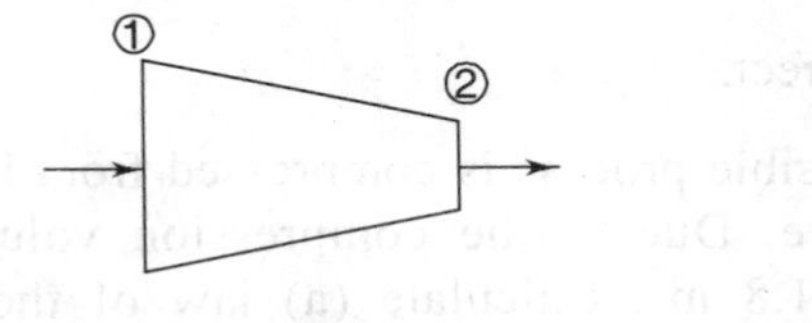

$$\frac{c_1^2}{2} + h_1 = \frac{c_2^2}{2} + h_2$$

Take $c_1 = 0$. Hence,

$$c_2 = \sqrt{2(h_1 - h_2)}$$

or $$c_2 = \sqrt{2 \times 0.8 \times 10^3} = 40 \text{ m/s}$$

Option (b) is correct.

28. Air ($c_p = 1$ kJ/kg, $v = 1.4$) enters a compressor at a temperature of 27°C. The compressor pressure ratio is 4. Assuming an efficiency of 80%. The compressor work required in kJ/kg is

(a) 160 (b) 172 (c) 182 (d) 225. (GATE: 1998)

Given: $\frac{P_2}{P_1} = 4$

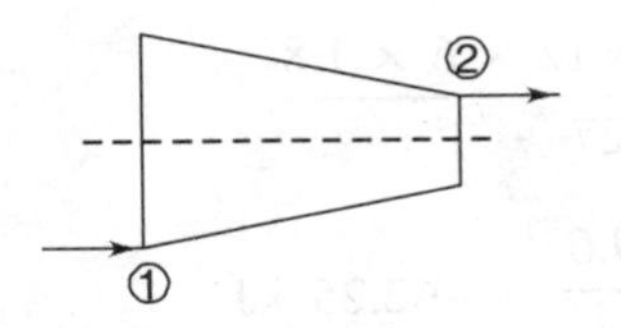

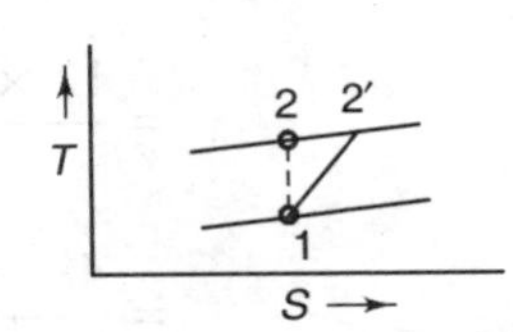

Now $$\frac{T_2}{T_1} = \left(\frac{P_2}{P_1}\right)^{\frac{\gamma-1}{\gamma}} = 4^{0.4/1.4} = 1.48$$

∴ $$T_2 = 1.48 \times 300 = 446 \text{ K}$$

$$\text{Work} = c_p(T_2 - T_1) \text{ for reversible}$$

$$\text{Work} = \frac{c_p(T_2 - T_1)}{\eta} \text{ for irreversible}$$

$$T_2' - T_1 = \frac{T_2 - T_1}{\eta} = \frac{446 - 300}{0.8} = 182.2$$

Compressor work = 1 × (182.2) = 182.2 kJ/kg

Option (c) is correct.

29. Air during a reversible process is compressed from initial process 12 kN/m^2 to 6 times the initial pressure. Due to the compression volume of air decreases from initial volume 4 m^3 to 1.8 m^3. Calculate (a) law of the process, and (b) work done in compressing the air. (UPTU: May 2008)

P_1 = 12 kN/m^2 P_2 = 12 × 6 kN/m^2

V_1 = 4 m^3 V_2 = 1.8 m^3

Applying $P_1V_1^n = P_2V_2^n$

$$\left(\frac{4}{1.8}\right)^n = \left(\frac{12 \times 6}{12}\right)$$

or $$(2.2)^n = 6$$

or $$n = 2.27$$

Hence the law of process is

$$PV^{2.27} = \text{constant}$$

$$\text{Work} = \frac{P_1V_1 - P_2V_2}{n - 1}$$

$$= \frac{12 \times 4 - 12 \times 6 \times 1.8}{2.27 - 1}$$

$$= \frac{48 - 129.6}{1.27} = -62.25 \text{ kJ}$$

OBJECTIVE TYPE QUESTIONS

Happiness comes from within your heart, not from your surroundings.

State True and False

1. The first law of thermodynamics is a form of the law of conservation of energy. *(True/False)*
2. Heat added to the system is considered negative. *(True/False)*
3. Work done on the system is considered positive. *(True/False)*
4. Free expansion work is considered negative. *(True/False)*
5. In a cyclic process, the area enclosed by the curve on a *PV* diagram is traced clockwise, then net work done is done by the system. *(True/False)*
6. If traced curve on a *PV* diagram is anticlockwise, then work is negative. *(True/False)*
7. (Q–W) value between two states is the same for all processes. *(True/False)*
8. Internal energy is a state function. *(True/False)*
9. If Q heat is given to a system, then it is used for increasing internal energy and in doing work. *(True/False)*
10. Change of internal energy in a cyclic process depends upon whether processes are reversible or irreversible processes. *(True/False)*
11. Internal energy of a system increases in an isothermal process. *(True/False)*
12. Internal energy is used by the system in doing work in an adiabatic process. *(True/False)*
13. In an isochoric process, heat given to the system is utilized completely in increasing internal energy. *(True/False)*
14. The internal energy of an isolated system remains constant. *(True/False)*
15. In a cyclic process, heat given to a system is equal to work done by the system. *(True/False)*
16. If in a process, the volume of the system decreases, then the work is done by the system. *(True/False)*
17. In an isobaric process, heat $Q = mR(T_1 - T_2)\left(\frac{1}{\gamma - 1} + 1\right)$ *(True/False)*
18. In an isochoric process, change of internal energy is $mc_v\,(T_2 - T_1)$, which is also equal to the total heat added to the system. *(True/False)*
19. In an isobaric process, heat given (Q) to the system is equal to $mc_p\,(T_2 - T_1)$, which is equal to increase in internal energy and work done by the system. *(True/False)*
20. If $n = 0$ for process Pv^n = constant, the process is isobaric. *(True/False)*

21. If $n = 1$ for a polytropic process, the process is isothermal. *(True/False)*
22. If $n = \alpha$ for a polytropic process, the process is isochoric. *(True/False)*
23. In free expansion, work done by a gas is not zero. *(True/False)*
24. Flow work is equal to the product of pressure and temperature. *(True/False)*
25. Enthalpy is the difference of internal energy and flow work. *(True/False)*
26. A turbine is a device in which fluid at high pressure and temperature is expanded to low pressure and temperature so that work can be done on the system. *(True/False)*
27. A turbine has a smaller inlet area and a bigger outlet area so that fluid may expand and gives out work. *(True/False)*
28. A compressor is a device to compress fluid to high pressure and temperature and doing so gives out work. *(True/False)*
29. A compressor has a large inlet area and a small outlet area so that fluid is compressed to high pressure and temperature when work is given to it. *(True/False)*
30. A compressor is used for vapour to increase pressure while a pump is used for liquid to increase pressure. *(True/False)*
31. A turbine and a compressor are suitable for liquid. *(True/False)*
32. A boiler is a device used for generation of steam at constant temperature. *(True/False)*
33. A condenser condenses vapour into liquid at constant pressure. Therefore, it is not a heat exchanger. *(True/False)*
34. A condenser extracts heat from vapour so as to condense vapour into liquid by dissipating heat to the surroundings. *(True/False)*
35. A nozzle is a device in which expansion of fluid takes place resulting in drop of pressure and increase of exit velocity. *(True/False)*
36. The velocity of fluid at the inlet of a nozzle is more than at the outlet of the nozzle. *(True/False)*
37. The diverging cross section for a nozzle is used for subsonic flow. *(True/False)*
38. A combustion chamber is used for adiabatic mixing of air and fuel. *(True/False)*
39. In adiabatic mixing of two fluids, heat is released. *(True/False)*
40. In throttling, fluid passes through a restricted opening resulting in increase of enthalpy. *(True/False)*
41. Heat is high grade energy and hence can be converted fully into work. *(True/False)*
42. Heat can flow from a low temperature body to a high temperature body without interaction of any other form of energy. *(True/False)*
43. Work can be fully converted into heat as work is high grade energy. *(True/False)*
44. Steady flow occurs when conditions do not change with time at any point. *(True/False)*
45. Is it correct to say that system contains heat? *(True/False)*
46. Work is a path function and not a state function or property of the system. *(True/False)*

47. A throttle is a device which can maintain high pressure at one side and low pressure at other side. (*True/False*)

48. Steam has to pass through a nozzle to convert steam energy into kinetic energy before driving shaft work from an impulse turbine. (*True/False*)

49. Rocket and jet engines have nozzles to obtain higher velocity for thrust. (*True/False*)

50. If pressure and volume increase, the work done by the system is positive and internal energy must increase. (*True/False*)

51. If initial pressure and volume are equal to final pressure and volume, then initial temperature must be equal to the final temperature and internal energy would remain unchanged. (*True/False*)

52. Pressure and volume can be increased by keeping temperature constant. (*True/False*)

53. When we switch on the fan we feel cool as internal energy of the room decreases. (*True/False*)

54. If a gas is compressed at constant temperature, the internal energy will remain the same. (*True/False*)

Multiple Choice Questions

1. Work done in free expansion is
   (a) zero (b) negative (c) positive

2. In a free expansion process
   (a) pressure is constant (b) work is constant (c) internal energy is constant

3. The following amounts of heat transfer occur during a cycle of four processes: +120 kJ, – 20 kJ, 16 kJ and 4 kJ. The work done during the cycle is
   (a) 100 kJ (b) 120 kJ (c) 130 kJ (d) 140 kJ

4. Throttling occurs when fluid passes through a
   (a) nozzle (b) diffuser
   (c) turbine (d) restricted opening

5. In the steady flow work done is generally
   (a) PV (b) $\int_1^2 PdV$
   (c) $-\int_1^2 PdV$ (d) $\frac{P_1V_1 - P_2V_2}{n-1}$

6. Steady flow energy equation when applied to a boiler is
   (a) $q = h_2 - h_1$ (b) $q = W$
   (c) $q = u_2 - u_1$ (d) $q = C_v(T_2 - T_1)$

7. The concept of internal energy is given by which of following laws of thermodynamics?
   (a) Zeroth law (b) First law (c) Second law (d) Third law

8. If a system undergoes an irreversible adiabatic change from state 1 to state 2, the work done by the system is given by

(a) $u_1 - u_2$ (b) $u_2 - u_1$ (c) $< u_1 - u_2$ (d) $> u_1 - u_2$

9. What is not given by the first law of thermodynamics

(a) $\oint dQ = \oint dW$

(b) $dE = dQ - dW$

(c) Energy of system is conserved

(d) $dQ - dW$ is constant

10. If a refrigerator is operated with the door open in an isolated room, then the temperature of the room

(a) decreases (b) remains constant (c) increases

11. In $Q = \Delta U + W$, where Q is heat given to system and W is work done by the system, then

(a) $Q > 0, W > 0$ (b) $Q < 0, W > 0$

(c) $Q > 0, W < 0$ (d) $Q < 0, W < 0$

12. We know that the density of water is more than that of ice. During melting of ice into water, we can say

(a) work done by ice on the atmosphere

(b) no work done

(c) work done by the atmosphere on ice

13. $Q - W = 0$ is true for

(a) a reversible process (b) an irreversible process

(c) a cyclic process

14. $W = -\Delta U$ is true for an

(a) isobaric process (b) isothermal process (c) adiabatic process

15. $Q = W$ is true for an

(a) isothermal process (b) isochoric process (c) adiabatic process

16. If heat is supplied to a gas in an isothermal process,

(a) internal energy will increase

(b) gas will do positive work

(c) gas will do negative work

(d) nothing can be said

17. A and B are two processes on the PV diagram. Let ΔQ_1 and ΔQ_2 be heat given in A and B processes. Then, the relation between ΔQ_1 and ΔQ_2 is

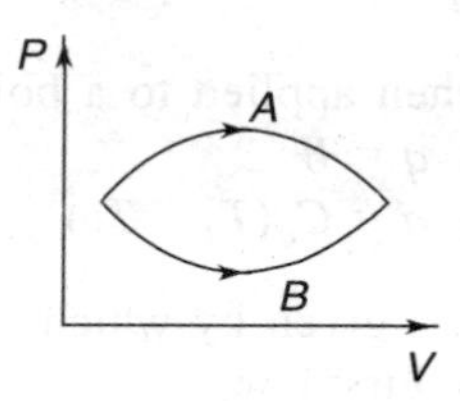

(a) $\Delta Q_1 > \Delta Q_2$ (b) $\Delta Q_1 = \Delta Q_2$ (c) $\Delta Q_1 < \Delta Q_2$ (d) $\Delta Q_1 \leq \Delta Q_2$

18. If ΔU_1 and ΔU_2 be change in internal energy in the processes A and B as shown in Question 17, then

(a) $\Delta U_1 > \Delta U_2$ (b) $\Delta U_1 = \Delta U_2$ (c) $\Delta U_1 < \Delta U_2$ (d) $\Delta U_1 \neq \Delta U_2$

19. In the process 1–2, the work done by the system

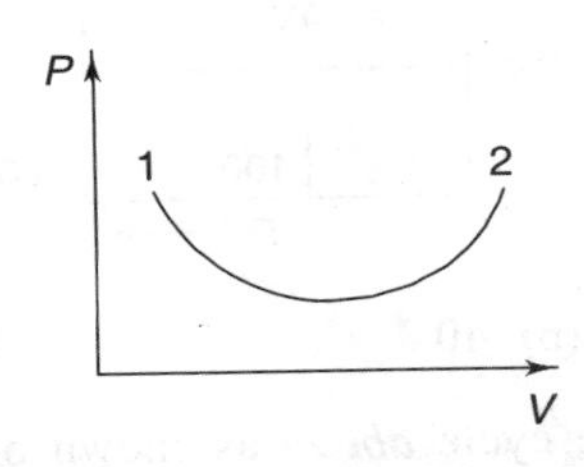

(a) continuously increases
(b) continuously decreases
(c) first decreases then increases
(d) first increases then decreases

20. A system moves from state A to state B as shown on the PT diagram. The work done by system is

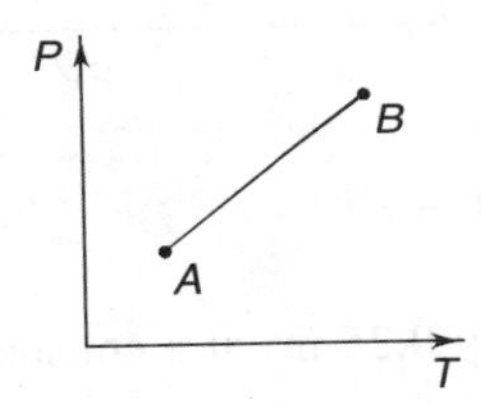

(a) positive (b) negative
(c) zero (d) cannot say about work

21. Processes A and B are as shown on the PT diagram. The initial and final volumes for both A and B processes are the same. If ΔW_1 and ΔW_2 are work done by the system in A and B processes, then

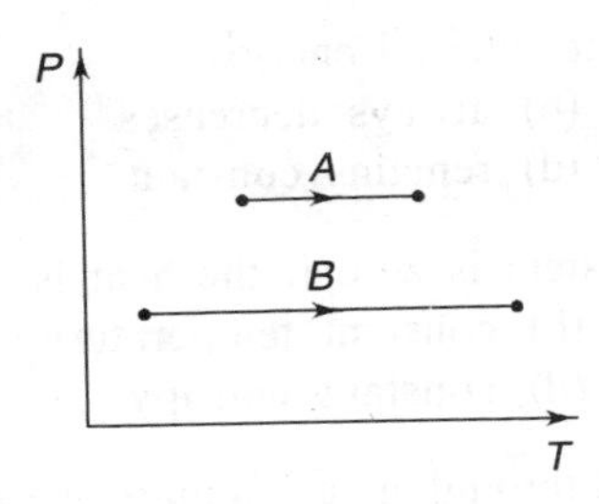

(a) $\Delta W_1 > \Delta W_2$ (b) $\Delta W_1 < \Delta W_2$ (c) $\Delta W_1 = \Delta W_2$

22. The heat absorbed by a system through a cyclic process as shown on the PV diagram is

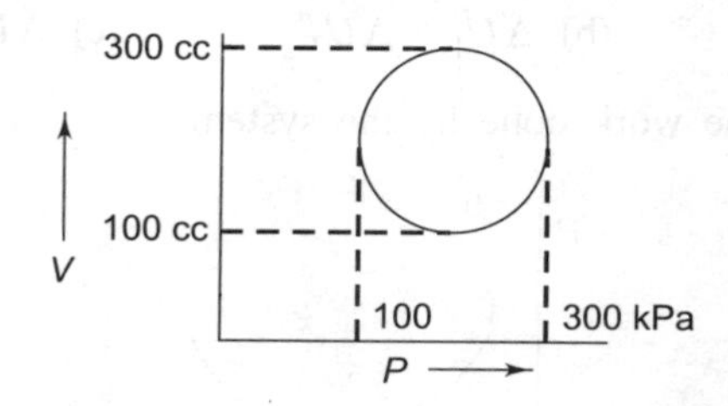

(a) 314 J (b) 40 J (c) 30 J

23. A system is taken round the cycle *abcda* as shown on the PV diagram. The work done during cycle is

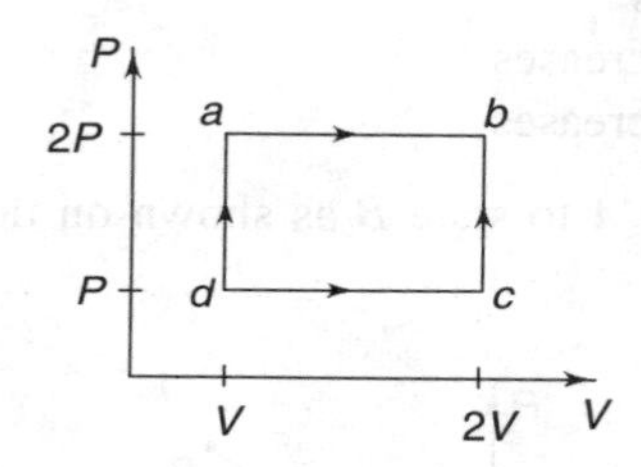

(a) PV (b) $-PV$ (c) $2PV$ (d) $\frac{1}{2}PV$

24. The volume of a gas expands by 0.25 m^3 at a constant pressure of 10^3 Pa. The work done is equal to
(a) 2.5 joules (b) 250 joules (c) 250 watts (d) 200 newtons

25. For a closed system, the difference between the heat added to the system and the work done by the system is equal to the change in
(a) enthalpy (b) entropy
(c) temperature (d) internal energy

26. In an isothermal process, the internal energy
(a) always increases (b) always decreases
(c) is zero (d) remains constant

27. Change in enthalpy of a system is zero if the heat is supplied at
(a) constant pressure (b) constant temperature
(c) constant volume (d) constant entropy

28. If a gas is allowed to expand through a very minute aperture, then such a process is known as
(a) free expansion (b) throttling process
(c) adiabatic expansion (d) parabolic expansion

29. The cyclic integral $(dQ - dW)$ for a process is equal to
(a) positive (b) negative (c) zero

30. If for a process both heat given to the system and work done by the system are equal, then the process is
(a) adiabatic (b) isobaric (c) isochoric (d) isothermal

31. If for a process both reduction of internal energy and work done by system are equal, then the process is
(a) adiabatic (b) isobaric (c) isochoric (d) isothermal

32. If for a process both heat supplied and increase of internal are equal, then the process is
(a) adiabatic (b) isobaric (c) isochoric (d) isothermal

Fill in the Blanks

1. During an adiabatic process work is done by the system using ________ energy.
(a) potential (b) internal

2. During an isothermal process, heat given to the system is used ________ for doing work.
(a) partially (b) fully

3. In a constant volume process, heat is used ________ for increasing internal energy.
(a) partially (b) fully

4. A turbine has a ________ cross-sectional area at the entrance as compared to the exit.
(a) larger (b) smaller

5. A compressor has a ________ cross-sectional area at the entrance as compared to the exit.
(a) larger (b) smaller

6. Positive work is done by a system, when the volume of the system ________.
(a) increases (b) decreases

7. Negative work is done by a system from state 1 to state 2 if ________.
(a) $V_2 > V_1$ (b) $V_2 < V_1$

8. A cyclic process performs positive work if the curve on a PV diagram traces ________.
(a) clockwise (b) anticlockwise

9. Throttling is a process which has constant ________.
(a) enthalpy (b) velocity

10. A perpetual motion machine violates ________.
(a) energy equation (b) first law of thermodynamics

11. A nozzle is a converging type if fluid has ________ flow.
(a) subsonic (b) supersonic

12. A nozzle is a diverging type if fluid has ________ flow.
(a) subsonic (b) supersonic

13. For adiabatic mixing of two fluids, the heat interaction is ________.
(a) unity (b) zero

14. Evaporation of water and freezing of water are examples of an ________ process.
(a) isochoric (b) isobaric

15. Internal energy remains constant during an ________ process.
(a) adiabatic (b) isothermal

16. $Q–W$ remains constant for ________ processes.
(a) reversible (b) all

17. Heat generation in a boiler takes place at constant ________.
(a) temperature (b) pressure

18. A pump is used to increase the pressure of ________.
(a) liquid (b) vapour

19. Enthalpy is the summation of internal energy and ________ energy.
(a) free (b) flow

20. At constant pressure, heat supplied is used for increasing ________.
(a) internal energy (b) enthalpy

ANSWERS

Great men tell you how to get where you're going, greater men take you there.

State True and False

1. True
2. False (Heat added is positive and heat rejected/released is negative.)
3. False (Workdone by the system is positive and work done on the system is negative.)
4. False (In free expansion, work is zero as the expansion of the system has not been carried out against atmospheric pressure.)
5. True (Clockwise is positive and anticlockwise is negative.)
6. True
7. True ($Q - W$ is equal to Δu $(u_2 - u_1)$ which is constant for a process undergoing change from state 1 to state 2.)
8. True
9. True ($Q = \Delta U + W$)
10. False (As the system in a cyclic process comes back to initial position, hence change of internal energy is zero for all processes.)
11. False (Internal energy depends upon temperature only. Hence internal energy remains constant for isothermal process.)
12. True ($Q = \Delta U + W$. As for adiabatic process $Q = 0$, therefore $W = -\Delta U$)
13. True ($Q = \Delta U + W$. As $W(=Pdv)$ is zero for an isochoric process, all the heat is used for increasing internal energy, i.e. $Q = \Delta U$)
14. True ($Q = \Delta U + W$. As for an isolated system $Q = 0$ and $W = 0$, internal energy remains constant, i.e. $\Delta U = 0$)
15. True ($Q = \Delta U + W$ and $\Delta U = 0$ for a cyclic process. Hence $Q = W$ for a cyclic process.)
16. False (If the volume of a system increases, work is done by the system and if volume decreases work is done on the system.)
17. True
18. True
19. True
20. True ($PV^0 = P =$ constant)
21. True ($PV^1 = \overline{R}_1 T =$ constant or temperature is constant)
22. True ($PV^\alpha = P^{\overline{\alpha}} V = V =$ constant since $P^{\overline{\alpha}} = P^\circ = 1$)
23. False (Free expansion is done in vacuum and hence work done is zero.)

24. False (Flow work = PV. It is energy required to take fluid inside and outside the system.)

25. False ($h = u + Pv$)

26. False (Fluid performs work which is available at the shaft of a turbine.)

27. True (Fluid expands to do work and the turbine has increasing cross section so that fluid can expand.)

28. False (Fluid is compressed to higher pressure and hence work is done on the fluid.)

29. True (Fluid is compressed and hence compressor has reducing area.)

30. True (A compressor can handle large volume of vapour which is not possible in a pump.)

31. False (Turbines and compressors are designed for vapours and gases only.)

32. False (For a boiler, heat (Q) is equal to increase of enthalpy of steam at constant pressure.)

33. False (A condensor is also a type of heat exchanger.)

34. True

35. True

36. False (Velocity at the outlet is higher than velocity at the inlet.)

37. False (Converging cross section is for subsonic flow and diverging for supersonic flow.)

38. False (Fuel is injected at high pressure in a combustion chamber having air at high pressure and temperature resulting into combustion.)

39. False (No heat is released in an adiabatic process.)

40. False (Enthalpy remains constant in a throttling process.)

41. False (Heat is low-grade energy and cannot be converted fully into work which is high-grade energy.)

42. False (Heat cannot flow from a low temperature body to a high temperature body without assistance of work.)

43. True

44. True

45. False (Heat is a transit energy. It is not a property of a system. It is a phenomenon which occurs at the boundary of the system.)

46. True

47. True

48. True

49. True

50. True

51. True

52. False

53. False (Internal energy remained the same as temperature of the room does not change.)

54. True (Internal energy depends on temperature only.)

Multiple Choice Questions

1. (a)
2. (c)
3. (b) ($\Sigma Q = 120 - 20 + 16 + 4 = 120 = \Sigma W$)
4. (d)
5. (b)
6. (a)
7. (b)
8. (a)
9. (b)
10. (c) (The refrigerator and the room form an isolated system and electrical energy generates heat.)
11. (a)
12. (c) (When ice melts, its volume decreases against the atmospheric pressure. Hence work is done by the atmosphere.)
13. (c) ($Q - W = \Delta U$ and internal energy is a state function. Hence $\Delta U = 0$ as the system comes back to the initial state after a cyclic process.)
14. (c) ($Q = \Delta U + W$ and $Q = 0$ for an adiabatic process which gives $W = -\Delta U$.)
15. (a) ($Q = \Delta U + W$ and $\Delta U = 0$ for an isothermal process which gives $Q = W$.)
16. (b) ($Q = \Delta U + W$. For an isothermal process $\Delta U = 0$ and which will give $Q = W$.)
17. (a) (More area under path A means more work and which requires more heat than along path B.)
18. (b) (Internal energy is a state function.)
19. (a) (Since volume is continuously increasing, work will continuously increase.)
20. (c) (Process AB is isochoric which will give $\Delta V = 0$. Hence $\int PdV = 0$)
21. (a) (ΔV is the same for processes A and B and pressure $P_A > P_B$. Hence $\Delta W_1 > \Delta W_2$ as work is $\int PdV$)
22. (a) (Work or heat is the area enclosed $= \pi r_1 r_2 = \pi \times \left(\frac{300-100}{2}\right) \times 10^{-6} \times \left(\frac{300-100}{2}\right) \times 10^3 = 3.14 \times 100 \times 100 \times 10^{-2} = 31.4$ J)
23. (d) [area = $(2P - P)(2V - V) = PV$]
24. (b)
25. (d) ($Q - W = \Delta U$)
26. (d)
27. (a)
28. (b)
29. (c) (For a cyclic process $Q - W = 0$.)
30. (d) ($Q = W$ for an isothermal process.)
31. (a) ($W = -\Delta U$ for an adiabatic process.)
32. (c) (For an isochoric process $W = 0$, therefore $Q = \Delta U$.)

Fill in the Blanks

1. (b) 2. (b) 3. (b)

4. (b) (Fluid for expansion requires increasing area.)

5. (a) (Fluid for compression requires decreasing area.)

6. (a) 7. (b) 8. (a) 9. (a)

10. (b) 11. (a) 12. (b) 13. (b)

14. (b) 15. (b) 16. (b) 17. (b)

18. (a) 19. (b) 20. (b)

CHAPTER 3

Second Law of Thermodynamics

If you've experienced the dark, you can better appreciate the light.

INTRODUCTION

The first law of thermodynamics cannot explain non-occurrence of certain processes as well as the direction of a process. Feasibility of a process, the direction of a process, and grades of energy (low and high) are clarified by the second law of thermodynamics. Other things like maximum possible efficiency of a heat engine; coefficient of performance of a heat pump and a refrigerator, and the concept of a temperature scale, which is independent of physical properties, are also explained by the second law of thermodynamics.

HEAT RESERVOIR

Heat reservoir is a system/body having extremely large heat capacity. It is capable of absorbing or rejecting finite amount of heat without any change in temperature. In this respect the atmosphere, rivers and seas are reservoirs from which we can extract or dump any amount of heat without changing temperature. Source is a heat reservoir at higher temperature from which heat is extracted without change of its temperature. The sun is a "source" heat reservoir. A sink is a heat reservoir capable to absorb any amount of heat without change of its temperature. The atmosphere or surroundings is a "sink" heat reservoir.

HEAT ENGINE

A heat engine is a device used for converting heat into work (Figure 3.1). It is possible to convert work into heat directly but the heat engine is required to convert heat into work.

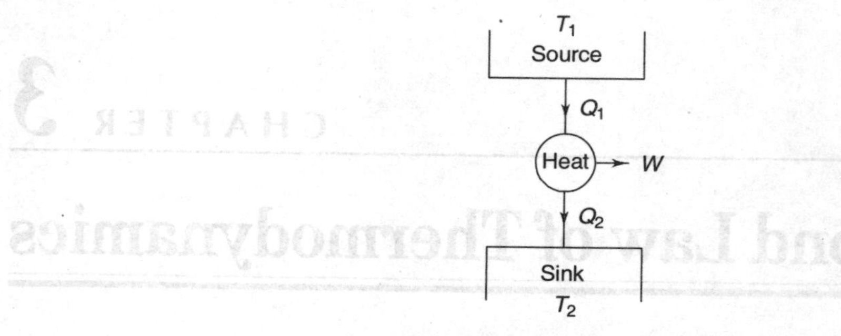

FIGURE 3.1 Heat engine.

A heat engine can be defined as a device operating in a cycle between a high temperature source and a low temperature sink and producing work. The heat engine receives heat (Q_1) from the source and transforms some portion of heat into work (W) and rejects balance heat (Q_2) to the sink.

Heat and work have been categorized as two forms of energy—low grade and high grade. Conversion of high-grade energy into low-grade energy can be carried out fully and spontaneously without aid of any device. However, complete conversion of low-grade energy into high-grade energy is impossible and non-spontaneous. We require a device like a heat engine or gas turbine plant to convert low-grade heat into high-grade energy (work). However, conversion from low-grade to high-grade energy cannot be achieved fully.

A gas turbine plant (Figure 3.2) consists of a boiler, a turbine, a heat exchanger and a compressor. Heat (Q_{add}) is added in the boiler, expansion takes place in the turbine producing work (W_e), the heat exchanger cools the fluid by extracting heat (Q_{rej}) and the compressor compresses the fluid when compression work (W_c) is given to it by external source.

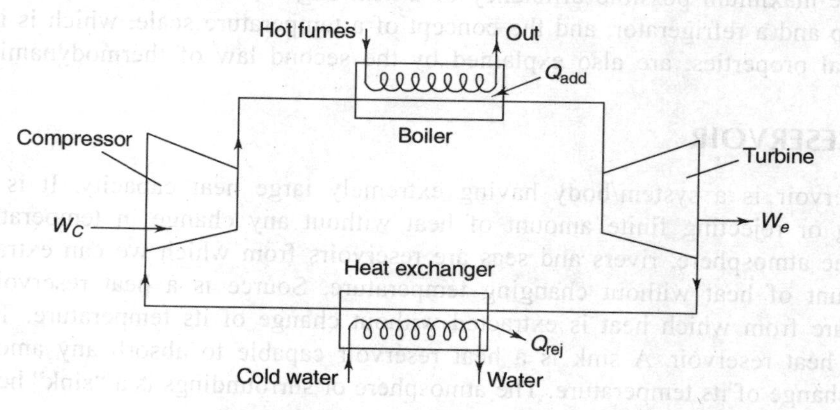

FIGURE 3.2 Gas turbine plant.

The efficiency of a heat engine is to convert as much heat into work as possible:

$$\text{Efficiency } \eta = \frac{\text{Net work output}}{\text{Heat supplied}} = \frac{W}{Q_{add}}$$

A gas turbine plant is also designed to extract as much work from the supplied heat.

$$W = W_e - W_c$$

where W_e = expansion work from turbine
W_c = compression work given to compressor.

Also
$$\Sigma W = \Sigma Q = Q_{\text{add}} - Q_{\text{rej}}$$

Therefore,
$$\eta = \frac{W_e - W_c}{Q_{\text{add}}} = \frac{Q_{\text{add}} - Q_{\text{rej}}}{Q_{\text{add}}}$$

$$\eta = 1 - \frac{Q_{\text{rej}}}{Q_{\text{add}}} \tag{3.1}$$

It can be seen from Eq. (3.1) that efficiency can be increased by reducing heat rejected.

HEAT PUMP

A heat pump is a device used for extracting heat from a low temperature body and sending it to a high temperature body while operating in a cycle (Figure 3.3(a)). Transfer of heat from a low temperature body to a high temperature body is a non-spontaneous process. However, it is possible with the help of a heat pump which uses external work supplied to it. A heat pump is used in cold regions where temperature of the surroundings is low and room temperature is to be kept at higher temperature (Refer to Figure 3.3(b)). A heat pump picks heat (Q_2) from the surroundings (low temperature T_2) and delivers heat (Q_1) to the room which is at high temperature (T_1) using external work (W).

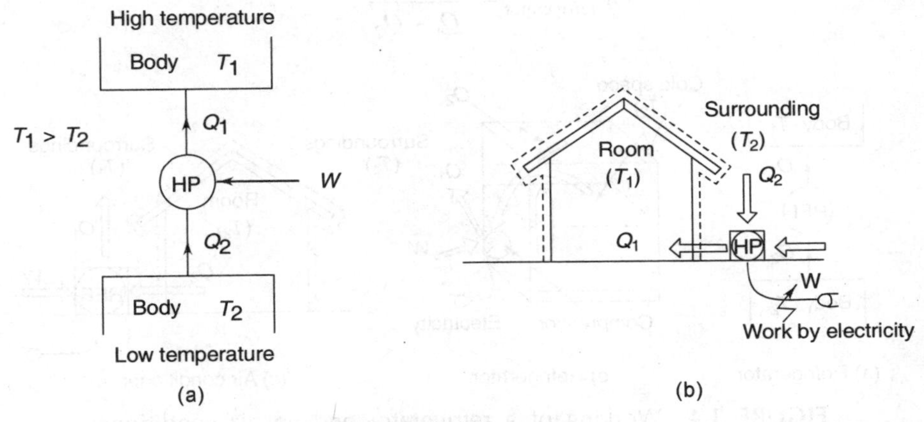

FIGURE 3.3 Heat pump.

Coefficient of performance: The coefficient of performance (COP) is defined as the ratio of desired effect to external work supplied for getting the desired effect:

$$\text{COP} = \frac{\text{Desired effect}}{\text{Work supplied}}$$

Desired effect for a heat pump is to supply heat Q_1 to the hot body or room.

$$\text{COP}_{\text{heat pump}} = \frac{Q_1}{W} \tag{3.2}$$

However

$$\Sigma Q = \Sigma W$$

$$Q_1 - Q_2 = W$$

Therefore, Eq. (3.2) becomes

$$\text{COP}_{\text{heat pump}} = \frac{Q_1}{Q_1 - Q_2} \tag{3.3}$$

REFRIGERATOR

A refrigerator is a device similar to a heat pump, but the desired effect is to extract heat as much as possible from the cold body/space and rejects to a high temperature body/surroundings. The desired effect of a refrigerator is heat (Q_2) removed from cold space.

In a domestic refrigerator and an air conditioner, heat (Q_2) is removed from the refrigerator or room as shown in Figure 3.4 by supplying work (W) and heat (Q_1) is rejected to the surroundings.

$$\text{COP}_{\text{refrigerator}} = \frac{Q_2}{W}$$

or

$$\text{COP}_{\text{refrigerator}} = \frac{Q_2}{Q_1 - Q_2} \tag{3.4}$$

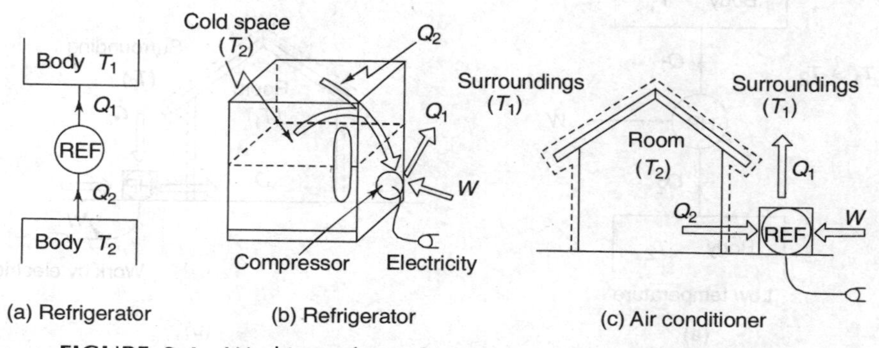

FIGURE 3.4 Working of a refrigerator and an air conditioner.

Since $Q_1 > Q_2$ (as $Q_1 = Q_2 + W$), COP of a heat pump is always greater than COP of a refrigerator.

$$\text{COP}_{\text{heat pump}} = \frac{Q_1}{Q_1 - Q_2} \qquad \text{[from Eq. (3.3)]}$$

$$\text{COP}_{\text{refrigerator}} = \frac{Q_2}{Q_1 - Q_2} \qquad \text{[from Eq. (3.4)]}$$

Therefore, $\text{COP}_{\text{heat pump}} - \text{COP}_{\text{refrigerator}} = \dfrac{Q_1 - Q_2}{Q_1 - Q_2} = 1$

or
$$\text{COP}_{\text{heat pump}} = 1 + \text{COP}_{\text{refrigerator}} \qquad (3.5)$$

STATEMENTS FOR THE SECOND LAW OF THERMODYNAMICS

Clausius statement: The Clausius statement for the second law of thermodynamics states that it is impossible to have a device that while operating in a cycle produces no effect other than transfer of heat from a body at low temperature to a body at higher temperature. A non-spontaneous process such as transfer of heat from a low temperature body to a high temperature body can be realised when some other effects such as external work requirement are bound to be there. Heat pumps and refrigerators use work (electrical energy) to transfer heat from low to high temperature surrounding.

Kelvin–Plank statement: The Kelvin–Plank statement for the second law of thermodynamics is that it is impossible for any device operating in a cycle to produce net work while exchanging heat with bodies at a single fixed temperature.

No cyclic engine can convert whole heat into work. It is impossible to build a heat engine which has 100% efficiency. There is degradation of energy in a cyclic heat engine as some heat has to be degraded or rejected to a low temperature body. There has to be atleast two heat reservoirs (source and sink) for a heat engine to perform.

There is an equivalence between the Kelvin–Plank statement and the Clausius statement. Any system based on violation of the Kelvin–Plank statement leads to violation of the Clausius statement and vice versa.

Violation of the Clausius statement leads to violation of the Kelvin–Plank statement. As shown in Figure 3.5(a), device A is violating the Clausius statement and it is tranferring heat Q_2 from the sink to the source without any work. If a heat engine is made to work in

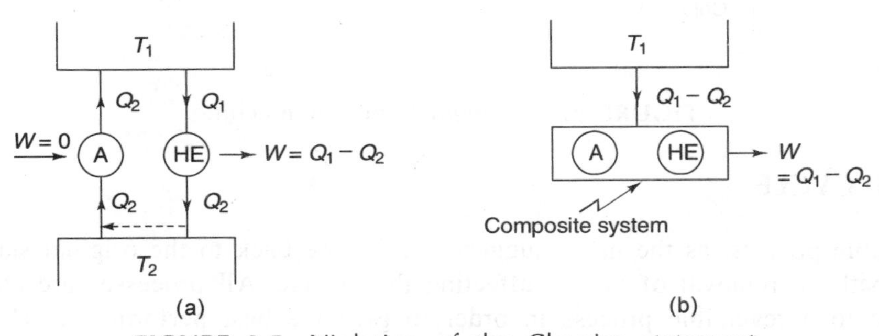

FIGURE 3.5 Violation of the Clausius statement.

parallel to it, we get a composite system (Figure 3.5(b)) which produces work interacting with one reservoir.

Violation of the Kelvin–Plank statement leads to violation of the Clausius statement. Refer to Figure 3.6. Device B is violating the Kelvin–Plank statement and it is producing work taking heat from one reservoir. If a heat pump (HP) is made to work using work output of device B, we get a composite system which extracts heat from the sink and delivers to the source without aid of external work. The composite system violates the Clausius statement.

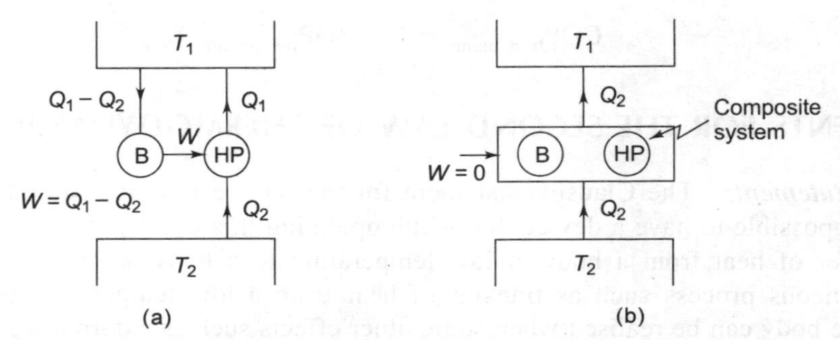

FIGURE 3.6 Violation of the Kelvin–Plank statement.

Perpetual Motion Machine: The Perpetual Motion Machine (PMM) of second kind is a machine which violates the Clausius or Kelvin–Plank statement of the second law of thermodynamics. Figure 3.7(a) shows PMM II, which extracts heat Q_2 from the cold body (T_2) and delivers to the hot body (T_1) without aid of work ($W = 0$). Figure 3.7(b) shows PMM II, which takes heat (Q) from the hot body (T_1) and converts fully into work (W) without any rejection of heat to the cold body.

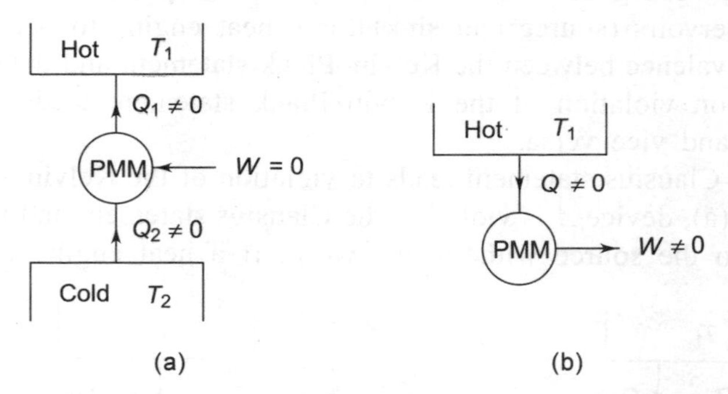

FIGURE 3.7 Perpetual motion machine.

CARNOT CYCLE

The reversible process, as the name suggests, can come back to the original state through the same path on removal of factors affecting the change. All processes are attempted to reach close to a reversible process in order to give the best performance. However, all

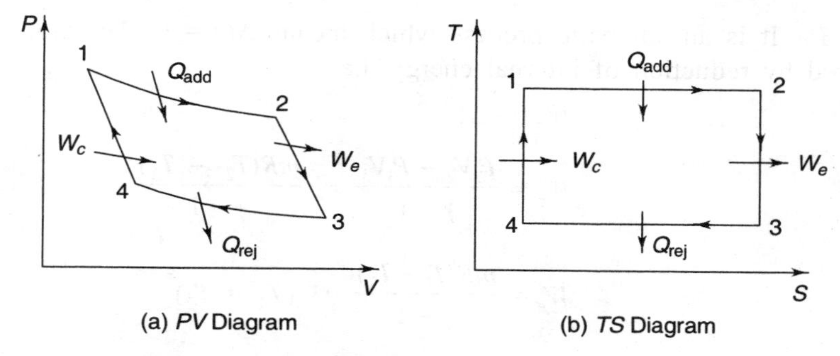

(a) *PV* Diagram (b) *TS* Diagram

FIGURE 3.8 Carnot cycle.

practical processes are irreversible as these cannot attain their original state or follow back the path due to friction or dissipation of energy.

The Carnot cycle is a reversible thermodynamics cycle comprising of four reversible processes (Figure 3.8):

1. Reversible isothermal heat addition (Q_{add})—process 1-2
2. Reversible adiabatic expansion process giving work output (W_e)—process 2-3.
3. Reversible isothermal heat rejection (Q_{rej})—process 3-4.
4. Reversible adiabatic compression using external work (W_c)—process 4-1.

$$\Sigma Q = \Sigma W \tag{3.6}$$

$$Q_{add} - Q_{rej} = W_e - W_c$$

$$\eta = \frac{\text{Net work}}{\text{Heat added}} = \frac{W_e - W_c}{Q_{add}}$$

Substituting Eq. (3.6) in the above equation,

$$\eta = \frac{Q_{add} - Q_{rej}}{Q_{add}} = 1 - \frac{Q_{rej}}{Q_{add}}$$

The analysis of each process of the Carnot cycle can be done for heat and work as given below:

Process 1-2: It is an isothermal process which means $\Delta U = 0$. Complete heat is converted into work.

$$Q_{add} = P_1V_1 \ln \frac{V_2}{V_1} = mRT_1 \ln \frac{V_2}{V_1}$$

$$= mRT_1 \ln r$$

where

$$r = \text{compression ratio} = \frac{V_2}{V_1}$$

Process 2–3: It is an adiabatic process which means $\Delta Q = 0$. Therefore, expansion work is achieved by reduction of internal energy, i.e.

$$W_e = -\Delta U$$

$$W_e = \frac{P_2V_2 - P_3V_3}{\gamma - 1} = \frac{mR(T_2 - T_3)}{\gamma - 1}$$

or

$$W_e = \frac{mR(T_1 - T_3)}{\gamma - 1} \quad \because (T_1 = T_2)$$

(c) *Process 3–4:* It is an isothermal process, i.e. $\Delta U = 0$. Therefore, $Q_{rej} = W_{34}$.

$$Q_{rej} = P_3V_3 \ln \frac{V_3}{V_4} = mRT_3 \ln \frac{V_2}{V_1}$$

$$= mRT_3 \ln r$$

(d) *Process 4–1:* It is an adiabatic, i.e. $\Delta Q = 0$. Therefore, $W_c = -\Delta U$.

$$W_c = \frac{P_1V_1 - P_4V_4}{\gamma - 1} = \frac{mR(T_1 - T_4)}{\gamma - 1} = \frac{mR(T_1 - T_3)}{\gamma - 1}$$

The Carnot cycle is not practical for the following reasons:

1. Frequent change of the cylinder head to make insulating for an adiabatic process and diathermic for an isothermal process.
2. It is practically impossible to achieve isothermal heat addition.
3. Reversible adiabatic expansion and compression are impossible.
4. Reversible isothermal processes are very slow processes while reversible adiabatic processes are fast processes. Speed fluctuation in the revolution of an engine is not possible.

The Carnot cycle can also be operated in reverse direction, i.e. anticlockwise. In that case, work is negative, i.e. work has to be done on the system to get the desired output (cooling of space or heating of room). The reversed Carnot cycle is also called *Carnot refrigeration cycle*.

CARNOT THEOREM

The Carnot theorem states that a Carnot heat engine has efficiency greater than that of any other heat engine operating between the same temperature limits.

To prove the Carnot theorem, consider heat engines *A* and *B* operating in parallel between two reservoirs at temperatures T_1 and T_2 (Figure 3.9(a)). Heat engine *A* is the Carnot heat engine while heat engine *B* operates irreversibly. Assume $\eta_B > \eta_A$ which will give $W_B > W_A$. Now we reverse heat engine *A* to operate as a heat pump (Figure 3.9(b)) and it can take work W_A from heat engine *B*. Now we get a composite system, which takes heat

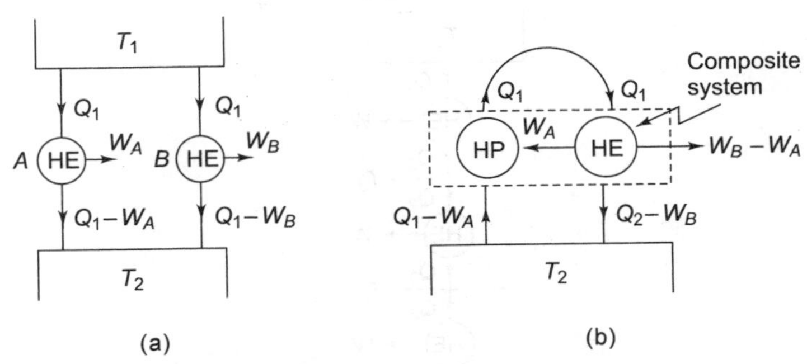

FIGURE 3.9 Carnot theorem.

from reservoir T_2 and it converts it fully to work output $(W_B - W_A)$. The composite system violates the Kelvin–Plank statement. Hence our assumption is wrong and the Carnot heat engine has efficiency greater than any other heat engines.

The corollaries (deductions) of the Carnot theorem are as follows:

1. Efficiency of all reversible engines operating between the same temperature limits is the same.
2. Efficiency of a reversible engine does not depend on the working fluid in the cycle.

THERMODYNAMIC TEMPERATURE SCALE

A temperature scale which is independent of the property of thermometric substance is defined as a *thermodynamic temperature scale*. The thermodynamic temperature scale is developed based on the fact that the efficiency of a reversible heat engine does not depend on the working medium, but it depends on the temperature of two reservoirs in which it is operating.

$$\eta_{\text{Carnot}} = 1 - \frac{Q_2}{Q_1} = \phi(T_1, T_2)$$

or

$$\frac{Q_1}{Q_2} = f(T_1, T_2) = \frac{\psi(T_1)}{\psi(T_2)}$$

By choosing suitable equivalent value of function, we may write as follows:

$$\frac{Q_1}{Q_2} = \frac{T_1}{T_2}$$

where T_1 and T_2 are in Kelvin.

Hence heat flow is proportional to temperature of the reservoir.

Now consider a series of reversible engines operating and producing equal work W while operating between a series of reservoirs (Figure 3.10). As work output is equal for each engine, we can write:

$$W = Q_1 - Q_2 = Q_2 - Q_3 = Q_3 - Q_4 = Q_4 - Q_5$$
$$= T_1 - T_2 = T_2 - T_3 = T_3 - T_4 = T_4 - T_5$$

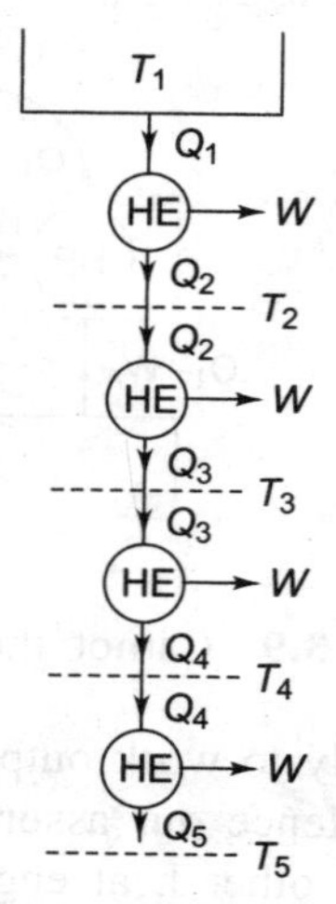

FIGURE 3.10 A series of reversible engines operating between a series of reservoirs.

The difference between the temperatures of successive reservoirs is equal, which can be made small or large depending upon the requirement of the temperature scale.

Heat interactions in a reversible engine are proportional to the absolute temperature of the source and the sink. Hence efficiency of a heat engine (HE) and COP of a heat pump (HP) can be written in terms of reservoir temperatures T_1 and T_2, i.e.

$$\frac{Q_1}{Q_2} = \frac{T_1}{T_2}$$

$$\eta_{HE} = 1 - \frac{Q_2}{Q_1} = 1 - \frac{T_2}{T_1}$$

$$COP_{HP} = \frac{Q_1}{Q_1 - Q_2} = \frac{T_1}{T_1 - T_2}$$

$$COP_{ref} = \frac{Q_2}{Q_1 - Q_2} = \frac{T_2}{T_1 - T_2}$$

CLAUSIUS INEQUALITY

For cyclic processes, let us see the ratio of change of heat to temperature:

In a reversible cycle (Figure 3.11)

$$AB = CD$$

or

$$\frac{Q_1}{T_1} = -\frac{Q_2}{T_2}$$

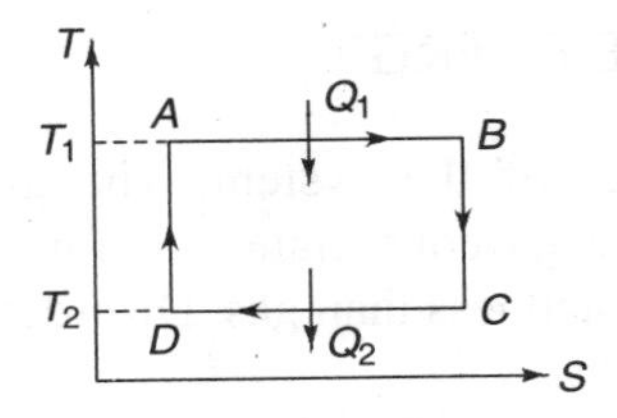

FIGURE 3.11 Reversible cycle.

or
$$\frac{Q_1}{T_1} + \frac{Q_2}{T_2} = 0$$

or
$$\oint \frac{dQ}{T} = 0$$

In an irreversible cycle (Figure 3.12)

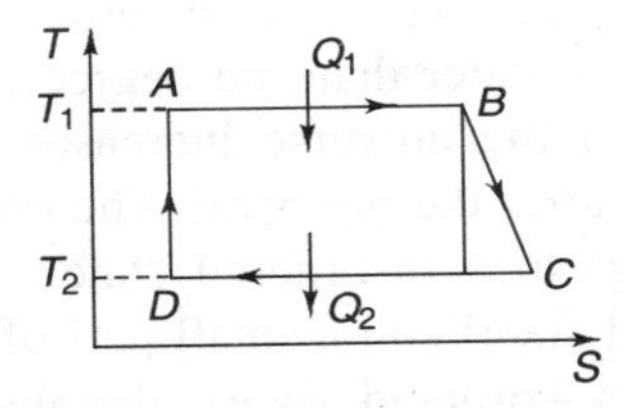

FIGURE 3.12 Irreversible cycle.

$$AB < CD$$

or
$$\frac{Q_1}{T_1} < -\frac{Q_2}{T_2}$$

or
$$\frac{Q_1}{T_1} + \frac{Q_2}{T_2} < 0$$

or
$$\oint \frac{dQ}{T} < 0$$

The Clausius inequality states that

1. $\oint \frac{dQ}{T} = 0$ for a reversible process
2. $\oint \frac{dQ}{T} < 0$ for an irreversible process
3. $\oint \frac{dQ}{T} > 0$ for an impossible process

ENTROPY AND AVAILABLE ENERGY

Entropy is a measure of disorder of the system. The greater is disorder, the higher is entorpy. Entropy is highest in the gaseous state and lowest in the crystalline solid state. Liquid has entropy more than solid and less than gas. Disorder also increases with irreversibility of the process.

A few examples of the increase of disorder or entropy are as follows:

1. A cup on a table (ordered state) falls on to the floor and breaks in small pieces (disordered state).
2. The pieces of the jigsaw game start off in a box in an ordered arrangement in which they form a picture. On shaking, the pieces will take up another arrangement in which the pieces do not form a proper picture as they are now in a disordered state.
3. The computer memory is in a disordered state in the beginning. When we feed information, computer memory changes from the disordered state to an ordered state. It is necessary to use a certain amount of electric energy to do so. The electric energy is also used for running a cooling fan and dissipation of heat from the computer which increases the degree of disorder in surroundings. The degree of disorder in surroundings is greater than the degree of order in the computer memory. Hence the total disorder of the universe increases.
4. The universe is expanding after the big bang. The universe is moving from an ordered state (before the big bang) to a disordered state.
5. We consume food (ordered state) and a small part of energy from it is used for useful purposes (thinking process—ordered state), thereby increasing the disorder.

All heats are not equally valuable for converting into work. Heat at higher temperatures has greater possibility of conversion into work than heat at lower temperatures.

The quantity $\int \frac{dQ}{T}$ is the same for all reversible processes between state 1 and state 2. It is independent of the path and hence it is a state function. Hence $\frac{dQ}{T}$ is an exact differential of entropy (s), i.e. ds. We can say $dQ = TdS$ which means entropy increases with heat addition and decreases with heat removal.

Entropy for a reversible process is defined by the following relation:

$$dS = \left(\frac{dQ}{T}\right)_{\text{reversible}}$$

To find entropy for a irreversible process, the actual process is substituted by an imaginary reversible process. The change of entropy from state 1 to state 2 for an imaginary reversible process and an actual irreversible process would be the same.

Entropy is a function of heat and temperature, which shows the possibility of conversion of that heat into work. The increase in entropy is small when heat is added at high temperature and greater when heat is added at lower temperatures. Therefore, heat having higher entropy has lower possibility for conversion into work. Similarly, heat having lower entropy has higher possibility for conversion into work.

Actually entropy is zero at absolute zero temperature, and it is impossible to achieve absolute zero temperature. Hence convenient temperature is selected at which entropy is given arbitrary value of zero. The selected temperature for steam is 0°C while for NH_3, Fe–12 and CO_2, it is – 40°C. We determine the change of entropy from these points. Therefore, we cannot measure absolute value of entropy. Entropy change for various systems is as follows:

1. Entropy change of an ideal gas for a closed system is:
   From the first law of thermodynamics

$$dQ = dU + dW$$

$$TdS = c_v\, dT + PdV \tag{3.7}$$

We know that

$$P = \frac{RT}{V}$$

Therefore Eq. (3.7) becomes

$$Tds = c_v\, dT + RT\,\frac{dV}{V}$$

$$\int_1^2 dS = \int_1^2 c_v \frac{dT}{T} + \int_1^2 R\frac{dV}{V}$$

$$S_2 - S_1 = c_v \ln \frac{T_2}{T_1} + R \ln \frac{V_2}{V_1} \tag{3.8}$$

For constant volume process $V_2 = V_1$. Therefore Eq. (3.8) becomes

$$S_2 - S_1 = c_v \ln \frac{T_2}{T_1} \tag{3.9}$$

2. Entropy change for an open system is:

$$h = u + Pv$$

$$dh = du + Pdv + vdP$$

$$= dQ + vdP$$

$$c_p dT = TdS - RT\,\frac{dP}{P} \quad (\because\ Pv = RT)$$

$$\int_1^2 dS = \int_1^2 c_p \frac{dT}{T} - \int_1^2 R\frac{dP}{P}$$

$$S_2 - S_1 = c_p \ln \frac{T_2}{T_1} - R \ln \frac{P_2}{P_1} \tag{3.10}$$

For constant pressure processes $P_2 = P_1$. Therefore, Eq. (3.10) becomes

$$S_2 - S_1 = c_p \ln \frac{T_2}{T_1}$$

3. Entropy change for a polytropic process is

$$S_2 - S_1 = c_v \ln \frac{T_2}{T_1} + R \ln \frac{V_2}{V_1} \tag{3.11}$$

From $PV = RT$ and PV^n = constant

$$\frac{V_2}{V_1} = \left(\frac{T_1}{T_2}\right)^{\frac{1}{n-1}}$$

Therefore Eq. (3.11) becomes

$$S_2 - S_1 = c_V \ln \frac{T_2}{T_1} + c_V (\gamma - 1) \ln \left(\frac{T_1}{T_2}\right)^{\frac{1}{n-1}} \qquad (\because R = c_v (\gamma - 1))$$

$$= c_v \ln \frac{T_2}{T_1}\left(\frac{n-\gamma}{n-1}\right)$$

4. Entropy change for an isoentropic process is zero while for an isothermal process, it is $\frac{Q}{T}$.

Entropy generation: For a reversible process, entropy change $(dS)_R$ is equal to $\frac{dQ}{T}$ and for an irreversible process, entropy change (ds) is more than reversible process. Hence for an irreversible process

$$dS > \frac{dQ}{T}$$

$$dS = \frac{dQ}{T} + S_G \tag{3.12}$$

where S_G = entropy generation

It is clear from Eq. (3.12) that $S_G = 0$ for a reversible process and $S_G > 0$ for an irreversible process.

Entropy increase: The principle of entropy increase states that the entropy of an isolated system will always increase, i.e. $\Delta S \geq 0$ for an isolated system. Our universe is also an isolated system in which all processes take place. If we take an individual system in the universe which receives heat ΔQ at temperature T from the surroundings (temperature T_s), then change of entropy of the universe is:

$$\Delta S_{\text{universe}} = \Delta S_{\text{system}} + \Delta S_{\text{surroundings}}$$

$$= \frac{\Delta Q}{T} - \frac{\Delta Q}{T_s}$$

(Change of entropy for a system is positive as heat ΔQ enters while change of entropy for the surrounding is negative as heat leaves).

$$\Delta S_{\text{universe}} = \Delta Q \left[\frac{1}{T} - \frac{1}{T_s}\right]$$

As $$T_s > T,$$

$$\Delta S_{\text{universe}} > 0$$

As change of entropy $dS = \left(\frac{dQ}{T}\right)_R$, heat interaction $dQ = Tds$ for a reversible process.

Hence $\int TdS$ is the area under curve on a TS diagram which is equal to the heat interaction for the process.

We know that in a cyclic process

$$\oint dQ = \oint dW$$

or $$\oint TdS = \oint pdv$$

We know that in a reversible adiabatic process, heat interaction is zero, i.e. $dQ = 0$ or $TdS = 0$ or $dS = 0$ or S = constant. Hence a reversible adiabatic process is also called isentropic (constant entropy) process. However, an isentropic process need not be an adiabatic process as the entropy during a particular process may be kept constant by heat transfer to or from the system.

We have already seen that $\Delta Q = TdS$. Also $dQ = c_v\, dT$ for constant volume and $dQ = c_p\, dT$ for constant pressure.

$$dQ = TdS = c_v\, dT$$

or $$\frac{T}{c_v} = \frac{dT}{dS} \quad \text{(for constant volume)}$$

Similarly $$dQ = TdS = c_p\, dT$$

or $$\frac{T}{c_p} = \frac{dT}{dS} \quad \text{(for constant pressure)}$$

Since $c_p > c_v$, the constant volume line has a greater slope than the constant pressure line on a TS diagram.

Third law: The entropy of pure substance approaches zero at absolute zero temperature. This is called the *third law of thermodynamics*.

Available energy: The second law of thermodynamics prohibits the complete conversion of low-grade energy into high-grade energy. A portion of energy that can be converted into work is called *available exergy* or *exergy* and the rest is unattainable energy or *anergy*. Therefore,

$$\text{Energy} = \text{exergy} + \text{anergy}$$

At each temperature (T), heat has two portions viz., available heat energy capable of doing work and unavailable heat energy which is rejected to the surroundings (T_0).

$$\eta = \frac{dW}{dQ} = 1 - \frac{T_0}{T_1}$$

or $$dW = dQ\left(1 - \frac{T_0}{T_1}\right)$$

$$= dQ - dQ\frac{T_0}{T_1}$$

or $$dQ = dW + dQ\frac{T_0}{T_1} \tag{3.13}$$

From Eq. (3.13), it is clear that heat rejection $= \dfrac{T_0}{T_1} \times dQ$.

However,

$$\frac{dQ}{T_1} = dS$$

Therefore, Heat rejection $= T_0\, dS$

Hence, Unavailable energy = anergy = heat rejection $= T_0\, dS$

As the surroundings temperature (T_0) is known, change of entropy (dS) is a measure of unavailable heat energy.

Entropy generation in a closed system is:

$$(\Delta S)_{\text{total}} = S_{\text{generation}} = (\Delta S)_{\text{system}} + (\Delta S)_{\text{surroundings}}$$

$$S_{\text{generation}} = m(S_2 - S_1) + \frac{Q_{\text{surroundings}}}{T_{\text{surroundings}}}$$

Entropy generation in an open system is:

$$(\Delta S)_{\text{total}} = S_{\text{generation}} = (\Delta S)_{\text{system}} + (\Delta S)_{\text{surroundings}}$$

$$S_{\text{generation}} = (S_2 - S_1) + (S_0 - S_1) + \frac{Q_{\text{surroundings}}}{T_{\text{surroundings}}}$$

For steady flow conditions, inside change of entropy ($S_2 - S_1$) is zero and only change of entropy at the inlet and the outlet ($S_0 - S_1$) will remain.

$$S_{generation} = (S_0 - S_1) + \frac{Q_{surroundings}}{T_{surroundings}}$$

Irreversibility: Entropy change (dS) is a point function and hence it is a thermodynamic property. However, entropy generation is not a thermodynamic property as it is dependent on the path which the system has followed. Irreversibility is

$$i = W^{rev} - W = T_0 \, S_{generation}$$

where W^{rev} is maximum work which can be extracted by a reversible engine.

Exergy is maximum possible theoretical work which can be extracted from a system and environment before they come to dead state. At the dead state, they have energy but no exergy.

Maximum work can be obtained when a process takes place in reversible manner. Almost all processes are irreversible. Exergy (availability) cannot be conserved like energy. Exergy gets destroyed due to irreversibility in process. Irreversibility is equal to the product of entropy generated and temperature. As irreversible processes are continuously increasing, it means unavailable energy is also gradually increasing. This is called *law of degradation of energy*.

Exergy or Availability of energy is a measure of the state of a system from the state of environment/surrounding. Therefore, exergy is an attribute of a system and environment together. If the environment is specified, a numerical value can be assigned to availability in terms of property value for the system only. Therefore, exergy can be regarded as the property of the system.

SOLVED PROBLEMS

1. A heat engine, a heat pump and a refrigerator are working between two reservoirs at temperatures of 600 K and 300 K. Find the efficiency of heat energy, COP of the heat pump and the refrigerator.

$$\eta \text{ of the heat engine} = 1 - \frac{T_2}{T_1}$$

$$= 1 - \frac{300}{600} = \frac{1}{2}$$

$$\text{COP of the heat pump} = \frac{T_1}{T_1 - T_2}$$

$$= \frac{600}{600 - 300}$$

$$= \frac{600}{300} = 2$$

$$\text{COP of the refrigerator} = \frac{T_2}{T_1 - T_2}$$

$$= \frac{300}{600 - 300} = \frac{300}{300} = 1$$

2. A heat engine, a heat pump and a refrigerator reject 50 kJ, 125 kJ and 150 kJ heat respectively. If each one get heat 100 kJ, find η of the heat engine and COP of the heat pump and the refrigerator.

Heat engine: $Q_1 = 100$ kJ and $Q_2 = 50$ kJ

$$W = Q_1 - Q_2 = 100 - 50 = 50$$

$$\eta = \frac{W}{Q_1} = \frac{50}{100} = 0.5$$

Heat pump: $Q_2 = 100$ and $Q_1 = 125$ kJ

$$W = Q_1 - Q_2 = 125 - 100 = 25 \text{ kJ}$$

$$\text{COP} = \frac{Q_1}{W} = \frac{125}{25} = 5$$

Refrigerator: $Q_2 = 100$ and $Q_1 = 150$ kJ

$$W = Q_1 - Q_2 = 150 - 100 = 50 \text{ kJ}$$

$$\text{COP} = \frac{Q_2}{W} = \frac{100}{50} = 2$$

3. A Carnot engine has an efficiency of 0.5. Find COP of a refrigerator working within the same temperature limit.

$$\eta \text{ of the heat engine} = 0.5 = 1 - \frac{T_2}{T_1}$$

or $$\frac{T_2}{T_1} = 0.5$$

or $$T_2 = 0.5\,T_1$$

$$\text{COP of the refrigerator} = \frac{T_2}{T_1 - T_2}$$

$$= \frac{0.5\,T_1}{T_1 - 0.5\,T_1} = 1$$

4. Determine the heat to be supplied to a Carnot engine operating between 800 and 400 K and producing 100 kJ of work.

$$\eta = 1 - \frac{T_2}{T_1}$$

$$= 1 - \frac{400}{800} = 0.5$$

$$\eta = \frac{W}{Q_{add}} = 0.5$$

$$\frac{100}{Q_{add}} = 0.5$$

or $$Q_{add} = \frac{100}{0.5} = 200 \text{ kJ}$$

5. A refrigerator operates on a reversed Carnot cycle between 900 and 300 K. If heat at the rate of 3 kJ/s is extracted from the low temperature space, find the power required to drive the refrigerator.

$$\frac{Q_1}{Q_2} = \frac{T_1}{T_2}$$

$$\frac{Q_1}{3} = \frac{900}{300}$$

or $$Q_1 = 9 \text{ kJ/s}$$

$$W = Q_1 - Q_2$$

$$= 9 - 3 = 6 \text{ kJ/s} = 6 \text{ kW}$$

Hence the power to derive the refrigerator is 6 kW.

6. A cold storage plant of 20 tonne of refrigerator capacity operates between 200 and 300 K. Determine the power required to run the plant if plant has half COP of a Carnot cycle. (Take 1 tonne refrigeration = 3.5 kW.)

$$\text{COP of the Carnot cycle} = \frac{T_2}{T_1 - T_2}$$

$$= \frac{200}{300 - 200} = 2$$

$$\text{COP of the plant} = \frac{1}{2} (\text{COP})_{\text{Carnot}}$$

$$= \frac{1}{2} \times 2 = 1$$

$$Q_2 = 20 \text{ tonne of refrigeration}$$

$$= 20 \times 3.5 = 70 \text{ kW}$$

$$\text{COP} = \frac{Q_2}{W} = \frac{70}{W} = 1$$

or $$W = 70 \text{ kW}$$

Hence, 70 kW power is required to run the plant.

7. 300 kJ/s of heat is supplied at a constant temperature of 500 K to a heat engine. The heat rejection takes place at 300 K. The following results are obtained:

(i) 210 kJ
(ii) 180 kJ
(iii) 150 kJ

Classify which of the result is reversible, irreversible or impossible result.

(i) $$\Sigma \frac{dQ}{T} = \frac{300}{500} - \frac{210}{300}$$

$$= 0.6 - 0.7 = -0.1 < 0$$

Hence the cycle is irreversible.

(ii) $$\Sigma \frac{dQ}{T} = \frac{300}{500} - \frac{180}{300}$$

$$= 0.6 - 0.6 = 0$$

Hence the cycle is reversible.

(iii) $$\Sigma \frac{dQ}{T} = \frac{300}{500} - \frac{150}{300}$$

$$= 0.6 - 0.5 = 0.1 > 0$$

Hence the cycle is impossible.

8. A steam turbine plant is as shown working in temperature range of 500 K and 300 K. Enthalpy at various states are as follows:

(a) State 1 = 700 kJ/kg
(b) State 2 = 2200 kJ/kg
(c) State 3 = 1500 kJ/kg
(d) State 4 = 500 kJ/kg

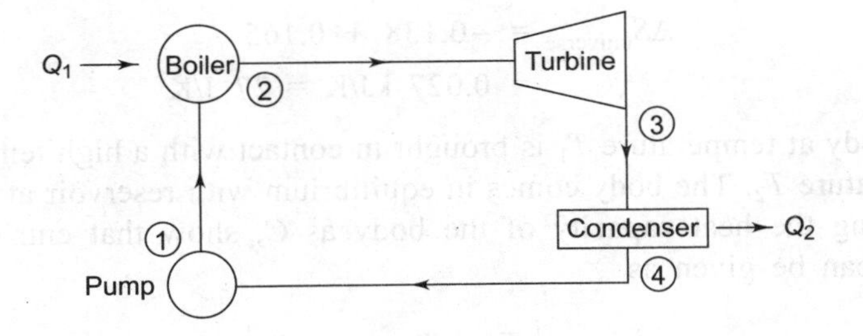

Verify the Clausius inequality.

$$Q_1 = h_2 - h_1 = 2200 - 700 = 1500 \text{ kJ/kg}$$
$$Q_2 = h_4 - h_3 = 1500 - 500 = 1000 \text{ kJ/kg}$$

$$\sum \frac{dQ}{T} = \frac{Q_1}{T_1} + \frac{Q_2}{T_2}$$
$$= \frac{1500}{500} - \frac{1000}{300}$$
$$= 3 - 3.33$$
$$= -0.33 \text{ kJ/kg K} < 0$$

Hence the Clausius inequality is proved.

9. Determine the change in entropy of the universe if a copper block of 1 kg at 150°C is placed in sea water at 25°C. The heat capacity of the copper block is 0.393 kJ/(kg K).

$$\Delta S_{\text{universe}} = \Delta S_{\text{block}} + \Delta S_{\text{water}}$$
$$T_1 = 150°\text{C} + 273 = 423 \text{ K}$$
$$T_2 = 25°\text{C} + 273 = 298 \text{ K}$$
$$\Delta S_{\text{block}} = mC \ln \frac{T_2}{T_1}$$
$$= 1 \times 0.393 \times \ln \frac{298}{423}$$
$$= -0.138 \text{ kJ/K}$$

$$\text{Heat given to water} = \text{Heat lost by the block}$$
$$= +1 \times 0.393 \times 125$$
$$= 49.13 \text{ kJ}$$

$$\Delta S_{\text{water}} = \frac{Q}{T_2} = \frac{49.13}{298} = 0.165 \text{ kJ/K}$$

$$\Delta S_{universe} = -0.138 + 0.165$$
$$= 0.027 \text{ kJ/K} = 27 \text{ J/K}$$

10. A cold body at temperature T_1 is brought in contact with a high temperature reservoir at temperature T_2. The body comes in equilibrium with reservoir at constant pressure. Considering the heat capacity of the body as C, show that entropy change of the universe can be given as

$$mC\left[\frac{T_1 - T_2}{T_2} - \ln\frac{T_1}{T_2}\right]$$

$$\Delta S_{universe} = \Delta S_{reservoir} + \Delta S_{body}$$

The body will heat up to temperature of T_2 from T_1.

$$\Delta S_{body} = mC \ln\frac{T_2}{T_1}$$

Heat lost by the reservoir = heat gained by body

$$= mC(T_2 - T_1)$$

$$\Delta S_{reservoir} = \frac{-mC(T_2 - T_1)}{T_2}$$

$$\Delta S_{universe} = mC \ln\frac{T_2}{T_1} - \frac{mC(T_2 - T_1)}{T_2}$$

$$= \frac{mC(T_1 - T_2)}{T_2} - mC \ln\frac{T_1}{T_2}$$

$$= mC\left(\frac{T_1 - T_2}{T_2} - \ln\frac{T_1}{T_2}\right)$$

11. A heat engine works between starting temperature limits of T_1 and T_2 of two bodies. Working fluid flows at the rate of "m" kg/s and the specific heat at constant pressure is c_p. Determine the maximum obtainable work till the bodies attain the same temperature.

Let final temperature = T_3

As the engine works, heat from the body is taken till its temperature falls to T_3.

$$\Delta S_{body1} = \int_{T_1}^{T_3} mc_p \frac{dT}{T} = mc_p \ln\frac{T_3}{T_1}$$

Similarly, the cold body will attain temperature T_3 from T_2.

$$\Delta S_{body2} = \int_{T_2}^{T_3} mc_p \frac{dT}{T} = mc_p \ln\frac{T_3}{T_2}$$

For the maximum work, the process must be reversible and entropy change is to be zero.

$$\Delta S_{\text{body1}} + \Delta S_{\text{body2}} = 0$$

$$mc_p \ln\frac{T_3}{T_1} + mc_p \ln\frac{T_3}{T_2} = 0$$

or $$\ln\frac{T_3}{T_1} + \ln\frac{T_3}{T_2} = 0$$

or $$\ln\frac{T_3}{T_1} \times \frac{T_3}{T_2} = 0\ (= \ln 1\ (\because \ln 1 = 0)$$

Therefore, $$\frac{T_3^2}{T_1 \times T_2} = 1$$

or $$T_3 = \sqrt{T_1 T_2}$$

Maximum work = heat given by the first body – heat taken by the second body

$$\text{Maximum work} = mc_p(T_1 - T_3) - mc_p(T_3 - T_1)$$

$$= mc_p(T_1 - 2T_3 + T_2)$$

$$= mc_p(T_1 - 2\sqrt{T_1 T_2} + T_2)$$

$$= mc_p(\sqrt{T_1} - \sqrt{T_2})^2$$

12. Determine anergy or unavailable energy if a heat engine is working between the temperature limit of 1000 and 300 K. Heat delivers to engine is 1000 J and work output is 400 J.

$$\Delta S = \frac{Q_1}{T_1} + \frac{Q_2}{T_2}$$

Here Q_1 = heat supplied to the engine

Q_2 = heat rejected by the engine

$$Q_1 - Q_2 = W$$

or $$Q_2 = Q_1 - W$$

$$= 1000 - 400 = 600 \text{ J}$$

$$\Delta S = \frac{-1000}{1000} + \frac{600}{300}$$

$$= 1 \text{ J/K}$$

$$\therefore \text{Unavailable energy} = T_2 \times \Delta S$$
$$= 300 \times 1$$
$$= 300 \text{ J}$$

13. A closed system executed a reversible cycle 1–2–3–4–5–6–1 consisting of six processes. During processes 1–2 and 3–4, the system receives 1000 kJ and 800 kJ of heat respectively at constant temperature of 500 K and 400 K respectively. Processes 2–3 and 4–5 are adiabatic expansions in which the temperature is reduced from 500 K to 400 K and from 400 K to 300 K respectively. During process 5–6, the system rejects heat at a temperature of 300 K. Process 6–1 is an adiabatic compression process. Determine the work done by the system during the cycle and thermal efficiency of the cycle.

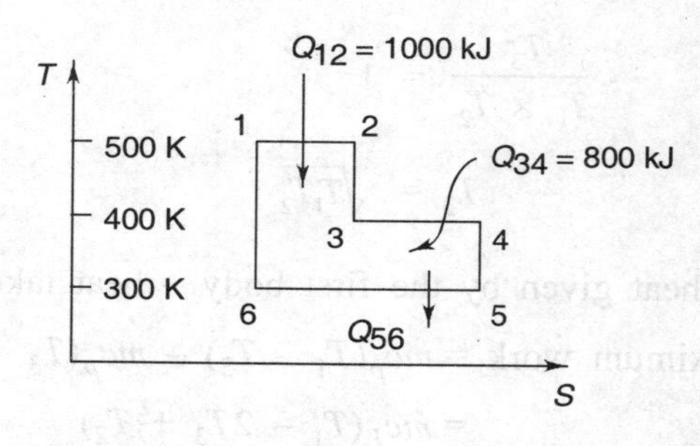

Process 1–2:
$$Q_{12} = T_1(S_2 - S_1)$$
$$1000 = 500(S_2 - S_1)$$
or
$$S_2 - S_1 = 2 \text{ kJ/K}$$

Process 3–4:
$$Q_{34} = T_3(S_4 - S_3)$$
$$800 = 400(S_4 - S_3)$$
or
$$S_4 - S_3 = 2 \text{ kJ/K}$$

Process 5–6:
$$Q_{56} = T_6(S_5 - S_6)$$
$$= 300[(S_2 - S_1) + (S_4 - S_3)]$$
$$= 300 \times 4 = 1200 \text{ kJ}$$

$$\text{Heat added} = Q_{12} + Q_{34} = 1000 + 800 = 1800 \text{ kJ}$$
$$\text{Heat rejected} = Q_{56} = 1200 \text{ kJ}$$
$$W = Q_{add} - Q_{rej}$$
$$= 1800 - 1200 = 600 \text{ kJ}$$

Therefore, $$\eta = \frac{W}{Q_{add}} = \frac{600}{1800} = \frac{1}{3}$$

$$= 33.34\%$$

14. A reversible heat engine operated between two reservoirs at temperatures of 600°C and 40°C. The engine drives a reversible refrigerator which operates between reservoirs at temperature of 40°C and –20°C. The heat transfer to the heat engine is 2000 kJ and net work output of the combined engine–refrigerator plant is 300 kJ. Evaluate the heat transfer to the refrigerator and the net heat transfer to the reservoir at 40°C.

(UPTU: Dec. 2005)

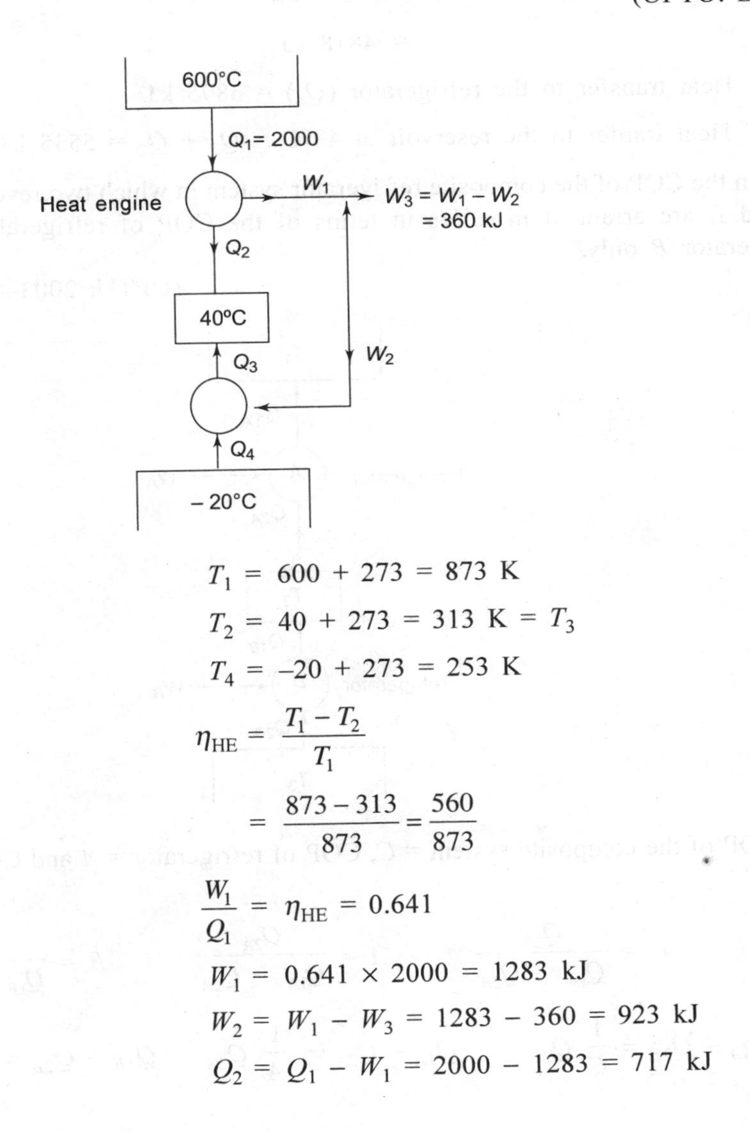

$$T_1 = 600 + 273 = 873 \text{ K}$$

$$T_2 = 40 + 273 = 313 \text{ K} = T_3$$

$$T_4 = -20 + 273 = 253 \text{ K}$$

$$\eta_{HE} = \frac{T_1 - T_2}{T_1}$$

$$= \frac{873 - 313}{873} = \frac{560}{873}$$

$$\frac{W_1}{Q_1} = \eta_{HE} = 0.641$$

$$\therefore \quad W_1 = 0.641 \times 2000 = 1283 \text{ kJ}$$

$$W_2 = W_1 - W_3 = 1283 - 360 = 923 \text{ kJ}$$

$$Q_2 = Q_1 - W_1 = 2000 - 1283 = 717 \text{ kJ}$$

$$\text{COP}_{\text{ref}} = \frac{T_4}{T_3 - T_4} = \frac{253}{313 - 253}$$

$$= \frac{253}{60} = 4.22$$

$$\text{COP} = \frac{Q_4}{W_2} = 4.22$$

$$\therefore \quad Q_4 = 923 \times 4.22 = 3895 \text{ kJ}$$

$$Q_3 = W_2 + Q_4 = 923 + 3895$$

$$= 4818 \text{ kJ}$$

$\therefore$ Heat transfer to the refrigerator (Q_4) = 3895 kJ

Heat tranfer to the reservoir at 40°C = $Q_3 + Q_2$ = 5535 kJ

15. Obtain the COP of the composite refrigerator system in which two reversible refrigerators *A* and *B* are arranged in series in terms of the COP of refrigerator *A* and COP of refrigerator *B* only.

(UPTU: 2003–04 and May 2008)

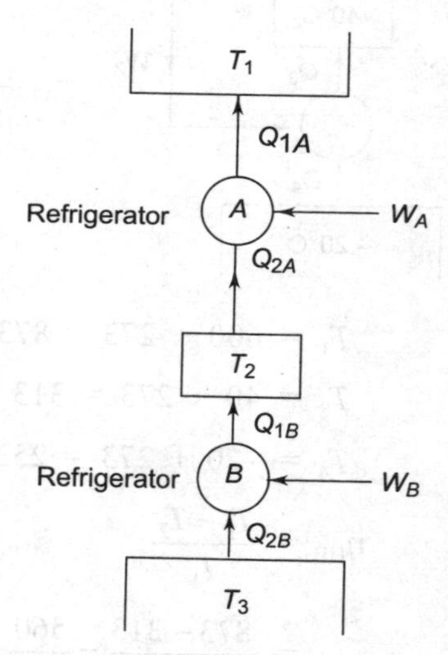

Let COP of the composite system = *C*, COP of refrigerator = *A* and COP of refrigerator = *B*.

$$C = \frac{Q_{2B}}{Q_{1A} - Q_{2B}} \qquad A = \frac{Q_{2A}}{Q_{1A} - Q_{2A}} \qquad B = \frac{Q_{2B}}{Q_{1B} - Q_{2B}}$$

$$\therefore \quad Q_{1A} - Q_{2B} = \frac{1}{C} Q_{2B} \qquad Q_{1A} - Q_{2A} = \frac{1}{A} Q_{2A} \qquad Q_{1B} - Q_{2B} = \frac{1}{B} Q_{2B}$$

or $$Q_{1A} = \left(\frac{1}{C}+1\right) Q_{2B} \qquad Q_{1A} = \left(\frac{1}{A}+1\right) Q_{2A} \qquad Q_{1B} = \left(\frac{1}{B}+1\right) Q_{2B}$$

Now $Q_{2A} = Q_{1B}$

Therefore, $$\frac{Q_{1A}}{\left(\frac{1}{A}+1\right)} = \left(\frac{1}{B}+1\right) Q_{2B}$$

or $$\left(\frac{\frac{1}{C}+1}{\frac{1}{A}+1}\right) Q_{2B} = \left(\frac{1}{B}+1\right) Q_{2B}$$

or $$\left(\frac{1}{C}+1\right) = \left(\frac{1}{A}+1\right)\left(\frac{1}{B}+1\right)$$

or $$\frac{1}{C}+1 = \frac{(1+A)(1+B)}{AB}$$

or $$\frac{1}{C} = \frac{1+AB+A+B-AB}{AB}$$

or $$C = \frac{AB}{A+B+1}$$

or $$(\text{COP})_{\text{composite}} = \frac{(\text{COP})_A (\text{COP})_B}{(\text{COP})_A + (\text{COP})_B + 1}$$

16. Which is more effective way to increase the efficiency of a reversible heat engine (i) to increase the source temperature T_1 while sink temperature T_2 kept constant or (ii) to decrease the sink temperature by the same amount while source temperature is constant.

(UPTU: 2006–2007)

$$\eta = 1 - \frac{T_2}{T_1}$$

Case 1: Reduce the temperature T_2 by dT:

$$\eta_{\text{now}} = 1 - \frac{T_2 - dT}{T_1}$$

$$= \left(1 - \frac{T_2}{T}\right) + \frac{dT}{T_1}$$

$$= \eta + \frac{dT}{T_1}$$

Case 2: Increase the temperature T_1 by dT

$$\eta_{new} = 1 - \frac{T_2}{T_1 + dT}$$

$$= 1 - \frac{T_2}{T_1\left(1 + \frac{dT}{T_1}\right)} = 1 - \frac{T_2}{T_1}\left(1 + \frac{dT}{T_1}\right)^{-1}$$

$$= 1 - \frac{T_2}{T_1}\left(1 - \frac{dT}{T_1}\right)$$

$$= \left(1 - \frac{T_2}{T_1}\right) + T_2 \times \frac{dT}{T_1}$$

$$= \eta + T_2\left(\frac{dT}{T_1}\right)$$

As
$$T_2\left(\frac{dT}{T_1}\right) > \left(\frac{dT}{T_1}\right)$$

Hence efficiency increases more when temperature of source is increased as compared to lowering temperature.

17. A metal block of 5 kg and 200°C is cooled in a surrounding of air which is at 30°C. If the specific heat of metal is 0.4 kJ/kgK, calculate the following to (a) entropy change of block and (b) entropy change of the surroundings and universe.

(UPTU: 2006–2007)

$$(\Delta S)_{block} = m \times C_p \times \int_{T_1}^{T_2} \frac{dT}{T}$$

$$= 5 \times 0.4 \times \log \frac{303}{473}$$

$$= -0.89 \text{ kJ/K}$$

Heat absorbed by the atmosphere is

$$Q = m \times c_p \times (T_1 - T_2)$$

$$= 5 \times 0.4 \times (473 - 303)$$

$$= 340 \text{ kV}$$

Now for surroundings

$$(\Delta S)_{\text{surroundings}} = \frac{Q}{T_2} = \frac{340}{303}$$

$$= 1.12 \text{ kJ/K}$$

$$(\Delta S)_{\text{universe}} = (\Delta S)_{\text{block}} + (\Delta S)_{\text{surroundings}}$$

$$= -0.89 + 1.12$$

$$= 0.23 \text{ kJ/K}$$

18. A cyclic heat engine operator between a source temperature of 800°C and a sink temperature of 30°C. What is the least rate of heat rejection per kW net output of the engine?

(UPTU: 2007–2008)

$$\eta = 1 - \frac{T_2}{T_1}$$

$$= 1 - \frac{303}{1017}$$

$$= 0.718$$

$$\eta = \frac{W}{Q_1} \quad \text{given } W = 1 \text{ kW} = 10^3 \text{ W}$$

$$0.718 = \frac{1 \times 10^3}{Q_1}$$

or

$$Q_1 = \frac{10^3}{0.718} = 1.4 \times 10^3$$

$$= 1.4 \text{ kW}$$

$$Q_2 = Q_1 - W$$

$$= 1.4 - 1 = 0.4 \text{ kW}$$

19. A fluid undergoes a reversible adiabatic compression from 0.5 MPa, 0.02 m^3 to 0.05 m^3 according to the law $pv^{1.3}$ = constant. Determine the change in enthalpy, internal energy, entropy and heat of work transfer during the process.

(UPTU: 2006–2007)

$$P_1 v_1^{1.3} = P_2 v_2^{1.3}$$

or
$$\left(\frac{v_1}{v_2}\right)^{1.3} = \frac{P_2}{P_1}$$

$$P_2 = 0.5 \times \left(\frac{0.02}{0.05}\right)^{1.3}$$

$$= 0.152 \text{ MPa}$$

(a) Work for adiabatic process

$$W = \frac{P_1 v_1 - P_2 v_2}{v - 1}$$

$$= \frac{1.52 \times 10^5 \times 0.05 - 5 \times 10^5 \times 0.02}{1.3 - 1}$$

$$= -8 \text{ kJ}$$

Negative sign means that work is done on the system.

(b) Change in internal energy

$$Q = \Delta U + W$$

For adiabatic process $Q = 0$. Therefore,

$$\Delta U = -W$$

$$= 8 \text{ kJ}$$

(c) Change in enthalpy

$$\Delta h = \Delta U + (P_2 v_2 - P_1 v_1)$$

$$= 8 \text{ kJ} + 2.4 \text{ kJ}$$

$$= 10.4 \text{ kJ}$$

(d) Change in entropy

$$Q = 0 \quad \text{and} \quad \Delta S = \frac{Q}{T}$$

$\therefore$
$$\Delta S = 0.$$

20. At a place where surroundings are at 1 bar 27°C, a closed rigid thermally insulated tank contains 2 kg air at 2 bar, 27°C. This air is then churned for a while, by a paddle wheel connected to an external motor. If it is given that the irreversibility of the process is 100 kJ, find the final temperature, and the increase in availability of air. Assume for air c_v = 6.718 kJ/kg K.

(GATE: 1997)

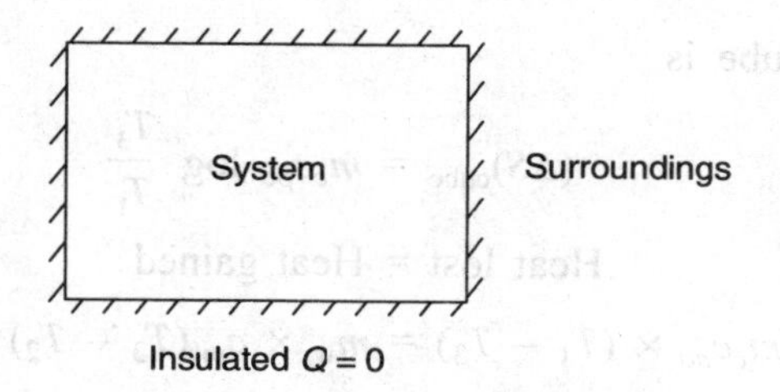

No heat flow from system and entropy change of surroundings is

$$(\Delta S)_{\text{surroundings}} = \frac{Q}{T_0} = 0$$

$$\text{Irreversibility } I = T_0(\Delta S_{\text{sys}} + \Delta S_{\text{sur}})$$

$$= T_0 \times \Delta S_{\text{sys}}$$

or

$$(\Delta S)_{\text{sys}} = \frac{I}{T_0} = \frac{100}{300} = 0.33 \text{ kJ/K}$$

But we know

$$(\Delta S)_{\text{sys}} = m\left[c_v \log \frac{T_2}{T_1} + R \log \frac{v_2}{v_1}\right]$$

Here $v_1 = v_2$

$\therefore$

$$(\Delta S)_{\text{sys}} = m\, c_v \log \frac{T_2}{T_1}$$

$$0.33 = 2 \times 0.718 \log \frac{T_2}{300}$$

$\therefore$

$$T_2 = 3783 \text{ K}$$

Increase of availability is

$$\Delta E_{\text{avail}} = m(\Delta E - T_0 \Delta S)$$

$$= 2[c_v(T_2 - T_1) - T_0 \Delta S]$$

$$= 2[0.718(3783 - 300) - T_0 \log \frac{3783}{300}]$$

$$= 12.54 \text{ kJ}$$

21. An iron cube at a temperature of 400°C is dropped into an insulated bath containing 10 kg water at 25°C. The water finally reaches a temperature of 50°C at steady state. Given that the specific heat of water is equal to 4186 J/kg K. Find the entropy change for the iron cube and the water. Is the process is irreversible? If so, why?

(GATE: 1996)

Entropy change of cube is

$$(\Delta S)_{\text{cube}} = m_c c_{pc} \log \frac{T_3}{T_1}$$

$$\text{Heat lost} = \text{Heat gained}$$

$$m_c c_{pc} \times (T_1 - T_3) = m_w \times c_{pw}(T_3 - T_2)$$

where T_1 = initial temperature of cube, T_2 = initial temperature of water and T_3 = final temperature $(m_c \times c_{pc})$ $(400 - 50) = 10 \times 4186 \times (50 - 25)$

$$\therefore \qquad m_c \times c_{pc} = \frac{10 \times 4186 \times 25}{350} = 2990 \text{ J/K}$$

$$(\Delta S)_{\text{cube}} = 2990 \log \frac{50 + 273}{400 + 273}$$

$$= -2990 \times 0.734$$

$$= -2195 \text{ kJ/K}$$

$$(\Delta S)_{\text{water}} = m_w\ c_{pw} \log \frac{T_3}{T_2}$$

$$= 10 \times 4186 \log \frac{323}{298}$$

$$= 3372 \text{ kJ/K}$$

$$(\Delta S)_{\text{total}} = (\Delta S)_{\text{cube}} + (\Delta S)_{\text{water}}$$

$$= -2195 + 3372$$

$$= 1177 \text{ kJ/K}$$

Since $(\Delta S)_{\text{total}} > 0$, the process is irreversible.

22. A certain mass of a pure substance undergoes an irreversible process from state 1 to state 2, the path of the process being a straight line on *TS* diagram. Calculate the work interaction. Some properties at the initial and final states are: T_1 = 330 K, T_2 = 440 K, U_1 = 170 kJ, U_2 = 190 kJ, H_1 = 220 kJ, H_2 = 24 kJ and S_1 = 0.23 kJ/K and S_2 = 0.3 kJ/K where *T*, *U*, *H* and *S* represent temperature, internal energy, enthalpy and entropy respectively.

(GATE: 2000)

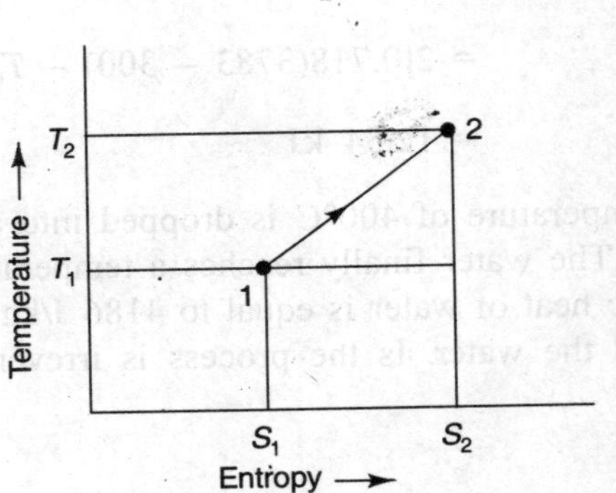

$$Q_{1-2} = \text{Area under line 1–2}$$

$$= T_1 \times (S_2 - S_1) + \frac{1}{2}(S_1 - S_2)(T_2 - T_1)$$

$$= 300(0.3 - 0.23) + \frac{1}{2}(0.3 - 0.23)(440 - 330)$$

$$= 26.95 \text{ kJ}$$

$$\Delta U_{1-2} = U_2 - U_1$$

$$= 190 - 170 = 20 \text{ kJ}$$

$$Q_{1-2} = \Delta U_{1-2} + W_{1-2}$$

or

$$W_{1-2} = Q_{1-2} - \Delta U_{1-2}$$

$$= 26.95 - 20$$

$$= 6.95 \text{ kJ}$$

23. One kilomole of an ideal gas is throttled from an initial pressure of 0.5 mPa to 0.1 mPa. The initial temperature is 300 K. The entropy change of universe is

(a) 13.38 kJ/K (b) 401.3 kJ/K (c) 0.0446 kJ/K (d) – 0.0446 kJ/K

(GATE: 1995)

$$S_2 - S_1 = c_p \log \frac{T_2}{T_1} - \bar{R} \log \frac{P_2}{P_1}$$

$$= -\bar{R} \log \frac{P_2}{P_1} \quad \text{as } T_2 = T_1$$

$$= +8.314 \log \frac{0.5}{0.1} = +13.38 \text{ kJ/K}$$

$$(\Delta S)_{\text{universe}} = (\Delta S)_{\text{sys}} + (\Delta S)_{\text{sur}}$$

$$= (\Delta S)_{\text{sys}} + 0$$

$$= 13.38$$

Option (a) is correct.

24. For two cycles coupled in series, the top cycle has efficiency of 30% and the bottom cycle has efficiency of 20%. The overall combined cycle efficiency is

(a) 50% (b) 44% (c) 38% (d) 55%

(GATE: 1996)

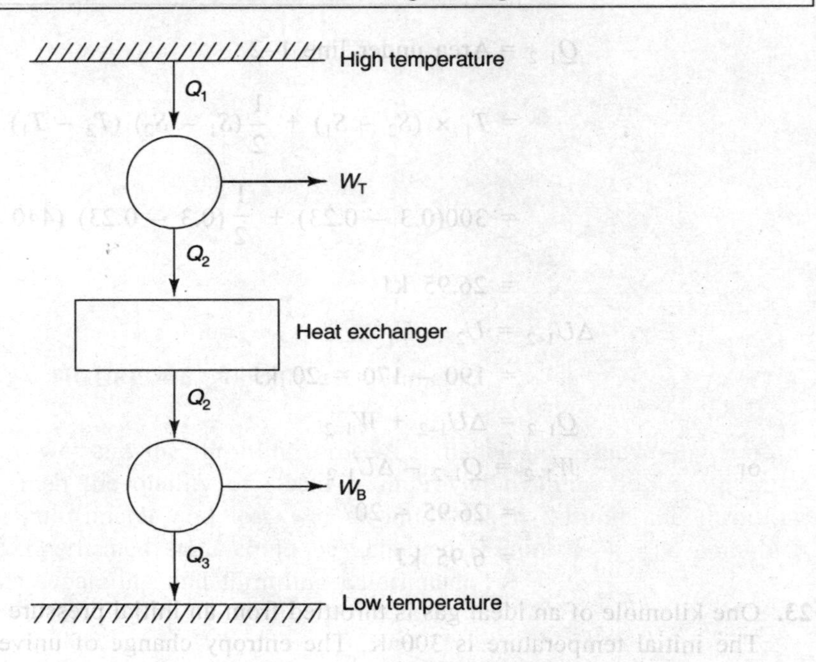

$$\eta_{top} = \frac{W_T}{Q_1} = \frac{Q_1 - Q_2}{Q_1} = 0.3$$

or $$\frac{Q_2}{Q_1} = 0.7$$

$$\eta_{bot} = \frac{W_B}{Q_2} = \frac{Q_2 - Q_3}{Q_2} = 0.2$$

or $\dfrac{Q_3}{Q_2} = 0.8$ or $Q_3 = 0.8,\ Q_2 = 0.8 \times 0.7,\ Q_1 = 0.56\ Q_1$

Combined $$\eta = \frac{W_T + Q_3}{Q_1} = \frac{Q_1 - Q_3}{Q_1}$$

$$= \frac{Q_1 - 0.56\ Q_1}{Q_1}$$

$$= 0.44 \text{ or } 44\%$$

Option (b) is correct.

25. A cyclic heat engine does 50 kJ of work per cycle. If the efficiency of the heat engine is 75%, the heat rejected per cycle is

(a) $16\frac{2}{3}$ kJ (b) $33\frac{1}{3}$ kJ (c) $37\frac{1}{2}$ kJ (d) $66\frac{2}{3}$ kJ

$$\eta = 0.75 = \frac{W}{Q_1} = \frac{50}{Q_1}$$

$$\therefore \quad Q_1 = \frac{50}{0.75} = \frac{50 \times 4}{3}$$

$$= \frac{200}{3} = 66^2/_3 \text{ kJ}$$

$$Q_2 = Q_1 - W$$

$$= \frac{200}{3} - 50$$

$$= \frac{200 - 150}{3}$$

$$= \frac{50}{3} = 16\frac{2}{3}$$

Option (a) is correct.

26. The operating temperature of a cold storage is –2°C. Heat leakage from the surrounding is 30 kW for the ambient temperature of 40°C. The actual COP of the refrigeration plant used is one fourth that of an ideal plant working between the same temperature. The power required to drive the plant is

(a) 1.86 kW (b) 7.72 kW (c) 7.44 kW (d) 18.6 kW

$$(\text{COP})_{\text{theoretical}} = \frac{271}{313 - 271} = \frac{271}{42}$$

$$(\text{COP})_{\text{actual}} = \frac{1}{4} \times \frac{271}{42} = \frac{Q_2}{W} = \frac{30}{W}$$

or

$$W = \frac{30 \times 4 \times 42}{271} = 18.6 \text{ kW}$$

Option (d) is correct.

27. If a heat engine gives an output of 3 kW when the input is 10,000 J/s, then the thermal efficiency of engine will be

(a) 20% (b) 30% (c) 70% (d) 76.7%

(Civil Services: 1995)

$$\eta = \frac{W}{Q} = \frac{3 \times 10^3}{10,000}$$

$$= 0.3 \text{ or } 30\%$$

Option (b) is correct.

28. **A heat engine is supplied with 250 kJ/c of heat at a constant fixed temperature of 227°C. The heat is rejected at 27°C. The cycle is reversible if the amount of heat rejected is**

(a) 273 kJ/s (b) 200 kJ/s (c) 180 kJ/s (d) 150 kJ/s.

(Civil Services: 1995)

$$\eta = \frac{T_1 - T_2}{T_1} = 1 - \frac{T_2}{T_1}$$

$$= 1 - \frac{300}{500} = 0.4$$

$$0.4 = \frac{W}{Q_1} = \frac{Q_1 - Q_2}{Q_1} = 1 - \frac{Q_2}{Q_1}$$

$$\frac{Q_2}{Q_1} = 0.6 \quad \text{but } Q_1 = 250 \text{ kJ/s}$$

$$\therefore \quad Q_2 = 0.6 \times 250 = 150 \text{ kJ}$$

Option (d) is correct.

29. **The efficiency of a reversible cycle process undergone by a substance as shown in the diagram is**

(a) 0.4 (b) 0.55 (c) 0.66 (d) 0.80.

(Civil Services: 1994)

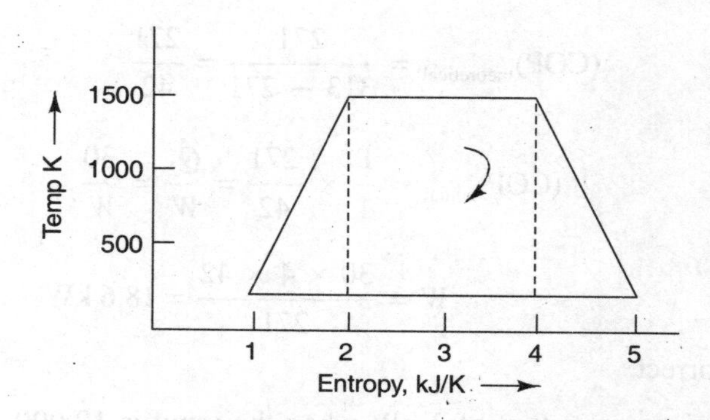

The substance is operating between 1500 K and 500 K. And the efficiency of a reversible cycle depends on only temperatures.

$$\eta = 1 - \frac{T_2}{T_1} = 1 - \frac{500}{1500} = \frac{2}{3}$$

$$= 0.6$$

Option (b) is correct.

30. Given that path 1-2-3, a system absorbs 100 kJ heat and does 60 kJ work while along the path 1-4-3 it does 20 kJ work. The heat absorbed during cycle 1-2-3:

(a) –140 kJ (b) –80 kJ (c) –40 kJ (d) 60 kJ.

(Civil Services: 1994)

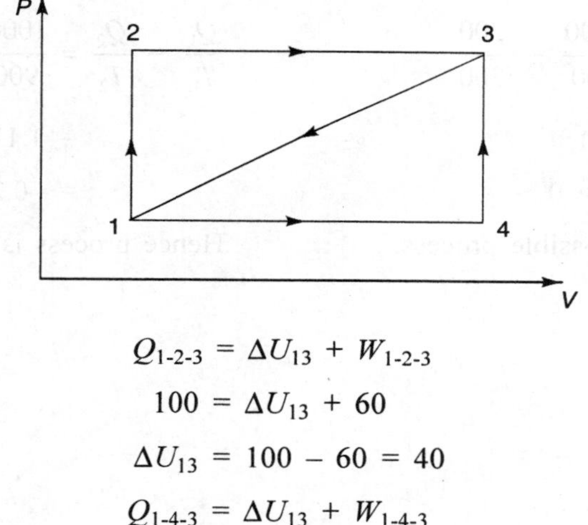

$$Q_{1\text{-}2\text{-}3} = \Delta U_{13} + W_{1\text{-}2\text{-}3}$$

$$100 = \Delta U_{13} + 60$$

$$\therefore \quad \Delta U_{13} = 100 - 60 = 40$$

$$Q_{1\text{-}4\text{-}3} = \Delta U_{13} + W_{1\text{-}4\text{-}3}$$

$$Q_{1\text{-}4\text{-}3} = 40 + 20 = 60 \text{ kJ}$$

Option (d) is correct.

31. The block diagrams of two systems are given below. Giving proper reasons indicate: (i) name of the system (i.e. HE, RE or HP) and (ii) type of cycle is possible or impossible and reversible or irreversible.

(UPTU: May 2008)

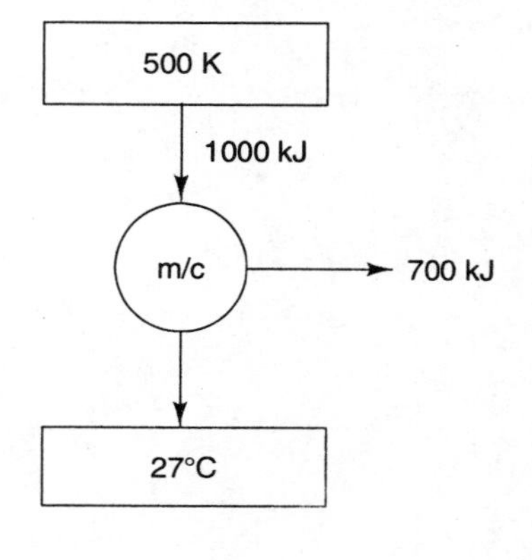

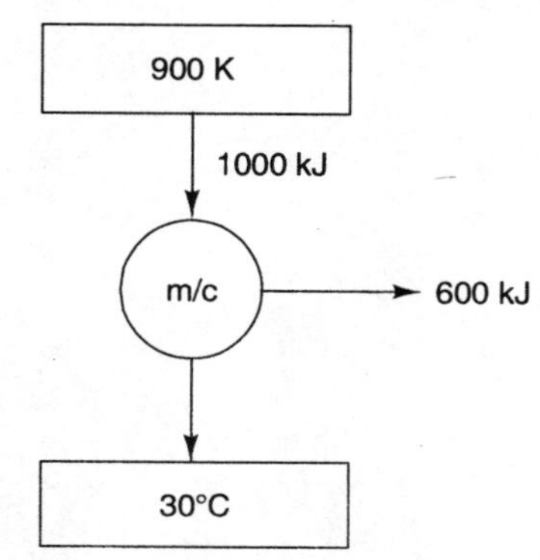

As machines are taking heat from one reservoir and rejecting part of heat to the second reservoir with output of work, machines are engines in both cases.

Machine 1

$Q_1 = 1000$ kJ

$T_1 = 500$ K

$Q_2 = 1000 - 700 = 300$ kJ

$T_2 = 27 + 273 = 300$

Applying Clausius Inequality:

$$\frac{Q_1}{T_1} - \frac{Q_2}{T_2} = \frac{1000}{500} - \frac{300}{300}$$

$$= 2–1$$

$$= 1 > 0$$

Hence it is impossible process.

Machine 2

$Q_1 = 1000$ kJ

$T_1 = 900$ K

$Q_2 = 1000 - 600 = 400$ kJ

$T_2 = 30 + 273 = 303$

Applying Clausius Inequality:

$$\frac{Q_1}{T_1} - \frac{Q_2}{T_2} = \frac{1000}{900} - \frac{400}{303}$$

$$= 1.11 - 1.32$$

$$= -0.21 < 0$$

Hence process is possible but irreversible.

OBJECTIVE TYPE QUESTIONS

Do you have troubles? Has some unexpected problem come up? Ask God to come to your aid. Put Him on the spot and Test Him. He can do the impossible.

State True or False

1. The first law of thermodynamics cannot explain non-occurrence of certain processes as well as the direction of processes. (*True/False*)
2. Heat and work are energy, and they have no difference in grade. (*True/False*)
3. High and low-grade energy is classified according to potential energy. (*True/False*)
4. All spontaneous processes proceed in one direction. (*True/False*)
5. A spontaneous process cannot be made to proceed in the reverse direction without aid. (*True/False*)
6. The second law of thermodynamics rules out the possibility of a spontaneous process proceeding in the reverse direction. (*True/False*)
7. A thermal reservoir at a low temperature to which heat is rejected is called source. (*True/False*)
8. A sink is a thermal reservoir to which a heat engine rejects heat. (*True/False*)
9. A thermal power plant is a heat engine. (*True/False*)
10. A heat engine is a device which receives heat from a source and rejects heat to a sink while undergoing a cyclic process and performs work. (*True/False*)
11. A heat reservoir has large heat capacity and can receive or reject heat without change of temperature. (*True/False*)
12. The surrounding is generally used as a sink for a heat engine. (*True/False*)
13. The surrounding is used as a low temperature body in a heat pump. (*True/False*)
14. Heat supplied to room is desired effect for a heat pump. (*True/False*)
15. Heat rejected to the surrounding is the desired effect for a refrigerator. (*True/False*)
16. COP of a refrigerator is greater than COP of a heat pump working within the same temperature limits. (*True/False*)
17. Performance of a heat engine can be given by COP. (*True/False*)
18. A heat pump is required to extract heat from room to bring down the temperature. (*True/False*)
19. A heat pump is also used during the warm season for heating a room. (*True/False*)
20. A heat pump is used in winter for heating room. (*True/False*)

21. The main purpose of a refrigerator is to extract heat from cold storage space. (*True/False*)
22. The surroundings are a high temperature reservoir for a domestic refrigerator. (*True/False*)
23. The thermal efficiency of a heat engine is the ratio of work output to the heat added. (*True/False*)
24. The second law gives the possibility of a spontaneous process to be reversed itself unaided. (*True/False*)
25. The second law of thermodynamics prohibits the possibility of designing of heat engine with 100% efficiency. (*True/False*)
26. In an isothermal process all heat is converted into work. (*True/False*)
27. The difference of COP of a heat pump and a refrigerator is unity. (*True/False*)
28. The Kelvin–Plank statement permits the possibility of a device producing work by drawing heat from a single source. (*True/False*)
29. A device can work as per the Clausius statement but violates the Kelvin–Plank statement. (*True/False*)
30. The Clausius and the Kelvin–Plank statements are equivalent. (*True/False*)
31. The Clausius statement permits heat to flow from a low temperature body to a high temperature body without interaction of any other energy. (*True/False*)
32. All spontaneous processes are irreversible. (*True/False*)
33. In a reversible cycle, all processes constituting the cycle are reversible. (*True/False*)
34. A process which can proceed forward or reverse direction without violating the second law is a reversible process. (*True/False*)
35. Violation of the Kelvin–Plank statement does not lead to violation of the Clausius statement of the second law of thermodynamics. (*True/False*)
36. A Carnot cycle must have more than one reservoir. (*True/False*)
37. A Carnot cycle has two adiabatic and two isobaric processes. (*True/False*)
38. A Carnot cycle is a hypothetical device and consists of only a reversible process. (*True/False*)
39. A Carnot heat pump and a refrigerator work on a reversed Carnot cycle or refrigeration cycle. (*True/False*)
40. The COP of a Carnot heat pump and refrigerator is the highest. (*True/False*)
41. The COP of a normal refrigerator is higher than that of a Carnot refrigerator. (*True/False*)
42. A Carnot refrigerator will require minimum energy for desired effect. (*True/False*)
43. The performance of other engines is compared with a Carnot heat engine as it is used as standard of perfection. (*True/False*)
44. The performance of an actual refrigerator and a heat pump cannot be compared with a Carnot heat pump and a refrigerator as standard of performance since these are hypothetical. (*True/False*)

45. As a heat pump and a refrigerator work on a reversed Carnot cycle, they cannot be used as replacement for each other with some modification. *(True/False)*
46. Desired effect for a heat pump will be more than that of a refrigerator as work input is also added up with heat removed from a cold body while rejecting to hot body. *(True/False)*
47. The Carnot engine has greater efficiency than any reversible engine between given temperature limits. *(True/False)*
48. The Carnot engine can have different efficiencies depending upon temperature limits. *(True/False)*
49. The Carnot engine can have different efficiencies depending upon working medium. *(True/False)*
50. All reversible engines have the same efficiency working between the same temperature limits. *(True/False)*
51. The Carnot engine will have greater efficiency with increase of source temperature. *(True/False)*
52. The Carnot engine will have increased efficiency with lowering of sink temperature. *(True/False)*
53. The efficiency of the Carnot engine depends upon source temperature and sink can have any lower temperatures. *(True/False)*
54. Lowering sink temperature will be a more effective way to increase the efficiency of the Carnot engine as compared to increasing source temperature. *(True/False)*
55. The COP of a refrigerator is the ratio of heat supplied to a hot body to work input. *(True/False)*
56. The COP of a heat pump is the ratio of heat extracted from a cold body to work input. *(True/False)*
57. A Carnot pump with COP = 4 is reversed to work as a heat engine. The efficiency of the heat engine will be 25%. *(True/False)*
58. If the COP of a refrigerator is 4, then the COP of a heat pump will be 5. *(True/False)*
59. If a refrigerator with COP = 5 is given 5 kW energy, then heat extracted from cold space will be 25 kW. *(True/False)*
60. The Clausius inequality is not based on the second law of thermodynamics. *(True/False)*
61. Considering temperatures T_1 and T_2 vary but the ratio of T_1 to T_2 remains the same. The efficiency of a Carnot engine will increase with lowering of T_2 (sink temperature). *(True/False)*
62. The COP of a refrigerator will slightly fall as surroundings temperature increases in day time. *(True/False)*
63. The difference of the COP of a heat pump and a refrigerator will remain unchanged even when temperatures of source and sink are changed. *(True/False)*
64. The efficiency of a Carnot engine is 33.34%. If its cycle is reversed to work as a heat pump, then the COP of heat pump will be 3. *(True/False)*

65. $\oint T ds = \oint P dv$ for a cyclic process. (*True/False*)

66. Entropy is not conserved. (*True/False*)

67. A reversible adiabatic process differs from an isentropic process. (*True/False*)

68. Spontaneous processes occur such as to decrease the entropy of the universe. (*True/False*)

69. The system having lower entropy is more prone to a spontaneous process. (*True/False*)

70. For a reversible engine $\frac{Q_1}{T_1} + \frac{Q_2}{T_2} = 0$, where T_1 and T_2 are source and sink temperatures and Q_1 and Q_2 are heat added and rejected. (*True/False*)

71. For an irreversible engine taking Q_1 heat from T_1 source and rejecting Q_2 heat from T_2 sink, then $\frac{Q_1}{T_1} + \frac{Q_2}{T_2} < 0$. (*True/False*)

72. Whenever a system undergoes a cyclic process $\oint \frac{dQ}{T} \geq 0$. (*True/False*)

73. $\oint \frac{dQ}{T}$ has the same value for all reversible processes. (*True/False*)

74. $\left(\frac{dQ}{T}\right)_{\text{reversible}}$ is an exact differential. (*True/False*)

75. $\frac{dQ}{T} = dS$ is true for all reversible processes. (*True/False*)

76. Change of entropy is a state function and does not depend upon the path. (*True/False*)

77. $\frac{dQ}{T} < dS$ for an irreversible process. (*True/False*)

78. Entropy change during melting is equal to the latent heat of freezing divided by 273 K. (*True/False*)

79. Entropy change during evaporation is equal to the latent heat of vaporization divided by 373 K. (*True/False*)

80. Entropy changes when a hot body is dropped in cold liquid. (*True/False*)

81. If two gases at the same temperature and pressure are mixed, entropy does not change. (*True/False*)

82. Entropy change between two states remains constant irrespective of the path (both reversible and irreversible). (*True/False*)

83. If a refrigerator is working in an isolated room where temperature is increasing, then the COP of the refrigerator will decrease. (*True/False*)

84. If two refrigerators with each having COP = 4, works in parallel, then the COP of such arrangement will remain the same. *(True/False)*

85. If a heat pump with COP = 5 and a refrigerator with COP = 4 are working in parallel, then the COP of both working as refrigerator will have COP = 4. *(True/False)*

86. If an engine is working between 800 and 400 K and supplied heat 100 kJ, then unavailable energy will be 40 kJ. *(True/False)*

87. If a system is working within the temperature limit of 600 and 300 K and change of entropy is 1.5, then unavailable energy will be 400 kJ. *(True/False)*

88. If a system has energy = 600 kJ and anergy = 200 kJ, then it has an exergy of 400 kJ. *(True/False)*

89. If the difference of unavailable energy between 300 K and 200 K is 100 J, then change of entropy of the system will be unity. *(True/False)*

90. Degree of irreversibility of a system will be 450 J, in case surroundings temperature is 300 K and entropy generation is 1.5 J/K. *(True/False)*

91. If work output of an engine = $0.3Q + a$ where Q = heat supplied and a is a constant, then the efficiency of engine will be 30%. *(True/False)*

Multiple Choice Questions

1. The second law of thermodynamics defines
(a) heat (b) work (c) enthalpy (d) entropy

2. For a reversible adiabatic process, change of entropy is
(a) minimum (b) zero (c) positive (d) negative

3. The net entropy change in any irreversible process is
(a) positive (b) zero (c) infinite (d) unity

4. For any reversible process, the change in entropy of the system and surroundings is
(a) unity (b) negative (c) positive (d) zero

5. Isentropic flow is
(a) irreversible adiabatic (b) reversible adiabatic
(c) isothermal (d) isoenthalpic

6. A Carnot cycle has two adiabatic and two other processes which are
(a) isochoric (b) isothermal (c) isobaric (d) polytropic

7. The efficiency of a Carnot cycle depends upon
(a) working substance (b) temperature of the sink only
(c) temperature of the source only (d) temperatures of the source and sink

8. The efficiency of a Carnot cycle depends upon
(a) work output only (b) amount of temperature rejected
(c) amount of heat rejected (d) ratio of work output to heat added

9. The efficiency of a Carnot engine is

(a) $\frac{T_1}{T_1 - T_2}$ (b) $\frac{T_2}{T_1 - T_2}$ (c) $\frac{T_1 + T_2}{T_2}$ (d) $\frac{T_1 - T_2}{T_1}$

10. The efficiency of a Carnot cycle can be 100% only if sink temperature can be

(a) 0°C (b) 0°F (c) – 200°C (d) 0 K

11. In a reversible cycle, the entropy of a system will

(a) increase (b) decrease
(c) remain constant (d) depend on working substance

12. The Kelvin–Plank statement of the second law of thermodynamics is about

(a) conservation of mass (b) conservation of heat
(c) conversion of heat into work (d) conversion of work into heat

13. In the device shown in the figure

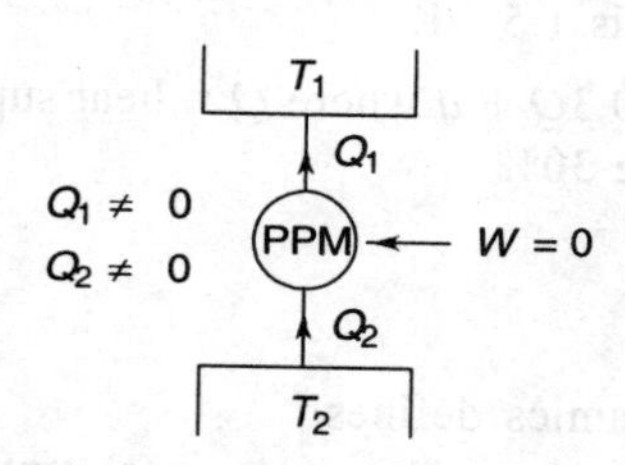

(a) PPM I violating the second law of thermodynamics
(b) PPM II violating the Kelvin–Plank statement
(c) PPM II violating the Clausius statement

14. In the device shown in the figure

(a) PPM I violating the second law of thermodynamics
(b) PPM II isolating the Kelvin–Plank statement
(c) PPM II isolating the Clausius statement

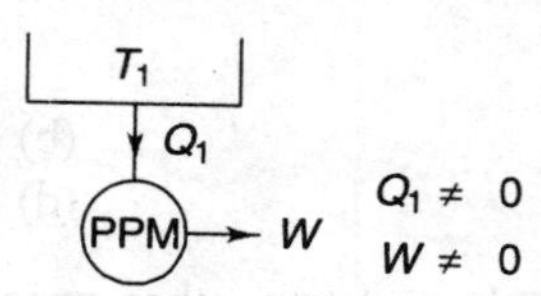

15. The Clausius statement is that

(a) PMM II can be devised
(b) PMM I can be devised
(c) heat cannot flow from cold body to hot body unaided

16. The Kelvin–Plank statement prohibits work output from a heat engine if the

(a) difference between source and sink temperatures < 200°C
(b) difference between source and sink temperatures < 300°C
(c) heat interaction is there with the source only

17. A heat engine is a device that
(a) generates heat from work
(b) converts full heat into work
(c) converts a portion of heat into work and rejects the remaining

18. A heat pump is a device that
(a) pumps heat out of a room
(b) supplies heat to a room extracting from surroundings with work supplied
(c) delivers heat to a room from surroundings without assistance

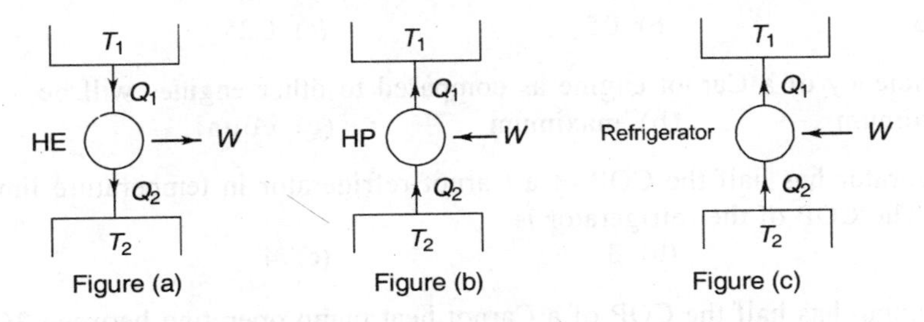

Figure (a) Figure (b) Figure (c)

19. The efficiency of heat engine shown in Figure (a) is

(a) $\dfrac{Q_1}{Q_1 - Q_2}$ (b) $\dfrac{Q_1 + Q_2}{Q_1}$ (c) $\dfrac{Q_1 - Q_2}{Q_1}$

20. The efficiency of the heat engine shown in Figure (a) can also be given as

(a) $\dfrac{T_1}{T_1 - T_2}$ (b) $\dfrac{T_1 - T_2}{T_1}$ (c) $\dfrac{T_1 + T_2}{T_1}$

21. The COP of the heat pump shown in Figure (b) is

(a) $\dfrac{Q_1 - Q_2}{Q_1}$ (b) $\dfrac{Q_2}{Q_1 - Q_2}$ (c) $\dfrac{Q_1}{Q_1 - Q_2}$

22. The COP of the heat pump shown in Figure (b) is

(a) $\dfrac{T_1}{T_1 - T_2}$ (b) $\dfrac{T_2}{T_1 - T_2}$ (c) $\dfrac{T_1 - T_2}{T_1}$

23. The COP of the refrigerator shown in Figure (c) is

(a) $\dfrac{Q_1 - Q_2}{Q_1}$ (b) $\dfrac{Q_2}{Q_1 - Q_2}$ (c) $\dfrac{Q_1}{Q_1 - Q_2}$

24. The COP of the refrigerator in Figure (c) can be given as

(a) $\dfrac{T_1}{T_1 - T_2}$ (b) $\dfrac{T_1 - T_2}{T_1}$ (c) $\dfrac{T_2}{T_1 - T_2}$

25. A heat engine will have greater efficiency in case source temperature (T_1) or sink temperature (T_2) is varied
(a) $T_1 + \Delta T$ (b) $T_1 - \Delta T$ (c) $T_2 + \Delta T$ (d) $T_2 - \Delta T$

26. Heat engines A and B are working on the Carnot cycle and have air and steam as working fluid respectively. Choose the correct answer.
(a) $\eta_A > \eta_B$ (b) $\eta_B > \eta_A$ (c) $\eta_A = \eta_B$

27. The efficiency of a heat engine having half the efficiency of a Carnot engine in temperature limits of 600 and 300 K is
(a) 0.75 (b) 0.5 (c) 0.25

28. The efficiency of a Carnot engine as compared to other engines will be
(a) minimum (b) maximum (c) equal

29. A refrigerator has half the COP of a Carnot refrigerator in temperature limit of 350 and 300 K. The COP of the refrigerator is
(a) 6 (b) 3 (c) 4

30. A heat pump has half the COP of a Carnot heat pump operating between 360 and 300 K. The COP of the heat pump is
(a) 3 (b) 4 (c) 6

31. The COP of a Carnot heat pump is 5. The COP of a Carnot refrigerator within the same temperature limit is
(a) 6 (b) 4 (c) 5

32. The efficiency of a Carnot engine is 25%. What is the COP of a heat pump working between the same temperature limit?
(a) 3 (b) 4 (c) 5

33. A reversible process can be replaced with a series of
(a) reversible adiabatic and isothermal processes
(b) reversible isobaric and isothermal processes
(c) reversible isochoric and isothermal processes

34. Two bodies of equal mass and material at T_1 and T_2 ($T_1 > T_2$) are used as a source and a sink for a Carnot heat engine. If T_3 is final temperature then

(a) $T_3 = \dfrac{T_1 - T_2}{2}$ (b) $T_3 = \dfrac{T_1}{T_2}$ (c) $T_3 = \sqrt{T_1 T_2}$

35. A Carnot engine has $Q_1 = 1000$ kJ, $T_1 = 1000$ K and $T_2 = 300$ K. Q_2 and W will be
(a) 300 kJ and 700 kJ
(b) 400 kJ and 600 kJ
(c) 200 kJ and 800 kJ

36. A Carnot engine takes 800 kJ and 700 kJ heat from two sources at 800 and 700 K respectively. If the sink is at 300 K, then heat rejected and efficiency are
(a) 400 kJ and 50% (b) 600 kJ and 60% (c) 500 kJ and 55%

37. A Carnot engine ($\eta = 0.5$) runs a Carnot refrigerator having COP = 5. The ratio of the heat added to a heat engine to the heat extracted from cold space by the refrigerator is
(a) 0.1 (b) 0.5 (c) 0.4

38. If a Carnot heat pump and a refrigerator are working within the same temperature limits, then
(a) $COP_{HP} = COP_{Ref}$
(b) $COP_{Ref} > COP_{HP}$
(c) $COP_{HP} > COP_{Ref}$

39. A Carnot heat pump and a refrigerator are working within the same temperature limit. If the COP of the heat pump is 6, then the COP of the refrigerator is
(a) 5 (b) 7 (c) 4

40. Energy, exergy and anergy are related as under
(a) Energy = Exergy + Anergy
(b) Energy = Exergy – Anergy
(c) Exergy = Energy + Anergy

41. For an adiabatic process, the change of entropy is
(a) $\Delta S \geq 0$ (b) $\Delta S \leq 0$ (c) $\Delta S = 0$

42. When a system undergoes a process, the entropy generation is
(a) $S_G = 0$ (b) $S_G \geq 0$ (c) $S_G < 0$

43. During throttling, the change of properties is
(a) $dh = 0, dS > 0$ (b) $dh = 1, dS > 0$ (c) $dh = 0, dS = 0$

44. When the entropies at the inlet and the outlet of an adiabatic turbine are S_1 and S_2, then
(a) $S_2 = S_1$ (b) $S_2 > S_1$ (c) $S_1 > S_2$

45. Combining $du = dQ - dW$ and $ds = \left(\frac{dQ}{T}\right)_R$, the relation is
(a) $du = Tds + pdv$ (b) $du = pdv - Tds$ (c) $du = Tds - pdv$

46. What will be the change of entropy of a reservoir if 1000 kJ of heat is supplied to it at 200 K?
(a) 3 kJ/K (b) 4 kJ/K (c) 5 kJ/K

47. What will be the degree of irreversibility of a system if S_g = 5 kJ/K and surroundings at 300 K?
(a) 2000 kJ (b) 1500 kJ (c) 1200 kJ

48. If two Carnot heat engines (work output each = ω) work in parallel within the same temperature limit, then total work will be
(a) ω (b) 2ω (c) $\omega/2$

49. If two Carnot heat engines are used in series with the temperatures of the source and the sink as T_1 and T_2 (individually work output of each is ω for T_1 and T_2), then output in series for each engine is:
(a) ω (b) 2ω (c) $\omega/2$

50. If 5 Carnot heat engines are operating in series within the source (1000 K) and the sink (500 K). The temperature difference across each heat engine having equal work output is
(a) 500 K (b) 100 K (c) 200 K

51. Four Carnot heat engines are operating in series in between 800 and 400 K. Temperature gradient is the same across each engine. If the first engine takes 80 J heat, then work output for each engine and total output are
(a) 10 J and 40 J (b) 20 J and 30 J (c) 9 J and 36 J

52. An inventer claims to develop a device which takes a stream of fluid (entropy = 4 kJ/K) and gives out two streams of fluid (entropy = 2 kJ/K and entropy = 3 kJ/K). Comment on his claim.
(a) claim is feasible (b) claim is unfeasible (c) insufficient data to comment

53. An inventor claims to develop a magic tube which takes two streams of fluid (S_1 = 3 and S_2 = 5 kJ/K) and merges the streams to one stream (S_3 = 6 kg/K). Comment on his claim.
(a) claim is feasible (b) claim is unfeasible (c) insufficient data to comment

54. The system at state *A* has entropy = 4 kJ/K and at state *B* has entropy = 3 kJ/K. Comment on the feasibility of the process from state *A* to state *B*.
(a) spontaneous (b) impossible (c) slow

55. Two Carnot heat engines are working in series (work output is equal) within the temperature limit of 600 and 200 K. If heat supplied by the source (600 K) is 30 J, then the temperature of the reservoir between engines and heat rejected at the sink (200 K) are
(a) 300 K and 25 joules
(b) 400 K and 10 joules
(c) 425 K and 30 joules

56. A Carnot heat engine is driving a refrigerator with a source = 800 K and a sink = 400 K. If the heat engine takes 40 J from the source and the refrigerator extracts 5 J from the sink, the net work output will be
(a) 20 J (b) 15 J (c) 10 J

57. If a system is working between 800 and 400 K and supplied heat = 200 kJ, the unavailable energy is
(a) 150 kJ (b) 125 kJ (c) 100 kJ

58. If a system works between 600 and 300 K and entropy change is 1.2 kJ/K, then unavailable energy is
(a) 360 kJ (b) 400 kJ (c) 300 kJ

59. If degree of irreversibility is 600 J at temperature of surroundings (300 K). The entropy generation is
(a) 2 kJ/K (b) 2.5 kJ/K (c) 3 kJ/K

Fill in the Blanks

1. Work is ________ grade energy.
(a) high (b) low

2. Heat is ________ grade energy.
(a) high (b) low

3. All spontaneous processes move in ________ directions.
(a) one (b) both

4. A source is a reservoir at ________ temperature.
(a) high (b) low

5. A sink is a reservoir at ________ temperature.
(a) high (b) low

6. Surroundings are an ideal ________.
(a) source (b) sink

7. $\oint dS$ for a system is ________.
(a) zero (b) > 0

8. $\oint \frac{dQ}{T}$ for a system is ________.
(a) zero (b) < 0

9. If Q_1 = heat supplied at T_1 and Q_2 = heat rejected at T_2 for a Carnot cycle, then Q_1 is equal to ________.
(a) $Q_2 \times \frac{T_1}{T_2}$ (b) $Q_2 \times \frac{T_2}{T_1}$

10. The efficiency of a Carnot cycle with source = T_1 and sink = T_2 is
(a) $1 - \frac{T_1}{T_2}$ (b) $1 - \frac{T_2}{T_1}$

11. A process which can proceed in a forward or reverse direction without violating the second law is ________ process.
(a) homogeneous (b) reversible

12. As per the Kelvin–Plank statement, a heat engine has to interact with ________ reservoir(s).
(a) one (b) atleast two

13. If a hypothetical heat engine can produce work interacting with one reservoir, then its efficiency is ________.
(a) infinity (b) 100%

14. If the temperature of a source is T_1 and that of a sink is T_2, then the COP of a heat engine is ________.

(a) $\frac{T_1}{T_1 - T_2}$ (b) $\frac{T_2}{T_1 - T_2}$

15. If the temperature of a source is T_1 and that of a sink is T_2, then the COP of a refrigerator will be ________.

(a) $\frac{T_1}{T_1 - T_2}$ (b) $\frac{T_2}{T_1 - T_2}$

16. If heat interaction = ΔQ at temperature T, then entropy change for a reversible process is ________.

(a) $T \times \Delta Q$ (b) $\frac{\Delta Q}{T}$

17. If S_g = entropy generation for a system interacting with surroundings at T_0, then degree of irreversibility is ________.

(a) $\frac{S_g}{T_0}$ (b) $T_0 \, S_g$

18. If ΔQ = heat interaction, ΔS = entropy charge and T = temperature, then their relation is ________.

(a) $\frac{\Delta Q}{T} = \Delta S$ (b) $\frac{\Delta Q}{T} < \Delta S$

19. For an impossible process, $\oint \frac{\Delta Q}{T}$ is ________.

(a) zero (b) > 0

20. The entropy of our universe will always ________.

(a) increase (b) decrease

21. The change of entropy of a heating body (C = heat capacity) from T_1 to T_2 is ________.

(a) $C \log \frac{T_2}{T_1}$ (b) $2C\frac{(T_2 - T_1)}{T_1 + T_2}$

22. The unavailable energy will ________ with lowering of sink temperature.

(a) increase (b) decrease

23. The COP of a refrigerator is ________ if the COP of a heat pump is 5, working within the same temperature limits.

(a) 4 (b) 6

24. The efficiency of a heat engine will ________ in case source temperature is increased.

(a) increase (b) decrease

25. The efficiency of a heat engine will ________ in case sink temperature is decreased.
(a) increase (b) decrease

26. If Q = latent heat at temperature T, then change of entropy is ________.
(a) $Q \times T$ (b) $\frac{Q}{T}$

27. If 600 kJ is rejected to surroundings at temperature 300 K, then change of entropy is ________.
(a) 2 (b) 0.5

28. PPM II is a device which ________ the second law of thermodynamics.
(a) violates (b) supports

29. The refrigeration cycle is ________ to a Carnot cycle.
(a) similar (b) reversed

30. A Carnot engine can have ________ depending upon the temperature limits.
(a) the same efficiency
(b) different efficiencies

31. Change of entropy is a ________ function.
(a) state (b) path

32. A portion of energy which can be converted into work is called ________.
(a) Anergy (b) Exergy

33. Exergy ________ be conserved like other energies.
(a) can (b) cannot

34. Exergy ________ be regarded as a property of the system.
(a) can (b) cannot

35. At the dead state, a system can have ________.
(a) exergy (b) energy

36. Entropy of steam is measured from ________.
(a) 0 K (b) 0°C

ANSWERS

You may not be able to choose your lot in life, but you can choose how to handle your lot.

State True or False

1. True
2. False (Energy is graded as high and low. Work is high-grade energy and heat is low-graded energy.)
3. False (High-grade energy can be converted fully into low-grade energy while low-grade energy cannot be converted fully into high-grade energy.)
4. True (All processes proceed to increase entropy.)
5. False (Interaction of other energy is required to reverse the process.)
6. False (Lays down the condition of interaction of other energy to reverse.)
7. False (Heat is rejected to the sink at low temperature.)
8. True
9. True (Both are devices to convert heat into work.)
10. True
11. True
12. True
13. True
14. True
15. False (Heat extraction from cold space is desired effect for the refrigerator. Heat rejected to surroundings has no significance to the refrigerator.)
16. False ($COP_{HE} > COP_{ref.}$)
17. False (A heat engine is required to convert heat into work. It has nothing to do with supply of heat or extraction of heat.)
18. False (A heat pump is required to supply heat to a hot body (room) by extracting heat from surroundings.)
19. False (A heat pump is meant for heating. In warm weather, heating is not required.)
20. True
21. True
22. True (For a refrigerator cold space is a low temperature body and the surrounding is a high temperature body.)

23. True $\left(\eta = \frac{W}{Q_{add}}\right)$

24. False (A spontaneous process cannot be reversed unaided.)

25. True

26. True (For an isothermal process $\Delta U = 0$. Therefore $Q = W$.)

27. True ($COP_{HP} = COP_{ref} + 1$)

28. False (A heat engine can work between the source and the sink. It cannot work with one reservoir. Such device is known as PMM-II.)

29. False (The Clausius and Kelvin–Plank statements are equivalent. Violation of one statement leads to violation of other.)

30. True

31. False (Any device working as violation of the Clausius statement is known as PMM-II.)

32. True

33. True (If even one process of a cycle is irreversible, then cycle becomes irreversible.)

34. True

35. False (Both statements are equivalent.)

36. True

37. False (A Carnot cycle has two adiabatic and two isothermal processes.)

38. True

39. True

40. True

41. False

42. True

43. True

44. False

45. False

46. True ($Q_1 = Q_2 + W$ where Q_2 is heat from cold body and Q_1 is heat rejected to hot body.)

47. False (All reversible engines have the same efficiency operating between the same temperature limit.)

48. True ($\eta = 1 - \frac{T_2}{T_1}$ and as T_1 and T_2 will vary, efficiency will also vary.)

49. False (Efficiency of a Carnot cycle does not depend on working medium.)

50. True

51. True ($\eta = 1 - \frac{T_2}{T_1}$. When T_1 increases, η will increase.)

52. True ($\eta = 1 - \frac{T_2}{T_1}$. When T_2 increases, η will increase.)

53. False

54. True [$\eta_1 = 1 - \frac{T_2 - \Delta T}{T_1} = \left(1 - \frac{T_2}{T_1}\right) + \frac{\Delta T}{T_1} = \eta + a$; $\eta_2 = 1 - \frac{T_2}{T_1 + \Delta T}$

$$= 1 - \frac{T_2}{T_1\left(1 + \frac{\Delta T}{T}\right)} = 1 - \frac{T_2}{T_1}\left(1 + \frac{\Delta T}{T_1}\right)^{-1} = 1 - \frac{T_2}{T_1}\left(1 - \frac{\Delta T}{T_1}\right)$$

$$= \left(1 - \frac{T_2}{T_1}\right) + \Delta T \times \frac{T_2}{T_1^2} = \eta + \frac{aT_2}{T_1}$$

Hence $\eta_1 > \eta_2$.]

55. False $\left(\text{COP}_{\text{ref}} = \frac{\text{Heat removed from cold body}}{\text{Work input}}\right)$

56. False $\left(\text{COP}_{\text{HP}} = \frac{\text{Heat supplied to hot body}}{\text{Work input}}\right)$

57. True ($\text{COP}_{\text{HE}} = \frac{T_1}{T_1 - T_2} = 4$. $\therefore T_2 = \frac{3}{4}T_1 \cdot \eta = 1 - \frac{T_2}{T_1} = 1 - \frac{3}{4} = \frac{1}{4}$)

58. True ($\text{COP}_{\text{HP}} = \text{COP}_{\text{ref}} + 1$)

59. True ($\text{COP}_{\text{ref}} = 5 = \frac{Q_2}{W} = \frac{Q_2}{5}$ $\therefore Q_2 = 25$)

60. False

61. False ($\eta = 1 - \frac{T_2}{T_1}$, hence η depends upon $\frac{T_2}{T_1}$ which is constant.)

62. True ($\text{COP}_{\text{ref}} = \frac{T_2}{T_1 - T_2}$ and hence COP will fall when T_1 increases.)

63. True ($\text{COP}_{\text{HP}} = \text{COP}_{\text{ref}} + 1$)

64. True ($\eta = \frac{T_1 - T_2}{T_1}$ and $\text{COP}_{\text{HP}} = \frac{T_1}{T_1 - T_2}$ which is reverse of η and equal to 3.)

65. True

66. True

67. True (A reversible adiabatic process is isentropic but isentropic may not be adiabatic.)

68. False (Spontaneous processes are irreversible and generate entropy. Entropy of the universe is continuously increasing.)

69. True (A system with lesser entropy has heat at higher temperature which is more valuable and prone to a spontaneous process.)

70. True ($\Sigma \frac{dQ}{T} = 0$ for reversible processes.)

71. True ($\Sigma \frac{dQ}{T} < 0$ for irreversible processes.)

72. False $\left(\oint \frac{dQ}{T} = \leq 0\right)$

73. True ($\oint \frac{dQ}{T}$ is a state function and does not depend upon the path.)

74. True $\left(\int dS = \int \left(\frac{dQ}{T}\right)_{\text{rev}}\right)$

75. False $\left(\left(\frac{dQ}{7}\right)_{\text{rev}} = ds\right)$

76. True

77. True

78. True $\left(\Delta S = \frac{h_{Sf}}{273}\right)$

79. True $\left(\Delta S = \frac{h_{fg}}{373}\right)$

80. True $[(\Delta S)_{\text{universe}} = (\Delta S)_{\text{body}} + (\Delta S)_{\text{liq}}]$

81. False (Entropy increases for all processes except for a reversible adiabatic process.)

82. True (Change of entropy is a state function.)

83. True ($\text{COP}_{\text{ref}} = \frac{T_2}{T_1 - T_2}$ with T_1 increasing COP will fall.)

84. True

85. True

86. False (Heat rejected = unavailable energy = $Q_1 \times \frac{T_2}{T_1} = 100 \times \frac{400}{800} = 50$ kJ)

87. False (Unavailable energy = $T_0\ \Delta S = 300 \times 1.5 = 450$ kJ)

88. True (Exergy = energy – anergy = 600 – 200 = 400)

89. True (Anergy $A_1 = 300 \times \Delta S$, $A_2 = 200 \times \Delta S$; $(A_1 - A_2) = (300 - 200)\Delta S = 100 \times 1 = 100$)

90. True (Degree of irreversibility = $T_0 \times \Delta S = 1.5 \times 300 = 450$ J)

91. True ($W = 0.3Q + a$, $\frac{dW}{dQ} = 0.3 = \eta$)

Multiple Choice Questions

1. (d) **2.** (b) **3.** (a) **4.** (c)

5. (b) **6.** (b) **7.** (d) $(\eta = 1 - \frac{T_2}{T_1})$

8. (d) $\left(\eta = \frac{W}{Q_{add}}\right)$ **9.** (d) $\left(\eta = \frac{T_1 - T_2}{T_1}\right)$ **10.** (d) $\left(\eta = \frac{T_1 - T_2}{T_1} = \frac{T_1 - 0}{T_1} = 1\right)$

11. (c) **12.** (c) $(Q_1 = W + Q_2)$ **13.** (c) **14.** (b)

15. (c) **16.** (c) **17.** (c) **18.** (b)

19. (c) **20.** (b) **21.** (c) **22.** (a)

23. (b) **24.** (c) **25.** (d)

26. (c) (η does not depend upon medium) **27.** (b) $\left(\eta = 1 - \frac{T_2}{T_1} = 1 - \frac{300}{600} = \frac{1}{2}\right)$

28. (b)

29. (b) ($COP_{\text{Carnot ref}} = \frac{T_2}{T_1 - T_2} = \frac{300}{50} = 6 \quad \therefore COP_{\text{actual}} = \frac{1}{2} \times 6 = 3$)

30. (a) ($COP_{HP} = \frac{T_1}{T_1 - T_2} = \frac{360}{360 - 300} = 6$, $COP_{\text{actual}} = \frac{1}{2} \times 6 = 3$)

31. (b) ($COP_{Ref} = COP_{HP} - 1 = 5 - 1 = 4$)

32. (b) $\left(\eta = \dfrac{T_1 - T_2}{T_1} \text{ and } COP_{HP} = \dfrac{T_1}{T_1 - T_2} = \dfrac{1}{0.25} = 4\right)$

33. (a)

34. (c) (See solved example of Question 11.)

35. (a) $(\eta = 1 - \dfrac{T_2}{T_1} = 1 - \dfrac{300}{1000} = 0.7 = \dfrac{W}{Q_1}$ or $W = 0.7 \times Q_1 = 0.7 \times 1000 = 700$; $Q_2 = Q_1 - W$ $= 1000 - 700 = 300$ kJ)

36. (b) $(Q_2' = Q_1' \times \dfrac{T_2}{T_1'} = 800 \times \dfrac{300}{800} = 300$ kJ, $Q_2'' = Q_1'' \times \dfrac{T_2}{T_1''} = 700 \times \dfrac{300}{700} = 300$ kJ)

$Q_2 = Q_2' + Q_2'' = 300 + 300 = 600$ kJ

$\eta = \dfrac{Q_1 - Q_2}{Q_1} = \dfrac{1500 - 600}{1500} = \dfrac{3}{5} = 60\%)$

37. (c) $(\eta = 0.5 = \dfrac{W}{Q_{add}}$ or $W = 0.5\, Q_{add}$, COP $= 5 = \dfrac{Q_2}{W} = \dfrac{Q_2}{0.5Q_{add}}$ $\therefore$ $\dfrac{Q_{add}}{Q_1} = 0.4)$

38. (c) **39.** (a) **40.** (a) **41.** (a)

42. (b) **43.** (a) **44.** (b) **45.** (c)

46. (c) $(\Delta S = \dfrac{dQ}{T} = \dfrac{1000}{200} = 5$ kJ/K)

47. (b) $(dQ = TdS = 300 \times 5 = 1500$ kJ)

48. (b)

49. (c) ($W = Q_1 - Q_2$. If one engine is operating and $W = (Q_1 - Q_3) + (Q_3 - Q_2)$ and if two engines are operating $W = W_1 + W_2$ but $W_1 = W_2$. Work output of each engine $= \dfrac{W}{2}$)

50. (b) ($T_1 - T_2 = T_2 - T_3 = T_3 - T_4 = T_4 - T_5 = T_5 - T_6$. The temperature difference $= \dfrac{1000 - 500}{5}$ $= 100$ K)

51. (a) $(T_1 - T_2 = T_2 - T_3 = T_2 - T_4 = T_5 - T_6 = \dfrac{800 - 400}{4} = 100$;

$\dfrac{W}{Q_1} = \dfrac{Q_1 - Q_2}{Q_1} = \dfrac{T_1 - T_2}{T_1} = \dfrac{100}{800}$ $\therefore W_1 = \dfrac{1}{8} \times 80 = 10$ J;

$W = 4\, W_1, = 4 \times 10 = 40$ J)

52. (a) ($\Delta S = S_2 - S_1 = (2 + 3) - 4 = 1 > 0$. Hence feasible)

53. (b) ($\Delta S = S_2 - S_1 = 6 - (3 + 5) = -2 < 0$. Hence impossible)

54. (b) [Since $S_B < S_A$, hence process A to process B is impossible, i.e. $S_B - S_A = 3 - 4 = -1$ kJ/K is impossible]

55. (b) $T_1 - T_2 = T_2 - T_3 = \dfrac{600-200}{2} = 200$ K

$T_2 = 600 - 200 = 400$ K. Now $\dfrac{Q_1}{T_1} = \dfrac{Q_2}{T_2} = \dfrac{Q_3}{T_3}$

or $\dfrac{30}{600} = \dfrac{Q_3}{200}$ or $Q_3 = 10$ J)

56. (b) ($\eta_{\text{HE}} = 1 - \dfrac{T_2}{T_1} = 1 - \dfrac{400}{800} = \dfrac{1}{2} = \dfrac{W_E}{Q} = \dfrac{W_E}{40}$

or $W_E = 20$ J; COP $= \dfrac{400}{800 - 400} = 1 = \dfrac{Q_2}{W_{\text{ref}}} = \dfrac{Q_2}{W_{\text{ref}}} = \dfrac{5}{W_{\text{ref}}}$ or $W_{\text{ref}} = 5$ J. Net work output $= W_E - W_{\text{ref}} = 20 - 5 = 15$ J)

57. (c) (unavailable energy $= Q_1 \times \dfrac{T_2}{T_1} = 200 \times \dfrac{400}{800} = 100$ kJ)

58. (a) (unavailable energy $= \Delta S \times T_2 = 1.2 \times 300 = 360$ kJ)

59. (a) (Degree of irreversibility $= \Delta S \times T_2 = 600$ or $\Delta S = 600/300 = 2$ kJ/K)

Fill in the Blanks

1. (a)	**2.** (b)	**3.** (a)	**4.** (a)
5. (b)	**6.** (b)	**7.** (a)	**8.** (b)
9. (a) $\left(\dfrac{Q_1}{T_1} = \dfrac{Q_2}{T_2}\right)$	**10.** (b)	**11.** (b)	**12.** (b)
13. (b)	**14.** (a)	**15.** (b)	**16.** (b)
17. (b)	**18.** (b)	**19.** (b)	**20.** (a)
21. (a)			

22. (b) $\left[\text{unavailable energy} = Q_1 \times \dfrac{T_2}{T_1} \text{ which will decrease on lowering of } T_2\right]$

23. (a) ($\text{COP}_{\text{ref}} = \text{COP}_{\text{HP}}^{-1}$)

24. (a) $\left[\eta = 1 - \dfrac{T_2}{T_1};\ \eta \text{ will increase with increase of } T_1\right]$

25. (a) $\left[\eta = 1 - \frac{T_2}{T_1}; \eta \text{ will increase with decrease of } T_2\right]$

26. (b)

27. (a) $\left[\theta_{rej} = \Delta s \times T_o, \text{ hence } \Delta s = \frac{600}{300} = 2\,\text{kJ/K}\right]$

28. (a) **29.** (b)

30. (b) $\left[\eta = 1 - \frac{T_2}{T_1}; \eta \text{ will be different for different } T_1 \text{ and } T_2\right]$

31. (a) **32.** (b) **33.** (b)

34. (b) **35.** (b) **36.** (b)

CHAPTER 4

Properties of Steam and Thermodynamics

A painting is made up of many different colours, each one important and necessary.

INTRODUCTION

Steam is the most common working substance employed as working fluid in stream engines, steam turbines and atomic power plants for power generation. It acts in these applications as a transport agent for energy and mass interactions. Steam is a pure substance. It can be easily converted into any of three states, i.e. solid, liquid and gas. It has capability to retain its chemical composition and homogeneity in liquid and gaseous phases.

A pure substance has chemical homogeneity and constant chemical composition. Water is a pure substance as it meets both the above requirements. A substance cannot be a pure substance if it undergoes a chemical change. In this chapter, we will learn about the properties of steam as a pure substance and how to change the properties of steam during a process to attain the desired effect, i.e. maximum work for a given heat intake.

DEFINITION

Pure substance: A pure substance is a substance which has constant chemical composition and chemical homogeneity. Water is a pure substance. Any substance which undergoes a chemical reaction, cannot be termed pure substance.

Sensible heating: Sensible heating of a substance is heating in a single phase. The heat is used for raising the temperature of the substance.

Latent heating: Latent heat of a substance is the heat that causes its phase change without raising its temperature.

Boiling point: Boiling point is the temperature at which vapour pressure is equal to atmospheric pressure and phase change takes place from liquid to gas.

Melting point: Melting point is the temperature at which phase changes from solid to liquid when heat supplied is equal to latent heat.

Saturation state: Saturation state is the state at which its phase transformation takes place without any change in pressure and temperature. Therefore, there can be saturated solid state, saturated liquid state and saturated vapour state.

Saturation pressure: Saturation pressure is the pressure for a given temperature at which a substance changes its phase. The saturation pressure for water at 100°C is one atmospheric pressure.

Saturation temperature: Saturation temperature for a given pressure is the temperature at which a substance changes its phase.

Triple point: The triple point of a substance is the state at which the substance can coexist in solid, liquid and gaseous phases in equilibrium. The triple point for water is 0.01°C.

Critical point: The critical point of a substance is the state at which the substance can co-exist in two phases for last time in equilibrium. It is the point of highest pressure and temperature at which water and vapour coexist for last time. The substance has vapour phase only above this point. The critical point of water has a pressure of 22.12 MPa and a temperature of 374.15°C.

Dryness fraction: Dryness fraction is the ratio of the mass of vapour to the mass of both liquid and vapour in any liquid–vapour mixture region.

The liquid having temperature less than saturation temperature corresponding to a given pressure is called *subcooled* or *compressed liquid.*

Steam having temperature more than saturation temperature corresponding to any given pressure is called *superheated steam*.

Refer to Figures 4.1 and 4.2. When ice is heated, ice will have sensible heating up to 0°C along *AB*. Here ice will start melting along *BC* at constant temperature. *CD* shows sensible heating of water. *DE* shows evaporation of water at constant temperature. *EF* shows sensible heating of steam.

The above temperature vs heat as well as temperature vs entropy diagrams have been drawn for one atmospheric pressure. On the similar line, *T–S* variation of water can be obtained for other pressures. If all points of *D* (saturation liquid side) and *E* (saturation vapour side) are joined, a bell type curve is obtained with critical point (CP) at top (Figure 4.3). The saturated liquid line and the saturated vapour line come closer as pressure

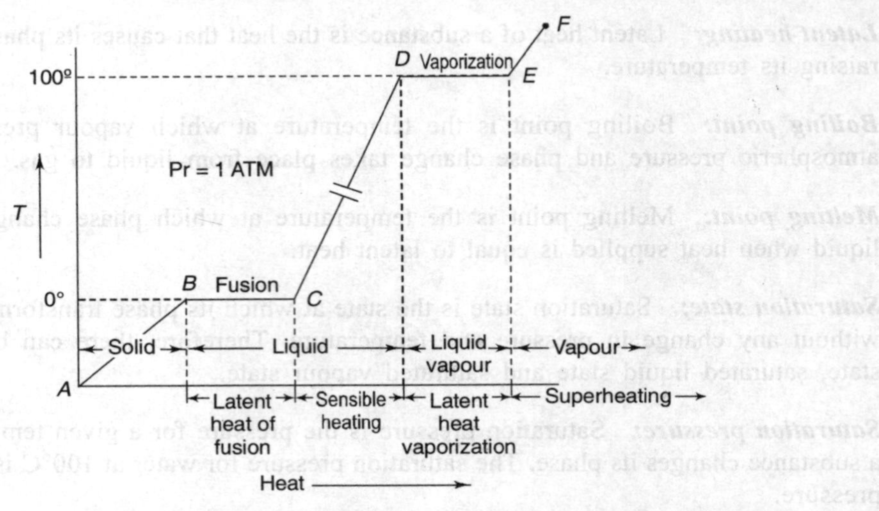

FIGURE 4.1 Temperature vs heat.

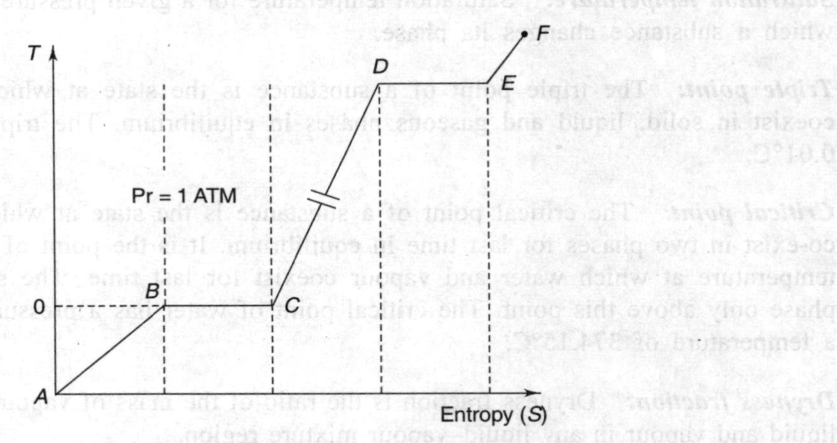

FIGURE 4.2 Temperature vs entropy.

and temperature increases. These lines meet at the critical point. The region enclosed between the saturated liquid line and the vapour line is wet steam (liquid–vapour) region. The region on the right of the saturation vapour line is the superheated region. The region left to the saturation liquid line is liquid (water) region.

The wet steam is the region between the saturated liquid line and the saturated vapour line. Dryness factor (x) is zero on the saturated liquid line and one on the saturated vapour line. Dryness at other point in the wet steam region is given by equal dryness lines. D_1E_1 shows heat of vaporization (h_{fg}) which decreases with increase of pressure. C_1D_1 shows sensible heating and ($T_{D1} - T_{C1}$) gives degree of subcooling at temperature C_1. Similarly, E_1F_1 shows sensible heating and ($T_{F1} - T_{E1}$) gives degree of superheating at temperature F_1.

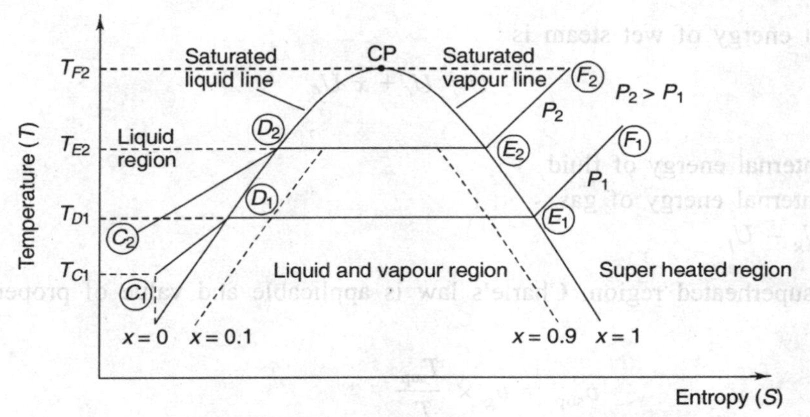

FIGURE 4.3 Temperature vs entropy.

PROPERTIES OF STEAM

Dryness fraction of steam is a factor to specify the quality of steam. It is the ratio of the mass of dry steam (m_g) to the mass of wet steam ($m_g + m_f$) where m_f is the mass of fluid (water) in wet steam.

$$x = \frac{m_g}{m_g + m_f}$$

The specific volume of wet steam is

$$v = v_f + xv_{fg}$$

where

v_f = specific volume of fluid (water)
v_g = specific volume of gas (steam)
$v_{fg} = v_g - v_f$

If $x = 0$, $v = v_f$
If $x = 1$, $v = v_g$

The enthalpy of wet steam is

$$h = h_f + x\ h_{fg}$$

where

h_f = enthalpy of fluid
h_g = enthalpy of gas
$h_{fg} = h_g - h_f$

The entropy of wet steam is

$$S = S_f + x\ S_{fg}$$

where

S_f = entropy of fluid
S_g = entropy of gas
$S_{fg} = S_g - S_f$

The internal energy of wet steam is

$$U = U_f + x\,U_{fg}$$

where

U_f = internal energy of fluid
U_g = internal energy of gas
$U_{fg} = U_g - U_f$

In the superheated region, Charle's law is applicable and value of properties can be found out:

$$v_{sup} = v_g \times \frac{T_{sup}}{T_g}$$

$$h_{sup} = h_g + c_p(T_{sup} - T_g)$$

$$S_{sup} = S_g + c_p \ln \frac{T_{sup}}{T_g}$$

$$U_{sup} = [h_g + c_p(T_{sup} - T_g) - Pv_{sup}]$$

STEAM TABLES AND MOLLIER DIAGRAM

Steam is a pure substance having constant properties at different pressure and temperature. These properties are estimated and tabulated in *steam table*. The table may be based on saturation pressure or temperature. The table gives enthalpy (h_f, h_g and h_{fg}), entropy (S_f, S_g and S_{fg},) specific volume (v_f, v_g, v_{fg}) and internal energy (U_f, U_g and U_{fg}).

Similarly, steam properties for superheated steam are also estimated and tabulated at some discrete pressure for varying degree of superheating in superheated steam tables.

Most of the devices in engineering operate on constant pressure. At constant pressure, heat supplied or removed increases or decreases the enthalpy of steam ($dQ = TdS$ and $dQ = dh - vdp$. As $dp = 0$, $dQ = dh$). Keeping this in view, the Mollier diagram is developed which is a diagram of enthalpy vs entropy (Figure 4.4). The Mollier diagram is widely used

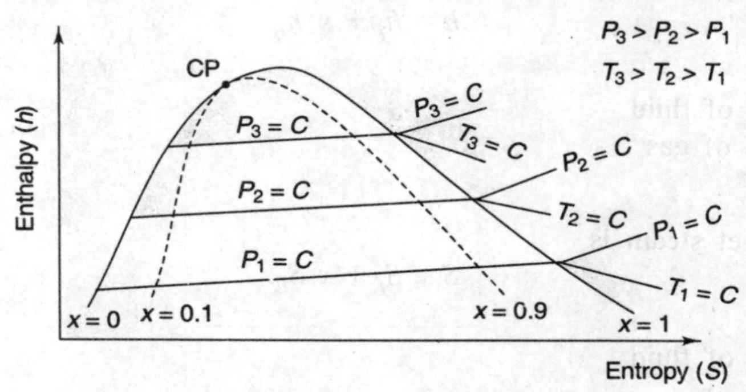

FIGURE 4.4 Mollier diagram.

since the amount of heat supplied or the change of enthalpy is readily available from it. The constant pressure line on the diagram has slope of temperature $\left[\left(\frac{dh}{dS}\right)_p = T\right]$. Isothermal lines in the wet region coincide with constant pressure lines. Every pressure has a definite saturation temperature in the wet region.

DRYNESS FACTOR MEASUREMENT

Dryness factor (x) is a basis parameter for fixing the state of a mixture in the wet region. It is zero at the saturated liquid line and one at the saturated vapour line. Generally, as far as possible dry steam is utilized in engineering applications. Therefore, dryness factor is measured to find the quality of steam. Dryness factor is measured by

1. Separating calorimeter
2. Throttling calorimeter
3. Separating and throttling calorimeter

In a separating calorimeter, water content of steam is separated in the calorimeter and the mass of water is found out (Figure 4.5).

$$\text{Dryness factor} = \frac{\text{total mass} - \text{mass of water}}{\text{total mass}}$$

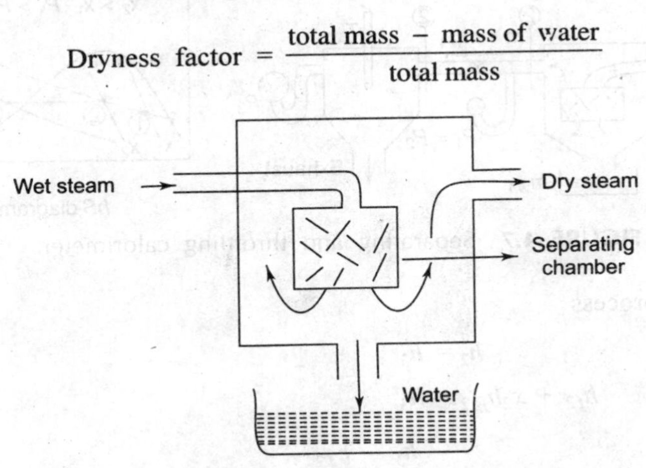

FIGURE 4.5 Separating calorimeter.

If wet steam is throttled (Figure 4.6), its enthalpy remains constant and a new state is adjusted to lower pressure in the superheated region which can be easily found out on the Mollier diagram ($h_1 = h_2$).

$$h_1 = h_{f1} + xh_{fg1} = h_2 \quad [h_2 \text{ at } P_2 \text{ and } T_2,\ h_1 \text{ at } P_1]$$

$$x = \frac{h_2 - h_{f1}}{h_{fg1}}$$

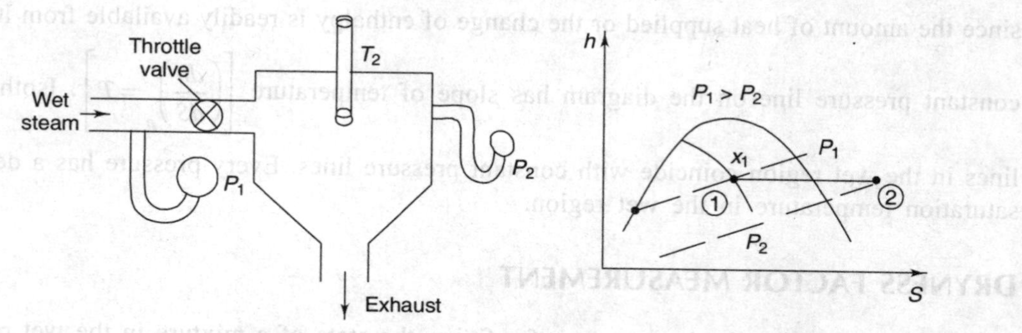

FIGURE 4.6 Throttling calorimeter.

If steam is extremely wet and the throttling process is unable to achieve any state in the superheated region, then the quality of steam is improved by removing some water content in a separating calorimeter and less wet steam is taken through the throttling calorimeter so that the superheated state could be achieved (Figure 4.7). The combined arrangement is known as separating and throttling calorimeter.

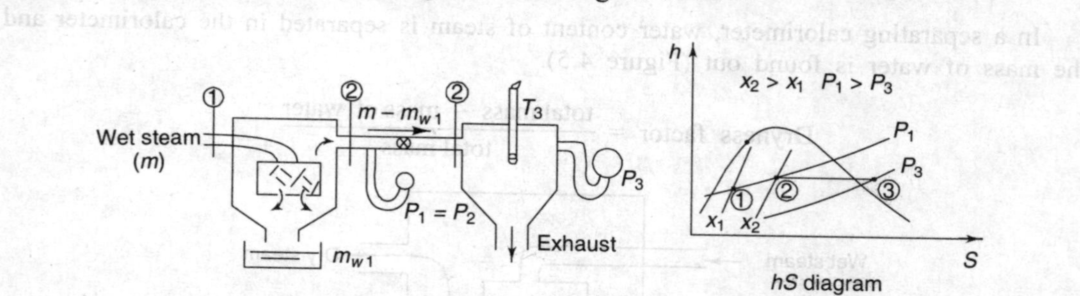

FIGURE 4.7 Separating and throttling calorimeter.

For throttling process

$$h_2 = h_3$$

$$h_{f2} + x_2 h_{fg2} = h_3$$

$$x_2 = \frac{h_3 - h_{f2}}{h_{fg2}}$$

Mass of vapour at state 2 is $x_2(m - m_{w1})$

$$x_1 = \frac{x_2(m - m_{w1})}{m}$$

where m_{w1} = water removed in the separating calorimeter

Total dryness factor

$$x = x_1 \, x_2$$

SOLVED PROBLEMS

1. Dry steam at 25 bar is throttled to lower pressure at 5 bar and then expanded adiabatically to 1 bar. Draw the process on the Mollier chart and find out dryness and temperature of the final state. Find also entropy change during throttling and total enthalpy change.

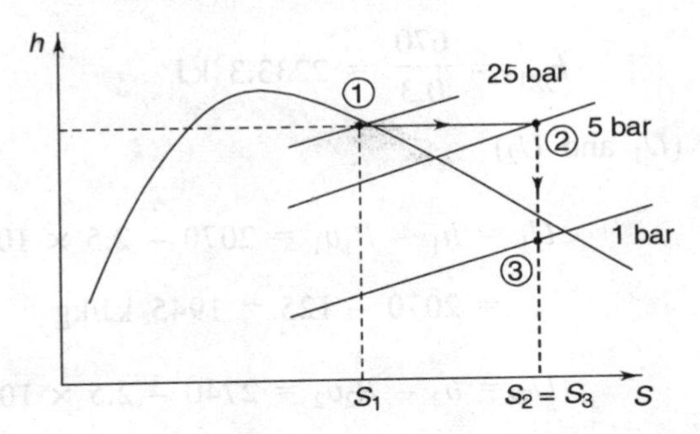

From the Mollier diagram:

State 1: Obtained from 25 bar line cutting saturation vapour line.
State 2: Obtained from constant enthalpy line from point 1 cutting 5 bar line.
State 3: Obtained from constant entropy line from state 2 cutting constant pressure line (1 bar)

From point 3, dryness factor (x_3) = 0.93 and temperature is 99.6°C

$$h_1 = 2800 \text{ kJ}$$

2. Calculate external work of evaporation, latent heat and internal energy for state 1 and state 2 as shown on the Mollier chart.

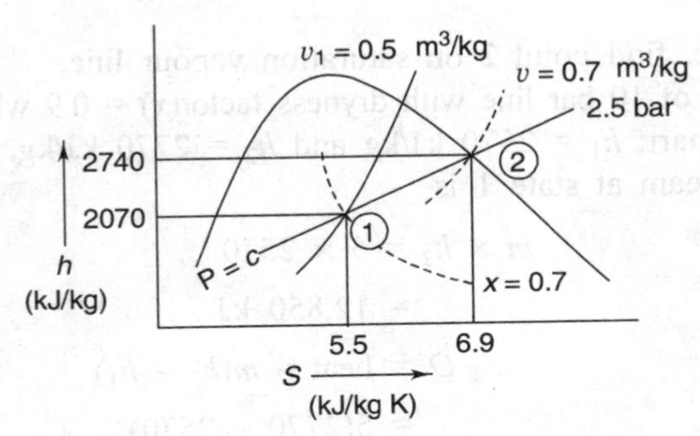

For state 1: $P_1 = 2.5 \text{ bar}, v_1 = 0.5 \text{ m}^3\text{/kg}, \quad h_1 = 2070$

For state 2: $P_2 = 2.5 \text{ bar}, v_2 = 0.7 \text{ m}^3\text{/kg}, \quad h_2 = 2740$

$$W_{1-2} = \text{Work} = P(v_2 - v_1) = 2.5 \times 10^5 \ (0.7 - 0.5)$$
$$= 50 \text{ kJ}$$

For latent heat (h_{fg1})

$$h_1 = h_f + x_1 h_{fg1} \quad \text{(i)}$$

$$h_2 = h_f + h_{fg1} \quad \text{(ii)}$$

$$(1 - x_1)h_{fg1} = h_2 - h_1 \quad [\because \text{Eq. (ii)–Eq. (i)}]$$

$$0.3 \times h_{fg1} = 2740 - 2070 = 670$$

$$h_{fg1} = \frac{670}{0.3} = 2233.3 \text{ kJ}$$

For internal energy (U_1 and U_2)

$$U_1 = h_1 - P_1v_1 = 2070 - 2.5 \times 10^5 \times 0.5 \times \frac{1}{10^3}$$

$$= 2070 - 125 = 1945 \text{ kJ/kg}$$

$$U_2 = h_2 - P_2v_2 = 2740 - 2.5 \times 10^5 \times 0.7 \times \frac{1}{10^3}$$

$$= 2740 - 175 = 2565 \text{ kJ/kg}$$

3. Calculate enthalpy of 5 kg of steam at 10 bar and dryness factor(x) = 0.9 from the Mollier chart. How much heat is to be given to convert it to full dry?

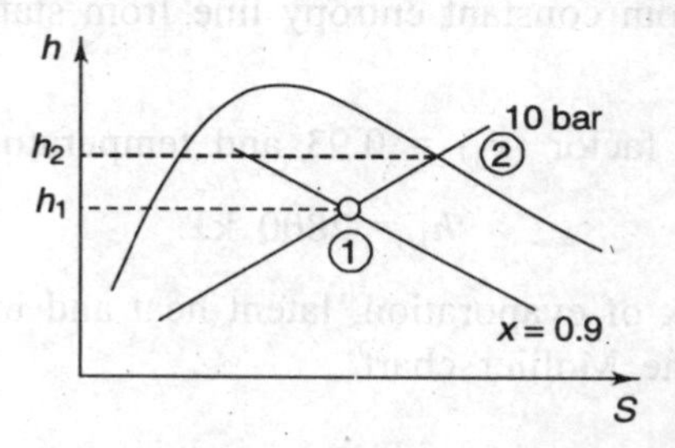

First for 10 bar line, find point 2 on saturation vapour line.
Find the intersection of 10 bar line with dryness factor(x) = 0.9 which will give point 1.
From the Mollier chart: h_1 = 2570 kJ/kg and h_2 = 2770 kJ/kg.
Total enthalpy of steam at state 1 is

$$m \times h_2 = 5 \times 2570$$

$$= 12{,}850 \text{ kJ}$$

$$Q = \text{heat} = m(h_2 - h_1)$$

$$= 5(2770 - 2570)$$

$$= 1000 \text{ kJ}$$

4. Calculate using the Mollier chart:

(i) Work done by 5 kg of steam when it is expanded from state 1 (h = 2900 kJ/kg and pressure 4 bar) to state 2 (pressure = 0.5 bar) isoentropically in a turbine.

(ii) Work done on 10 kg of steam to compress isentropically from state 1 ($P_1 = 0.5$ bar, $x_1 = 0.85$) to state 2 ($P_2 = 12$ bar) in a compressor.

(iii) Heat removed by a condensor in cooling 5 kg of steam from state 1 ($P_1 = 1.5$ bar, $T_1 = 350°C$) to saturation temperature isobarically.

(iv) Pressure drop and superheat from state 1 ($v_1 = 0.5 \frac{m^3}{kg}$ and $x_1 = 0.95$) to state 2($P_2 = 0.2$ bar).

(v) Superheat if enthalpy changes from 2700 kJ/kg to 3170 kJ/K at 1.5 bar.

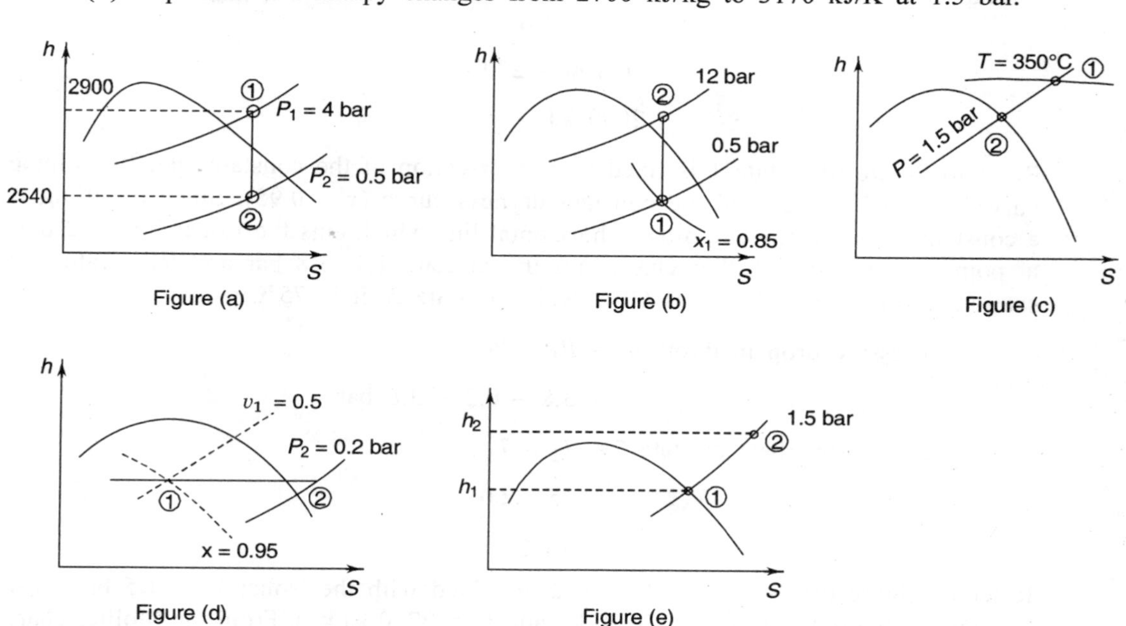

Refer to Figure (a). Expansion is isentropically from state 1 to state 2. State 1 is obtained by value of h_1(2900) and isobar line (4 bar). State 2 is obtained by drawing a vertical line to the isobar line (0.5 bar) from state 1.

$$h_2 = 2540 \text{ kJ/kg}$$

$$W_{1\text{-}2} = m(h_1 - h_2)$$
$$= 5(2900 - 2540)$$
$$= 1800 \text{ kJ}$$

Refer to Figure (b). State 1 is obtained by intersection of the isobar line (0.5 bar) with the constant dryness curve ($x = 0.85$). State 2 is obtained by drawing a vertical line from state 1 to the isobar line (12 bar).

$$h_1 = 2300,\ h_2 = 2840 \text{ from the Mollier chart}$$

$$W_{1\text{-}2} = m(h_2 - h_1) = 10(2840 - 2300)$$
$$= 5400 \text{ kJ}$$

Refer to Figure (c). State 1 is obtained by intersection of the isothermal line (350°C) with the isobar line (1.5 bar). Cooling is a constant pressure process. State 2 is obtained at the intersection of the isobar line (1.5 bar) with the saturation vapour curve. Temperature and enthalpy of state 2 are read from the Mollier chart.

$$T_2 = 125°\text{C}, \ h_2 = 2700 \text{ kJ/kg}, \ h_1 = 3100 \text{ kJ/kg}$$

Heat removed

$$Q = m(h_1 - h_2)$$

$$= 5(3100 - 2700)$$

$$= 2000 \text{ kJ}$$

Refer to Figure (d). State 1 is fixed with intersection of the constant specific volume curve (v = 0.5 m^3/kg) and the constant dryness curve (x = 0.95). Since throttling is a constant enthalpy process, draw a horizontal line which cuts the isobar line (0.2 bar) at point 2. From the Mollier chart, pressure at state 1 is 3.8 bar and temperature at saturation pressure (0.2 bar) is 60°C while at state 2, it is 75°C.

$$\text{Pressure drop in throttling} = P_1 - P_2$$

$$= 3.8 - 0.2 = 3.6 \text{ bar}$$

$$\text{Superheat at state 2} = T_2 - T_S$$

$$= 75 - 60$$

$$= 15°\text{C}$$

Refer to Figure (e). State 1 and State 2 are fixed with the isobar line (1.5 bar) with constant enthalpy lines (h_2 = 3170 kJ/kg and h_1 = 2700 kJ/kg). From the Mollier chart:

$$T_1 = 120°\text{C and } T_2 = 350°\text{C}$$

$$\text{Superheat} = T_2 - T_1 = 350 - 120 = 230°\text{C}$$

5. Calculate the state of steam, i.e. wet, dry or superheat for given conditions

(i) Pressure 15 bar and specific volume 0.125 m^3/kg
(ii) Pressure 10 bar and temperature 225°C
(iii) 5016 kJ heat given to 2 kg of water to generate steam at 30 bar

Let us do this problem by steam table instead of the Mollier chart. We use saturated steam (pressure) table as follows:

Absolute pressure	*Temperature*	*Specific volume (m³/kg)*		*Specific enthalpy*		
(*bar*)	(°C)	v_f	v_g	h_f	h_{fg}	h_g
10	179.9	0.001127	0.194	762.6	2013.6	2776.2
15	198.3	0.001154	0.132	844.7	1945.2	2789.9
30	233.8	0.001216	0.0666	1008.4	1793.9	2802.3

(i) At 15 bar:

$$v_g = 0.132 \text{ m}^3/\text{kg}, \quad v_f = .001154 \text{ m}^3/\text{kg}$$

Given: $v = 0.125 \text{ m}^3/\text{kg}$

Since $v_g > v$ hence steam is wet:

$$v = v_f + x(v_g - v_f)$$

$$0.125 = 0.0011 + x(0.132 - 0.0011)$$

$$x = \frac{0.124}{0.131} = 0.946$$

(ii) At 10 bar, $T_{sat} = 179.9°C$.

Since temperature of steam is 225°C (>179.9°C) hence the steam is in the superheat region. Therefore

$$\text{Superheat} = 225 - 179 = 46°C$$

(iii) At 30 bar, $h_g = 2803.2$, $h_f = 1008.4$ and $h_{fg} = 1793.9$.

$$\text{Actual enthalpy} = \text{heat given per kg}$$

$$= \frac{5016}{2} = 2508 \text{ kJ/kg}$$

Since h_g > actual enthalpy, hence steam is wet.

$$h = h_f + x\, h_{fg}$$

$$2508 = 1008 + x \times 1793.9$$

$$x = \frac{1500}{1794} = 0.836$$

6. A container having a volume of 0.200 m^3 contains equal volume of saturated dry steam and water at 300°C. Calculate the volume of steam and water.

Since saturation temperature is given, here we use the saturated steam table (temperature).

Temperature (°C)	*Saturated pressure*	*Specific volume* (m^3/kg)		*Specific enthalpy*		
		v_f	v_g	h_f	h_{fg}	h_g
300	8.581	0.001404	0.02167	1349	1404.9	2749

At saturation temperature (300°C)

$$v_f = .001404 \quad v_g = 0.02167$$

The volumes of steam and water are equal. Hence the volume of steam is $\frac{0.200}{2} = 0.100 \text{ m}^3$

$$\text{Mass of steam} = \frac{\text{volume}}{v_g} = \frac{0.100}{0.02167} = 4.6 \text{ kg}$$

$$\text{Mass of water} = \frac{\text{volume}}{v_f} = \frac{0.100}{0.001404} = 71.22 \text{ kg}$$

7. Find the specific volume enthalpy, entropy and internal energy of wet steam at 10.5 bar with dryness factor of 0.8.
Use saturation steam (pressure) table.

Pr (bar)	*Temp* (°C)	*Enthalpy*			*Entropy*			*Internal energy*		*Specific vol.*	
		h_f	h_{fg}	h_g	S_f	S_{fg}	S_g	U_f	U_{fg}	v_f	v_g
10.5	182	772	2056	2778	2.159	4.407	6.566	770	1814	0.001130	0.185

$$v = v_f + x(v_g - v_f)$$
$$= 0.001177 + 0.8(0.185 - 0.0011)$$
$$= 0.001 + 0.8 \times 0.184 = 0.148 \text{ m}^3/\text{kg}$$
$$h = h_f + xh_{fg}$$
$$= 772 + 0.8 \times 2056$$
$$= 2416.8 \text{ kJ/kg}$$
$$U = U_f + xU_{fg}$$
$$= 770 + 0.8 \times 1814$$
$$= 2221 \text{ kJ/kg}$$
$$S = S_f + xS_{fg}$$
$$= 2.159 + 0.8 \times 4.407$$
$$= 5.685 \text{ kJ/(kg K)}$$

8. Find the internal energy if steam has enthalpy, pressure and specific volume as 2848 kJ/kg, 120 bar and 0.017 m^3.

$$h = U + pv$$

or
$$U = h - pv$$
$$U = 2848 - 120 \times 10^2 \times 0.017$$
$$U = 2644 \text{ kJ/kg}$$

9. In a piston-cylinder arrangement, 1 kg of steam at 1.5 MPa and 200°C is cooled till 2/3 kg of dry steam is left.

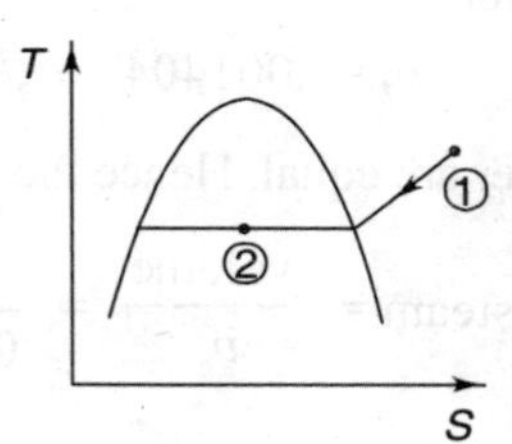

Here we use the superheated steam table.

P (bar)	Superheat	200	250	300	350	400
15 sat temp (198.32)	v U h_S	0.132	0.152	0.169	0.187	0.203

$$\text{Here superheat } (t) = 200°\text{C}$$

$$v_1 = 0.132$$

Cooling is taken place at constant pressure and volume becomes $\frac{2}{3}$ of state 1. Therefore,

$$v_2 = \frac{2}{3} v_1 = \frac{2}{3} \times 0.132 = 0.0883$$

From saturated steam (pressure) table as given in earlier problem (4) v_g at 15 bar = 0.132

Since $v_g > v_2$, steam at state 2 is in the wet region. Also T-saturation = 198.32°C.

10. In a steam power plant, the steam of 0.1 bar and 0.95 dry enters the condenser and leaves as saturated liquid at 0.1 bar and 45°C. Cooling water enters the condenser at 20°C and leaves at 35°C. Determine the mass flow rate of cooling water to the condenser.

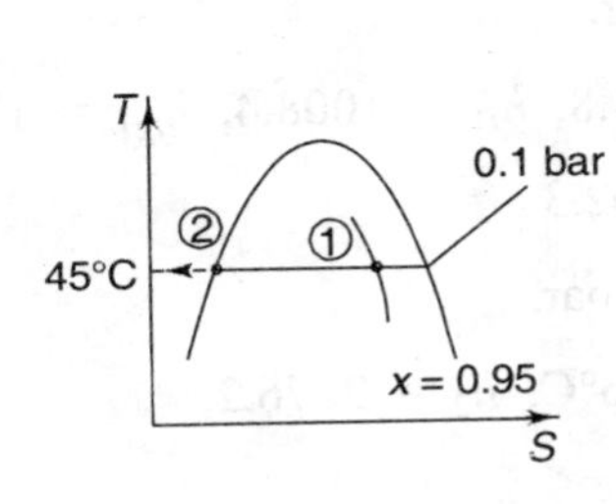

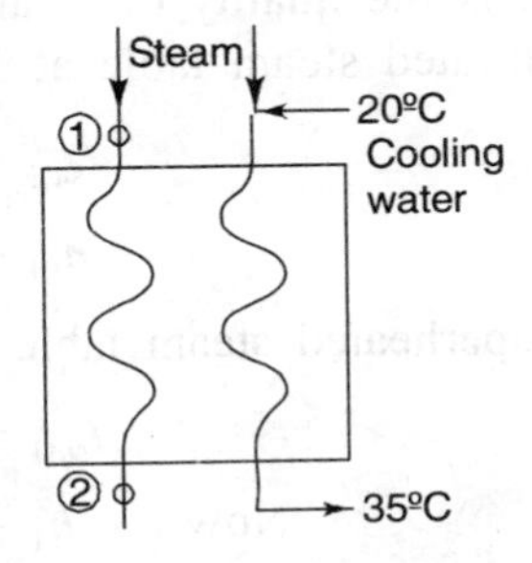

Using the saturated steam table (pressure), for 0.1 bar:

$$h_{f1} = 191.81 \text{ and } h_{fg1} = 2393$$

$$h_1 = h_{f1} + x\, h_{fg1}$$

$$= 191.81 + 0.95 \times 2393$$

$$= 2465 \text{ kJ/kg}$$

$$h_2 = h_{f1} = 191.81 \text{ kJ/kg}$$

Heat gained by cooling water is $m_w \times 4.18 \times (35 - 20)$

$$= m_w \times 4.18 \times 15$$

Heat lost by steam is $m_s (h_1 - h_2)$

$$= m_s (2465 - 191.81)$$

$$= m_s \times 2273$$

$$m_s \times 2273 = m_w \times 4.18 \times 15$$

$$\frac{m_w}{m_s} = \frac{2273}{4.18 \times 15} = 36.25$$

11. Steam enters a throttling calorimeter at 10 MPa and comes out at 0.05 KPa and 100°C. Calculate dryness factor of steam.

Using the saturated steam (pressure) table for 10 MPa,

$$h_{f1} = 1408 \text{ kJ/kg } h_{fg1} = 1317 \text{ kJ/kg}$$

$$h_1 = h_{f1} + xh_{fg1}$$

$$= 1408 + x \times 1317$$

Using the superheated steam (pressure) table for pressure = 0.5 bar and T = 100°C,

$$h_2 = 2682.5 \text{ kJ/kg}$$

$$1408 + 1317 \times x = 2682.5$$

$$x = 0.968$$

12. For estimating the quality of steam of 3 MPa in a boiler, its sample is throttled in a throttling calorimeter where its pressure and temperature are found to be 1 bar and 140°C. What is the quality of steam. (UPTU carry over: Dec. 2005)

From the saturated steam table at 30 bar,

$$t_{sat1} = 233.8,\ h_{f1} = 1008.4,\ h_{fg1} = 1793.9$$

and

$$h_{g1} = 2802.3$$

From the superheated steam table at 1 bar,

$$t_{sat1} = 99.6°C,\ h_2 = 2676.2$$

Now $\quad h_1 = h_2$

$$h_{f1} + x\, h_{fg1} = h_2 = 2676.2$$

$$1008.4 + x \times 1793.9 = 2676.2$$

or

$$x = 0.929$$

13. Draw (1) TS diagram, (2) TV diagram, (3) Th diagram, (4) PV diagram and (5) hS diagram when water is heated at one atmosphere.

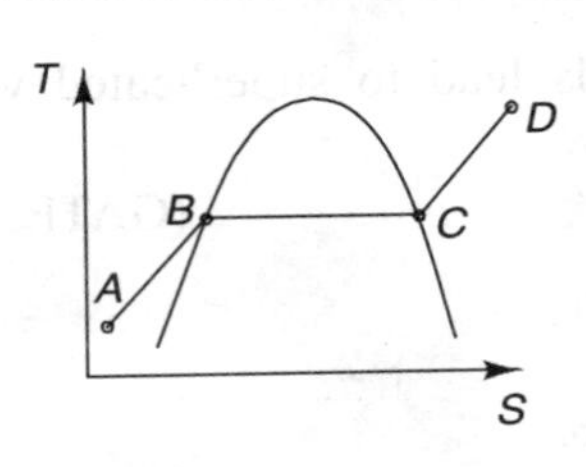

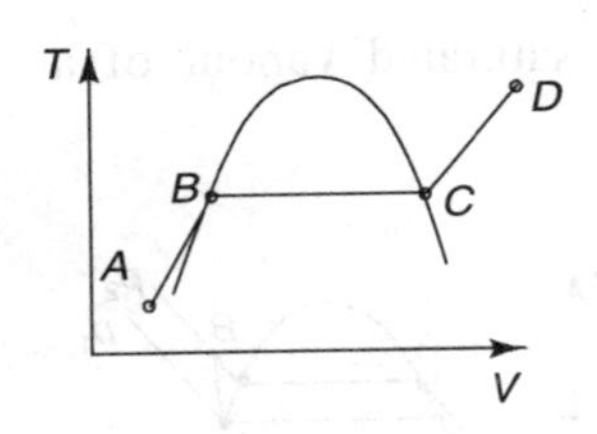

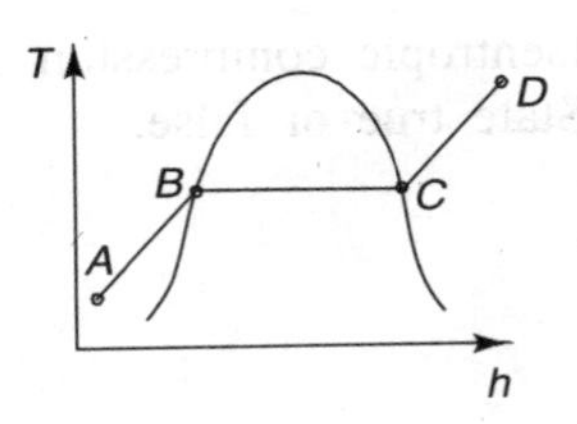

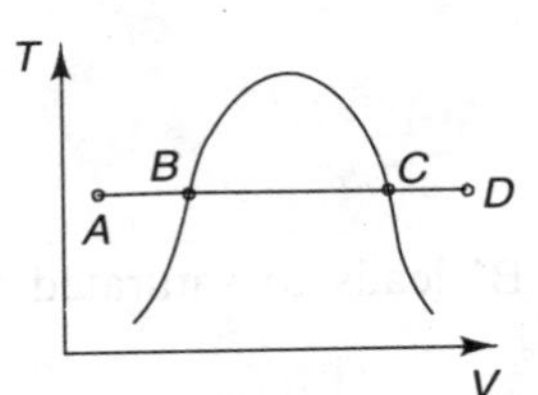

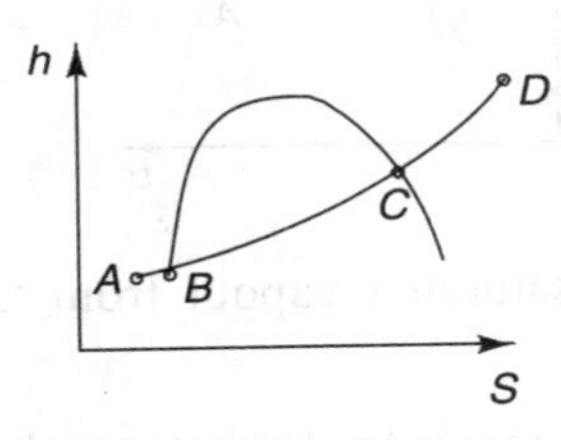

14. Draw the triple and critical points of water on *PT* diagram.

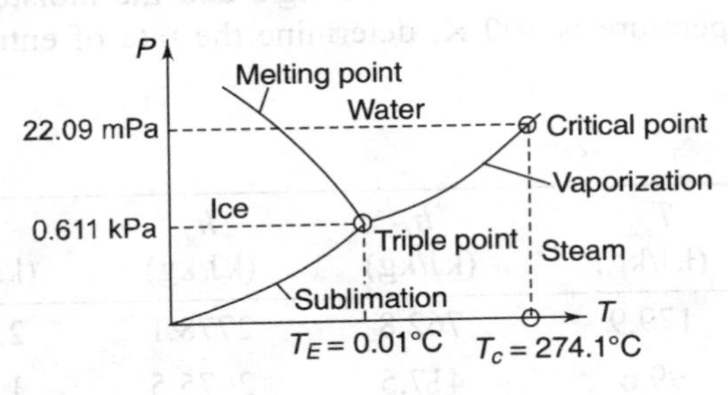

15. The equivalent evaporation (kg/hr) of a boiler producing 2000 kg/hr of steam with enthalpy content of 2426 kJ/kg from feed water at temperature 40°C (liquid enthalpy = 168 kJ/kg, enthalpy of evaporation of water at 100°C = 2258 kJ/kg) is

(a) 2000 (b) 2149 (c) 1682 (d) 1649.

(GATE: 1993)

$$h_1 = h_{f1} = 168 \text{ kJ/kg}, \; h_2 = h_{v_2} = 2426 \text{ kJ/kg}$$

$$\text{Enthalpy of evaporation} = \frac{m(h_2 - h_1)}{(h_{fg})_{100°C}}$$

$$= \frac{2000 \times (2426 - 168)}{2258}$$

$$= 2000 \text{ kg/hr}$$

16. Isentropic compression of saturated vapour of all fluids lead to superheated vapour. State true or false.

(GATE: 1994)

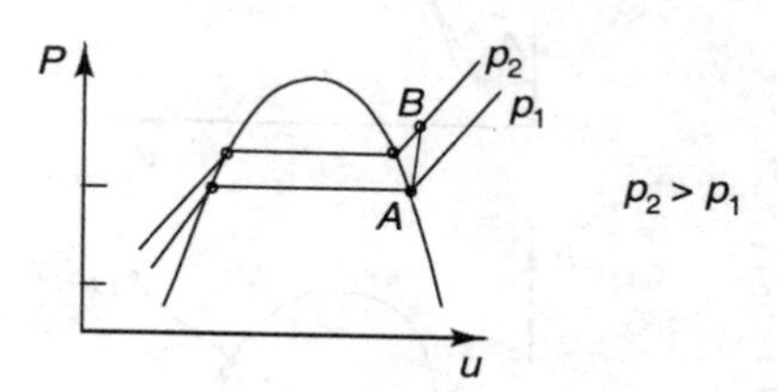

Isentropic compression of saturated vapout from 'A' to 'B' leads to saturated vapour. Hence statement is true.

17. An adiabatic steam turbine receives dry saturated steam at 1.0 MN/m^2 and discharge it at 0.1 MN/m^2. The steam flow rate is 3 kg/s and the moisture at exit is negligible. If the ambient temperature is 300 K, determine the rate of entropy production and the lost power.

Steam properties:

P (MN/m^2)	T_{sat} (kJ/kg)	h_f (kJ/kg)	h_g (kJ/kg)	S_f (kJ/kg)	S_g (kJ/kg)
1.0	179.9	762.8	2778.1	2.139	6.586
0.1	99.6	417.5	2675.5	1.303	7.359

(GATE: 1999)

Given: moisture at exit is nil, i.e. steam is saturated at exit of the turbine.

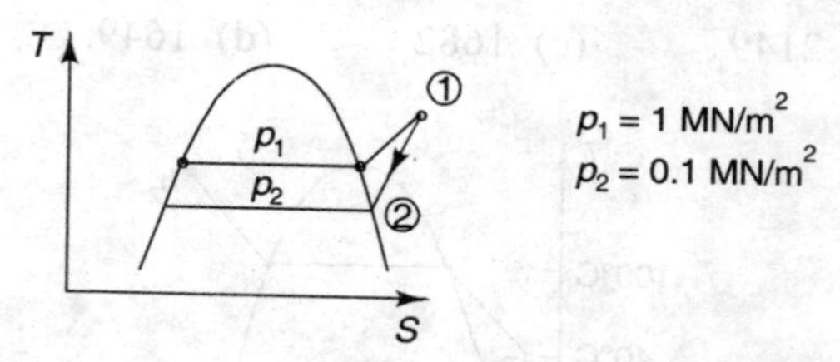

From table, $S_1 = (S_g)_{p_1} = 6.586$

and $S_2 = (S_g)_{p_2} = 7.359$

$\therefore$ Rate of entropy production $= m[(S_g)_{p_2} - (S_g)_{p_1}]$

$= 3 \times [7.359 - 6.586)$

$= 2.32$ kJ/s K

Lost power $= m \times T_0 \times \Delta S = 3 \times 300 \times (7.359 - 6.568)$

$= 686.7$ kJ/s.

18. A spherical shell of a boiler of 40 cm in radius contains saturated steam and water at 80°C. Calculate the mass of each if their volumes are equal.

$$\text{Total volume of the shell} = \frac{4}{3}\pi r^3$$

$$= \frac{4}{3}\pi(0.4)^3$$

$$= 0.268 \text{ m}^3$$

$$\text{Volume of steam} = \text{volume of water} = \frac{0.268}{2} = 0.134 \text{ m}^3$$

At saturation temperature 300°C from the steam table, we have

$$v_f = 1.404 \times 10^{-3} \text{ m}^3/\text{kg}$$

$$v_g = 2.167 \times 10^{-2} \text{ m}^3/\text{kg}$$

$$\therefore \text{ mass of water} = \frac{\text{Volume of water}}{v_f}$$

$$= \frac{0.134}{1.404 \times 10^{-3}} = 95.44 \text{ kg}$$

$$\therefore \text{ mass of steam} = \frac{V_g}{v_g} = \frac{0.134}{2.167 \times 10^{-2}} = 6.184 \text{ kg}$$

19. In a boiler, feed water supplied per hour is 205 kg while coal fired per hour is 23 kg. Net enthalpy rise per kg of water is 145 kV for conversion to steam. If the calorific value of coal is 2050 kJ/kg, then the boiler efficiency is

(a) 78% (b) 74% (c) 62% (d) 59%

$$\text{Boiler efficiency} = \frac{\text{heat utilized}}{\text{heat supplied by coal}}$$

$$= \frac{205 \times 145}{23 \times 2050}$$

$$= 0.62$$

Option (c) is correct.

OBJECTIVE TYPE QUESTIONS

When the problems mount so high that you cannot see anything else, it pays to step back from your work so you can see the big picture.

State True or False

1. The saturation temperature of water at one atmospheric pressure is 100°C. *(True/False)*
2. The saturation temperatures at pressure P_1 and P_2 are T_1 and T_2. If $P_1 > P_2$, then $T_1 > T_2$. *(True/False)*
3. The saturation temperature does not increase above 100°C if pressure is increased above 1 atmosphere. *(True/False)*
4. Latent heat decreases as temperature and pressure increase. *(True/False)*
5. At the critical temperature, latent heat becomes zero. *(True/False)*
6. The internal energy at the saturated liquid line is comparable to the internal energy at the saturated vapour line as internal energy depends upon temperature. *(True/False)*
7. The specific volume at the saturated liquid line is too small as compared to specific volume at the saturated vapour line. *(True/False)*
8. The change of enthalpy (h_{fg}) is the difference of enthalpy at saturated vapour line (h_g) and saturated liquid line (h_f). *(True/False)*
9. Sensible heating is heating of a substance in two phases. *(True/False)*
10. Latent heating is heating of a substance which transforms its phase without raising its temperature. *(True/False)*
11. At the boiling point, vapour pressure is equal to atmospheric pressure. *(True/False)*
12. To melt ice, heat equal to latent heat of fusion is to be given at 0°C. *(True/False)*
13. To evaporate water, heat equal to latent heat of evaporation is to be given to water at 100°C. *(True/False)*
14. During phase transformation, pressure and temperature may change. *(True/False)*
15. Saturated solid state and saturated liquid state are two possible saturation states. *(True/False)*
16. Corresponding to every temperature, there is a saturation pressure at which transformation takes place. *(True/False)*
17. The triple point is a state at which a substance can coexist in solid, liquid and gaseous phases in equilibrium. *(True/False)*
18. The critical point is the last state (highest temperature and pressure) of a substance in which liquid and vapour can coexist. *(True/False)*

19. Dryness factor is the ratio of the weight of water in wet steam to total mass. (*True/False*)

20. Subcooled liquid has temperature lower than saturation temperature for a given pressure. (*True/False*)

21. Superheated steam has temperature lower than saturation temperature for a given pressure. (*True/False*)

22. In the wet region on a *TS* diagram, temperature and pressure lines coincoide and are horizontal. (*True/False*)

23. In the wet region on the Mollier chart, temperature and pressure lines coincide and slanting with a slope equal to temperature. (*True/False*)

24. The Mollier chart is a *TS* diagram. (*True/False*)

25. The dryness increases in the wet region as the state moves towards the saturated vapour line. (*True/False*)

26. On the saturated vapour line, dryness factor is unity. (*True/False*)

27. On the saturated liquid line dryness factor is zero. (*True/False*)

28. The change of entropy in the wet region is $\frac{x \times h_{fg}}{T_s}$ where x = dryness, h_{fg} = heat of evaporation and T_s = saturation temperature. (*True/False*)

29. Subcooling is the difference between saturation temperature and actual temperature of water. (*True/False*)

30. Superheating is the difference between actual temperature and saturation temperature. (*True/False*)

31. In a separating calorimeter, the state of wet steam to dry state is achieved by reducing pressure. (*True/False*)

32. In a throttling calorimeter, water from wet steam is removed by sudden change of direction of flow. (*True/False*)

33. In a throttling process, the enthalpy of steam remains constant. (*True/False*)

34. A separating and throttling calorimeter is used for very wet steam to determine dryness factor. (*True/False*)

35. Heat capacity (C) = Mass × specific heat and sensible heating = $C \times \Delta T$ where ΔT = temperature rise. (*True/False*)

36. In the wet region, sensible heating takes place when heat is supplied. (*True/False*)

37. During vaporization, entropy remains constant. (*True/False*)

38. In the wet region, entropy changes without change of temperature. (*True/False*)

39. In the liquid region, entropy increases with increase of temperature. (*True/False*)

40. In the superheated region, entropy changes with temperature. (*True/False*)

41. In the superheated region, entropy change = $c \ln \frac{T_{\text{actual}}}{T_{\text{saturation}}}$ where c = heat capacity. (*True/False*)

42. In subcooled region, entropy change = $c \ln \dfrac{T_{\text{actual}}}{T_{\text{saturation}}}$ where c = heat capacity. *(True/False)*

43. On a *hS* diagram and in the wet region, the slope of the isobar line will increase with temperature rise. *(True/False)*

44. On a *TS* diagram, the slope of the isochoric line is more than the slope of the isobaric line. *(True/False)*

45. The enthalpy of water is measured from 0 K. *(True/False)*

46. Heat removal in a condensor is change of enthalpy at the inlet and the outlet of the condensor. *(True/False)*

47. The change of internal energy of stream at the inlet and the outlet of a turbine is work done by steam. *(True/False)*

48. Steam has more enthalpy at the outlet than at the inlet of a turbine. *(True/False)*

49. Steam has more enthalpy at the outlet than at the inlet of a compressor. *(True/False)*

50. Enthalpy of steam is amount of heat required to convert 1 kg of water from freezing point into dry and saturated steam. *(True/False)*

51. Enthalpy of vaporization (h_{fg}) is amount of heat required to convert 1 kg of saturated water into dry and saturated steam. *(True/False)*

52. The Mollier diagram is most commonly used to obtain steam properties for steam turbines and nozzles. *(True/False)*

53. The constant temperature lines in the Mollier diagram are sloping straight in the wet region and horizontal in the superheated region. *(True/False)*

54. The constant pressure lines do not change slope in the wet region and the superheated region on the *hS* diagram. *(True/False)*

55. On the *TS* diagram, constant pressure and constant temperature lines are horizontal in the wet region. *(True/False)*

56. In the superheated region on a *TS* diagram, the constant pressure lines are steeper than constant volume lines. *(True/False)*

57. A pure substance has constant chemical composition. *(True/False)*

Multiple Choice Questions

1. Process 1–2 on the *hS* diagram is
   (a) heating in a boiler (b) cooling in a condenser (c) compression in a compressor

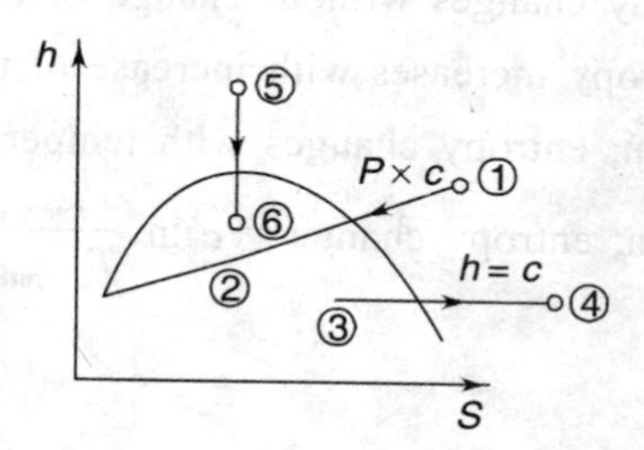

2. Process 3–4 on the hS diagram in Q.1 is
   (a) throttling (b) free expansion (c) heating

3. Process 5–6 on the hS diagram in Q.1 is
   (a) isentropic compression in a compressor
   (b) isentropic expansion in a turbine
   (c) isobaric cooling in a condenser

4. Work done during evaporation if P = pressure, v_f = specific volume of liquid, v_g = specific volume of gas and v_{fg} = change of specific volume from liquid to gas is
   (a) Pv_f (b) Pv_{fg} (c) Pv_g

5. The value of work in process 1–2 on the PV diagram is
   (a) 150 kJ (b) 300 kJ (c) 400 kJ

6. The value of heat in process 1–2 on the TS diagram is

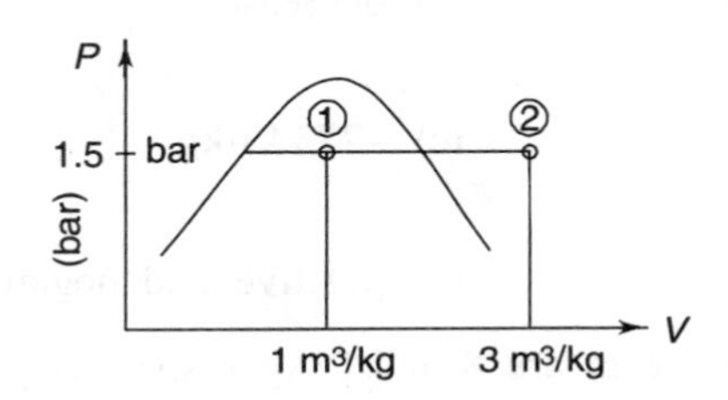

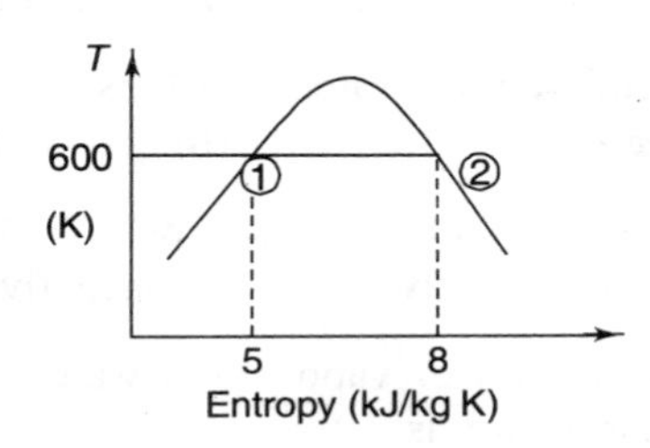

   (a) 3000 kJ (b) 1800 kJ (c) 4800 kJ

7. Three processes have been shown on the hS diagrams. Identify the process where heat is equal to the area under the curve of

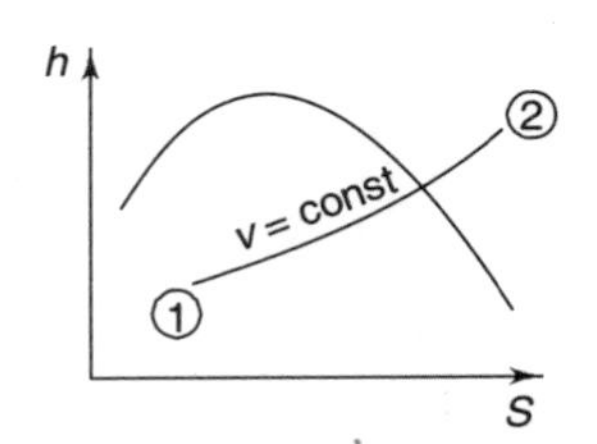

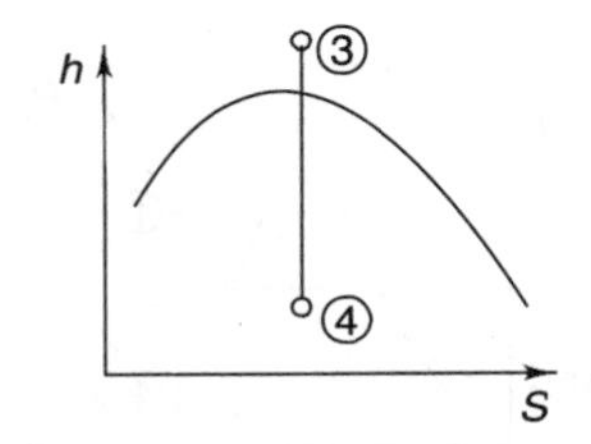

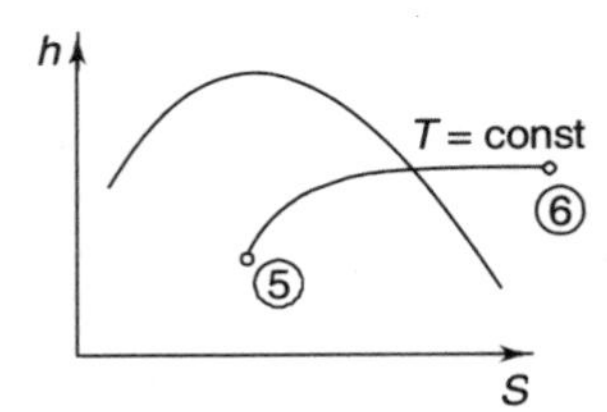

   (a) process 5–6 (b) process 1–2 (c) process 3–4

8. Out of the above three processes on the hS diagrams (Q.7), identify the process which will give change of internal energy as change of enthalpy.
   (a) process 5–6 (b) process 1–2 (c) process 3–4

9. Identify the process from the three processes Q.7, which will give work done by change of enthalpy.
   (a) process 3–4 (b) process 1–2 (c) process 5–6

10. If dryness factor = 0.8, the mass of dry steam in 5 kg of wet steam is
    (a) 3 kg (b) 5 kg (c) 4 kg (d) 1 kg

11. As pressure increases, the latent heat of steam will
(a) remain constant (b) decrease (c) increase

12. The critical point of water is
(a) 221.297 bar and 374.15°C
(b) 230.297 bar and 374.15°C
(c) 219.297 bar and 374.15°C

13. On a *TS* diagram, the change of state during transformation from saturated liquid to saturated vapour on heating is a
(a) horizontal line (b) vertical line (c) slant line

14. After throttling of wet steam in a throttling calorimeter, the state of steam must be in the
(a) superheated state (b) wet state (c) dry and saturated state

15. If pressure is increased, the specific volume of steam will
(a) remain constant (b) increase (c) decrease

16. The enthalpy of water at 0°C is
(a) zero (b) 273 kJ/kg (c) –273 kJ/kg

17. The enthalpy of water can be
(a) positive only (b) negative only (c) positive and negative

18. If the masses of vapour and water in wet steam are 8 and 2 kg respectively. Then the dryness factor is
(a) 0.4 (b) 0.5 (c) 0.8

19. What is the amount of work lost due to irreversibility during an adiabatic expansion (process 1–2′ in the figure) if T_2 = 300 K and $\Delta S = (S_{2'} - S_2)$ = 0.5 kJ(kg K)?
(a) 300 kJ (b) 200 kJ (c) 150 kJ

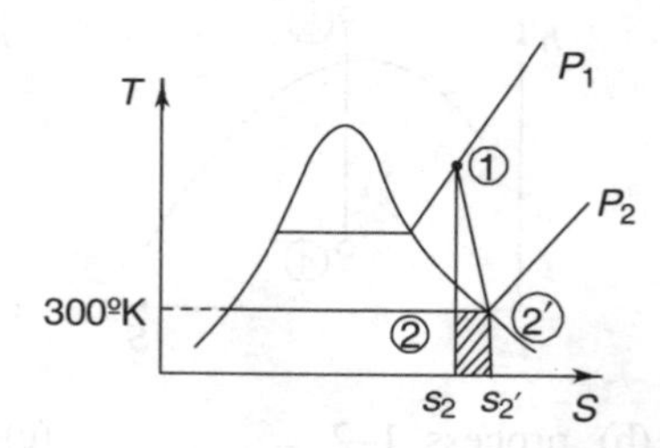

20. What is the amount of work lost due to irreversibility for adiabatic expansion (process 1–2′ on the *hS* diagram) if h_2 = 2500 kJ/kg h_2' = 2600 and mass of steam = 5 kg?

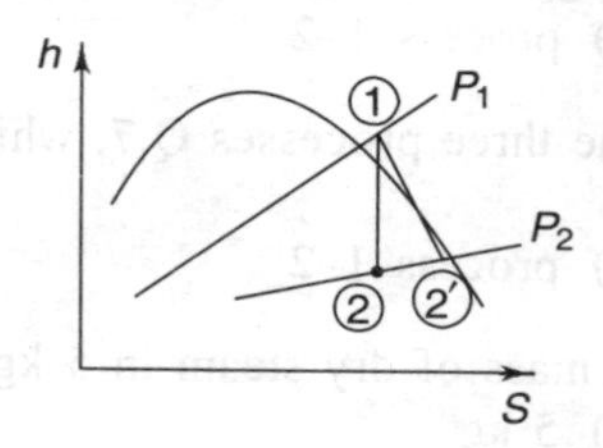

(a) 100 kJ (b) 500 kJ (c) 400 kJ

21. The degree of cooling given by state A in the figure is

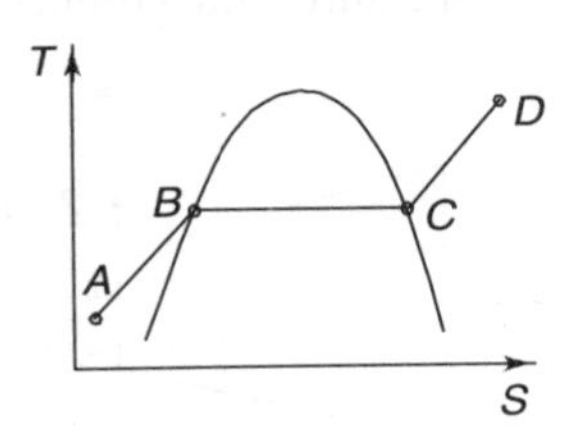

(a) $(T_B - T_A)$ (b) $(T_D - T_C)$ (c) $(T_C - T_A)$

22. The degree of superheating given by state D in the figure of Q.21 is
(a) $(T_B - T_A)$ (b) $(T_D - T_C)$ (c) $(T_D - T_A)$

23. In the PV diagram of water, identify the correct set.

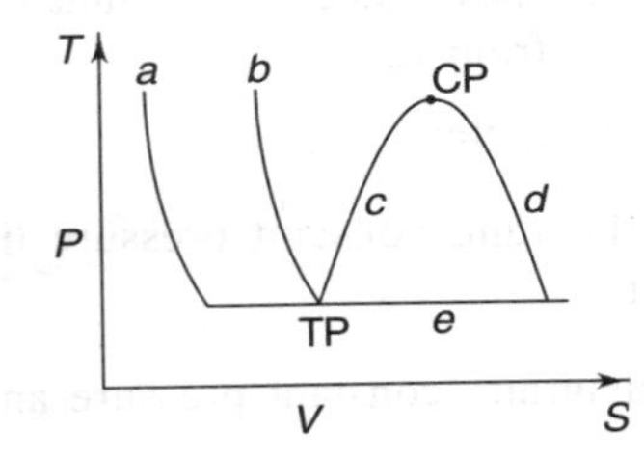

(a) Set 1 – a = saturated solid line, b and c = saturated liquid line, d = saturated vapour line and c = triple point line
(b) Set 2 – a and b saturated solid line, c = saturated liquid line, d = saturated vapour line and c = triple point line

24. Steam enters the turbine with enthalpy h_1 and leaves with enthalpy h_2. If m is the mass of steam, then work done by steam is

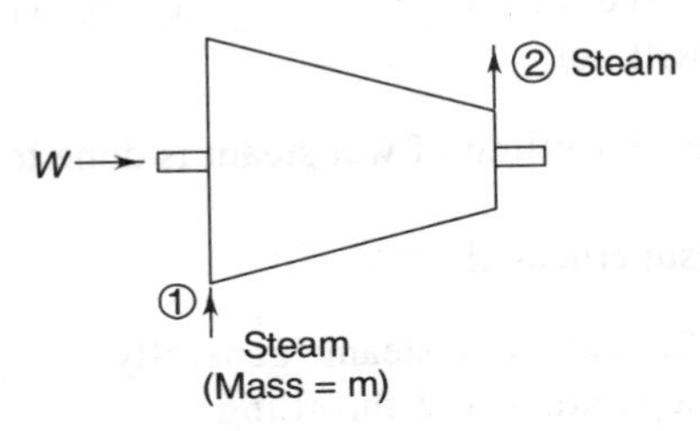

(a) $m(h_2 - h_1)$ (b) $m(h_1 - h_2)$ (c) $\dfrac{h_1 - h_2}{m}$

25. Steam enters the compressor with temperature T_1 and leaves it at temperature T_2. If c_p and c_v are specific heat at constant pressure and volume, the work done in compression per kg of steam is

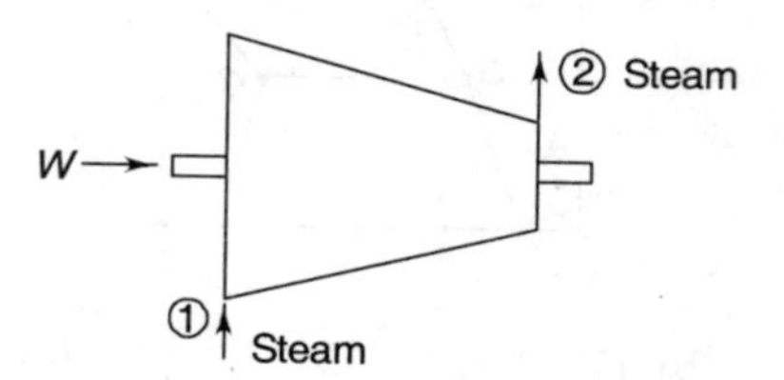

(a) $c_p(T_2 - T_1)$ (b) $c_p(T_1 - T_2)$ (c) $c_v(T_2 - T_1)$

Fill in the Blanks

1. The Mollier diagram is widely used since the amount of heat supplied or the change of enthalpy is readily _________ from it.
   (a) available (b) not available
2. In the wet region on a *hS* diagram, constant pressure lines are _________.
   (a) horizontal (b) slant
3. In the wet region on a *TS* diagram, constant pressure and constant temperature lines are _________.
   (a) horizontal (b) slant
4. The enthalpy of water is measured from _________.
   (a) 0°C (b) –273°C
5. If temperature and pressure increase, the latent heat of vaporization will _________.
   (a) decrease (b) increase
6. Above the critical point, the steam _________ liquefy when it is cooled.
   (a) will (b) will not
7. In a throttling calorimeter, throttling of wet steam is done to get the state to the _________ region.
   (a) saturated (b) superheated
8. For measuring dryness for very wet steam, generally a _________ calorimeter is used.
   (a) throttling (b) separating and throttling
9. The latent heat of fusion for water is _________.
   (a) 330 kJ/kg (b) 335 kJ/kg
10. The latent heat of vaporization for water is _________.
    (a) 2256.9 kJ/kg (b) 2260 kJ/kg

11. The enthalpy of water is ________ at 0°C.
(a) zero (b) 335 kJ/kg

12. The enthalpy of ice at 0°C is ________.
(a) 335 kJ/kg (b) –335 kJ/kg

13. In a separating and throttle calorimeter, dryness factor (x_1) = 0.8 in the separator part and it is 0.9 in the throttling part. Then dryness is ________.
(a) 0.72 (b) 0.9

14. If in a process change of enthalpy = 7200 kJ/kg, maximum possible work output = 4800 kJ/kg and change of entropy = 8 kJ/(kg K), then temperature is ________.
(a) 300 K (b) 200 K

15. If during evaporation, latent heat = 2193 kJ/kg at 23°C, then change of entropy is ________.
(a) 7.31 kJ/kg K (b) 17.84 kJ/kg K

ANSWERS

Try to replace critical thoughts with positive ones, and you'll be amazed what a difference it will make in your day, your spirit, and your life.

State True or False

1. True
2. True (The higher is saturation pressure, the higher is saturation temperature.)
3. False (Saturation temperature will be more than 160°C in case saturation pressure is more than atmospheric pressure.)
4. True (Latent heat decreases with increase of temperature and pressure. At critical point, latent heat is zero.)
5. True
6. False (Internal energy of the saturated vapour line is more than that of the saturated liquid line. $h_g - h_f = h_{fg}$ = latent heat.)
7. True (As $v_g >> v_f$, we can neglect v_f)
8. True
9. False (Tensible heating is in one phase. It is latent heating which is heating in two phases.)
10. True
11. True (Boiling takes place when vapour pressure is equal to pressure applied.)
12. True
13. True
14. False (During phase transformation, pressure and temperature remain constant.)
15. False (At the triple point all saturation states, viz saturation solid state, saturation liquid state and saturation vapour state coexist.)
16. True (If saturation pressure is higher, saturation temperature is also higher.)
17. True
18. True
19. False (Dryness factor is the ratio of the weight of vapour to total mass.)
20. True (Subcooling = saturation temperature – actual temperature)
21. False (Superheating is actual temperature of steam – temperature at saturation)
22. True
23. True

24. False (The Mollier chart is an enthalpy–entropy chart.)

25. True

26. True

27. True

28. True (For constant temperature, charge of entropy = $\frac{\Delta Q}{T}$)

29. True

30. True

31. False (Dryness is achieved by separating water from wet steam by sudden change in direction of flow.)

32. False (Steam is taken to superheated state by throttling resulting into lower pressure of steam which is now superheated at this state.)

33. True ($h_1 = h_2$)

34. True

35. True

36. False (Heat is used at constant temperature for phase transformation.)

37. False (Vaporization takes place at constant temperature. Entropy change = $\frac{\text{Latent heat}}{T_{sat}}$)

38. True

39. True ($\Delta S = m_w C_w \ln \frac{T_2}{T_1}$)

40. True ($\Delta S = m_s \times C_{sup} \ln \frac{T_2}{T_1}$)

41. True

42. False ($\Delta S = C \ln \frac{T_S}{T_{actual}}$)

43. True

44. True

45. False (Enthalpy of water is measured from 0°C.)

46. True ($Q = m_s(h_1 - h_2)$)

47. False ($w = m_s(h_1 - h_2)) \neq m_s(u_1 - u_2)$)

48. False (Enthalpy at the inlet > enthalpy at the outlet and the difference is work.)

49. False (Enthalpy at the outlet > enthalpy at the inlet. The difference is work input into a compressor for compression.)

50. True

51. True

52. True

53. True

54. True

55. True

56. Tru

57. True

Multiple Choice Questions

1. (b) (constant pressure)

2. (a) (constant enthalpy)

3. (b) ($\Delta S = 0$)

4. (b) [$W = Pdv = P(v_g - v_f) = p v_{fg}$]

5. (b) [$W = P \times dV = 1.5 \times 10^2 \times (3 - 1)$ kJ = 300 kJ]

6. (b) [$Q = T \times \Delta S = 600 \times (8 - 5) = 1800$ kJ]

7. (a) [$Q = T \times \Delta S$ which is constant temperature process (5–6)]

8. (b) ($dh = \Delta U + Pdv$ and for constant volume process $Pdv = 0$ and $dh = dU$. Hence process 1–2 will give change of internal energy as change of enthalpy.)

9. (a) [$dQ = dU + pdv$ and $dQ = T\Delta S$. If $\Delta S = 0$ (Isentropic process) $dQ = 0$ and $-dU = W(= Pdv)$. Process 3–4 is an isentropic process and will give change of enthalpy as work output.]

10. (c) ($0.8 = \dfrac{m_g}{m_{total}}$ or $m_g = 4$ kg)

11. (b) **12.** (a) **13.** (a) **14.** (a)

15. (c) **16.** (a)

17. (c) (Enthalpy of water above 0°C is positive and below 0°C it is negative.)

18. (c) (Dryness $= \dfrac{m_g}{m_g + m_g} = \dfrac{8}{2+8} = 0.8$)

19. (c) (Degree of irreversibility = $T \times \Delta S = 300 \times 0.5 = 150$ kJ)

20. (b) (Work lost = $(h_2' - h_2) \times$ mass = $100 \times 5 = 500$ kg)

21. (a) **22.** (b) **23.** (a)

24. (b) ($h_1 > h_2$)

25. (a) ($T_2 > T_1$ and process is at constant pressure.)

Fill in the Blanks

1. (a) **2.** (b) **3.** (a) **4.** (a)

5. (a) **6.** (b) **7.** (b) **8.** (b)

9. (b) **10.** (a) **11.** (a) **12.** (b)

13. (a) [In a separating and throttling calorimeter, the total dryness factor of steam (x) = dryness in separating portion (x_1) × dryness in throttling portion (x_2)

$$x = x_1 \times x_2 = 0.8 \times 0.9 = 0.72]$$

14. (a) Unavailable energy = 7200 – 4800 = 2400 kJ/kg, $T_0 \times \Delta S = 2400$ or $T_0 = \dfrac{2400}{8} = 300$ K

15. (a) $\Delta Q = 2193$ kJ/kg at 300 K. Hence $\Delta S = \dfrac{\Delta Q}{T} = \dfrac{2193}{300} = 75.31$ kJ/(kg K)

CHAPTER 5

Vapour Cycles

A pat on the back and a sympathetic ear are valuable gifts you can give to those you work and live with. Times of crisis only increase their value.

INTRODUCTION

Vapour cycles with vapour as working fluid are classified based on utility as:

1. Heat engine or power cycle
2. Refrigeration or heat pump cycle

In a vapour power cycle, the working fluid changes from liquid to vapour and back to its original state after performing work. The vapour power cycle is essentially a closed cycle, i.e. the working substance undergoes a series of processes and is always brought back to the initial state. However, some of the power cycles operates in an open system which means that the working substance is taken into the unit from the atmosphere and is discharged after undergoing a series of processes. In petrol and diesel engines, air and fuel are taken into the cylinder and the products of combustion are exhausted into the atmosphere. Similarly, in a steam engine water is taken into the boiler and steam after expansion in the cylinder is exhausted into the atmosphere. Such devices actually do not form a cycle. However, they are analyzed by considering that working substance after undergoing all processes is brought back to the inlet state and thus completing a cycle.

A vapour cycle comprises a series of steady flow processes. Each process is carried out in a separate piece of equipment specially designed for the purpose. The whole plant forms a closed system but each piece of equipment separately forms an open system. If

KE and PE remain constant for each piece of equipment, the steady flow energy equation is applicable as

$$q = (h_2 - h_1) + w$$

The working fluid employed in a vapour power cycle is water/stream. Water is a pure substance as it is chemically stable, cheapest and readily available. The common working substances used in vapour cycle refrigeration are ammonia, freon 11, kreon 12 and freon 22.

CARNOT VAPOUR CYCLE

An ideal vapour power cycle would be the Carnot vapour cycle comprised of two reversible isothermal and two reversible adiabatic processes. As working substance changes phase in a vapour cycle, the two isothermal processes are easily attainable as internally reversible in the forms of boiling of the liquid and condensation of the vapour. But heat transfer from a high temperature reservoir as well as from condensing vapour to a low temperature reservoir will remain externally irreversible. The Carnot vapour cycle is shown on a *TS* diagram as in Figure 5.1.

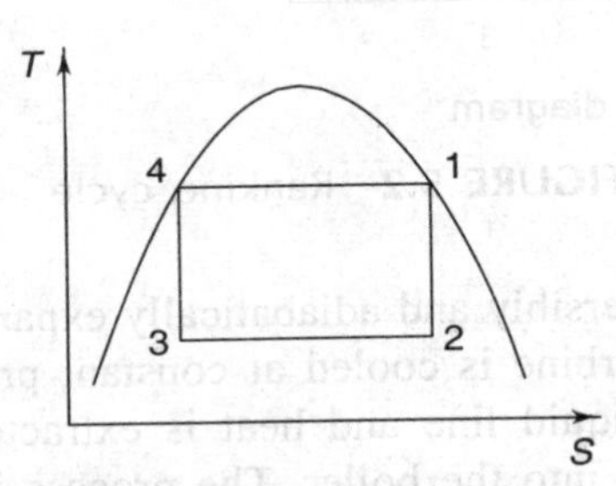

FIGURE 5.1 *TS* diagram.

In process 4–1, heat is added to water isothermally. Steam is expanded adiabatically in the turbine in process 1–2. Heat is removed in condensor in process 2–3. Wet steam is compressed adiabatically in the compressor in process 3–4.

$$\eta = \frac{W}{Q_{\text{add}}} = \frac{Q_{\text{add}} - Q_{\text{rej}}}{Q_{\text{add}}}$$

$$= 1 - \frac{Q_{\text{rej}}}{Q_{\text{add}}} = 1 - \frac{h_2 - h_3}{h_1 - h_4}$$

Also $$\eta = 1 - \frac{T_2}{T_1}$$

A few practical difficulties are experienced in the application of Carnot vapour cycle:

1. Outlet condition of wet steam from the condensor (state 3) cannot be controlled.

2. The size of the compressor has to be large to handle mixture of water and vapour corresponding to state 3.
3. Compression work is large resulting in a low cycle efficiency.
4. The wet steam (state 2) from the outlet of the turbine contains water droplets at high velocity which can pit and corrode turbine blades.

RANKINE CYCLE

The Rankine cycle is based on the modified Carnot vapour cycle to overcome its limitations. It consists of four steady flow processes as shown on the flow diagram and *TS* diagram in Figure 5.2.

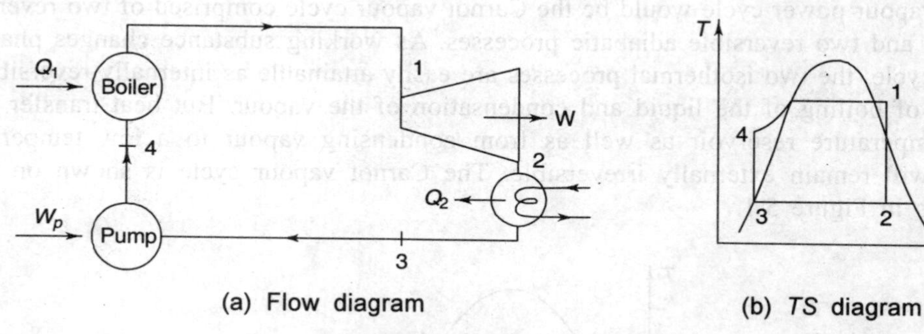

FIGURE 5.2 Rankine cycle.

In process 1–2, steam is reversibly and adiabatically expanded in the turbine. In process 2–3 steam coming out of the turbine is cooled at constant pressure in a condensor so that point 3 lies on the saturation liquid line and heat is extracted in the process. In process 3–4, water at state 3 is pumped into the boiler. The process is reversible and adiabatic. In process 4–1, the water is heated at constant pressure to get saturation vapour line.

Energy interaction in the above four processes are as follows:

1. *Process 1–2:* Turbine work output, $W_e = h_1 - h_2$
2. *Process 2–3:* Heat rejection, $Q_{rej} = h_2 - h_3$
3. *Process 3–4:* Pumping work, $W_p = h_4 - h_3 = v_{3f}(P_1 - P_2)$
4. *Process 4–1:* Heat is added in the boiler $Q_{add} = h_1 - h_4$.

$$\eta \text{ of the Rankine engine} = \frac{\text{network}}{\text{heat added}}$$

$$= \frac{W_e - W_p}{Q_{add}}$$

$$= \frac{(h_1 - h_2) - (h_4 - h_3)}{h_1 - h_4}$$

In case steam is superheated before delivering to the turbine (state 1′ instead of state 1), the work output and heat added increase by area 11′ 2′ 21 (Figure 5.3). This gives more efficiency.

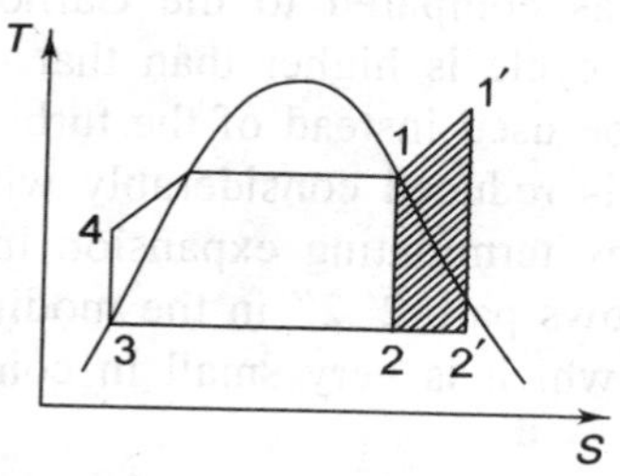

FIGURE 5.3 *TS* diagram when steam is superheated.

In case the boiler pressure is increased from P_1 to P_1' and condensor pressure is unchanged and also maximum temperature T_1' is kept equal T_1 (Figure 5.4), the work output is almost the same [Area (a) – Area (b) = 0] but heat rejection is reduced [by area (c)].

As $\eta = 1 - \dfrac{Q_{rej}}{Q_{add}}$, efficiency increases.

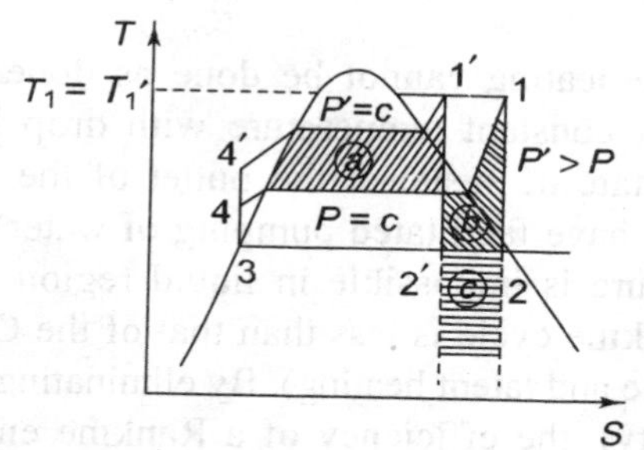

FIGURE 5.4 *TS* diagram when the boiler pressure is increased.

If the condensor pressure is reduced, the net work is increased by 2–2′–3′–4′–4–3–2 while heat added is increased by area *a*–4′–4–*b* (Figure 5.5). The net effect is increased in the efficiency of the cycle as Q_{rej} is also decreased.

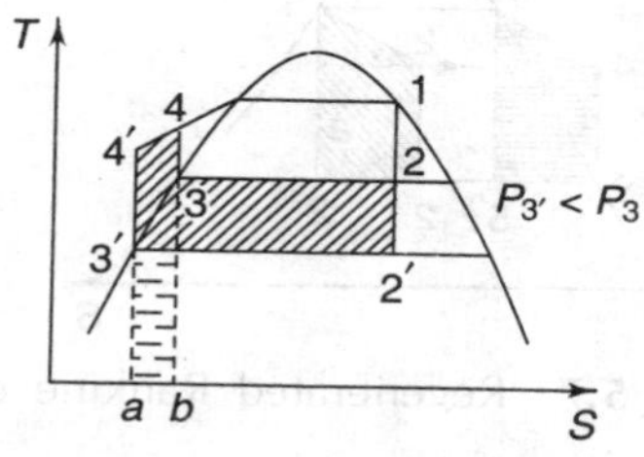

FIGURE 5.5 *TS* diagram when the condensor pressure is reduced.

During heating of water in a boiler, part of heat is used for sensible heating of water to boiling point. Hence heat addition in the Rankine cycle does not take place at the maximum temperature, i.e. boiling point. The average temperature at which heat is added is lower in the Rankine cycle as compared to the Carnot vapour cycle. Therefore, the efficiency of the Carnot vapour cycle is higher than that of the Rankine cycle.

A reciprocating engine can be used instead of the turbine in a power cycle. The stroke length and size of the cylinder is reduced considerably with sacrifice of small amount of work output. This is achieved by terminating expansion in the reciprocating engine at 2′ instead of 2 and the system follows path 2′ 2″ in tne modified Rankine cycle (Figure 5.6). The work lost is area 2′ 22″2′ which is very small in comparison to advantage gained.

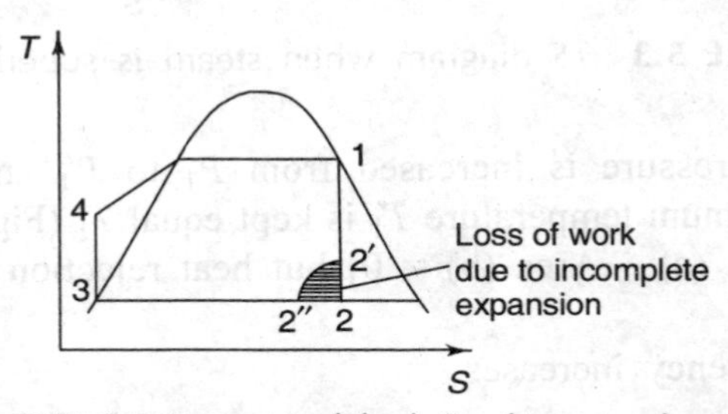

FIGURE 5.6 Modified Rankine cycle.

In the Carnot cycle, superheating cannot be done as done in the Rankine cycle. The reason is that superheating at constant temperature with drop in pressure cannot be done in the boiler. Similarly, the state of steam at the outlet of the condensor cannot be at the saturated liquid line (it would have facilitated pumping of water) in the Carnot cycle as heat addition at constant temperature is impossible in liquid region.

The efficiency of the Rankine cycle is less than that of the Carnot cycle as heat is added at average temperature (sensible and latent heating). By eliminating (not possible) or minimizing sensible heating (irreversibility), the efficiency of a Rankine engine can be increased. This is done by sensible heating of water by some other arrangement before it enters the boiler. Regenerative feed heating (Figure 5.7) is one of such arrangements. The expanding steam of the turbine is used to heat the feed water.

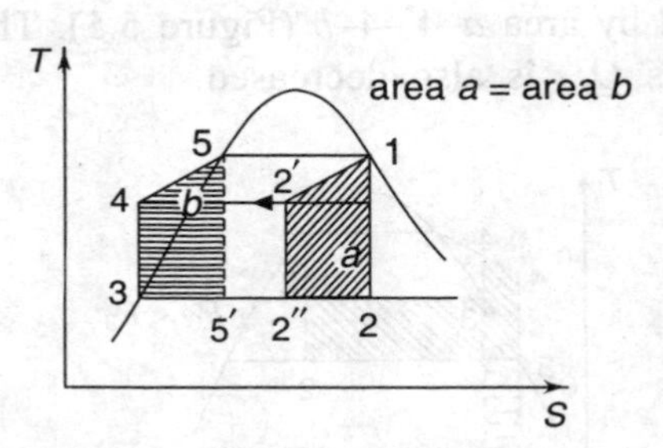

FIGURE 5.7 Regenerated Rankine cycle.

Steam becomes wet as it expands in the turbine and water droplets erode and corrode the turbine blades. To avoid this, steam is reheated. Reheating (Figure 5.8) is done by taking out the whole steam from the turbine at a suitable point before it becomes wet and steam is reheated. Reheated steam is again expanded in the turbine to condensor pressure.

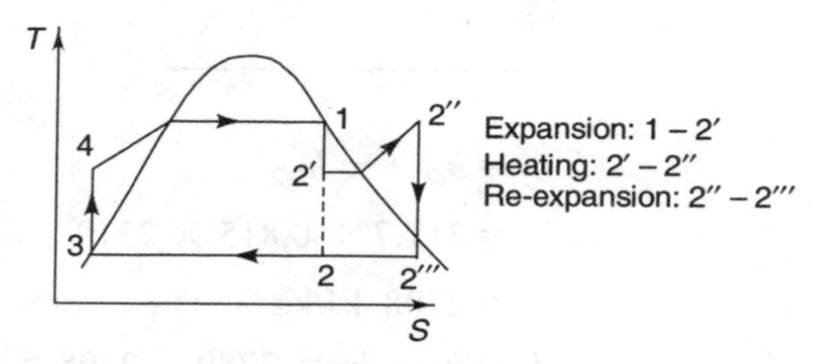

FIGURE 5.8 Reheat Rankine cycle.

SOLVED PROBLEMS

1. In a stream power cycle, the steam supply is at 15 bar fully dry and saturated. The condensor pressure is 0.4 bar. Calculate the Carnot and Rankine efficiencies of the cycle neglecting pump work.

Using steam table (pressure) we get properties at 15 bar and 0.4 bar:

Pressure (bar)	*Temperature* (°C)	*Specific enthalpy*			*Specific entropy*		
		h_f	h_{fg}	h_g	s_f	s_{fg}	s_g
15	198.3	844.7	1945	2789	2314	4.126	6.440
0.4	75.9	317.7	2.319	2637	1.026	6.645	7.671

$$T_1 = 198.3 + 273 = 471.3, \; T_2 = 75.9 + 273 = 348.9$$

$$\eta_{\text{Carnot}} = \frac{T_1 - T_2}{T_1} = \frac{471.3 - 348.9}{471.3}$$

$$= 0.259 \text{ or } 25.9\%$$

State 2 has to be fixed by finding dryness factor by equating entropy at state 1 = state 2.

$$s_1 = s_2 = s_{f2} + x\, s_{fg2}$$

$$6.440 = 1.026 + x \times 6.645$$

$$x = \frac{6.440 - 1.026}{6.645} = 0.815$$

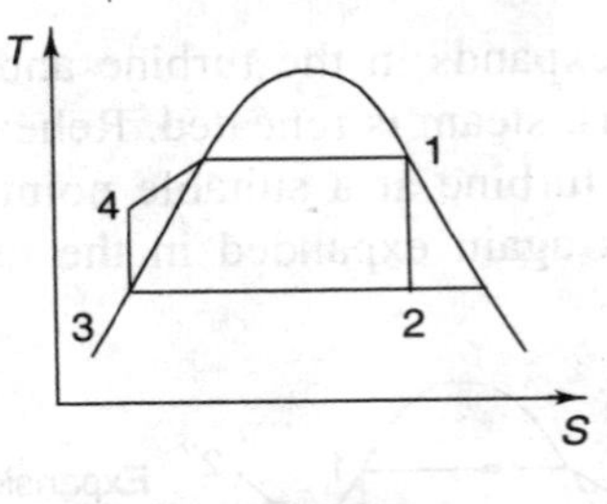

Now $h_2 = h_{f2} + x\ h_{fg2}$

$$= 317.7 + 0.815 \times 2319$$

$$= 2208 \text{ kJ/kg}$$

$$W_e = h_1 - h_2 = 2789 - 2208 = 581$$

$$Q_{add} = h_1 - h_4$$

if $h_3 = h_4$ (given),

$$Q_{add} = h_1 - h_3$$

$$= 2789 - 317.7 = 2471.3 \frac{\text{kJ}}{\text{kg}}$$

$$\eta_{Rankine} = \frac{W_e}{Q_{add}} = \frac{581}{2471.3} = 0.235 = 23.5\%$$

It can be seen that $\eta_{Carnot} > \eta_{Rankine}$.

2. A steam power plant is to be designed with superheated 350°C at the inlet and an exhaust pressure of 0.08 bar. The moisture content is designed not to exceed 15% after expansion in the turbine. Determine the steam pressure at the turbine inlet and the Rankine cycle efficiency neglecting pump work.

From the steam table (pressure) at a saturation pressure of 0.08 bar:

$$h_{f2} \approx 174,\ h_{fg2} \approx 2403$$

$$S_{f2} \approx 0.59,\ S_{fg2} = 7.64$$

$$x = 0.85 \text{ (given)}$$

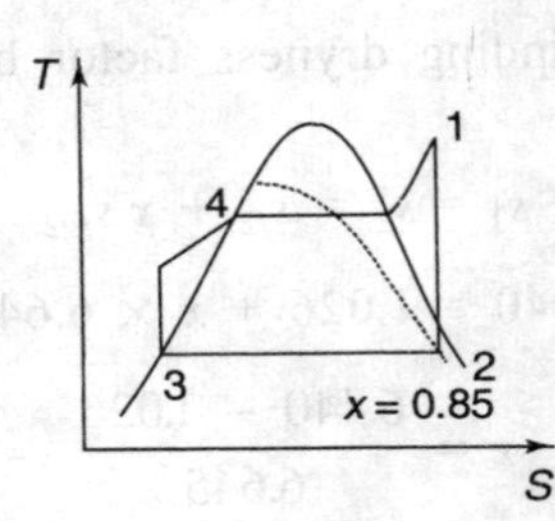

$$h_2 = h_{f2} + x\ h_{fg2}$$
$$= 174 + 0.85 \times 2403$$
$$= 2217 \text{ kJ/kg}$$
$$S_2 = S_{f2} + x\ S_{fg2} = 0.59 + 0.85 \times 7.64$$
$$= 7.085 \text{ kJ/(kg K)}$$

Now $$S_1 = S_2 = 7.085 \text{ kJ/(kg K)}$$

From the superheated steam table for $S_1 = 7.085$ kJ/(kg K) and temperature 350°C, pressure = 16 bar and $h_1 = 3147$ kJ/kg.

$$\eta = \frac{h_1 - h_2}{h_1 - h_3} = \frac{3147 - 2217}{3147 - 174} = 0.313 = 31.3\%$$

3. A steam power plant working on a Rankine cycle has steam parameter at the turbine inlet as 100 bar and 550°C and condensor pressure = 0.05 bar. Determine the cycle efficiency.

The solution to this problem is simpler if the Mollier chart is used. State 2 and state 4 can be easily determined as process 1–2 and process 3–4 are isentropic. The cycle on the hS diagram is shown.

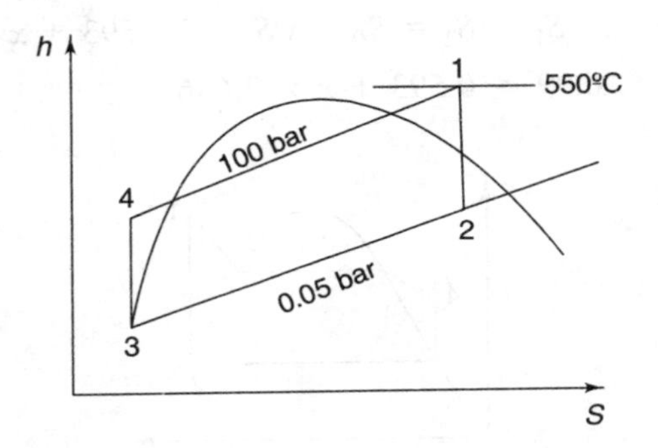

Point 1 is got at an intersection of 100 bar line with 550°C line. Point 2 is got vertically moving down to 0.05 bar line. Point 3 is 0.05 bar line with the saturated liquid line. Point 4 is at an intersection of 100 bar line and isentropic line from point 3.

$$h_1 = 3500 \text{ kJ/kg}$$
$$h_2 = 2050 \text{ kJ/kg}$$
$$h_3 = 140 \text{ kJ/kg}$$
$$h_4 = 150 \text{ kJ/kg}$$

$$\eta = \frac{W_e - W_p}{Q_{add}}$$

$$= \frac{(h_1 - h_2) - (h_4 - h_3)}{h_1 - h_4}$$

$$= \frac{(3500 - 2050) - (150 - 140)}{3500 - 150}$$

$$= \frac{1440}{3350} = 0.429 \text{ or } 42.9\%$$

4. Steam at 20 bar and 360°C is expanded in a turbine to 0.08 bar. Steam is condensed to the saturated liquid line and then pumped to the boiler. Find work output and efficiency.

From steam tables at 20 bar and 360°C:

$$h_1 = 3159 \text{ kJ/kg}, S_1 = 6.99 \text{ kJ/(kg K)}, T_{sat} = 212.4$$

At 0.08 bar,

$$h_3 = h_{f3} = 174 \text{ kJ/kg}, h_{fg3} = 2403$$

$$S_{f3} = 0.593, S_{fg3} = 7.636, v_{f3} = 0.001$$

$$S_1 = S_2 = S_{f3} + x\,S_{fg3} = 0.593 + x\,S_{fg3}$$

$$6.99 = 0.593 + x \times 7.636$$

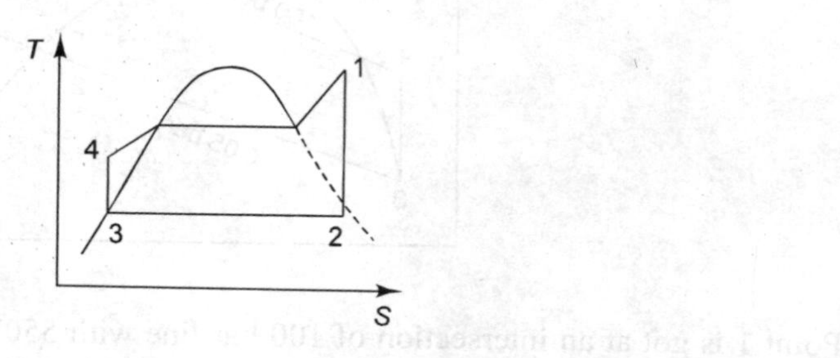

or $$x = \frac{6.40}{7.64} = 0.84$$

$\therefore$ $$h_2 = h_{f3} + xh_{fg3} = 174 + 0.84 \times 2403$$

$$= 2192.5 \text{ kJ/kg}$$

$$W_{pump} = h_{f4} - h_{f3} = v_{f3}(P_1 - P_2)$$

$$= 0.001\,(20 - 0.08) \times 10^2 \text{ kJ/kg}$$

$$= 1.992 \text{ kJ/kg} \approx 2 \text{ kJ/kg}$$

$$W_{\text{turbine}} = h_1 - h_2 = 3159 - 2192.5 = 966.5 \text{ kJ/kg}$$

$$Q_{\text{add}} = h_1 - h_4 = 3159 - (174 + 2)$$

$$= 2983 \text{ kJ/kg}$$

$$\eta_{\text{cycle}} = \frac{W_{\text{net}}}{Q_{\text{add}}} = \frac{W_{\text{turbine}} - W_{\text{pump}}}{Q_{\text{add}}}$$

$$= \frac{966.5 - 2}{2983} = 0.323 \text{ or } 32.3\%$$

5. Find net output and thermal efficiency of a theoretical Rankine cycle in which the boiler is at 40 bar and it is generating steam at 300°C. Condensor pressure is 0.1 bar. (UPTU: Dec. 2005)

$$W_{\text{turbine}} = h_1 - h_2$$

$$W_{\text{pump}} = v_{f3} (P_1 - P_2)$$

$$Q_{\text{add}} = h_1 - h_4$$

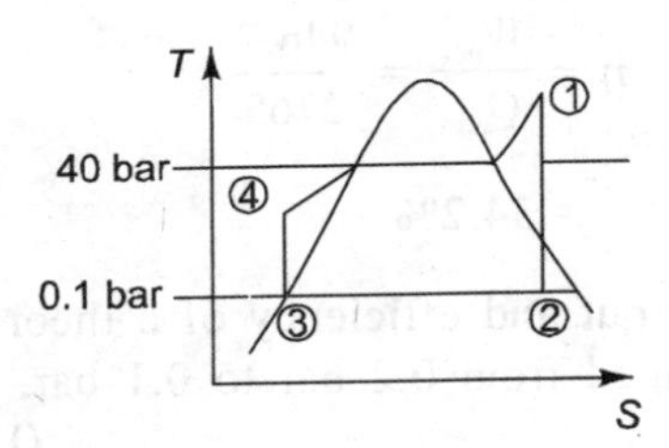

For 40 bar, state 1 is in the superheated region for temperature = 300°C

$$h_1 = 2960.7, \ S_1 = 6.362, \ v_1 = 0.0706$$

For 0.1 bar, state 2 is in the wet steam region.

$$h_{f2} = 191.8, \ h_{fg2} = 2392.8, \ h_{g2} = 2584.7 \text{ kJ/kg}$$

$$S_{f2} = 0.649, \ S_{fg2} = 7.501, \ S_{g2} = 8.150 \text{ kJ/kg°C}$$

$$v_{f3} = 0.001008 \text{ m}^3\text{/kg}$$

Process 1–2 is isentropic, therefore

$$S_1 = S_2$$

$$S_1 = S_{f_2} + x \, S_{fg_2} \quad \text{(where } x = \text{dryness factor)}$$

$$6.362 = 0.649 + x \times 7.501$$

$$x = \frac{5.713}{7.501} = 0.76$$

Now
$$h_2 = h_{f_2} + x\ h_{fg2}$$
$$= 191.8 + 0.76 \times 2392.8$$
$$= 191.8 + 1818.5 \approx 2010$$
$$W_{pump} = v_{f3}\ (P_2 - P_1) = 0.001008\ (40 - 0.1) \times 10^2$$
$$= 4 \text{ kJ}$$
$$h_4 = h_3 + W_{pump} = 191.8 + 4 = 195.8$$
$$W_{net} = W_{turbine} - W_{pump}$$
$$= (h_1 - h_2) - 4$$
$$= (2960.7 - 2010) - 4$$
$$= 950.7 - 4 = 946.7 \text{ kJ}$$
$$Q_{add} = h_1 - h_4$$
$$= 2960.7 - 195.8$$
$$= 2765 \text{ kJ}$$
$$\eta = \frac{W_{net}}{Q_{add}} = \frac{946.7}{2765}$$
$$= 34.2\%$$

6. Calculate the change in output and efficiency of a theoretical Rankine cycle when its condenser pressure is changed from 0.2 bar to 0.1 bar. The inlet condition is 40 bar and 400°C. (UPTU carry over: Aug. 2005)

At 40 bar and 400°C: State 1 in superheated region
$$h_1 = 2960.7,\ S_1 = 6.362$$

At 0.2 bar:
$$h_{f_2} = 251.5,\ h_{fg_2} = 2358.4,\ h_{g_2} = 2609.9$$
$$S_{f_2} = 0.8321,\ S_{fg_2} = 7.0773,\ S_{fg_2} = 7.909$$
$$S_1 = S_{f_2} + xS_{fg_2}$$

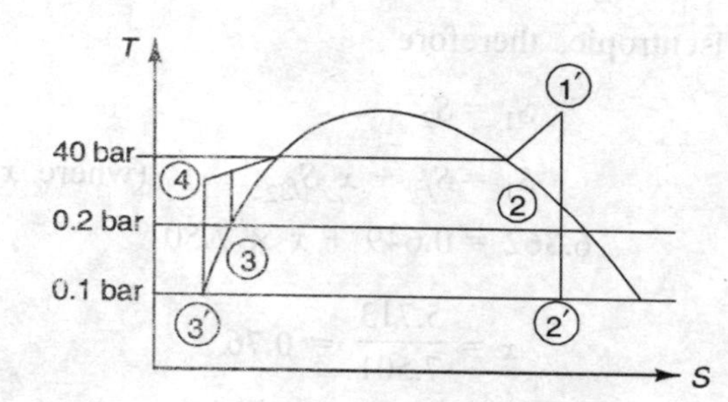

$$6.362 = 0.8321 + x \times 7.909$$

$$x = \frac{5.5299}{7.909} = 0.699$$

$$h_2 = h_{f_2} + x \times h_{fg_2}$$

$$= 251.5 + 0.699 \times 2358.4$$

$$= 251.5 + 1648.5$$

$$= 1900 \text{ kJ/kg}$$

At 0.1 bar,

$$h_{f_2'} = 191.8 \text{ and } h_{fg_2'} = 2392.8$$

$$h_2' = h_{f_2'} + x\, h_{fg_2'}$$

$$= 191.8 + 0.699 \times 2392.8$$

$$= 1864.4 \text{ kJ/kg}$$

$$W_{\text{pump}} = v_f\,(P_1 - P_2) = 1.007 \times 10^{-3}\,(40 - 0.2) \times 10^2$$

$$\approx 4 \text{ kJ/kg}$$

$$W_{\text{turbine}} = h_1 - h_2 = 2960.7 - 1900 = 1060.7 \text{ kJ/kg}$$

Now

$$W'_{\text{turbine}} = h_1 - h_{2'} = 2960.7 - 1864.4 = 1096.3 \text{ kJ/kg}$$

$$\Delta W = W' - W = 1096.3 - 1060.7$$

$$= 35.6 \text{ kJ/kg}$$

$$\Delta\eta = \frac{\Delta W}{Q_{\text{add}}} = \frac{35.6}{2960.7 - (h_{f_2'} + w_p)}$$

$$= \frac{35.6}{2960.7 - (251.5 + 4)}$$

$$= \frac{35.6}{2705} = 0.013 = 1.3\%$$

7. In a steam plant, steam is supplied to the turbine at 36 bar and 410°C. The condenser pressure is 0.075 bar. If the turbine develops a power of 12 mW, calculate for a theoretical cycle, (a) mass flow rate of the steam, (b) heat addition and heat rejection, (c) pump work and (d) thermal efficiency.

(UPTU: May 2008)

The cycle is

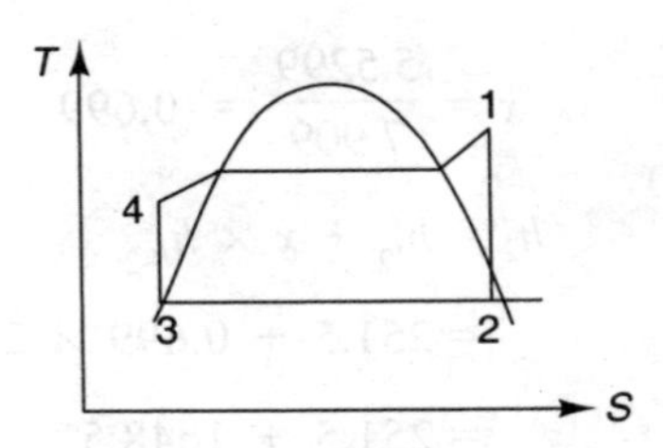

Using steam tables, for point 1

$$h_1 = 3241 \text{ kJ/kg} \quad \text{and} \quad s_1 = 6.920 \text{ kJ/kg}$$

For point 2, we have

$$s_{2f} = 0.574 \text{ kJ/kg}, \; s_{2g} = 7.778 \text{ kJ/kg}$$

$$h_{2f} = 168.65 \text{ kJ/kg}, \; h_{2fg} = 2406.2 \text{ kJ/kg}$$

$$v_{2f} = 0.0010085 \text{ m}^3\text{/kg}$$

Now

$$s_1 = s_{2f} + x.s_{2fg}$$

$$6.920 = 0.574 + x \cdot 7.778$$

or

$$x = 0.815$$

Therefore,

$$h_2 = h_{2f} + x \cdot h_{2fg}$$

$$= 168.85 + 0.815 \times 2406.2$$

$$= 2129.54 \text{ kJ}$$

Also

$$h_3 = h_{2f} = 168.85 \text{ kJ/kg}$$

Now

$$h_4 = h_3 + v_f(p_2 - p_1)$$

$$168.85 + 0.0010085 \times (36 - 0.075) \times 10^2$$

$$h_4 = 168.85 + 36.23$$

$$= 215.08 \text{ kJ/kg}$$

$$w_{\text{exp}} = h_1 - h_2 = 3241 - 2129.54$$

$$= 1011.46 \text{ kJ/kg}$$

$$w_{\text{pump}} = 36.23 \text{ kJ/kg}$$

$$w = w_{\text{exp}} - w_{\text{pump}}$$

$$= 1011.46 - 36.23$$

$$= 975.23 \text{ kJ/kg}$$

Now $\quad m \times w = 12 \times 10^3$ kJ [where m = mass flow rate of steam] or $$m = \frac{12 \times 10^3}{975.23} = 12.3 \text{ kg/s}$$

$$Q_1 = \text{heat addition} = h_1 - h_4$$
$$= 3241 - 215.08$$
$$= 3025.92 \text{ kJ/kg}$$
$$Q_2 = \text{heat rejection} = h_2 - h_3$$
$$= 2129.54 - 168.85$$
$$= 1960.69 \text{ kJ/kg}$$
$$\eta = 1 - \frac{Q_2}{Q_1}$$
$$= 1 - \frac{1960.69}{3025.92}$$
$$= 1 - 0.65$$
$$= 0.35$$

OBJECTIVE TYPE QUESTIONS

Dare to be different. Dare to take a stand for what you know is right.

State True or False

1. A vapour power cycle is essentially a closed cycle. (*True/False*)
2. In a vapour power cycle, each piece of equipment of a power cycle separately forms an open system. (*True/False*)
3. Water is used as working fluid in a vapour refrigeration cycle. (*True/False*)
4. Water is a pure substance, cheap and readily available. It is used as working fluid in a vapour power cycle. (*True/False*)
5. The ideal vapour cycle is a Carnot vapour cycle. (*True/False*)
6. Heat addition to vaporizing water in a boiler and heat rejection from condensing steam in a condensor can be easily done reversibly. (*True/False*)
7. The compression of wet steam requires a large compressor and compression work. (*True/False*)
8. Pumping of water from a condensor to a boiler requires small work. (*True/False*)
9. The average temperature of heat addition is lower in the Rankine cycle as compared to the Carnot cycle for the same temperature limits. (*True/False*)
10. The steam at the exit of a turbine should be as dry as possible to avoid pitting and erosion of turbine blades. (*True/False*)
11. The thermal efficiency of the Rankine cycle reduces with superheating of inlet steam. (*True/False*)
12. The thermal efficiency of the Rankine cycle can be increased by subcooling of water coming out of a condensor. (*True/False*)
13. The thermal efficiency of the Rankine cycle can be improved by lowering of condensor pressure. (*True/False*)
14. The thermal efficiency of the Rankine cycle can be improved by increasing the boiler pressure. (*True/False*)
15. The condensor of the Carnot cycle is difficult to design to match specific state of wet steam at the outlet of the condensor. (*True/False*)
16. The Carnot vapour cycle has two reversible isothermal processes and two reversible adiabatic processes. (*True/False*)

17. The Rankine cycle has two reversible isobaric and two adiabatic processes. (*True/False*)
18. The modified Rankine cycle can be used for a turbine. (*True/False*)
19. Vapour cycle has vapour as working fluid. (*True/False*)
20. Superheating in Carnot cycle is possible. (*True/False*)
21. Wet steam in condensor can be cooled to saturation liquid in Carnot cycle. (*True/False*)
22. If boiler pressure is increased in the Rankine cycle without changing temperature, the dryness fraction of steam after expansion will increase. (*True/False*)
23. Reheating in the Rankine cycle is done to avoid pitting and erosion of turbine blades. (*True/False*)
24. The regenerated Rankine cycle gives improved average temperature at which heat is added. (*True/False*)
25. In the boiler of the Rankine cycle, latent heating of water at constant pressure takes place. (*True/False*)
26. Plant working on the Rankine cycle has a boiler, a turbine, a condensor and a compressor. (*True/False*)

Multiple Choice Questions

1. In the Carnot vapour cycle, heat addition and rejection is
   (a) isothermal (b) isobaric (c) isochoric
2. In the Rankine cycle, heat addition and rejection is
   (a) isothermal (b) isobaric (c) isochoric
3. The efficiency of the Carnot cycle is
   (a) more than that of the Rankine cycle
   (b) less than that of the Rankine cycle
   (c) equal to the Rankine cycle
4. If boiler pressure in the Rankine cycle is reduced, then efficiency
   (a) increases (b) decreases (c) remains constant
5. If superheating of steam is done in the boiler, then efficiency of the Rankine cycle
   (a) increases (b) decreases (c) remains constant
6. If condensor pressure of the Rankine cycle is increased, then efficiency
   (a) increases (b) decreases (c) remains constant
7. A Carnot vapour cycle is impracticable as
   (a) heat addition and rejection at constant temperature is not feasible
   (b) turbine work is irreversible
   (c) pump work is irreversible

8. The compressor in the Carnot cycle has to be large as the
   (a) state is saturated liquid
   (b) state is wet steam
   (c) state is liquid region

9. The compressor in the Carnot cycle is difficult to design as inlet steam has to have a
   (a) definite dryness factor
   (b) dryness factor of unity
   (c) dryness factor of zero

10. The Carnot vapour cycle is impractical though it may have the highest efficiency, because it has
    (a) high heat addition (b) low heat rejection (c) low net work output

11. The work output from the turbine in the Rankine cycle is given by
    (a) change of enthalpy between the inlet and the outlet
    (b) change of internal energy between the inlet and the outlet
    (c) change of entropy between the inlet and the outlet

12. The efficiency of the Rankine cycle will improve
    (a) in summer as average temperature of heat supplied increases
    (b) in winter as it decreases the temperature at which heat added is improved
    (c) none of the above

13. The modified Rankine cycle is used
    (a) when a plant has a turbine
    (b) when a plant has a reciprocating engine
    (c) when steam is reheated in a turbine

14. In the reheat Rankine cycle
    (a) water is heated after pumping
    (b) steam is heated after expansion in the turbine
    (c) steam is taken out in between expansion in the turbine and reexpanded after reheat

15. The regenerated Rankine engine has
    (a) low average temperature of heat addition
    (b) high average temperature of heat addition
    (c) large heat rejection

16. The state of steam at the outlet of the condensor in the Rankine cycle has dryness as
    (a) zero (b) unity (c) 0.5

Fill in the Blanks

1. The Rankine cycle consists of a turbine, a condensor, a boiler and a ________.
   (a) compressor (b) pump

2. A Carnot cycle consists of a boiler, a condensor, a turbine and a ________.
   (a) pump (b) compressor

3. Net work output due to large size of a compressor in the Carnot cycle is ________
(a) small (b) large

4. High pressure in the boiler will give ________ efficiency in the Rankine cycle.
(a) lower (b) higher

5. Low pressure in the condensor will give ________ efficiency in the Rankine cycle.
(a) lower (b) higher

6. Superheat in the boiler will lead to ________ efficiency in the Rankine cycle.
(a) lower (b) higher

7. Higher dryness of steam is preferred at the exit of a turbine to avoid ________.
(a) corrosion of blades (b) lower efficiency

8. Each unit of a vapour power cycle operates as ________ system.
(a) open (b) closed

9. A vapour power cycle has ________ as working fluid.
(a) water (b) freon

10. A power cycle and a refrigeration cycle are ________ cycles.
(a) gas (b) vapour

11. The modified Rankine cycle helps to ________ the size of the cylinder of the reciprocating engine.
(a) increase (b) decrease

12. The regenerated Rankine cycle is to ________ heat to water before it enters the boiler.
(a) extract (b) add

13. The regenerated Rankine cycle has ________ average temperature of source as compared to the simple Rankine cycle.
(a) low (b) high

14. In the reheat cycle, steam is heated when it is at ________.
(a) outlet (b) mid expansion

15. Superheating of steam in the Carnot cycle is ________.
(a) possible (b) impossible

16. Steam cannot be liquefied to the saturation liquid line in the Carnot cycle as water cannot be heated at ________.
(a) constant temperature (b) constant volume

ANSWERS

It's how you handle your problems and troubles that counts, not the troubles themselves.

State True or False

1. True
2. True
3. False (Refrigerants like ammonia and freons are used as working fluids.)
4. True
5. True (Carnot vapour cycle has the maximum efficiency.)
6. False (Heat addition and heat rejection are irreversible at constant pressure.)
7. True
8. True
9. True (The average of sensible heating of water and latent heating is lower.)
10. True (Water content increases as steam becomes wet and damages the turbine blade.)
11. False (Superheating increases the efficiency of the Rankine cycle.)
12. False (Temperature of water at the inlet of the boiler should be as high as possible so that average temperature of heat addition is high which gives higher efficiency. Temperature of water at the inlet of the boiler is increased in the regenerative Rankine cycle.)
13. True
14. True
15. True
16. True
17. True
18. False (The modified Rankine cycle can be used for reciprocating engine to reduce the size of the engine.)
19. True
20. False (Heating of steam at constant temperature with falling pressure cannot be achieved.)
21. False (Steam cannot be cooled to water state in the condensor as further water cannot be heated at constant temperature.)

22. False (Refer to *TS* diagram.)

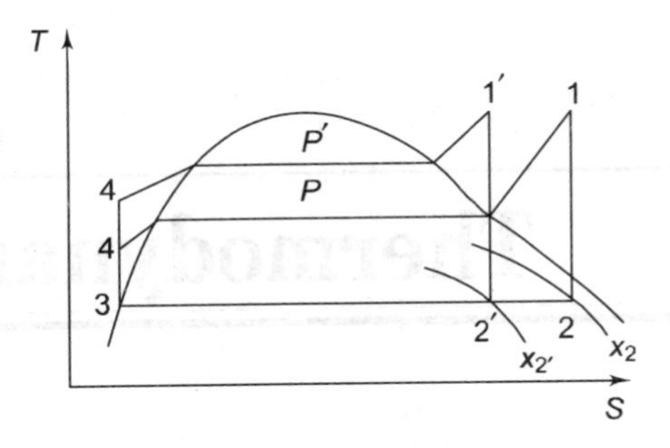

$P' > P$ and $T_1 = T_1'$. It is evident that dryness $x_2 > x_2'$.

23. True

24. True

25. False (Sensible heating and latent heating take place.)

26. False (Rankine cycle has a pump instead of a compressor.)

Multiple Choice Questions

1. (a) **2.** (b) **3.** (a) **4.** (b)

5. (a)

6. (b) (If condensor pressure is decreased, then only efficiency increases.)

7. (a) **8.** (b)

9. (a) (State must have entropy matching the entropy of saturated liquid line at high (boiler) pressure.)

10. (c) (Wet steam has water and dry steam mixture and the specific volume is large resulting large compression work and low net work output.)

11. (a)

12. (a) (Higher inlet temperature to the boiler will give higher efficiency.)

13. (b) **14.** (c)

15. (b) **16.** (a)

Fill in the Blanks

1. (b) **2.** (b) **3.** (a) **4.** (b)

5. (b) **6.** (b) **7.** (a) **8.** (a)

9. (a) **10.** (b) **11.** (b) **12.** (b)

13. (b) **14.** (b) **15.** (b) **16.** (a)

CHAPTER 6

Thermodynamic Cycles

Acquiring knowledge is not an end in itself, but only a means to an end, knowledge without purpose is a destructive tool.

INTRODUCTION

Thermodynamic cycles can be classified based on their utility as power cycles and refrigeration cycles. Power cycles can be vapour power cycles and gas power cycle/air cycles. Vapour power cycles have been explained in Chapter 5. In this chapter, we will study air-standard cycles.

An air-standard cycle is a thermodynamic cycle working with air as the working substance with certain assumptions. The assumptions are as follows:

1. Air as working substance behaves as a perfect gas.
2. The mass and composition of working substance do not change.
3. All processes are reversible.
4. Heat transfer from combustion is considered from an external source.
5. The engine operates as a closed system.
6. Specific heats remain constant.
7. KE and PE remain constant.
8. Heat loss to surroundings is nil.

Otto and Diesel cycles are gas power cycles with air as working substance. Otto and Diesel cycles are modified forms of the Carnot cycle in order to make the cycles realistic. Engines are generally designed based on Otto and Diesel cycles.

Certain terms used for the analysis of Otto and Diesel cycles are as follows:

1. ***Top dead centre (TDC)*:** It is the highest point to which the piston can reach during a stroke (Figure 6.1).
2. ***Bottom dead centre (BDC)*:** It is the lowest position to which the piston can reach during a stroke (Figure 6.1).
3. ***Swept length* (l_s):** It is the distance between TDC and BDC.
4. ***Clearance length* (l_c):** It is the distance between TDC and the cylinder top.
5. ***Clearance volume* (V_c):** It is the product of the area of the piston and the clearance length.
6. ***Swept volume* (V_s):** It is the product of the area of the piston and the swept length.
7. ***Compression ratio* (r):** It is the ratio of the sum of the swept volume (V_s) and clearance volume (V_c) to the clearance volume (V_c):

$$r = \frac{V_s + V_c}{V_c} = 1 + \frac{V_s}{V_c} = 1 + \frac{l_s}{l_c}$$

The compression ratio (r) for SI (spark ignition) is in the range of 5 : 1 to 11 : 1 while for compression ignition (CI), it is in the range of 12 : 1 to 25 : 1.

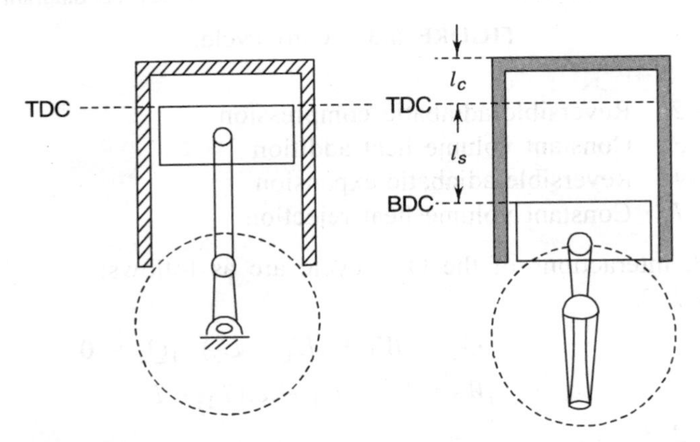

FIGURE 6.1 Piston positions in TDC and BDC.

The cylinder head block and the engine block are joined together as illustrated in Figure 6.2. These blocks get worn out (shaded portion) over a period of time and need to be machined resulting in reduction of clearance length (l_c). The compression ratio (r) increases due to machining of the cylinder head.

OTTO AND DIESEL CYCLES

The Otto cycle consists of two isochoric and two adiabatic processes. The Otto cycle is also known as a *constant volume cycle*. Reciprocating spark ignition engines are based on this cycle. Four processes of the Otto cycle are shown in Figure 6.3.

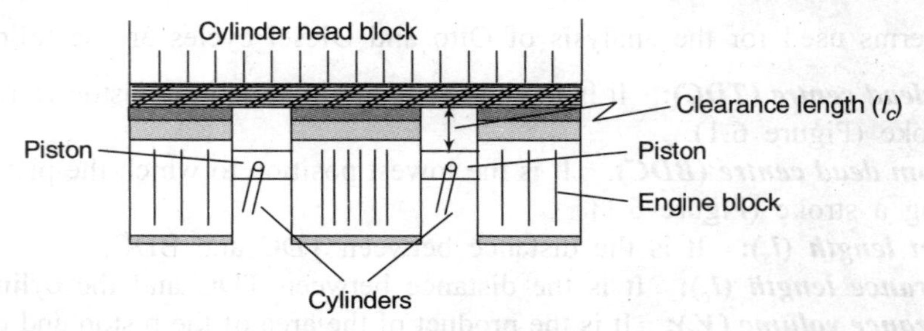

FIGURE 6.2 Cylinder head block joined with engine block.

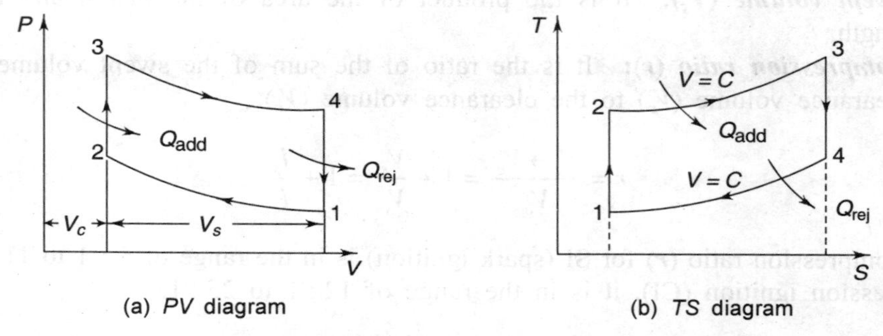

FIGURE 6.3: Otto cycle.

1. *Process 1–2:* Reversible adiabatic compression
2. *Process 2–3:* Constant volume heat addition
3. *Process 3–4:* Reversible adiabatic expansion
4. *Process 4–1:* Constant volume heat rejection

Heat and work interactions of the Otto cycle are as follows:

1. *Process 1–2:*

$$_1Q_2 = {}_1W_2 + (U_1 - U_2),\ {}_1Q_2 = 0$$
$$_1W_2 = U_2 - U_1 = c_v(T_2 - T_1)$$

2. *Process 2–3:*

$$_2Q_3 = (U_3 - U_2)\ (\text{As } {}_2W_3 = 0) = c_v(T_3 - T_2)$$

3. *Process 3–4:*

$$_3W_4 = (U_3 - U_4)\ (\text{As } {}_3Q_4 = 0) = c_v(T_3 - T_4)$$

4. *Process 4–1:*

$$_4Q_1 = (U_1 - U_4)\ (\text{As } {}_4W_3 = 0) = c_v(T_1 - T_4)$$

The efficiency of the Otto cycle can be worked out as follows:

$$\eta = \frac{\text{net work}}{\text{heat added}} = \frac{Q_{add} - Q_{rej}}{Q_{add}}$$

$$= \frac{c_v(T_3 - T_2) - c_v(T_4 - T_1)}{c_v(T_3 - T_2)}$$

$$= 1 - \frac{T_4 - T_1}{T_3 - T_2}$$

For an adiabatic process, we have $Pv^\gamma = c$. Combining this with $\frac{Pv}{T} = R$, we get

$$T_2 = T_1 r^{\gamma - 1} \quad \text{where } r = \frac{v_1}{v_2} = \frac{v_4}{v_3} = \text{compression ratio}$$

$$T_3 = T_4 r^{\gamma - 1}$$

$$\eta = 1 - \frac{(T_4 - T_1)}{r^{\gamma-1}(T_4 - T_1)} = 1 - \frac{1}{r^{\gamma-1}}$$

Therefore the efficiency of the Otto cycle depends upon the compression ratio (r) and the adiabatic index (γ). Efficiency increases if the compression ratio increases (see Figure 6.4).

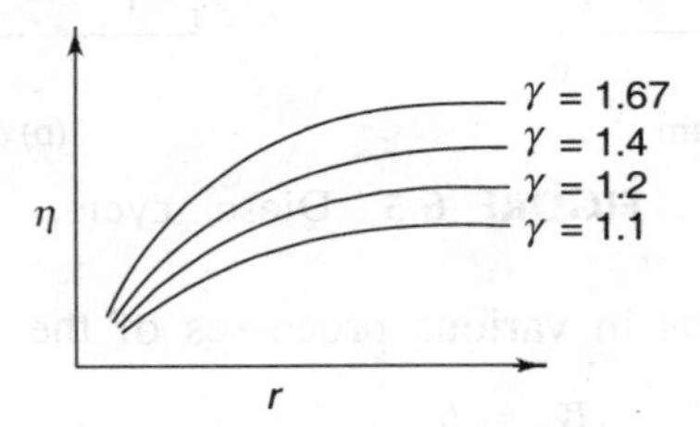

FIGURE 6.4 Variation of η with r.

The maximum work from the Otto cycle depends upon the compression ratio at the highest temperature (T_3) and lowest temperature (T_1):

$$W = Q_{add} - Q_{rej}$$

$$= c_v(T_3 - T_2) - c_v(T_4 - T_1)$$

$$= c_v\left(T_3 - T_1 \times r_r^{\gamma-1} - \frac{T_3}{r^{\gamma-1}} + T_1\right)$$

$\frac{dW}{dr} = 0$ for maximum work w.r.t. compression ratio. On solving, we get

$$r = \left(\frac{T_3}{T_1}\right)^{\frac{1}{2(\gamma-1)}}$$

Also $\quad T_2 = T_4 = \sqrt{T_1 T_3}$

The Diesel cycle consists of two reversible adiabatic process, one constant pressure process and one constant volume process. The Diesel cycle has four processes shown in Figure 6.5. Let us discuss these processes one by one.

1. *Process 1–2:* Reversible adiabatic compression
2. *Process 2–3:* Constant pressure heat addition

   Heat addition is stopped at point 3 which is known as cutoff point. Cutoff ratio $(\beta) = V_3/V_2$.

3. *Process 3–4:* Reversible adiabatic expansion
4. *Process 4–1:* Constant-volume heat rejection

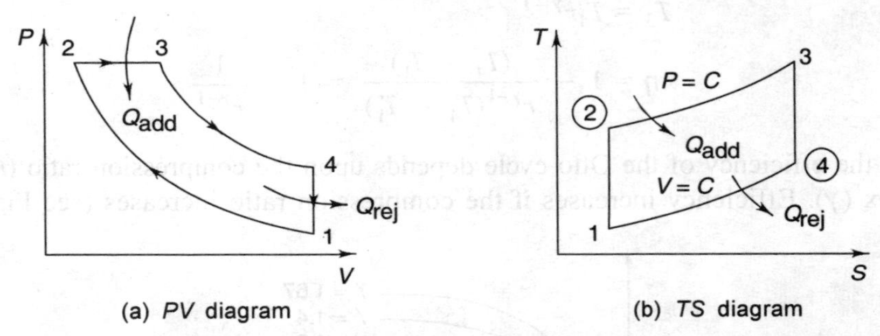

FIGURE 6.5 Diesel cycle.

Heat and work interactions in various processes of the Diesel cycle are as follows:

1. *Process 1–2:* $_1W_2 = U_2 - U_1$

   As $Q_2 = 0$

   $W_2 = c_v(T_2 - T_1)$

   Also $T_2 = T_1 r^{\gamma-1}$ where $r = \dfrac{V_1}{V_2}$ = compression ratio

2. *Process 2–3:* $_2Q_1 = \int dU + \int Pdv$

$$= \int d(U + Pv)$$

$$= \int dh = c_p(T_3 - T_2)$$

   Also $T_3 = T_1 r^{\gamma-1}\beta$ where $\beta = \dfrac{V_3}{V_2} = \dfrac{T_3}{T_2}$ = cutoff ratio

3. *Process 3–4:* $_3W_4 = U_3 - U_4$

   As $_3Q_4 = 0,$

$$= c_v(T_3 - T_4)$$

4. *Process 4–1:*

$${}_4Q_1 = U_1 - U_4$$
$$= c_v(T_1 - T_4)$$

Also $T_4 = T_1\beta^\gamma$ and $\dfrac{v_4}{v_3} = \dfrac{v_1}{v_2} \times \dfrac{v_2}{v_3} = r \times \dfrac{1}{\beta}$

The efficiency of the Diesel cycle can be found out as follows:

$$\eta = \frac{\text{net work}}{Q_{add}} = \frac{Q_{add} - Q_{rej}}{Q_{add}}$$

$$= 1 - \frac{Q_{rej}}{Q_{add}}$$

$$= 1 - \frac{c_v(T_4 - T_1)}{c_p(T_3 - T_2)}$$

$$= 1 - \frac{1}{\gamma}\frac{(T_4 - T_1)}{(T_3 - T_2)}$$

$$= 1 - \frac{1}{\gamma}\frac{(T_1\beta^\gamma - T_1)}{(T_1\beta r^{\gamma-1} - T_1 r^{\gamma-1})}$$

$$= 1 - \frac{1}{r^{\gamma-1}}\left[\frac{\beta^\gamma - 1}{\gamma(\beta - 1)}\right]$$

$$= 1 - \frac{1}{r^{\gamma-1}} \times k$$

where

$$k = \frac{\beta^\gamma - 1}{\gamma(\beta - 1)}.$$

As the cutoff ratio (β) increases, the volume of the factor k will also increase. The efficiency of the Diesel cycle will decrease with increase of the factor k. However, it is obvious that the value of the factor k is always greater than unity. Therefore, the efficiency of the Diesel cycle is always less efficient than a corresponding Otto cycle having the same compression ratio (r). The variation of the efficiency with the compression ratio is shown in Figure 6.6.

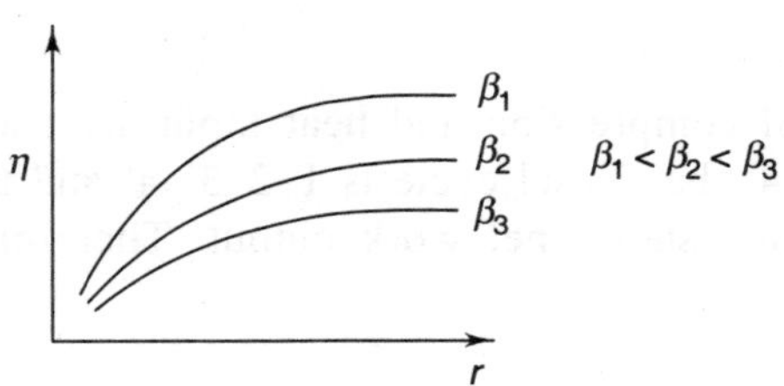

FIGURE 6.6 Variation of η with r.

Determination of Temperatures

Otto cycle: If initial temperature T_1, compression ratio (r), c_v and γ are given, then other temperatures can be found out as explained

(a) $T_2 = T_1 r^{\gamma-1}$

(b) $T_3 = T_2 + \dfrac{\text{heat added}}{c_v}$

(c) $T_4 = T_2 = \dfrac{T_3}{r^{\gamma-1}}$ or $T_4 = T_1 + \dfrac{\text{heat rejected}}{c_v}$

Diesel cycle: If initial temperature T_1, compression ratio (r), cutoff ratio (β) and γ are given, then other temperatures can be found out

(a) $T_2 = T_1 \cdot r^{\gamma-1}$

(b) $T_3 = \beta \cdot T_2$

(c) $T_4 = \dfrac{T_3}{\left(\dfrac{r}{\beta}\right)^{\gamma-1}}$

It is difficult to achieve either constant volume or constant pressure heat addition in SI and CI engines due to appreciable amount of time required for the completion of the combustion process. In actual condition, a part of heat addition takes place at constant volume and the rest at constant pressure. Such cycle having heat addition partly at constant volume and the rest at constant pressure is called *dual cycle* (Figure 6.7).

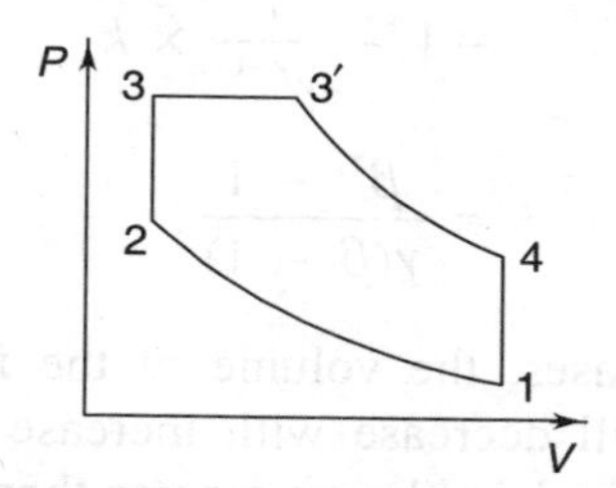

FIGURE 6.7 Dual cycle.

The efficiency of Otto, dual and Diesel cycles can be compared in many ways as follows:

1. Cycles for equal compression and heat input are shown in Figure 6.8. The Otto cycle is 1–2–3–4, the Diesel cycle is 1–2–3′–4′ and the dual cycle is 1–2″–3″–4″. The area enclosed shows net work output. Therefore $\eta_{\text{Otto}} > \eta_{\text{Dual}} > \eta_{\text{Diesel}}$.

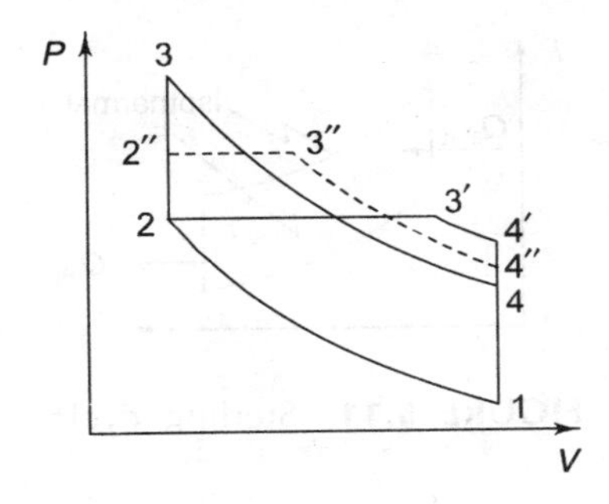

FIGURE 6.8 Equal compression and heat input.

2. Cycles for equal maximum pressure and heat input are shown in Figure 6.9. The Otto cycle is 1–2–3–4, the Dual cycle is 1–2″–2‴–3″–4″ and the Diesel cycle is 1–2′–3′–4′. As per the area enclosed, we can say $\eta_{\text{Diesel}} > \eta_{\text{Dual}} > \eta_{\text{Otto}}$.

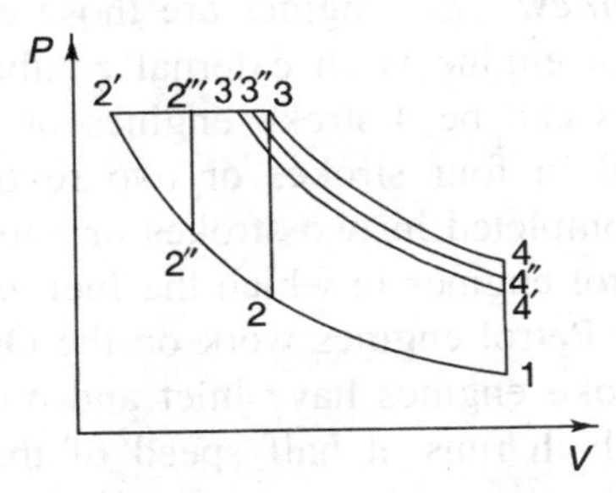

FIGURE 6.9 Equal maximum pressure and heat input.

3. Cycles for maximum pressure and temperature are shown in Figure 6.10. Otto cycle is 1–2–3–4, the dual cycle is 2″–2‴–3–4 and the Diesel cycle is 1–2′–3′–4. As per area enclosed in the cycles, we can say $\eta_{\text{Diesel}} > \eta_{\text{Dual}} > \eta_{\text{Otto}}$.

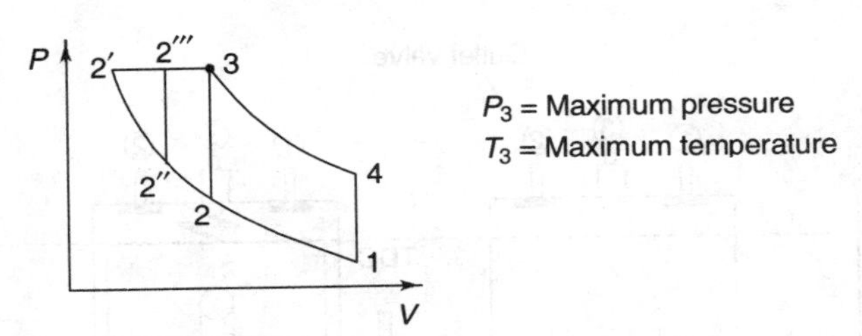

FIGURE 6.10 Equal maximum pressure and temperature.

In the Sterling cycle (regenerative cycle), a regenerator is used, which stores the rejected heat energy during the heat rejection process and supplies the same during the heat addition process. It consists of two isothermal and two constant volume processes as shown in Figure 6.11.

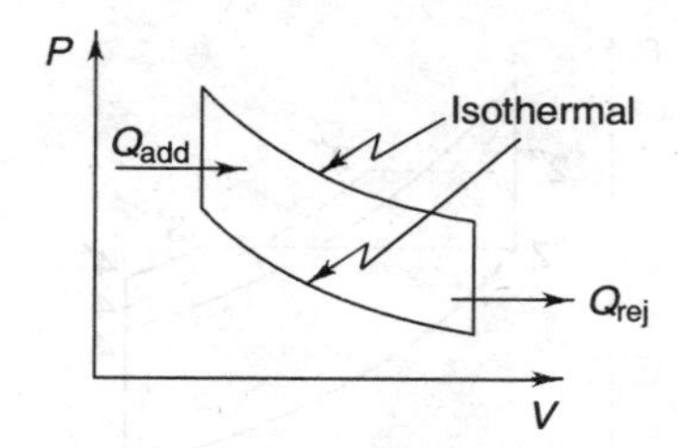

FIGURE 6.11 Sterling cycle.

ENGINES

Internal combustion (IC) engines: IC engines are those engines in which fuel is burnt inside a cylinder. Petrol and diesel engines are internal combustion engines as fuel is burnt inside the cylinder of these engines.

External combustion (EC) engines: EC engines are those engines in which fuel is burnt outside the cylinders. The steam engine is an external combustion engine.

Internal combustion engines can be 4-stroke engines or 2-stroke engines. In 4-stroke engines, the cycle is completed in four strokes or two revolutions of the crankshaft. In 2-stroke engines, the cycle is completed in two strokes or one revolution of the crankshaft.

4-stroke engines can be petrol engines in which the fuel used is petrol or diesel engines in which the fuel used is diesel. Petrol engines work on the Otto cycle while diesel engines work on the Diesel cycle. 4-stroke engines have inlet and outlet valves which are opened and closed by the camshaft, which runs at half speed of the crankshaft.

Spark ignition (SI) engine has a carburettor for making air-fuel mixture and a spark plug to ignite the mixture. Four strokes of SI engines (Figures 6.12 and 6.13) are as follows:

1. *Suction stroke* (Process 0–1): The piston moves from TDS to BDC creating vacuum inside the cylinder. The inlet valve opens and air fuel mixture enters into cylinder.

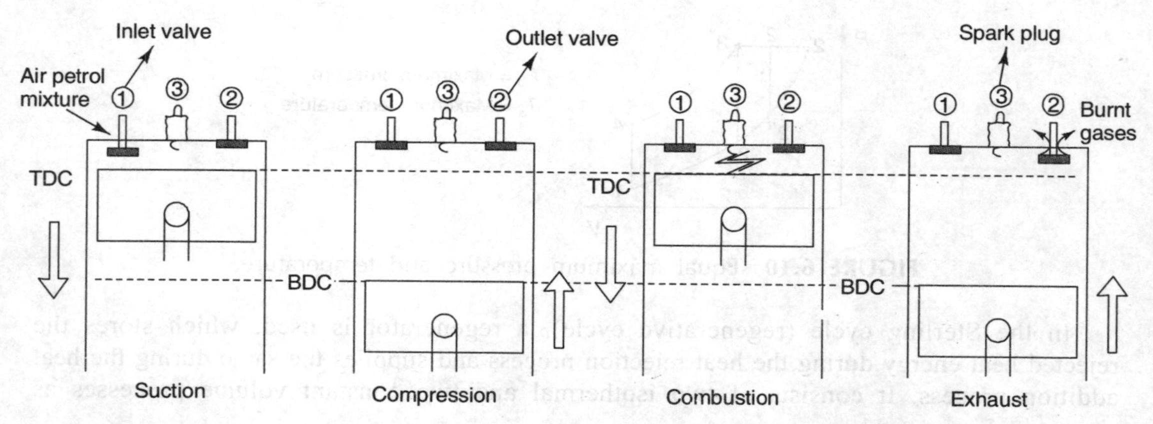

FIGURE 6.12 Four strokes of SI engine.

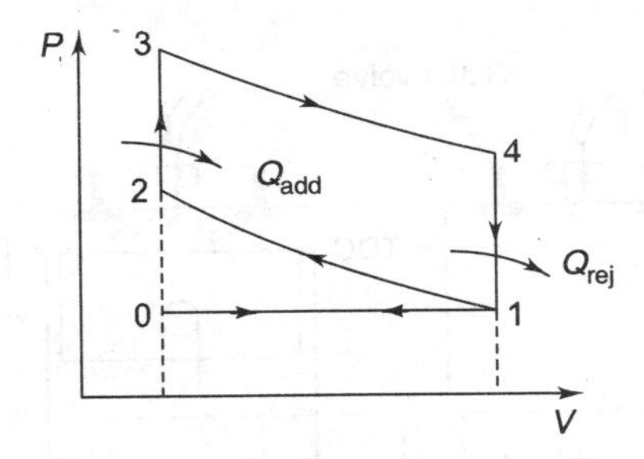

FIGURE 6.13 *PV* diagram for 4-stroke SI engine.

2. *Compression stroke* (Process 1–2): The piston moves from BDC to TDC. Both valves are closed. Air-fuel mixture is compressed.
3. *Combustion and power stroke* (Processes 2–3 and 3–4): The air-fuel mixture is ignited before the piston reaching TDS. Both valves remain closed. There is a rise of temperature and pressure due to combustion at constant volume. The temperature is around 1800–2000°C and pressure is around 30–40 bar. Due to high pressure, a force acts on the piston and the piston moves from TDS to BDS. Power is obtained from this stroke. At the end of this stroke, heat is rejected to surroundings at constant volume (Process 4–1).
4. *Exhaust stroke* (Process 1–0): The inlet valve remains closed and the exhaust valve opens. Burnt gases are pushed out through the exhaust valve.

Compression ignition (CI): Engine has a fuel injection pump to inject fuel at high pressure through the injector in the cylinder. It has no spark plug or carburettor as combustion takes place by compressing air (no air-fuel mixture) at high pressure and then injecting diesel through the injector at very high pressure so that spontaneous combustion takes place. CI engine works on the diesel cycle. The four strokes of a CI engine (Figures 6.14 and 6.15) are as follows:

1. *Suction stroke* (Process 0–1): The piston is at TDC, the inlet valve is open and outlet valve is closed. The piston moves from TDC to BDC creating vacuum in the cylinder. The air is sucked in.
2. *Compression stroke* (Process 1–2): Both inlet and outlet valves are closed. The piston moves from TDC to BDC compressing the air. The temperature and pressure of air increase. The compression ratio reaches 12–25. The pressure of air is around 60 bar and the temperature is about 600°C. Temperature is sufficient for auto ignition of fuel. Diesel is injected by the fuel injection pump at high pressure when the piston is about to reach TDC.
3. *Combustion and power stroke* (Processes 2–3 and 3–4): The combustion of fuel takes place at constant pressure. The fuel enters at point 2 and fuel is cut off at point 3. Hot gases now expand pushing the piston downwards towards BDC. Power is obtained in this stroke. Both valves are closed during the stroke. At BDC heat is rejected at constant volume (Process 4–1).

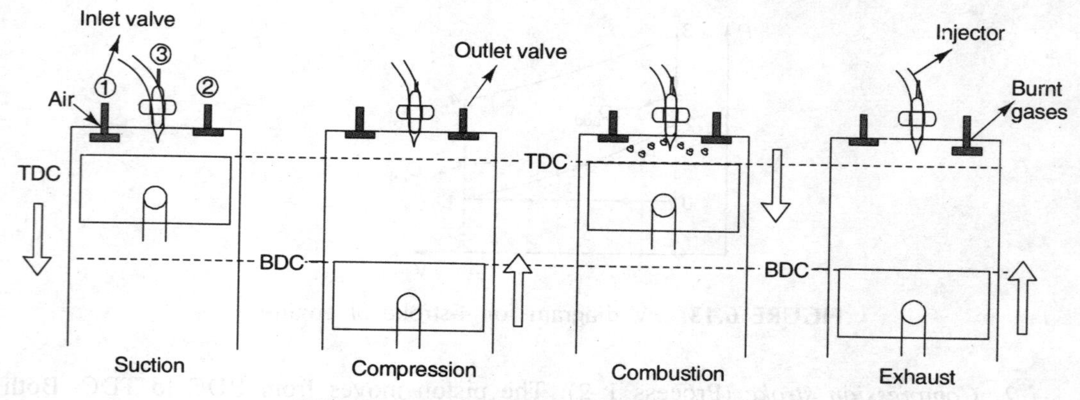

FIGURE 6.14 Four strokes of CI engine.

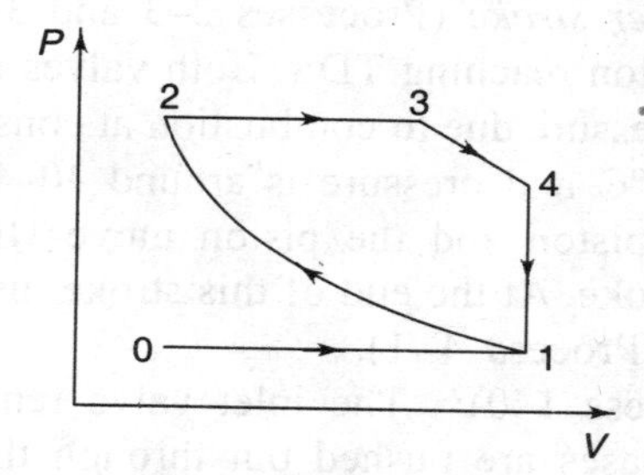

FIGURE 6.15 *PV* diagram for four strokes of CI engine.

4. *Exhaust stroke* (Process 1–0): Inlet valve remain closed while exhaust valve opens. The piston moves towards TDC pushing the burnt gases out of cylinder through exhaust valve.

In a 2-stroke petrol engine, the cycle is completed in two strokes of the piston or one revolution of the crankshaft, the shape of the piston is having deflector-like shape at top to deflect the incoming air fuel mixture to top thereby, preventing its escape through the exhaust port (Figure 6.16). The cylinder body has three ports (*Note:* No valves as in a 4-stroke engine) which are opened in turn by the piston. The inlet port is opened when the piston is at TDC. When the piston moves down, it first opens the exhaust port and closes the inlet port. Later on the piston opens the transport port. The cycle is completed (Figures 6.17 and 6.18) as follows:

1. *Ignition and induction:* The piston is almost at TDC towards the end of the compression stroke. The exhaust port and transfer port are covered by the piston while the inlet port is uncovered. The compressed charge is ignited by spark. Combustion of air fuel mixture takes place. Temperature and pressure increase in the combustion chamber of the cylinder. Vacuum is simultaneously created below the piston in the crankcase and fresh charge is inducted in the crankcase through the inlet port.

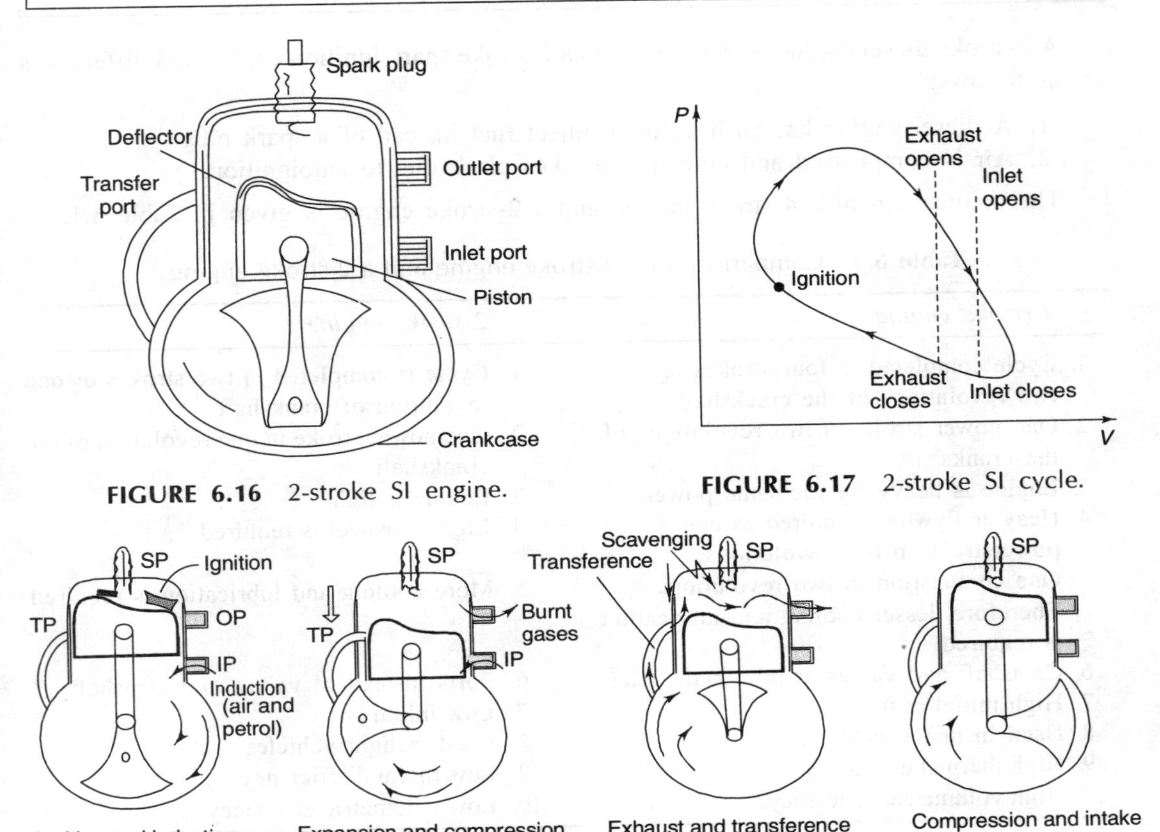

FIGURE 6.16 2-stroke SI engine.

FIGURE 6.17 2-stroke SI cycle.

FIGURE 6.18 2-stroke cycle.

2. *Expansion and compression:* It is expansion of high pressure and temperature gases after combustion and compression of fresh charge in the crankcase. The piston moves down due to expansion of gases after combustion and power is developed. The downward movement of the piston closes the inlet port and compresses fresh charge in the crankcase to about 1.4 bar. The moving piston uncovers the exhaust port when about 80% expansion has taken place. The burnt out gases being higher in pressure than atmospheric pressure escape out from the cylinder through the exhaust port.
3. *Exhaust and transference:* It is exhaust of burnt out gases and transference of compressed charge from the crankcase to the cylinder when the piston is almost at BDC. At this point, the piston uncovers the transfer port and transfers the slightly compressed charge from the crankcase to the cylinder through the transfer port. The deflector like shape on the piston helps in pushing out the burnt out gases through the exhaust port. This process is known as *scavenging*.
4. *Compression and intake:* When piston moves upwards from BDC, it first covers the transfer port to stop transfer of fresh charge. A little later it covers the exhaust port to stop exhaust and to start compression. The compression is complete near TDC, and the inlet port is uncovered for fresh intake in the crankcase.

A 2-stroke diesel engine works similar to a 2-stroke spark ignition engine and differences are as follows:

1. A diesel engine has an injector to inject fuel instead of a spark plug.
2. Air is compressed and combustion takes place due to autoignition.

The comparison of a 4-stroke engine and a 2-stroke engine is given in Table 6.1.

Table 6.1 Comparison of a 4-stroke engine and a 2-stroke engine

4-stroke engine	*2-stroke engine*
1. Cycle completed in four strokes or two revolutions of the crackshaft	1. Cycle is completed in two strokes or one revolution of crankshaft
2. One power stroke in two revolutions of the crankshaft	2. One power stroke in one revolution of the crankshaft
3. Engine is heavy for the same power	3. Engine is light
4. Heavier flywheel required as one powerstroke in two revolutions	4. Light flywheel is required
5. One combustion in two revolutions Therefore, lesser cooling and lubrication is required	5. More cooling and lubrication is required
6. Camshaft and valves at inlet and outlet	6. Ports instead of valves and camshaft
7. High initial cost	7. Low initial cost
8. Used in heavy vehicles	8. Used in light vehicles
9. High thermal efficiency	9. Low thermal efficiency
10. High volumetric efficiency	10. Low volumetric efficiency

The comparison of a spark ignition engine and a compression ignition engine is given in Table 6.2.

Table 6.2 Comparison of a spark ignition engine and a compression ignition engine

SI engine	*CI engine*
1. Based on the Otto cycle	1. Based on the Diesel cycle
2. Low compression ratio (range 5 to 11)	2. High compression ratio (range 12 to 25)
3. Petrol used as fuel	3. Diesel used as fuel
4. Air-fuel mixture is compressed in the cylinder	4. Only air is compressed
5. Carburettor makes air-fuel mixture	5. Fuel injection pump is required
6. Ignition by a spark plug	6. High pressure fuel from the injection pump is injected through the injector in the cylinder and autoignition takes place
7. Combustion is isochoric	7. Combustion is isobaric
8. Low compression and engine is light	8. High compression and engine is heavy
9. High engine speed (8000–6000 rpm)	9. Low speed (400–3500 rpm)
10. Low thermal efficiency	10. High thermal efficiency
11. Low maintenance cost and high running cost	11. High maintenance cost and low running cost
12. Used in light vehicles	12. Used in heavy vehicles

INDICATED, BRAKE AND FRICTION POWER

Combustion takes place in the cylinder and power is taken from crankshaft. The shaft work available is less than total energy released inside the cylinder due to frictional and others losses. *Indicated power* is the power available inside the cylinder and provided to the piston. Indicated power is measured from an indicator diagram (similar to *PV* diagram) obtained using an indicator mechanism.

Indicated power = Energy in fuel – Energy loss in exhaust, coolant and radiation

Brake power (BP) is the power available at crankshaft. Brake power is measured by dynamometers.

Brake power = Indicated power – Energy loss in friction and pumping

$$= \frac{2\pi NT}{60}; \quad \text{where } T \text{ is Torque and } N \text{ is rpm}$$

Friction power (FP) is the power lost due to friction and other reasons.

Friction power = Indicated power – Brake power

Mean effective pressure (mep or P_m) is the average pressure per stroke and it can be obtained from an indicator diagram. The indicator diagram indicates displacement of the piston on *x*-axis and cylinder pressure on *y*-axis.

$$P_m = \text{mep} = \frac{\text{Enclosed area of indicator diagram}}{\text{Length of diagram}} \times \text{Indicator spring constant}$$

Indicated power (IP) is the power generated per unit time:

$$\text{IP} = \frac{n \times P_m \times A \times L \times N \times K}{60}$$

where n = number of cylinders

P_m = mean effective pressure

A = Area of the piston

L = Length of the stroke

N = rpm

K = 1 for a 2-stroke engine

$= \frac{1}{2}$ for a 4-stroke engine

EFFICIENCIES

Mechanical efficiency (η_{mech}) is the ratio of brake power (BP) to indicated power (IP):

$$\eta_{mech} = \frac{BP}{IP} = \frac{IP - FP}{IP}$$

Volumetric efficiency (h_v) is the ratio of the actual volume of the charge (V_{actual}) admitted during suction stroke reduced to NTP to the swept volume of the piston (V_{swept}):

$$\eta_v = \frac{V_{actual}}{V_{swept}}$$

Brake thermal efficiency is the ratio of brake power (BP) generated to fuel energy used:

$$\eta_{brake\ thermal} = \frac{BP}{m_f \times c}$$

where m_f = mass of fuel

c = calorific value of fuel

Indicated thermal efficiency can be given as follows:

$$\eta_{indicated\ thermal} = \frac{IP}{m_f \times c}$$

The knocking in an SI engine is ignition of air-fuel mixture before spark reaches it. Isooctane content in fuel for SI engines retards autoignition while normal heptane accelerates autoignition. The knocking in an SI engine increases with increase in the compression ratio and decrease in speed. The ignition quality of fuels in SI engines is determined by octane number rating. Higher octane of fuel will decrease tendency of knocking. The knocking tendency in a CI engine is increased with decrease of the compression ratio.

In an SI engine, an ignition coil is used to generate high voltage for the spark plug. It can be appreciated that in a multicylinder SI engine, spark at particular order is given to each cylinder so that power obtained is smooth in revolution of the crankshaft. Hence a distributor is used to obtain required firing order of spark plugs so that spark is obtained at right moment in each cylinder.

SOLVED PROBLEMS

1. An engine cylinder has a piston of 0.12 m^2 and contains gas at a pressure of 1.5 MPa. The gas expands according to a process which is represented by a straight line on a pressure–volume diagram. The final pressure is 0.15 MPa. Calculate the work done by the gas on the piston if the stroke is 0.3 m. (UPTU: 2005)

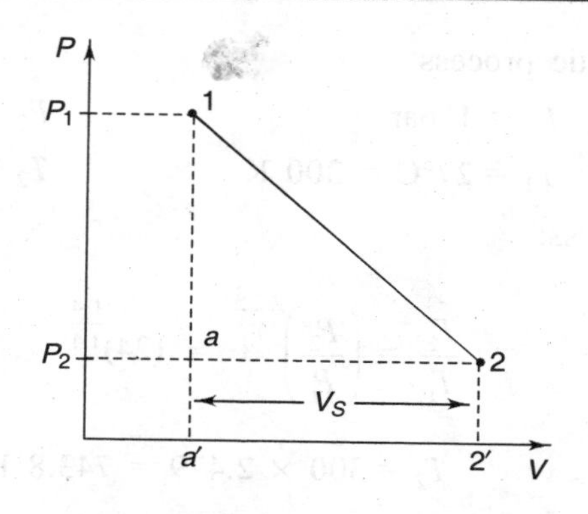

$$V_s = \text{Swept volume} = A \times l_{\text{stroke}} = 0.12 \times 0.3 = 0.036$$

$$\text{Work done} = \text{area of triangle } 12a + \text{area of rectangle } 2aa'2'$$

$$= \frac{1}{2} \times (1500 - 150) \times 10^3 \times (0.036) + 150 \times 10^3 \times 0.036$$

$$= (24300 + 5400) \text{ kJ}$$

$$= 29700 \text{ kJ}$$

2. Suction pressure and temperature for a theoretical diesel cycle are 1 bar and 27°C. The pressure at the end of the compression stroke found to be 24 bar. The maximum temperature of the cycle is limited to 1200°C. Determine

(a) cutoff ratio
(b) net output of the cycle
(c) thermal efficiency (UPTU: 2005)

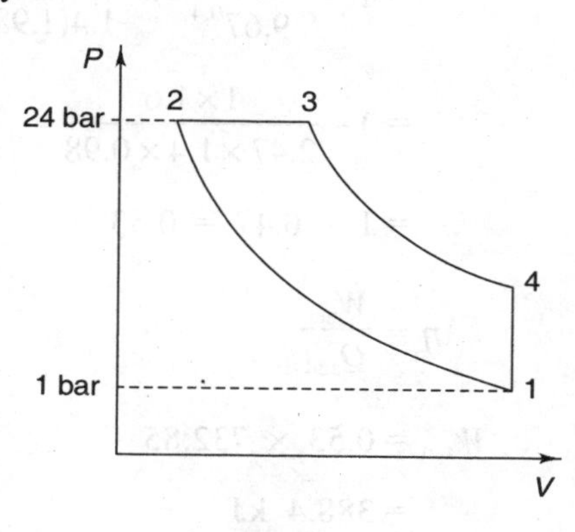

For air:

$$R = 0.287 \text{ kJ/(kg K)}$$

$$c_p = 1.005 \text{ kJ/(kg K)}$$

$$c_v = 0.718 \text{ kJ/(kg K)}$$

$$\gamma = 1.4$$

Process 1–2: Adiabatic process

$P_1 = 1$ bar $\qquad P_2 = 24$ bar

$T_1 = 27°C = 300$ K $\qquad T_2 = ?$

For an adiabatic process:

$$\frac{T_2}{T_1} = \left(\frac{P_2}{P_1}\right)^{\frac{\gamma-1}{\gamma}} = (24)^{\frac{0.4}{1.4}}$$

or $$T_2 = 300 \times 2.479 = 743.8 \text{ K}$$

Process 2–3: Isobaric process

$$\frac{V_3}{V_2} = \beta = \frac{T_3}{T_2} = \frac{1200 + 273}{743.8}$$

$\therefore$ $$\beta = 1.98$$

$$Q_{add} = C_p(T_3 - T_2) = 1.005 \times (1473 - 743.8)$$

$$= 1.005 \times 729.2$$

$$= 732.85$$

$$\eta_{Diesel} = 1 - \frac{1}{r^{\gamma-1}} \times \frac{(\beta^{\gamma} - 1)}{\gamma(\beta - 1)}$$

$$= 1 - \frac{1}{9.67^{0.4}} \times \frac{(1.98^{1.4} - 1)}{1.4(1.98 - 1)}$$

$$= 1 - \frac{1 \times 1.6}{2.47 \times 1.4 \times 0.98}$$

$$= 1 - 0.47 = 0.53$$

Now $$\eta = \frac{W_{net}}{Q_{add}}$$

Therefore, $$W_{net} = 0.53 \times 732.85$$

$$= 388.4 \text{ kJ}$$

Note: Similar problems can be solved for the Otto cycle with only difference is heat add $= c_v(T_3 - T_2)$ and $\eta = 1 - \frac{1}{r^{\gamma-1}}$.

Also in case mep is to be calculated, then

$$\text{mep} = \frac{W_{net}}{V_s}$$

where V_s = Swept volume.

3. An engine of 250 mm bore and 375 mm stroke works on the Otto cycle. The clearance volume is 0.00263 m³. The initial temperature and pressure are 50°C and 1 bar. If the maximum pressure is limited to 25 bar. Find (a) the air standard efficiency of the cycle and (b) the mean effective pressure for the cycle. (UPTU: 2000)

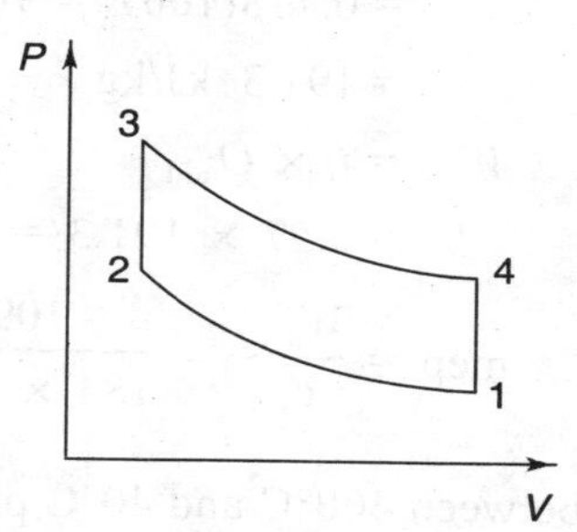

Given: $P_1 = 1$ bar, $T_1 = 50°\text{C}$ and $P_3 = 25$ bar

$L = 375$ mm, $D = 250$ mm, $V_c = 26.3 \times 10^{-4}$ m³

$$V_s = \frac{\pi D^2 \times L}{4}$$

or

$$V_s = 184 \times 10^{-4} \text{ m}^3$$

$$\text{Compression ratio } (r) = \frac{V_s + V_c}{V_c} = \frac{(184 + 26.3) \times 10^{-4}}{26.3 \times 10^{-4}}$$

$$= 8$$

$$P_2 = P_1 \left(\frac{V_2}{V_1}\right)^{\gamma} = 1 \times r^{\gamma} = 1 \times 8^{1.4} = 18.38 \text{ bar}$$

$$\eta = 1 - \frac{1}{r^{\gamma-1}} = 1 - \frac{1}{(8)^{0.4}} = 1 - \frac{1}{2.29}$$

$$= 1 - 0.43 = 0.57$$

Also

$$\frac{T_2}{T_1} = \left(\frac{P_2}{P_1}\right)^{\frac{\gamma-1}{\gamma}} = (18.38)^{\frac{0.4}{1.4}}$$

$$T_2 = 323 \times (18.38)^{.285} \text{ as } (\because T_1 = 273 + 50°\text{C} = 323 \text{ K})$$
$$= 323 \times 2.29$$
$$= 740.5 \text{ K}$$

$$T_3 = T_2\left(\frac{P_3}{P_2}\right) = 740.5 \times \left(\frac{25}{18.38}\right)$$
$$= 1007 \text{ K}$$
$$Q_{add} = c_v(T_3 - T_2)$$
$$= 0.718(1007 - 740.5)$$
$$= 191.3 \text{ kJ/kg}$$
$$\therefore \quad W_{net} = \eta \times Q_{add}$$
$$= 0.57 \times 191.3 = 109 \text{ kJ/kg}$$
$$\text{mep} = \frac{W_{net}}{V_s} = \frac{109}{184 \times 10^{-4}} = 0.59 \times 10^4 \text{ kPa}$$

4. A Carnot engine working between 400°C and 40°C produces 130 kJ of work. Determine the following:
(a) Thermal efficiency
(b) Heat added
(c) Entropy change during heat rejection

$$T_1 = 400 + 273 = 673 \text{ K}$$
$$T_2 = 40 + 273 = 313 \text{ K}$$
$$\therefore \quad \eta = 1 - \frac{T_1}{T_2} = 1 - \frac{313}{673}$$
$$= 0.535$$
$$\eta = \frac{W_{net}}{Q_{add}} = \frac{130}{Q_{add}} = 0.535$$
$$\therefore \quad Q_{add} = \frac{130}{0.535} = 243 \text{ kJ}$$
$$\text{Heat rejected} = Q_{rej} = Q_{add} - W_{net}$$
$$= 243 - 130$$
$$= 113 \text{ kJ}$$
$$Q_{rej} = T_2 \times \Delta S$$
$$113 = 313 \times \Delta S$$
$$\therefore \quad \Delta S = \frac{113}{313} = 0.361 \text{ kJ/(kg K)}$$

5. If an engine works on the Otto cycle between temperature limits 1450 K and 310 K, find the maximum power developed by the engine assuming the circulation of air per minute as 0.38 kg.

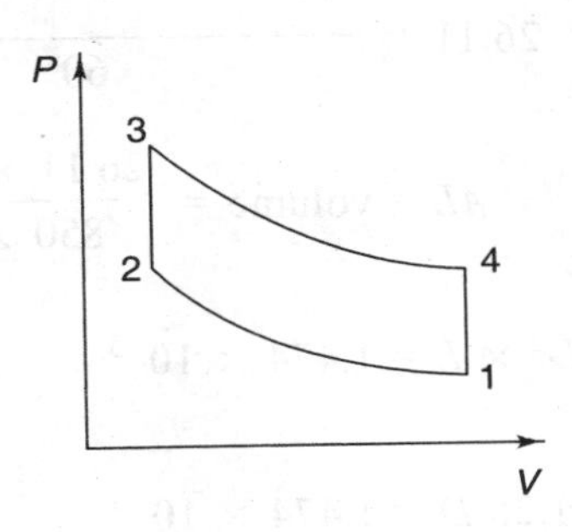

For maximum power $T_2 = T_4 = \sqrt{T_1 T_3}$

where T_3 = maximum temperature

and T_1 = minimum temperature

$$\therefore \quad T_2 = T_4 = \sqrt{1450 \times 310} = 670.4 \text{ K}$$

$$W = Q_{\text{add}} - Q_{\text{rej}}$$

$$W = c_v[(T_3 - T_2) - (T_4 - T_1)]$$

$$c_v = 0.718[(1450 - 670.4) - (670.4 - 310)]$$

$$= 0.718(779.6 - 360.4)$$

$$= 0.718(419.2) = 301 \text{ kJ/kg}$$

$$W = m \times W = \frac{0.38}{60} \times 301 \text{ kW}$$

$$= 1.9 \text{ kW}$$

6. A 4-stroke diesel engine has L/A ratio of 1.25. The mean effective pressure is found with the help of an indicator equal to 0.85 MPa. If the engine produces indicated process of 35 HP while it is running at 2500 rpm, find the dimension of the engine. (UPTU: 2004)

$$P_m = 0.85 \times 10^3 \text{ kPa}$$

$$\text{IP} = 35 \times 0.746 \text{ kW} = 26.11 \text{ kW}$$

$$= \frac{n \times P_m \times A \times L \times N \times K}{60}$$

$n = 1$ for 1-cylinder and $K = 1/2$ for 4-stroke. Therefore,

$$26.11 = \frac{850 \times A \times L \times 2500 \times \frac{1}{2}}{60}$$

$$AL = \text{volume} = \frac{26.11 \times 60 \times 2}{850 \times 2500} = 1.474 \times 10^{-3}$$

$$\frac{\pi}{4} \times D^2 \times L = 1.474 \times 10^{-3}$$

$$\frac{\pi}{4} \times D^2 \times 1.25\, D = 1.474 \times 10^{-3}$$

$$D^3 = \frac{1.474 \times 4}{\pi \times 1.25} \times 10^{-3} = 1.502 \times 10^{-3}$$

$$D = 1.145 \times 10^{-1} \text{ m} = 11.45 \text{ cm}$$

$$\frac{L}{D} = 1.25$$

$$\therefore \quad L = 1.25 \times 11.45 = 14.31 \text{ cm}$$

7. A 4-cylinder diesel engine of 4-stroke type has stroke to bore ratio as 1.2 and the cylinder diameter is 12 cm. Estimate indicated power of the engine using the indicator diagram arrangement. The indicator card shows the diagram having an area of 30 cm^2 and length as half of the stroke. The indicator spring constant is 20×10^3 Nm/m^2 and the engine is running at 2000 rpm. Also find out the mechanical efficiency of the engine if 10% of power is lost in friction and other losses.

Given: $L/D = 1.2$, $D = 12$ cm

Therefore, $L = 1.2 \times 12 = 14.4$ cm

$$\text{Length of the indicator diagram} = \frac{1}{2} \times \text{stroke}$$

$$= \frac{1}{2} \times 14.4 = 7.2 \text{ cm}$$

$$P_m = \text{mep} = \frac{\text{Area of the indicator diagram}}{\text{Length of the indicator diagram}} \times \text{Spring constant}$$

$$= \left(\frac{30}{7.2} \times \frac{1}{10^2}\right) \times (20 \times 10^3 \times 10^3)$$

$$= 8.333 \times 10^5 \text{ N/m}^2$$

$$\text{IP} = \frac{n \times P_m \times A \times L \times N \times k}{60}$$

$$= \frac{4 \times 8.333 \times 10^5 \times \frac{\pi}{4} \times (0.12)^2 \times 0.144 \times \frac{1}{2} \times 2000}{60}$$

$$= 90{,}400 \text{ watts}$$

$$\text{Friction power loss} = 0.1 \times 90{,}400 = 9040$$

$$\eta_{\text{mechanical}} = \frac{\text{IP} - \text{FP}}{\text{IP}} = \frac{90{,}400 - 9040}{90{,}400}$$

$$= 0.899 \text{ or } 89.9\%$$

8. An engine with bore 7.5 cm and stroke 10 cm has a compression ratio of 6 to 1. To increase the compression ratio, 5 mm is machined off from the cylinder head face. Calculate the new compression ratio.

$$\text{Compression ratio } (r) = \frac{V_c + V_s}{V_c} = 1 + \frac{V_s}{V_c}$$

$$6.1 = 1 + \frac{l_s}{l_c}$$

or
$$\frac{l_s}{l_c} = 5.1$$

or
$$l_c = \frac{0.1}{5.1} \text{ m} = 19.6 \text{ mm}$$

$$l'_c = \text{Clear length after machining} = 19.6 - 5 = 14.6 \text{ mm}$$

$$\text{New compression ratio} = 1 + \frac{l_s}{l'_c}$$

$$= 1 + \frac{100}{14.6}$$

$$= 1 + 6.85$$

$$= 7.85$$

9. The power output of an IC engine is measured by a rope dynamometer. The diameter of the brake pulley is 700 mm and the rope diameter is 25 mm. The load on the tight side of the rope is 50 kg mass and the spring balance reads 50 N. The engine running at 900 rev/min consumes the fuel, of Calorific value 40,000 kJ/kg at a rate of 4 kg/hr. Assume g = 9.81 m/s^2. Calculate (a) brake specific fuel consumption, and (b) brake thermal efficiency.

(GATE: 1997)

$$\text{Brake power} = \frac{(W - S)\pi(D + d)N}{6000}$$

where W = weight on tight side, S = spring balance reading, D = diameter of the pulley, d = diameter of the rope, N = rpm.

$$\text{Brake power} = \frac{(50 \times 9.81 - 50)\pi(0.7 + 0.025)900}{60,000}$$

$$= 15.03 \text{ kW}$$

$$\text{Brake specific fuel consumption} = \frac{\text{kg of fuel/hr}}{\text{BHP(kW)}}$$

$$= \frac{4}{15.03} = 0.26 \text{ kg/kW hr}$$

$$\text{Brake thermal } \eta = \frac{\text{BHP}}{m_f \times \text{calorific value}}$$

$$= \frac{15.03}{\frac{4}{3600} \times 44,000} = \frac{15.03 \times 3600}{4 \times 44 \times 10^3}$$

$$= 0.31 \quad \text{or} \quad 31\%.$$

10. The minimum pressure and temperature in an Otto cycle are 100 kPa and 27°C. The amount of heat added to air per cycle is 1500 kJ/kg. Determine the pressure and temperature at all points of the air standard Otto cycle. Also calculate the specific work and the thermal efficiency of the cycle for a compression ratio of 8 : 1 (Take c_v(air) = 0.72 kJ/kgK and c_p/c_v = 1.4).

(GATE: 1998)

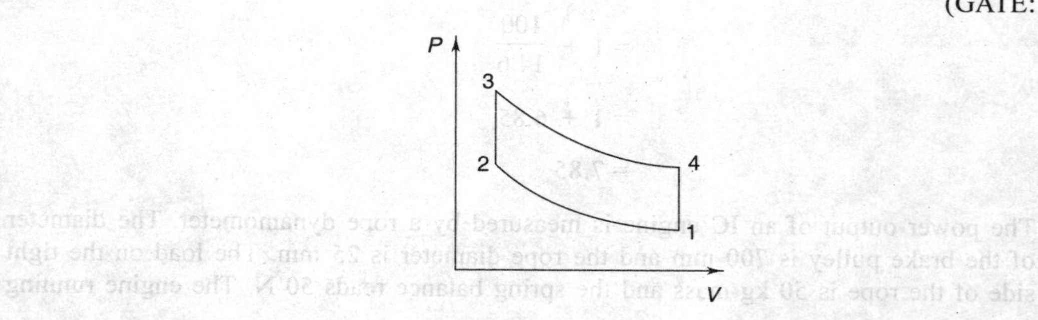

Process 1–2:

$$\frac{P_2}{P_1} = \left(\frac{V_1}{V_2}\right)^{\gamma} = (r)^{\gamma}$$

$$P_2 = P_1(8)^{1.4} = 100 \times (8)^{1.4}$$

$$= 1838 \text{ kPa}$$

Also

$$T_2 = T_1\left(\frac{P_2}{P_1}\right)^{\frac{\gamma-1}{\gamma}}$$

$$= 300 \times \left(\frac{1838}{100}\right)^{\frac{0.4}{1.4}}$$

$$= 689 \text{ K}$$

Process 2–3:

$$\frac{P_3}{P_2} = \frac{T_3}{T_2} \text{ and } Q_{\text{add}} = c_v(T_3 - T_2)$$

or

$$T_3 = 689 + \frac{Q_{\text{add}}}{c_v} = 689 + \frac{1500}{0.72}$$

$$= 2773 \text{ K}$$

∴

$$P_3 = P_2 \times \frac{T_3}{T_2} = 1838 \times \frac{2773}{689}$$

$$= 7394 \text{ kPa}$$

Process 3–4:

$$\frac{P_4}{P_3} = \left(\frac{V_3}{V_4}\right)^{\gamma} = \left(\frac{1}{8}\right)^{1.4}$$

$$P_4 = 7394 \times \left(\frac{1}{8}\right)^{1.4}$$

$$= 402 \text{ kPa}$$

$$\frac{P_4}{P_3} = \left(\frac{T_4}{T_3}\right)^{\frac{\gamma}{\gamma-1}}$$

$$T_4 = T_3 \times \left(\frac{P_4}{P_3}\right)^{\frac{\gamma-1}{\gamma}}$$

$$= 2773 \times \left(\frac{402}{7394}\right)^{\frac{0.4}{1.4}}$$

$$= 1207 \text{ K}$$

$$\text{Heat rejected } Q_{rej} = c_v(T_4 - T_1)$$

$$= 0.72(1207 - 300)$$

$$= 653 \text{ kJ/kg}$$

$$\text{Work} = Q_{add} - Q_{rej}$$

$$= 1500 - 653$$

$$= 847 \text{ kJ/kg}$$

$$\eta = \frac{W}{Q_{add}} = \frac{847}{1500} = 0.565$$

11. A large diesel engine runs on a stroke cycle at 2000 rpm. The engine has a displacement of 25 litres and a brake mean affective pressure of 0.6 mn/m^2. It consumes 0.018 kg/s (calorific value = 42,000 kJ/kg). Determine the brake power and the brake thermal efficiency.

(GATE: 1999)

$$\text{Break power} = \frac{(p_m)(l \times A) \times N}{60} \times n$$

$$l \times A = \text{displacement volume}$$

$$= 25 \times 10^{-3} = 0.025 \text{ m}^3$$

$$n = 1/2 \text{ for 4 stroke}$$

$$\text{Break power} = \frac{0.6 \times 10^3 \times 0.025 \times 2000}{2 \times 60}$$

$$= 250 \text{ kW}$$

$$\text{Break efficiency } \eta = \frac{\text{Break power}}{\text{Heat supplied}}$$

$$= \frac{250}{0.018 \times 42{,}000}$$

$$= 0.331 \quad \text{or} \quad 33.1\%$$

12. A diesel engine develops a brake power of 45 kW. Its indicated thermal efficiency is 30% and the mechanical efficiency is 85%. Take the calorific value of the fuel as 40,000 kJ/kg and calculate (a) the fuel consumption in kg/hr and (b) the indicated specific fuel consumption.

(GATE: 2000)

Given: $\eta_{thermal} = 0.3$ and $\eta_{mech} = 0.85$, BP = 4.5, $\eta_{mech} = \frac{BP}{IP}$

or
$$IP = \frac{BP}{\eta_{mech}} = \frac{4.5}{0.85} = 5.29 \text{ kW}$$

$$\eta_{thermal} = \frac{IP}{\text{Heat given}} = \frac{IP}{m_f \times c_f}$$

$$0.3 = \frac{5.29}{m_f \times 40 \times 10^6}$$

$$m_f = 0.44 \times 10^{-3} \text{ kg/s}$$

$$= 0.44 \times 10^{-3} \times 3600 = 1.58 \text{ kg/hr}$$

Indicated specific fuel consumption

$$(m_f)_{indicated} = \frac{m_f}{IP}$$

$$= \frac{1.58}{5.29} = 0.298 \text{ kg/kW hr}$$

13. In a spark ignition engine working on the ideal Otto cycle, the compression rated is 5.5. The work output per cycle (i.e. area of *PV* diagram) is equal to $23.625 \times 10^5 \times V_c$ joules where V_c is clearance volume in m^3. The indicated mean effective pressure is
(a) 4.295 bar (b) 5.25 bar (c) 86.87 bar (d) 106.3 bar

(GATE: 2001)

Compression ratio $r = \frac{V}{V_c} = 5.5$

or
$$V = 5.5\ V_c$$

$$\text{Work} = P_{mean}\ (V - V_c)$$

$$= 23.625 \times 10^{15} \times V_c$$

$$\therefore \quad P_{\text{mean}} = \frac{23.625 \times 10^5 \times V_c}{V - V_c}$$

$$= \frac{23.625 \times 10^5 \times V_c}{5.5\,V_c - V_c}$$

$$= \frac{23.625 \times 10^5}{4.5}$$

$$= 5.25 \text{ bar}$$

Option (b) is correct.

14. An ideal air standard Otto cycle has a compression ratio of 8.5. If the ratio of the specific heats of air (ν) is 1.4, what is the thermal efficiency (in percentage) of the Otto cycle?

(a) 57.5 (b) 45.7 (c) 52.5 (d) 95

(GATE: 2002)

$$\eta = 1 - \frac{1}{r^{\nu-1}}$$

$$= 1 - \frac{1}{(8.5)^{1.4-1}}$$

$$= 1 - \frac{1}{(8.5)^{0.4}}$$

$$= 1 - 0.425$$

$$= 0.575 \text{ or } 57.5\%$$

Option (a) is correct.

15. An engine working on air standard Otto cycle has a cylinder diameter of 10 cm and stroke length of 15 cm. The ratio of specific heats for air is 1.4. If the clearance volume is 196.3 cc and the heat supplied per kg of air per cycle is 1800 kJ/kg, the work output per cycle per kg of air is

(a) 879.1 kJ (b) 890.2 kJ (c) 895.3 kJ (d) 973.5 kJ

(GATE: 2004)

$$\text{Volume swept } V_s = \frac{\pi}{4} \times d^2 \times L$$

$$= \frac{\pi}{4} \times (10)^2 \times 15$$

$$= 1178 \text{ cc}$$

$$r = \text{Compression ratio} = \frac{V_c + V_s}{V_c}$$

$$= \frac{196.3 + 1178}{196.3} = 7$$

$$\eta_{\text{Otto}} = 1 - \frac{1}{r^{\gamma-1}}$$

$$= 1 - \frac{1}{7^{1.4-1}}$$

$$= 0.541 \quad \text{or} \quad 54.1\%$$

$$\eta = \frac{W}{Q}$$

or

$$W = \eta \times Q$$

$$= 0.541 \times 1800$$

$$= 973.5 \text{ kJ}$$

Option (d) is correct.

16. A 4-cylinder petrol engine has a swept volume of 2000 cm^3 and the clearance volume in each cylinder is 60 cm^3. If the pressure and temperature at the beginning of compression are 1 bar and 24°C and the maximum cycle temperature is 1500°C, the air standard efficiency will be

(a) 58% (b) 59% (c) 61% (d) 63%

(GATE: 2005)

$$\text{Compression ratio } r = \frac{V_s + V_c}{V_c} = \frac{2000 + 60 \times 4}{4 \times 60}$$

$$= 9.333$$

$$\eta_{\text{Otto}} = 1 - \frac{1}{r^{\gamma-1}}$$

$$= 1 - \frac{1}{(9.333)^{1.4-1}}$$

$$= 0.59 \quad \text{or} \quad 59\%$$

Option (b) is correct.

17. For the previous problem of the petrol engine the mean effective pressure will be
(a) 4.83 bar (b) 5.83 bar (c) 6.83 bar (d) 8.83 bar

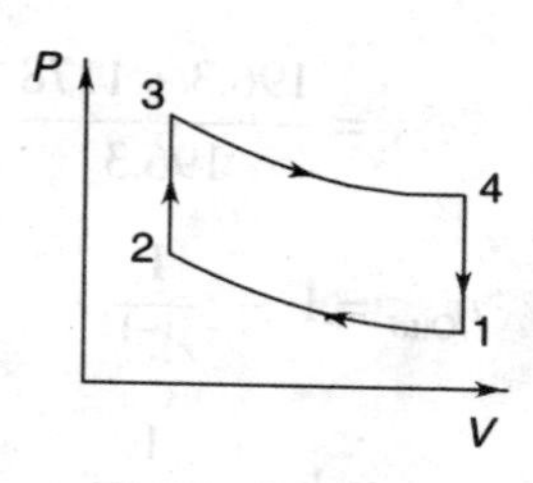

$$T_1 = 24°C = 307 \text{ K}$$

$$T_2 = T_1\left(\frac{V_2}{V_1}\right)^{\gamma-1} = 307 \times r^{\gamma-1}$$

$$= 307 \times (9.333)^{0.4}$$

$$= 726 \text{ K}$$

$$T_3 = 1500°C \text{ (Given)}$$

$$= 1500 + 273$$

$$= 1773 \text{ K}$$

$$a = \frac{T_3}{T_2} = \frac{1773}{726} = 2.44$$

$$P_m = P_1 \times r \times \frac{a-1}{\gamma-1}\left[\frac{r^{\gamma-1}-1}{r-1}\right]$$

$$= 1 \times 9.33 \times \frac{2.44-1}{1.4-1}\left[\frac{(9.33)^{1.4-1}-1}{9.33-1}\right]$$

$$= 5.83 \text{ bar}$$

Option (b) is correct.

18. A diesel cycle takes air at 1.0 bar and 300 K and compresses it to 16 bar. Heat is added till its temperature becomes 1700 K. Calculate (a) work from cycle, and (b) air standard efficiency.

(UPTU: 2007–2008)

Process 1–2:

$$\text{Compression ratio} = r = \frac{V_1}{V_2} = \left(\frac{P_2}{P_1}\right)^{1/\gamma} = (16)^{1/1.4} = 7.24$$

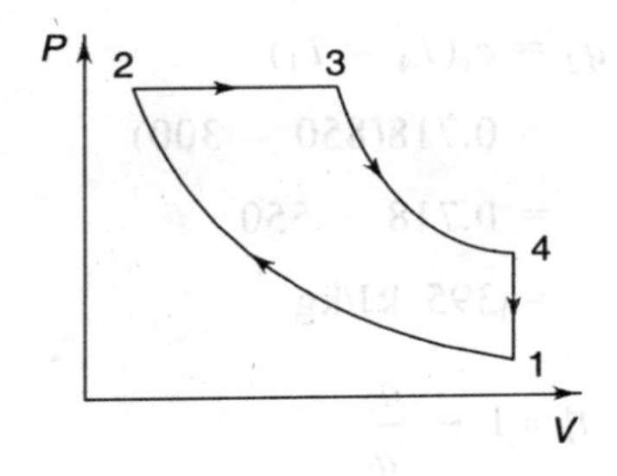

Also $$\frac{T_2}{T_1} = \left(\frac{P_2}{P_1}\right)^{\frac{\nu-1}{\nu}}$$

$$T_2 = 300 \times (16)^{0.4/1.4}$$

$$= 616 \text{ K}$$

Process 2–3:
Given $T_3 = 1700$ K

$$\therefore \quad \frac{V_3}{V_2} = \frac{T_3}{T_2} = \frac{1700}{616} = 2.78 = \beta$$

Process 3–4:

$$\frac{T_3}{T_4} = \left(\frac{V_4}{V_3}\right)^{\nu-1}$$

$$= \left(\frac{V_1}{V_2} \cdot \frac{V_2}{V_3}\right)^{\nu-1} \qquad \text{as } V_4 = V_1$$

$$= \left(r \cdot \frac{1}{\beta}\right)^{\nu-1}$$

$$= \left(16 \cdot \frac{1}{2.78}\right)^{\nu-1} = 2$$

$$\therefore \quad T_4 = \frac{1700}{2} = 850 \text{ K}$$

$$q_1 = c_p(T_3 - T_2)$$
$$= 1.005(1700 - 616)$$
$$= 1089.4 \text{ kJ/kg}$$

$$q_2 = c_v(T_4 - T_1)$$
$$= 0.718(850 - 300)$$
$$= 0.718 \times 550$$
$$= 395 \text{ kJ/kg}$$

$$\eta = 1 - \frac{q_2}{q_1}$$
$$= 1 - \frac{395}{1089.4}$$
$$= 1 - 0.363$$
$$= 0.637$$

$$w = q_1 - q_2$$
$$= 1089.4 - 395$$
$$= 694.4 \text{ kJ/kg}$$

19. An engine working on a direct cycle has air intake conditions of 1 bar and 310 K and compression ratio of 17. Heat added at high pressure is 1250 kJ/kg. Make calculation for the maximum temperature of the cycle, net power output and thermal efficiency. (UPTU: 2006–2007)

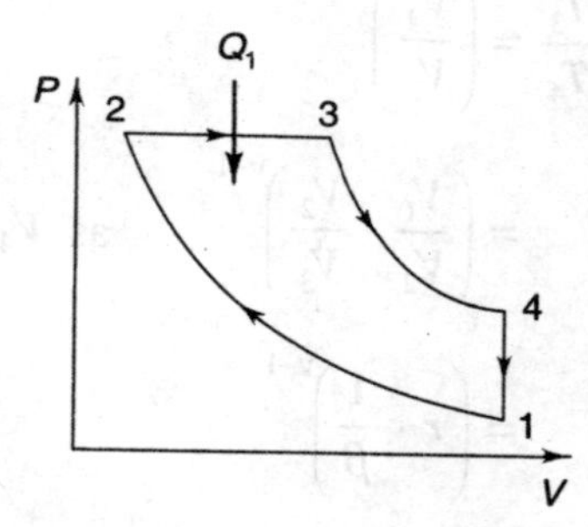

$$q_1 = 1250 \text{ kJ/kg}$$

Process 1–2:

For adiabatic process: $\frac{T_2}{T_1} = \left(\frac{V_1}{V_2}\right)^{\gamma-1} = 17^{0.4} = 3.1$

$\therefore$ $T_2 = 3.1 \times 310 = 963 \text{ K}$

Process 2–3:

$$q_1 = 1250 = c_p(T_3 - T_2)$$
$$= 1.005(T_3 - 963)$$

$\therefore \quad T_3 = 2007 \text{ K}$

$$\beta = \text{cutoff ratio} = \frac{V_3}{V_2} = \frac{T_3}{T_2} = \frac{2007}{963} = 2.3$$

Process 3–4:

For adiabatic process: $$\frac{T_3}{T_4} = \left(\frac{V_4}{V_3}\right)^{\nu-1}$$

$$= \left(\frac{V_1}{V_2} \cdot \frac{V_2}{V_3}\right)^{\nu-1} \quad [\text{as } V_4 = V_1]$$

$$= \left(\frac{r}{\beta}\right)^{\nu-1}$$

$$= \left(\frac{17}{2.3}\right)^{0.4} = 2.17$$

or $$T_4 = \frac{2207}{2.17} = 1017$$

$$q_2 = c_v(T_4 - T_1) = 0.718(1017 - 310)$$
$$= 507.5 \text{ kJ/kg}$$

$$\eta = 1 - \frac{q_2}{q_1} = 1 - \frac{507.5}{1250}$$
$$= 0.7084$$

$$w = \eta \times q_1$$
$$= 0.7084 \times 1250$$
$$= 885.6 \text{ kJ/kg}$$

20. In a Diesel cycle, the compression ratio is 10 and cutoff ratio is 3. If the initial temperature is 300 K, find other temperaturs. Assumes $\nu = 1.4$.
T_2 is retated to T_1 by compression ratio, i.e. $T_2 = T_1\, r^{\nu-1}$. T_2 is related to T_3 by cut-off ratio, i.e. $T_3 = \beta T_2$. Also T_4 is related to T_3 by relation $T_3 = T_4\left(\frac{r}{\beta}\right)^{\nu-1}$

$$T_2 = 300 \times 10^{1.4-1} = 300 \times 10^{0.4}$$
$$= 753.6 \text{ K}$$

$$T_3 = \beta \times T_2 = 3 \times 753.6$$

$$= 2260 \text{ K}$$

$$T_4 = \frac{T_3}{\left(\frac{r}{\beta}\right)^{\gamma-1}} = \frac{2260}{\left(\frac{10}{3}\right)^{0.4}}$$

$$= \frac{2260}{1.618} = 1396 \text{ K}$$

21. For an Otto cycle shown below has $T_1 = 300$ K, $T_2 = 800$ K, $T_3 = 2100$ K and $T_4 = 900$ K. If $c_v = 0.718$, find work and efficiency.

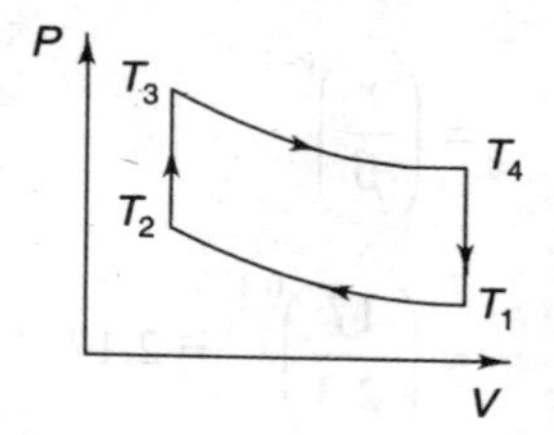

$$\text{Work} = q_1 - q_2$$

$$= c_v(T_3 - T_2) - C_V(T_4 - T_1)$$

$$= c_v[(T_3 - T_2) - (T_4 - T_1)]$$

$$= 0.718[(2100 - 800) - (900 - 300)]$$

$$= 0.718(1300 - 600)$$

$$= 502.6 \text{ J/kg}$$

$$q_1 = 0.718 \times 1300$$

$$= 933.4 \text{ kJ/kg}$$

$$\eta = \frac{w}{q_1} = \frac{502.6}{933.4} = 0.54$$

22. If in an Otto cycle, the product of pressure and volume at each point starting from initial is: (a) $p_1v_1 = 700$ kJ/kg, (b) $p_2v_2 = 1200$ kj/kg, (c) $p_3v_3 = 4200$ kJ and (d) $p_4v_4 = 1400$ kJ. Find work. Assume $R = 0.286$ kJ/kg K and $c_v = 0.718$ kJ/kg K

$$\text{Work} = c_v[(T_3 - T_4) - (T_2 - T_1)]$$

But $$PV = RT$$

$\therefore$ $$\text{Work} = \frac{C_V}{R}[(P_3v_3 - P_4v_4) - (P_2v_2 - P_1v_1)]$$

$$= \frac{0.718}{0.286}[(4200 - 1400) - (1200 - 700)]$$

$$= \frac{0.718}{0.286}[1800 - 500]$$

$$= 3.26 \text{ kJ/kg}$$

23. For a Diesel cycle, the following data were observed. Air inlet pressure and temperature = 1 bar and 300 K. Compression ratio = 20, cutoff ratio = 2. Calculate the temperatures at all points and cycle, net power output and thermal efficiency of the cycle.

(UPTU: May 2008)

From initial temperature T_1, we can find out other temperatures as under:

(a) $T_2 = T_1(r)^{\nu-1}$

(b) $T_3 = T_2 \times \beta$

(c) $T_4 = \dfrac{T_3}{\left(\dfrac{r}{\beta}\right)^{\nu-1}}$

Now T_1 = 300 K, hence we have

$$T_2 = T_1 \times (r)^{\nu-1}$$

$$= 300 \times (20)^{1.4-1}$$

$$= 994.3 \text{ K}$$

also

$$T_3 = \beta T_2 = 2 \times 994.3$$

$$= 1988.6 \text{ K}$$

also

$$T_4 = \frac{T_3}{\left(\frac{r}{\beta}\right)^{\nu-1}} = \frac{1988.6}{\left(\frac{20}{2}\right)^{0.4}}$$

$$= 791.64 \text{ K}$$

Now

$$q_{add} = q_1 = C_p(T_3 - T_2) = 1.005(1988.6 - 994)$$

$$= 999.6 \text{ kJ/kg}$$

$$q_{rej} = q_2 = c_v(T_4 - T_1) = 0.718(791.64 - 300)$$

$$= 353 \text{ kJ/kg}$$

$$w = q_1 - q_2 = 999.6 - 353 = 966.6$$

$$\eta = 1 - \frac{q_2}{q_1} = 1 - \frac{353}{999.6} = 0.647$$

OBJECTIVE TYPE QUESTIONS

Enthusiasm is the baking powder of life. Without it, you're flat. With it, you rise.

State True or False

1. An air-standard cycle with assumptions is a gas power cycle (True/False)
2. An air-standard cycle is assumed to be a closed air cycle. (True/False)
3. Heat transferred is assumed to be through combustion in the cylinder. (True/False)
4. If the cylinder head is machined off, the compression ratio decreases. (True/False)
5. The Otto cycle consists of two isobaric and two adiabatic processes. (True/False)
6. The Diesel cycle consists of one isochoric, one isobaric and two adiabatic processes. (True/False)
7. Spark ignition engines are based on Diesel cycle. (True/False)
8. Compression ignition engines use diesel as fuel. (True/False)
9. Diesel has high autoignition temperature. (True/False)
10. Petrol engines are based on the Otto cycle. (True/False)
11. Petrol has low self ignition temperature. (True/False)
12. The efficiency of the Otto cycle decreases with increase of compression ratio. (True/False)
13. For maximum work from the Otto cycle with the highest temperature = T_3 and the lowest temperature = T_1, T_2 and T_4 are equal and the value is $\sqrt{T_1 T_3}$. (True/False)
14. The maximum work from the Otto cycle is possible if $r = \left(\frac{T_3}{T}\right)^{\frac{1}{2(\gamma-1)}}$ where T_1 and T_3 are the highest and lowest temperatures in the cycle. (True/False)
15. The efficiency of the Diesel cycle is higher than Otto cycle. (True/False)
16. The Otto and Diesel cycles are used in external combustion engines. (True/False)
17. The cutoff ratio of the Diesel cycle is $\beta = \frac{V_3}{V_2}$ given in the following figure. (True/False)

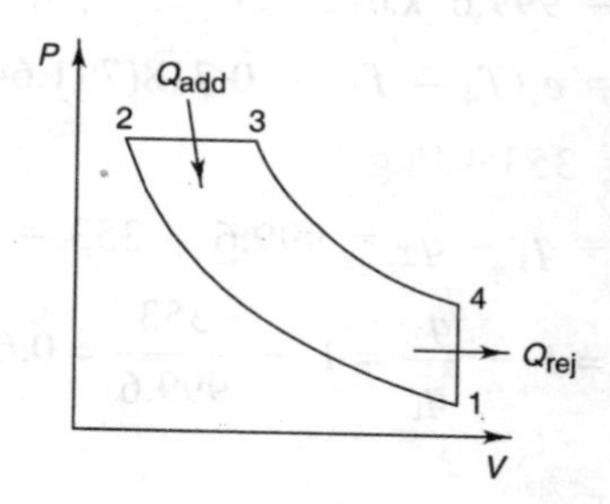

18. The efficiency of the Diesel cycle increases with increase of the cutoff ratio. (*True/False*)
19. The efficiency of the Diesel cycle increases with increase of the compression ratio. (*True/False*)
20. A fuel injection pump is used in diesel engines. (*True/False*)
21. A carburettor makes air-fuel mixture in SI engines. (*True/False*)
22. The compression ratio is the same in compression and spark ignition engines. (*True/False*)
23. The flywheel is heavy in a 4-stroke engine as compared to a 2-stroke engine. (*True/False*)
24. A 2-stroke engine can produce more power per stroke as compared to a 4-stroke engine of the same size. (*True/False*)
25. Valves and cam arrangement controls suction and exhaust in a 2-stroke engine. (*True/False*)
26. Air-fuel mixture is precompressed in the crankcase in 2-stroke engines. (*True/False*)
27. The transfer port is provided for transference of air or air-fuel mixture from the crankcase to the cylinder. (*True/False*)
28. A camshaft operates inlet and outlet valves. (*True/False*)
29. A camshaft runs at the same speed as a crankshaft. (*True/False*)
30. A fuel injector pump with a fuel injector is used for spraying fuel in the cylinder in a CI engine. (*True/False*)
31. Brake horse power has lesser value than that of indicated horse power and their ratio is called mechanical efficiency. (*True/False*)
32. Indicated horse power is the shaft power available at a crankshaft. (*True/False*)
33. The volumetric efficiency of a 2-stroke engine is higher than that of a 4-stroke engine. (*True/False*)
34. The brake thermal efficiency is the ratio of brake power to fuel energy used. (*True/False*)
35. The mean effective pressure can be found out by dividing the area of the indicated diagram by the length of the diagram if spring mechanism constant is unity (N/m^3). (*True/False*)
36. Deflector-like shape of the piston in a 2-stroke engine helps in scavenging out the burnout gases. (*True/False*)
37. The combustion in the Diesel cycle is isobaric and heat rejection is isochoric. (*True/False*)
38. Both combustion and heat rejection are isochoric in the Otto cycle. (*True/False*)
39. Both expansion and compression are reversible adiabatic in both the Otto and Diesel cycles. (*True/False*)
40. Engine speed of a CI engine is higher as compared to an SI engine. (*True/False*)
41. CI engines are used in light vehicles. (*True/False*)
42. The efficiency of the Carnot cycle is higher than that of the Otto and Diesel cycles. (*True/False*)
43. The difference of brake power and indicated power is friction power. (*True/False*)

44. The extreme positions of the piston are called dead centres. *(True/False)*

45. The piston reverses its direction of motion at the dead centre. *(True/False)*

46. The volume between the cylinder head and TDC is called clearance volume. *(True/False)*

47. The volume between TDC and BDC is called swept volume. *(True/False)*

48. Mean effective pressure is the ratio of net work done to the swept volume. *(True/False)*

49. The ports in a 2-stroke engine are opened and closed by external mechanism. *(True/False)*

50. Fuel is lost in scavenging in a 2-stroke engine. *(True/False)*

51. Knocking in an SI engine is self ignition of air-fuel mixture before spark ignites the mixture. *(True/False)*

52. An outlet port is nearest to TDC and an inlet port is nearest to BDC in a 2-stroke engine. *(True/False)*

53. Lubricating oil is mixed in petrol for a 2-stroke SI engine. *(True/False)*

54. For the same compression ratio, the efficiency of the Otto cycle is higher than that of the Diesel cycle. *(True/False)*

Multiple Choice Questions

1. If swept volume (V_s) and clearance volume (V_c) are given, then the compression ratio is

(a) $\frac{V_s}{V_c}$ (b) $1 + \frac{V_s}{V_c}$ (c) $1 + \frac{V_c}{V_s}$

2. The Otto cycle has two adiabatic processes and other two are

(a) one isochoric and one isobaric process
(b) both isochoric processes
(c) both isobaric processes

3. The Diesel cycle has two adiabatic processes and other two are

(a) both isochoric (b) both isobaric (c) one isochoric and one isobaric

4. In a 4-stroke IC engine, the cycle is completed in *n* number of revolutions of the crankshaft. The *n* is

(a) one (b) two (c) three

5. In a 2-stroke IC engine, the cycle is completed in *n* revolution of the crankshaft. Then the value of *n* is

(a) one (b) two (c) three

6. During suction stroke of an SI engine, the engine intake in the cylinder is

(a) air (b) air fuel mixture (c) fuel

7. During suction stroke of a CI engine, the engine intake in the cylinder is

(a) air (b) air fuel mixture (c) fuel

8. Air-fuel mixture in an SI engine is prepared by
(a) a feel injection pump (b) carburettor (c) an injector

9. In a CI engine, fuel is pumped through the injector into the cylinder by
(a) a carburettor (b) a fuel pump (c) an injector pump

10. In a 4-stroke engine, the ratio of speed of the crankshaft to that of the camshaft is
(a) 1 : 1 (b) 2 : 1 (c) 1 : 2

11. In a 2-stroke engine, the inlet port, transference port and output port are controlled by
(a) valves mechanism by the crankshaft
(b) valves mechanism by the camshaft
(c) the piston covering and uncovering the ports

12. A spark ignition engine is also known as
(a) petrol engine (b) gas engine (c) diesel engine

13. A compression ignition engine is also known as
(a) petrol engine (b) gas engine (c) diesel engine

14. Engines having very high speed are
(a) SI engines (b) CI engines (c) gas engines

15. In an SI engine, high voltage for the spark plug is generated by
(a) a battery (b) an alternator (c) an ignition coil

16. The desired firing order of spark plugs in a multicylinder engine is ensured by
(a) battery (b) ignition coil (c) distributor

17. In an SI engine, correct air-fuel mixture is ensured by
(a) a distributor (b) a fuel pump (c) a carburettor

18. The compression ratio in a CI engine is in the range of
(a) 5–10 (b) 8–20 (c) 12–25

19. In an SI engine, the compression ratio is in the range of
(a) 5–11 (b) 8–20 (c) 12–25

20. The thermal efficiency of a 4-stroke engine as comparison to a 2-stroke engine is
(a) low (b) equal (c) high

21. The volumetric efficiency of a 4-stroke engine as compared to a 2-stroke engine is
(a) low (b) equal (c) high

22. The thermal efficiency of a CI engine as compared to an SI engine for the same compression ratio is
(a) low (b) equal (c) high

23. The effect on autoignition by the content of isooctane in fuel in an SI engine is to
(a) accelerate (b) retard (c) no effect

24. The content of normal heptane is not preferred in the fuel of an SI engine as it can
(a) retard autoignition (b) corrode the piston (c) accelerate autoignition

25. Knocking in an SI engine increases with
(a) high compression ratio
(b) low compression ratio
(c) high voltage at the spark plug

26. The compression ratio in an SI engine is generally not increased beyond 10 or 11 due to
(a) high wear and tear of the engine
(b) high tendency of knocking
(c) low tendency of knocking

27. The knocking tendency in an SI engine increases with
(a) low speed (b) high speed (c) low compression ratio

28. In a CI engine, knocking tendency increases with
(a) high compression ratio
(b) high surrounding temperature
(c) low compression ratio

29. If an engine has half load, the friction load will be correspondingly
(a) half (b) full (c) double

30. The actual volume of fresh charge in a 4-stroke SI engine is
(a) more than the stroke volume
(b) equal to the stroke volume
(c) less than the stroke volume

31. In the Diesel cycle shown in the figure the compression ratio and cutoff ratio are
(a) 9 and 2 (b) 8 and 2 (c) 10 and 15

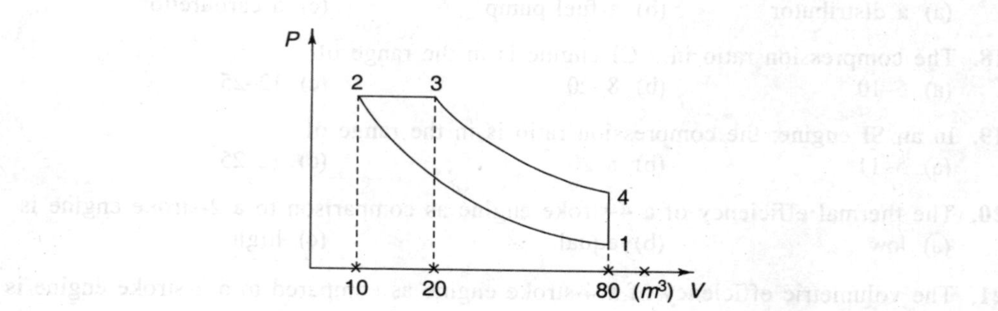

32. The efficiency of the Otto cycle, if $r = 0.43$ where r is compression ratio is
(a) 33% (b) 67% (c) 50%

33. In which engine, does the charge consist of mixture of air, fuel and lubricating oil?
(a) 4-stroke SI engine (b) 2-stroke diesel engine (c) 2-stroke SI engine

34. The reduced knocking in an SI engine is generally observed at
(a) increased atmospheric humidity
(b) reduced atmospheric humidity
(c) increased exhaust pressure

35. For minimum knocking tendency in an SI engine, the spark plug should be located
(a) near the inlet valve
(b) near the exhaust valve
(c) midway between the inlet and exhaust valves

36. The Otto cycle operates with volumes of 40 cm^3 and 400 cm^3 at TDC and BDC. If power output is 100 kW, then heat input (kJ/s) for γ = 1.4 will be
(a) 162 (b) 245 (c) 93

37. In the Diesel cycle, the volumes in the cylinder are 30 cm^3 and 45 cm^3 at the time of start and stop of fuel injection in the cylinder, then the cutoff ratio is
(a) 2 (b) 1.5 (c) 3

38. Reduced knocking is observed in an SI engine with
(a) increased rate of burning
(b) rich air-fuel mixture
(c) increased charge density

39. A Diesel cycle is shown on the *TS* diagram and if T_2 = 500 K and T_3 = 800 K, then heat added is (c_p = 1 kJ/(kg K) and c_v = 0.72 kJ/kg)
(a) 300 kJ/kg (b) 216 kJ/kg (c) 84 kJ/kg

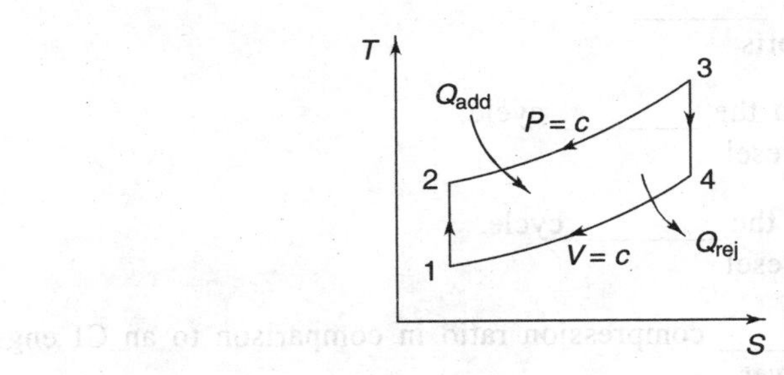

40. In the above question if T_1 = 300 K and T_4 = 400 K, then heat rejection is
(a) 100 kJ/kg (b) 72 kJ/kg (c) 120 kJ/kg

41. In the above question, the net work output is
(a) 228 kJ/kg (b) 240 kJ/kg (c) 200 kJ/kg

42. In a petrol engine, the tendency of knocking decreases with
(a) increase of isooctane of fuel
(b) increase of compression ratio
(c) increase of cylinder diameter

43. A fuel has 65 parts isooctane and 35 parts *n*-heptane by volume. The octane number of fuels is
(a) 35 (b) 50 (c) 65

44. Which engine will have a heavier flywheel?
(a) 40 HP, 4-stroke SI engine
(b) 40 HP, 2-stroke SI engine
(c) 40 HP, 2-stroke CI engine

45. For the same maximum pressure and temperature, the efficiency of
(a) Otto cycle > Diesel cycle
(b) Diesel cycle > Otto cycle
(c) Otto cycle = Diesel cycle

46. An SI engine working on the Otto cycle has the compression ratio 5.5, the work output/cycle on a *PV* diagram is $23.625 \times 10^3 \times V_c$ joules (V_c = clearance volume). The indicated mcp is
(a) 4295 bar (b) 5250 bar (c) 106.3 bar

Fill in the Blanks

1. In the Otto cycle, heat addition and rejection is an ________ process.
(a) isochoric (b) isobaric

2. In the Diesel cycle, heat is added by an ________ process.
(a) isobaric (b) Isochoric

3. A 2-stroke IC engine has ________ .
(a) valves (b) ports

4. A 4-stroke IC engine has ________ .
(a) valves (b) ports

5. An SI engine is based on the ________ cycle.
(a) Otto (b) Diesel

6. A CI engine is based on the ________ cycle.
(a) Otto (b) Diesel

7. An SI engine has a ________ compression ratio in comparison to an CI engine.
(a) higher (b) lower

8. An SI engine has ________ speed in comparison to CI engine.
(a) higher (b) lower

9. The intake of an SI engine is ________ during the suction stroke.
(a) air (b) air-fuel mixture

10. The intake of a CI engine is ________ during the suction stroke.
(a) air (b) air-fuel mixture

11. A 2-stroke IC engine has an inlet port, a ________ port and an exhaust port.
(a) transfer (b) midport

12. The exhaust port is located ________ to TDC in a 2-stroke IC engine.
(a) nearest (b) farthest

13. The inlet port is located ________ to BDC in a 2-stroke IC engine.
(a) nearest (b) farthest

14. The transfer port is located below the ________ port in a 2-stroke IC engine.
(a) exhaust (b) inlet

15. A crank also performs the operation of ________ of charge in the crankcase before its transfer to the cylinder in 2-stroke CI engine.
(a) transfer (b) precompression

16. High octane SI fuel has ________ tendency of knocking.
(a) high (b) low

17. For the same compression, the Otto cycle has a ________ efficiency than that of the Diesel cycle.
(a) higher (b) lower

18. A Carnot engine has a ________ efficiency than that of the Otto or Diesel cycle.
(a) lower (b) higher

19. The efficiency of the Otto cycle ________ with increase of the compression ratio.
(a) increases (b) decreases

20. The efficiency of the Diesel cycle ________ with increase of the cutoff ratio.
(a) increases (b) decreases

21. Knocking in an SI engine is ________ by reduced turbulence of air-fuel mixture in the cylinder.
(a) encouraged (b) discouraged

22. For the same heat input and maximum pressure, the thermal efficiency of the Otto cycle is ________ than that of the Diesel cycle.
(a) higher (b) lower

ANSWERS

Let us look behind us with understanding, before us with faith, and around us with love.

State True or False

1. True
2. True
3. False (Heat transfer is considered from an external source instead of combustion in the cylinder.)
4. False (In an actual engine, the cylinder head block and engine block are joined together to form the cylinder of the engine. Machining of any of two reduces the clearance length.)
5. True
6. True
7. False (An SI engine uses petrol only as petrol gives readily air-fuel mixture which has high self ignition temperature and can be ignited by spark.)
8. True (Diesel has lower self-ignition temperature and undergoes spontaneous combustion when diesel is sprayed in highly compressed and hot air.)
9. True
10. True
11. False (Petrol has high self-ignition temperature to avoid any self ignition of mixture before spark reaches it, otherwise knocking will take place if any self ignition starts.)
12. False (The efficiency of the Otto cycle increases with increase of compression ratio as $\eta = 1 - \frac{1}{r^{\gamma - 1}}$)
13. True
14. True
15. False (Diesel cycle efficiency $= 1 - \frac{1}{r^{\gamma - 1}} k$ and k factor > 1)
16. False (Both are used in IC engines.)
17. True
18. False ($\eta = 1 - \frac{1}{r^{\gamma - 1}} k$ and $k = \frac{\beta^{\gamma} - 1}{\gamma(\beta - 1)}$. As k increases with cutoff (β), η decreases.)

19. True ($\eta = 1 - \frac{1}{r^{\gamma - 1}} \times k$ and as r increases, the negative quantity decreases.)

20. True (High pressure for fuel is required to be injected in compressed air.)

21. True (Fuel-air mixture is taken in the suction stroke.)

22. False (Compression ratio is more in a CI engine. As for self ignition of diesel, air is to be at high pressure and temperature.)

23. True (There is one power stroke in 2 revolutions of the crankshaft. Energy is stored in power stroke which is to be used in remaining 3 strokes. Hence a heavy flywheel is to store more energy.)

24. True (A 2-stroke engine has one power stroke per revolution while a 4-stroke engine has one power stroke per two revolutions.)

25. False (A 2-stroke engine has ports which are covered and uncovered by the piston side.)

26. True (Precompression of charge takes place in the crankcase when the piston moves down from TDC during the power stroke.)

27. True (The transfer port connects the crankcase to the cylinder at a place between the inlet port and the outlet port. The precompressed charge from the crankcase is transferred to the cylinder when the piston is moving down in power stroke and uncovers the transfer port.)

28. True (Inlet and outlet valves are to be opened once in two revolutions of the crankshaft. The camshaft is designed to run at half speed of the crankshaft to operate inlet and outlet valve mechanism.)

29. False (The camshaft runs at half speed of the crankshaft.)

30. True

31. False (IHP > BHP and IHP-friction horse power = BHP. Also $\eta_{\text{mechanical}}$ = BHP/IHP)

32. False (IHP is power actually generated in the cylinder while power available at the crankshaft is BHP = (IHP–FHP))

33. False (Volumetric efficiency in a 2-stroke engine is lower as charge is lost in scavenging, i.e. pushing out the burnt gases out of the cylinder through the exhaust port.)

34. True

35. True

36. True (The piston has deflector-like shape and pushes incoming precompressed charge through the transfer port towards top of the cylinder, thereby preventing charge rushing out of the exhaust port.)

37. True (Combustion cannot be isochoric as the piston moves down as diesel is being sprayed and spontaneously undergoing combustion.)

38. True (Spark is given to the compressed fixed volume of air and fuel mixture which burns spontaneously at constant volume).

39. True

40. False (As compression is higher in a CI engine as compared to an SI engine, the stroke length and pressure against which the piston is moving. TDC is higher resulting into lower speed.)

41. False (Since stroke length is more, the size of the cylinder of a CI engine is more. Higher compression means more pressure requiring thicker wall cylinder, piston and other parts. Therefore, a CI engine is heavy and can be used for heavy vehicles.)

42. True (The average temperature at which heat can be added to the system is higher in the Carnot cycle as compared to the Otto and Diesel cycles. Therefore, the efficiency of the Carnot cycle is maximum.)

43. False (IP = BP + FP)

44. True (Extreme positions are called TDC and BDC.)

45. True

46. True

47. True

48. True

49. False (Ports are covered and uncovered by the piston while moving from TDC to BDC and vice versa.)

50. Tree (Due to the volumetric efficiency of a 4-stroke engine is higher than that of a 2-stroke engine.)

51. True [In case mixture gets ignited before spark reaches the mixture. (Piston is yet to reach TDC) power will be used firstly to stop the upward movement of the piston abruptly and remaining power for acceleration of the piston towards BDC. The effect will be knocking (jerking) and wastage of power.]

52. True

53. True

54. True

Multiple Choice Questions

1. (b) **2.** (b) **3.** (c) **4.** (b)

5. (a) **6.** (b) **7.** (a) **8.** (b)

9. (c)

10. (b) (Inlet and outlet are to be opened once in 2 revolutions of the crankshaft.)

11. (c) **12.** (a) **13.** (c)

14. (a) (Lesser compression means shorter piston movement resulting higher speed.)

15. (c) **16.** (c) **17.** (c) **18.** (c)

19. (a)

20. (c) (Work output depends upon the heat added. Since fuel is lost to some extent in scavenging, heat added is smaller in a 2-stroke engine in comparison to a 4-stroke engine for work output resulting into lower efficiency of a 2-stroke engine.)

21. (c) (Some charge in a 2-stroke engine is wasted in scavenging.)

22. (a) ($\eta_{\text{Otto}} = 1 - \frac{1}{r^{\gamma-1}}$ and $\eta_{\text{Diesel}} = 1 - \frac{1}{r^{\gamma-1}} \times k$ where $k > 1$. Therefore, for the same r (compression ratio) $\eta_{\text{Otto}} > \eta_{\text{Diesel}}$)

23. (b) **24.** (c) **25.** (a) **26.** (b)

27. (a) **28.** (c)

29. (b) (Friction power remains constant.)

30. (c) **31.** (b) $\left[\beta = \frac{v_2}{v_1} = \frac{20}{10} = 2,\quad r = \frac{v_4}{v_2} = \frac{80}{10} = 8\right]$

32. (b) ($\eta = 1 - \frac{1}{r^{\gamma-1}} = \left(1 - \frac{1}{r^{0.4}}\right) = 1 - 0.33 = 0.67$)

33. (c) **34.** (a) **35.** (b)

36. (a) ($r = 1 + \frac{V_s}{V_c} = 1 + \frac{400}{40} = 1 + 10 = 11$

$$\eta = 1 - \frac{1}{r^{\gamma-1}} = 1 - \frac{1}{11^{0.4}} = 1 - 0.383 = 0.617 = \frac{W}{Q_{\text{add}}}$$

$$\therefore Q_{\text{add}} = \frac{100}{0.616} = 162 \text{ kW})$$

37. (b) ($\beta = \frac{v_3}{v_2} = \frac{45}{30} = 1.5$)

38. (a)

39. (a) ($Q_{\text{add}} = c_p(T_3 - T_2) = 1 \times (800 - 500) = 300$ kJ/kg)

40. (b) ($Q_{\text{rej}} = c_v(T_4 - T_1) = 0.72(400 - 300) = 72$ kJ/kg)

41. (a) ($W = Q_{\text{add}} - Q_{\text{rej}} = 300 - 72 = 228$ kJ/K)

42. (a)

43. (c) (The octane number is parts of isooctane in 100 parts of fuel.)

44. (a) (A 4-stroke engine will require a heavier flywheel as there is one power stroke in 2 revolutions of the crankshaft.)

45. (b) (Work of the Diesel cycle is higher as shown in the diagram by the shaded portion.)

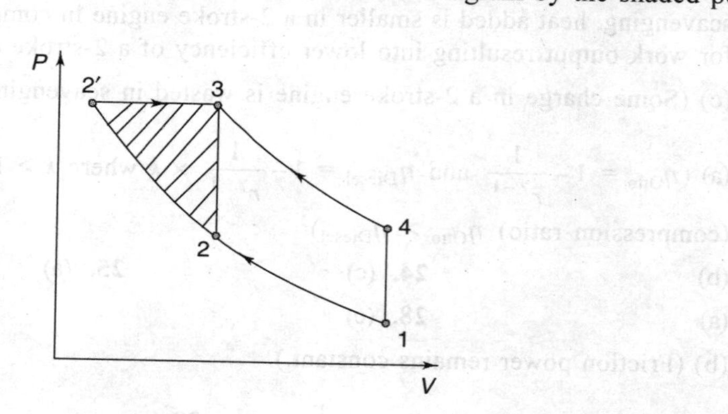

46. (a) (mep $= \dfrac{\text{area}}{\text{swept vol}} = \dfrac{23.625 \times 10^5 \times V_c}{V_s}$ and $1+\dfrac{V_s}{V_c} = 5.5$ or $\dfrac{V_c}{V_s} = \dfrac{1}{4.5}$

$\therefore$ mep $= \dfrac{23.625 \times 10^5}{4.5} = 5250$ bar)

Fill in the Blanks

1. (a) **2.** (a) **3.** (b) **4.** (a)

5. (a) **6.** (b) **7.** (b) **8.** (a)

9. (b) **10.** (a) **11.** (a) **12.** (a)

13. (a) **14.** (a) **15.** (b) **16.** (b)

17. (a) ($\eta_{\text{Otto}} = 1 - \dfrac{1}{r^{\gamma-1}}$, and $\eta_{\text{Diesel}} = 1 - \dfrac{1}{r^{\gamma-1}} k$ where $k > 1$ $\therefore \eta_{\text{Otto}} > \eta_{\text{Diesel}}$)

18. (b)

19. (a) **20.** (b) **21.** (a) **22.** (a)

CHAPTER 7

Mechanism and Simple Machines

Books are man's best friend.

INTRODUCTION

Man has invented devices to augment his abilities. The development of mechanisms and machines has been the outcome of this endeavour. Initially, the primary engineering activities were restricted to construction works which demanded the shifting and lifting of heavy stones and other items of the construction. This necessity led to the development of lever mechanisms, pulleys and wedges, etc. As other new activities such as irrigation, mining, shipping and water supply started, the use of wind power and water power for running pump and many other activities became more common. This was possible as new ideas and concepts emerged in transferring and transforming power and motion by using different types of machines and mechanisms.

The theory of machines and mechanisms is an applied science which helps us to understand the relationship between the geometry and motion of the parts of a machine or mechanism and the forces producing these motions. A mechanism is a set of machine elements or components or parts which are arranged in specific order to produce a desired motion. A machine in simple term can be defined as a contrivance which receives energy in some available form and uses it to do some particular kind of work. For example, a crowbar with its fulcrum also forms a machine as it can transform the muscular energy of a man in raising a heavy stone. Similarly a petrol engine is also a machine which transfers the heat energy of the fuel into power for propelling a vehicle. The theory of machines comprises of the study of the relative motion between the parts of a machine and the study of the forces which act on these parts. The study of the relative motion between the parts without considering the forces causing the motion is called the sciences of kinematics. The sciences of kinetics deal with the inertia forces arising from the combined effect of the mass and the motion of the parts. The designing of a machine involves (i) determination of kinematic chain, (ii) determination of forces, and (iii) proportioning of the parts.

KINEMATIC LINK OR ELEMENT

A kinematic link is a resistant body or an assembly of resistant bodies which are utilized to connect a part or parts of a machine with other parts which have relative motions with it. A kinematic link can be an assembly of parts forming one unit but these parts cannot have any relative motion with respect to one another. A link need not necessarily be a rigid body. However, it must be a resistant body which means that it must be capable of transmitting the required force with negligible deformation. Such links are (i) liquids which are resistant to compressive force and they are used as links in hydraulic machines and (ii) flexible links such as chains, belts and ropes which are resistant to tensile force and therefore they are used to transmit motion and force.

Classification of Links

Depending upon its ends on which revolute or turning pairs can be placed, links can be classified as (i) binary link having two vertices, (ii) ternary link having three vertices, (iii) quaternary link having four vertices and so on (Figure 7.1).

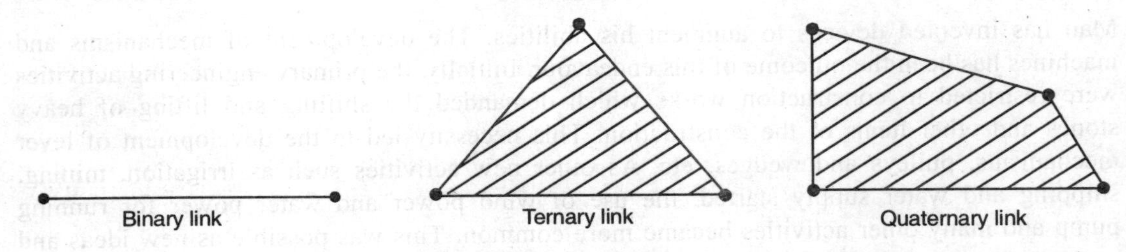

FIGURE 7.1 Classification of links.

Types of Links

The links can be:

(a) *Rigid link:* A rigid link can transmit motion and force without undergoing any deformation. A connecting rod, a crank and a tappet rod of a valve are rigid links.

(b) *Flexible link:* Flexible links can transmit motion and force without any deformation in the desired direction only as they are resistant to such forces and motion in that direction. Belts, ropes and chains are resistant to tensile forces but not to compressive forces. Hence, they can transmit tensile forces only.

(c) *Fluid link:* Fluids are resistant to compressive force. A fluid link is used to transmit motion and force through the fluid by pressure. A hydraulic press, a hydraulic jack and a fluid brake are examples of fluid links.

(d) *Floating link:* As the name suggests, it is a link which is not connected to the frame of the machine.

KINEMATIC PAIR

A kinematic pair consists of two links of a machine which are in contact with each other and they have a relative motion between them. For example, a slider-crank mechanism of an IC engine consists of four links, viz. (i) link 1 = frame; (ii) link 2 = crank; (iii) link 3 = connecting rod; and (iv) link 4 = slider or piston. Hence a four-link kinematic chain has four constituted kinematic pairs of (i) links 1 and 2; (ii) links 2 and 3; (iii) links 3 and 4; and (iv) links 4 and 1 as shown in Figure 7.2.

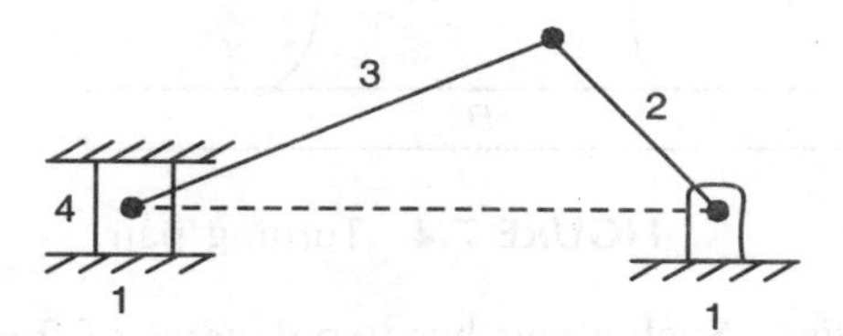

FIGURE 7.2 Slider crank mechanism: 4 links and 4 kinematic pairs.

Classification of Kinematic Pairs

Kinematic pairs can be classified according to (i) type of relative motion, (ii) type of contact with each other, and (iii) type of mechanical restraint.

Kinematic pairs according to the relative motion can be classified as:

(a) *Sliding pair:* The links in a sliding pair are constrained to have sliding motion relative to each other (Figue 7.3). This is also called *prismatic pair*. The examples of such pairs are: (i) piston and cylinder, (ii) tailstock on lathe bed, and (iii) ram and guides on the column of a shaper. The relative motion between elements A and B can be expressed by a single coordinate S and thus such a pair has one degree of freedom.

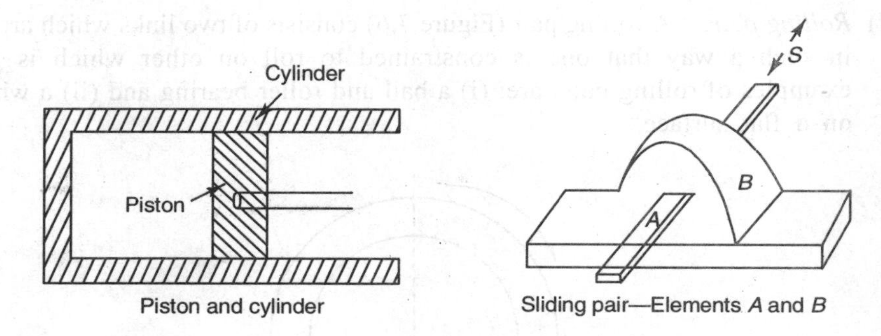

FIGURE 7.3 Sliding pair.

(b) *Turning pair:* It is also called a revolute or hinged pair. The turning pair consists of two links which are connected in such a manner that one link is constrained to turn or revolve about a fixed axis of another link. Examples of turning pair are: (i) turning of crankshaft in a bearing, (ii) revolving of a cycle wheel over its axle, and (iii) a shaft with a collar at both ends revolving in a circular hole. Such a pair allows only relative motion of rotation which can be expressed by a single coordinate θ (Figure 7.4). Hence a turning pair has a single degree of freedom.

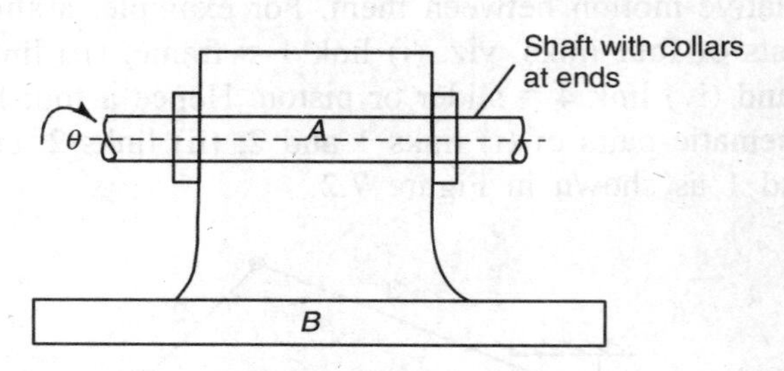

FIGURE 7.4 Turning pair.

(c) *Cylindrical pair:* Such a pair has two degrees of freedom, namely that of rotation and translation parallel to the axis of rotation. These two motions have no relationship with each other. If a shaft has no collar at ends, the motion between elements *A* and *B* will be both of sliding and turning. These relative motions of rotation and translation can be expressed by coordinates θ and *S* respectively (Figure 7.5).

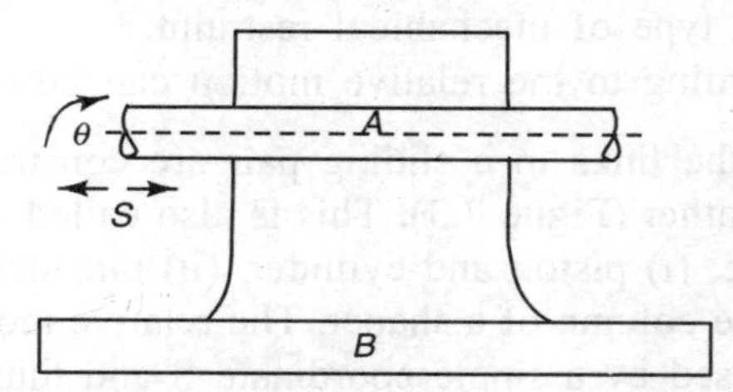

FIGURE 7.5 Cylinder pair: Translation and rotation.

(d) *Rolling pair:* A rolling pair (Figure 7.6) consists of two links which are connected in such a way that one is constrained to roll on other which is fixed. The examples of rolling pairs are: (i) a ball and roller bearing and (ii) a wheel rolling on a flat surface.

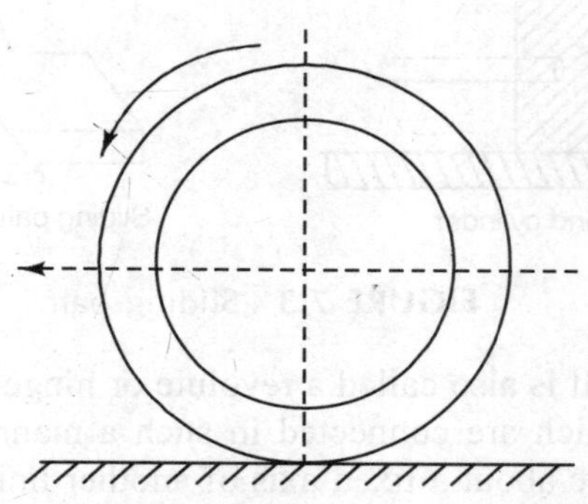

FIGURE 7.6 Rolling wheel on ground: Rolling pair.

(e) *Screw pair:* This is also called a helical pair. The pair consists of two links in which one link turns about the other link by means of threads. The motion in this pair consists of a combination of sliding and turning motion. However, a screw pair has one degree of freedom as the relative movement between the links can be expressed by a single coordinate θ or s (Figure 7.7). These two coordinates are related by the relation of $\frac{\Delta\theta}{2\pi} = \frac{\Delta s}{L}$ where L = load of the thread. Examples of screw pairs are: (i) a half nut and lead screw of the lathe machine, (ii) bolt and its nut, and (iii) a screw and nut of a lifting jack.

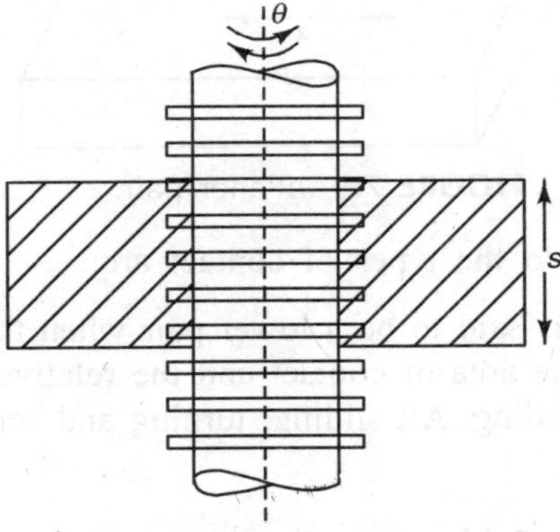

FIGURE 7.7 Screw pair.

(f) *Spheric pair:* A spheric pair consists of two elements in which one element is in the form of a ball which turns about the other fixed element having the form of a socket. Examples are: (i) a ball and socket joint of arm with shoulder, and (ii) ball and socket joint of a pen stand. This connection has three degrees of freedom as three coordinates are required to describe the relative movement between the connected elements. α and β coordinates are required to specify the position of the axis OA and the third coordinate θ is used to describe rotation about the axis OA (Figure 7.8).

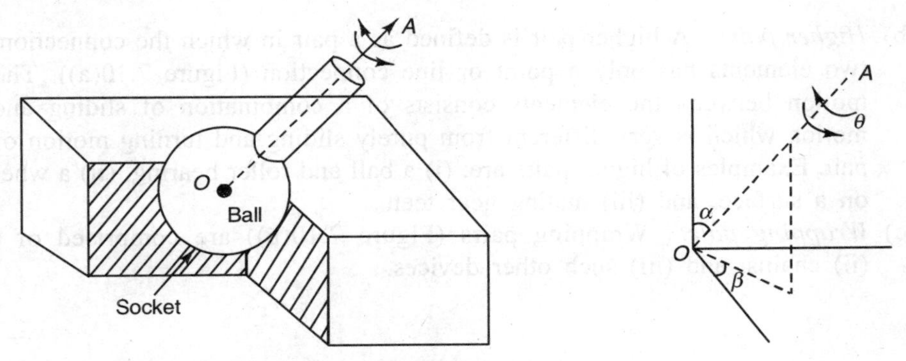

FIGURE 7.8 Spheric pair: Ball and socket joint.

(g) *Planer pair:* A planer pair as the name suggests consists of two elements which can move in a plane. A planer pair has three degrees of freedom. Two coordinates x and y describe the relative translation in the xy plane and the third coordinate θ describes the relative rotation about the z-axis (Figure 7.9).

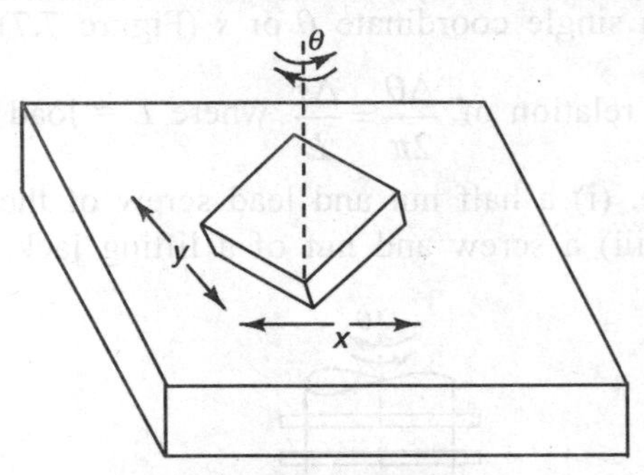

FIGURE 7.9 Planer pair.

Kinematic pairs according to the types of contact are:

(a) *Lower pair:* A pair is said to be a lower pair when the connection between two elements is through the area of contact and the relative motion between elements is purely turning or sliding. All sliding, turning and screw pairs form lower pairs as listed in Table 7.1.

Table 7.1 The lower pairs

	Pair	*Pair variable*	*Degree of freedom*	*Relative motion*
1.	Revolute/Turning	$\Delta\theta$	1	Circular
2.	Sliding/Prismatic	ΔS	1	Linear
3.	Screw	$\Delta\theta$ or ΔS	1	Helical
4.	Cylinder	$\Delta\theta$ and ΔS	2	Cylindric
5.	Spheric	$\Delta\theta$, $\Delta\alpha$, $\Delta\beta$	3	Spheric
6.	Flat	Δx, Δy, $\Delta\theta$	3	Planer

(b) *Higher pair:* A higher pair is defined as a pair in which the connection between two elements has only a point or line connection (Figure 7.10(a)). The relative motion between the elements consists of a combination of sliding and turning motion which is very different from purely sliding and turning motion of a lower pair. Examples of higher pairs are: (i) a ball and roller bearing, (ii) a wheel rolling on a surface, and (iii) mating gear teeth.

(c) *Wrapping pair:* Wrapping pairs (Figure 7.10(b)) are comprised of (i) belts, (ii) chains, and (iii) such other devices.

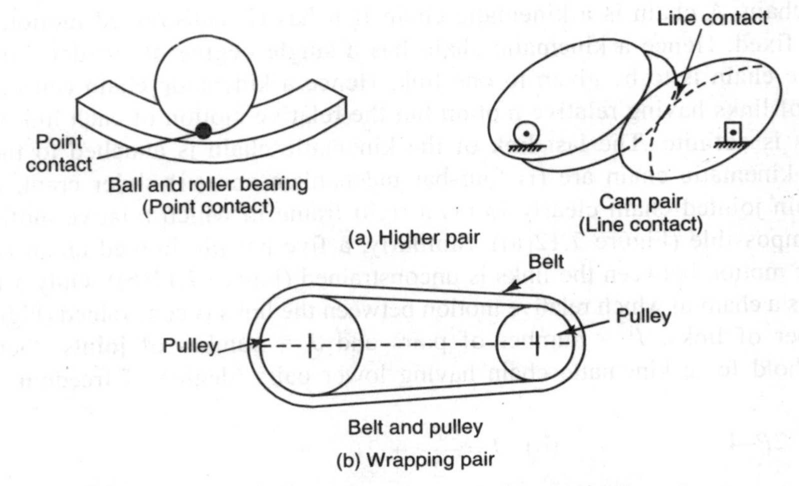

FIGURE 7.10 Kinematic pairs: Higher and Wrapping.

Kinematic pairs according to the type of mechanical constraint are:

(a) *Closed pair:* A closed pair or form closed (Figure 7.11(a)) has its elements held together mechanically by geometrically enclosing one element in the other in such a manner that the desired type of relative motion can only take place. All lower pairs and few higher pairs such as enclosed cam and follower are closed pairs.

(b) *Unclosed pair:* This is also called *force closed* as the contact between two elements can be maintained with the help of an externally applied force (Figure 7.11(b)).

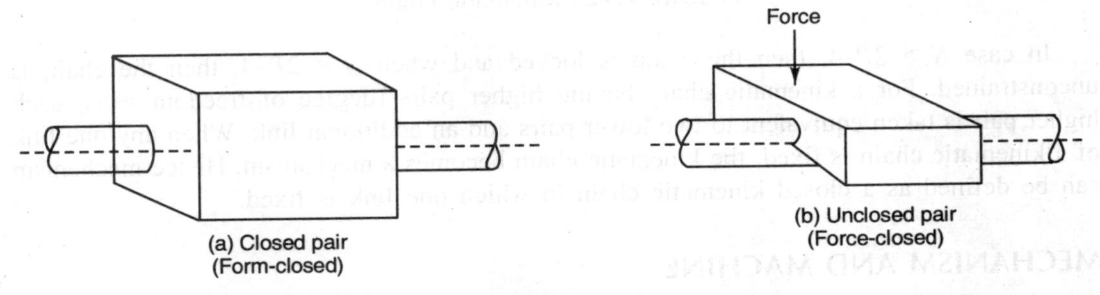

FIGURE 7.11 Kinematic pairs according to type of mechanical constraint.

KINEMATIC CHAIN

When a number of links are connected in space in such a way that the relative motion of any point on a link with respect to any other point follows a law or in other words their relative motion is constrained, then the links are said to form a kinematic chain. A constrained motion is a motion in which the position of a point on a link can be predicted at any instant as against an unconstrained motion in which a point can have any of the possible infinite motions or positions at any instant. A chain having unconstrained motion cannot be called a kinematic chain. A chain is a kinematic chain if it has (i) constrained motion and (ii) no link of its is fixed. Hence a kinematic chain has a single degree of freedom. Input motion to a kinematic chain is to be given to one link. Hence a kinematic chain can be defined as an assembly of links having relative motion but the relative motion of each link with respect to other links is definite. The last link of the kinematic chain is attached to the first link. Examples of kinematic chain are (i) four-bar mechanism and (ii) slider crank mechanism. A three-bar pin jointed chain clearly forms a rigid frame in which relative motion between the links is impossible (Figure 7.12(a)). Similarly, a five-bar pin jointed chain is a chain in which relative motion between the links is unconstrained (Figure 7.12(b)). Only a four-bar pin jointed chain is a chain in which relative motion between the links is constrained (Figure 7.12(c)). If N = number of links, P = number of pairs and J = number of joints, then following relationships hold for a kinematic chain having lower pairs (degree of freedom = 1) only:

(a) $N = 2P - 4$ (b) $J = \dfrac{3N}{2} - 2$

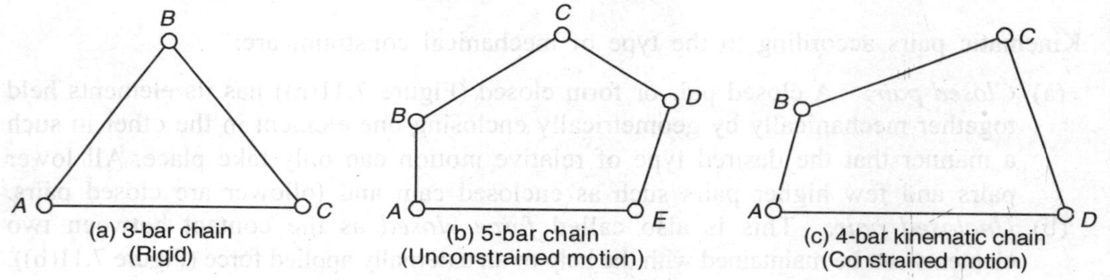

FIGURE 7.12 Kinematic chain.

In case $N > 2P - 4$, then the chain is locked and when $N < 2P - 4$, then the chain is unconstrained. For a kinematic chain having higher pairs (degree of freedom > 1), each higher pair is taken equivalent to two lower pairs and an additional link. When any one link of a kinematic chain is fixed, the kinematic chain becomes a mechanism. Hence mechanism can be defined as a closed kinematic chain in which one link is fixed.

MECHANISM AND MACHINE

If one of the links of a kinematic chain is fixed or made unmovable, the resulting kinematic chain is called a mechanism. The mechanism is primarily used to transmit or to modify motion. On the other hand, a machine is also basically made of kinematic chains and it is primarily used to transmit motions as well as forces. If a kinematic chain is considered purely from the point of view of motion modifier or transmitter, then it is called or referred to as mechanism. In case a kinematic chain is considered as a mean for applying or modifying mechanical work (force × distance), then it is called or referred to as a machine. In order to form a simple closed chain, at least three links with three kinematic pairs are required. However, if one of these three links is fixed, there cannot be any relative motion between any of the links and so such a closed chain does not form a mechanism. Such a type of chain is called a structure which is completely rigid. Hence, the simplest mechanism consists of four binary links which are connected to each other by a kinematic pair of revolute type and this mechanism is called a four-bar mechanism.

When a mechanism is required to transmit power or to perform some particular kind of work, the various links or elements have to be designed so as to carry safely the forces to which they are subjected to. The arrangement of such designed links into a mechanism becomes a machine. A mechanism may therefore be considered as a machine, in which each part is reduced to the simplest form which is necessary in order to transmit the desired motion. Hence, a machine can be defined as a combination of resistant bodies, with successfully constrained relative motions, which is used for transmitting or transforming input power so as to do some particular kind of work. The difference between a machine and a mechanism is tabulated in Table 7.2.

Table 7.2 Difference between a machine and a mechanism

Mechanism	*Machine*
1. A mechanism transmits and modifies input motion.	1. A machine modifies and transmits input work or power.
2. The simplest mechanism is a kinematic chain which has a minimum of four binary links which are connected to each other by a kinematic pair of revolute type and one link is fixed.	2. A machine may have many mechanisms.
3. When a kinematic chain is analyzed as a mechanism, no special consideration is given to the form or size or cross section of any link. The main consideration is their lengths and their assembly locations.	3. Cross-sectorial areas and proportional lengths of all links forming a kinematic chain are considered to provide strength, stiffness and clearances to the links so that they can transmit safely power and motion.
4. Type-writers and clock work are some examples of mechanisms as they are required to transmit motion only.	4. Lathes, shapers and milling machines are required to receive power which is suitably converted to perform metal cutting.

Types of Mechanisms

Mechanisms can be plane, spherical and spatial. The type of the mechanism depends upon the characteristics of the motions of the links. A planer mechanism is one on which all particles describe plane curves in space and all these curves have to lie in parallel planes. In other words, all points on the mechanism have the loci which are plane curves parallel to a single common plane. This characteristic helps in representing the locus of any point of a planer mechanism in its true size and shape on a single drawing. In spherical and spatial mechanisms, the links of mechanism lie in different planes.

The mechanisms can also be classified as:

(a) *Simple mechanism:* A simple mechanism has four links only.
(b) *Compound mechanism:* A compound mechanism has more than four links.
(c) *Complex mechanism:* Simple and compound mechanisms are formed by binary links while a complex mechanism is formed by the inclusion of ternary or higher order floating links.

Mobility and Kutzbach Criterion

The degree of freedom of a mechanism is a measure of the mobility of the device. The mobility of a mechanism indicates the number of inputs or kinematic pairs which must be controlled independently in order to bring the mechanism into a particular position. The mobility of a mechanism depends upon (i) the number of links and (ii) the number and types of joints forming the mechanism.

A link in a planer mechanism has three degrees of freedom, i.e. x, y and θ. If there are n links and one link is fixed, then a n-link planer mechanism has $3(n - 1)$ degrees of freedom before any of the joints is connected. A revolute joint has one degree of freedom, i.e. only rotation (θ and not x and y), thereby a revolute pair provides two constraints per joint. If a two-degree of freedom pair is connected, it provides one constraint. When the constraints for all joints are subtracted from the total freedoms of the unconnected links, the resulting mobility of the connected mechanism can be found out. The resulting mobility m of a planer n-link mechanism can be given as

$$m = 3(n - 1) - 2J_1 - J_2$$

where J_1 = single degree of freedom pair and J_2 = two degree of freedom pairs.

The above equation is called the Kutzbach criterion for the mobility of a planer mechanism as shown in Figure 7.13. If the Kutzbach criterion yields $m > 0$, the mechanism has m degrees of freedom. If $m = 1$, the mechanism can be driven by a single input motion. If $m = 2$, then two separate input motions are required to produce constraint motion for the mechanism. If the Kutzbach criterion yields $m = 0$, the motion is impossible and the chain forms a rigid structure. If $m = -1$ or loss, it means the chain has redundant constraints, indicating the chain is a statically indeterminate structure. The Kutzbach criterion can also

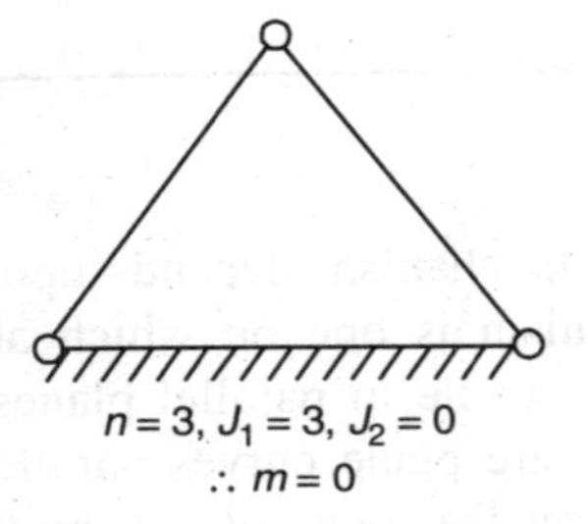

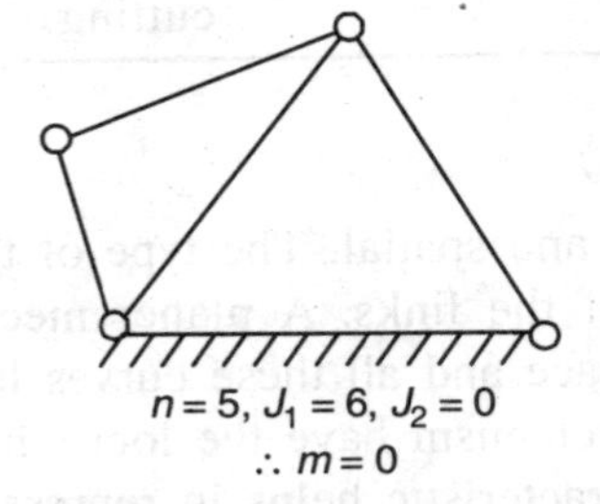

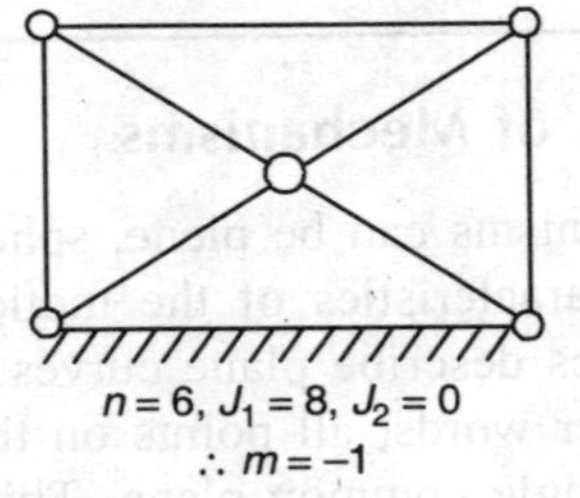

Rigid and statically indeterminate structure

FIGURE 7.13 (Contd.)

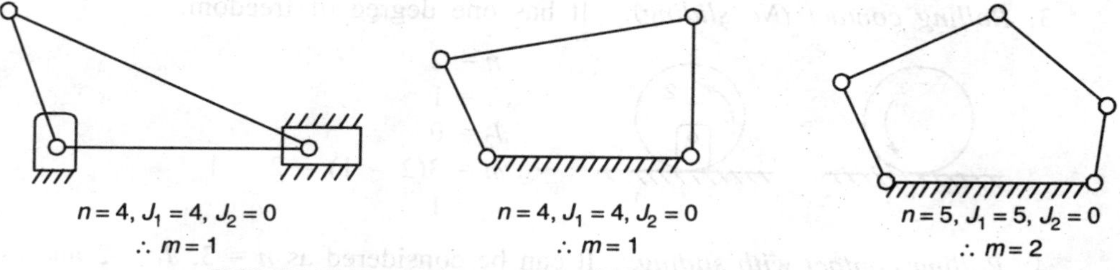

FIGURE 7.13 Degree of freedom: Kutzbach criterion.

be applied to a mechanism having kinematic pairs with two degrees of freedom. A cam and follower has one higher pair at the contact of the cam and follower. Similarly, a rotating wheel with slippage with a fixed link has a higher pair. They have, however, mobility as one and two as shown in Figure 7.1.4.

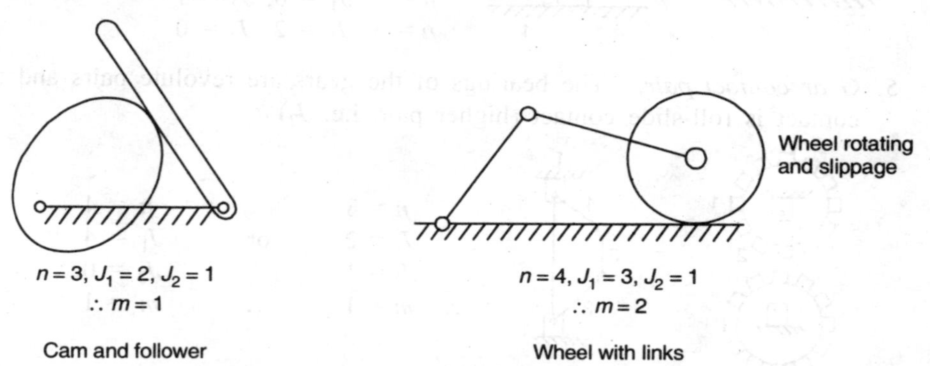

FIGURE 7.14 Mechanism with higher kinematic pairs.

Example 7.1 Find the degrees of freedom of lower and higher pairs.

1. *Revolute pair:* It has only rotating constraint.

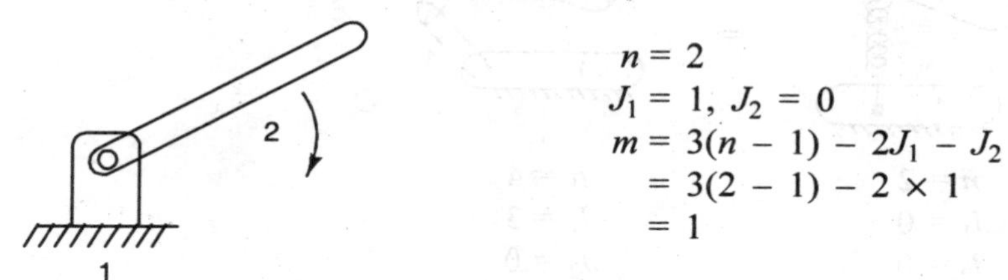

$$n = 2$$
$$J_1 = 1,\ J_2 = 0$$
$$m = 3(n - 1) - 2J_1 - J_2$$
$$= 3(2 - 1) - 2 \times 1$$
$$= 1$$

2. *Slider (Prismatic) pair:* It has constraint motion of sliding. It can also be considered as a revolute pair with a fixed link at the radius of infinity.

$$n = 2$$
$$J_1 = 1$$
$$J_2 = 0$$
$$m = 3(2 - 1) - 2 \times 1 = 1$$

3. *Rolling contact (No sliding):* It has one degree of freedom.

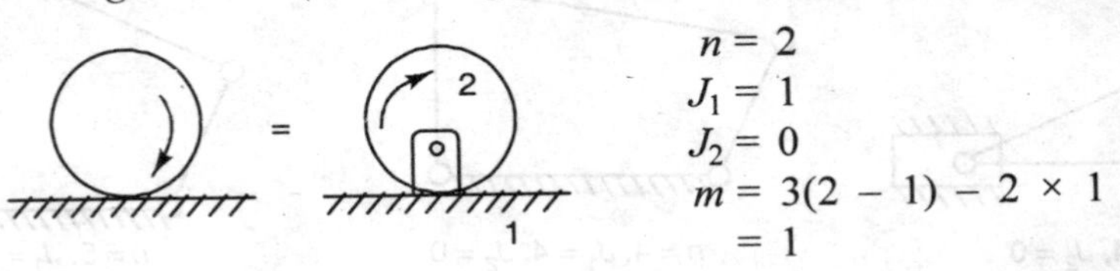

$n = 2$
$J_1 = 1$
$J_2 = 0$
$m = 3(2 - 1) - 2 \times 1$
$= 1$

4. *Rolling contact with sliding:* It can be considered as $n = 3$, $J_1 = 2$ and $J_2 = 0$ or $n = 2$, $J_1 = 0$ and $J_2 = 1$

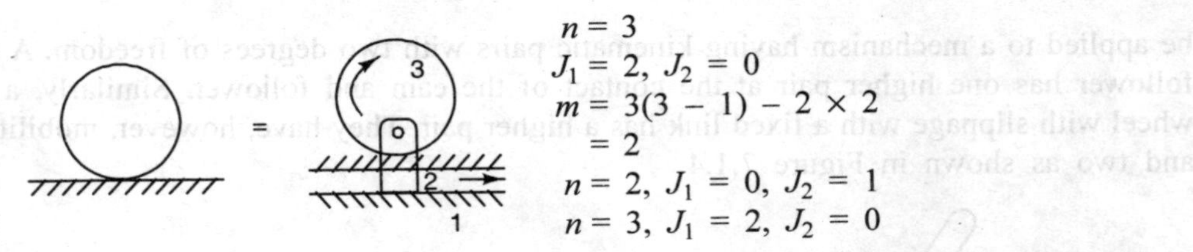

$n = 3$
$J_1 = 2,\ J_2 = 0$
$m = 3(3 - 1) - 2 \times 2$
$= 2$
$n = 2,\ J_1 = 0,\ J_2 = 1$
$n = 3,\ J_1 = 2,\ J_2 = 0$

5. *Gear contact pair:* The bearings of the gears are revolute pairs and the teeth contact is roll-slide contact (higher pair, i.e. J_2)

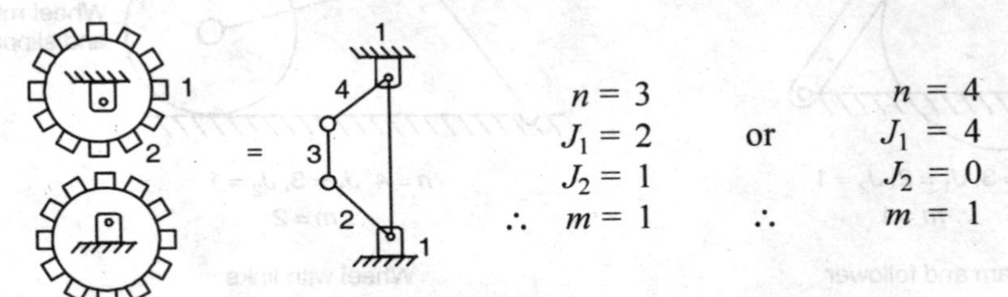

$n = 3$ or $n = 4$
$J_1 = 2$ $J_1 = 4$
$J_2 = 1$ $J_2 = 0$
$\therefore\ m = 1$ $\therefore\ m = 1$

6. *Spring connection:* The spring connection does not constraint the relative motion between two links.

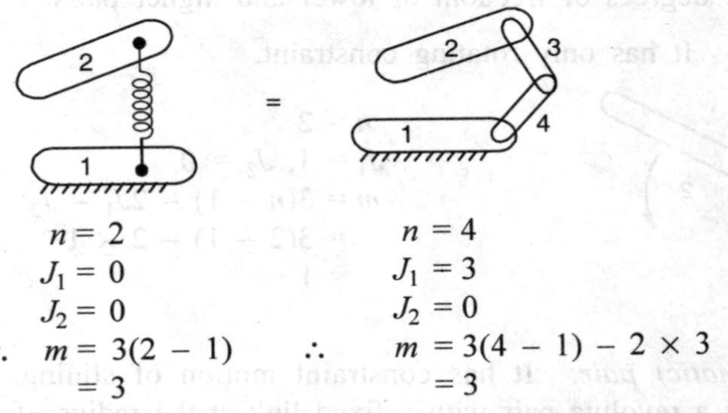

$n = 2$ $n = 4$
$J_1 = 0$ $J_1 = 3$
$J_2 = 0$ $J_2 = 0$
$\therefore\ m = 3(2 - 1)$ $\therefore\ m = 3(4 - 1) - 2 \times 3$
$= 3$ $= 3$

7. *Belt and pulley:* A ternary link with three revolute pairs is equivalent to a six-link mechanism. The pulley is rolling on the belt without sliding and hence it is a rolling pair.

$n = 2$ $n = 4$

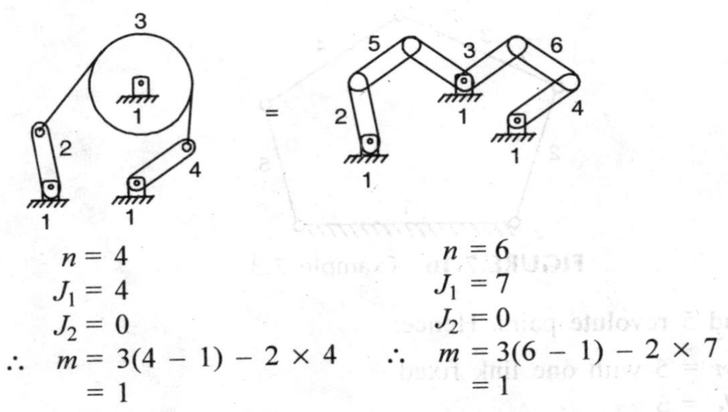

$n = 4$
$J_1 = 4$
$J_2 = 0$
$\therefore \quad m = 3(4 - 1) - 2 \times 4$
$= 1$

$n = 6$
$J_1 = 7$
$J_2 = 0$
$\therefore \quad m = 3(6 - 1) - 2 \times 7$
$= 1$

8. *Chain and spare kit:* Similar to belt and pulley mechanism and it has one degree of freedom.

Example 7.2 The two-links system shown in Figure 7.15 is constrained to move with planer motion. It possesses:

(a) 2 degrees of freedom
(b) 3 degrees of freedom
(c) 4 degrees of freedom
(d) 6 degrees of freedom

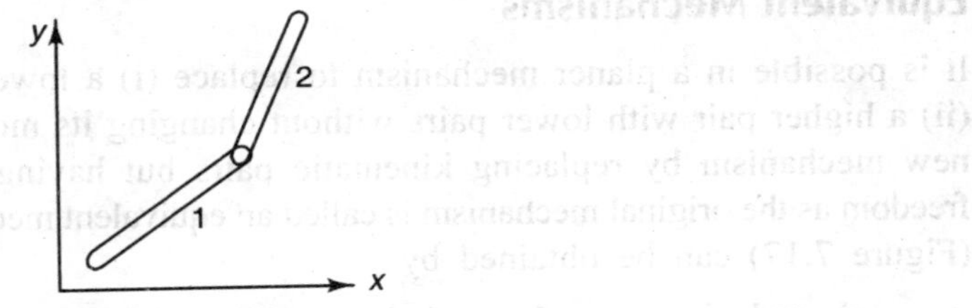

FIGURE 7.15 Example 7.2.

(IES: 1994)

There are two links (none of link is fixed) and one revolute pair. Hence:

$n = 2$ with no link is fixed
$J_1 = 1$
$\therefore \quad m = 3(n - 0) - 2J_1$
$= 3(2 - 0) - 2 \times 1$
$= 6 - 2$
$= 4$

Answer c is correct

Example 7.3 The number of degrees of freedom of a five links plane mechanism with five revolute pairs as shown in Figure 7.16 is

(a) 3 (b) 4 (c) 2 (d) 1

(GATE: 1993)

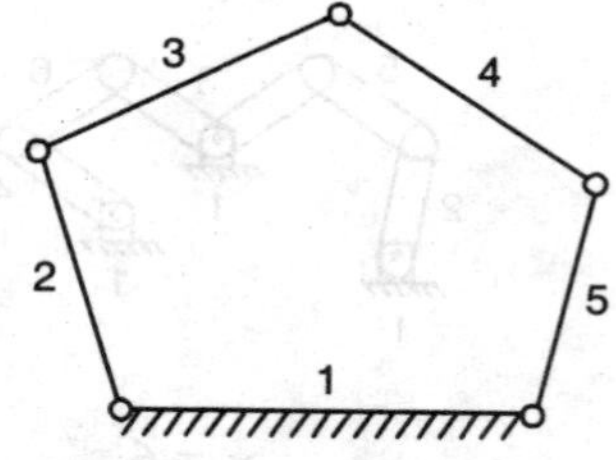

FIGURE 7.16 Example 7.3.

There are 5 links and 5 revolute pairs. Hence:

$$
\begin{aligned}
n &= 5 \text{ with one link fixed} \\
J_1 &= 5 \\
J_2 &= 0 \\
m &= 3(n - 1) - 2J_1 - J_2 \\
&= 3(5 - 1) - 2 \times 5 \\
&= 12 - 10 \\
&= 2
\end{aligned}
$$

Option (c) is correct.

Equivalent Mechanisms

It is possible in a planer mechanism to replace (i) a lower pair by another lower pair, and (ii) a higher pair with lower pairs without changing its mobility or degrees of freedom. The new mechanism by replacing kinematic pairs but having the same number of degrees of freedom as the original mechanism is called an equivalent mechanism. The equivalent mechanism (Figure 7.17) can be obtained by

(a) replacing a turning pair by a sliding pair or vice versa
(b) replacing a spring by two binary links
(c) replacing a cam by one binary link with two revolute pairs at each end.

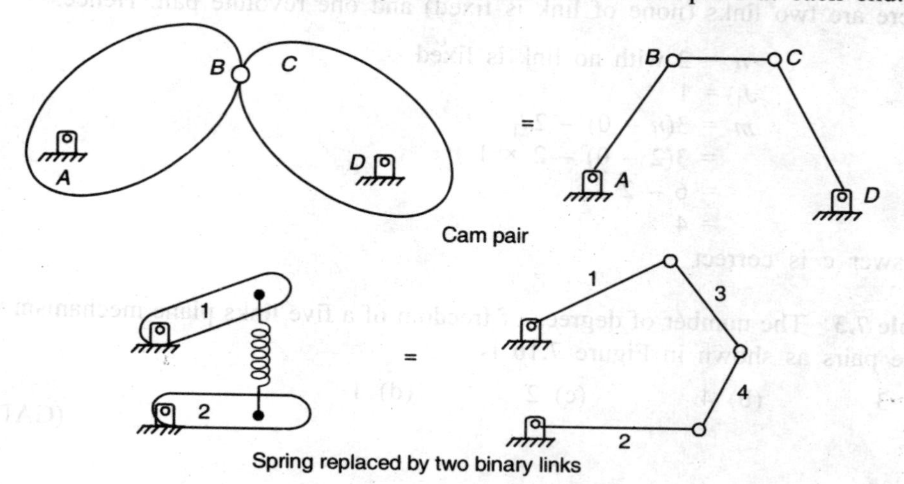

FIGURE 7.17 Equivalent mechanisms.

INVERSION

A mechanism is obtained by fixing one link of a kinematic chain. If there are n links in a kinematic chain, then n different mechanisms can be obtained by fixing each of its link of the kinematic chain in turn. The mechanisms obtained in this way may be very different in their appearance and in the purpose for which they can be used. Each mechanism is termed inversion of the original kinematic chain.

The Four-bar Chain (4 Turning Pairs)

The four-bar chain is one of the most important kinematic chains with four lower kinematic pairs (either sliding or turning) from the practical point of view. It is found that many complicated machines are based on the combinations of different inversions of the four-bar chain. The four links may be of different lengths and the use of the various inversions of the mechanism is dependent solely on the relative lengths of the links. There are a few practical mechanisms obtained from the four-bar chain:

(a) *The mechanism of the coupling rod of a locomotive:* In this inversion, the opposite links are equal, i.e. $AB = CD$ and $AD = BC$ (Figure 7.18). Equal velocity is imparted to wheels.

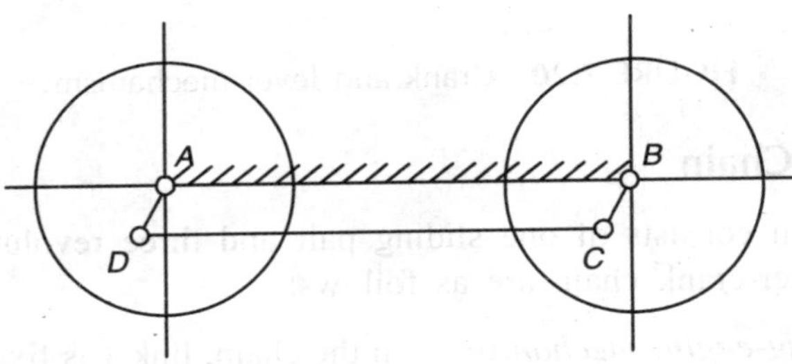

FIGURE 7.18 Mechanism of a coupling rod.

(b) *The mechanism of the Ackermann—Steering gear:* In this inversion of the four-bar chain, two short links are equal while the long links AB and CD are unequal in length (Figure 7.19). When car is moving along a straight path the two long links AB and CD remain parallel. However, when a car moves along a curved path, the links move up to such a position due to the lengths of the links that the axes

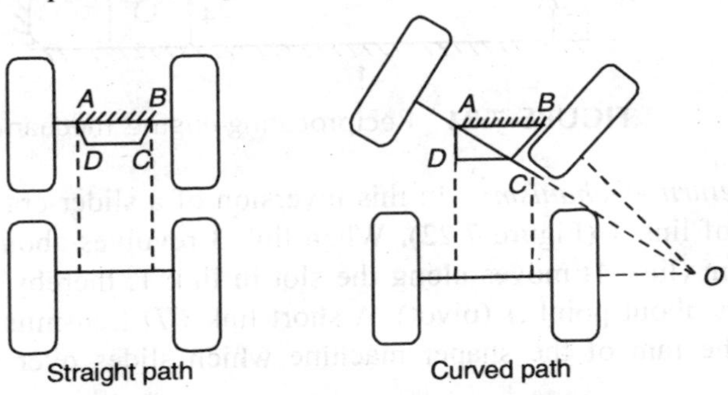

FIGURE 7.19 Ackermann steering gear.

of all four wheels intersect at the point O as shown in the figure. The mechanism thus ensures that the relative motion between the tyres and the road surface remains pure rolling.

(c) *Beam engine mechanism:* This is also called a crank and lever mechanism. When the crank AB rotates about the fixed centre A, the beam CDE oscillates about the fixed centre D and the vertical reciprocating motion of the piston is transmitted to end E of the beam (Figure 7.20). The reciprocating motion at point E is converted into rotating motion of the crank AB.

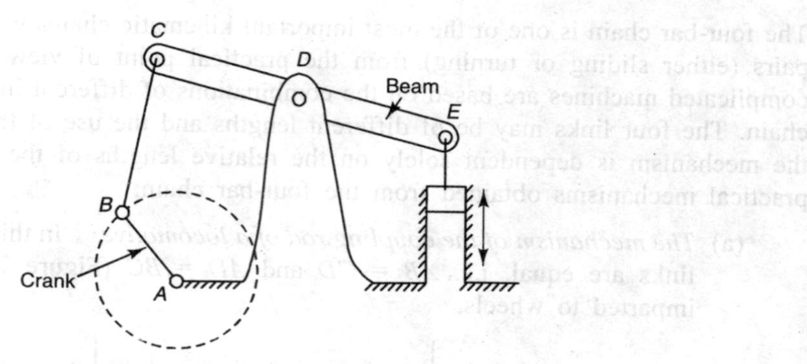

FIGURE 7.20 Crank and lever mechanism.

The Slider-Crank Chain

The slider-crank chain consists of one sliding pair and three revolute pairs. The possible inversions of the slider-crank chain are as follows:

(a) *Reciprocating-engine mechanism:* In the chain, link 1 is fixed, link 2 revolves and link 4 slides. The reciprocating motion of link 4 is converted into revolving motion of link 1. See Figure 7.21.

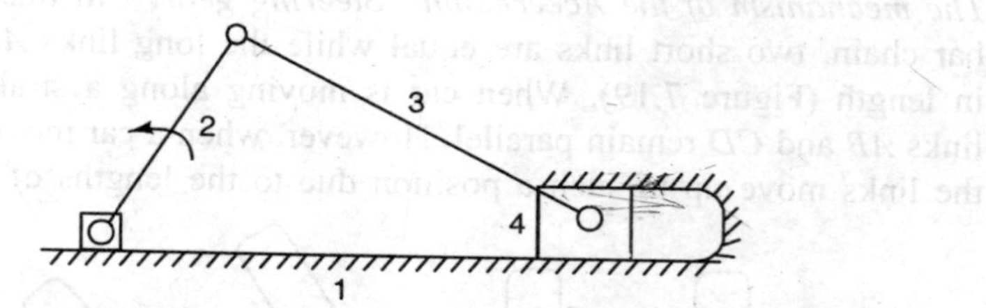

FIGURE 7.21 Reciprocating-engine mechanism.

(b) *Quick-return mechanism:* In this inversion of a slider-crank chain, link 2 is fixed instead of link 1 (Figure 7.22). When link 3 revolves about an axis about point B, the piston (link 4) moves along the slot in link 1, thereby slotted link AC (link 1) oscillates about point A (pivot). A short link CD transmits this motion of the link AC to the ram of the shaper machine which slides over the guides.

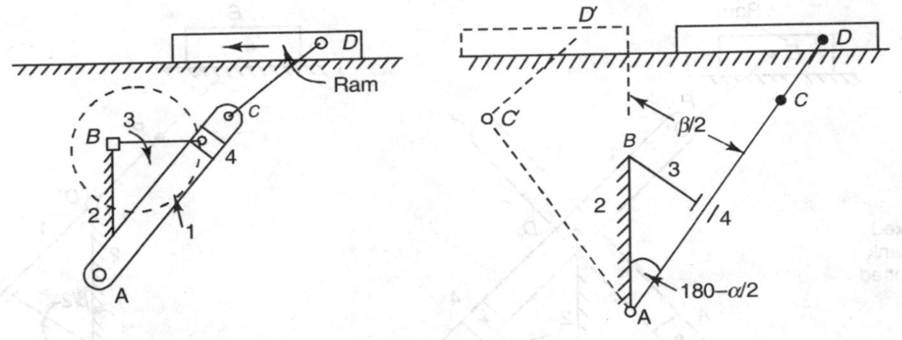

FIGURE 7.22 Quick-return mechanism.

Cutting stroke = motion of AC to AC'

$= \omega \times t_1$

where ω = angular speed and t_1 = cutting time.

return stroke = $\omega \times t_2$ where t_2 = return time

$$t_1 \propto \text{angle } \alpha$$

$$t_2 \propto \text{angle } \beta$$

Therefore, since cutting stroke = return stroke,

$$\frac{t_1}{t_2} = \frac{\alpha}{\beta}$$

Since angle $\alpha >$ angle β, $t_1 > t_2$, i.e. cutting time is more than return time.

$$\text{Length of stroke} = \frac{2 \times \text{length of link 3}}{\text{length of link 2}} \times \text{length of link 1}$$

(c) *Whitworth quick-return mechanism:* In this inversion of a slider crank mechanism which is similar to a quick-return mechanism as link 2 is fixed. The only difference between a quick-return mechanism and a Whitworth quick-return mechanism lies in the different length proportions adopted for driving the crank (link 3) and the fixed link (link 2). The slotted link oscillates about point D as shown in Figure 7.23. The connecting rod PE is used to move the ram of the shaper machine on the guides. The ram thus reciprocates on the horizontal guides. The cutting time in the motion is achieved more than idle or return stroke as explained below:

$$\frac{\text{Cutting time}}{\text{Idle/return time}} = \frac{\alpha}{\beta} \text{ and } \alpha > \beta$$

Also length of stroke = $2 \times PD$

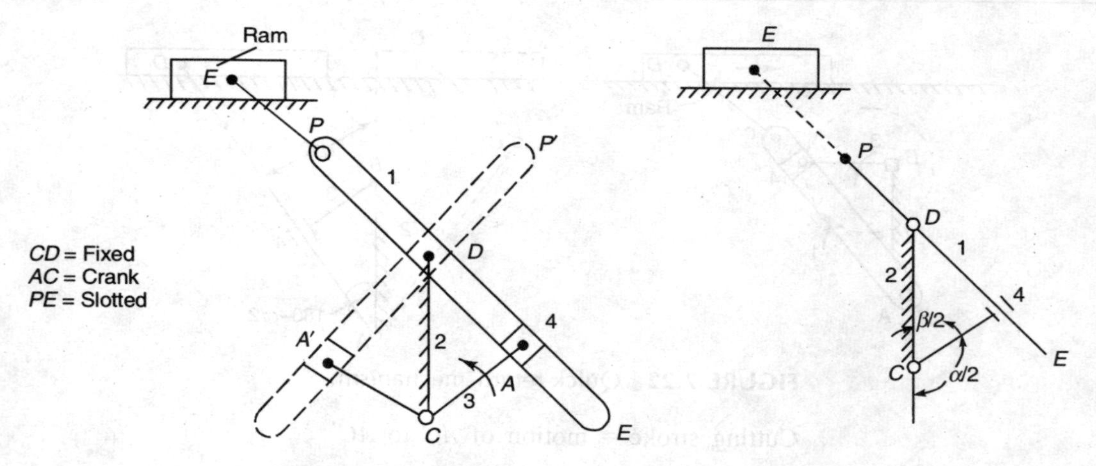

FIGURE 7.23 Whitworth quick-return mechanism.

(d) *Oscillating-cylinder engine mechanism:* In this inversion of a slider-crank chain, link 3 is fixed as shown in Figure 7.24. When link 2 rotates, link 4 slides in slotted link 1 (cylinder) up and down, making slotted link 1 (cylinder) to oscillate about point *A*. This is the reason why this mechanism is called oscillating cylinder engine.

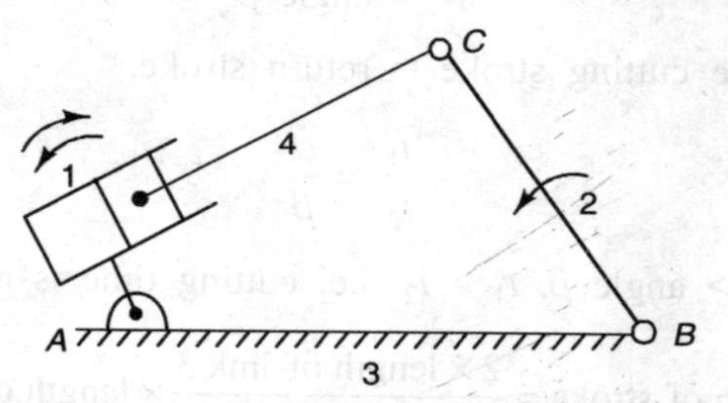

FIGURE 7.24 Oscillating-cylinder engine mechanism.

(e) *Hand pump mechanism:* In this inversion of a slider-crank chain, the cylinder (link 4) is fixed while link 1 (piston) slides in link 4 as shown in Figure 7.25. Link 2 is rotated to move the piston up and down.

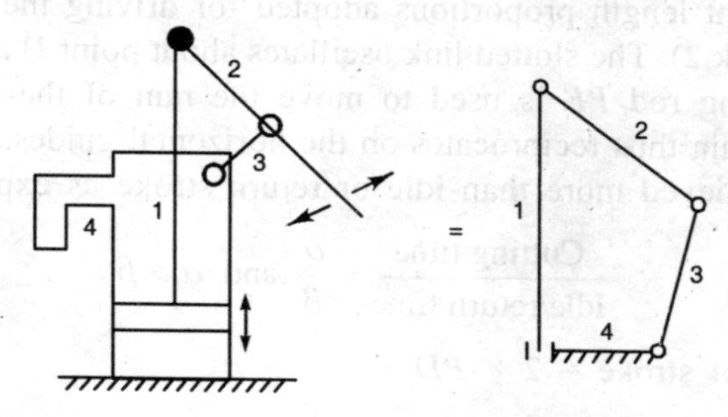

FIGURE 7.25 Hand pump mechanism.

The Double Slider-Crank Chain

The double slider-crank kinematic chain consists of two turning and two sliding pairs. Two slider blocks slide along slots in a frame, and pins P and Q on the slider blocks are connected by the link PQ. Each of the slider blocks constitutes a sliding pair with the frame and a furning pair with the link PQ. Three inversion are possible which are obtained by fixing different links as explained below:

(a) *Elliptical trammel:* In this mechanism (Figure 7.26), the slotted frame (link 4) is fixed. Any point S on the link PQ (link 2) will trace out an ellipse. This can be verified as under:

$$x_s = SQ \cos \theta, \qquad \text{or} \qquad \cos \theta = \frac{x_s}{SQ}$$

$$y_s = PS \sin \theta, \qquad \text{or} \qquad \sin \theta = \frac{y_s}{PS}$$

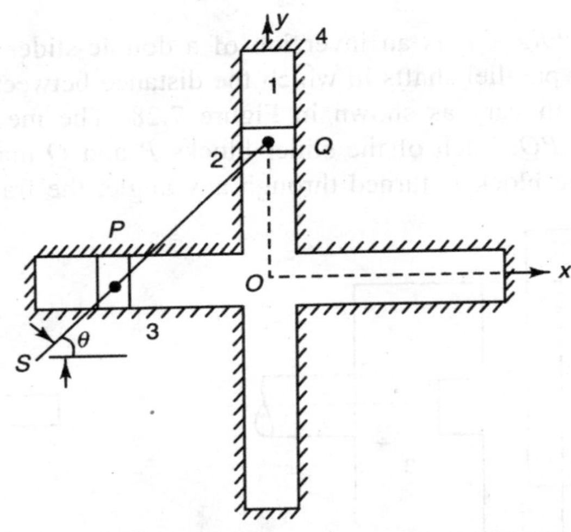

FIGURE 7.26 Elliptical trammel.

On squaring and adding we get

$$\cos^2\theta + \sin^2\theta = \left(\frac{x_s}{SQ}\right)^2 + \left(\frac{y_s}{PS}\right)^2 = 1$$

The above equation is for an ellipse. Clearly, length QS and length PS are semi-major and semi-minor axes of the ellipse.

(b) *Scotch yoke:* It is the inversion of a double-slider crank chain. It is used for converting rotary motion into reciprocating motion. In this inversion, the slider block P is fixed so that link PQ (link 2) can rotate about P as centre and thus causes the frame to reciprocate (Figure 7.27). The fixed block P guides the frame to reciprocate.

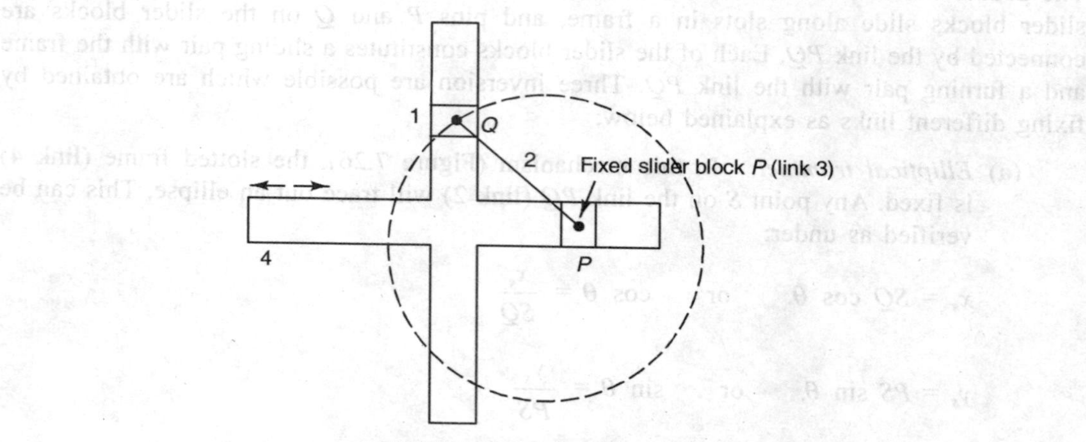

FIGURE 7.27 Scotch yoke.

(c) *Oldham's coupling:* It is an inversion of a double-slider crank chain. It is used to connect two parallel shafts in which the distance between two axes is small and this distance can vary as shown in Figure 7.28. The mechanism is obtained by fixing the link PQ. Each of the slider blocks P and Q may rotate about the pins P and Q. If one block is turned through any angle, the frame and the other block

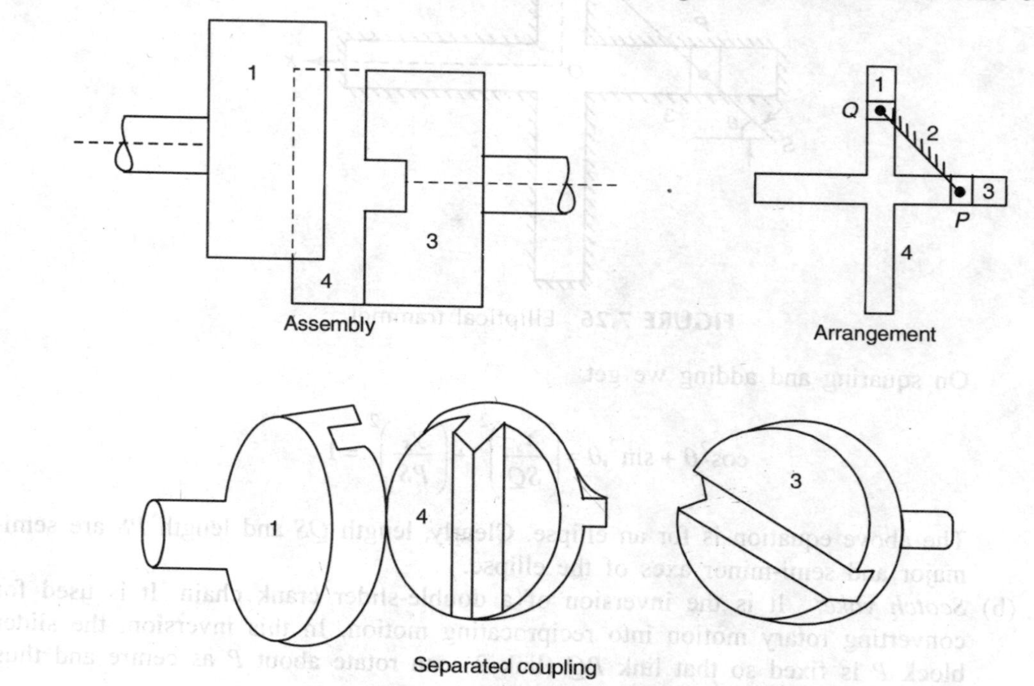

FIGURE 7.28 Oldham's coupling.

must turn through the same angle. As rotation takes place, the frame has to slide relative to each of the two blocks. In Oldham's coupling, the shafts have flanges at the end in which slots or grooves are cut diameterically. An intermediate circular piece has tongues at right angles on its opposite side which can fit between the slots on the two flanges. The tongues can also slide in the slots of the flanges. The intermediate piece or circular disc forms link 4 and it can slide or reciprocate in link 1 and link 3 (flanges of the shafts). So long as the shafts remain parallel to each other, their distance may vary while the shafts are in motion without affecting the transmission of uniform motion from one shaft to the other. If the shafts are at a constant distance apart, the centre of the disc has to describe a circular path with this distance as the diameter. The maximum speed of sliding of each tongue of the disc along the grooves on the flanges is equal to the peripheral velocity of the centre of the disc along its circular path. It can be expressed in terms of the distance apart of the two shafts and the angular velocity of the rotation of the shafts.

Example 7.4 The distance between two parallel shafts connected by Oldham's coupling is 20 mm. The driving shaft is rotating at 100 rpm. Find the maximum sliding speed of the tongue of the central disc along the grooves of the flanges.

The maximum sliding speed of each tongue of the disc along the grooves on the flanges is equal to the peripheral velocity of the centre of disc along its circular path

$$\text{Maximum velocity of sliding} = \omega \times d$$

where ω = angular velocity of the disc which is the same as that of driving shaft

$= 100$ rpm

$= \dfrac{2\pi \times 100}{60} = 10.5$ rod/s

$\therefore$ Velocity $(v) = 10.5 \times 0.02 = 0.21$ m/s

Example 7.5 In a quick-return mechanism made out of the slider-crank chain, the length of the crank is 500 mm and the time ratio of cutting to the idle stroke is 3. Find (a) distance between the fixed centres and (b) the length of the slotted link. See Figure 7.29.

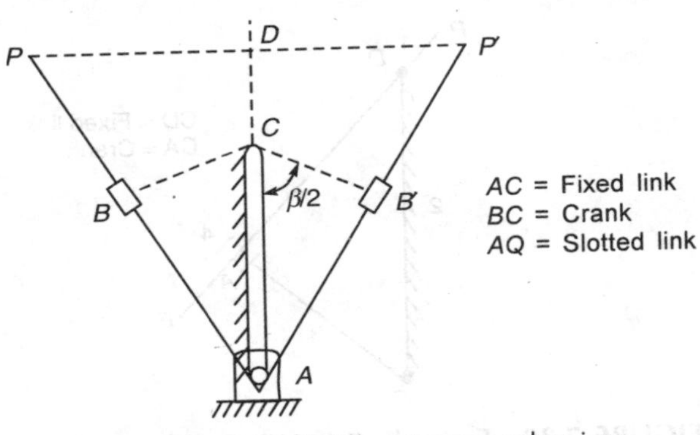

FIGURE 7.29 Example 7.5: Quick-return mechanism.

$$\frac{\text{Cutting time}}{\text{Idle time}} = \frac{360 - \beta}{\beta} = 3$$

or
$$4\beta = 360$$
or
$$\beta = 90°$$
Now in ΔABC, we have

$$\cos\left(\frac{\beta}{2}\right) = \frac{BC}{AC}$$

where $\frac{\beta}{2} = 45°$.

or
$$AC = \frac{BC}{\cos\left(\frac{\beta}{2}\right)} = \sqrt{2}\,BC = \sqrt{2} \times 500 = 707.11 \text{ mm}$$

Now the length of the stroke is $= 2PD = 200$

$\therefore$
$$PD = 100 \text{ mm}$$

Now in the triangle APD, we have:

$$\sin\left(90 - \frac{\beta}{2}\right) = \frac{PD}{AP}$$

$\therefore$
$$AP = \frac{PD}{\text{sm } 45} = \sqrt{2} \times 100$$
$$= 141.42 \text{ mm}$$

Example 7.6 In a Whitworth quick-return mechanism, the distance between the fixed centres is 60 mm and the length of the driving crank is 80 mm. The length of slotted link is 160 mm and connecting rod is 140 mm as shown in Figure 7.30. Find the ratio of the cutting to the idle time.

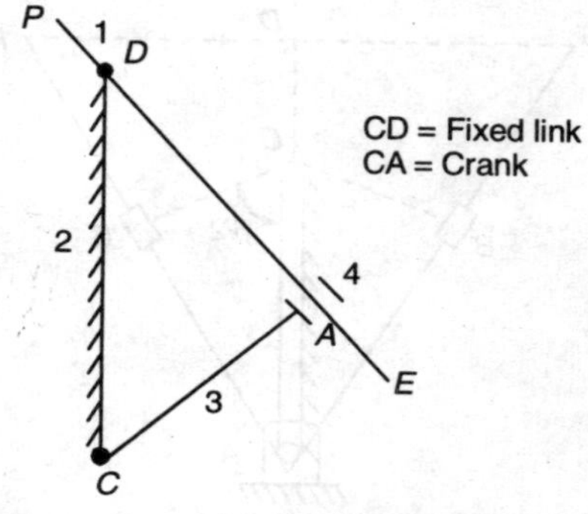

FIGURE 7.30 Example 7.6: Whitworth quick-return mechanism.

Given: $CD = 60$ mm, $CA = 80$ mm

$$\cos\frac{\beta}{2} = \frac{CD}{CA} = \frac{60}{80}$$

$$= 0.75$$

$$\therefore \quad \beta = 82.8°$$

$$\frac{\text{Time for cutting stroke}}{\text{Time for idle stroke}} = \frac{360 - \beta}{\beta}$$

$$= \frac{360 - 82.8}{82.8}$$

$$= 3.35$$

Grashof's Law

A minimum of four kinematic pairs are required so that a kinematic chain can transmit motion according to a definite law. A chain consisting of four links having revolute pairs at the ends forms a four-bar chain or a quadric cycle chain. Various mechanisms can be obtained from a four-bar chain depending upon some relationship involving the lengths of industrial links. Another important consideration while designing a motor driven four-bar mechanism is to ensure that the input link can make a complete revolution about its hinged point.

Grashof's law states that in a planer four-bar kinematic chain (Figure 7.31), the sum of the shortest and the longest length cannot be greater than the sum of the remaining two link lengths, if there has to be a continuous relative motion between two members.

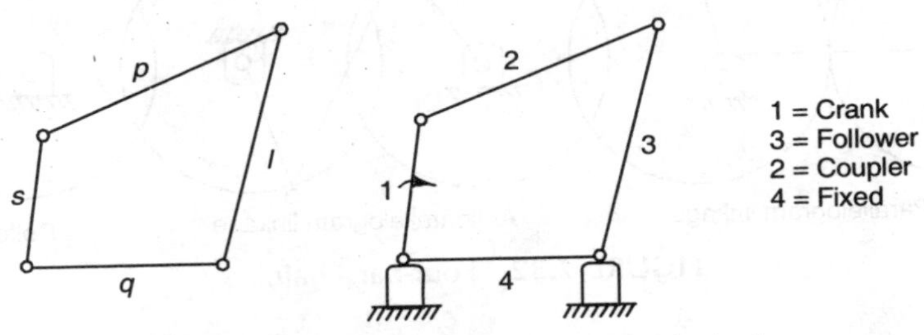

FIGURE 7.31 Four-bar linkage: Grashof's law.

If the longest link is l, the shortest link is s and the remaining two links have lengths of p and q, then as per Grashof' law:

$$l + s \leq p + q$$

The above relation ensures that a four-bar linkage having the simplest possible pin-jointed mechanism would have a single degree freedom controlled motion. In a four-bar linkage, the link having no connection to the frame is called the coupler while the two links hinged to the frame are called the crank and follower. Three different mechanisms possible depending upon lengths of links which are (i) the double-crank or drag-link mechanism in which both the crank and follower can make complete notation, (ii) the crank-rocker mechanism in which the complete rotation of one link (crank) causes an oscillation of the follower (rocker), and (iii) the double-rocker mechanism in which both the driver and driven links only oscillate, i.e. none of the driver and follower make a complete rotation. When $l + s < p + q$, the linkage is called Grashof's linkage. Grashof's linkage gives three mechanism as inversions which are: (i) a double-crank mechanism when the shortest link s is a frame, (ii) two different crank-rocker mechanisms when the shortest link s is the crank and any one of the adjacent links is the frame and (iii) one double-rocker mechanism when the shortest link s is the coupler.

If $l + s > p + q$, then four triple rocker mechanisms are possible depending upon the link selected to be fixed. Similarly, if $l + s = p + q$, the four inversions are obtained similar to those which are obtained when $l + s < p + q$, but these have difficulties of dead centres. To overcome this, the links must be guided in power direction using the inertia of the links to cross dead centres. The situation where $l + s = p + q$ and the linkage having two pairs of equal lengths, gives (i) the parallelogram and antiparallelogram linkage in which equal links are not kept adjacent, and (ii) the deltoid linkage in which the equal links are kept adjacent (Figure 7.32). The parallelogram linkage is quite useful as it can exactly duplicate the rotary motion of the driver crank by the driven crank. One common use of this mechanism is to couple the output of the two wipers to cover the width of the windshield of an automobile.

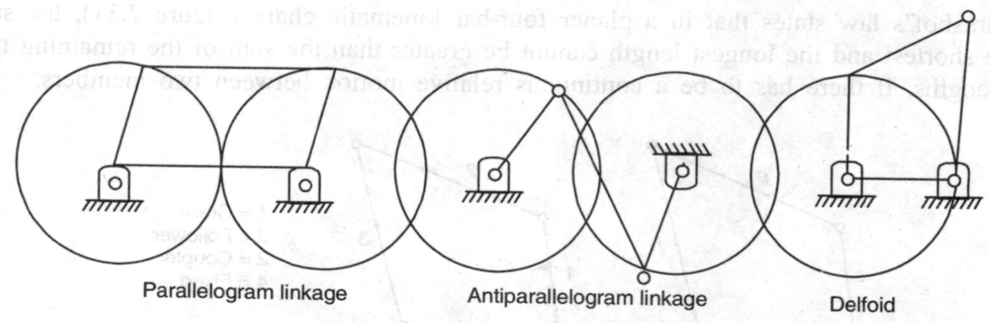

FIGURE 7.32 Four-bar chain.

Pantograph

A pantograph is a mechanism which is a kinematic linkage comprising lower pairs and it is used to enlarge or to reduce the input movements. Therefore, pantographs are mechanisms to reproduce drawings to a different scale. They are also used in guiding tools to cut the part as per the sample. It is infact a four-bar linkage, and four links AB, BC, CD and DA form a parallelogram in which link AB and link CD are equal and parallel. Similarly, link BC and AD are equal and parallel. Link CD is extended to point P and link CB is extended to O such that the points O and P lie in a straight line through a point at the turning pair A

as shown in Figure 7.33. Thus OAP is a straight line. Make point O as a pivot point. This arrangement works now as a pantograph mechanism with point P tracing the same path as described by point A. To verify, consider the triangles OAB and OPC which are similar as $\angle AOB = \angle BOC$ (included angle), $\angle OBA = \angle OCB$ (as BA is parallel to CP) and $\angle OBA = \angle OPC$. Hence, we have

$$\frac{OB}{OC} = \frac{OA}{OP} = \frac{BA}{CP}$$

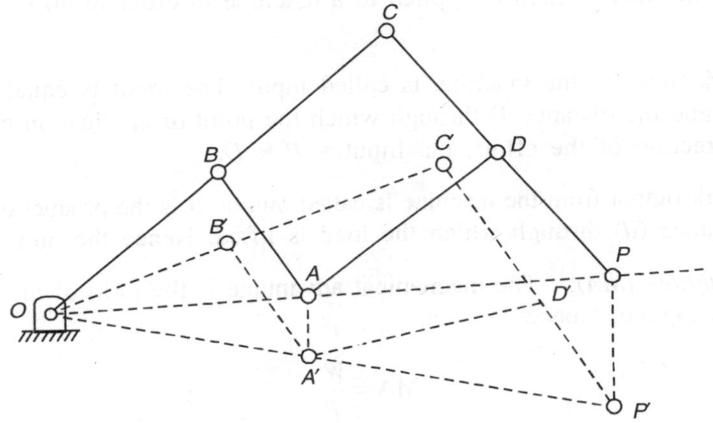

FIGURE 7.33 Pantograph.

Now the pantograph is moved from point A to A' so that P moves to P'. It can be seen that triangles $A'OB'$ and $P'OC'$ are again similar. Hence, we have

$$\frac{OB'}{OC'} = \frac{OA'}{OP'} = \frac{B'A'}{C'P'}$$

or

$$\frac{OA}{OP} = \frac{OA'}{OP'}$$

Thus the ratio of length $OA{:}OP$ remains constant as links are moved. As this is true for all positions, the point P traces out the same path as point A.

CLASSIFICATION OF MACHINES

The machines can be classified as follows:

(a) *Simple machine:* A simple machine is a machine in which there is only one point for the application of effort and one point for the load to be lifted. Levers, screw jacks, bicycles and inclined planes are examples of a simple machine.

(b) *Compound machine:* A compound machine is a machine which has more than one point for the application of the effort as well as the load to be lifted. Lathe machines, grinding machines, shapers, slotter and milling machines are examples of a compound machine.

Terms Used with Lifting Machines

Machine. A machine is defined as a device which is capable of doing some useful work or lifting load or overcoming resistance at some desirable point by the application of an effort at any desired point.

Load or Resistance (W). It is the amount of load to be lifted by the machine or the amount of the frictional resistance which has to be overcome.

Effort (P). It is the force which is applied to a machine in order to do some some useful work.

Input. The work input to the machine is called input. The input is equal to the product of the effort (P) and the distance D through which the point of application of the effort has moved in the direction of the effort, i.e. Input = $P \times D$.

Output. The work output from the machine is called output. It is the product of the load lifted (W) and the distance (d) through which the load is lifted. Hence the output is = $W \times d$.

Mechanical advantage (MA). The mechanical advantage is the ratio of the load (W) lifted to the effort (P) applied. Hence

$$\text{MA} = \frac{W}{P}$$

Velocity ratio (VR). The velocity ratio is the ratio of the distance (D) moved by the effort (P) to the distance (d) moved by the lifted load (W) or the resistance overcome. Hence

$$\text{VR} = \text{Velocity ratio} = \frac{D}{d}$$

Efficiency (η). The efficiency is the ratio of the mechanical advantage (MA) to the velocity ratio. Hence

$$\eta = \frac{\text{MA}}{\text{VR}} = \frac{W/P}{D/d} = \frac{W \times d}{P \times D} = \frac{\text{Output}}{\text{Input}}$$

Hence, the efficiency can also be defined as the ratio of the output of the machine to the input of the machine.

Lifting machine. A machine which is mainly used for lifting of load only is called a lifting machine.

Ideal machine. An ideal machine is a machine which has 100% efficiency. No machine can be ideal but efforts are made to achieve efficiency as close to an ideal machine as possible. For an ideal machine

$$\text{Mechanical advantage} = \text{Velocity ratio}$$

Reversible machine. A reversible machine is a machine which is capable to perform some work in the reverse direction when effort is removed. In other words, the removal of the effort while lifting results in lowering of the load in such machines. The efficiency of such a machine has to be more than 50%.

Irreversible or self locking machine. When the effort is removed, a machine which is incapable to perform any work in the reverse direction is called an irreversible or self-locking machine. In other words, the removal of the effort while lifting does not result into the lowering of the load in such machines. The efficiency of such a machine has to be less than 50%.

Law of a Machine

The law of a machine is given by the relationship between the effort applied and the load lifted by the machine. If the readings of applied effort (P) and load (W) lifted are noted, a graph is obtained as shown in Figure 7.34. The effort for zero load for an ideal machine is zero. However, an actual machine requires some effort (C) even at zero load. If 'θ' is the slope of the line, then $\tan\theta = m$ and the law machine can be written as

$$P = mW + C$$

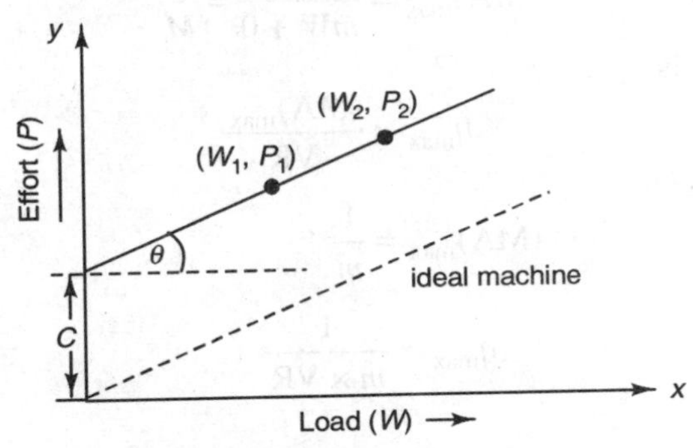

FIGURE 7.34 Law of machine.

The above is the equation of a straight line and the slope 'm' can be given as:

$$m = \frac{\Delta P}{\Delta W} = \frac{P_2 - P_1}{W_2 - W_1}$$

Maximum Mechanical Advantage

The relationship between applied effort (P) and lifted load (W) is given by the law of a machine as given below:

$$P = mW + C$$

But the mechanical advantage is

$$\text{MA} = \frac{W}{P}$$

$$= \frac{W}{mW + C}$$

The maximum MA can be obtained by differentiating and equating to zero, i.e.

$$\frac{d(\text{MA})}{dW} = 0$$

$\therefore$

$$\frac{d}{dW}\left(\frac{W}{mW + C}\right) = 0$$

or

$$\frac{(mW + C) \times 1 - W \times m}{(mW + C)^2} = 0$$

or

$$C = 0$$

$\therefore$

$$(\text{MA})_{\max} = \frac{W}{mW + 0} = \frac{1}{M}$$

The maximum efficiency is

$$\eta_{\max} = \frac{(\text{MA})_{\max}}{\text{VR}}$$

But

$$(\text{MA})_{\max} = \frac{1}{m}$$

$\therefore$

$$\eta_{\max} = \frac{1}{m \times \text{VR}}$$

Variation of Mechanical Advantage

The mechanical advantage is given by

$$\text{MA} = \frac{W}{P}$$

$$= \frac{W}{mW + C}$$

$$= \frac{1}{m + \frac{C}{W}}$$

As the load increases, C/W decreases and the mechanical advantage increases. In the limiting case, when W tends to infinity, then $C/W = 0$ which gives the mechanical advantage equal to $1/m$. The variation of the mechanical advantage with respect to load is as shown in Figure 7.35.

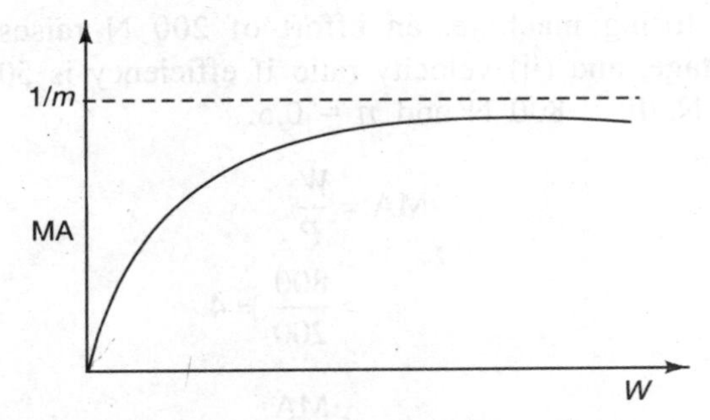

FIGURE 7.35 Variation of mechanical advantage.

Variation of Efficiency

The effeciency of the machine is given by

$$h = \frac{\text{MA}}{\text{VR}}$$

$$= \frac{W/P}{\text{VR}}$$

$$= \frac{\dfrac{W}{mW + C}}{\text{VR}}$$

$$= \frac{1}{\text{VR}}\left(\frac{1}{m + C/W}\right)$$

The efficiency increases as load (W) increase. When load (W) approaches infinity, the value of C/W becomes equal to zero (Figure 7.36). Hence, the maximum efficiency approaches to $\frac{1}{\text{VR}} \times \frac{1}{m}$.

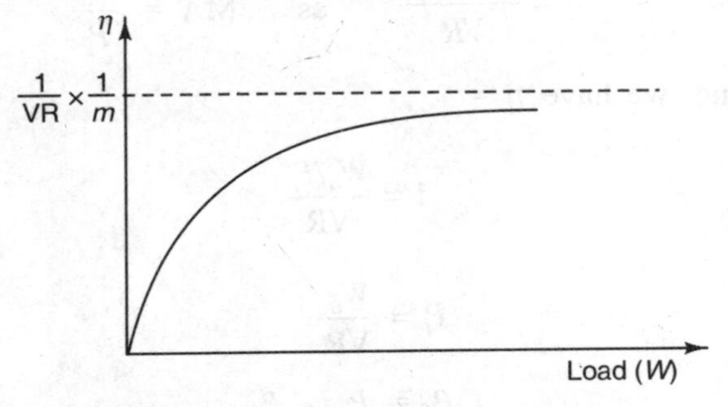

FIGURE 7.36 Variation of efficiency.

Example 7.7 In a lifting machine, an effort of 200 N raises a load of 800 N. Find (i) mechanical advantage, and (ii) velocity ratio if efficiency is 50%.

Given: P = 200 N, W = 800 N and η = 0.5.

$$\text{MA} = \frac{W}{P}$$

$$= \frac{800}{200} = 4$$

Now $$\eta = \frac{\text{MA}}{\text{VR}}$$

$\therefore$ $$\text{VR} = \frac{\text{MA}}{\eta}$$

$$= \frac{4}{0.5} = 8$$

Loss of Effort in Friction

In actual machines, there is a loss of effort due to friction. More effort is required in actual machines to lift a load. Similarly, lesser load is lifted by an effort in an actual machine as compared to an ideal machine. Consider the following:

P_a = actual effort applied
P_i = ideal effort required
W_a = actual load lifted
W_i = ideal load lifted

Now we can find out:

Effort lost due to friction $(P_f) = P_a - P_i$
Loss in lifting load due to friction = $(W_f) = W_i - W_a$

$$\eta = \frac{(W_a/P_i)}{VR} \quad \text{as} \quad \text{MA} = \frac{W_a}{P_i}$$

For an ideal machine, we have η = 1.

$$1 = \frac{W_a/P_i}{\text{VR}}$$

$$P_i = \frac{W_a}{\text{VR}}$$

Now $$P_f = P_a - P_i$$

$$= P_a - \frac{W_a}{\text{VR}}$$

$$= \frac{P_a \times \text{VR} - W_a}{\text{VR}}$$

or $$P_f \times \text{VR} = P_a \times \text{VR} - W_a \quad \text{(i)}$$

Similarly $$\eta = \frac{W_i/P_a}{\text{VR}}$$

If $\eta = 1$, then $$\frac{W_i}{P_a} = \text{VR}$$

or $$W_i = P_a \times \text{VR}$$

Now $$W_f = W_i - W_a$$

$$= P_a \times \text{VR} - W_a \quad \text{(ii)}$$

From Eqs. (i) and (ii), we get

$$W_f = P_f \times \text{VR}$$

Now $$\eta = \frac{P_i}{P_a} = \frac{W_a}{W_i}$$

Example 7.8 In a lifting machine, an effort of 400 N is to be moved by a distance of 20 m to raise a load of 8000 N by a distance of 0.8 m. Find (i) mechanical advantage, (ii) velocity ratio and (iii) efficiency. Also find (iv) ideal effort, (v) effort lost in friction, (vi) ideal load and (vii) frictional resistance.

$$\text{MA} = \frac{W}{P}$$

$$= \frac{8000}{400} = 20$$

$$\text{VR} = \frac{D}{d}$$

$$= \frac{20}{0.8} = 25$$

$$\eta = \frac{\text{MA}}{\text{VR}}$$

$$= \frac{20}{25} = 0.8$$

$$\text{Ideal effort } (P_i) = \frac{W}{\text{VR}}$$

$$= \frac{8000}{25} = 320 \text{ N}$$

$$\text{Effort lost in friction} = P - P_i$$

$$= 400 - 320$$

$$= 80 \text{ N}$$

$$\text{Ideal load } (W_i) = P \times \text{VR}$$

$$= 400 \times 25$$

$$= 10{,}000 \text{ N}$$

$$\text{Frictional resistance} = W_i - W$$

$$= 10{,}000 - 8000$$

$$= 2000 \text{ N}$$

Example 7.9 In a lifting machine, the following are the readings:

Effort (P)	*Load (W)*
1000 N	4000 N
1250 N	7500 N

Find (i) the maximum mechanical advantage, and (ii) maximum efficiency if velocity ratio is 5.

$$m = \frac{P_2 - P_1}{W_2 - W_1}$$

$$= \frac{1250 - 1000}{7500 - 4000}$$

$$= \frac{250}{3500} = \frac{1}{14}$$

$$(\text{MA})_{\text{max}} = \frac{1}{m}$$

$$= \frac{1}{1/14} = 14$$

$$\eta_{max} = \frac{1}{m \times VR} \times 100\%$$

$$= \frac{1}{14 \times 5} \times 100\%$$

$$= \frac{1}{70} \times 100\% = 1.43\%$$

Example 7.10 A lifting machine has the following readings:

Effort (P)	*Load (W)*
300 N	9000 N
500 N	14,000 N

Find the law of the machine.

The law of machine is

$$P = mW + C$$

$$m = \frac{P_2 - P_1}{W_2 - W_1}$$

$$= \frac{500 - 300}{14000 - 9000}$$

But

$$= \frac{200}{5000} = 0.02$$

Puffing the value of m, P and w in the law of machine, we get

$$300 = 0.02 \times 9000 + C$$

$$300 = 180 + C$$

$\therefore$ $$C = 120$$

Hence, the law of a machine is

$$P = 0.02\,W + 20$$

Example 7.11 In a simple machine, an effort of 200 N is just sufficient to lift a load of 1200 N. The velocity ratio is 20. Find (i) the efficiency, (ii) the loss of effort due to friction, and (iii) the loss of load due to friction.

$$\eta = \frac{\text{MA}}{\text{VR}} = \frac{W}{P \times \text{VR}}$$

$$= \frac{1200}{200 \times 20} = 0.3$$

Loss of effort due to friction is:

$$P_f = P - \frac{W}{\text{VR}}$$

$$= 200 - \frac{1200}{20}$$

$$= 200 - 60 = 140 \text{ N}$$

Loss of load due to friction is

$$W_f = P \times \text{VR} - W$$

$$= 200 \times 20 - 1200$$

$$= 4000 - 1200$$

$$= 2800 \text{ N}$$

Reversibility of a Machine

In case the removal of effort while lifting of a load results into the lowering of the load, the machine is called a reversible machine. On the other hand, if the removal of the effort does not result into the lowering of the load, the machine is said to be a self-locking or irreversible lifting machine. A lifting jack is an irreversible or self-locking machine as lifting jack keeps on holding the vehicle even when effort is removed. However, while lifting water from a well, the pail of water falls back into the well if the effort is removed. Hence, lifting water with the rope constitutes a reversible lifting machine.

A simple lifting machine can be reversible or irreversible depending upon its efficiency. It is seen that a lifting machine is reversible if its efficiency is greater than 50%. If efficiency is less than 50%, then the machine is self-locking type. It can be proved as given below:

$$\text{Input} = \text{Effort} \times \text{Distance}$$

$$= P \times D$$

$$\text{Output} = \text{Load} \times \text{Distance}$$

$$= W \times d$$

$$\text{Work lost in friction} = \text{Input} - \text{Output}$$

$$= P \times D - W \times d$$

When effort is removed, the load has to overcome the frictional resistance in order to start moving down. Hence the condition for reversibility is

$$W \times d > \text{frictional resistance}$$

or

$$W \times d > p \times d - W \times D$$

or

$$2W \times d > P \times D$$

or
$$\frac{W \times d}{P \times D} > \frac{1}{2}$$

or
$$\left(\frac{W}{P}\right) \times \left(\frac{d}{D}\right) > \frac{1}{2}$$

or
$$\text{MA} \times \frac{1}{\text{VR}} > \frac{1}{2}$$

or
$$\eta > \frac{1}{2}$$

or
$$\eta > 50\%$$

Therefore, a lifting machine is reversible in case its efficiency is greater than 50%. In case efficiency is less than 50%, then the lifting machine is self-locking.

Example 7.12 A lifting machine has velocity ratio as 30 and it is lifting a load of 5000 N with an effort of 400 N. Find whether the machine is self locking. Also determine its frictional resistance.

Given: VR = 30, W = 5000 N, P = 400 N

Hence
$$\text{MA} = \frac{W}{P} = \frac{5000}{400} = 12.5$$

Now
$$\eta = \frac{\text{MA}}{\text{VR}} = \frac{12.5}{30} = 0.4167 = 41.67\%$$

Since η < 50%, the machine is self locking.

$$\text{Ideal load } W_i = P \times \text{VR}$$
$$= 400 \times 30 = 12{,}000 \text{ N}$$

$\therefore$
$$\text{Frictional resistance} = W_i - W$$
$$= 12000 - 5000$$
$$= 7000 \text{ N}$$

Single Pulley

A pulley can be used as a single unit for lifting load (Figure 7.37). Consider a single pulley system in which load is at one end and effert is applied at other end. Let P be the tension in the rope. As there is equilibrium existing, we will have

$$T = W$$

and
$$T = P$$

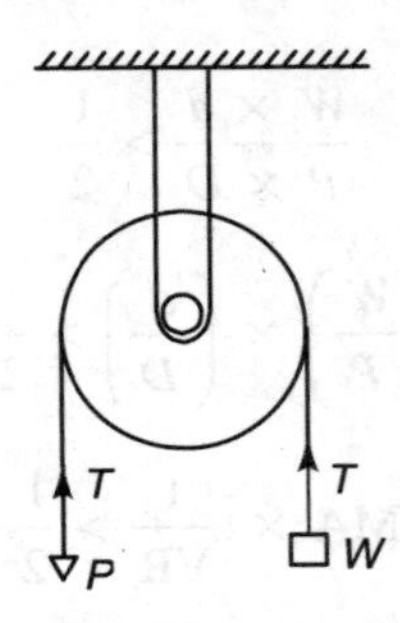

FIGURE 7.37 Single pulley.

Hence
$$W = P$$

$$\therefore \quad \text{MA} = \frac{W}{P} = 1$$

Since distance d moved by the load (W) is the same as the distance D moved by the effort, hence $D = d$ which gives:

$$\text{VR} = 1$$

$$\therefore \quad \eta = \frac{\text{MA}}{\text{VR}} = 1 \text{ or } 100\%$$

System of Pulleys

In order to have a higher mechanical advantage, a system consisting of several pulleys is often used. Whenever two or more pulleys are combined together to obtain a high mechanical advantage, it is called a system of pulleys. There are three systems of pulleys commonly used which are:

(a) First-order system of pulleys
(b) Second-order system of pulleys
(c) Third-order system of pulleys

The following assumptions are made while finding their mechanical advantage and velocity ratio:

(a) The weight of a pulley is small and negligible.
(b) The friction between a pulley and the rope is negligible. Hence tension in the rope remains constant throughout the length of the rope.
(c) The bearings of the pulleys are frictionless.

First-Order System of Pulleys

In the first-order system of pulleys, there are as many fixed ropes as the number of movable pulleys. Consider a first-order system of pulleys consisting of three fixed ropes and three movable pulleys numbered from 1 to 3 as shown in Figure 7.38. Finally, there is a fixed pulley which is numbered as 4. The tensions in the various parts of the ropes are as shown in the figure. It can be seen that the tension in the rope keeps on reducing to half from the first movable pulley to the last fixed pulley.

$$MA = \frac{W}{P}$$

$$= \frac{W}{W/8} = 8 = 2^3$$

The above is the equation for 3 movable pulley system. In case the first-order pulley system has n movable pulleys, then we have

$$\text{Mechanical advantage (MA)} = 2^n$$

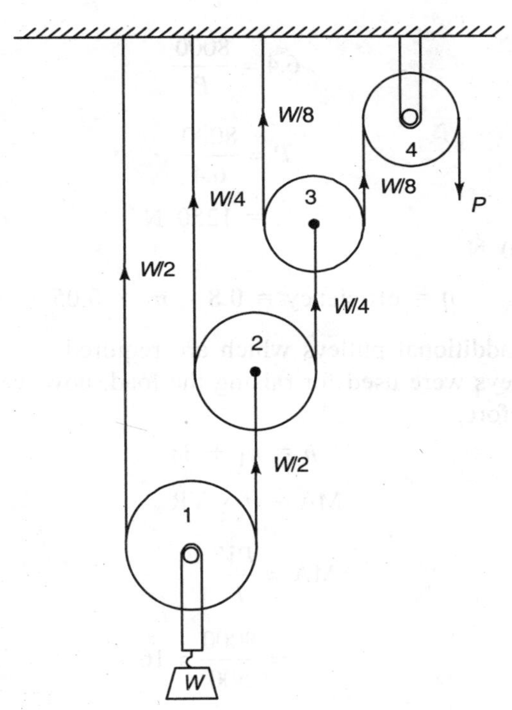

FIGURE 7.38 First-order system of pulleys.

If the pulley system is considered ideal, the efficiency is 100%. Hence, we have

$$MA = VR$$

or

$$VR = 2^n$$

Example 7.13 In a first-order system of pulleys, there are three movable pulleys. What is the effort required to raise a load of 8000 N? The efficiency of the system is 80%. In case the same load is to be lifted using effort of 500 N, find the number of movable pulleys that are necessary. Assume a reduction of efficiency of 5% for each additional pulley used in the system.

Case 1:

$$\text{VR} = 2^n$$

$$= 2^3 = 8$$

But

$$\eta = \frac{\text{MA}}{\text{VR}} = 0.8$$

$\therefore$

$$\text{MA} = 0.8 \times 8$$

$$= 6.4$$

But

$$\text{MA} = \frac{W}{P}$$

$$6.4 = \frac{8000}{P}$$

or

$$P = \frac{8000}{6.4}$$

$$= 1250 \text{ N}$$

Case 2: Now $P = 600$ N

$$\eta = \text{efficiency} = 0.8 - n_1 \times 0.05$$

where n_1 is number of additional pulleys which are required.

As earlier three pulleys were used for raising the load, now we require $(n_1 + 3)$ pulleys to raise the load. Therefore,

$$n = n_1 + 3$$

$\therefore$

$$\text{MA} = n \times \text{VR}$$

But

$$\text{MA} = \frac{PV}{P}$$

$$= \frac{8000}{500} = 16$$

$\therefore$

$$16 = n \times 2^n \quad (\text{as VR} = 2^n)$$

But

$$\eta = 0.8 - n_1 \times 0.05$$

$$= 0.8 - (n - 3) \times 0.05$$

$\therefore$

$$16 = (0.8 - (n - 3) \times 0.05) \times 2^n$$

Now if $n = 4$, RHS $= (0.8 - 0.05) \times 24 = 12 < 16$

$n = 5$, RHS $= (0.8 - 0.1) \times 25 = 22.4 > 16$

Therefore the number of pulleys required = 5.

Second-Order System of Pulleys

The second-order system of pulleys consists of two pulley blocks, one fixed and the other movable. The top pulley block is fixed in position to the top support, whereas the bottom pulley block can move vertically with the load which is attached to it. The number of pulleys in the fixed block may be either equal or one more than the number of pulleys in the movable block. If the number of pulleys in both blocks is equal, then the end of the rope is fixed to the movable block as shown in Figure 7.39. In case, the number of pulleys is one more in the fixed block, then the end of the rope is tied to the fixed block as shown in the figure. The tension in the rope all along the length is P.

If there are n number of pulleys in the fixed and movable block, then considering equilibrium condition when section is taken along A–A in between the fixed and movable blocks for the movable block:

$$2n \times P = W$$

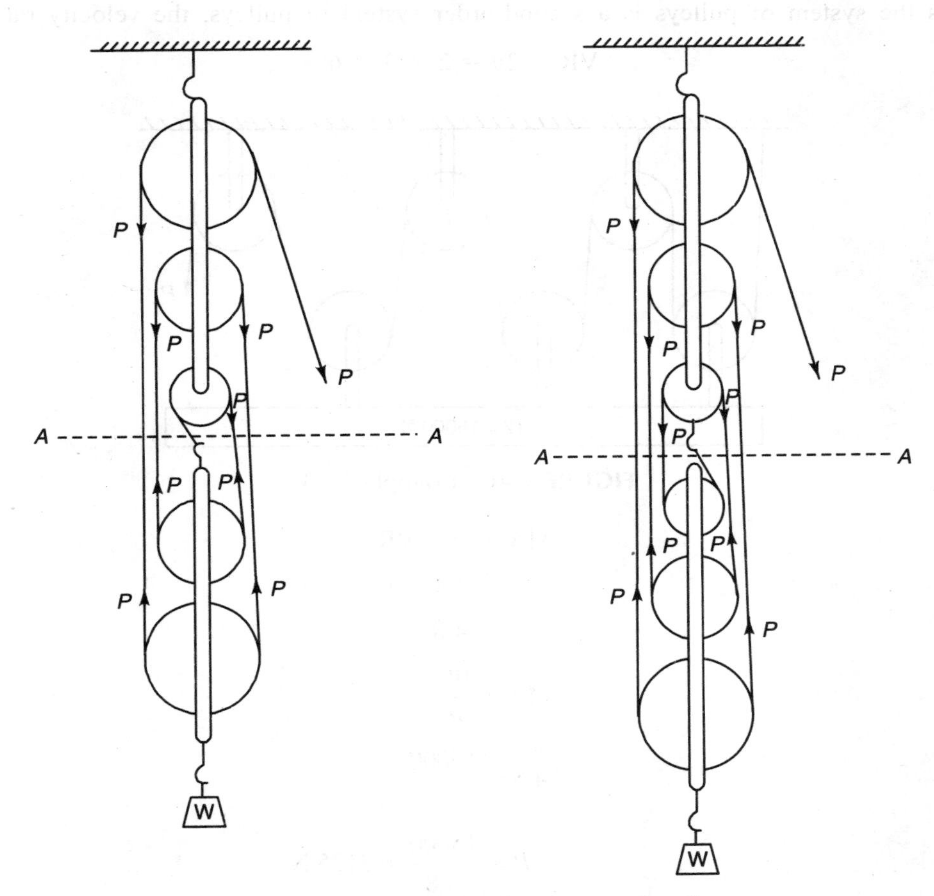

FIGURE 7.39 Second-order system of pulleys.

Now $$\text{MA} = \frac{W}{P}$$

$$= \frac{2n \times P}{P} = 2n$$

For an ideal system, efficiency $\eta = 1$.

$$\text{MA} = \eta \times \text{VR} = 1 \times \text{VR}$$

$\therefore$ $$\text{VR} = 2n$$

Example 7.14 A load of 15000 N is being lifted by an effort P using a second-order system of pulleys as shown in Figure 7.40. Find P assuming the efficiency of the system is 80%.

As the system of pulleys is a second-order system of pulleys, the velocity ratio is

$$\text{VR} = 2n = 2 \times 3 = 6$$

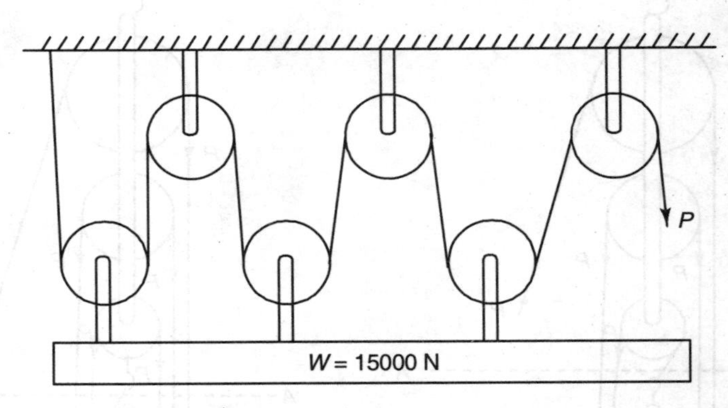

FIGURE 7.40 Example 7.14.

Now $$\text{MA} = \eta \times \text{VR}$$

$$= 0.8 \times 6$$

$$= 4.8$$

But $$\text{MA} = \frac{W}{P}$$

$$4.8 = \frac{15000}{P}$$

$$P = \frac{15000}{4.8} = 3125 \text{ N}$$

Example 7.15 Find the pull required to lift the load (w) as shown in Figure 7.41. Assume the efficiency of the system is 80%.

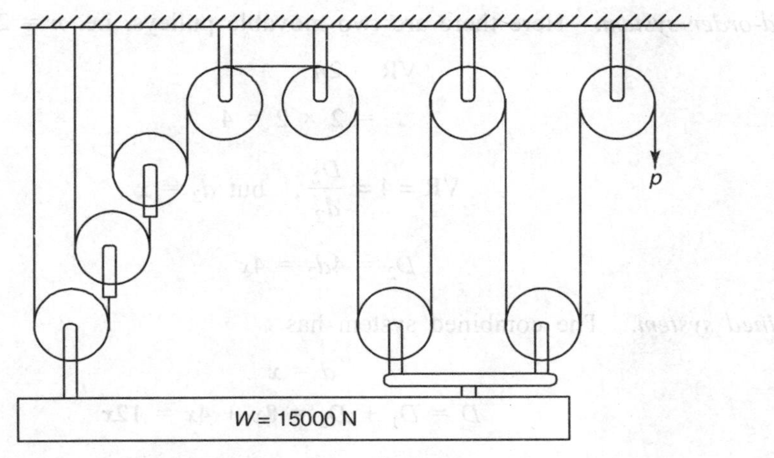

FIGURE 7.41 Example 7.15.

The pulley system as shown in the figure can be considered to be constituted of one first-order system and one second-order system as shown in Figure 7.42.

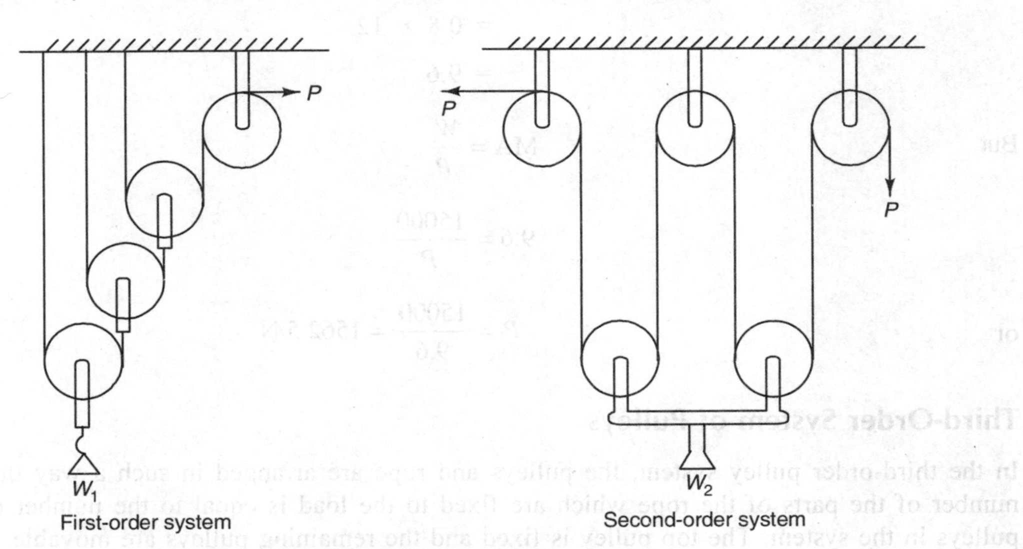

FIGURE 7.42 Example 7.15.

First-order system. As shown in the figure, the system has three movable and one fixed pulley. Hence, velocity ratio is

$$\text{VR} = 2^n = 2^3 = 8$$

Now if load W moves by a distance x, then we have

$$\text{VR} = \frac{D_1}{d_1} = \frac{D_1}{x} = 8$$

or $$D_1 = 8x, \; D_1 = \text{distance moved by effort.}$$

Second-order system. Here there are two movable pulleys, i.e. $n = 2$.

$$VR = 2n$$
$$= 2 \times 2 = 4$$
$$VR = 4 = \frac{D_2}{d_2}, \quad \text{but } d_2 = x$$

or
$$D_2 = 4d_2 = 4x$$

Combined system. The combined system has

$$d = x$$
$$D = D_1 + D_2 = 8x + 4x = 12x$$

$\therefore$
$$VR = \frac{D}{d} = \frac{12x}{x} = 12$$

Now
$$MA = \eta \times VR$$
$$= 0.8 \times 12$$
$$= 9.6$$

But
$$MA = \frac{W}{P}$$
$$9.6 = \frac{15000}{P}$$

or
$$P = \frac{15000}{9.6} = 1562.5 \text{ N}$$

Third-Order System of Pulleys

In the third-order pulley system, the pulleys and rope are arranged in such a way that the number of the parts of the rope which are fixed to the load is equal to the number of the pulleys in the system. The top pulley is fixed and the remaining pulleys are movable. If we take a section A–A as shown in Figure 7.43, then the equilibrium of the system is

$$W = T_1 + T_2 + T_3 + T_4$$

But the equilibrium of each pulley gives

$$T_4 = P$$
$$T_3 = 2T_4 = 2P = 2^1P$$
$$T_2 = 2T_3 = 2 \times 2P = 4P = 2^2P$$
$$T_1 = 2T_2 = 2 \times 4P = 8P = 2^3P$$

In case we have n pulleys instead of four pulleys as in this case, then we have

$$W = P + 2^1P + 2^2P + 2^3P \ldots 2^{n-1}P$$
$$= P(1 + 2^1 + 2^2 + 2^3 \ldots 2^{n-1})$$
$$= P(2^n - 1)$$

$$\therefore \quad \text{MA} = \frac{W}{P}$$
$$= \frac{P(2^n - 1)}{P}$$
$$= 2^n - 1$$

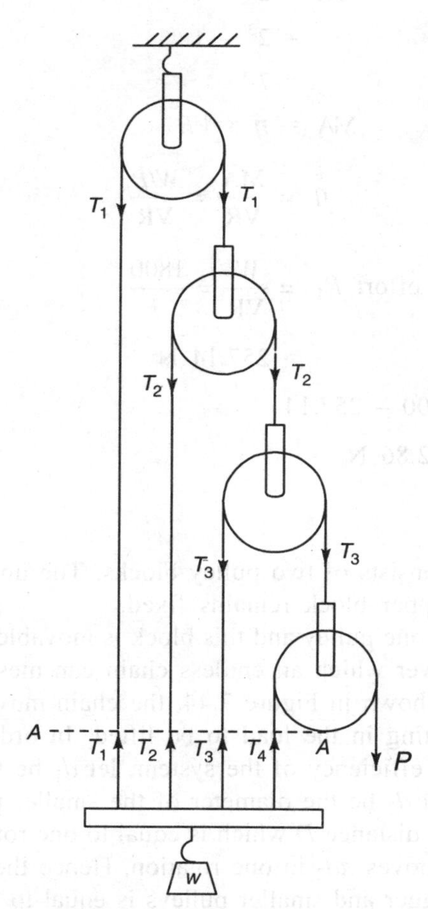

FIGURE 7.43 Third-order system of pulleys.

For an ideal system, $\eta = 100\%$.

$$\eta = \frac{\text{MA}}{\text{VR}} = 1$$

or

$$\text{MA} = \text{VR}$$

$\therefore$

$$\text{VR} = 2^n - 1$$

Example 7.16 A lifting machine consists of a third-order system of pulleys. There are three pulleys in the system. A load of 1800 N is lifted by an effort of 300 N. Find the efficiency of the pulley system and the effort lost in friction.

In the third-order system of pulley, the velocity ratio in term of number of pulleys is

$$\text{VR} = 2^n - 1$$

$$= 2^3 - 1$$

$$= 7$$

Now

$$\text{MA} = \eta \times VR$$

or

$$\eta = \frac{\text{MA}}{\text{VR}} = \frac{W/P}{\text{VR}}$$

$$\text{Ideal effort } P_1 = \frac{W}{\text{VR}} = \frac{1800}{7}$$

$$= 257.14 \text{ N}$$

$$\therefore \text{ Effort lost in friction} = 300 - 257.14$$

$$= 42.86 \text{ N}$$

Differential Pulley Block

A differential pulley block consists of two pulley blocks. The upper block has two pulleys which are concentric. The upper block remains fixed.

The lower block has only one pulley and this block is movable. These pulleys have teeth on their grooved periphery over which an endless chain can mesh and move. When effort P is applied to the chain as shown in Figure 7.44, the chain moves as indicated by arrows numbered from 1 to 4, resulting in the load to be lifted. In order to find the mechanical advantage, velocity ratio and efficiency of the system, let d_1 be the diameter of the bigger pulley in the upper block and d_2 be the diameter of the smaller pulley in the upper block. Consider effort is moved by a distance D which is equal to one rotation of the bigger pulley, i.e. πd_1. The smaller pulley moves πd_2 in one rotation. Hence the change in the movement length of the chain on the bigger and smaller pulleys is equal to $\pi(d_1 - d_2)$. The difference in the movement of the chain results into the lifting of load by distance $d = \dfrac{\pi(d_1 - d_2)}{2}$.

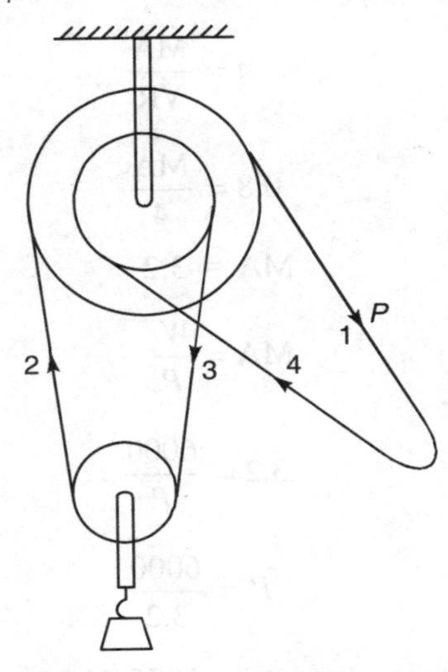

FIGURE 7.44 Differential pulley block.

$$\text{Now, velocity ratio} = \frac{D}{d}$$

$$= \frac{\pi d_1}{\pi/2(d_1 - d_2)}$$

$$= \frac{2d_1}{(d_1 - d_2)}$$

$$\text{Mechanical advantage} = \frac{W}{P}$$

$$\text{Efficiency} = \frac{\text{MA}}{\text{VR}}$$

$$= \frac{W}{P}\left(\frac{d_1 - d_2}{2d_1}\right)$$

Example 7.17 A differential pulley block has bigger and smaller diameters as 60 cm and 30 cm respectively. Find effort required if a load of 6000 N is to be lifted. Assume the efficiency as 80%.

$$\text{VR} = \frac{2d_1}{d_1 - d_2}$$

$$= \frac{2 \times 60}{60 - 30} = 4$$

$$\eta = \frac{\text{MA}}{\text{VR}}$$

$$0.8 = \frac{\text{MA}}{4}$$

$\therefore$ $$\text{MA} = 3.2$$

Now $$\text{MA} = \frac{W}{P}$$

or $$3.2 = \frac{6000}{P}$$

or $$P = \frac{6000}{3.2}$$

$$= 1875 \text{ N}$$

Wheel and Axle

The machine consists of an axle having the smaller diameter (d_a) and a wheel having the bigger diamter (d_w) which are coaxially jointed and mounted on a bearing as shown in Figure 7.45. The axle and wheel rotate as one piece. A rope is wound round the wheel whose one end is fixed to the wheel itself and other end is used for the application of the effort (P). Another rope is wound on the axle in an opposite direction to the wheel such that when the rope on the wheel unwinds, the rope of the axle winds. The other end of the

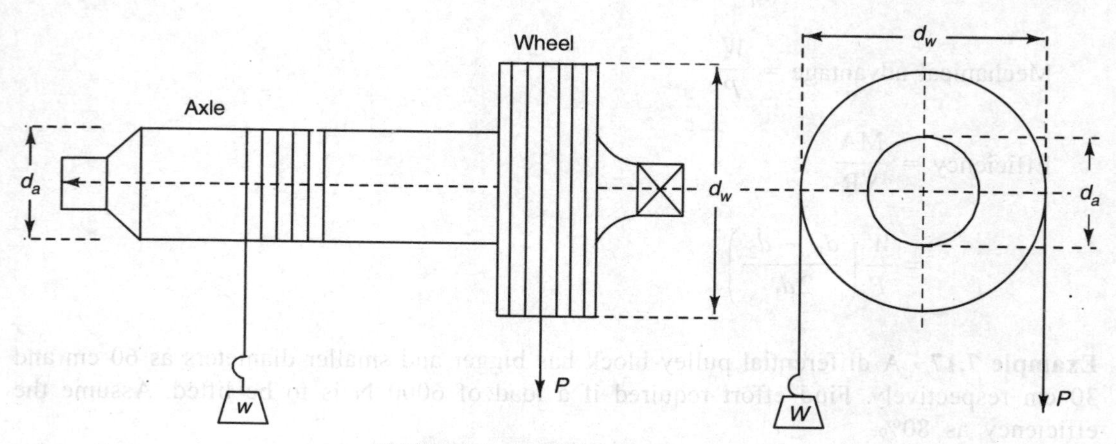

FIGURE 7.45 Wheel and axle.

rope of the axle is attached to the load which is lifted when the rope on the axle winds. Consider now one rotation of the wheel due to applied effort P. The effort P has moved by a distance $D = \pi d_w$ while the load W has moved by a distance of $d = \pi d_a$. The velocity ratio is

$$\text{VR} = \frac{D}{d} = \frac{\pi d_w}{\pi d_a}$$

$$= \frac{d_w}{d_a}$$

$$\text{MA} = \frac{W}{P}$$

$$\text{Efficiency} = \frac{\text{MA}}{\text{VR}} = \frac{W}{P} \times \frac{d_a}{d_w}$$

Example 7.18 A load of 800 N is to be lifted by a wheel and axle machine. The diameter of the wheel is 600 mm while that of the axle is 200 mm. If the efficiency of the machine is 80%, find the effort P required to lift the load.

$$\text{VR} = \frac{d_w}{d_a}$$

$$= \frac{600}{200} = 3$$

$$\eta = \frac{\text{MA}}{\text{VR}}$$

$$0.8 = \frac{\text{MA}}{3}$$

$$\therefore \quad \text{MA} = 2.4$$

$$\text{MA} = \frac{W}{P}$$

$$2.4 = \frac{600}{P}$$

$$\therefore \quad P = \frac{600}{2.4}$$

$$= 250 \text{ N}$$

Wheel and Differential Axle

The wheel and defferential axle machine is an improvement over the wheel and axle machine (Figure 7.46). A rope is wound on the wheel of diameter d_w whose one end is fixed to the wheel itself and to the other end an effort P is applied. Another rope is wound over the

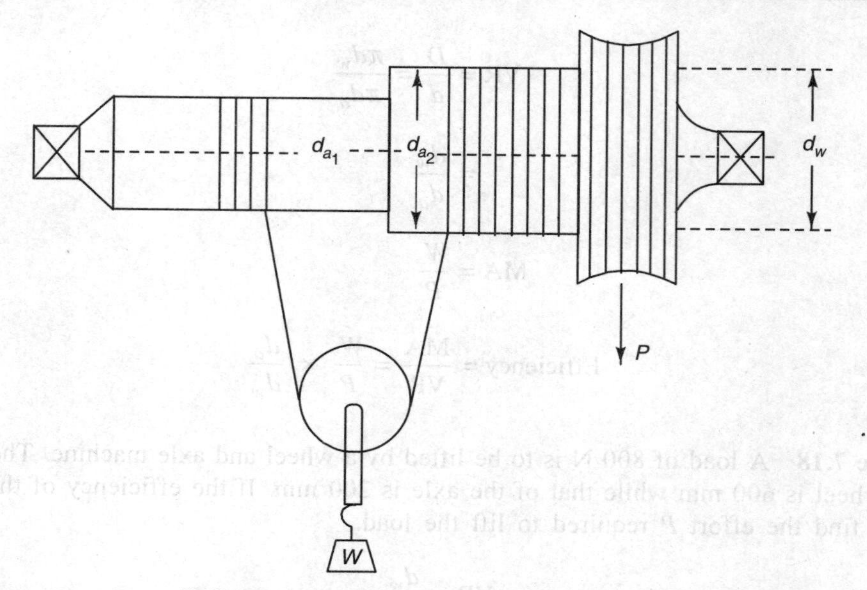

FIGURE 7.46 Wheel and differential axle.

smaller-sized axle with diameter d_{a_2} which firstly passes over a hanging pulley and then it is wound round the bigger-sized axle with diameter d_{w_1} in the opposite direction to the direction of wound on d_{a_2}. It is ensured that the direction of winding of the rope on d_w and d_{a_1} should be the same. Consider the effort $\angle P$ is made to move by the distance of one rotation of wheel, i.e. $D = \pi d_w$. Distance d moved by the load $\angle W$ is $\pi/2(d_{a_1} - d_{a_2})$. Hence, velocity ratio is

$$\text{VR} = \frac{D}{d}$$

$$= \frac{\pi d_w}{\pi/2(d_{a_1} - d_{a_2})}$$

$$= \frac{2d_w}{d_{a_1} - d_{a_2}}$$

Now
$$\text{MA} = \frac{W}{P}$$

and
$$\eta = \frac{\text{MA}}{\text{VR}}$$

$$= \frac{W(d_{a_1} - d_{a_2})}{P \times 2d_w}$$

Example 7.19 A differential wheel and axle machine has the wheel diameter of 600 mm and axle diameters of 200 and 150 mm. Find the load which can be lifted by an effort of 300 N if the efficiency of the machine is 80%.

$$\eta = \frac{W(d_{a_1} - d_{a_2})}{P \times 2d_w}$$

$$0.8 = \frac{W(200 - 150)}{300 \times 2 \times 600}$$

$$= \frac{50\,W}{300 \times 2 \times 600}$$

or

$$W = \frac{0.8 \times 300 \times 2 \times 600}{50}$$

$$= 5760 \text{ N}$$

Worm and Worm Wheel

The worm and worm wheel (Figure 7.47) machine consists of a horizontal shaft on which a worm is provided. A wheel is provided on this shaft and the shaft is supported at both ends on bearings. A rope is wound on the wheel which has one end fixed to the wheel and the other end is used to apply effort. The worm meshes with the worm gear which is provided with a load drum. A rope is attached to the load drum whose other end is attached to load.

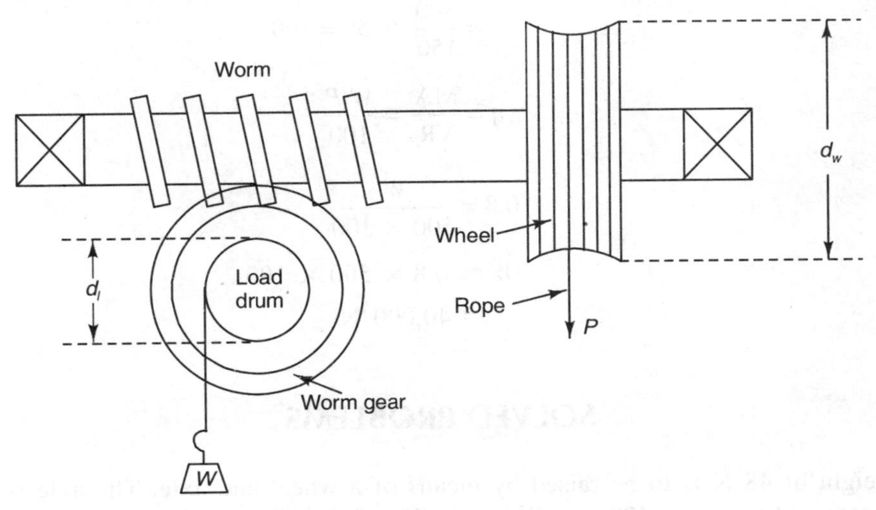

FIGURE 7.47 Worm and worm wheel.

Let d_w = diameter of the wheel

d_l = diameter of the load drum

z = number of feeth on the worm gear

Distance moved by the effort in one revolution of the wheel, $D = \pi d_w$

Distance moved by the load, $d = \frac{\pi d_l}{z}$

$$\therefore \quad \text{Velocity ratio} = \text{VR} = \frac{D}{d}$$

$$= \frac{\pi d_w}{\pi dl/z}$$

$$= \frac{d_w}{dl} \times z$$

$$\text{MA} = \frac{W}{P}$$

$$\eta = \frac{W}{P} \times \frac{d_l}{z \times d_w}$$

Example 7.20 A worm and wheel machine has a worm gear with 50 teeth. The diameter of the effort wheel is 300 mm while the diameter of the load wheel is 150 mm. Find (i) velocity ratio and (ii) load which can be lifted with an effort of 500 N if efficiency is 80%.

$$\text{VR} = \frac{d_w}{d_l} \times z$$

$$= \frac{300}{150} \times 50 = 100$$

$$\eta = \frac{\text{MA}}{\text{VR}} = \frac{W/P}{100}$$

$$0.8 = \frac{W}{500 \times 100}$$

$$\therefore \quad W = 0.8 \times 500 \times 100$$

$$= 40{,}000 \text{ N}$$

SOLVED PROBLEMS

1. A weight of 48 N is to be raised by means of a wheel and axle. The axle is 100 mm diameter and wheel is 400 mm diameter. If a force of 16 N has to be applied to the wheel find,
   (i) Mechanical advantage
   (ii) Velocity ratio
   (iii) Efficiency of the machine

(PTU: 2007–2008)

Here we have

$$W = 48 \text{ N}$$
$$d_a = 100 \text{ mm}$$
$$d_w = 400 \text{ mm}$$
$$P = 16 \text{ N}$$

Now $$\text{MA} = \frac{W}{P} = \frac{48}{16} = 3$$

and $$\text{VR} = \frac{d_w}{d_a} = \frac{400}{100} = 4$$

and $$\eta = \frac{\text{MA}}{\text{VR}} \times 100$$
$$= \frac{3}{4} \times 100 = 75\%$$

2. A simple lifting machine raised a load of 360 N through a distance of 200 mm. The effort a force of 60 N moved 1.8 m during the process. Calculate the velocity ratio, mechnaical advantage and efficiency of the machine.

(PTU: 2004–2005)

$$\text{Load } W = 360 \text{ N}$$
$$\text{Effort } P = 60 \text{ N}$$
$$D = \text{distance moved by effort}$$
$$= 1.8 \text{ m}$$
$$d = \text{distance moved by load}$$
$$= 0.2 \text{ m}$$

Now $$\text{MA} = \frac{W}{P}$$

and $$\text{MA} = \frac{360}{60} = 6$$

and $$\text{VR} = \frac{D}{d}$$
$$= \frac{1.8}{0.2} = 9$$

and $$\eta = \frac{\text{MA}}{\text{VR}} \times 100$$
$$= \frac{6}{9} \times 100 = 66.7\%$$

3. In a lifting machine, an effort of 30 N is required to raise a load of 1 kN. If a efficiency of the machine is 0.75, what is the velocity ratio? If on this machine an effort of 59 N raised a load of 2 kN, what is now the efficiency? What will the effort be required to raise a load of 6 kN.

(PTU: 2004–2005)

Case 1:

$$\text{Load } W = 1000 \text{ N}$$

$$\text{Effort } P = 30 \text{ N}$$

Now
$$\text{MA} = \frac{W}{P} = \frac{1000}{30} = 33.34$$

and
$$\eta = \frac{\text{MA}}{\text{VR}} = 0.75$$

∴
$$\text{VR} = \frac{\text{MA}}{0.75} = \frac{33.34}{0.75} = 44.45$$

Case 2:

$$W = 2000 \text{ N}$$

$$P = 59 \text{ N}$$

∴
$$\text{MA} = \frac{W}{P} = \frac{2000}{59} = 33.9$$

and
$$\eta = \frac{\text{MA}}{\text{VR}} = \frac{33.9}{44.45} = 0.76\%$$

Effort required to raise load of 6 kN

$$\text{VR} = 44.5 \text{ and } W = 6000 \text{ N}$$

As
$$\eta = \frac{\text{MA}}{\text{VR}} = \frac{W/P}{VR} = \frac{6000}{P \times \text{VR}} = 0.76$$

∴
$$P = \frac{6000}{0.76 \times 44.45} = 177 \text{ N}$$

4. For a differential wheel and axle, the diameter of the wheel is 25 cm. The larger and smaller diameters of the differential axle are 10 cm and 9 cm respectively. An effort of 30 N is applied to lift a load of 900 N. Determine the efficiency of the differential wheel and axle.

(PTU: 2004–2005)

$$\text{Wheel diameter } D = 25 \text{ cm}$$

$$\text{Axle larger diameter } d_1 = 10 \text{ cm}$$

Axle smaller diameter $d_2 = 9$ cm

$$\text{VR} = \frac{2D}{d_1 - d_2} = \frac{2 \times 25}{10 - 9} = 50$$

$$W = 900 \text{ N}$$

$$P = 30 \text{ N}$$

$$\text{MA} = \frac{W}{P} = \frac{900}{30} = 30$$

$$\eta = \frac{\text{MA}}{\text{VR}} = \frac{30}{50} = 60\%$$

OBJECTIVE TYPE QUESTIONS

One drop of blood can save thousand lives.

State True or False

1. A floating link is connected to the frame. (*True/False*)
2. A spring can be replaced by two binary links. (*True/False*)
3. A cam and spring loaded follower is an example of a closed pair. (*True/False*)
4. A ball and socket joint constitutes a spherical pair. (*True/False*)
5. The lead screw of a lathe with a half nut constitutes a helical pair. (*True/False*)
6. A crank turning on a hinged joint constitutes a rolling pair. (*True/False*)
7. A slider-crank mechanism is used to convert rotatory motion into reciprocating motion. (*True/False*)
8. The Whitworth-quick return mechanism is an inversion of a slider-crank chain. (*True/False*)
9. The Scotch-yoke mechanism is an inversion of a double slider-crank chain. (*True/False*)
10. Oldham's coupling mechanism is an inversion of a single slider-crank chain. (*True/False*)
11. Mechanical advantage is the ratio of effort appllied to the load lifted. (*True/False*)
12. Efficiency is the ratio of output to input. (*True/False*)
13. Velocity ratio is the ratio of distance moved by the load to the distance moved by the effort. (*True/False*)
14. The efficiency of a reversible machine is more than 50% while of a self-locking machine is less than 50%. (*True/False*)
15. The velocity ratio of a pulley system varies with effort and load. (*True/False*)
16. The velocity ratio of the first-order system of pulleys is 2^n where n is the number of movable pulleys. (*True/False*)
17. The velocity ratio has to be more than one in case load has to be lifted by lessor effort. (*True/False*)
18. The velocity ratio and mechanical advantage are equal for an ideal machine. (*True/False*)
19. The law of motion for a lifting machine is given by a line on the load and effort axes. (*True/False*)
20. A bicylce is a simple machine as compared to a milling machine which is a compound machine. (*True/False*)
21. A lifting jack is not a self-locking machine. (*True/False*)
22. A mechanism is a kinematic chain in which one link is fixed. (*True/False*)
23. A lifting machine tends to become an ideal machine when load tends to become infinity. (*True/False*)
24. The velocity ratio of a second-order lifting machine is $2n$ where n is the number of pulleys in the movable block. (*True/False*)

Multiple Choice Questions

1. A kinematic chain requires at least n links and p kinematic pairs where
   (a) $n = 2, p = 3$ (b) $n = 3, p = 4$ (c) $n = 4, p = 4$ (d) $n = 4, p = 3$
2. Which one of the following is a lower pair?
   (a) ball and socket (b) piston and cylinder (c) cam and follower (d) (a) and (b)
3. A slider-crank mechanism consists of r revolute pairs and p prismatic paris whre r and p are given as
   (a) 3, 1 (b) 2, 2 (c) 1, 3 (d) 1, 3
4. The number of links in a pantograph mechanism is equal to
   (a) 2 (b) 4 (c) 3 (d) 5
5. A rigid chain has no mobility and the number of links in a rigid chain is
   (a) 5 (b) 4 (c) 3 (d) 6
6. Which one of the given kinematic pair is a higher pair?
   (a) sliding pair (b) revolute pair (c) belt and pulley (d) screw pair
7. The degree of freedom (m) of a kinematic chain in terms of number of links (n), single degree of freedom pairs (J_1) and two degrees of freedom of pair (J_2) is given by
   (a) $m = 3(n - 1) - 2J_1 - J_2$ (b) $m = 3(n + 1) - 2J_1 - J_2$
   (c) $m = 3(n - 1) - J_1 - J_2$ (d) $m = 3(n + 3) - J_1 - J_2$
8. If the longest link = l, shortest link = s and p and q are lengths of other links in a four-bar chain, then Grashof's law to ensure the single degree of freedom is given by
   (a) $l + s > p + q$ (b) $l + s > 2(p + q)$
   (c) $2(l + s) < p + q$ (d) $l + s < p + q$
9. Ackermann sterring mechanism is an inversion of a
   (a) single slider-crank chain (b) double slider-crank chain
   (c) 3-bar chain (d) 4-bar chain
10. A quick-return mechanism is an inversion of a
    (a) 4-bar chain (b) single slider-crank chain
    (c) double slider-crank chain (d) 5-bar chain
11. The law of machine in a lifting machine is given by
    (a) $W = mP - c$ (b) $P = mW + C$ (c) $W = mP + C$ (d) $W = P + C/m$
12. The velocity ratio of a first-order system of pulleys is
    (a) 2^{n-1} (b) 2^n (c) 2^{n+1} (d) $2^{n-1} + 1$
13. The ratio of mechanical advantage to the velocity ratio in a machine is
    (a) > 1 (b) ≥ 1 (c) < 1 (d) $= 1$
14. In an ideal machine, the relation of MA and VR is
    (a) MA > VR (b) MA < VR (c) MA = VR (d) 2MA = VR
15. In a self-locking machine, the efficiency is
    (a) 25% (b) > 25% (c) < 50% (d) > 50%
16. In a reversible machine, the efficiency is
    (a) > 50% (b) < 50% (c) = 50% (d) = 55%

17. The velocity ratio of a second-order of system of pulleys is
(a) $3n$ (b) $2n$ (c) n (d) $n/2$

18. The velocity ratio of a third-order system of pulleys is
(a) $2n$ (b) 2^{n-1} (c) 2^n-1 (d) 2^n+1

19. The efficiency of a lifting machine is given by
(a) $\frac{VR}{MA}$ (b) $\frac{VR + MA}{MA}$ (c) $\frac{MA}{VR}$ (d) $\frac{VR + MA}{MA}$

20. In a single pulley, velcoty ratio is
(a) 0.5 (b) 1 (c) 1.5 (d) 2

Fill in the Blanks

1. A kinematic chain becomes a machanism when one of its link ________ .
(a) fixed (b) floating

2. A belt is a ________ link.
(a) sliding (b) flexible

3. A link not connected to the frame is called ________ link.
(a) sliding (b) floating

4. A bolt with a nut is ________ pair.
(a) sliding (b) screw

5. The tailstock of a lathe constitutes a ________ pair with the lathe bed.
(a) floating (b) sliding

6. ________ pairs have surface contact while in motion.
(a) Lower (b) Higher

7. ________ pairs have point or line contact while in motion.
(a) Higher (b) Lower

8. A screw jack is a ________ machine.
(a) simple (b) compound

9. A milling machine is a ________ machine.
(a) simple (b) compound

10. In an actual machine the velocity ratio is ________ than mechanical advantage.
(a) lesser (b) greater

11. In an ideal machine, mechanical advantage is ________ an ideal machine.
(a) lesser than (b) equal to

12. In a first-order pulley system, there are three movable pulleys, therefore velocity ratio is ________.
(a) 4 (b) 8

13. In a second-order pulley system there are two pulleys in the movable block. Hence the velocity ratio is ________.
(a) 2 (b) 4

14. In a third-order system of pulleys, there are 4 pulleys. The velocity ratio is ________.
(a) 16 (b) 15

ANSWERS

Do not be afraid to take a big step in your work as a ditch cannot be crossed in two small steps.

State True or False

1. False. Links connected to the frame are driver and driven cranks and link not connected to the frame is a floating link or coupler.
2. True
3. False. It is a forced pair or unclosed pair
4. True
5. False. It is a sliding pair.
6. False. It is a revolute pair.
7. True
8. True
9. True
10. False. It is an inversion of a double slider-crank chain.
11. False. MA = W/P
12. True
13. False $VR = \dfrac{\text{Distance moved by the effort}}{\text{Distance moved by the load}}$
14. True
15. False. Velocity ratio remains constant.
16. True
17. True. MA = VR if η = 100% and $W/P > 1$ if VR > 1
18. True
19. True. $W = mW + C$
20. True. A bicycle is given effort at one point while a milling machine is given effort at various points.
21. False. It is a self-locking machine as effort can be removed without any danger of load moving down.
22. True
23. Ture. $\eta = \dfrac{W}{P} = \dfrac{W}{mW + C} = \dfrac{1}{1 + C/W} = 1$ when W tends to be infinity.
24. True

Multiple Choice Questions

1. (c)	**2.** (d)	**3.** (a)	**4.** (b)
5. (c)	**6.** (c)	**7.** (a)	**8.** (d)
9. (d)	**10.** (b)	**11.** (b)	**12.** (b)
13. (c)	**14.** (c)	**15.** (c)	**16.** (a)
17. (b)	**18.** (b)	**19.** (c)	**20.** (b)

Fill in the Blanks

1. (a)	**2.** (b)	**3.** (b)	**4.** (b)
5. (b)	**6.** (a)	**7.** (a)	**8.** (a)
9. (b)	**10.** (b)	**11.** (b)	**12.** (b)
13. (b)	**14.** (b)		

CHAPTER 8

Force System and Analysis

If you just take a step by faith to do the right thing, then God will do the rest— the things you can't do.

INTRODUCTION

Mechanics is a science which deals with the state of rest or the state of motion of a body under the action of forces. The application of this science to actual problems is called *applied mechanics*. Statics is a branch of mechanics which relates to bodies at rest. Dynamics is a branch of mechanics which deals with bodies in motion. The analysis of force system on bodies is based on some of basic laws which are fundamental laws of mechanics. There are six fundamental laws of mechanics.

FUNDAMENTAL LAWS OF MECHANICS

The *first law of motion* states that a body tends to stay in the state of rest or of uniform motion unless an external force is applied.

The *second law of motion* states that the rate of change of momentum of a body is directly proportional to the applied force and is in the same direction.

$$\text{Force (F)} \propto \text{rate of change of momentum} = m \times \left(\frac{v-u}{t}\right)$$

$$= m \times a$$

where

m = mass
v = final velocity
u = initial velocity
t = time
a = acceleration

The *third law of motion* states that for every action, there is an equal and opposite reaction. A roller is resting on the ground (Figure 8.1a). If we remove the ground surface, we have to exert force R to hold the roller in place to counteract its weight acting downwards. Hence the ground is exerting reaction R on the roller which is equal to its weight. Similarly, if we remove the floor and wall on which the ladder is resting, we have to apply reactions R_f and μR_f for the floor, and R_w and μR_w for the wall (Figure 8.1b).

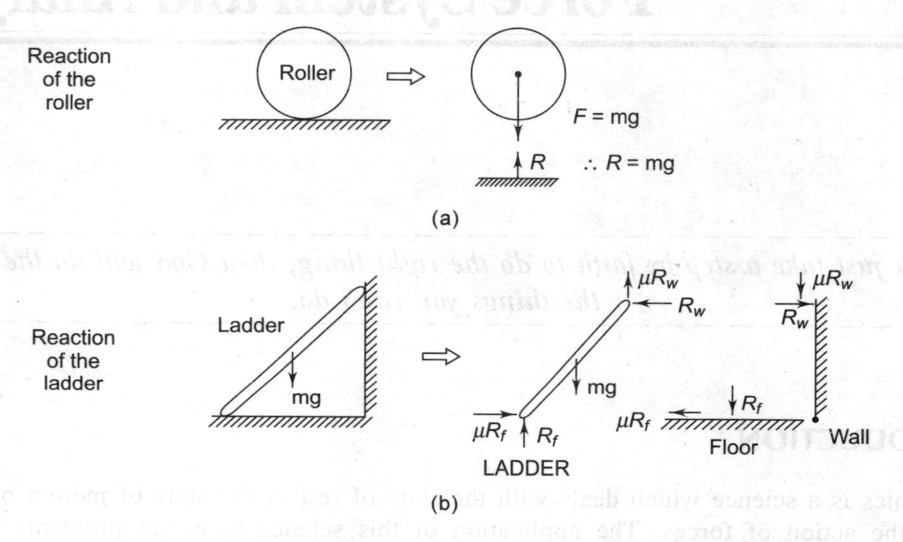

FIGURE 8.1 Third law of motion.

Newton's *law of gravitation* states that the force of attraction between two bodies is directly proportional to their masses and inversely proportional to the square of the distance between them:

$$F = \frac{GM_1M_2}{d^2}$$

$$g = \frac{GM_{\text{earth}}}{R^2_{\text{earth}}} = 9.81 \text{ m/s}^2$$

The value of g increases as a star starts collapsing. As the star collapses and becomes more and more dense, the force of gravity on its surface becomes stronger and stronger as per the relation given above. Such a collapsing star forms a black hole and nothing can pass through a black hole (Figure 8.2). Even light cannot pass through a black hole due to very strong gravity force.

It takes eight minutes for light to travel from the Sun to the Earth and hence we see the sun at any moment as it existed eight minutes ago. The length of an object will not have the same length when moving as compared to whom it is at rest as per relativity theory.

FIGURE 8.2 Black hole.

The length depends on its motion relative to the observers and it changes with relative velocity. The object seems to contract in the direction of its motion. Similarly, a clock in motion runs slower or time slows down during motion. Even a human heartbeat slows down if he is in motion. If one of two forming a twin goes on a fast trip in a rocket in space (Figure 8.3), he will be younger than the other one when he comes back after some years because his heartbeat will slow down during the space travel. This is called *twin paradox*.

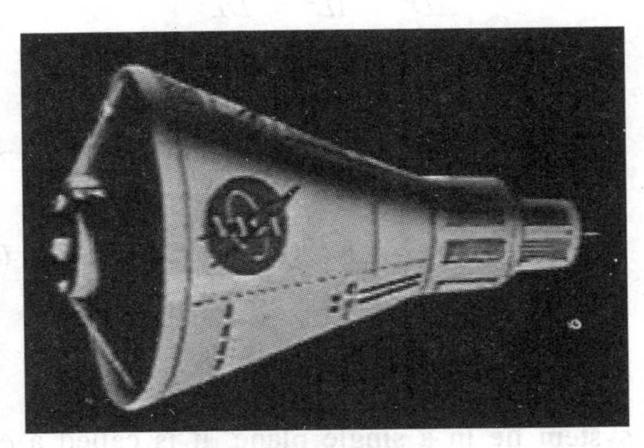

FIGURE 8.3 Rocket.

The *law of transmissibility of force* states that the state of a rigid body (rest or motion) is unaltered if a force acting on the body is replaced by another force of the same magnitude and direction but acting anywhere along the line of action of replaced force. Consider a force (F) acting at point 1 of the body (Figure 8.4). The force will have the same effect if it is acting at point 2 as point 2 lies on the line of action of the force (F).

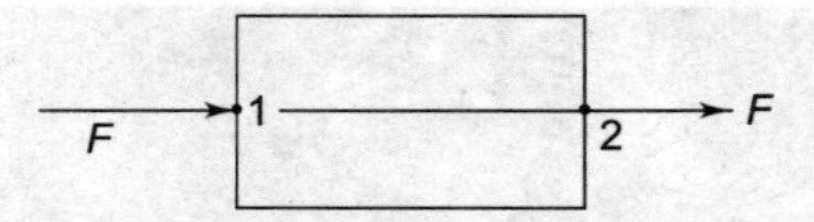

FIGURE 8.4 Law of transmissibility.

The parallelogram law of forces states that if two forces are acting on a body at a point and represented in magnitude and direction by two adjacent sides of a parallelogram, their resultant is represented in magnitude and direction by the diagonal of the parallelogram which passes through the point of intersection (Figure 8.5).

In Figure 8.5, forces F_1 and F_2 are acting at point A with magnitude and direction as $\overrightarrow{AC}$ and $\overrightarrow{AB}$. If parallelogram $ABDC$ is drawn with sides AB and AC, then resultant $\vec{R}$ will be given by diagonal $\overrightarrow{AD}$. Extend AB to E so that ED is perpendicular.

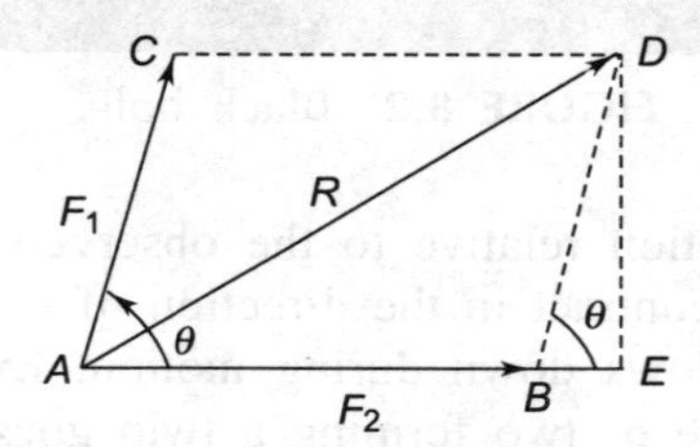

FIGURE 8.5 Parallelogram law of forces.

In ΔAED

$$AD^2 = AE^2 + DE^2$$

$$R^2 = (F_2 + BE)^2 + DE^2$$

In ΔDBE

$$BE = F_1 \cos\theta \text{ and } DE = F_1 \sin\theta$$

$$\therefore \quad R^2 = (F_2 + F_1 \cos\theta)^2 + (F_1 \sin\theta)^2$$

$$= F_1^2 + F_2^2 + 2F_1F_2 \cos\theta$$

FORCE SYSTEM

If all the forces in a system lie in a single plane, it is called a *coplanar force system*.

If the line of action of all forces lies along a single line then it is called a *collinear force system*.

The coplanar force system can be

1. coplanar parallel forces
2. coplanar like parallel forces
3. coplanar concurrent forces
4. coplanar non-concurrent forces

Concurrent forces can be

1. coplanar concurrent forces
2. non-coplanar concurrent forces

Non-concurrent forces can be

1. coplanar non-concurrent forces
2. non-coplanar non-concurrent forces

Moment of a force about a point is the measure of its rotational effect. It is the product of the magnitude of the force and the perpendicular distance of the point from the line of action of the force. The point from where the moment is taken is called *moment centre* and the perpendicular distance of the point from the line of action of the force is called *moment arm* (Figure 8.6). The moment can be clockwise or anticlockwise.

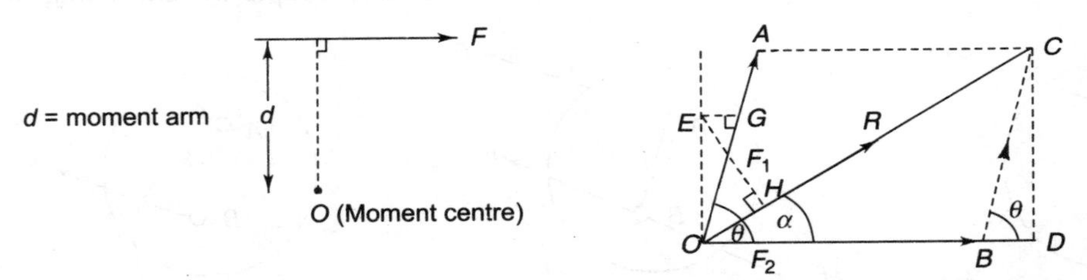

FIGURE 8.6 Moment of a force and Varignon's theorem.

Moment from point O is $F \times d$ and it is clockwise.

Varignon's theorem (Principle of moments) states that the algebraic sum of the moments of a system of coplanar forces about a moment centre in their plane is equal to the moment of their resultant force about the same moment centre. To prove, draw perpendiculars EG, EO and EH on forces F_1, F_2 and their resultant R as shown in Figure 8.6.

Now the moment of F_1 and F_2 from point $E = F_1 \times EG + F_2 \times EO$

$$= F_1 \times EO \cos \theta + F_2 \times EO = EO(F_1 \cos \theta + F_2)$$

$$= EO \times (OB + BD) = EO \times OD$$

Similarly the moment of R from point E is

$$EH \times R = EO \cos \alpha \times R = EO \times OD$$

Hence, the moment of resultant is equal to the moment of forces when taken from point E.

A *couple* is formed by two parallel forces equal in magnitude and opposite in direction and separated by a definite distance. The translatory effect of a couple is zero and it has only rotational effect.

Two unlike parallel and equal magnitude forces F_1 separated by a distance of d forms a couple with only effect of moment equal to $F_1 \times d$ (Figure 8.7). The couple remains unchanged if:

1. it is rotated through an angle
2. moment centre is shifted to another position
3. it is replaced by another pair of forces whose rotational effect is the same

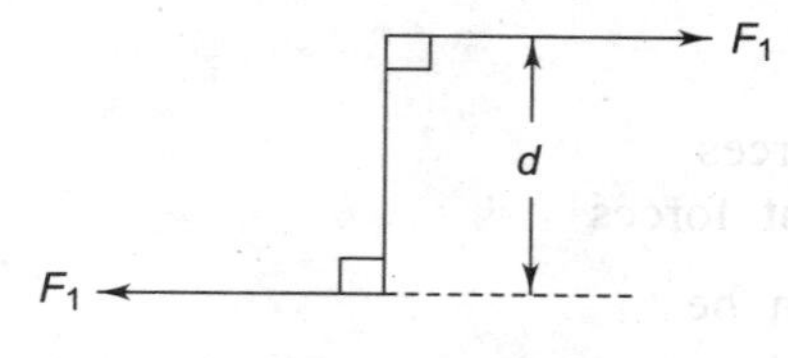

FIGURE 8.7 A couple.

Transfer of a force to a parallel force at another point is possible. A force acting at any point on a body can be replaced by a parallel force acting at the some other point and a couple.

A force F is acting at a point A on the body (Figure 8.8). We went to shift the force to point B. To do so, we apply force F in equal magnitude and direction at B. To balance the extra force F, at B, we also apply equal force in the opposite direction at B. The force E at A and the opposite force at B forms couple. Hence we have a force and couple system acting at B.

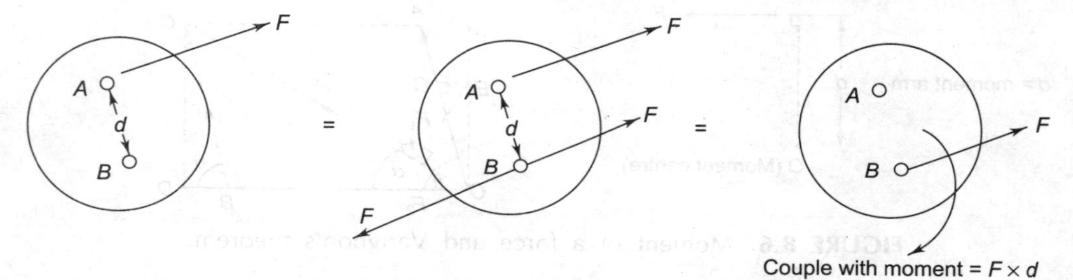

FIGURE 8.8 Transfer of a force to a parallel force.

The resultant of coplanar concurrent force system can be found out by

1. graphical method
2. analytical method

In the graphical method, each force is drawn with a magnitude and direction to find the resultant. The law of parallelogram of forces, triangle law of forces and polygon law of forces are used to find the resultant by the graphical method. According to the triangular law of forces if two forces act on a body are represented by the sides of a triangle taken in direction, their resultant is represented by the closing side of the triangle taken in the opposite direction (Figure 8.9a). The polygon law of forces states that if a number of concurrent forces acting on a body are represented in a magnitude and direction by the sides of a polygon taken in order, then the resultant is represented in a magnitude and direction by the closing sides of the polygon taken in the opposite direction (Figure 8.9b).

The analytical method of finding the resultant of the coplanar concurrent force system consists of finding components of each force in two mutually perpendicular direction (x and y direction) and then combining these components in each direction (ΣP_x and ΣP_y). These two components that are mutually perpendicular are combined to get the resultant ($R = \sqrt{(\Sigma P_x)^2 + (\Sigma P_y)^2}$). Finding the component of a force P as P_x in x direction and P_y in y

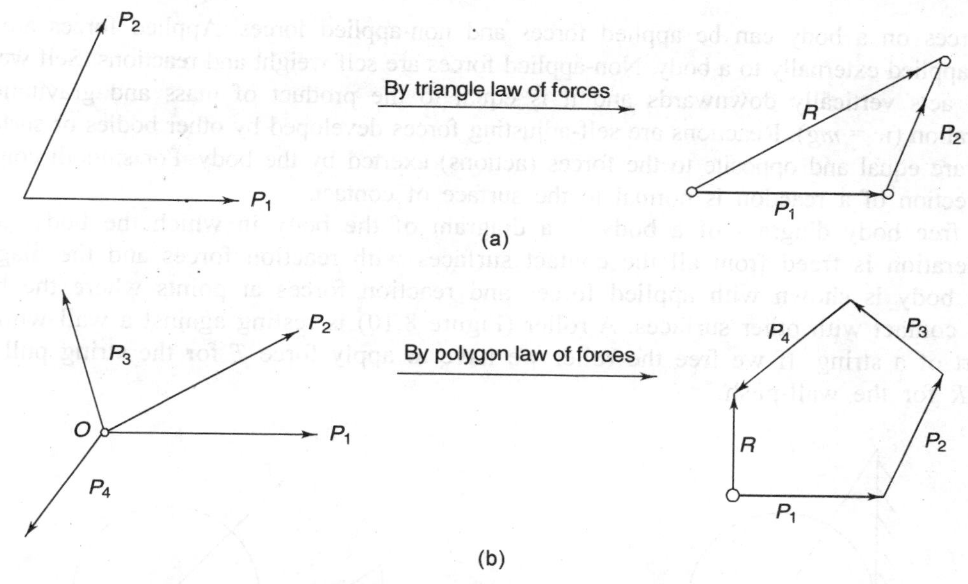

FIGURE 8.9 Graphical method.

direction is called *resolution of force* where $P = \sqrt{P_x^2 + P_y^2}$ and the angle of inclination of the resultant (R) to x-axis is given by $\tan^{-1} \dfrac{\Sigma P_y}{\Sigma P_x}$.

The analytical method of finding the resultant of the coplanar nonconcurrent force system replaces each force by a force of the same magnitude and direction acting at the reference point (0) and a moment about point (0). The coplanar nonconcurrent force system is thereby converted into a coplanar concurrent force system and a moment ΣM_0. The resultant (R) of the concurrent force system can be found out. The force R and moment ΣM_0 can be replaced by a single force R acting at a distance d from point 0 such that the moment $R \times d$ is equal to ΣM_0.

$$\Sigma P_x = P_1 x + P_2 x + \cdots$$

$$\Sigma P_y = P_1 y + P_2 y + \cdots$$

$$R^2 = (\Sigma P_x)^2 + (\Sigma P_y)^2$$

$$R \times d = \Sigma M_0$$

If the resultant (R) is at distance x and y from point 0, then

$$x = \frac{\Sigma M_0}{R_y} = \frac{\Sigma M_0}{\Sigma P_y}$$

$$y = \frac{\Sigma M_0}{R_x} = \frac{\Sigma M_0}{\Sigma P_x}$$

Forces on a body can be applied forces and non-applied forces. Applied forces are the forces applied externally to a body. Non-applied forces are self weight and reactions. Self weight always acts vertically downwards and it is equal to the product of mass and gravitational acceleration ($w = mg$). Reactions are self-adjusting forces developed by other bodies or surfaces which are equal and opposite to the forces (actions) exerted by the body. For smooth contact, the direction of a reaction is normal to the surface of contact.

A free body diagram of a body is a diagram of the body in which the body under consideration is freed from all the contact surfaces with reaction forces and the diagram of the body is shown with applied forces and reaction forces at points where the body makes contact with other surfaces. A roller (Figure 8.10) is resting against a wall with the support of a string. If we free the roller we have to apply force T for the string pull and force R for the wall-push.

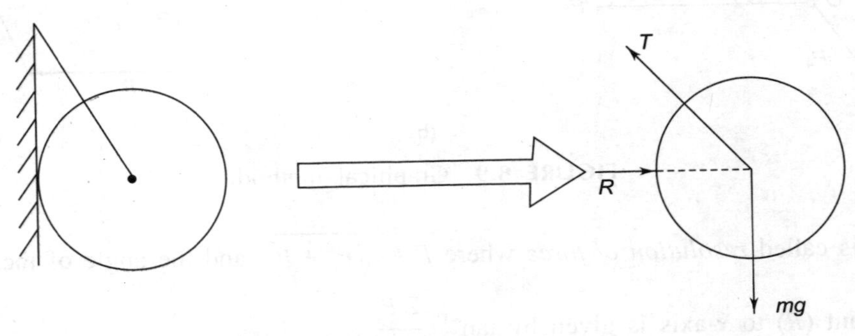

FIGURE 8.10 Free body diagram.

A body is said to be in equilibrium when it is at rest or in uniform motion. It means that the resultant of all forces acting is zero. Mathematically,

$$\Sigma P_x = 0, \ \Sigma P_y = 0, \ \Sigma M = 0$$

When a body is in equilibrium under a concurrent force system, then $\Sigma P_x = 0$ and $\Sigma P_y = 0$. If a body is in equilibrium under the action of three forces, then Lami's theorem can be applied for quick analysis.

According to Lami's theorem if a body is in equilibrium under the action of three forces, (P_1, P_2, P_3), then each force is proportional to the sine of the angle between other two forces (Figure 8.11):

$$\frac{P_1}{\sin \alpha_{23}} = \frac{P_2}{\sin \alpha_{13}} = \frac{P_3}{\sin \alpha_{12}}$$

When two or more bodies are in contact with one another, the system appears as though it is a non-concurrent force system. However, when each body is considered separately, we will find that it is two or more concurrent force systems.

For example, in Figure 8.12, A and B bodies are attached with strings and they are further tied to supports E, C and D. Forces P_1 and P_2 are applied to them. They form a

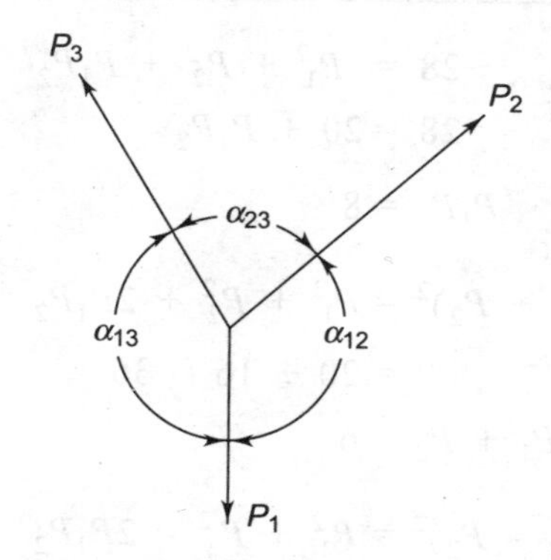

FIGURE 8.11 Lami's theorem.

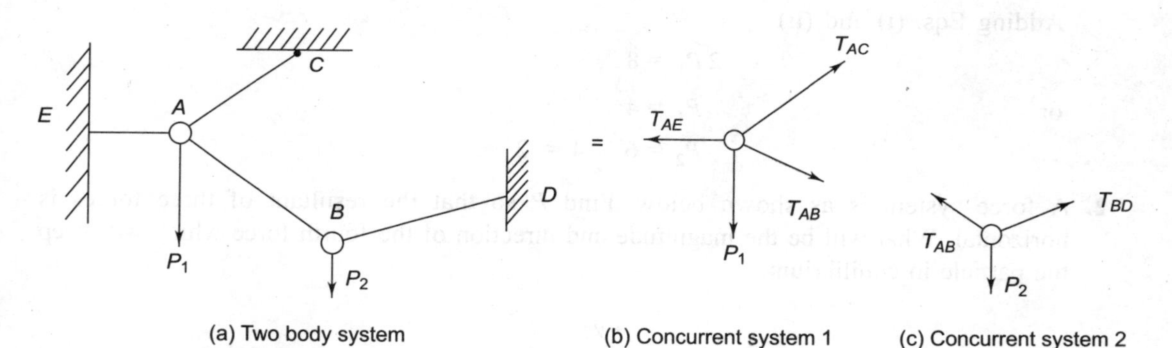

FIGURE 8.12 Force systems.

coplanar force system which is difficult to solve. However, consideration of each body separately gives concurrent force systems which are simpler to solve.

SOLVED PROBLEMS

1. The magnitude of two forces is such that when acting at a right angle produce a resultant force of $\sqrt{20}$ and when acting at 60° produce a resultant equal to $\sqrt{28}$. Find the magnitude of the forces.

If P_1 and P_2 are forces and the angle between them is θ,

$$R^2 = P_1^2 + P_2^2 + 2P_1P_2 \cos \theta$$

Putting $\theta = 90°$ and $R = \sqrt{20}$ in the above equation,

$$20 = P_1^2 + P_2^2$$

Now putting $\theta = 60$ and $R = \sqrt{28}$,

$$28 = P_1^2 + P_2^2 + P_1P_2$$

$$28 = 20 + P_1P_2$$

or $$P_1P_2 = 8$$

Now

$$(P_1 + P_2)^2 = P_1^2 + P_2^2 + 2P_1P_2$$

$$= 20 + 16 = 36 \quad \text{(i)}$$

$\therefore$ $$P_1 + P_2 = 6$$

Also

$$(P_1 - P_2)^2 = P_1^2 + P_2^2 - 2P_1P_2$$

$$= 20 - 16 = 4 \quad \therefore \quad P_1 - P_2 = 2 \quad \text{(ii)}$$

Adding Eqs. (i) and (ii)

$$2P_1 = 8$$

or $$P_1 = 4$$

$\therefore$ $$P_2 = 6 - 4 = 2$$

2. A force system is as shown below. Find P so that the resultant of three forces is horizontal. What will be the magnitude and direction of the fourth force which will keep the particle in equilibrium.

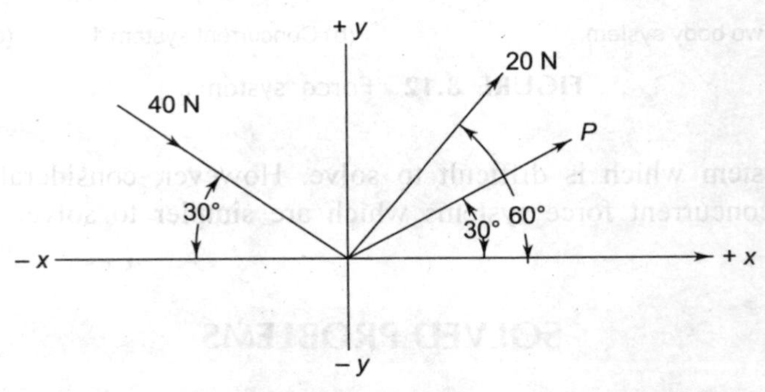

For finding the value of P, it has been given that the resultant is horizontal which gives us the condition $\Sigma P_y = 0$.

Applying $\Sigma P_y = 0$

$$+20 \sin 60 - 40 \sin 30 + P \sin 30 = 0$$

or $$20\sqrt{3} - 40 + P = 0$$

$$P = 5.36 \text{ N}$$

Hence if $P = 5.36$ N, the resultant will be horizontal. To find the resultant $\Sigma P_x = R$.

$$R = 40 \cos 30 + 20 \cos 60 + 5.36 \cos 30$$

$$= 39.28 + 10 = 49.28 \text{ N}$$

The resultant has the value of 49.28 N acting towards $+x$ direction. To make the force system in equilibrium, the fourth force has to be 49.28 N and acts towards $-x$ direction.

3. A rigid bar is subjected to a system of parallel forces as shown in the figure. Reduce the system to (a) a single force system and (b) a single force moment system at B.

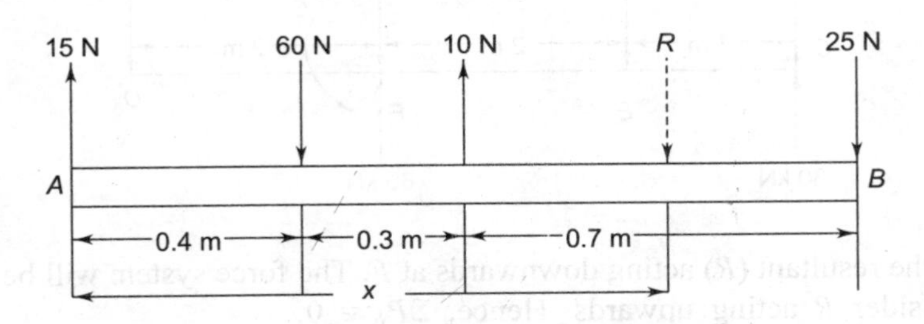

Let R be the resultant acting x metre from A. The force system will be in equilibrium if we consider R acting in the reverse direction, i.e. $\Sigma P_y = 0$.

$$R = 60 - 15 - 10 + 25 = 60 \text{ N}$$

$$\Sigma M_A = 0$$

$$-R \times x + 25 \times 1.4 - 10 \times 0.7 + 60 \times 0.4 = 0$$

$$60x = 35 - 7 + 24 = 52$$

$$x = 0.866$$

Hence the single force system can be shown as follows:

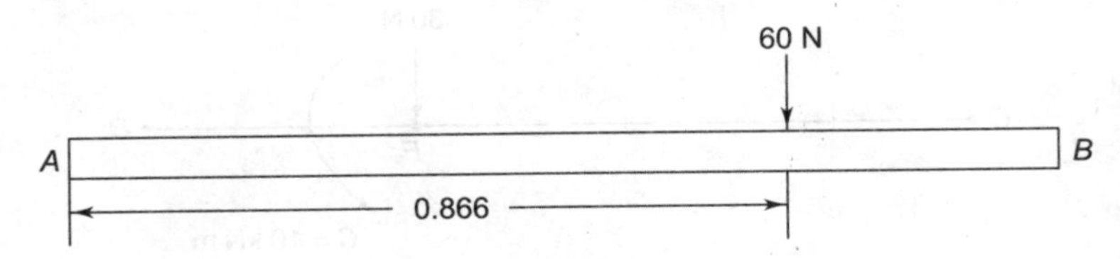

Now to convert it to a single force moment system at B, we will have a force = 60 N and moment equal to force × distance from B, i.e.

$$M_B = 60 \times (1.4 - 0.866)$$

$$= 32.04 \text{ N m (anticlockwise)}$$

The system is as shown below:

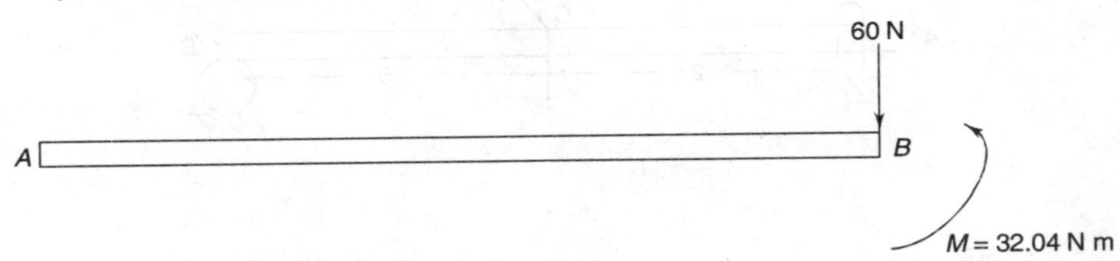

4. A rigid bar CD is subjected to a system of parallel forces as shown in the figure. Reduce the given system of forces to an equivalent force couple system at F.

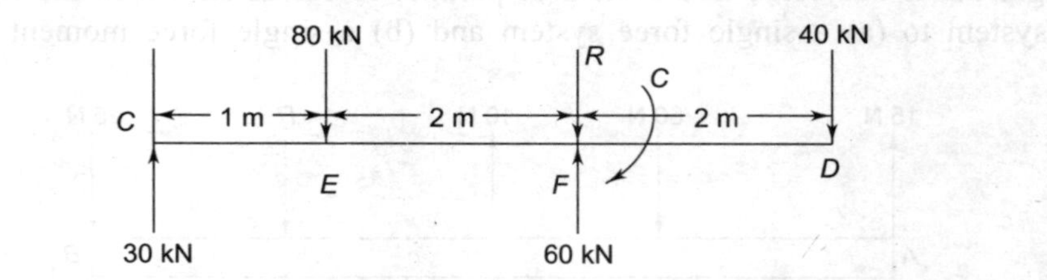

Consider the resultant (R) acting downwards at F. The force system will be in equilibrium if we consider R acting upwards. Hence, $\Sigma P_y = 0$.

$$30 - 80 + 60 - 40 + R = 0$$

$$R = 30 \text{ kN}$$

Consider couple C acting at F.

$$\Sigma M_F = 0$$

$$30 \times 3 - 80 \times 2 + 40 \times 2 - C = 0$$

$$C = 10 \text{ kN m (clockwise)}$$

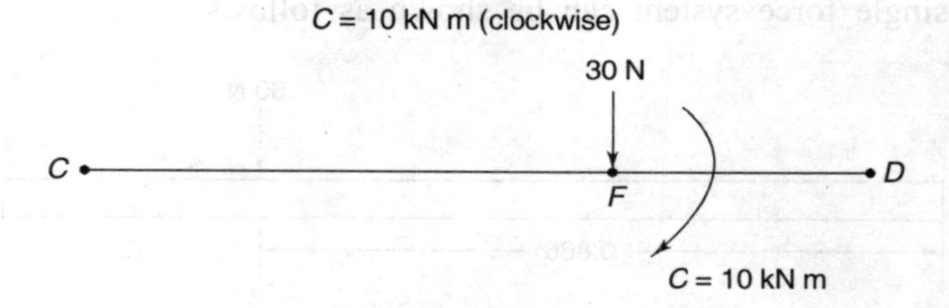

5. A force $P = 5000$ N is applied at the centre C of the beam AB of length 5 m as shown below. Find the reaction at the hinge and roller support.

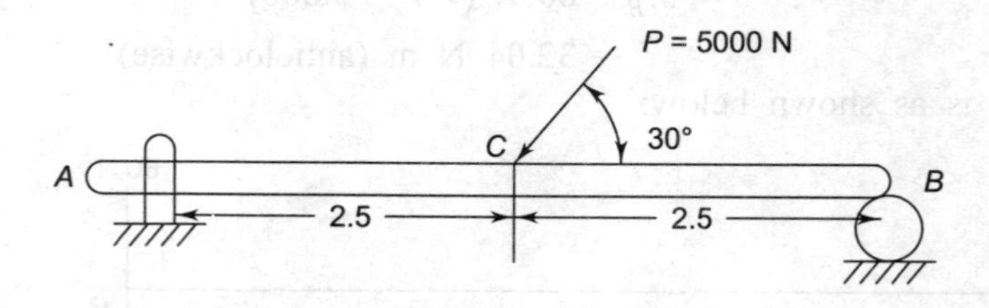

Draw free body diagram of beam:

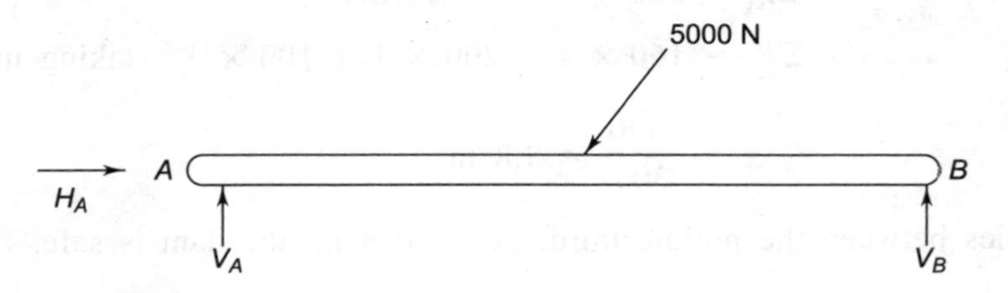

Since the system is in equilibrium, $\Sigma P_x = 0$ and $\Sigma P_y = 0$.

$$\Sigma P_x = 0$$

$$H_A = 5000 \cos 30 = 4330.13$$

$$\Sigma P_y = 0$$

$$V_A + V_B = 5000 \sin 30 = 2500$$

$$\Sigma M_A = 0,\ 5000 \sin 30 \times 2.5 - V_A \times 5 = 0$$

$$\therefore \quad V_B = 1250 \text{ N}$$

$$\therefore \quad V_A = 2500 - 1250 = 1250 \text{ N}$$

$$R_A = \sqrt{(H_A)^2 + (V_A)^2} = \sqrt{(4330.13)^2 + (1250)^2}$$

$$= 4507 \text{ N}$$

6. The forces acting on a dam is as shown in the figure. The dam is safe if the resultant passes through the middle third of the base. Find if the dam is safe.

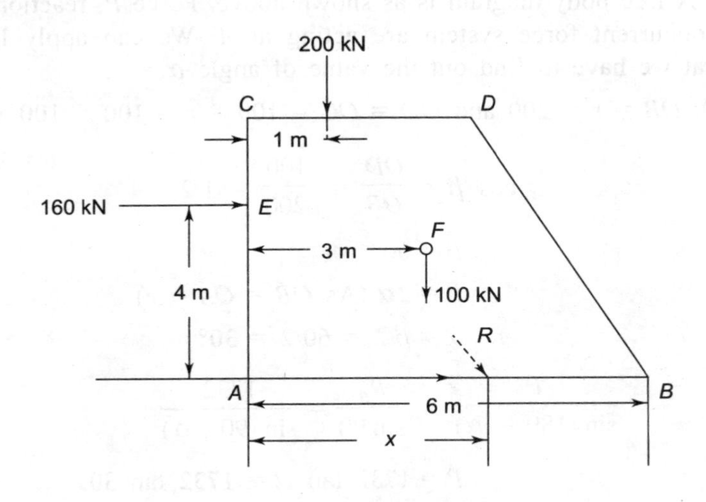

Let x be distance of the resultant from edge AC.

$$\Sigma P_x = 160 \text{ kN}$$

$$\Sigma P_y = 200 + 100 = 300 \text{ kN}$$

$$x \times \Sigma P_y = 160 \times 4 + 200 \times 1 + 100 \times 3 \quad \text{(taking moment at } A)$$

$$x = \frac{1140}{300} = 3.8 \text{ m}$$

As x lies between the middle third, i.e. 2 to 4 m, the dam is safe.

7. A roller of radius r = 200 mm and weight 1732 N is to be pulled over a curb of height 100 mm by a horizontal force P applied to the end of string wound tightly around the circumference of the roller. Find the magnitude of P required to start the roller move over the curb. Also find the least pull P through the centre O the wheel to just turn the roller over the curb.

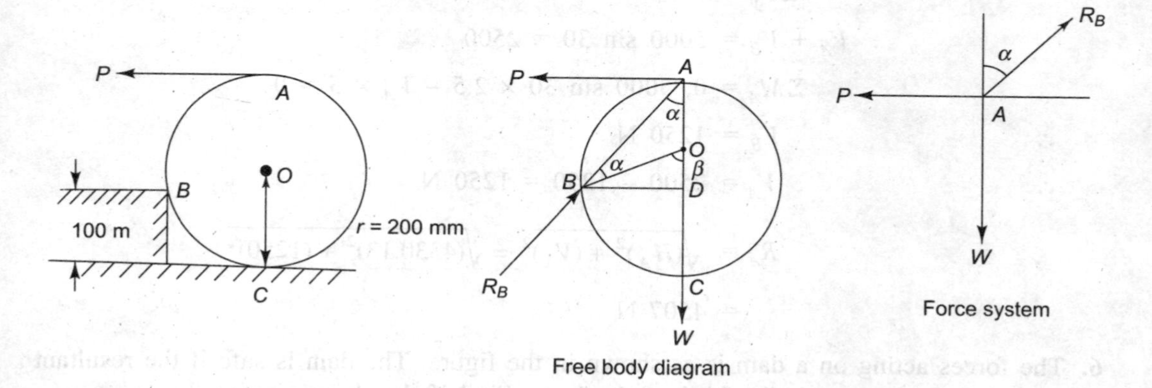

Case 1: A free body diagram is as shown above. Force P, reaction R_B and weight form a concurrent force system are acting at A. We can apply Lami's theorem. Before that we have to find out the value of angle α.

In ΔOBD, $OB = r = 200$ and $OD = OC - 100 = r - 100 = 100$

Hence $\cos \beta = \dfrac{OD}{OB} = \dfrac{100}{200} = 1/2$

$\therefore$ $\beta = 60°$

Now $\beta = 2\alpha$ (As $OB = OA = r$)

$\therefore$ $\alpha = \beta/2 = 60/2 = 30°$

Now $$\frac{P}{\sin (180 - \alpha)} = \frac{R_B}{\sin 90} = \frac{1732}{\sin (90 + \alpha)}$$

$\therefore$ $P = 1732 \tan \alpha = 1732 \tan 30$

$= 1000$ N

Case 2:

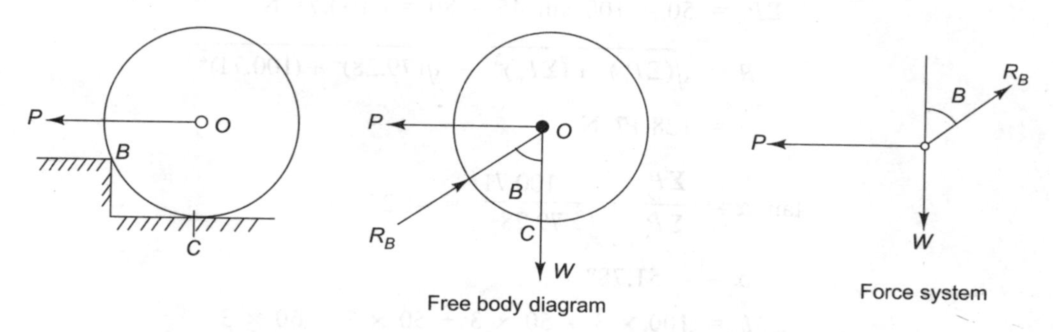

The free body diagram is as shown above and reaction R_B will now pass through point O. Now we have a concurrent force system at point O. Applying Lami's theorem (β = 60°),

$$\frac{P}{\sin(180 - \beta)} = \frac{1732}{\sin(90 + \beta)} = \frac{R_B}{\sin 90}$$

$$P = 1732 \tan 60 = 3000 \text{ N}$$

8. Determine the horizontal force P to be applied to a body (weight = 2000 N) to hold it in position on a smooth inclined plane AB which makes an angle of 45° with the horizontal.

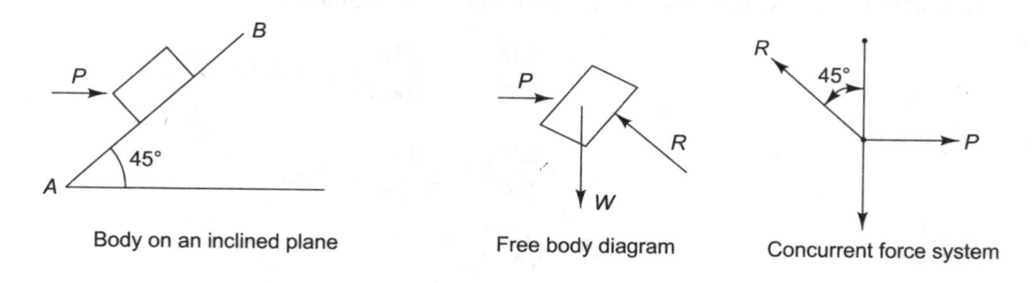

Body on an inclined plane

Free body diagram

Concurrent force system

A free body diagram of the body on an inclined surface is as shown above. We have three forces (concurrent) acting at a point. Applying Lami's theorem

$$\frac{P}{\sin(180 - 45)} = \frac{W}{\sin(90 + 45)} = \frac{R}{\sin 90}$$

$$\therefore \quad P = W \tan 45 = 2000$$

9. Determine the resultant of the forces acting tangential to circle of radius 3 m as shown in the figure. What will be the location with respect to the centre of the circle?

$$\Sigma P_x = 150 - 100 \cos 45 = 79.28 \text{ N}$$

$$\Sigma P_y = 50 - 100 \sin 45 - 80 = -100.71 \text{ N}$$

$$R = \sqrt{(\Sigma P_x)^2 + (\Sigma P_y)^2} = \sqrt{(79.28)^2 + (100.71)^2}$$

$$= 128.17 \text{ N}$$

$$\tan \alpha = \frac{\Sigma P_y}{\Sigma P_x} = \frac{-100.71}{79.28} = -1.27$$

$$\alpha = -51.78°$$

$$\Sigma M_0 = 100 \times 3 - 50 \times 3 + 50 \times 3 - 60 \times 3$$

$$= 210 \text{ N m}$$

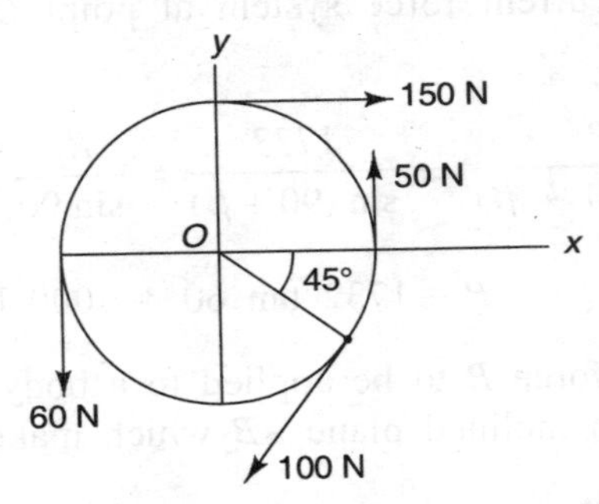

If x and y are the distance of the resultant from point O,

$$x = \frac{\Sigma M_0}{\Sigma P_y} = \frac{210}{-100.71} = -2.09$$

$$y = \frac{\Sigma M_0}{\Sigma P_x} = \frac{210}{79.28} = 2.65$$

$$d = \sqrt{x^2 + y^2}$$

$$= \sqrt{(2.09)^2 + (2.65)^2}$$

$$= 3.37$$

10. A bracket is subjected to a force of 141 kN as shown in the figure. Find (a) an equivalent force-couple system at A and (b) an equivalent force couple system at C.

(a) If force is converted into a force couple system at point A, then it will consist of force 141 N and a clockwise couple is

$$141 \times AD = 141 \times \sqrt{4+1} = 315.3 \text{ N m}$$

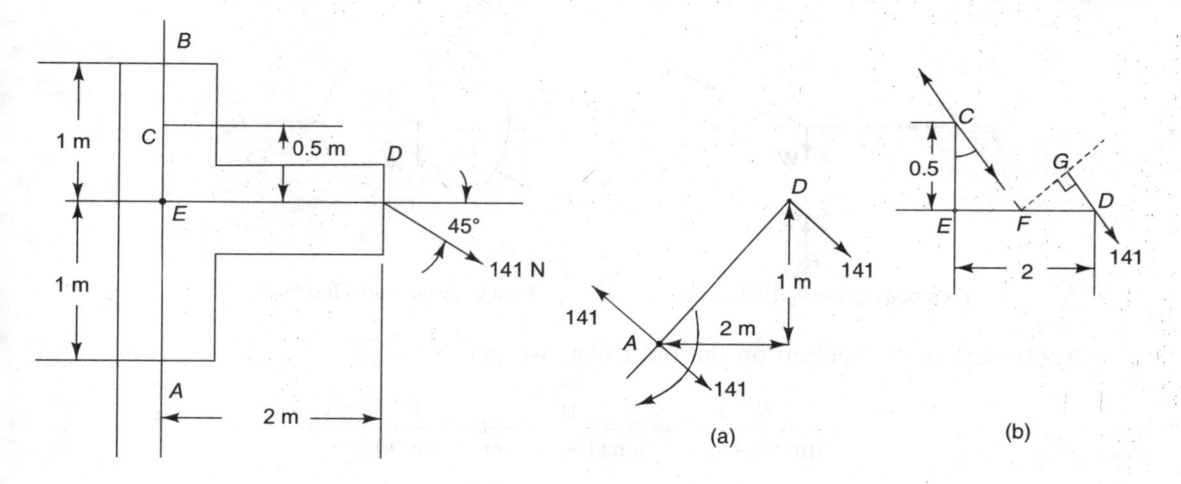

(b) If force is converted into a force couple system at C, then it will consist of force 141 N and a clockwise couple is

$$141 \times FD \cos 45 = 141 \times 1.5 \times \frac{1}{\sqrt{2}} = 150 \text{ kN}$$

11. Two smooth spheres each of weight w and each of radius r are in equilibrium in a horizontal channel of width b ($b < 4r$) and vertical sides as shown in the figure.

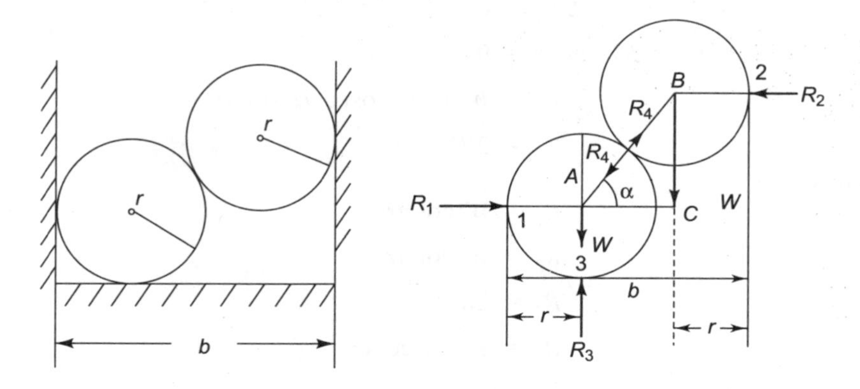

Find the three reactions from the sides of the channel which are all smooth. Also find the force exerted by each sphere on the other. (UPTU: 2005)

$$\text{In } \Delta ABC, \; AC = b - 2r \quad \text{and} \quad AB = 2r$$

$$\cos \alpha = \frac{AC}{AB} = \frac{b - 2r}{2r}$$

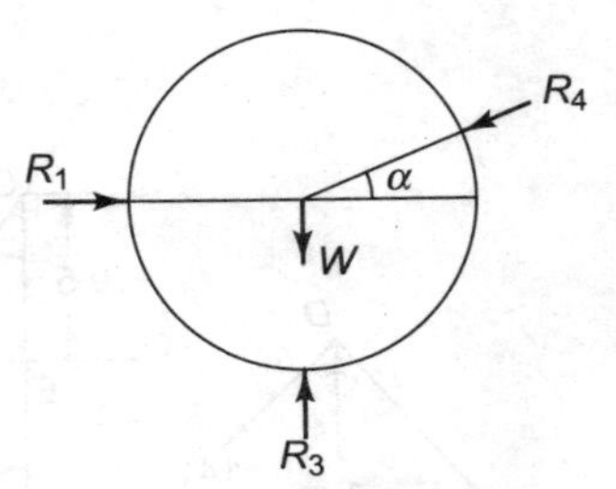

Free body diagram (Bottom roller)

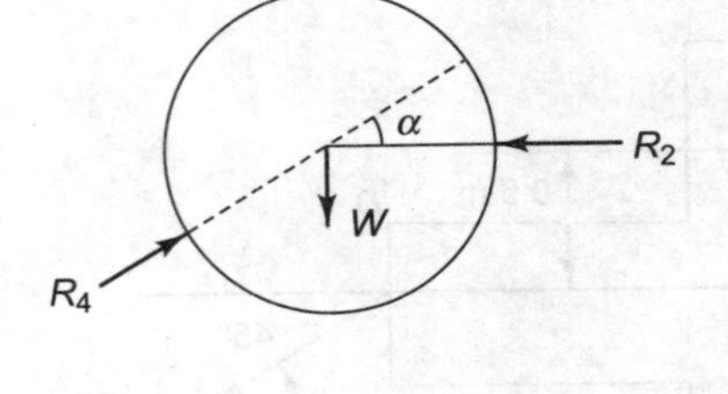

Free body diagram (Top roller)

Applying Lami's theorem on the top roller, we get

$$\frac{R_2}{\sin(90+\alpha)} = \frac{W}{\sin(180 - \alpha)} = \frac{R_4}{\sin 90}$$

$$R_2 = W \cot \alpha \quad \text{and} \quad R_4 = W \operatorname{cosec} \alpha$$

Now applying $\Sigma P_x = 0$ and $\Sigma P_y = 0$ on the bottom roller, we get

$$\Sigma P_x = 0$$

$$R_1 - R_4 \cos \alpha = 0$$

or

$$R_1 = R_4 \cos \alpha = W \operatorname{cosec} \alpha \times \cos \alpha$$

$$= W \cot \alpha$$

$$\Sigma P_y = 0$$

$$W - R_3 + R_4 \sin \alpha = 0$$

or

$$R_3 = W + W \operatorname{cosec} \alpha \sin \alpha$$

$$= 2W$$

Therefore

$$R_1 = W \cot \alpha$$

$$R_2 = W \cot \alpha$$

$$R_3 = 2W$$

$$R_4 = W \operatorname{cosec} \alpha$$

where

$$\alpha = \cos^{-1} \frac{b - 2r}{2r}$$

12. A 12-m boom AB weighs 1 kN, the distance of the centre of gravity G being 6 m from A. For the position shown, determine the tension T in the cable and the reaction at A. (UPTU: 2003–2004)

$$\Sigma M_A = 0$$

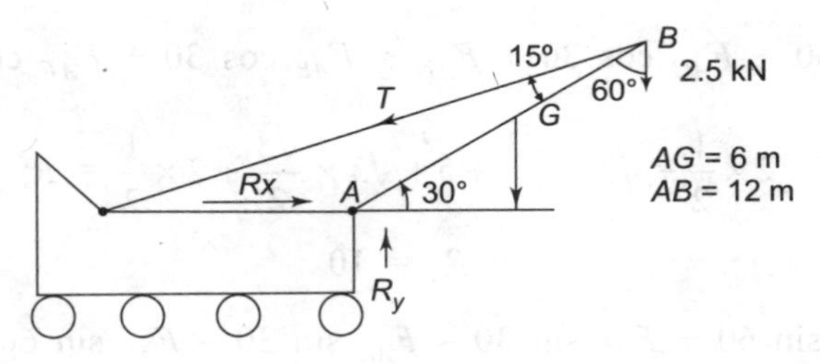

$$T \cos 75 \times 12 \cos 30 - T \sin 75 \times 12 \sin 30 + 2.5 \times 12 \cos 30 + 1 \times \frac{1}{2} \times 12 \cos 30 = 0$$

or $$0.259T \times 12 \times 0.866 - T \times 0.966 \times 12 \times \frac{1}{2} + 2.5 \times 12 \times 0.866 + 6 \times 0.866 = 0$$

or $$2.69T - 5.796T + 25.98 = 0$$

or $$T = \frac{25.98}{3.106} = 8.36 \text{ kN}$$

$$\Sigma P_x = 0, \; R_x = T \sin 75 = 0$$

$$R_x = 8.36 \times 0.966 = 8.08 \text{ kN}$$

$$\Sigma P_y = 0$$

$$R_y - 1 - 2.5 - 8.36 \cos 75 = 0$$

$$R_y = 5.65 \text{ kN}$$

$$R = \sqrt{R_x^2 + R_y^2} = \sqrt{65.2 + 31.9}$$

$$= 9.85 \text{ kN}$$

13. Forces 2, $\sqrt{3}$, 5, $\sqrt{3}$ and 2 kN respectively act at one of the angular point of a regular hexagon towards five other angular points. Determine the magnitude and direction of the resultant force. (UPTU Carryover: Aug. 2005)

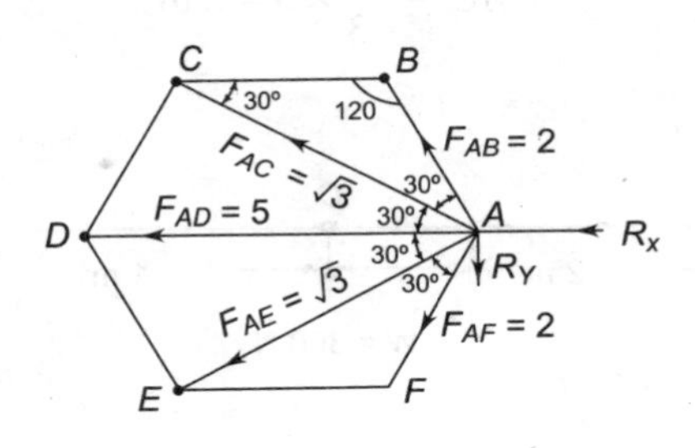

Let R_x and R_y be resultants. Take reverse direction forces for equilibrium.

$$F_{AB} \cos 60 + F_{AC} \cos 30 + F_{AD} + F_{AE} \cos 30 + F_{AF} \cos 60 = R_x$$

$$2 \times \frac{1}{2} + \sqrt{3} \times \frac{\sqrt{3}}{2} + 5 + \sqrt{3} \times \frac{\sqrt{3}}{2} + 2 \times \frac{1}{2} = R_x$$

or $$R_x = 10$$

$$F_{AB} \sin 60 + F_{AC} \sin 30 - F_{AE} \sin 30 - F_{AF} \sin 60 = R_y$$

or $$2 \sin 60 + \sqrt{3} \sin 30 - \sqrt{3} \sin 30 - 2 \sin 60 = R_y$$

or $$R_y = 0$$

$$R = \sqrt{R_x^2 + R_y^2} = 10$$

The magnitude of the resultant is 10 kN along negative x direction.

14. A beam supports a distributed load as shown below. Find the resultant of this loading and reactions at support A and B.

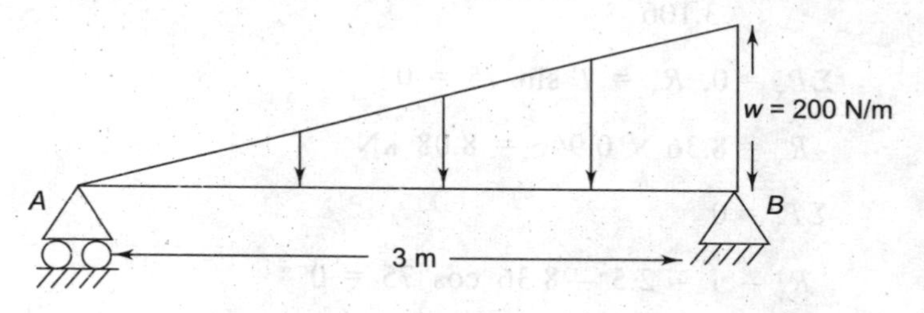

Total load for UVL (w = 200 N/m) is

$$W = \text{area of triangle} = \frac{1}{2} \times wL = \frac{1}{2} \times 200 \times 3$$

$$= 300 \text{ N}$$

Total load will act at $\frac{2}{3}$ length from point A,

$$AC = \frac{2}{3} \times 3 = 2 \text{ m}$$

Resultant loading is

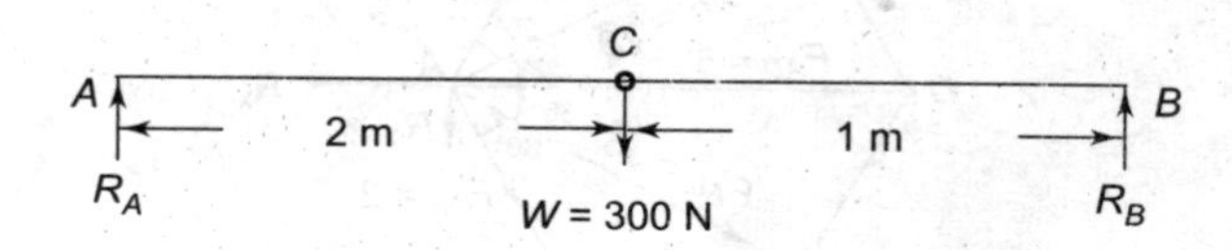

Now

$$\Sigma F_y = 0,$$

$$R_A + R_B = 300$$

Also

$$\Sigma M_A = 0,$$

$$300 \times 2 = R_B \times 3$$

or $$R_B = 200 \text{ N and } R_A = 100 \text{ N}$$

15. Two beams AB and CD are arranged and supported as shown below. Find the reaction at D due to a force of 1000 N acting at B.

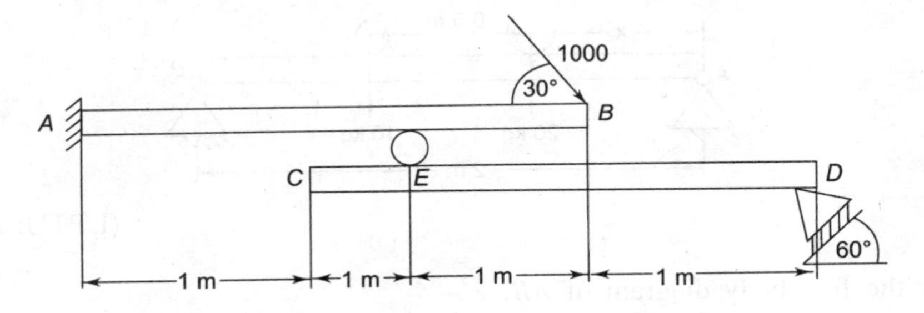

(a) Draw the free body diagram of AB:

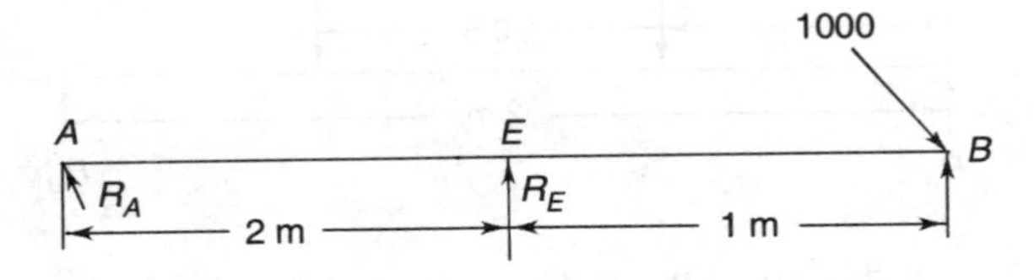

Now

$$\Sigma M_A = 0, \quad R_E \times 2 - 1000 \sin 30 \times 3 = 0$$

or $$R_E = \frac{1000 \times \frac{1}{2} \times 3}{2} = 750 \text{ N}$$

(b) Draw the free body diagram of CD:

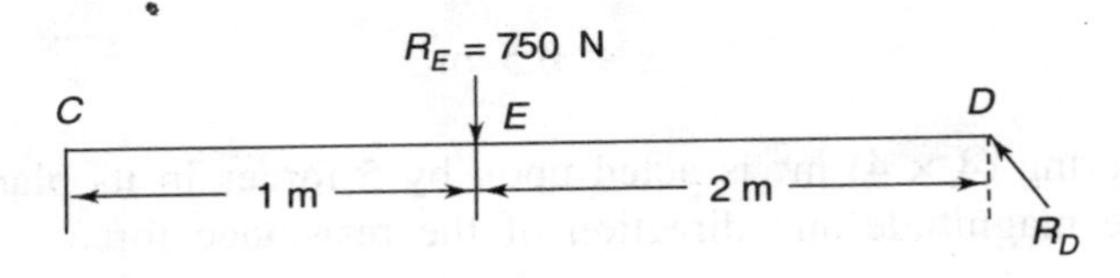

Now

$$\Sigma M_C = 0,$$

$$R_E \times 1 - R_D \sin 30 \times 3 = 0$$

or $$R_D = \frac{750}{3 \times \frac{1}{2}} = 500 \text{ N}$$

16. Two rollers of mass 20 kg and 10 kg rest on a horizontal beam as shown in the figure below with a massless wire fixing the two centres. Determine the distance x of the load 20 kg from the support A, if the reaction R_A is twice the support reaction R_B. The length of the beam is 2 m and the length of the connecting wire is 0.5 m. Neglect the weight of the beam. Assume the rollers to the point masses neglecting their dimensions.

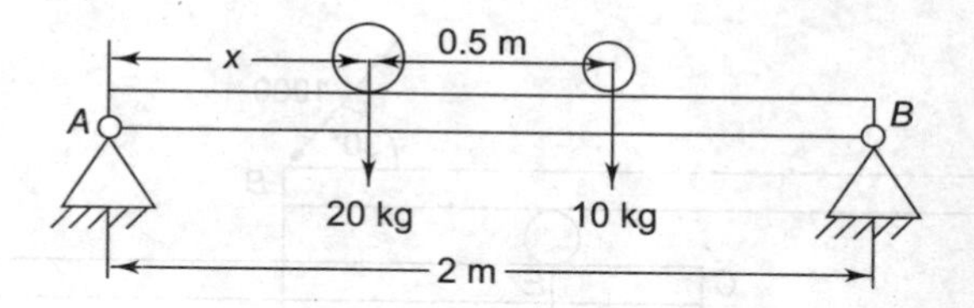

(UPTU: 2007–2008)

Draw the free body diagram of AB:

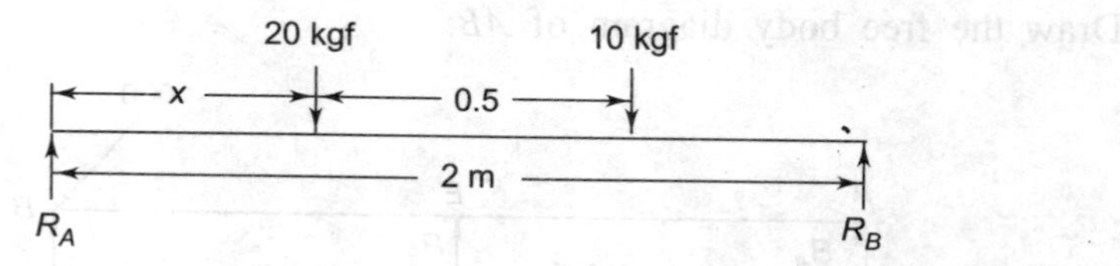

$$\Sigma F_y = 0, \quad R_A + R_B = 20 + 10 = 30 \text{ kgf}$$

Given $$R_A = 2R_B$$

$\therefore$ $$2R_B + R_B = 30$$

or $$R_B = 10 \text{ kgf}$$

$$\Sigma M_A = 0, \quad 20 \times x + 10(x + 0.5) - 10 \times 2 = 0$$

$$30x - 15 = 0$$

or $$x = 0.5 \text{ m}$$

17. A plate measuring (4×4) m^2 is acted upon by 5 forces in its plane as shown below. Determine the magnitude and direction of the resistance force.

(UPTU: 2006–2007)

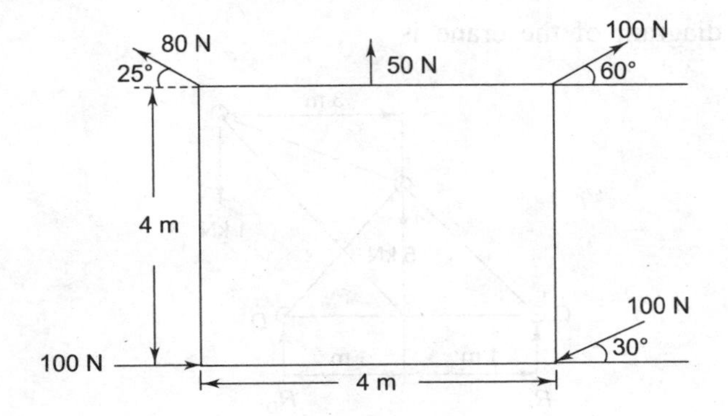

Let us find out

$$\Sigma F_y = R_y = 50 + 80 \sin 25 + 100 \sin 60 - 100 \sin 30$$
$$= 50 + 33.81 + 86.6 - 50$$
$$= 120.41$$

$$\Sigma F_x = R_x = -80 \cos 25 + 100 \cos 60 - 100 \cos 30 + 100$$
$$= -72.5 + 50 - 86.6 + 100$$
$$= -9.1$$

$$R = \sqrt{R_x^2 + R_y^2}$$

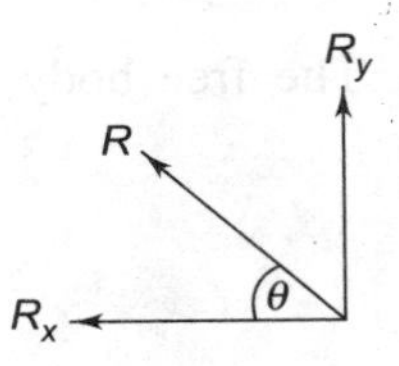

$$= \sqrt{(-9.1)^2 + (120.41)^2}$$
$$= 120.75$$

$$\tan \theta = \frac{R_y}{R_x} = \frac{120.41}{9.1} = 13.27$$

$$\theta = 85.7°$$

18. Find the support reactions in the beam shown below.

(UPTU: 2000–2001)

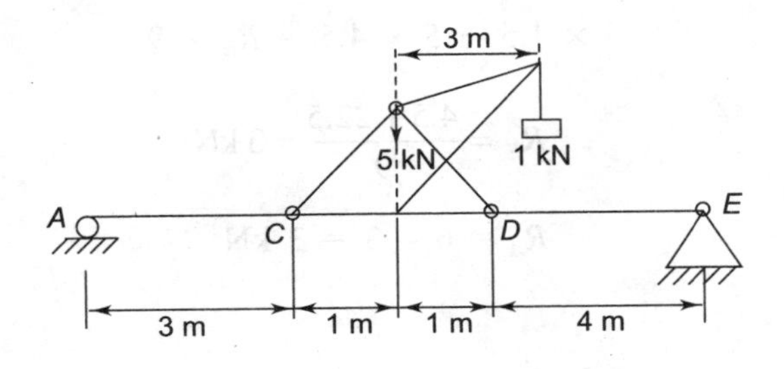

The free body diagram of the crane is

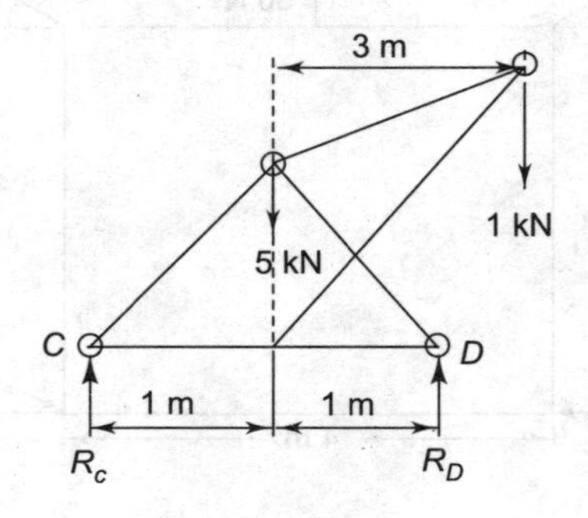

Now

$$\Sigma F_y = 0,$$

$$R_C + R_D = 5 + 1 = 6$$

$$\Sigma M_C = 0,$$

$$R_D \times 2 = 5 \times 1 + 1 \times 4 = 9$$

$$R_D = 4.5 \text{ kN}$$

$\therefore$

$$R_C = 6 - 4.5 = 1.5 \text{ kN}$$

The free body diagram of the beam AB is

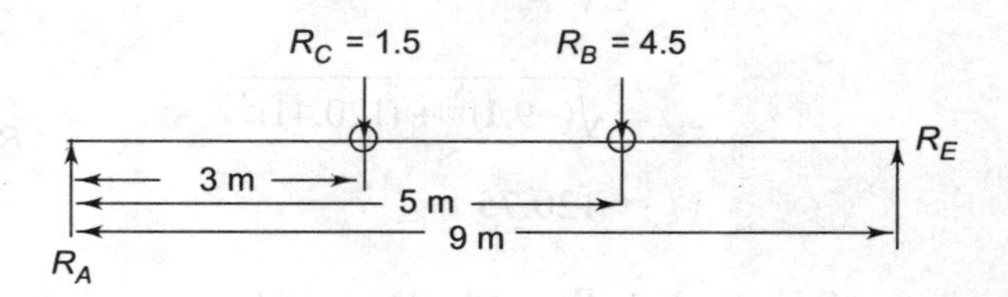

Now

$$\Sigma F_y = 0,$$

$$R_A + R_E = 6$$

Also

$$\Sigma M_A = 0,$$

$$3 \times 1.5 + 5 \times 4.5 = R_E \times 9$$

or

$$R_E = \frac{4.5 + 22.5}{9} = 3 \text{ kN}$$

$\therefore$

$$R_A = 6 - 3 = 3 \text{ kN}$$

19. Find the reactions at A and B for the beam shown below.

(UPTU: 2006–2007)

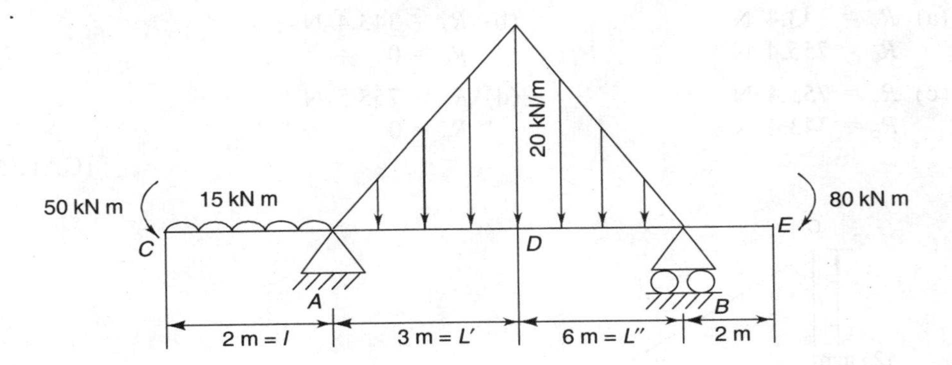

The equivalent loading is

$$wl = 15 \times 2 = 30 \text{ N (equivalent of UDL)}$$

$$\frac{1}{2} L' \times w = \frac{1}{2} \times 3 \times 20 = 30 \text{ N} \quad \text{(equivalent of UVL–left triangle)}$$

$$\frac{1}{2} L'' \times w = \frac{1}{2} \times 6 \times 20 = 60 \text{ N} \quad \text{(equivalent of UVL–right triangle)}$$

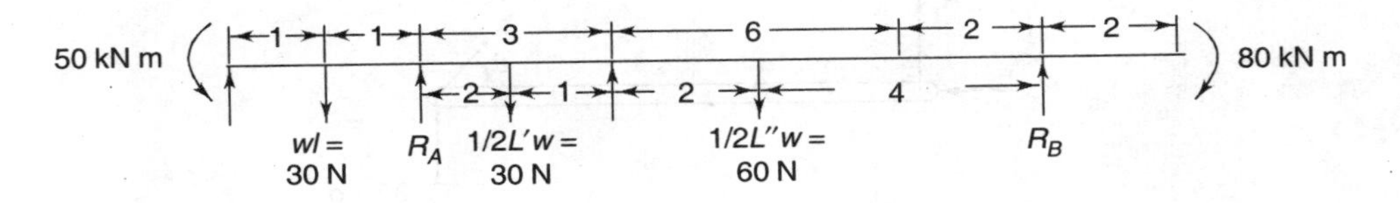

Now, we have

$$\Sigma F_y = 0,$$

$$-30 + R_A - 30 - 60 + R_B = 0$$

or

$$R_A + R_B = 120$$

$$\Sigma M_A = 0,$$

$$-50 - 30 \times 1 + 30 \times 2 + 60 \times 5 - R_B \times 9 + 80 = 0$$

$$R_B = \frac{360}{9} = 40 \text{ N}$$

$\therefore$

$$R_A = 120 - 40 = 80 \text{ N}$$

20. A mass of 35 kg is suspended from a weightless bar AB which as supported by a cable CB and a pin at A as shown in the figure. The pin reaction at A on the bar AB are:

(a) $R_x = 343.4$ N
$R_y = 755.4$ N

(b) $R_x = 343.4$ N
$R_y = 0$

(c) $R_x = 755.4$ N
$R_y = 343.4$ N

(d) $R_x = 755.5$ N
$R_y = 0$

(GATE: 1997)

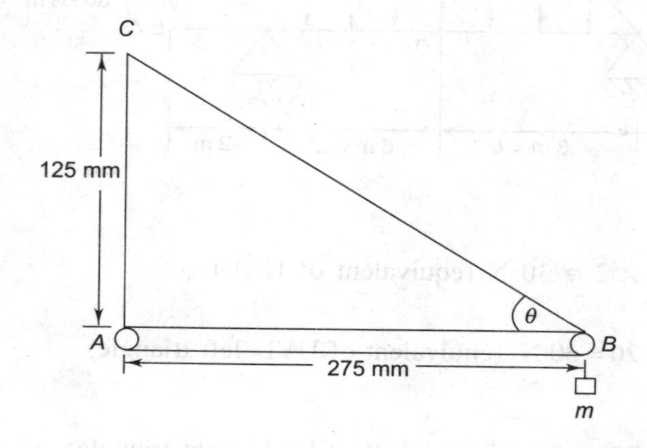

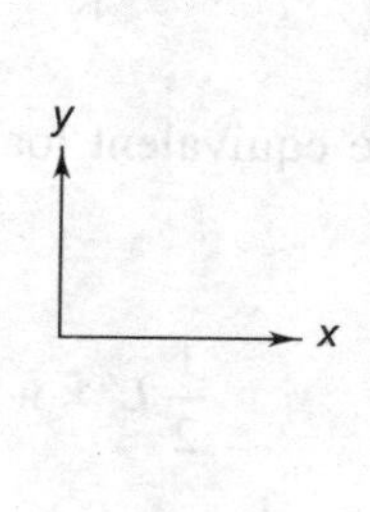

The free body diagram of the bar:

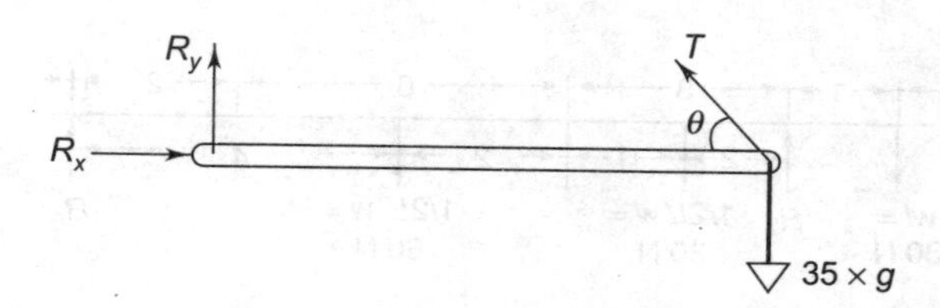

$$\cos\theta = \frac{AB}{BC} = \frac{275}{\sqrt{275^2 + 125^2}}$$

$$\theta = 24.5°$$

$$\Sigma F_y = 0,$$

$$R_y + T\sin\theta = 35 \times 9.81$$

$$\Sigma F_x = 0,$$

$$R_x = T\cos\theta$$

However, $R_y = 0$. Therefore,

$$T = \frac{35 \times 9.81}{\sin 24.5} = 830 \text{ N}$$

and $$R_x = 830 \cos 24.5$$
$$= 755.4 \text{ N}$$

Hence, option (d) is correct.

21. The following figure shows a rigid bar hinged at A and supported in a horizontal by two vertical identical wires. Neglect the weight of the beam. The tension T_1 and T_2 induced in these wires by a vertical load P applied as shown are

(a) $$T_1 = T_2 = \frac{P}{2}$$

(b) $$T_1 = T_2 = P$$

(c) $$T_1 = \frac{Pal}{a^2 + b^2}, T_2 = \frac{Pbl}{a^2 + b^2}$$

(d) $$T_1 = \frac{Pbl}{2(a^2 + b^2)}$$
$$T_2 = \frac{Pal}{2(a^2 + b^2)}$$

(GATE: 1994)

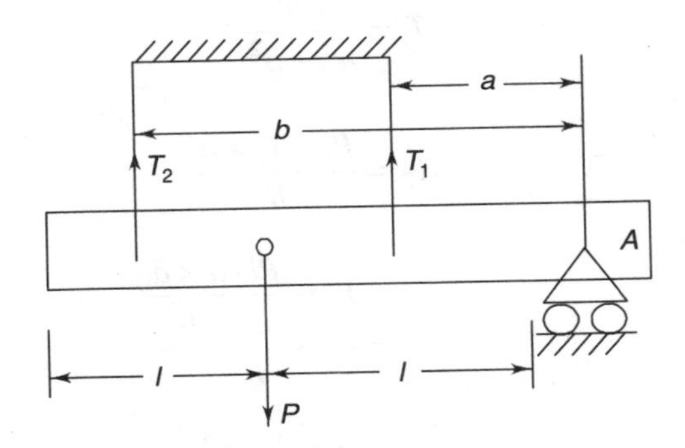

Taking moment about A,

$$T_2 \times b + T_1 \times a = P \times l \quad \text{(i)}$$

Now elongation in wire ∂l_1 and ∂l_2

$$\partial l_1 = \frac{T_1 \times L}{AE} \text{ and } \partial l_2 = \frac{T_2 L}{AE}$$

$$\therefore \quad \frac{\partial l_1}{\partial l_2} = \frac{T_1}{T_2} = \frac{a}{b}$$

or $$\frac{T_2}{b} = \frac{T_1}{a}$$

or $$T_2 = \frac{b}{a} \times T_1$$

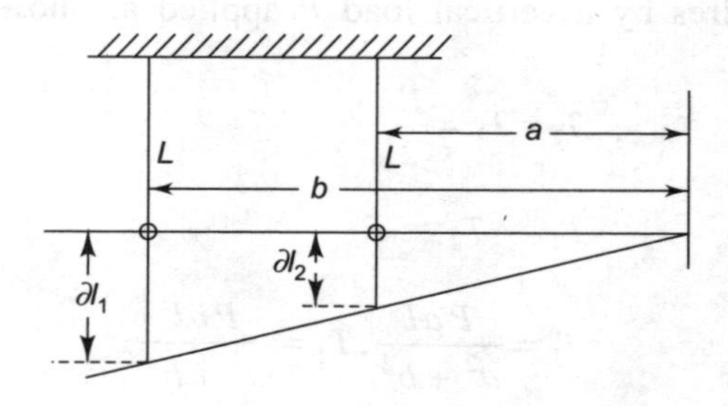

From Eq. (i)

$$T_1 \times \frac{b^2}{a} + T_1 a = P \times l$$

or $$T_1(b^2 + a^2) = Pal$$

or $$T_1 = \frac{Pal}{a^2 + b^2}$$

$$T_2 = \frac{Pl - T_1 \times a}{b}$$

$$= \frac{Pl - \dfrac{Pl \cdot a \times a}{a^2 + b^2}}{b}$$

$$= \frac{Pl(a^2 + b^2 - a^2)}{b(a^2 + b^2)}$$

$$= \frac{Plb}{a^2 + b^2}$$

Hence, option (c) is correct.

22. A roller of weight W is rolled over the wooden block as shown in the figure. The pull F required to just cause the said motion is

(a) $\frac{W}{2}$ (b) W (c) $\sqrt{3}W$ (d) $2W$

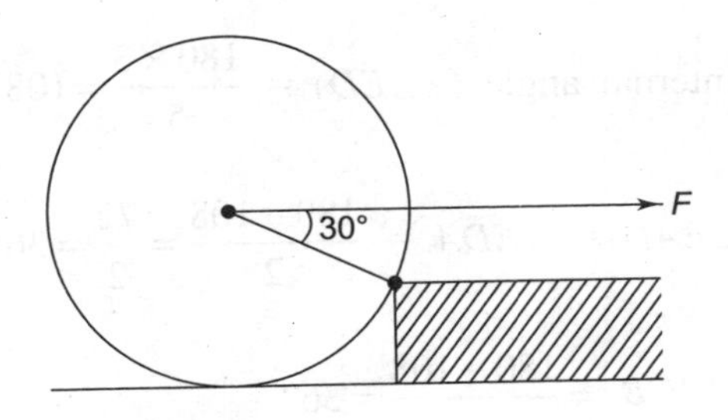

The free body diagram of the roller is

$$\Sigma F_y = 0, \quad R \cos 60 = W$$

or $$R = 2W$$

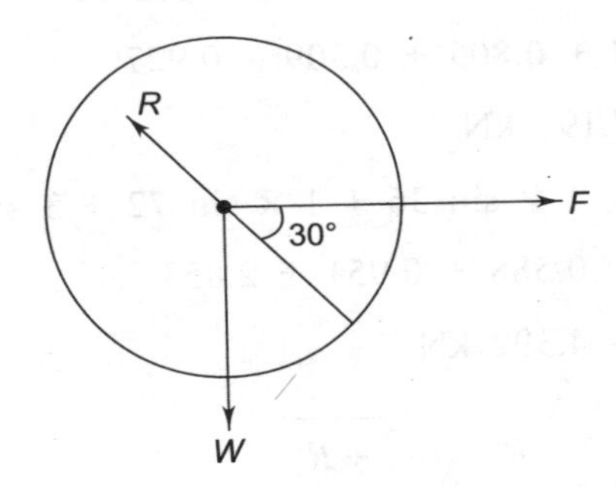

$$\Sigma F_x = 0, \quad F = R \sin 60$$

or $$F = 2W \times \frac{\sqrt{3}}{2}$$

$$= \sqrt{3}\,W$$

Therefore, option (c) is correct.

23. Forces 7, 1, 1 and 3 kN act at one of the angular points of a regular pentagon towards four other angular points taken in order. Obtain the resultant of this force system. What is the direction?

(UPTU: May 2008)

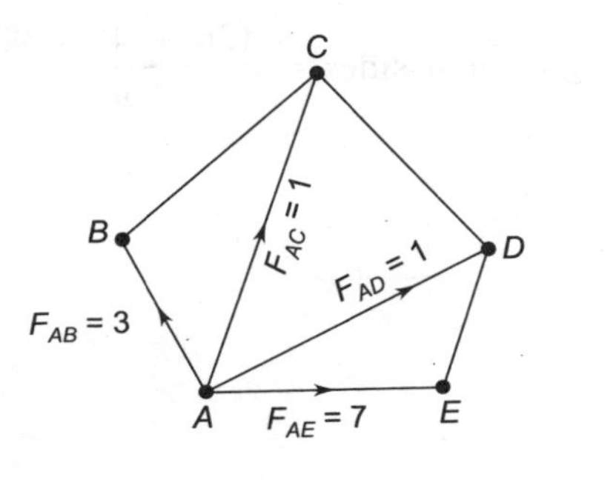

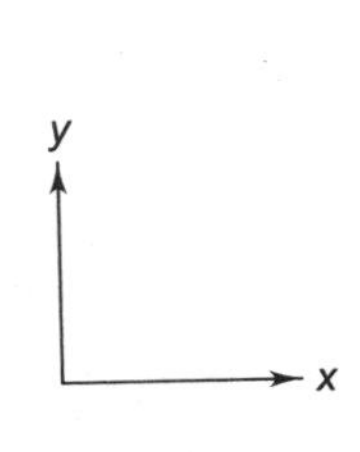

$$\text{Internal angle } (\angle AED) = \frac{180 \times 3}{5} = 108°$$

$$\angle EAD = \angle EDA = \frac{180 - 108}{2} = \frac{72}{2} = 36°$$

Similarly, $\angle CAB = \angle ACB = \dfrac{180 - 108}{2} = 36°$

$\therefore$ $\angle DAC = 108 - 36 - 36$

$= 36°$

$$\Sigma R_x = 7 + 1 \times \cos 36 + 1 \times \cos 72 - 3 \cos 72$$

$$= 7 + 0.809 + 0.309 - 0.927$$

$$= 7.191 \text{ kN}$$

$$\Sigma R_y = 1 \times \sin 36 + 1 \times \sin 72 + 3 \sin 72$$

$$= 0.588 + 0.951 + 2.853$$

$$= 4.392 \text{ kN}$$

$\therefore$

$$R = \sqrt{R_x^2 + R_y^2}$$

$$= \sqrt{7.191^2 + 4.392^2}$$

$$= \sqrt{51.71 + 19.289} = 8.426 \text{ kN}$$

Now $\tan\theta = \dfrac{R_y}{R_x} = \dfrac{4.392}{7.191}$

or $\theta = 31.42°$

Hence the resultant is 8.426 kN acting at 31.42° at point *A*.

Note: The interior angle of the polygon of n sides $= \dfrac{(2n - 4) \times 90}{n}$

OBJECTIVE TYPE QUESTIONS

Great men are not born great. God uses troubles and trials to make them great.

State True or False

1. Mechanics is science which deals with state of rest or state of motion of bodies under the action of forces. *(True/False)*

2. Statics is the branch of mechanics which relates to bodies at rest. *(True/False)*

3. Dynamics is the branch of mechanics which deals with bodies in motion. *(True/False)*

4. There are five fundamental laws of mechanics which are called basic laws. *(True/False)*

5. The weight of a body (mass = m) in an accelerating lift (a m/s^2) will be $m(a + g)$ when moving up and $m(g - a)$ when moving down where g is the acceleration due to gravity. *(True/False)*

6. The weight of a free-falling body of mass (m) will be zero. *(True/False)*

7. The state of a body is unaltered if force is replaced by unlike parallel forces of the same magnitude acting along the line of action of replaced force. *(True/False)*

8. If two forces in magnitude and direction are given by the sides of a parallelogram, the diagonal will give in magnitude and direction the resultant passing through their point of contact. *(True/False)*

9. If F_1 and F_2 are two forces with in between angle θ, then the resultant (R) is $R^2 = F_1^2 + F_2^2 + F_1F_2 \cos \theta$. *(True/False)*

10. If two forces are represented by the sides of a triangle in magnitude and direction, then the closing side will show the resultant in magnitude but in the opposite direction. *(True/False)*

11. If a number of concurrent forces acting on a body are represented by the sides of a polygon, then the resultant will be represented by the closing side in the reverse order. *(True/False)*

12. If a coplanar concurrent force system forms a closing polygon, then resultant is zero. *(True/False)*

13. Coplanar forces pass through a single point. *(True/False)*

14. Concurrent forces are required to be in one plane. *(True/False)*

15. Coplanar non-concurrent forces are collinear. *(True/False)*

16. Moment tends to rotate a body, i.e. rotational motion. *(True/False)*

17. Force tends to move a body, i.e. translatory motion. *(True/False)*

18. A body cannot have both translatory and rotational motion. *(True/False)*

19. Moment is the product of magnitude of the force with the moment arm. *(True/False)*

20. The value of moment remains unaltered if the direction of the force is reversed. (*True/False*)

21. The value of two moments of the same magnitude but opposite direction is zero. (*True/False*)

22. Two unlike and equal forces separated by any distance will form a couple. (*True/False*)

23. Couples of 10 and 20 N m can be replaced with a single couple of 30 N m. (*True/False*)

24. Couples of 50 N m (clockwise) and 30 Nm (anticlockwise) can be replaced with a couple of 20 N m (clockwise). (*True/False*)

25. A couple will have changed value when moment centre is changed. (*True/False*)

26. A force acting at a point on a body can be replaced by a parallel force acting at some other point and a couple. (*True/False*)

27. Weight is a non-applied force which acts vertically downwards. (*True/False*)

28. Reactions are non-applied forces which are equal and opposite to forces exerted by the body on the surfaces. (*True/False*)

29. For smooth surface, the direction of reaction is normal to the surface of contact. (*True/False*)

30. For a coplanar concurrent force system, the resultant is equal to $\sqrt{(\Sigma P_x)^2 + (\Sigma P_y)^2}$. (*True/False*)

31. If a coplanar concurrent force system is in equilibrium, then $\Sigma P_x = 0$, $\Sigma P_y = 0$. (*True/False*)

32. The resultant of a coplanar force system will intercept *x*-axis at $x = \dfrac{\Sigma M_0}{\Sigma P_y}$. (*True/False*)

33. If a body is in equilibrium under a force system consisting of three forces, then each force is proportional to sine of the angle between other two forces. (*True/False*)

34. The condition for equilibrium of a noncurrent coplanar force system is $\Sigma P_x = 0$, $\Sigma P_y = 0$. (*True/False*)

35. Three unknown forces can be found out by Lami's theorem. (*True/False*)

Multiple Choice Questions

1. A body can be in equilibrium under the action of
   (a) two unequal and opposite forces
   (b) two equal and opposite forces
   (c) two equal and opposite parallel forces

2. A couple can be balanced by
   (a) an equal and opposite couple
   (b) another couple
   (c) a torque

3. What do we apply while opening the cap of a bottle with an opener?
(a) force (b) torque (c) moment

4. What is applied while screwing a cap on a bottle
(a) force (b) torque (c) couple

5. What do we apply to start a home generator with a rope-pully arrangement?
(a) force (b) torque (c) couple

6. The condition for equilibrium for a coplanar non-current force system is
(a) $\Sigma P_x = 0, \Sigma P_y = 0$ (b) $\Sigma P_x = 0, \Sigma M = 0$ (c) $\Sigma P_x = 0, \Sigma P_y = 0, \Sigma M = 0$

7. The condition for equilibrium for a coplanar concurrent force system is
(a) $\Sigma P_x = 0, \Sigma P_y = 0$ (b) $\Sigma P_x = 0, \Sigma M = 0$ (c) $\Sigma P_x = 0, \Sigma P_y = 0, \Sigma M = 0$

8. The effect of a given force remains unaltered along the line of action. This is according to
(a) resolution (b) law of motion (c) law of transmissibility

9. The algebraic sum of the moments of a coplanar force system is equal to the moment of the resultant about the same moment centre is as per
(a) Varignon's theorem (b) law of resolution (c) Newton's law

10. The third unknown force of a coplanar force system in equilibrium will be given by
(a) Varignon's theorem (b) triangle of forces (c) parallelogram law of forces

11. If a coplanar concurrent force system has $\Sigma P_x = 0$ but $\Sigma P_y \neq 0$, then the resultant will be
(a) a force acting upwards or downwards
(b) a force having magnitude zero
(c) a force inclined to x-axis

12. What do we apply while we cycle?
(a) moment (b) torque (c) couple

13. The resultant of two forces (each with a magnitude of $P/2$) acting at a right angle is
(a) $P/2$ (b) $\frac{P}{\sqrt{2}}$ (c) $\sqrt{2}\,P$

14. The resultant of two forces (each P) acting at an angle of 60° is
(a) $2P$ (b) $\sqrt{3}\,P$ (c) $2/3\ P$

15. The resultant of two forces P_1 and P_2 is R. If P is doubled and the new resultant becomes perpendicular to P_2, then
(a) $P_1 = P_2$ (b) $P_2 = R$ (c) $P_1 = R$

16. If two forces of 7 N and 8 N act at 60°, then the resultant will be
(a) 10 N (b) 15 N (c) 13 N

17. If two forces of magnitude P acts at angle of θ. Then resultant will be
(a) $2P \cos \theta$ (b) $P \cos 2\theta$ (c) $2P \cos \theta/2$

18. If the resultant of two equal forces has the same magnitude then, the angle between them is
(a) 120° (b) 60° (c) 90°

19. The angles between two forces, when the resultant is maximum and minimum, are
(a) 180° and 0° (b) 90° and 0° (c) 0° and 180°

20. Three forces acting on a rigid body is represented by the three sides of a triangle. The forces are equivalent to a couple whose moment is equal to
(a) the area of the triangle
(b) half the area of the triangle
(c) twice the area of the triangle

21. If three coplanar and concurrent force systems as shown are in equilibrium, then

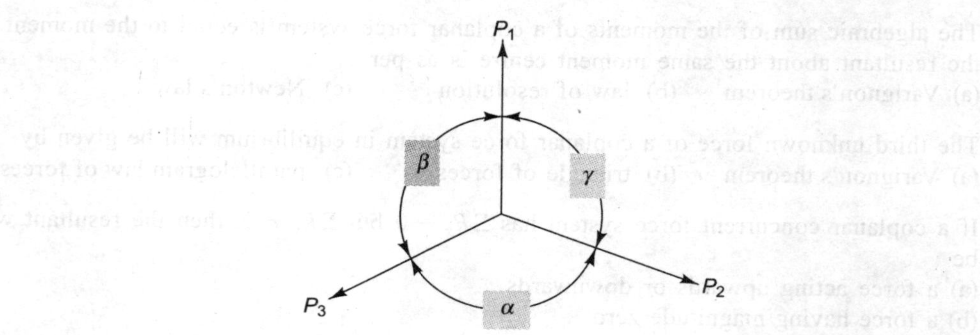

(a) $\dfrac{P_1}{\cos \alpha} = \dfrac{P_2}{\cos \beta} = \dfrac{P_3}{\cos \gamma}$

(b) $\dfrac{P_1}{\sin \gamma} = \dfrac{P_2}{\sin \alpha} = \dfrac{P_3}{\sin \beta}$

(c) $\dfrac{P_1}{\sin \alpha} = \dfrac{P_2}{\sin \beta} = \dfrac{P_3}{\sin \gamma}$

22. If a coplanar non-concurrent force system acts on a body and $\Sigma P_x = 0$ and $\Sigma P_y = 0$, then it may be
(a) at rest (b) moving in one direction (c) rotating about itself

23. According to which of the following, if three coplanar and concurrent forces are in equilibrium, then each force is proportional to the sine angle between other two?
(a) Varignon's theorem (b) Lami's theorem (c) Law of transmissibility

24. An automobile of weight W is as shown. A pull P is applied horizontally. The reaction at the front wheel is

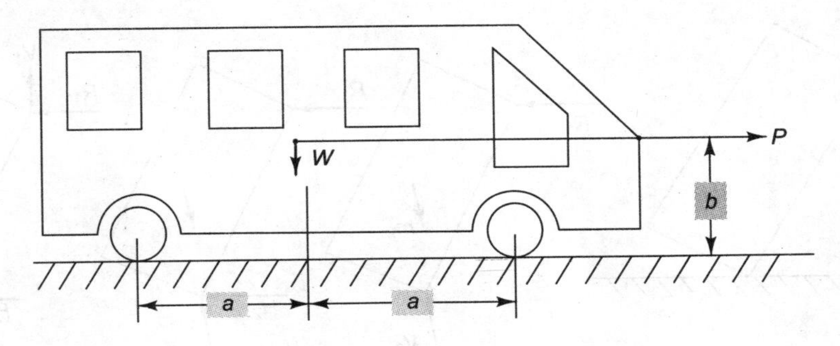

(a) $\dfrac{W}{2} + \dfrac{Pb}{2a}$ (b) $\dfrac{W}{2} - \dfrac{Pb}{2a}$ (c) $\dfrac{W}{2} - \dfrac{P_a}{2b}$

25. A simply supported beam carries a load P through a bracket as shown. The relation R_a at A will be ___

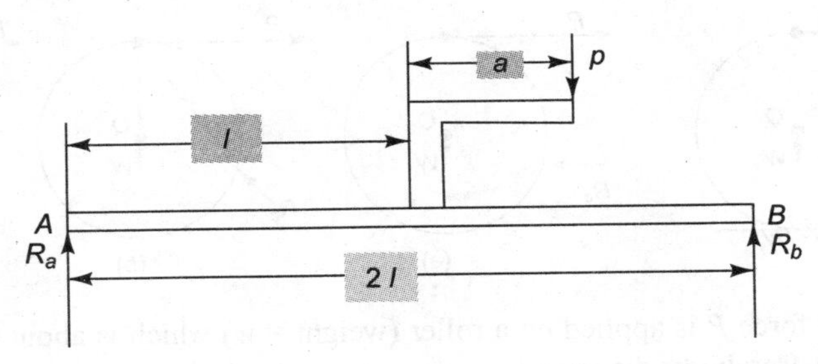

(a) $R_a = \dfrac{P(l-a)}{l}$ and $R_b > R_a$

(b) $R_a = \dfrac{P(l+a)}{l}$ and $R_a > R_b$

(c) $R_a = \dfrac{2P(l-a)}{l}$ and $R_a > R_b$

26. A roller is hold against a wall with a beam- and string-arrangement. If R_w = reaction of wall, R_b = reaction of beam and w = weight of the roller, then which figure shows a correct free body diagram of the roller?

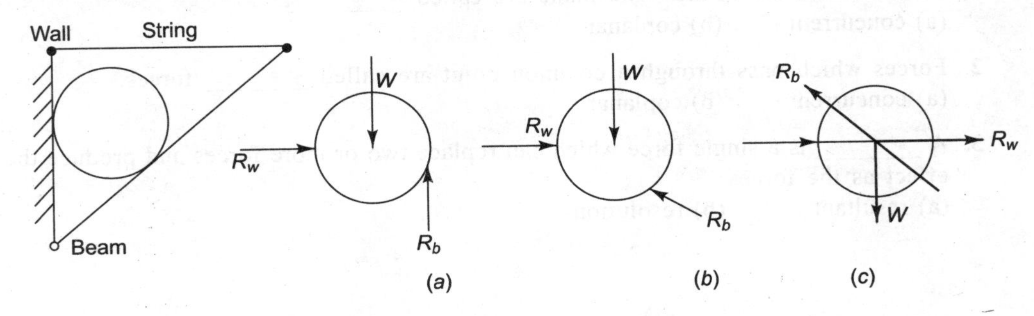

27. A rectangular crate of mass m rests against a smooth wall and a rough floor as shown in the figure. The free body diagram for the crate is

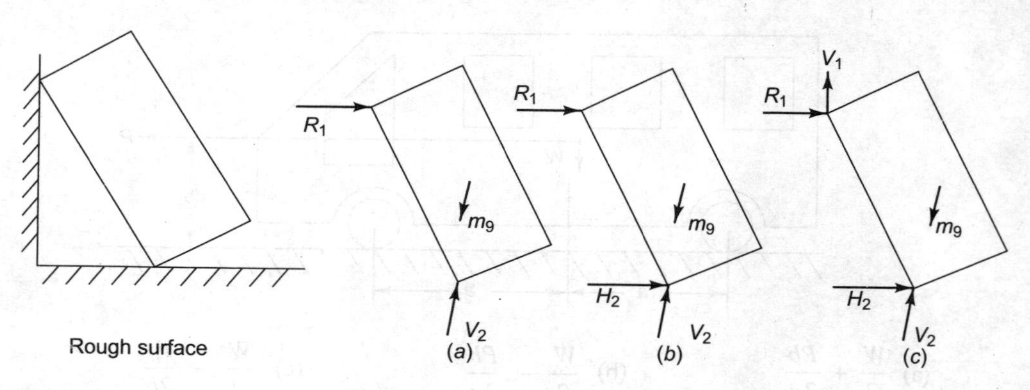

28. If a roller (radius = r) is about to climb the curb (height = $r/2$), then the free body diagram is

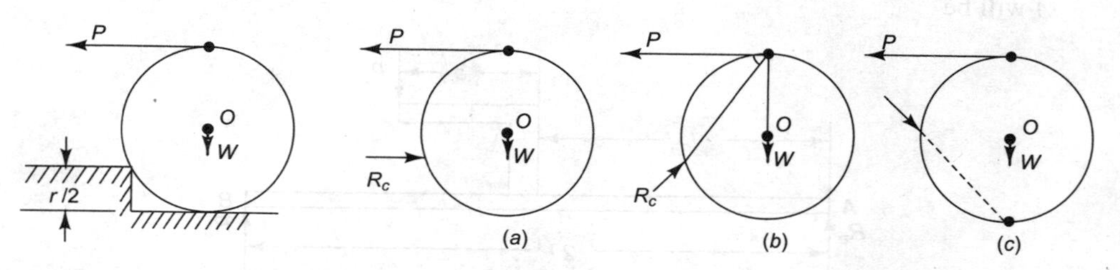

29. If a horizontal force P is applied on a roller (weight = w) which is about to climb curb (height = r_2), then the free body diagram is

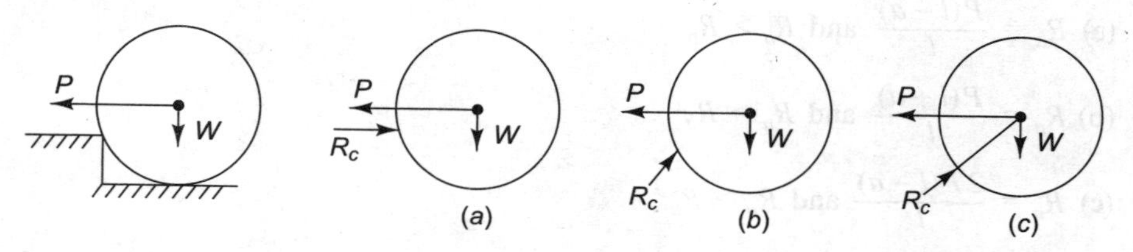

Fill in the Blanks

1. Forces which are in the same plane are called ________.
   (a) concurrent (b) coplanar
2. Forces which pass through a common point are called ________ forces.
   (a) concurrent (b) coplanar
3. A ________ is a single force which can replace two or more forces and produce the same effect as the forces.
   (a) resultant (b) resolution

4. The splitting of a force into two perpendicular directions without changing its effect is called
   (a) resultant (b) resolution

5. The square of the resultant of forces P_1 and P_2 with a θ angle between them is ________.
   (a) $P_1^2 + P_2^2 + 2P_1P_2$
   (b) $P_1^2 + P_2^2 + 2P_1P_2 \cos \theta$

6. If coplanar and concurrent forces are in equilibrium, then the polygon drawn of these forces will be ________.
   (a) open (b) closed

7. If coplanar and concurrent forces are in equilibrium, then the condition for equilibrium will be ________.
   (a) $\Sigma P_x = 0$, $\Sigma P_y = 0$
   (b) $\Sigma P_x = 0$, $\Sigma P_y = 0$ and $\Sigma M = 0$

8. A body isolated from other bodies which are connected with it, and is subjected to all applied and non-applied forces is called ________ diagram.
   (a) an isolated body (b) a free body

9. If the resultant is equal to both the forces, then the angle between the forces is________.
   (a) 90° (b) 20°

10. The ________ is a rotational tendency of a force.
    (a) moment (b) resolution

11. The moment of two parallel and equal forces is a ________ and equal to force multiplied by the distance between forces.
    (a) torque (b) couple

12. Magnitude, direction, sense and ________ are characteristics of a force.
    (a) point of application (b) active

13. Lami's theorem can be applied for three forces which are ________.
    (a) coplanar (b) concurrent

14. The steering wheel of a car is an example of ________.
    (a) torque (b) couple

15. When we crank a car to start it, then we apply a ________.
    (a) moment (b) couple

16. A ________ is applied while opening a water tap.
    (a) couple (b) moment

17. Conventionally a clockwise moment is taken as ________ moment.
    (a) negative (b) positive

ANSWERS

To succeed in a new field, make a new discovery, or do anything new in life, you have to be willing to fail.

State True or False

1. True
2. True
3. True
4. False (There are six fundamental laws of mechanics)
5. True
6. True (weight = $m(g - a)$. For free falling body $a = g$ and hence weight = $m \times 0 = 0$)
7. False (It is to be like and in line force.)
8. True
9. False ($R^2 = F_1^2 + F_2^2 + 2F_1F_2 \cos\theta$)
10. True
11. True
12. True
13. False
14. False (Concurrent forces can be coplanar or non-coplanar.)
15. False
16. True
17. True
18. False (A body can have both translatory and rotational motion, e.g. a wheel of a car.)
19. True
20. False (The value becomes negative.)
21. True
22. True
23. True ($\Sigma M = 10 + 20 = 30$ N m)
24. True ($\Sigma M = 50 - 30 = 20$ N m)
25. False (A couple does not depend upon moment centre but the shortest distance between the forces.)
26. True
27. True
28. True
29. True
30. True
31. True
32. True
33. True (Lami's theorem)
34. False (Conditions are $\Sigma P_x = 0$, $\Sigma P_y = 0$ and $\Sigma M_y = 0$)
35. False (Only two unknown forces).

Multiple Choice Questions

1. (b)
2. (a)
3. (a)
4. (c)
5. (b)
6. (c)
7. (a)
8. (c)
9. (a)
10. (b)
11. (a)
12. (c)
13. (b) $\left[R^2 = \left(\frac{P}{2}\right)^2 + \left(\frac{P}{2}\right)^2 + 2 \times \frac{P}{2} \times \frac{P}{2} \times 0 \text{ or } R = \frac{P}{\sqrt{2}}\right]$

14. (b) ($R^2 = P^2 + P^2 + 2P \cdot P \cdot \frac{1}{2} = 3P^2$ or $R = \sqrt{3}P$)

15. (c) (If AB is extended to D so that $AB = BD$ and CD becomes perpendicular to AC. Hence points A, C and D are lying on the circle. Therefore $P_1 = R_1$ = radius)

16. (c) ($R^2 = 7^2 + 8^2 + 2 \times 7 \times 8 \times \cos 60 = 49 + 64 + 56 = 169$ or $R = 13$)

17. (c) ($R^2 = P^2 + P^2 + 2P^2 \cos\theta = 2P^2 (1 + \cos\theta) = 2P^2 \times 2\cos^2\theta/2 = 4P^2\cos^2\theta/2$ or $R = 2P\cos\theta/2$

18. (a) ($R^2 = 4P^2\cos^2\theta/2 = P^2$, $\cos\theta/2 = \pm 1/2$ or $\theta = 120°$)

19. (c) (Maximum when both forces are collinear and acting in the same direction. Minimum when both forces are collinear but in the opposite direction.)

20. (c)

21. (c)

22. (c) (Unless $\Sigma M = 0$, the body will have rotational motion.)

23. (b)

24. (a) $\left(\Sigma M_{\text{backwheel}} = 0,\ R \times 2a = W \times a + P \times b \text{ or } R = \frac{W}{2} + P \times \frac{b}{2a}\right)$

25. (a) $\left[\Sigma M_b = 0,\ R_a \times 2l - P(l - a) = 0 \text{ or } R_a = \frac{P(l-a)}{zl}\right]$

26. (c)

27. (b)

28. (b) (Forces will form a concurrent force system.)

29. (c) (Forces will form a concurrent force system.)

Fill in the Blanks

1. (b)	**2.** (a)	**3.** (b)	**4.** (b)
5. (b)	**6.** (b)	**7.** (a)	**8.** (b)
9. (b)	**10.** (a)	**11.** (b)	**12.** (a)
13. (b)	**14.** (b)	**15.** (a)	**16.** (a)
17. (a)			

CHAPTER 9
Friction

> *Procrastination will be overcome when you take little steps toward the goal. With each step you gain momentum, and with consistency you'll be crossing the finish line before you know it.*

INTRODUCTION

When two surfaces are in contact, burr and roughness get interlocked thereby making movement difficult.

When a body moves or tends to move over another body, a force opposing this motion is developed at the contact surface. The force that opposes the movement or tendency of movement is called *frictional force* or *friction*.

Frictional force has a property of adjusting its magnitude to the force trying to produce motion to the body so that motion is prevented. The magnitude of friction cannot be increased beyond a limit which is known as *limiting* or *maximum friction force*. If applied force is more than limiting friction, the body will move. The static friction is friction force till the body is stationary. Dynamic friction is friction force acting when body is moving.

A body (weight = W) is being pulled with a force P (Figure 9.1). The body is stationary till force P increases to limiting friction (F). The coefficient of friction (μ) is defined as the ratio of limiting friction force (F) to normal reaction of the body.

$$\mu = \frac{F}{N}$$

where, N = normal reaction = W

If λ = friction angle

$$\tan \lambda = \mu = \frac{F}{N}$$

A body is stationary if $F > P$ and the body starts moving if $P > F$.

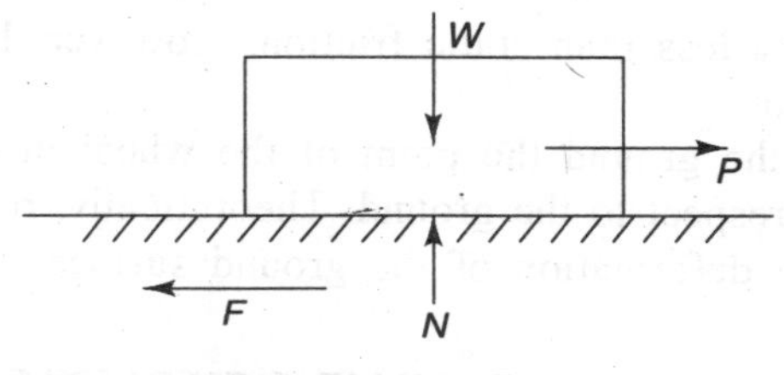

FIGURE 9.1 Friction.

If we draw a diagram (Figure 9.2) of applied force (P) against friction force (F), the friction force will increase linearly with the increase of the applied force (OA). The friction force cannot increase after it reaches the value of the limiting friction. Now if applied force (P) is increased, the body starts moving and even friction force falls (AB). The body moves with constant dynamic friction force along BC.

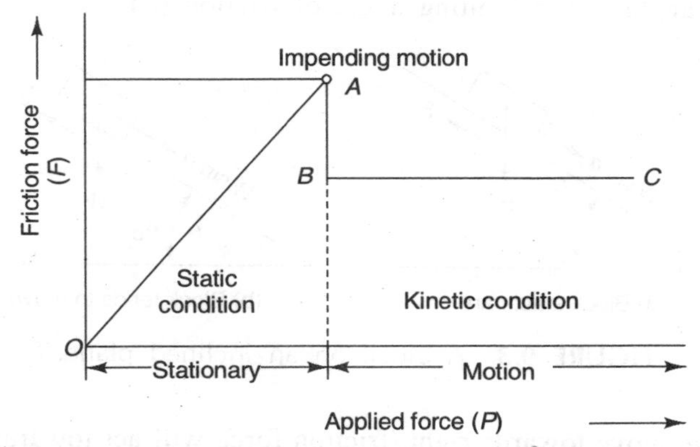

FIGURE 9.2 Friction force (F) vs applied force (P).

COULOMB'S LAW OF FRICTION

Coulomb's law of friction encompasses that

1. The force of friction always acts in a direction opposite to the direction in which the body tends to move.
2. The magnitude of friction force is equal to the applied force till it remains stationary.
3. The limiting friction bears a constant ratio to the reaction force between the two surfaces.
4. The friction force depends upon the roughness/smoothness between the surfaces.
5. The force of friction does not depend on the area of contact.
6. Dynamic friction has a lower value than that of the limiting friction.

Dynamic friction force is given as $F = \mu_k N$ where N is normal reaction force and μ_k is coefficient of dynamic friction which is always less than static friction (μ). Dynamic friction is about 20% to 25% less than static friction. However, laws of dynamic friction are the same as static friction.

When a wheel rolls on the ground the point of the wheel in contact with the ground has no relative motion with respect to the ground. Theoretically, rolling friction is zero but rolling friction exists due to deformation of the ground surface.

THE ANGLE OF REPOSE AND THE CONE OF FRICTION

If a block of weight W is put on an inclined plane and it is stationary, this means friction force (F) is greater than sliding force ($W \sin \alpha$) acting along the plane due to weight of the block (Figure 9.3a). However, on increasing the inclination angle to a critical angle, the sliding force ($W \sin \alpha$) will become greater than friction force and the body tends to slide (Figure 9.3b). The steepest angle of inclination (α) for which the block remains in equilibrium is called the *angle of repose*. The block can repose (sleep) up to the repose angle. It can be seen that repose angle (α) = limiting angle of friction (λ)

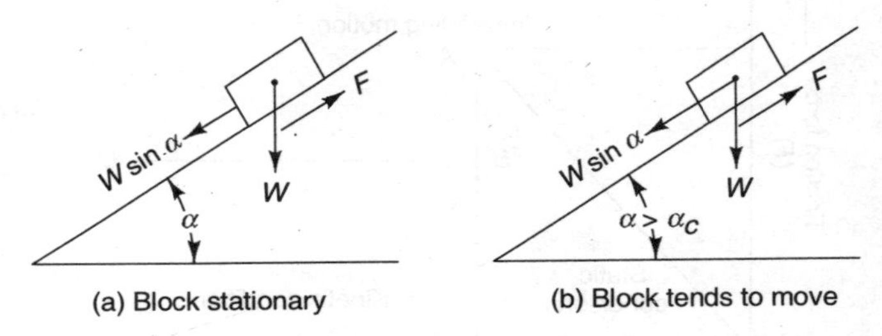

FIGURE 9.3 A block on an inclined plane.

If a body tends to move towards right, friction force will act towards left. The resultant (R) of friction force with the normal will make an angle λ with the normal towards left (Figure 9.4a). If now the body tries to move towards left, the resultant will act towards right (Figure 9.4b). Similarly, it will be in all directions and the resultant will seem to be moving and generating a cone surface (Figure 9.4c). The *cone of friction* is defined as the surface generated by the direction of resultant when a body is moved in different directions.

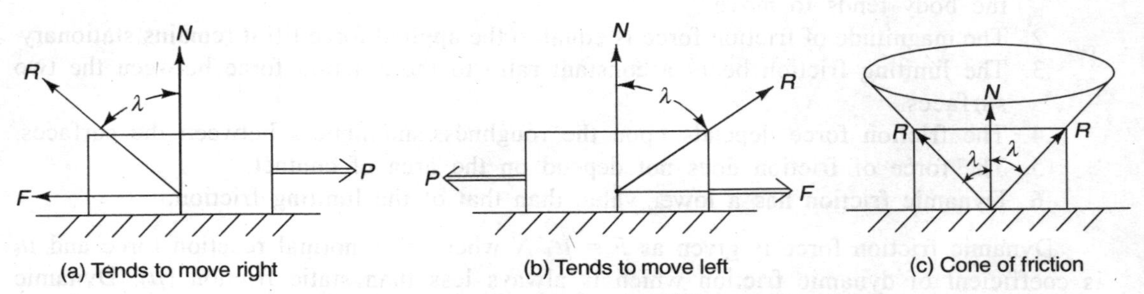

FIGURE 9.4 A body moving in different directions.

EQUILIBRIUM: BLOCK, WEDGE AND LADDER

During equilibrium of a block (Figure 9.5), the system of forces including friction may be treated as the concurrent force system. Hence using the equilibrium equations for concurrent forces ($\Sigma P_x = 0$, $\Sigma P_y = 0$) and the laws of friction, the force system can be analyzed.

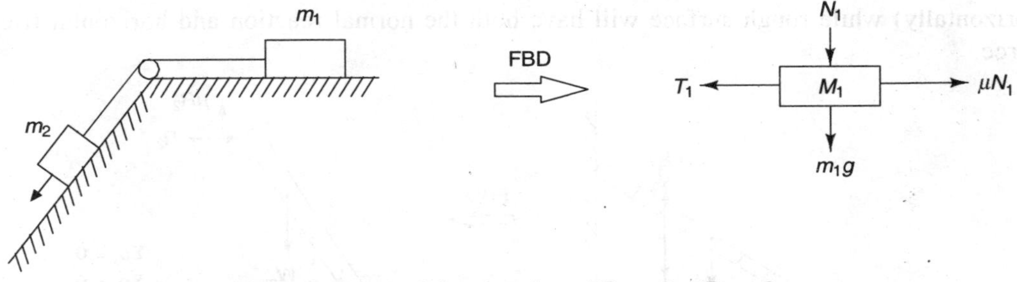

(a) One concurrent force system: $\Sigma P_x = 0$, $\Sigma P_y = 0$

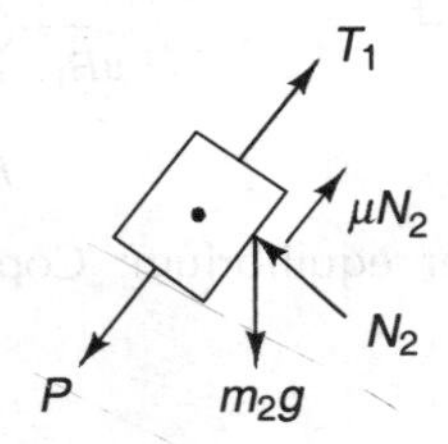

(b) Second concurrent force system: $\Sigma P_x = 0$, $\Sigma P_y = 0$

FIGURE 9.5 Blocks equilibrium: Two concurrent force systems.

Wedges are small pieces of material with two of their opposite surfaces are not parallel (Figure 9.6). They are used to lift heavy objects like beams for final alignment or to make a space for inserting lifting devices like lifting jack. In analysis of the equilibrium of wedges, the weight of wedges is neglected as it is very small. Friction force and the normal are

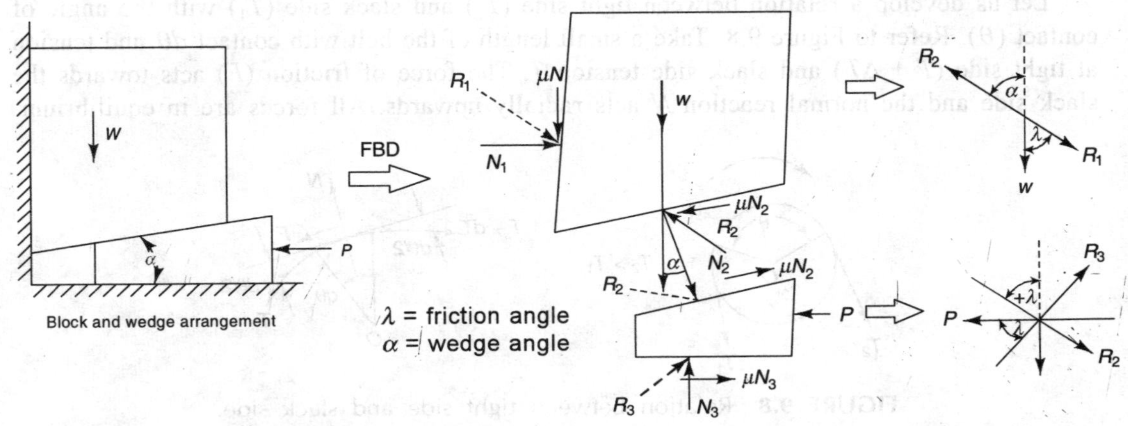

FIGURE 9.6 Wedge equilibrium: Two concurrent force systems.

combined to get the resultant which is used on a free body diagram to simplify the analysis of force system on the wedge.

The system of forces acting on a ladder constitutes non-concurrent system. Therefore, in analysis of the force system for a ladder, laws of friction and equations of equilibrium of non-concurrent forces ($\Sigma P_x = 0$, $\Sigma P_y = 0$, $\Sigma M = 0$) have to be used (Figure 9.7). The smooth surface will have only normal reaction (no friction force which otherwise acts horizontally) while rough surface will have both the normal reaction and horizontal friction force.

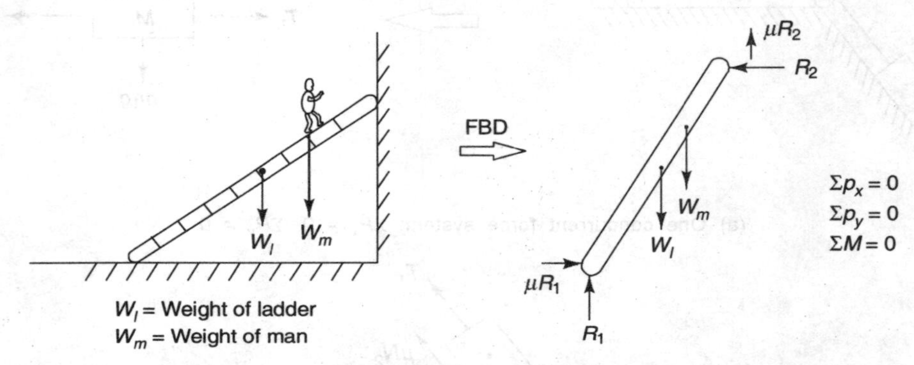

FIGURE 9.7 Ladder equilibrium: Coplanar force system.

POWER TRANSMITTED

The power can be transmitted by means of a belt as friction exists between the pulley (wheel) and the belt. Friction is also used in band brakes to stop the rotating wheel.

The power transmitted by a belt-pulley arrangement depends upon the angle of contact of the belt with the pulley. If the angle of contact increases, more power can be transmitted. The tension in the belt is more on the side it is pulled as it has to overcome friction force and this side of the belt is called *tight side*. The other side having less tension is called *slack side*.

Let us develop a relation between tight side (T_2) and slack side (T_1) with the angle of contact (θ). Refer to Figure 9.8. Take a small length of the belt with contact $d\theta$ and tension at tight side ($T + \Delta T$) and slack side tension T. The force of friction (F) acts towards the slack side and the normal reaction N acts radially upwards. All forces are in equilibrium.

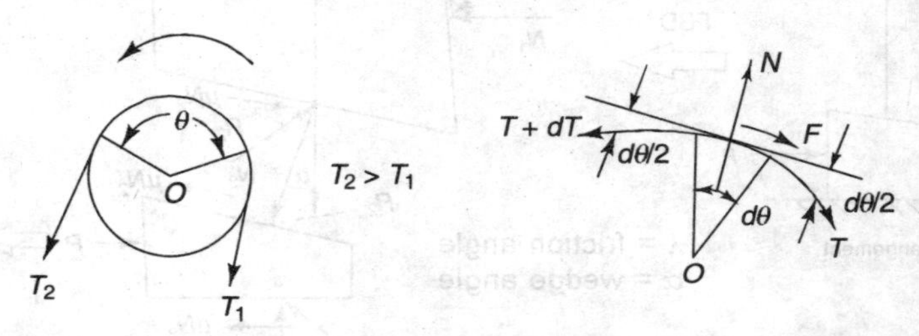

FIGURE 9.8 Relation between tight side and slack side.

$$\Sigma \text{ Forces in the radial side} = 0$$

$$N - T \sin\frac{d\theta}{2} - (T + dT) \sin\frac{d\theta}{2} = 0$$

As $d\theta$ is small,

$$\sin\frac{d\theta}{2} = \frac{d\theta}{2}$$

Therefore,

$$N - T\frac{d\theta}{2} - T\frac{d\theta}{2} - dT\frac{d\theta}{2} = 0$$

or

$$N = Td\theta \quad \left(\text{as } dT\frac{d\theta}{2} \approx 0\right)$$

Σ Forces on the tangential side = 0

$$(T + dT) \cos\frac{d\theta}{2} - F - T \cos\frac{d\theta}{2} = 0$$

As $d\theta$ is small,
$$\cos\frac{d\theta}{2} = 1$$

Therefore,

$$T + dT - F = T = 0$$

$$F = dT$$

$$\mu N = dT \quad (\text{as } F = \mu N)$$

$$\mu T\, d\theta = dT \quad (\text{as } N = Td\theta)$$

$$\int_{T_1}^{T_2} \frac{dT}{T} = \mu \int_0^{\theta} d\theta$$

$$T_2 = T_1 e^{u\theta} \quad (\theta \text{ is in radians})$$

The value of angle of contact or lap θ depends upon the arrangement of a belt wrapping over pulleys (Figure 9.9). If the sizes of the driver pulley and the driven pulley are the same, then the lap angle at both the driver and the driven pulley is equal $(=\pi)$.

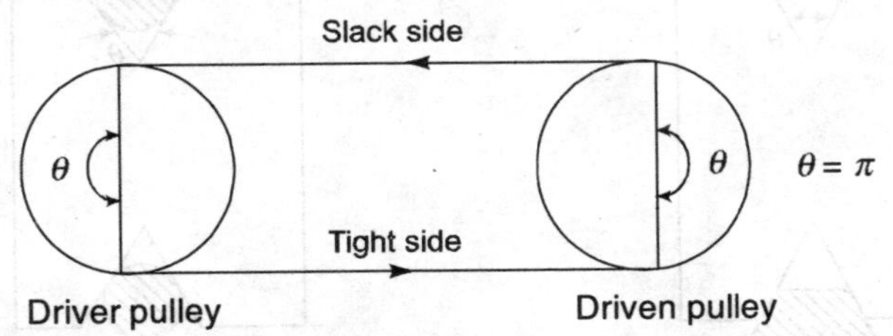

FIGURE 9.9 Arrangement of a belt and pulleys.

Now refer to Figure 9.10. If the diameter of the driver pulley (D_1) is more than that of the driven pulley (D_2), the lap angle on the driver pulley is $ADC(\pi + 2\alpha)$ and the driven pulley is $(\pi - 2\alpha)$ where $\sin \alpha = \dfrac{D_1 - D_2}{2l}$. As the lap angle on a smaller pulley is small, the maximum power before the belt starts slipping will be limited by the small pulley. Therefore, the smaller lap angle $(\pi - 2\alpha)$ is considered while calculating maximum power transmission. This problem can be overcome by a cross belt arrangement. The lap angles on both big and small pulleys become equal to $(\pi + 2\alpha)$ where $\sin \alpha = \dfrac{r_1 + r_2}{l}$. The lap angle can also be increased by increasing the turns of rope on pulleys. If rope has n turns on the pulley then the wrap angle will be $2\pi n$.

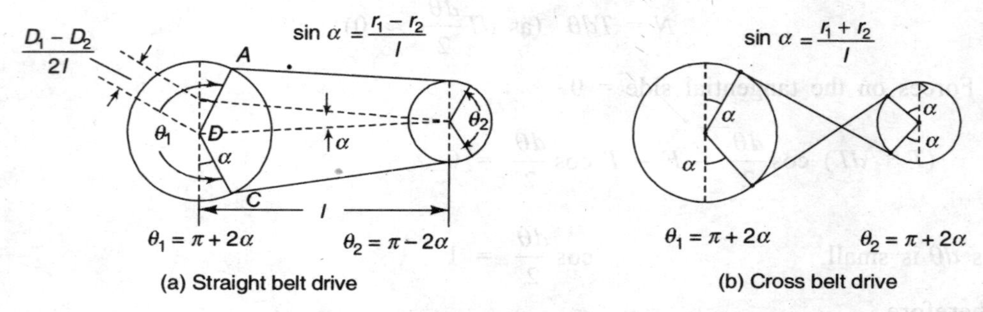

(a) Straight belt drive (b) Cross belt drive

FIGURE 9.10 Different belt arrangements.

The length of the belt is given by

(a) Straight belt drive $= \pi(r_1 + r_2) + \dfrac{(r_1 - r_2)^2}{l} + 2l$

(b) Cross belt drive $= \pi(r_1 + r_2) + \dfrac{(r_1 + r_2)^2}{l} + 2l$

In a *V*-belt drive and a rope drive (Figures 9.11 and 9.12) $T_2 = T_1 \times e^{\mu\beta \operatorname{cosec} \theta/2}$ where θ is the groove angle. As cosec $\theta/2 > 1$,

$$e^{\mu\beta \operatorname{cosec} \theta/2} > e^{\mu\beta}$$

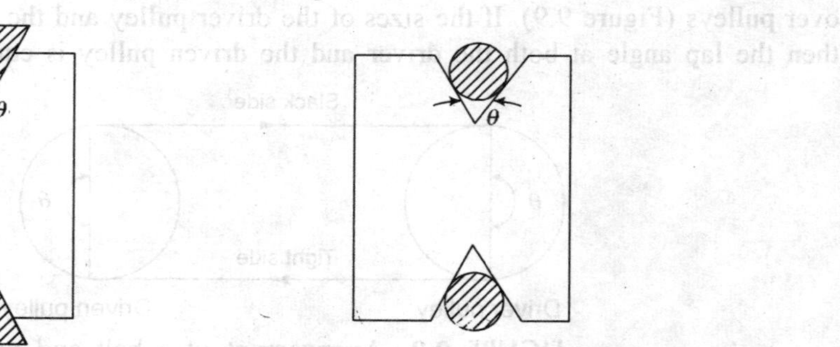

FIGURE 9.11 V-belt drive. **FIGURE 9.12** Rope drive.

As the more frictional resistance is provided by the V-belt and rope drives, the more power can be transmitted by these belts.

Power transmitted by a belt-pulley arrangement can be found out (Figure 9.13).

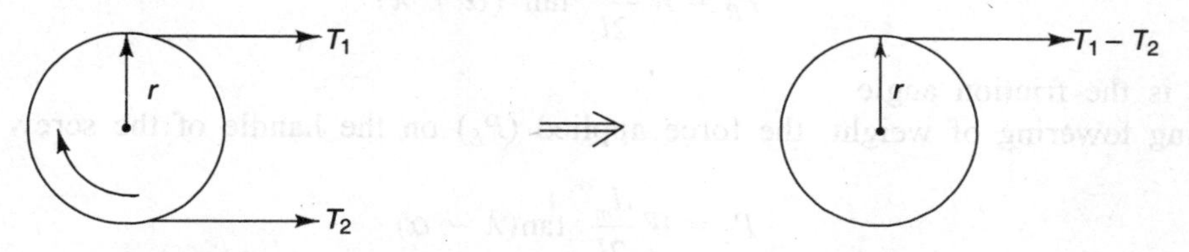

FIGURE 9.13 Power generated by a belt-pulley arrangement.

$$\text{Net force on pulley} = T_1 - T_2$$
$$\text{Turning moment} = (T_1 - T_2)r$$
$$\text{Work done/second} = (T_1 - T_2) \times r \times \omega \quad (\omega = \text{angular velocity})$$
$$= (T_1 - T_2)v \quad (v = \text{belt velocity})$$
$$= (T_1 - T_2)\frac{\pi DN}{60}$$
$$N = \text{rpm of pulley}$$

If D_1 and N_1 are the diameter and the speed of a driver and D_2 and N_2 are the diameter and the speed of a driven pulley, and if t is thickness of the belt, then

$$\frac{N_1}{N_2} = \frac{D_1 + t}{D_2 + t}$$

SCREW JACK

Mechanical advantage is the ratio of weight (W) lifted to force (F) used, i.e.

$$\text{Mechanical advantage} = \frac{W}{F}$$

The mechanical advantage should be more than one.

A screw jack (square thread type) is a device used for lifting heavy loads by applying less effort. It works on the principle of an inclined plane. The inclination of thread (α) is given by

$$\tan^{-1}\frac{P}{\pi\, d_m}$$

where P is the pitch (distance between two consecutive threads) and d_m is the mean diameter of threads.

During raising of a weight (W), P_R horizontal force applied at the handle (length R) of the screw jack is

$$P_R = W\frac{d_m}{2l}\tan(\alpha + \lambda)$$

where λ is the friction angle.

During lowering of weight, the force applied (P_L) on the handle of the screw jack is

$$P_L = W\frac{d_m}{2l}\tan(\lambda - \alpha)$$

If $\alpha = \lambda$, then $P_L = 0$. Also if $\alpha > \lambda$ then P_L is negative and the weight will start moving downwards by itself and P_L force has to be applied to hold the weight from coming down. Such a screw jack is not a self-locking type. To guard against this undesirable effect in the screw, the screw angle (α) is always kept less than the friction angle (λ).

The efficiency of a screw jack is given by the ratio of ideal effort (when friction is zero) to actual effort, i.e.

$$\eta = \frac{\tan\alpha}{\tan(\alpha + \lambda)}$$

The efficiency of a screw jack is independent of weight being lifted. For maximum efficiency $\left(\frac{d\eta}{d\alpha} = 0\right)$,

$$\text{the angle of inclination } (\alpha) = 45° - \frac{\lambda}{2}$$

and

$$\eta_{\max} = \frac{1 - \sin\lambda}{1 + \sin\lambda}$$

If η is more than 50%, then the screw jack is not a self-locking type.

Square threads take load perpendicular to thread while V-threads do not take load perpendicular to the surface of thread (Figures 9.14, 9.15 and 9.16). Therefore in V-threads, the coefficient of friction can be considered as $\mu/\cos\beta(2\beta$ is the angle between two sides

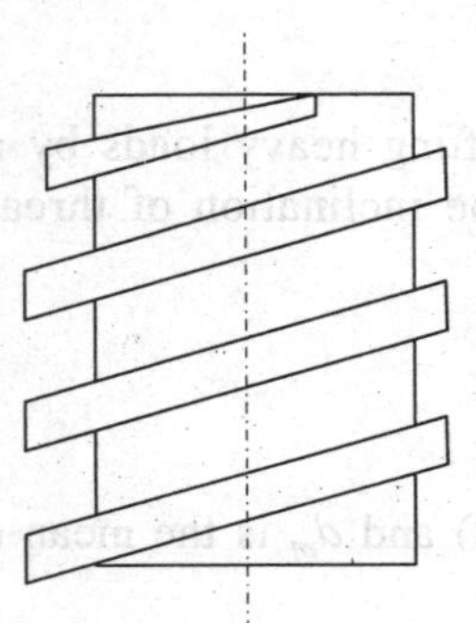

FIGURE 9.14 Square thread.

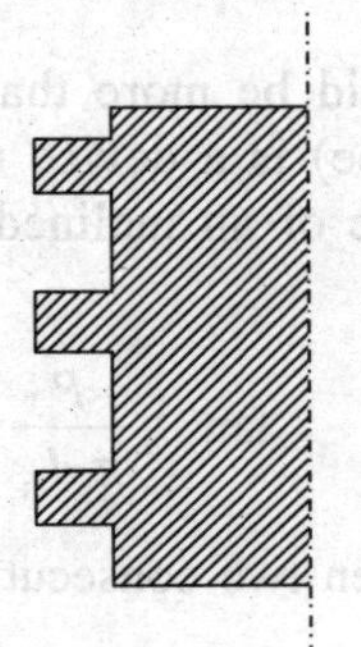

FIGURE 9.15 Square thread (Sectional view).

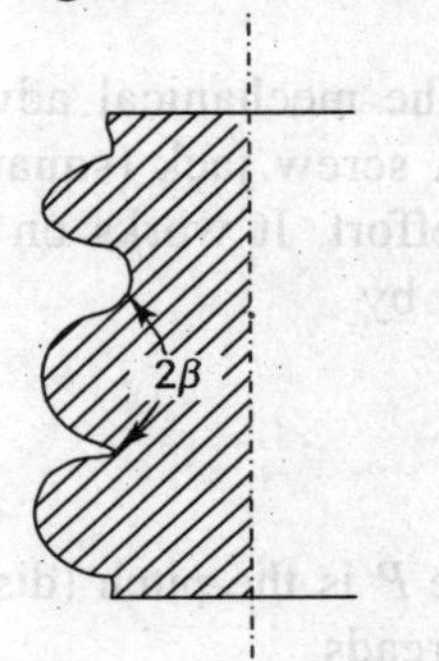

FIGURE 9.16 V-thread (Sectional view).

of thread) in finding effort by relations given above for square threads. As $\mu/\cos\beta > \mu$, force required to lift load with a V-thread is more than a square thread.

Square threads are generally used for transmission of power in machines like lathes and milling machines. Square threads are also used for transmitting power without any side thrusts. However, square threads are difficult to be manufactured and they are difficult to be cleaned. They cannot be used with split or half nut applications on account of difficulty in disengagement. On the other hand, ACME threads, though not so efficient as square threads, are easier to be manufactured and they also permit easier applications in split nut.

SOLVED PROBLEMS

1. Block A weighing 1000 N rests over block B, which weighs 2000 N. Block A is tied to wall with a horizontal string. Find the value of P to move Block B if the coefficient of friction between A and B is 0.5 and the coefficient between B and the floor is 0.33.

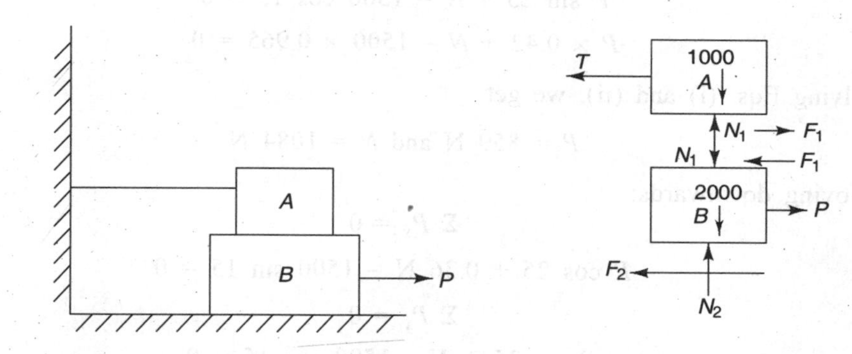

Draw a free body diagram of blocks A and B. Considering equilibrium equations for block A(limiting friction),

$$\Sigma P_y = 0, \quad 1000 = N_1$$

$$\Sigma P_x = 0, \quad T = F_1 = \mu N_1 = 1000 \times 0.25 = 250 \text{ N}$$

Now considering equilibrium of Block B,

$$\Sigma P_y = 0, \quad N_2 = 2000 + 1000 = 3000 \text{ N}$$

$$\Sigma P_x = 0, \quad P = F_1 + F_2 = \mu_2 N_2 + 250$$

$$= 0.33 \times 3000 + 250$$

$$= 990 + 250 = 1240 \text{ N}$$

2. A block of weight 1500 N is lying on a plane inclined 15° to the horizontal. The angle of friction is 20°. An effort inclined at 25° with the plane is applied on the block. Determine the effort required to move the block (i) upwards and (ii) downwards.

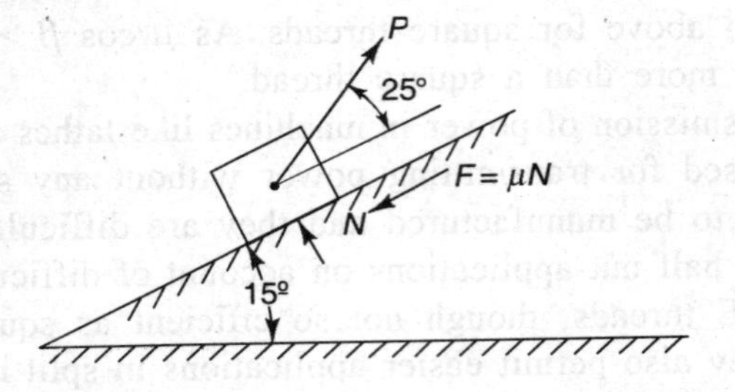

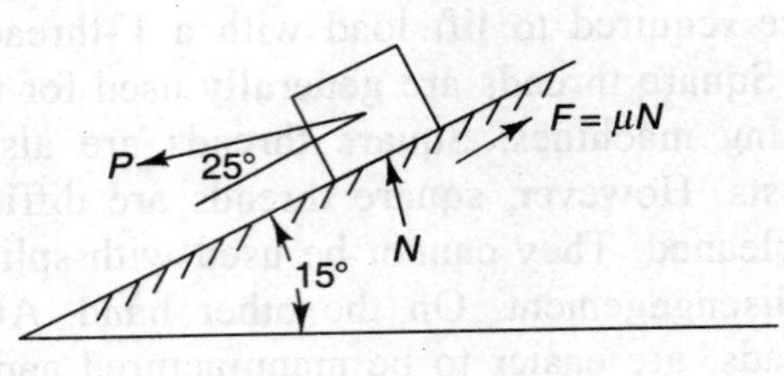

(i) Moving upwards (x along the plane):

$$\Sigma P_x = 0, P \cos 25 - \mu N - 1500 \sin 15 = 0$$

$$\mu = \tan \lambda = \tan 20 = 0.36$$

Therefore,

$$P \times 0.9 - 0.36 N - 1500 \times 0.258 = 0 \quad \text{(i)}$$

$$\Sigma P_y = 0,$$

$$P \sin 25 + N - 1500 \cos 15 = 0$$

$$P \times 0.42 + N - 1500 \times 0.965 = 0 \quad \text{(ii)}$$

On solving Eqs. (i) and (ii), we get

$$P = 859 \text{ N and } N = 1084 \text{ N}$$

(ii) Moving downwards:

$$\Sigma P_x = 0$$

$$-P \cos 25 + 0.36 \text{ N} - 1500 \sin 15 = 0 \quad \text{(iii)}$$

$$\Sigma P_y = 0$$

$$-P \sin 25 + N - 1500 \cos 15 = 0 \quad \text{(iv)}$$

On solving Eqs. (iii) and (iv), we get

$$P = 175.8 \text{ N and } N = 4467 \text{ N}$$

3. **P force is applied at 30° to the horizontal to a 2-block (500 and 750 N) system. Assume the pulley is smooth and the coefficient of friction is 0.20. Find P which will move the arrangement upwards.**

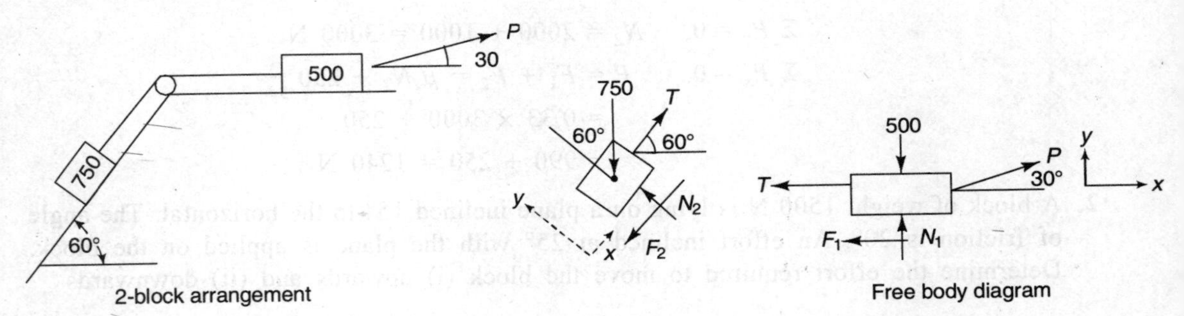

Considering the equilibrium of the block 750 N,

$$\Sigma P_y = 0$$

$$N_2 - 750 \cos 60 = 0$$

or $$N_2 = 375 \text{ N}$$

$$\Sigma P_x = 0$$

$$F_2 = T - 750 \sin 60$$

But

$$\mu N_2 = F_2 = 0.2 \times 375 = 75$$

Therefore,

$$T = 75 + 750 \times \frac{\sqrt{3}}{2} = 724.5 \text{ N}$$

Considering equilibrium of the block 500 N,

$$\Sigma P_y = 0$$

$$N_1 - 500 + P \sin 30 = 0$$

$$N_1 + 0.5P = 500 \qquad \text{(i)}$$

$$\Sigma P_x = 0, P \cos 30 - T - F_1 = 0$$

$$P \cos 30 - 724.5 - 0.2 N_1 = 0 \qquad (\because F_1 = 0.2 \times N_1) \qquad \text{(ii)}$$

Solving Eqs. (i) and (ii), we get

$$P = 853.5 \text{ N}$$

4. A block (10 kN) is to be raised by a wedge (30°) as shown by a horizontal force P. Determine the minimum value of P and mechanical advantage of the system ($\mu = 0.36$ for all).

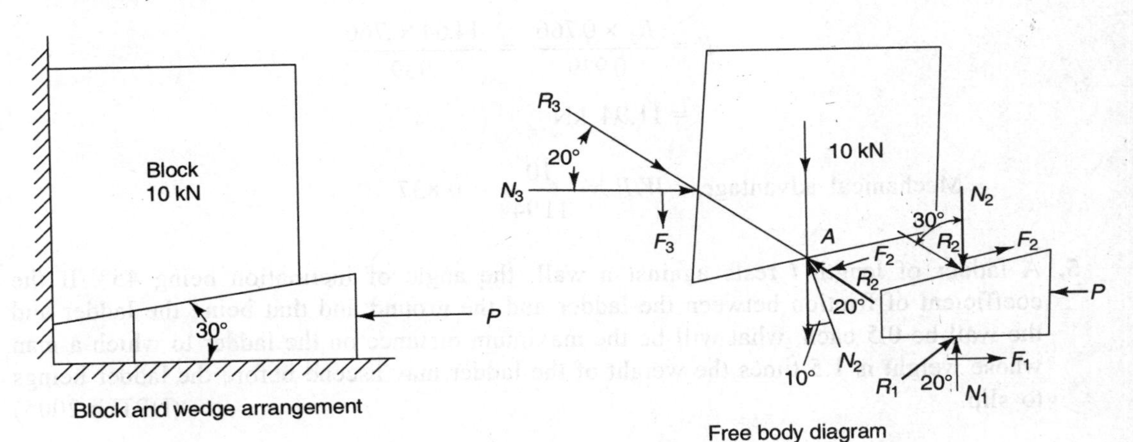

If $\mu = 0.36$,

$$\tan \lambda = 0.36$$

or $\lambda = 20°$

Considering the equilibrium of the block, we have three forces, viz. R_3, 10 kN and R_2 meeting at point. Applying Lami's theorem,

$$\frac{R_3}{\sin(180 - 30)} = \frac{R_2}{\sin(90 - 20)} = \frac{10}{\sin(20 + 90 + 30)}$$

$$\therefore \quad \frac{R_3}{\sin 30} = \frac{R_2}{\cos 20} = \frac{10}{\sin 40}$$

$$\frac{R_3}{0.5} = \frac{R_2}{0.94} = \frac{10}{0.642}$$

$$R_3 = 7.79 \quad \text{and} \quad R_2 = 14.64$$

R2
30°
20°
R3
10 kN

Now considering the equilibrium of wedge, we have three forces, viz R_2, P and R_1 meeting at a point. Applying Lami's theorem,

$$\frac{R_1}{\sin(90 + 30)} = \frac{R_2}{\sin(90 + 20)} = \frac{P}{\sin(180 - 20 - 30)}$$

$$\frac{R_1}{\cos 30} = \frac{R_2}{\cos 20} = \frac{P}{\sin 50}$$

$$\frac{R_1}{0.866} = \frac{R_2}{0.939} = \frac{P}{0.766}$$

R1
20°
P
30°
R2

Therefore,

$$P = \frac{R_2 \times 0.766}{0.939} = \frac{14.64 \times .766}{.939}$$

$$= 11.94 \text{ kN}$$

$$\text{Mechanical advantage} = W/P = \frac{10}{11.94} = 0.837$$

5. A ladder of length l rests against a wall, the angle of inclination being 45°. If the coefficient of friction between the ladder and the ground and that being the ladder and the wall be 0.5 each, what will be the maximum distance on the ladder to which a man whose weight is 1.5 times the weight of the ladder may ascend before the ladder beings to slip. (UPTU: 2005)

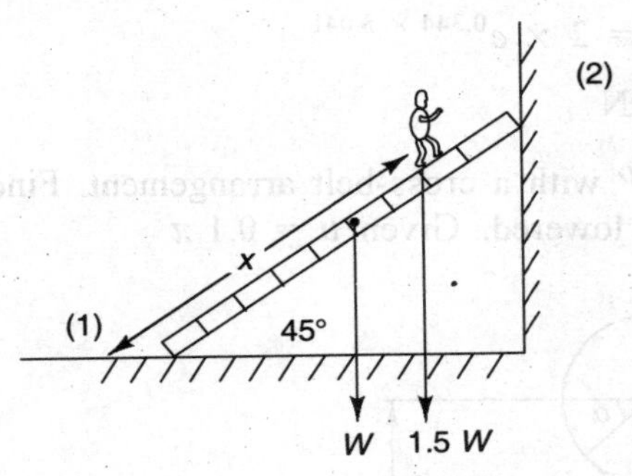

Ladder and man

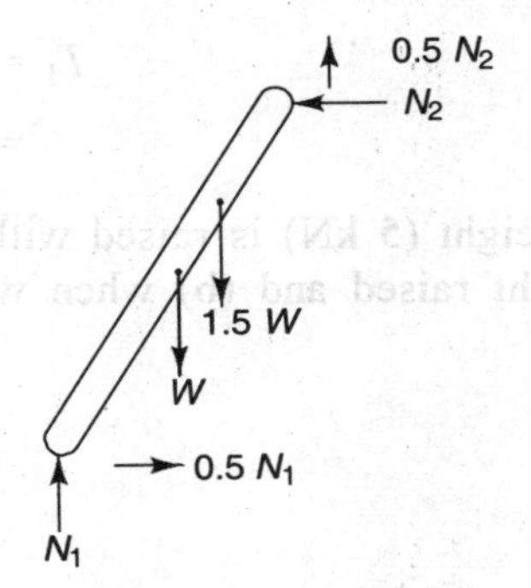

Free body diagram

For equilibrium when the ladder is about to slide and the man is at distance x from the base of the ladder,

$$\Sigma P_x = 0$$

$$N_2 = 0.5\ N_1 \quad \text{or} \quad N_1 = 2\ N_2$$

$$\Sigma P_y = 0$$

$$W + 1.5W - N_1 - 0.5N_2 = 0$$

$$2.5W = 2.5N_2 \quad \text{or} \quad N_2 = W$$

$$\Sigma M_1 = 0$$

$$W \times \frac{l}{2} \times \cos 45 + 1.5 \times Wx \cos 45 - N_2 \times l \sin 45 - 0.5N_2 \times l \cos 45 = 0$$

or
$$W \times \frac{l}{2} + 1.5 \times Wx - Wl - 0.5\ Wl = 0$$

or
$$1.5 \times Wx = Wl$$

or
$$x = \frac{2}{3} \times l$$

6. An open belt drive transmits power from one pulley to another located 3 m apart. The larger pulley has diameter = 1 m and smaller has diameter = 0.7 m. If λ = 19°, find the maximum tension if tension in the slack side is 2 kN.

$$\mu = \tan \lambda = \tan 19 = 0.344$$

$$\sin \alpha = \frac{D_1 - D_2}{2l} = \frac{1 - 0.7}{2 \times 3} = \frac{0.3}{6} = 0.05$$

$$\therefore \quad \alpha = 2.86°$$

$$\text{Lap angle} \quad \beta = \pi - 2\alpha = \pi - \frac{2 \times 2.86 \times \pi}{180}$$

$$= 3.041 \text{ rad}$$

$$T_1 = T_2\, e^{\mu\beta} = 2 \times e^{0.344 \times 3.041}$$
$$= 5.693 \text{ kN}$$

7. A weight (5 kN) is raised with effort P with a cross-belt arrangement. Find (a) when weight raised and (b) when weight is lowered. Given $\mu = 0.1\,\pi$

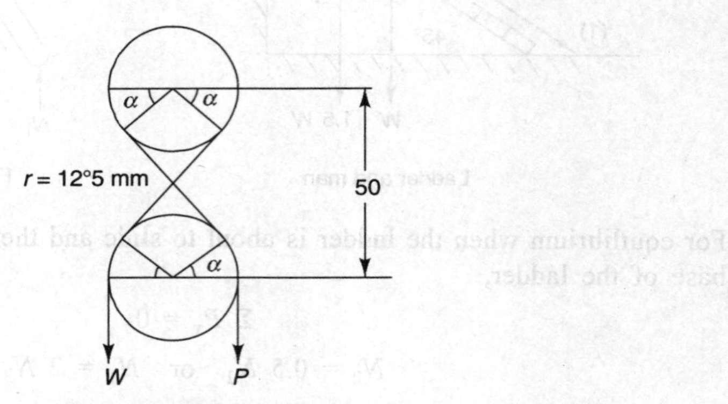

Cross-belt arrangement

$$\sin\alpha = \frac{r_1 + r_2}{l} = \frac{12.5 + 12.5}{50} = 0.5$$

$$\alpha = 30°$$

Lap angle $\beta = 180 + 4 \times 30 = 300°$

(2α of the top pulley and 2α of the bottom pulley)

$$\beta = 5.24 \text{ rad}$$

(a) Weight is being raised. It means the weight side is the slack side. Therefore,

$$P = we^{\mu\beta} = 5 \times e^{5.24 \times 0.1\,\pi}$$
$$= 5 \times e^{1.646} = 5 \times 5.18$$
$$= 25.9 \text{ N}$$

(b) Weight is being lowered.

Now $w = Pe^{\mu\beta}$

or $$P = \frac{5}{5.18} = 0.965 \text{ kN}$$

8. A rope drive transmits power with a pulley (diameter = 1.25 m, lap angle = 180°, groove angle = 45°, maximum tension = 1200 N, μ = 0.3). Find the torque on the pulley.

$$T_1 = T_2\, e^{\mu\beta \text{ cosec } \theta/2}$$

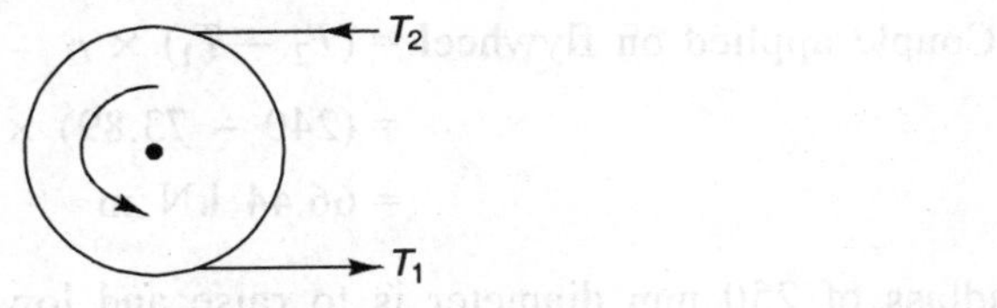

$$1200 = T_2 \, e^{0.3 \times \pi \text{ cosec } 45/2}$$

$$= T_2 \, e^{2.463} = T_2 \times 11.732$$

$$\therefore \quad T_2 = \frac{1200}{11.732} = 102.2$$

$$\text{Torque} = (T_1 - T_2)r = (1200 - 102.2) \times 0.625$$

$$= 686 \text{ N m}$$

9. The speed of a flywheel is controlled by a band brake as shown $\mu_s = 0.3$ and $\mu_k = 0.25$. Find the couple to be applied to keep it rotating at constant speed when P = 60 kN.

$$\Sigma M_A = 0$$

$$P \times 4 = T_2 \times 1$$

$$T_2 = 4 \times 60 = 240 \text{ kN}$$

$$\text{Lap angle } \beta = 270° = 1.5\pi$$

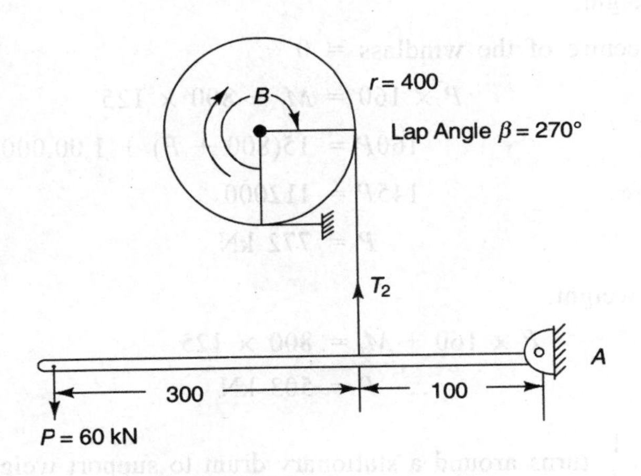

Here we have to use the coefficient of dynamic friction, i.e. $\mu_k = 0.25$.

$$T_2 = T_1 \, e^{\mu k \times \beta} = T_1 \times e^{0.25 \times 1.5\pi}$$

$$240 = T_1 \times 3.248$$

$$T_1 = 73.89 \text{ kN}$$

Couple applied on flywheel $= (T_2 - T_1) \times r$

$= (240 - 73.89) \times 0.4$

$= 66.44$ kN m

10. A windlass of 250 mm diameter is to raise and lower 800 N block. The windlass is supported by poorly lubricated bearing ($\mu_s = 0.5$) of 60 mm diameter. Find effort (P) for (a) raising and (b) lowering load.

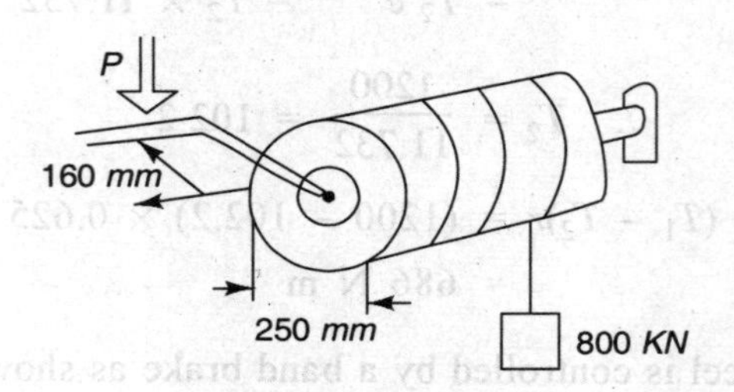

Reaction on the bearing $= P + 800$

Friction due to reaction $= \mu(P + 800)$

$= 0.5(P + 800)$

M_f = Friction moment = Friction force × rad of the bearing

$= (800 + P) \times 0.5 \times 30$

(a) Raising weight:

ΣM at the centre of the windlass = 0

$$P \times 160 = M_f + 800 \times 125$$

$$160P = 15(800 + P) + 1,00,000$$

$$145P = 112000$$

$$P = 772 \text{ kN}$$

(b) Lowering weight:

$$P \times 160 + M_f = 800 \times 125$$

$$\therefore \quad P = 503 \text{ kN}$$

11. A rope makes $1\frac{1}{4}$ turns around a stationary drum to support weight W. If $\mu = 0.3$ and $P = 600$ N, find weight (a) while raising and (b) while lowering.

$$\text{Lap angle} = 2\pi + \frac{2\pi}{4} = 2.5\pi$$

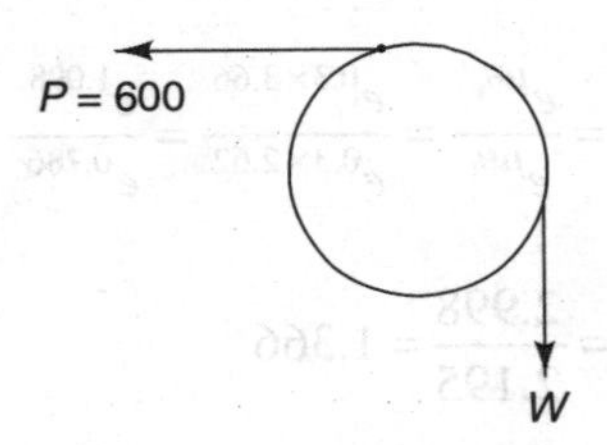

(a) Raising:

$$T_2 = T_1 e^{\mu\beta}$$

$$T_2 = 600, \quad \beta = 2.5\pi \quad \text{and} \quad \mu = 0.3$$

$$T_1 = \frac{600}{e^{0.75\pi}} = \frac{600}{10.55} = 56.87 \text{ N}$$

Therefore,

$$W = 56.87 \text{ N}$$

(b) Lowering:

$$T_2 = 600 \times e^{0.75}\pi = 600 \times 10.55 = 6330 \text{ N}$$

12. The percentage improvement in power capacity of a flat bolt drive, when the wrap angle at the driving pulley is increased from 150° to 210° by an idler arrangement for a friction coefficient of 0.3, is

(a) 25.21 (b) 33.92
(c) 40.17 (d) 67.85 (GATE: 1997)

$$\text{Power} = (T_1 - T_2)\frac{\pi\, dx}{60}$$

$$\propto (T_1 - T_2)$$

$$\frac{T_1}{T_2} = e^{\mu\theta}$$

or

$$T_1 \propto e^{\mu\theta}$$

or

$$P_1 \propto e^{\mu\theta}$$

$$\theta_1 = 150 \times \frac{\pi}{180} = 2.62 \text{ rad}$$

$$\theta_1' = 210 \times \frac{\pi}{180} = 3.66 \text{ rad}$$

$$\frac{P_1'}{P_1} = \frac{e^{\mu\theta_1'}}{e^{\mu\theta_1}} = \frac{e^{0.3\times3.66}}{e^{0.3\times2.62}} = \frac{e^{1.098}}{e^{0.786}}$$

$$= \frac{2.998}{2.195} = 1.366$$

The percentage improvement is 36.6%. Therefore, option (b) is correct.

13. A felt-belt open drive transmits 1.5 kW of power. The coefficient of friction between the belt and the drive pulley is 0.25 and the lap angle is 159°. The drive pulley is rotating in the clockwise direction with a linear velocity of 3.75 mk. Determine the x and y component of the reaction force on the drive pulley shaft.

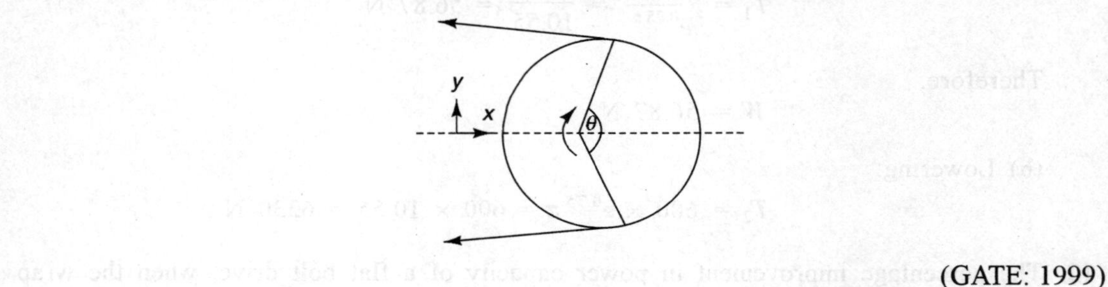

(GATE: 1999)

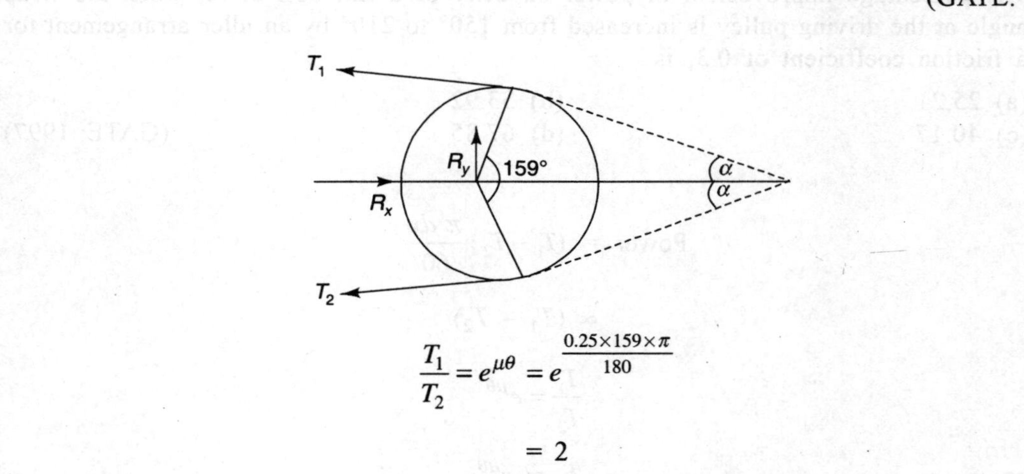

$$\frac{T_1}{T_2} = e^{\mu\theta} = e^{\frac{0.25\times159\times\pi}{180}}$$

$$= 2$$

$$T_1 = 2T_2$$

$$P = (T_1 - T_2)v$$

where P = Power, v = Velocity of belt

$$1.5 \times 10^3 = (2T_2 - T_2) \times 3.75$$

$\therefore$
$$T_2 = \frac{1.5\times10^3}{3.75} = 400 \text{ N}$$

$$\therefore \quad T_1 = 800 \text{ N}$$

$$\theta = 180 - \frac{159}{2} - 90$$

$$= 10.5$$

$$R_x = -T_1 \cos\theta - T_2 \cos\theta$$

$$= -(800 + 400) \cos 10.5 = -1179.9 \text{ N}$$

$$R_y = 800 \sin 10.5 - 400 \sin 10.5$$

$$= 72.89$$

14. A belt drive shown in the figure has an angle of wrap 160° on the smaller pulley. Adding an idler as shown in the figure increases the wrap angle to 200°. The slack side tension is the same in both cases and the centrifugal force is negligible. By what percentage is the torque capacity of the bolt drive increased by adding the idler? (use $\mu = 0.3$). (GATE: 2001)

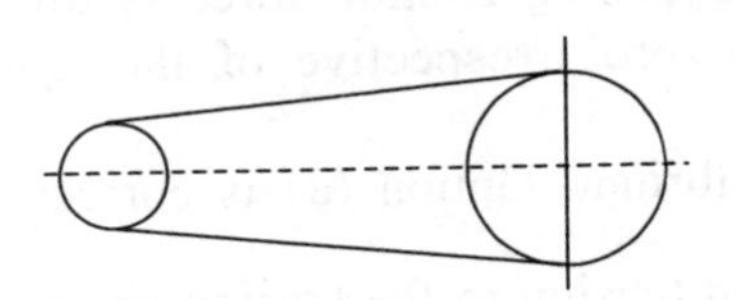

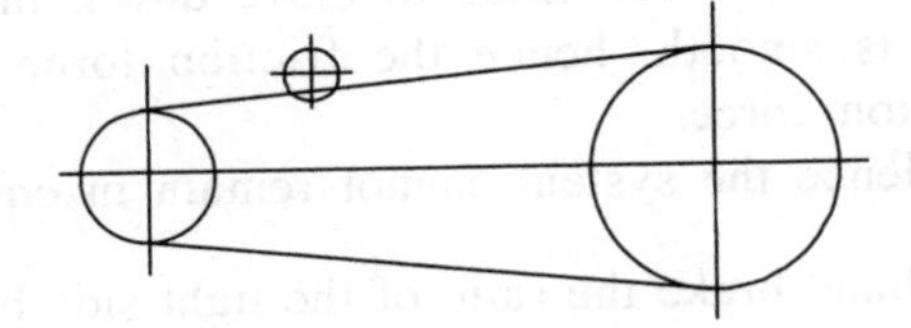

$$\text{Torque} = (T_1 - T_2)r \text{ and } \frac{T_1}{T_2} = e^{\mu\theta}$$

T_1 = Tension without the idler and T_1' = Tension with the idler

$$T_1 = T_2 e^{0.3\times\frac{160}{180}\times\pi}$$

$$= T_2 e^{0.837} = 2.31\ T_2$$

Now,

$$T_1' = T_2 e^{0.3\times\frac{200}{180}\times\pi} = T_2 e^{1.047} = 2.848T_2$$

$$\frac{(\text{Torque})'}{(\text{Torque})} = \frac{(T_1' - T_2)\times r}{(T_1 - T_2)\times r} = \frac{T_2\,(2.848-1)}{T_2\,(2.317-1)} = \frac{1.848}{1.317} = 1.403$$

Hence the torque has increased by 40.3%.

15. Consider the situation shown in the figure. The wall is smooth but the surface A and B in contact are rough. The friction on B due to A in equilibrium is

(a) upward (b) downward

(c) zero (d) system cannot remain in equilibrium

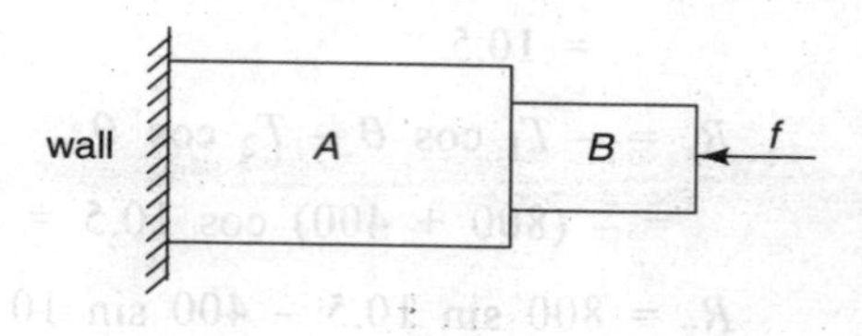

Consider the wall:

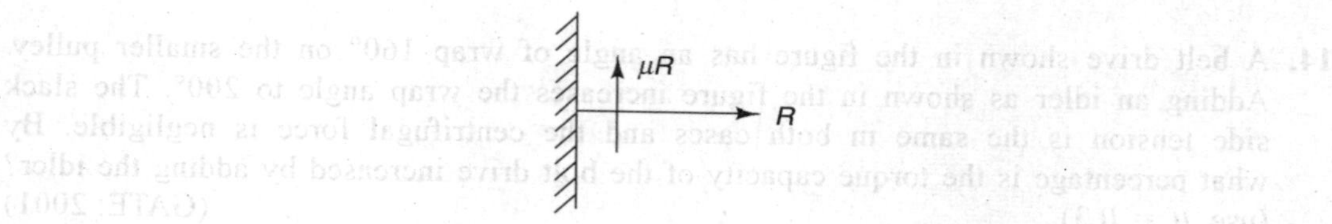

In case the block tends to move down, the opposing friction force is μR. Since the wall is smooth, hence the friction force is zero irrespective of the magnitude of reaction force.

Hence the system cannot remain in equilibrium. Option (d) is correct.

16. In a band brake the ratio of the tight side band tension to the tension on the slack side is 3. If the angle of overlap of the band on the drum is 180°, the coefficient of friction required between the drum and the band is

(a) 0.2 (b) 0.25

(c) 0.3 (d) 0.35

(GATE: 2003)

$$\frac{T_1}{T_2} = e^{\mu\theta} = e^{\mu \times \pi} = 3$$

$\therefore$ $$\mu \times 3.14 = 1.0986$$

or $$\mu = 0.35$$

The option (d) is correct.

17. A uniform ladder weighing 300 N rests against a smooth vertical wall and on a rough horizontal floor making an angle 60° with the horizontal. Find the force of friction at the floor ($\mu_f = 0.3$).

(UPTU: 2001–2002)

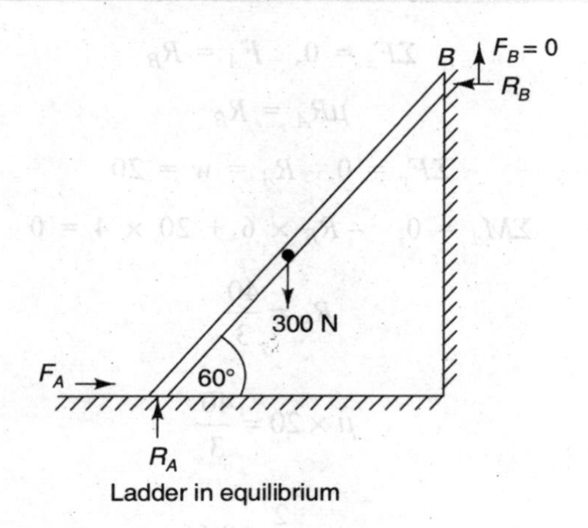

Ladder in equilibrium

As the wall is smooth, $F_B = 0$

$$\Sigma F_x = 0, \quad F_A = R_B$$

$$\mu_f R_A = R_B$$

$$\Sigma F_y = 0, \quad R_A = w = 300 \text{ N}$$

$$F_A = \mu_f R_A = \mu_f \times 300$$

$$= 0.3 \times 300 = 90 \text{ N}$$

18. A uniform ladder of length 10 m and weighing 20 N is placed against a smooth vertical wall with its lower and 8 m from the wall. In this position the ladder is just to slip. Determine:

(i) the coefficient of friction between the ladder and the floor.

(ii) the frictional force acting on the ladder at the point of contact between the ladder and the floor. (UPTU: 2004–2005)

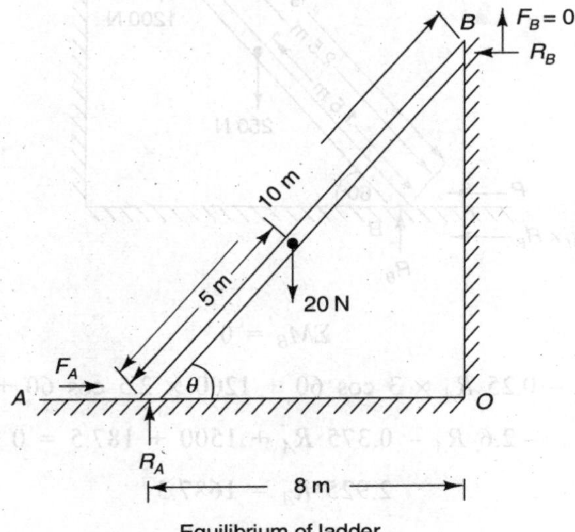

Equilibrium of ladder

$$\Sigma F_x = 0, \quad F_A = R_B$$

or

$$\mu R_A = R_B$$

$$\Sigma F_y = 0, \quad R_A = w = 20$$

$$\Sigma M_A = 0, \quad -R_B \times 6 + 20 \times 4 = 0$$

$$R_B = \frac{40}{3}$$

$$\mu \times 20 = \frac{40}{3}$$

$\therefore$

$$\mu = \frac{2}{3} = 0.66$$

19. A ladder 3 m long and weighing 250 N is placed against a wall with end *B* at the floor level and *A* on the wall. In addition to self weight, the ladder supports a man weighing 1200 N at 2.5 m from *B* on the ladder. If the coefficient of friction at the wall is 0.25 and at the floor is 0.35 and if the ladder makes an angle 60° with the floor, find the minimum horizontal force which if applied at *B* will prevent the slipping of the ladder. (UPTU: 2006–2007)

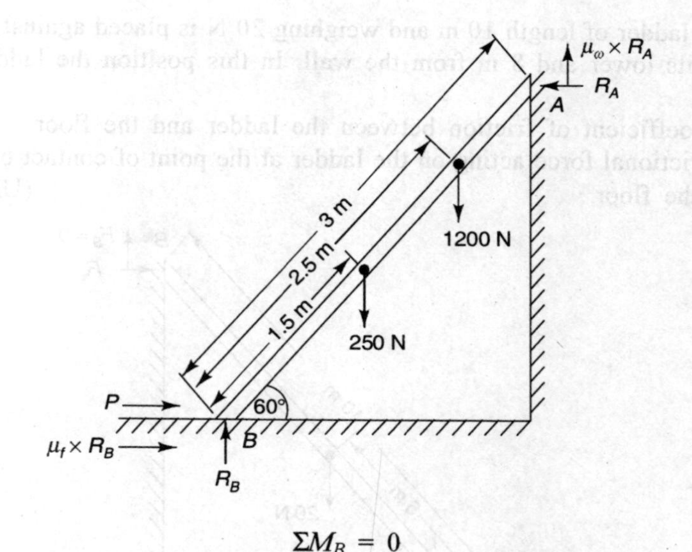

$$\Sigma M_B = 0$$

$$-R_A \times 3 \sin 60 - 0.25 R_A \times 3 \cos 60 + 1200 \times 2.5 \cos 60 + 250 \times 1.5 \cos 60 = 0$$

$$-2.6 R_A - 0.375 R_A + 1500 + 187.5 = 0$$

$$2.925 R_A = 1687.5$$

or $$R_A = 567.23 \text{ N}$$

$$\Sigma F_x = 0,$$

$$P + 0.35 \times R_B = R_A = 567.23 \qquad \text{(i)}$$

$$\Sigma F_y = 0,$$

$$R_B = 1200 + 250 - 0.25 \times 567.23$$

$$= 1308.2$$

Therefore, from Eq. (i),

$$P = 567.23 - 0.35 \times 1308.2$$

$$= 109.36$$

20. A fan belt running at a speed of 500 m/min drives a pulley. Determine the power transmitted by the belt, if the maximum tension on the tight side of the belt is 1200 N. Neglect the centrifugal tension effect. The angle of lap is 160° and the coefficient of friction between the belt and the pulley material is 0.35.

(UPTU: 2006–2007)

Lap angle $\theta = 160 \times \dfrac{\pi}{180} = 2.79$ rad

$$\frac{T_1}{T_2} = e^{\mu\theta}$$

$$\frac{1200}{T_2} = e^{0.35 \times 2.79}$$

$$\therefore \qquad T_2 = 452$$

Power transmitted

$$P = (T_1 - T_2)v$$

where v = Velocity of belt.
Therefore,

$$P = (1200 - 452)\frac{500}{60}$$

$$= 6.241 \text{ kW}$$

21. A block of stone weighing 50 kN rests on a horizontal floor. If the coefficient of friction between the floor and the block is 0.3 and if a man pulls the block through a string which makes an angle α with the horizontal, find for what value of the force necessary to move the block will be minimum. Find the force also.

(UPTU: May 2008)

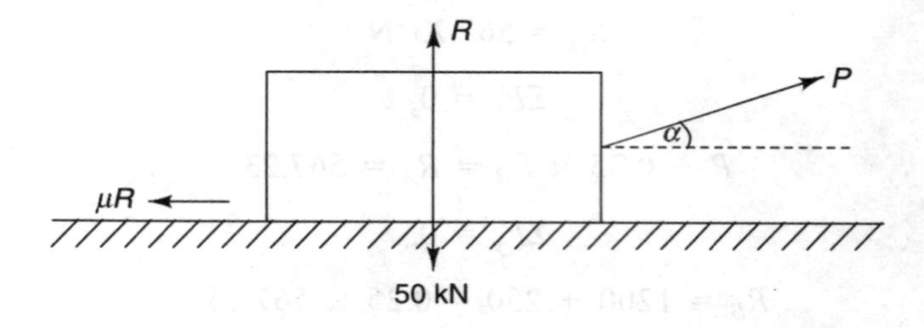

Now $\Sigma F_y = 0$,

$$P \sin \alpha + R = 50 \quad \text{(i)}$$

also $\Sigma F_x = 0$,

$$P \cos \alpha = 0.3R$$

or

$$R = \frac{P \cos \alpha}{0.3} = 3.33P \cos \alpha$$

Putting the value of R in Eq. (i),

$$P \sin \alpha + 3.33P \cos \alpha = 50$$

$$P (\sin \alpha + 3.33 \cos \alpha) = 50 \quad \text{(ii)}$$

Now $\frac{dP}{d\alpha} = 0$ for minimum P.

$$\frac{dP}{d\alpha}(\sin \alpha + 3.33 \cos \alpha) + P(\cos \alpha - 3.33x \sin \alpha) = 0$$

$\therefore$

$$\cos \alpha = 3.33 \sin \alpha$$

or

$$\tan \alpha = \frac{1}{3.33} = 0.3$$

or

$$\alpha = 16.715$$

Putting the value of α' in Eq. (ii)

$$P(\sin 16.715 + 3.33 \cos 16.715) = 50$$

or

$$P(0.288 + 3.189) = 50$$

or

$$P = \frac{50}{3.477}$$

$$= 14.38 \text{ kN}$$

OBJECTIVE TYPE QUESTIONS

If you want to be a winner, you have to be willing to give it your all.

State True or False

1. In engineering applications, friction force is both desirable and undesirable. (*True/False*)
2. Frictional force is self adjusting to applied force. (*True/False*)
3. Frictional force is a reactive force. (*True/False*)
4. Static friction is less than dynamic friction. (*True/False*)
5. Kinetic friction and dynamic friction differ and depend upon the area of contact. (*True/False*)
6. Friction force does not depend upon the area of contact and the shape of contacting surface. (*True/False*)
7. Rolling friction has a theoretical value of zero. (*True/False*)
8. The coefficient of friction is the ratio of frictional force to normal reaction. (*True/False*)
9. Static friction force follows Coulomb's laws of friction. (*True/False*)
10. Dynamic friction force does not follow laws of friction. (*True/False*)
11. The friction angle is measured from the normal. (*True/False*)
12. The angle of repose is equal to friction angle when the body tends to move. (*True/False*)
13. No force is required to move down the body on an inclined plane if the angle of plane is less than friction angle. (*True/False*)
14. The screw jack works on the principle of an inclined plane. (*True/False*)
15. Lesser effort is required to lift weight by a screw jack if it has V threads instead of square threads. (*True/False*)
16. A screw jack is not a self-locking type in case efficiency is more than 50%. (*True/False*)
17. The maximum efficiency of a screw jack is $\dfrac{1 - \sin \lambda}{1 + \sin \lambda}$. (*True/False*)
18. The cone of friction is the surface generated by the resultant of the normal and friction force when a body is moved in different directions. (*True/False*)
19. Wedges are small pieces with two opposite sides not parallel. (*True/False*)
20. The weights of wedges are considered while analyzing a force system. (*True/False*)
21. The resultant of a normal and friction force is not taken to make the force system concurrent in analysis of the force system acting on wedges. (*True/False*)
22. The force system acting on a ladder is concurrent. (*True/False*)

23. Smooth surface will have normal reaction only in the free body diagram of a ladder. *(True/False)*

24. Rough surface will have normal reaction and frictional force in analysis of a ladder. *(True/False)*

25. Wedges are used for final alignment of a heavy block or lifting heavy machine so that a lifting jack can be inserted. *(True/False)*

26. Power can be transmitted by a belt-and-pulley system. *(True/False)*

27. The tension in a belt-and-pulley arrangement is less in the direction it is pulled. *(True/False)*

28. The higher tension side in a belt-pulley arrangement is called a tight side. *(True/False)*

29. The lower tension side in a belt-pulley arrangement is called a slack side. *(True/False)*

30. The belt moves from the slack side to the tight side in a belt-pulley arrangement. *(True/False)*

31. The value of a lap angle does not depend upon the arrangement of a belt wrapping over the pulley. *(True/False)*

32. The lap angle will be the same if driver and driven pulleys have the same size. *(True/False)*

33. The lap angle will be more for a larger pulley than that of a small pulley in a straight belt arrangement. *(True/False)*

34. In a cross belt arrangement, the lap angle is equal irrespective of the size of driver and driven pulley. *(True/False)*

35. The lap angle cannot be increased by increasing the turns of rope on a pulley. *(True/False)*

36. If T_1 is tension in the tight side and T_2 is tension in the slack side, then $T_1 = T_2\, e^{\mu\beta}$ where μ = coefficient of friction and β = lap angle. *(True/False)*

37. The belt will slip first on a small pulley in a straight belt drive. *(True/False)*

38. In a V-belt drive, friction coefficient (μ) is taken as μ cosec $\theta/2$ where θ is a groove angle. *(True/False)*

39. A flat belt can transmit more power than a V-belt. *(True/False)*

40. Turning moment on a pulley will be equal to $(T_1 - T_2) \times r$ where T_1 and T_2 are tensions in the tight and slack sides and r is the radius of the pulley. *(True/False)*

41. The velocity of a belt is given by $\dfrac{\pi DN}{60}$ where D = diameter of the pulley and N = rpm. *(True/False)*

42. Power transmitted by belt is $(T_1 - T_2)v$ where T_1 and T_2 are tensions in belt and v is velocity of belt. *(True/False)*

43. Mechanical advantage is the ratio of weight to effort applied (W/E). *(True/False)*

44. Mechanical advantage should be less than one. *(True/False)*

45. Square threads are used for transmission of power. (True/False)

46. ACME threads permit the use of split nut. (True/False)

47. ACME threads are preferred as they are easier to be manufactured. (True/False)

48. Square threads are not simpler to be manufactured. (True/False)

49. Square threads do not permit the use of split nut. (True/False)

50. The pitch is the distance between two consecutive threads. (True/False)

51. For a self-locking lifting jack, the inclination angle (α) of the screw should be less than the friction angle (λ). (True/False)

52. The lap angle can be increased for a belt with an idler pulley. (True/False)

53. No force is required for a body to slide down even when the angle of an inclined plane is less than the repose angle. (True/False)

54. A band brake stops a rotating wheel due to the friction force developed between the band and the wheel. (True/False)

55. Bicycle wheels are circular as we want to reduce effort by working against rolling friction instead of sliding friction. (True/False)

Multiple Choice Questions

1. The maximum value of friction force which comes into play when a body tends to move on a surface is called
   (a) sliding friction (b) limiting friction (c) milling friction

2. The ratio of static friction to dynamic friction is
   (a) less than 1 (b) equal to 1 (c) greater than 1

3. The angle of friction is equal to the
   (a) ratio of friction to the normal
   (b) angle of an inclined plane when a body tends to slide down
   (c) angle of an inclined plane when a body is sliding.

4. The coefficient of friction depends upon
   (a) area of contact (b) shape of the body (c) nature of contact surfaces

5. Kinetic friction is
   (a) limiting friction
   (b) friction when a body is moving
   (c) friction when a body is stationary

6. The force required to move a body up an inclined plane will be least when the angle of inclination is
   (a) equal to friction angle (b) greater than friction angle (c) less than friction angle

7. Dynamic friction as compared to static friction is
   (a) less (b) equal (c) greater

8. Friction resistance depends upon
   (a) weight of the body and nature of contacting surface
   (b) speed of the body
   (c) shape of the body

9. We can walk or run as
   (a) friction on a foot is equal to forward thrust
   (b) friction is greater than forward thrust
   (c) friction is less than forward thrust

10. We slide on a wet surface as the coefficient of friction of a wet surface as compared to a dry surface is
    (a) more (b) less (c) equal

11. The maximum frictional force which comes into play when a body just begins to slide over the surface of another body is known as
    (a) sliding friction (b) rolling friction (c) limiting friction

12. The angle of friction is
    (a) the angle between the normal and the resultant of normals and limiting friction force
    (b) the ratio of friction force to normal reaction
    (c) the angle between the horizontal and the resultant of normals and friction force

13. The angle of inclination of an inclined plane when a body is about to slide down is called an
    (a) angle of friction (b) angle of repose (c) angle of kinetic friction

14. A block of weight 600 N is placed on a horizontal surface which tends to move when a horizontal force of $200\sqrt{3}$ N is applied. The angle of friction is
    (a) 30° (b) 45° (c) 60°

15. When a ladder is resting on a smooth ground and leaning against a rough vertical wall, then the force of friction acts
    (a) towards the wall at its upper end
    (b) downwards at its upper end
    (c) upwards at its upper end

16. A ladder resting on a rough ground and leaning against a smooth vertical wall will experience a force of friction
    (a) zero at its upper end
    (b) towards the wall at its upper end
    (c) downwards at its upper end

17. If the angle of inclination of a plane is less than the friction angle, then we require a force
    (a) to move the body upwards only
    (b) to move the body downwards only
    (c) to move the body upwards and downwards both

18. If the friction angle is λ, the efficiency of a screw jack is
    (a) $\dfrac{1 + \sin \lambda}{1 - \sin \lambda}$ (b) $\dfrac{1 - \sin \lambda}{1 + \sin \lambda}$ (c) $\dfrac{\sin \lambda}{1 + \sin \lambda}$

19. A cube rests on a rough horizontal surface. If the cube is gradually tilted by tilting the plane, then sliding will occur without toppling of the cube if the coefficient of friction (μ) is
(a) μ greater than 1 (b) μ equal to 1 (c) μ less than 1

20. Mechanical advantage should be
(a) unity (b) less than unity (c) greater than unity

21. If a weight W is lifted up y distance and a force P is applied corresponding to x movement and friction of the device is F, then mechanical advantage is
(a) $\dfrac{W+F}{P}$ (b) $\dfrac{W}{P+F}$ (c) $\dfrac{W}{P}$

22. Friction force (F), normal reaction and coefficient friction (μ) are related as
(a) $N = \mu F$ (b) $F = \mu N$ (c) $\dfrac{F}{N} = \mu$

23. The angle of repose α is related to the friction angle λ by which of the following relations when a body tends to slide?
(a) $\alpha = 2\lambda$ (b) $\alpha = \lambda$ (c) $\lambda = 2\alpha$

24. In a screw jack, the effort (P) required to lift the load (w) is given by which of the following relations if α = angle of the screw and λ = friction angle?
(a) $P = w \tan(\alpha - \lambda)$ (b) $P = w \tan(\alpha + \lambda)$ (c) $w = P \tan(\alpha + \lambda)$

25. What is the mechanical advantage of a screw jack if α = angle of inclination and λ = friction angle?
(a) $\tan(\alpha + \lambda)$ (b) $\dfrac{1}{\tan(\alpha + \lambda)}$ (c) $\tan(\alpha - \lambda)$

26. In case of a screw jack, the condition of the maximum efficiency is given by in which of the following if α = angle of the screw and λ = angle of friction?
(a) $\alpha = 45 + \lambda/2$ (b) $\alpha = 45 - \lambda/2$ (c) $\alpha = \lambda/2 - 45$

27. If a driver pulley is bigger than a driven pulley, then the belt will slide off on increasing transmission power from
(a) the bigger pulley
(b) the smaller pulley
(c) both the pulleys simultaneously

28. The lap angle considered for transmission of power in a straight-belt drive is the
(a) angle of contact of the smaller pulley
(b) angle of contact of the bigger pulley
(c) average angle of contact of both pulleys

29. The lap angle of a driver pulley and a driven pulley of equal size is
(a) 2π (b) $2\pi/3$ (c) π

30. If rope is wrapped round a pillar for five turns, the lap angle is
(a) 5π (b) 10π (c) 15π

31. The percentage improvement in power capacity of a flat-belt drive when the wrap angle at the driving pulley is increased from 150° to 210° by an idler arrangement for a friction coefficient of 0.3 is
(a) 47.2 (b) 36.97 (c) 77.15

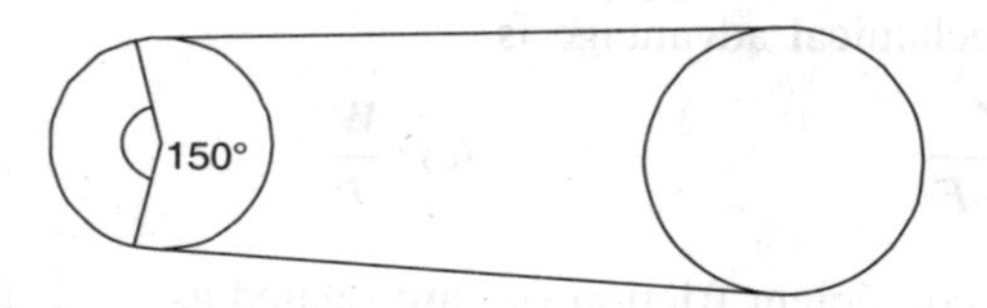

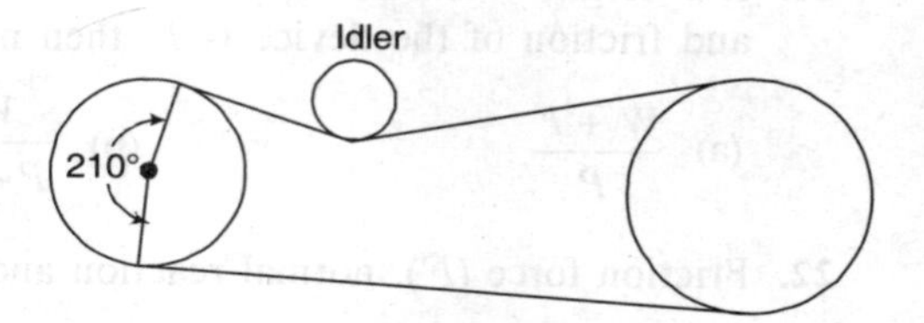

32. A belt-pulley arrangement has wrap angles = 160° and 200° on the large pulley. Putting an idler which has contact angle (α) = 40°, then lap angle on small pulley will be
(a) 200 (b) 160 (c) 240

33. Wrap angles of the belt are 180° and 2α (α = 30°) on pulleys A and B respectively. The lap angle for the arrangement is
(a) 180° (b) 240° (c) 200°

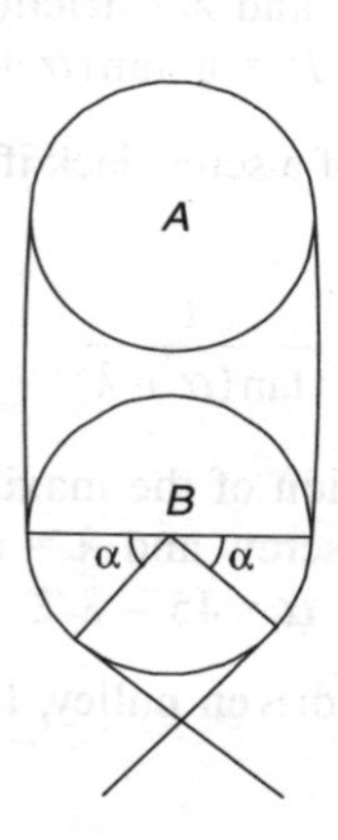

34. If the difference of tensions in a belt is 10 N and the velocity of the belt is 60 m/min, then power transmitted is
(a) 10 watts (b) 60 watts (c) 15 watts

35. If 30 N and 10 N are tensions in a belt with rpm = 60 and diameter = 7/22, then power transmitted is
(a) 30 watts (b) 15 watts (c) 20 watts

36. Friction on the wheel of a cycle acts
(a) forwards (b) backwards (c) upwards

37. The cycle stops when wheels are stopped rolling with the help of the brake as
(a) rolling of the wheels decreases
(b) power to the wheels stops
(c) sliding friction is much higher than rolling friction which acts against motion

38. Sand is spread on an ice-covered road to
(a) harden the ice
(b) soften the ice
(c) increase the friction of the surface

39. What is the maximum acceleration of the conveyor belt which permits a man standing stationary on it if the weight of the man is 500 N, mass 50 kg and the coefficient of friction (μ) = 0.1.
(a) 1 m/s^2 (b) 2 m/s^2 (c) 0.5 m/s^2

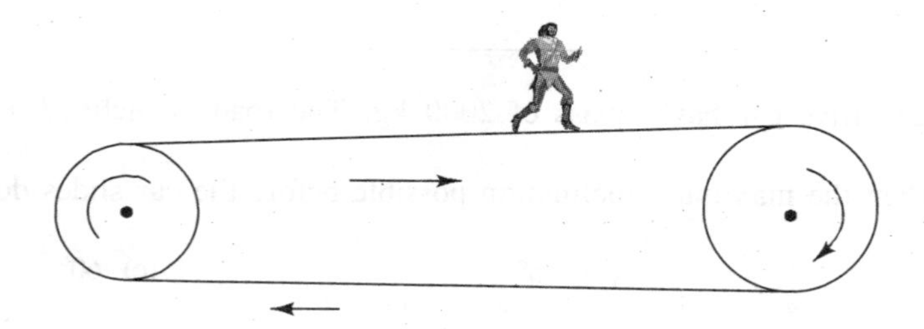

40. Can anyone get off a friction less horizontal surface by jumping?
(a) yes (b) no (c) depends on gravity

41. If $\mu = \frac{1}{\sqrt{3}}$, then the angle of friction (λ) is
(a) 45° (b) 30° (c) 60°

42. A ball bearing is used to
(a) reduce sliding friction
(b) reduce the coefficient of friction
(c) convert sliding friction to very low rolling friction

43. A ladder of weight 300 N rests on a smooth wall and a horizontal floor (μ = 0.2) making an angle of 60° with the horizontal. The force of friction is
(a) 300 N (b) 60 N (c) 150 N

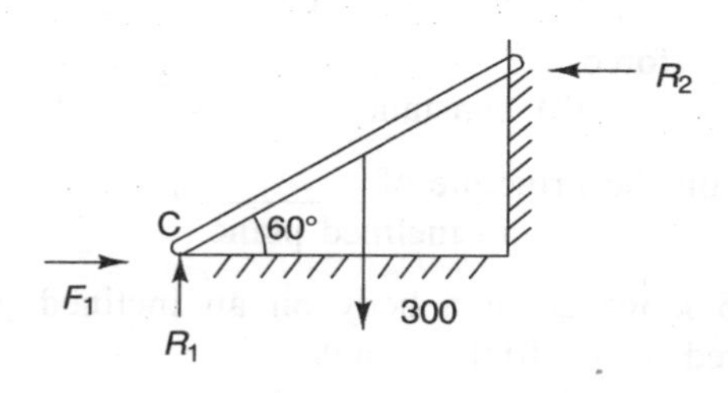

44. A body of weight 50 N is pulled with $10\sqrt{2}$ N force at 45° to the horizontal. Find the coefficient of friction if the body is about to move.
(a) 0.3 (b) 0.25 (c) 0.2

45. The ratio of tension in the tight side and the slack side is 1.2. The maximum tension permissible is 240 kN. If the belt has a velocity of 2 m/s, then the power transmitted
(a) 70 kW (b) 80 kW (c) 60 kW

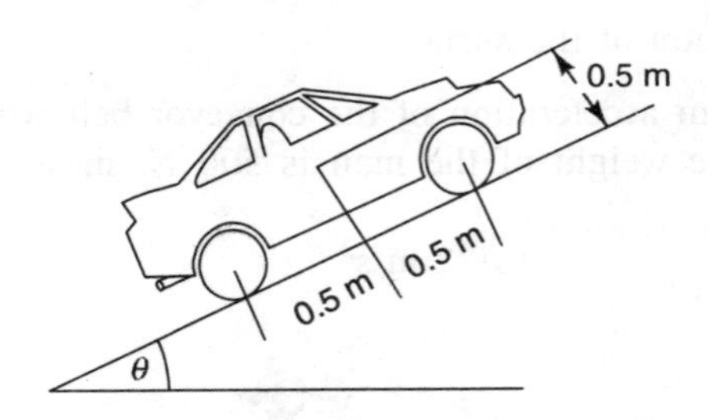

46. A four-wheel drive car has a mass of 2000 kg. The road is inclined at an angle θ. If $\mu = \frac{1}{\sqrt{3}}$, then the maximum inclination possible before the car slides down is
(a) 30° (b) 45° (c) 60°

Fill in the Blanks

1. Static friction is ________ as compared to dynamic friction.
(a) greater (b) lesser
2. Friction force depends upon ________ .
(a) roughness of surface (b) shape of the body
3. A wet surface will have ________ friction as compared to dry friction.
(a) lesser (b) greater
4. A person is more prone to slip on ice due to ________ .
(a) less friction (b) hardness of ice
5. The friction angle is an angle formed by the resultant with the ________ .
(a) normal (b) horizontal
6. Friction is a ________ force.
(a) reactive (b) applied
7. Friction is a ________ force.
(a) self-adjusting (b) constant
8. A screw jack works on the principle of ________ .
(a) friction (b) an inclined plane
9. Force is required to move down a body on an inclined plane if the repose angle is ________ as compared to the friction angle.
(a) less (b) greater

10. The effort required to lift a weight by a screw jack will be ________ if it has V-threads instead of square threads.
    (a) less (b) more

11. If the efficiency of a screw jack is more than 50%, then the screw jack will be ________.
    (a) self locking (b) not self locking

12. If the screw angle is less than the friction angle, the screw jack will be ________.
    (a) self locking (b) not self locking

13. The weights of wedges are ________ in analysis of the equilibrium of the wedges.
    (a) considered (b) not considered

14. The wedge has two opposite sides which are ________.
    (a) parallel (b) unparallel

15. The cone of friction is the surface generated by the ________.
    (a) resultant (b) friction force

16. The tight side has ________ tension as compared to the slack side.
    (a) more (b) less

17. The difference of tensions in the tight side and the slack side will depend upon the ________.
    (a) lap angle (b) width of the belt

18. The power transmitted is the product of the difference of tensions in two sides and the ________ of the belt.
    (a) velocity (b) friction

19. The belt will slip first on a ________ pulley.
    (a) large (b) small

20. If W is weight and P is effort, then the mechanical advantage is ________.
    (a) $W \times P$ (b) $\frac{W}{P}$

21. The coefficient of friction will be________ in case weight is doubled.
    (a) double (b) same

22. When an axle rotates in a sleeve, the type of friction acting is ________.
    (a) rolling friction (b) sliding friction

23. Rubber tyres are preferred to steel tyres as the coefficient of friction of a rubber tyre is ________ as compared to a steel tyre.
    (a) less (b) more

24. Car tyres are treaded to ________ friction.
    (a) increase (b) decrease

25. V-belt can transmit ________ power as compared to a flat belt.
    (a) less (b) more

26. A car moves ahead due to ________ friction on wheels.
    (a) forward (b) backward

ANSWERS

Life is a school. Those who learn to love and help others graduate with honours.

State True or False

1. True
2. True
3. True
4. False (static friction > dynamic friction)
5. False (Laws of dynamic and static friction are similar. Neither depends upon the area of contact.)
6. True
7. True (As no sliding takes place on the line of contact)
8. True ($\mu = F/N$)
9. True
10. False (Dynamic friction also follows Coulomb's law of friction.)
11. True
12. True ($\alpha_{repose} = \lambda$)
13. False (Force equal to mg sin $(\theta - \lambda)$ is required to move the body upon the plane where θ = inclination angle and λ = friction angle. If $\theta = \lambda$, then no force is required.)
14. True
15. False (Effort is proportional to tan $(\alpha + \lambda)$ where $\lambda = \tan^{-1} \mu$. As μ increases, λ increases.

    Since $\dfrac{\mu}{\cos \beta} > \mu$, λ for V-thread > λ for square thread. Therefore, more effort required for a V-thread screw jack.)
16. True
17. True
18. True
19. True
20. False (Weight of wedges is neglected.)
21. False (Analysis is simplified by considering the resultant instead of normal reaction and friction force individually as then we get a concurrent force system.)
22. False (A force system acting on a ladder is a coplanar non-concurrent force system.)
23. True
24. True
25. True
26. True
27. False (Tension in the belt is more in the direction it is moving as it is overcoming the friction force acting against its movement on the pulley.)
28. True
29. True
30. True
31. False (A lap angle is the angle of contact and it depends upon wrapping.)
32. True
33. True
34. True
35. False (A lap angle increases with the number of turns. One turn gives a lap angle of 2π. If there are n turns, then lap angle becomes $2\pi n$.)

36. True

37. True (The lap angle on the small pulley in a straight belt drive is smaller than that of the bigger pulley. Hence, the belt will slip first on the smaller pulley).

38. True

39. False ($\frac{T_1}{T_2} = e^{\mu\beta}$ for a flat belt while it is $\frac{T_1}{T_2} = e^{m \text{cosec} \frac{\theta}{2} \times b}$ for a V-belt where θ is the groove angle. As cosec $\theta/2 \geq 1$, the ratio of $\frac{T_1}{T_2}$ is more for a V-belt resulting into higher power transmission)

40. True
41. True
42. True
43. False
44. True
45. True
46. True
47. True
48. True
49. True
50. True
51. True
52. True

53. False (For sliding, inclined angle ≥ repose angle where repose angle is equal to friction angle λ.)

54. True
55. True

Multiple Choice Questions

1. (b)
2. (c)
3. (b)
4. (c)
5. (b)
6. (c) (The effort required to move the body up is $mg(\sin\theta + \mu\cos\theta)$ which will be least when θ is less.)
7. (a)
8. (a)
9. (b)
10. (b)
11. (c)
12. (a)
13. (b)
14. (a) ($\mu = \tan\lambda = \frac{200\sqrt{3}}{600} = \frac{1}{\sqrt{3}}$ or $\lambda = 30°$)
15. (c) (Friction is between the ladder and the wall. Also the ladder slides down, and friction will act up.)
16. (a) (The wall is smooth and will have zero friction when the ladder slides down.)
17. (c) (Zero effort to move down when angle of repose = angle of friction)
18. (b)
19. (c) (The cube will topple when the inclination angle is 45°. If the angle of repose or angle of friction is less than 45°, the cube will slide first, therefore $\mu = \tan\lambda < \tan 45$ or $\mu < 1$.)
20. (c)
21. (c)
22. (b)
23. (b)
24. (b)

25. (b) (For a screw jack, $P = W \tan(\alpha + \lambda)$. Therefore, W/P = Mech advantage = $\dfrac{1}{\tan(\alpha + \lambda)}$)

26. (b) **27.** (b)

28. (a) (The belt slips first on the small pulley which has a lower lap angle.)

29. (c)

30. (b) (Lap angle = $2\pi n$. For $n = 5$, lap angle = 10π)

31. (b) (Percentage improvement is $e^{\mu(\theta_1 - \theta_2)} \times 100 = e^{0.3 \times 60 \times \pi/180} \times 100 = 36.97\%$)

32. (a)

33. (b) (Lap angle = $180 + 2\alpha = 180 + 60 = 240°$)

34. (a) (Power = $(T_1 - T_2)$ velocity = $10 \times \dfrac{60}{60} = 10$ watts)

35. (c) (Power = $(T_1 - T_2)\dfrac{\pi DN}{60} = (30 - 10)\dfrac{\pi \times \dfrac{7}{22} \times 60}{60} = 20$ watts)

36. (a)

Friction force

37. (c) **38.** (c)

39. (a) (Friction force = $\mu W = 0.1 \times 500 = 50$ N. Force acting towards the front = $m \times \alpha = 50 \times \alpha$. The man will not fall till force acting is equal to the friction force. Therefore, $50 \times \alpha = 50$ or $\alpha = 1$ m/s)

40. (b) (No reaction is available as friction force is absent.)

41. (b) ($\tan \lambda = \dfrac{1}{\sqrt{3}}$ or $\lambda = 30°$)

42. (c)

43. (b) (Friction force on the ground = $\mu N = 0.2 \times 300 = 60$ N)

44. (b) ($\Sigma P_y = 0$, $N - 50 - 10\sqrt{2} \times \dfrac{1}{\sqrt{2}} = 0$ or $N = 40$, also $\Sigma P_x = 0$, therefore $10\sqrt{2}\dfrac{1}{\sqrt{2}} - 40\mu = 0$, or $\mu = 0.25$)

45. (b) ($\dfrac{T_1}{T_2} = 1.2$, or $T_2 = \dfrac{240}{1.2} = 200$ kN. Now $T_1 - T_2 = 240 - 200 = 40$ kN. Power = $(T_1 - T_2)v = 40 \times 2 = 80$ kW]

46. (a) (The slide will start irrespective of the weight of the body at the angle of repose of an inclined plane, which is equal to the angle of friction (λ). $\tan \lambda = \frac{1}{\sqrt{3}}$ or $\lambda = 30°$)

Fill in the Blanks

1. (a)
2. (a)
3. (a)
4. (a)
5. (a)
6. (a)
7. (a)
8. (b)
9. (a)
10. (b) (The apparent friction in V-thread is $\frac{\mu}{\cos \beta}$ and $\cos \beta \leq 1$)
11. (b)
12. (a)
13. (b)
14. (a)
15. (a)
16. (a)
17. (a)
18. (a)
19. (b)
20. (b)
21. (b)
22. (b)
23. (a)
24. (a)
25. (b)
26. (a)

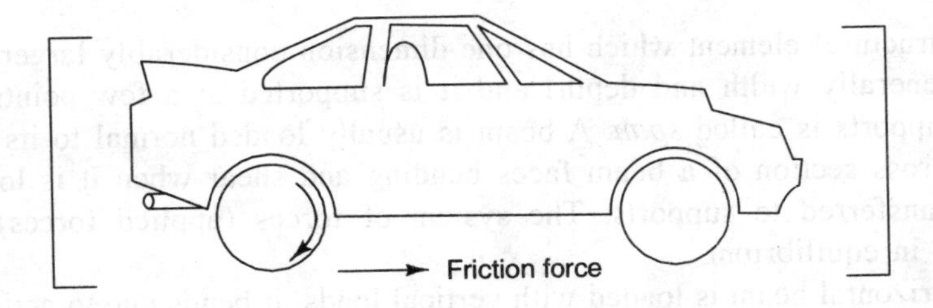

CHAPTER 10

Analysis of Beams

Don't let your mind become cluttered with worry. It leaves less room for the good stuff.

INTRODUCTION

A beam is a structural element which has one dimension considerably larger than other two dimensions (generally width and depth) and it is supported at a few points. The distance between the supports is called *span*. A beam is usually loaded normal to its cross-sectional areas. Every cross section of a beam faces bending and shear when it is loaded. The load finally gets transferred to supports. The system of forces (applied forces) and reactions keep the beam in equilibrium.

When a horizontal beam is loaded with vertical loads, it bends due to action of the loads. The internal shear stress and bending moment are developed to resist bending. The amount of bending in the beam depends upon the amount and type of the loads, length of the beam, elasticity of the beam and dimensions of the beam. The best way of studying the deflection or any other effect is to draw and analyze the shear force diagram (SFD) and the bending moment diagram (BMD) of the beam.

TYPES OF BEAMS

Beams can be any of the following types (Figure 10.1):

1. Simply supported
2. Cantilever type
3. Overhanging
4. Hinged and roller supported
5. Fixed
6. Continuous (having more than two supports)

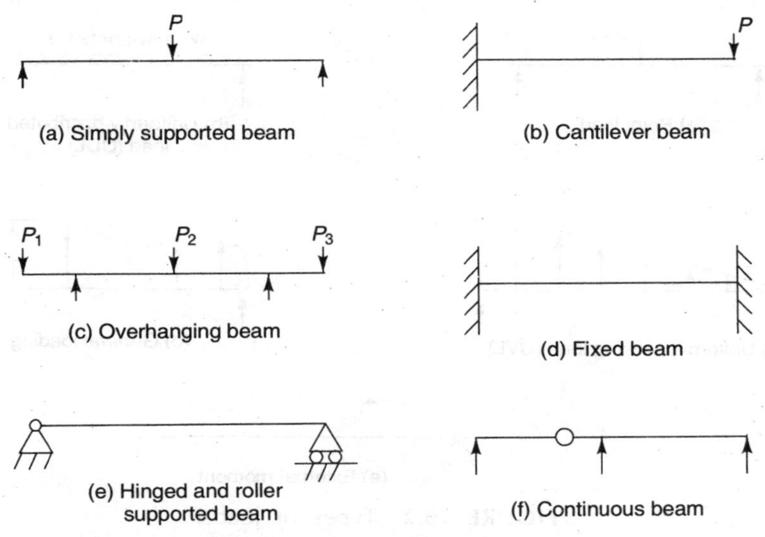

FIGURE 10.1 Types of beams.

TYPES OF SUPPORTS

Beams can have the following supports:

1. ***Simple support:*** When a beam rests on a simple support, the reaction is at a right angle to the support. The beam is free to move in the direction of an axis (along the length) and it can also rotate about its axis.
2. ***Roller support:*** When a beam rests on a roller support, the reaction is normal to the support. The beam is free to move along the axis.
3. ***Hinged support:*** It keeps the end of the beam stationary. The reaction has both the horizontal and vertical components.

TYPES OF LOADS

Types of loads which a beam may have are the following (Figure 10.2):

1. Point load or concentrated load
2. Uniformly distributed load (UDL)
3. Uniformly varying load (UVL)
4. General loading
5. External moment

The load applied on a beam gets transferred to its supports. Every section of a beam experiences the following:

Shear force: Shear force tries to shear off the section. It is obtained as the algebraic sum of all the forces acting normal to the axis of the beam either to the left or to the right of the

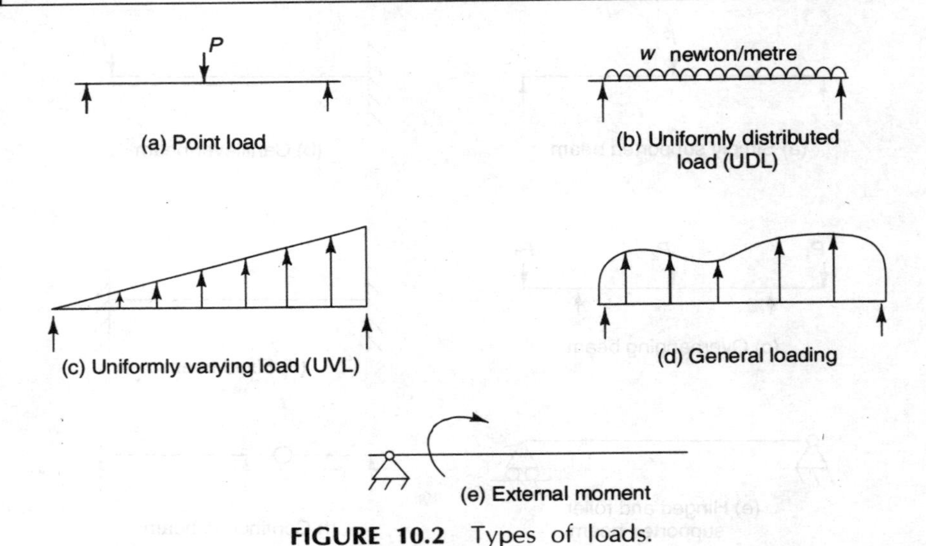

FIGURE 10.2 Types of loads.

section. The shear force that tends to move the left portion upwards relative to the right portion is taken as a positive shear force (Figure 10.3a). Otherwise it is negative (Figure 10.3b). The following sign convention ensures that the shear force diagram has the same shape whether it is drawn from the right ro left side of the beam.

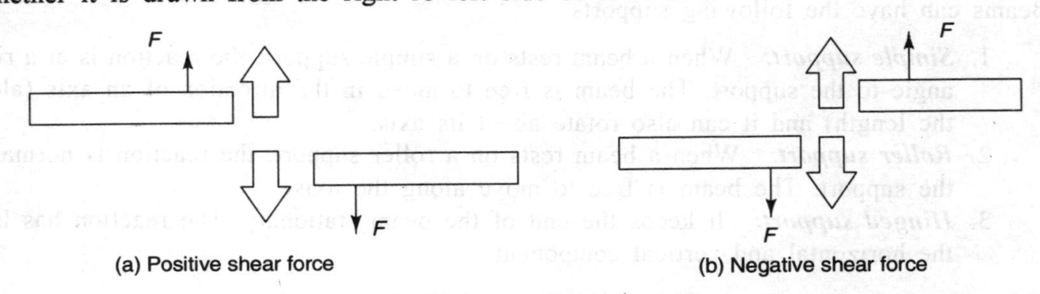

FIGURE 10.3 Shear force.

Bending moment: A bending moment (BM) is the moment of a section of the beam which tries to bend it. It is obtained as the algebraic sum of moments of all the forces about the section acting either to the left or to the right of the section. The bending moment which tends to sag a beam (beam bending upwards) is taken as a positive bending moment (Figure 10.4a). The bending moment which tries to hog a beam (beam bending downwards) is taken as a negative bending moment (Figure 10.4b).

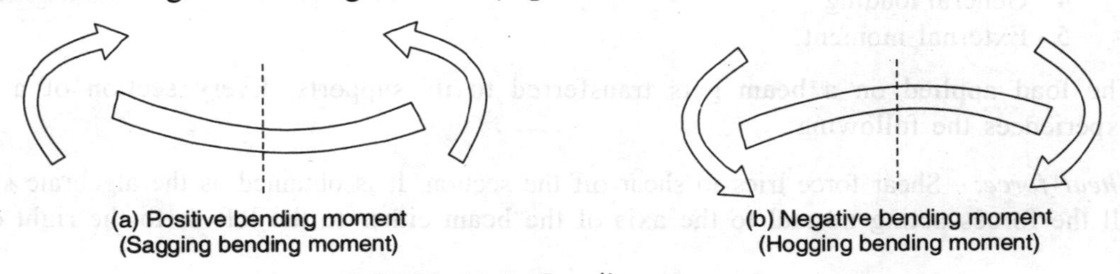

FIGURE 10.4 Bending moment.

RELATIONSHIP: LOAD INTENSITY, SHEAR FORCE AND BM

The shear force and bending moment in a beam vary from section to section along its length. If w is the load per unit length on the beam, F is the shearing force and M is the bending moment (Figure 10.5), then

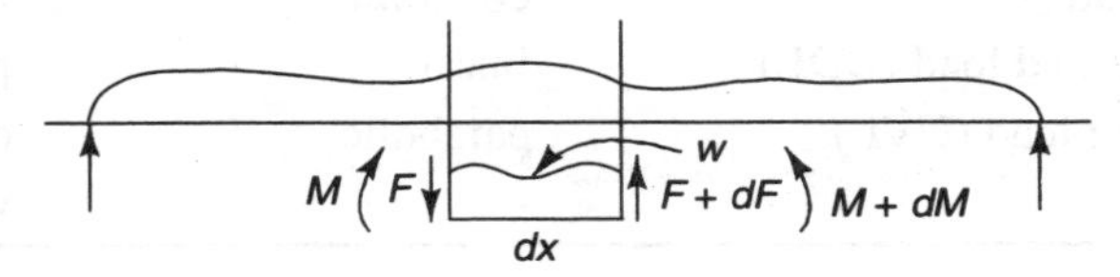

FIGURE 10.5 Relationship between load intensity, shear force and BM.

Now

$$\Sigma Fy = 0, \; -F + F + dF - w\,dx = 0$$

$$dF = w\,dx$$

or

$$\frac{dF}{dx} = w$$

Now taking the moment from point of action of $F + dF$,

$$M - Fdx - (M + dM) = 0$$

or

$$\frac{dM}{dx} = +F$$

$$M = \int dM = \int F dx$$

The above equations give the following useful deductions:

1. Bending moment is maximum or minimum where shear force is zero.
2. The inflection or contraflexure point on a beam lies at BM = 0.
3. The area of a shear force diagram at any point on a beam from a support gives bending moment.

The variation of shear force along the length of a beam can be represented on a graph in which an ordinate represents shear force. Such a shear force graph is called shear force diagram (SFD). Positive shear force is plotted above the beam and negative force below the beam in a shear force diagram.

The variation of bending moment along the length of a beam can be represented graphically. Such a graph is called bending moment diagram (BMD). Positive bending moment is plotted above the beam and negative bending moment below the beam. Bending moment changes its sign at the point of contraflexure. Bending moment is zero at the point of contraflexure.

Shear force diagrams and bending moment diagrams depend upon types of loading on the beam. The nature of variation of shear force diagrams and bending moment diagrams with types of load is given in Table 10.1.

Rules for drawing shear force diagrams and bending moment diagrams are as follows:

1. Wherever a point load (including support reaction) acts, there is a sudden change in value of shear force. The change is equal to the load and in the direction of the load.

Table 10.1 Nature of variation of shear force diagrams and bending moment diagrams with types of load.

Load	*Shear force diagram*	*Bending moment diagram*
(a) Point load/No load	constant	linear
(b) Uniformly distributed load (UDL)	linear	parabolic
(c) Uniformly varying load (UVL)	parabolic	cubic
(d) External moment	—	vertical change

2. The bending moment is the algebraic sum of the area of the SFD at that point from either support. The slope of the lines between two points is equal to shear force between them.
3. The bending moment is maximum or minimum where shear force is zero.
4. If an external moment is acting at a point on a beam, there is a sudden change in value of bending moment. The change is numerically equal to the external moment.
5. Inflection or contraflexure points occur where bending moment is zero.
6. Bending moment is zero at supports at both ends of a simply supported beam.
7. Bending moment is zero at the free end of cantilever and overhanging beams.
8. A shear force diagram and a bending moment diagram must be started from the free end of a cantilever as the fixed end has bending moment and shear force which are unknown.
9. Shear force diagrams and bending moment diagrams of a simply supported beam can be drawn from either end.

SFDs and BMDs for standard cases are as follows:

1. Simply supported beam

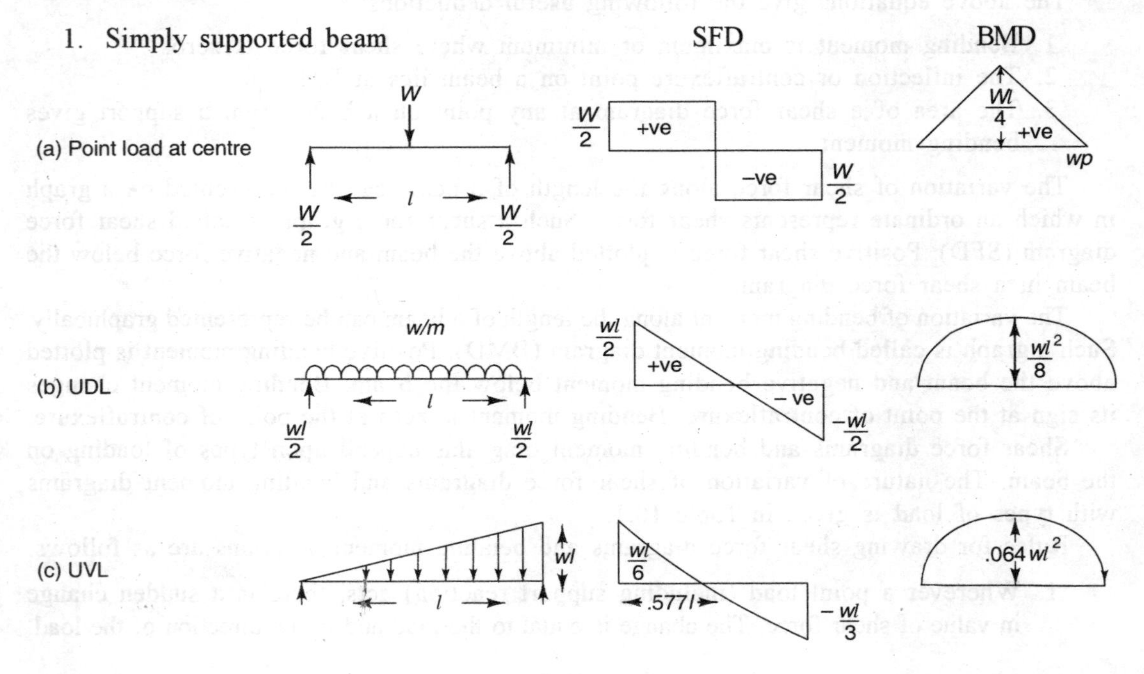

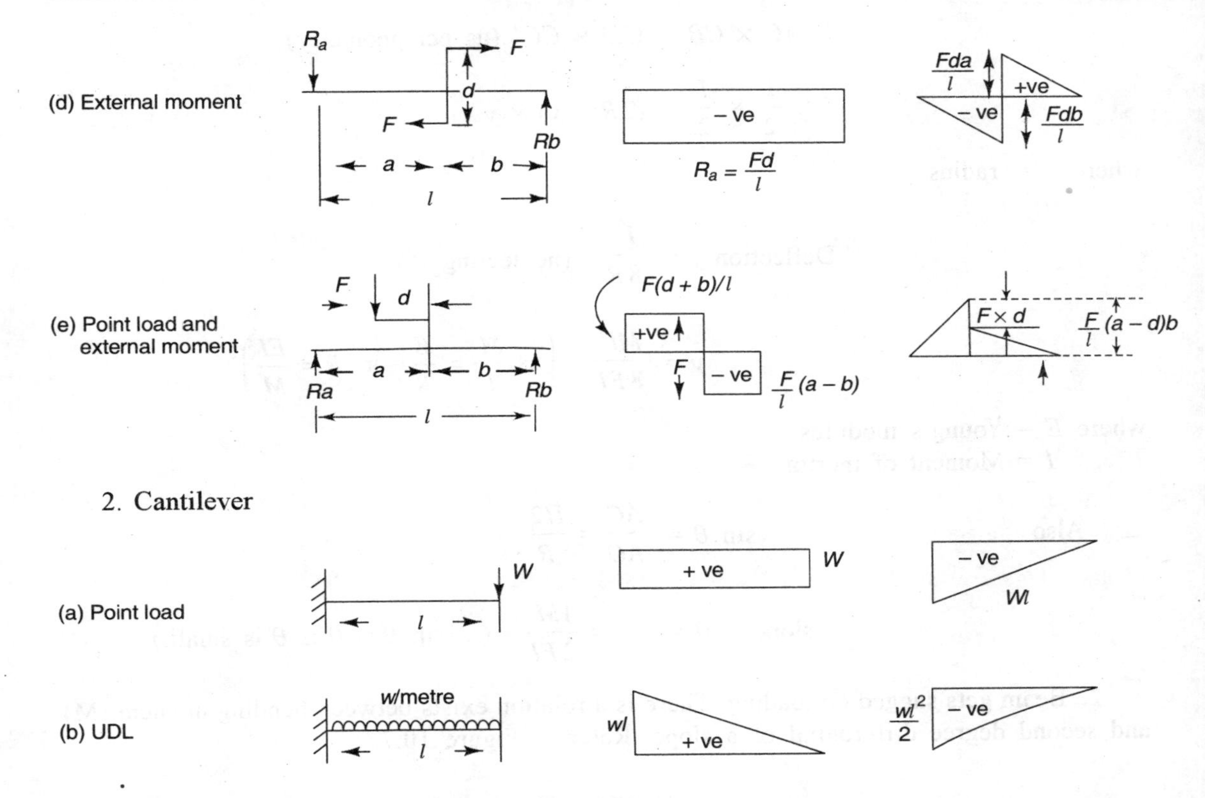

In the design of a beam, we are usually not only interested in stresses produced by the loads but also in the deflection produced by these loads so that deflections should not exceed a permissible limit. Though deflections and slope are generally covered in strength of materials, we are explaining, them in this chapter as both are linked with bending moment diagrams. Three methods of finding slope and deflections have been covered as under:

1. On loading, the mid point of a beam is sagged/deflected (y) and also there is a slope (θ) at the end. In Figure 10.6 *ACB* beam is sagged to shape *AC′B*.

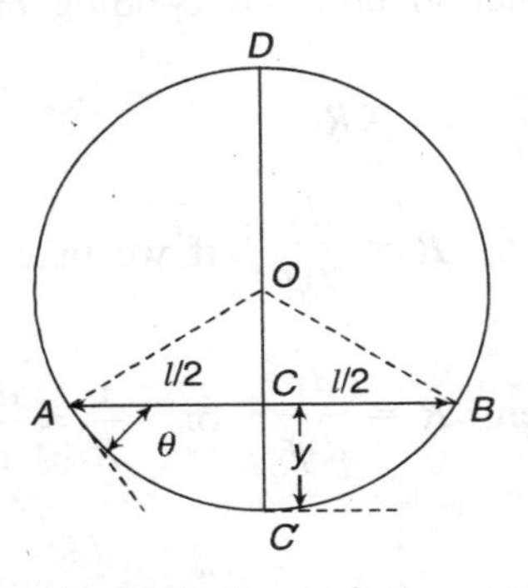

FIGURE 10.6 A beam being sagged.

$$AC \times CB = CD \times CC' \text{ (as per geometry)}$$

$$\frac{l}{2} \times \frac{l}{2} = (2R - y) \times y$$

where R = radius

$$\text{Deflection } y = \frac{l^2}{8R} \quad \text{(neglecting } y^2\text{)}$$

$$y = \frac{Ml^2}{8EI} \quad \left(\text{as } \frac{M}{l} = \frac{E}{R} \text{ or } R = \frac{EI}{M}\right)$$

where E = Young's modulus
I = Moment of inertia

Also

$$\sin \theta = \frac{AC}{AO} = \frac{l/2}{R}$$

$$\text{slope} = \theta = \frac{l}{2R} = \frac{1M}{2EI} \quad (\because \sin \theta = \theta \text{ if } \theta \text{ is small.})$$

2. Beam gets sagged on loading. There is a relation exists between bending moment (M) and second degree differential of a slope. Refer to Figure 10.7.

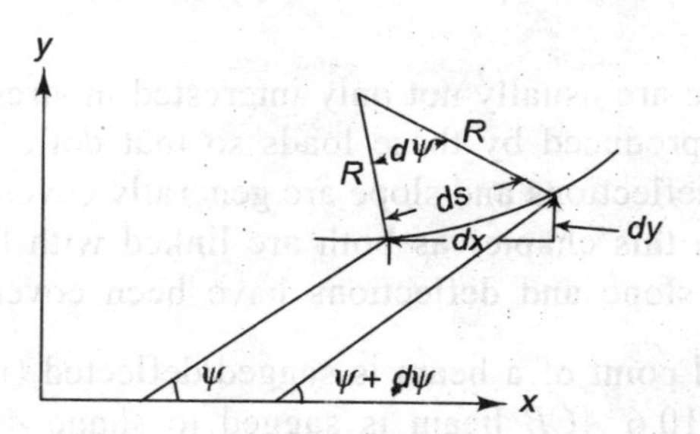

FIGURE 10.7 Relation between bending moment and slope.

$$ds = Rd\psi$$

$$\therefore \quad R = \frac{ds}{d\psi} \quad \text{if we take } ds \approx dx$$

$$\text{then } R = \frac{dx}{d\psi} \quad \text{or} \quad \frac{1}{R} = \frac{d\psi}{dx}$$

However,

$$\tan \psi = \psi = \frac{dy}{dx}$$

$$\therefore \quad \frac{1}{R} = \frac{d\psi}{dx} = \frac{d^2 y}{dx^2}$$

We know

$$\frac{M}{I} = \frac{E}{R}$$

or

$$\frac{1}{R} = \frac{M}{EI}$$

$$\therefore \quad M = EI\frac{d^2 y}{dx^2}$$

The equations of the slope and deflection for some particular cases by this method are as follows:

(a) A simply supported beam with point load (W) at the centre. Refer to Figure 10.8.

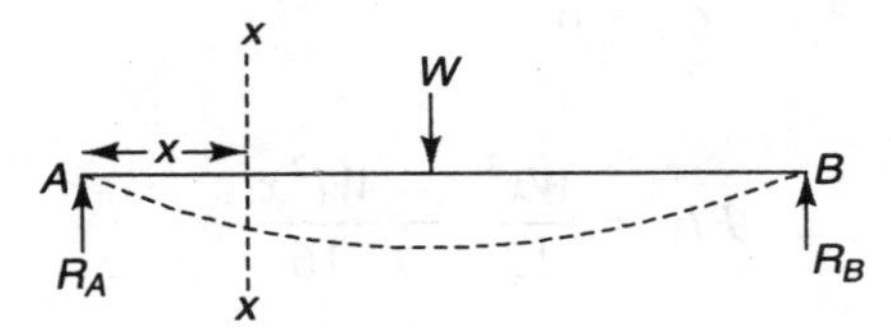

FIGURE 10.8 A simply supported beam with a point load.

$$M_x = R_A \times x = \frac{W}{2}x$$

Now

$$M_x = EI\frac{d^2 y}{dx^2}$$

Therefore,

$$EI\frac{d^2 y}{dx^2} = \frac{W}{2}x \tag{10.1}$$

Integrating Eq. (10.1)

$$EI\frac{dy}{dx} = \frac{Wx^2}{4} + C_1$$

Condition: At $x = \frac{l}{2}$

$$\frac{dy}{dx} = \text{slope} = 0$$

Therefore,

$$C_1 = -\frac{Wl^2}{16}$$

So,

$$EI\frac{dy}{dx} = \frac{Wx^2}{4} - \frac{Wl^2}{16} \tag{10.2}$$

Slope at $x = 0$,

$$\left(\frac{dy}{dx}\right)_{x=0} = \frac{Wl^2}{16\,EI}$$

Integrating Eq. (10.2),

$$EIy = \frac{Wx^3}{12} - \frac{Wl^2 x}{16} + C_2$$

Condition: $x = 0$, $y = 0$. Therefore,

$$C_2 = 0$$

So,

$$EIy = \frac{Wx^3}{12} - \frac{Wl^2 x}{16}$$

Now at $x = l/2$,

$$y = -\frac{1}{EI} \times \frac{Wl^3}{48}$$

(b) Cantilever: Refer to Figure 10.9.

$$M_x = -Wx$$

$$EI\frac{d^2y}{dx^2} = -Wx \tag{10.3}$$

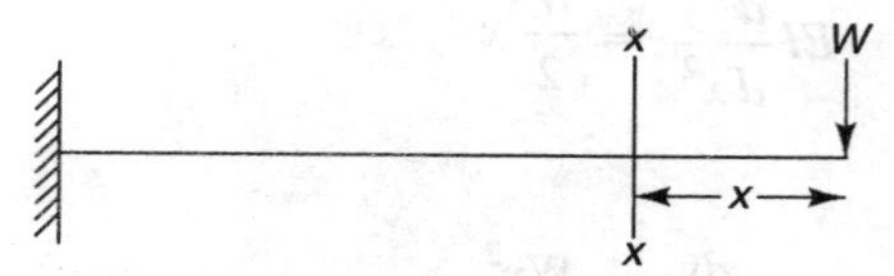

FIGURE 10.9 A cantilever type beam with a load at the end.

Integrating Eq. (10.3), we get

$$EI\frac{dy}{dx} = -\frac{Wx^2}{2} + C_1 \tag{10.4}$$

At $x = l$

$$\frac{dy}{dx} = 0$$

Therefore,

$$C_1 = + \frac{Wl^2}{2}$$

So Eq. (10.4) becomes $$EI\frac{dy}{dx} = -\frac{Wx^2}{2} + \frac{Wl^2}{2} \quad (10.5)$$

For maximum slope (at $x = 0$),

$$\left(\frac{dy}{dx}\right)_{x=0} = \frac{Wl^2}{2EI}$$

Integrating Eq. (10.5),

$$EIy = \frac{-Wx^3}{6} + \frac{Wl^2x}{2} + C_2 \quad (10.6)$$

At $x = l$, $y = 0$. Therefore,

$$C_2 = \frac{-Wl^3}{3}$$

So Eq. (10.6) becomes

$$EIy = \frac{Wl^2x}{2} - \frac{Wx^3}{6} - \frac{Wl^3}{3}$$

Maximum deflection (at $x = 0$)

$$y = \frac{-Wl^3}{3EI}$$

3. Mohr's theorem (Moment Area Method): The change of the slope between any two points on an elastic curve is equal to the net area of the bending moment diagram between these points divided by *EI*. The intercepts taken on the vertical reference line of tangents at any two points on an elastic curve is equal to the moment taken from the reference point of the area of the BM diagram between these two points divided by E1. This is called *Mohr's theorem* or *moment area method*. Refer to Figure 10.10. Showing a simply supported beam loaded at the centre BM diagrams are also shown for reaction R_b (= $W/2$) and load (W).

We want to find slope and deflection between B and C.

$$\text{Area } A = \text{Area } C \text{ to } B = \frac{1}{2} \times \frac{Wl}{4} \times \frac{l}{2} = \frac{Wl^2}{16}$$

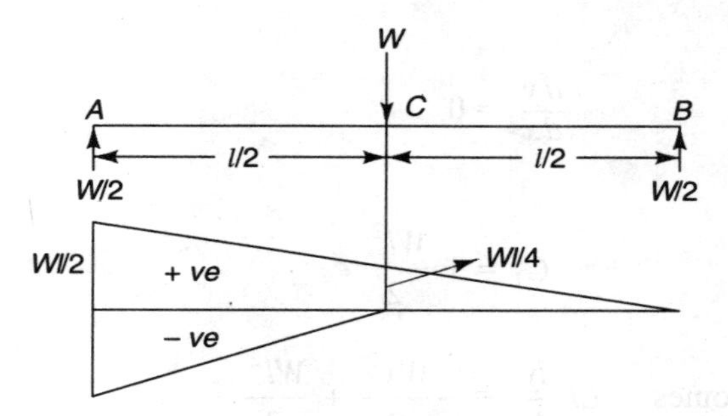

FIGURE 10.10 Mohr's theorem.

$$\text{Slope at } C = \frac{\text{Area } A}{EI} = \frac{Wl^2}{16EI}$$

$$\bar{x} = \text{Centre of gravity of bending moment area between } BC$$

$$= \frac{2}{3} \times \frac{l}{2} = \frac{l}{3}$$

Deflection at C = Moment of area from point B

$$y_c = \frac{A}{EI} \times \bar{x} = \frac{Wl^2}{16EI} \times \frac{l}{3} = \frac{Wl^3}{48EI}$$

SOLVED PROBLEMS

1. A simply supported beam of span L carries a concentrated load W at the mid span. Draw the SFD and BMD.

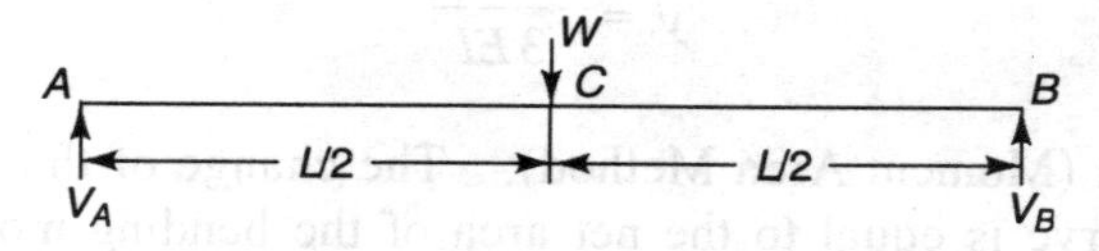

Reactions

Since loading is symmetrical $V_A = V_B = W/2$

SFD

$$F_A = V_A = \frac{W}{2}$$

$$F_C = V_A - W = \frac{W}{2} - W = -\frac{W}{2}$$

$$F_B = V_B = -\frac{W}{2}$$

BMD

$$M_x = \frac{W}{2}x$$

At $x = 0$

$$M_A = 0$$

At $x = \frac{l}{2}$

$$M_C = \frac{Wl}{4}$$

At $x = l$

$$M_B = 0$$

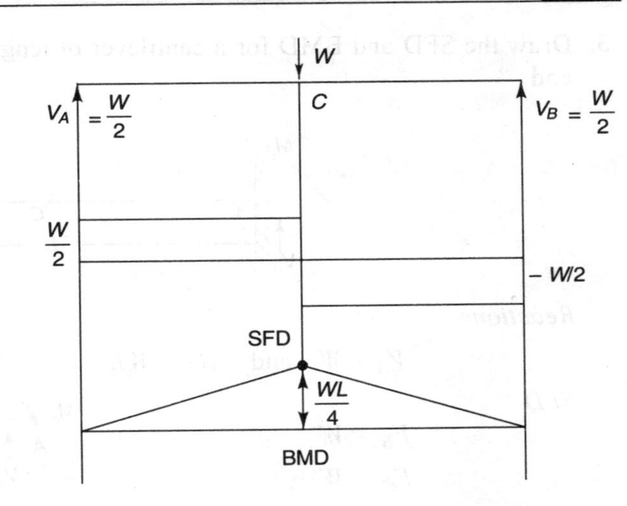

2. A simply supported beam of span L carries a uniformly distributed load as shown. Draw the SFD and BMD.

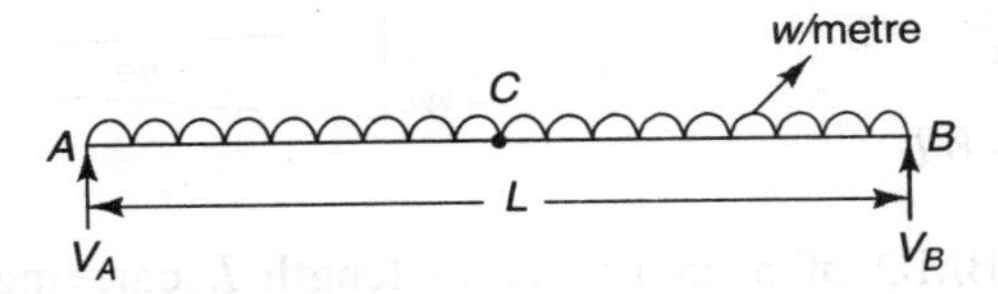

Reactions

$$V_A = V_B = \frac{wL}{2}$$

SFD

$$F_A = V_A = \frac{wL}{2}$$

$$F_C = V_A - \frac{wL}{2} = 0$$

$$F_B = \frac{-wL}{2}$$

BMD

$$M_x = -wx \times \frac{x}{2} + V_A \times x$$

$$M_0 = 0$$

$$M_{L/2} = \frac{-wL^2}{8} + \frac{wL^2}{4} = \frac{wL^2}{8}$$

$$M_L = 0$$

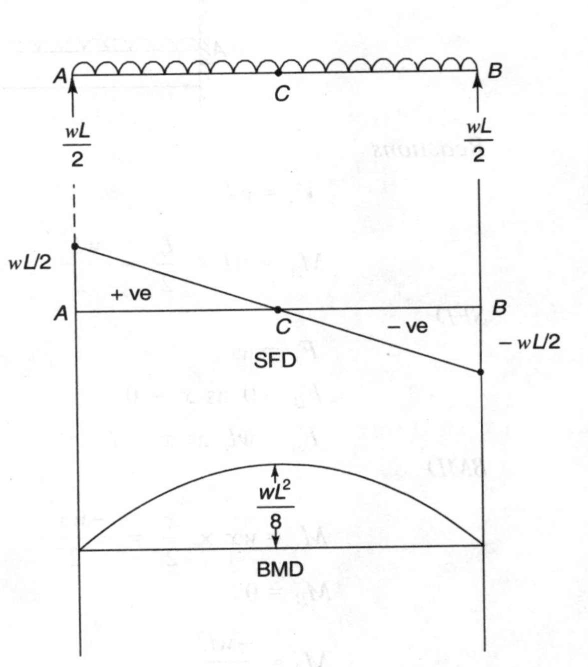

3. Draw the SFD and BMD for a cantilever of length L carrying a point load W at the free end.

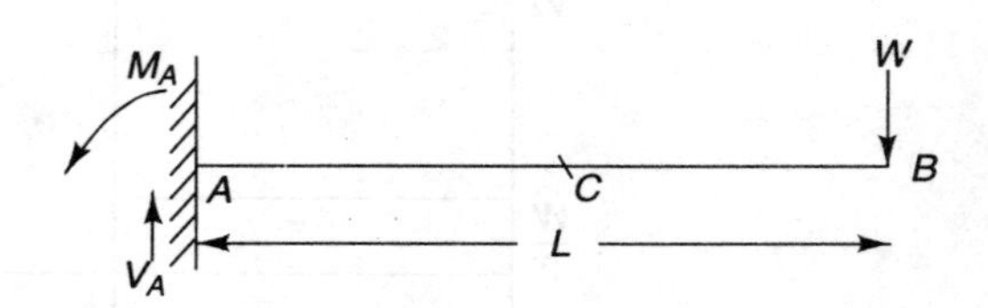

Reactions

$$V_A = W \quad \text{and} \quad M_A = WL$$

SFD

$$F_B = W$$
$$F_C = W$$
$$F_A = W$$

BMD

$$M_x = -W \times x$$
$$M_B = 0$$
$$M_A = -WL$$

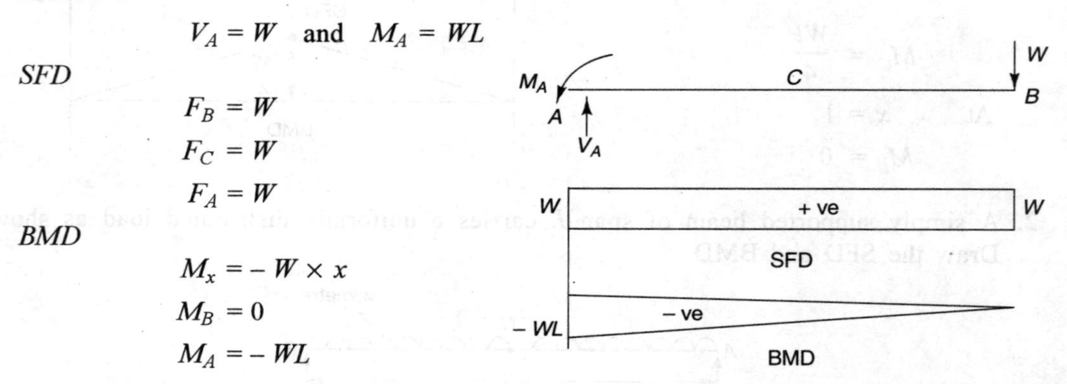

4. Draw the SFD and BMD of a cantilever of length L carrying UDF = w/metre.

Reactions

$$V_A = wL$$

$$M_A = wL \times \frac{L}{2} = \frac{wL^2}{2}$$

SFD

$$F_x = wx$$
$$F_B = 0 \text{ as } x = 0$$
$$F_A = wL \text{ as } x = L$$

BMD

$$M_x = wx \times \frac{x}{2} = \frac{-wx^3}{2}$$
$$M_B = 0$$
$$M_A = \frac{-wL^3}{2}$$

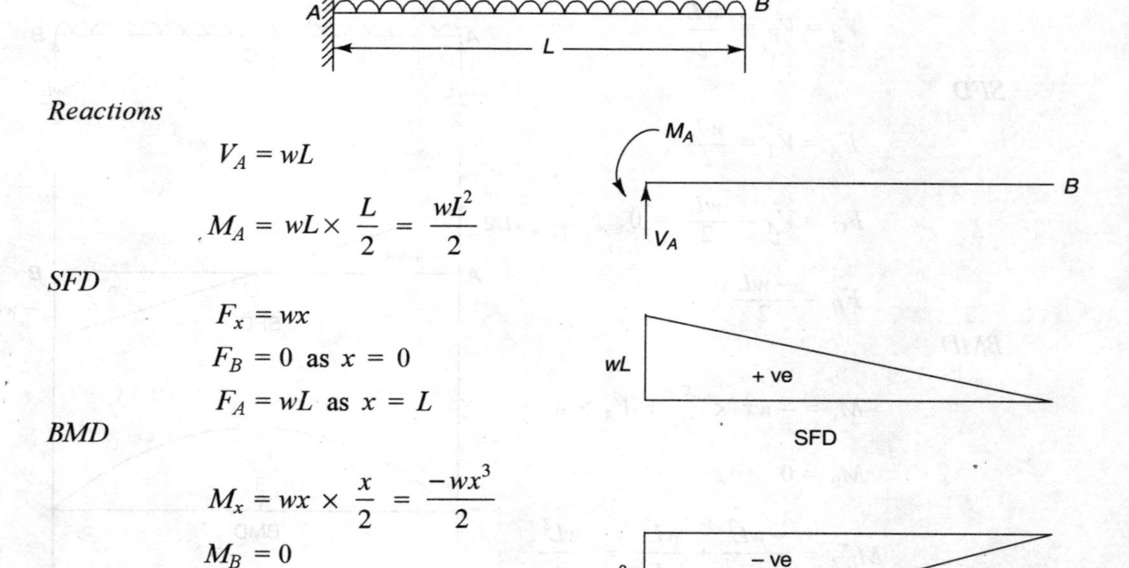

5. Draw the SFD and BMD for the beam shown below.(UPTU: 2001–2002)

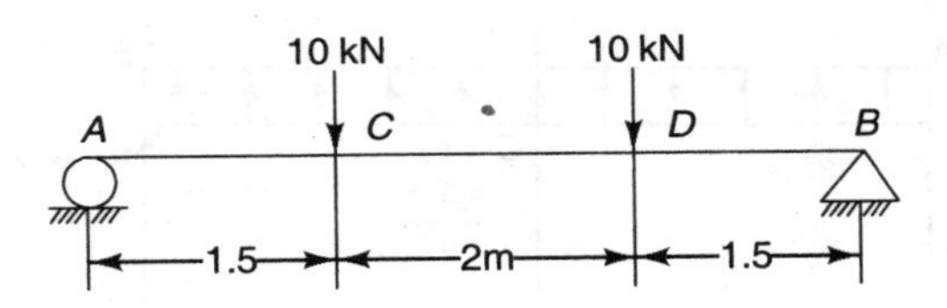

Reactions

Since the beam is symmetrically loaded,

$$V_A = V_B = 10 \text{ kN}$$

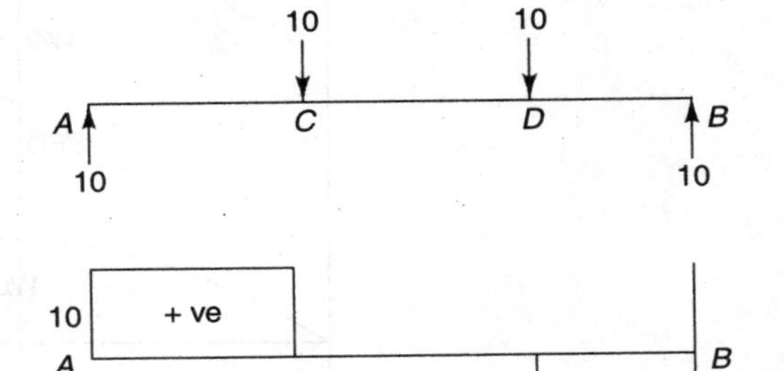

SFD

$$F_A = V_A = 10 \text{ kN}$$

$$F_C = V_A - 10 = 0$$

$$F_D = -10$$

$$F_B = -10 = V_B$$

BMD

$$M_x = V_A \times x \text{ (up to } C)$$

$$M_A = 0$$

$$M_C = 10 \times 1.5 = 15$$

$$M_x = V_A x - 10x \text{ (up to } D)$$

$$M_D = 10 \times 3.5 - 20 = 15$$

$$M_B = 0$$

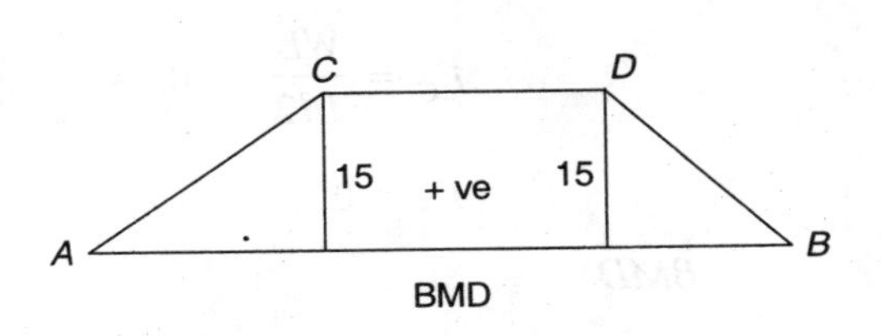

6. A log of wood is floating in water with a weight W placed at its middle as shown. Neglecting the weight of the log, draw the SFD and BMD of the log.
(UPTU: 2002–2003)

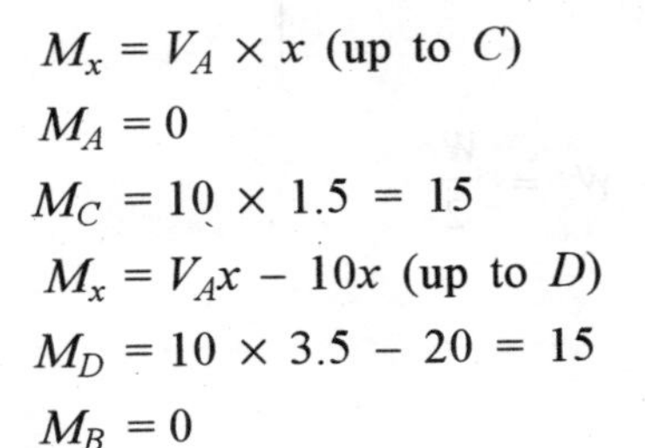
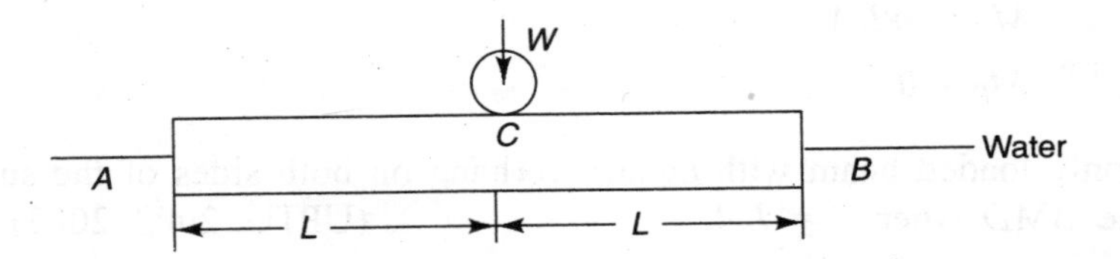

Reactions

$$\text{UDL} = w = \frac{W}{2L}$$

SFD

$$F_x = wx = \frac{W}{2L}x$$

$$F_A = 0$$

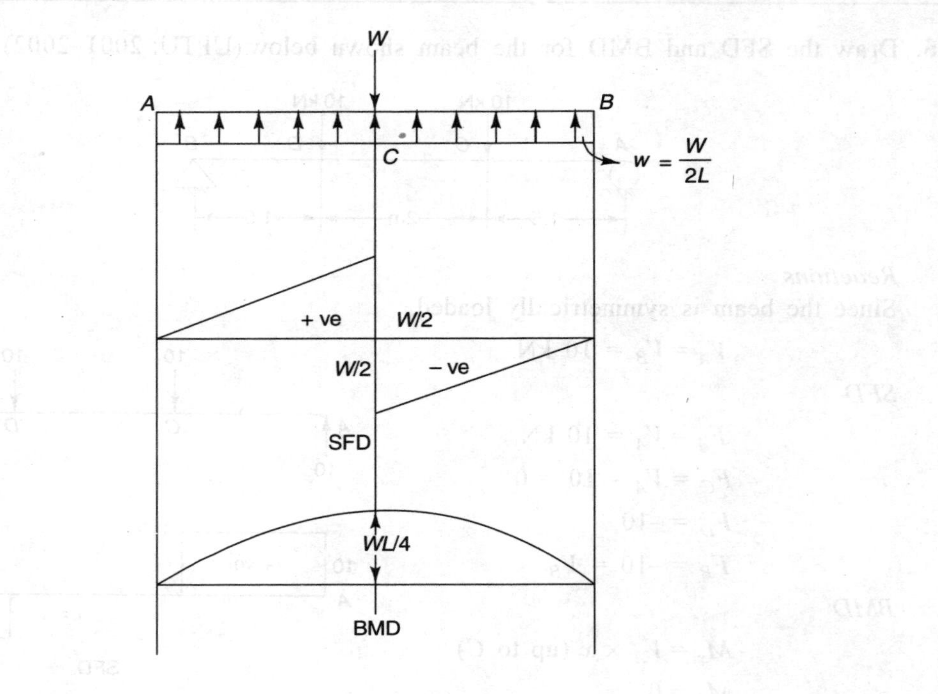

$$F_C = \frac{WL}{L/2} - W = \frac{W}{2} - W = \frac{W}{2}$$

$$F_B = 0$$

BMD

$$M_x = wx \times \frac{x}{2} = \frac{W}{2L} \times \frac{x^2}{2} = \frac{Wx^2}{4L}$$

$$M_A = 0$$

$$M_C = WL/4$$

$$M_B = 0$$

7. A uniformly loaded beam with equal overhang on both sides of the support is shown. Draw the BMD when $a = L/4$. (UPTU: 2002–2003)

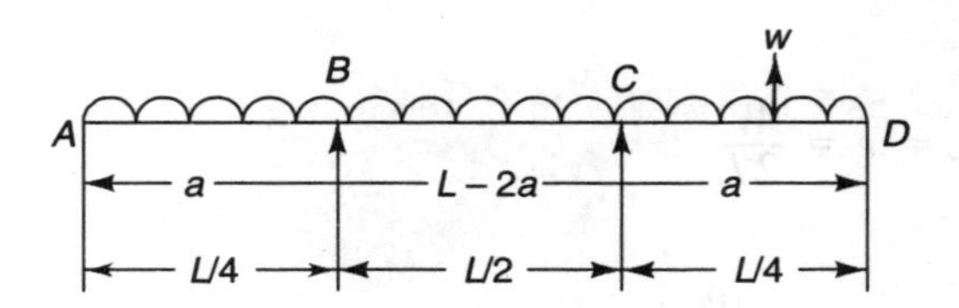

Reactions

Symmetrical loading gives:

$$V_B = V_C = \frac{wL}{2}$$

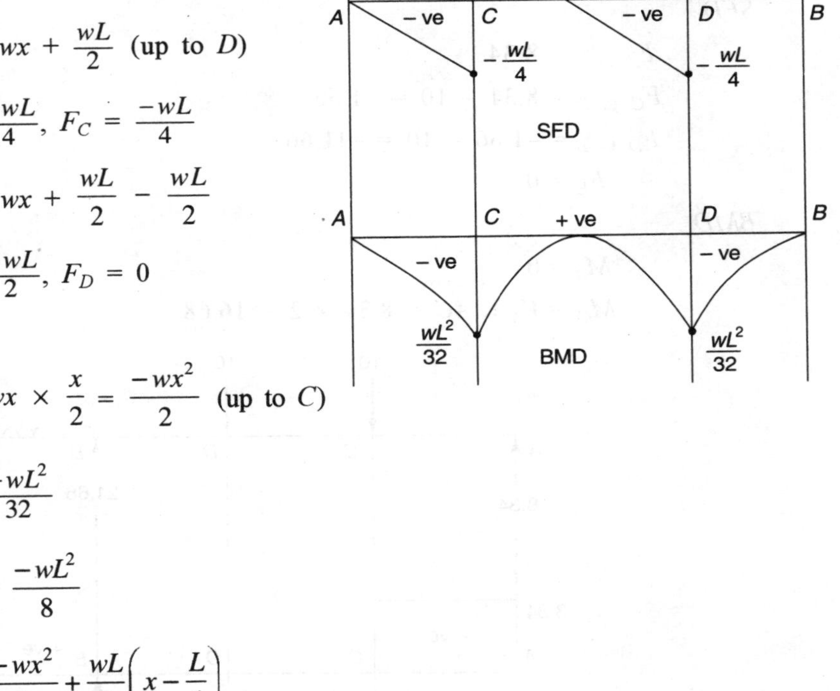

SFD

$$F_x = -wx \text{ (up to } C)$$

$$F_A = 0 \text{ and } F_B = \frac{-wL}{4}$$

$$F_x = -wx + \frac{wL}{2} \text{ (up to } D)$$

$$F_B = \frac{+wL}{4},\ F_C = \frac{-wL}{4}$$

$$F_x = -wx + \frac{wL}{2} - \frac{wL}{2}$$

$$F_c = \frac{+wL}{2},\ F_D = 0$$

BMD

$$M_x = wx \times \frac{x}{2} = \frac{-wx^2}{2} \text{ (up to } C)$$

$$M_C = \frac{-wL^2}{32}$$

Similarly, $M_D = \dfrac{-wL^2}{8}$

$$M_x = \frac{-wx^2}{2} + \frac{wL}{2}\left(x - \frac{L}{4}\right)$$

$$M_{L/2} = \frac{-wL^2}{8} + \frac{wL}{2} \times \frac{L}{4}$$

$$= 0$$

8. Draw the shear force and bending moment diagrams for the beam shown. (UPTU: 2003–2004)

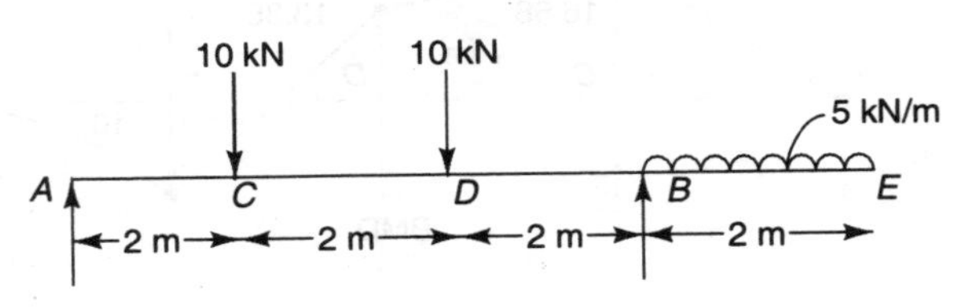

Reactions

$$V_A + V_B = 10 + 10 + 5 \times 2 = 30$$

$$M_A = 0,$$

$$= 10 \times 2 + 10 \times 4 - V_B \times 6 + 2 \times 5 \times 7 = 0$$

$$V_B = \frac{130}{6} = 21.66 \text{ kN}$$

$$V_A = 30 - 21.66 = 8.34 \text{ kN}$$

SFD

$$F_{A \text{ to } C} = 8.34$$

$$F_{C \text{ to } D} = 8.34 - 10 = -1.66$$

$$F_{D \text{ to } B} = -1.66 - 10 = -11.66$$

$$F_E = 0$$

BMD

$$M_A = 0$$

$$M_C = V_A \times AC = 8.34 \times 2 = 16.68$$

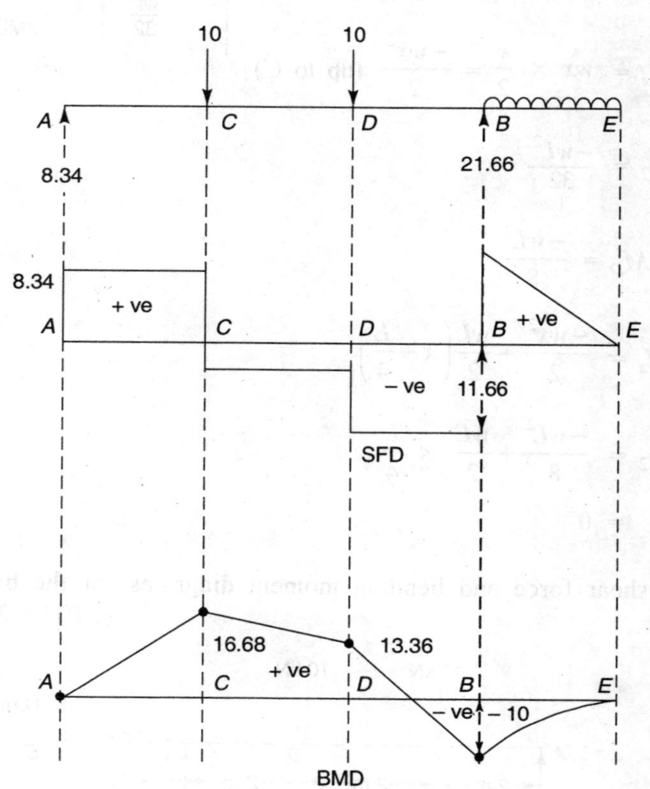

$$M_D = V_A \times AD - 10 \times 2 = 13.36$$
$$M_B = V_A \times AB - 10 \times 4 - 10 \times 2 = -10$$
$$M_E = 0$$

9. A simply supported beam is subjected to various loadings as shown in the figure. Sketch the SFD and BMD showing their values at significant locations. (UPTU: 2005)

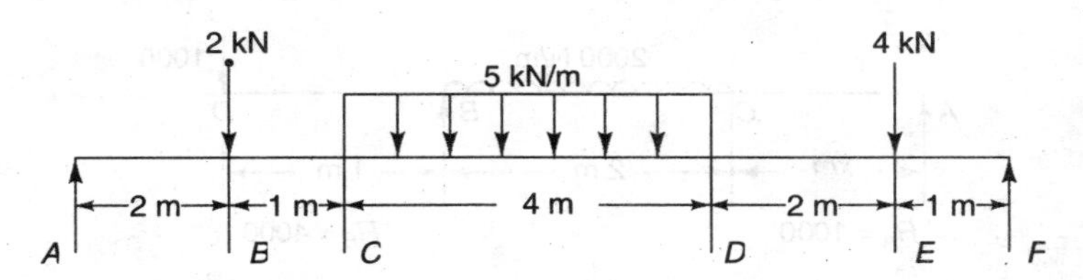

Reactions

$$V_A + V_F = 2 + 5 \times 4 + 4 = 26$$
$$\Sigma M_A = 0,\ 2 \times 2 + 20 \times 5 + 4 \times 9 - V_F \times 10 = 0$$

or $$V_F = \frac{4 + 100 + 36}{10} = 14 \text{ kN}$$

$\therefore$ $$V_A = 26 - 14 = 12 \text{ kN}$$

SFD

$$F_A = +12$$
$$F_B = 12 - 2 = 10$$
$$F_C = 10$$
$$F_D = 10 - 20 = -10$$
$$F_E = -10 - 4 = -14$$
$$F_F = -14$$

BMD

$$M_A = 0$$
$$M_B = 12 \times 2 = 24$$
$$M_C = 12 \times 3 - 2 \times 1 = 34$$
$$M_D = 12 \times 7 - 2 \times 5 - 20 \times 2$$
$$= 84 - 10 - 40 = 34$$
$$M_E = V_P \times 1 = 14$$
$$M_D = 14 \times 3 - 4 \times 2$$
$$= 42 - 8 = 34$$

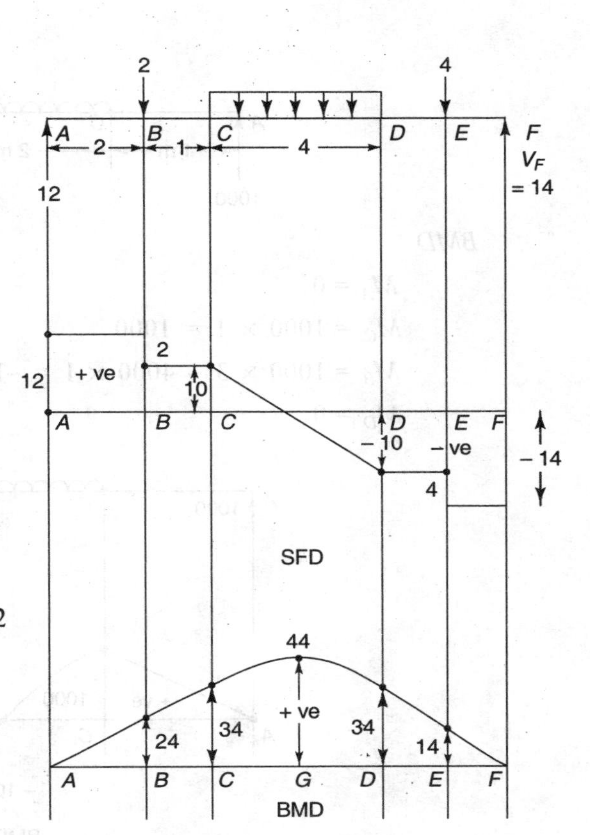

$$M_G = M_C + \frac{1}{2} \times 10 \times 2$$
$$= 34 + 10 = 44$$

10. Find the value of x and draw the BMD for the beam shown below.
(UPTU: 2000–2001)

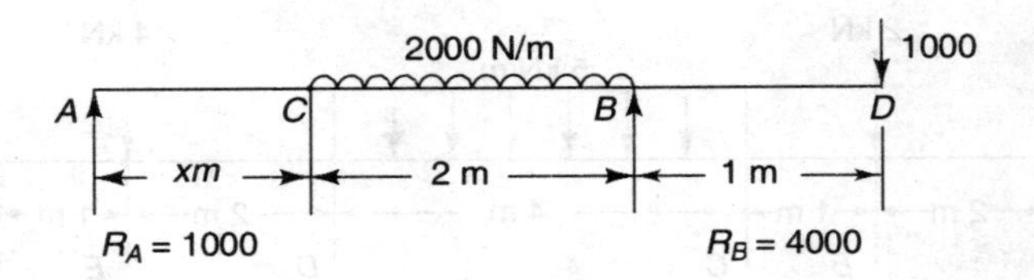

Reactions

$$R_A + R_B = 1000 + 2000 \times 2 = 5000$$
$$\Sigma M_A = 0,\ 4000\ (x + 1) - 4000\ (2 + x) + 1000\ (3 + x) = 0$$
or $$x = 1 \text{ m}$$

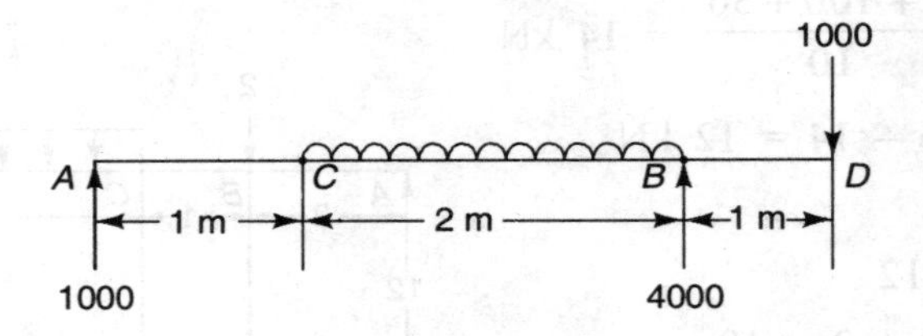

BMD

$$M_A = 0$$
$$M_C = 1000 \times 1 = 1000$$
$$M_B = 1000 \times 3 - 4000 \times 1 = -1000$$
$$M_D = 0$$

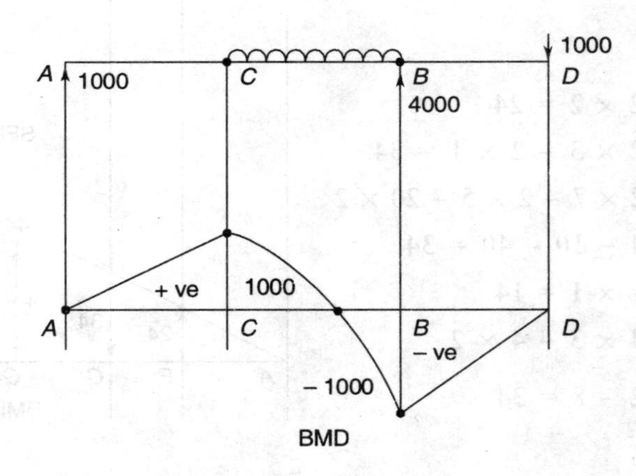

11. Draw the SFD of the beam shown and indicate the maximum shear force value in the SFD.

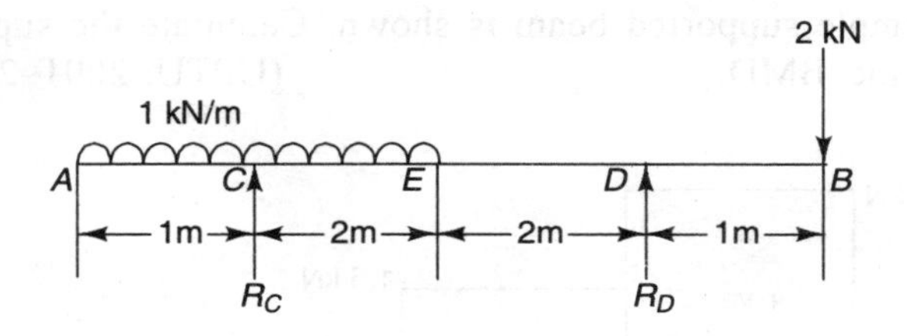

Reactions

$R_C + R_D = 2 + 1 \times 3 = 5$

$\Sigma M_C = 0,$

$$= -1 \times \frac{1}{2} + 2 \times 1 - R_D \times 4 + 2 \times 5 = 0$$

$$\therefore \quad R_D = \frac{-0.5 + 2 + 10}{4} = \frac{11.5}{4} = 2.875 \text{ kN}$$

$R_C = 5 - 2.875 = 2.125$ kN

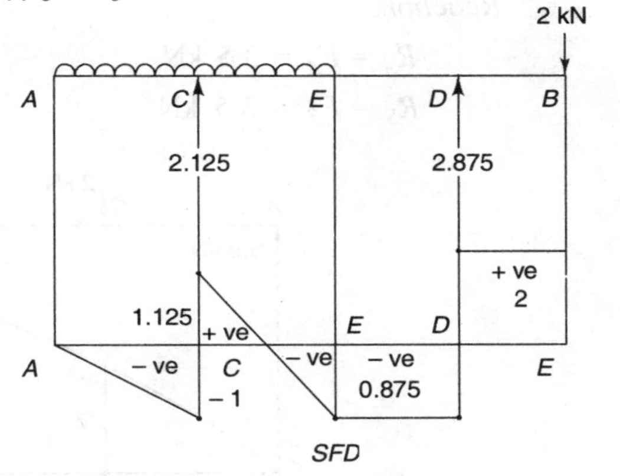

SFD

$F_A = 0$

$F_C = +1.125$

$F_E = 11.125 - 2 = 0.875$

$F_D = +2.875 - 0.875 = 2$

12. The BMD of a simple supported beam is shown. Calculate the support reactions of the beam. (UPTU: 2000–2001)

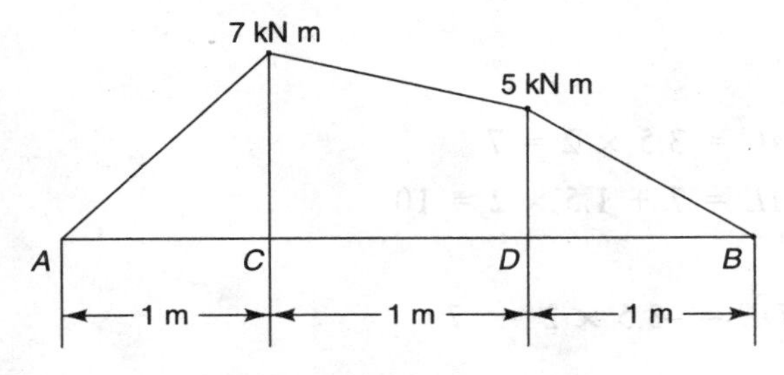

If reaction is R_A at A, then moment about C,

$$M_C = R_A \times AC = R_A \times 1$$

$$7 = 1 \times R_A \quad \text{(ordinate at point } C\text{)}$$

$$\therefore \quad R_A = 7 \text{ kN}$$

Similarly, $M_D = R_B \times 1 = 5$ (ordinate at point D)

$\therefore \quad R_B = 5$ kN

13. **The SFD of a simple supported beam is shown. Calculate the support reactions of the beam and draw the BMD. (UPTU: 2001–2002)**

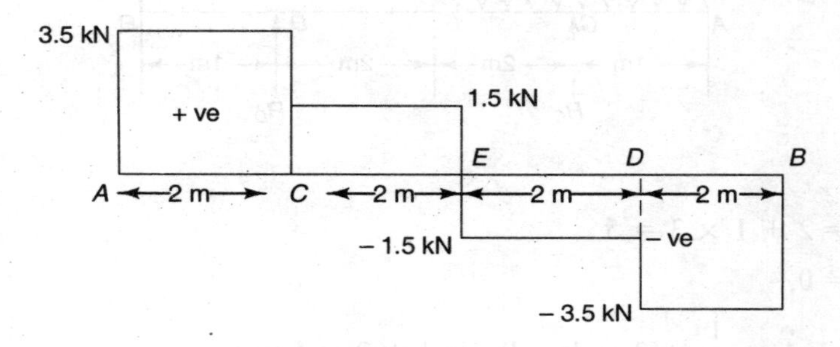

Reactions

$R_A = F_A = 3.5$ kN

$R_B = F_B = 3.5$ kN

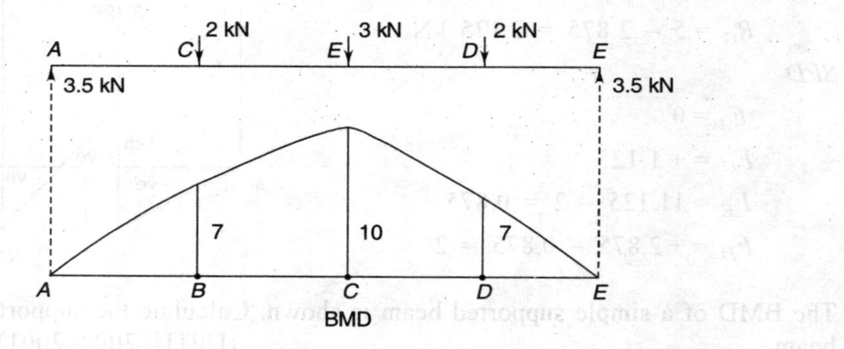

BMD

$M_A = 0$

$M_C =$ Area $AC = 3.5 \times 2 = 7$

$M_E =$ Area $AE = 7 + 1.5 \times 2 = 10$

$M_B = 0$

$M_D =$ Area $DE = -3.5 \times 2 = -7$

14. **The SFD of a simply supported beam at A and B is given. Calculate the support reactions and draw the BMD. (UPTU: 2002–2003)**

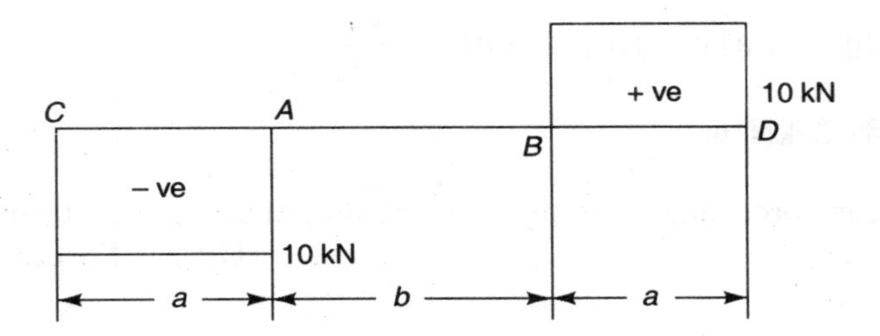

Reactions

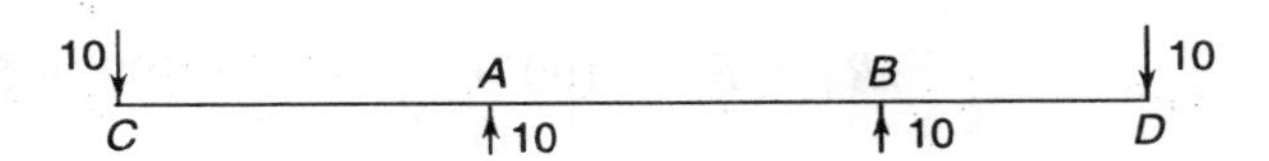

$$R_A = -10,\ R_B = +10$$

BMD

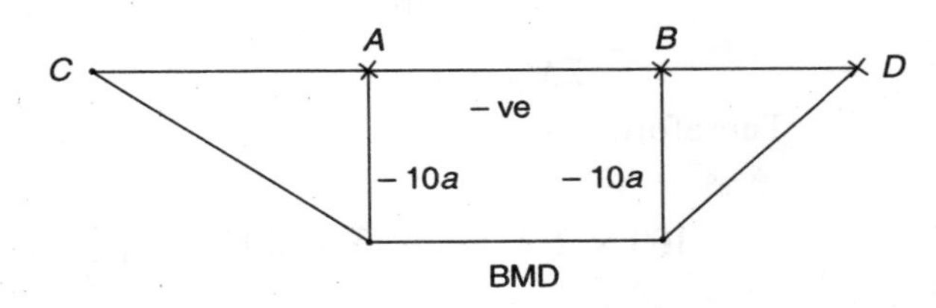

$$M_C = 0$$

$$M_A = \text{Area } AC = -10 \times a$$

$$M_B = -10a$$

$$M_D = -10a + 10a = 0$$

15. Draw the load diagram for the SFD shown for a simply supported beam. Calculate the maximum bending moment and its location. (UPTU: 2001–2002)

Reactions

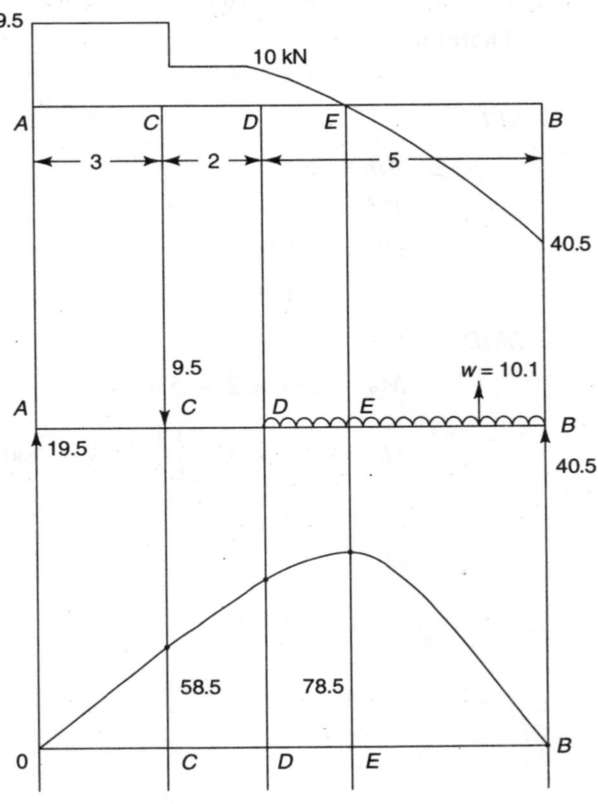

$$R_A = 19.5$$

$$R_B = 40.5$$

$$\Sigma P_y = 0,\ 19.5 + 40.5$$

$$= 9.5 + w \times 5$$

$$w = \frac{60 - 9.5}{5} = 10.1$$

BMD

$$M_A = 0$$

$$M_C = 19.5 \times 3 = 58.5$$

$$M_D = 19.5 \times 5 - 9.5 \times 2$$

$$= 97.5 - 19 = 78.5$$

Maximum *BM* will be at point *E* where *SF* = 0.

Now

$$DE \times 40.5 = BE \times 10$$

$$= (5 - DE) \times 10$$

$$\therefore\ DE = 0.99 \text{ mm}$$

$$\therefore \quad M_E = 40.5\ (4.01) - 10.1 \times 4.01 \times \frac{4.01}{2}$$

$$= 81.2 \text{ kN m}$$

16. Draw the shear force and bending moment diagrams for the beam as shown below. (UPTU: Sp carry over Aug. 2005)

Reactions

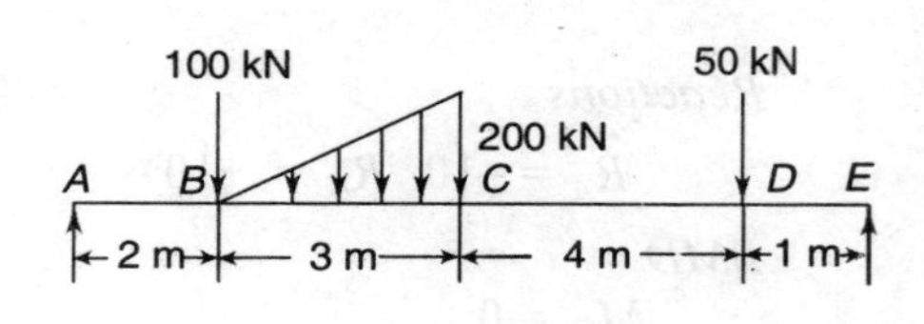

$$R_A + R_E = 100 + \frac{1}{2} \times 3 \times 200 + 50$$

$$= 450$$

$$\Sigma M_A = 0$$

Therefore,

$$100 \times 2 + \left(\frac{1}{2} \times 3 \times 200\right) \times \left(\frac{2}{3} \times 3 + 2\right) + 50 \times 9 - R_E \times 10 = 0$$

or $\quad 200 + 1200 + 450 - 10\ R_E = 0$

or $\quad R_E = 185$

Therefore, $\quad R_A = 265$

SFD

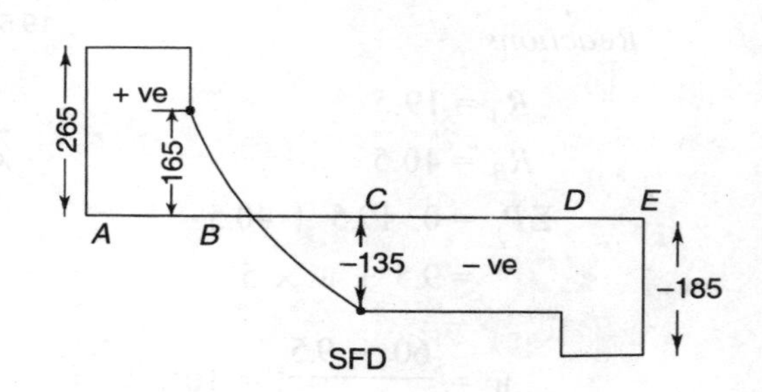

$AB = 265$

$BC = 165 \text{ to } - 135$

$CD = -135$

$DE = -155$

BMD

$$M_B = 265 \times 2 = 530$$

$$M_C = 265 \times 5 - \left(\frac{1}{2} \times 3 \times 200\right)\left(\frac{1}{3} \times 3\right) - 100 \times 3$$

$$= 725$$

$$M_D = 185$$

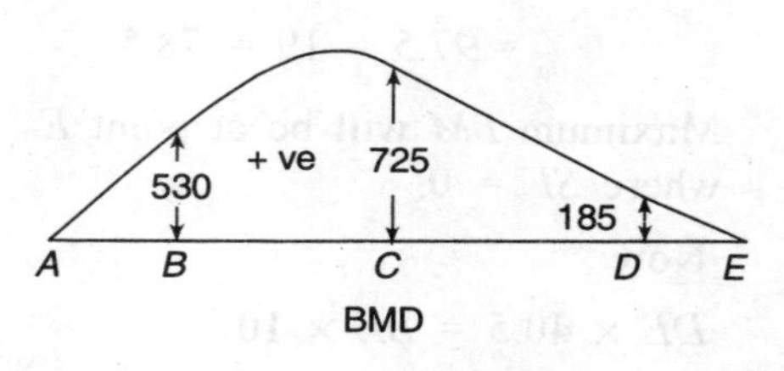

17. A beam 5 m long, hinged at both the ends is subjected to a moment $M = 60$ kN m at a point 3 m from end A as shown below. Draw the shear force and bending moment diagram. (UPTU: 2006–2007)

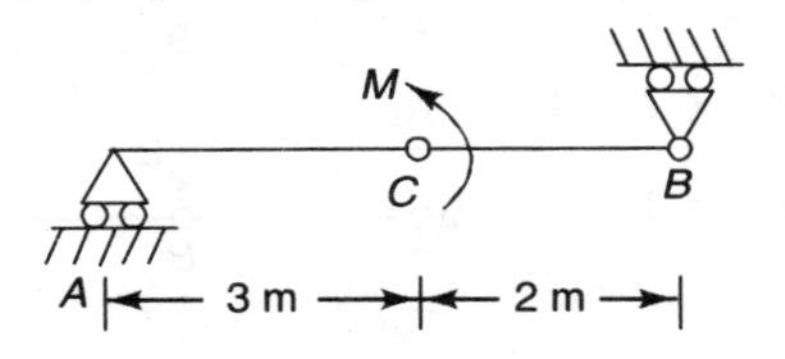

The free body diagram of the beam is

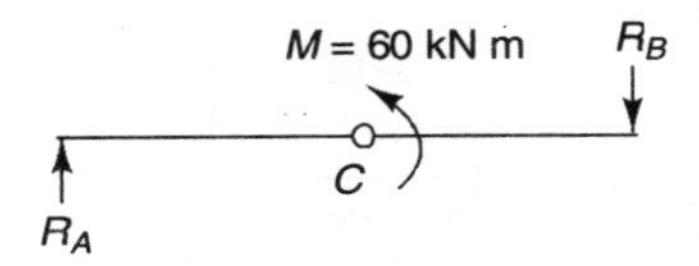

Now

$$\Sigma F_y = 0, \quad R_A = R_B$$

Also

$$\Sigma M_A = 0, \quad R_B = \frac{60}{5} = 12 \text{ kN}$$

Hence,

$$R_A = 12 \text{ kN.}$$

SFD

$$F_A = +\ 12 \text{ kN}$$
$$F_B = -\ 12 \text{ kN}$$

BMD

$$M_A = 0$$
$$M_C = 12 \times 3 \text{ (little left of } C)$$
$$= 36$$
$$M_C = 12 \times 3 - 60$$
$$= -24 \text{ (little right of } C)$$
$$M_B = 0$$

18. Draw the bending moment diagram of the beam shown below. (UPTU: 2006–2007)

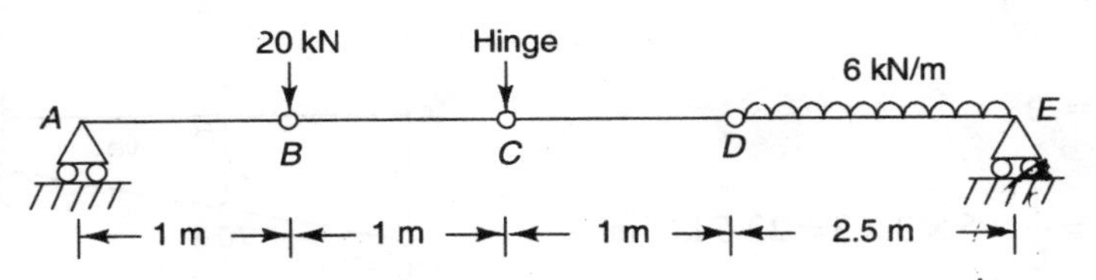

Beam can be considered to be of two parts, viz. AC and CE.

Reactions (Part AC)

$$R_A + R_C = 2$$

$$\Sigma M_A = 0, \quad 2 \times 1 - R_C \times 2 = 0$$

or $\quad R_C = 1$ and $R_A = 1$

Reaction (Part CE)

$$-R_C - 6 \times 2.5 + R_E = 0$$

or $\quad R_E = 1 + 15 = 16$

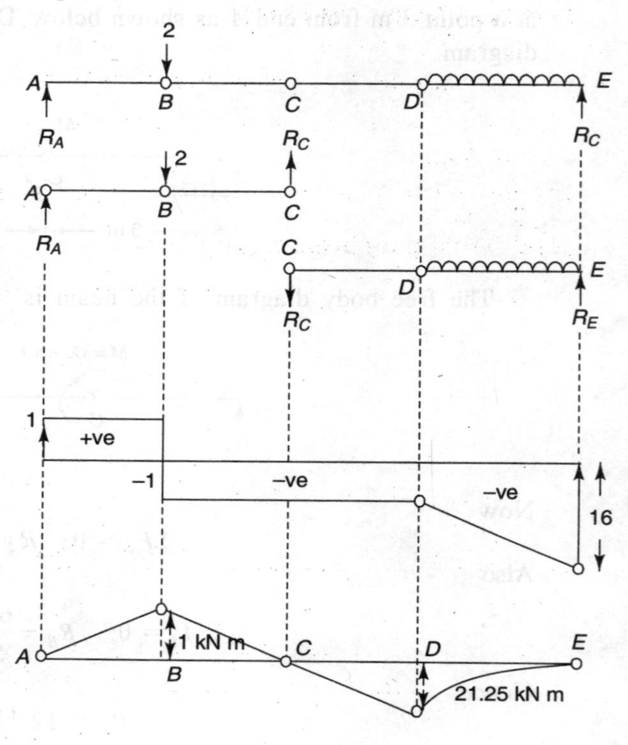

SFD

$$F_A = 1,\ F_B = -1,\ F_E = -16$$

BMD

$$M_A = 0$$
$$M_B = +1$$
$$M_C = 0$$
$$M_E = 0$$

$$M_D = -16 \times 2.5 + 6 \times 2.5 \times \frac{2.5}{2}$$
$$= -40 + 18.75 = -21.25$$

19. Draw the SF and BM diagram for the beam as shown below.

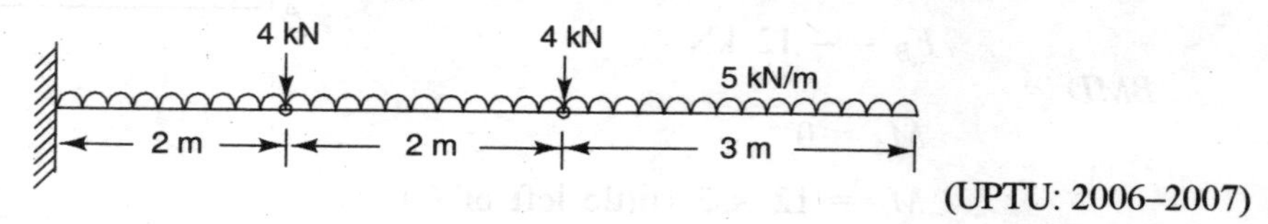

(UPTU: 2006–2007)

SFD

$$F_D = 0$$

$$F_C = +5 \times 3 = 15 \text{ (just right of } C)$$
$$= 15 + 4 = 19 \text{ (just left of } C)$$
$$F_B = 5 \times 5 + 4 = 29 \text{ (just right of } B)$$
$$= 29 + 4 = 33 \text{ (just left of } B)$$
$$F_A = 33 + 2 \times 5 = 43$$

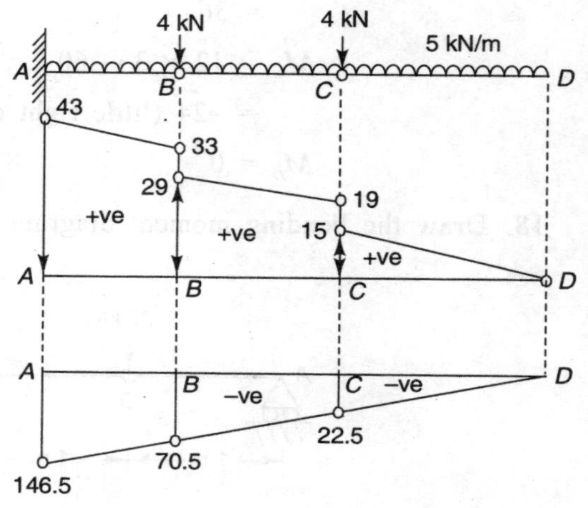

BMD

$$M_D = 0$$

$$M_C = \frac{1}{2} \times 5 \times 3 \times 3 = 22.5 \text{ kN m}$$

$$M_B = \frac{1}{2} \times 5 \times 5 \times 5 + 4 \times 2$$

$$= 70.5 \text{ kN m}$$

$$M_A = \frac{1}{2} \times 5 \times 7 \times 7 + 4 \times 4 + 4 \times 2$$

$$= 122.5 + 8 + 16 = 146.5 \text{ kN m}$$

20. **A simply supported beam carries a load P through a bracket as shown in the figure. The maximum bending moment in the beam is**

(a) $\dfrac{Pl}{2}$ (b) $\dfrac{Pl}{2} + \dfrac{aP}{2}$

(c) $\dfrac{Pl}{2} + aP$ (d) $\dfrac{Pl}{2} - aP$ (GATE: 2000)

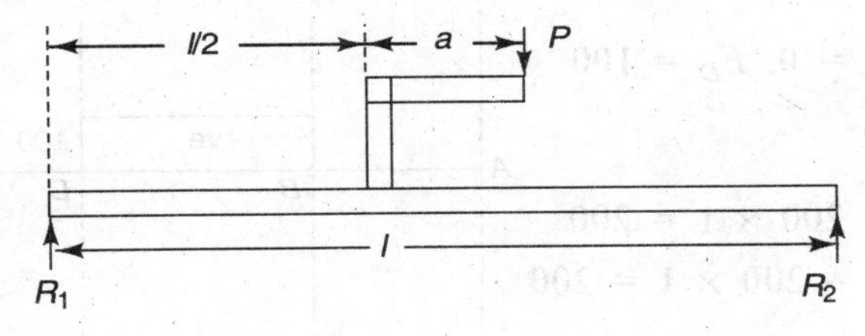

The free body diagram of the beam is

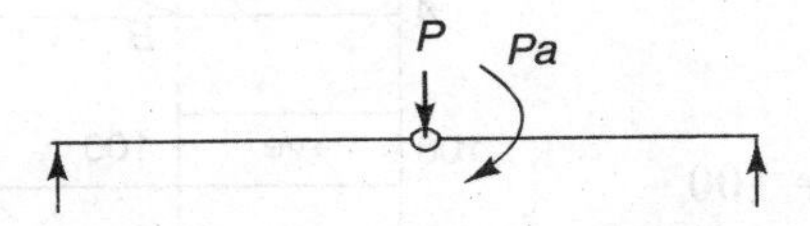

The bending moment can be considered to be consisted of point loading P and moment of Pa.

Hence the total BM at centre is

$$M = \frac{P \times l}{2} + Pa$$

Hence the option (c) is correct.

21. Two bars AB and BC are connected by a frictionless hinge at B. The assembly is supported and loaded as shown. Draw the SFD and BMD for the combined beam AC, clearly labelling the important values. Also indicate your sign convention.

(GATE: 1996)

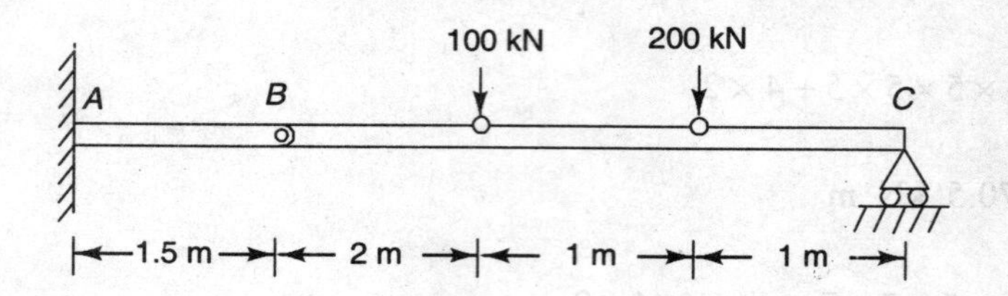

Beam can be considered as two parts, viz. *AB* and *BC*. Taking part *BC*:

Reactions

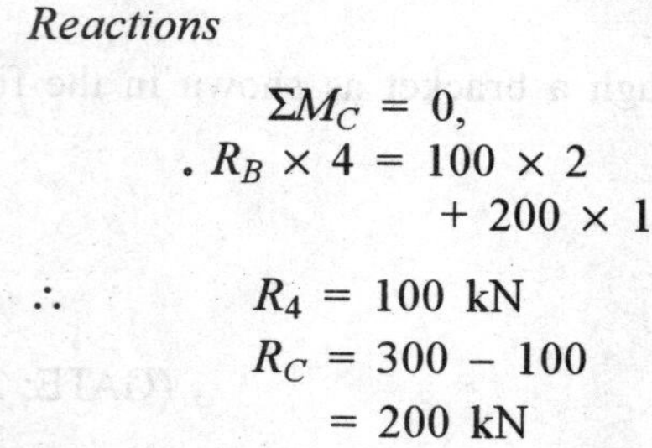

$$\Sigma M_C = 0,$$

$$R_B \times 4 = 100 \times 2 + 200 \times 1$$

$$\therefore \quad R_4 = 100 \text{ kN}$$

$$R_C = 300 - 100 = 200 \text{ kN}$$

SFD

$$F_C = -200,\ F_E = 0,\ F_D = 100$$

BMD

$$M_C = 0,\ M_E = 200 \times 1 = 200$$

$$M_D = 200 \times 2 - 200 \times 1 = 200$$

$$M_B = 0$$

Now take part *AB*:

SFD

$$F_B = 100 \text{ and } F_A = 100$$

BMD

$$M_B = 0,$$

$$M_A = -100 \times 1.5 = -150 \text{ kN m}$$

Now combining SFD

Now combining BMD

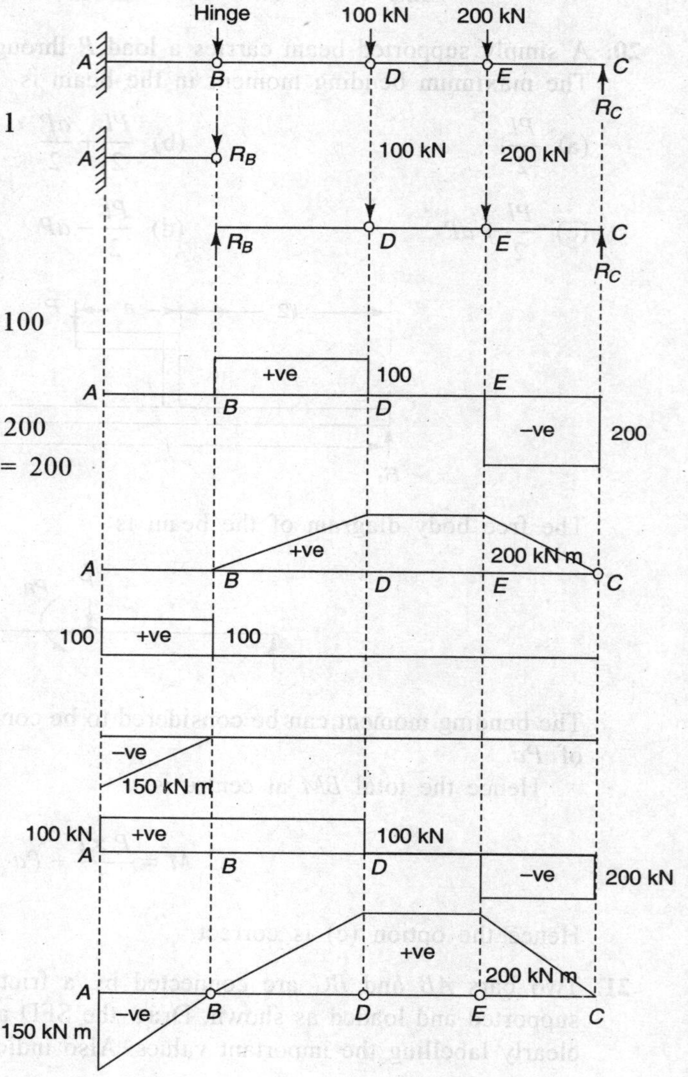

22. The shear force diagram of a loaded beam is shown in the figure. The maximum bending moment is
(a) 16 kN m (b) 11 kN m
(c) 28 kN m (d) 8 kN m (GATE: 2001)

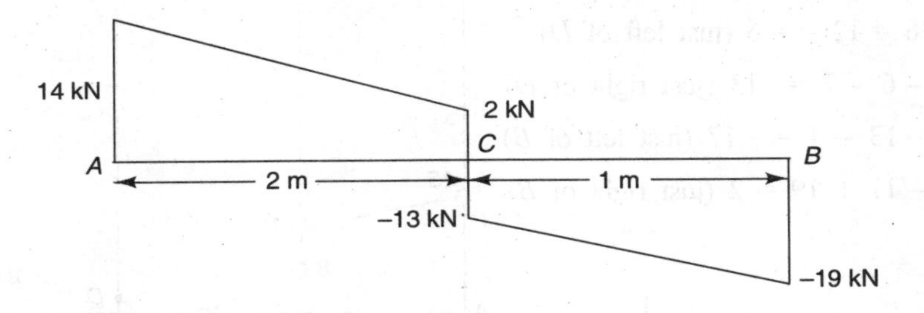

Bending moment at any point is the area of the shear force diagram from one end to that point.

$$\text{Area at point } C = \frac{1}{2} \times 2 \times (14 + 2)$$

$$= 16 \text{ kN m}$$

Hence the option (a) is correct.

23. Draw SF and BM diagrams for the beam shown in the figure. Find the location and magnitude of maximum bending moment. Also determine the location of any point of contraflexure. (UPTU: 2007–2008)

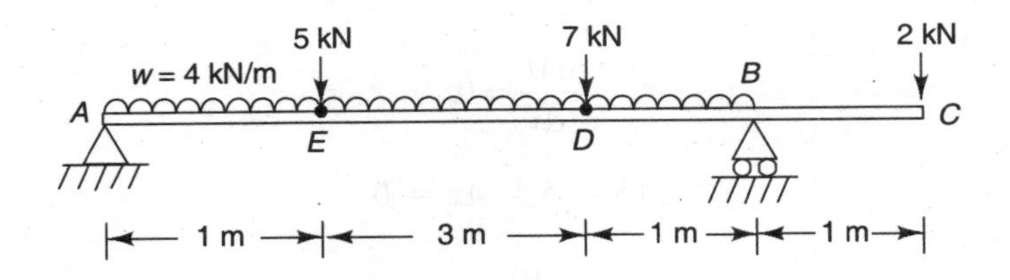

Reactions

$$\Sigma F_y = 0, \quad R_A + R_B = 14 + 20 = 34$$

$$\Sigma M_A = 0, \quad 5 \times 1 + 7 \times 4 - R_B \times 5 + 2 \times 6 + 20 \times \frac{5}{2} = 0$$

$$R_B = \frac{5 + 28 + 12 + 50}{5} = 19 \text{ kN}$$

$$R_A = 34 - 19 = 15 \text{ kN}$$

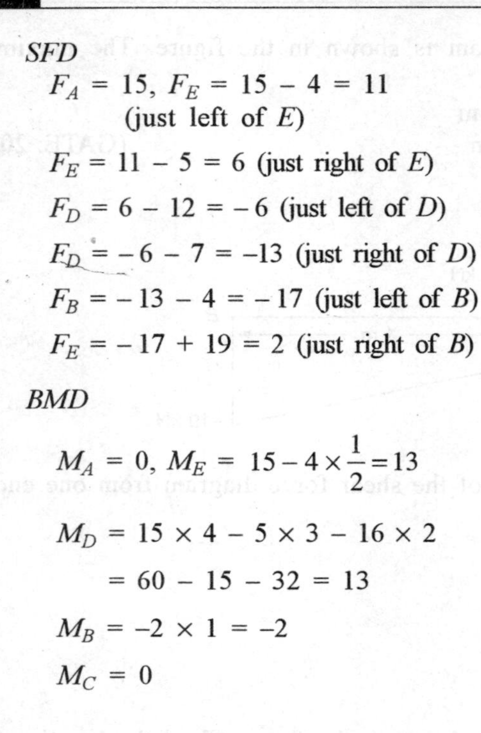

SFD

$F_A = 15,\ F_E = 15 - 4 = 11$ (just left of E)

$F_E = 11 - 5 = 6$ (just right of E)

$F_D = 6 - 12 = -6$ (just left of D)

$F_D = -6 - 7 = -13$ (just right of D)

$F_B = -13 - 4 = -17$ (just left of B)

$F_E = -17 + 19 = 2$ (just right of B)

BMD

$M_A = 0,\ M_E = 15 - 4 \times \frac{1}{2} = 13$

$M_D = 15 \times 4 - 5 \times 3 - 16 \times 2$

$= 60 - 15 - 32 = 13$

$M_B = -2 \times 1 = -2$

$M_C = 0$

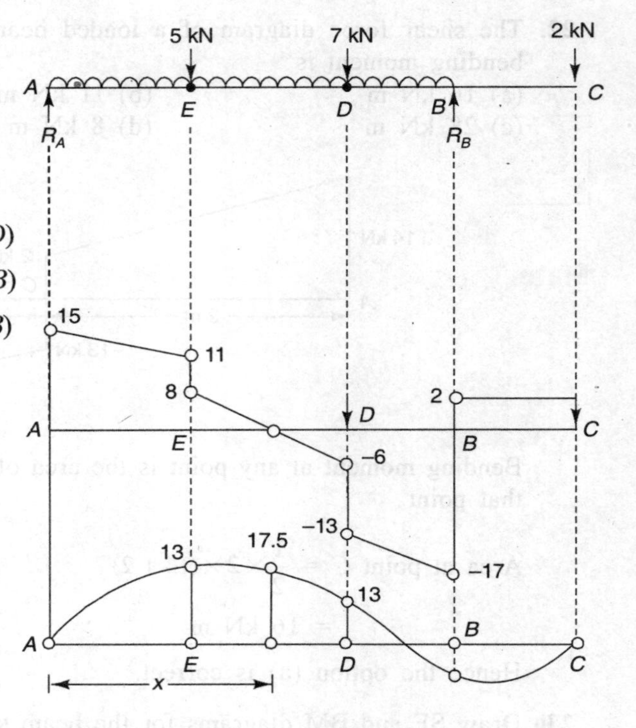

The maximum bending moment will be x distance from A

$$M = 15 \times (x) - 5 \times (x - 1) - x \times 4 \times \frac{x}{2}$$

$$\frac{\partial M}{\partial x} = 0,$$

$$15 - 5 - 4x = 0$$

or
$$x = \frac{10}{4} = 2.5$$

$$M_{max} = 15 \times 2.5 - 5 \times 1.5 - 2 \times (2.5)^2$$

$$= 37.5 - 7.5 - 12.5$$

$$= 17.5 \text{ kN}$$

OBJECTIVE TYPE QUESTIONS

It's not always what you say that makes the difference; sometimes it's the way you say it.

State True or False

1. A beam is a structural element which has length considerably larger than width and depth. (*True/False*)
2. Load on a beam acts at a right angle to the axis. (*True/False*)
3. A cantilever has one end hinged. (*True/False*)
4. A fixed beam has one end fixed. (*True/False*)
5. An overhanging beam has both ends supported. (*True/False*)
6. The load on a beam is finally transferred to supports. (*True/False*)
7. The system of applied forces and reactions keep a beam in equilibrium. (*True/False*)
8. A continuous beam has two supports. (*True/False*)
9. Shear force tries to shear off the section of a beam. (*True/False*)
10. Shear force that tends to move the left portion downwards relative to the right portion is taken as positive. (*True/False*)
11. Binding moment is the moment at the section of a beam, which tries to bend it. (*True/False*)
12. Bending moment that tries to sag a beam is taken as negative. (*True/False*)
13. BM is maximum where SF is maximum. (*True/False*)
14. The contraflexure is a point on a beam where shear force is zero. (*True/False*)
15. The slope of a BMD is equal to shear force between two points. (*True/False*)
16. The area of an SFD from end to any point gives BM at that point. (*True/False*)
17. A cantilever has BM only at the fixed support. (*True/False*)
18. A uniform distributed load (UDL) on a beam gives a parabolic curve on the BMD. (*True/False*)
19. A uniformly varying load (UVL) on a beam gives a cubic curve on the BMD. (*True/False*)
20. A UDL on a beam gives a constant line on the SFD. (*True/False*)
21. A UVL on a beam gives a linear line on the SFD. (*True/False*)
22. A point load gives a constant line on the BMD. (*True/False*)

23. The bending moment at point C in the figure is 50 kN m. *(True/False)*

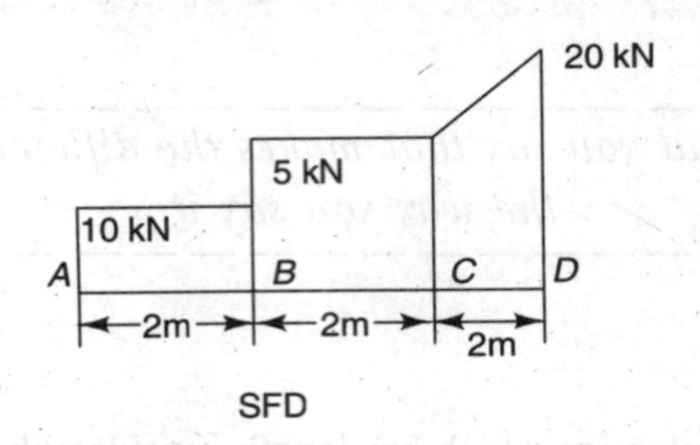

24. The bending moment at point D in the above figure is 42 kN m. *(True/False)*

25. The shear force in the figure at point B is 5 kN. *(True/False)*

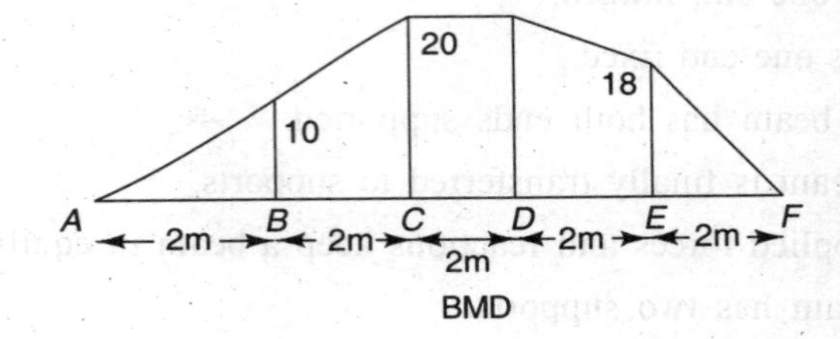

26. The reaction at the support for the beam at point A of the figure of Q. 25 is 5 kN. *(True/False)*

27. A cantilever has a point load at its free end. The maximum slope on bending is at the free end. *(True/False)*

Multiple Choice Questions

1. A cantilever is a beam whose
(a) one end is fixed and other end is free
(b) both ends simply supported
(c) both ends are fixed

2. The beam having its ends not supported is called
(a) a cantilever beam (b) a simply supported beam (c) an overhanging beam

3. A continuous beam is one which has
(a) less than two supports (b) two supports only (c) more than two supports

4. A UDL on a cantilever gives on the SFD a
(a) linear line (b) cubic curve (c) parabolic curve

5. A UVL on a cantilever gives on the SFD a
(a) linear line (b) parabolic curve (c) cubic curve

6. A UVL on a cantilever gives on the BMD a
(a) cubic curve (b) parabolic curve (c) linear line

7. Bending moment is maximum where shear force is
(a) zero (b) maximum (c) constant

8. The point of contraflexure on a beam is the point where bending is
(a) maximum (b) zero (c) constant

9. If the distance between two points is 2 m and the difference between the heights of ordinates is 10 kN m on the BMD, then the shear force at first point is
(a) 10 kN (b) 20 kN (c) 5 kN

10. For a UVL on a cantilever (0 at the free end to w at the fixed end) of length l, the maximum bending moment is
(a) $\frac{wl^2}{3}$ (b) $\frac{wl^2}{6}$ (c) $\frac{wl^2}{12}$

11. For a simply supported beam, the bending moment at supports is
(a) less than unity (b) more than unity (c) zero

12. The maximum bending moment at the centre of a simply supported beam (length = l) with point load (w) at the centre is
(a) $\frac{wl}{4}$ (b) $\frac{wl}{2}$ (c) $\frac{wl}{3}$

13. Shear force at the centre of a simply supported beam with UDL = w/meter and length l is
(a) wl (b) $\frac{wl}{2}$ (c) zero

14. The maximum bending moment of a simply supported beam (length = l) and UDL = w/meter is
(a) $\frac{wl^2}{2}$ (b) $\frac{wl^2}{4}$ (c) $\frac{wl^2}{8}$

15. In a simply supported beam carrying UVL = w/meter, the maximum BM is
(a) $\frac{wl^2}{4}$ (b) $\frac{wl^2}{8}$ (c) $\frac{wl^2}{12}$

16. The maximum deflection of a cantilever of span l carrying a point load of w at its free end is
(a) $\frac{wl}{2EI}$ (b) $\frac{wl^3}{3EI}$ (c) $\frac{wl^3}{8EI}$ (d) $\frac{wl^3}{16EI}$

17. The maximum slope of a cantilever carrying a point load at its free end is at the
(a) fixed end (b) centre of spine
(c) free end (d) none of these

18. A cantilever of span l carries a uniform distributed load on the entire span. The maximum slope of the cantilever is

(a) $\frac{wl^2}{3EI}$ (b) $\frac{wl^4}{3EI}$ (c) $\frac{wl^2}{6EI}$ (d) $\frac{wl^4}{16EI}$

19. The maximum deflection of a cantilever which is uniformly loaded (w) is

(a) $\frac{wl^4}{2EI}$ (b) $\frac{wl^4}{3EI}$ (c) $\frac{wl^4}{8EI}$ (d) $\frac{wl^4}{16EI}$

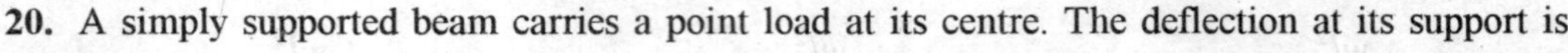

20. A simply supported beam carries a point load at its centre. The deflection at its support is

(a) $\frac{wl^2}{16EI}$ (b) $\frac{wl^3}{16EI}$ (c) $\frac{wl^2}{48EI}$ (d) $\frac{wl^3}{48EI}$

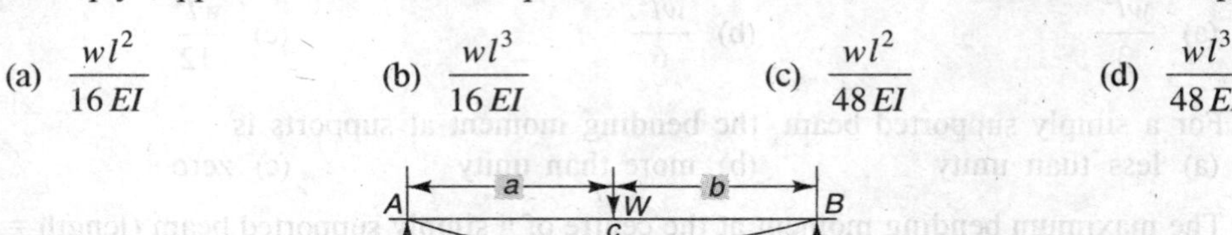

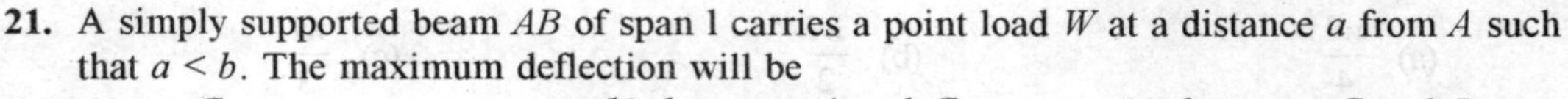

21. A simply supported beam AB of span 1 carries a point load W at a distance a from A such that $a < b$. The maximum deflection will be

(a) at C (b) between A and C (c) between C and B

22. A simply supported beam of span l has a uniformly distributed load (w). The maximum deflection at the centre of the beam is

(a) $\frac{5wl^4}{48EI}$ (b) $\frac{5wl^4}{96EI}$ (c) $\frac{5wl^4}{192EI}$ (d) $\frac{5wl^4}{384EI}$

23. Two simply supported beams of the same span carry the same load but one has a point load and other a UDL. The maximum slope of the first to the second is

(a) 1 : 1 (b) 1 : 1.5 (c) 1.5 to 1 (d) 2 : 1

24. Sagging moment is

(a) positive (b) negative (c) constant

25. Hogging moment is

(a) negative (b) positive (c) constant

Fill in the Blanks

1. The distance between supports of a beam is called ________.
   (a) span (b) overhang
2. A beam is generally loaded ________ to its axis.
   (a) horizontal (b) normal
3. A beam is free to move along the ________ in a roller support.
   (a) axis (b) section

4. A beam is free to ________ along the axis in a simple support.
(a) rotate (b) relocate

5. Sagging bending moment is ________.
(a) positive (b) negative

6. Hogging bending moment is ________.
(a) positive (b) negative

7. A fixed support has ________ and shear force.
(a) bending moment (b) couple

8. The area of the SFD from a point to the support gives ________ at that point.
(a) total load (b) bending moment

9. The slope of a BMD between two points gives ________.
(a) shear force (b) bending moment

10. The maximum bending moment is at a point where shear force is ________.
(a) zero (b) maximum

11. The point of contraflexure is at a point when bending moment is ________.
(a) maximum (b) zero

12. The bending moment at supports of a simply supported beam is ________.
(a) zero (b) maximum

13. The bending moment at the centre of a simply supported beam, which has a uniform distributed load on its span is ________.
(a) zero (b) maximum

14. The bending moment of a cantilever having a point load at the free end is maximum at the ________.
(a) free end (b) fixed support

15. For a point load on a beam, the SFD shows ________.
(a) an inclined straight line (b) a horizontal straight line

16. For a point load on a beam, the BMD shows ________.
(a) an inclined straight line (b) a horizontal straight line

17. For a UDL on a beam, the SFD shows ________ straight line.
(a) an inclined (b) horizontal

18. For a UDL on a beam, the BMD shows ________.
(a) an inclined line (b) a parabolic curve

19. At a hinge in a beam, the bending moment is ________.
(a) zero (b) maximum

20. An external moments on a BMD gives ________ change.
(a) uniform (b) abrupt

21. A continuous beam has more than ________ support(s).
(a) one (b) two

ANSWERS

You are a wise man today if you have learned from yesterday's blunders.

State True or False

1. True
2. True
3. False (A cantilever has one free end and one fixed end.)
4. False (A fixed beam has both ends fixed.)
5. False (An overhanging beam can have one or both ends free.)
6. True
7. True
8. False (A continuous beam has more than two supports.)
9. True
10. False (Shear force that tries to move the left portion of the beam upwards relative to the right portion is positive and vice versa.)
11. True
12. False (Sagging BM is positive.)
13. False (The maximum BM is where shear force is zero.)
14. False (Contraflexure is a point on a beam where bending moment is zero.)
15. True ($\frac{dM}{dx}$ = shear force on the BMD)
16. True ($\int dM = \int F dx$)
17. False (It has both BM and shear force.)
18. True ($M_x = \frac{w}{2}x^2$ which is parabolic.)
19. True ($M_x = \frac{wx^3}{6}$ which is cubic)
20. False ($F_x \propto wx$. Hence it will give a linear/inclined line on the SFD.)
21. False ($F_x \propto \frac{w}{2}x^2$. Hence it will give a parabolic line on the SFD.)
22. False ($M = Fx$, hence linear line)

23. True (Area on $AC = 10 \times 2 + 15 \times 2 = 50$)

24. False (Area on $AD = 50 + 2 \times 15 + \frac{1}{2} \times 2 \times 5 = 85$)

25. True (Shear force is the slope on $BC = \frac{20 - 10}{2} = 5$)

26. True (Reaction = slope on $AB = \frac{10 - 0}{2} = 5$)

27. True

Multiple Choice Questions

1. (a) **2.** (c) **3.** (c) **4.** (a)
5. (b) **6.** (a) **7.** (a) **8.** (b)
9. (c) (Shear force = slope = $\frac{10}{2} = 5$) **10.** (b) **11.** (c)
12. (b) **13.** (c) **14.** (b) **15.** (c)
16. (b) **17.** (c) **18.** (c) **19.** (c)
20. (a) **21.** (c) **22.** (c) **23.** (c)
24. (a) **25.** (a)

Fill in the Blanks

1. (a) **2.** (b) **3.** (a) **4.** (a)
5. (a) **6.** (b) **7.** (a) **8.** (b)
9. (a) **10.** (a) **11.** (b) **12.** (a)
13. (b) **14.** (b) **15.** (b) **16.** (a)
17. (a) **18.** (b) **19.** (a) **20.** (b)
21. (b)

CHAPTER 11
Trusses

If you believe even when you cannot see, your reward will be one which you cannot imagine.

INTRODUCTION

A truss is a structure made of slender members which are assumed to be pin connected at ends. It is capable of taking loads at joints. It is designed to resist geometrical distortion under any applied system of loading.

A truss can be a plane truss or space truss. In a plane truss, all members lie in a single plane. The force system acting on the truss is coplanar and it is in the plane of the truss. Bridge trusses and roof trusses are examples of plane trusses. A truss in which all members do not lie in the same plane is called a space truss. A transmission tower is an example of a space truss.

TYPES OF PLANE TRUSS

A triangular structure made up of three members joined by pins is stable and it is the basic element of a plane truss. Trusses formed by four or more members in the shape of a polygon of as many sides are non-rigid or unstable. However, they can be made stable by adding diagonal members which convert a polygon into a group of triangle forms. If we add one diagonal to a non-rigid four members truss, we get a rigid and stable truss having two triagular forms.

A truss is said to be perfect if the number of members is just sufficient to prevent its distortion when it is subjected to an external force system. For a perfect truss, the correlation between the number of joints (J) and the number of members (m) is given by the equation $m = 2J - 3$. The truss in Figure 11.1(a) is a perfect truss as it has 3 members & 3 joints ($m = 3 = 2J - 3 = 3$). A perfect truss is statically determinable, i.e. equations of static equilibrium are sufficient to determine the forces in its members.

A truss is termed an imperfect truss if the number of members in it is less than what is required for a perfect truss. Figure 11.1(b) shows an imperfect truss as it has four members, instead of five members, which are required for a perfect truss ($m < 2J - 3$). An imperfect truss cannot prevent geometrical distortion when loaded. It is also statically indeterminable.

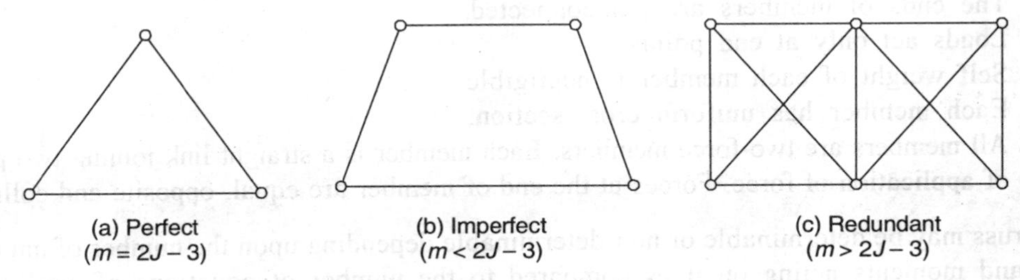

FIGURE 11.1 Types of plane truss.

A redundant truss has members more than that is required for a perfect truss. A redundant truss is shown in Figure 11.1(c). It has 6 joints and 11 members ($m > 2J - 3$) while the truss needs only 9 members. A redundant truss is over rigid and statically indeterminable.

SUPPORTS

A perfect truss has a support at both ends. One end is generally resting on the roller support and the other end is hinged. The roller support is frictionless and provides a reaction at a right angle to the roller base as shown in Figure 11.2(a). For the hinged support, the direction of the reaction depends upon the load system on the truss. It depends upon the net horizontal (H_R) and vertical forces (V_R) acting at the hinged support as shown in Figure 11.2(b). Resultant, $R = \sqrt{(H_R)^2 + (V_R)^2}$ and $\tan\theta = V_R/H_R$. If a truss has a roller support at both ends, the truss can slide and fall under horizontal load such as wind load.

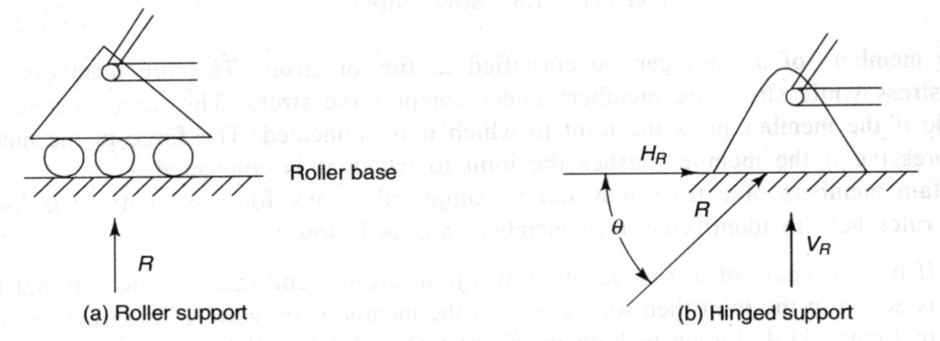

FIGURE 11.2 Supports.

Also if truss has a hinged support at both ends, the movement of the truss under varying temperature is impossible, which may lead to failure of the truss.

ANALYSIS OF PLANE TRUSS

The analysis of a perfect and plane truss involves the determination of the reactions at supports and internal stresses induced in the members due to external loads. For analysis, following assumptions are made:

1. The ends of members are pin connected.
2. Loads act only at end points.
3. Self weight of each member is negligible.
4. Each member has uniform cross-section.
5. All members are two-force members. Each member is a straight link joining two points of application of force. Forces at the end of member are equal, opposite and collinear.

A truss may be determinable or non-determinable depending upon the number of unknown forces and moments acting on it as compared to the number of equations of equilibrium which can be formed. A determinate truss can be analyzed with three equations of statical equilibrium. Since three unknowns can be solved with three equations, therefore the truss having up to three unknowns at a joint can be determined.

Bow's notation is used in the graphical solution of a truss. It is a method of designating a force by placing capital letters on either side of force. As shown in Figure 11.3, forces P_1, P_2 and P_3 can be designated as AB, BC and CA.

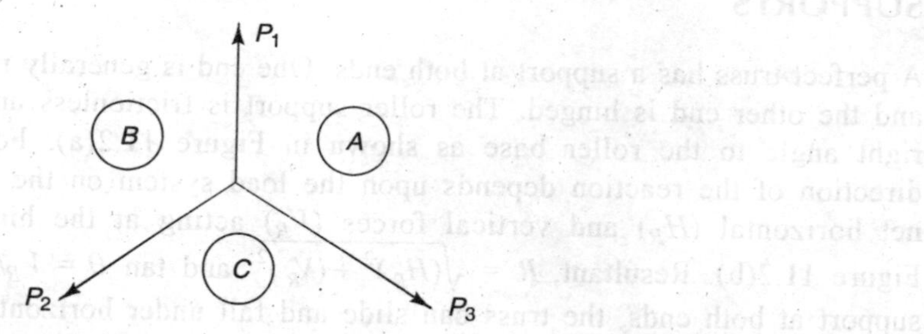

FIGURE 11.3 Bow's notation.

The members of a truss can be classified as ties or struts. Ties are members under tensile stress while struts are members under compressive stress. The force in a member is tensile if the member pulls the joint to which it is connected. The force in the member is compressive if the member pushes the joint to which it is connected.

Certain members of a truss may not be subjected to any force when truss is loaded. Certain rules help in identifying such members are as follows:

1. If two members of a truss meeting at a joint are not collinear and no external force is acting at the joint then forces in both the members are zero. In the truss as shown in Figure 11.4, forces in both members CD and DE will be zero.
2. When three members meeting at a joint with load and two members from three are collinear, then force in the third member is zero. In the figure two members AF and FE are collinear and the third member BF is meeting them at joint F, hence the force in member BF will be zero.

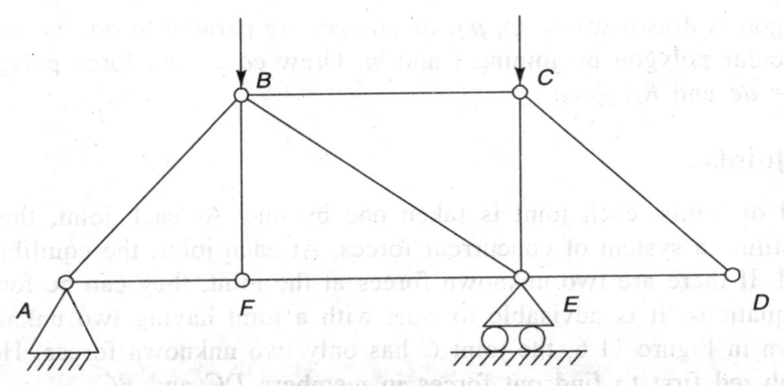

FIGURE 11.4 Members having no force.

3. A single force cannot form a system in equilibrium. If a single force is acting in equilibrium during analysis, then this force is zero.

Graphical Method

The graphical method of analysis consists of drawing *space diagram*, *force polygon* and *funicular polygon* (Figure 11.5). A force polygon is drawn by drawing forces in order on a small scale on a vertical line ($P_1 = ab$, $P_2 = bc$ and $P_3 = cd$). A pole is selected (point O) and it is joined to point a, b, c, and d as shown in the figure. Using a force polygon, a

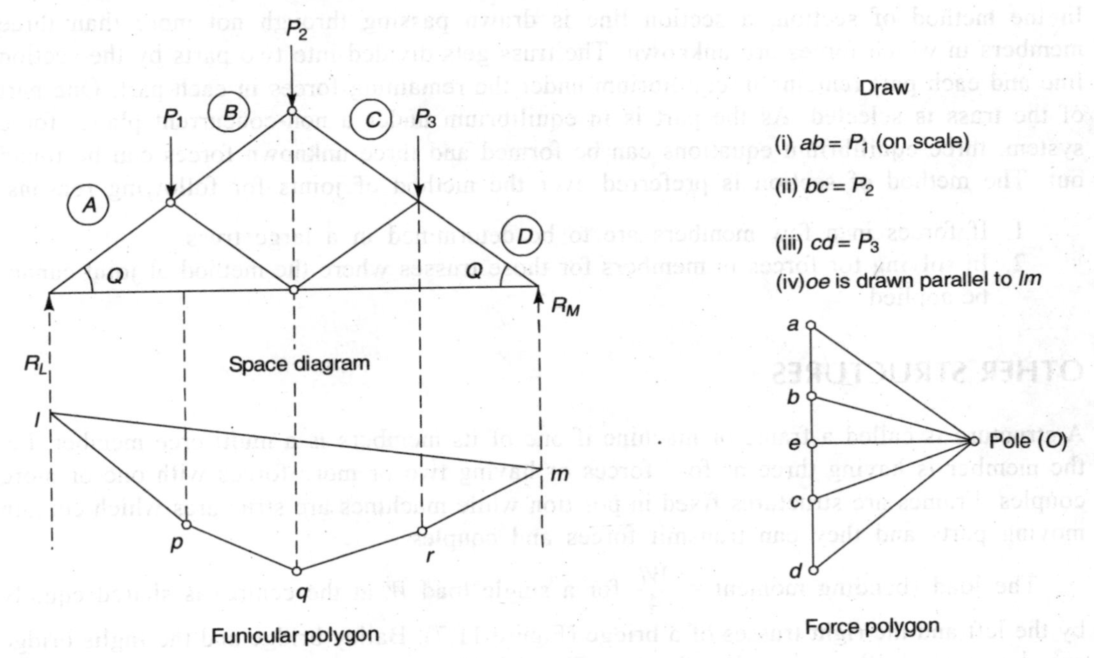

FIGURE 11.5 Graphical method.

funicular polygon is drawn where lp, pq, qr and rm are parallel to ao, bo, co and do. Now close the funicular polygon by joining l and m. Draw eo on the force polygon parallel to lm. Then $R_L = de$ and $R_M = ea$.

Method of Joints

In the method of joints, each joint is taken one by one. At each joint, the forces in the members constitute a system of concurrent forces. At each joint, the equilibrium equations can be formed. If there are two unknown forces at the joint, they can be found from these equilibrium equations. It is advisable to start with a joint having two unknown forces. In the truss shown in Figure 11.6, the joint C has only two unknown forces. Hence this joint should be analyzed first to find out forces in members DC and BC.

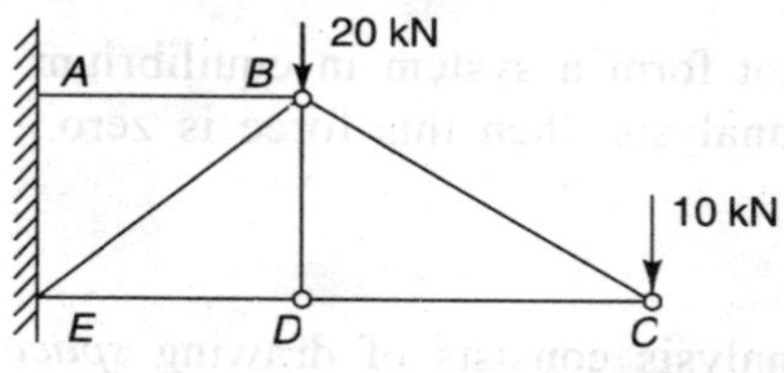

FIGURE 11.6 Method of joints.

Method of Section

In the method of section, a section line is drawn passing through not more than three members in which forces are unknown. The truss gets divided into two parts by the section line and each part remains in equilibrium under the remaining forces in each part. One part of the truss is selected. As the part is in equilibrium under a non-concurrent planer force system, three equilibrium equations can be formed and three unknown forces can be found out. The method of section is preferred over the method of joints for following reasons:

1. If forces in a few members are to be determined in a large truss
2. In solving for forces in members for those trusses where the method of joint cannot be applied

OTHER STRUCTURES

A structure is called a frame or machine if one of its members is a multiforce member, i.e. the member is having three or four forces or having two or more forces with one or more couples. Frames are structures fixed in position while machines are structures which contain moving parts and they can transmit forces and couples.

The load (bending moment $= \frac{Wl}{4}$ for a single load W at the centre) is shared equally by the left and the right trusses of a bridge (Figure 11.7). Bailey bridge and the Inglis bridge are designed on the principle of truss. These bridges are mechanical bridges in which

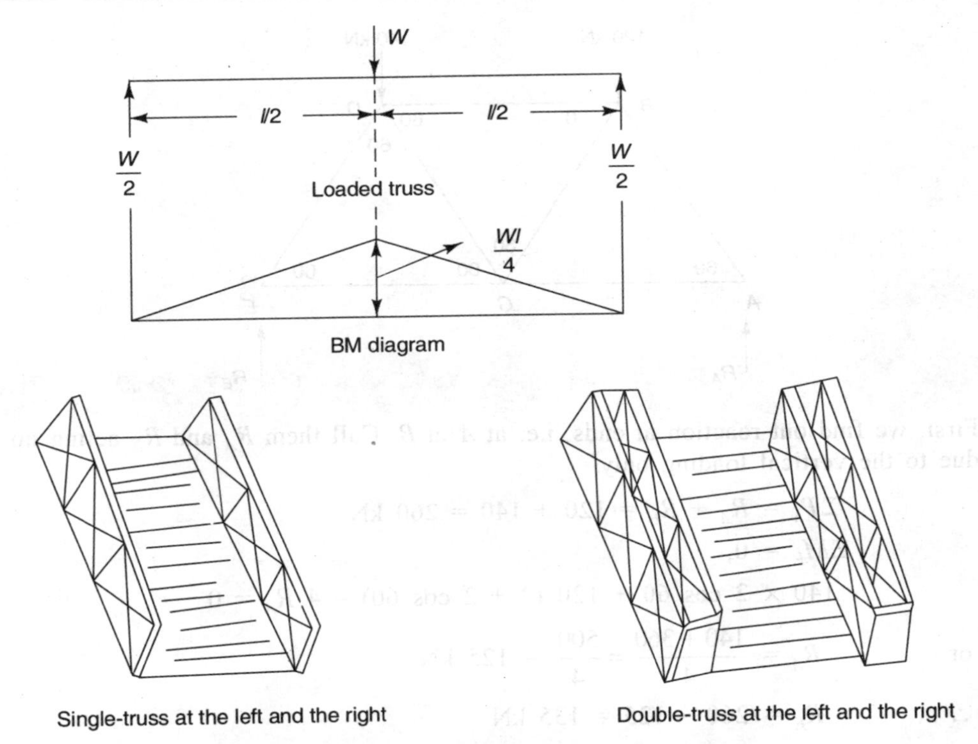

FIGURE 11.7 Different structures of truss.

trusses are formed by inter-connecting tubular members in the Inglis bridge and panels in the Bailey bridge. Left-hand and right-hand trusses can be single or double as shown in Figure 11.7. A truss can be also single-storeyed or double-storeyed. The load carrying capacity can be increased by adding a truss side by side or by increasing the storey in each truss. Adding storey is more effective way of increasing strength of a beam (strength depends on moment of inertia which is equal to $\frac{bd^3}{12}$ where b = width and d = depth). Adding a truss side by side is equivalent to increasing width (b) while adding storey is equivalent to increasing depth (d) and hence it is many times more effective. Bridges are designed with maximum possible height of a truss at the centre of a bridge.

SOLVED PROBLEMS

1. Each member of a truss as given in the figure is 2 m long. The truss is simply supported at the ends. Determine the forces in all members clearly showing whether they are in tension or compression. (UPTU: Dec. 2005)

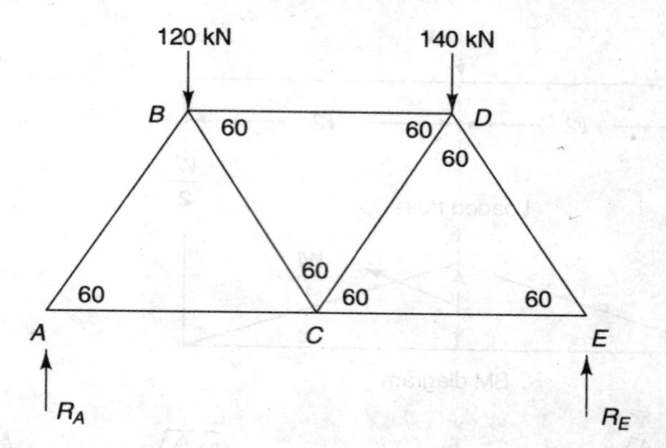

First, we find out reaction at ends, i.e. at A at B. Call them R_A and R_E acting up only due to the vertical loading only.

$$\Sigma P_y = R_A + R_E = 120 + 140 = 260 \text{ kN}$$

$$\Sigma M_E = 0,$$

$$140 \times 2 \cos 60 + 120 (2 + 2 \cos 60) - 4 R_A = 0$$

or $$R_A = \frac{140 + 360}{4} = \frac{500}{4} = 125 \text{ kN}$$

$\therefore$ $$R_E = 260 - 125 = 135 \text{ kN}$$

Now draw the stresses in each member.

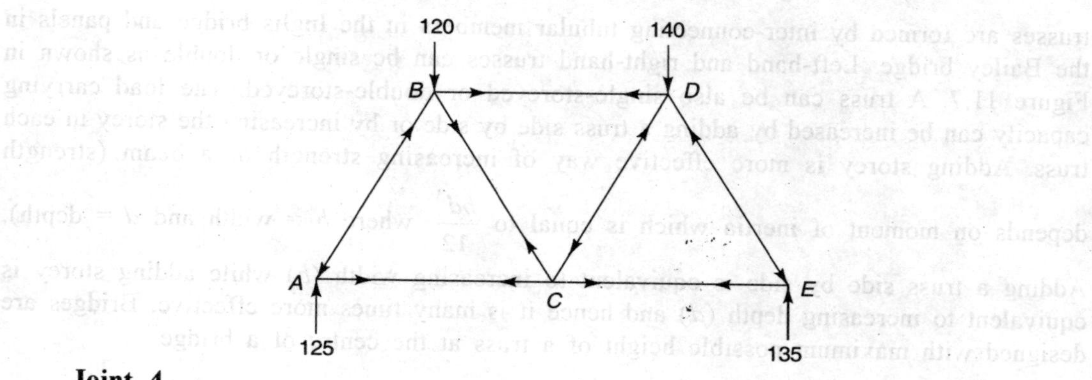

Joint A

$$\Sigma P_y = 0,$$

$$F_{BA} \sin 60 = 125$$

or $$F_{BA} = 144.3 \text{ kN}$$

$$\Sigma P_x = 0,$$

$$F_{BA} \cos 60 = F_{AC}$$

or $$F_{AC} = 72.2 \text{ kN}$$

Joint *B*

$$\Sigma P_y = 0$$

$$120 - F_{AB} \sin 60 + F_{CB} \sin 60 = 0$$

or $$120 - 144.3 \times 0.866 + F_{CB} \times 0.866 = 0$$

or $$F_{CB} = -5.73 \text{ kN}$$

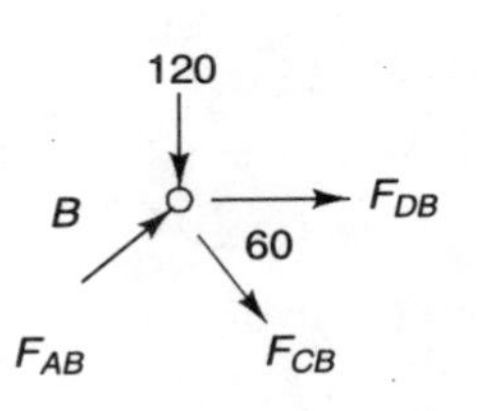

$$\Sigma P_x = 0$$

$$F_{AB} \cos 60 + F_{CB} \cos 60 - F_{DB} = 0$$

$$F_{DB} = (144.3 + 5.73) \times \frac{1}{2}$$

$$= 75 \text{ kN}$$

Joint *C*

$$\Sigma P_y = 0$$

$$F_{CB} \sin 60 = F_{CD} \sin 60$$

Therefore,

$$F_{CD} = F_{CB} = 5.73$$

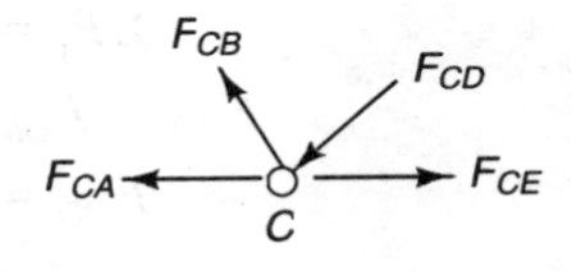

$$\Sigma P_x = 0$$

$$F_{CE} - F_{CB} \cos 60 - F_{CD} \cos 60 = 0$$

or $$F_{CE} = 72.2 + (5.73 + 5.73) \times \frac{1}{2}$$

$$= 72.2 + 5.73$$

$$= 78 \text{ kN}$$

Joint *E*

$$\Sigma p_y = 0$$

$$F_{ED} \sin 60 = 135$$

or $$F_{ED} = 155.9$$

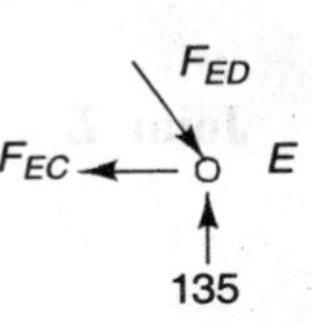

The following table list all the forces and show whether they are tensile or compressive.

S.No.	*Member*	*Force* (kN)	*Type*
1	*AB*	144.3	Compressive
2	*AC*	72.2	Tensile
3	*BC*	5.73	Compressive
4	*BD*	75	Tensile
5	*CD*	5.73	Compressive
6	*CE*	78	Tensile
7	*DE*	155.9	Compressive

Note: Members having forces pulling the joints are in tensile while members having forces pushing the joints are in compression.

2. A cantilever truss is loaded as shown below. Find the forces in each member.

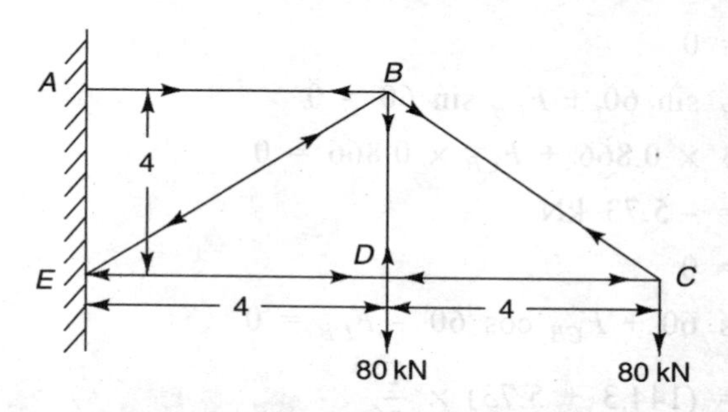

Joint *C*

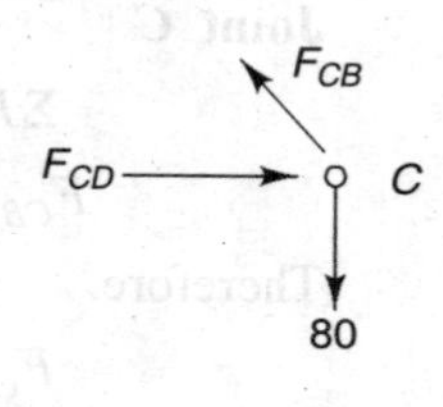

$$\Sigma P_x = 0$$
$$F_{CB} \sin 45 - 80 = 0$$
$$F_{CB} = 113 \text{ kN}$$
$$\Sigma P_y = 0,$$
$$F_{CD} - F_{CB} \cos 45 = 0$$
$$F_{CD} = 80 \text{ kN}$$

Joint *D*

$$\Sigma P_x = 0$$
$$F_{ED} = 80$$
$$\Sigma P_y = 0$$
$$F_{DB} = 80 \text{ kN}$$

Joint *B*

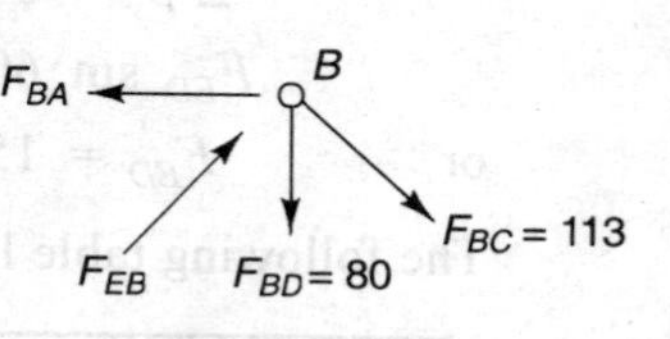

$$\Sigma P_x = 0$$
$$-F_{BA} + F_{BC} \cos 45 + F_{BE} \cos 45 = 0$$
$$\Sigma P_y = 0,\ F_{BE} \sin 45 - 80 - 113 \sin 45 = 0$$
$$F_{BE} = 226 \text{ kN}$$

Also $F_{BA} = 240$ kN

The following table lists all the forces.

S.No.	*Member*	*Force* (kN)	*Type*
1	*BC*	113	Tensile
2	*CD*	80	Compressive
3	*BD*	80	Tensile
4	*AB*	240	Tensile
5	*BE*	226	Compressive
6	*DE*	80	Compressive

3. A truss as shown below is supported at both ends. Using the sectional method, determine the forces in members *FH*, *HG* and *GI*. The truss has members connected in seven equilateral triangles with side 4 m each and a load of 10 kN is put on top of each equilateral triangle.

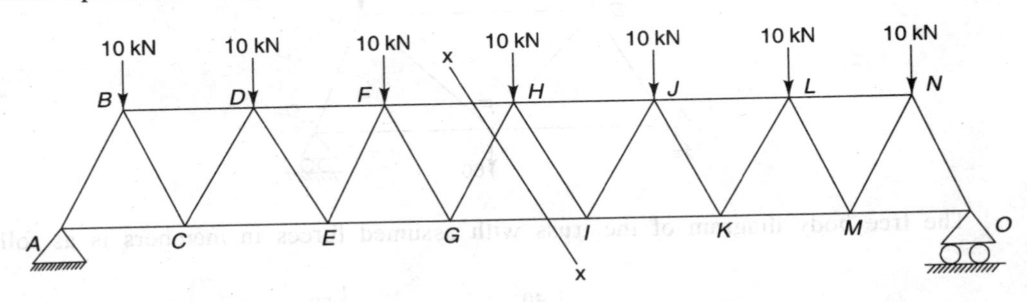

Due to symmetry

$$R_A = R_0 = \frac{1}{2} \times 10 \times 7 = 35 \text{ kN}$$

Since force in members *FH*, *HG* and *GI* is to be determined, take section x–x cutting these three members which divide the truss into two parts. We can take any part. Let us take the left part of the truss.

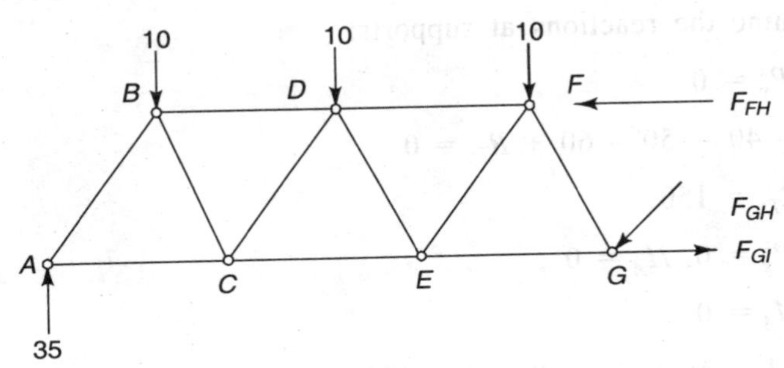

Now

$$\Sigma M_G = 0$$

$$-F_{FH} \times 4 \sin 60 + 35 \times 12 - 10 \times 10 - 10 \times 6 - 10 \times 2 = 0$$

$$\therefore \quad F_{FH} = 69.2 \text{ kN}$$

$$\Sigma P_y = 0$$

$$35 - F_{GH} \sin 60 - 10 - 10 - 10 = 0$$

$$\text{or} \quad F_{GH} = 5.7 \text{ kN}$$

$$\Sigma P_x = 0,$$

$$F_{GI} - F_{GH} \cos 60 - F_{FH} = 0$$

$$\text{or} \quad F_{GI} = 72.05 \text{ kN}$$

4. Using the joint method, determine the forces in all members of the truss and indicate the magnitude and notion of forces. All members are 2 m long.

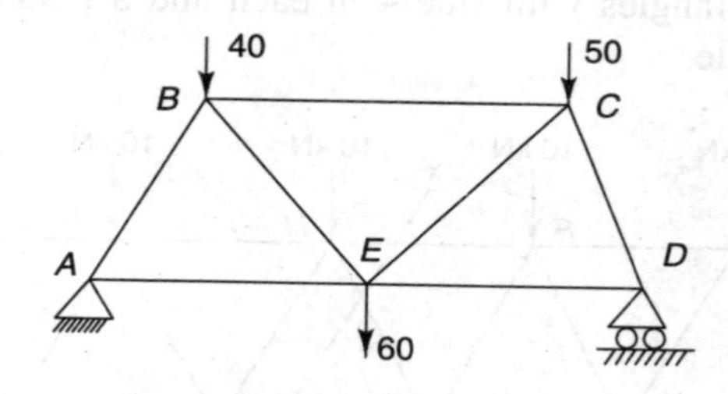

The free body diagram of the truss with assumed forces in members is as follows:

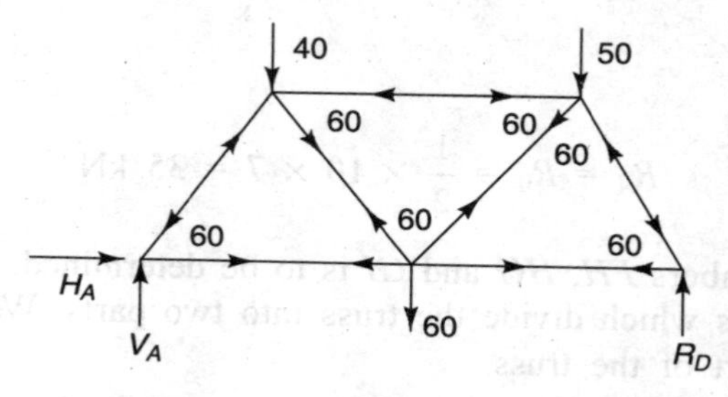

Let us determine the reactions at supports:

$$\Sigma P_y = 0$$

$$V_A - 40 - 50 - 60 + R_D = 0$$

or $$V_A + R_D = 150$$

$$\Sigma P_x = 0, H_A = 0$$

$$\Sigma M_A = 0$$

$$R_D \times 4 - 50 \times 3 - 60 \times 2 - 40 \times 1 = 0$$

or $$R_D = 77.5 \text{ kN}$$

Therefore,

$$V_A = 72.5 \text{ kN}$$

Joint *A*

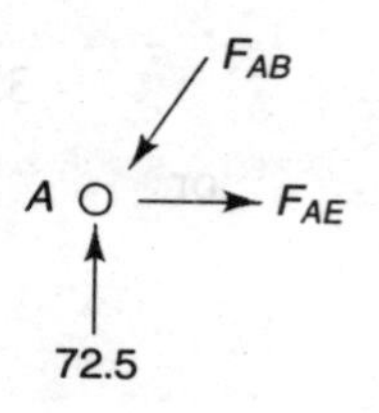

$$\Sigma P_y = 0$$

$$F_{AB} \sin 60 - 72.5 = 0$$

or $$F_{AB} = 83.71 \text{ kN}$$

$$\Sigma P_x = 0$$

$$F_{AE} = F_{AB} \cos 60 = 41.85 \text{ kN}$$

Joint D

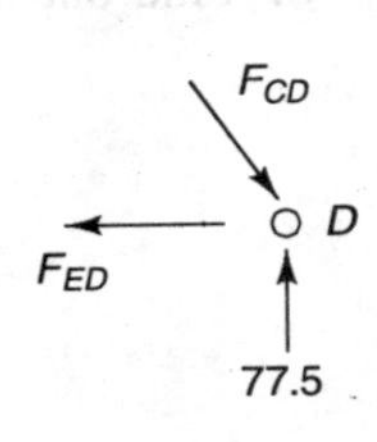

$$\Sigma P_y = 0$$

$$-F_{CD} \sin 60 + 77.5 = 0$$

or $$F_{CD} = 89.48 \text{ kN}$$

$$\Sigma P_x = 0$$

$$-F_{ED} + F_{CD} \cos 60 = 0$$

or $$F_{ED} = 44.74 \text{ kN}$$

Joint E

$$\Sigma P_x = 0$$

$$F_{ED} + F_{EC} \cos 60 - F_{AE} - F_{BE} \cos 60 = 0$$

or $$F_{EC} - F_{BE} = -5.78 \qquad \text{(i)}$$

$$\Sigma P_y = 0$$

$$F_{EC} \sin 60 + F_{BE} \sin 60 - 60 = 0$$

or $$F_{EC} + F_{BE} = 69.28 \qquad \text{(ii)}$$

Now solving Eq. (i) and (ii) we get

$$F_{EC} = 31.75 \text{ kN}$$

$$F_{BE} = 37.53 \text{ kN}$$

Joint B

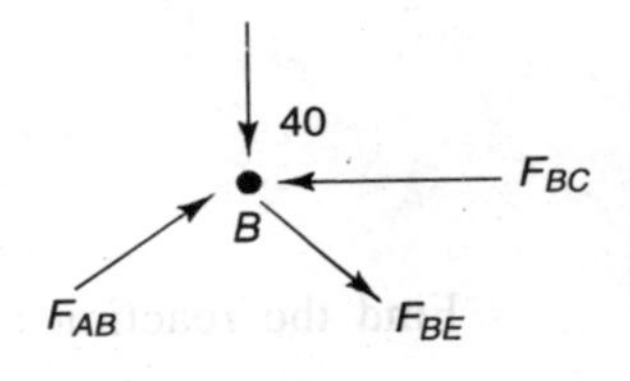

$$\Sigma P_x = 0$$

$$-F_{BC} + F_{BE} \cos 60 + F_{AB} \cos 60 = 0$$

$$F_{BC} = 60.61 \text{ kN}$$

The following table list the forces, the magnitude and nature of the forces.

S.No.	*Member*	*Force* (kN)	*Nature*
1	*AB*	83.71	Compressive
2	*BC*	60.61	Compressive
3	*CD*	89.48	Compressive
4	*ED*	44.74	Tensile
5	*EC*	31.75	Tensile
6	*BE*	37.52	Tensile
7	*AE*	41.85	Tensile

5. Find out the axial forces in all members of a truss with loading as shown in the figure. (UPTU: Dec. 2001)

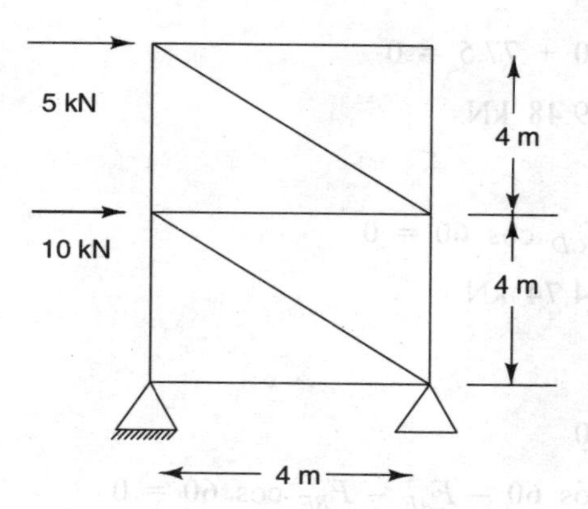

Draw the free body diagram of the truss:

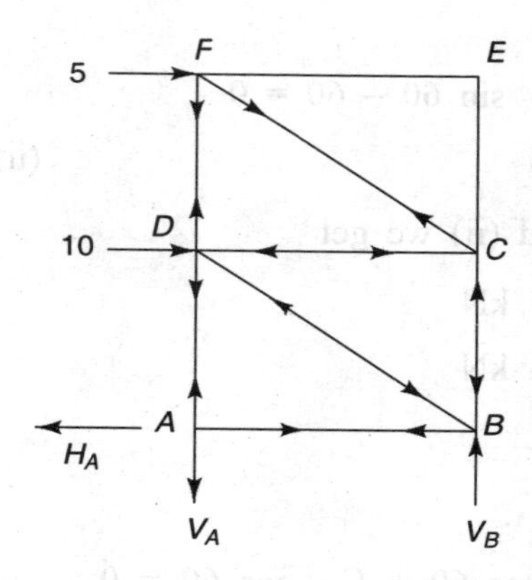

Find the reactions:

$$\Sigma P_x = 0$$

$$H_A = 10 + 5 = 15$$

$$\Sigma P_y = 0$$

$$-V_A + V_B = 0 \text{ or } V_A = V_B$$

$$\Sigma M_A = 0, -V_B \times 4 + 10 \times 4 + 5 \times 8 = 0$$

or $\quad V_B = 20$

Therefore, $\quad V_A = 20$

Joint *A*

F_{AD}, 15, F_{AB}, A, 20

$$\Sigma P_y = 0, F_{AD} = 20 \text{ kN}$$

$$\Sigma P_x = 0, F_{AB} = 15 \text{ kN}$$

Joint B

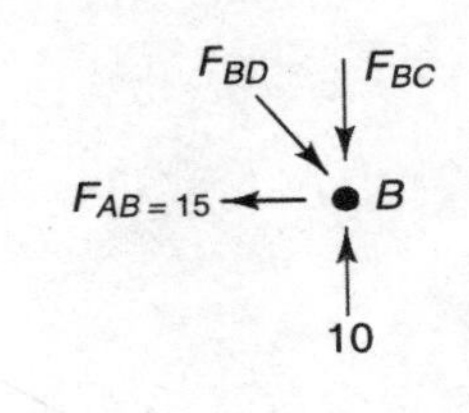

$$\Sigma P_x = 0$$

$$F_{AB} = F_{BD} \cos 45$$

or $$F_{BD} = 15 \times \sqrt{2} = 21.2 \text{ kN}$$

$$\Sigma P_y = 0$$

$$F_{BC} = 10 + 21.2 \cos 45$$

or $$F_{BC} = 25 \text{ kN}$$

Joint D

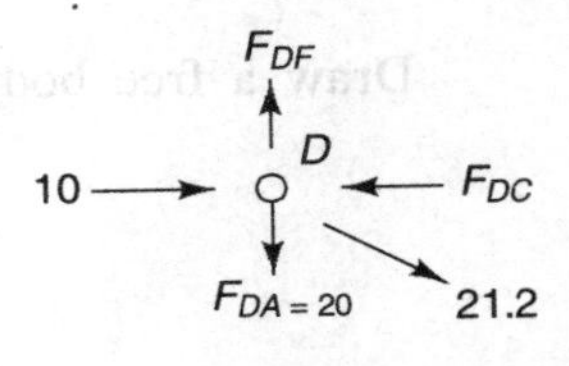

$$\Sigma P_x = 0$$

$$F_{DC} = 10 + 21.2 \cos 45 = 25 \text{ kN}$$

$$\Sigma P_y = 0$$

$$F_{DF} = 20 + 21.2 \sin 45 = 35 \text{ kN}$$

Joint C

F_{FC} F_{EC} $F_{DC} = 25$ C $F_{BC} = 25$

$$\Sigma P_x = 0,$$

$$F_{DC} = F_{FC} \cos 45$$

or $$F_{FC} = 25 \times \sqrt{2}$$

$$= 35.35 \text{ kN}$$

$$\Sigma P_y = 0$$

$$F_{FC} \cos 45 + 25 - F_{EC} = 0$$

$$F_{EC} = 25 - 25 = 0$$

Similarly, $F_{EF} = 0$ at point E.

The following table lists the forces and their nature.

S.No.	*Member*	*Force* (kN)	*Nature*
1	*AB*	15	Tensile
2	*AD*	20	Tensile
3	*BD*	21.2	Compressive
4	*BC*	25	Compressive
5	*DC*	25	Compressive
6	*DF*	35	Tensile
7	*CF*	35.35	Tensile
8	*EF*	0	—
9	*CE*	0	—

6. For the simply supported truss shown in the figure, find the force in the members *BD*, *DE*, *EG*, *BE* and *CE* using the method of sections. (UPTU: Dec. 2003)

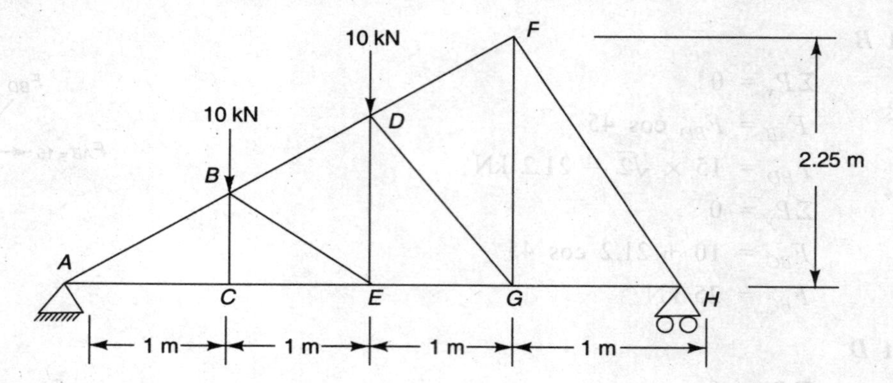

Draw a free body diagram of the truss.

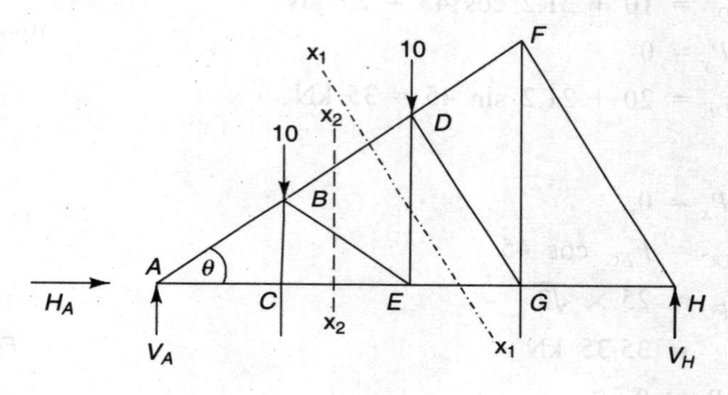

Find reactions:

$$\Sigma P_x = 0,\ H_A = 0$$

$$\Sigma P_y = 0$$

$$V_A + V_H = 20 \text{ kN}$$

$$\Sigma M_A = 0$$

$$10 \times 1 + 10 \times 2 - V_H \times 4 = 0$$

or $$V_H = \frac{30}{4} = 7.5 \text{ kN}$$

$\therefore$ $$V_A = 20 - 7.5 = 12.5 \text{ kN}$$

Take section x_1–x_1, which divide the truss into two parts. Take the left part.

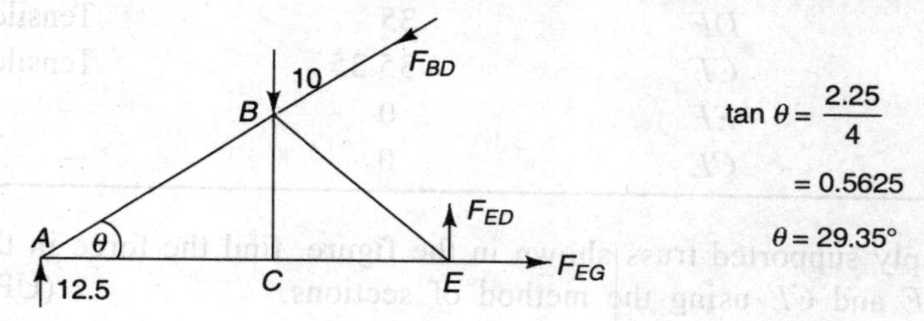

$$\tan\theta = \frac{2.25}{4}$$

$$= 0.5625$$

$$\theta = 29.35°$$

$$\Sigma M_E = 0$$

$$12.5 \times 2 - 10 \times 1 - F_{BD} \times 2 \sin \theta = 0$$

$$15 = F_{BD} \times 2 \times 0.49$$

$$F_{BD} = 15.3 \text{ kN}$$

$$\Sigma P_x = 0$$

$$F_{EG} = F_{BD} \cos \theta = 13.34 \text{ kN}$$

$$\Sigma P_y = 0$$

$$F_{ED} = 10 + F_{BD} \sin \theta - 12.5$$

$$= 10 + 7.5 - 12.5 = 5 \text{ kN}$$

Now take section x_2–x_2 and take the left part.

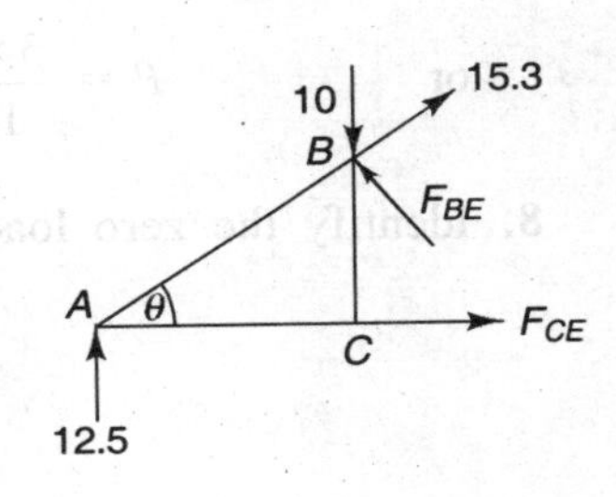

$$\Sigma M_A = 0$$

$$10 \times 1 - F_{BE} \times 2 \cos \theta = 0$$

$$F_{BE} = \frac{10}{2 \times 0.871} = 5.74 \text{ kN}$$

$$\Sigma M_B = 0$$

$$F_{CE} \times 1 \times \tan \theta - 12.5 \times 1 = 0$$

$$F_{CE} = \frac{12.5}{0.5623} = 22.23 \text{ kN}$$

7. A cantilever truss is loaded and supported as shown. Find the value of load P which would produce an axial force of magnitude 3 kN in the member AC using the method of section. (UPTU: 2002–2003)

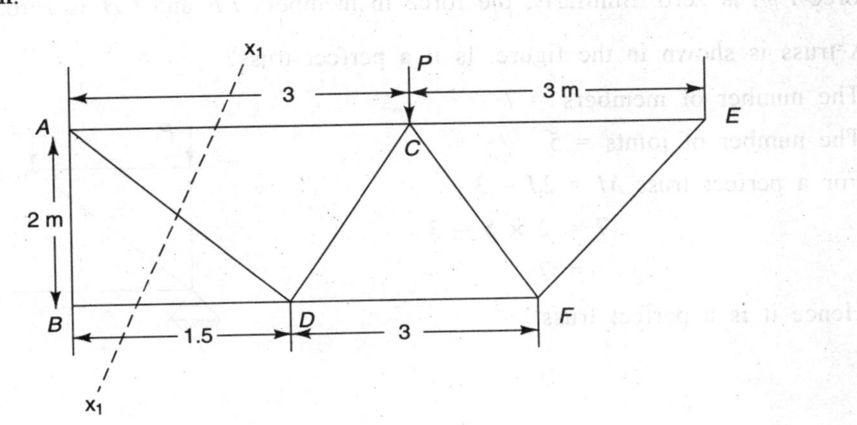

Take a section x_1–x_1 cutting members *AC*, *CD* and *BD*. Take the right-hand part.

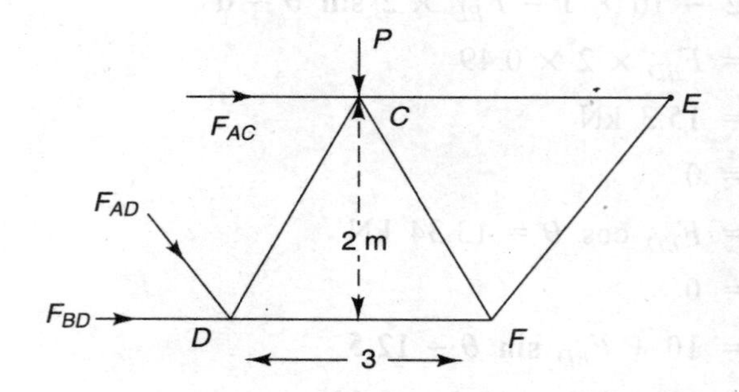

$$\Sigma M_D = 0,\ P \times 1.5 = F_{AC} \times 2$$

or $$P = \frac{3 \times 2}{1.5} = 4 \text{ kN}$$

8. Identify the zero load member in the truss shown.

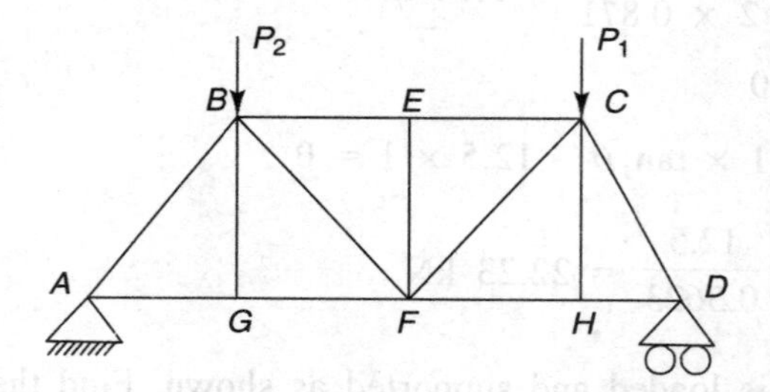

Three members are joining at point *G* and members *AG* and *GF* are collinear. Hence force F_{BG} is zero. Similarly, the force in members *EF* and *CH* is zero.

9. A truss is shown in the figure. Is it a perfect truss?

The number of members = 7

The number of joints = 5

For a perfect truss $M = 2J - 3$

$$7 = 2 \times 5 - 3$$
$$= 7$$

Hence it is a perfect truss.

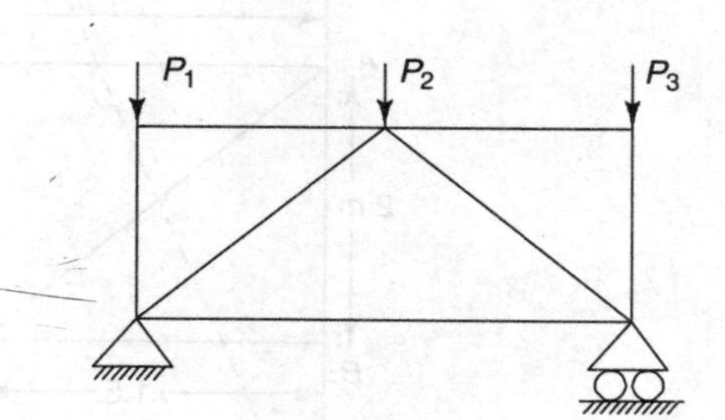

10. A cantilever truss is shown below. Identify zero load members.

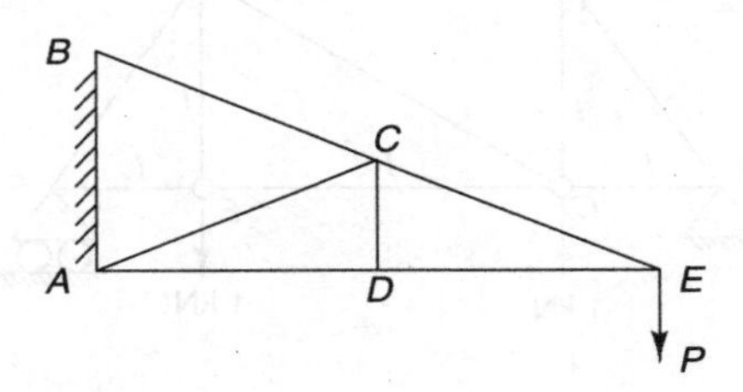

Three members are joining at D and members AD and DE are collinear. Hence the force in member CD is zero.

Now the three members (member CD is redundant having zero load) are joining at C and members BC and CE are collinear. Hence the load in member AC is zero.

Hence members AC and CD have zero load.

11. A truss is shown in the figure. Identify zero load members.
Here reactions are

$$R_H = P \text{ and } R_G = 0$$

At point G, members EG and GH are joining. Hence load is zero in both the members.
At point H, $F_{HF} = R_H$ and $F_{EH} = 0$.
At point E, the load in members EF and CE is zero.
At point F, $F_{FD} = P$ and $F_{CF} = 0$.
At point C, the load in members CD and CA is zero.
At point A, the load in members AC and AB is zero.

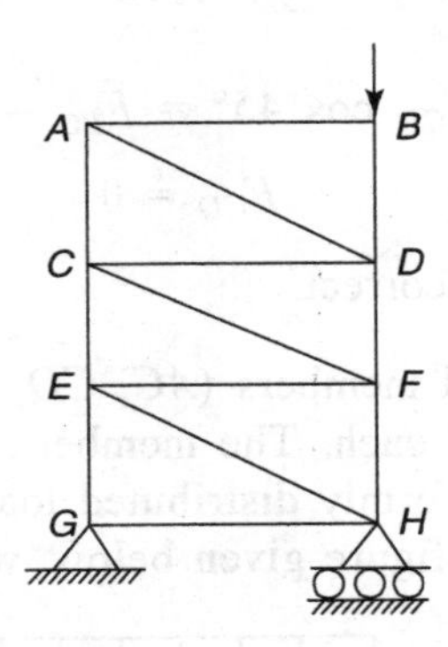

12. For the loading on the truss shown in the figure, the force in members CDs

(a) 0 (b) 1 kN

(c) $\sqrt{2}$ kN (d) $\dfrac{1}{\sqrt{2}}$ kN (GATE: 2001)

$R_A = R_B = 1$ as per symmetry.

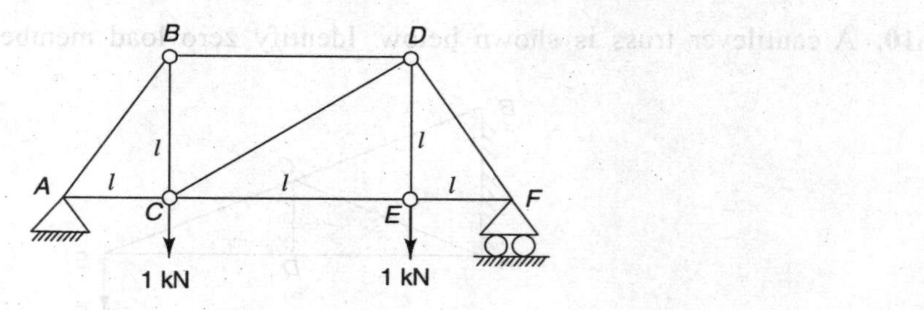

Joint *A*

$$\tan\theta = \frac{l}{l} = 1$$

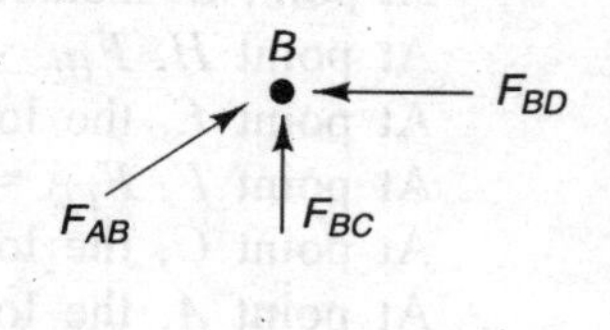

or
$$\theta = 45°$$

$$F_{AB} \sin 45 = 1$$

or
$$F_{AB} = \sqrt{2}$$

$$F_{AC} = F_{AB} \cos 45 = \sqrt{2} \times \frac{1}{\sqrt{2}} = 1$$

Joint *B*

$$F_{BC} = F_{AB} \cos 45°$$

$$= \sqrt{2} \times \frac{1}{\sqrt{2}} = 1$$

Joint *C*

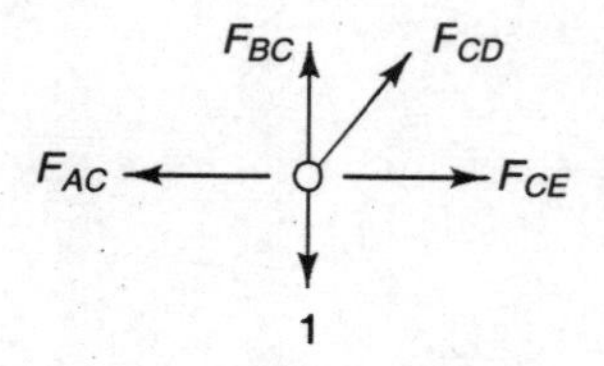

$$F_{CD} \cos 45° = F_{BC} - 1$$

$$F_{CD} = 0$$

Therefore, the option (a) is correct.

13. A truss consists of horizontal members (*AC*, *CD*, *DB* and *EF*) and vertical members (*CE* and *DE*) having length *l* each. The member *AE*, *DE* and *BF* are inclined at 45° to the horizontal. For the uniformly distributed load *P* per unit length on the member *EF* of the truss shown in the figure given below, what is the force in the member *CD* is?

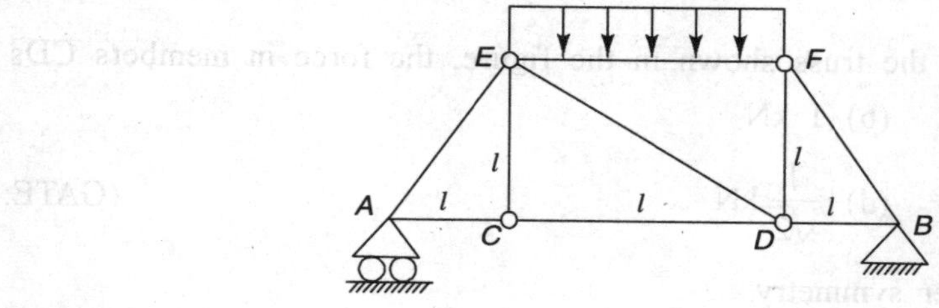

(GATE: 2003)

$$\Sigma F_y = 0,$$

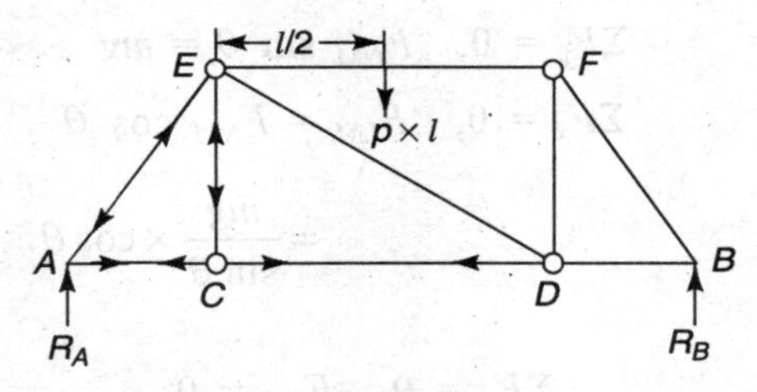

$$R_A + R_B = P \times l$$

$$\Sigma M_A = 0, \quad Pl \times \frac{3l}{2} = R_B \times 3l$$

or
$$R_B = \frac{Pl}{2}$$

Joint *A*

$$\Sigma F_y = 0,$$

$$F_{AE} \cos 45 = \frac{Pl}{2}$$

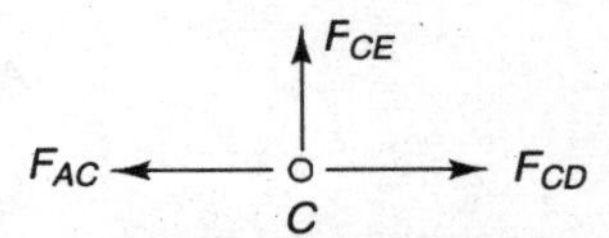

$$F_{AE} = \frac{Pl}{\sqrt{2}}$$

$$\Sigma F_x = 0,$$

$$F_{AE} \sin 45 = F_{AC}$$

$$F_{AC} = \frac{Pl}{\sqrt{2}} \times \frac{1}{\sqrt{2}} = \frac{Pl}{2}$$

Joint *C*

$$\Sigma F_x = 0, F_{AC} = F_{CD} = \frac{Pl}{2}$$

14. The figure shows a pin-jointed plane truss loaded at the point M by hanging a mass of 100 kg. The member LN of the truss is subjected to a load of

(a) 0 (b) – 490

(c) – 981 (d) +981 (GATE: 2004)

Joint *M*

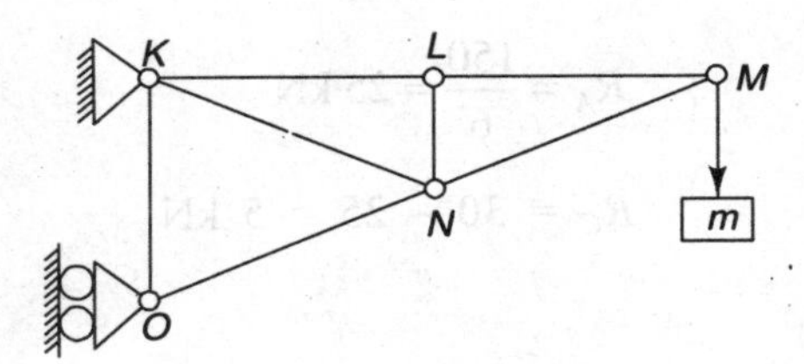

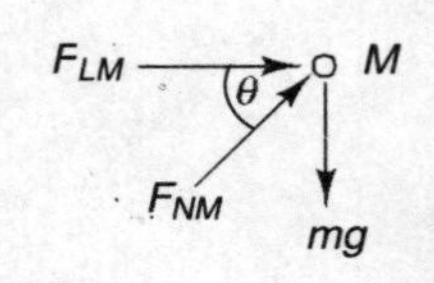

$$\Sigma F_y = 0, \quad F_{NM} \sin \theta = mg$$

$$\Sigma F_x = 0, \quad F_{LM} = F_{NM} \cos \theta$$

$$= \frac{mg}{\sin \theta} \times \cos \theta$$

Joint *L*

$$\Sigma F_y = 0, \quad F_{LN} = 0$$

F_{KL} L F_{LM} F_{LN}

Therefore, the option (a) is correct.

15. Determine the magnitude and nature of forces with the members of truss shown in the figure.

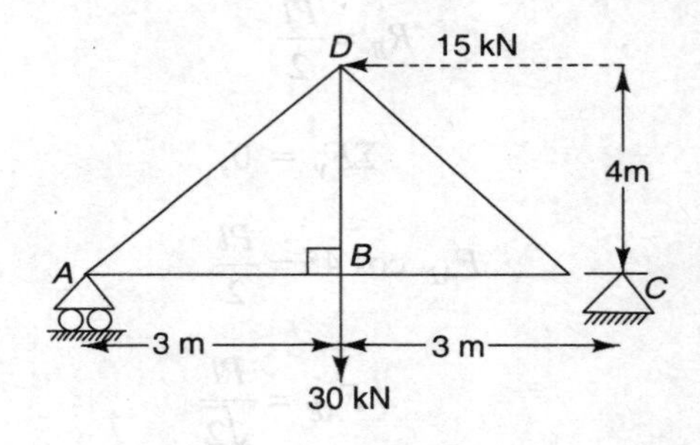

(UPTU: 2006–2007)

The free body diagram of the truss is:

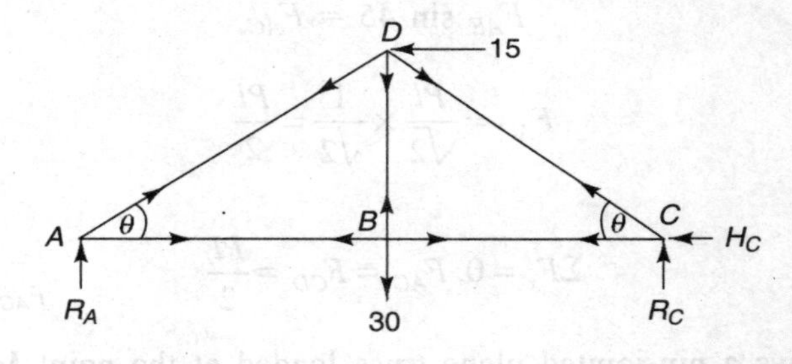

Reactions

$$\Sigma F_y = 0, \quad R_A + R_C = 30 \text{ kN}$$

$$\Sigma F_x = 0, \quad H_C = 15 \text{ kN}$$

$$\Sigma M_C = 0, \quad R_A \times 6 - 30 \times 3 - 15 \times 4 = 0$$

$$R_A = \frac{150}{6} = 25 \text{ kN}$$

$$\therefore \qquad R_C = 30 - 25 = 5 \text{ kN}$$

Now $$\tan \theta = \frac{4}{3}$$

or $$\theta = 53.12°$$

Joint *A*

$$\Sigma F_y = 0, \quad F_{AB} = \frac{25}{\sin \theta} = 31.25 \text{ kN}$$

$$\Sigma F_x = 0, \quad F_{AB} = F_{AD} \cos \theta$$

$$= 18.76 \text{ kN}$$

Joint *B*

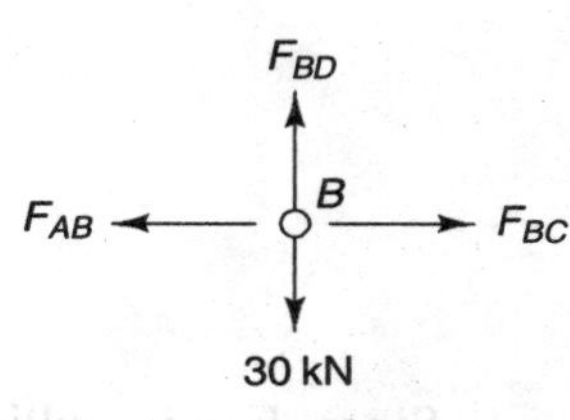

$$\Sigma F_y = 0, \quad F_{BD} = 30 \text{ kN}$$

$$\Sigma F_x = 0, \quad F_{BC} = F_{AB} = 18.76 \text{ kN}$$

Joint *C*

$$\Sigma F_y = 0, \quad F_{BC} \sin \theta = 5$$

$$F_{DC} = \frac{5}{\sin \theta} = 6.25 \text{ kN}$$

The following table list the forces and their types.

S.No.	*Member*	*Force* (kN)	*Type*
1	*AD*	31.25	Compressive
2	*AB*	18.76	Tensile
3	*BD*	30	Tensile
4	*CD*	5	Compressive
5	*BC*	18.76	Tensile

16. The force in member *FD* in the given truss in the figure is

(a) –50 kN (b) +50 kN

(c) +150 kN (d) –100 kN

(Civil Services: 1996)

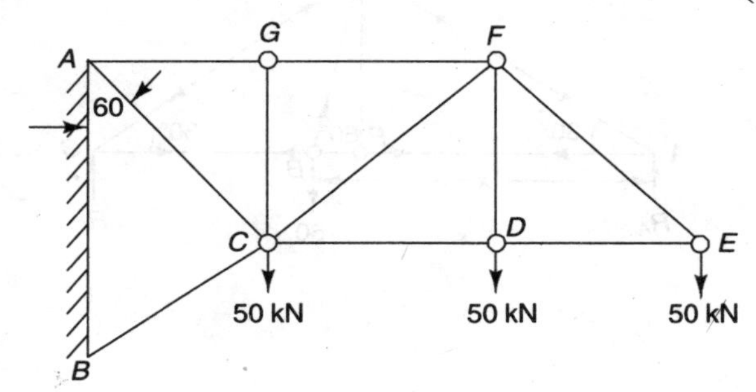

Taking a section $x - x$ of the truss as shown and taking the right portion of the truss. Now taking moment w.r.t. point E

$$\Sigma M_E = 0, \quad F_{FD} \times DE - 50 \times DE = 0$$

$$\therefore \qquad F_{FD} = 50 \text{ kN}$$

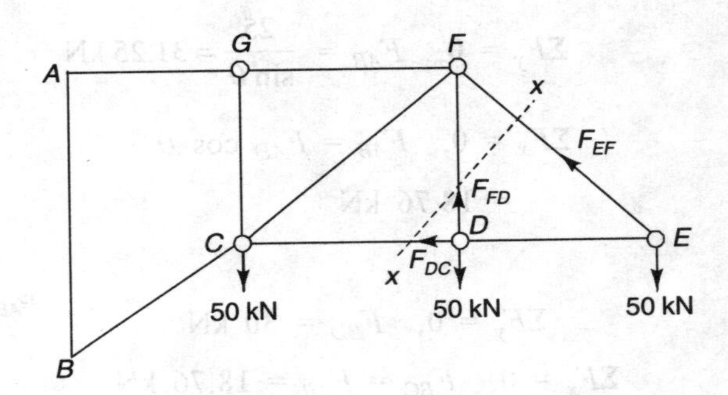

Since F_{FD} is pulling the joint D, hence the force F_{FD} is tensile, i.e. positive. Hence, the option (b) is correct.

17. Determine the forces and their nature in each member of truss loaded as shown in the figure.

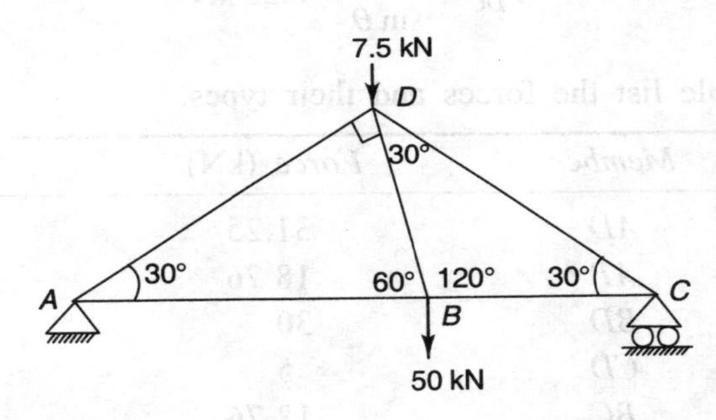

(UPTU: May 2008)

Draw firstly the free body diagram as shown below:

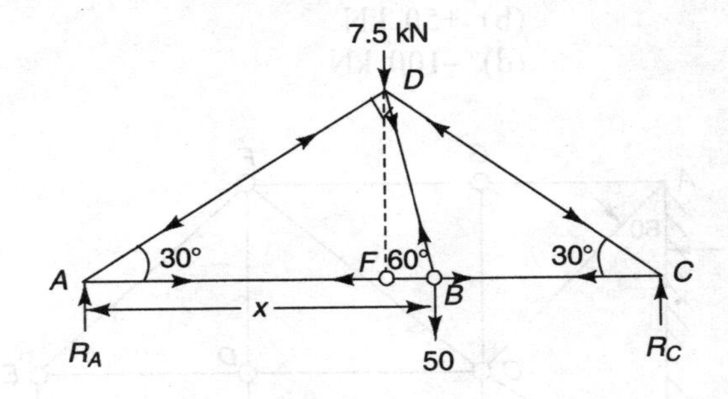

Take $AB = x$ in the right angled triangle ABD. Therefore,

$$AD = x \cos 30 \quad \text{and} \quad BD = x \sin 30$$

Now in the equilateral triangle BDC,

$$BD = DC = x \sin 30$$

Also $AF = AD \cos 30 = x \cos^2 30 = \dfrac{3x}{4}$

Now

$$\Sigma F_x = 0,$$

$$R_A + R_B = 7.5 + 50 = 57.5$$

$$\Sigma M_A = 0, \quad 7.5 \times AF + 50 \times x = R_C \times (x + x \sin 30)$$

or $$7.5 \times x \times \frac{3}{4} + 50 \times x = R_C \left(1 + \frac{1}{2}\right) \times x$$

or $$R_C = \frac{5.625 + 50}{1.5} = 37.08 \text{ kN}$$

∴ $$R_A = 57.5 - 37.08$$

$$= 20.42$$

Joint *A*

$$\Sigma F_y = 0, \quad F_{AD} \sin 30 = 20.42$$

or $$F_{AD} = \frac{20.42}{0.5} = 40.84 \text{ kN}$$

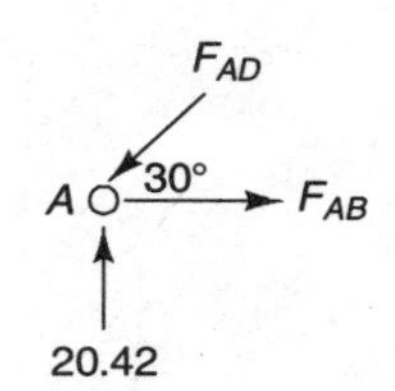

$$\Sigma F_x = 0,$$

$$F_{AD} = \cos 30 = F_{AB}$$

or $$F_{AB} = \frac{40.84 \times \sqrt{3}}{2}$$

$$= 35.37$$

Joint *C*

$$\Sigma F_y = 0, \quad F_{DC} \sin 30 = 37.08$$

or $$F_{DC} = 37.08 \times 2$$

$$= 74.16 \text{ kN}$$

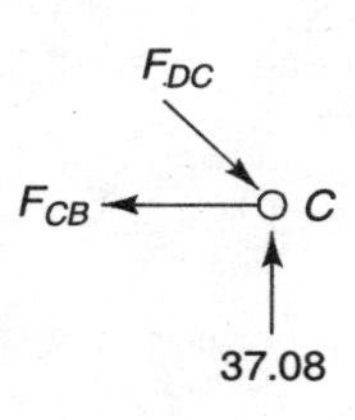

$$\Sigma F_x = 0, \quad F_{CB} = F_{DC} \cos 30$$

$$= 74.16 \times 0.866$$

$$= 64.22 \text{ kN}$$

Joint B

$$\Sigma F_y = 0, \quad F_{BD} \sin 60 = 50$$

or

$$F_{BD} = \frac{50}{\sin 60} = \frac{50}{0.866}$$

$$= 57.74 \text{ kN}$$

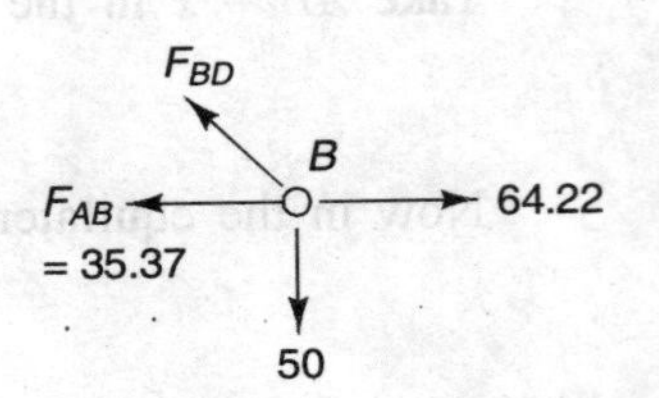

The following are the list of forces and their nature.

S.No.	*Component*	*Force* (kN)	*Nature*
1	*AD*	40.84	Compressive
2	*AB*	35.37	Tensile
3	*DC*	74.16	Compressive
4	*BC*	64.22	Tensile
5	*BD*	57.74	Tensile

OBJECTIVE TYPE QUESTIONS

Are you sure you're right, or are you just being stubborn?

State True or False

1. A truss is a structure mode of slender members pin connected at ends. *(True/False)*
2. A truss is capable to take loads at any point on a member. *(True/False)*
3. A truss is designed to resist geometrical distortion under any applied system of loading. *(True/False)*
4. A roof truss and a bridge truss are examples of a plane truss. *(True/False)*
5. All members of a plane truss lie in a single plane. *(True/False)*
6. Members of a space truss can be anywhere in space. *(True/False)*
7. A transmission tower is a space truss. *(True/False)*
8. A member of trusses under tension is called a strut. *(True/False)*
9. A member of trusses under compression is a called tie. *(True/False)*
10. A truss has generally a roller support at one end and its other end is hinged. *(True/False)*
11. A roller support is frictionless and provides a reaction at a right angle to the roller base of the support. *(True/False)*
12. The direction of reaction of a hinged support depends upon the load system on the truss. *(True/False)*
13. An imperfect truss has more members than what is required for a perfect truss. *(True/False)*
14. A perfect truss has more members than a redundant truss. *(True/False)*
15. The number of members (m) of a perfect truss is given by $m = 2J - 3$ where J is the number of joints. *(True/False)*
16. A perfect truss is statically determinate. *(True/False)*
17. An imperfect and redundant trusses are indeterminate. *(True/False)*
18. The weight of a member of a truss is considered depending on the length of the member. *(True/False)*
19. Load is considered to be acting at the joint of a truss. *(True/False)*
20. The cross section of a member of a truss is considered to change uniformly. *(True/False)*
21. The member is under tensile force in case force pulls the joint. *(True/False)*
22. The member is under compressive force if force pushes the joint. *(True/False)*
23. If three members join at a joint and two members are collinear, then force in the third member is zero. *(True/False)*

24. If two non-collinear members meet at a joint, then force in each member is zero. *(True/False)*

25. A graphical method is used to find out reactions at supports. *(True/False)*

26. The forces in members of a truss constitute a system of concurrence forces at each joint. *(True/False)*

27. Three equations of equilibrium can be formed at each joint in the method of joint for force analysis of a truss. *(True/False)*

28. A joint having two unknown forces is selected in a method of joint for force analysis of a truss. *(True/False)*

29. A section line passing through not more than three members of a truss is drawn in the method of section for force analysis of a truss. *(True/False)*

30. We divide a truss into two parts in the method of section so as to get one part of the truss which is a non-concurrent force system and the other part is a concurrent force system. *(True/False)*

31. In the method of section, the section is taken so that the least number of members is cut. *(True/False)*

32. Three equations of equilibrium can be formed in the method of section. *(True/False)*

33. Bow's notation is a method of designating a force by placing capital letters on either side of the force. *(True/False)*

34. If there are six joints and nine members in a truss, then the truss is a perfect truss. *(True/False)*

35. Frames do not differ from trusses. *(True/False)*

36. Machines do not differ from frames. *(True/False)*

37. A single-storeyed double truss bridge can take twice the load as taken by a single-storyed single truss bridge. *(True/False)*

38. A double-storeyed single truss bridge can take eight times the load as taken by a single-storeyed single truss bridge. *(True/False)*

39. Forces at the end of members are equal, opposite and collinear. *(True/False)*

40. A four-member truss can be made rigid by adding a diagonal member. *(True/False)*

Multiple Choice Questions

1. If a perfect truss has J points, then the number of members is
(a) $2J$ (b) $2J - 3$ (c) $2J - 4$

2. If J is the number of joints and $2J$ is the number of members, then the truss is
(a) perfect (b) imperfect (c) redundant

3. A truss is determinate if it is
(a) an imperfect truss (b) a perfect truss (c) a redundant truss

4. A perfect truss can have
   (a) one end on the roller support and the other hinged
   (b) both ends on the roller support
   (c) both ends hinged

5. The force polygon representing a system of forces in equilibrium is
   (a) a triangle (b) a closed polygon (c) an open polygon

6. A smooth surface support always develops a
   (a) vertical reaction (b) reaction normal to it (c) horizontal reaction

7. The free body diagram of a joint should satisfy which of the following equilibrium equations?
   (a) $\Sigma H = 0$, $\Sigma M = 0$ (b) $\Sigma V = 0$, $\Sigma M = 0$ (c) $\Sigma H = 0$, $\Sigma V = 0$

8. The support which can develop a reaction and moment together is
   (a) hinged (b) fixed support (c) roller support

9. A truss subjected to wind load must have
   (a) both ends at the roller support
   (b) both ends hinged support
   (c) one at the roller support and other hinged

10. A system of coplanar forces is in equilibrium when
   (a) a funicular polygon closes
   (b) force polygon closes
   (c) both force and funicular polygons close

11. The diagram showing the point of application and line of action of forces in their plane is known as
   (a) vector diagram (b) force diagram (c) space diagram

12. The ordinate in a funicular polygon represents
   (a) shear force (b) load (c) bending moment

13. The number of funicular polygons which can be drawn to pass through two specified points in a space diagram is
   (a) zero (b) infinite (c) two

14. The pole distance is measured on the
   (a) force scale (b) mass scale (c) distance scale

15. If a perfect truss has five joints, then the number of members is
   (a) 6 (b) 7 (c) 8

16. If a redundant truss has four joints, then the number of members is
   (a) 5 (b) 4 (c) more than 5

17. If an imperfect truss has seven joints, then the number of members is
   (a) 11 (b) 12 (c) less than 11

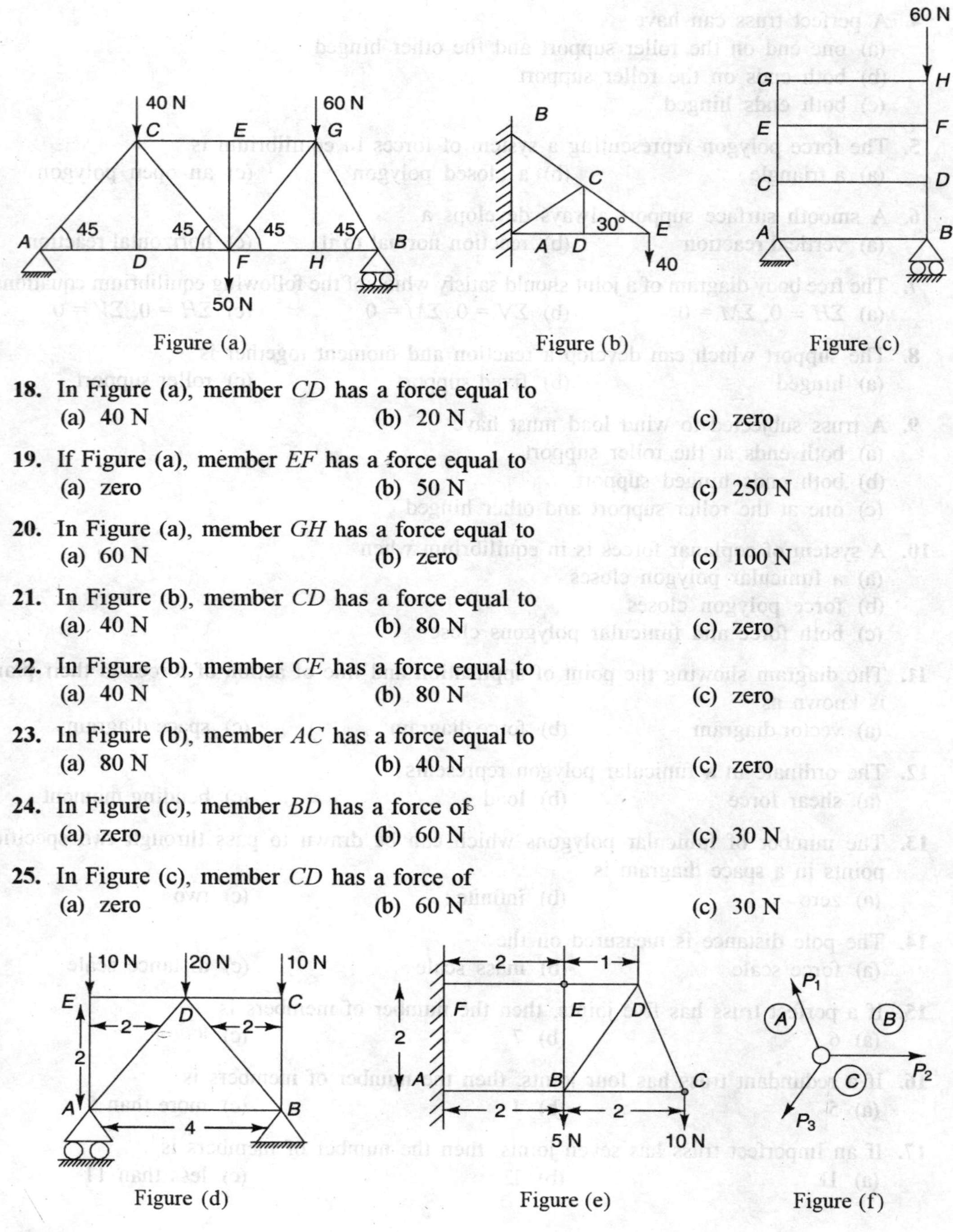

Figure (a) Figure (b) Figure (c)

18. In Figure (a), member *CD* has a force equal to
(a) 40 N (b) 20 N (c) zero

19. If Figure (a), member *EF* has a force equal to
(a) zero (b) 50 N (c) 250 N

20. In Figure (a), member *GH* has a force equal to
(a) 60 N (b) zero (c) 100 N

21. In Figure (b), member *CD* has a force equal to
(a) 40 N (b) 80 N (c) zero

22. In Figure (b), member *CE* has a force equal to
(a) 40 N (b) 80 N (c) zero

23. In Figure (b), member *AC* has a force equal to
(a) 80 N (b) 40 N (c) zero

24. In Figure (c), member *BD* has a force of
(a) zero (b) 60 N (c) 30 N

25. In Figure (c), member *CD* has a force of
(a) zero (b) 60 N (c) 30 N

Figure (d) Figure (e) Figure (f)

26. In Figure (d), the truss is
(a) perfect (b) imperfect (c) redundant

27. In Figure (d), the reaction R_A is
(a) 40 N (b) 20 N (c) 80 N

28. In Figure (e), the force in member *EF* is
(a) 10 N (b) 15 N (c) 5 N

29. In Figure (f), the force P_2 is represented in Bow's notation as
(a) *AB* (b) *BC* (c) *AC*

30. A member of a truss under tension is called
(a) tie (b) strut (c) column

31. A member of a truss under compression is called
(a) tie (b) strut (c) frame

32. A truss having the number of members less than 2*J*–3 is called
(a) redundant truss (b) imperfect truss (c) perfect truss

33. A perfect truss must obey the relation
(a) $m = 2J - 2$ (b) $m = 2J - 4$ (c) $m = 2J - 3$

34. A truss in which the number of members is more than $2J - 3$ is called
(a) perfect truss (b) redundant truss (c) imperfect truss

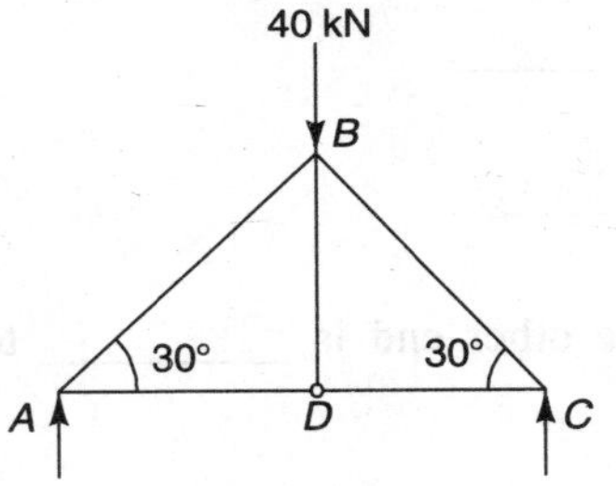

Figure (g) 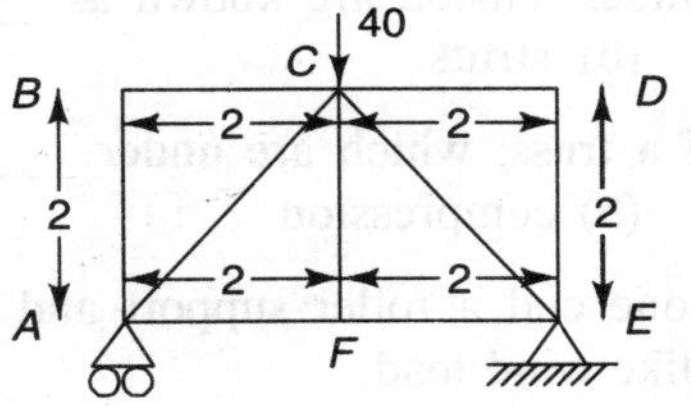 Figure (h) 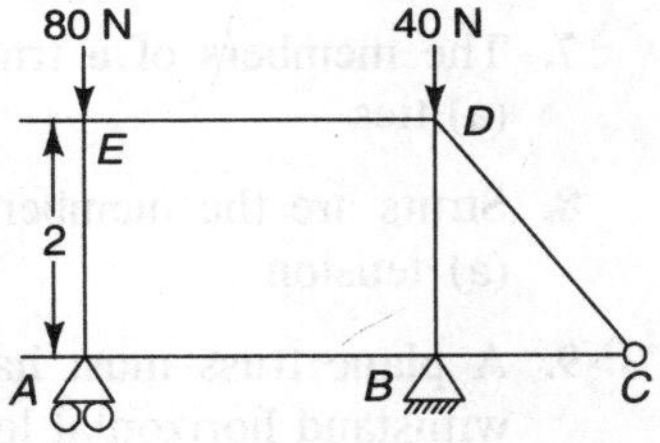 Figure (i)

35. In Figure (g), the force in member *AB* is
(a) 30 kN (b) 40 kN (c) 25 kN

36. In Figure (h), the force in member *AC* is
(a) 40 kN (b) 120 kN (c) $20/\sqrt{2}$ kN

37. In Figure (i), the force in member *BC* is
(a) 20 N (b) 40 N (c) zero

38. In Figure (i), the force in member *CD* is
(a) 20 N (b) 40 N (c) zero

39. In a bridge with a single truss, each side takes *W* load. If one truss is added to each side, then the bridge can take a load of
(a) *W* (b) 3*W* (c) 2*W*

40. In a bridge with a single truss, each side takes a load of W. If the truss is made double the original height, then the bridge can take a load of
(a) $2W$ (b) $8W$ (c) $4W$

Fill in the Blanks

1. A truss having members = $2J - 3$ is __________ truss.
(a) an imperfect (b) a perfect

2. If a truss has more members than what is required, it is called __________ truss.
(a) an imperfect (b) a redundant

3. If a truss has less members than required, then it is called an __________ truss.
(a) economical (b) imperfect

4. A truss is a structure made of __________ members.
(a) solid (b) slender

5. A truss is designed to take load at __________ of the members.
(a) midpoint (b) joint

6. The member of a plane truss must lie in __________ planes.
(a) one (b) two

7. The members of a truss under tension are known as __________.
(a) ties (b) struts

8. Struts are the members of a truss, which are under __________.
(a) tension (b) compression

9. A plane truss must have one end a roller support and the other end is __________ to withstand horizontal load like wind load.
(a) fixed (b) hinged

10. In order to compensate elongation and contraction, one end of a truss is to be __________.
(a) fixed (b) roller supported

11. If two members meet at a point and they are not collinear, then the force in each member is __________.
(a) equal (b) zero

12. If two collinear members meet a third member at a joint, the force in the third member is __________.
(a) half (b) zero

13. In the graphical method for drawing a space diagram, __________ notation is used for designating forces.
(a) Arrow's (b) Bow's

14. In the joint method of analysis of a truss, the first selected joint cannot have more than __________ unknown forces.
(a) two (b) three

15. In the section method of analysis of a truss, the section line must not cut more than __________ members.
(a) three (b) four

16. When the method of section is used, we get __________ parts of a truss.
(a) two (b) three

17. In the graphical method __________ polygons are drawn in that order.
(a) force and funicular (b) funicular and force

18. A railway bridge is a __________ truss.
(a) space (b) plane

19. A roof truss is a __________ truss.
(a) plane (b) space

20. The weight of the members of a plane truss is taken as __________
(a) zero (b) actual

21. The ends of members are considered as __________ connected.
(a) pin (b) rivet

22. A railway bridge has __________ trusses joined by transoms and stringers.
(a) front and back (b) left and right

23. If the height of a truss is doubled, it can take __________ times of load.
(a) 4 (b) 8

24. If the number of trusses is doubled, it can take __________ times of load.
(a) 4 (b) 2

25. A perfect truss is __________.
(a) determinate (b) indeterminate

26. Three members joined by a pin is always stable and it is a __________ element of a plane truss.
(a) solid (b) basic

27. All members of a truss are __________ force members.
(a) One (b) two

ANSWERS

Nothing great was ever achieved by only sitting still.

State True or False

1. True **2.** False (load at joints only) **3.** True

4. True **5.** True

6. True (Members in three-dimensional space)

7. True

8. False (A member under tension is called tie.)

9. False (A member under compression is called strut.)

10. True (A roller support to compensate for elongation and contraction, and a hinged support to prevent sliding off during horizontal loading like wood load)

11. True

12. True

13. False (An imperfect truss has less members than required from stability consideration.)

14. False (A redundant truss has more than required from stability point of view.)

15. True

16. True

17. True (Equilibrium equations are not sufficient to find unknown forces.)

18. False (The weight of members is neglected.)

19. True

20. False (Members are considered to have uniform cross section)

21. True (joint —→ ←— joint = tensile)

22. True (joint ←— —→ joint = compressive)

23. True **24.** True **25.** True **26.** True

27. False (Only two equations of equilibrium can be formed as each point gives a force systems of collinear forces.)

28. True (Each joint gives two equations of equilibrium which can solve two unknown forces.)

29. True (The coplanar system formed in the section method gives three equations of equilibrium which can solve three unknowns.)

30. False (Two parts of a truss with each part a force system of coplanar forces)

31. True

32. True

33. True

34. True ($m = 2J - 3$ or $m = 2 \times 6 - 3 = 9$)

35. False (A frame has at least one multiforce member)

36. False (A machine has moving members.)

37. True

38. True (Moment of inertia is proportional to (height)^3.)

39. True

40. True

Multiple Choice Questions

1. (b)
2. (c)
3. (b)
4. (a)
5. (b)
6. (b)
7. (c)
8. (b)
9. (c)
10. (c)
11. (c)
12. (c)
13. (b)
14. (a)
15. (b) ($M = 2J - 3 = 2 \times 5 - 3 = 7$)
16. (c) ($m > 2J - 3 > 2 \times 4 - 3 > 5$)
17. (c) ($m < 2J - 3 < 2 \times 7 - 3 < 1$)
18. (c) (If two collinear members joined by the third member, then the force member is zero. *CD* is a member meeting collinear members *AD* and *DF*.)
19. (a)
20. (b)
21. (c) (Members *AD* and *DE* are collinear and third member *CD* is joining them.)
22. (b) (F_{CE} sin 30 = 40 or $F_{CE} = 80$)
23. (c) (Members *BC* and *CE* are collinear and member *AC* is joining them. Whole force in member *CD* is meeting them.)
24. (b) ($R_B = 60$, $F_{BD} = R_B = 60$)
25. (a) (Members *BD* and *DF* are collinear and member *CD* is meeting them.)
26. (a) ($m = 7, J = 5 \therefore m = 2J - 3 = 7 = 2 \times 5 - 3 = 7$)
27. (b) (Due to symmetry $V_A = V_B = \frac{1}{2}(10 + 20 + 10) = 20$)
28. (a)

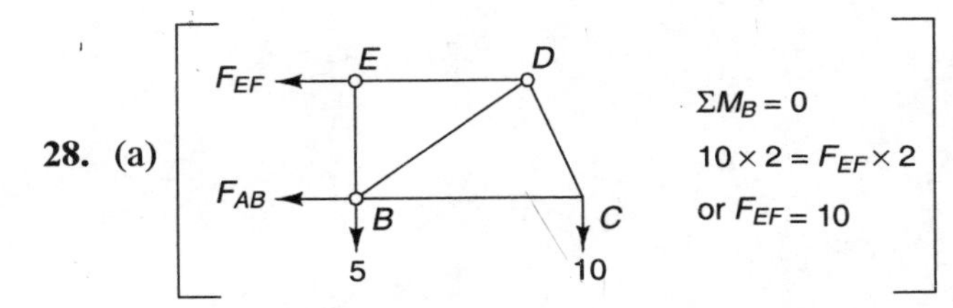

29. (b)
30. (a)
31. (b)

32. (b) **33.** (c) **34.** (b)

35. (a) (F_{AB} sin 30 = R_A = 20 or F_{AB} = 40)

36. (c) (Members AB and BC are non-collinear. Force is zero in them: $F_{AC} = 20 \cos 45 = 20/\sqrt{2}$)

37. (c) (Non-collinear member)

38. (c)

39. (b) (Moment of inertia ∝ (height)^3)

40. (b) [I = moment of inertia = $\dfrac{bd^3}{12}$ $\quad\therefore\quad \dfrac{I_2}{I_1} = 8$]

Fill in the Blanks

1. (b) **2.** (b) **3.** (b) **4.** (b)
5. (b) **6.** (a) **7.** (a) **8.** (b)
9. (b) **10.** (b) **11.** (b) **12.** (b)
13. (b) **14.** (a) **15.** (a) **16.** (a)
17. (a) **18.** (b) **19.** (a) **20.** (a)
21. (a) **22.** (b) **23.** (b) **24.** (b)
25. (a) **26.** (b) **27.** (b)

CHAPTER 12

Centroid and Moment of Inertia

Walk a mile in others' shoes before you say no to their request for a new pair.

INTRODUCTION

The centre of the mass of a body is a very special point where its mass is concentrated as a particle. The motion of the body is just like the motion of a single particle placed at the centre of mass having the same mass. Every body has only one point at which the whole weight of the body can be considered to be concentrated. This point is called *centre of gravity* (CG) of the body. The plane bodies have only areas. The point at which the total area is considered to be concentrated is called *centroid of the plane area.* The centre of gravity and centroid are located at the same point for a plane body.

According to Newton's first law of motion, a body continues in its state of rest or uniform translatory motion unless some external force acts upon it to change its present state. The property by virtue of which any body opposes any change in its present state is called *inertia.* In the same way, when a body rotates about an axis, then it has a tendency to oppose any change in its state. This property by virtue of which a body opposes any change in its state of rotation about an axis is called the *moment of inertia* of the body about that axis.

CENTRE OF MASS

Consider a body consisting of N particles as shown in Figure 12.1 having total mass M. Let its ith particle has m_i mass and it is located at distance of x_i, y_i and z_i from the origin along x, y and z directions. The centre of mass of this body can be given as follows:

$$\bar{x} = \frac{1}{M} \sum_{i=1}^{i=N} m_i x_i$$

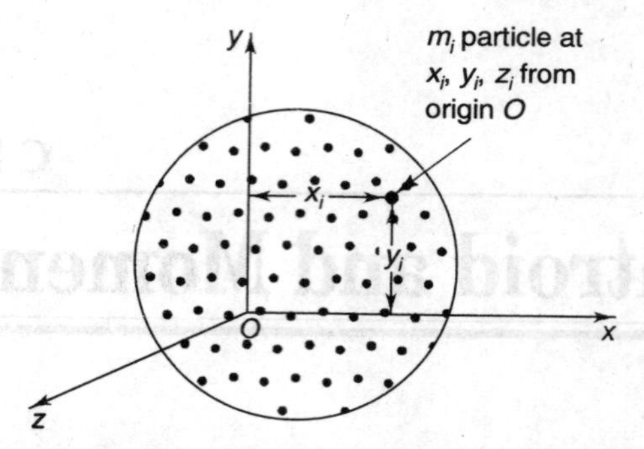

FIGURE 12.1 Centre of mass.

$$\overline{y} = \frac{1}{M}\sum_{i=1}^{i=N} m_i y_i$$

and

$$\overline{z} = \frac{1}{M}\sum_{i=1}^{i=N} m_i z_i$$

Centroid

The centroid is infact the first moment of a plane surface of area A in the xy-plane is as shown in Figure 12.2. We can define the first moment of area A about the x-axis as follows:

$$M_x = \int_A y \cdot dA$$

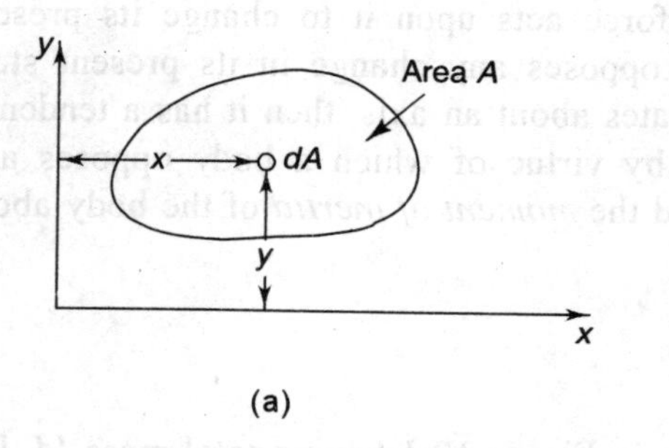

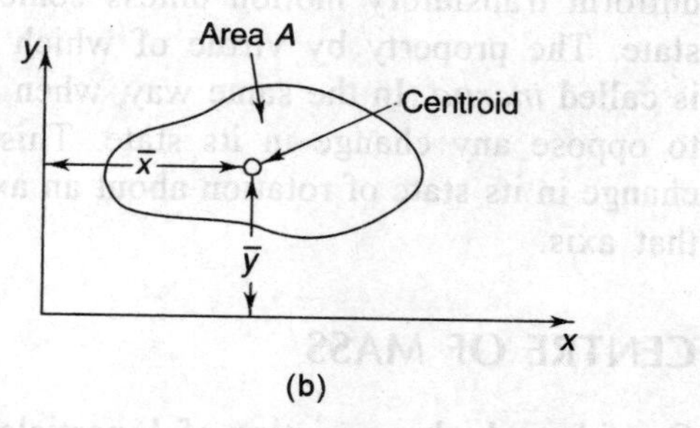

FIGURE 12.2 (a) Plane area and (b) Centroid coordinates.

Similarly, the first moment about the y-axis is

$$M_y = \int_A x \cdot dA$$

The above two quantities convey a definite knowledge about the shape, size and orientation of the area which are useful in many analysis of mechanics. In case we concentrate the entire area A at a point known as centroid having position as $(\bar{x}, \bar{y})$, the new arrangement is equivalent to original distribution. Now in order to compute $\bar{x}$ and $\bar{y}$, it is simply to equate moments of the distributed area with that of the concentrated area about both axes as it is done below:

$$A \cdot \bar{x} = \int_A x\, dA$$

or

$$\bar{x} = \int_A \frac{x\, dA}{A}$$

Similarly,

$$A \cdot \bar{y} = \int_A y\, dA$$

or

$$\bar{y} = \frac{\int_A y\, dA}{A}$$

If the axes x and y have their origin at the centroid, then these axes are called *centroidal axes.* It is evident that the first moments about centroidal axes must be zero.

Example 12.1 A plane surface as shown in Figure 12.3 is bounded by the x-axis, curve $y^2 = 10x$ and a line parallel to the y-axis. What are the first moments of area about the x- and y-axes? Find also the centroidal coordinates.

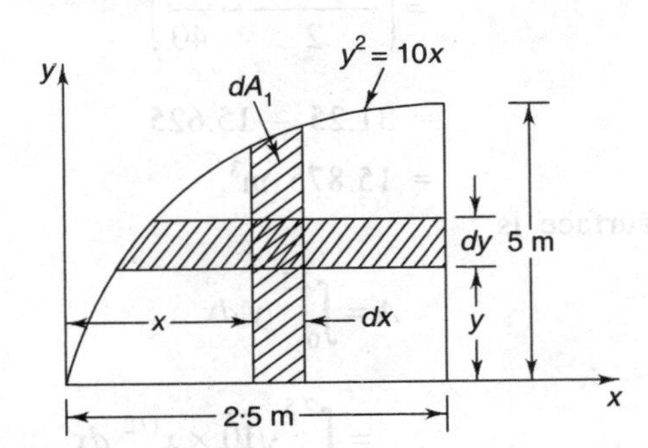

FIGURE 12.3 Example 12.1: Plane surface.

The first moment of area about the y-axis is

$$M_y = \int_A x\, dA_1 = \int_A x(y\, dx)$$

$$= \int_0^{2.5} x(\sqrt{10} \times \sqrt{x})\, dx$$

$$= \int_0^{2.5} \sqrt{10} \times x^{3/2} \times dx$$

$$= \frac{\sqrt{10}}{5/2} \times \left[x^{5/2} \right]_0^{2.5}$$

$$= \frac{\sqrt{10} \times 2}{5} \times 9.88 = 12.5 \text{ m}^3$$

Similarly,

$$M_x = \int_0^5 y(2.5 - x)\, dy$$

$$= \int_0^5 y\left(2.5 - \frac{y^2}{10}\right) dy$$

$$= \int_0^5 \left(2.5y - \frac{y^3}{10}\right) dy$$

$$= \left[2.5 \times \frac{y^2}{2} - \frac{y^4}{40}\right]_0^5$$

$$= \left(\frac{2.5 \times 25}{2} - \frac{5^4}{40}\right)$$

$$= 31.25 - 15.625$$

$$= 15.875 \text{ m}^3$$

Now the area A of the surface is

$$A = \int_0^{2.5} y\, dx$$

$$= \int_0^{2.5} \sqrt{10} \times x^{1/2}\, dx$$

$$= \sqrt{10} \times \frac{(2.5)^{3/2}}{3/2}$$

$$= \frac{2 \times 3.16 \times 3.95}{3}$$

$$= 8.32 \text{ m}^2$$

Hence the centroid coordinates are as follows:

$$\bar{x} = \frac{M_y}{A} = \frac{12.5}{8.32}$$

$$= 1.502 \text{ m}$$

$$\bar{y} = \frac{M_x}{A} = \frac{15.875}{8.32}$$

$$= 1.9 \text{ m}$$

PLANE AREA WITH AN AXIS OF SYMMETRY

If a plane area has its axis of symmetry about the y-axis, then $\bar{x} = 0$. Similarly, if the axis of symmetry is about the x-axis, then $\bar{y} = 0$.

Consider a plane area with an axis of symmetry as shown in Figure 12.4 and the y-axis is collinear with the axis of symmetry.

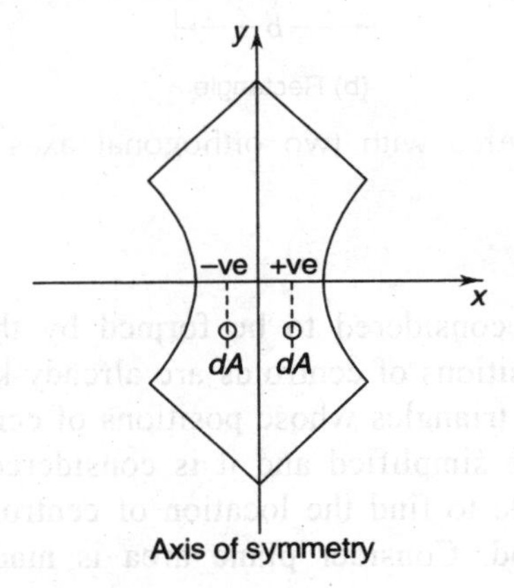

FIGURE 12.4 Area with one axis of symmetry.

Now $\bar{x}$ for the centroid of the body can be given as

$$\bar{x} = \frac{1}{A}\int_A x\, dA$$

In case we take elemental areas (dA) in symmetric pairs as shown in the figure, these elemental areas are mirror images of each other about the y-axis (axis of symmetry). The first moment of such a pair about the axis of symmetry has to be zero. The entire area of the plane can be considered to be composed of such pairs. Therefore, the integral $\int_A x\, dA$

has to be zero and it gives $\overline{x} = 0$. Another valuable deduction from this is about the location of the centroid. The centroid of an area with one axis of symmetry must lie somewhere along the axis of symmetry. The axis of symmetry coincides with the centroidal axis.

AREA WITH TWO ORTHOGONAL AXES OF SYMMETRY

Area with two orthogonal axes of symmetry must have its centroid at the intersection of these axes. The areas such as squares, rectangles and circles must have centroids at the intersections of their centroidal axes as shown in Figure 12.5.

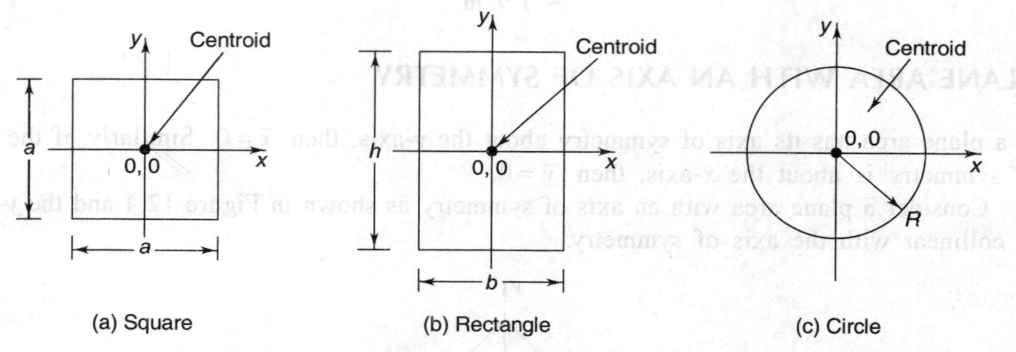

FIGURE 12.5 Area with two orthogonal axes of symmetry.

COMPOSITE AREAS

A complex plane area can be considered to be formed by the addition or subtraction of simple familiar areas whose positions of centroids are already known. The familiar areas are squares, rectangles, circles and triangles whose positions of centroids are commonly known. The complex plane area can be simplified and it is considered to be made of such simple and familiar areas. It is possible to find the location of centroid of the complex plane area by this composite areas method. Consider plane area is made of N familiar areas, then centroid of the plane area is

$$\overline{x} = \frac{\sum_{i=1}^{i=N} A_i \overline{x}_i}{A}$$

$$\overline{y} = \frac{\sum_{i=1}^{i=N} A_i \overline{y}_i}{A}$$

where A_i is area of ith familiar area $\overline{x}_i$ & $\overline{y}_i$ are the coordinates of the centroid of this area. A is the total area of the plane body.

Example 12.2 From a circular plate of diameter 100 mm, a circular part is cut out whose diameter is 50 mm (see Figure 12.6). Find the centroid of the remainder.

(UPTU: 2002–2003)

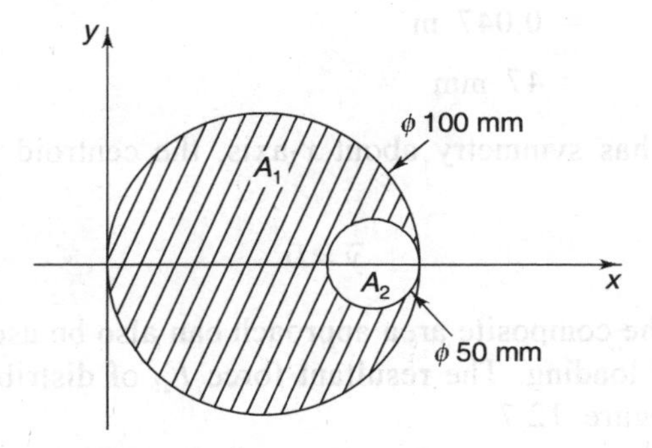

FIGURE 12.6 Example 12.2.

Guidance. It is to be solved with the composite area approach with $N = 2$. First moment of area for cut portion is to be taken as negative. The centroid is

$$\bar{x} = \frac{\sum_{i=1}^{i=N} A_i \bar{x}_i}{A}$$

$$= \frac{A_1 \bar{x}_1 - A_2 x_2}{A_1 - A_2}$$

$$A_1 = \frac{\pi}{4} \times (0.1)^2 \text{ m}^2$$

$$A_2 = \frac{\pi}{4} \times (0.05)^2 \text{ m}^2$$

$$A = A_1 - A_2 = \frac{\pi}{4}(0.1^2 - 0.05^2) \text{ m}^2$$

$$\bar{x}_1 = 0.05 \text{ m}$$

$$\bar{x}_2 = 0.1 - 0.25 = 0.075 \text{ m}$$

$$\therefore \quad \bar{x} = \frac{\frac{\pi}{4} \times 0.1^2 \times 0.05 - \frac{\pi}{4} \times 0.05^2 \times 0.075}{\frac{\pi}{4}(0.1^2 - 0.05^2)}$$

$$= \frac{5\times10^{-4} - 1.875\times10^{-4}}{(100-25)\times10^{-4}}$$

$$= 0.047 \text{ m}$$

$$= 47 \text{ mm}$$

Since circular plate has symmetry about x-axis, the centroid will lie on x-axis. Hence we have

$$\bar{y} = 0.$$

Distributed Loading: The composite area approach can also be used for finding the simplest resultant of a distributed loading. The resultant force F_R of distributed loading $w(x)$ can be depicted as shown in Figure 12.7.

$$F_R = \int_0^l w(x)\, dx$$

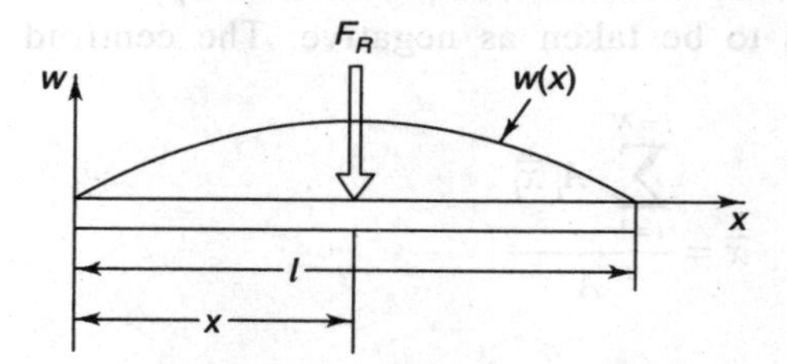

FIGURE 12.7 Composite area.

The position of the resultant for the distributed loading can be found out by equating

$$F_R \times \bar{x} = \int_0^l x\, w(x)\, dx$$

or

$$\bar{x} = \frac{\int_0^l x\, w(x)\, dx}{F_R}$$

where $\bar{x}$ is the centroid of the area under the loading. The resultant force of a distributed loading always acts at the centroid of the area under the loading curve. This helps in finding the position of the resultant of distributed loading such as UDL and UVL on beams. The resultant of a triangular loading as shown in Figure 12.8 will be at $\frac{2}{3}l$ from end A which has been worked out using the composite area approach.

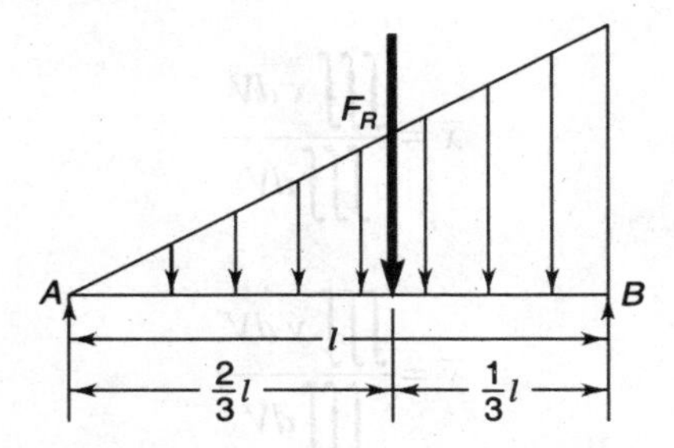

FIGURE 12.8 Triangular loading.

MOMENT OF VOLUME

Similar to a plane area, the concept of moments and centroids can also be used for three dimensional bodies. Consider a body with volume V (see Figure 12.9). The first moment of the body about a point O can be given as follows:

$$\text{Moment vector of volume} \equiv \iiint_V r \cdot dV$$

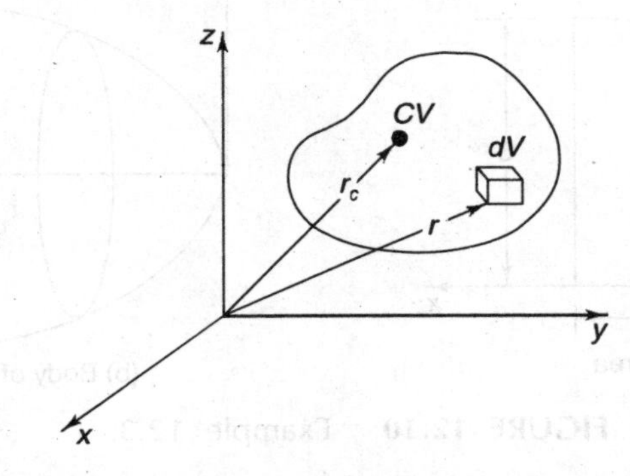

FIGURE 12.9 Body with centre of volume (CV).

The centre of volume (r_c) can be equated as

$$Vr_c = \iiint_V r\,dV$$

$$\therefore \qquad r_c = \frac{1}{V}\iiint_V r\,dV$$

The centre of volume (r_c) can be defined as a point at which the entire volume of the body can be considered to be concentrated for the purpose of computing the first moment of the volume of the body. The centroidal distances of volume in $\overline{x}, \overline{y}$ and $\overline{z}$ can be given as

$$\bar{x} = \frac{\iiint x\,dV}{\iiint dV}$$

$$\bar{y} = \frac{\iiint y\,dV}{\iiint dV}$$

$$\bar{z} = \frac{\iiint z\,dV}{\iiint dV}$$

Example 12.3 A volume of revolution is formed by revolving the area as shown in Figure 12.10 about the x-axis. The volume generated is also shown. Determine the centroidal distance $\bar{x}$.

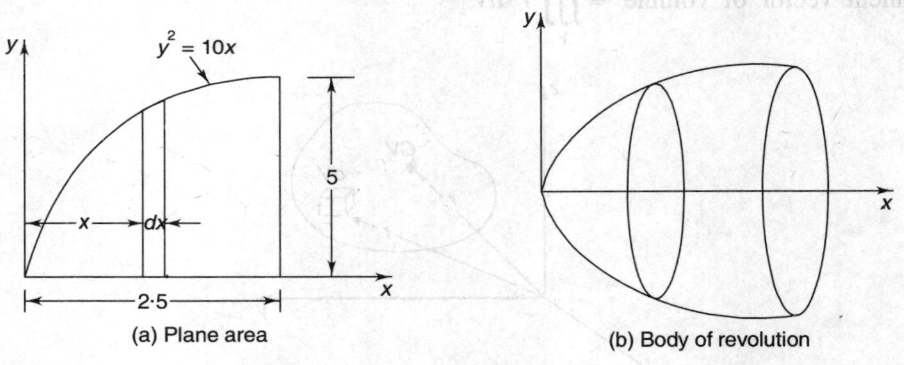

FIGURE 12.10 Example 12.3.

The centroid can be given as

$$\bar{x} = \frac{\iiint x\,dV}{\iiint dV}$$

$$V = \iiint dV = \iiint (\pi r^2)\,dx = \int_0^{2.5} \pi\,(10x)\,dx \qquad \text{as } r^2 = y^2 = 10x$$

or $$V = 10 \times \pi \times \int_0^{2.5} x\,dx$$

$$= 10\pi \left[\frac{x^2}{2} \right]_0^{2.5}$$

$$= \frac{10 \times \pi \times 2.5^2}{2} = 98.125 \text{ m}^3$$

Now

$$\bar{x} = \frac{1}{V} \int_0^{2.5} x \times (\pi r^2)\, dx$$

$$= \frac{1}{98.125} \times \pi \times \int_0^{2.5} 10x^2\, dx \quad \text{as } r^2 = y^2 = 10x$$

$$= \frac{10 \times \pi}{98.125} \times \left(\frac{x^3}{3} \right)_0^{2.5}$$

$$= \frac{10 \times \pi \times 2.5^3}{98.125 \times 3}$$

$$= 0.667 \text{ m}$$

COMPOSITE VOLUMES

Any complex volume can be considered to be composed of simple and familiar volumic shapes whose centres of volume are known. Such volumes can be called composite volumes. In order to determine the centroid of a complex volume, the known centroids of the composite parts can be used. The centroid of a composite body having volume V can be given as

$$\bar{x} = \frac{\Sigma \bar{x}_i \times V_i}{V}$$

$$\bar{y} = \frac{\Sigma \bar{y}_i \times V_i}{V}$$

$$\bar{z} = \frac{\Sigma z_i V_i}{V}$$

Example 12.4 Find the centroid $\bar{x}$ for a volume of a body of revolution as shown in Figure 12.11. The left end has a cone cut-out while at the right end contains a hemispherical region.

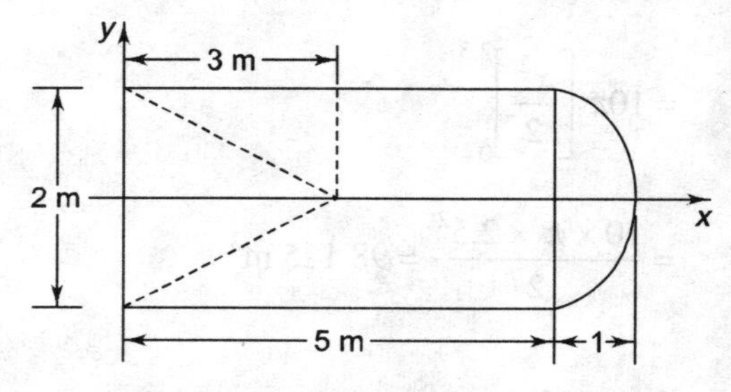

FIGURE 12.11 Example 12.4.

The above composite body consists of three familiar body volumes, viz. the cone, the cylinder and the hemisphere. The following table lists the volumes of item.

S.No.	*Familiar body*	*Volume* (V_i)	*Centroid* $\bar{x}_i$	$V_i \bar{x}_i$
1	Cone	$-\frac{1}{3}\times\pi\times 1^2\times 3 = -3.14$	$\frac{3}{4}$	– 2.25
2	Cylinder	$\pi \times 1^2 \times 5 = 15.7$	$\frac{5}{2}$	39.25
3	Hemisphere	$\frac{1}{2}\times\frac{4}{3}\pi\times 1^3 = 2.09$	$5+\frac{3}{8}\times 1$	11.23
		$V = \Sigma V_i = 14.65$		$\Sigma V_i \bar{x}_i = 50.23$

Now

$$\bar{x} = \frac{\Sigma V_i \times \bar{x}_i}{V}$$

$$= \frac{50.23}{14.65}$$

$$= 3.428 \text{ m}$$

CENTRE OF MASS

The centre of volume is given by

$$r_c = \frac{1}{V}\iiint_V r\,dV$$

Now if we replace dV by dm where $dm = \rho \times dV$ and ρ = mass density, then the centre of mass is given by

$$r_c = \frac{1}{M} \iiint_V \rho\, r\, dV$$

or

$$\bar{x} = \frac{\iiint x\,\rho\, dV}{\iiint \rho\, dV}, \bar{y} = \frac{\iiint y\,\rho\, dV}{\iiint \rho\, dV}, \bar{z} = \frac{\iiint z\,\rho\, dV}{\iiint \rho\, dV}$$

The centre of mass can also be expressed in case we consider the mass consisting of n particles as shown in Figure 12.12.

$$r_c = \frac{\sum_{i=1}^{i=N} m_i \times r_i}{M}$$

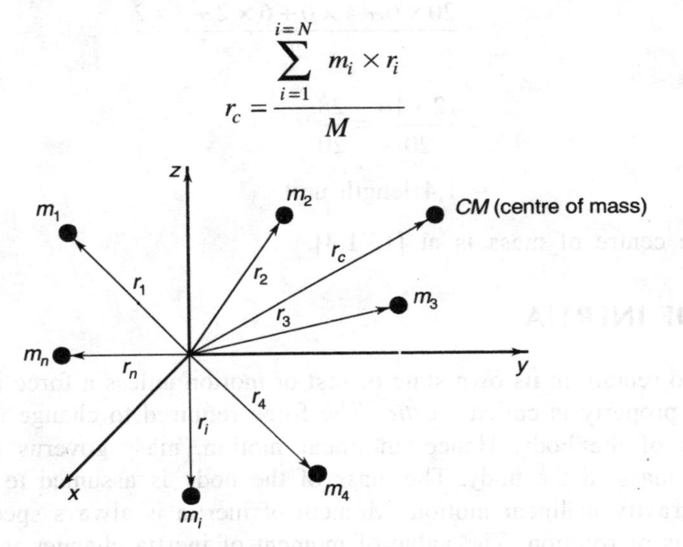

FIGURE 12.12 Mass consisting of n particles.

Example 12.5 Consider four particles A, B, C and D having masses 2, 4, 6 and 8 mass unit respectively are placed at the corner of a square of side 2 length unit as shown in Figure 12.13. Find the centre of mass.

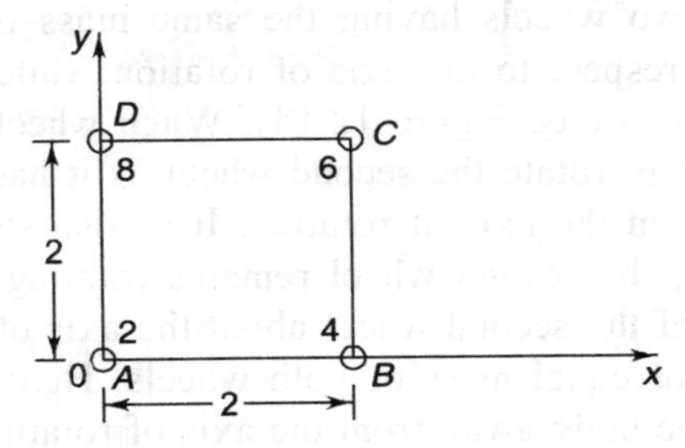

FIGURE 12.13 Example 12.5.

$$\bar{x} = \frac{\Sigma m_i \times x_i}{M}$$

$$= \frac{2 \times 0 + 4 \times 2 + 6 \times 2 + 8 \times 0}{2 + 4 + 6 + 8}$$

$$= \frac{8 + 12}{20} = \frac{20}{20} = 1 \text{ length unit}$$

$$\bar{y} = \frac{\Sigma m_i \times y_i}{M}$$

$$= \frac{20 \times 0 + 4 \times 0 + 6 \times 2 + 8 \times 2}{20}$$

$$= \frac{12 + 16}{20} = \frac{28}{20}$$

$$= 1.4 \text{ length unit}$$

Hence, the centre of mass is at [1, 1.4].

MOMENT OF INERTIA

A body tends to remain in its own state of rest or motion unless a force is applied to change the state. This property is called *inertia*. The force required to change to the state depends upon the mass of the body. Hence, in linear motion, mass governs inertia and not the distribution of mass in the body. The mass of the body is assumed to be concentrated at the centre of gravity in linear motion. Moment of inertia is always specified in relation to a particular axis of rotation. The value of moment of inertia changes whenever the axis of rotation of the body is changed. Moment of inertia is specified for an area while mass moment of inertia is given for the mass of the body. Moment of inertia is also called the second moment of area or mass.

The moment of inertia of a body about an axis not only depends upon the mass of the body but also upon the distribution of the mass of the body about the axis of rotation. To understand this, consider two wheels having the same mass but the first wheel has mass uniformly distributed with respect to the axis of rotation while the second wheel has most of its mass situated at the rim (see Figure 12.14). When wheels are rotated, it is seen that a greater torque is required to rotate the second wheel as it has a greater inertia due to the mass located more away from the axis of rotation. It is also seen that once the two wheels are set in rotation and left, the second wheel remains rotating for a longer time. It means that the moment of inertia of the second wheel about the axis of rotation is greater than that of the first wheel in spite of equal mass of both wheels. Hence it is clear that the greater is the part of the mass of the body away from the axis of rotation, the greater is the moment

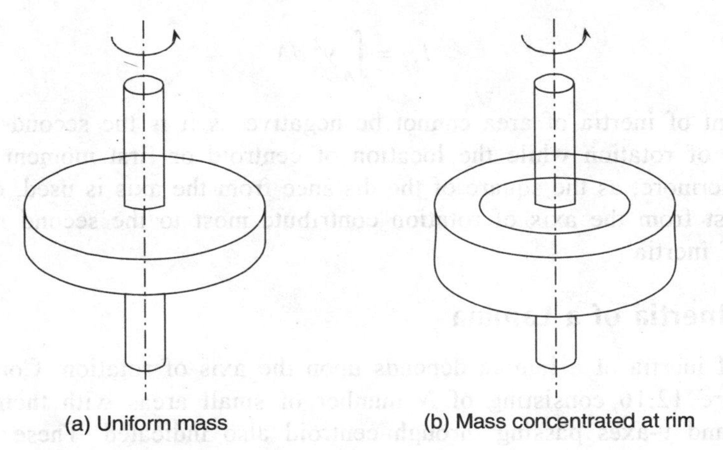

FIGURE 12.14 Two wheels with the same mass.

of inertia of the body about that axis. An important use of this property is made in stationary engines. The torque rotating the shaft of the engine changes periodically so that the shaft cannot rotate uniformly. In order to make the rotation uniform, a large heavy wheel is connected to the shaft which is called flywheel having a large moment of inertia. When engine starts, shaft rotates with the flywheel. Due to its large moment of inertia, flywheel with the shaft continues to rotate almost uniformly in spite of fluctuating torque of the engine. Same concept is also used for normal bi-wheel cycle. The moment of inertia of the wheel of a bi-wheel cycle is increased by concentrating most of wheel mass at the rim of the wheel and connecting the rim to the axle of the wheel through slender spokes. It results into a large moment of inertia of the wheel, resulting the wheel of the cycle continues rotating at same speed for sometime even when the cyclist is not pedalling.

Moment of inertia is the second moment of area. The second moments of the area A about the x- and y-axes as shown in Figure 12.15 are denoted as I_{xx} and I_{yy} respectively and they are defined as

$$I_{xx} = \int_A y^2 \, dA$$

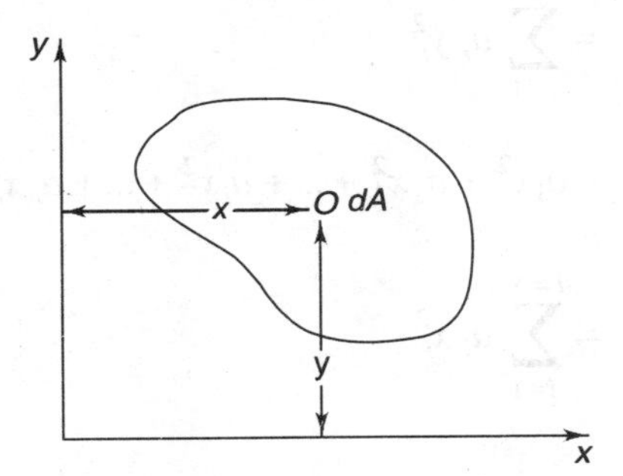

FIGURE 12.15 Plane surface.

$$I_{yy} = \int_A y^2\, dA$$

The moment of inertia of area cannot be negative as it is the second moment of area about the axis of rotation while the location of centroid or first moment of area can be negative. Furthermore, as the square of the distance from the axis is used, elements of area that are farthest from the axis of rotation contribute most to the second moment of area, i.e. moment of inertia.

Moment of Inertia of a Lamina

The moment of inertia of a lamina depends upon the axis of rotation. Consider lamina as shown in Figure 12.16 consisting of N number of small areas with their locations with respect to x- and y-axes passing through centroid also indicated. These axes are called *centroidal axes*.

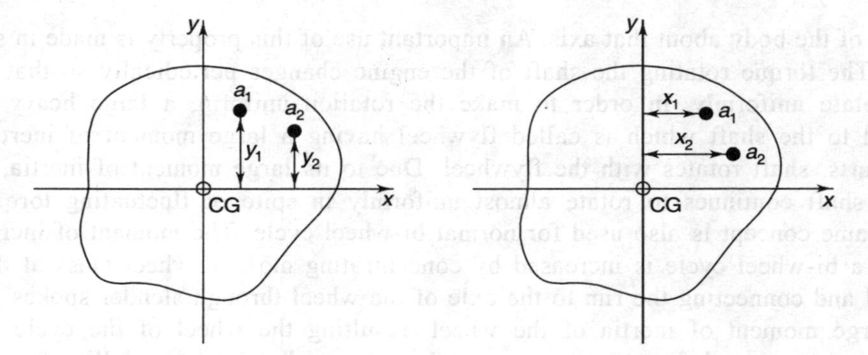

FIGURE 12.16 Moment of inertia of lamina.

Now moment of inertia about x-x and y-y-axes are follows:

$$I_{xx} = a_1 y_1^2 + a_2 y_2^2 + \ldots + a_i y_i^2 + \ldots + a_n y_n^2$$

$$= \sum_{i=1}^{i=N} a_i\, y_i^2$$

$$I_{yy} = a_1 x_1^2 + a_2 x_2^2 + \ldots + a_i x_i^2 + \ldots + a_n x_n^2$$

$$= \sum_{i=1}^{i=N} a_i\, x_i^2$$

Parallel Axis Theorem

In order to find out moment of inertia about any axis other than the centroidal axes, the parallel axis theorem is applied. Consider a lamina with *x-x* and *y-y*-axes passing through centroid as shown in Figure 12.17. We want to find out the moments of inertia about *AA* and *BB* axes which are at distances *h* and *j* from *y-y* and *x-x*-axes respectively.

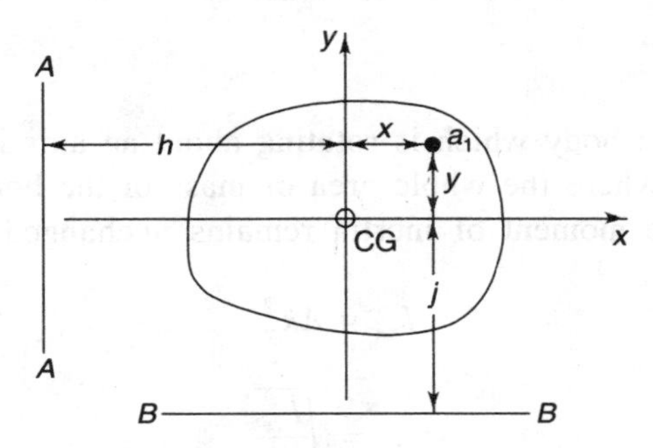

FIGURE 12.17 Parallel axis theorem.

It is possible to find out the value of I_{AA} and I_{BB} of the lamina in terms of I_{xx} and I_{yy}. As per the definition of moment of inertia, we have

$$I_{AA} = \Sigma a_i(x + h)^2$$
$$= \Sigma a_i(x^2 + 2hx + h^2)$$
$$= \Sigma a_i x^2 + 2h\Sigma a_i x + h^2\Sigma a_i$$
$$= I_{yy} + 0 + Ah^2$$
$$= I_{yy} + Ah^2$$

Similarly, we have

$$I_{BB} = \Sigma a_i(y + j)^2$$
$$= \Sigma a_i(y^2 + 2yj + j^2)$$
$$= \Sigma a_i y^2 + 2j\Sigma a_i y + j^2\Sigma a_i$$
$$= I_{xx} + 0 + j^2 A$$
$$= I_{xx} + Aj^2$$

Theorem of the Perpendicular Axis

The moment of inertia about the *z-z*-axis, which is called polar axis, is given by

$$I_{zz} = \Sigma az^2$$

where the polar axis is perpendicular to *x-y* plane.

But

$$z^2 = x^2 + y^2$$

$$\therefore \quad I_{zz} = \Sigma a_i(x^2 + y^2)$$

$$= \Sigma a_i x^2 + \Sigma a_i y^2$$

$$= I_{yy} + I_{xx}$$

Radius of Gyration

The radius of gyration of a body which is rotating about an axis is defined as the distance from the axis of rotation where the whole area or mass of the body can be assumed to be concentrated such that the moment of inertia remains unchanged. Hence we have

$$I_{xx} = Ak_y^2$$

or

$$k_y = \sqrt{\frac{I_{xx}}{A}}$$

where I_{xx} is moment of inertia and k_y is the radius of gyration.

Similarly, we have

$$k_x = \sqrt{\frac{I_{yy}}{A}}$$

The distances k_x and k_y are called the radii of gyration. This radius of gyration will have a position that depends not only on the shape of the area but also on the position of the reference axis or the axis of rotation. This situation is unlike the centroid whose location is independent of the position of the reference axis.

Product of Area

The product of area (Figure 12.18) relates an area directly to a set of axes. It is defined as

$$I_{xy} = \int_A x\,y\,dA$$

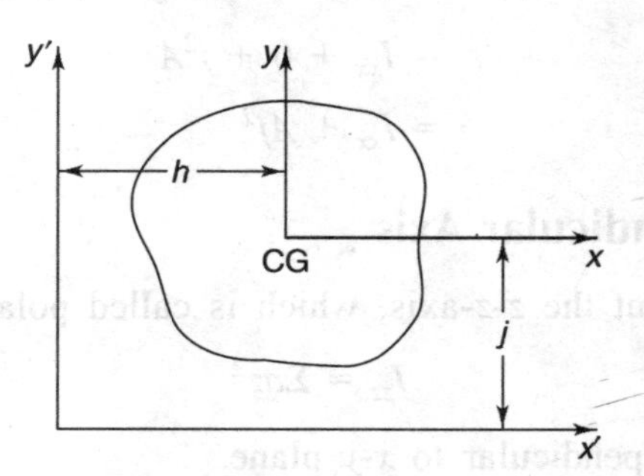

FIGURE 12.18 Product of area.

The product of area can have a negative value unlike the second moment of area or moment of inertia. Secondly, the product area must be zero if any of the axis out of the set of orthogonal axis is also the axis of symmetry of the body.

Let us now see what happens to the product of area when axes of reference are shifted by j and h distances from centroidal axes. By applying definition of product of area with respect to new axes, we have

$$I_{x'y'} = \int_A x' \cdot y' \, dA$$

$$= \int_A (x+h) \cdot (y+j) \, dA$$

$$= \int_A x\,y\,dA + j\int_A x\,dA + h\int_A y\,dA + jh\int_A dA$$

$$= I_{xy} + j \times 0 + h \times 0 + jhA$$

$$= I_{xy} + jhA$$

Moment of Inertia of a Rectangular Section

Consider a rectangular lamina of height h and width b as shown in Figure 12.19. We are going to find moment of inertia about its centroidal axis x and y. Take a strip of thickness dy at a distance y from the x-axis as shown in the figure.

$$\text{Area of the strip} = b \times dy$$

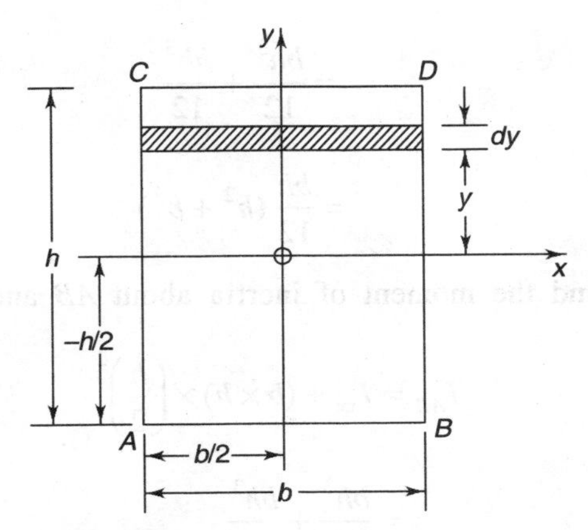

FIGURE 12.19 Rectangular section.

Moment of inertia of the strip with respect to the x-x axis is

$$dI_{xx} = y^2 \times (b \times dy)$$

Total moment of inertia of the lamina can be found out by integrating the above from $-\frac{h}{2}$ to $\frac{h}{2}$.

$$I_{xx} = \int_{-h/2}^{h/2} b \times y^2 \, dy$$

$$= 2 \times b \int_{0}^{h/2} y^2 \, dy$$

$$= 2b\left[\frac{y^3}{3}\right]_0^{h/2}$$

$$= \frac{bh^3}{12}$$

Similarly, we can find out moment of inertia about the *y*-axis which is

$$I_{yy} = \frac{hb^3}{12}$$

Now we can apply the perpendicular axis theorem to find the moment of inertia about the *z*-axis which is

$$I_{zz} = I_{xx} + I_{yy}$$

$$= \frac{bh^3}{12} + \frac{hb^3}{12}$$

$$= \frac{bh}{12}(h^2 + b^2)$$

In case we like to find the moment of inertia about *AB* and *AC* axes, then we have

$$I_{AB} = I_{xx} + (b \times h) \times \left(\frac{h}{2}\right)^2$$

$$= \frac{bh^3}{12} + \frac{bh^3}{4}$$

$$= \frac{bh^3}{3}$$

It is to be understood that $I_{AB} > I_{xx}$ as more area is situated at a greater distance from the *AB*-axis as compared to the *x-x* axis.

Similarly, moment of inertia about the AC-axis is

$$I_{AC} = I_{yy} + (b \times h)\left(\frac{b}{2}\right)^2$$

$$= \frac{hb^3}{12} + \frac{hb^3}{4}$$

$$= \frac{hb^3}{4}$$

The radius of gyration about the x-x axis is

$$k_{xx} = \sqrt{\frac{I_{xx}}{A}}$$

$$= \sqrt{\frac{bh^3}{hb \times 12}}$$

$$= \frac{h}{2\sqrt{3}}$$

Similarly, the radius of gyration about the y-y axis is

$$k_{yy} = \frac{h}{2\sqrt{3}}$$

The product of area about the x and y-axes has to be zero due to symmetry. Hence $I_{xy} = 0$.

Moment of Inertia of a Circular Section

Consider a circular lamina of diameter D and centroidal axes x and y are passing through its centre O. Take a round strip of thickness dr at a distance r from the centre O as shown in Figure 12.20.

$$\text{Area of the strip} = 2\pi r \cdot dr$$

Polar moment of area of the strip $= r^2 \times (2\pi r \cdot dr)$

Now we integrate the above from 0 to R to find out the moment of inertia of the complete lamina. We have

$$I_{zz} = \int_0^R r^2 \times (2\pi r \,.\, dr)$$

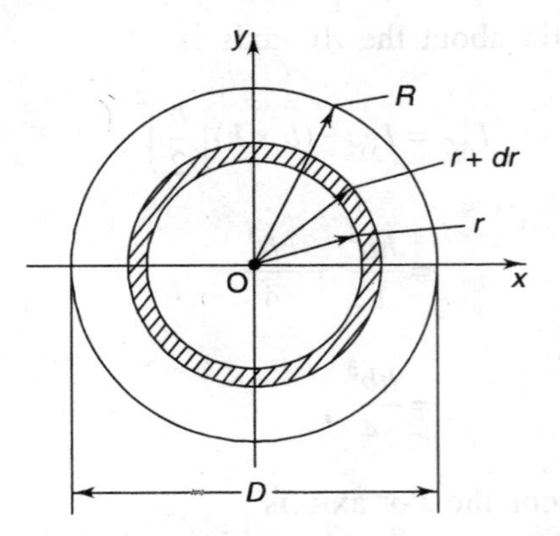

FIGURE 12.20 Circular section.

$$= \int_0^R 2\pi r^3 \, . dr$$

$$= 2\pi \left[\frac{r^4}{4} \right]_0^R$$

$$= \frac{\pi R^4}{2}$$

$$= \frac{\pi D^4}{32}$$

In order to find moment of inertia about the x- and y-axes, we have

$$I_{zz} = I_{xx} + I_{yy}$$

But $I_{xx} = I_{yy}$ as per symmetry. Therefore,

$$I_{xx} = I_{yy} = \frac{I_{zz}}{2} = \frac{\pi D^4}{2 \times 32}$$

$$= \frac{\pi D^4}{64}$$

Moment of Inertia of a Hollow Rectangular Section

Consider a hollow rectangular lamina with h_1 and h_2 as outer and inner heights while b_1 and b_2 as outer and inner widths. The centroidal axes are as shown in Figure 12.21.

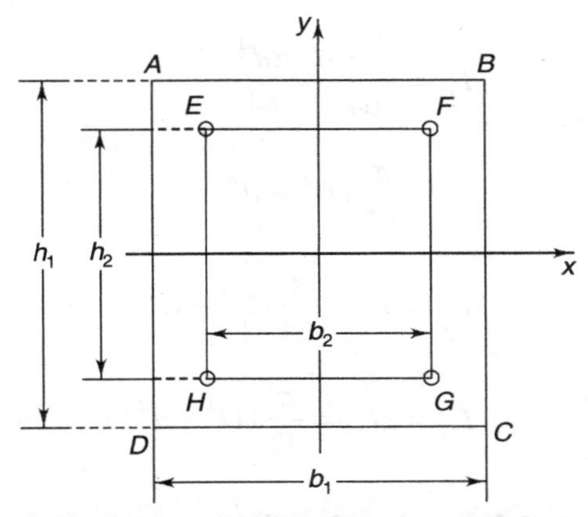

FIGURE 12.21 Hollow rectangular section.

$$I_{xx} = I_{ABCD} - I_{EFGH}$$

$$= \frac{b_1 \times h_1^3}{12} - \frac{b_2 \times h_2^3}{12}$$

Also

$$I_{yy} = \frac{h_1 \times b_1^3}{12} - \frac{h_2 \times b_2^3}{12}$$

Similarly,

$$I_{zz} = I_{xx} + I_{yy}$$

$$= \left(\frac{b_1 h_1^3 + h_1 b_1^3}{12}\right) - \left(\frac{b_2 h_2^3 + h_2 b_2^3}{12}\right)$$

Moment of Inertia of a Hollow Circular Section

Consider a hollow circular lamina with the outside diameter D and the inside diameter d. The centroidal axes are as shown in Figure 12.22.

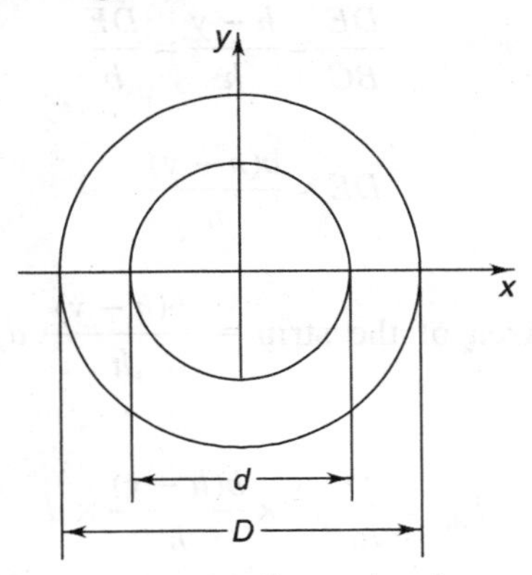

FIGURE 12.22 Hollow circular section.

$$I_{xx} = \frac{\pi D^4}{64} - \frac{\pi d^4}{64}$$

$$= \frac{\pi}{64}(D^4 - d^4)$$

$$I_{yy} = I_{xx} = \frac{\pi}{64}(D^4 - d^4)$$

$$I_{zz} = 2I_{xx} = \frac{\pi}{32}(D^4 - d^4)$$

Moment of Inertia of a Triangular Section

Consider a triangular lamina *ABC* with the reference axis *BC* as shown in Figure 12.23. Take a strip *DE* of thickness *dy* as shown in the figure.

$$\text{Area of the strip} = DE \cdot dy$$

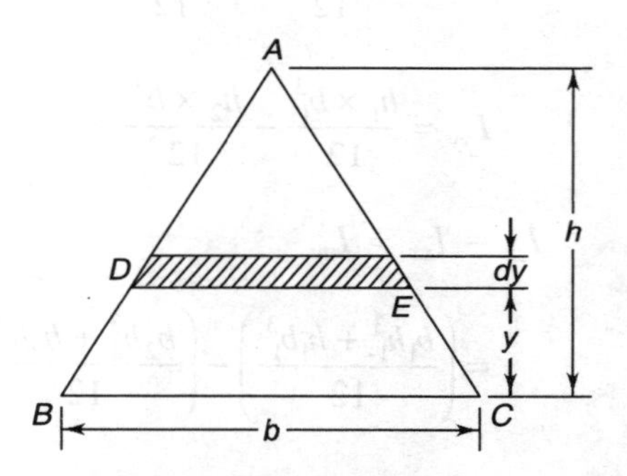

FIGURE 12.23 Triangular section.

Now

$$\frac{DE}{BC} = \frac{h-y}{h} = \frac{DE}{b}$$

∴
$$DE = \frac{b(h-y)}{h}$$

∴
$$\text{Area of the strip} = \frac{b(h-y)}{h}.dy$$

Now

$$I_{BC} = \int_0^h y^2 \times \frac{b(h-y)}{h} \times dy$$

$$= b\int_0^h y^2\,dy - \frac{b}{h}\int_0^h y^3\,dy$$

$$= b\left[\frac{y^3}{3}\right]_0^h - \frac{b}{h}\left[\frac{y^4}{4}\right]_0^h$$

$$= b \times \frac{h^3}{3} - \frac{bh^3}{4}$$

$$= bh^3\left[\frac{4-3}{12}\right]$$

$$= \frac{bh^3}{12}$$

In case the axis of moment of inertia is passing through the centroid of the triangular section, which is also parallel to base BC, then the centroid of the triangular section is at height of $\frac{h}{3}$ from the base as shown in Figure 12.24.

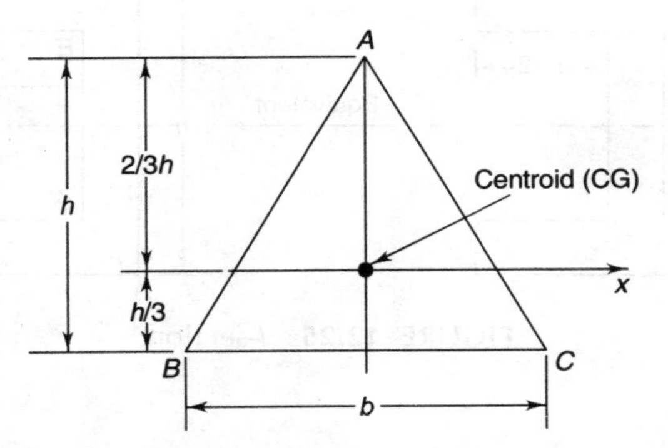

FIGURE 12.24 Triangular section and centroid axis.

$$I_{BC} = I_{CG} + \text{Area} \times \left(\frac{h}{3}\right)^2$$

or
$$I_{CG} = \frac{bh^3}{12} - \left(\frac{b \times h}{2}\right)\left(\frac{h}{3}\right)^2$$

$$= \frac{bh^3}{12} - \frac{bh^3}{18}$$

$$= bh^3\left[\frac{3-2}{36}\right]$$

$$= \frac{bh^3}{36}$$

Moment of Inertia of *I*-Section

The moment of inertia of an *I*-section is to be found about the centroidal axis. Consider an *I*-section as shown in Figure 12.25. In order to simplify, the mid vertical portion can be shifted to left to obtain an equivalent C-section as shown in the figure. Consider now C-section which consists of outer rectangle *ABCD* with missing inner rectangle *EFGH*. Therefore, we have

$$I_{xx} = I_{ABCD} - I_{EFGH}$$

$$= \frac{bh^3}{12} - \frac{b_1h_1^3}{12}$$

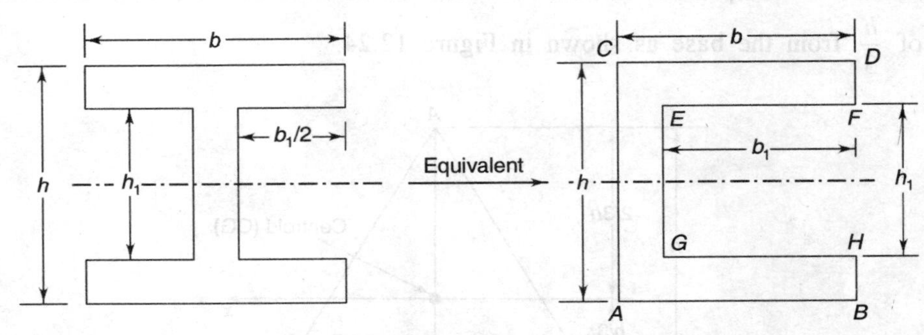

FIGURE 12.25 *I*-section.

CENTRE OF MASS

It has already been explained that the centre of mass is the location at which the mass of the body can be assumed to be concentrated. If we consider the body to have continuous distribution of matter, the summation in the formula about the centre of mass can be replaced by integration. Hence the centre of mass is

$$\bar{x} = \frac{1}{M}\int x\, dm$$

$$\bar{y} = \frac{1}{M}\int y\, dm$$

$$\bar{z} = \frac{1}{M}\int z\, dm$$

where M = mass of the body and dm is the mass of small element of the body.

Centre of Mass of a Uniform Straight Rod

Consider a uniform straight rod having mass M and length L (Figure 12.26).

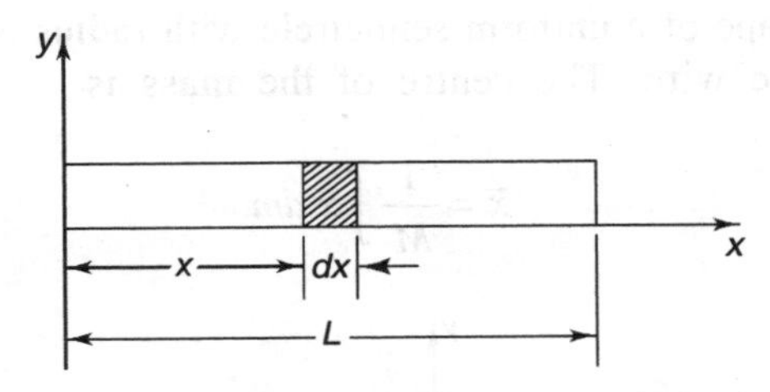

FIGURE 12.26 Uniform straight rod.

Consider an element dx at a distance x from the left end. The coordinate of the centre of mass is

$$\bar{x} = \frac{1}{M}\int_0^L x\, dm$$

But

$$dm = \frac{M}{L} \times dx$$

$\therefore$

$$\bar{x} = \frac{1}{M}\int_0^L x \cdot \frac{M}{L} \times dx$$

$$= \frac{1}{L}\int_0^L x\, dx$$

$$= \frac{1}{L}\left[\frac{x^2}{2}\right]_0^L$$

$$= \frac{L^2}{2L} = \frac{L}{2}$$

$$\bar{y} = \frac{1}{M}\int y\, dm = 0$$

$$\overline{z} = \frac{1}{M}\int z\, dm = 0$$

Hence the centre of mass is at $\left(\frac{L}{2}, 0, 0\right)$.

Centre of Mass of a Uniform Semicircle Wire

Consider a wire having shape of a uniform semicircle with radius R as shown in Figure 12.27. Let M be the mass of the wire. The centre of the mass is

$$\overline{x} = \frac{1}{M}\int x\, dm$$

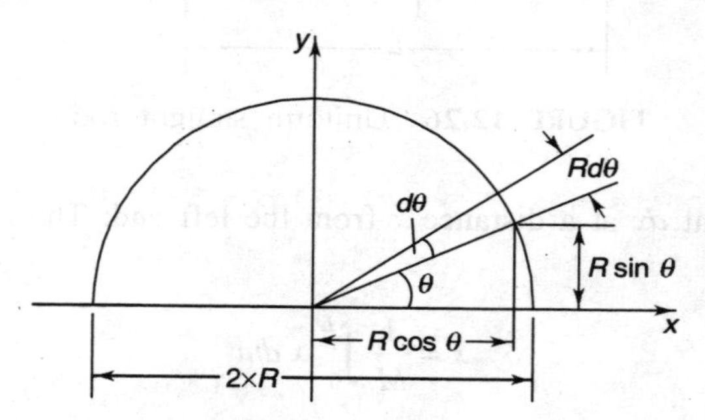

FIGURE 12.27 Uniform semicircle wire.

But

$$dm = \frac{M}{\pi R} \times (R\, d\theta)$$

and

$$x = R \cos \theta$$

$\therefore$

$$\overline{x} = \frac{1}{M}\int_0^{\pi} (R\cos\theta) \times \frac{M}{\pi R} \times (R \times d\theta)$$

$$= \frac{R}{\pi}\int_0^{\pi} \cos\theta\, d\theta$$

$$= \frac{R}{\pi}[\sin\theta]_0^{\pi}$$

$$= 0$$

$$\bar{y} = \frac{1}{M}\int_0^{\pi} y\, dm$$

$$= \frac{1}{M}\int_0^{\pi} (R\sin\theta)\left(\frac{M}{\pi}\right) d\theta$$

$$= \frac{R}{\pi}\int_0^{\pi} \sin\theta\, d\theta$$

$$= \frac{R}{\pi}[\cos\theta]_{\pi}^{0}$$

$$= \frac{2R}{\pi}$$

The centre of mass is at $\left[0, \dfrac{2R}{\pi}\right]$.

Centre of Mass of a Uniform Semicircular Plate

Consider a uniform semicircular plate of radius R and mass M (Figure 12.28). Take a strip at radius r and thickness dr.

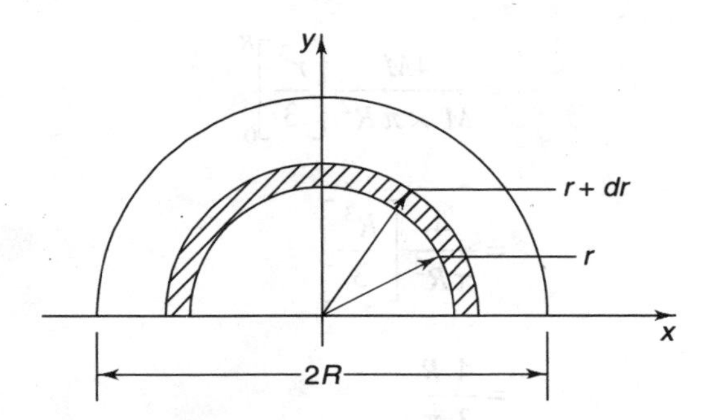

FIGURE 12.28 Semicircular plate.

The centre of mass is

$$\bar{x} = \frac{1}{M}\int x\, dm$$

$$\bar{y} = \frac{1}{M}\int y\, dm$$

The area of semicircular plate is $\frac{\pi R^2}{2}$. As the plate is uniform, the mass per unit area is $\frac{M}{\pi R^2/2}$. The mass of the strip is

$$dm = \frac{M}{\pi R^2/2} \times [\pi r\, dr]$$

$$= \frac{2M \cdot r \cdot dr}{R^2}$$

The strip chosen is equivalent to the uniform semicircular wire and we have found out that the mass can be considered to be located as a point of mass at $\left(0, \frac{2r}{\pi}\right)$. Hence, the centre of mass of semicircular plate is

$$\therefore \quad \bar{x} = \frac{1}{M}\int_0^R 0 \times \frac{2M \cdot r \cdot dr}{R^2}$$

$$= 0$$

$$\bar{y} = \frac{1}{M}\int_0^R \frac{2r}{\pi} \times \frac{2M \cdot r \cdot dr}{R^2}$$

$$= \frac{4M}{M \times \pi R^2}\left[\frac{r^3}{3}\right]_0^R$$

$$= \frac{4}{\pi R^2}\left[\frac{R^3}{3}\right]$$

$$= \frac{4}{3}\frac{R}{\pi}$$

MASS MOMENT OF INERTIA

In case the body is assumed to be continuous, it is possible to use the technique of integration in obtaining its moment of inertia about an axis. Let a small element has mass dm and its perpendicular distance from a reference axis is r, then the mass moment of inertia is

$$I = \int r^2\, dm$$

Mass Moment of Inertia of Uniform Rod

Consider a uniform rod of mass M and length L as shown in Figure 12.29. The moment of inertia is to be found out about a vertical axis passing its centre, taken as the y-axis. Consider an element of the rod at x distance from the centre with thickness dx as shown in the figure. The mass of rod per unit length is $\frac{M}{L}$. The moment of inertia is

$$I = \int r^2 \, dm$$

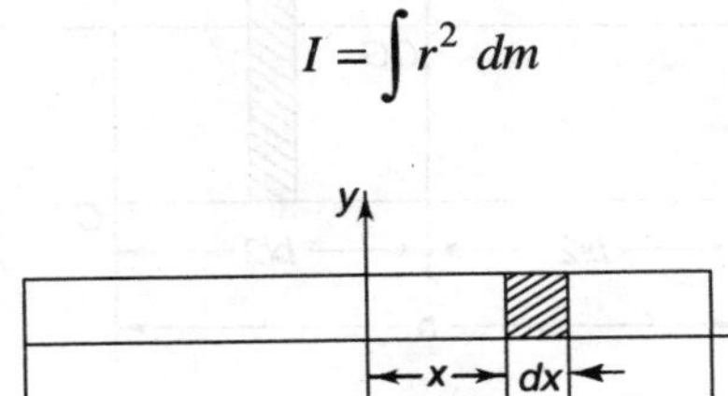

FIGURE 12.29 Uniform rod.

$\therefore$
$$I_{yy} = \int_0^{L/2} x^2 \, dm$$

$$dm = \frac{M}{L} \times dx$$

$\therefore$
$$I_{yy} = \int_{-L/2}^{L/2} x^2 \times \frac{M}{L} \times dx$$

$$= \frac{M}{L} \int_{-L/2}^{L/2} x^2 \, dx$$

$$= \frac{M}{L} \left[\frac{x^3}{3} \right]_{-L/2}^{L/2}$$

$$= \frac{M}{L} \times \frac{L^3}{12}$$

$$= \frac{ML^2}{12}$$

Mass Moment of Inertia of a Rectangular Plate

Consider a rectangular plate of height h and width b having mass M as shown in Figure 12.30.

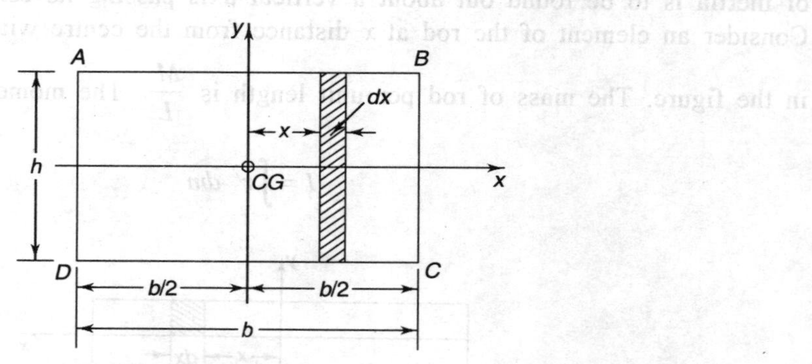

FIGURE 12.30 Rectangular plate.

Take a strip dx at distance x from the centre. The mass moment of inertia about the y-axis is

$$I_{yy} = \int x^2 \, dm$$

$$\text{Mass per unit area} = \frac{M}{b \times h}$$

and

$$\text{Area of strip} = h \times dx$$

Now

$$dm = \frac{M}{b \cdot h} \times (h \times dx)$$

$\therefore$

$$I_{yy} = \int_{-b/2}^{b/2} x^2 \times \frac{M}{h} \times dx$$

$$= \frac{M}{b} \left[\frac{x^3}{3} \right]_{-b/2}^{b/2}$$

$$= \frac{M}{b} \times 2 \times \frac{1}{3} \times \left[\left(\frac{b}{2} \right)^3 \right]$$

$$= \frac{Mb^2}{12}$$

In case the moment of inertia has to be worked out about axis BC instead of mid-point (y-axis), then we have

$$I_{BC} = I_{yy} + M \times \left(\frac{b}{2}\right)^2 \quad \text{(as per parallel theorem)}$$

$$\therefore \quad I_{BC} = \frac{Mb^2}{12} + M \times \left(\frac{b}{2}\right)^2$$

$$= \frac{Mb^2}{12} + \frac{Mb^2}{4}$$

$$= \frac{Mb^2}{12}(1 + 3)$$

$$= \frac{Mb^2}{3}$$

The mass moment of inertia about an centroidal axis passing parallel to the base of the rectangular body is

$$I_{xx} = \frac{Mh^2}{12}$$

Mass Moment of Inertia of a Uniform Circular Ring

Consider a circular ring of radius R and mass M (Figure 12.31). The moment of inertia is

$$I = \int r^2 \, dM$$

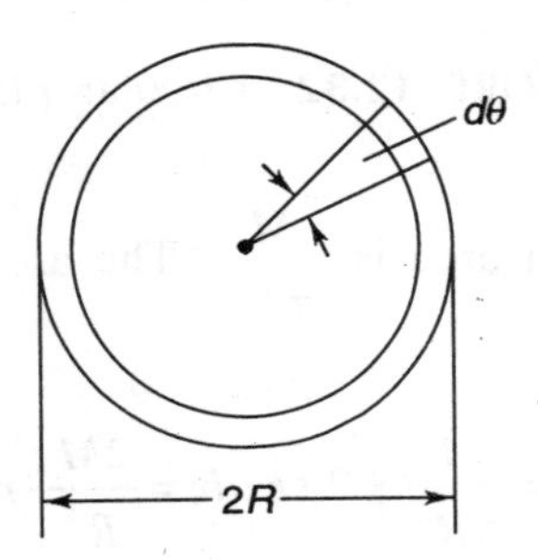

FIGURE 12.31 Circular ring.

Now

$$dm = \frac{M}{2\pi R} \times R d\theta$$

$\therefore$

$$I = \int_0^{2\pi} \frac{M}{2\pi R} \times R d\theta \times R^2$$

$$= \frac{M}{2\pi} \times R^2 \int_0^{2\pi} d\theta$$

$$= \frac{MR^2}{2\pi} \times [(\theta)]_0^{2\pi}$$

$$= MR^2$$

Mass Moment of Inertia of a Uniform Circular Plate

Consider a plate with mass M and radius R. Select a circular strip at radius r and thickness dr as shown in Figure 12.32.

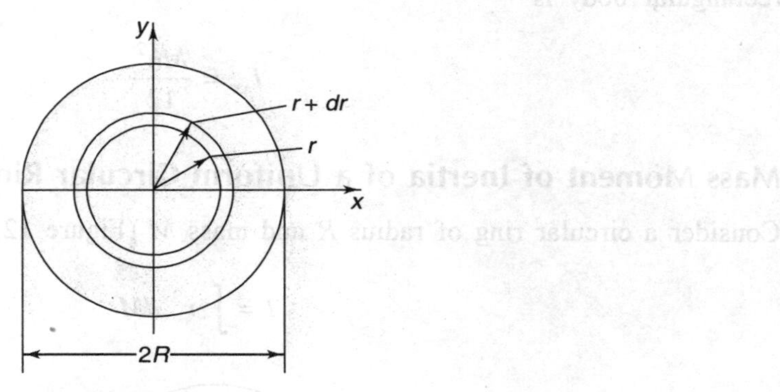

FIGURE 12.32 Circular plate.

The mass of the plate per unit area is $\frac{M}{\pi R^2}$. The area of the selected strip is $2\pi r \cdot dr$. Hence, the mass of the strip is

$$dM = \frac{M}{\pi R^2} \times 2\pi r \cdot dr = \frac{2M}{R^2} \cdot r \cdot dr$$

The moment of inertia of the plate is

$$I_{zz} = \int r^2 \, dm$$

$$= \int_0^R r^2 \frac{2M}{R^2} \cdot r \cdot dr$$

$$= \frac{2M}{R^2} \int_0^R r^3 \, dr$$

$$= \frac{2M}{R^2} \left[\frac{r^4}{4} \right]_0^R$$

$$= \frac{M}{2} . R^2$$

Mass Moment of Inertia of a Uniform Solid Cylinder

Consider a solid cylinder of mass M, radius R and height h. Consider reference axes passing its centre of mass as shown in Figure 12.33.

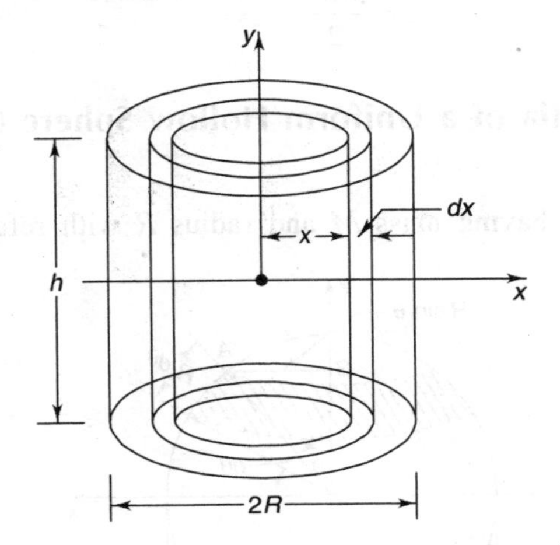

FIGURE 12.33 Solid cylinder.

Consider a cylindrical strip at radius x and thickness dx as shown in the figure.

$$I_{yy} = \int x^2 \, dM$$

where yy is the axis of symmetry.

Now mass per unit volume $= \dfrac{M}{\pi R^2 h}$

Volume of cylindrical strip $= 2\pi x h \cdot dx$

$$\therefore \quad dm = \frac{M}{\pi R^2 h} \times 2\pi x h \, dx$$

$$= \frac{2Mx}{R^2} dx$$

$$I_{yy} = \int_0^R x^2 \frac{2Mx}{R^2} dx$$

$$= \frac{2M}{R^2} \int_0^R x^3 \, dx$$

$$= \frac{2M}{R^2} \left[\frac{x^4}{4} \right]_0^R$$

$$= \frac{MR^2}{2}$$

Mass Moment of Inertia of a Uniform Hollow Sphere (Thin Thickness Sphere)

Consider a hollow sphere having mass M and radius R with reference axes as shown in

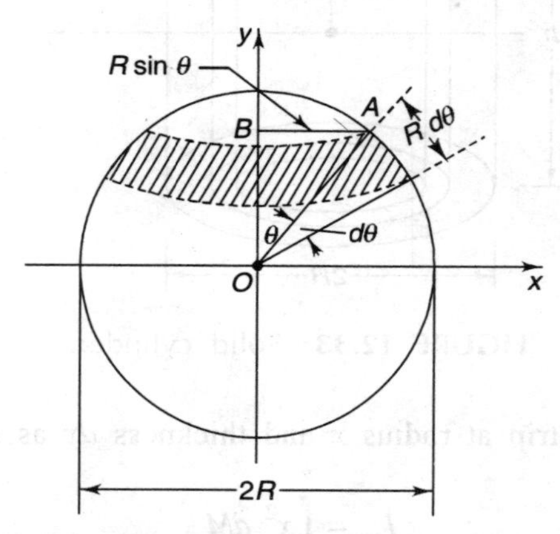

FIGURE 12.34 Thin-walled sphere.

Figure 12.34. The mass of the sphere is mainly spread over the surface of the sphere.

Now consider a radius OA at an angle θ with the axis oy. Taking BA as the radius, trace a circle about axis-oy on the sphere. Now change the angle from θ to $\theta + d\theta$ and draw another circle on the sphere with larger radius as shown in the figure. The part of the sphere between these circles has been shown as shaded in the figure. This part of the sphere can be taken as a ring of radius $R \sin\theta$ having width as $Rd\theta$. The periphery of the ring is $2\pi R \sin\theta$.

Area of the ring $= (2\pi R \sin\theta)\,(Rd\theta)$

$$\text{Mass per unit area} = \frac{M}{4\pi R^2}$$

$$\text{Mass of the ring} = (2\pi R \sin\theta)\,(Rd\theta) \times \frac{M}{4\pi R^2}$$

or

$$dM = \frac{M}{2}\sin\theta \cdot d\theta$$

Now the moment of inertia of the ring about the *y-y* axis is

$$\int dI = \int r^2\, dM$$

where $r = R \sin\theta$.

$$I = \int_0^{\pi} (R\sin\theta)^2 \times \frac{M}{2}\sin\theta\, d\theta$$

$$= \frac{MR^2}{2}\int_0^{\pi} \sin^3\theta\, d\theta$$

$$= \frac{MR^2}{2}\int_0^{\pi} (1-\cos^2\theta)\sin\theta\, d\theta$$

$$= -\frac{MR^2}{2}\int_0^{\pi} (1-\cos^2\theta)\, d(\cos\theta)$$

$$= -\frac{MR^2}{2}\left[\cos\theta - \frac{\cos^3\theta}{3}\right]_0^{\pi}$$

$$= \frac{2}{3}MR^2$$

Mass Moment of Inertia of a Uniform Solid Sphere

Consider a uniform solid sphere with mass M and radius R. Now consider a thin spherical elemental volume with radius r and thickness dr as shown in Figure 12.35. The mass per unit volume of the solid sphere is $\frac{M}{4/3\pi R^3}$. The thin elemental sphere has surface area $4\pi r^2$ and thickness dr. Its volume is $4\pi r^2.dr$ and therefore it has mass, given as follows.

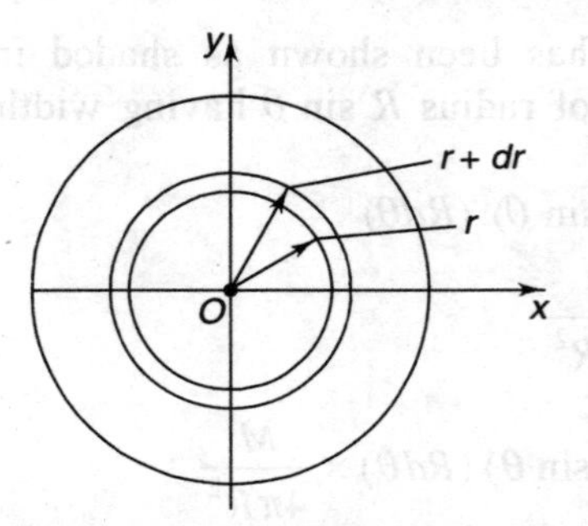

FIGURE 12.35 Solid sphere.

$$dM = \left(\frac{3M}{4\pi R^3}\right)(4\pi r^2\ dr)$$

$$= \frac{3M}{R^3} \cdot r^2 \cdot dr$$

Now the moment of inertia of a thin spherical sphere as worked out earlier is

$$dI = \frac{2}{3} \times dM \times r^2$$

$$= \frac{2}{3} \times r^2 \times \frac{3M}{R^3} \cdot r^2 \cdot dr$$

$$= \frac{2M}{R^3} \cdot r^4 \cdot dr$$

$$\therefore \qquad \int dI = \int_0^R \frac{2M}{R^3} \cdot r^4 \cdot dr$$

$$I = \frac{2M}{R^3}\left[\frac{r^5}{5}\right]_0^R$$

$$= \frac{2}{5} MR^2$$

Mass Moment of Inertia of a Uniform Solid Cone

Consider a cone of mass M, base $2R$ and height h as shown in Figure 12.36. Consider an elemental strip on a conical strip at a distance y from the base having radius x and

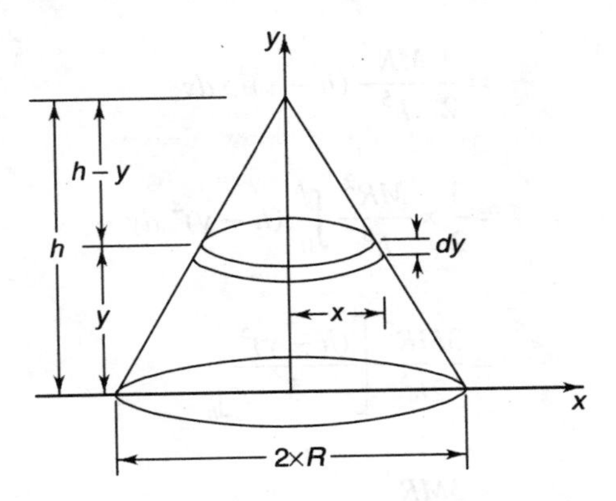

FIGURE 12.36 Solid cone.

thickness dy as shown in the figure. The mass per unit volume is $\dfrac{M}{1/3\pi R^2 h}$. The area of the strip is

$$dA = \pi \cdot x^2$$

But

$$\frac{x}{h-y} = \frac{R}{h}$$

or

$$x = \frac{R(h-y)}{h}$$

Now the elemental volume of the disc = $dV = dA \times dy = \pi \cdot x^2 \cdot dy$

$$\therefore \qquad dV = \pi \cdot \frac{R^2(h-y)^2}{h^2} \cdot dy$$

$$\therefore \qquad dM = \frac{3M}{\pi R^2 h} \times \frac{\pi R^2 (h-y)^2}{h^2} \times dy$$

$$= \frac{3M}{h^3}(h-y)^2\, dy$$

Now for this disc, moment of inertia is

$$dI = \frac{dM \cdot x^2}{2}$$

$$= \frac{1}{2} \times \frac{3M}{h^3}(h-y)^2 \cdot \frac{R^2(h-y)^2}{h^2} \cdot dy$$

$$= \frac{3}{2} \frac{MR^2}{h^5} (h - y)^4 \cdot dy$$

$\therefore$
$$I = \frac{3}{2} \times \frac{MR^2}{h^5} \int_0^h (h - y)^4 \, dy$$

$$= \frac{3MR^2}{2h^5} \left[\frac{(h - y)^5}{5} \right]_h^0$$

$$= \frac{3MR^2}{10h^5} [h^5]$$

$$= \frac{3}{10} MR^2$$

ROTATION OF AXES

Consider an area whose second moments and product of area are known about the reference axes x and y. Now the reference axes are rotated by angle α in the counter clockwise direction to x' and y' as shown in Figure 12.37.

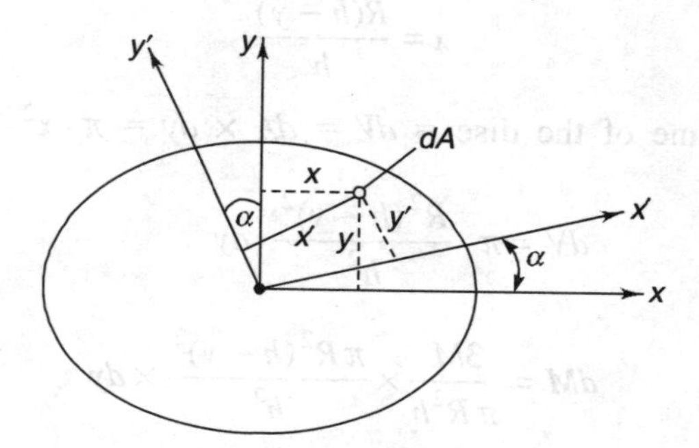

FIGURE 12.37 Rotation of axes.

We have

$$x' = x \cos \alpha + y \sin \alpha$$
$$y' = -x \sin \alpha + y \cos \alpha$$

Now

$$I_{x'x'} = \int_A (y')^2 \, dA$$

$$= \int_A (-x \sin \alpha + y \cos \alpha)^2 \, dA$$

$$= \sin^2 \alpha \int x^2\, dA + \cos^2 \alpha \int y^2\, dA - 2 \cos \alpha \sin \alpha \int xy\, dA$$

$$= I_{yy} \sin^2 \alpha + I_{xx} \cos^2 \alpha - 2I_{xy} \cos \alpha \sin \alpha$$

or

$$I_{x'x'} = \frac{I_{xx} + I_{yy}}{2} + \frac{I_{xx} - I_{yy}}{2} \cos 2\alpha - I_{xy} \sin 2\alpha$$

Similarly, it can be found out

$$I_{y'y'} = \frac{I_{xx} + I_{yy}}{2} - \frac{I_{xx} - I_{yy}}{2} \cos 2\alpha + I_{xy} \sin 2\alpha$$

Also, it can be given

$$I_{x'y'} = \frac{I_{xx} - I_{yy}}{2} \sin 2\alpha + I_{yy} \cos 2\alpha$$

PRINCIPAL AXES

The second moments of area ($I_{x'x'}$ and $I_{y'y'}$) have different values depending up the value of the angle of rotation α. However, the sum of second moments remains constant for all reference axes which are orthogonal, i.e., for all x and y reference axes irrespective of angle α. It means that the maximum moment of area about any axis say the x-axis will have corresponding minimum moment of area about the y-axis and vice versa. Since the second moment of area is a function of angle α as shown below:

$$I_{x'x'} = \frac{I_{xx} + I_{yy}}{2} + \frac{I_{xx} - I_{yy}}{2} \cos 2\alpha - I_{xy} \sin 2\alpha$$

For maximum moment of area, put $\frac{\partial I_{x'x'}}{\partial \alpha} = 0$. Therefore,

$$\frac{\partial I_{x'x'}}{\partial \alpha} = (I_{xx} - I_{yy})(-\sin 2\alpha) - 2I_{xy} \cos 2\alpha = 0$$

or

$$\tan 2\alpha = \frac{2I_{xy}}{I_{yy} - I_{xx}}$$

There are two values of α which will give maximum moments of area. However, the product of area for this value of α is zero. Therefore, we can define the principal axes as those axes about which the product of area is zero. The principal axes give maximum and minimum moments of area but the sum of the moments of area remain the same as for other reference axes.

Example 12.6 An area has $I_{xx} = 113\ \text{cm}^4$, $I_{yy} = 32\ \text{cm}^6$ and $I_{xy} = -42\ \text{cm}^4$. Find the principal second moment of area.

For the principal second moments of area, we have

$$\tan 2\alpha = \frac{2I_{xy}}{I_{yy} - I_{xx}}$$

$$= \frac{2(-42)}{32 - 113}$$

$$= \frac{84}{81} = 1.04$$

or

$$2\alpha = \tan^{-1} 1.04$$

$$= 46.12° \text{ and } (46.12 + 180)$$

$$2\alpha = 46.12° \text{ and } 226.12°$$

For $2\alpha = 46.12°$, the principal second moment of area

$$I_1 = \frac{113 + 32}{2} + \frac{113 - 32}{2} \cos 46.12 - (-42) \sin (46.12)$$

$$I_1 = 72.5 + 40.5 \times 0.693 - (-42) \times 0.72$$

$$= 72.5 + 28.07 + 30.27$$

$$= 130.84\ \text{cm}^4$$

For $2\alpha = 226.12°$, the principal second moment of area is

$$I_2 = 72.5 - 28.07 - 30.27$$

$$= 14.16\ \text{cm}^4$$

Now we have

$$I_1 + I_2 = 130.84 + 14.16$$

$$= 145.00\ \text{cm}^4$$

While we have for the x- and y-axes

$$I_{xx} + I_{yy} = 113 + 32$$

$$= 145$$

Hence

$$I_1 + I_2 = I_{xx} + I_{yy}$$

It means the sum of moments of area remains constant about any pair of orthogonal axes. Also, we have

(a) Maximum moment of area $I_1 = 130.84 \text{ cm}^4$
(b) Minimum moment of area $I_2 = 14.16 \text{ cm}^4$

Table 12.1 lists centroids of some geometrical shapes.

Table 12.1 Centroids of geometrical shapes

Description	Shape	Area	$\bar{x}$	$\bar{y}$
1. Rectangle	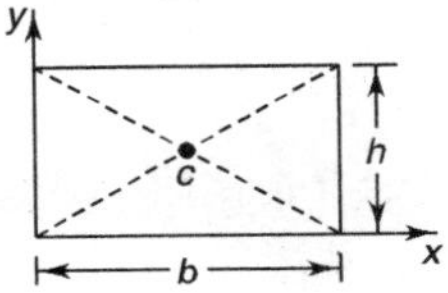	bh	$\dfrac{b}{2}$	$\dfrac{h}{2}$
2. Square	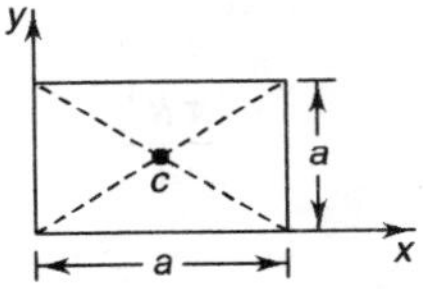	a^2	$\dfrac{a}{2}$	$\dfrac{a}{2}$
3. Parallelogram	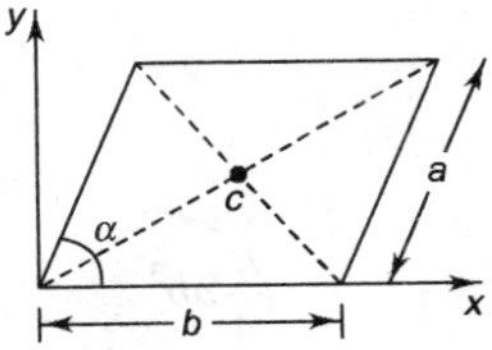	$ab \sin \alpha$	$\dfrac{b + a\cos\alpha}{2}$	$a \sin \alpha$
4. Triangle	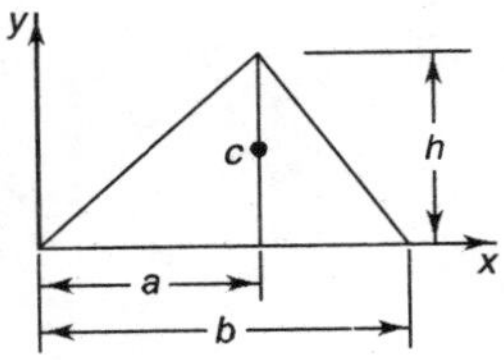	$\dfrac{1}{2}bh$	$\dfrac{1}{3}(a+b)$	$\dfrac{h}{3}$
5. Semi circle	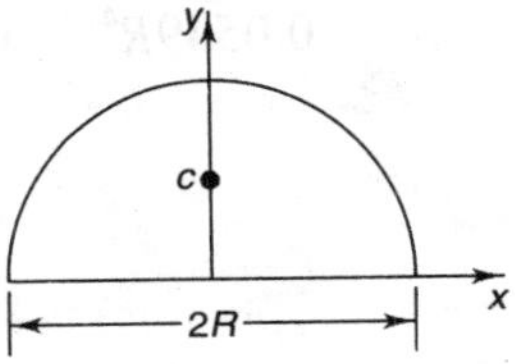	$\dfrac{\pi R^2}{2}$	0	$\dfrac{4R}{3\pi}$
6. Quarter circle	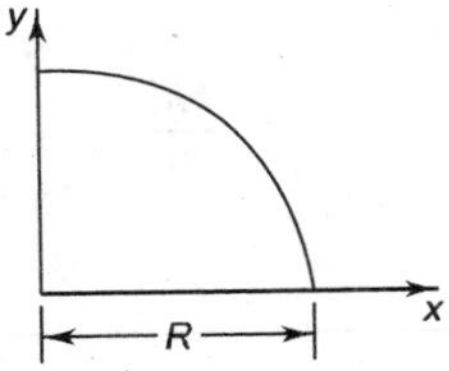	$\dfrac{\pi R^2}{4}$	$\dfrac{4R}{3\pi}$	$\dfrac{4R}{3\pi}$

Table 12.2 shows moments and products of inertia of some plane figures.

Table 12.2 Moments and products of inertia of plane figures

Description	*Figure*	I_{xx}	I_{yy}	I_{xy}
1. Rectangle	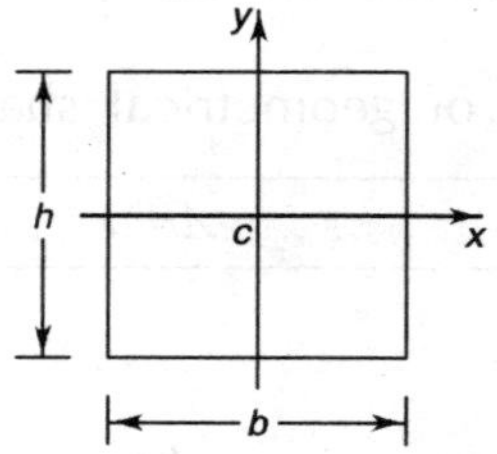	$\frac{bh^3}{12}$	$\frac{hb^3}{12}$	0
2. Circle	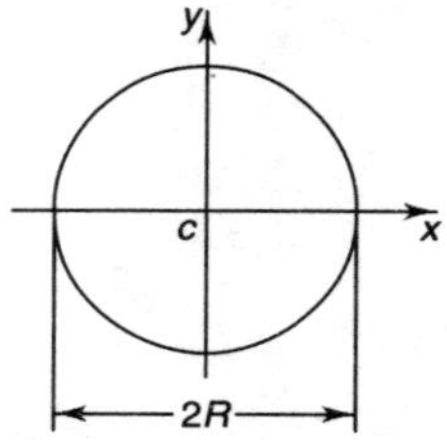	$\frac{\pi R^4}{4}$	$\frac{\pi R^4}{4}$	0
3. Ellipse	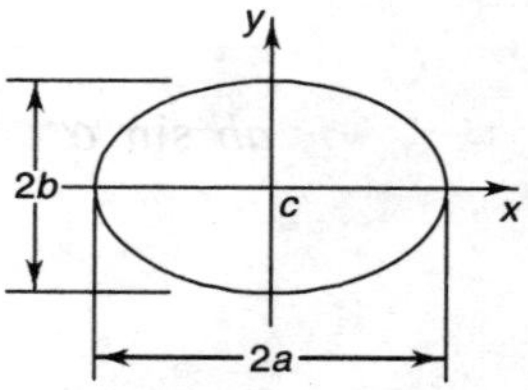	$\frac{\pi}{4}ab^3$	$\frac{\pi}{4}ba^3$	0
4. Quarter circle	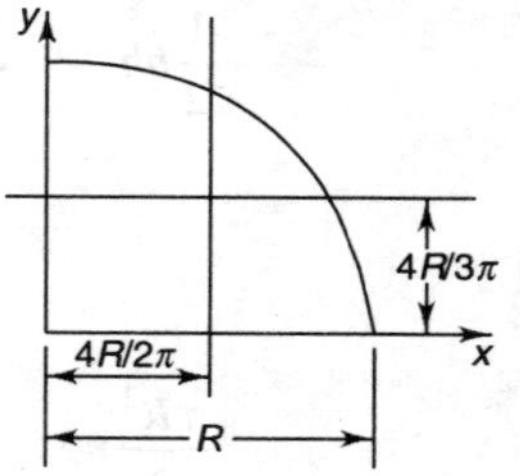	$0.0549R^4$	$0.0549R^4$	$-0.0163R^4$
5. Triangle	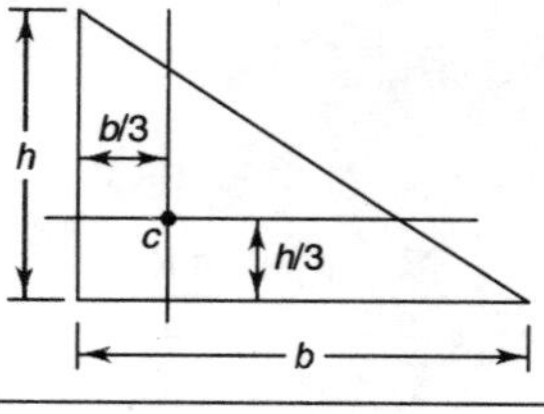	$\frac{bh^3}{36}$	$\frac{hb^3}{36}$	$-\frac{b^2h^2}{72}$

Table 12.3 depicts mass moments of inertia of some bodies.

Table 12.3 Mass moments of inertia of bodies

Description	*Figure*	I_y	I_x	I_z
1. Uniform rod	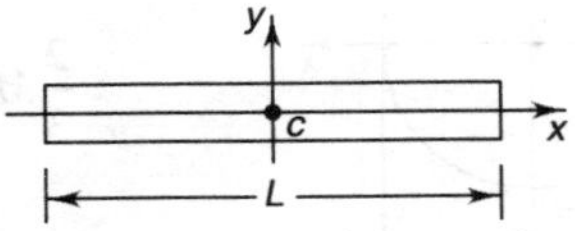	$\frac{ML^2}{12}$	–	–
2. Rectangular plate	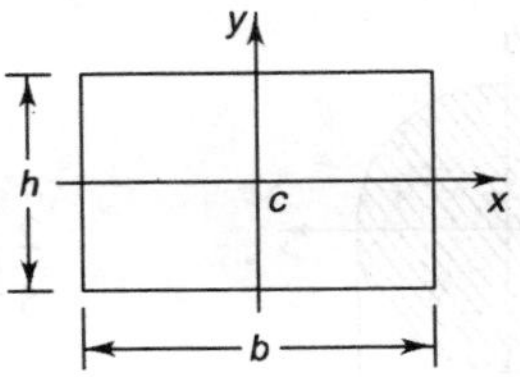	$\frac{Mb^2}{12}$	$\frac{Mh^2}{12}$	$\frac{M(b^2+h^2)}{12}$
3. Uniform circular ring	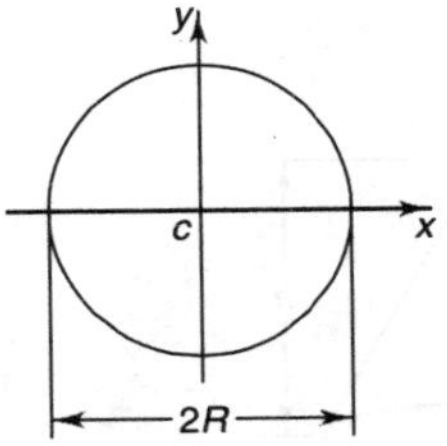	MR^2	MR^2	$2MR^2$
4. Uniform circular plate	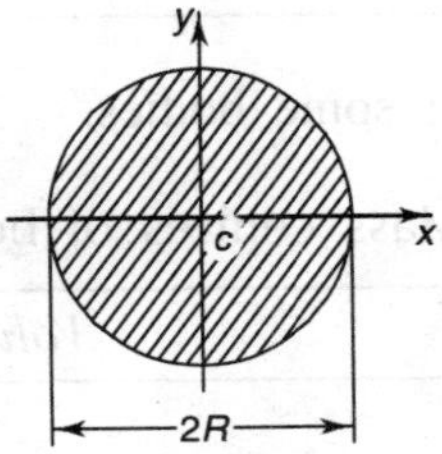	$\frac{MR^2}{2}$	$\frac{MR^2}{2}$	MR^2
5. Cylinder solid	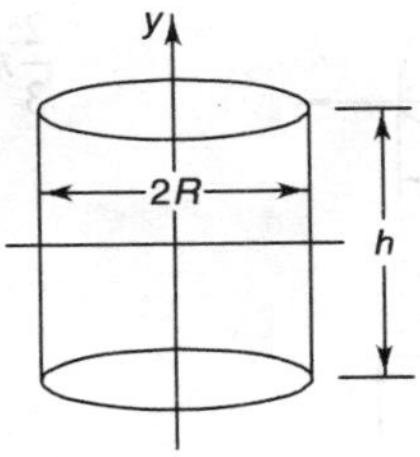	$\frac{1}{2}MR^2$		

(*Contd.*)

Table 12.3 Mass moments of inertia (Contd.)

Description	*Figure*	I_y	I_x	I_z
6. Hollow sphere (Thin walled)	y, x, 2R	$\frac{2}{3}MR^2$	$\frac{2}{3}MR^2$	—
7. Solid sphere	y, x, 2R	$\frac{2}{5}MR^2$	$\frac{2}{5}MR^2$	—
8. Solid cone	y, h, 2R	$\frac{3}{10}MR^2$	—	—

Table 12.4 describes mass centres of some bodies.

Table 12.4 Mass centres of bodies

Description	*Shape*	*Volume*	*Mass centre*
1. Hemisphere	y, c, 2R	$\frac{2}{3}\pi R^3$	$\overline{y} = \frac{3R}{8}$ from the base
2. Right circular cone	y, h, 2R	$\frac{1}{3}\pi R^2 h$	$\overline{y} = \frac{h}{4}$ from the base

(*Contd.*)

Table 12.4 Mass centres of bodies (Contd.)

Description	*Shape*	*Volume*	*Mass centre*
3. Cube		l^3	$\frac{l}{2}$ from every face
4. Sphere		$\frac{4}{3}\pi R^3$	R from every point on the surface
5. Cylinder		$\pi R^2 h$	$\frac{h}{2}$ from the base

SOLVED PROBLEMS

1. Determine the coordinates $\bar{x}$ and $\bar{y}$ of the centre of a 100 mm diameter circular hole cut in a thin plate so that this point will be the centroid of the remaining shaded areas as shown in the figure.

(UPTU: 2001–2002)

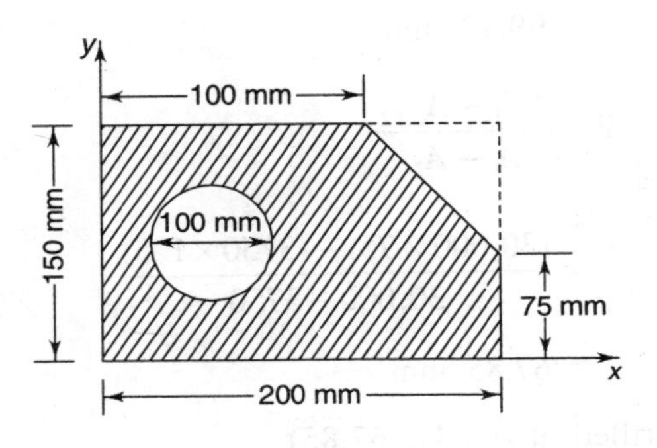

In case the circular hole is cut in the thin plate where the centroid is located, its centroid will remain unchanged even after the drilling of the hole. The problem reduces to finding the centroid of the thin plate without any consideration to the hole. The thin plate can be taken as a composite body of area (1) and area (2) as shown in the figure.

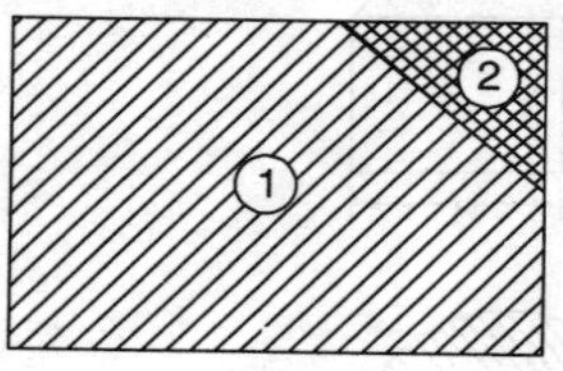

$$A_1 = 200 \times 150 = 30{,}000 \text{ mm}^2$$

$$A_2 = \frac{1}{2} \times 100 \times 75 = 3750 \text{ mm}^2$$

$$\bar{x}_1 = 100 \text{ mm}$$

$$\bar{y}_1 = 75 \text{ mm}$$

$$\bar{x}_2 = 100 + \frac{2}{3} \times 100 = 166.67 \text{ mm}$$

$$\bar{y}_2 = 75 + \frac{2}{3} \times 75 = 125 \text{ mm}$$

For a composite body, we have the centroid as

$$\bar{x} = \frac{A_1\bar{x}_1 - A_2\bar{x}_2}{A_1 - A_2}$$

$$= \frac{(30{,}000 \times 100) - (3750 \times 166.67)}{30{,}000 - 3750}$$

$$= 90.47 \text{ mm}$$

$$\bar{y} = \frac{A_1\bar{y}_1 - A_2\bar{y}_2}{A_1 - A_2}$$

$$= \frac{(30{,}000 \times 75) - (3750 \times 125)}{30{,}000 - 3750}$$

$$= 67.85 \text{ mm}$$

The hole is to be drilled at (90.47, 67.85).

2. Find the centroid of the Z-section as shown in the figure.

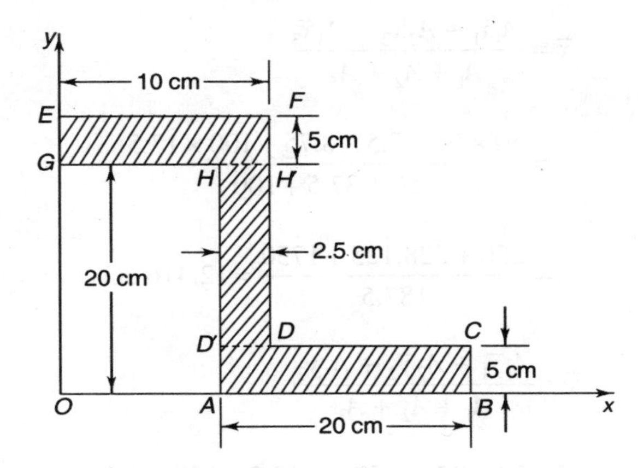

The Z-section can be considered to be a composite body of three areas.

$$A_1 = \text{rectangle } EFH'G$$
$$= 10 \times 5 = 50 \text{ cm}^2$$
$$A_2 = \text{rectangle } HH'DD'$$
$$= 15 \times 2.5 = 37.5 \text{ cm}^2$$
$$A_3 = \text{rectangle } AD'CB$$
$$= 20 \times 5 = 100 \text{ cm}^2$$

Now the centroid of these areas are:

$$\bar{x}_1 = 5 \text{ cm}$$

$$\bar{y}_1 = 20 + 2.5 = 22.5 \text{ cm}$$

$$\bar{x}_2 = 10 - \frac{2.5}{2} = 8.75 \text{ cm}$$

$$\bar{y}_2 = 5 + \frac{15}{2} = 12.5 \text{ cm}$$

$$\bar{x}_3 = 10 - 2.5 + \frac{20}{2} = 17.5 \text{ cm}$$

$$\bar{y}_3 = \frac{5}{2} = 2.5 \text{ cm}$$

Now the centroid of the Z-section is

$$\bar{x} = \frac{A_1\bar{x}_1 + A_2\bar{x}_2 + A_3\bar{x}_3}{A_1 + A_2 + A_3}$$

$$= \frac{50 \times 5 + 37.5 \times 8.75 + 100 \times 17.5}{50 + 37.5 + 100}$$

$$= \frac{250 + 328.125 + 1750}{187.5} = 12.416 \text{ cm}$$

$$\bar{y} = \frac{A_1\bar{y}_1 + A_2\bar{y}_2 + A_3\bar{y}_3}{A_1 + A_2 + A_3}$$

$$= \frac{50 \times 22.5 + 37.5 \times 12.5 + 100 \times 2.5}{50 + 37.5 + 100}$$

$$= \frac{1125 + 468.75 + 250}{187.5}$$

$$= \frac{1843.75}{187.5}$$

$$= 9.833$$

The centroid of the *Z*-section is (12.416, 9.833).

3. A body consists of a right circular solid cone of height 120 mm and radius 100 mm placed on a solid hemisphere of radius 100 mm of the same material. Find the position of centre of gravity.

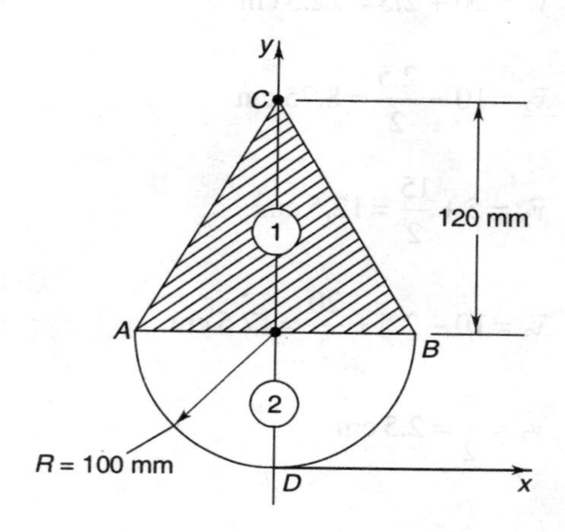

Guidance. The body is a composite body consisting of two volumes, viz. (1) cone and (2) hemisphere.

$$V_1 = \frac{1}{3} \times \pi \times (100)^2 \times 120$$

$$= 125.84 \times 10^4 \text{ mm}^3$$

$$V_2 = \frac{1}{2} \times \frac{4}{3} \times \pi \times (100)^3$$

$$= 2.1 \times 10^6 \text{ mm}^3$$

$$\bar{y}_1 = \left(100 - \frac{3}{8} \times 100\right) \text{ as CG of hemisphere is } \frac{3}{8} \times R \text{ from the base}$$

$$= 62.5 \text{ mm}$$

$$\bar{y}_2 = 100 + \frac{1}{3} \times 120 \text{ as CG is at } \frac{1}{3} \times h$$

$$= 140 \text{ mm}$$

Now CG is

$$\bar{y} = \frac{V_1 \times \bar{y}_1 + V_2 \bar{y}_2}{V_1 + V_2}$$

$$= \frac{125.84 \times 10^4 \times 62.5 + 210 \times 10^4 \times 140}{125.84 \times 10^4 + 210 \times 10^4}$$

$$= \frac{7865 + 29400}{335.84}$$

$$= \frac{37265}{335.84}$$

$$\approx 111 \text{ mm}$$

4. For the shaded area shown below, find the moment of inertia about the lines AA' and AB'.

(UPTU: 2003–2004)

$I_{AB'}$ = (I of rectangle $AA'CB$ about the centroid + area × h^2)
+ (I of $\Delta BB'C$ about the base) – (I of semicircle about the centroid + area × R^2)

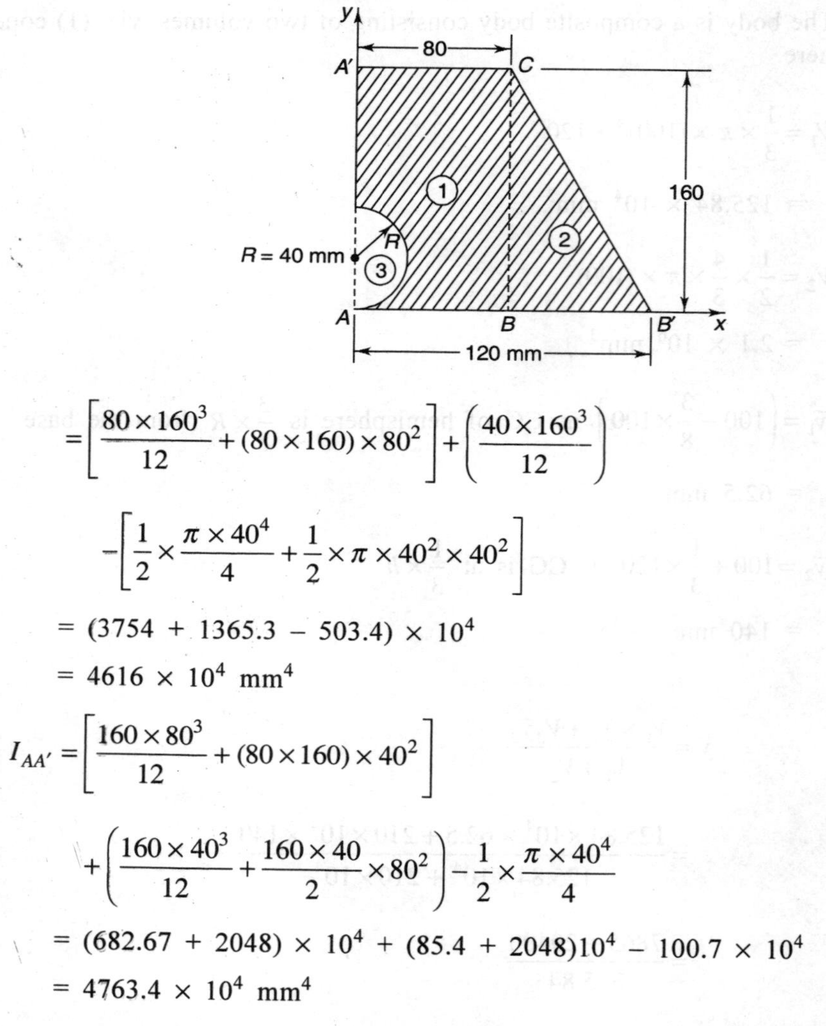

$$= \left[\frac{80 \times 160^3}{12} + (80 \times 160) \times 80^2\right] + \left(\frac{40 \times 160^3}{12}\right)$$

$$- \left[\frac{1}{2} \times \frac{\pi \times 40^4}{4} + \frac{1}{2} \times \pi \times 40^2 \times 40^2\right]$$

$$= (3754 + 1365.3 - 503.4) \times 10^4$$

$$= 4616 \times 10^4 \text{ mm}^4$$

$$I_{AA'} = \left[\frac{160 \times 80^3}{12} + (80 \times 160) \times 40^2\right]$$

$$+ \left(\frac{160 \times 40^3}{12} + \frac{160 \times 40}{2} \times 80^2\right) - \frac{1}{2} \times \frac{\pi \times 40^4}{4}$$

$$= (682.67 + 2048) \times 10^4 + (85.4 + 2048)10^4 - 100.7 \times 10^4$$

$$= 4763.4 \times 10^4 \text{ mm}^4$$

5. Determine the moment of inertia of the *I*-section of the following dimensions about an axis passing through the centroid and parallel to the flange.

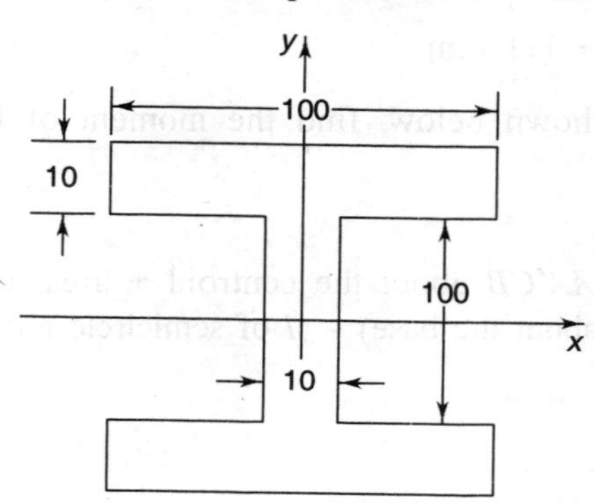

The *I*-section is equivalent to the *C*-section as shown below. It reduces to a composite area consisting of two rectangles.

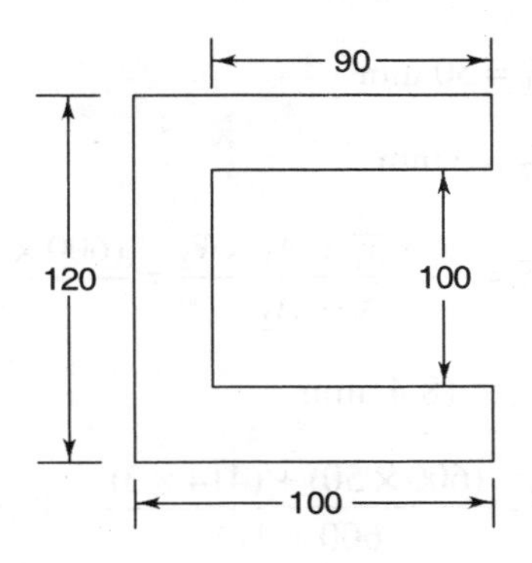

$$I_{xx} = \frac{100 \times 120^3}{12} - \frac{90 \times 100^3}{12}$$

$$= 144 \times 10^5 - 75 \times 10^5$$

$$= 69 \times 10^5 \text{ mm}^4$$

6. Find the moment of inertia of ISA 100 × 75 × 6 about the centroidal *x-x* and *y-y* axes.
(UPTU: 2001–2002)

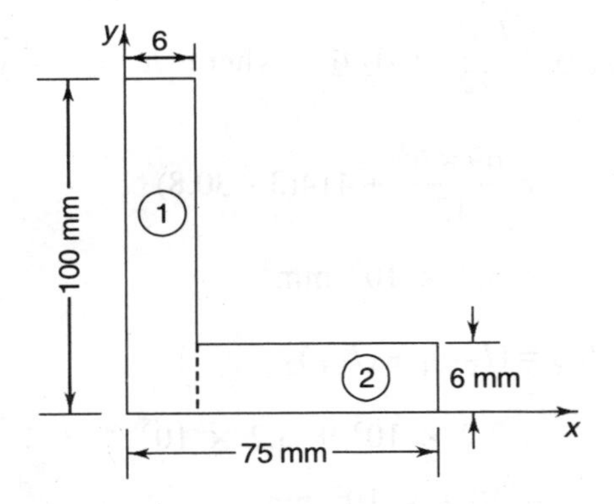

First we have to find the centroid about reference axes as shown and later moment of inertia found about *x-x* and *y-y* axes to be modified as per the parallel theorem.

$$A_1 = 100 \times 6 = 600 \text{ mm}^2$$

$$A_2 = (75 - 6) \times 6 = 414 \text{ mm}^2$$

$$\bar{x}_1 = 3 \text{ mm}$$

$$\bar{x}_2 = 6 + \frac{69}{2} = 40.5 \text{ mm}$$

$$\bar{y}_1 = 50 \text{ mm}$$

$$\bar{y}_2 = 3 \text{ mm}$$

$$\bar{x} = \frac{A_1 \times \bar{x}_1 + A_2 \times \bar{x}_2}{A_1 + A_2} = \frac{(600 \times 3) + (414 \times 40.5)}{600 + 414}$$

$$= 18.4 \text{ mm}$$

$$\bar{y} = \frac{(600 \times 50) + (414 \times 3)}{600 + 414} = 30.8 \text{ mm}$$

The moment of inertia about the centroidal axes are

$$(I_{\bar{x}\bar{x}})_1 = \frac{b_1 d_1^3}{12} + A_1 h_1^2 \quad \text{where } h_1 = \bar{y}_1 - \bar{y}$$

$$= \frac{6 \times 100^3}{12} + 600(50 - 30.8)$$

$$= 72 \times 10^5 \text{ mm}^4$$

$$(I_{\bar{x}\bar{x}})_2 = \frac{b_2 d_2^3}{12} + A_2 h_2^2 \quad \text{where } h_2 = \bar{y}_2 - \bar{y}$$

$$= \frac{69 \times 6^3}{12} + 414(3 - 30.8)$$

$$= 3.2 \times 10^5 \text{ mm}^4$$

$$\therefore \quad I_{\bar{x}\bar{x}} = (I_{\bar{x}\bar{x}})_1 + (I_{\bar{x}\bar{x}})_2$$

$$= 7.2 \times 10^5 + 3.2 \times 10^5$$

$$= 10.4 \times 10^5 \text{ mm}^4$$

Now about the $\bar{y}$-$\bar{y}$ axis, we have

$$(I_{\bar{y}\bar{y}})_1 = \frac{d_1 b_1^3}{12} + A_1 J_1^2 \quad \text{where } J_1 = \bar{x}_1 - \bar{x}$$

$$= \frac{100 \times 6^3}{12} + 600(3 - 18.4)^2$$

$$= 1.4 \times 10^5 \text{ mm}^4$$

$$(I_{\bar{y}\bar{y}})_2 = \frac{d_2 b_2^3}{12} + A_2 J_2^2 \quad \text{where} \quad J_2 = \bar{x}_2 - \bar{x}$$

$$= \frac{6 \times (60)^3}{12} + 414(40.5 - 18.4)^2$$

$$= 3.7 \times 10^5 \text{ mm}^4$$

$$\therefore \qquad I_{\bar{y}\bar{y}} = (I_{\bar{y}\bar{y}})_1 + (I_{\bar{y}\bar{y}})_2$$

$$I_{\bar{y}\bar{y}} = 1.4 \times 10^5 + 3.7 \times 10^5$$

$$= 5.1 \times 10^5 \text{ mm}^4$$

7. A square hole is punched out of a circular lamina, the diagonal of the square being the radius of the circle as shown in the figure. Find the centroid of the remainder if R is the radius of the circle.

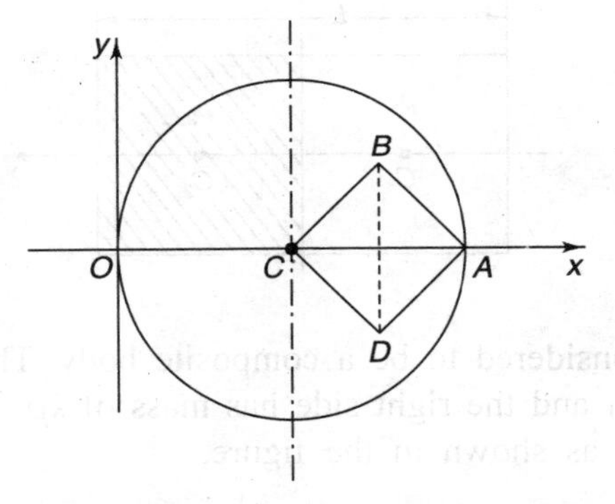

It can be seen that the circular lamina with the punched square has symmetry about the diameter OA. Hence the centroid will lie on the diameter OA. The lamina is also a composite body. The centroid is

$$\bar{x} = \frac{A_1 x_1 - A_2 x_2}{A_1 - A_2}$$

$$A_1 = \pi R^2$$

$$A_2 = \frac{R \times R}{2} = \frac{R^2}{2}$$

$$x_1 = R$$

$$x_2 = R + 0.5R = 1.5R$$

$$\therefore \quad \bar{x} = \frac{(\pi R^2) \times R - \left(\frac{R^2}{2}\right) \times 1.5R}{\pi R^2 - \frac{R^2}{2}}$$

$$= \frac{R(3.14 - 0.75)}{(3.14 - 0.5)}$$

$$= \frac{2.39}{2.64} R$$

$$= 0.905\, R$$

8. A rectangular plate of length L has half length made of material of density ρ_1 and the other half of density ρ_2. Find the centre of mass of the plate.

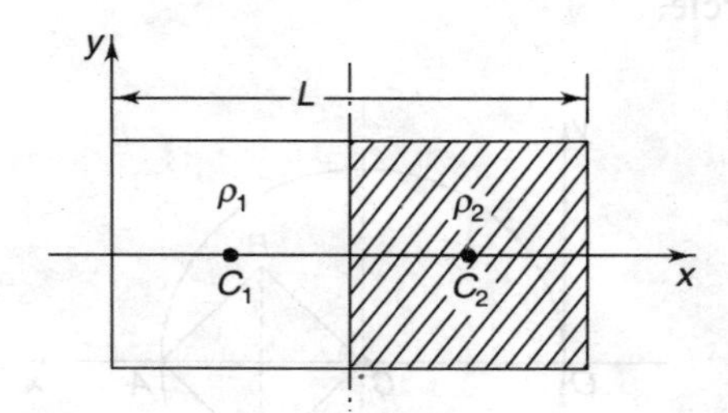

The plate can be considered to be a composite body. The mass of left is $k\rho_1$ and it is concentrated at C_1 and the right side has mass of $k\rho_2$ which is concentrated at C_2. Take reference axes as shown in the figure.

$$x_1 = \frac{L}{4}$$

$$x_2 = \frac{L}{2} + \frac{L}{4} = \frac{3L}{4}$$

Hence

$$\bar{x} = \frac{(k\rho_1) \times \frac{L}{4} + (k\rho_2) \times \frac{3L}{4}}{k\rho_1 + k\rho_2}$$

$$= \frac{\rho_1 + 3\rho_2}{\rho_1 + \rho_2} \times \frac{L}{4}$$

9. The density of a linear rod of length L varies as $\rho = A + B \cdot x$ where x is the distance from the left end. Find the centre of mass.

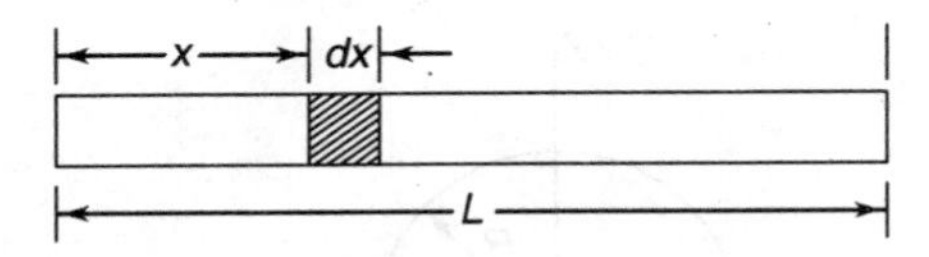

The centre of mass is

$$\bar{x} = \frac{\int_0^L x \cdot dm}{\int_0^L dm}$$

$$= \frac{\int_0^L x \cdot k\rho \cdot dx}{\int_0^L k\rho \cdot dx}$$

$$= \frac{\int_0^L x \cdot k \cdot (A + B \cdot x)\, dx}{\int_0^L k\,(A + Bx)\, dx}$$

$$= \frac{\int_0^L (Ax + Bx^2)\, dx}{\int_0^L (A + Bx)\, dx}$$

$$= \frac{\left[\frac{Ax^2}{2} + \frac{Bx^3}{3}\right]_0^L}{\left[Ax + \frac{Bx^2}{2}\right]_0^L}$$

$$= \frac{\frac{AL^2}{2} + \frac{BL^3}{3}}{AL + \frac{BL^2}{2}}$$

$$= \frac{(3A + 2BL)L^2}{3(2A + BL)L}$$

$$= \frac{(3A + 2BL)L}{3(2A + BL)}$$

10. Find the mass moment of inertia of a uniform ring of mass M and radius R about the diameter.

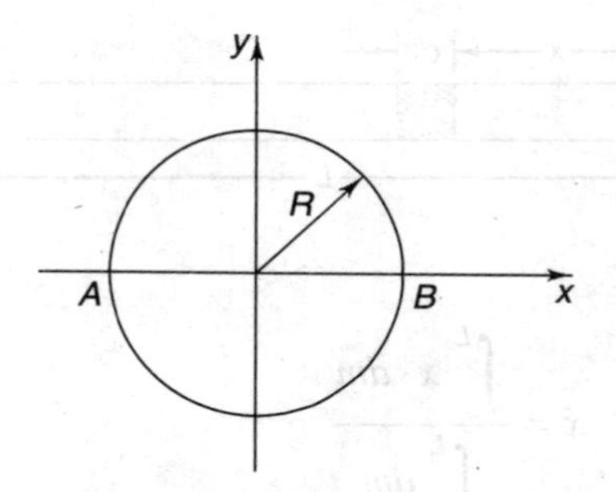

The mass moment of inertia of the ring about the Z-axis has been found out as

$$I_{ZZ} = MR^2$$

As per the perpendicular axis theorem, we have

$$I_{zz} = I_{xx} + I_{yy}$$

Due to symmetry, we have

$$I_{xx} = I_{yy}$$

Hence,

$$I_{xx} = I_{yy} = \frac{I_{zz}}{2} = \frac{MR^2}{2}$$

11. Find the moment of inertia of a solid cylinder of mass M and radius R about a line parallel to the axis of the cylinder and on the surface of the cylinder. Also find the mass moment of inertia of a sphere (radius = R, mass = M) about its tangent.

The mass moment of inertia of a cylinder about its centroidal axis has been calculated to be equal to $\frac{MR^2}{2}$.

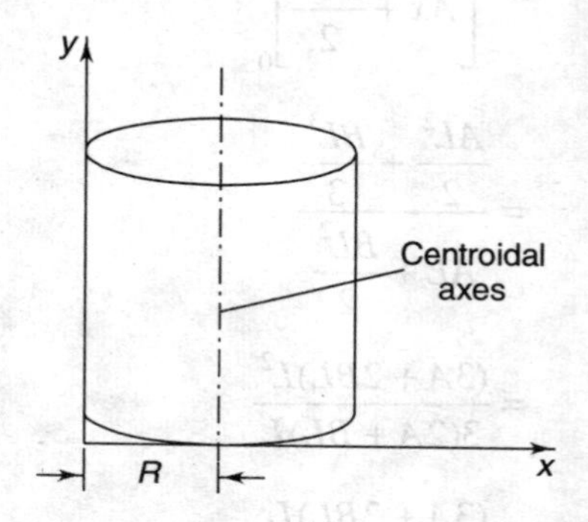

Now we have to find out I_{yy} which can be found out by using the parallel axes theorem.

$$I_{yy} = \frac{MR^2}{2} + M \times R^2$$

$$= \frac{3}{2} MR^2$$

The mass moment of inertia of a sphere about centroidal axes is $\frac{2}{5}MR^2$.

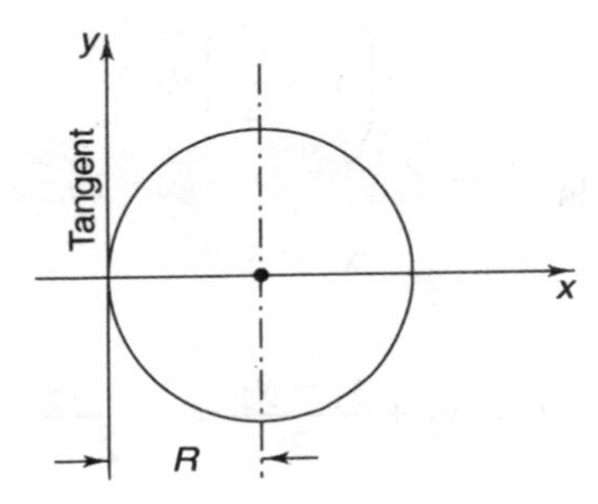

Now using the parallel axis theorem, the moment of inertia about the tangent is

$$I_{xx} = \frac{2}{5}MR^2 + MR^2$$

$$= \frac{7}{5}MR^2$$

12. Determine the coordinates of the centroid of the shaded area enclosed by parabola $4y = x^2$ and the straight line $x - y = 0$.

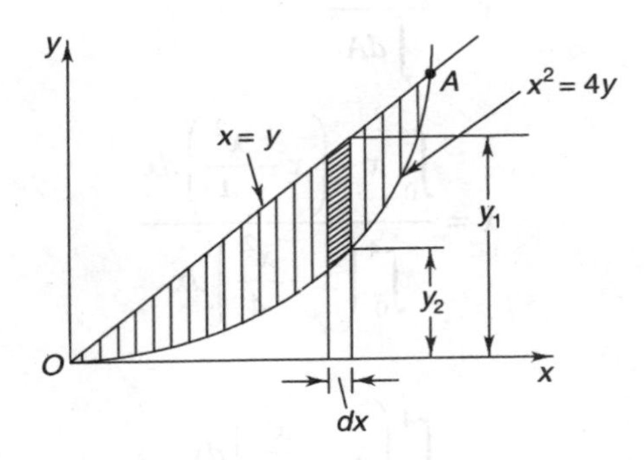

The coordinates of point O and A are to be found out by putting $x = y$ in equation $x^2 = 4y$. Therefore,

$$x^2 = 4x$$

or

$$x(x - 4) = 0$$

$\therefore$ $$x_1 = 0, y_1 = 0$$

and

$$x_2 = 4, y_2 = 4$$

The shaded portion can be considered to be consisting of a number of strips like one is selected and marked. The area of strip is

$$dA = (y_1 - y_2)\, dx$$

$$= \left(x - \frac{x^2}{4} \right) dx$$

The centroid of area dA is

$$x' = x$$

and

$$y' = y_2 + \frac{y_1 - y_2}{2} = \frac{y_1 + y_2}{2}$$

$$= \frac{1}{2}\left(x + \frac{x^2}{4} \right)$$

Now the centroid of the shaded region is

$$\bar{x} = \frac{\int x'\, dA}{\int dA}$$

and

$$\bar{y} = \frac{\int y'\, dA}{\int dA}$$

$\therefore$ $$\bar{x} = \frac{\int_0^4 x - \left(x - \frac{x^2}{4} \right) dx}{\int_0^4 \left(x - \frac{x^2}{4} \right) dx}$$

$$= \frac{\int_0^4 \left(x^2 - \frac{x^3}{4} \right) dx}{\int_0^4 \left(x - \frac{x^2}{4} \right) dx}$$

$$= \frac{\left[\frac{x^3}{3} - \frac{x^4}{16}\right]_0^4}{\left[\frac{x^2}{2} - \frac{x^3}{12}\right]_0^4}$$

$$= \frac{\left(\frac{4^3}{3} - \frac{4^4}{16}\right)}{\left(\frac{4^2}{2} - \frac{4^3}{12}\right)}$$

$$= \frac{4^2(64-48)\times 12}{3\times 16\times 4^2(6-8)}$$

$$= \frac{16}{8}$$

$$= 2$$

Now we have

$$\bar{y} = \frac{\int_0^4 y\,dA}{\int_0^4 dA}$$

$$= \frac{\int_0^4 \frac{1}{2}\left(x + \frac{x^2}{4}\right)\left(x - \frac{x^2}{4}\right)dx}{\int_0^4 \left(x - \frac{x^2}{4}\right)dx}$$

$$= \frac{\frac{1}{2}\int_0^4 \left(x^2 - \frac{x^4}{16}\right)dx}{\int_0^4 \left(x - \frac{x^2}{4}\right)dx}$$

$$= \frac{\frac{1}{2}\times\left[\frac{x^3}{3} - \frac{x^5}{5\times 16}\right]_0^4}{\left[\frac{x^2}{2} - \frac{x^3}{12}\right]_0^4}$$

$$= \frac{\frac{1}{2} \times \left(\frac{4^3}{3} - \frac{4^5}{80}\right)}{\left(\frac{4^2}{2} - \frac{4^3}{12}\right)}$$

$$= \frac{\frac{1}{2} \times 4^2 (320 - 192) \times 6}{\frac{1}{2} \times 4^2 (6 - 4) \times 80 \times 3}$$

$$= \frac{128}{2 \times 40} = \frac{8}{5}$$

13. A frustum of a solid right circular cone has an axial hole of 50 cm diameter. Determine centre of gravity.

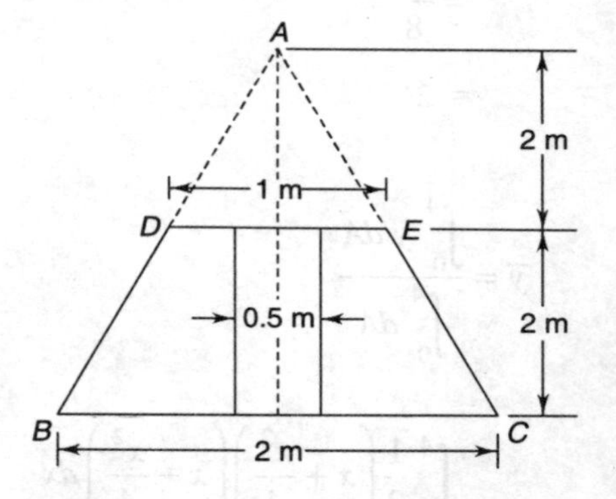

The problem can be solved considering three volumes, viz. (i) the complete cone of height 4 m, (ii) the cylinder of height 2 m and (iii) the top cone of height 2 m.

$$V_1 = \frac{1}{3}\pi r_1^2 h_1 = \frac{1}{3}\pi \times 1^2 \times 4 = 1.33\pi$$

$$V_2 = \pi r_2^2 h_2 = \pi \times (0.25)^2 \times 2 = 0.125\pi$$

$$V_3 = \frac{1}{3}\pi r_3^2 h_3 = \frac{1}{3} \times \pi \times (0.5)^2 \times 2 = 0.167\pi$$

The centroid will lie on the y-axis which is the axis is symmetry. It can be given

$$\bar{y} = \frac{V_1 y_1 - V_2 y_2 - V_3 y_3}{V_1 - V_2 - V_3}$$

$$y_1 = 1,\ y_2 = 1,\ y_3 = 2 + \frac{2}{4} = 2.5$$

$$\therefore \quad \bar{y} = \frac{1.33\pi \times 1 - 0.125\pi \times 1 - 0.167\pi \times 2.5}{1.33\pi - 0.125\pi - 0.167\pi}$$

$$\bar{y} = \frac{1.33 - 0.125 - 0.418}{1.04}$$

$$= \frac{0.787}{1.04}$$

$$= 0.756 \text{ mm}$$

14. Find the centroid of the *I*-section as shown in the figure.

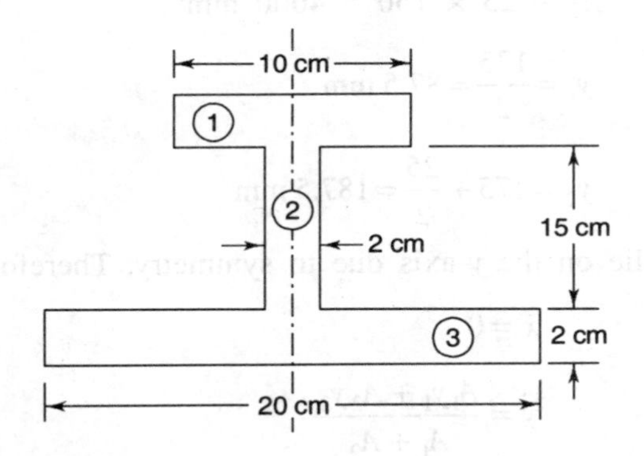

I-section can be considered to consist of three areas as shown in the figure.

$$\bar{y} = \frac{A_1 y_1 + A_2 y_2 + A_3 y_3}{A_1 + A_2 + A_3}$$

$$= \frac{(10 \times 2)\left(17 + \frac{2}{2}\right) + (15 \times 2)\left(2 + \frac{15}{2}\right) + (20 \times 2)\left(\frac{2}{2}\right)}{10 \times 2 + 15 \times 2 + 20 \times 2}$$

$$= \frac{360 + 285 + 40}{90}$$

$$= \frac{685}{90}$$

$$= 7.6 \text{ cm}$$

15. Find the centroid of a *T*-section as shown in the figure.

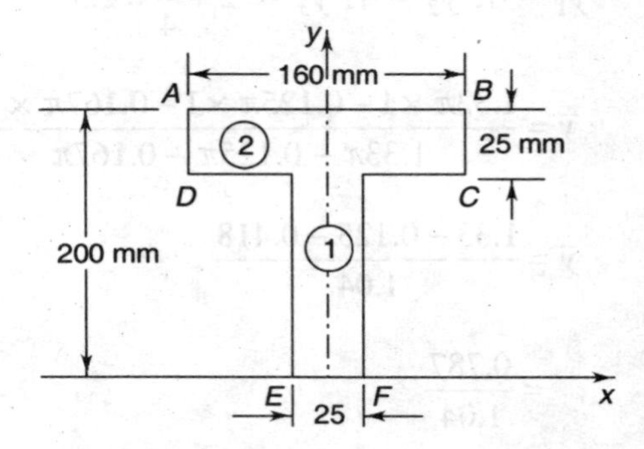

T-section can be considered to consist of two areas as shown.

$$A_1 = 25 \times (200 - 25) = 4375 \text{ mm}^2$$
$$A_2 = 25 \times 160 = 4000 \text{ mm}^2$$

$$y_1 = \frac{175}{2} = 87.5 \text{ mm}$$

$$y_2 = 175 + \frac{25}{2} = 187.5 \text{ mm}$$

The centroid will lie on the *y*-axis due to symmetry. Therefore,

$$\bar{x} = 0$$

$$\bar{y} = \frac{A_1 y_1 + A_2 y_2}{A_1 + A_2}$$

$$= \frac{4375 \times 87.5 + 4000 \times 187.5}{4375 + 4000}$$

$$= \frac{383 \times 10^3 + 750 \times 10^3}{8375}$$

$$= 135.3 \text{ mm}$$

16. Find the centroid of an angle section shown in the figure.

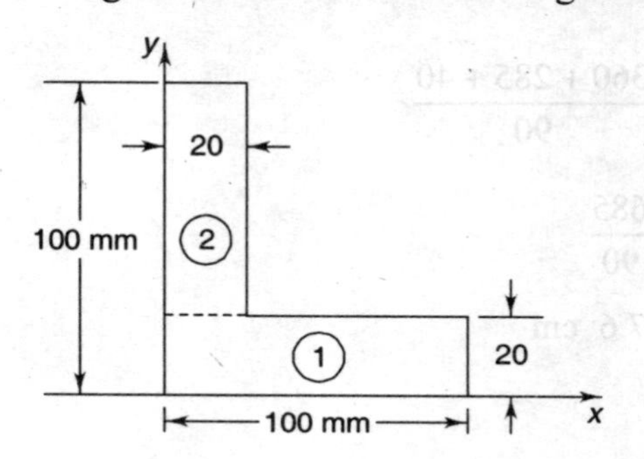

L-section consists of two areas as shown.

$$A_1 = 100 \times 20 = 200 \text{ mm}^2$$

$$A_2 = 80 \times 20 = 160 \text{ mm}^2$$

$$y_1 = 10 \text{ mm}$$

$$y_2 = 20 + \frac{80}{2} = 60 \text{ mm}$$

$$x_1 = 50 \text{ mm}$$

$$x_2 = 10 \text{ mm}$$

$$\bar{x} = \frac{A_1x_1 + A_2x_2}{A_1 + A_2}$$

$$= \frac{200 \times 50 + 160 \times 10}{200 + 160}$$

$$= \frac{10 \times 10^3 + 1.6 \times 10^3}{360}$$

$$= \frac{116 \times 10^2}{360}$$

$$= 32.2 \text{ mm}$$

$$\bar{y} = \frac{A_1y_1 + A_2y_2}{A_1 + A_2} = \frac{200 \times 10 + 160 \times 60}{360}$$

$$= \frac{11600}{360}$$

$$= 32.2 \text{ mm}$$

17. Using the analytical method, determine the centre of gravity of the plane uniform lamina as shown.

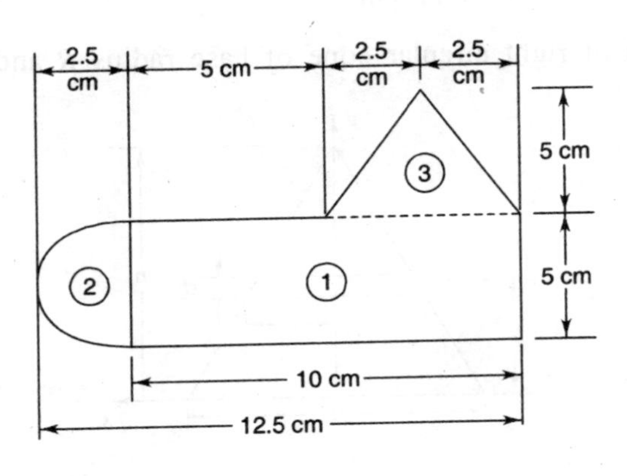

(AMIE–75)

$$A_1 = 10 \times 5 = 50 \text{ cm}^2$$

$$y_1 = \frac{5}{2} = 2.5 \text{ cm} \text{ and } x_1 = 2.5 + \frac{10}{2} = 7.5 \text{ cm}$$

$$A_2 = \frac{\pi r^2}{2} = \frac{\pi \times 2.5^2}{2} = 9.82 \text{ cm}^2$$

$$y_2 = \frac{5}{2} = 2.5 \text{ cm} \text{ and } x_2 = 2.5 - \frac{4r}{3\pi} = 1.44 \text{ cm}$$

$$A_3 = \frac{5 \times 5}{2} = 12.5 \text{ cm}^2$$

$$y_3 = 5 + \frac{5}{3} = 6.67 \text{ cm} \text{ and } x_3 = 2.5 + 5 + 2.5 = 10 \text{ cm}$$

Now for this composite area, we have

$$\bar{y} = \frac{A_1 y_1 + A_2 y_2 + A_3 y_3}{A_1 + A_2 + A_3}$$

$$= \frac{50 \times 2.5 + 9.82 \times 2.5 + 12.5 \times 6.67}{50 + 9.82 + 12.5}$$

$$= \frac{233}{72.3} = 3.23$$

Similarly, we have

$$\bar{x} = \frac{A_1 x_1 + A_2 x_2 + A_3 x_3}{A_1 + A_2 + A_3}$$

$$= \frac{50 \times 7.5 + 9.82 \times 1.44 + 12.5 \times 10}{72.3}$$

$$= 7.11 \text{ cm}$$

18. Find the centroid of right circular cone of base radius R and height h.

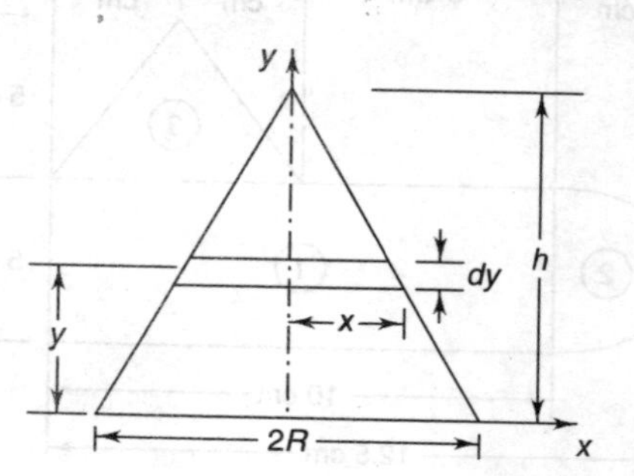

As the body has symmetry about the y-axis, the centroid will lie on this axis. Consider a disc of thickness dy at a distance y from the base, then the volume of the disc is

$$dV = (\pi x^2)\ dy$$

$$\bar{y} = \frac{\int y\, dV}{\int dV}$$

$$\bar{y} = \frac{\int_0^h y\,(\pi x^2)\, dy}{\int_0^h dV}$$

$$\int_0^h dV = \frac{1}{3}\pi R^2 h$$

and

$$\frac{x}{R} = \frac{h-y}{h}$$

$$\therefore \qquad \bar{y} = \frac{\pi \int_0^h y \left(\frac{h-y}{h}\right)^2 R^2\, dy}{\frac{1}{3}\pi R^2 h}$$

$$= \frac{3}{h^3} \int_0^h y(h^2 + y^2 - 2hy)\, dy$$

$$= \frac{3}{h^3} \left[\frac{h^2 y^2}{2} + \frac{y^4}{4} - \frac{2hy^3}{3} \right]_0^h$$

$$= \frac{3}{h^3} \left(\frac{h^4}{2} + \frac{h^4}{4} - \frac{2h^4}{3} \right)$$

$$= \frac{3h}{12}(6 + 3 - 8)$$

$$= \frac{h}{4}$$

19. A right circular cone of base radius R and height h is attached to a hemisphere of radius R as shown in the figure. Find the ratio h/R so that the centroid of the composite volume is located in the plane between the cone and hemisphere.

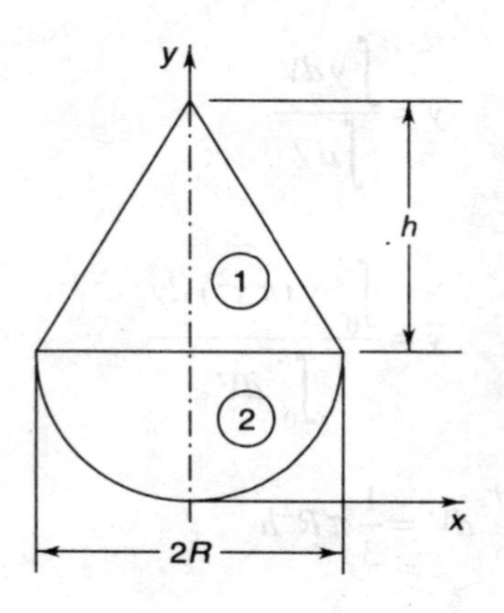

$$V_1 = \frac{1}{3}\pi R^2 h$$

$$y_1 = R + \frac{h}{4}$$

$$V_2 = \frac{2}{3}\pi R^3$$

$$y_2 = R - \frac{3}{8}R = \frac{5}{8}R$$

$$\therefore \quad \bar{y} = \frac{V_1 y_1 + V_2 y_2}{V_1 + V_2}$$

$$= \frac{\left(\frac{1}{3}\pi R^2 h\right) \times \left(R + \frac{h}{4}\right) + \left(\frac{2}{3}\pi R^3\right)\left(\frac{5}{8}R\right)}{\frac{1}{3}\pi R^2 h + \frac{2}{3}\pi R^3}$$

$$= \frac{h\left(R + \frac{h}{4}\right) + 2R\left(\frac{5}{8}R\right)}{h + 2R}$$

Now it is given $\bar{y} = R$. Therefore,

$$R = \frac{h\left(R + \frac{h}{4}\right) + \frac{10}{8}R^2}{h + 2R}$$

or $$8(hR + 2R^2) = 8h\left(R + \frac{h}{4}\right) + 5R^2$$

or $$8hR + 16R^2 = 8hR + 2h^2 + 10R^2$$

or $$6R^2 = 2h^2$$

or $$\frac{h}{R} = \sqrt{3}$$

20. Determine the centre of mass of a homogeneous solid body of revolution as shown in the figure.

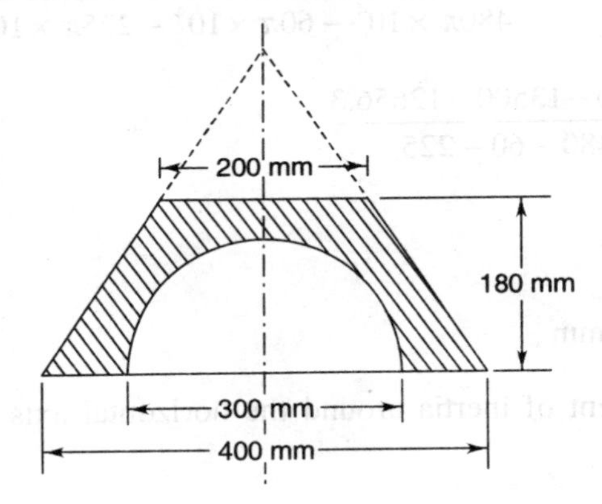

The composite volume generated on revolution (i) a right-angled cone of height 360 mm and base 400 mm minus, (ii) a right-angled cone of height 180 mm and base 200 mm (minus) and (iii) a hemisphere of radius 150 mm (minus).

$$V_1 = \frac{1}{3}\pi r_1^2 h_1 = \frac{1}{3} \times \pi \times 200^2 \times 360$$

$$= 480 \times \pi \times 10^4 \text{ mm}^3$$

$$y_1 = \frac{360}{4} = 90 \text{ mm}$$

$$V_2 = \frac{1}{3}\pi r_2^2 h_2 = \frac{1}{3} \times \pi \times 100^2 \times 180$$

$$= 60\pi \times 10^4 \text{ mm}^3$$

$$y_2 = 180 + \frac{180}{4} = 180 + 45 = 225 \text{ mm}$$

$$V_3 = \frac{2}{3}\pi r_3^3 = \frac{2}{3}\pi \times 150^3$$

$$= 225\pi \times 10^4 \text{ mm}^3$$

$$y_3 = \frac{3}{8} \times 150 = 56.25 \text{ mm}$$

Now

$$\bar{y} = \frac{V_1 y_1 - V_2 y_2 - V_3 y_3}{V_1 - V_2 - V_3}$$

$$= \frac{(480\pi \times 10^4) \times 90 - (60\pi \times 10^4) \times 225 - (225\pi \times 10^4) \times 56.25}{480\pi \times 10^4 - 60\pi \times 10^4 - 225\pi \times 10^4}$$

$$= \frac{43200 - 13500 - 12656.3}{480 - 60 - 225}$$

$$= \frac{17044}{195}$$

$$= 87.4 \text{ mm}$$

21. Determine the moment of inertia around the horizontal axis for the area as shown in the figure.

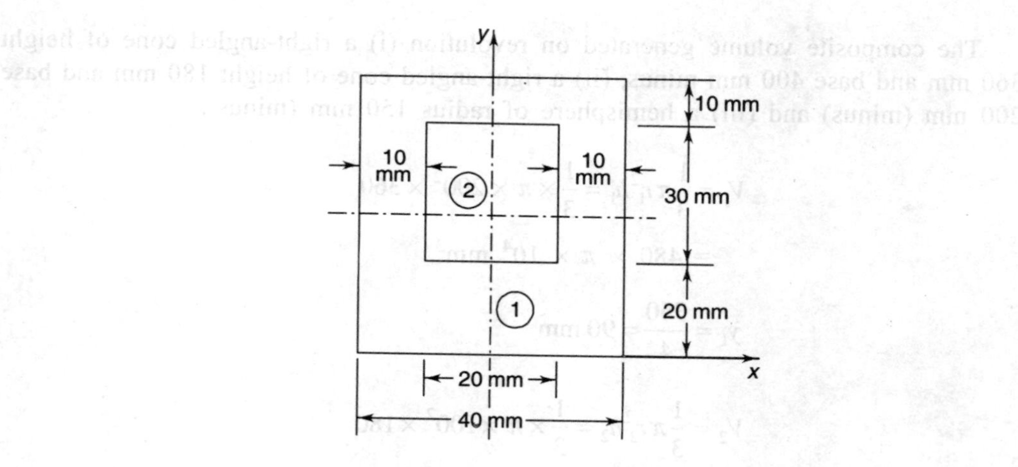

As there exists symmetry about the *y*-axis, the centroid of the lamina will lie on the *y*-axis. Composite areas are (i) a rectangle 40 × 60 mm and (ii) a rectangle 20 × 30 mm.

$$\bar{y} = \frac{A_1 y_1 - A_2 y_2}{A_1 - A_2}$$

$$= \frac{(60 \times 40) \times 30 - (20 \times 30) \times 35}{60 \times 40 - 20 \times 30}$$

$$= 28.3 \text{ mm}$$

Now, the moment of inertia about the horizontal axis passing through centroid is

$$(I_{xx})_1 = (I_1 + A_1 h_1^2)$$

$$= \frac{40 \times 60^3}{12} + (40 \times 60)\,(30 - 28.3)^2$$

$$= 72.7 \times 10^4 \text{ mm}^4$$

$$(I_{xx})_2 = I_2 + A_2 h_2^2$$

$$= \frac{20 \times 30^3}{12} + (20 \times 30)(35 - 28.3)^2$$

$$= 7.2 \times 10^4 \text{ mm}^4$$

$$I_{xx} = (I_{xx})_1 - (I_{xx})_2$$

$$= 72.7 \times 10^4 - 7.2 \times 10^4$$

$$= 65.5 \times 10^4 \text{ mm}^4$$

OBJECTIVE TYPE QUESTIONS

Enthusiasm is the fuel of life, it helps you get where you are going.

State True or False

1. Centre of mass is a point at which complete mass of the body can be assumed to be concentrated. (*True*/*False*)
2. A centroid is a point at which the complete area of the lamina can be assume to be concentrated. (*True*/*False*)
3. The flywheel of an engine has lower value of moment of inertia. (*True*/*False*)
4. Moment of inertia of a man increases in case he spreads his arms and legs. (*True*/*False*)
5. If a body has maximum moment of inertia about the x-axis, then it will have maximum moment of inertia about the y-axis also. (*True*/*False*)
6. The principal axes have zero product of area. (*True*/*False*)
7. The sum of moment of inertia about a pair of orthogonal axes remains constant irrespective or the rotation of the reference axes. (*True*/*False*)
8. Centre of mass is governed by

$$\bar{x} = \frac{1}{M} \Sigma m_i x_i.$$ (*True*/*False*)

9. The centroid of a lamina is given by

$$\bar{x} = \frac{1}{A} \int_A x \, dA.$$ (*True*/*False*)

10. The centroid of a volume is given by

$$\bar{x} = \frac{\iiint x \, dV}{\iiint dV}$$ (*True*/*False*)

11. Moment of inertia depends upon mass only. (*True*/*False*)
12. Moment of inertia depends upon mass and distribution of mass about the axis of rotation. (*True*/*False*)
13. As per the parallel axis theorem, the moment of inertia does not change in case the axis of rotation is shifted to a distance parallel to the original position. (*True*/*False*)
14. The centroid of a lamina will always lie on the axis of symmetry. (*True*/*False*)
15. A circular lamina will have equal value of moment of inertia about the x-x and y-y axes if located at the centroid. (*True*/*False*)

16. The radius of gyration $k_{xx} = \sqrt{\frac{I_{yy}}{A}}$ is a point where whole mass of body is assumed to be concentrated to give same moment of inertia. (*True/False*)
17. Moment of inertia can be negative. (*True/False*)
18. Product of area is always positive. (*True/False*)
19. Product of area is zero if body has one axis of symmetry. (*True/False*)
20. A wheel of a cycle has a rim and spoke system to increase the moment of inertia of the wheel. (*True/False*)

Multiple Choice Questions

1. The mass moment of inertia of a uniform rod is

   (a) $\frac{ML^2}{6}$ (b) $\frac{ML^2}{16}$ (c) $\frac{ML^2}{12}$

2. The mass moment of inertia of a rectangular plate is

   (a) $\frac{Mb^2}{6}$ (b) $\frac{Mb^2}{12}$ (c) $\frac{Mb^2}{16}$

3. The mass moment of inertia of a uniform circular ring is

   (a) MR^2 (b) $2MR^2$ (c) $\frac{1}{2}MR^2$

4. A uniform circular plate has mass moment of inertia given by

   (a) MR^2 (b) $\frac{MR^2}{2}$ (c) $\frac{MR^2}{4}$

5. A solid cylinder has mass moment inertia as

   (a) $\frac{1}{3}MR^2$ (b) $\frac{4}{3}MR^2$ (c) $\frac{2}{3}MR^2$

6. A hallow sphere has mass moment of inertia given by

   (a) $\frac{2}{3}MR^2$ (b) $\frac{1}{3}MR^2$ (c) MR^2

7. The mass moment of inertia of a solid sphere is

   (a) $\frac{1}{5}MR^2$ (b) $\frac{2}{5}MR^2$ (c) $\frac{2}{3}MR^2$

8. A solid right-angled cone has mass moment of inertia given by

   (a) $\frac{3}{5}MR^2$ (b) $\frac{3}{8}MR^2$ (c) $\frac{3}{10}MR^2$

9. The centre of mass of a hemisphere from its base is

(a) $\frac{3R}{8}$ (b) $\frac{2R}{7}$ (c) $\frac{5R}{8}$

10. The centre of mass of a right circular cone from its base is at

(a) $\frac{h}{5}$ (b) $\frac{h}{4}$ (c) $\frac{h}{3}$

11. The centre of mass of a cylinder from its base is at

(a) $\frac{h}{3}$ (b) $\frac{h}{4}$ (c) $\frac{h}{2}$

12. The moment of inertia of a rectangle about its centroid axis (which is parallel to its base) is

(a) $\frac{bh^3}{12}$ (b) $\frac{bh^2}{6}$ (c) $\frac{bh^3}{36}$

13. The moment of inertia of a circle about its diameter is

(a) $\frac{\pi R^2}{2}$ (b) $\frac{\pi R^2}{4}$ (c) $\frac{\pi R^2}{6}$

14. The moment of inertia of a triangle about its base is

(a) $\frac{bh^3}{3}$ (b) $\frac{bh^3}{6}$ (c) $\frac{bh^3}{12}$

15. The centroid of a semicircle located at a distance from its diameter is

(a) $\frac{4R}{3\pi}$ (b) $\frac{2R}{3\pi}$ (c) $\frac{4R}{\pi}$

16. The value of product of area about its principal axes is

(a) maximum (b) zero (c) minimum

Fill in the Blanks

1. The centre of mass is a point where the __________ of the body is concentrated.

(a) mass (b) area

2. The centroid of a plane area is a point where the __________ of the plane area is concentrated.

(a) mass (b) area

3. The axes of rotation passing through the centroid are called __________ axes.

(a) centroidal (b) principal

4. The axes about which a plane area has maximum and minimum moments of inertia are called __________ axes.

(a) centroidal (b) principal

5. The product of area is __________ about principal axes.
(a) zero (b) maximum

6. The sum of moments of inertia __________ when the reference axes are rotated.
(a) changes (b) remains constant

7. The moment of inertia of a wheel __________ when more mass is located at the rim.
(a) increases (b) decreases

8. The property of a body which opposes a change of its rotation is called __________.
(a) continuum (b) moment of inertia

9. A quarter circle plane area has its centroid from its base at

(a) $\frac{4R}{3\pi}$ (b) $\frac{2R}{3\pi}$

10. The polar moment of inertia is __________ of I_{xx} and I_{yy}.
(a) sum (b) difference

ANSWERS

The optimist fell through twelve storeys
And at each window bar
He yelled to his friends who were frightened below
Well I'm all right so far

State True or False

1. True
2. True
3. False. The moment of inertia of the flywheel has to be large to oppose any change in the rotational speed.
4. True. When a man spreads his arms and legs, there will be more mass of his body away from his axis of symmetry, thereby increasing his moment of inertia.
5. False. If the x-axis has maximum moment of inertia, then the y-axis will have minimum moment of inertia so that the sum of moments of inertia remains constant.
6. True
7. True
8. True
9. True
10. True
11. False. Moment of inertia depends upon mass and distribution of mass about the axis of rotation.
12. True
13. False. The moment of inertia changes as per the parallel axis theorem, i.e. $I_{x'x'} = I_{xx} + A \times h^2$
14. True
15. True. $I_{xx} = I_{yy} = \dfrac{\pi R^4}{4}$
16. True.
17. False. Moment of inertia is the second moment of area and it cannot be negative.
18. False. Product of area can be negative or positive.
19. True. $I_{xy} = \int xy\, dA$ will be zero if the area has the axis of symmetry.
20. True.

Multiple Choice Questions

1. (a) **2.** (b) **3.** (c) **4.** (b)
5. (c) **6.** (a) **7.** (b) **8.** (c)
9. (a) **10.** (b) **11.** (c) **12.** (a)
13. (b) **14.** (c) **15.** (a) **16.** (b)

Fill in the Blanks

1. (a) **2.** (b) **3.** (a) **4.** (b)
5. (a) **6.** (b) **7.** (a) **8.** (b)
9. (a) **10.** (a)

CHAPTER 13

Kinematics of Rigid Body

A word of encouragement is as refreshing as a cold drink on a hot summer day.

INTRODUCTION

A body is called a rigid body when the body does not produce any displacement of its particles relative to each other on the application of any external force. No real body is perfectly rigid. However, the relative displacement by external force is extremely small and negligible in most of the solid bodies. Therefore the solid bodies can be idealised as rigid. In simple words, we can say that distance between any pairs of points within the rigid body remains constant.

Kinematics of rigid body involves the study of the velocity and acceleration of the rigid body without taking into the consideration the forces causing the motion. The motion of a rigid body can be classified as pure translation, pure rotation, plane motion and space motion. The motion of a rigid body is said to be a plane motion if all points of the rigid body stay in the same parallel planes. To understand it, a plane motion can be considered to be composed of translation and rotation. Further, this combined motion of translation and rotation of the rigid body at any instant of time may be assumed to be a motion of entirely rotation about a certain point which is called instantaneous centre of rotation. The position of instantaneous centre of rotation can be easily determined by graphical method, thereby helping in the determination of velocity and acceleration at different points on the rigid body.

MOTION AND FRAME OF REFERENCE

If a body changes its position with time, it is said to be moving. A glass placed on a table remains on the table and we can say that the glass is at rest. However, if we locate ourself in space, the whole earth is changing its position including the room, table and glass. Hence, the glass is at rest if it is viewed from the room, but it is moving if it is viewed from the space. Motion is a combined property of the object under study and the observer. The rest or motion has no meaning without the viewer. Nothing is in absolute rest or in absolute motion.

A train is moving with respect to a man standing on the platform. Similarly, the man standing on the platform is moving with respect to the train. Similarly, the passengers on the train are at rest with respect to the train but are moving with respect to the man standing on the platform. To locate the position of a body, we need a frame of reference. A convenient method of fixing a frame of reference is to select three mutually perpendicular axes and name them x, y and z axes. The coordinates x, y, z of any point on the body then specify the position of the point with respect to that frame. If there is no change in x, y and z with time, the point is at rest otherwise the point or body is moving with respect to the frame.

MOTIONS REFERRED TO MOVING FRAME OF REFERENCE

In case a point is moving in a moving frame of reference, the motion of the point can be described about inertial frame of reference (application of Newton's laws are possible for motions about inertial frame of reference only) by first ascertaining the motion in relation to a moving frame and then relating it to the inertial frame. The moving frame can have translation or rotation motion. It can also accelerate linearly or angularly. The relation of motion about a moving frame and a fixed frame is explained as follows:

(a) *Motion in a fixed frame.* Consider two points P_1 and P_2 moving with v_1 and v_2 velocities and a_1 and a_2 accelerations with respect to a fixed frame of reference (Figure 13.1). Then relative velocity of P_1 with respect to P_1 is given by

$$v_{12} = v_1 - v_2$$

Similarly, the acceleration of P_1 with respect to P_2 is

$$a_{12} = a_1 - a_2$$

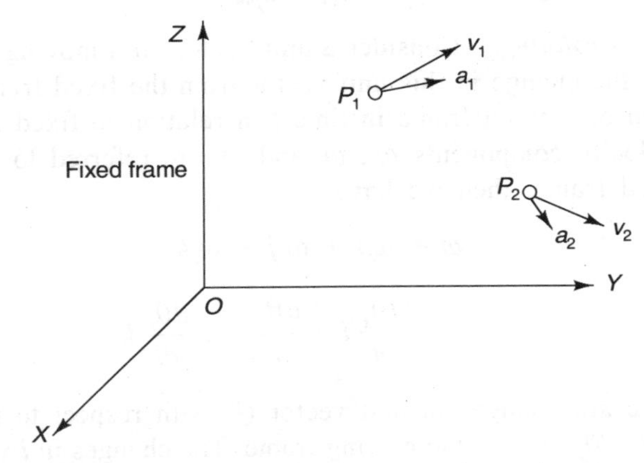

FIGURE 13.1 Relative velocity and acceleration.

(b) *Moving frame is translating with respect to fixed frame.* A moving frame of reference x, y and z is translating with respect to a fixed frame X, Y, Z as shown in Figure 13.2. If the velocity of the moving frame is v_0 and the velocity of a point P with respect to the moving frame is v_{pm}, the velocity of the point with respect to fixed frame can be given as

$$v_{pf} = v_0 + v_{pm}$$

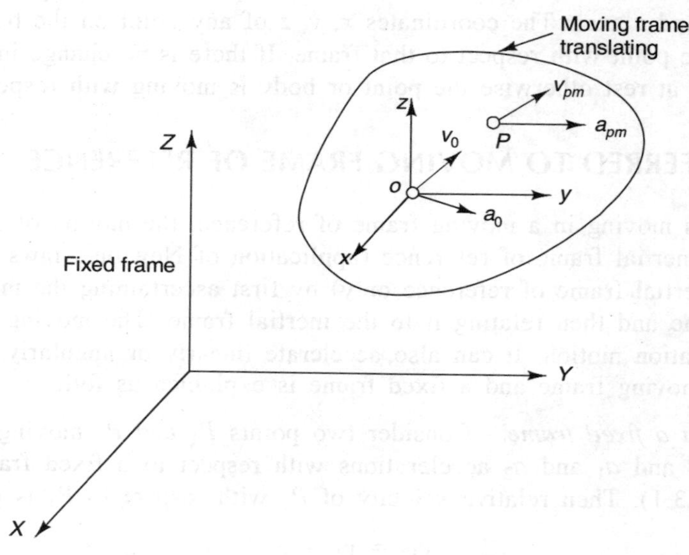

FIGURE 13.2 Moving frame translating.

Similarly, the relation between accelerations is

$$a_{pf} = a_0 + a_{pm}$$

(c) *Moving frame is rotating.* Consider a unit vector on a moving reference frame and let us observe the change in this unit vector from the fixed frame. Let $d\theta_x$, $d\theta_y$ and $d\theta_z$ be rotation of moving frame in time t in relation to fixed frame (Figure 13.3). If angular velocity components ω_x, ω_y and ω_z are referred to the rates of rotation about the fixed frame, then we have

$$\omega = \omega_x \boldsymbol{i} + \omega_y \boldsymbol{j} + \omega_z \boldsymbol{k}$$

$$= \frac{d\theta_x}{dt}\boldsymbol{i} + \frac{d\theta_y}{dt}\boldsymbol{j} + \frac{d\theta_z}{dt}\boldsymbol{k}$$

However, there are changes in unit vector ($\boldsymbol{i}$) with respect to the fixed frame due to rotation ($d\theta_x$, $d\theta_y$, $d\theta_z$) of the moving frame. The changes in $\boldsymbol{i}$ vector are as follows:

(a) $d\theta_x$ produces no change in $\boldsymbol{i}$
(b) $d\theta_y$ produces a change in $\boldsymbol{i} = -\boldsymbol{k}\, d\theta_y$
(c) $d\theta_z$ produces a change in $\boldsymbol{i} = +\boldsymbol{j}\, d\theta_z$

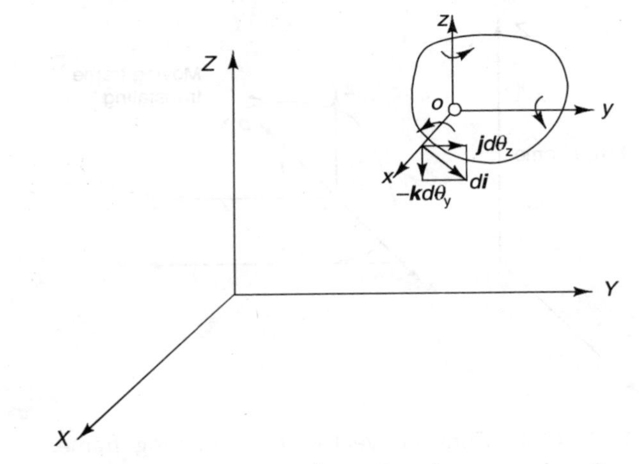

FIGURE 13.3 Rotation of moving frame and unit vector.

Total changes in vector $\boldsymbol{i}$ with respect to the fixed frame is

$$(di)_f = \boldsymbol{j}\ d\theta_z - \boldsymbol{k}\ d\theta_y$$

Therefore rate of change of unit vector $\boldsymbol{i}$ is

$$\left(\frac{di}{dt}\right)_f = \boldsymbol{j}\frac{d\theta_z}{dt} - \boldsymbol{k}\frac{d\theta_y}{dt}$$

$$= \boldsymbol{j}\ \omega_z - \boldsymbol{k}\ \omega_y$$

But we know

$$\omega \times \boldsymbol{i} = (\omega_x \boldsymbol{i} + \omega_y \boldsymbol{j} + \omega_z \boldsymbol{k}) \times i$$

$$= \omega_x \times 0 - \boldsymbol{k}\ \omega_y + \boldsymbol{j}\ \omega_z$$

$\therefore$

$$\left(\frac{di}{dt}\right)_f = \omega \times \boldsymbol{i}$$

Similarly, we have

$$\left(\frac{dj}{dt}\right)_f = \omega \times \boldsymbol{j}$$

and

$$\left(\frac{dk}{dt}\right)_f = \omega \times \boldsymbol{k}$$

(d) *Derivative of a constant vector (r) in moving frame.* Consider a fixed position vector $\boldsymbol{r}$ of a point P located on the moving frame which is rotating with an angular velocity ω as shown in Figure 13.4.

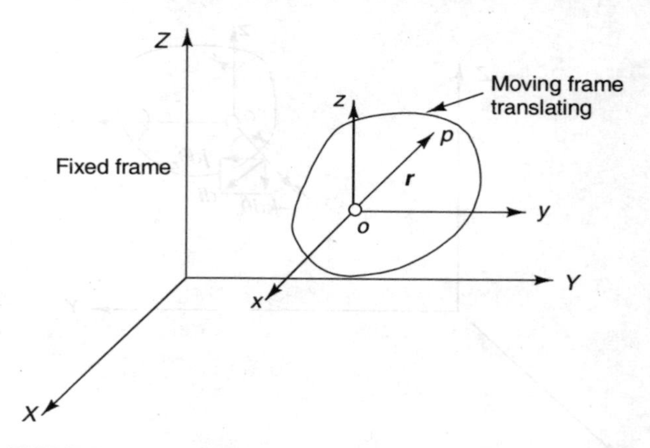

FIGURE 13.4 Constant vector ($\boldsymbol{r}$) in moving frame.

Now,

$$\boldsymbol{r} = x\boldsymbol{i} + y\boldsymbol{j} + z\boldsymbol{k}$$

$$\therefore \quad \left(\frac{dr}{dt}\right)_f = \frac{dx}{dt}\boldsymbol{i} + \frac{dy}{dt}\boldsymbol{j} + \frac{dz}{dt}\boldsymbol{k} + x\left(\frac{di}{dt}\right)_f + y\left(\frac{di}{dt}\right)_f + z\left(\frac{di}{dt}\right)_f$$

$$= \frac{dx}{dt}\boldsymbol{i} + \frac{dy}{dt}\boldsymbol{j} + \frac{dz}{dt}\boldsymbol{k} + x(\omega \times \boldsymbol{i}) + y(\omega \times \boldsymbol{j}) + z(\omega \times \boldsymbol{k})$$

Also

$$\left(\frac{dr}{dt}\right)_m = \frac{dx}{dt}\boldsymbol{i} + \frac{dy}{dt}\boldsymbol{j} + \frac{dz}{dt}\boldsymbol{k}$$

$$= 0 \text{ (as there is no change)}$$

$$\therefore \quad \left(\frac{dr}{dt}\right)_f = x(\omega \times \boldsymbol{i}) + y(\omega \times \boldsymbol{j}) + z(\omega \times \boldsymbol{k})$$

$$= \omega \times (x\boldsymbol{i} + y\boldsymbol{j} + z\boldsymbol{k})$$

$$= \omega \times \boldsymbol{r}$$

(e) *Derivative of a vector **A** located on a moving frame.* Consider a vector A located on a moving frame which is rotating with angular velocity ω. If the position vectors specifying the start and the end of the vector $\boldsymbol{A}$ are $\boldsymbol{r}$ and $\boldsymbol{r}'$ as shown in Figure 13.5, then we have

$$\boldsymbol{A} = \boldsymbol{r} - \boldsymbol{r}'$$

Now

$$\left(\frac{dA}{dt}\right)_f = \left(\frac{dr}{dt}\right)_f - \left(\frac{dr'}{dt}\right)_f$$

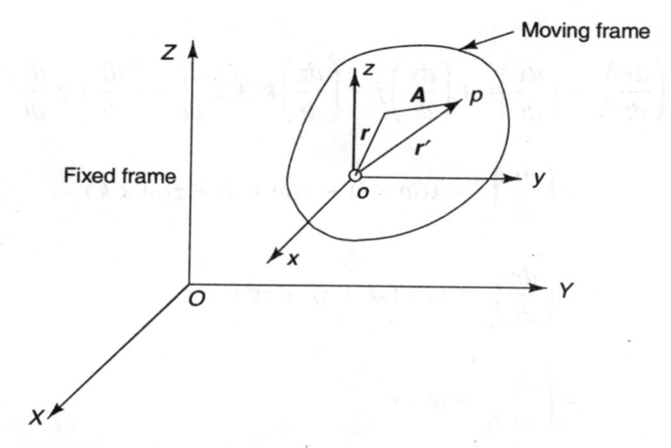

FIGURE 13.5 A vector located on a moving frame.

$$= \omega \times \boldsymbol{r} - \omega \times \boldsymbol{r}'$$

$$= \omega \times (\boldsymbol{r} - \boldsymbol{r}')$$

(f) *Derivative of a positional vector ($\boldsymbol{r}$) located in moving frame and being referred to fixed frame.* Consider the positional vector ($\boldsymbol{r}$) of point P located in a moving frame. The position vector of point P is R from the fixed frame as shown in Figure 13.6. It is evident from the figure

$$\boldsymbol{r} = x\boldsymbol{i} + y\boldsymbol{j} + z\boldsymbol{k}$$

and

$$R = R_0 + \boldsymbol{r}$$

Now

$$\left(\frac{dr}{dt}\right)_m = \left(\frac{dx}{dt}\right)\boldsymbol{i} + \left(\frac{dy}{dt}\right)\boldsymbol{j} + \left(\frac{dz}{dt}\right)\boldsymbol{k}$$

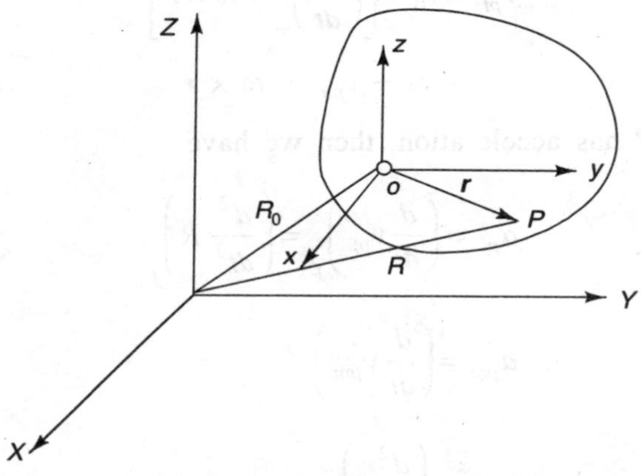

FIGURE 13.6 Derivative of positive vector from fixed frame.

and

$$\left(\frac{dr}{dt}\right)_f = \left(\frac{dx}{dt}\right)\boldsymbol{i} + \left(\frac{dy}{dt}\right)\boldsymbol{j} + \left(\frac{dz}{dt}\right)\boldsymbol{k} + x\frac{di}{dt} + y\frac{di}{dt} + z\frac{dz}{dt}$$

$$= \left(\frac{dr}{dt}\right)_m + x(\omega \times \boldsymbol{i}) + y(\omega \times \boldsymbol{j}) + z(\omega \times \boldsymbol{k})$$

$$= \left(\frac{dr}{dt}\right)_m + \omega \times (x\boldsymbol{i} + y\boldsymbol{j} + z\boldsymbol{k})$$

$$= \left(\frac{dr}{dt}\right)_m + \omega \times \boldsymbol{r}$$

Now in case point P has velocity v_{pm}, then

$$v_{pm} = \left(\frac{dr}{dt}\right)_m$$

and

$$v_{pf} = \left(\frac{dR}{dt}\right)_f$$

But

$$R = R_0 + \boldsymbol{r}$$

$$\therefore \qquad \left(\frac{dR}{dt}\right)_f = \left(\frac{dR_0}{dt}\right)_f + \left(\frac{dr}{dt}\right)_f$$

$$v_{pf} = v_0 + \left[\left(\frac{dv}{dt}\right)_m + \omega \times \boldsymbol{r}\right]$$

$$= v_0 + v_{pm} + \omega \times \boldsymbol{r}$$

In case point P has acceleration, then we have

$$a_{pf} = \left(\frac{d}{dt}v_{pf}\right)_f = \left(\frac{d^2}{dt^2}R\right)_f$$

$$a_{pm} = \left(\frac{d}{dt}v_{pm}\right)_m$$

$$= \left(\frac{d^2 r}{dt^2}\right)_m$$

Now

$$a_{pf} = \left(\frac{d^2}{dt^2}R\right)_f = \left[\frac{d^2}{dt^2}(R_0 + r)\right]_f$$

$$= \left(\frac{d}{dt}v_0\right)_f + \left[\frac{d^2}{dt^2}(r)\right]_f$$

$$= a_0 + \left[\frac{d}{dt}(v_{pm} + \omega \times r)\right]_f$$

$$= a_0 + \left[\frac{d}{dt}v_{pm}\right]_f + \left[\frac{d}{dt}\omega \times r\right]_f$$

$$= a_0 + \left\{\left[\frac{d}{dt}v_{pm}\right]_m + \omega \times v_{pm}\right\} + \left\{\omega \times \frac{dr}{dt} + \frac{d\omega}{dt} \times r\right\}_f$$

$$= a_0 + a_{pm} + \omega \times v_{pm} + \left\{\omega\left[\left(\frac{dr}{dt}\right)_m + \omega \times r\right] + a \times r\right\}$$

$$= a_0 + a_{pm} + \omega \times v_{pm} + \omega \times v_{pm} + \omega \times (\omega \times r) + a \times r$$

$$= a_0 + a_{pm} + a \times r + 2\omega \times v_{pm} + \omega \times (\omega \times r)$$

TRANSLATION MOTION

A rigid body is said to be carrying out translation motion if the linear displacement of every point in the rigid body is the same. When a rigid body is in translation, all points or particles of the body have the same velocity and same acceleration at any particular instant. The translation can be (i) rectilinear and (ii) curvilinear. In rectilinear translation, a point P on the rigid body moves in straight line PP_1P' and a straight line PQ moves straightly to $P'Q'$ through P_1Q_1 as shown in Figure 13.7.

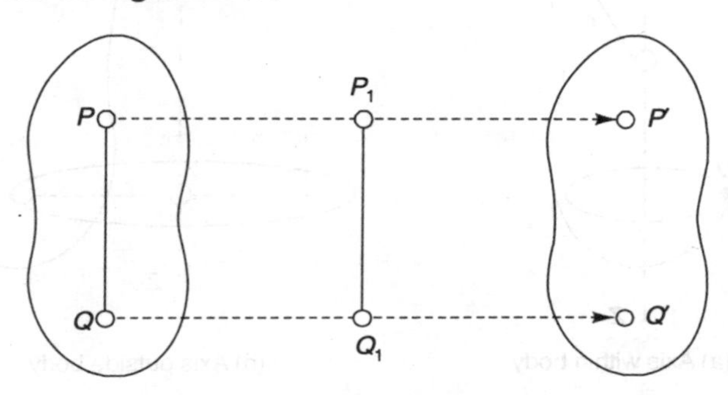

FIGURE 13.7 Rectilinear translation.

In curvilinear translation, a selected point P may move in plane or space as it trace curve PP_1P'. Also line PQ while moving is parallel to P_1Q_1 and $P'Q'$ as shown in Figure 13.8. The features of translatory motion are as follows:

(a) The curve traced by each point of the rigid body is identical.
(b) All points of the rigid body have the same linear displacement, same velocity and same acceleration at a given instant.

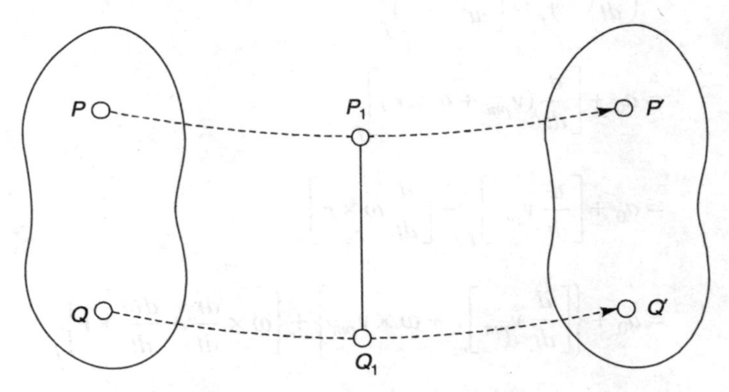

FIGURE 13.8 Curvilinear translation.

ROTATIONAL MOTION

Rotational motion is characterized by the same angular displacement of all points of the rigid body. If a body is rotated about an axis (axis can be within or outside the body), the angular displacement, angular velocity and angular acceleration at a given instant of time are the same for all points in the rigid body (see Figure 13.9). The trajectory of the movement of each point on the rigid body in rotation is a circle with centre on the axis of rotation. In

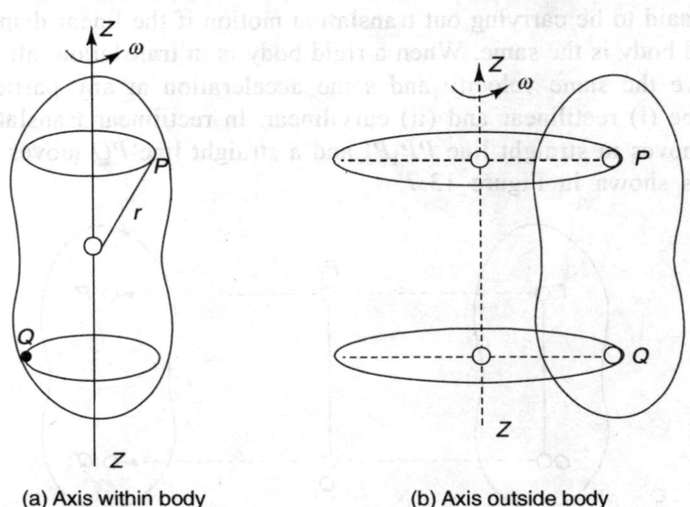

FIGURE 13.9 Rotational motion.

case, rotation is ω and acceleration is a, then the velocity and acceleration of point P having positional vector $\boldsymbol{r}$ are

$$v_p = \omega \times \boldsymbol{r}$$

$$a_p = a \times \boldsymbol{r} + \omega \times (\omega \times \boldsymbol{r})$$

PLANE MOTION

As the name suggests, the plane motion of a rigid body is a type of motion in which all points in the body stay in the same parallel planes. A plane motion may be (i) pure translation, (ii) pure rotation and (iii) a combination of translation and rotation.
The examples of plane motion are as follows:

(a) Linear translation of a rigid body.
(b) Rotation of a rigid body about a fixed axis.
(c) Rolling of a cylinder (Figure 13.10) or disc on a flat surface.

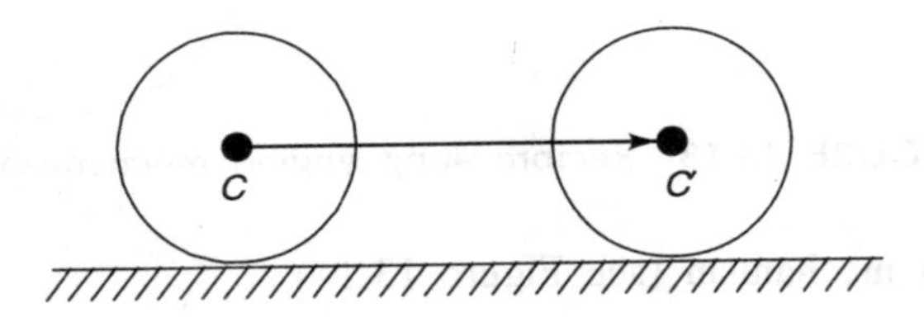

FIGURE 13.10 Rolling of a cylinder.

(d) Sliding of a ladder or a slender bar (see Figure 13.11).

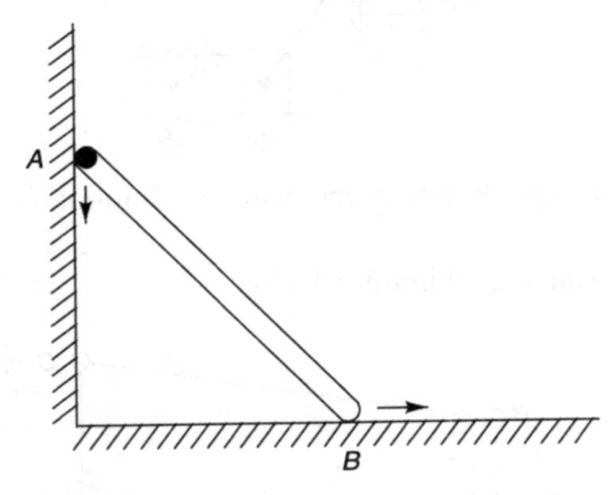

FIGURE 13.11 Sliding of a slender bar.

(e) Sliding of a straight bar in a semi-cylindrical trough (see Figure 13.12).

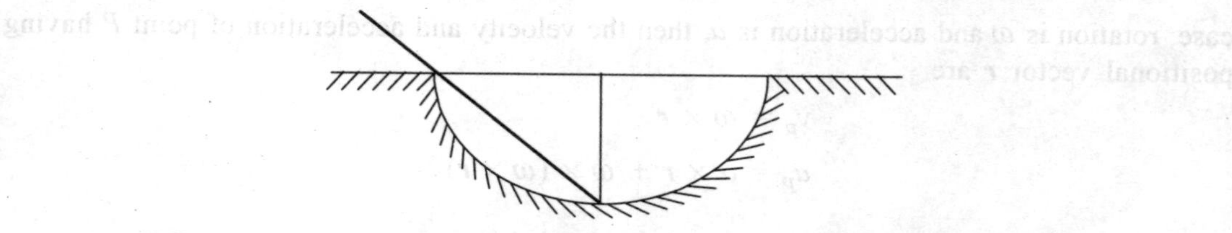

FIGURE 13.12 Sliding of a straight bar in a semi-cylindrical trough.

(f) A reciprocating engine mechanism (see Figure 13.13).

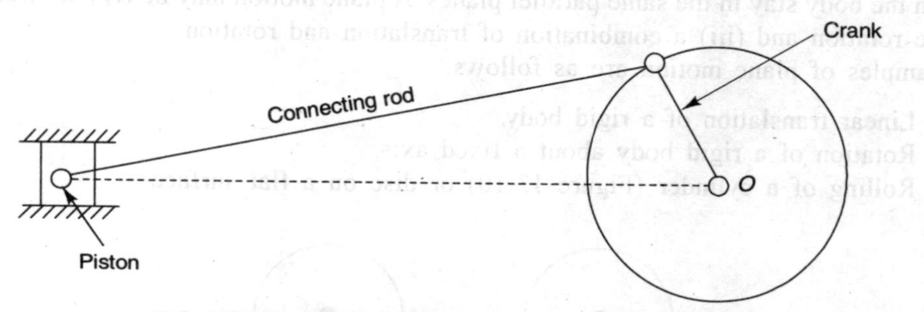

FIGURE 13.13 Reciprocating engine mechanism.

(g) A quick return mechanism (see Figure 13.14).

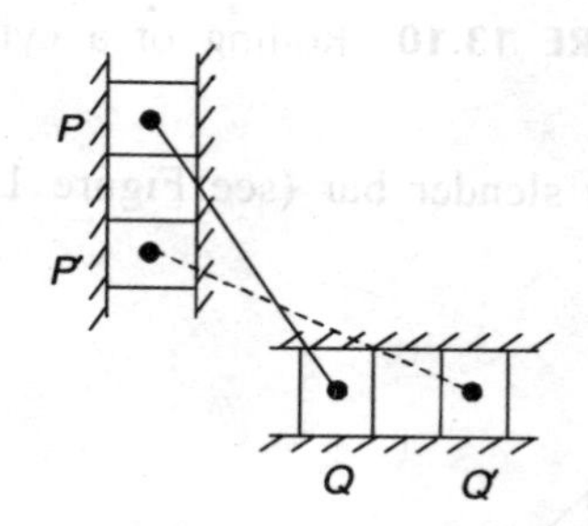

FIGURE 13.14 Quick return mechanism: Translation and Rotation.

(h) Four bar mechanism (see Figure 13.15).

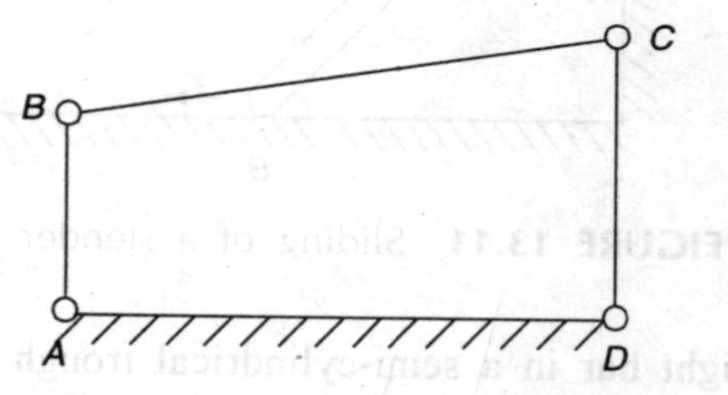

FIGURE 13.15 Four bar mechanism: Translation and Rotation.

(i) A thin plate hanging by two equal strings and performing plane curvilinear plane translation (see Figure 13.16).

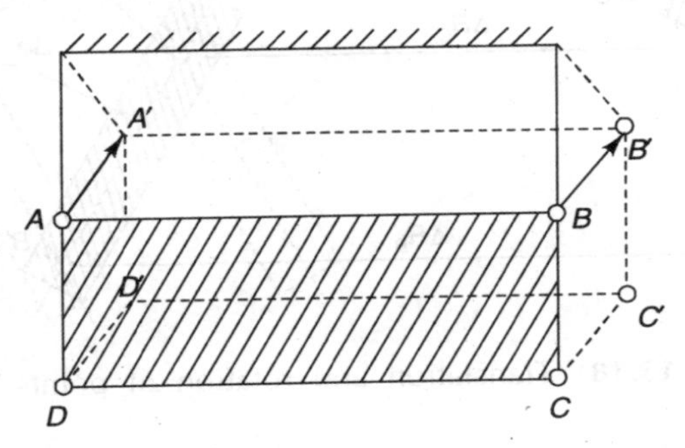

FIGURE 13.16 Curvilinear translation.

Chasles Theorem

Two simple motions of a body are pure translation and pure rotation. It can be shown that any motion of any rigid body can be considered to be made by the superposition of a translation and a rotational motion. Consider a body moving in a plane and its positions shown in Figure 13.17 at time t and $t + \Delta t$. Choose a point B on the body. The point B of the body displaces to point B in time Δt and displacement vector for the translation is shown as ΔR_B. In order to achieve the correct orientation of the body for $(t + \Delta t)$, the body

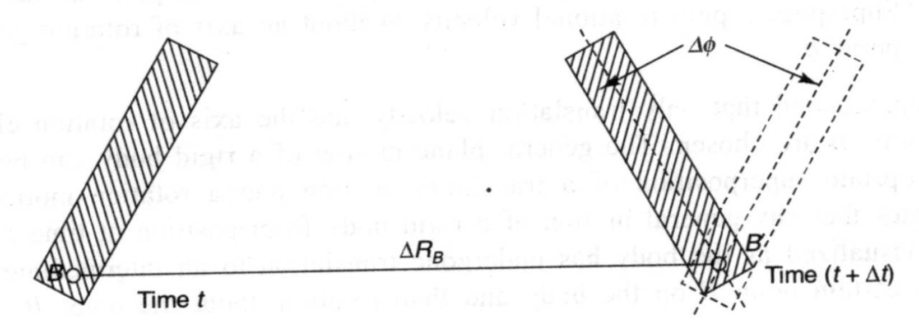

FIGURE 13.17 Translation and rotation of point B.

has to be rotated by an angle of $\Delta\phi$ about an axis which is normal to the plane and it passes through point B. Consider another point C on the body which moves to point C in time Δt. The body has to be rotated by angle $\Delta\phi$ to achieve the correct final orientation of the body (Figure 13.18). The displacement vector for translation in this case is ΔR_C. The displacement ΔR_C differs from ΔR_B but there is no difference in the amount of rotation which is $\Delta\phi$ for both cases. The following are deductions:

(a) ΔR and the axis of rotation depend on the point selected on the rigid body.
(b) The amount of rotation $\Delta\phi$ is the same for all points for the rigid body.

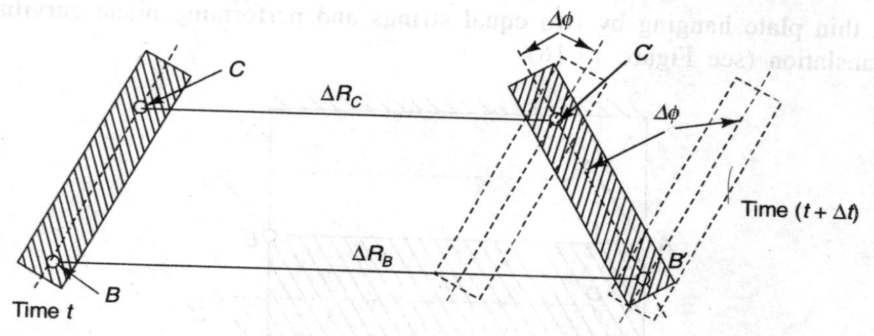

FIGURE 13.18 Translation and rotation of points B and C.

(c) $\dfrac{\Delta R}{\Delta t}$ with limit $\Delta t \to 0$ gives instantaneous translation velocity which is varying from point to point on the rigid body. The translation velocity of a point depends upon instantaneous translation velocity of the point.

(d) $\dfrac{\Delta \phi}{\Delta t}$ with limit $\Delta t \to 0$ is the angular velocity ω which is the same for all points on the rigid body. According to Chasles' theorem

(i) Select any point B in the body and assume all points or particles on the body have the same velocity which is equal to velocity of point B, i.e. v_B.

(ii) Superpose a pure rotational velocity ω about an axis of rotation going through point B.

It can be appreciated that only translation velocity and the axis of rotation change when different points B are chosen. The general plane motion of a rigid body can be considered as an appropriate superposition of a translation motion and a rotation motion. Chasles' theorem states that any general motion of a rigid body from position at time t to time $t + \Delta t$ can be visualized as the body has undergone translation to an intermediate position in regard to a certain point B on the body and than rotation about the point B.

INSTANTANEOUS CENTRE OF ROTATION

The general plane motion of a rigid body can be considered as the sum of a plane translation motion and a rotational motion about an axis perpendicular to the plane of motion. The velocity of a rigid body can therefore be completely specified by specifying the translation velocity of point $P(v_P)$ and the rotational velocity (ω) about an axis through the point as shown in Figure 13.19.

The rotational velocity (ω) is the same for every point of the body and the displacement or translation is different for different points which suggests that a point can be considered to exist on the body with respect to which the body can be considered to rotate about an axis passing through this point at that instant. Such a point is called *instantaneous centre*

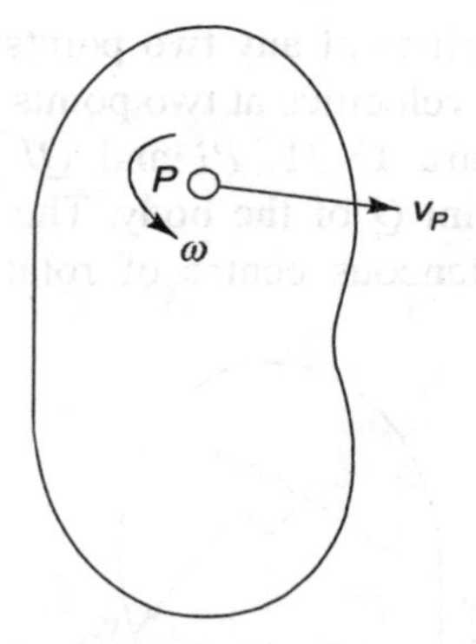

FIGURE 13.19 Translation and rotation for specifying general plane motion.

of rotation. The point of instantaneous centre of rotation has zero velocity at that instant. The point of instantaneous centre can be located on the body or outside the body. The velocity of other points in the body can be found out by comparing the normal distances of the points from the instantaneous centre of rotation as shown in Figure 13.20.

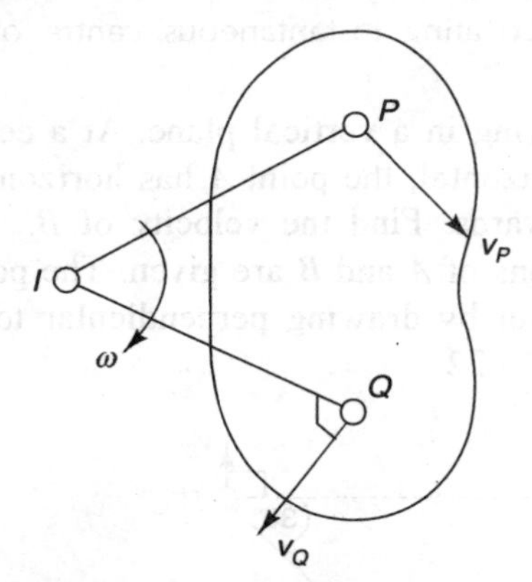

FIGURE 13.20 Instantaneous centre of rotation.

The velocities at two points P and Q of the body are

$$v_P = (IP) \times \omega$$

$$v_Q = (IQ) \times \omega$$

or

$$\frac{v_Q}{v_P} = \frac{IQ}{IP}$$

or

$$v_Q = v_P \times \frac{IQ}{IP}$$

Hence, it is possible to find v_Q in case velocity v_P is known and distances IQ and IP are measured. This fact provides another method of locating the instantaneous centre of rotation (point I) if the directions of the velocities at any two points on a rigid body are known. The perpendiculars to the directions of the velocities at two points will intersect at the instantaneous centre of rotation as shown in Figure 13.21. PI and QI are drawn perpendicular to the velocities v_P at point P and v_Q at point Q of the body. The perpendiculars intersect at point I which is now the point of instantaneous centre of rotation.

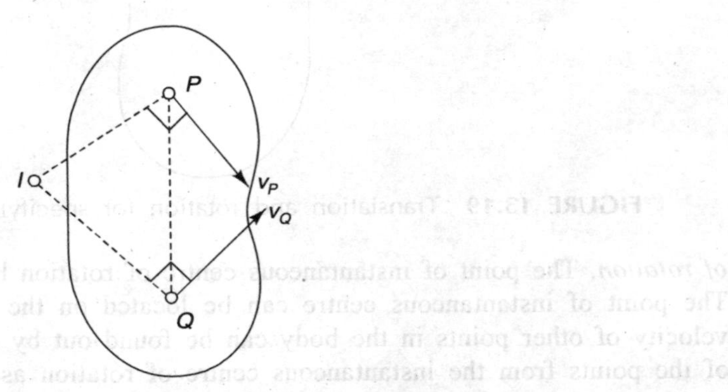

FIGURE 13.21 Locating instantaneous centre of rotation.

Example 13.1 A link AB is moving in a vertical plane. At a certain instant when the link is inclined at angle 30° to the horizontal, the point A has horizontal velocity of 8 m/s while point B has vertical velocity upwards. Find the velocity of B.

The actual directions of motions of A and B are given. The position of the instantaneous centre of rotation can be found out by drawing perpendicular to the directions of motions at A and B as shown in Figure 13.22.

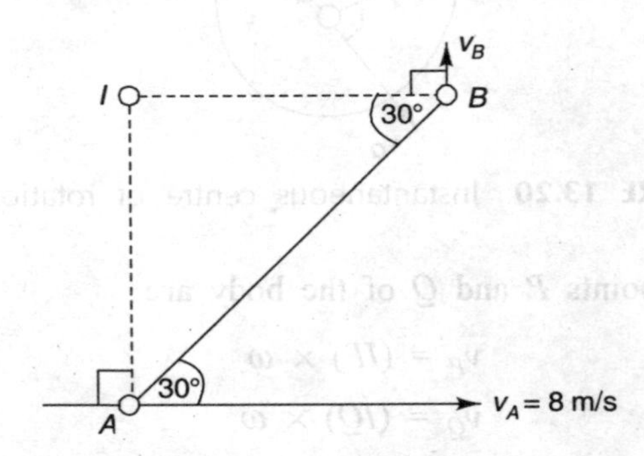

FIGURE 13.22 Example 13.1.

Draw AI perpendicular to v_A and BI perpendicular to v_B. The perpendicular intersects at I which is the instantaneous centre of rotation. Hence we have

$$\frac{v_A}{v_B} = \frac{AI}{BI} = \tan 30$$

$$\therefore \qquad \frac{8}{v_B} = \frac{1}{\sqrt{3}}$$

or

$$v_B = 8 \times \sqrt{3}$$

$$= 8 \times 1.732$$

$$= 13.856 \text{ m/s}$$

Example 13.2 A wheel of radius 0.4 m rolls without slipping down on an inclined plane making an angle of 30° from the horizontal plane as shown in Figure 13.23. At this instant of time, the velocity of the centre of wheel is 8 m/s. Find the location of the instantaneous centre of rotation and velocities of the points at *A*, *B* and *C* on the periphery.

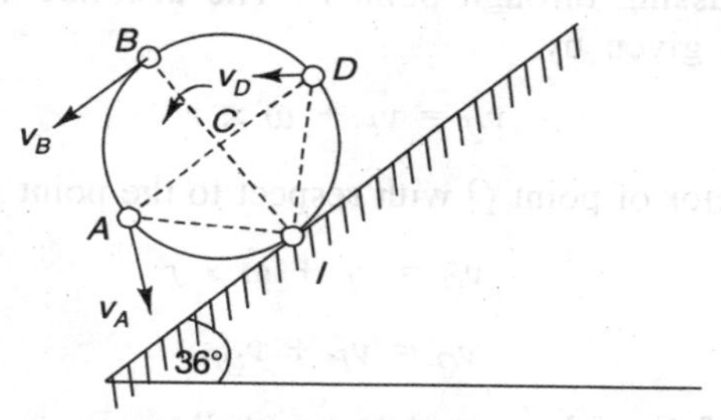

FIGURE 13.23 Example 13.2.

As the wheel is rolling without slippage, the point of contact of the wheel with the ground at point *I* has zero velocity. Hence, point *I* is the instantaneous centre of rotation of the wheel. This gives us

$$v_C = IC \times \omega$$

$$v_A = IA \times \omega$$

$$v_B = IB \times \omega$$

and

$$v_D = I_D \times \omega$$

Given:

$$v_C = 8 \text{ m/s and } IC = R = 0.4 \text{ m}$$

$$\therefore \qquad \omega = \frac{v_C}{IC} = \frac{v_C}{R} = \frac{8}{0.4} = 20 \text{ m/s}$$

Now

$$IA = \sqrt{2} \times R = \sqrt{2} \times 0.4 = 0.566 \text{ m}$$

$$ID = \sqrt{2} \times R = \sqrt{2} \times 0.4 = 0.566 \text{ m}$$

$$IB = 2R = 2 \times 0.4 = 0.8 \text{ m}$$

Hence, we have

$$v_A = IA \times \omega$$
$$= 0.566 \times 20$$
$$= 11.32 \text{ m/s}$$
$$v_D = v_A = 11.32 \text{ m/s}$$
$$v_B = 0.8 \times 20$$
$$= 16 \text{ m/s}$$

Relative Velocity and Acceleration for Points on a Rigid Body

The general plane motion is made up by the combination of a translation of point P and a rotation about the axis passing through point P. The absolute velocity of another point Q on the rigid body can be given by

$$v_Q = v_P + \omega \times \boldsymbol{r}$$

where $\boldsymbol{r}$ is the position vector of point Q with respect to the point P as shown in Figure 13.24.

$$v_Q = v_P + \omega \times r$$

or

$$v_Q = v_P + v_{QP}$$

where v_{QP} = Velocity of Q with respect to point P

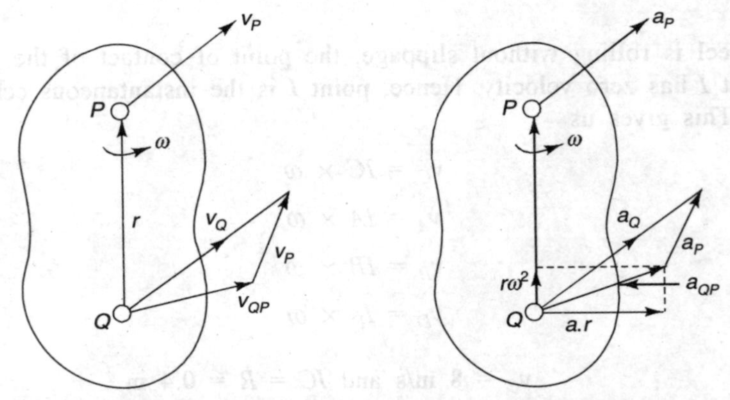

FIGURE 13.24 Relative velocity and acceleration.

As $v_{QP} = \omega \times \boldsymbol{r}$ which is a cross product vector which means that vector v_{QP} will lie in a plane which is perpendicular to the plane containing ω and $\boldsymbol{r}$. Hence v_{QP} has to lie in the plane of the motion and it is acting in the direction perpendicular to the line joining P and Q as shown in the figure.

The absolute acceleration of a point Q in terms of the acceleration of point P can be given similarly as

$$a_Q = a_P + a_{QP}$$

The acceleration a_{QP} is consisted of

(a) tangential component $= a \times \boldsymbol{r}$
(b) normal component $= \omega \times (\omega \times \boldsymbol{r})$

The tangential component $(a \times \boldsymbol{r})$ will act perpendicular to line PQ while normal component $[\omega \times (\omega \times \boldsymbol{r})]$ will act parallel to the line PQ.

Example 13.3 A straight rigid link AB is 40 cm long. At a given instant, end B is moving along a line OX with velocity 0.8 m/s and acceleration 4 m/s². The other end is moving along YO. Find the velocity and acceleration of the end A as well as of the mid-point C of the link when it is inclined 30° with the OX-axis (see Figure 13.25).

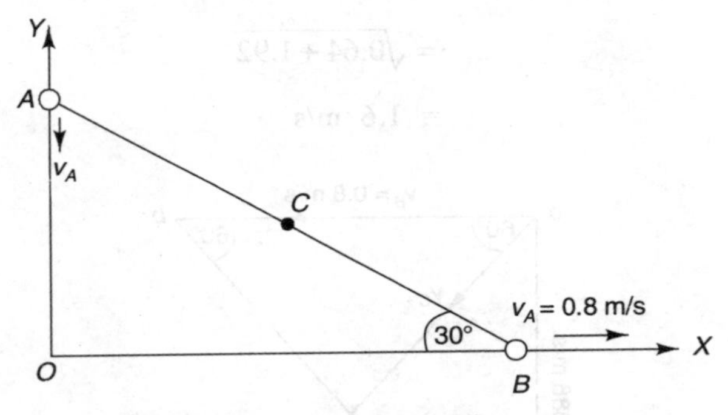

FIGURE 13.25 Example 13.3.

The velocity of end A can be found out by finding the point of instantaneous centre of rotation by drawing perpendiculars to velocities v_A and v_B (see Figure 13.26).

$$\frac{v_A}{v_B} = \frac{AI}{BI} = \frac{OB}{OA} = \frac{1}{\tan 30} = \sqrt{3}$$

or
$$v_A = \sqrt{3} \times v_B$$

$$= \sqrt{3} \times 0.8 = 1.386 \text{ m/s}$$

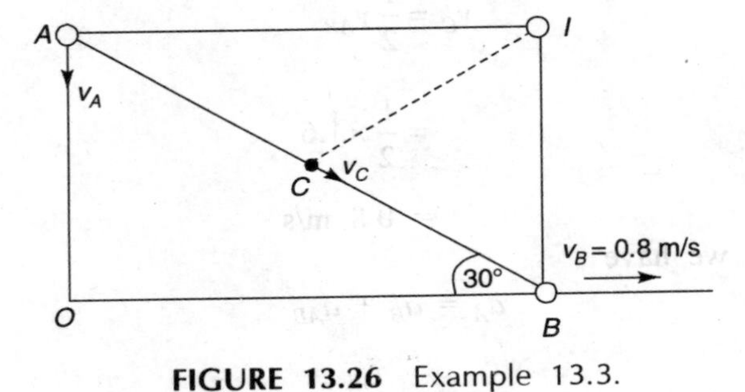

FIGURE 13.26 Example 13.3.

Now

$$v_A = v_B + v_{BA}$$

Draw the velocity diagram as shown below (see Figure 13.27). Line Ob is drawn parallel to the OX-axis with length equal to $v_B = 0.8$ m/s. Line Oa is drawn parallel to the OY-axis with length equal to $v_A = 1.386$ m/s. Join ab which represents v_{AB}. It is to understand that line ab is perpendicular to link AB.

$$v_{AB} = \sqrt{v_A^2 + v_B^2}$$

$$= \sqrt{0.8^2 + 1.386^2}$$

$$= \sqrt{0.64 + 1.92}$$

$$= 1.6 \text{ m/s}$$

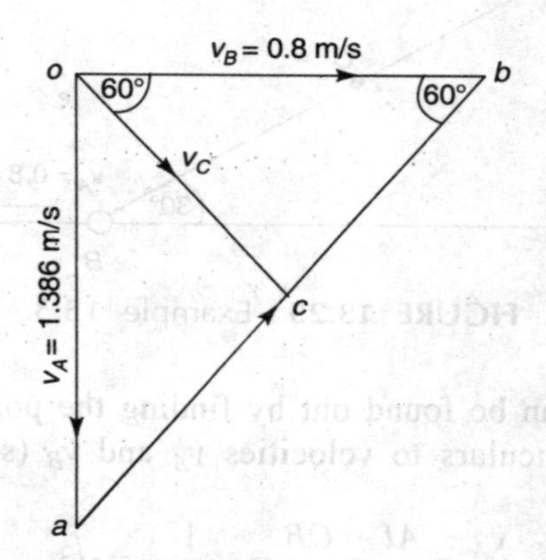

FIGURE 13.27 Example 13.3.

The relative motion of link AB is given by line ab. Now join point O to mid-point of ab which is point C. Now we have v_C which is half of the velocity v_{AB} and perpendicular to link AB. Hence, we have

$$v_C = \frac{1}{2} v_{AB}$$

$$= \frac{1}{2} \times 1.6$$

$$= 0.8 \text{ m/s}$$

Now for acceleration, we have

$$a_A = a_B + a_{AB}$$

Now a_B = 4 m/s^2 along OX. The acceleration a_{AB} consists of

(a) normal component $= \dfrac{v_{AB}^2}{AB}$

$$= \frac{1.6^2}{0.4}$$

$$= 6.4 \text{ m/s}^2$$

It is directed along link AB. Draw ba' = 6.4 m/s^2 (see Figure 13.28).

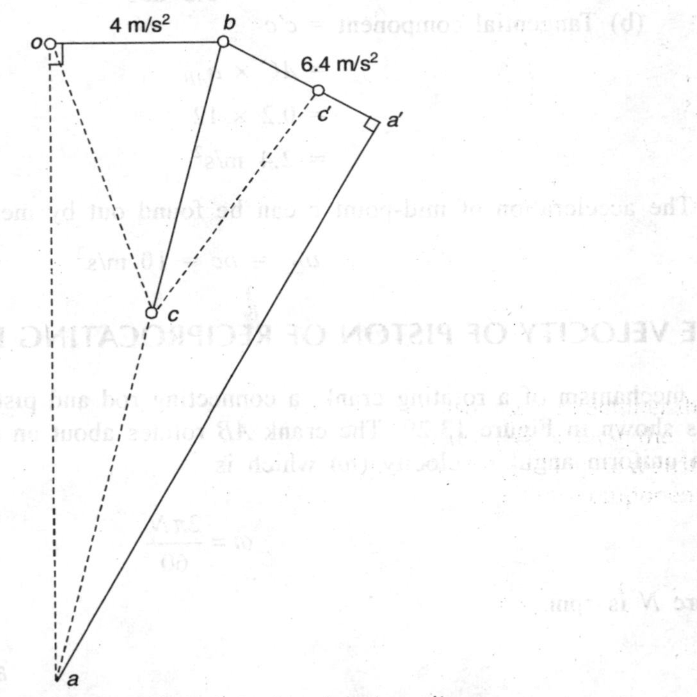

FIGURE 13.28 Example 13.3: Acceleration diagram.

(b) Tangential component = $a_{AB} \times \boldsymbol{r}$. It is directed perpendicular to link AB or line ba' and it cuts vertical line oa at point a. Then $a'a = AB \times a_{AB}$.
From the diagram aa' = 4.8 m/s^2
Hence

$$a_{AB} = \frac{aa'}{AB}$$

$$= \frac{4.8}{0.4}$$

$$= 12 \text{ m/s}^2$$

and

$$a_A = oa = 20 \text{ m/s}^2 \text{ (by measurement)}$$

Now the acceleration for mid-point C is

$$a_C = a_B + a_{CB}$$

The components of a_{CB} are as follows:

(a) Centripetal component $= bc' = 0.2 \times \omega^2$

$$= 0.2 \times \left(\frac{1.6}{0.4}\right)^2$$

$$= 3.2 \text{ m/s}^2$$

(b) Tangential component $= c'c$

$$= AC \times a_{AB}$$

$$= 0.2 \times 12$$

$$= 2.4 \text{ m/s}^2$$

The acceleration of mid-point c can be found out by measuring oc.

$$a_C = oc = 10 \text{ m/s}^2$$

THE VELOCITY OF PISTON OF RECIPROCATING ENGINE

The mechanism of a rotating crank, a connecting rod and piston of a reciprocating engine is as shown in Figure 13.29. The crank AB rotates about an axis passing through point A with uniform angular velocity (ω) which is

$$\omega = \frac{2\pi N}{60}$$

where N is rpm.

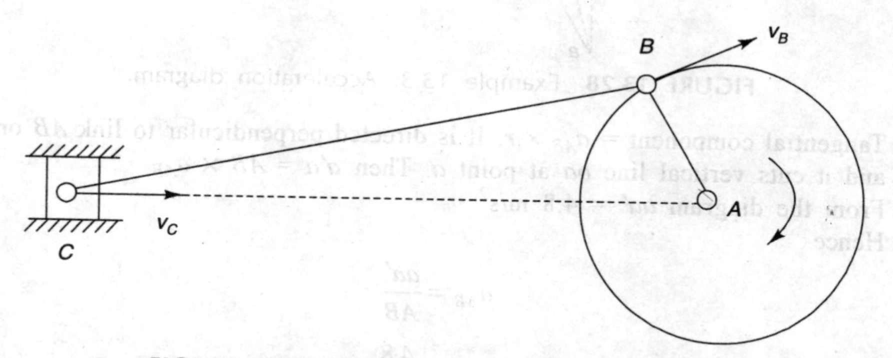

FIGURE 13.29 Reciprocating engine mechanism.

Piston C can make to and fro motion. The connecting rod connects piston C to the crank BA. The velocity of point B is v_B which is perpendicular to crank BA. The velocity of piston is v_C which is acting along line CA. The velocity v_B is known and it is required to find piston velocity v_C. The velocity v_C can be found out by three methods:

(a) graphical method
(b) analytical method
(c) velocity diagram

Graphical method. The construction (Figure 13.30) involves the following steps:

(a) Draw a circle of radius equal to crank length.
(b) Draw crank position for given angle θ. AB is crank position.
(c) Draw a horizontal line from A and cut BC equal to connecting rod length.
(d) Velocity at B is tangential to circle while velocity at C along horizontal line connecting C to A. Mark these velocities v_B and v_C as shown in the figure.
(e) Draw perpendicular lines to v_B and v_C which meet at point I. Point I is the point of instantaneous centre of rotation. Measure the lengths IB and IC.
(f) The velocity of v_C is

$$v_C = \frac{v_B \times IC}{IB}$$

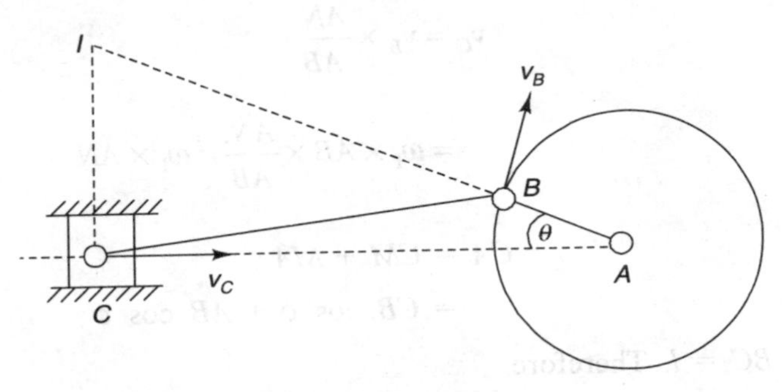

FIGURE 13.30 Graphical method.

Analytical method. Consider the mechanism of a reciprocating engine in which AB is a crank, BC is a connecting rod and C is a piston. Let I be the instantaneous centre of rotation of the connecting rod BC. Extent line CB to meet the vertical line through A at N as shown in Figure 13.31. From B, draw a perpendicular to CA which is BM. Triangles ICB and ABN are similar as per the geometrical construction. Hence, we have

$$\frac{IC}{IB} = \frac{AN}{AB}$$

If ω_1 is the angular velocity of the crank and ω_2 is the angular velocity of the connecting rod with respect to the instantaneous centre of rotation, then we have

$$v_C = \omega_2 \times IC$$

where I is instantaneous centre

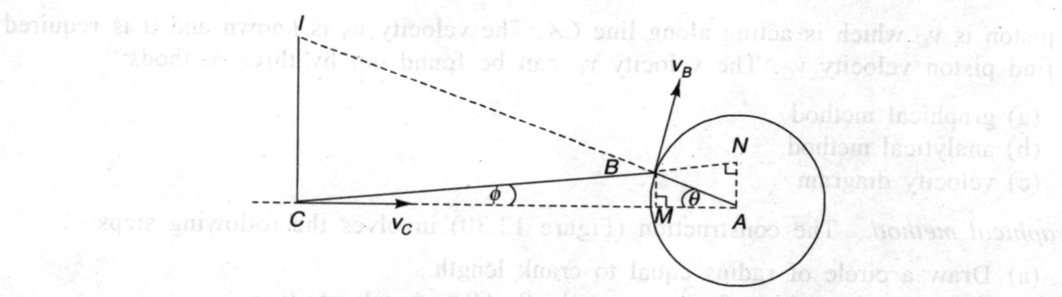

FIGURE 13.31 Analytical method.

$$v_B = \omega_2 \times IB$$

But v_B is also equal to $\omega_1 \times AB$

Now

$$\frac{v_C}{v_B} = \frac{IC}{IB} = \frac{AN}{AB}$$

or

$$v_C = v_B \times \frac{AN}{AB}$$

$$= \omega_1 \times AB \times \frac{AN}{AB} = \omega_1 \times AN$$

Now

$$CA = CM + MA$$

$$= CB \cos \phi + AB \cos \phi$$

Take $AB = r$ and $BC = l$. Therefore,

$$CA = l \cos \phi + r \cos \theta$$

Also

$$AN = CA \tan \phi$$

$$= (l \cos \phi + r \cos \theta) \tan \phi$$

$$= l \sin \phi + r \cos \theta \tan \phi$$

But

$$v_C = \omega_1 \times AN$$

$$= \omega_1 \times (l \sin \phi + r \cos \theta \tan \phi)$$

Velocity diagram. The following is the method of construction of velocity diagram (Figure 13.32) as explained

(a) Draw a horizontal line from a point a.
(b) Draw a line ab parallel to v_B or a right angle to crank AB.

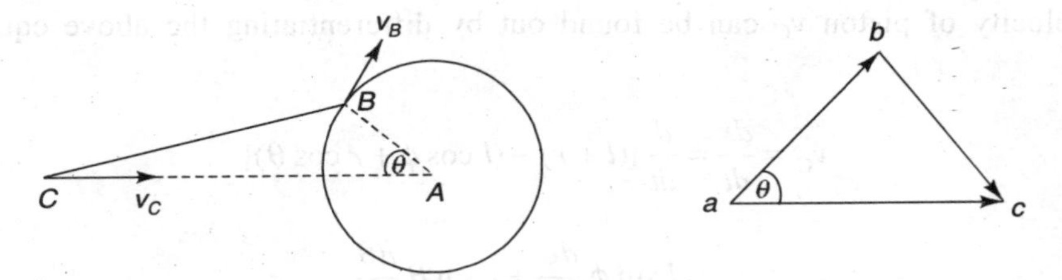

FIGURE 13.32 Velocity diagram.

(c) Length ab on some scale is equal to v_B ($v_B = \omega r$ where ω = angular velocity and r is the length of the crank AB)

(d) Through point b on the velocity diagram draw line bc which is perpendicular to the connecting rod BC.

(e) Line ac can be measured which is equal to the velocity of piston (v_C).

ACCELERATION OF RECIPROCATING PISTON

Figure 13.33 shows the reciprocating engine mechanism. The crank moves from B to B' (top dead centre), thereby the piston moves by distance x as shown in the figure.

$$
\begin{aligned}
x &= AC' - AC \\
&= (AB' + B'C') - (AM + CM) \\
&= (l + r) - (l \cos \phi + r \cos \theta)
\end{aligned}
$$

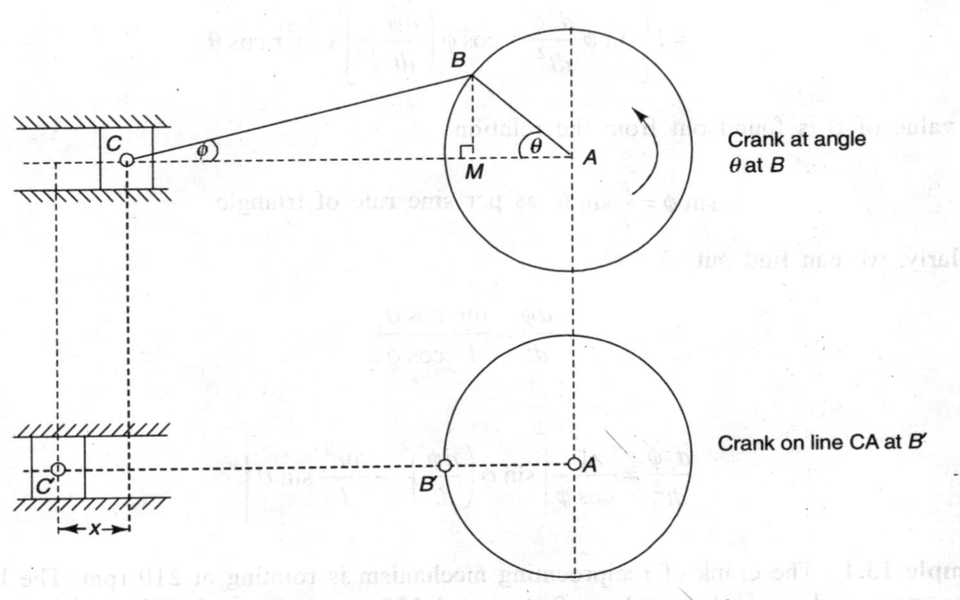

FIGURE 13.33 Acceleration of reciprocating piston.

The velocity of piston v_C can be found out by differentiating the above equation. We get

$$v_C = \frac{dx}{dt} = \frac{d}{dt}[(l + r) - (l \cos \phi + r \cos \theta)]$$

$$= l \sin \phi \frac{d\phi}{dt} + r \sin \theta \frac{d\theta}{dt}$$

But we know

$$\frac{d\theta}{dt} = \omega = \text{angular velocity of crank}$$

$$\therefore \qquad v_C = l \sin \phi \frac{d\phi}{dt} + r\omega \sin \theta$$

The acceleration of piston a_C can be found out by differentiating v_C.

$$a_C = \frac{dv_C}{dt} = \frac{d}{dt}\left(l \sin \phi \frac{d\phi}{dt} + r\omega \sin \theta\right)$$

$$= l\left[\sin \phi \frac{d^2\phi}{dt^2} + \cos \phi \left(\frac{d\phi}{dt}\right)^2\right] + \omega r \times \cos \theta \frac{d\theta}{dt}$$

$$= l\left[\sin \phi \frac{d^2\phi}{dt^2} + \cos \phi \left(\frac{d\phi}{dt}\right)^2\right] + \omega^2 r \cos \theta$$

The value of ϕ is found out from the relation

$$\sin \phi = \frac{r}{l} \sin \theta \quad \text{as per sine rule of triangle}$$

Similarly, we can find out

$$\frac{d\phi}{dt} = \frac{\omega r}{l} \frac{\cos \theta}{\cos \phi}$$

and

$$\frac{d^2\phi}{dt^2} = \frac{1}{\cos \phi}\left[\sin \phi \left(\frac{d\phi}{dt}\right)^2 - \frac{\omega r^2}{l} \sin \theta\right]$$

Example 13.4 The crank of reciprocating mechanism is rotating at 210 rpm. The lengths of the crank and connecting rod are 20 cm and 100 cm respectively. Find the velocity of the piston when the crank is making an angle of 45° to the horizontal.

The velocity of the crank is given, which is v_B. The piston velocity is along CA. Hence, the instantaneous centre of rotation can be found out by drawing perpendiculars to v_B and v_C (Figure 13.34).

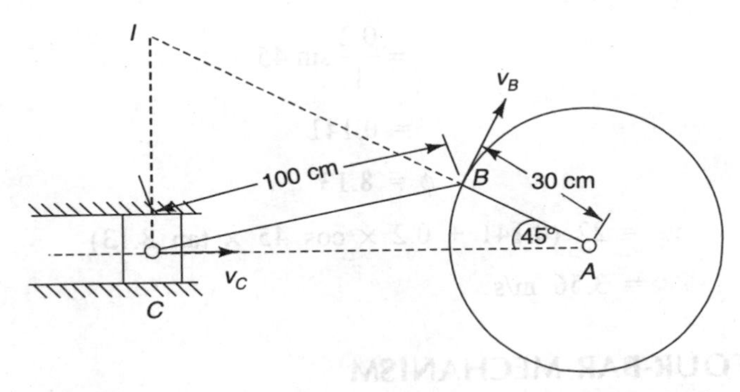

FIGURE 13.34 Example 13.4: Graphical method.

As $N = 120$ rpm,

$$\omega = \frac{2\pi N}{60}$$

$$= \frac{2\pi \times 120}{60}$$

$$= 22 \text{ rad/s}$$

$$\therefore \quad v_B = \omega \times r$$

$$= 22 \times 0.2$$

$$= 4.4 \text{ m/s}$$

length $CI = 1.15$ m (from the diagram)
and $BI = 1.41$ m (from the diagram)

$$v_C = v_B \times \frac{CI}{BI}$$

$$= 4.4 \times \frac{1.15}{1.41}$$

$$= 3.58 \text{ m/s}$$

Analytical method. The piston velocity is

$$v_C = \omega \, (l \sin \phi + r \cos \theta \times \tan \phi)$$

However,

$$\sin\phi = \frac{r}{l}\sin\theta$$

$$= \frac{0.2}{1}\sin 45$$

$$= 0.141$$

$$\phi = 8.13°$$

$\therefore$

$$v_C = 22\ (0.141 + 0.2 \times \cos 45 \times \tan 8.13)$$

$$= 3.56 \text{ m/s}$$

ANALYSIS OF FOUR-BAR MECHANISM

Figure 13.35 shows a four-bar mechanism consisting of a fixed link *AD*, two movable links *AB* and *CD* rotating about points *A* and *D* respectively, and a connecting link *BC*. If *AB* is rotating at uniform angular velocity, then the velocity of link *BC* and *CD* can be found out. Link *AB* and *CD* have rotary motion while link *BC* has a combination of rotary and translation motion. The velocity at *B* and *C* will be perpendicular to link *AB* and *CD* as shown by v_B and v_C in the figure. To find the instantaneous centre of rotation of link *BC*, we have to draw perpendiculars to velocity v_B and v_C which are along length *AB* and *DC* respectively. If we extend *AB* and *DC*, we get the instantaneous centre of rotation which is point *I*. Now we have

$$v_B = \omega_1 \times AB = \omega_0 \times IB$$

$$v_C = \omega_2 \times AB = \omega_0 \times IC$$

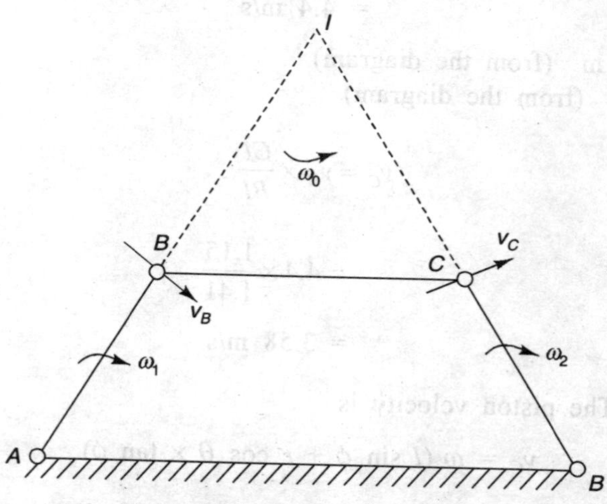

FIGURE 13.35 Four-bar link.

From the above, we get

$$\omega_0 = \frac{\omega_1 \times AB}{IB}$$

$$\omega_2 = \frac{\omega_0 \times IC}{AB}$$

$$= \frac{\omega_1 \times IC}{IB}$$

Example 13.5 A four-bar mechanism is as shown in Figure 13.36. If link AB revolves with 50 rpm, find (i) angular velocity of link CD and (ii) angular velocity of link BC.

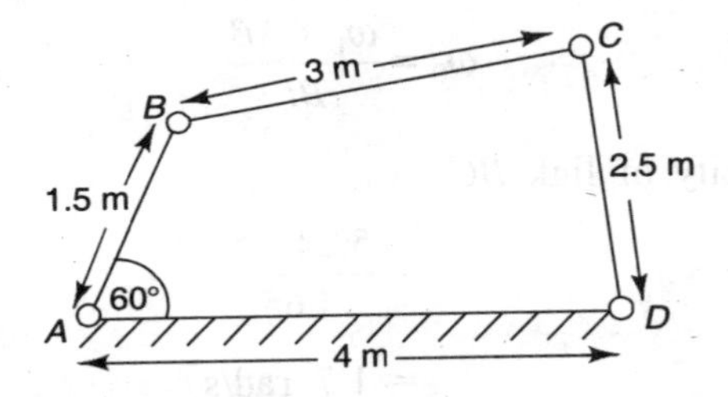

FIGURE 13.36 Example 13.5.

The velocities v_B and v_C at B and C are perpendicular to AB and DC. The instantaneous centre of rotation can be located by drawing perpendiculars to v_B and v_C or just extending AB and DC.

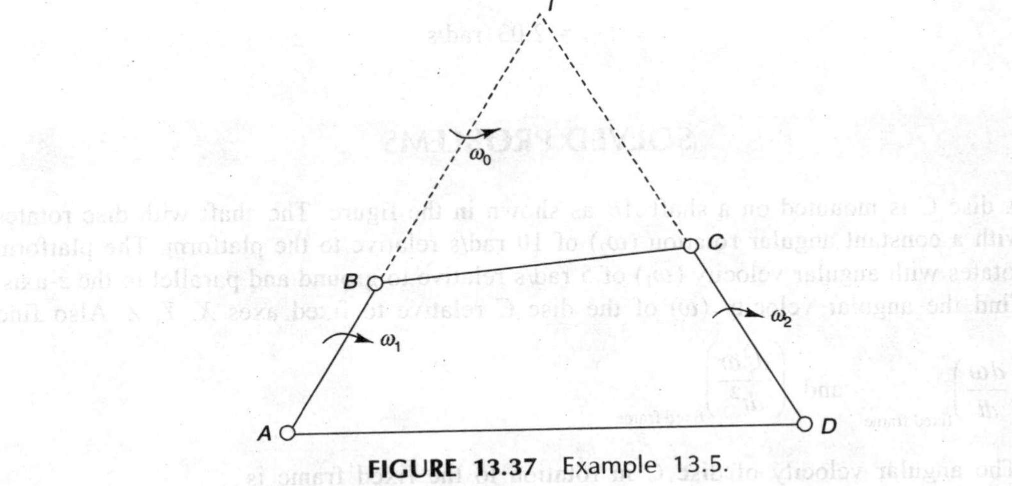

FIGURE 13.37 Example 13.5.

Extend AB and DC, and locate point I. Measure BI and CI. We get

$$BI = 4.65 \text{ m}$$

and
$$CI = 3 \text{ m}$$

Now

$$\omega_1 = \frac{2\pi N}{60}$$

$$= \frac{2\pi \times 50}{60}$$

$$= 5.24 \text{ rad/s}$$

$\therefore$
$$\omega_0 = \frac{\omega_1 \times AB}{BI}$$

where ω_0 = angular velocity of link BC

$$= \frac{5.24 \times 5}{4.65}$$

$$= 1.7 \text{ rad/s}$$

$$\omega_2 = \frac{\omega_0 \times CI}{CD}$$

$$= \frac{1.7 \times 3}{2.5}$$

$$= 2.03 \text{ rad/s}$$

SOLVED PROBLEMS

1. A disc C is mounted on a shaft AB as shown in the figure. The shaft with disc rotates with a constant angular rotation (ω_2) of 10 rad/s relative to the platform. The platform rotates with angular velocity (ω_1) of 5 rad/s relative to ground and parallel to the z-axis. Find the angular velocity (ω) of the disc C relative to fixed axes X, Y, Z. Also find

$$\left(\frac{d\omega}{dt}\right)_{\text{fixed frame}} \quad \text{and} \quad \left(\frac{d^2\omega}{dt^2}\right)_{\text{fixed frame}}$$

The angular velocity of disc C in rotation to the fixed frame is

$$\omega = \omega_1 + \omega_2 \cdot$$

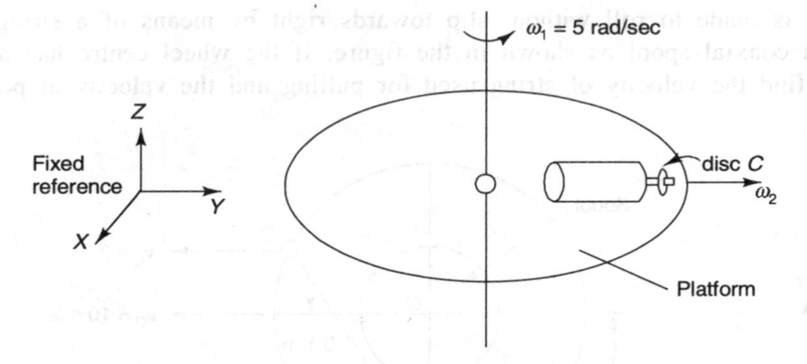

The platform is rotating about an axis parallel to the z-axis with 5 rad/s. The velocity vector is

$$\omega_1 = 5\boldsymbol{k}$$

The disc is rotating about an axis parallel to the y-axis with 10 rad/s at that instant of time. The velocity vector is

$$\omega_2 = 10\boldsymbol{j}$$

$\therefore$ $$\omega = 5\boldsymbol{k} + 10\boldsymbol{j}$$

Again we have

$$\omega = \omega_1 + \omega_2$$

On differentiation with respect to time, we have

$$\dot{\omega} = \dot{\omega}_1 + \dot{\omega}_2$$

As ω_1 is constant, hence $\dot{\omega}_1 = 0$. However, ω_2 is fixed to the platform which has angular velocity ω_1 relative to the fixed frame, we have

$$\dot{\omega}_2 = \omega_1 \times \omega_2$$

$\therefore$ $$\dot{\omega} = \omega_1 \times \omega_2$$

$$= 5\boldsymbol{k} \times 10\boldsymbol{j}$$

$$= -50\boldsymbol{i} \text{ rad/s}$$

Now as

$$\dot{\omega} = \omega_1 \times \omega_2$$

Differentiating again, we get

$$\ddot{\omega} = \dot{\omega}_1 \times \omega_2 + \omega_1 \times \dot{\omega}_2$$

$$= 0 + \omega_1 \times (\omega_1 \times \omega_2)$$

$$= 5\boldsymbol{k} \times (5\boldsymbol{k} \times 10\boldsymbol{j})$$

$$= 5\boldsymbol{k} \times (-50\boldsymbol{i})$$

$$= -250\boldsymbol{j}.$$

2. A wheel is made to roll without slip towards right by means of a string wrapped around a coaxial spool as shown in the figure. If the wheel centre has a speed of 20 m/s, find the velocity of string used for pulling and the velocity at periphery at point C:

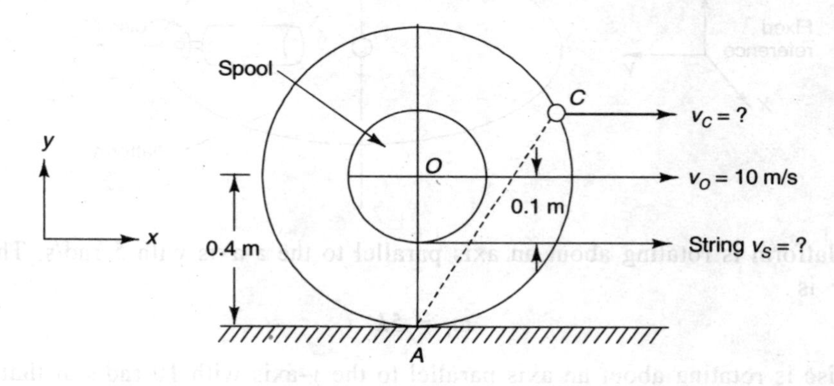

The wheel has instantaneous centre of rotation at point A where it touches the ground. Hence,

$$v_A = 0$$

The velocity at centre O is 10 m/s

$$\therefore \qquad v_O = 10\boldsymbol{i}$$

The velocity of O in relation to A is

$$v_{OA} = v_O + v_A$$

$$= 10\boldsymbol{i}$$

However $v_{OA} = \omega \times r_{OA}$
where $r_{OA} = 0.4\boldsymbol{j}$ and ω is about the z-axis.

$$\therefore \qquad 10\boldsymbol{i} = \omega\boldsymbol{k} \times 0.4\boldsymbol{j}$$

$$= -0.4 \times \omega\boldsymbol{i} \quad (\because \text{ vector } \boldsymbol{k} \times \boldsymbol{j} = -\boldsymbol{i})$$

$$\therefore \qquad \omega = -\frac{10}{0.4} = -25 \text{ rad/s}$$

Now the velocity of string is

$$v_S = v_A + \omega \times r_{SA}$$

$$= 0 + (-25\boldsymbol{k}) \times (+0.3\boldsymbol{j})$$

$$= 7.5\boldsymbol{i} \text{ m/s}$$

Now the velocity of point C is

$$v_C = v_A + \omega \times r_{CA}$$

$$= 0 + (-25\boldsymbol{k}) \times 0.4[\cos 30\boldsymbol{i} + (1 + \sin 30)\boldsymbol{j}]$$

$$= 10(-0.866j + 1.5i)$$
$$= 15i - 8.66j$$

3. A straight bar AB is placed in a semi-cylindrical trough of radius 20 cm and it is released to slide in the trough in such a way that end A slides inside the trough as shown in the figure. The bar can touch and slip at corner O of the trough. At an instant of time, bar makes 45° with the horizontal and A has been found to move at a velocity of 6 m/s horizontally. Find the velocity of sliding at point P of the bar.

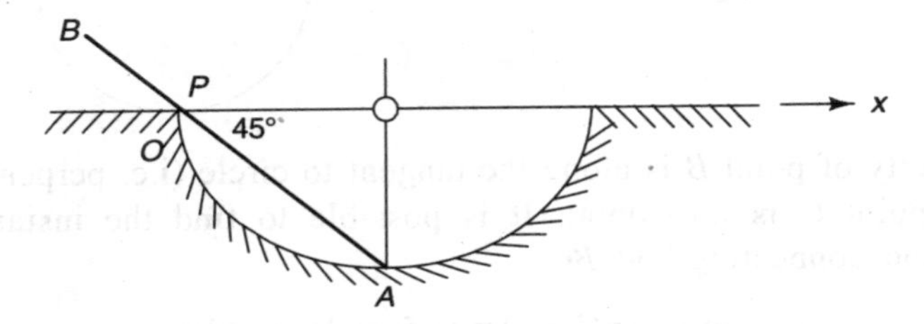

The velocity at point A is in horizontal direction and velocity at point B is along PA (45° to horizontal). The instantaneous centre of rotation can be located by drawing perpendicular to velocities v_A and v_P.

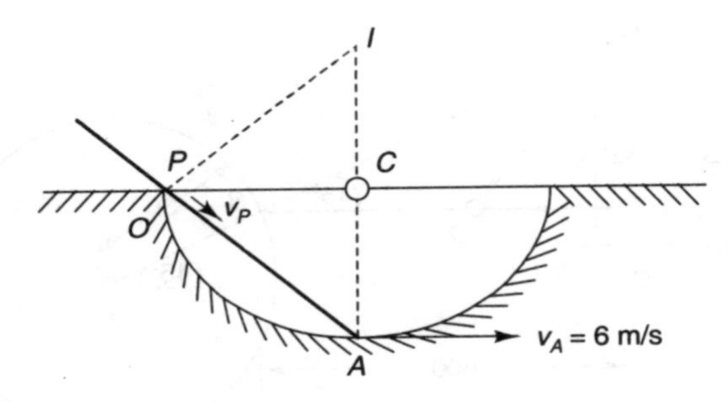

As per geometry, we have

$$IC = CA = r$$

$$\therefore \quad PI = \sqrt{r^2 + r^2}$$

$$= r \times \sqrt{2}$$

$$AI = 2r$$

$$\therefore \quad v_P = v_B \times \frac{PI}{AI}$$

$$= 6 \times \frac{r \times \sqrt{2}}{2r}$$

$$= \frac{6}{\sqrt{2}} = 4.24 \text{ m/s}$$

4. The crank AB is 300 mm long and it rotates at 5 rev/s. The link CB is 600 mm long. The piston C can move horizontally only. When crank is making 45° to horizontal, find (i) the velocity of piston C, (ii) the angular velocity of connecting rod BC and (iii) the velocity of a point D at the centre of connecting rod.

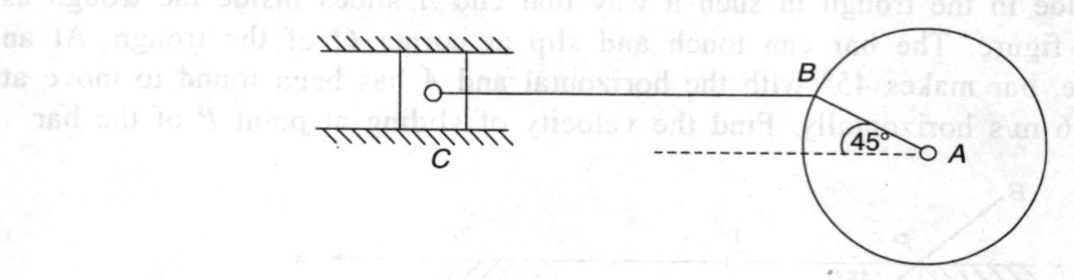

The velocity of point B is along the tangent to circle (i.e. perpendicular to AB) and velocity of point C is horizontal. It is possible to find the instantaneous centre of rotation of the connecting rod BC.

$$\omega = 2\pi N = 2\pi \times 5 = 10\pi \text{ rad/s}$$

$$v_B = \omega \cdot r = 10\pi \times 0.3$$

$$= 9.4 \text{ m/s}$$

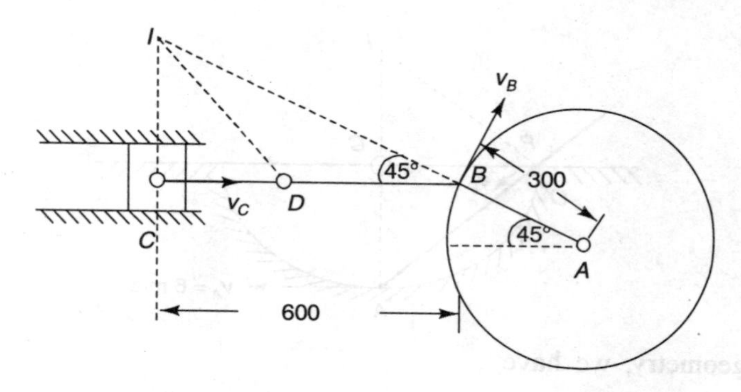

Now

$$v_C = v_B \times \frac{IC}{IB}$$

$$= v_B \times \sin 45$$

$$= 9.4 \times \frac{1}{\sqrt{2}}$$

$$= 6.65 \text{ m/s}$$

If ω_0 is angular velocity of the connecting rod, then we have

$$\omega_0 \times IB = v_B = 9.4$$

or
$$\omega_0 = \frac{9.4}{IB}$$

$$= \frac{9.4}{\frac{0.6}{\cos 45}}$$

$$= \frac{9.4}{0.6} \times \frac{1}{\sqrt{2}}$$

$$= 11.08 \text{ rad/s}$$

Now
$$v_D = \omega_0 \times ID$$

But $ID = 0.67$ m by measurement.

Hence
$$v_D = 11.08 \times 0.67$$
$$= 7.42 \text{ m/s}$$

5. Three links are hinged together to form a triangle *ABC* as shown in the figure. At a certain instant, the point *A* is moving towards the mid-point of *BC* with a velocity of 10 m/s and *B* is moving in perpendicular direction to *AC*. Find the velocity of *C*.

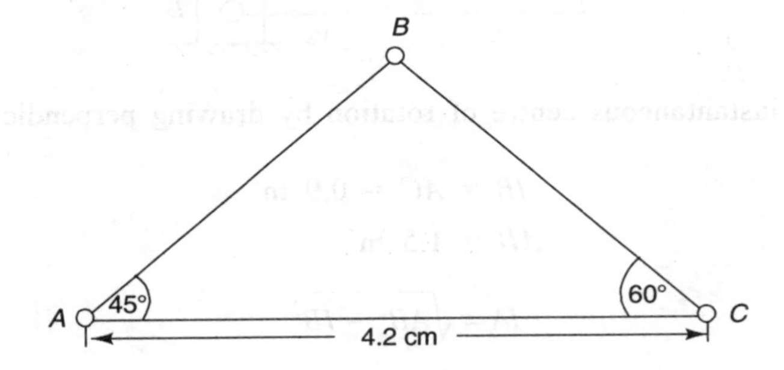

The instantaneous centre of the rotation of the structure *ABC* can be found out by drawing perpendicular to velocity v_A at *A* and velocity v_B at *B* as shown in the figure.

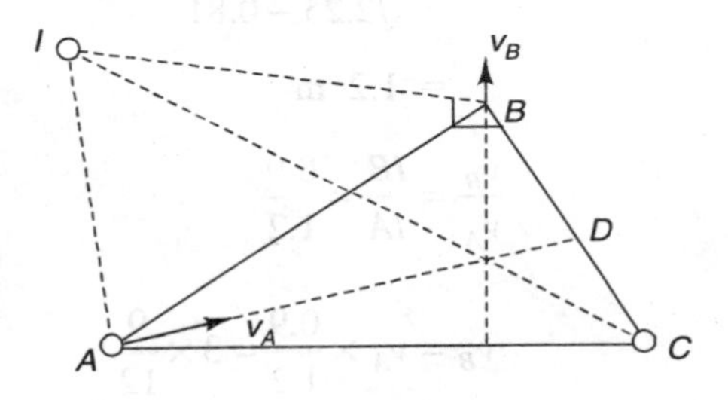

The instantaneous centre of rotation of the structure ABC is found out at I by drawing perpendiculars to v_A and v_B. On measurement, we have IA = 2.6 cm and IC = 5.4 cm.

Now

$$\frac{v_C}{v_A} = \frac{IC}{IA} = \frac{5.4}{2.6}$$

or

$$v_C = v_A \times \frac{5.4}{2.6} = 10 \times \frac{5.4}{2.6}$$

$$= 20.8 \text{ m/s}$$

6. The ends A and B of a link 1.5 m long are constrained to move in the vertical direction in horizontal and vertical guides as shown in the figure. At a given instant, when A is 0.9 m above C, it was moving at 3 m/s upwards. Find the velocity of B at this instant.

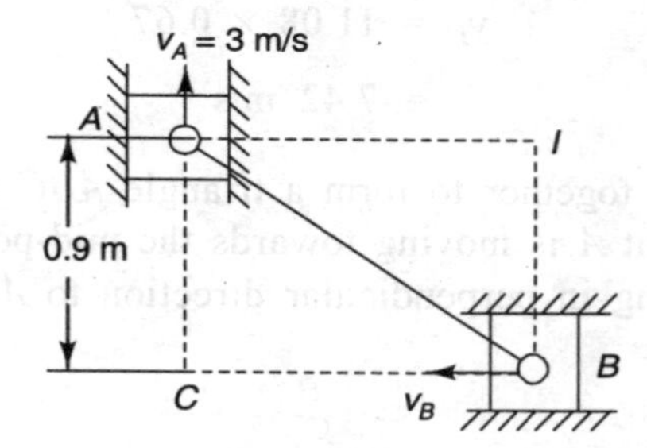

Find the instantaneous centre of rotation by drawing perpendiculars to v_A and v_B.

Now

$$IB = AC = 0.9 \text{ m}$$

$$AB = 1.5 \text{ m}$$

$\therefore$

$$IA = \sqrt{AB^2 - IB^2}$$

$$= \sqrt{(1.5)^2 - 0.9^2}$$

$$= \sqrt{2.25 - 0.81}$$

$$= 1.2 \text{ m}$$

Now

$$\frac{v_B}{v_A} = \frac{IB}{IA} = \frac{0.9}{1.2}$$

$\therefore$

$$v_B = v_A \times \frac{0.9}{1.2} = 3 \times \frac{9}{12}$$

$$= 2.25 \text{ m/s}$$

7. A steam engine has a crank radius of length 15 cm and a connecting rod of length 75 cm as shown in the figure. The crank CQ rotates in a clockwise direction with constant speed of 300 rpm. Calculate the velocity and acceleration of the piston P at the instant when angle $\theta = 30°$. (AMIE: 1980)

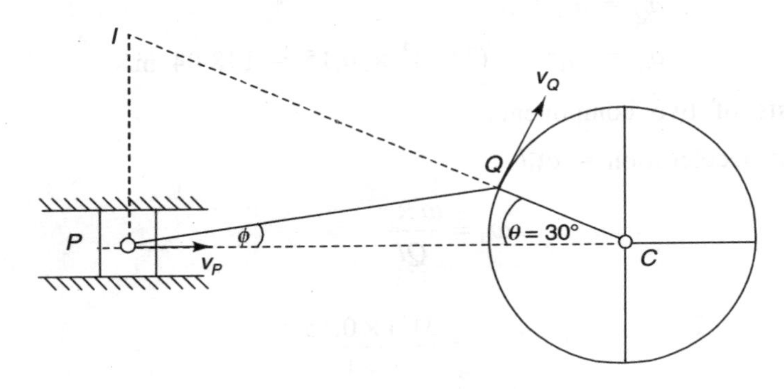

Graphical method: The velocity v_Q is normal to QC and velocity v_P is along PC. Draw perpendiculars to v_Q and v_P which meet at point of instantaneous centre of rotation. On measurement, we get $IP = 0.495$ cm and $IQ = 0.84$ cm.

Now

$$\frac{v_P}{v_Q} = \frac{IP}{IQ}$$

$$\omega_1 = \frac{2\pi N}{60} = \frac{2\pi \times 300}{60}$$

$$= 31.4 \text{ rad/s}$$

$$v_Q = r \times \omega_1 = 0.15 \times 31.4$$

$$= 4.71 \text{ m/s}$$

$\therefore$

$$v_P = v_Q \times \frac{0.495}{0.84} = 2.77 \text{ m/s}$$

Analytical method

$$v_P = \omega_1 \,(l \sin \phi + r \cos \theta \tan \phi)$$

$$\sin \phi = \frac{CQ}{PQ} \sin \theta$$

$$= \frac{0.15}{0.75} \sin 30$$

$$= 0.1$$

or

$$\phi = 5.74°$$

$$\therefore \quad v_P = 31.4\ (0.75 \sin 5.74 + 0.15 \sin 30 \tan 5.74)$$
$$= 2.77 \text{ m/s}$$

Acceleration (a_P)

$$a_Q = a_Q + a_{PQ}$$
$$a_Q = \omega_1{}^2 r = (31.4)^2 \times 0.15 = 148.04 \text{ m/s}^2$$

a_{PQ} consists of two components.

(a) Normal acceleration $= \omega_0^2 l$

$$\omega_0 = \frac{\omega \times r}{QI}$$

$$= \frac{31.4 \times 0.15}{0.84}$$

$$= 5.6 \text{ rad/s}$$

$$(a_{PQ})_{nor} = \omega_0^2 l = (5.6)^2 \times 0.75 = 23.52 \text{ m/s}^2$$

(b) Tangential acceleration $= (a_{PQ})_{tan}$

Now we can draw the acceleration diagram as described below:

(a) Draw a line oa parallel to QC and length $oa = 148.04$ m/s^2 on the same scale.
(b) From point a, draw a line $ab = 23.52$ m/s^2 on the selected scale which is parallel to PQ.
(c) Draw a normal to line ab through point b which cuts a line from o which is horizontal at a point c.
(d) The distance oc is acceleration at P.

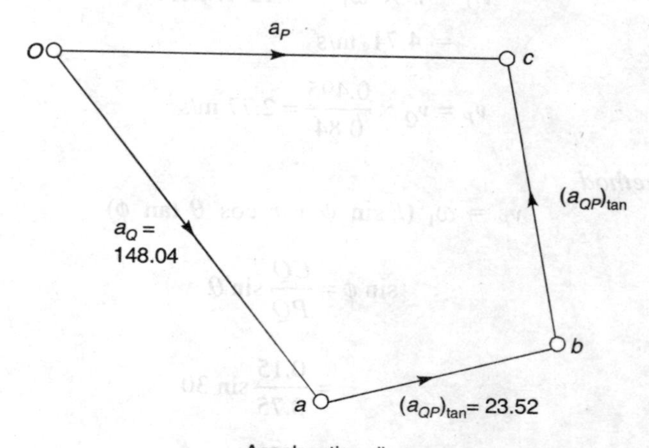

Acceleration diagram

By measurement of oc, the acceleration of piston $a_P = 143.5$ m^2/s.

8. In a four bar mechanism *ABCD*, points *A* and *C* are fixed points 30 cm apart and *AB*, *CD* are bars 60 cm and 70 cm long respectively, which are connected by a rod *BD* which is 50 cm long. If *AB* rotates with a uniform speed of 60 rpm. Determine (i) the velocity of *D* when *AB* is perpendicular to *AC* and also when it makes 10° on either side of the perpendicular and (ii) the instantaneous centre of the bar *BD* and its angular velocities in the three positions. (AMIE: 1960)

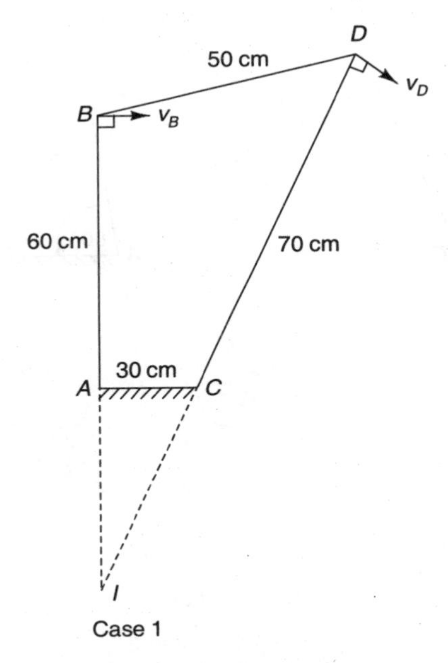

Case 1

Case 1: The velocity v_B is normal to *AB* and velocity v_D is normal to *CD*. Hence, the instantaneous centre can be located by drawing perpendiculars to v_B and v_D. By measurement, we get, *ID* = 1.77 m and *IB* = 1.63 m.

$$\text{Angular velocity of } AB = \omega_1 = \frac{2\pi N}{60}$$

or
$$\omega_1 = \frac{2\pi \times 60}{60} = 2\pi \text{ rad/s}$$

∴
$$v_B = AB \times \omega_1$$
$$= 0.6 \times 2\pi$$
$$= 3.768 \text{ m/s}$$

Now we have

$$\frac{v_D}{v_B} = \frac{ID}{IO} = \frac{1.77}{1.63}$$

∴
$$v_D = 3.768 \times \frac{1.77}{1.63}$$
$$= 4.09 \text{ m/s}$$

Case 2: The angle between AB and AC is 100°. Draw again link mechanism as shown below.

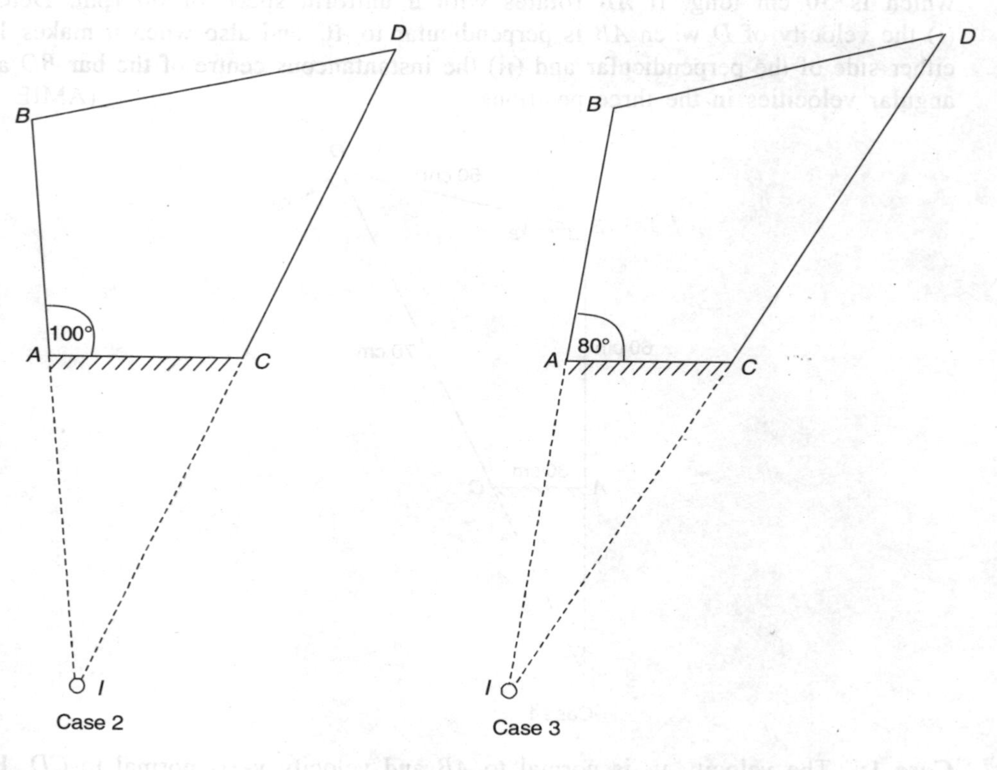

The instantaneous centre I is obtained by extending BA and DC. By measurement, we get

$$IB = 1.58 \text{ m}$$

$$ID = 1.67 \text{ m}$$

Now

$$v_D = v_B \times \frac{ID}{IB}$$

$$= 3.768 \times \frac{1.67}{1.58}$$

$$= 3.98 \text{ m/s}$$

Case 3: The angle between AB and AC is 80°. Draw the link mechanism and obtain intantaneous centre I by extending BA and DC. By measurement, we get

$$IB = 1.6 \text{ m}$$

$$ID = 1.79 \text{ m}$$

$$\therefore \quad v_B = v_B \times \frac{ID}{IB}$$

$$= 3.768 \times \frac{1.79}{1.6}$$

$$= 4.215 \text{ m/s}$$

9. A link AB is moving in a vertical plane. At a certain instant when the link is inclined at 60° to the horizontal, the point A is moving horizontally at 5 m/s while B is moving vertically upwards. Find velocity of B.

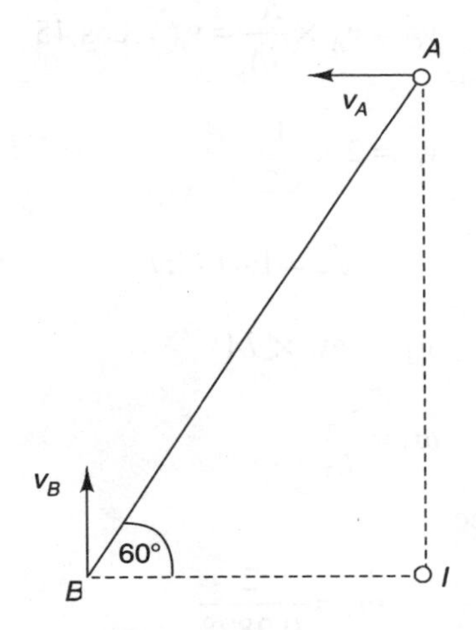

To obtain the instantaneous centre, draw perpendiculars to v_A and v_B velocities. Now

$$\frac{v_B}{v_A} = \frac{IB}{IA}$$

or

$$v_B = v_A \times \frac{1}{\tan 60}$$

$$= \frac{5}{\sqrt{3}} = 2.88 \text{ m/s}$$

10. A slender bar AB slides down a circular surface and on a horizontal surface as shown in the figure. At an instant when $\theta = 45$, the velocity of the end A is 2 m/s. Determine the angular velocity of the bar and the velocity of the point of contact on the circular surface.

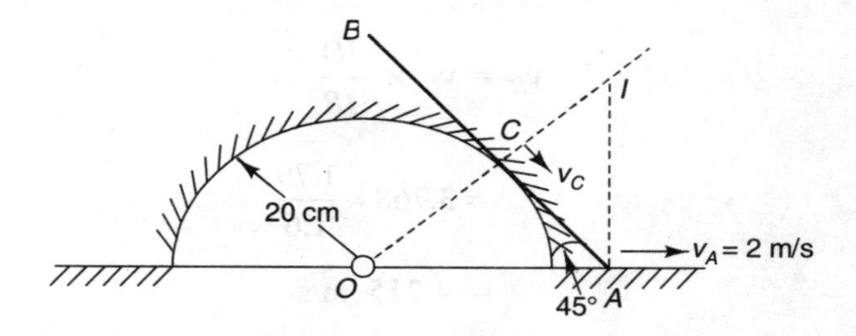

Instantaneous centre I is obtained by drawing perpendiculars on v_A and v_C. Now

$$v_C = v_A \times \frac{IC}{IA} = v_A \times \cos 45$$

$\therefore$
$$v_C = 2 \times \frac{1}{\sqrt{2}}$$

$$= \sqrt{2} = 1.414 \text{ m/s}$$

Also

$$v_A = \omega_0 \times IA$$

$\therefore$
$$\omega_0 = \frac{v_A}{IA}$$

But $IA = 20\sqrt{2}$ cm. Hence

$$\omega_0 = \frac{2}{0.2828}$$

$$= 7.072 \text{ rad/s}$$

11. An airplane moving at 100 m/s is undergoing a roll of 6 rad/min. When the plane is horizontal, an antenna is moving out at a speed of 2 m/s relative to the plane and is at a position 4 m from the centreline of the plane. Assuming the axis of roll about centreline, find the velocity of the antenna end relative to the ground when the plane is horizontal.

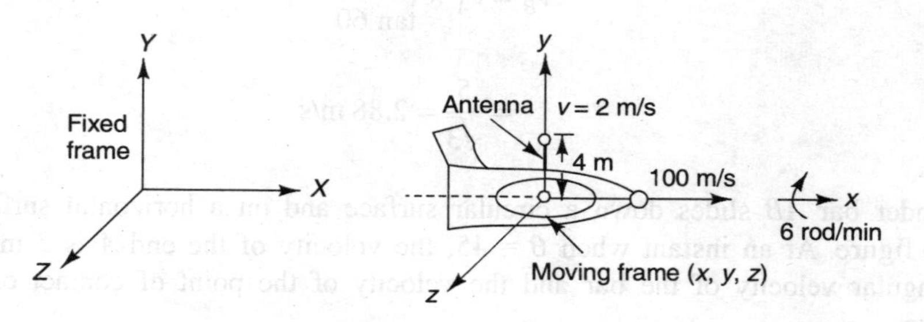

Let P be the position vector showing the end of the antenna relative to moving frame. The position of the antenna end is 4 m from the origin along the y-axis. Hence we have

$$P = 4\boldsymbol{j}$$

Also the velocity of the antenna end is 2 m/s along the y-axis. Hence we have

$$v_m = \text{velocity relative to moving frame}$$
$$= 2\boldsymbol{j}$$

Motion of moving frame (with airplane) relative to the fixed frame is along the x-axis.

$$R = 100\boldsymbol{i}$$

Also

$$\omega = -\frac{6}{60}\boldsymbol{i} = -0.1\boldsymbol{i}$$

Now we know

$$v_f = v_m + \dot{R} + \omega \times \rho$$
$$= 2\boldsymbol{j} + 100\boldsymbol{i} + (-0.1\boldsymbol{i} \times 2\boldsymbol{j})$$
$$= 2\boldsymbol{j} + 100\boldsymbol{i} - 0.2\boldsymbol{k}$$
$$= 100\boldsymbol{i} + 2\boldsymbol{j} - 0.2\boldsymbol{k}$$

12. A right circular cylinder rolls without slipping along a horizontal AB surface and its centre has at a certain instant a velocity v_C as shown in the figure. Find the velocities at the same instant at the point D and E on the periphery of the cylinder.

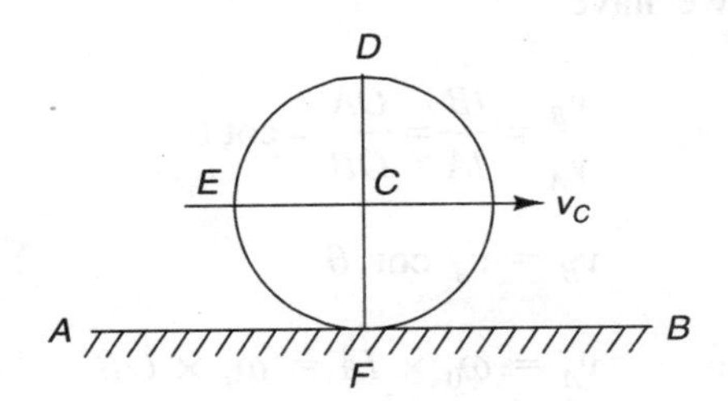

Since the cylinder is rolling without slipping, the velocity at F of the roller in contact with ground is zero. Hence F is the point of instantaneous rotation of the roller. Hence

$$v_C = \omega_0 \times r$$

or

$$\omega_0 = \frac{v_C}{r}$$

Now

$$FD = 2r$$

$$\therefore \quad v_D = \omega_0 \times 2r$$
$$= 2(\omega_0 \times r)$$
$$= 2v_C$$

Also $\quad FE = \sqrt{2}r$

$$\therefore \quad v_E = \omega_0 \times \sqrt{2} \times r$$
$$= \sqrt{2}\,(\omega_0 \times r)$$
$$= \sqrt{2}\,v_C$$

13. A prismatic bar *AB* has its end *A* and *B* are constrained to move horizontally and vertically as shown in the figure. If the end *A* of the bar-moves with constant velocity v_A, find the angular velocity ω_0 of the bar and velocity v_B of the end *B* for the instant when the axis of the bar makes an angle θ with the horizontal *x*-axis.

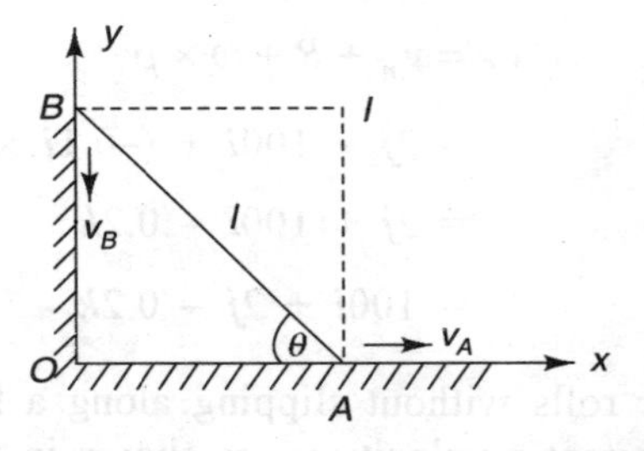

The point of instantaneous centre of rotation can be found out by drawing perpendicular to v_A and v_B. Hence we have

$$\frac{v_B}{v_A} = \frac{IB}{IA} = \frac{OA}{OB} = \cot\theta$$

$$\therefore \quad v_B = v_A \cot\theta$$

Now

$$v_A = \omega_0 \times IA = \omega_0 \times OB$$
$$= \omega_0 \times l \times \sin\theta$$

$$\therefore \quad \omega_0 = \frac{v_A}{l \sin\theta} = \frac{v_A}{l} \operatorname{cosec}\theta$$

14. A circular roller of radius 12 cm is contacted at top and bottom points of its circumference by two conveyor belts *AA* and *BB* as shown in the figure. If the belts run with uniform speed v_1 = 6 m/s and v_2 = 4 m/s, find the linear velocity v_C of the roller and also its angular velocity ω_0.

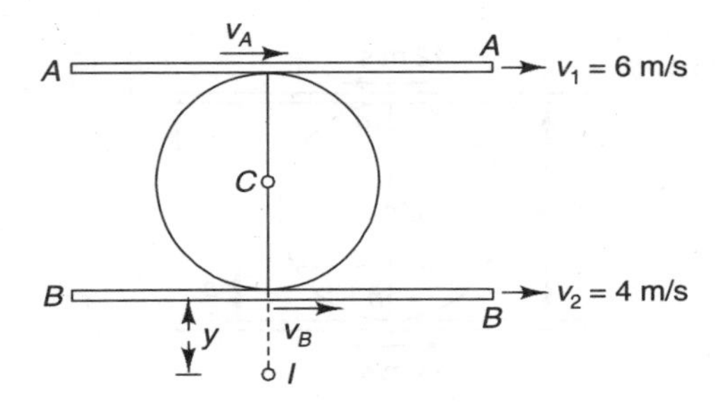

The instantaneous centre is located at the perpendiculars to v_A and v_B and it will therefore lie on the vertical diameter at a point outside the roller as shown in the figure.

Now

$$\frac{v_B}{y} = \frac{v_A}{2R + y}$$

or

$$\frac{4}{y} = \frac{6}{0.24 + y}$$

or

$$0.48 + 2y = 3y$$

or

$$y = 0.48 \text{ m}$$

If ω_0 is angular velocity, then we have

$$v_B = y \times \omega_0$$

or

$$\omega_0 = \frac{v}{y}$$

$$= \frac{4}{0.48}$$

$$= 8.33 \text{ rad/s}$$

Now

$$v_C = \omega_0 \times (y + R)$$

$$= 8.33 \times (0.48 + 0.12)$$

$$= 8.33 \times 0.6$$

$$= 4.998$$

$$\approx 5 \text{ m/s}$$

15. A roller of radius $r = 10$ cm rides between two horizontal bars moving in opposite directions as shown in the figure. Calculate the distance a defining the position of the horizontal path of the instantaneous centre of rotation of the roller. Assume that there is no slip at the points of contact m and n.

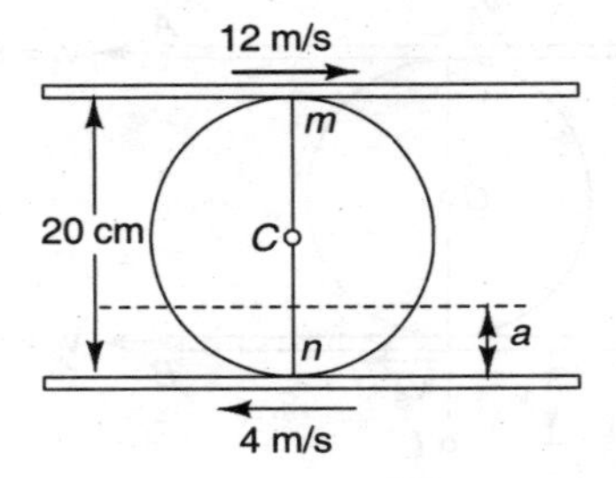

As the instantaneous centre is given we have

$$v_n = \omega_0 \times \frac{a}{100} = 4 \tag{i}$$

$$v_m = \omega_0 \times \frac{20 - a}{100} = 12 \tag{ii}$$

Dividing Eq. (ii) by Eq. (i), we get

$$\frac{20 - a}{a} = \frac{12}{4} = 3$$

or $$4a = 20$$

or $$a = 5 \text{ cm}$$

16. Find the ratio of the angular velocities $\dot{\theta}_1$ and $\dot{\theta}_2$ of the crank O_1A and O_2B of the system shown in the figure for the instantaneous positions shown.

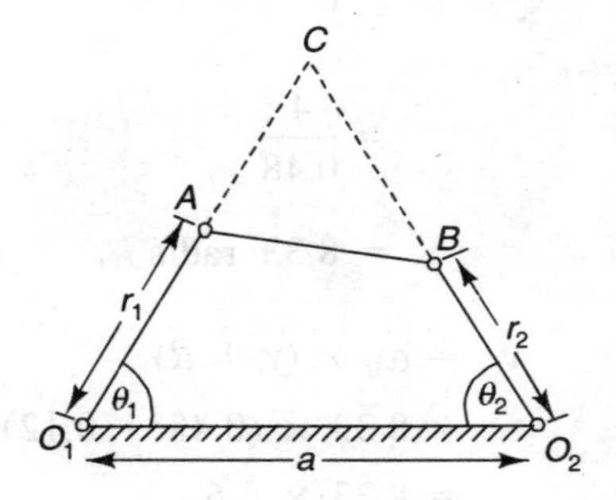

It is obvious that the instantaneous centre of the bar AB will be located at point C.

Velocity at point A is

$$v_A = r_1 \times \dot{\theta}_1 = \omega_0 \times AC \tag{i}$$

where ω_0 is angular velocity of bar AB.

Similarly, velocity at point B is

$$v_B = r_2 \times \dot{\theta}_2 = \omega_0 \times BC \tag{ii}$$

Hence we have from Eqs. (i) and (ii),

$$\frac{r_1 \times \dot{\theta}_1}{r_2 \times \dot{\theta}_2} = \frac{AC}{BC}$$

or

$$\frac{\dot{\theta}_1}{\dot{\theta}_2} = \frac{r_2 AC}{r_1 BC}$$

17. The ends A and B of a slender bar of length l are constrained to follow the straight lines OA and OB with an exterior angle as shown in the figure. Prove that, for motion of the bar in the plane of the figure, the instantaneous centre of rotation describes a circle of radius $\dfrac{l}{\sin \alpha}$ with centre at O.

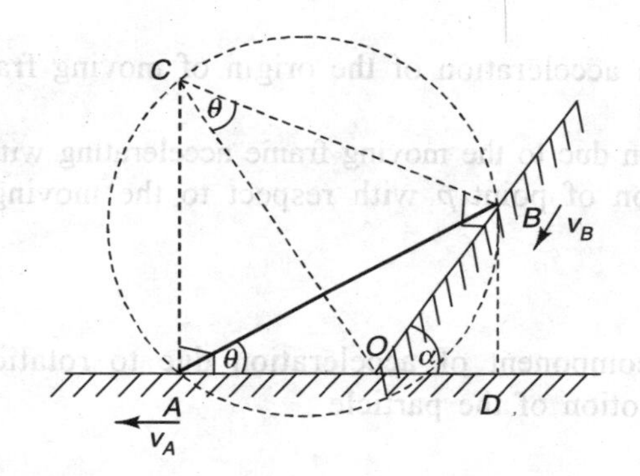

The velocities at points A and B of slender bar AB are as shown in the figure. Now draw a normal to these velocities or a normal to surface OB and OA to obtain the instantaneous centre of rotation of the slender bar which is point C as shown in the figure. Now angle $CBO = 90°$ and angle $CAO = 90°$. Hence points C, A, O and B lie on a circle which has CO as the diameter. Now $\angle OAB$ and $\angle OCB$ on periphery are located on the same chord BO of the circle, hence they are equal. Therefore we have

$$BD = l \sin \theta = d \sin \theta \sin \alpha$$

where $d = CO$.

$$\therefore \qquad d = \frac{l}{\sin \alpha}$$

As d is independent of angle θ which gives the instantaneous position of the slender bar, it means that the distance of the instantaneous centre from the point O remains constant. Hence the instantaneous centre of rotation describes a circle of radius $\dfrac{l}{\sin \alpha}$ with centre at O.

18. A particle p moves with constant relative velocity v_r along the circumference of a circular disk of radius r while the disk rotates with uniform angular velocity ω in a opposite direction. Find the absolute acceleration a_p of the particle.

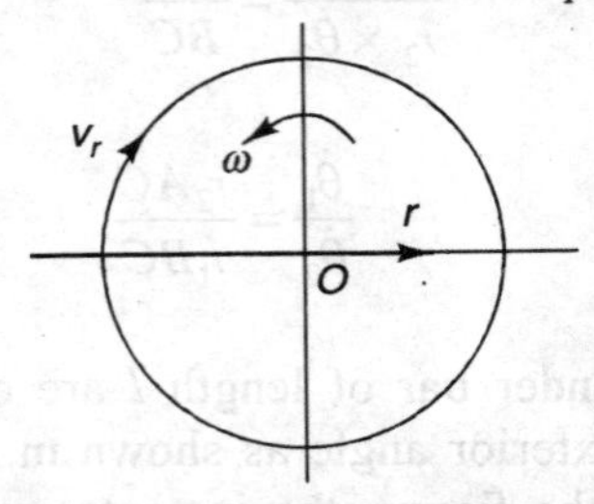

We have already seen that the acceleration of a particle is

$$a_{pf} = a_{pm} + a_0 + a \times r + 2\omega \times v_{pm} + \omega \times (\omega \times r)$$

Here we have

a_0 = translation acceleration of the origin of moving frame with respect to fixed origin = 0

$a \times r$ = acceleration due to the moving frame accelerating with angular acceleration = 0

a_{pm} = acceleration of point p with respect to the moving frame of reference

$$= \frac{v_r^2}{r}$$

$2\omega v_{pm}$ = Coriolis component of acceleration due to rotation of moving frame and relative motion of the particle

$$= 2\omega v_r$$

$$\omega \times (\omega \times r) = \text{centripetal acceleration}$$

$$= \omega^2 r$$

$$\therefore \quad a_{pf} = \frac{vr^2}{r} + \omega^2 r - 2\omega v_r$$

19. The disk shown in the figure rolls without slip with a constant speed of 1.0 m/s. Find the velocities v_D and v_E at points D and E at the top and bottom of the rim.

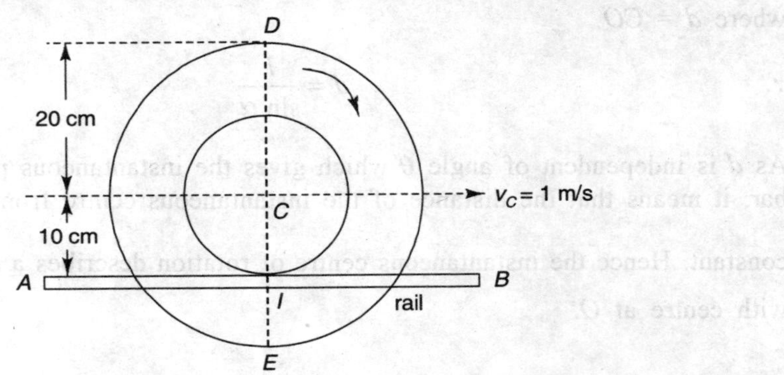

The instantaneous centre will be located at I. Hence we have

$$v_C = \omega_0 \times CI$$

or $$1 = 0.1 \times \omega_0$$

or $$\omega_0 = 10 \text{ rad/s}$$

Now velocity at D is

$$v_D = \omega_0 \times ID$$

$$= 10 \times \frac{30}{100}$$

$$= 3 \text{ m/s}$$

Also velocity at E is

$$v_E = -10 \times \frac{10}{100}$$

$$= -1 \text{ m/s}$$

20. A link OAR rotates anticlockwise at an angular velocity of 2 rad/s. Another link BCS at right angles to it has collar at B which slides over OAR at 2 m/s^2 and decelerates at 4 m/s^2 with respect to OAR as shown in the figure. A collar D slides over BCS with a velocity of 6 m/s and decelerates at 8 m/s^2 with respect to BCS. Find the velocity and acceleration of collar D with respect to the ground reference at the instant of interest.

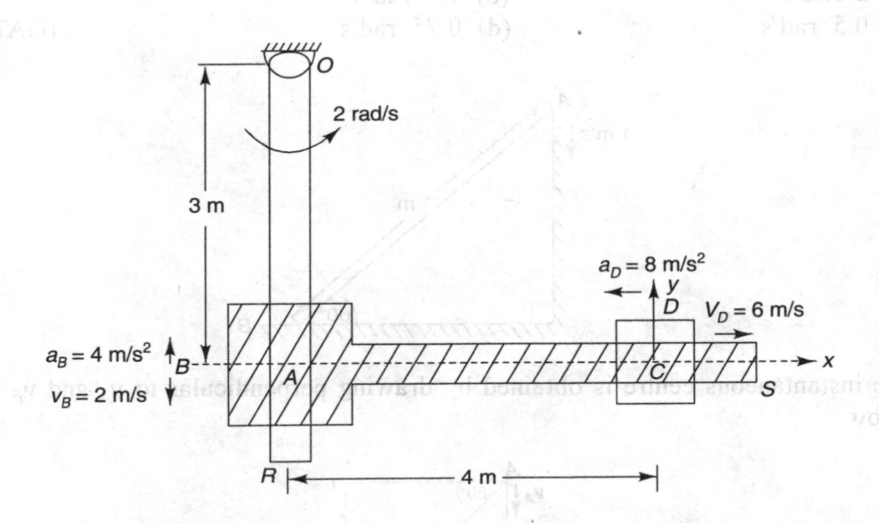

Take the moving frame about point C on the horizontal link BCS. Collar velocity is 6 m along the x-axis and 2 m/s along the y-axis. Hence, we have

$$v_{Dm} = 6\boldsymbol{i} - 2\boldsymbol{j}$$

$$\omega = 2\boldsymbol{k} \text{ rad/s (along } z\text{-axis)}$$

$$v_C = 0$$

$$\therefore \quad v_{Df} = v_{Dm} + v_C + \omega \times r_D$$

$$= (6\boldsymbol{i} - 2\boldsymbol{j}) + 0 + 2\boldsymbol{k} \times (4\boldsymbol{i} - 3\boldsymbol{j})$$

$$= 6\boldsymbol{i} - 2\boldsymbol{j} + 8\boldsymbol{j} + 6\boldsymbol{i}$$

$$= 12\boldsymbol{i} + 6\boldsymbol{j}$$

Similarly we have

$$a_{Dm} = -8\boldsymbol{i} + 4\boldsymbol{j}$$

$$\therefore \quad a_{Df} = a_{Dm} + a_0 \times r_D + 2\omega v_{Dm} + \omega \times (\omega \times r_D)$$

$$= (-8\boldsymbol{i} + 4\boldsymbol{j}) + 0 + 0 + 2\boldsymbol{k} \times 2\boldsymbol{k} \times (6\boldsymbol{i} - 2\boldsymbol{j})$$

$$+ 2\boldsymbol{k} \times [2\boldsymbol{k} \times (4\boldsymbol{i} - 3\boldsymbol{j})]$$

$$= -8\boldsymbol{i} + 4\boldsymbol{j} + 24\boldsymbol{j} + 8\boldsymbol{i} + 2\boldsymbol{k} \times (8\boldsymbol{j} + 6\boldsymbol{i})$$

$$= 28\boldsymbol{j} - 16\boldsymbol{i} + 12\boldsymbol{j}$$

$$= -16\boldsymbol{i} + 40\boldsymbol{j} \text{ m/s}^2$$

21. A rod of length 1 m is sliding in a corner as shown in the figure. At an instant when rod makes an angle of 60° with horizontal plane, the velocity of point A on the rod is 1 m/s. The angular velocity of the rod at this instant is

(a) 2 rad/s (b) 1.5 rad/s
(c) 0.5 rad/s (d) 0.75 rad/s (GATE: 1996)

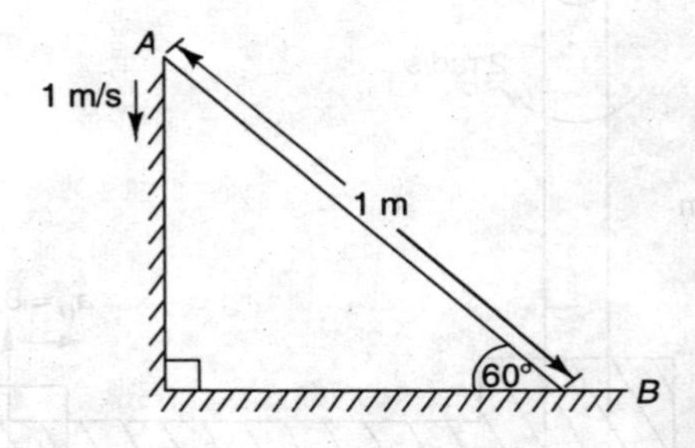

The instantaneous centre is obtained by drawing perpendicular to v_A and v_B as shown below.

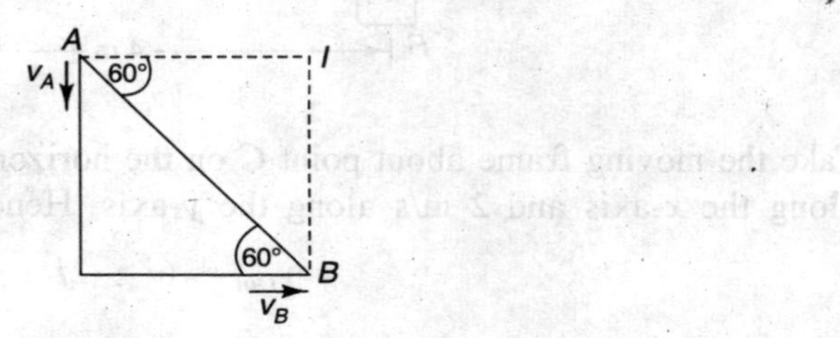

Now

$$\frac{IA}{AB} = \cos 60 = \frac{1}{2}$$

or

$$IA = \frac{AB}{2} = 0.5 \text{ m}$$

Now

$$\omega_{AB} = \frac{v_A}{IA} = \frac{1}{0.5} = 2 \text{ rad/s}$$

Option (a) is correct.

OBJECTIVE TYPE QUESTIONS

Happiness comes from within your heart, not from your surroundings

State True or False

1. A cone rolling on a flat surface is a plane motion. [*True/False*]
2. A cone sliding on a flat surface is a translatory motion. [*True/False*]
3. A spherical ball rolling down an inclined plane is rotary motion. [*True/False*]
4. A compound pendulum oscillating about an fulcrum is a rotary motion. [*True/False*]
5. A ladder sliding from a wall has a plane motion. [*True/False*]
6. The connecting rod of a reciprocating engine has translatory motion. [*True/False*]
7. The instantaneous centre of motion exists for all plane motions. [*True/False*]
8. Plane motion is a combination of translatory and rotary motion. [*True/False*]
9. The number of independent coordinates required to specify a rigid body in space is equal to the number of degrees of freedom of the body. [*True/False*]
10. Linear translation of a body has a point of instantaneous centre. [*True/False*]
11. Rotary motion of a body is a plane motion. [*True/False*]
12. Space motion of a rigid body is a general type of motion but has less than 6 degrees of freedom. [*True/False*]
13. Space motion of the body is a general motion which is not constrained in any direction. [*True/False*]
14. If a body moves such a way that all its particles move in parallel planes and travel in the same distance, then the body is having motion of translation. [*True/False*]
15. If a body rotates about a fixed point in such a way that all its particles move in circular paths, then the body is having motion of rotation. [*True/False*]
16. If a body is having a combined motion of translation and rotation, then the body is assumed to be rotating about a certain point which is known as instantaneous centre of rotation. [*True/False*]
17. The translatory motion can also have instantaneous centre of rotation. [*True/False*]
18. The instantaneous centre can be found out by drawing perpendiculars to the velocities at two points of the body. [*True/False*]
19. The point of the body at a greater distance from the instantaneous centre will have lesser velocity. [*True/False*]
20. The ratio of the velocities of two points depends upon the ratio of their distances from the instantaneous centre. [*True/False*]

Multiple Choice Questions

1. Figure shows an instantaneous centre of rotation. If $AI = 0.4$ m, $BI = 0.2$ m, and $v_A = 10$ m/s, the velocity v_B is

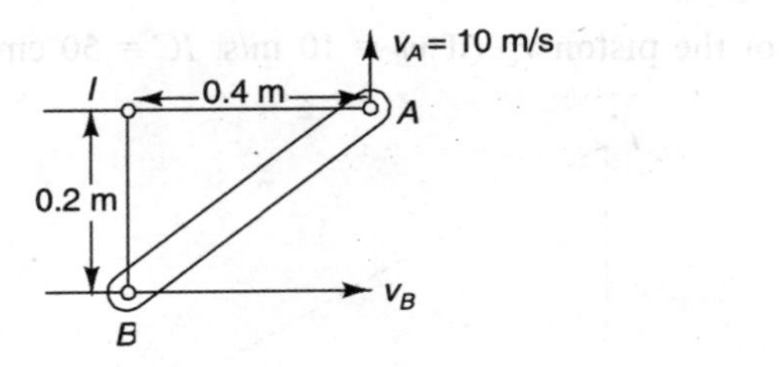

(a) 5 m/s (b) 4 m/s
(c) 20 m/s (d) 15 m/s

2. In the figure shown, the relative velocity of link 1 to link 2 is 12 m/s. Link 2 rotates at a constant speed of 120 rpm. The magnitude of coriolis component of acceleration of link 1 is

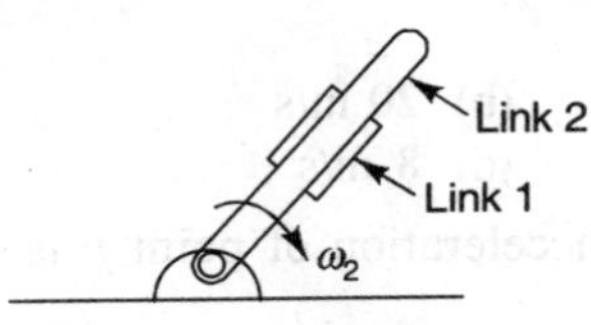

(a) 302 m/s (b) 604 m/s^2
(c) 906 m/s^2 (d) 1208 m/s^2 (GATE: 2004)

3. The figure below shows a planer mechanism with single degree of freedom. The instantaneous centre for the given configuration is located at

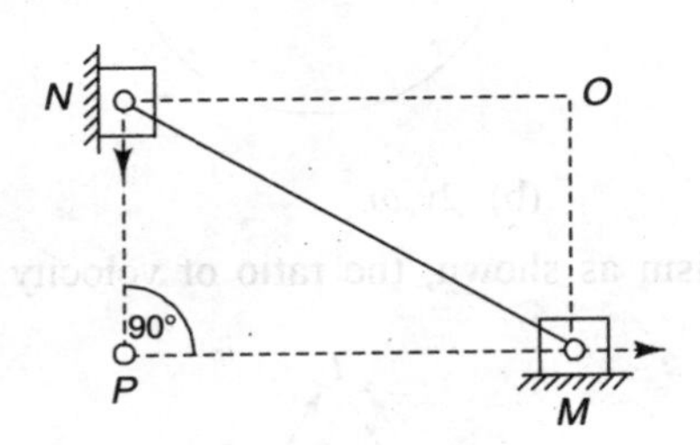

(a) N (b) P
(c) M (d) O

4. The degree of freedom of a rigid body imply the
   (a) angles that it may turn through
   (b) angular motion the body can have
   (c) constraint to its motion
   (d) total number of modes of displacement.

5. The instantaneous centre of rotation within the body
   (a) can exist for any space motion
   (b) can exist for any plane motion
   (c) can exist for any translatory motion.

6. A rigid body in translation
   (a) can move in straight line
   (b) may move in straight or curved path
   (c) may undergo plane motion.

7. The velocity of the piston v_C (if v_B = 10 m/s, IC = 50 cm and IB = 100 cm) is equal to

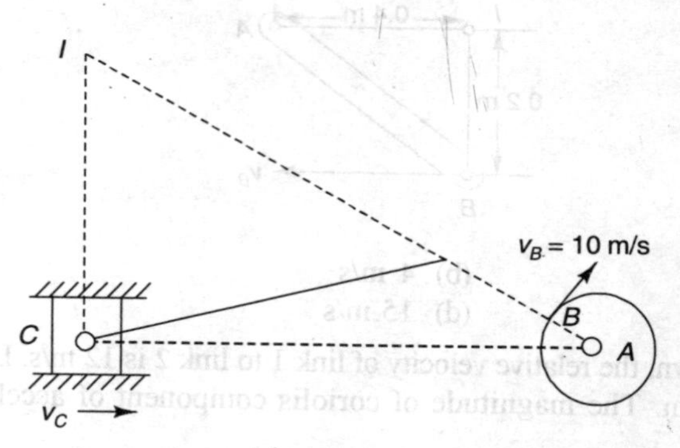

   (a) 5 m/s
   (b) 20 m/s
   (c) 15 m/s
   (d) 8 m/s

8. The coriolis component of acceleration of point p is

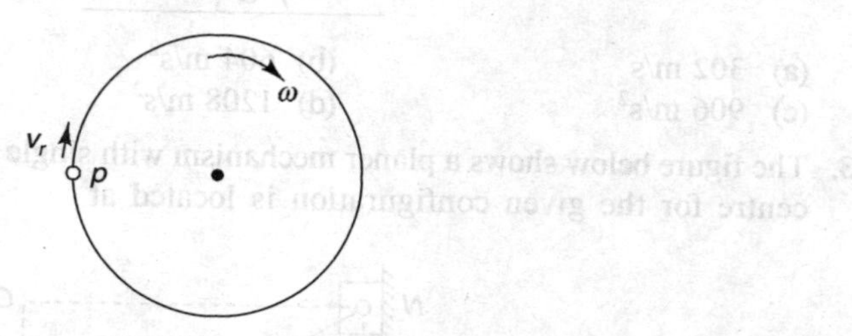

   (a) $v_r\omega$
   (b) $2v_r\omega$
   (c) $3v_r\omega$

9. For a four bar mechanism as shown, the ratio of velocity at B and C is

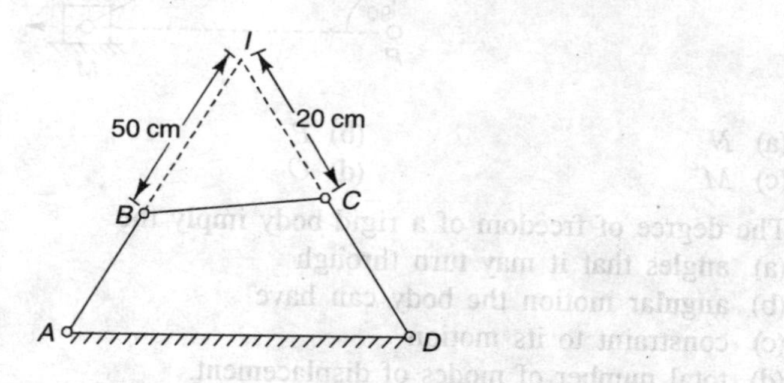

   (a) 1
   (b) 2
   (c) 1/2
   (d) 2.5

10. For a four-bar mechanism as shown, the velocity at C is given by

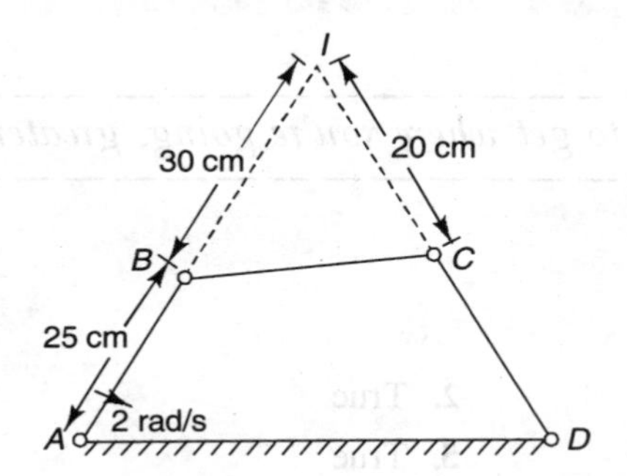

(a) $\frac{1}{3}$ m/s (b) $\frac{1}{2}$ m/s (c) 1 m/s

Fill in the Blanks

1. All plane motions have a point of __________ centre.
   (a) rotary (b) instantaneous
2. Piston in reciprocating mechanism has __________ motion
   (a) translatory (b) plane
3. The connecting rod of the reciprocating mechanism has __________ motion.
   (a) translatory (b) plane
4. The crank of the reciprocating mechanism has __________ motion.
   (a) rotary (b) plane
5. The sliding ladder has __________ motion.
   (a) translatory (b) plane
6. If a rotating disc with ω rad/s has a particle with velocity v_r, then the Coriolis component of acceleration is given as __________.
   (a) ωv_r (b) $2\omega v_r$
7. As the distance of a point increases from the instantaneous centre of rotation, its velocity __________.
   (a) increases (b) decreases
8. The velocity of instantaneous centre of rotation is __________.
   (a) zero (b) infinity
9. The instantaneous centre of rotation of a wheel is located at __________.
   (a) centre (b) point of contact at ground

ANSWERS

Great men tell you how to get where you're going, greater men take you there.

State True or False

1. True **2.** True **3.** False. It is plane motion.

4. True **5.** True **6.** False. It has plane motion.

7. True **8.** True **9.** True

10. True. A translatory motion is also a plane with instantaneous centre is located at infinity.

11. True. Pure translatory and rotation are also examples of plane motions.

12. False. It has six degrees of freedom.

13. True **14.** True **15.** True

16. True

17. True but instantaneous centre is located at infinity.

18. True

19. False. It will have greater velocity.

20. True. $\frac{v_A}{v_B} = \frac{IA}{IB}$ where IA and IB are distances from the instantaneous centre.

Multiple Choice Questions

1. (a) $$\frac{v_B}{v_A} = \frac{IB}{IA}$$

or $$v_B = v_A \times \frac{IB}{IA}$$

$$= 10 \times \frac{0.2}{0.4}$$

$$= 5 \text{ m/s}$$

2. (a) $$v_{1-2} = 12 \text{ m/s}$$

Also $$\omega = \frac{2\pi N}{60} = \frac{2\pi \times 120}{60}$$

$$= 4\pi \text{ rad/s}$$

The Coriolis component of acceleration is

$$= 2v_{1-2} \times \omega$$
$$= 2 \times 12 \times 4\pi$$
$$= 302 \text{ m/s}^2$$

3. (d). The perpendiculars on velocities at point N and point M meet at point O.

4. (d)

5. (b). Instantaneous centre for translatory motion exists at infinity.

6. (b)

7. (a)
$$v_C = v_B \times \frac{IC}{IB}$$
$$= 10 \times \frac{50}{100}$$
$$= 5 \text{ m/s}$$

8. (b)

9. (d)
$$\frac{v_B}{v_C} = \frac{IB}{IC}$$
$$= \frac{50}{20}$$
$$= 2.5$$

10. (a)
$$\frac{v_C}{v_B} = \frac{IC}{IB} = \frac{20}{30}$$

or
$$v_C = \frac{2}{3} v_B$$

But
$$v_B = \omega \times AB = 0.25 \times 2 = 0.5 \text{ m/s}$$

$\therefore$
$$v_C = \frac{2 \times 0.5}{3} = \frac{1}{3} \text{ m/s}$$

Fill in the Blanks

1. (b) **2.** (a) **3.** (b) **4.** (a)

5. (b) **6.** (b) **7.** (a) **8.** (a)

9. (b)

CHAPTER 14

Kinetics of Rigid Body

A pat on the back and a sympathetic ear are valuable gifts you can give to those you work with and live with. Times of crisis only increase their value.

INTRODUCTION

For convenience, dynamics is divided into two fields, known as *kinematics* and *kinetics*. In kinematics, we study motion of a body without concerning the forces that cause the motion. In kinetics, we are concerned with finding the kind of motion that a given body or system of bodies will have under the action of given forces, or with what forces must be applied to produce a prescribed motion. A plane motion of a rigid body is infact a superposition of translatory motion and a rotational motion. To specify the translatory and rotationary motion of a rigid body, it is convenient to select a centre of mass. The translatory motion is then specified using particles dynamics by Newton's laws of motion and similarly rotational motion using particles dynamics by Euler's equation. The mass of the body is assumed to be concentrated in the centre of the mass and all particles are assumed to be rotating around the centre of mass.

The equations of motions as obtained from Newton's laws of motion and Euler's equation can be integrated with respect to time to form an alternate set of equations for studying the dynamics of rigid body. They are called *linear impulse-momentum equation* and *angular impulse-momentum equation* which are obtained from the centre of mass and the axis of rotation passing through the centre of mass respectively. These equations are much simpler to apply and to evaluate the motion of a body specially when external forces and moments are varying with time. The principle of conservation of momentum is also obtained from it when no external force or moment is acting on the body.

Work-energy equation is a scalar equation as both work and energy are scaler quantities. It provides a simpler method to find velocities whenever work can be evaluated easily. The principle of conservation of energy is obtained from it when work done is zero.

D'Alembert's principle helps in changing the dynamic equilibrium of the body into an equivalent static equilibrium. The concept of D'Alembert is to introduce the inertia forces and inertia moments in addition to real forces and moments so that equations of motion can be written as equilibrium equations, thereby helping in solving the problems related to the dynamics of rigid body.

FORCE, MASS AND ACCELERATION

Force is defined as any action that tends to change the state of a body to which it is applied. The force is defined by (i) its magnitude, (ii) its point of application and (iii) its direction. There are many kinds of force, such as gravity force, the simple push or pull that we can exert upon a body with our hands.

The gravity force W if acting alone produces an acceleration of the particle equal to g. If now a force F acts on a body instead of gravity force, then from second law it follows that the acceleration G is produced by the force which will be in the direction of the force. This acceleration a will have same ratio to the acceleration g due to gravity as the force F is to the gravity force w. Therefore we have

$$\frac{a}{g} = \frac{F}{W}$$

or

$$F = \frac{w}{g}.a$$

The factor w/g measures the degree of sluggishness with which the particle yields to the action of an applied force and it is a measure of the inertia of the particle. It is called the *mass of the particle* and it is generally denoted by m. Hence we have

$$m = \frac{W}{g}$$

$$F = ma$$

ROTATORY MOTION OF A RIGID BODY

When a body rotates about a fixed axis, this motion is called a *rotatory motion*. The axis about which the body rotates is called the axis of *rotation*. In rotatory motion, every particle of the body, moves in circle and the centres of all these circles lie at the axis of rotation. The number of revolutions per second is called *angular velocity* ω. If the angular velocity changes from ω_1 to ω_2 in t seconds, then angular acceleration $\alpha = \frac{\omega_2 - \omega_2}{t}$. The angular velocity ω and angular acceleration α are related to linear velocity v and linear acceleration 'a' as given below

$$\omega = \frac{v}{r}$$

where r = radius

$$\alpha = \frac{a}{r}$$

When an external force acting on a body has a tendency to rotate the body about an axis, then the force is said to exert a torque about that axis. The torque or moment of force about an axis of rotation (see Figure 14.1) is equal to the product of the magnitude of the force and the perpendicular distance of the line of action of the force from the axis of rotation. The torque or moment for force F acting at a distance r from the axis of rotation is

$$M = T = F \times r$$

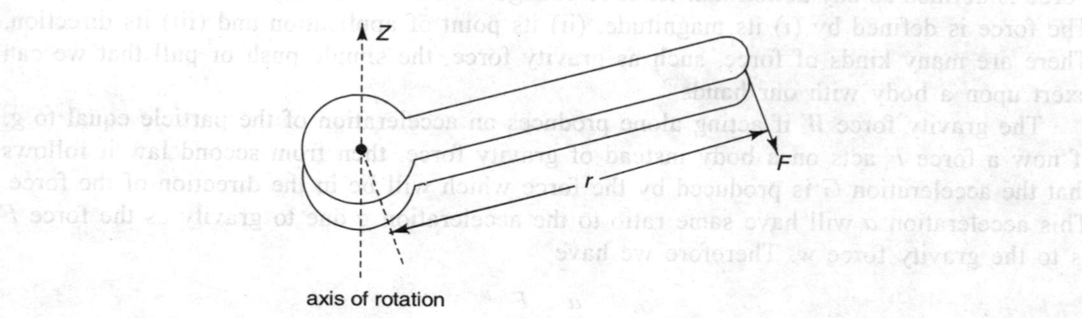

FIGURE 14.1 Torque about an axis of rotation.

Relation between Torque and Moment of Inertia

Suppose a body is acted upon by a torque T and it is rotating about an axis passing through a fixed point O. The body is rotating with constant angular acceleration α. All particles will have same angular acceleration but their linear accelerations will differ. Consider a particle m_1 and its distance from the axis is r_1. Then we have

$$\text{Linear acceleration } a_1 = \alpha \times r_1$$

$\therefore$ Force F_1 acting on it = mass × acceleration

or
$$F_1 = m_1 \times a_1$$
$$= m_1 \times \alpha \times r_1$$

Now torque or moment of this force is

$$T_1 = \text{Force} \times \text{distance}$$
$$= F_1 \times r_1$$
$$= m_1 \times \alpha \times r_1^2$$

There total torque or moment of all particles is

$$M = T = m_1 \times \alpha \times r_1^2 + m_2 \times \alpha \times r_2^2 + m_3 \times \alpha \times r_2^3 \ldots$$

$$= (m_1 r_1^2 + m_2 r_2^2 + m_3 r_3 \ldots) \times \alpha$$

$$= (\Sigma m r^2) \times \alpha$$

But Σmr^2 = moment of inertia

$$T = M = I \times \alpha$$

Equations of angular motion. The equations of linear motion are

$$v = u + at$$

$$S = ut + \frac{1}{2}at^2$$

$$v^2 = u^2 + 2as$$

The equations of angular motion can be given as

$$\omega_2 = \omega_1 + \alpha t$$

$$\theta = \omega_1 t + \frac{1}{2}\alpha t^2$$

$$\omega_2^2 = \omega_1^2 + 2\alpha\theta$$

Kinetic energy of rotation. The kinetic energy of a particle m_1 having velocity v_1 is given by

$$k_1 = \frac{1}{2}m_1 v_1^2$$

But $v_1 = r_1\omega$

$$\therefore \quad k_1 = \frac{1}{2}m_1 r_1^2 \omega^2$$

Hence kinetic energy of all particles of the rigid body is

$$k = \frac{1}{2}m_1 r_1^2 \omega^2 + \frac{1}{2}m_2 r_2^2 \omega^2 + \frac{1}{2}m_3 r_3^2 \omega^2 \ldots$$

$$= \frac{1}{2}(m_1 r_1^2 + m_2 r_2^2 + m_3 r_3^2 \ldots)\,\omega^2$$

$$= \frac{1}{2}(\Sigma mr^2)\,\omega^2$$

$$= \frac{1}{2}I\omega^2$$

Incase a body rotating about an axis is simultaneously moving along a straight line, then total kinetic energy is

$$k = \frac{1}{2}I\omega^2 + \frac{1}{2}mv^2$$

Angular momentum. If a body is rotating about an axis, then the sum of the moments of linear momentums of all the particles about the given axis is called angular momentum of the body about that axis.

Moment of momentum of m_1 = momentum × distance

$$= m_1 v_1 \times r_1$$

$$= m_1 \times (\omega \times r_1) r_1$$

$$= m_1 r_1^2 \omega$$

Total moment of momentum of all particles or angular momentum of the body is

$$J = m_1 r_1^2 \omega + m_2 r_2^2 \omega + m_3 r_3^2 \omega \ldots$$

$$= (m_1 r_1^2 + m_2 r_2^2 + m_3 r_3^2 \ldots) \omega$$

$$= (\Sigma m r^2) \times \omega$$

$$= I \times \omega$$

Relation between Torque and Angular Momentum

The body is rotating about an axis, then torque T is

$$T = I \times \alpha$$

$$= I \times \frac{d\omega}{dt}$$

The angular momentum is given by

$$J = I\omega$$

$$\therefore \quad \frac{dJ}{dt} = I \frac{d\omega}{dt} = I\alpha$$

$$= M \text{ or } T$$

Therefore, the rate of change of angular momentum of a body is equal to external moment or torque acting upon the body.

Example 14.1 A wheel of radius 10 cm rotates freely about its centre as shown in the Figure 14.2. A string is wrapped over its rim and is pulled by a force of 10 N. It is found that torque produces an angular acceleration of 4 rad/s^2. in the wheel. Find the moment of inertia of the wheel.

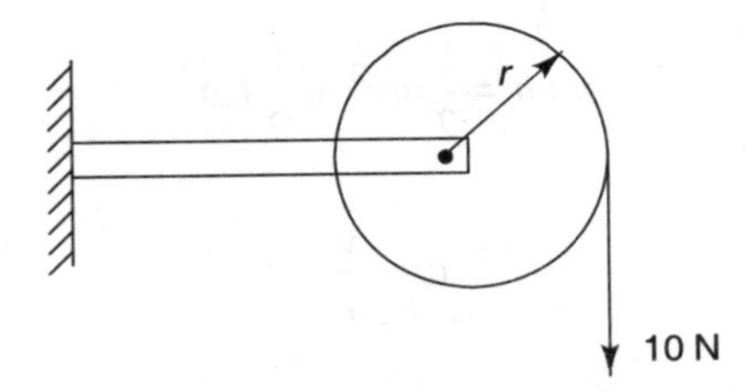

FIGURE 14.2 Example 14.1.

Torque $T = F \times r = 10 \times 0.1 = 1$ Nm

But $T = I \times \alpha$

or
$$I = \frac{T}{\alpha}$$

$$= \frac{1}{4}$$

$$= 0.25 \text{ kg m}^2$$

Conservation of angular momentum. If no external torque is acting upon a body which is rotating about an axis, then the angular momentum of the body remains constant. Hence we have

$$J = I \times \omega = \text{Constant}$$

The above is called the *law of conservation of angular momentum*. If moment of inertia I decreases, the angular velocity of the body will increase.

Example 14.2 A wheel of moment of inertia I and radius r is free to rotate about its centre as shown in Figure 14.3. A string is wrapped over its rim and a block of mass m is attached to the free end of the string. The system is released from rest. Find the speed of the block as it descends through a height h.

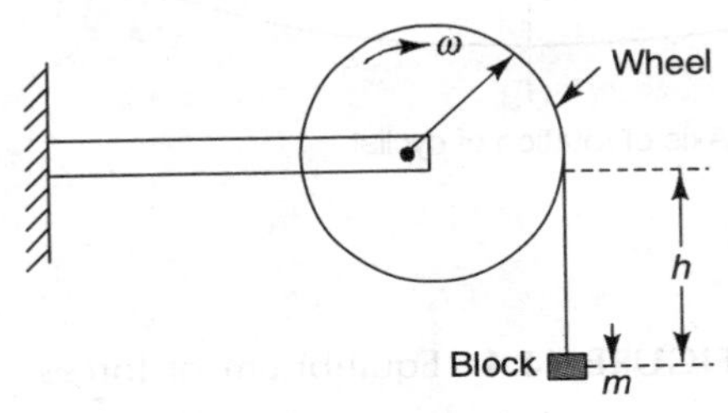

FIGURE 14.3 Example 14.2.

Consider the speed of the block be v when it descends through a height h. Here the gravitational potential energy is lost by the block must be equal to the kinetic energy gained by the block and the wheel. Hence we have

$$\text{PE of block} = \text{KE of block} + \text{KE of wheel}$$

$$mgh = \frac{1}{2}mv^2 + \frac{1}{2}I\omega^2$$

But

$$\omega = \frac{v}{r}$$

$\therefore$

$$mgh = \frac{1}{2}mv^2 + \frac{1}{2}I\frac{v^2}{r^2}$$

$\therefore$

$$v = \left[\frac{2mg}{m + \dfrac{I}{r^2}}\right]^{1/2}$$

Bending of a cyclist on a horizontal turn. Consider a cyclist is going at speed v on a circular horizontal road of radius r which is not banked. The cycle and rider has centre of mass at C as shown in Figure 14.4 and centre of mass is travelling in a circle of radius r with centre at O.

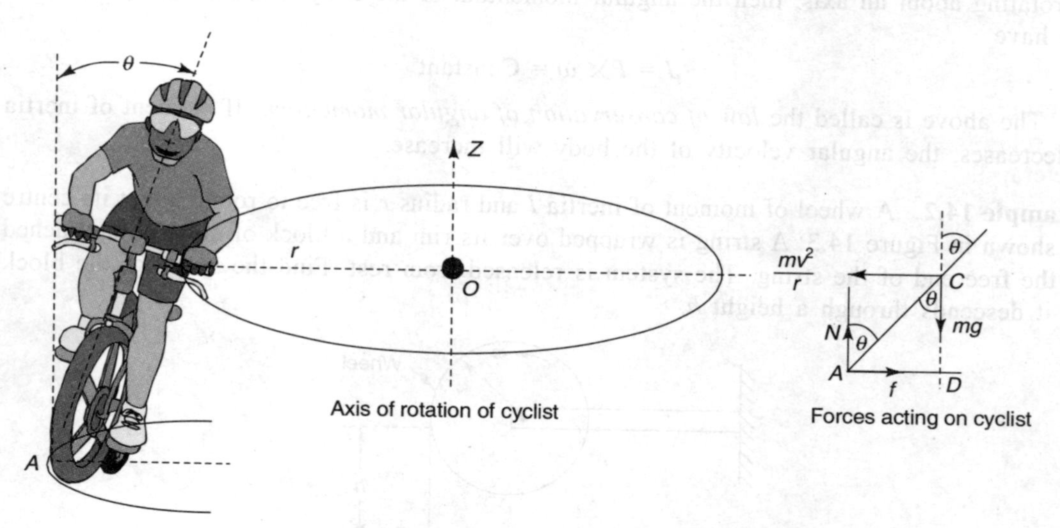

FIGURE 14.4 Equilibrium of forces.

The forces acting on a cyclist when he takes a horizontal turn are:

(a) Centrifugal force $\frac{mv^2}{r}$ acting away from the centre O.

(b) Friction force f acting towards the centre O.

(c) Weight *mg* acting down.
(d) Normal *N* acting upwards.

Incase of the equilibrium, total external forces and total external torques must be zero. Taking moment about point *A* where wheel touches the ground, we have

$$mg \times AD = \frac{mv^2}{r} \times CD$$

Now

$$\tan\theta = \frac{AD}{CD} = \frac{\frac{mv^2}{r}}{mg}$$

or

$$\tan\theta = \frac{v^2}{r \cdot g}$$

or

$$\theta = \tan^{-1}\left(\frac{v^2}{rg}\right)$$

The cyclist has to bend inside at angle of $\tan^{-1}\left(\frac{v^2}{rg}\right)$ while negotiating a horizontal turn.

Example 14.3 A force *F* acts tangentially at the highest point of a sphere of mass *m* kept on a rough horizontal plane (see Figure 14.5). If the sphere rolls without slipping, find the acceleration of the centre of the sphere.

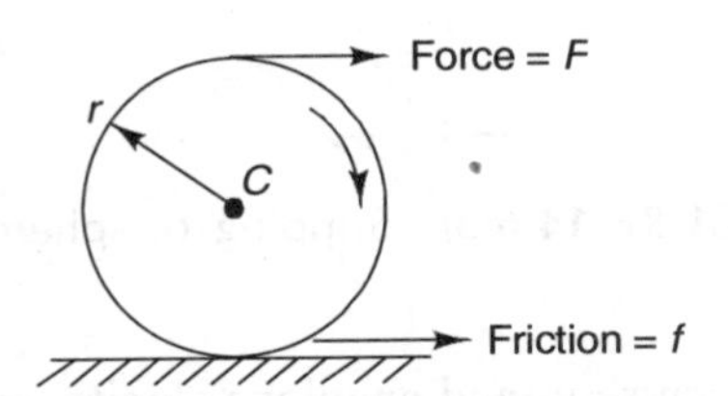

FIGURE 14.5 Rolling of sphere.

When *F* acts on the sphere, the friction force *f* will also act in the direction of the force *F* as shown in the figure. If linear acceleration is *a*, then we have

$$F + f = m \cdot a \quad \text{for linear motion}$$

and

$$F \cdot r - f \cdot r = I\alpha \quad \text{for angular motion}$$

$$I \text{ for sphere } = \frac{2}{5} mr^2$$

and angular acceleration $\alpha = \dfrac{a}{r}$

$$\therefore \qquad F \cdot r - f \cdot r = \frac{2}{5} mr^2 \times \frac{a}{r}$$

or
$$F - f = \frac{2}{5} ma \qquad \text{(i)}$$

Also
$$F + f = ma \qquad \text{(ii)}$$

Adding Eqs. (i) and (ii), we have

$$2F = \frac{7}{5} ma$$

or
$$a = \frac{10}{7} \frac{F}{m}$$

Example 14.4 A sphere of mass m has radius r slips on a rough horizontal plane as shown in Figure 14.6(a). It has translation velocity v_0 and rotational velocity $\dfrac{v_0}{2r}$ at the some instant. Determine translation velocity when the sphere starts pure rolling.

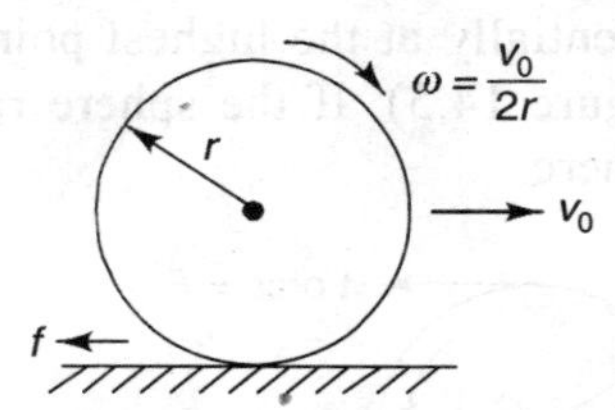

FIGURE 14.6(a) Slipping of sphere.

The sphere has velocity at centre v_0 and angular velocity $\omega = \dfrac{v_0}{2r}$. As $v_0 > \omega r$, the sphere will keep on slipping forwards till translation velocity v_0 decreases such that $v(t) = \omega r$. During slipping, the friction force f will act against the motion. If a is deceleration, then we have

$$f = -m \times a$$

or
$$a = -\frac{f}{m}$$

Now velocity of sphere is

$$v(t) = v_0 - at$$

$$= v_0 - \frac{f}{m} \times t \qquad \text{(i)}$$

The friction force forms a friction torque about the centre of the sphere given by $f \times r$ (see Figure 14.6(b)) which is also acting clockwise. Hence the angular acceleration about centre will be

$$I\alpha = f \times r$$

where I = moment of inertia

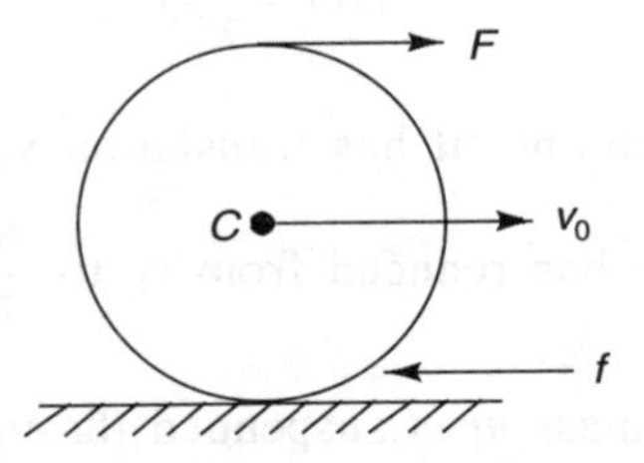

FIGURE 14.6(b) Slipping of sphere and equilibrium of moments.

But I of sphere $= \frac{2}{5} mr^2$

$$\therefore \qquad \alpha = \frac{f \times r}{I} = \frac{f \times r}{\frac{2}{5} m \times r^2}$$

$$= \frac{5f}{2mr}$$

Due to angular acceleration, the angular velocity of sphere after time t is

$$\omega(t) = \omega_0 + \alpha t$$

$$= \omega_0 + \frac{5f}{2mr} \times t$$

$$= \frac{v_0}{2r} + \frac{5f}{2mr} \times t$$

Now the pure rolling takes place when $v(t) = r \times \omega(t)$. Hence, we have

$$v(t) = \left(\frac{v_0}{2r} + \frac{5f}{2mr} \times t \right) r$$

$$2v(t) = v_0 + \frac{5f}{m} \times t$$

or

$$\frac{2}{5} v(t) = \frac{v_0}{5} + \frac{f}{m} \times t \qquad \text{(ii)}$$

Adding Eqs. (i) and (ii), we have

$$\frac{7}{5}v(t) = \frac{6}{5}v_0$$

or
$$v(t) = \frac{6}{7}v_0$$

Hence the sphere when starts rolling, it has translation velocity of $\frac{6}{7}v_0$.

Note: The translation velocity has reduced from v_0 to $\frac{6}{7}v_0$ when pure rolling starts.

Example 14.5 A cylinder of mass m is suspended through two strings wrapped around it as shown in Figure 14.7. Determine (i) tension T in the string and (ii) the speed of the cylinder when it falls by a distance h.

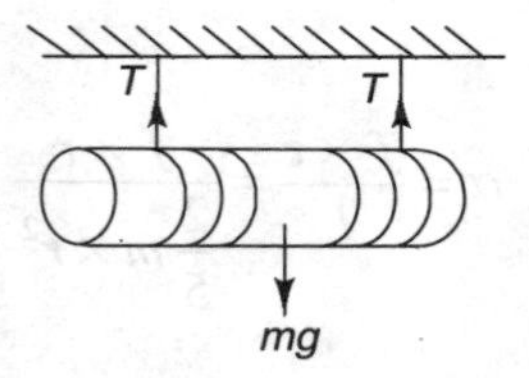

FIGURE 14.7 Example 14.5.

Let the cylinder falls with linear acceleration of a. Then we have

$$mg - 2 \times T = m \cdot a \tag{i}$$

Torque of all forces about the centre of mass must be zero as no external torque is applied.

$$2 \times T \times r - I \times \alpha = 0$$

For cylinder,
$$I = \frac{1}{2}mr^2 \text{ and } \alpha = \frac{a}{r}$$

$\therefore$
$$2 \times T \times r - \frac{1}{2}mr^2 \times \frac{a}{r} = 0$$

or
$$2T = \frac{ma}{2} \tag{ii}$$

Adding Eqs. (i) and (ii), we have

$$mg = \frac{3}{2}ma$$

or
$$a = \frac{2}{3}g$$

As initial velocity $u = 0$, then velocity v when cylinder has fallen through distance h is

$$v^2 = u^2 + 2ah$$

$$= 0 + 2 \times \frac{2}{3} g \times h$$

or

$$v = \sqrt{\frac{4gh}{3}}$$

Example 14.6 A cylinder is released from rest from the top of an inclined surface having inclination θ and length l (see Figure 14.8). If the cylinder rolls without slipping, find the speed when it reaches the bottom?

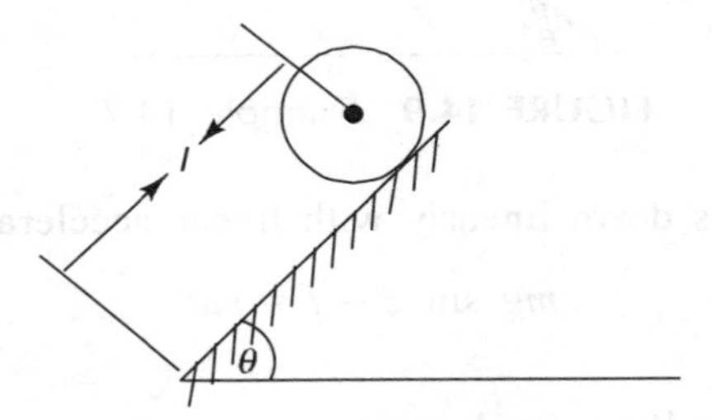

FIGURE 14.8 Example 14.6.

Guidance: The problem is about conservation of energy, i.e. PE is converted into KE of the cylinder.

$$\text{PE} = mgl \sin \theta$$

$$\text{KE} = \frac{1}{2} mv^2 + \frac{1}{2} I\omega^2$$

$$= \frac{1}{2} mv^2 + \frac{1}{2} \times \left(\frac{1}{2} mr^2 \right) \times \left(\frac{v}{r} \right)^2$$

$$= \frac{1}{2} mv^2 + \frac{1}{4} mv^2$$

$$= \frac{3}{4} mv^2$$

$$\text{PE} = \text{KE}$$

$$mgl \sin \theta = \frac{3}{4} mv^2$$

or
$$v = \sqrt{\frac{4}{3} gl \sin \theta}$$

Example 14.7 A sphere of mass m rolls without slipping on an inclined plane of inclination θ (see Figure 14.9). Find the linear acceleration of the sphere and the force of friction acting on it. What should be minimum coefficient of static friction to support pure rolling?

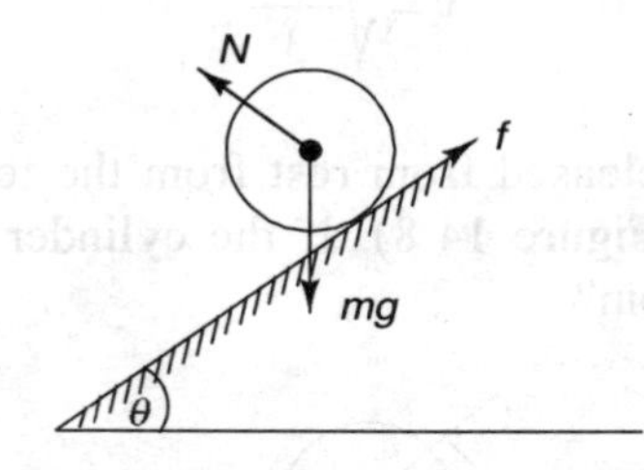

FIGURE 14.9 Example 14.7.

Initially the sphere slides down linearly with linear acceleration a. Hence

$$mg \sin \theta - f = ma \qquad \text{(i)}$$

where f = friction force.

When the sphere starts rolling, we have

$$f \times r = I \times \alpha$$

where α = angular acceleration

$$f \times r = \left(\frac{2}{5} mr^2\right) \times \frac{a}{r}$$

$\therefore$
$$f = \frac{2}{5} ma \qquad \text{(ii)}$$

Adding Eqs. (i) and (ii), we have

$$mg \sin \theta = \frac{7}{5} ma$$

or
$$a = \frac{5}{7} g \sin \theta$$

and
$$f = \frac{2}{7} mg \sin \theta$$

It means that $f = \frac{2}{7} mg \sin \theta$ when rolling starts. However, maximum friction that can act on sphere is $N \times \mu = mg \cos \theta \times \mu$.

Hence for pure rolling to the start, we must have

$$mg \cos \theta \times \mu > \frac{2}{7} mg \sin \theta$$

or
$$\mu > \frac{2}{7} \tan \theta$$

WORK-ENERGY PRINCIPLE

Work is the product of force and distance. The distance should be in the direction of the force. Consider a force F acting on a particle or on a mass centre which is displaced from r_1 to r_2 in the direction of force. The work done by the force would be

$$W = \int_{r_1}^{r_2} F \cdot dr$$

$$= \int_{x_1}^{x_2} F_x \cdot dx + \int_{y_1}^{y_2} F_y \cdot dy + \int_{z_1}^{z_2} F_z \cdot dz$$

As per Newton's second law of motion, the force F is

$$F = ma$$

$$= m \cdot \frac{dv}{dt}$$

$$\int_{r_1}^{r_2} F \cdot dr = m \int_{r_1}^{r_2} \frac{dv}{dt} \cdot dr$$

or
$$W = m \int_{t_1}^{t_2} \frac{dv}{dt} \cdot \frac{dx}{dt} \cdot dt$$

$$= m \int_{t_1}^{t_2} v \frac{dv}{dt} dt$$

$$= m \int_{r_1}^{r_2} \frac{d}{dt} \left(\frac{v^2}{2} \right) dt$$

$$= \left[\frac{1}{2} mv^2 \right]_{v_1}^{v_2}$$

$$= \frac{1}{2} mv_2^2 - \frac{1}{2} mv_1^2$$

$$= \text{KE}_2 - \text{KE}_1$$

The above is called *work-energy equation* which states that the work done on a particle must be equal to the change of kinetic energy of the particle. Similarly, the work done by the resultant force which is acting at the mass centre, is equal to the change in the translational kinetic energy of the body. The work done on a particle may be due to varying or constant force. The advantages of work-energy equations are:

(a) work-energy equation is a scalar equation as both works and kinetic energy are scalar quantities. There is no need to use Newton's law which involves force and acceleration which are vector quantities.
(b) work-energy equation is a path integration of the equation of motion. Hence it is a solution of the second order differential equation of the motion.
(c) it provides a simpler method to find velocities whenever work can be evaluated easily.

Conservation of Mechanical Energy

When a particle is moved from position 1 to 2 in a conservative force field, then work done is

$$W = \text{PE}_1 - \text{PE}_2$$

The change in potential energy leads to change in kinetic energy. Hence we have

$$W = \text{KE}_2 - \text{KE}_1$$

$\therefore$
$$\text{PE}_1 - \text{PE}_2 = \text{KE}_2 - \text{KE}_1 \quad \text{if } W = 0$$

or
$$(\text{PE} + \text{KE})_1 = (\text{PE} + \text{KE})_2$$

Work Done Against Spring Force

Consider a block is moved against spring force (spring constant k) from $x = 0$ to $x = x_1$ as shown in Figure 14.10.

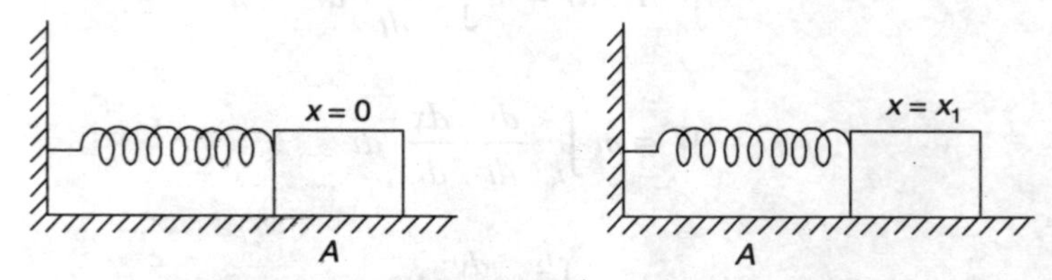

FIGURE 14.10 Work done against spring force.

The work done is

$$W = \int_0^{x_1} -(kx) \times dx$$

$$= \left[-\frac{kx^2}{2}\right]_0^{x_1}$$

$$= -\frac{kx_1^2}{2}$$

Example 14.8 A block of mass m slides along a frictionless surface as shown in Figure 14.11. If block is released from rest from point A, find the speed at B.

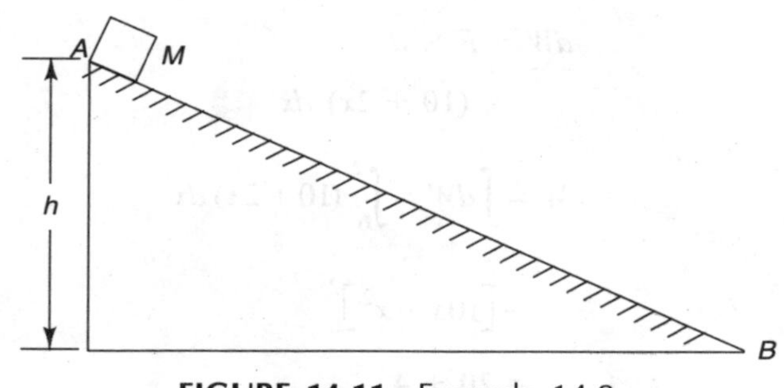

FIGURE 14.11 Example 14.8.

As no external force is applied, the work done is zero. There is a conservation of energy i.e. PE is converted into KE. Hence we have

$$mgh = \frac{1}{2} mv_b^2$$

or

$$v_b = \sqrt{2gh}$$

Example 14.9 A block of mass m attached to a spring of spring constant k, oscillate on a smooth horizontal table. Other end of spring is fixed to a wall (see Figure 14.12). If the block has a speed of v when the spring is at its natural length, find distance x travelled by the block before it comes to rest.

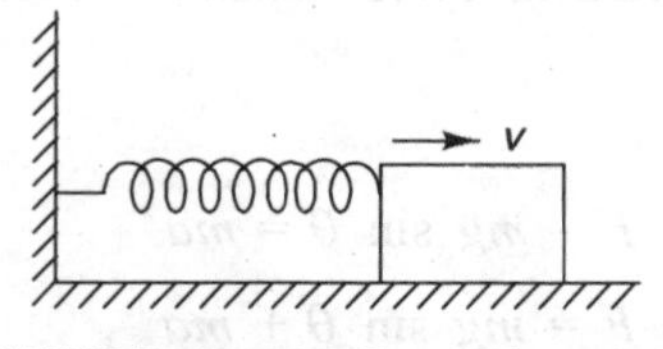

FIGURE 14.12 Example 14.9.

Here energy of the system is conserved as no work is done by any external force:

$$\text{KE} = \text{PE}$$

$$\frac{1}{2} mv^2 = \frac{1}{2} kx^2$$

or

$$x = v\sqrt{\frac{m}{k}}$$

Example 14.10 A force $F = (10 + 2x)$ acts on a body in x direction where F is in newton and x is in metre. If displacement is from $x = 0$ to $x = 2$ m, determine the work done by the force.

$$dW = F \times dx$$

$$= (10 + 2x)\ dx$$

$$\therefore \qquad W = \int dW = \int_0^2 (10 + 2x)\, dx$$

$$= \left[10x + x^2\right]_0^2$$

$$= 20 + 4$$

$$= 24 \text{ J}$$

Example 14.11 A block of a mass of 2 kg is pulled up on a smooth incline of angle 30° with horizontal (see Figure 14.13). If block moves with an acceleration of 1.0 m/s^2, find the work done by the pulling force in time $t = 4$ sec after the motion starts.

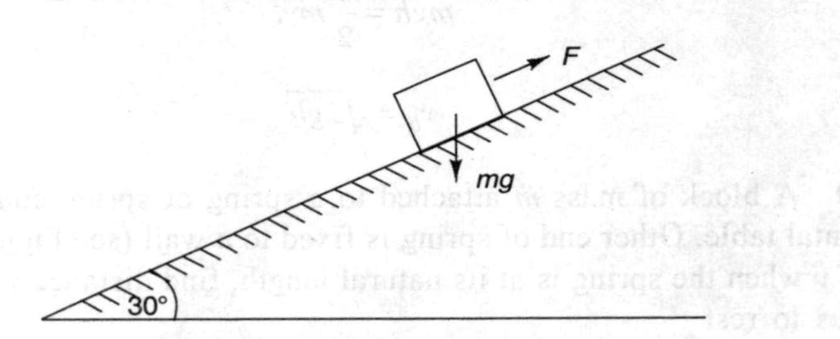

FIGURE 14.13 Example 14.11.

Forces acting on the block is

$$F - mg \sin \theta = ma$$

or

$$F = mg \sin \theta + ma$$

$$= 2 \times 9.81 \times \sin 30 + 2 \times 1$$

$$= 11.8 \text{ N}$$

Displacement d in direction of force is

$$d = \frac{1}{2} at^2$$

$$= \frac{1}{2} \times 1 \times 4^2$$

$$= 8 \text{ m}$$

Hence, work done is

$$W = F \times d$$

$$= 11.8 \times 8 = 94.4 \text{ J}$$

Example 14.12 A car is moving at 15 m/s when driver puts on his brakes, thereby car skids in the direction of motion (see Figure 14.14). Car weighs 500 kg and dynamic coefficient of friction is 0.6. How far will car moves before it stops?

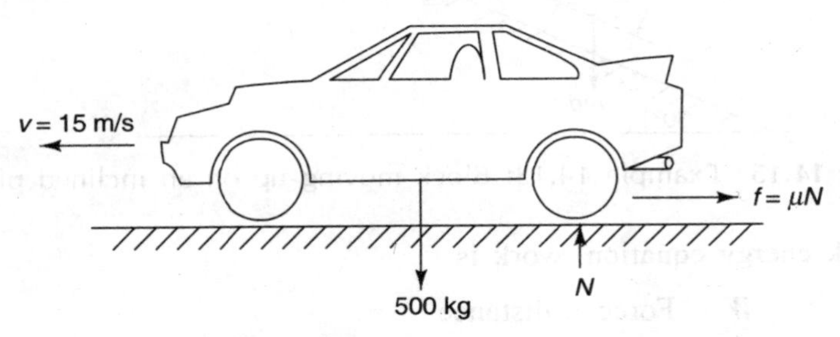

FIGURE 14.14 Example 14.12.

Guidance: The problem can be easily solved by using work energy equation

$$\text{The change of KE} = \frac{1}{2}m(v_2^2 - v_1^2)$$

$$= \frac{1}{2} \times 500 \times (0 - 15^2)$$

$$= 56.25 \times 10^3 \text{ N m}$$

Work done = $f \times$ distance

But f = friction = $(2000 \times g \times \mu)$

Hence $W = (500 \times 9.81 \times 0.6) \times l$ where l = distance

$$= 2943 \times l \text{ N m}$$

$$W = \text{change in KE}$$

$$2943 \times l = 56.25 \times 10^3$$

or

$$l = \frac{56.25 \times 10}{2943}$$

$$= 19.1 \text{ m}$$

Example 14.13 A weight of 400 N is placed on an inclined plane making 30° with the horizontal (see Figure 14.15). What value of force P is required to get a velocity of 10 m/s in distance of 1.0 m? Neglect friction.

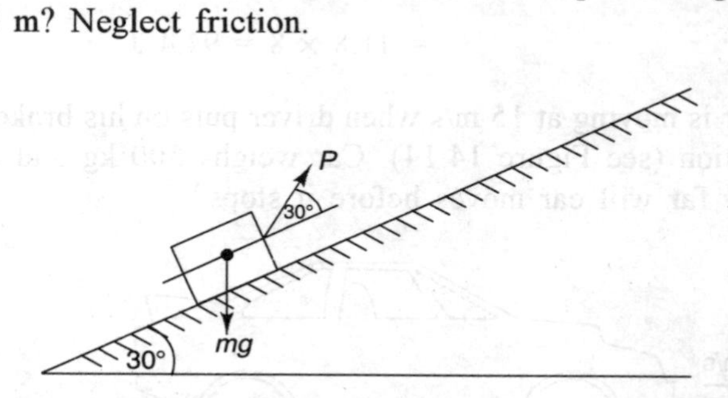

FIGURE 14.15 Example 14.13: Block moving up on an inclined plane.

Applying work energy equation, work is

$$W = \text{Force} \times \text{distance}$$

$$= (P \cos 30 + 400 \sin 30) \times 1 = P \cos 30 + 200$$

$$\Delta PE = -mg\Delta h = -400 \times 1 \times \sin 30$$

$$= -200 \text{ J}$$

$$\Delta KE = \frac{1}{2} \times \frac{400}{9.81} \times 10^2$$

$$= 2038.7 \text{ J}$$

$\therefore$ $$W = \Delta PE + \Delta KE$$

$$= -200 + 2038.7 = 1838.7$$

$\therefore$ $$P \cos 30 + 200 = 1838.7$$

or $$P = 1892.2 \text{ N}$$

Kinetic Energy-Based on Centre of Mass

Consider a system of n particles as shown in Figure 14.16. The total kinetic energy of the system of particles in relation to x, y, z axes.

$$KE = \sum_{i=1}^{i=n} \frac{1}{2} m_i v_i^2$$

The position vector of the centre of mass (CM) is r_c and the position vector of the particle m_i is r_i. The ρ_{ci} is the displacement vector from CM to m_i particle. Therefore we have

$$r_i = r_c + \rho_{ci}$$

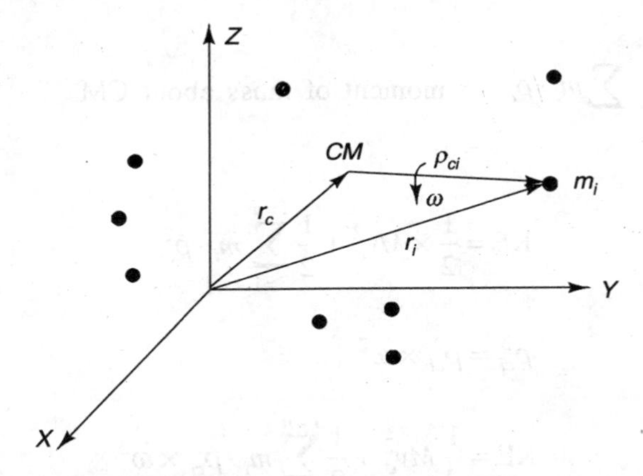

FIGURE 14.16 System of particles with centre mass.

On differentiation, we have

$$\dot{r}_i = \dot{r}_c + \dot{\rho}_{ci}$$

or

$$v_i = v_c + \dot{\rho}_{ci}$$

$\therefore$

$$\text{KE} = \sum_{i=1}^{i=n} \frac{1}{2} m_i v_i^2$$

$$= \sum_{i=1}^{i=n} \frac{1}{2} m_i \, (v_c + \dot{\rho}_{ci}) \cdot (v_c + \dot{\rho}_{ci})$$

$$= \frac{1}{2} \sum_{i=1}^{i=n} m_i v_c^2 + \sum_{i=1}^{i=n} m_1 v_c \cdot \dot{\rho}_{ci} + \frac{1}{2} \sum m_i \dot{\rho}_{ci}^2$$

$$= \frac{1}{2} \left(\sum_{i=1}^{i=n} m_i \right) v_c^2 + v_c \cdot \left(\sum_{i=1}^{i=n} m_i \cdot \dot{\rho}_{ci} \right) + \frac{1}{2} \left(\sum_{i=1}^{i=n} m_i \cdot \dot{\rho}_{ci}^2 \right)$$

But

$$\sum_{i=1}^{i=n} m_i = M$$

and

$$\sum_{i=1}^{i=n} m_i \dot{\rho}_{ci} = \frac{d}{dt} \sum_{i=1}^{i=n} m_i \cdot \rho_{ci}$$

But

$$\sum m_i \cdot \rho_{ci} = \text{moment of mass about CM}$$

$$= 0$$

$\therefore$

$$\text{KE} = \frac{1}{2} \times M v_c^2 + \frac{1}{2} \sum_{i=1}^{i=n} m_i \cdot \dot{\rho}_{ci}^2$$

But

$$\dot{\rho}_{ci}^2 = \rho_{ci}^2 \times \omega^2$$

$\therefore$

$$\text{KE} = \frac{1}{2} M v_c^2 + \frac{1}{2} \sum_{i=1}^{i=n} m_i \cdot \rho_{ci}^2 \times \omega^2$$

$$= \frac{1}{2} M v_c^2 + \frac{1}{2} I_c \omega^2$$

Hence kinetic energy of a body for some reference is composed of

(a) kinetic energy of total mass moving relative to that reference with velocity of mass centre.
(b) kinetic energy of the mass of the particles relative to the mass centre.

Work-Energy Equations for a Rigid Body

The work-energy equations for a rigid body may be obtained from Newton's law and Euler's equation as done for a particle.

$$f = m a_c \qquad (a_c = \text{acceleration of centre of mass})$$

$$= m \frac{dv_c}{dr_c} \cdot \frac{dr_c}{dt}$$

$$\int F \cdot dr_c = \int_{r_1}^{r_2} m \times \frac{1}{2} \frac{d}{dt} v_c^2$$

$$W = \frac{1}{2} m (v_{c_2}^2 - v_{c_1}^2) = \text{KE}_2 - \text{KE}_1$$

The left-hand expression shows that net work is done if net force is acting at the centre of mass. Similarly right hand shows the change in kinetic energy of total mass m of the body as if it is concentrated at the centre of mass. This is the reason why this equation is called the work-energy equation applied to centre of mass.

When Euler's equation is applied to the centre of mass of the rigid body, we have moment as

$$M_c = I_c\alpha$$

where I_c = moment of inertia of body at centre of mass and α = angular acceleration.

or
$$M_c = I_c \frac{d\omega}{dt}$$

$$= I_c \frac{d\omega}{d\theta}.\frac{d\theta}{dt} = I_c\omega.\frac{d\omega}{d\theta}$$

$$= I_c \times \frac{1}{2}\frac{d\omega^2}{d\theta}$$

$\therefore$
$$W = \int_{\theta_1}^{\theta_2} M_c \cdot d\theta = \frac{1}{2} I_c (\omega_2^2 - \omega_1^2)$$

$$= \text{KE}_2 - \text{KE}_1$$

The left-hand expression stands for the work done by net moment acting on the body about the centre of mass which brings angular displacement from θ_1 to θ_2. Similarly right hand side expression stands for the change in rotational kinetic energy in relation to the centre of mass.

Total work done by action of forces and moments acting on a body is the sum of the work done by the net force and moment both acting at the centre of mass.

$$W_{\text{total}} = \int_{r_1}^{r_2} F \cdot dr_c + \int_{\theta_1}^{\theta_2} M_c \, d\theta$$

$$= \frac{1}{2} m(v_{c_2}^2 - v_{c_1}^2) + \frac{1}{2} I_c (\omega_2^2 - \omega_1^2)$$

Example 14.14 A cylinder with mass 50 kg is released from rest on an incline as shown in Figure 14.17. The diameter of the cylinder is 2 m. If the cylinder rolls without slipping,

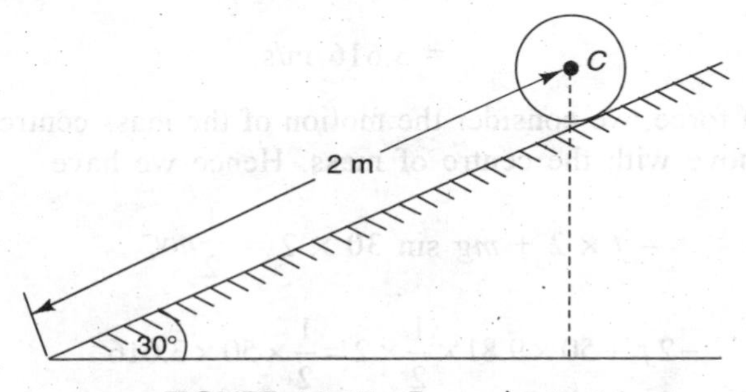

FIGURE 14.17 Example 14.14.

then compute (i) the speed of centre point C after it has moved 2 m along the inclined plane and (ii) friction force acting on the cylinder.

Work-energy equation can be used which states

$$W = \Delta KE + \Delta PE$$

However, as no force acting, hence $W = 0$

$$\Delta PE_{1\text{-}2} = -mg\Delta h$$

$$= -50 \times 9.81 \times 2 \sin 30$$

$$= -490.5 \text{ J}$$

$$\Delta KE_{1-2} = \left(\frac{1}{2}mv_c^2 + \frac{1}{2}I\omega^2\right)_2 - 0$$

$$= \frac{1}{2} \times 50 \times v_c^2 + \frac{1}{2}\left(\frac{1}{2}mr^2\right)\omega^2$$

But

$$\omega = \frac{v_c}{r} = \frac{v_c}{1} = v_c$$

$\therefore$

$$\Delta KE_{1-2} = 25v_c^2 + \frac{1}{4} \times 50 \times (1)^2 \times v_c^2$$

$$= (25 + 12.5)v_c^2$$

$$= 37.5v_c^2$$

$$(\Delta PE)_{1-2} + (\Delta KE)_{1-2} = 0$$

$$-490.5 + 37.5v_c^2 = 0$$

or

$$v_c = \sqrt{\frac{490.5}{37.5}}$$

$$= 3.616 \text{ m/s}$$

To find out friction force, we consider the motion of the mass centre of the cylinder. All external forces must move with the centre of mass. Hence we have

$$-f \times 2 + mg \sin 30 \times 2 = \frac{1}{2}mv_c^2$$

$$-2f + 50 \times 9.81 \times \frac{1}{2} \times 2 = \frac{1}{2} \times 50 \times 3.616^2$$

$$-2f + 490.5 = 326.88$$

or
$$f = \frac{490.5 - 326.88}{2}$$

$$= 81.81 \text{ J}$$

Example 14.15 A flywheel of diameter 1 m is made to rotate by means of a suspended mass of 200 kg by a string wound round a concentric drum of 0.6 m diameter as shown in Figure 14.18. Find the moment of inertia of the flywheel if velocity of the suspended mass is 0.6 m/s after a fall of 1 m. Take friction moment as 100 Nm at bearing centre.

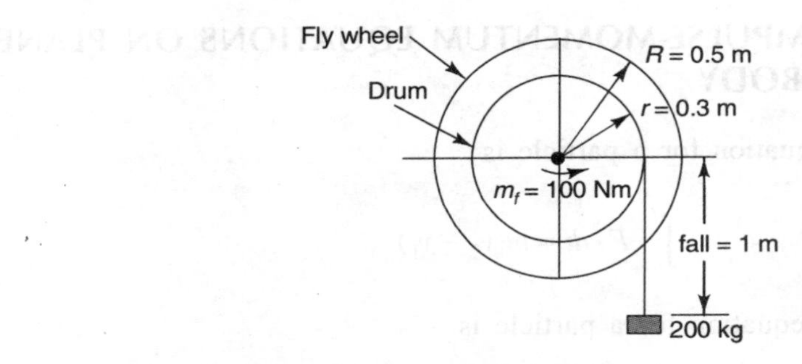

FIGURE 14.18 Example 14.15.

Work-energy equation can be applied on the system.

$$W_{\text{total}} = (W)_{\text{mass}} - (W)_{\text{friction}}$$

$$(\text{Work})_{\text{total}} = \text{Force} \times \text{distance}$$

$$= (mg) \times h = 200 \times 9.8 \times 1 = 1962 \text{ J}$$

$$(\text{Work})_{\text{friction}} = M_f \times \theta = 100 \times \frac{\text{fall}}{r} = 100 \times \frac{1}{0.3} = 333.3 \text{ Nm}$$

Change in energy $\Delta\text{KE} = (\text{KE}_2 - \text{KE}_1)_{\text{total}}$

$\text{KE}_1 = 0$ as both mass and flywheel are at rest

$$\text{KE}_2 = \text{KE of mass} + \text{KE of flywheel}$$

$$= \frac{1}{2}Mv^2 + \frac{1}{2}I\omega^2$$

$$= \frac{1}{2} \times 200 \times 0.6^2 + \frac{1}{2} \times I \times \left(\frac{0.6}{0.3}\right)^2$$

$$= 36 + 2 \times I$$

$$\therefore \quad \Delta KE = (36 + 2I) - 0$$

$$= 36 + 2I$$

$$W_{total} = \Delta KE$$

$$1962 - 333.33 = 36 + 2I$$

or
$$I_f = \frac{1592.67}{2}$$

$$= 796.335 \text{ kg m}^2$$

APPLICATIONS OF IMPULSE-MOMENTUM EQUATIONS ON PLANE MOTION OF RIGID BODY

Linear impulse moment equation for a particle is

$$\int_{t_1}^{t_2} F \cdot dt = m(v_2 - v_1)$$

Angular impulse moment equation for a particle is

$$\int_{t_1}^{t_2} M_c \, dt = I(\omega_2 - \omega_1)$$

In case of rigid body, the linear impulse moment equation is derived in relation of centre of mass and it can be expressed as

$$\int_{t_1}^{t} F_c \cdot dt = m_c (v_{c_2} - v_{c_1})$$

The expression at left side is the linear impulse due to net external force acting at the centre of mass over a period of time whereas the right side expression is the change in the linear momentum of the centre of mass over the same interval of time.

Similarly, the angular impulse-momentum equation of a rigid body can be given as

$$\int_{t_1}^{t_2} M_c \cdot dt = I_c (\omega_1 - \omega_2)$$

The left-hand side expression is the angular impulse due to the net moment M_c acting about the centre of mass over a period of time, whereas the right side expression is the change in angular momentum about the centre of mass for same time interval.

Example 14.16 Find the velocity v_c that the right circular cylinder of weight W and radius r will acquire after falling from rest through a vertical distance h (Figure 14.19).

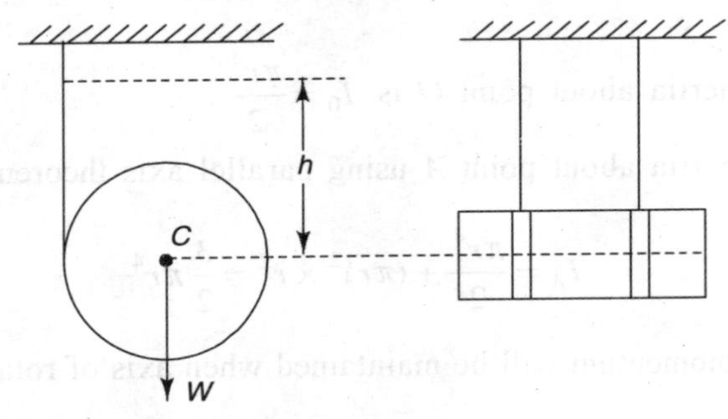

FIGURE 14.19 Example 14.16.

The problem can be solved with work-energy equation.

$$\text{Work} = W \times h \tag{i}$$

Energy changed is

$$\frac{1}{2}\frac{W}{g}v_c^2 + \frac{1}{2}\left(\frac{1}{2}\frac{W}{g} \times r^2\right)\left(\frac{v_c}{r}\right)^2 \tag{ii}$$

Equating Eqs. (i) and (ii)

$$W \times h = \frac{1}{2}\frac{W}{g} \times v_c^2 + \frac{1}{4}\frac{W}{g}v_c^2$$

$$= \frac{3}{4}\frac{W}{g}v_c^2$$

or

$$v_c = \sqrt{\frac{4gh}{3}}$$

Example 14.17 A circular disc of radius r is supported by a perfectly smooth horizontable table. The disc spins with angular velocity ω about its vertical geometric axis as shown in Figure 14.20. What new angular velocity ω will the disk have if a point A on its circumference is suddenly pinned to the table?

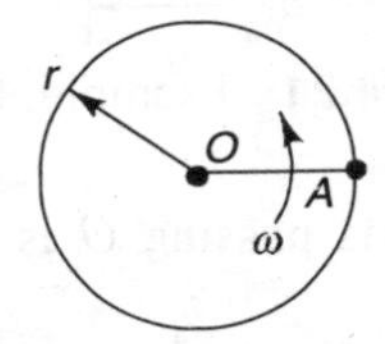

FIGURE 14.20 Example 14.17.

The moment of inertia about point O is $I_0 = \dfrac{\pi r^4}{2}$.

The moment of inertia about point A using parallel axis theorem is

$$I_A = \frac{\pi r^4}{2} + (\pi r)^2 \times r^2 = \frac{3}{2}\pi r^4$$

Now the angular momentum will be maintained when axis of rotation is changed. Hence we have

$$I_0 \times \omega = I_A \times \omega'$$

or

$$\omega' = \frac{I_0 \times \omega}{I_A}$$

$$= \frac{\frac{\pi r^4}{2}}{\frac{3}{2}\pi r^4} \times \omega$$

$$= \frac{\omega}{3}$$

Example 14.18 If a square plate (see Figure 14.21) initially spinning about a vertical axis through its centre of gravity, then find new angular speed if one corner is suddenly pinned down.

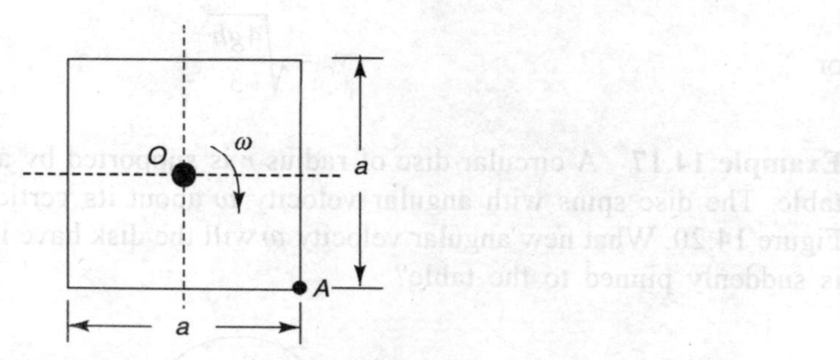

FIGURE 14.21 Example 14.18.

Moment of inertia about vertical axis passing O is

$$I_0 = \frac{a^4}{6}$$

Moment of inertia about vertical axis passing A is

$$I_A = \frac{a^4}{6} + (a^2)\left(\frac{a}{\sqrt{2}}\right)^2$$

$$= \frac{4}{6}a^6$$

When axis of rotation is changed, the angular momentum will be conserved. Hence we have

$$I_0 \times \omega = I_A \times \omega'$$

or

$$\omega' = \frac{I_0}{I_A}\omega$$

$$= \frac{a^4/6}{4a^4/6} \times \omega$$

$$= \frac{\omega}{4}$$

Example 14.19 At what height h above the table surface should a billiard ball of radius c be struck by a horizontal impact F in order to have no sliding at the point of contact O (see Figure 14.22).

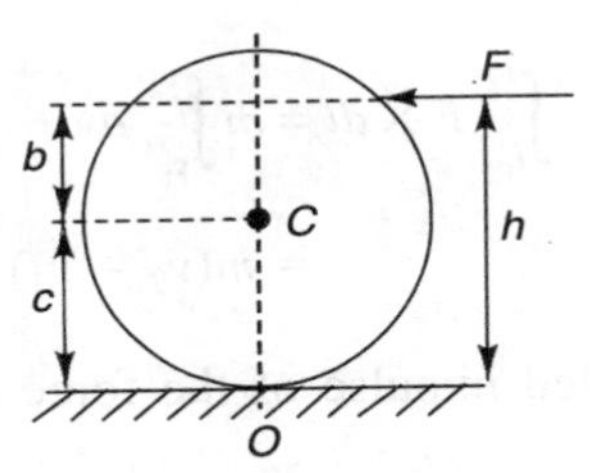

FIGURE 14.22 Example 14.19.

The moment of inertia of the ball about point O is

$$I_0 = I_c + m \times c^2$$

$$= \frac{2}{5}mc^2 + mc^2 = \frac{7}{5}mc^2$$

Due to impact, the force is

$$F = ma$$

where a is linear acceleration

$$\therefore \qquad \text{Moment} = F \times h$$

$$= ma \times h$$

For equilibrium, we have

$$I_0 \times \alpha = mah$$

where α = angular acceleration

$$\frac{7}{5} mc^2 \times \frac{a}{c} = mah$$

or

$$h = \frac{7}{5} c$$

LINEAR IMPULSE-MOMENTUM PRINCIPLE

As per Newton's first law of motion, we have

$$F = m \times a \qquad m = \text{mass and } a = \text{acceleration}$$

$$F = m \times \frac{dv}{dt} \qquad v = \text{velocity, } t = \text{time}$$

or

$$F \times dt = m \times dv$$

Suppose body has initial velocity v_1 at t_1 and final velocity v_2 at t_2. Then if we integrate the above equation, we have

$$\int_{t_1}^{t_2} F \times dt = m \int_{v_1}^{v_2} dv$$

$$= m(v_2 - v_1)$$

The integral $\int_{t_1}^{t_2} F \times dt$ is called impulse of the force during the time interval $(t_2 - t_1)$ and term $m(v_2 - v_1)$ is the change of the linear momentum of the body. Therefore Impulse-Momentum principle states that impulse acting on a body over a time interval is equal to the change in linear momentum of the body during that time interval. The force F may vary or remain constant in the time interval specified. In case force is varying, then impulse can be ascertained from the area under a force versus time curve (see Figure 14.23).

For example, the impulse of the force F in time interval from t_1 to t_2 is equal to area under the curve as shaded $\left(\int_{t_1}^{t_2} F \cdot dt\right)$.

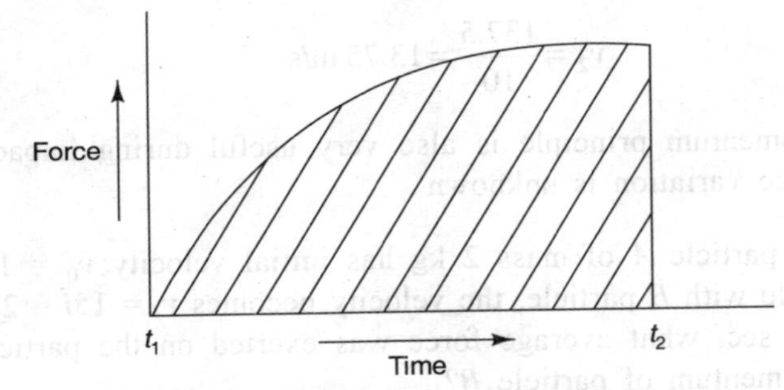

FIGURE 14.23 Force versus time graph.

Example 14.20 Find the impulse, if a force acting over an interval of 20*s* is as shown in Figure 14.24.

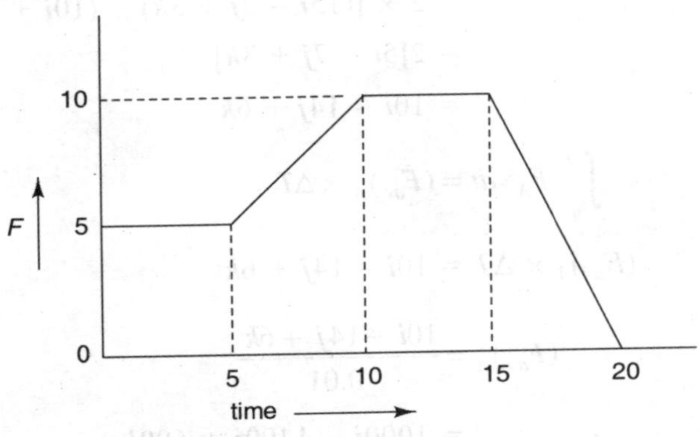

FIGURE 14.24 Example 14.20.

The impulse is equal to area under curve. Hence area can be found out as shown below

$$\text{area} = (5\times5)+\left(5\times5+\frac{1}{2}\times5\times5\right)+5\times10+\frac{1}{2}\times10\times5$$

$$= 25 + 25 + 12.5 + 50 + 25$$

$$= 137.5 \text{ N s}$$

Hence impulse = 137.5 N s.

Now if the mass of body is 10 kg, then we can easily find the velocity of the body after 20 sec from stationary position if this impulse is applied on the body. It is possible due to impulse-momentum principle otherwise finding final velocity with changing force with time is a difficult exercise. If we apply the above impulse, then we have

$$\text{Impulse} = \text{Change of linear momentum}$$

$$137.5 = 10(v_2 - v_1)$$

$$= 10(v_2 - 0)$$

or
$$v_2 = \frac{137.5}{10} = 13.75 \text{ m/s}$$

The impulse-momentum principle is also very useful during impact of two particles where the exact force variation is unknown.

Example 14.21 A particle A of mass 2 kg has initial velocity $v_0 = 10\boldsymbol{i} + 5\boldsymbol{j}$ m/s. After collision of A particle with B particle, the velocity becomes $v_f = 15\boldsymbol{i} - 2\boldsymbol{j} + 3\boldsymbol{k}$ m/s. If time of collision of 0.01 sec, what average force was exerted on the particle A? What is the change of linear momentum of particle B?

$$\int_{t_0}^{t_f} F_A \cdot dt = \text{change of momentum}$$

$$= m(v_f - v_0)$$

$$= 2 \times [(15\boldsymbol{i} - 2\boldsymbol{j} + 3\boldsymbol{k}) - (10\boldsymbol{i} + 5\boldsymbol{j})]$$

$$= 2[5\boldsymbol{i} - 7\boldsymbol{j} + 3\boldsymbol{k}]$$

$$= 10\boldsymbol{i} - 14\boldsymbol{j} + 6\boldsymbol{k}$$

But
$$\int_{t_0}^{t_f} F_A \cdot dt = (F_{av})_A \times \Delta T$$

$\therefore$
$$(F_{av})_A \times \Delta T = 10\boldsymbol{i} - 14\boldsymbol{j} + 6\boldsymbol{k}$$

or
$$(F_{av})_A = \frac{10\boldsymbol{i} - 14\boldsymbol{j} + 6\boldsymbol{k}}{0.01}$$

$$= 1000\boldsymbol{i} - 1400\boldsymbol{j} + 600\boldsymbol{k}$$

Using Newton's third law that action equals reaction, we find the impulse of particle B which is negative of impulse of particle A. The change of momentum of particle B is

$$\int_{t_0}^{t_f} F_B \times dt = -\int_{t_0}^{t_f} F_A \times dt = -(10\boldsymbol{i} - 14\boldsymbol{j} + 6\boldsymbol{k})$$

$$= -10\boldsymbol{i} + 14\boldsymbol{j} - 6\boldsymbol{k}$$

The impulse-momentum equation can also be usefully applied in computing the velocity of the body at any time t when the body reaches the end of the inclined plane as shown in Figure 14.25. It shows two bodies W_1 and W_2 are connected by a cord and body W_1 on the inclined surface at an angle θ with the horizontal and μ_d is the dynamic coefficient of friction.

For body W_1, we have for equilibrium in vertical direction

$$\Sigma F_y = 0, N_1 = W_1 \cos\theta$$

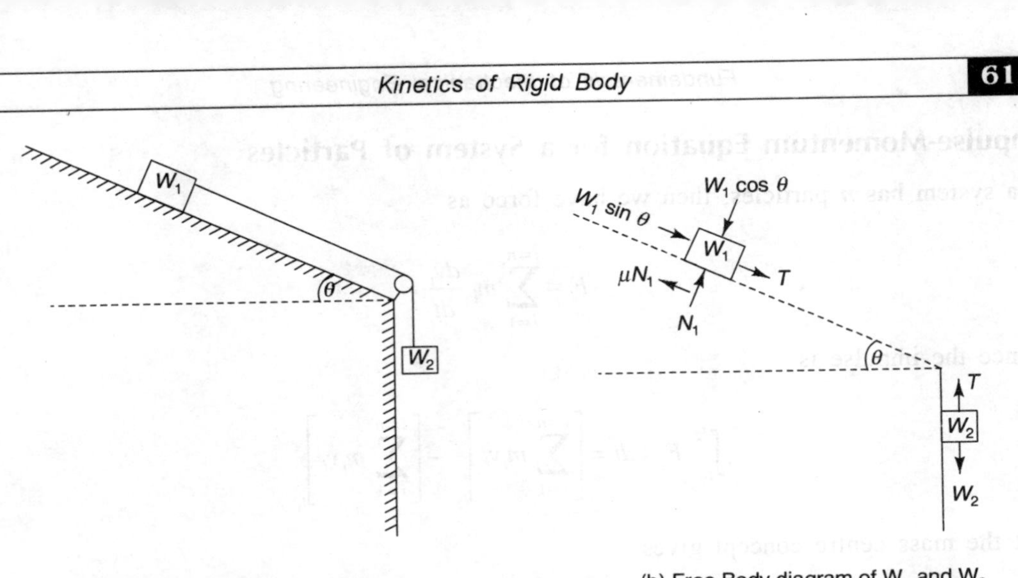

FIGURE 14.25 Example 14.21.

Now apply the impulse-momentum equation on body W_1 along the surface, we have

$$= \int_0^t (W_1 \sin\theta + T - \mu_d W_1 \cos\theta)\, dt = \frac{W_1}{g}(v - 0)$$

or
$$(W_1 \sin\theta + T - \mu_d W_1 \cos\theta)t = \frac{W_1}{g} v \qquad \text{(i)}$$

Now apply the impulse-moment equation for body W_2, we have

$$\int_0^t (W_2 - T)\, dt = \frac{W_2}{g}(v - 0)$$

or
$$(W_2 - T)t = \frac{W_2}{g} v \qquad \text{(ii)}$$

Now on adding Eqs. (i) and (ii), we have

$$(W_1 \sin\theta + W_2 - \mu_d W_1 \cos\theta) \times t = \frac{v}{g}(W_1 + W_2)$$

$$\therefore \qquad v = \frac{g \times t}{W_1 \times W_2}(W_1 \sin\theta + W_2 - \mu_d W_1 \cos\theta)$$

Impulse-Momentum Equation for a System of Particles

If a system has n particles, then we have force as

$$F = \sum_{i=1}^{i=n} m_i \frac{dv_i}{dt}$$

Hence the impulse is

$$\int_{t_1}^{t_2} F \times dt = \left[\sum_{i=1}^{i=n} m_i v_i\right]_{t_2} - \left[\sum_{i=1}^{i=n} m_i v_i\right]_{t_1}$$

But the mass centre concept gives

$$Mr_c = \sum_{i=1}^{i=n} m_i r_i$$

where M is mass at mass centre.

On differentiation, we get

$$Mv_c = \sum_{i=1}^{i=n} m_i v_i$$

Hence impulse is

$$\int_{t_1}^{t_2} F \cdot dt = M(v_c)_{t_2} - M(v_c)_{t_1}$$

The total impulse in a system of particles equals to the change in linear momentum of a hypothetical particle having the mass of the entire aggregate of the particles and moving with the mass centre.

Angular Impulse-Momentum

As per Euler's equation, if M_c is the net moment acting about the centre of mass, then it is equal to angular momentum of the body.

$$M_c = \text{angular momentum}$$

$$= I_c \times \alpha \qquad \text{(where } \alpha = \text{angular acceleration)}$$

$$= I_c \times \frac{d\omega}{dt}$$

where ω = angular velocity

If external moment is acting from time t_1 to t_2, then we have

$$\int_{t_1}^{t_2} M_c \times dt = \int_{\omega_1}^{\omega_2} I_c \times d\omega$$

or

$$\int_{t_1}^{t_2} M_c \, dt = I_c(\omega_2 - \omega_1)$$

The above is angular impulse momentum equation which states that angular impulse $\left[\int_{t_1}^{t_2} M_c \, dt\right]$ is equal to the change of angular momentum for a time interval. It must be noted that for a plane motion of a rigid body, the angular impulse and angular momentum about the centre of mass must be about the axis of rotation passing through the centre of mass. The principles of linear impulse momentum and angular impulse momentum are mutually independent principles and therefore both can be applied to a rigid body for finding its dynamic condition in just the same way as Newton's law and Euler's equations are used. Impulse-momentum equations are generally preferred over Newton's law and Euler's equation as these give gross effect of the change in linear and angular velocities over a period of time. Secondly, there is another advantage of the usage of impulse moment equations. When the net external force or moment applied on a body is zero, the corresponding impulse momentum equation reduces to momentum conservation and this can be used more conveniently. Whenever two bodies impact upon, they have no external force or moment acting on them. Hence momentum conservation equations can be applied in finding final velocities of these bodies. The momentum conservation equations are

(a) linear momentum conservation equation as given

$$m_1 v_{c_1} + m_2 v_{c_2} = m_2 v'_{c_1} + m_2 v'_{c_2}$$

(b) angular momentum conservation equation as given

$$I_{c_1}\omega_1 + I_{c_2}\omega_2 = I_{c_1}\omega_1' + I_{c_2}\omega_2'$$

Example 14.22 A wheel is rotating at an angular speed about its axis which is kept vertical. An identical wheel initially at rest is gently dropped into the same axle and two wheels start rotating with a common angular speed. Find the common angular speed.

As no external torque or moment is applied on this two-wheeled system, the angular momentum will remain unchanged. Hence we have

$$\therefore \; I\omega_1 + 0 = I\omega_2 + I\omega_2$$

or

$$\omega_2 = \frac{\omega_1}{2}$$

Final angular speed is one-half of the initial angular speed.

Example 14.23 A man of mass 60 kg stands at one end of 6 m long floating boat of mass 300 kg (see Figure 14.26). If the man starts walking to other end of the boat at 2 m/s, find (i) absolute velocity of the boat, (ii) distance of shifting of the boat, (iii) velocity of the boat when man reaches other end of the boat and (iv) velocity of the boat if man falls in water while walking.

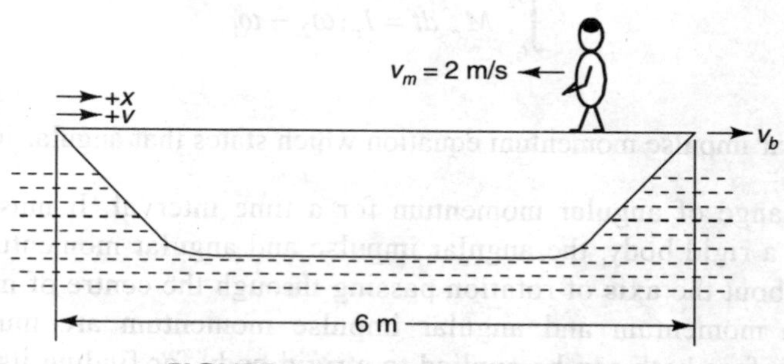

FIGURE 14.26 Example 14.23.

Initially the man and boat are at rest, hence their momentum is zero. As no external force is applied, the momentum will be conserved.

1. *Absolute velocity of boat.* When man moves to the left with velocity $v_m = 2$ m/s, the boat will move towards the right with velocity of v_b. The absolute velocity of the man is

$$v_{ma} = -2 + v_b$$

The momentum of the system is

$$m_m \times v_{ma} + m_b \times v_b = (-2 + v_b) \times 60 + 300 \times v_b$$

$$= 360v_b - 120$$

As the initial momentum is zero, we have

$$\therefore \qquad 360v_b - 120 = 0$$

or

$$v_b = \frac{120}{360} = 0.334 \text{ m/s}$$

$$\therefore \qquad v_{ma} = -2 + 0.334$$

$$= -1.666 \text{ m/s}$$

Hence the absolute velocities of the man and boat are –1.666 m/s and 0.334 m/s.

2. *Distance of boat shifting.* The time taken by the man to travel 6 m of the boat at speed of 2 m/s is

$$t = \frac{6}{2} = 3 \text{ s}$$

The boat has a velocity of 0.334 m/s, hence the boat shifting is

$$S = 0.334 \times 3$$

$$= 1 \text{ m}$$

3. *Velocity of boat when man reaches other end.* When man reaches the other end, his velocity is again zero, resulting the velocity of the boat also becoming zero.

4. *Man falls in water while walking.* The man has certain momentum before falling which is equal to

$$m_m \times v_{m_a} = 60 \times v_{m_a}$$

$$= 60 \times (-1.666)$$

$$= -99.96 \text{ N s}$$

Hence the boat will gain this momentum. If boat has now new velocity as v_{b_2}, then

$$m_b \times v_{b_2} = 99.96$$

or

$$v_{b_2} = \frac{99.96}{300}$$

$$= 0.334 \text{ m/s}$$

Example 14.24 A particle of mass 4 kg is tied to a string and rotated with angular velocity of 20 rad/s in a circle of 1 m radius over a smooth horizontal table surface as shown in Figure 14.27. It is possible to reduce the radius of circle by pulling the string through a slot in the table surface at a speed of 5 m/s. Find the speed of the particle when particle is circling at radius of 0.5 m.

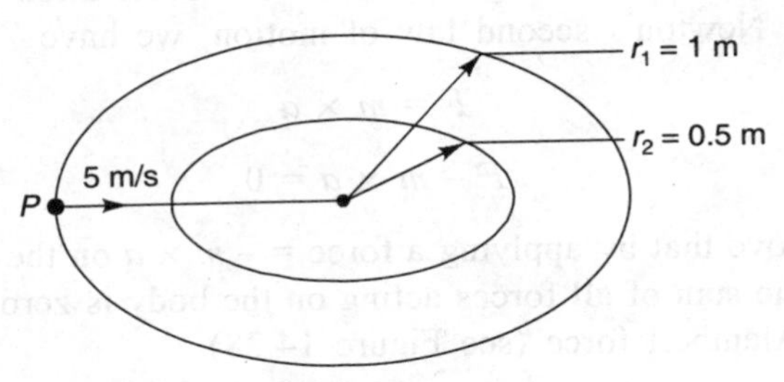

FIGURE 14.27 Example 14.24.

As there is no external torque applied, the moment of momentum of the particle is conserved.

$$\text{Initial moment of momentum} = mv_1 \times r_1$$

$$= m\omega_1 \times r_1^2$$

$$= 4 \times 20 \times 1^2$$

$$= 80 \text{ N ms}$$

$$\text{Final moment of momentum} = m\omega_2 \times r_2^2$$
$$= 4 \times \omega_2 \times 0.5^2$$
$$= \omega_2 \text{ N ms}$$

As initial moment of momentum is equal to final moment of momentum, we have

$$\omega_2 = 80 \text{ rad/s}$$

$$\therefore \quad \text{Tangential velocity } (v_2)_{tan} = \omega_2 \times r$$
$$= 80 \times 0.5$$
$$= 40 \text{ m/s}$$

$$\text{Radial velocity } (v_2)_{rad} = 0.5 \text{ m/s}$$

$$\therefore \quad \text{Total velocity } v_2 = \sqrt{(v_2)_{tan}^2 + (v_2)_{rad}^2}$$
$$= \sqrt{40^2 + 0.5^2}$$
$$= 40.3 \text{ m/s}$$

D'ALEMBERT'S PRINCIPLE

D'Alembert's principle states that a moving body having a dynamic equilibrium can be brought to a static equilibrium by applying an imaginary inertia force of the same magnitude as that of the accelerating force but in the opposite direction. D'Alembert's principle helps in changing the dynamic equilibrium of the body into a static equilibrium.

Let a body of mass m be moving with uniform acceleration (a) under the action of external force. As per Newton's second law of motion, we have

$$F = m \times a$$

or

$$F - m \times a = 0$$

It is clear from above that by applying a force = $- m \times a$ on the body, the body has now static equilibrium as the sum of all forces acting on the body is zero. The inertia force given by $- ma$ is called D'Alembert force (see Figure 14.28).

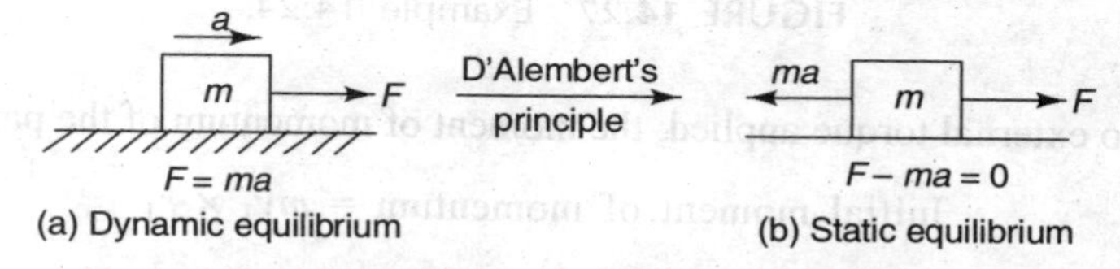

FIGURE 14.28 D'Alembert force.

Inertia force ($m \times a$) acts in the opposite direction of motion of the body and it passes through the centre of gravity of the body. It is simpler to solve static equilibrium as compared to dynamic equilibrium. D'Alembert's principle also helps in changing the dynamic equilibrium of fluid mass into static equilibrium. An imaginary inertia force of same magnitude but opposite in direction in place of actual acceleration force is applied to the moving fluid so that it can be brought to static equilibrium (see Figure 14.29).

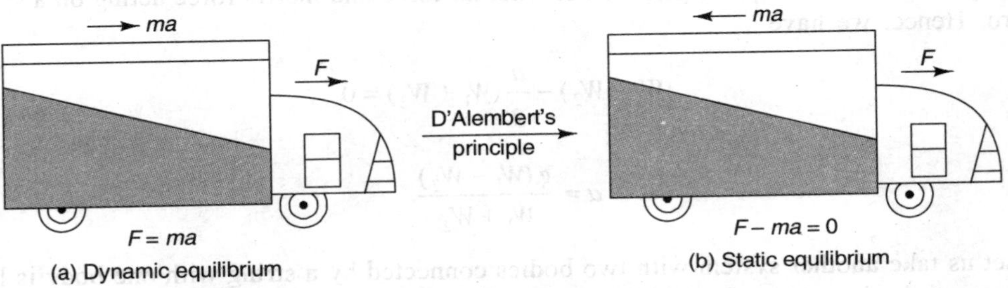

FIGURE 14.29 Change of dynamic equilibrium.

D'Alembert's principle provides a simpler method for the analysis of two bodies connected by a string. Consider two bodies of weights W_1 and W_2 connected to the two ends of a string passing over a smooth pulley as shown in Figure 14.30. The weight W_1 is more than weight W_2 and the net motion is downwards with acceleration a.

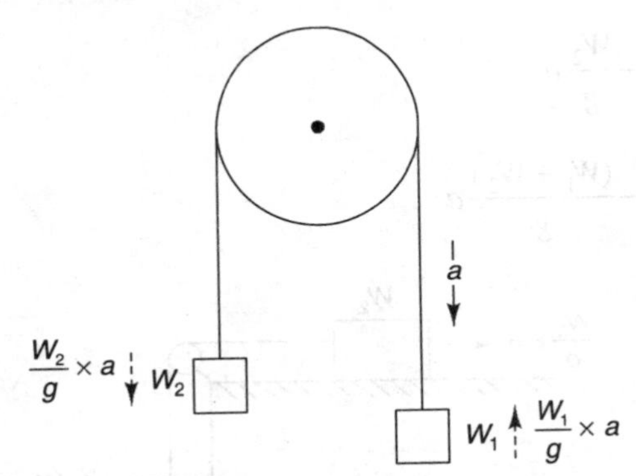

FIGURE 14.30 Application of D'Alembert's principle.

Net force $\Sigma F = W_1 - W_2$

Inertia force of $W_1 = -\dfrac{W_1}{g} \times a$

Inertia force of $W_2 = +\dfrac{W_2}{g} \times a$

$$\text{Total inertia force} = -\frac{W_1}{g}a - \frac{W_2}{g}a$$

$$= -\frac{a}{g}(W_1 + W_2)$$

As per D'Alembert's principle, the net external force and inertia force acting on a system is zero. Hence, we have

$$(W_1 - W_2) - \frac{a}{g}(W_1 + W_2) = 0$$

or

$$a = \frac{g\,(W_1 - W_2)}{W_1 + W_2}$$

Let us take another system with two bodies connected by a string with one body is lying on a smooth horizontal surface while other is hanging free as shown Figure 14.31. Weight W_1 is net force and both weights are moving with acceleration a. Now we have

$$\text{Net force acting} = W_1$$

$$\text{Inertia force of } W_1 = -\frac{W_1}{g}a$$

$$\text{Inertia force of } W_2 = -\frac{W_2}{g}a$$

$$\text{Total inertia force} = -\frac{(W_1 + W_2)}{g}a$$

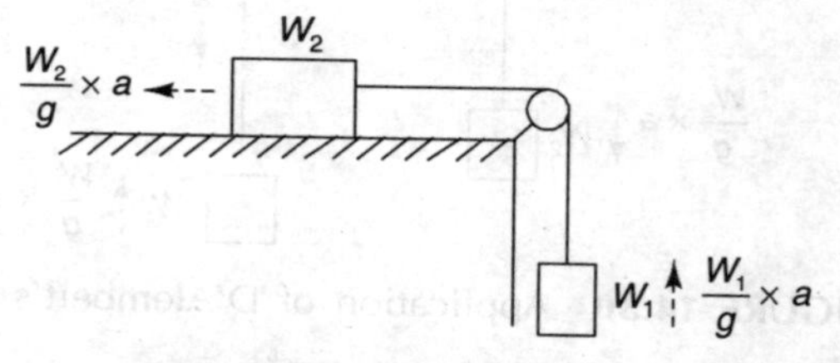

FIGURE 14.31 Application of D'Alembert's principle.

Applying D'Alembert's principle that the sum of net force and inertia force is zero, we have

$$+W_1 - \frac{(W_1 + W_2)}{g} \times a = 0$$

or

$$a = \frac{W_1 g}{W_1 + W_2}$$

Rotary Motion and D'Alembert's Principle

D'Alembert's principle can also be applied for rotary motion. When external torque is applied to a system having rotating motion then algebraic sum of all torques acting on the system due to external forces and inertia forces is zero. Consider a rotation due to weight W attached to one end of a string which is passing over a pulley having weight W_0 as shown in Figure 14.32. When weight W moves down, the pulley rotates clockwise. Let

a = linear acceleration,
α = angular acceleration of the pulley,
R = radius of the pulley
I = moment of inertia of the pulley.

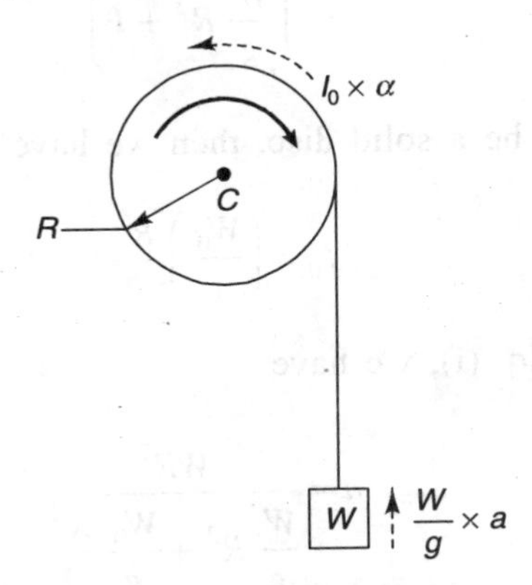

FIGURE 14.32 Rotary motion and D' Alembert's principle.

$$\text{Net torque acting on the system} = \text{force} \times \text{distance} = W \times R$$

$$\text{Inertia force due to } W = -\frac{W}{g} \times a$$

$$\therefore \quad \text{Torque due this inertial force} = -\frac{W}{g} \times a \times R$$

$$\text{Inertia torque due to pulley} = -\text{moment of inertia} \times \text{angular acceleration}$$

$$= -I_0 \times \alpha$$

$$= -I_0 \times \frac{a}{R}$$

as $$\alpha = \frac{a}{R}$$

As per D'Alembert's principle, the algebraic sum of torques acting on the system due to external forces and inertia forces should be zero.

$$\therefore \qquad W \times R - \frac{W}{g} aR - I \times \frac{a}{R} = 0$$

or
$$WR^2 - \frac{W}{g} aR^2 - I \times a = 0$$

or
$$a = \frac{WR^2}{\left(\frac{W}{g} R^2 + I\right)} \qquad \text{(i)}$$

If the pulley is assumed to be a solid disc, then we have

$$I_0 = \left(\frac{W_0}{g}\right)\frac{R^2}{2}$$

Putting the value of I_0 in Eq. (i), we have

$$a = \frac{WR^2}{\frac{W}{g} R^2 + \frac{W_0}{g}\frac{R^2}{2}}$$

$$= \frac{W \cdot g}{\left(W + \frac{W_0}{2}\right)}$$

Let us consider another system in which weights are attached to two ends of a string which passes over a rough pulley of weight W_0 as shown in Figure 14.33. Consider weight W_1 is more than W_2, resulting clockwise rotation of the pulley. Now torques acting are

(a) Torque due to resultant force = $(W_1 - W_2)R$

(b) Torque due to inertia forces

(i) Weight $W_1 = -\frac{W_1}{g} \times a \times R$

(ii) Weight $W_2 = -\frac{W_2}{g} \times a \times R$

(iii) Pulley = $-I \times \alpha$

$$= -I \times \frac{a}{R}$$

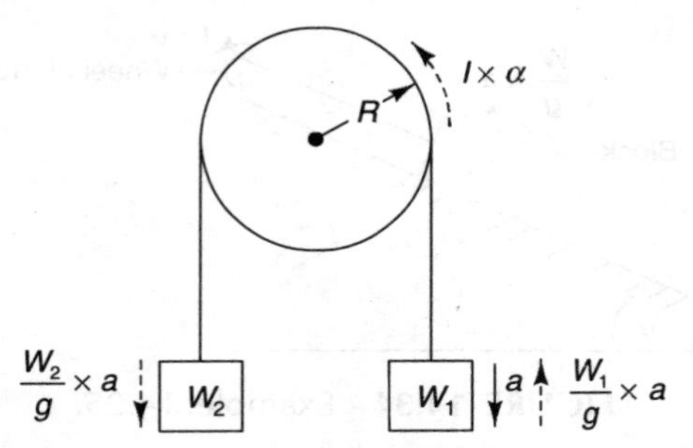

FIGURE 14.33 Clockwise rotation and application of D′ Alembert's principle.

Now applying D'Alembert's principle, we have

$$(W_1 - W_2)R - \frac{W_1}{g} \times a \times R - \frac{W_2}{g} \times a \times R - I \times \frac{a}{R} = 0$$

or
$$(W_1 - W_2)R^2 - a\left(\frac{W_1R^2}{g} + \frac{W_2R^2}{g} + I\right) = 0$$

or
$$a = \frac{(W_1 - W_2)R^2}{\frac{W_1R^2}{g} + \frac{W_2R^2}{g} + I}$$

In case the pulley is taken as a solid disc, then we have

$$I = \frac{W_0}{g} \times \frac{R^2}{2}$$

$\therefore$
$$a = \frac{(W_1 - W_2)R^2}{\frac{W_1}{g}R^2 + \frac{W_2}{g}R^2 + \frac{W_0}{g} \times \frac{R^2}{2}}$$

$$= \frac{(W_1 - W_2)g}{\left(W_1 + W_2 + \frac{W_0}{2}\right)}$$

Example 14.25 A wheel of radius R and moment of inertia I about its axis is fixed at the top of an inclined plane of inclination θ as shown in Figure 14.34. A string is wrapped round the wheel and its free end supports a block of weight W which can slide on the plane. Find the acceleration of the block when it slides down.

Net force acting is = $W \sin \theta$

Torque due to net force on pulley = $W \sin \theta \times R$

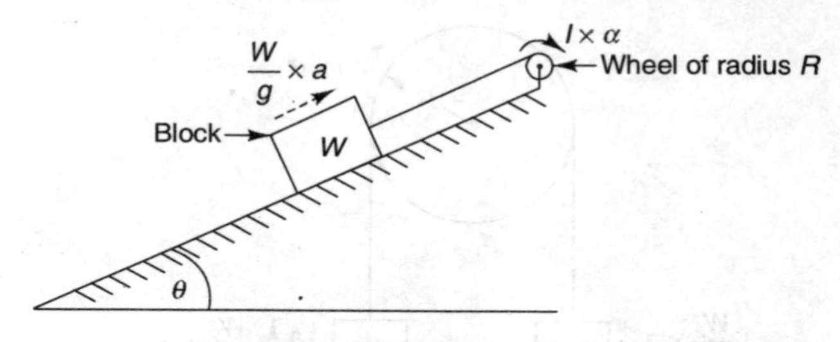

FIGURE 14.34 Example 14.25.

$$\text{Inertia force} = -\frac{W}{g} \times a$$

$$\text{Inertia torque} = -\frac{W}{g} \times a \times R$$

Inertia torque of the pulley due to rotation $= -I \times \dfrac{a}{R}$

Applying D'Alembert's principle, we have

$$W \sin\theta \times R - \frac{W}{g} \times a \times R - I \times \frac{a}{R} = 0$$

or

$$a = \frac{W \cdot \sin\theta \cdot R^2}{\frac{W}{g} R^2 + I}$$

In case we have two weights on inclined planes as shown in Figure 14.35, then we have

$$\text{Net force acting} = (W_1 - W_2) \sin\theta$$

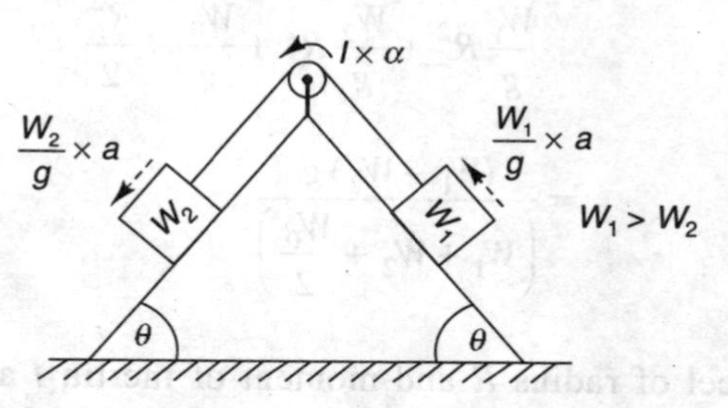

FIGURE 14.35 Example 14.25.

Now torques acting are:

1. Net torque due to net force $= (W_1 - W_2) \cdot \sin\theta \cdot R$
2. Net torque due to inertia forces from W_1 and $W_2 = -\left(\dfrac{W_1}{g} + \dfrac{W_2}{g}\right) \times a \times R$

3. Net inertia torque due to rotation of pulley $= -I \times \dfrac{a}{R}$

Applying D'Alembert's principle, we have

$$(W_1 - W_2)\sin\theta \cdot R - \left(\frac{W_1}{g} + \frac{W_2}{g}\right) \times a \times R - I\frac{a}{R} = 0$$

or

$$a = \frac{(W_1 - W_2)\sin\theta \cdot R^2}{\left(\dfrac{W_1}{g}R^2 + \dfrac{W}{g}R^2 + I\right)}$$

Let $\theta = 45$, $R = 10$ cm, $I = 0.5$ kg m^2, $\dfrac{W_1}{g} = 4$ kg, $\dfrac{W_2}{g} = 2$ kg, then we have

$$a = \frac{(m_1 - m_2)\sin 45 \times g}{m_1 + m_2 + \dfrac{I}{R^2}}$$

$$= \frac{(4-2) \times \dfrac{1}{\sqrt{2}} \times 9.81}{4 + 2 + \dfrac{0.5}{(0.1)^2}} = \frac{2 \times 0.707 \times 9.81}{6 + 50}$$

$$= 0.248 \text{ m/s}^2$$

Example 14.26 By means of a rope, a man of weight W is climbing vertically upwards with a constant acceleration a. Find tension T in the rope (see Figure 14.36).

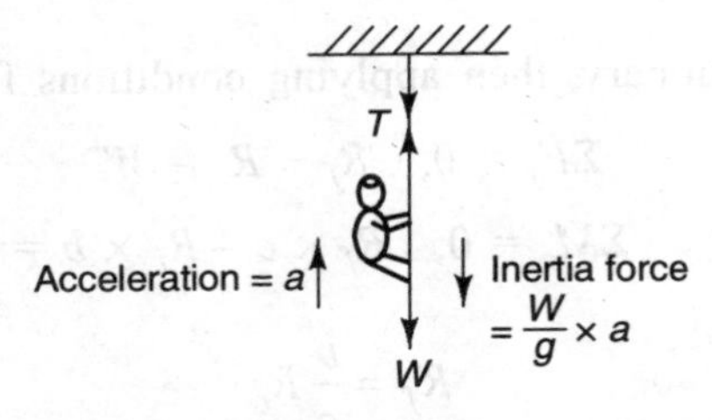

FIGURE 14.36 Example 14.26.

As the man is climbing with acceleration a, the inertia force $= \dfrac{W}{g} \times a$ will be acting upwards as per D'Alembert's principle.

Hence we have

$$T - W - \frac{W}{g}a = 0$$

or

$$T = W + \frac{W}{g} \times a$$

$$= W\left(1 + \frac{a}{g}\right)$$

As the value of acceleration increases, the tension T increases. If the man climbs down, the tension T will decrease. This is the reason why the rope is likely to break down while the man is climbing.

Example 14.27 Find the maximum acceleration along a level road that the rear-wheel drive automobile can attain if the coefficient of friction between the tyre and road surface is μ (see Figure 14.37).

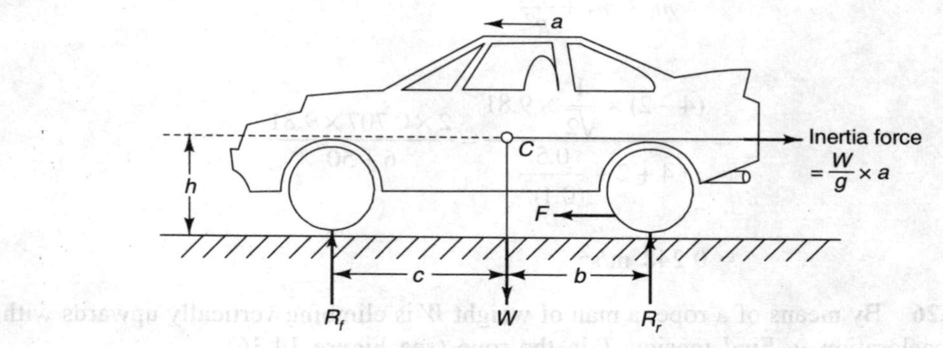

FIGURE 14.37 Example 14.27.

When automobile is stationary, then applying conditions for equilibrium, we have

$$\Sigma F_y = 0, \quad R_f + R_r = W \tag{i}$$

$$\Sigma M_c = 0, \quad R_f \times c - R_r \times b = 0$$

or

$$R_f = \frac{b}{c} R_r \tag{ii}$$

From Eqs. (i) and (ii) we have

$$R_f = W\left(\frac{b}{b+c}\right)$$

$$R_r = W\left(\frac{c}{b+c}\right)$$

Now when the automobile is moving with acceleration a, then inertia force = $\frac{W}{g} \times a$ can be applied at point C (which is CG of the automobile) as per D'Alembert's principle. If we take moment about rear wheel, then we have

$$R_f = W \frac{b}{b+c} - \frac{W}{g} \frac{ah}{b+c}$$

Similarly, taking moment about front wheel, we have

$$R_r = W\left(\frac{c}{b+c}\right) + \frac{W}{g} \frac{a \times h}{b+c}$$

It can be seen that R_f is decreasing as acceleration a is increasing while R_r is increasing.

The driving force $\left(F = \frac{W}{g} \times a\right)$ is dependent upon having a large reaction on the rear wheel.

The condition when rear wheel is about to slip is the limiting condition to which acceleration can be increased.

As $F = \mu R_r$, then the limit condition is

$$\frac{W}{g} \times a = \mu \times \left[\frac{Wgc + W \times a \times h}{g(b+c)}\right]$$

or
$$a = \mu \frac{(gc + ah)}{b+c}$$

or
$$a(b + c) = \mu gc + \mu ah$$

or
$$a(b + c - \mu ah) = \mu gc$$

or
$$a = \frac{\mu gc}{b + c - \mu ah}$$

SOLVED PROBLEMS

1. A uniform homogeneous cylinder rolls without slip along a horizontal levelled surface with a translation velocity of 20 cm/s. If the weight is 0.1 N and its radius is 10 cm, what is its total kinetic energy? (AMIE: 1976)

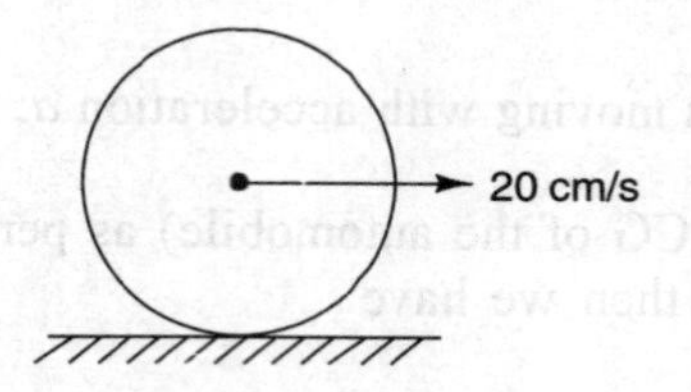

$$\text{Total KE} = \frac{1}{2}mv^2 + \frac{1}{2}I\omega^2$$

$$= \frac{1}{2} \times \frac{0.1}{9.81} \times 0.2^2 + \frac{1}{2} \times \left(\frac{i}{2} \times \frac{0.1}{9.81} \times r^2\right)\omega^2$$

$$= 204 \times 10^{-6} + 102 \times 10^{-6}$$

$$= 306 \times 10^{-6} \text{ N m}$$

2. Determine the work done by electric motor in winding up a uniform cable which hangs from a hoisting drum if its free length is 10 m and it weighs 500 N. The drum is rotated by the motor. (AMIE: 1984)

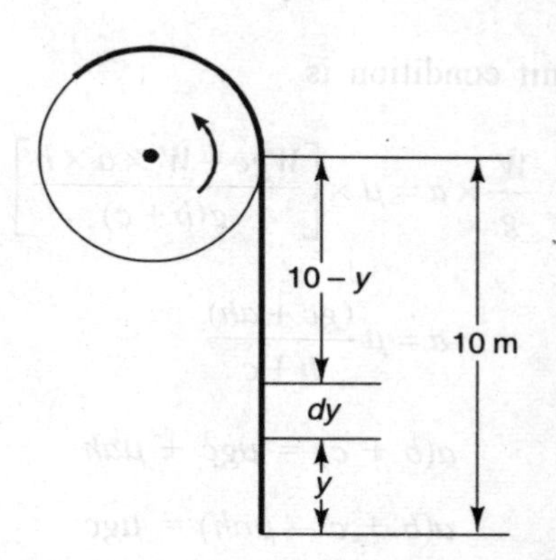

Consider an elemental length *dy* of cable at height *y* from its end.

$$\text{Weight of } dy \text{ length of cable} = \frac{500}{10} \times dy$$

$$= 50\, dy \text{ N}$$

Work done to hoist this *dy* length of cable to remaining full height, i.e. $(10 - y)$ is

$$dW = \text{force} \times \text{distance}$$

$$= (50\, dy) \times (10 - y)$$

$$= 50(10 - y)\, dy \text{ N m}$$

Total work done for the complete cable is

$$W = \int_0^{10} dW = \int_0^{10} 50(10 - y)\, dy$$

$$= \left[500y - 25y^2\right]_0^{10}$$

$$= 5000 - 2500$$

$$= 2500 \text{ N m}$$

3. A truck of weight W is moving at a constant speed v while being loaded with coal at a constant rate of k kg per second as shown in the figure. Find force F necessary to sustain the constant speed.

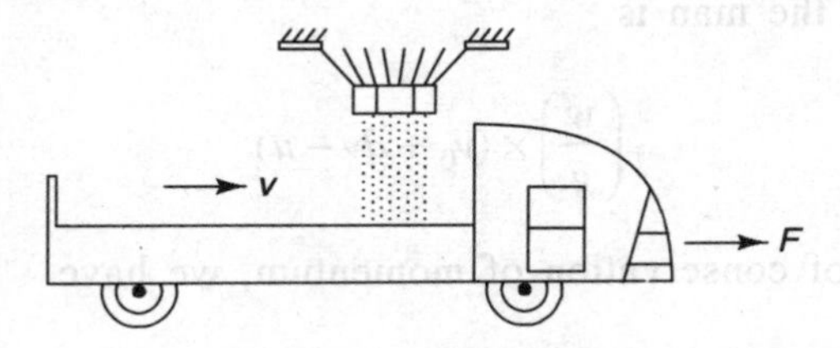

As the coal is being loaded, the mass of truck and loaded coal is gradually increasing, thereby momentum at constant speed is increasing. We have to use impulse-momentum equation to determine F. Consider it takes t time to load the truck. Then the change of momentum is

$$\left[\frac{W}{g} + k \times t\right] v - \frac{W}{g} \times v$$

$$= k \times t \times v$$

$$\text{Impulse} = F \times t$$

As per impulse-momentum equation, we have

$$\therefore \qquad F \times t = k \times t \times v$$

or $$F = kv \text{ Newtons}$$

4. A truck having weight W can roll without resistance along a horizontal road as shown in the figure. Initially the truck with a man of weight w moves to right with speed v_0. What increment of velocity Δv will the truck obtain if the man runs with speed u relative to the floor of the truck and jumps out of it from its rear end.

The problem is based on conservation of momentum as no external force is acting. Initial momentum of man and truck is

$$\left(\frac{W + w}{g}\right) \times v_0$$

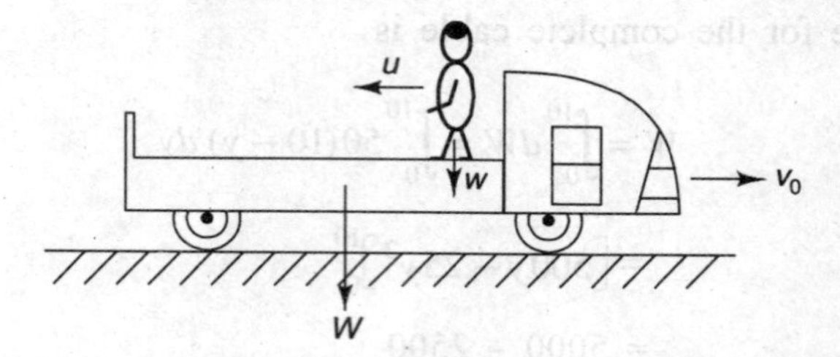

Final momentum of truck when velocity increases by dv is

$$\left(\frac{W+w}{g}\right)\times(v_0+dv)$$

Final momentum of the man is

$$\left(\frac{w}{g}\right)\times(v_0+dv-u)$$

Now applying law of conservation of momentum, we have

$$\left(\frac{W+w}{g}\right)v_0=\left(\frac{W+w}{g}\right)(v_0+dv)+\frac{w}{g}(v_0+dv-u)$$

or $$W\times dv - wu + w\,dv = 0$$

$\therefore$ $$dv=\frac{wu}{W+w}$$

5. A man weighing 800 N stands in a boat so that he is at 10 m from a pier as shown. He walks 5 m in the boat towards the pier and then stops. How far from the pier will be man at the end of this time? The boat weighs 1200 N. Assume no friction between boat and water.

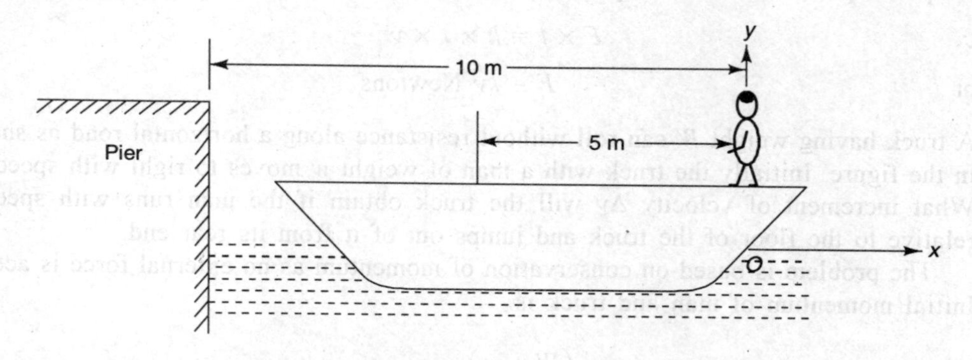

As no force is acting on the boat and man system, we can apply the principle of conservation of momentum. Take reference axes about point O on the right end of the boat as shown. Suppose man moves in time t a distance of 5 m towards the left. The boat say moves x distance toward right in time t. The absolute velocity of the boat is $= \frac{x}{t}$. The relative velocity of the man relative to the boat is $= -\frac{5}{t}$. Hence the absolute velocity of the man is $= -\frac{5}{t} + \frac{x}{t}$.

The momentum when the man has moved is

$$\frac{800}{g}\left(-\frac{5}{t}+\frac{x}{t}\right)+\frac{1200}{g}\left(\frac{x}{t}\right)$$

This must zero as initial momentum is zero. Hence we have

$$\frac{800}{g}\left(-\frac{5}{t}+\frac{x}{t}\right)+\frac{1200}{g}\left(\frac{x}{t}\right)=0$$

or $$-4000 + 800x + 1200x = 0$$

or $$x = \frac{4000}{2000} = 2 \text{ m}$$

The boat shifts 2 m to right. Hence the distance of the man from the pier is

$$10 - 5 + 2 = 7 \text{ m}.$$

6. A right circular cylinder of radius r and weight W is suspended by a cord that is wound round its surface as shown in the figure. If the cylinder is allowed to fall, prove that its centre of gravity C will follow a vertical rectilinear path and find the acceleration a along this path. Determine also tension T in the cord.

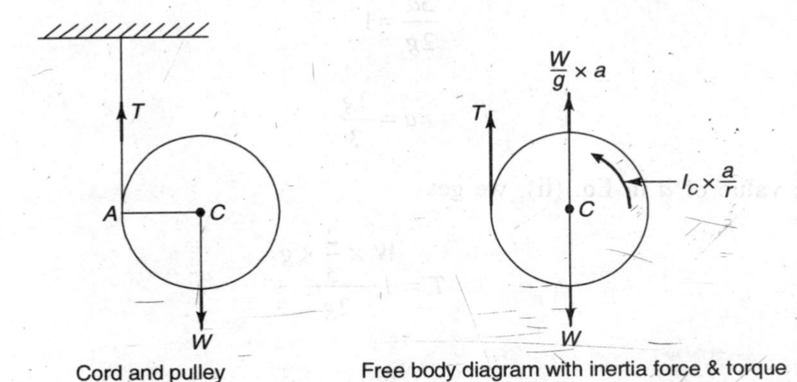

Cord and pulley　　Free body diagram with inertia force & torque

As the cylinder unwinds, the vertical cord will always remain in unchanged position in space and cylinder will rotate about point A which will be its instantaneous centre of rotation. The distance of point C is always equal to r from cord and it will follow rectilinear path.

Draw the free diagram of the pulley. Apply D'Alembert's principle and put the inertia force $= \frac{W}{g} \times a$ and inertia moment $= I_c \times \frac{a}{r}$ as shown in the free body diagram.

For equilibrium, we have

$$\Sigma F_y = 0, \quad T + \frac{W}{g} a = W \qquad \text{(i)}$$

$$\Sigma M_c = 0, \quad T \times r = I_c \frac{a}{r}$$

For cylinder $I_c = \frac{1}{2} mr^2$

$$\therefore \qquad T \times r = \frac{1}{2} \frac{W}{g} r^2 \times \frac{a}{r}$$

or
$$T = \frac{Wa}{2g} \qquad \text{(ii)}$$

From Eqs. (i) and (ii), we have

$$\frac{Wa}{2g} + \frac{W}{g} a = W$$

$$\frac{Wa}{2g}(1+2) = W$$

or
$$\frac{3a}{2g} = 1$$

or
$$a = \frac{2g}{3}$$

Put the value of a in Eq. (ii), we get

$$\therefore \qquad T = \frac{W \times \frac{2}{3} \times g}{2g}$$

$$= \frac{W}{3}$$

7. Two identical right circular disks are arranged in a vertical plane as shown in the figure. Neglecting friction, find the acceleration a of the centre C of the falling disk.

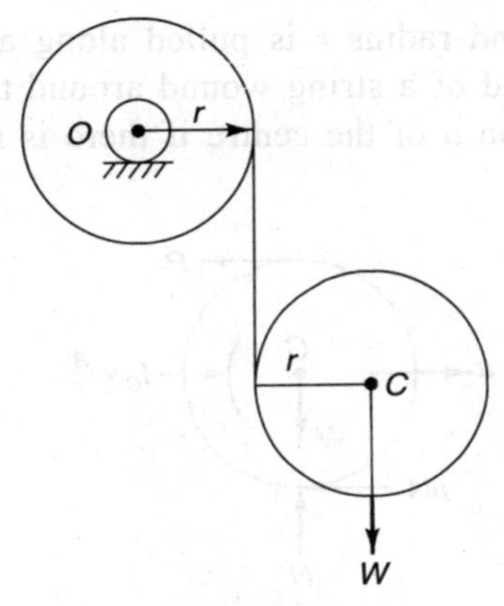

Circular discs

Free body diagram with inertia force & torque

The mass centre is moving down due to unwinding of the cord by both the cylinders. Hence the linear acceleration a is the double of angular acceleration of each disc.

$$\frac{a}{r} = 2\alpha$$

Consider the free body diagram of the bottom disk and apply inertia force and inertia moments as shown. Using D'Alembert's principle, we have

$$\Sigma F_y = 0, \quad T + \frac{W}{g}a = W \tag{i}$$

$$\Sigma M_c = 0, \quad T \times r = I_c \times \frac{a}{2r}$$

But $$I_c = \frac{1}{2}\frac{W}{g} \times r^2$$

$\therefore$ $$T \times r = \frac{1}{2} \times \frac{W}{g} r^2 \times \frac{a}{2r}$$

or $$T = \frac{Wa}{4g} \tag{ii}$$

From Eqs. (i) and (ii) we have

$$\frac{Wa}{4g} + \frac{Wa}{g} = W$$

$$\frac{a}{4g}(1+4) = 1$$

$$a = \frac{4g}{5}$$

8. A solid right circular cylinder of weight W and radius r is pulled along a horizontal plane by a horizontal force P applied to the end of a string wound around the cylinder as shown in the figure. Find (i) the acceleration a of the centre if there is no slippage and (ii) value of μ to prevent any slippage.

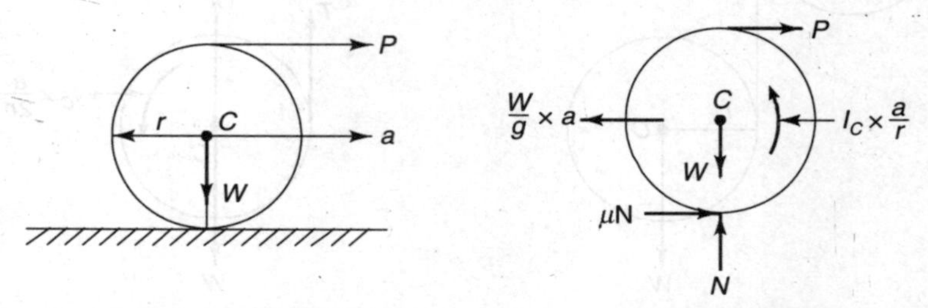

The free body diagram of the cylinder is as shown with inertia force and inertia moment as per D'Alembert's principle. For equilibrium, we have

$$\Sigma F_y = 0, \quad W = N \tag{i}$$

$$\Sigma F_x = 0, \quad P - \frac{W}{g} \times a + \mu N = 0 \tag{ii}$$

$$\Sigma M_c = 0, \quad P \times r + \mu N \times r = I_c \times \frac{a}{r}$$

But $I_c = \dfrac{1}{2} \times \dfrac{W}{g} r^2$

$$\therefore \qquad P - \mu N = \frac{Wa}{2g} \tag{iii}$$

Using $W = N$, Eqs. (ii) and (iii) reduces to

$$P - \frac{Wa}{g} + \mu W = 0 \tag{iv}$$

$$P - \mu W = \frac{Wa}{2g} \tag{v}$$

From Eqs. (iv) and (v), we have

$$2P - \frac{Wa}{g} = \frac{Wa}{2g}$$

or
$$2P = \frac{Wa}{g}\left(1 + \frac{1}{2}\right) = \frac{3}{2}\frac{Wa}{g}$$

or
$$a = \frac{4Pg}{3W}$$

Putting the value of a in Eq. (v), we have

$$\mu W = -\frac{W}{2g} \times \frac{4Pg}{3W} + P$$

$$\mu W = -\frac{2}{3}P + P$$

$$= \frac{P}{3}$$

or $$\mu = \frac{P}{3W}$$

$\therefore$ $$\mu \geq \frac{P}{3W}$$

9. A prismatic timber of weight W rests on two rollers, each of weight $\frac{W}{2}$ and radius r and is pulled along a horizontal plane by a force P as shown in the figure. Assuming that there no slippage, find the acceleration a of the timber.

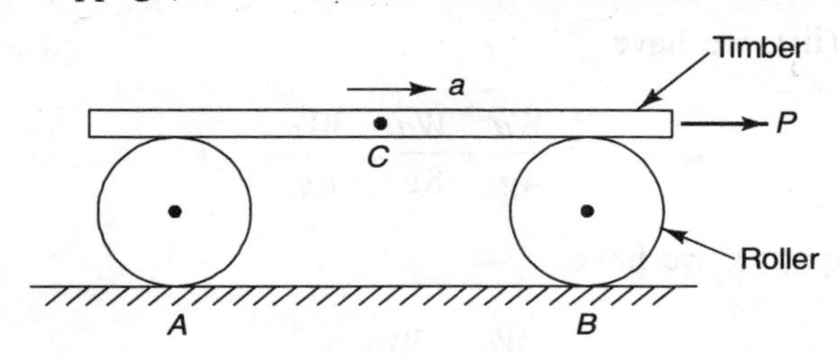

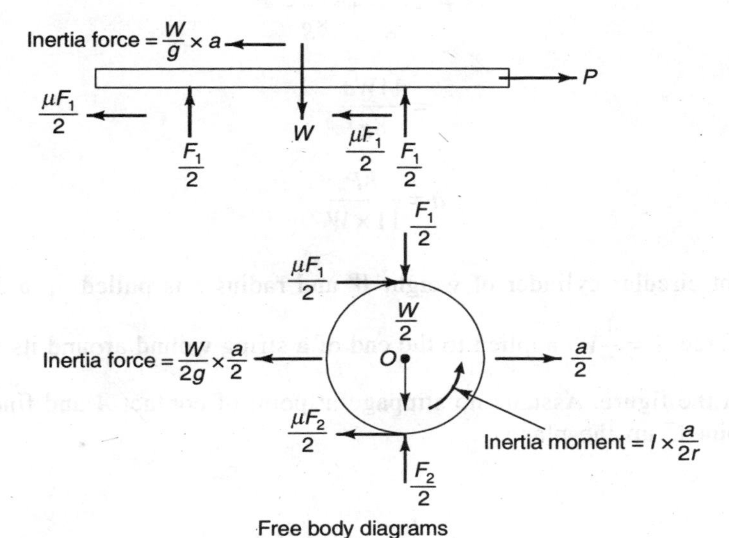

Free body diagrams

The free body diagrams of the timber and roller are as shown. The centre of mass of the roller will have half acceleration as that of timber due to geometry (instantaneous centre is at point B). $\frac{F_1}{2}$ and $\frac{F_2}{2}$ are normal reaction and $\mu\frac{F_1}{2}$ and $\mu\frac{F_2}{2}$ are friction forces as indicated in the free body diagrams.

Now for timber, we have

$$\Sigma F_x = 0, \quad \frac{W}{g}a = P - 2 \times \frac{\mu F_1}{2} \quad \text{where } \frac{W}{g}a = \text{inertia force.} \tag{i}$$

Now for roller, we have

$$\Sigma F_x = 0, \quad \frac{W}{2g} \times \frac{a}{2} = \frac{\mu F_1}{2} - \frac{\mu F_2}{2} \tag{ii}$$

and

$$\Sigma M_O = 0, \quad \frac{W}{2g} \times \frac{r^2}{2} \times \frac{a}{2r} = \left(\frac{F_1}{2} + \frac{F_2}{2}\right) \times \mu \times r$$

or

$$\mu \times \left(\frac{F_1}{2} + \frac{F_2}{2}\right) = \frac{Wa}{8g} \tag{iii}$$

From Eqs. (ii) and (iii), we have

$$\mu \times F_1 = \frac{Wa}{4g} + \frac{Wa}{8g} = \frac{3Wa}{8g} \tag{iv}$$

From Eq. (i) and Eq. (iv), we have

$$P = \frac{Wa}{g} + \frac{3Wa}{8g}$$

$$= \frac{11Wa}{8g}$$

$$\therefore \quad a = \frac{8P}{11 \times W}$$

10. A solid right circular cylinder of weight W and radius r is pulled up a 30° incline by a constant force $P = \frac{1}{2}W$ applied to the end of a string wound around its circumference as shown in the figure. Assume no slippage at point of contact A and find acceleration a of the point C up the plane.

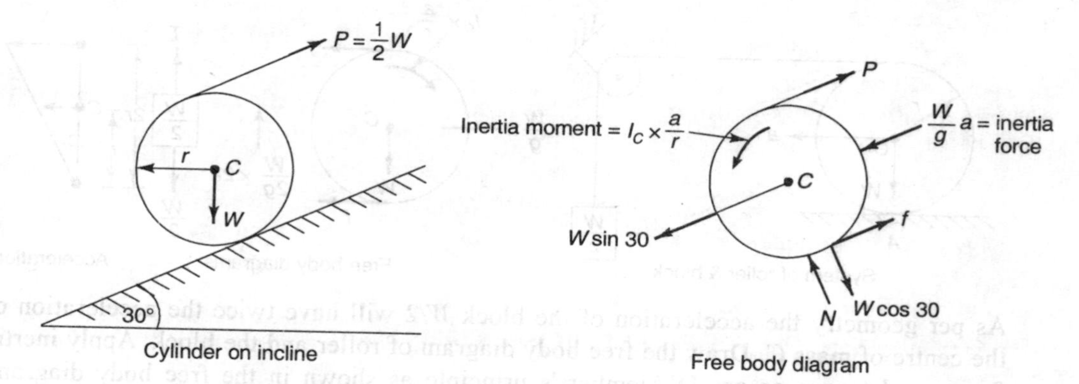

The free body diagram of the cylinder is as shown in the figure. Inertia force and inertia moment are applied as per D'Alembert's principle to have static equilibrium. Hence, we have

Now along surface $\Sigma F = 0,\ P + f - \dfrac{W}{2} - \dfrac{W}{g}a = 0$ (i)

$$\Sigma M_c = 0, \quad P \times r - f \times r = I_c \times \frac{a}{r}$$

$$= \frac{1}{2}\frac{W}{g} \times r^2 \times \frac{a}{r}$$

or
$$P - f = \frac{Wa}{2g} \tag{ii}$$

Adding Eqs. (i) and (ii), we have

$$2P = \frac{W}{2} + \frac{3Wa}{2g}$$

But
$$P = \frac{W}{2}$$

$\therefore$
$$\frac{3Wa}{2g} = \frac{W}{2}$$

or
$$a = \frac{g}{3}$$

11. Find the acceleration a of the right circular roller of weight W which is pulled along a horizontal plane by means of a weight $W/2$ on the end of a string wound round the circumference of the roller if friction at A is sufficient to prevent slipping.

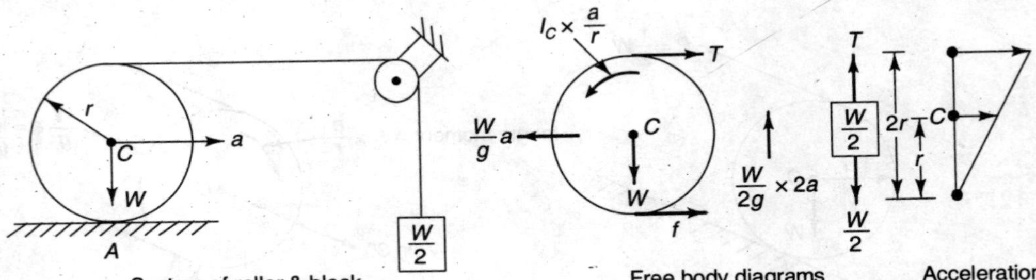

As per geometry the acceleration of the block $W/2$ will have twice the acceleration of the centre of mass C. Draw the free body diagram of roller and the block. Apply inertia forces and torque as per D'Alembert's principle as shown in the free body diagram. Taking block first, we have

$$\Sigma F_y = 0, \quad T + \frac{W}{2g} \times 2a = \frac{W}{2} \tag{i}$$

Now take roller, we have

$$\Sigma F_v = 0, \quad T + f = \frac{W}{g} \times a \tag{ii}$$

$$\Sigma M_c = 0, \quad T \times r - f \times r = \frac{2}{5} \times \frac{W}{g} \times r^2 \times \frac{a}{r}$$

or
$$T - f = \frac{2}{5} \frac{Wa}{g} \tag{iii}$$

Adding Eqs. (ii) and (iii), we have

$$2T = \frac{Wa}{g} + \frac{2Wa}{5g}$$

$$= \frac{7Wa}{5g}$$

or
$$T = \frac{7}{10} \frac{Wa}{g}$$

Putting the value of T in Eq. (i), we have

$$\frac{7}{10} \frac{Wa}{g} - \frac{Wa}{g} = \frac{W}{2}$$

or $$\frac{17}{10} \times \frac{Wa}{g} = \frac{W}{2}$$

or $$a = \frac{5}{17} g$$

12. An 890 N rowboat containing a 668 N man is pushed off the pier by an 800 N man. The speed that is imparted to the boat is 0.3 m/s by this push. The man then leaps into boat from the pier with a speed of 0.6 m/s relative to pier in the direction of motion of the boat. When the two men have settled down in the boat and before rowing commences, what is the speed of the boat? Neglect water resistances.

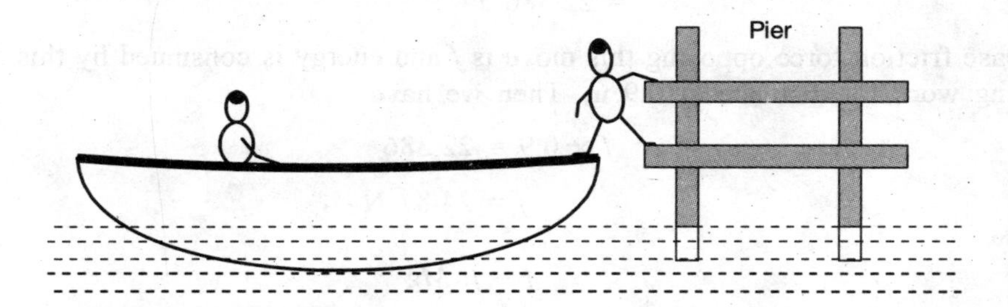

The problem can be solved by the principle of conservation of momentum.

Initial momentum is $\frac{(890+668)}{g} \times 0.3 + \frac{800}{g} \times 0.6$

Final momentum is $\left(\frac{890+668+800}{g}\right) v$

Equating both, we get

$$\frac{1558}{g} \times 0.3 + \frac{800 \times 0.6}{g} = \frac{2358}{g} \times v$$

or $$v = \frac{467.4 + 480}{2358}$$

$$= 0.402 \text{ m/s}$$

13. A bullet of mass 25 gm moving with velocity of 600 m/s horizontally and it strikes a wooden block of mass 5 kg resting on a rough horizontal surface. The bullet after striking the block remains buried in the block and they both move 90 cm before coming to rest. Find (i) average resistance between block and horizontal surface (ii) coefficient of friction between block and horizontal surface.

Total momentum before impact = total momentum after impact

or $$m_1u_1 + m_2u_2 = m_1v_1 + m_2v_2$$

$$= (m_1 + m_2)\, v \text{ as } v_1 = v_2 = v$$

or $$0.25 \times 600 + 5 \times 0 = (5 + 0.25)\, v$$

or $$v = 2.98 \text{ m/s}$$

Now block and bullet move with velocity $v = 2.98$ m/s till they come to rest.

$\therefore$ $$\Delta KE = \frac{1}{2} \times 5.025 (2.98^2 - 0)$$

$$= 22.386 \text{ J}$$

Incase friction force opposing this move is f and energy is consumed by this f force by doing work for distance of 0.9 m. Then we have

$\therefore$ $$f \times 0.9 = 22.386$$

or $$f = 24.87 \text{ N}$$

Now

$$f = \mu Mg$$

$\therefore$ $$\mu = \frac{24.87}{5.025 \times 9.81}$$

$$= 0.5$$

14. A bullet of 10 gm gets embedded in the block of 1 kg with some horizontal velocity. The block with bullet is displaced on a rough horizontal surface ($\mu = 0.2$) for a distance of 1 m. What was the velocity of bullet?

Equating momentum before and after impact, we have

$$m_1u_1 + m_2u_2 = M \times v$$

$$(0.01 \times u_1 + 1 \times 0) = 1.01 \times v$$

or $$v = \frac{0.01u_1}{1.01}$$

As block & bullet moves 1 m, then work done by the friction force f is

$$W_f = f \times 1$$

$$= (M \times g \times \mu) \times 1$$

$$= 1.01 \times 9.81 \times 0.2$$

$$= 1.98 \text{ J}$$

Now change of kinetic energy of block with bullet is

$$\Delta KE = \frac{1}{2} \times 1.01 \times v^2$$

$$= 0.505v^2 \text{ Joules}$$

Now $\Delta KE = W_f$

$$0.505v^2 = 1.98$$

or $$v^2 = \frac{1.98}{0.505}$$

or $$\left(\frac{0.01u_1}{1.01}\right)^2 = \frac{1.98}{0.505}$$

or $$u_1 \approx 200 \text{ m/s.}$$

15. A vehicle of mass m is negotiating a circular road of radius r with a constant speed v. Find (i) speed at which the vehicle will tend to overturn and (ii) speed at which the vehicle will tend to slip sideway?

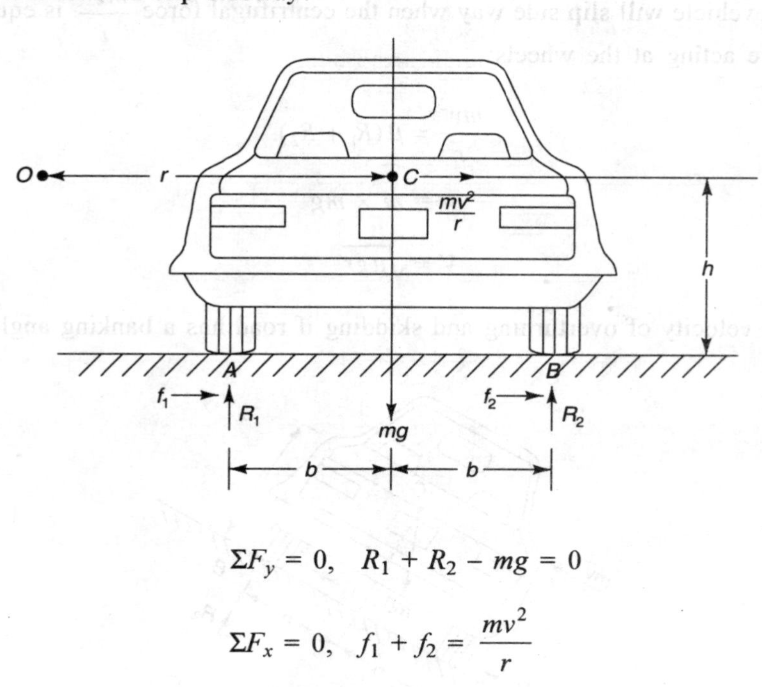

$$\Sigma F_y = 0, \quad R_1 + R_2 - mg = 0$$

$$\Sigma F_x = 0, \quad f_1 + f_2 = \frac{mv^2}{r}$$

Taking moment about point A at the inner wheel

$$\Sigma M_A = 0, \quad R_2 \times 2b - mg \cdot b = m\frac{v^2}{r} \times h$$

Solving the above equations, we get

$$R_1 = \frac{mg}{2} - \frac{h}{2b} \cdot m\frac{v^2}{r}$$

$$R_2 = \frac{mg}{2} + \frac{h}{2b} \cdot m\frac{v^2}{r}$$

When the reaction at inner wheel will become zero, the entire weight of the vehicle is shifted to outer wheel and the vehicle tends to overturn toward right

$$\therefore \qquad R_1 = 0$$

or
$$\frac{mg}{2} - \frac{h}{2b} \times m \times \frac{v^2}{r} = 0$$

or
$$v = \sqrt{\frac{b}{h} \times gr}$$

Now the vehicle will slip side way when the centrifugal force $\frac{mv^2}{r}$ is equal to the total resistance acting at the wheels:

$$\therefore \qquad \frac{mv^2}{r} = \mu(R_1 + R_2)$$

$$= m \times mg$$

or
$$V = \sqrt{\mu g r}$$

16. Find the velocity of overturning and skidding if road has a banking angle θ as shown.

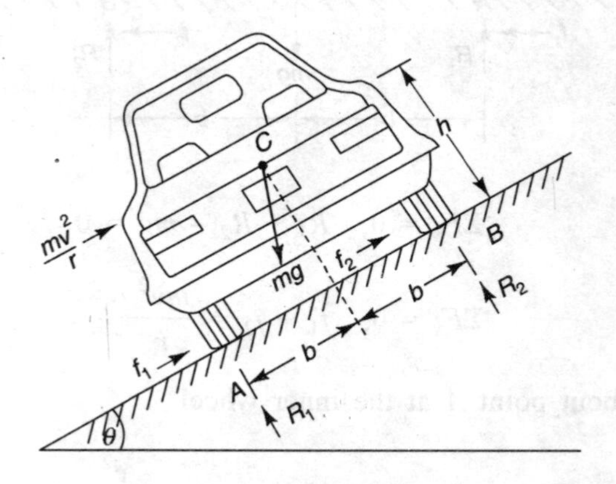

$$\Sigma F_x = 0, \quad f_1 + f_2 + mg \sin \theta = \frac{mv^2}{r} \cos \theta$$

$$\Sigma F_y = 0, \quad R_1 + R_2 - mg \cos \theta = \frac{mv^2}{r} \sin \theta$$

$$\Sigma M_A = 0, \quad R_2 \cdot 2b - mg\,(b \cos \theta - h \sin \theta) = \frac{mv^2}{r}(b \sin \theta + h \cos \theta)$$

$$\therefore \qquad R_2 = \frac{\frac{mv^2}{r}(b \sin \theta + h \cos \theta) + mg\,(b \cos \theta - h \sin \theta)}{2b}$$

and $$R_1 = \frac{mv^2}{r} \sin \theta + mg \cos \theta - \frac{\frac{mv^2}{r}(b \sin \theta + h \cos \theta) + mg\,(b \cos \theta - h \sin \theta)}{2b}$$

The vehicle will overturn if $R_1 = 0$

$$\therefore \qquad v = \sqrt{\frac{1 + h/b \tan \theta}{1 - b/h \tan \theta} \times \frac{b}{h} \times gr}$$

$$= \sqrt{\frac{b}{h} \times gr} \times \sqrt{\frac{1 + h/b \tan \theta}{1 - b/h \tan \theta}}$$

When a vehicle is about to skid, we have

$$f_1 + f_2 = \mu (R_1 + R_2) = \frac{mv^2}{r} \cos \theta$$

or $$\frac{mv^2}{r} \cos \theta = \mu mg \sin \theta$$

or $$\tan \theta = \frac{mv^2}{r \cdot \mu \cdot mg}$$

$$v^2 = mgr \tan \theta$$

$$v = \sqrt{\mu\, gr \tan \theta}$$

$$= \sqrt{\mu\, gr} \cdot \sqrt{\tan \theta}$$

17. A string is wrapped around a thin disc of radius of 0.5 m and mass 10 kg as shown. Find (i) acceleration of centre of mass, (ii) angular acceleration of the disc when the string is pulled up with a force of 250 N.

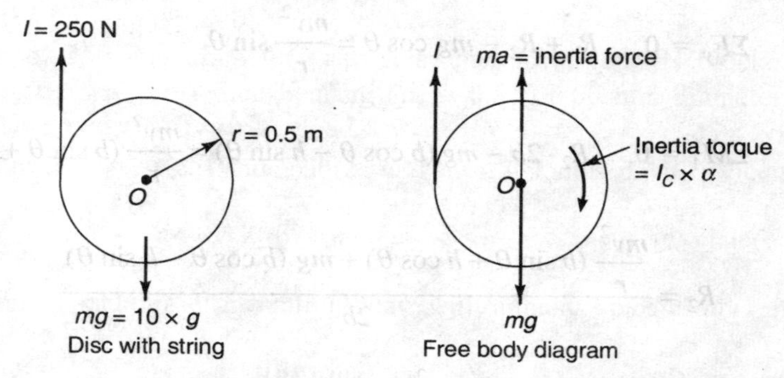

It is apparent that linear acceleration will take place in vertical direction as no force is acting in horizontal direction. Applying inertia force and inertia torque as per D'Alembert's principle, we have

$$\Sigma F_y = 0, \quad T + ma = mg \tag{i}$$

$$\Sigma M_O = 0, \quad T \times r = I_c \times \alpha$$

$$= \left(\frac{1}{2} mr^2\right) \times \alpha \tag{ii}$$

$$T = \frac{1}{2} mr\alpha$$

From Eq. (i), we have

$$250 + 10 \times a = 10 \times 9.81$$

or

$$a = \frac{250 - 98.1}{10}$$

$$= 15.19 \text{ m/s}^2$$

From Eq. (ii), we have

$$\alpha = \frac{2T}{mr} = \frac{2 \times 250}{10 \times 0.5} = 100 \text{ rad/s}^2$$

18. A carpet of mass M and it is rolled in the form of a cylinder of radius R. It is now unrolled with a small push. Find the linear velocity at centre of mass when the carpet roll has remaining radius of $\frac{R}{2}$ as shown in the figure.

The problem can be solved by interchanging of potential energy into kinetic energy, i.e. conservation of energy.

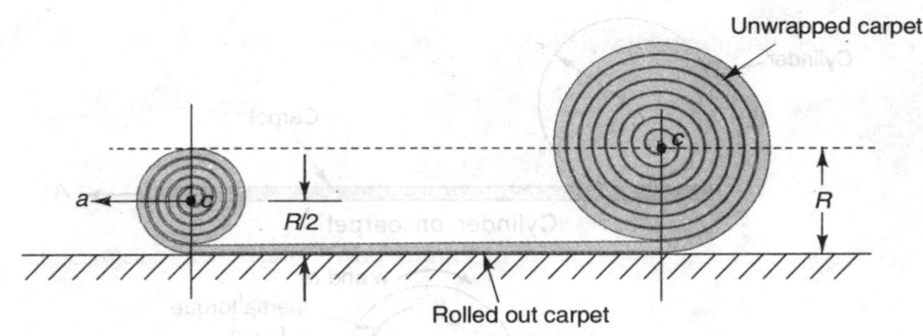

Unrolling out of the carpet

$$\Delta\text{PE} = MgR - \frac{M \times 4\pi \left(\frac{R}{2}\right)^2}{4\pi R^2} \times g \times \frac{R}{2}$$

$$= \frac{7}{8} MgR$$

$$\Delta\text{KE} = \frac{1}{2}\left(\frac{M}{4}\right)v_c^2 + \frac{1}{2} Iw^2$$

$$= \frac{Mv_c^2}{8} + \frac{1}{2}\left(\frac{1}{2} \times \frac{M}{4} \times \left(\frac{R}{2}\right)^2\right) \times \left(\frac{v_c}{R/2}\right)^2$$

$$= \frac{Mv_c^2}{8} + \frac{Mv_c^2}{16} = \frac{3Mv_c^2}{16}$$

Now $\Delta\text{PE} = \Delta\text{KE}$

$$\frac{7}{8} MgR = \frac{3}{16} Mv_c^2$$

or
$$v_c = \sqrt{\frac{14}{3} gR}$$

19. A cylinder of radius *r* and mass *m* rests on a horizontal rug. If the rug is pulled with an acceleration of *A* horizontally as shown in figure, find the (i) linear acceleration *a* at centre of mass and (ii) angular acceleration α of the cylinder.

The free body diagram of the cylinder with forces and moments acting on it are as shown below:

$$\Sigma F_x = 0, \quad f = ma \quad \text{(i)}$$

$$\Sigma F_y = 0, \quad N = mg \quad \text{and} \quad f = \mu N = \mu mg$$

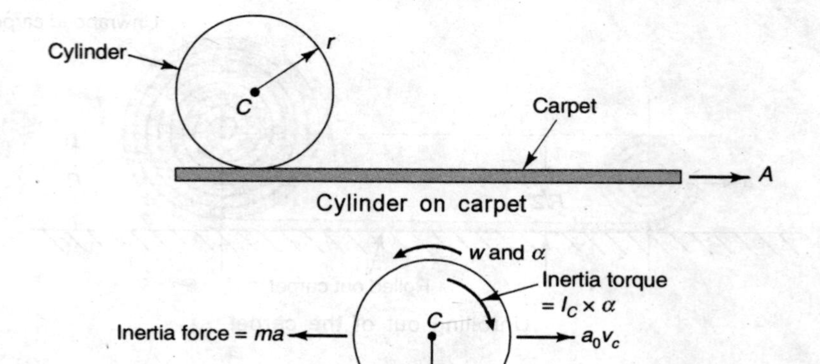

$$\Sigma M_c = 0, \quad f \times r = I_c \times \alpha$$

$$= \frac{1}{2} mr^2 \times \alpha$$

or $$f = \frac{1}{2} mr \times \alpha \qquad \text{(ii)}$$

For no slippage, we must have

$$a + r\alpha = A \qquad \text{(iii)}$$

From Eqs. (i) and (ii), we have

$$a = \frac{1}{2} r\alpha \qquad \text{(iv)}$$

From Eqs. (iii) and (iv), we have

$$a + 2a = A$$

or $$a = \frac{A}{3}$$

$\therefore$ $$\alpha = \frac{2a}{r} = \frac{2 \times A}{3r}$$

$$= \frac{2A}{3r}$$

20. At what distance P should the horizontal force F be applied to homogeneous bar, cylinder and sphere so that the horizontal component of the reaction at the point of suspension is zero.

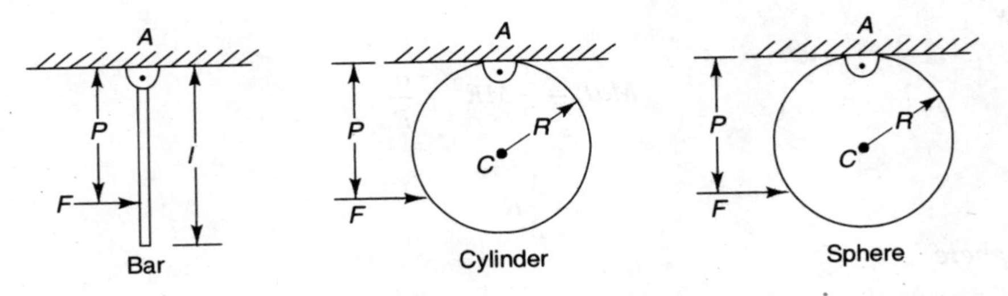

The horizontal component of the reaction at the point of suspension (A) is zero if the body starts rotating about point A, i.e. A is instantaneous centre of rotation.

1. *Bar*

$$(I_c)_{z\text{-axis}} = \frac{Ml^2}{12}$$

where C is the centre of mass

$$\therefore \quad I_A = \frac{Ml^2}{12} + M \times \left(\frac{l}{2}\right)^2$$

$$= Ml^2\left(\frac{1}{12} + \frac{1}{4}\right)$$

$$= \frac{1}{3} Ml^2$$

Now ΣM_A is

$$F \times P = \frac{1}{3} \times Ml^2 \times \alpha$$

But $F = M \times a$ and $\alpha = \dfrac{a}{l/2}$

$$\therefore \quad MaP = \frac{1}{3} Ml^2 \times \frac{a}{l/2}$$

or

$$P = \frac{2}{3} \times l$$

2. *Cylinder*

$$I_c = \frac{1}{2} MR^2$$

$$\therefore \quad I_A = \frac{1}{2} MR^2 + MR^2$$

$$= \frac{3}{2} MR^2$$

Now

$$MaP = \frac{3}{2} MR^2 \times \frac{a}{R}$$

or

$$P = \frac{3}{2} R$$

3. *Sphere*

$$I_c = \frac{2}{5} MR^2$$

$$I_A = \frac{2}{5} MR^2 + MR^2$$

$$= \frac{7}{5} MR^2$$

Now

$$MaP = \frac{7}{5} MR^2 \times \frac{a}{R}$$

or

$$P = \frac{7}{5} R$$

21. A tractor weighs 4000 N including the driver. The larger driver wheels each weigh 200 N with a radius of 0.8 m and radius of gyration of 0.7 m. The small wheel weighs 100 N each with radius of 0.3 m and radius of gyration 0.2 m. The tractor is dragging a load of 300 N. The coefficient of friction between load and ground is 0.2. What torque is required to drive wheels to increase the speed from 3 m/s to 6 m/s in 30 s. Assume no slippage.

The linear impulse momentum equation initially is

$$(f_2 - f_1 - f_3) \times 30 = (4000 + 300)\,(6 - 3)$$

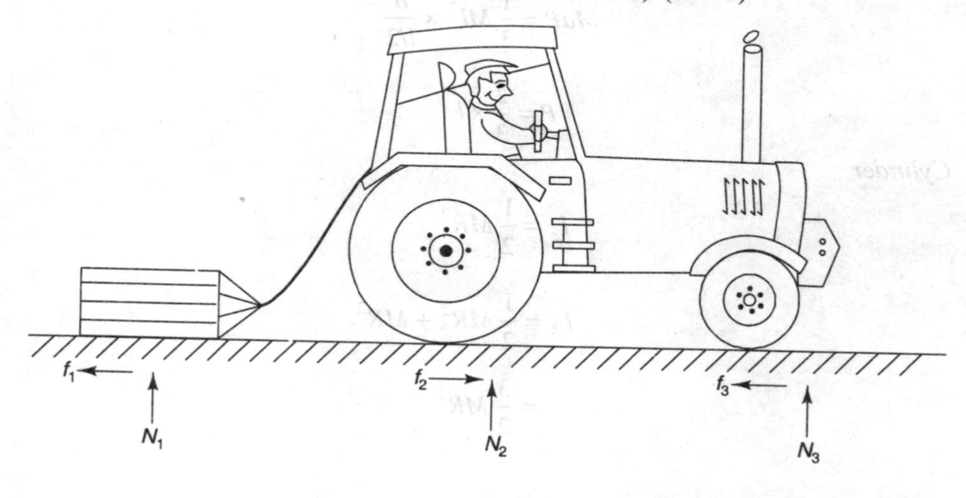

But $f_1 = 300 \times \mu = 300 \times 0.2 = 60$

$$\therefore \quad f_2 - f_3 = \frac{4300 \times 3}{30} + 60$$

$$= 430 + 60 = 490 \text{ N} \qquad \text{(i)}$$

If torque given to rear wheels is T Nm, then impulse and angular momentum is

$$(-T + 0.8 f_2)30 = \frac{400}{g} \times (0.7)^2 [(\omega_r)_2 - (\omega_r)_1]$$

$$(\omega_r)_2 - (\omega_r)_1 = \frac{(v_2 - v_1)}{r_r} = \frac{6-3}{0.8} = \frac{3}{0.8} = 2.75 \text{ rad/s}$$

or
$$-T + 0.8 f_2 = 1.83 \qquad \text{(ii)}$$

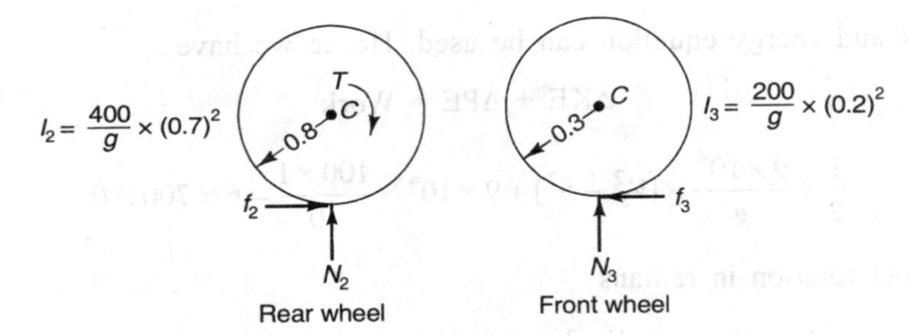

Now angular impulse momentum equation for front wheel is

$$30 \times (-f_3 \times 0.3) = \frac{200}{g} (0.2)^2 [(\omega_f)_2 - (\omega_f)_1]$$

$$= \frac{200}{g} (0.2)^2 \left[\frac{v_2 - v_1}{0.3}\right]$$

or
$$f_3 = \frac{200 \times 4 \times 10^{-2} \times 3}{g \times 30 \times 0.3 \times 0.3}$$

$$= 2.72 \text{ N} \qquad \text{(iii)}$$

From Eqs. (i) and (iii), we get

$$f_2 = 490 + 2.72$$

$$= 492.72 \text{ N}$$

Putting value of f_2 in Eq. (ii), we get

$$-T + 0.8 \times 492.72 = 1.83$$

or $$T = 394.18 - 1.83$$
$$= 392.35 \text{ N m}$$

22. A train moves up on a 1 in 10 inclined track. The train has 6 pairs of drive wheels and each pair developing 700 N m torque. The train has an initial velocity of 5 m/s. Find the speed after train has moved 100 m. Train weighs 90 kN and the diameter of wheel is 0.6 m. Neglect the rotational energy of the drive wheels.

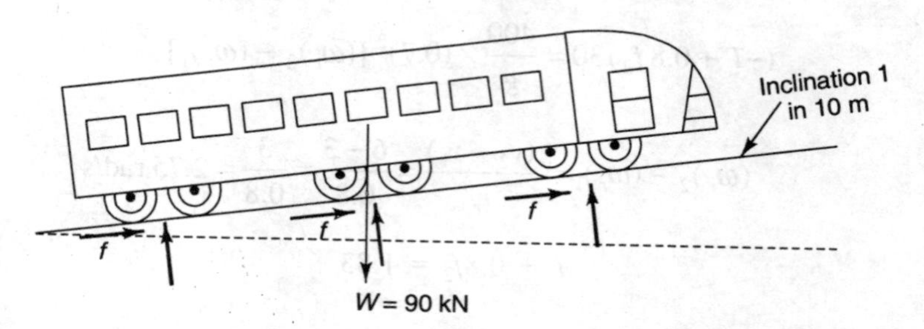

The work and energy equation can be used. Hence we have

$$\Delta \text{KE} + \Delta \text{PE} = \text{Work}$$

$$\frac{1}{2} \times \frac{9 \times 10^4}{g} \times [V_2^2 - 5^2] + 9 \times 10^4 \times \frac{100 \times 1}{10} = 6 \times 700 \times \theta$$

where θ = wheel rotation in radians

$$0.459 \times 10^4 (V_2^2 - 25) + 900000 = 4200 \times \theta$$

$$\theta = \frac{\text{distance}}{2\pi r} \times 2\pi = \frac{100}{2\pi r} \times 2\pi \text{ rad/s}$$

$$= \frac{100}{0.3} = 333.3 \text{ rad/s}$$

$$4590 \times (V_2^2 - 25) + 900 \times 10^3 = 4200 \times 333.3$$
$$= 1400 \times 10^3$$

or $$4590\,(V_2^2 - 25) = 1400 \times 10^3 - 900 \times 10^3$$
$$= 500 \times 10^3$$

or $$V_2^2 - 25 = \frac{500 \times 10^3}{4590}$$
$$= 101$$

or $$V_2^2 = 126$$

or $$V_2 = 11.22 \text{ m/s}$$

OBJECTIVE TYPE QUESTIONS

Great men are not born great. God uses troubles and trials to make them great.

State True or False

1. The factor W/g of a body is called mass. [*True/False*]
2. In rotary motion, every particle of the body moves in circle and centres of all these circles lie at the axis of rotation. [*True/False*]
3. Number of revolutions per min. is called angular velocity. [*True/False*]
4. The angular velocity is related to linear velocity by relation $\omega/r = v$. [*True/False*]
5. The angular acceleration is related to linear acceleration by relation $\alpha = a/r$. [*True/False*]
6. Moment of a force F from the axis of rotation at distance r is given by $M = F \times r$. [*True/False*]
7. The external moment (M) develops in a body an angular acceleration given by the relation $M = I \times a$. [*True/False*]
8. Kinetic energy of a rotating body is given by the relation $k = I\omega^2$. [*True/False*]
9. Moment of momentum is called angular momentum. [*True/False*]
10. The change in the momentum developed in a body is equal to the impulse acting on it. [*True/False*]
11. A cyclist has to bend outward while negotiating a horizontal road turn. [*True/False*]
12. As per work-energy equation, the work done by a force is equal to the change of kinetic energy in a body. [*True/False*]
13. If m is mass and a is acceleration, then $m \times a$ is inertia force as per D'Alembert's principle. [*True/False*]
14. If I = moment of inertia and α is acceleration, then $I \times \alpha$ is the inertia moment acting on the body as per D'Alembert's principle. [*True/False*]
15. D'Alembert's principle helps in converting dynamic equilibrium into a static equilibrium. [*True/False*]

Multiple Choice Questions

1. The rotational mass of a body is
   (a) inertial moment (b) rotational moment
   (c) internal moment (d) moment of inertia

2. If a body has mass = m velocity at centre of mass = v_c, moment of inertia = I_c and rotational velocity = ω, then total kinetic energy is

(a) $\frac{1}{2} m v_c^2$ (b) $\frac{1}{2} I_c \omega^2$

(c) $\frac{1}{2} I_c \omega^2 - \frac{1}{2} m v_c^2$ (d) $\frac{1}{2} I_c \omega_c^2 + \frac{1}{2} m v_c^2$

3. Bending inward of a cyclist on a horizontal turn of radius r with velocity v is

(a) $\tan \theta = \frac{v^2}{rg}$ (b) $\tan \theta = \frac{v}{rg}$ (c) $\tan \theta = \frac{v}{r^2 g}$

4. The angular momentum of a rotating body, with angular velocity ω is given by

(a) $\frac{1}{2} \times I\omega$ (b) $I\omega$ (c) $\frac{1}{2} \times I\omega^2$

5. The kinetic energy of a uniform disc with mass = m and radius r rotating with angular velocity ω is

(a) $\frac{1}{2} mr^2\omega^2$ (b) $\frac{1}{3} mr^2\omega^2$ (c) $\frac{1}{4} mr^2\omega^2$

6. The kinetic energy of a sphere of mass = m, radius = r and angular velocity = ω is

(a) $\frac{1}{5} mr^2\omega^2$ (b) $\frac{2}{5} mr^2\omega^2$ (c) $\frac{3}{5} mr^2\omega^2$

7. The kinetic energy of a cylinder of mass = m, radius = r and angular velocity = ω is

(a) $\frac{1}{2} mr^2\omega^2$ (b) $\frac{1}{4} mr^2\omega^2$ (c) $mr^2\omega^2$

8. A chain is placed on a frictionless table such that its 1/5 part is hanging down over the edge of the table. If the length of the chain be l and mass be m, then how much work will be done in pulling up the hanging part of the chain

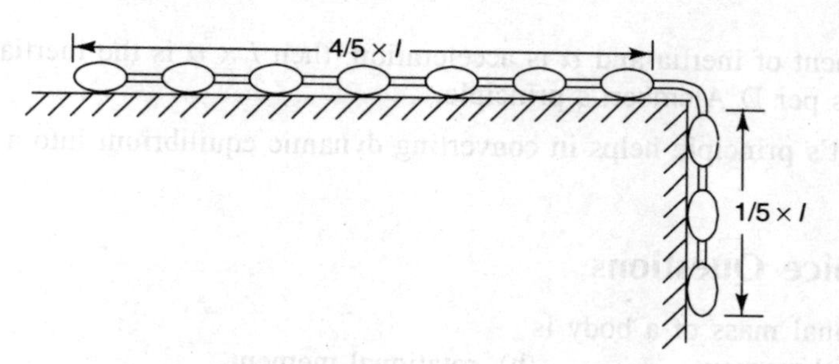

(a) $W = \frac{mgl}{50}$ (b) $W = \frac{mgl}{25}$ (c) $W = \frac{mgl}{75}$

9. A body has mass = m and momentum = P, then its kinetic energy is

(a) $\frac{P^2}{2m}$ (b) $\frac{P^2}{m}$ (c) Pm

10. Two balls of different masses have the same kinetic energy. The ball having greater momentum will be

(a) heavier one (b) lighter one (c) both having equal

11. A bullet is fired from rifle. If the rifle recoils freely, the kinetic energy of the rifle in comparison to that of the bullet is

(a) less (b) greater (c) equal

12. The angular velocity of an ice skater as he shortens the radius of skating will

(a) decrease (b) increase (c) remain same

13. The impulse acting on a body for $t = 0$ to $t = 4$ sec is

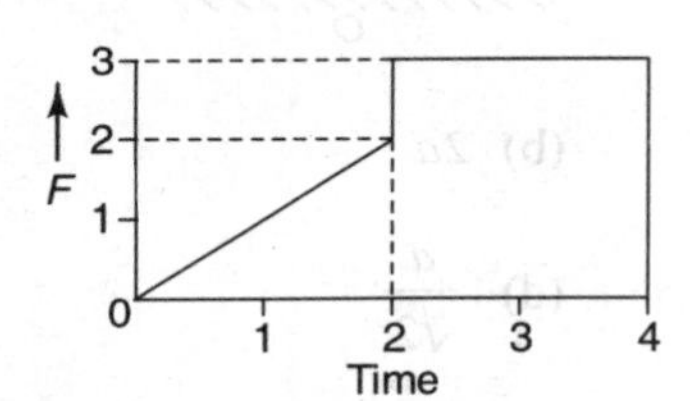

(a) 8 (b) 6
(c) 10 (d) 7

14. If a body of mass = 10 kg has its mass centre linearly accelerating with 4 m/s, then inertia force as per D'Alembert's principle is

(a) 20 N (b) – 40 N
(c) 80 N (d) – 80 N

15. If a body has moment of inertia = 10 kg m^2 and angular acceleration = 2 rad/s^2, then inertia moment of the body as per D'Alembert's principle is

(a) 20 N m (b) 10 N m (c) –20 N m

16. The velocity of the ball at point A is
c = centre of curvature; m = mass of ball; r = radius = 4m

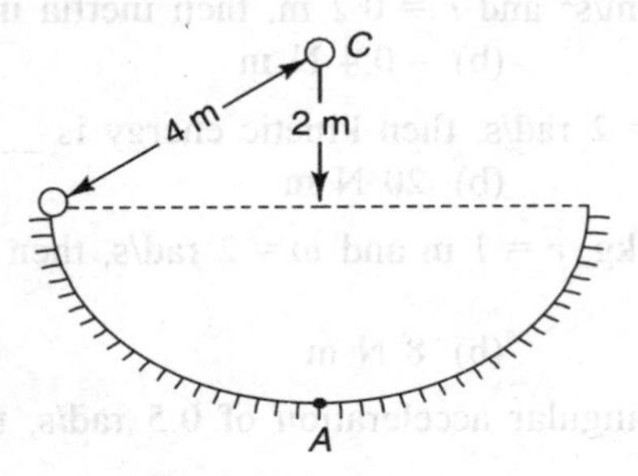

(a) $\sqrt{mg}$ (b) $3\sqrt{mg}$
(c) $\sqrt{2mg}$ (d) $2\sqrt{g}$

17. As per work-energy principle, the relation between work (W) and kinetic energy (KE) is

(a) W = KE (b) $W = \Delta$KE (c) $W \times$ KE = constant

18. As per impulse momentum equation, the relation between impulse (I) and momentum (M) is

(a) $I = M$ (b) $I = \Delta M$ (c) $I \times M$ = constant

19. If a disc is rolling on a horizontal surface as shown below and has linear acceleration at mass centre = a, then linear acceleration at point A on periphery is

(a) $\frac{a}{2}$ (b) $2a$

(c) $\sqrt{2}a$ (d) $\frac{a}{\sqrt{2}}$

20. The linear acceleration in above problem at point O is

(a) $a/2$ (b) $2a$

(c) zero (d) $\sqrt{2}a$

21. The linear acceleration in above problem at point B is

(a) $a/2$ (b) $\sqrt{2}a$

(c) zero (d) $2a$

Fill in the Blanks

1. If m = 2 kg, v = 4 m/s, a = 1 m/s^2, then inertia force is __________.

(a) – 8 N (b) – 2N

2. If I = 4 kg m^2, a = 0.2 m/s^2 and r = 0.2 m, then inertia moment is __________.

(a) – 4 N m (b) – 0.4 N m

3. If I = 10 kg m^2 and ω = 2 rad/s, then kinetic energy is __________.

(a) 40 N m (b) 20 N m

4. If a roller has mass = 5 kg, r = 1 m and ω = 2 rad/s, then the kinetic energy of the roller is __________.

(a) 4 N m (b) 8 N m

5. If the above roller has angular acceleration of 0.5 rad/s, then the angular momentum is __________.

(a) 4.0 N s (b) 2.0 N s

6. A vehicle of mass m is negotiating a circular road (friction coefficient = μ) of radius r with constant speed v. The vehicle will slip sideway if speed v is __________.

(a) $\sqrt{\mu gr}$ (b) $\sqrt{2\mu gr}$

7. In the above problem, the vehicle will overturn if reaction at __________ wheels is zero.

(a) outer (b) inner

8. In rear wheel drive, the friction force will act in the __________ direction of the vehicle movement.

(a) opposite (b) same

9. If earth contracts to half its radius, the length of day will be __________.

(a) 12 hours (b) 6 hours

10. When a diver jumps into water, he pulls his arms and legs towards his centre of body as shown below then rotation of his body __________.

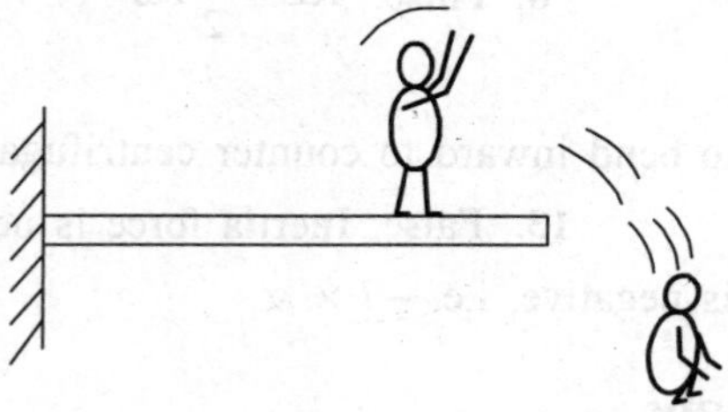

(a) decreases (b) increases

ANSWERS

It's how you handle your problems and troubles that counts, not the troubles themselves.

State True or False

1. True
2. True
3. False. Angular velocity is the revolutions per second and not per min.
4. False. $v = \omega \times r$
5. True
6. True
7. True
8. False. $KE = \frac{1}{2} I\omega^2$
9. True
10. True
11. False. The cyclist has to bend inward to counter centrifugal force.
12. True
13. False. Inertia force is negative, i.e. $-ma$.
14. False. Inertia moment is negative, i.e. $-I \times \alpha$
15. True

Multiple Choice Questions

1. (a)
2. (d)
3. (a)
4. (b)
5. (c) Kinetic energy $= \frac{1}{2} I\omega^2$

$$= \frac{1}{2} \times \left(\frac{1}{2} mr^2\right) \omega^2$$

$$= \frac{1}{4} mr^2\omega^2$$

6. (a) Kinetic energy $= \frac{1}{2} I\omega^2$

$$= \frac{1}{2} \times \left(\frac{2}{5} mr^2\right) \omega^2$$

$$= mr^2\omega^2$$

7. (b) Kinetic energy $= \frac{1}{2} I\omega^2$

$$= \frac{1}{2} \times \left(\frac{1}{2} mr^2\right)\omega^2$$

$$= \frac{1}{4} mr^2\omega^2$$

8. (a) Force acting on chain = mass × g

$$= \frac{m}{5} \times g$$

Work = Force × distance

$$= \frac{m}{5} \times g \times \frac{l/5}{2} \quad (CG \text{ at centre of hanging part})$$

$$= \frac{mgl}{50}$$

9. (a) Momentum $= P = m \times v$

$$\therefore \quad v = \frac{P}{m}$$

$$\therefore \quad \text{KE} = \frac{1}{2} mv^2 = \frac{1}{2} m \times \left(\frac{P}{m}\right)^2$$

$$= \frac{1}{2} \frac{P^2}{m}$$

10. (a) $\text{KE} = \frac{1}{2} m_1 v_1^2 = \frac{1}{2} m_2 v_2^2$

$$= \frac{1}{2} \times \frac{P_1^2}{m_1} = \frac{1}{2} \frac{P_2^2}{m_2}$$

$\therefore$ Angular momentum $P \propto \sqrt{m}$

$\therefore$ Angular momentum is more if mass is more.

11. (a) $M_g \times v_g = m_b \times v_b$

or $\quad M_g^2 v_g^2 = m_b^2 \times v_b^2$

or $\quad 2 \times M_g \times (\text{KE}_g) = 2m_b\,(\text{KE}_b)$

As $M_g >> m_b$, hence $\text{KE}_g << \text{KE}_b$

12. (a)

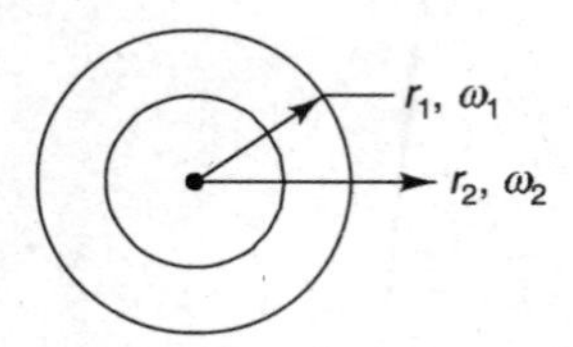

Angular momentum of the skater is maintained as he shortens the radius. Hence

$$I_1\omega_1 = I_2\omega_2$$

$$(mr_1^2)\,\omega_1 = (mr_2^2)\,\omega_2$$

where m = mass of skater

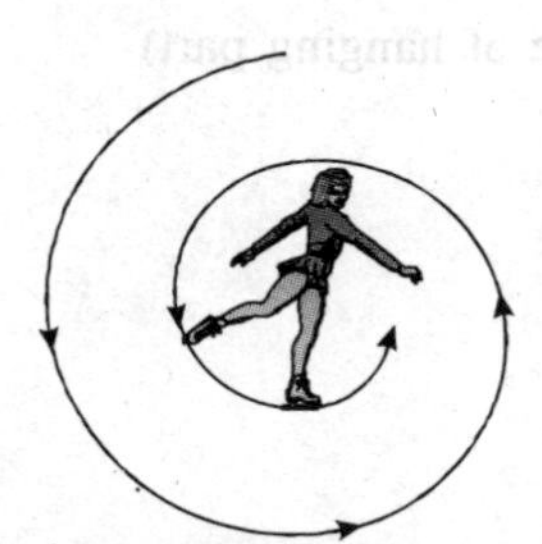

Angular velocity increasing

$$\therefore \quad \omega_2 = \left(\frac{r_1}{r_2}\right)^2 \times \omega_1$$

as $r_1 >> r_2$, hence $\omega_1 >> \omega_2$

13. (b) Impulse is equal to area of the curve on force-time diagram

$$\text{Area} = \frac{1}{2} \times 2 \times 2 + 3 \times 2 = 2 + 6 = 8 \text{ N s}$$

14. (b) Inertia force = $-ma$

$= -10 \times 4$

$= -40$ N

15. (c) Inertia moment = $-I \times \alpha$

$= -10 \times 2$

$= -20$ N m

16. (d) At point A, we will have

$$\Delta\text{KE} = \Delta\text{PE}$$

$$\frac{1}{2}mv^2 = mg\,\Delta h$$

$$v^2 = 2g \times (4 - 2)$$
$$= 4g$$

or $$v = 2\sqrt{g}$$

17. (b)

18. (b)

19. (b) Point O is instantaneous centre of rotation of the disc.

20. (d)

21. (c)

Fill in the Blanks

1. (b) Inertia force = $-m \times a$

2. (a) Inertia torque = $-I \times \alpha$

$$= -I \times \frac{a}{r}$$
$$= -4 \times \frac{0.2}{0.2}$$
$$= -4 \text{ N m}$$

3. (b) $\text{KE} = \frac{1}{2} I\omega^2$

$$= \frac{1}{2} \times 10 \times 2^2$$
$$= 20 \text{ N m}$$

4. (a) $\text{KE} = \frac{1}{2} I\omega^2$

$$= \frac{1}{2}\left(\frac{2}{5} mr^2\right)\omega^2$$
$$= \frac{1}{5} \times 5 \times 1^2 \times 2^2$$
$$= 4 \text{ N m}$$

5. (a) Angular momentum = $I\omega$

$$= \left(\frac{2}{5} mr^2\right) \times \omega$$
$$= \frac{2}{5} \times 5 \times 1^2 \times 2$$
$$= 4 \text{ N s}$$

6. (a)
7. (a) Centrifugal force reduces the reaction at inner wheel, thereby counteracting friction force at the inner wheels vanishes.
8. (b)

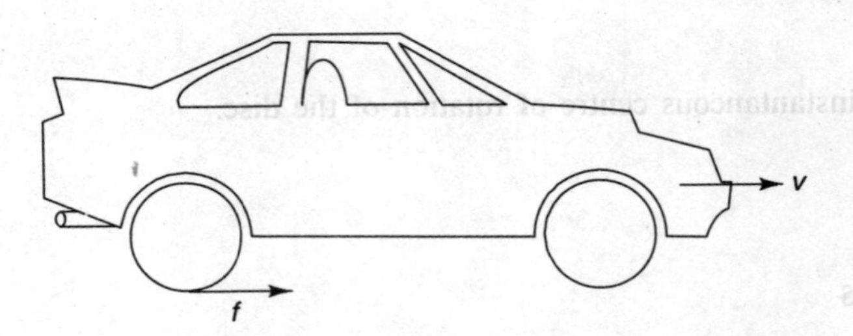

9. (b) The angular momentum is maintained. Hence we have

$$I_1\omega_1 = I_2\omega_2$$

$$\left(\frac{2}{5} M \times r_1^2\right)\omega_1 = \left(\frac{2}{5} \times M \times r_2^2\right)\omega_2$$

But $r_2 = \frac{r_1}{2}$

$\therefore \quad \omega_2 = 4\omega_1$

But $T = \frac{2\pi}{\omega}$

$\therefore \quad T_2 = \frac{T_1}{4}$

$= \frac{24}{4}$

$= 6$ hours

10. (b) When diver pulls his arms and legs, he reduces his body's moment of inertia. The angular momentum is $I \times \omega$. When moment of inertia reduces angular velocity increases, thereby helping the diver to execute acrobatic turns in the air before landing in water.

CHAPTER 15

Stress and Strain Analysis

Even if all you can do is crawl across the floor, it's better than just sitting there doing nothing.

INTRODUCTION

Whenever a load is attached to a thin hanging wire, the wire elongates. The elongation depends upon the magnitude of the load applied to the wire. It also depends upon the material of the wire and the thickness of the wire. The resistance to the elongation is developed in the wire due to the cohesive force between the molecules of the wire which tend to stay in the resting position. The force of resistance increases with the deformation as the applied load is increased. The deformation stops when the force of the resistance is equal to the applied force. However, there is a limit to which the force of the resistance can be increased. Beyond this limit the force of the resistance cannot increased and the deformation continues until failure takes place. The resistive force per unit area is stress and elongation per unit length is strain.

We will learn in this chapter, types of materials and loads. We will understand the types of stresses and strains; the relationship between stress and strain and their diagrams for different materials; and stresses developed due to own weight and due to change in temperature.

TYPES OF MATERIALS

Materials can be classified as follows:

1. Elastic
2. Elastoplastic
3. Plastic
4. Ductile
5. Brittle

Elastic materials: Elastic materials are those materials which undergo deformation when subjected to load, but their deformation disappears on removal of load.

Elastoplastic materials: Elastoplastic materials are those materials which like elastic materials undergo deformation when subjected to load, but deformation disappears partially on removal of load.

Plastic materials: Plastic materials undergo deformation when subjected to load, but deformation does not disappear at all after removal of load.

Ductile materials: Ductile materials can be drawn to a smaller section as they have a lower elastic limit as well as they can sustain large strain before rupture. Mild steel, aluminium and copper are ductile materials as they have ability to withstand large elongation or bending. Ductile materials do not snap off without giving sufficient warning by elongation.

Brittle materials: Brittle materials are those materials which can have little or zero deformation before they fail or rupture on loading. Glass and cast iron are brittle materials. Lack of ductility is brittleness. When body breaks easily when subjected to shocks, it is said to be brittle.

TYPES OF LOADS

Loads can be of various types. A *static load* is a gradually applied load which reaches equilibrium in a short time. A *sustained load* is a constant load applied over a long period of time. The weight of a structure is a sustained load. A *concentrated load* is a load applied at a point or a small area of a large-size member. A distributed load is applied over a large area with uniform or non-uniform distribution. A load passing through the centroid of the resisting section is called a *centric load* while a load not passing through the centroid is called an *eccentric load.*

Loads can also be identified by the effects produced by them. A *torsional load* causes a twist of the shaft. A *bending load* is applied transverse to the longitudinal axis of a beam which tends to bend the beam. The combination of axial and torsional or axial and bending or bending and torsion or any other combination of loads are called *combined loads*. A load applied rapidly for short duration is called *impact load*. The loads which are applied and removed repeatedly are called *repeated loads*.

Loads can be a pulling type (tensile) or pushing type (compressive). Hence resistive reactions to these applied loads can be tensile or compressive forces. Therefore, *direct stresses* are resistive forces per unit area. Direct stresses can be tensile or compressive depending on resistive forces. *Tensile stresses* are generally taken as positive and compressive stresses are taken as negative. Since stress is resistive force per area, it is more where the area of the member is small and less where the area of the member is large so as to ensure resistive force at any section of the member has constant magnitude of applied force.

Members which are used to take a tensile load are known as *tension members* or *ties* while members which are used to take compressive loads are known as *columns* or *struts*.

STRESS AND STRAIN

A simple stress is developed by one directional forces while compound stresses are developed by more than one directional force.

Normal stresses are developed due to forces which act normal to the cross section. Shear stresses are developed due to forces which act tangential to the cross section.

A longitudinal strain is defined as change in length to the original length (Figure 15.1(a)). Strain is dimensionless.

$$\varepsilon_{\text{long}} = \frac{\text{change in length } (\delta)}{\text{original length } (l)} = \frac{\delta}{l}$$

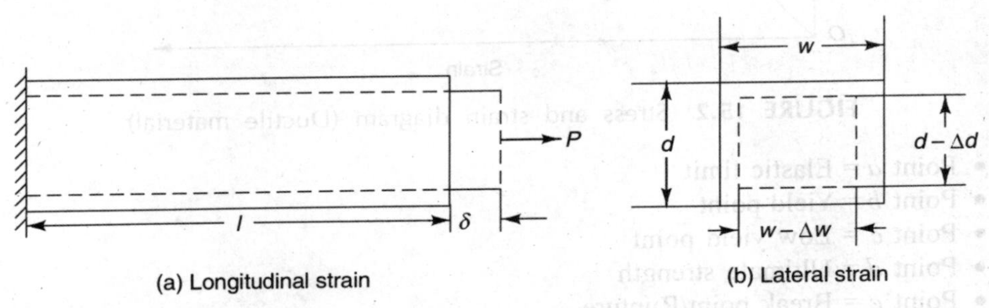

FIGURE 15.1 Strain.

Now refer to Figure 15.1(b) a member is subjected to axial load (P). The length of the member increases by δ. However, the width and depth decrease when the length increases. Let Δw and Δd are reduction in width (w) and depth (d). The lateral strains are

$$\varepsilon_{\text{lateral}} = -\frac{\Delta w}{w}$$

Also $$\varepsilon_{\text{lateral}} = \frac{\Delta d}{d}$$

Hooke's Law

When material is loaded within the elastic limit, stress is proportional to strain, i.e.

$$E = \frac{\text{stress}}{\text{strain}}$$

where E = modulus of elasticity or Young's modulus.

If a stress and strain diagram for ductile material (mild steel, aluminium and copper) is plotted, it can be seen that material yields (permanent deformation starts) after critical stress. A large deformation takes place after yield point and before material ruptures with a small increase of load. Important points and stretches on the diagram (Figure 15.2) are as follows:

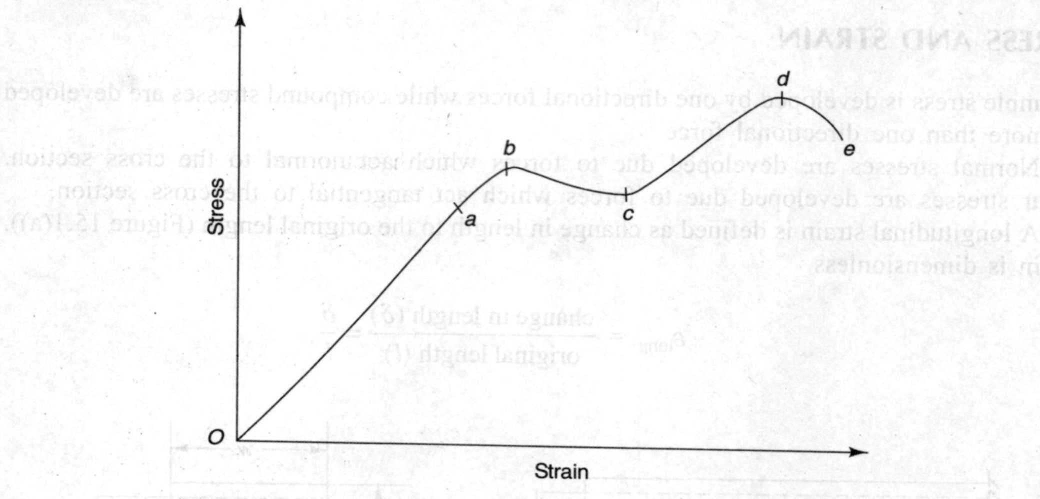

FIGURE 15.2 Stress and strain diagram (Ductile material).

- Point a = Elastic limit
- Point b = Yield point
- Point c = Low yield point
- Point d = Ultimate strength
- Point e = Break point/Rupture
- Stretch oa = Material is elastic and obeys Hooke's law
- Stretch ab = Rapid deformation
- Stretch bc = Yielding
- Stretch cd = Strain hardening
- Stretch de = Necking

It is apparent that strain within the elastic limit is very small as compared to plastic strain (Elastic strain is in stretch oa and plastic strain is in stretch be). The elastic strain is generally 0.1 to 2% of total strain before rupture.

If a stress and strain diagram for brittle material (cast iron) is plotted, it can be seen that material ruptures without any noticeable deformation after critical stress (Figure 15.3). In other words, brittle material does not have a yield point.

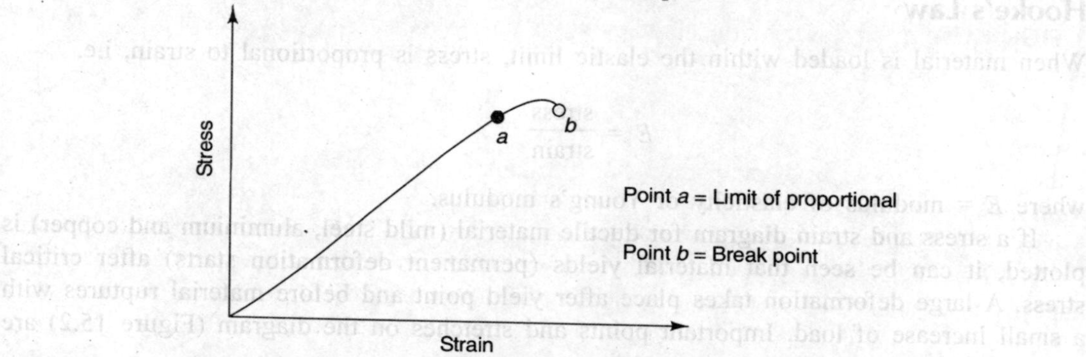

FIGURE 15.3 Stress and strain diagram (Brittle material).

If force P is applied on a specimen (length = l) and elongation is δ, then

$$\text{Stress } (\sigma) = \frac{\text{Force}}{\text{Area}} = \frac{P}{A}$$

$$\text{Strain } (\varepsilon) = \frac{\delta}{l}$$

From Hooke's law:

$$E = \text{Young's modulus} = \frac{\sigma}{\varepsilon}$$

or
$$E = \frac{P \times l}{A \times \delta}$$

or
$$\delta = \text{elongation} = \frac{Pl}{AE}$$

The compressive stress-strain diagram can be plotted on minus strain and minus stress axes. The diagram of cast iron is as indicated (Figure 15.4). Cast iron has much higher strength in compression than in tension.

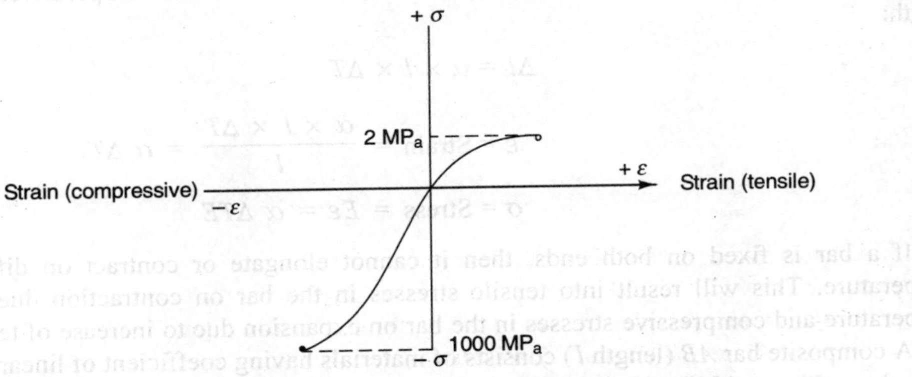

FIGURE 15.4 Comparison of compressive and tensile stress and strain in cast iron.

If material is loaded beyond elastic limit (point a) to stress b, then permanent strain oc is set in (Figure 15.5). If material is again loaded, it follows the path cb. As stress $b >$ stress a, the elastic limit of the specimen has increased after plastic deformation. This is called *strain hardening*. The point of rupture will remain the same. However, ductility has decreased as it is now measured from point c.

A shear stress and shear strain diagram (τ vs $\varepsilon_{\text{shear}}$) for ductile material has the same shape as a normal stress and strain diagram has. In the elastic limit,

$$\text{shear stress } (\tau) = G \text{ (modulus of rigidity)} \times \text{shear strain } (\varepsilon_{\text{shear}})$$

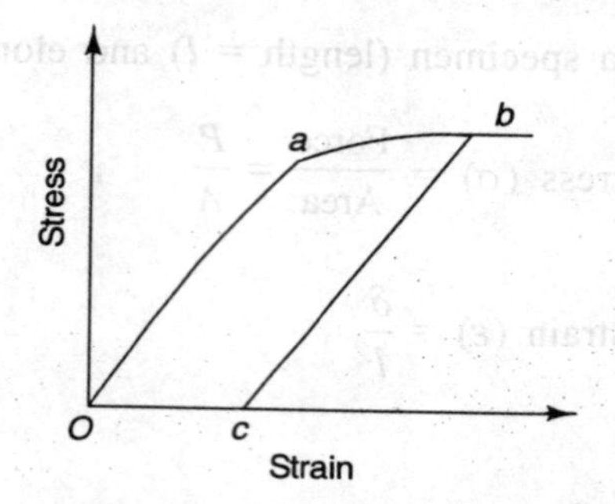

FIGURE 15.5 Strain hardening.

However, yield shear stress will be half of the normal stress, i.e.

$$\tau_{max} = \frac{\sigma_{max}}{2}$$

THERMAL STRESSES

Stress and strain are also developed in a bar due to temperature variation. A bar elongates on increase of temperature and contracts on fall of temperature. Elongation or contraction depends on the linear coefficient of expansion (α), difference of temperatures (ΔT) and length:

$$\Delta l = \alpha \times l \times \Delta T$$

$$\varepsilon = \text{Strain} = \frac{\alpha \times l \times \Delta T}{l} = \alpha\, \Delta T$$

$$\sigma = \text{Stress} = E\varepsilon = \alpha\, \Delta T E$$

If a bar is fixed on both ends, then it cannot elongate or contract on difference of temperature. This will result into tensile stresses in the bar on contraction due to fall of temperature and compressive stresses in the bar on expansion due to increase of temperature.

A composite bar AB (length l) consists of materials having coefficient of linear expansion α_1 and α_2 (Figure 15.6). On heating, the extension of composite bar is Δl. If $\alpha_1 > \alpha_2$, then extention in material-1 would have been more than actual Δl, i.e.

$$\Delta l + x_1 = l\alpha_1 \Delta T$$

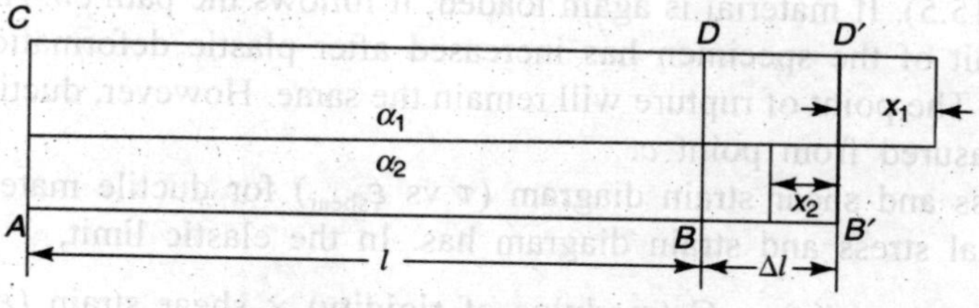

FIGURE 15.6 Composite bar.

while the extension in material-2 would have been less than actual Δl, i.e.

$$\Delta l - x_2 = l\alpha_2\Delta T$$

In other words material-1 has been contracted by x_1 $(\alpha_1\, l\Delta T - \Delta l)$ and material-2 has been elongated by $x_2(\Delta l - \alpha_2\, l\Delta T)$. The total contraction and elongation is

$$x_1 + x_2 = (\alpha_1 - \alpha_2)\, l\Delta T$$

The total strain will be

$$\varepsilon_1 + \varepsilon_2 = (\alpha_1 - \alpha_2)\,\Delta T$$

From $\dfrac{\sigma_1}{E_1} + \dfrac{\sigma_2}{E_2} = (\alpha_1 - \alpha_2)\,\Delta T$ and $\sigma_1 A_1 = \sigma_2 A_2$, we can find out σ_1 and σ_2.

DEFORMATION UNDER OWN WEIGHT

A body elongates due to its own weight when it is hanging from one end (Figure 15.7). If it has a uniform cross section, the load on dx thick part due to the weight of x part of the body is

$$\begin{aligned} \Delta P &= \text{weight of } x \text{ part of the body} \\ &= \text{density} \times \text{volume} \times g \\ &= \rho \times x \times A \times g \end{aligned}$$

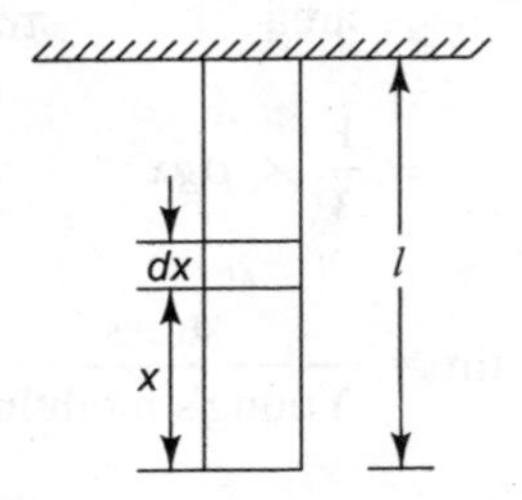

FIGURE 15.7 Uniform cross section bar.

$$\text{Stress} = \frac{\Delta P}{A} = \rho x g$$

$$\text{Strain } \varepsilon = \frac{\text{stress}}{E} = \frac{\rho x g}{E}$$

$$\text{Also strain} = \frac{\delta}{dx} = \frac{\rho x g}{E}$$

$$\delta = \frac{\rho x g}{E}\, dx$$

$$\text{Total elongation} = \frac{\rho g l}{E}\int_0^l x\,dx = \frac{\rho g l^2}{2E}$$

Elongation in a body due to its self weight is half of the elongation which can be achieved in case full weight is acting at the end.

If a tapered body is hanged as shown in Figure 15.8, there is an elongation in part *dx* due to the weight of *x* part of the body acting on it.

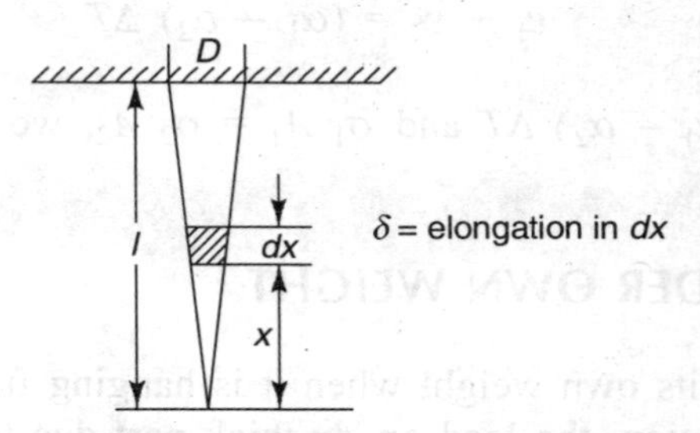

FIGURE 15.8 Tapered round bar.

$$\text{Self weight acting on } dx = \frac{1}{3} \times \pi d^2 x \times \rho \times g$$

$$\text{Stress} = \frac{\text{load}}{\text{area}} = \frac{1/3 \times \pi d^2 x \rho g \times 4}{\pi d^2 \times 4}$$

$$= \frac{1}{3} \times \rho g x$$

$$\text{Strain} = \frac{\text{stress}}{\text{Young's modulus}} = \frac{1}{3} \times \frac{\rho g x}{E}$$

$$\frac{\delta}{dx} = \frac{1}{3} \times \frac{\rho g}{E} x$$

$$\int \delta = \frac{1}{3} \times \frac{\rho g}{E}\int_0^l x$$

$$\text{Elongation} = \frac{1}{6} \times \frac{\rho g}{E} l^2$$

DEFORMATION UNDER EXTERNAL LOAD

Principle of super position: If the cross section of a bar varies in steps, elongation can be found out by the principle of super position. If P_1', P_2' and P_3' are forces acting in l_1, l_2

and l_3 parts of a bar as shown in Figure 15.9. The principle of superposition states that the total elongation is the sum total elongation of each part having a constant cross section.

$$\delta_1 = \frac{P'_1}{A_1E}\,l_1 \quad \delta_2 = \frac{P'_2}{A_2E}\,l_2 \quad \delta_3 = \frac{P'_3}{A_3E}\,l_3$$

$$\therefore \qquad \delta = \delta_1 + \delta_2 + \delta_3 = \frac{P'_1}{A_1E}\,l_1 + \frac{P'_2}{A_2E}\,l_2 + \frac{P'_3}{A_3E}\,l_3$$

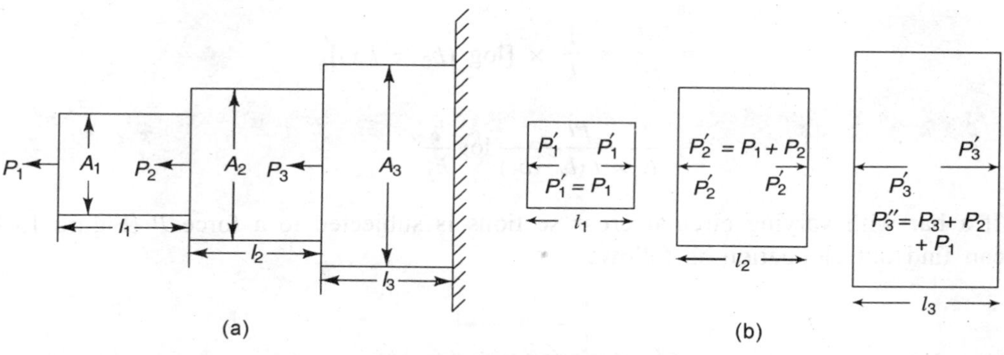

FIGURE 15.9 (a) Bar with cross section varying in steps and (b) forces in different parts of a bar.

A bar having a constant thickness (t) and varying breadths (b_1 to b_2) is subject to force P as shown in Figure 15.10. Let us find out its elongation.

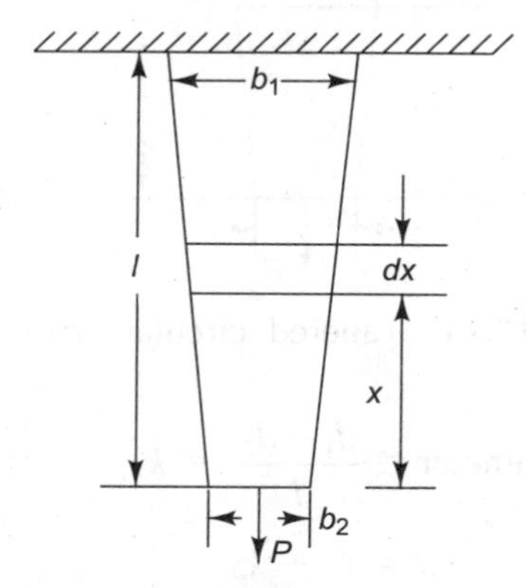

FIGURE 15.10 A bar with constant thickness (t) and varying breadth (b).

$$\text{Breadth } b \text{ at } x \text{ distance} = b_2 + \frac{b_1 - b_2}{l}\,x$$

$$= b_2 + kx$$

where $k = \dfrac{b_1 - b_2}{l}$ = breadth reduction constant

Cross sectional area at $x = b \times t$

Elongation in dx is

$$\delta = \frac{P}{AE}dx = \frac{P}{t\,(b_2 + kx)\,E}\,dx$$

$$\text{Elongation} = \int_0^l \frac{Pdx}{tE\,(b_2 + kx)} = \frac{P}{tE}\int_0^l \frac{dx}{(b_2 + kx)}$$

$$= \frac{P}{tE} \times \frac{1}{k} \times [\log\,(b_2 + kx)]_0^l$$

$$= \frac{Pl}{E \times t\,(b_1 - b_2)} \log \frac{b_1}{b_2}$$

If a bar with varying circular cross sections is subjected to a force P (Figure 15.11), we can find out elongation as follows:

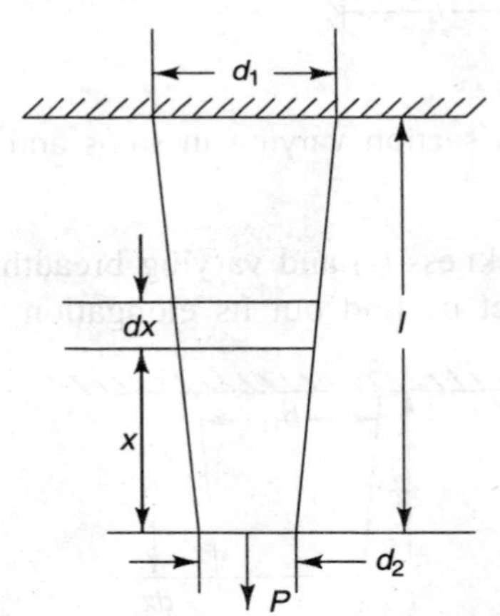

FIGURE 15.11 Tapered circular cross section.

$$\text{Rate of change of diameter} = \frac{d_1 - d_2}{l} = k$$

$$d = d_2 + kx$$

Take an element dx at distance x.

$$\text{Stress} = \frac{P}{\frac{\pi}{4}d^2} = \frac{P}{\frac{\pi}{4}(d_2 + kx)^2}$$

$$\delta = \text{Elongation in } dx = \frac{Pdx}{E \times \pi/4\,(d_2 + kx)^2}$$

$$\text{Total elongation} = \int_0^l \frac{P\,dx}{E \times \pi/4\,(d_2 + kx)^2}$$

$$= \frac{4\,Pl}{\pi E d_1 d_2}$$

A composite bar consists of two materials with Young's modulus as E_1 and E_2 as shown in Figure 15.12. Let us find out the total elongation. P_1 and P_2 are loads shared by two materials.

$$P = P_1 + P_2$$

As

$$P_1 = \sigma_1 A_1 \text{ and } P_2 = \sigma_2 A_2$$

$$P = \sigma_1 A_1 + \sigma_2 A_2$$

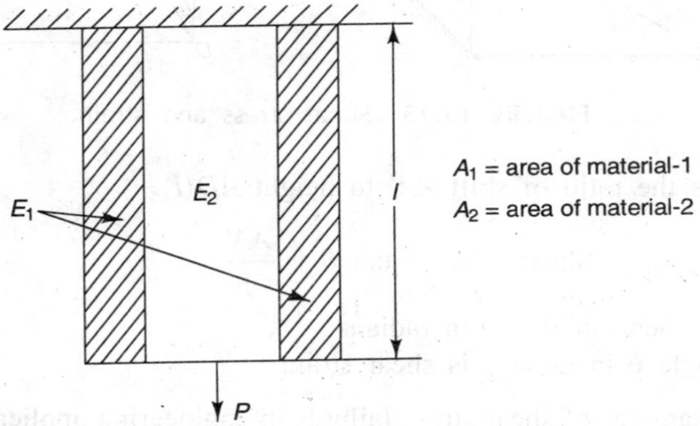

A_1 = area of material-1

A_2 = area of material-2

FIGURE 15.12 Composite bar.

The strain in material-1 and material-2 is the same, i.e.

$$\varepsilon_1 = \varepsilon_2$$

$$\therefore \quad \frac{\sigma_1}{E_1} = \frac{\sigma_2}{E_2}$$

$$P_1 = \delta \frac{A_1 E_1}{l}, \qquad P_2 = \delta \frac{A_2 E_2}{l}$$

Therefore,

$$P = P_1 + P_2 = \frac{\delta}{l}\,(A_1 E_1 + A_2 E_2)$$

or

$$\delta = \frac{Pl}{A_1 E_1 + A_2 E_2}$$

SHEAR STRESS AND STRAIN

If a rectangular block (Figure 15.13) is rigidly held and a force P is applied on the upper surface, the block gets deformed, i.e. rectangle face $ABCD$ will change to $A'B'CD$. There is a sliding tendency between the upper layer where force is acting and the lower layer which is rigidly held. The resistance to the sliding is provided by the shear stress developed in the material. The shear stress is directly proportional to the applied force P and inversely proportional to area $ABFE$.

$$\tau = \text{shear stress} = \frac{P}{\text{Area } ABFE}$$

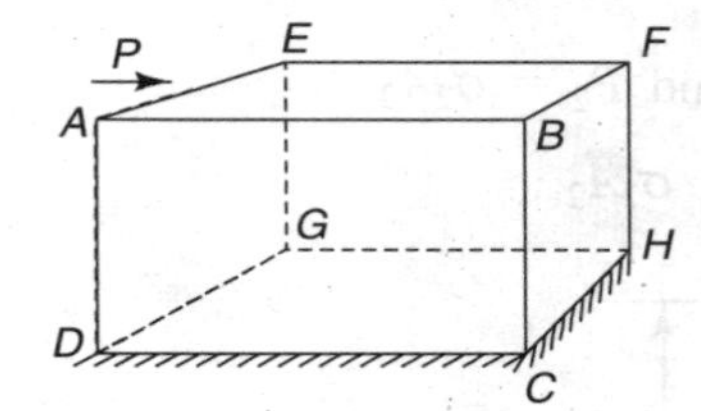

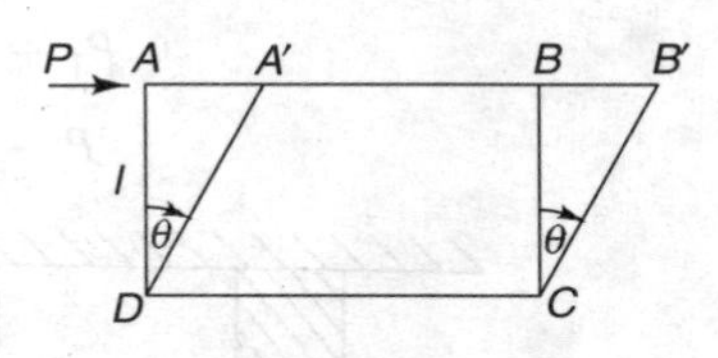

FIGURE 15.13 Shear stress and strain.

Shear strain is the ratio of shift AA' to height $AD(l)$.

Therefore $$\text{Shear strain} = \tan\theta = \frac{AA'}{l}$$

If θ is small, then $\tan\theta = \theta$ in radians.
Hence the angle θ in radians is shear strain.

There are many examples of shear stress failures in engineering applications (Figure 15.14):

- Formation of metal chip during machining (Lathe and milling)
- Failure of single-rivet or double-rivet joint (Figures 15.14(a) and (b)). A single rivet joining two strips under shear force P will generate shear stress equal to P divided by cross-sectional area ($A = \pi d^2/4$). In the case of double rivets shear stress in each rivet is equal to $P/2A$.
- Failure of bolts in a flange connecting two shafts (Figure 15.14(c)). When two shafts are connected to a large with bolts, the bolts are subjected to shear stress, which is equal to the torque transmitted divided by rnA (where n = number of bolts, A = area of each bolt and r = distance of the centre of the bolt from the centre of the shaft.)
- Punching of a hole (Figure 15.14(d)). When a hole is punched, shear stress (τ) is equal to the load (P) of punch divided by πdt (where d = diameter of the punch, t = thickness of plate) which must be higher than ultimate shear stress for punching.

Like normal stress and strain, shear stress and strain have also a constant ratio. Shear modulus (G) is the ratio of shear stress to shear strain.

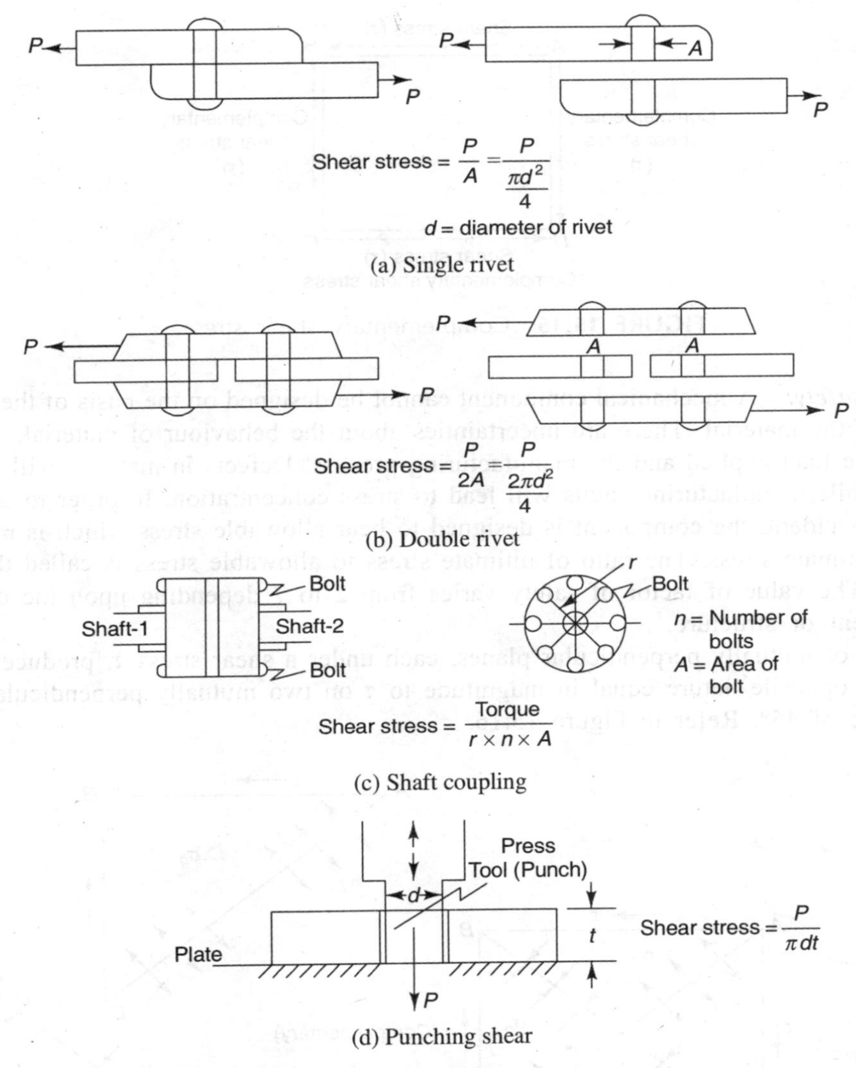

FIGURE 15.14 Applications of shear stress.

$$\text{Modulus of rigidity} = \frac{\text{shear stress } (\tau)}{\text{shear strain } (\varepsilon_s)}$$

If a set of shear stresses acts on a body, then a set of balancing shear stresses of the same value but normal to shear planes is developed. If τ is acting on planes AB and CD, then complementary shear stresses having the same value τ on planes AD and BC are developed (Figure 15.15). However, the moment of main shear forces is opposite to the moment of complementary shear forces. The moment of shear stresses is $\tau \times l^2 \times l$ which is in the clockwise direction but the moment of complementary stresses is also $\tau \times l^2 \times l$ which is in the anticlockwise direction. Here $AB = AD = AE = l$.

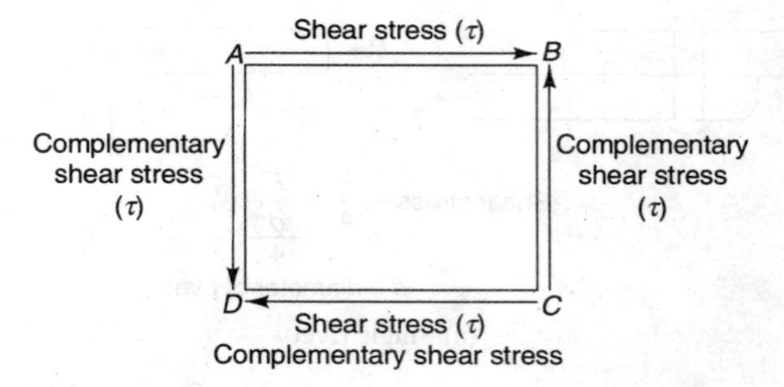

FIGURE 15.15 Complementary shear stress.

Factor of safety: A mechanical component cannot be designed on the basis of the ultimate strength of the material. There are uncertainties about the behaviour of material, the exact value of the load applied and the manufacturing process. Defects in material will give less strength while manufacturing faults will lead to stress concentration. In order to avoid any failure or accident, the component is designed to bear allowable stress which is much less than the ultimate stress. The ratio of ultimate stress to allowable stress is called the *factor of safety*. The value of factor of safety varies from 2 to 7 depending upon the design of a component or structure.

A pair of mutually perpendicular planes, each under a shear stress τ, produces normal stresses of opposite nature equal in magnitude to τ on two mutually perpendicular planes at an angle of 45°. Refer to Figure 15.16.

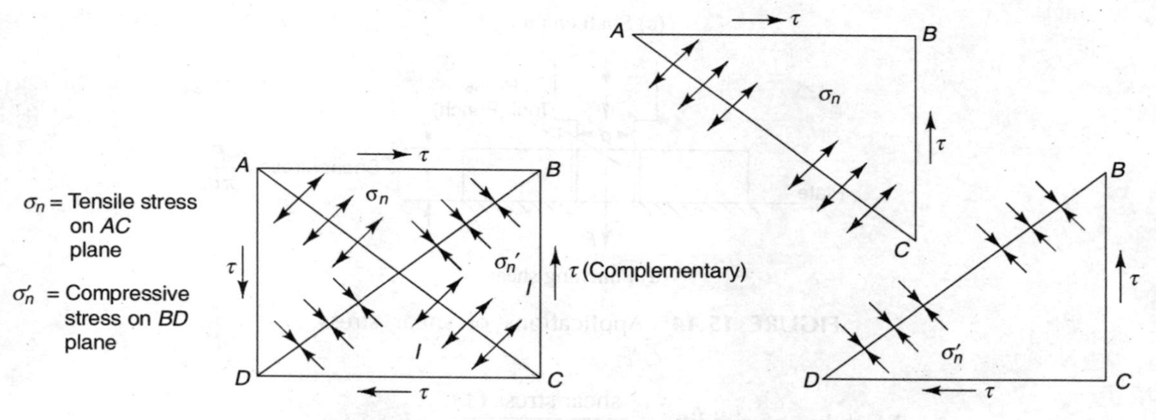

FIGURE 15.16 Normal stresses due to shear stress.

$$\text{Shear force on plane } BC = \tau \times l^2$$

$$\text{Shear force on plane } CD = \tau \times l^2$$

$$\text{Resultant of two} = \sqrt{2}\,\tau \times l^2 \text{ normal to plane } AC$$

Normal stress σ_n on $AC = \dfrac{\sqrt{2} \times \tau \times l^2}{\sqrt{2}\, l^2}$ ($\because$ Area of plane $AC = \sqrt{2}\, l^2$)

$$= \tau$$

$$\therefore \quad \sigma_n = \sigma'_n = \tau$$

The breaking of a piece of chalk and a piece of cast iron (brittle material) is due to normal (tensile) stresses induced by the shear stresses at an angle of 45°. Brittle materials fail due to tension.

As we have seen that there are normal stresses due to shear stresses, similarly there are shear stresses due to a normal load (Figure 15.17).

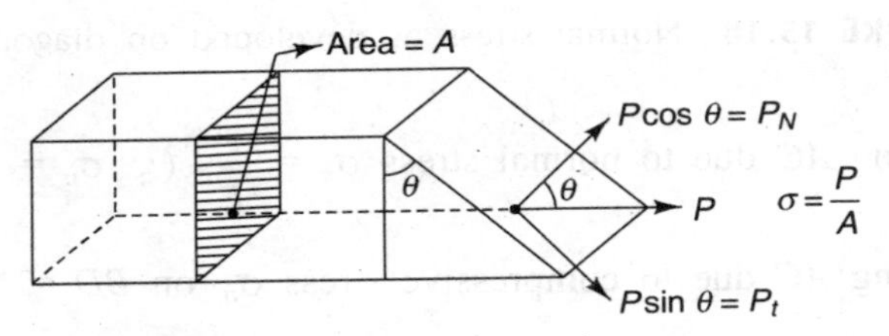

FIGURE 15.17 Shear stress due to normal load.

Force P is acting along the axis of the specimen on a plane inclined θ with the perpendicular plane. Now force $P \cos \theta$ is acting normal and $P \sin \theta$ is acting tangential to plane. Therefore, the normal stress on the plane,

$$\sigma_\theta = \frac{P \cos \theta}{A / \cos \theta} = \sigma \cos^2 \theta$$

or

$$\sigma_\theta = \frac{\sigma}{2}\ (1 + \cos 2\theta)$$

Shear stress

$$\tau_\theta = \frac{P \sin \theta}{A / \cos \theta} = \frac{\sigma}{2} \sin 2\theta$$

Normal stress is maximum when $\theta = 0°$. Shear stress is maximum when $\theta = 45°$.

$$\tau_{\max} = \frac{\sigma}{2}$$

Poisson's ratio: When a bar is loaded, its length increases but its width and thickness decrease. The lateral or transverse strain in the width and thickness has a definite relationship with the longitudinal strain.

$$\text{Poisson's ratio} = \gamma = \frac{\text{transverse strain}}{\text{longitudinal strain}}$$

Due to applied shear stress, normal stresses are developed on diagonals AC and BD (Figure 15.18). Let us find diagonal strain due to these stresses.

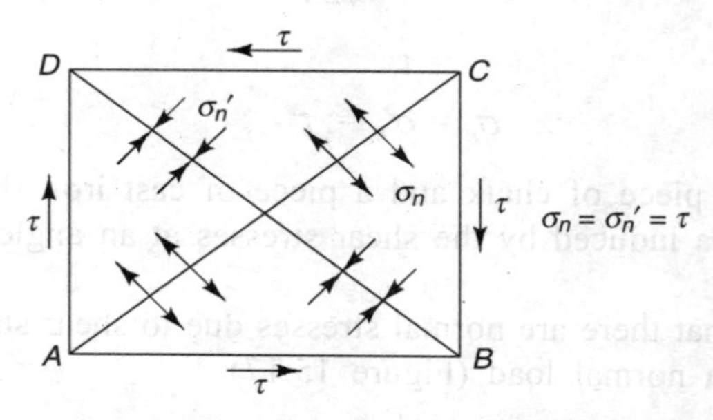

FIGURE 15.18 Normal stress as developed on diagonals.

Tensile strain along AC due to normal stress $\sigma_n = \dfrac{\tau}{E}$. ($\because\ \sigma_n = \tau$)

Tensile strain along AC due to compressive stress $\sigma_{n'}$ on $BD = \gamma \dfrac{\tau}{E}$

Total tensile strain along $AC = \dfrac{\tau}{E}\ (1 + \gamma)$

Similarly, the total compressive strain along $BD = \dfrac{\tau}{E}\ (1 + \gamma)$

VOLUMETRIC STRAIN, BULK MODULUS AND ELASTIC CONSTANTS

Volumetric strain: When a member is subjected to a single force, all sides undergo changes resulting into a net change in the volume of the member. The ratio of change in volume to the original volume is called volumetric strain (ε_v).

$$\varepsilon_v = \frac{\delta V}{V}$$

where δV = change in volume

V = original volume

The volumetric strain for a rectangular section due to axial force (P) can be found out (Figure 15.19).

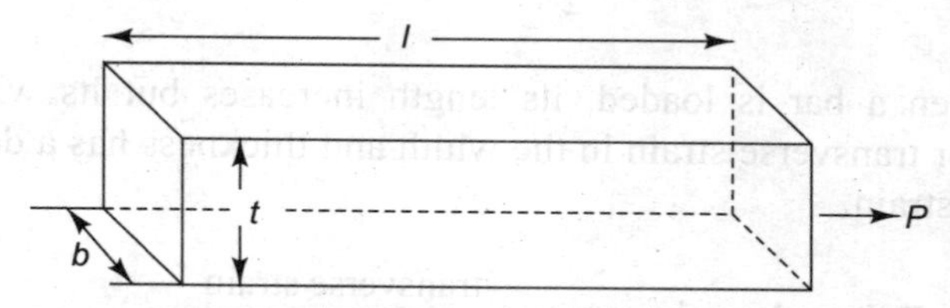

FIGURE 15.19 Volumetric strain for a rectangular section.

$$V_1 = \text{Initial volume} = l \times b \times t$$

$$V_2 = \text{Final volume} = (l + \delta l)\ (b - \delta b)\ (t - \delta t)$$

where l = length
b = breadth
t = thickness
δl = change in length
δb = change in breadth
σt = change in thickness

Neglecting $\sigma l \times \delta b$ and δt being very small, we get

$$V_2 = lbt + bt\delta l - bl\delta t - lt\delta b$$

$$\delta V = V_2 - V_1 = bt\delta l - bl\delta t - lt\delta b$$

$$\frac{\delta V}{V} = \frac{\delta l}{l} - \frac{\delta t}{t} - \frac{\delta b}{b}$$

Volumetric strain

$$\varepsilon_v = \varepsilon - \varepsilon_l - \varepsilon_l$$
$$= \varepsilon - \gamma\varepsilon - \gamma\varepsilon$$
$$= \varepsilon(1 - 2\gamma)$$

where ε = longitudinal strain
ε_l = lateral strain

The volumetric strain for a cylindrical rod (diameter = d and length = l) subjected to the axial load can be found out as follows:

$$V_2 - V_1 = \frac{\pi}{4}\ (d - \delta d)^2\ (1 + \delta l) - \frac{\pi}{4}\ d^2 l$$

$$\frac{\delta V}{V} = \frac{\delta l}{l} - \frac{2\delta d}{d} \quad (\because \text{ neglecting } \delta d \times \delta l \text{ being small})$$

$$\varepsilon_v = \varepsilon - 2\varepsilon_l$$
$$= \varepsilon - 2\gamma\varepsilon$$
$$= \varepsilon(1 - 2\gamma)$$

Bulk modulus of elasticity: The bulk modulus of elasticity (k) is the ratio of the normal stress (σ) to the volumetric strain (ε_v), i.e.

$$k = \frac{\sigma}{\varepsilon_v}$$

The bulk modulus is related to Young's modulus. When a cube is subjected to normal stress in x, y and z directions, then

$$\varepsilon_x = \text{strain in } x\text{-direction} = \frac{\sigma}{E}(1 - 2\gamma)$$

$$\varepsilon_y = \text{strain in } y\text{-direction} = \frac{\sigma}{E}(1 - 2\gamma)$$

$$\varepsilon_z = \text{strain in } z\text{-direction} = \frac{\sigma}{E}(1 - 2\gamma)$$

The volumetric strain $\varepsilon_v = \varepsilon_x + \varepsilon_y + \varepsilon_z$

$$= \frac{3\sigma}{E}(1 - 2\gamma)$$

$$k = \text{Bulk modulus} = \frac{\sigma}{\varepsilon_v} = \frac{\sigma}{\frac{3\sigma}{E}(1 - 2\gamma)} = \frac{E}{3(1 - 2\gamma)}$$

$$E = 3k(1 - 2\gamma)$$

The modulus of rigidity (G) is related to Young's modulus which is derived as follows. Refer to Figure 15.20.

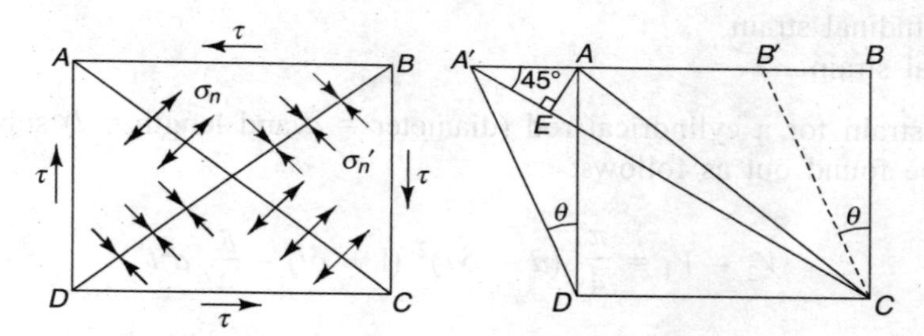

FIGURE 15.20 Relationship between modulus of rigidity and Young's modulus.

$$\text{Longitudinal strain in } AC = \frac{A'C - AC}{AC} = \frac{A'E}{AC}$$

$$= \frac{A'A\cos 45}{AD\sqrt{2}} = \frac{A'A}{2AD}$$

$$= \frac{\tan\theta}{2} = \frac{\theta}{2} = \frac{\tau}{2G} \qquad \left(\theta = \text{strain} = \frac{\tau}{G}\right)$$

The longitudinal strain due to the compressive stress in BD and the longitudinal stress in AC is

$$\frac{\sigma_n'}{E} + \frac{\sigma_n}{E}\gamma$$

But

$$\sigma_n = \sigma_{n'} = \tau$$

Therefore, the longitudinal strain is

$$\frac{\tau}{E} + \frac{\tau}{E}\gamma = \frac{\tau}{E}(1+\gamma)$$

Equating the strains found in both methods, we get

$$\frac{\tau}{2G} = \frac{\tau}{E}(1 + \gamma)$$

or

$$E = 2G(1 + \gamma)$$

Now the relation between the bulk modulus (k), Young's modulus (E) and the modulus of rigidity (G) is as follows:

$$E = 3k(1 - 2\gamma) = 2G(1 + \gamma)$$

Now putting the value $\gamma = \frac{E}{2G} - 1$ in the equation $E = 3k(1 - 2\gamma)$, we get

$$E = 3k\left[1 - 2\left(\frac{E}{2G} + 1\right)\right] = 3k\left[3 - \frac{E}{G}\right]$$

or

$$E = \frac{9\,kG}{G + 3k}$$

STRAIN ENERGY AND RESILENCE

When an elastic body is subjected to an external load, it gets deformed. As the point of the load application is displaced, work is done by the applied force. An internal resistance is developed in the body and work done by the applied force is stored in the body as strain energy, which brings back the body to the original shape on removal of the applied force.

Strain energy depends upon the following ways of loading:

(a) Slowly applied force (Figure 15.21(a))
(b) Suddenly applied force (Figure 15.21(b))

We know that

$$\text{Strain energy} = \text{force} \times \text{displacement}$$

In the case of slowly applied force,

$$\text{Strain energy} = \frac{P}{2} \times \delta l = \frac{P}{2} \times \frac{Pl}{AE}$$

$$= \frac{P^2 l}{2AE}$$

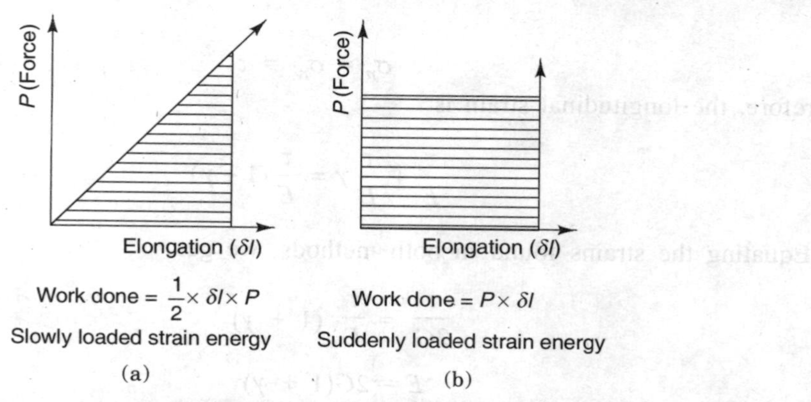

FIGURE 15.21 Strain energy.

In the case of suddenly applied force,

$$\text{Strain energy} = P \times \delta l$$

$$= \frac{P^2 L}{AE}$$

Resilence is strain energy per unit volume stored by body within the elastic limit when loaded externally. The maximum energy stored within the elastic limit is called *proof resilence*. It is the capacity of material to bear shock. Proof resilence per unit volume is called the *modulus of resilence*.

$$\text{Strain energy} = \frac{P^2 l}{2AE}$$

$$\text{Modulus resilence} = \frac{P^2 l}{2AE} \bigg/ (A \times l) = \left(\frac{P}{A}\right)^2 \times \frac{1}{2E}$$

$$= \frac{\sigma_P^2}{2E}$$

where σ_P is proof or maximum stress.

Now we will find strain energy due to shear force P. Refer to Figure 15.22.

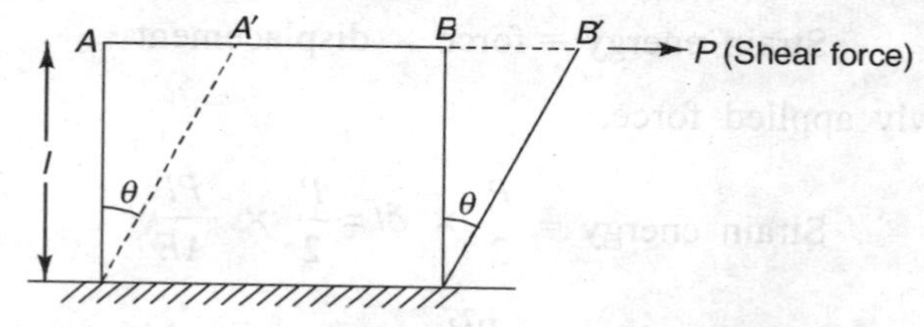

FIGURE 15.22 Strain energy due to shear force.

$$\text{Average shear force} = \frac{0+P}{2} = \frac{P}{2}$$

$$\text{Strain energy} = \frac{P}{2} \times AA' \qquad (\because \text{ Work done} = \text{force} \times \text{distance})$$

$$= \frac{1}{2} \times \left(\frac{P}{l^2}\right) \times l^2 \times (\theta l) \qquad \left[\because \frac{P}{l^2} = \frac{P}{\text{area}} = \tau\right]$$

$$= \frac{1}{2} \tau \times \theta l^3$$

$$= \frac{1}{2} \times \frac{\tau^2}{G} V \quad \left[\because \theta = \frac{\tau}{G}\right]$$

where V is volume.

$$\text{Modulus of resilence} = \frac{\text{strain energy}}{\text{volume}} = \frac{1}{2} \times \frac{\tau^2}{G}$$

COMPOUND STRESSES (2-D SYSTEM)

Let normal stresses σ_x and σ_y and shear stress τ are acting as shown in Figure 15.23. We will find out normal stress σ_θ and shear stress τ_θ on an inclined plane DE making an angle θ with plane CD.

Equating normal forces to plane DE, we get

$$\sigma_\theta \times 1 \times DE = \sigma_x \times \cos\theta \times 1 \times CD + \sigma_y \sin\theta \times 1 \times CE + \tau \sin\theta \times 1 \times CD + \tau \cos\theta \times 1 \times CE$$

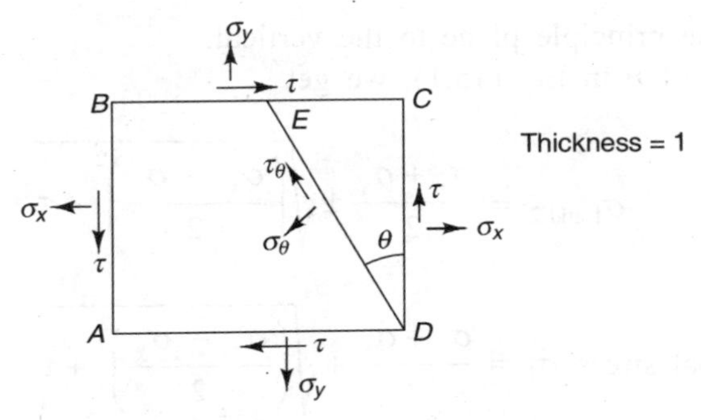

FIGURE 15.23 2-D system of compound stresses.

$$\sigma_\theta = \sigma_x \cdot \frac{CD}{DE} \cos\theta + \sigma_y \cdot \frac{CE}{DE} \sin\theta + \tau \cdot \frac{CD}{DE} \sin\theta + \tau \cdot \frac{CE}{DE} \cos\theta$$

$$= \sigma_x \cos^2\theta + \sigma_y \sin^2\theta + 2\tau \sin\theta \cos\theta$$

$$= \sigma_x\left(\frac{1+\cos 2\theta}{2}\right) + \sigma_y \frac{(1 - \cos 2\theta)}{2} + \tau \sin 2\theta$$

$$\therefore \quad \sigma_\theta = \frac{\sigma_x + \sigma_y}{2} + \frac{\sigma_x - \sigma_y}{2} \cos 2\theta + \tau \sin 2\theta \qquad (15.1)$$

Equating tangential forces to plane DE, we get

$$\tau_\theta \times 1 \times DE = \sigma_y \sin\theta \times 1 \times CD - \sigma_y \cos\theta \times 1 \times CE - \tau \cos\theta \times 1 \times CD + \tau \sin\theta \times 1 \times CE$$

$$\tau_\theta = \sigma_x \frac{CD}{DE} \sin\theta - \sigma_y \frac{CE}{DE} \cos\theta - \tau \frac{CD}{DE} \cos\theta + \tau \frac{CD}{DE} \sin\theta$$

$$= \sigma_x \cos\theta \sin\theta - \sigma_y \cos\theta \sin\theta - \tau(\cos^2\theta - \sin^2\theta)$$

$$\therefore \quad \tau_\theta = \left(\frac{\sigma_x - \sigma_y}{2}\right) \sin 2\theta - \tau \cos 2\theta \qquad (15.2)$$

Principal Planes and Principal Stresses

Principal planes are the planes where shear stress is zero. Principal planes have only normal stresses which are called principal stresses.

For principal planes $\tau_\theta = 0$. Therefore, substituting this value in Eq. (15.2),

$$\tan 2\theta = \frac{2\tau}{\sigma_x - \sigma_y}$$

where θ = angle of the principle plane to the vertical.

Putting the value of θ in Eq. (15.1), we get

$$\sigma_{1 \text{ and } 2} = \frac{\sigma_x + \sigma_y}{2} \pm \sqrt{\left(\frac{\sigma_x - \sigma_y}{2}\right)^2 + \tau^2}$$

$$\text{Major principal stress } \sigma_1 = \frac{\sigma_x + \sigma_y}{2} + \sqrt{\left(\frac{\sigma_x - \sigma_y}{2}\right)^2 + \tau^2}$$

$$\text{Minor principal stress } \sigma_2 = \frac{\sigma_x + \sigma_y}{2} - \sqrt{\left(\frac{\sigma_x - \sigma_y}{2}\right)^2 + \tau^2}$$

Now we can find out maximum shear stress.

$$\tau_\theta = \left(\frac{\sigma_x - \sigma_y}{2}\right) \sin 2\theta - \tau \cos 2\theta$$

For maximum shear stress, $\dfrac{d\tau_\theta}{d\theta} = 0.$

Therefore,

$$\left(\frac{\sigma_x - \sigma_y}{2}\right) \times 2 \cos 2\theta + 2\tau \sin 2\theta = 0$$

or
$$\tan 2\theta = -\left(\frac{\sigma_x - \sigma_y}{2\tau}\right)$$

$\therefore$
$$\tau_{\max} = \frac{1}{2}(\sigma_1 - \sigma_2)$$

Hence maximum shear stress is half the difference between principal stresses, i.e. $\frac{1}{2}(\sigma_1 - \sigma_2)$. Also maximum shear stress occurs on planes at 45° to the principal planes.

MOHR'S CIRCLE

Mohr's circle is a graphical method for finding normal and shear stresses on any plane when conditions of normal and shear stresses on any other plane are given. Normal stresses are drawn on x-axis, i.e. tensile stress on the positive x-axis and compressive stress on the negative x-axis. Two points of Mohr's circle cutting x-axis are principal planes P_1 and P_2. Shear stresses are drawn on y-axis, i.e. clockwise shear stress is on the positive y-axis and anticlockwise shear stress is negative.

If σ_x, σ_y and τ are given, we will draw Mohr's circle (Figure 15.24) as follows:

1. Locate point A on (σ, τ) plane with $OA' = \sigma_x$ and $AA' = \tau$.
2. Similarly, locate point B with $OB' = \sigma_y$ and $BB' = -\tau$.

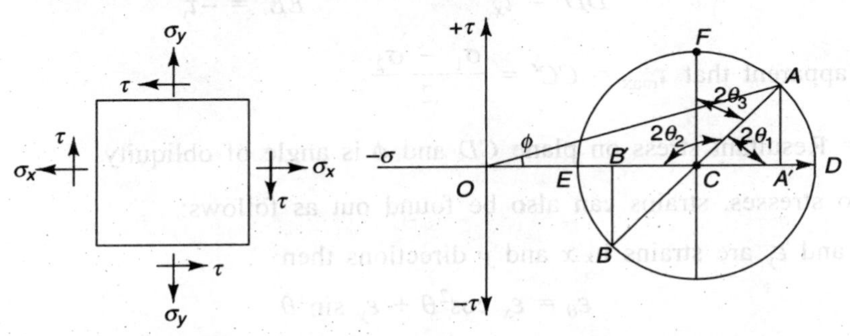

FIGURE 15.24 Mohr's circle when σ_x, σ_y and τ are given.

3. Join A and B which cuts x-axis at point C.
4. With point C as the centre and radius AC or BC, draw Mohr's circle.
5. Mohr's circle cuts x-axis at points D and E where shear stress is zero. Hence major principal stress $\sigma_1 = OD$ and minor principal stress $\sigma_2 = OE$.
6. Maximum shear stress is CF (perpendicular to x-axis).
7. Angle θ_1 between the major principal plane and the actual plane is given by $1/2 \angle ACD$.
8. Angle θ_2 between the minor principal plane and the actual plane is given by $1/2 \angle ACE$.
9. Angle θ_3 between the actual plane and maximum shear stress plane is given by $1/2 \angle ACF$
10. Resultant of forces on actual plane is given by OA and ϕ is the angle of obliquity.

If principal stresses σ_1 and σ_2 are given, we can draw Mohr's circle (Figure 15.25) as per the method given as follows:

1. As principal stresses are given, we have two points on Mohr's circle as $(\sigma_1, 0)$ and $(\sigma_2, 0)$ because shear stress on principal planes is zero.
2. Draw $OA = \sigma_1$ and $OB = \sigma_2$.
3. Divide AB and get its centre as point C.
4. With C as the centre and radius CA or CB, draw Mohr's circle.

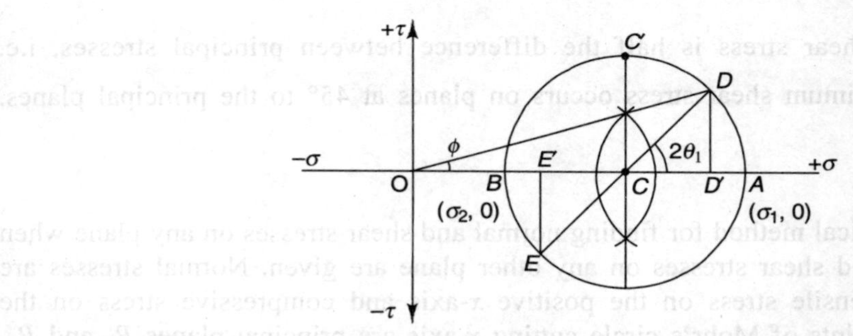

FIGURE 15.25 Mohr's circle when principal stresses (σ_1 and σ_2) are given.

5. To find stresses on a plane making an angle θ_1 with σ_1 plane, locate point D such that $\angle DCA = 2\theta_1$. Now join DC and extend it to point E

$$OD' = \sigma_x \qquad OE' = \sigma_y$$
$$DD' = \tau_x \qquad EE' = -\tau_y$$

It is apparent that $\tau_{max} = CC' = \dfrac{\sigma_1 - \sigma_2}{2}$

6. OD = Resultant stress on plane CD and ϕ is angle of obliquity.

Similar to stresses, strains can also be found out as follows:

1. If ε_x and ε_y are strains in x and y directions then

$$\varepsilon_\theta = \varepsilon_x \cos^2\theta + \varepsilon_y \sin^2\theta$$

2. If ε_s be shear strain, the direct strain in any direction is

$$\varepsilon_\theta = \varepsilon_s \frac{\sin 2\theta}{2}$$

3. Principal strain due to direct strain and shear strain are

$$\varepsilon_{1 \text{ and } 2} = \frac{\varepsilon_x + \varepsilon_y}{2} \pm \frac{1}{2}\sqrt{(\varepsilon_x - \varepsilon_y)^2 + \varepsilon_s^2}$$

Mohr's circle for strains can be drawn as it is done for stresses. For this either ε_x, ε_y and ε_s are given which give points (ε_x, ε_s) and (ε_y, $-\varepsilon_s$) on Mohr's circle or ε_1 and ε_2 are given, which give points (ε_1, 0) and (ε_2, 0) on Mohr's circle.

PROPERTIES OF METAL

Some important properties of metal are as follows:

Malleability: Malleability is the property by virtue of which the material may be hammered or rolled into thin sheets without breaking and cracking. This property generally increases with increase in temperature. It represents the ability of the material to allow expansion in all lateral directions under compression loading.

Toughness: Toughness is the strength with which the material opposes rupture. Molecules of material have attraction for each other giving the material the power to resist tearing apart. The area under stress and strain diagrams indicates the toughness.

Hardness: Hardness is resistance of material to penetration. Hard material resists scratches or wearing out by another body on rubbing. The converse of hardeness is softness.

Fatigue: Under repeated loads, the material tends to fracture under loads lower than the maximum strength of the material.

Creep: Under prolonged loading usually at high temperatures, the material is likely to fail at stresses much lower than maximum stresses that the material can normally withstand.

Impact testing is carried out on an impact testing machine by subjecting the components to impact loads. The stresses induced in these components are much greater than the stresses product by gradual loading. The value is an indication of shock absorbing capacity of materials. The capacities are expressed as rupture energy, modulus of ruptures and notch impact strength. The Charpy test and Izod test are two types of notch impact tests which are commonly conducted. In both tests, the standard specimen is in the form of a *V*-shaped notched beam. In the Charpy test, the specimen is placed as a simply supported beam while it is kept as a cantilever beam in the case of the Izod test. The specimens are provided with a 45° V-shaped notch and the notch is kept on the tension side during testing (facing swinging hammer).

Hardness can be tested on the Brinell cum Rockwell testing machine. In the Brinell hardness test (up to 500 BHN) an indentor (ball with diameter 2.5 mm, 5 mm or 10 mm) is used to make identation on the surface of specimen under load for 30 seconds.

$$\text{BHN} = \frac{\text{Load applied (kg)}}{\text{Spherical surface area of identation (mm}^2\text{)}}$$

Rockwell hardness test is the most common method as Rockwell hardness can be read directly on the dial. Different indentors are used depending upon scales (*A*, *B*, *C*, *D*, *E*, *M*, *R*, etc.). *B*-scale and *C*-scale are commonly used and hardness read on these scales are specified as HRB and HRC respectively. The depth of identification is read on the scale.

SOLVED PROBLEMS

1. A tensile axial load of 20 kN is applied to a 4-metre long mild steel member having a cross-sectional area of 4 cm^2. Find (i) stress, (ii) strain, (iii) elongation and (iv) work done if $E = 2 \times 10^5$ N/mm^2.

(i) Stress $= \sigma = \dfrac{P}{A} = \dfrac{20{,}000}{4 \times 100} = 50$ N/mm^2

(ii) Strain $= \varepsilon = \dfrac{\text{stress}}{E} = \dfrac{50}{2 \times 10^5} = 25 \times 10^{-5}$ mm

(iii) Elongation = strain × length = $25 \times 4000 \times 10^{-5} = 1$ mm

(iv) Work done $= \dfrac{1}{2} P \times$ elongation

$$= \frac{1}{2} \times 20{,}000 \times 1 = 10{,}000 \text{ Nmm}$$

2. Calculate the stress at – 6°C if any contraction is not permitted. Rails are 27 m long and they are laid so that there is no stress at 20°C. Assume $E = 210$ N/mm^2 and $\alpha = 12 \times 10^{-6}$ per °C. If there is an allowance for contraction of 6 mm per rail, then find stress at –6°C.

Case 1: *No allowance for contraction*

$$\varepsilon = \text{strain} = \alpha\,\Delta T$$

$$\sigma = \text{stress} = \alpha\,\Delta T \times E$$

$$= 12 \times 10^{-6} \times [20 - (-6)] \times 210 \times 10^9 = 65.55 \text{ MN/m}^2$$

$$\delta l = l\,\alpha\,\Delta T = 27 \times 12 \times 10^{-6} \times 26 = 8.42 \text{ mm}$$

Case 2: *6 mm is permitted*

$$\text{Net contraction} = 8.42 - 6 = 2.42 \text{ mm}$$

$$\therefore \qquad \sigma = \frac{2.47}{27} \times E = \frac{2.47}{27} \times 210 \times 10^9 = 18.86 \text{ MN/m}^2$$

3. A steel bar is subjected to loads as shown. If Young's modulus for the bar material is 200 kN/mm², determine the change in length of the bar. The bar is 200 mm in diameter. (UPTU: Dec. 2005)

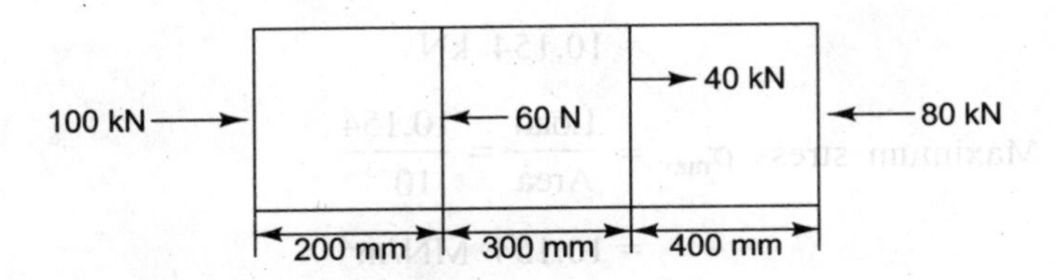

Load-wise diagrams of each length are as follows:

100 kN 100 kN 40 kN 40 kN 80 kN 80 kN

$$\delta L = \frac{1}{E}\left[\frac{P_1L_1}{A_1} + \frac{P_2L_2}{A_2} + \frac{P_3L_3}{A_3}\right]$$

Here $\qquad A_1 = A_2 = A_3 = \dfrac{\pi(200)^2}{4} = 3.14 \times 10^4 \text{ mm}^2$

$$\delta L = \frac{1}{200}\left[\frac{100 \times 200 + 40 \times 300 + 80 \times 400}{3.14 \times 10^4}\right]$$

$$= \frac{1}{200 \times 3.14}(2 + 1.2 + 3.2) = \frac{6.4}{200 \times 3.14} = 0.0102 \text{ mm}$$

4. A rectangular bar of uniform cross section 4 cm × 2.5 cm and of length 2 m is hanging vertically from a rigid support. It is subjected to an axial tensile loading of 10 kN. Take the density of steel as 7850 kg/m³ and E = 200 GN/m². Find the maximum stress and elongation of the bar. (UPTU: 2003–04)

$$\text{Area of cross section} = \frac{4}{100} \times \frac{2.5}{100} = 10^{-3} \text{ m}^2$$

$$\text{Length } (L) = 2 \text{ m}$$

$$\text{Force } (P) = 20 \text{ kN}$$

$$E = 200 \text{ GN/m}^2 = 200 \times 10^6 \text{ kN/m}^2$$

$$\text{Weight of the bar} = AL\rho g$$

$$= 10^{-3} \times 2 \times 7850 \times 9.81$$

$$= 154 \text{ N} = 0.154 \text{ kN}$$

$$\text{Total Axial load} = \text{Force } (P) + \text{weight}$$

$$= 10 + 0.154$$

$$= 10.154 \text{ kN}$$

$$\text{Maximum stress } \sigma_{max} = \frac{\text{Load}}{\text{Area}} = \frac{10.154}{10^{-3}}$$

$$= 10.154 \text{ MN/m}^2$$

$$\text{Elongation} = \frac{PL}{AE} + \frac{W}{2AE}L$$

$$= \left(\frac{10 \times 2}{10^{-3} \times 200 \times 10^6} + \frac{0.154 \times 2}{2 \times 10^{-3} \times 200 \times 6}\right) 10^3 \text{ mm}$$

$$= 0.1 + 0.00077 = 0.10077 \text{ mm}$$

5. Three bars of equal length and having cross-sectional area in ratio 1 : 2 : 4 are all subjected to equal load. Compare their strain energy. (UPTU: 2003–04)

$$\text{Strain energy} = \frac{P^2 L}{2AE}$$

$$\left.\begin{aligned} A_1 &= A_1 \\ A_2 &= 2A_1 \\ A_3 &= 4A_1 \end{aligned}\right\} \text{Area in ratio } 1:2:4$$

$$U_1 = \frac{P^2 L}{ZA_1 E}$$

$$U_2 = \frac{P^2 L}{2A_2 E} = \frac{P^2 L}{2 \times 2A_1 \times E}$$

$$U_3 = \frac{P^2 L}{2A_3 E} = \frac{P^2 L}{2 \times 4A_1 \times E}$$

$$U_1 : U_2 : U_3 = 1 : \frac{1}{2} : \frac{1}{4}$$

6. A load of 100 kg is supported upon rods A and C each of 10 mm diameter and another rod B of 15 mm diameter as shown in the figure. Find stresses in rods A, B and C.

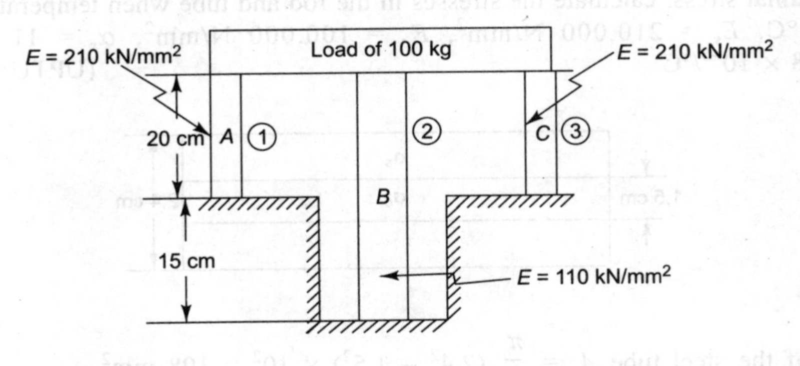

$$\text{Force } P = 100 \times 9.81 = 981 \text{ N}$$

$$A_3 = A_1 = \frac{\pi}{4} D^2 = \frac{\pi}{4} \times 10^2 = 78.54$$

$$L_3 = L_1 = 200 \text{ mm } E_3 = E_1 = 2.1 \times 10^5 \text{ N/mm}^2$$

$$A_2 = \frac{\pi}{4} \times 15^2 = 176.7 \text{ mm}^2$$

$$L_2 = 350 \text{ mm}$$

$$E_2 = 1.1 \times 10^5 \text{ N/mm}^2$$

Now

$$P = P_1 + P_2 + P_3$$

$$981 = \sigma_1 A_1 + \sigma_2 A_2 + \sigma_3 A_3$$

$$= 2\sigma_1 A_1 + \sigma_2 A_2 \qquad (\because \sigma_1 A_1 = \sigma_3 A_3)$$

$$\delta L_1 = \delta L_2 = \delta L_3$$

$$\frac{\sigma_1 L_1}{E_1} = \frac{\sigma_2 L_2}{E_2} = \frac{\sigma_3 L_3}{E_3}$$

$\therefore$

$$\sigma_1 = \frac{E_1}{E_2} \times \frac{L_2}{L_1} \sigma_2 = \frac{2.1 \times 10^5}{1.1 \times 10^5} \times \frac{350}{200} \sigma_2$$

$$= 3.34 \sigma_2$$

$$981 = 2 \times 3.34 \sigma_2 \times 78.54 + \sigma_2 \times 176.7$$

$$= 701.35 \sigma_2$$

$$\sigma_2 = \frac{981}{701.35} = 1.4 \text{ N/mm}^2$$

$$\sigma_1 = \sigma_3 = 3.34 \times 1.4 = 4.7 \text{ N/mm}^2$$

7. A steel tube with 2.4 cm external diameter encloses a copper rod of 1.5 cm diameter to which it is rigidly joined at each end. If, at a temperature of 10°C, there is no longitudinal stress, calculate the stresses in the rod and tube when temperature is raised to 200°C. E_s = 210,000 N/mm², E_c = 100,000 N/mm², $\alpha_s = 11 \times 10^{-6}$/°C, $\alpha_c = 18 \times 10^{-6}$/°C (UPTU: 2001–2002)

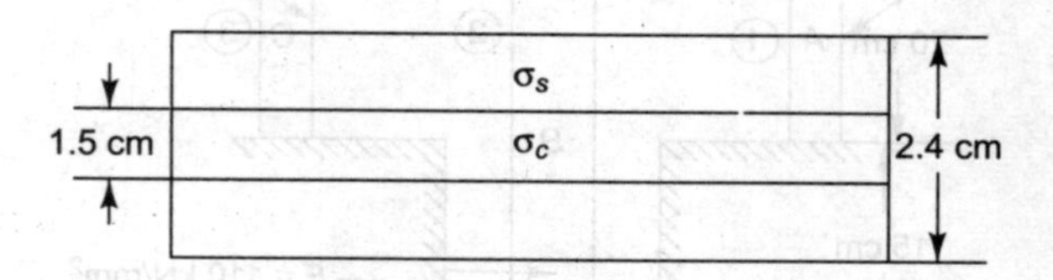

Area of the steel tube $A_s = \dfrac{\pi}{4}(2.4^2 - 1.5^2) \times 10^2 = 198 \text{ mm}^2$

Area of the copper rod $A_c = \dfrac{\pi}{4} \times 15^2 = 177 \text{ mm}^2$

Total tension in the steel tube = Total compression in the copper bar

$$\sigma_s A_s = \sigma_c A_c$$

$$\sigma_s = \frac{A_c}{A_s} \times \sigma_c = \frac{177}{198} \times \sigma_c = 0.89\ \sigma_c$$

Actual expansion of the steel tube = actual expansion of the copper rod

$$L \times \alpha_s\ \Delta T + \frac{\sigma_s}{E_s} \times L = \alpha_c\ \Delta T \times L - \frac{\sigma_c}{E_c} \times L$$

or
$$\alpha_s\ \Delta T + \frac{\sigma_s}{E_s} = \alpha_c\ \Delta T - \frac{\sigma_c}{E_c}$$

$$1.1 \times 10^{-5} \times (200 - 10) + \frac{0.89}{2.1 \times 10^5}\sigma_c = 1.8 \times 10^{-5} \times (200 - 10) - \frac{\sigma_c}{10^5}$$

$\therefore$ $\sigma_c = 93.35$ N/mm² (Compression)

$\sigma_s = 0.89 \times 93.35 = 83.25$ N/mm² (Tension)

8. A 500-mm long bar has a rectangular cross section of 20 mm × 40 mm. The bar is subjected to

(i) 40 kN tensile force on 20 mm × 40 mm face
(ii) 200 kN compressive force on 20 mm × 500 mm face
(iii) 300 kN tensile force on 40 mm × 500 mm face

Find the change in dimensions and column if $E = 2 \times 10^{15}$ N/mm² and $\gamma = 0.3$.

$$\sigma_x = \frac{40 \times 10^3}{20 \times 40} = 50 \text{ N/mm}^2 \qquad \text{(Tension)}$$

$$\sigma_y = \frac{200 \times 10^3}{20 \times 500} = 20 \text{ N/mm}^2 \qquad \text{(Compression)}$$

$$\sigma_z = \frac{300 \times 10^3}{500 \times 40} = 15 \text{ N/mm}^2 \qquad \text{(Tension)}$$

$$\varepsilon_x = \frac{\sigma_x}{E} - \gamma\frac{\sigma_y}{E} - \gamma\frac{\sigma_z}{E}$$

$$= \frac{50}{2 \times 10^5} + 0.3\left(\frac{20}{2 \times 10^5} - \frac{15}{2 \times 10^5}\right)$$

$$= \frac{1}{2 \times 10^5}\ [50 + 0.3\ (20 - 15)]$$

$$= +\frac{51.5}{2 \times 10^5}$$

$$\varepsilon_y = \frac{\sigma_y}{E} - \gamma\frac{\sigma_x}{E} - \gamma\frac{\sigma_z}{E}$$

$$= \frac{1}{2 \times 10^5}\ [-20 - 0.3 \times 50 - 0.3 \times 15)$$

$$= -\frac{39.5}{2 \times 10^5}$$

$$\varepsilon_z = \frac{1}{E}\ [\sigma_z - \gamma\sigma_x - \gamma\sigma_y]$$

$$= \frac{1}{2 \times 10^5}\ [15 - 0.3 \times 50 + 0.3 \times 20]$$

$$= \frac{6}{2 \times 10^5}$$

$$\varepsilon_v = \varepsilon_x + \varepsilon_y + \varepsilon_z = \frac{51.5 - 39.5 + 6}{2 \times 10^5} = 9 \times 10^{-5}$$

$$\text{Volume} = L \times b \times t = 500 \times 40 \times 20$$

$$= 4 \times 10^5 \text{ mm}^3$$

$$\frac{\delta V}{V} = \varepsilon_V = 9 \times 10^{-5}$$

$$\therefore \quad \delta V = 4 \times 10^5 \times 9 \times 10^{-5} = 36$$

$$\delta L = \varepsilon_x \times L = \frac{51.5}{2 \times 10^5} \times 500 = +0.129 \text{ mm}$$

$$\delta b = \varepsilon_y \times b = -\frac{39.5}{2 \times 10^5} \times 40 = -0.0079 \text{ mm}$$

$$\delta t = \varepsilon_z t = \frac{6}{2 \times 10^5} \times 20 = +0.0006 \text{ mm}$$

9. A load of 300 kN is applied on a short concrete column 250 × 250 mm. The column is reinforced by steel bars of total area 5600 mm^2. If the modulus of elasticity of steel is 15 times that of concrete, find the stresses in concrete and steel. The stress in the concrete should not exceed 4 N/mm^2. Find the area of the required so that the column may support a load of 600 kN. (UPTU: 2001–2002)

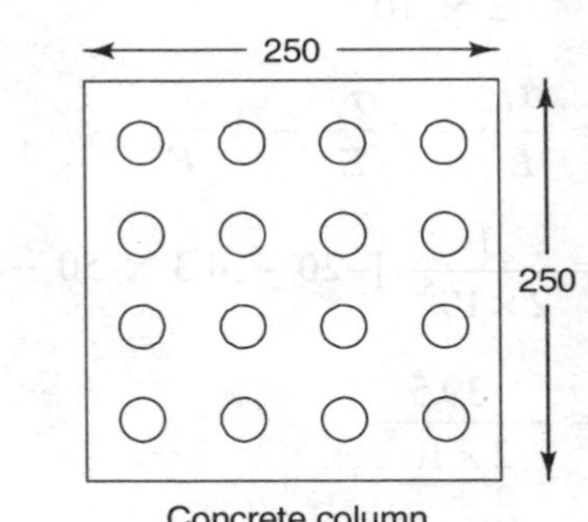

Concrete column with steel bars

Case 1: The area of steel bars = 5600 mm^2

The area of the column = 250 × 250 = 62500 mm^2

The net area of concrete = $62.5 \times 10^3 - 5.6 \times 10^3$

$= 56.9 \times 10^3$ mm^2

Strain of steel bars = Strain of concrete

$$\frac{\sigma_s}{E_s} = \frac{\sigma_c}{E_c}$$

$$P = P_s + P_c = \sigma_s A_s + \sigma_c A_c$$

$$300000 = \sigma_s \times 5600 + \frac{\sigma_s}{15} \times 56900$$

$$\therefore \quad \sigma_s = 31.94 \text{ N/mm}^2$$

$$\sigma_c = \frac{31.94}{15} = 2.13 \text{ N/mm}^2$$

Case 2:

$$\sigma_c = 4 \text{ N/mm}^2$$

$$\sigma_s = 15\ \sigma_c = 15 \times 4 = 60$$

$$P = \sigma_s A_s + 4 \times A_c$$

$$= 60\ A_s + 4\ (62500 - A_s)$$

$$600000 = 56\ A_s + 4 \times 62500$$

$$\therefore \quad A_s = 6250 \text{ mm}^2$$

10. A steel punch can sustain a compressive stress of 800 N/mm². Find the least diameter of the hole which can be punched through a steel plate of 10 mm thickness if the ultimate shear stress is 350 N/mm².

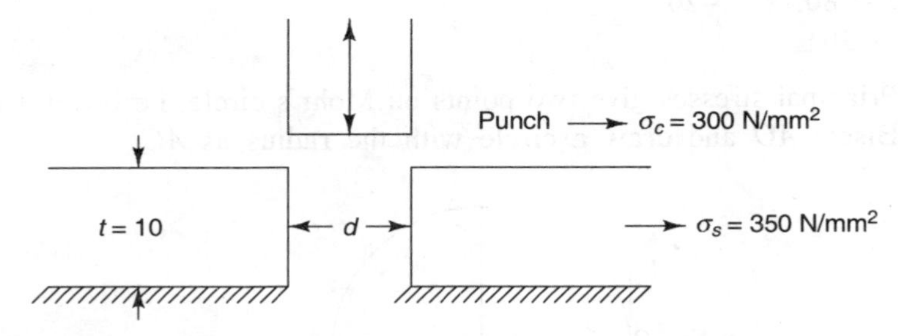

Compressive force of the punch = Shear force of the hole

$$\sigma_c \times \frac{\pi d^2}{4} = \sigma_s \times \pi d t$$

$$800 \times \frac{d}{4} = 350 \times 10$$

or

$$d = \frac{3500}{200} = 17.5 \text{ mm}$$

11. Calculate the strain energy stored in a bar which is 250 cm long, 5 cm wide and 4 cm thick, when it is subjected to a tensile load of 6 kN. Given $E = 2 \times 10^5$ N/mm².

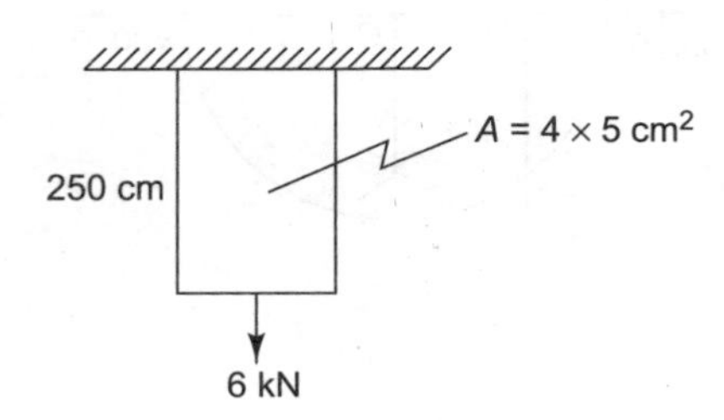

$$\sigma = \text{Stress} = \frac{6 \times 10^3}{20 \times 10^2} = 3 \text{ N/mm}^2$$

$$\text{Strain energy} = \frac{\sigma^2 AL}{2E}$$

$$= \frac{9 \times 250 \times 20 \times 10^3}{2 \times 2 \times 10^5}$$

$$= 112.5 \text{ Nmm}$$

12. Draw Mohr's circle for the following:

Case 1: Principal stresses $\sigma_1 = 80, \sigma_2 = 0$
Case 2: Principal stress $\sigma_1 = 80, \sigma_2 = -20$
Case 3: $\sigma_x = 60, \sigma_y = 10, \tau = 30$
Case 4: $\sigma_x = 80, \tau = -20$
Case 5: $\tau = 30$

Case 1: Principal stresses give two points on Mohr's circle, i.e. $A(80, 0)$ and $O(0, 0)$. Bisect AO and draw a circle with the radius as AC.

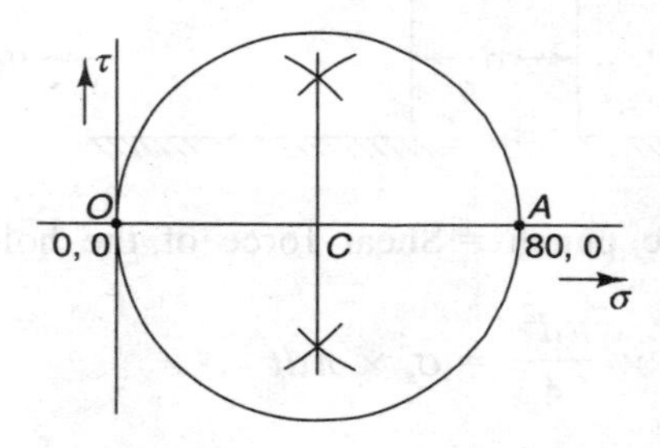

Case 2: Principal stresses give two points, i.e. $A(80, 0)$ and $B(-20, 0)$. Bisect AB and draw a circle from point C with the radius as AC.

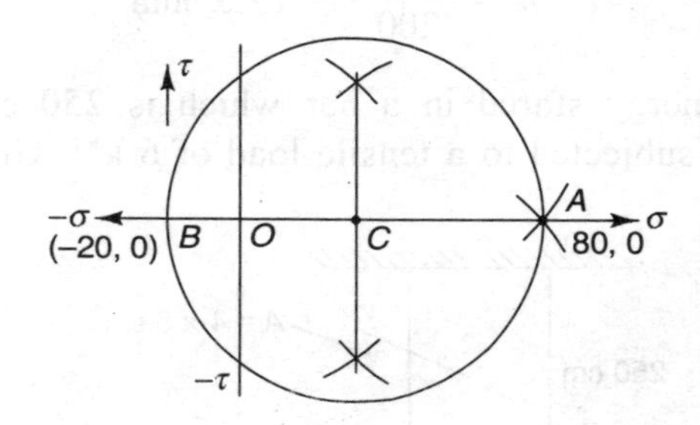

Case 3: Two points $A(60, -30)$ and $B(10, 30)$ will lie on Mohr's circle. Plot the points and join them. Line AB will cut x-axis at C. Draw a circle with the radius as AC.

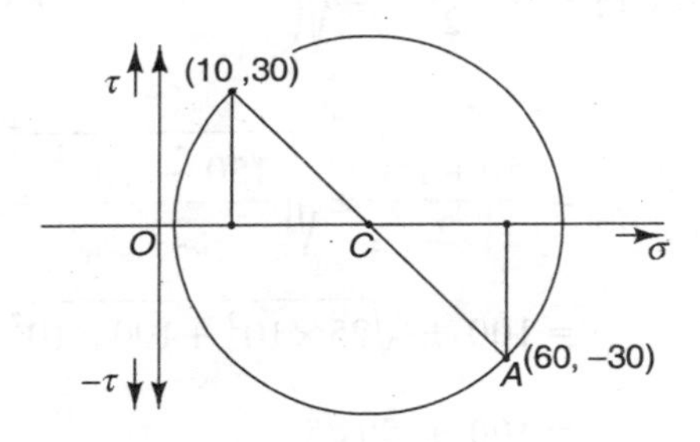

Case 4: Since σ_y is not given, $\sigma_y = 0$. Two points on Mohr's circle are $A(80, -20)$ and $B(0, 20)$. Join AB wheel.

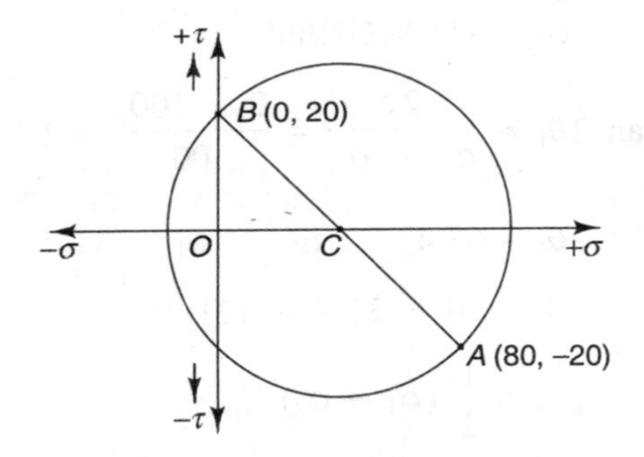

Case 5: Two points on Mohr's circle are $A(0, +30)$ and $B(0, -30)$. Draw a circle with the radius as AO.

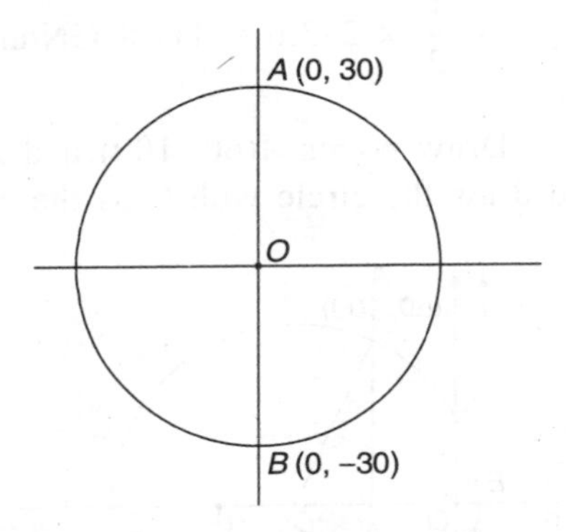

13. The state of stress at a point in a loaded component is found to be $\sigma_x = 60$ GN/m^2 (tensile), $\sigma_y = 150$ GN/m^2 (tensile) and shear force 100 GN/m^2 (clockwise). Determine the principal stresses and their planes. Also find out the maximum shear.

(UPTU: Dec. 2005)

Case 1: *Analytical method*

$$\sigma_{1 \text{ and } 2} = \frac{\sigma_x + \sigma_y}{2} \pm \sqrt{\left(\frac{\sigma_x - \sigma_y}{2}\right)^2 + \tau^2}$$

$$= \frac{50 + 150}{2} \pm \sqrt{\left(\frac{150 - 50}{2}\right)^2 + 100^2}$$

$$= 100 \pm \sqrt{25 \times 10^2 + 100 \times 10^2}$$

$$= 100 \pm 50\sqrt{5}$$

$$= 100 \pm 111.8$$

$$\therefore \quad \sigma_1 = 211.8 \text{ GN/m}^2$$

and $$\sigma_2 = -11.8 \text{ GN/m}^2$$

$$\tan 2\theta_1 = \frac{2\tau}{\sigma_x - \sigma_y} = \frac{2 \times 100}{100} = 2$$

$$2\theta_1 = 63.4° \quad \text{or} \quad \theta_1 = 31.7°$$

and $$\theta_2 = 90 + 31.7 = 121.7°$$

$$\tau_{max} = \frac{1}{2}(\sigma_1 - \sigma_2)$$

$$= \frac{1}{2}(211.8 + 11.8)$$

$$= \frac{1}{2} \times 223.6 = 111.8 \text{ GN/m}^2$$

Case 2: *Mohr's circle:* Draw points $A(60, 100)$ and $B(150, -100)$. (Scale 1 cm = 25 GN/m^2.) Join AB and draw the circle with C as the centre and radius = BC = AC.

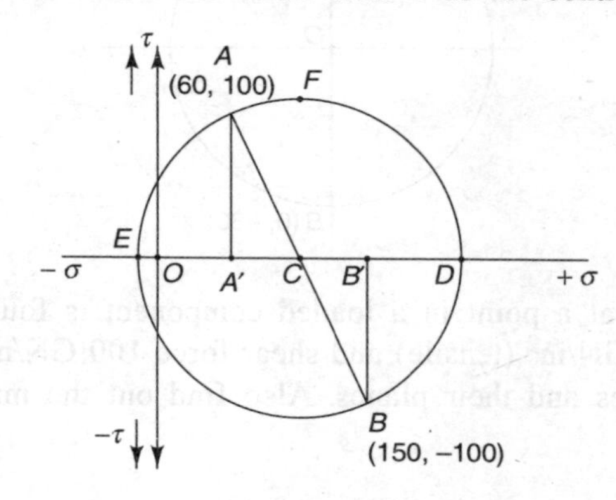

$$\text{Major principle stress} = OD = 8.472 \times 25 = 211.8 \text{ GN/m}^2$$

$$\text{Minor principle stress} = -OE = -0.472 \times 25 = -11.8 \text{ GN/m}^2$$

$$2\theta_1 = \angle ACA' = 63.4$$

$$\therefore \quad \theta_1 = 31.7°$$

$$2\theta_2 = \angle BCB' = 180 + 63.4$$

$$\therefore \quad \theta_2 = 121.7°$$

$$\tau_{max} = CF = 4.472 \times 25 = 111.8 \text{ GN/m}^2$$

14. In an elastic material the direct stresses 100 MN/m^2 and 80 MN/m^2 are applied at a certain point on planes at a right angle to each other in tension and compression respectively. Estimate the shear stress to which the material can be subjected if maximum principal stress is 130 MN/m^2. Also find the magnitude of other principal stress and its inclination to 100 MN/m^2 stress. (UPTU: Dec. 2005)

Given $\sigma_1 = 130, \sigma_x = 100, \sigma_y = -80$

$$\sigma_1 = \frac{\sigma_x + \sigma_y}{2} + \sqrt{\left(\frac{\sigma_x - \sigma_y}{2}\right)^2 + \tau^2}$$

$$130 = \frac{100 - 80}{2} + \sqrt{\left(\frac{100 + 80}{2}\right)^2 + \tau^2}$$

$$130 = 10 + \sqrt{(90)^2 + \tau^2}$$

$$\tau^2 + 90^2 = 120^2$$

or
$$\tau^2 = 120^2 - 90^2 = 210 \times 30$$

or
$$\tau = 79.37 \text{ MN/m}^2$$

$$\sigma_2 = \frac{\sigma_x + \sigma_y}{2} - \sqrt{\left(\frac{\sigma_x - \sigma_y}{2}\right)^2 + \tau^2}$$

$$= 10 - 120 = -110 \text{ MN/m}^2$$

$$\tan \theta_1 = \frac{2 \times \tau}{\sigma_x - \sigma_y} = \frac{2 \times 79.37}{20} = 7.937$$

or
$$\theta_1 = 81.8°$$

Therefore, θ_2 (inclination of the minor Principal plane) = 90 + 81.8 = 171.8.

Note: The formulas for compound stresses can be easily remembered with Mohr's circle.

(a) The centre of Mohr's circle $= \dfrac{\sigma_x + \sigma_y}{2}$

(b) Radius of Mohr's circle $= \sqrt{\left(\dfrac{\sigma_x - \sigma_y}{2}\right)^2 + \tau^2}$

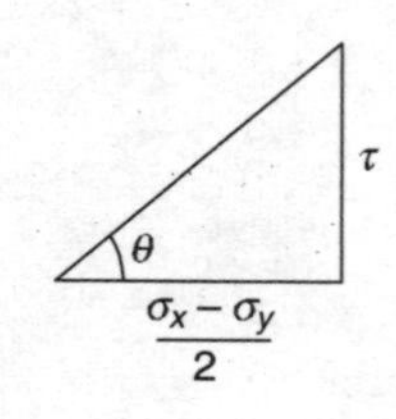

(c) Principal stresses = centre ± radius

$$= \frac{\sigma_x - \sigma_y}{2} \pm \sqrt{\left(\frac{\sigma_x - \sigma_y}{2}\right) + \tau^2}$$

(d) $\tan 2\theta_1 = \dfrac{\tau}{\left(\dfrac{\sigma_x - \sigma_y}{2}\right)}$

15. A piece of steel plate is subjected to a perpendicular stress of 50 N/mm² tensile and compression as shown in the figure. Calculate the normal and shear stresses at a plane making 45°.

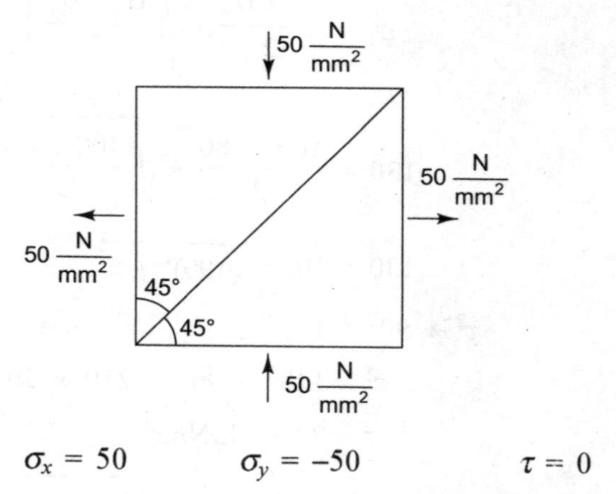

$\sigma_x = 50 \qquad \sigma_y = -50 \qquad \tau = 0$

Draw Mohr's circle with point $A(50, 0)$ and $B(-50, 0)$. Take scale 1 cm = 25 N/mm² and draw Mohr's circle with radius = OA.

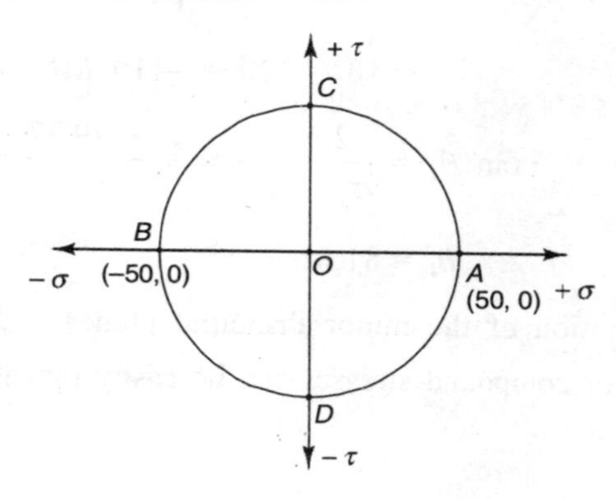

Plane 45° to σ_y is OD and stresses are as follows:
(a) Normal stress = 0
(b) Shear stress = –50 N/mm^2

16. The principal stresses at a point in a strained material are P_1 and P_2. Show that the resultant stress P_r is given by

$$P_r = \left[\frac{P_1^2 + P_2^2}{2}\right]^{\frac{1}{2}}$$

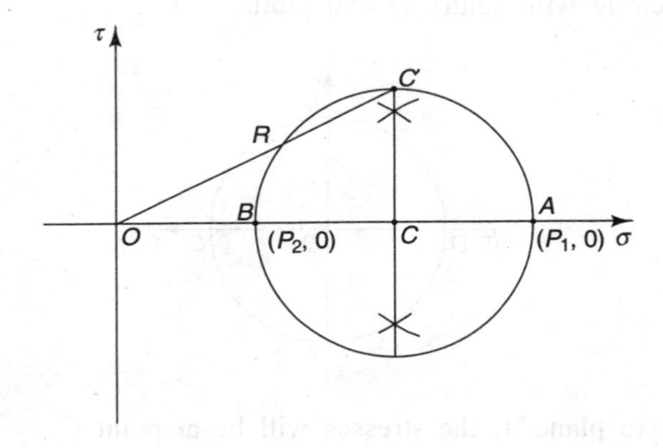

Draw a Mohr's circle with points $A(P_1, 0)$ and $B(P_2, 0)$. Centre $= \dfrac{P_1 + P_2}{2}$ and radius $= \dfrac{P_1 - P_2}{2}$.

$$OC = \frac{P_1 + P_2}{2}$$

$$CC' = \text{radius} = \frac{P_1 - P_2}{2}$$

$$\therefore \qquad R^2 = OC^2 + CC' = \left(\frac{P_1 + P_2}{2}\right)^2 + \left(\frac{P_1 - P_2}{2}\right)^2$$

$$P_r = \sqrt{\frac{P_1^2 + P_2^2}{2}}$$

17. Using Mohr's circle, derive the expression for normal and tangential stress on a diagonal plane of a material subjected to pure shear. (UPTU: May 2001)

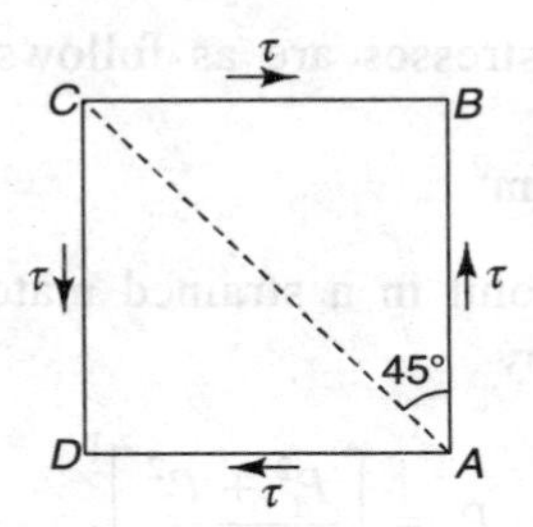

Here $\sigma_x = 0$ and $\sigma_y = 0$. Therefore, two point $A(0, -\tau)$ and $B(0, +\tau)$ lie on Mohr's circle. Draw a circle with centre O and radius $= \tau$.

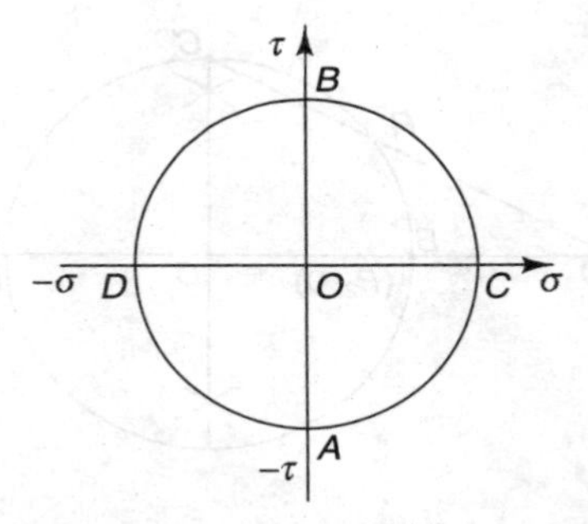

For a plane 45° to plane A, the stresses will be at point C.

(a) Normal stress $= OC = \tau$

(b) Shear stress $= 0$

18. A uniform steel bar of 2 cm × 2 cm is subjected to an axial pull of 4000 kg. Calculate the intensity of normal stress, shear stress and resultant stress on a plane, the normal to the plane being inclined at 30° to the axis of the bar. Solve the problem graphically by drawing Mohr's circle.

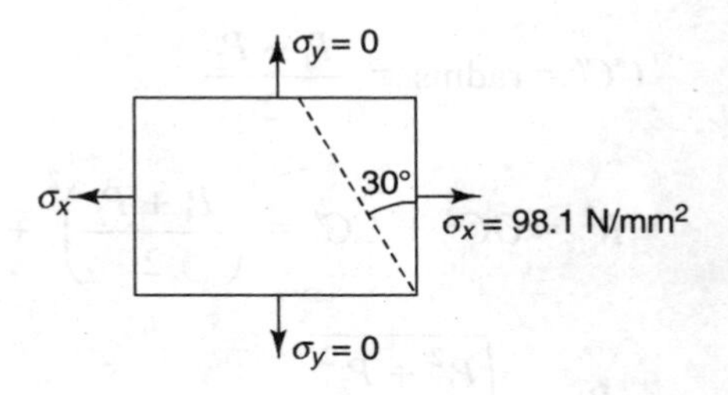

$$\sigma_x = \frac{4000 \times 9.81}{2 \times 2 \times 100} = 10 \times 9.81 = 98.1 \text{ N/mm}^2$$

Draw Mohr's circle with points $A(98.1, 0)$ and $B(0, 0)$. Centre $= \dfrac{98.1}{2} = 49.05$ and radius $= 49.05$. Take scale 1 cm $= 20$ N/mm^2.

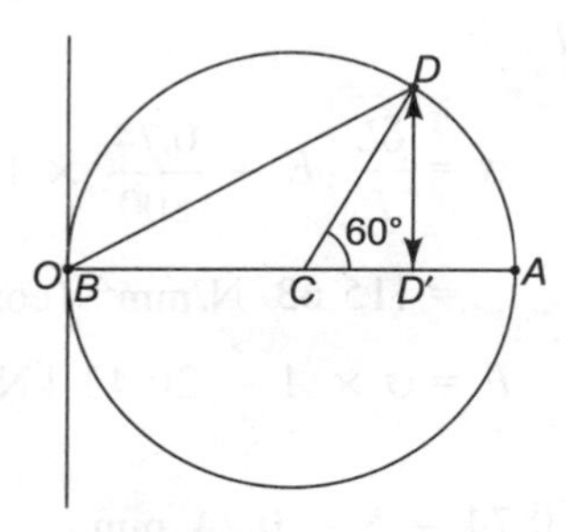

The plane inclined at 30° to the plane at A is on 60° line from CA at point D. Then stresses at D are:

(a) Normal = OD' = 3.7 cm = 74 N/mm^2
(b) Shear = DD' = 2.1 cm = 42 N/mm^2
(c) Resultant = OD = 4.3 cm = 86 N/mm^2

19. Find the principal stress for the state of stress given below:

$\sigma_x = 100$ N/mm^2, $\sigma_y = 0$ and $\tau_{xy} = 50$ N/mm^2 (UPTU: July 2002)

Draw Mohr's circle with points $A(100, 50)$ and $B(0, -50)$. Take scale 1 cm = 20 N/mm^2.

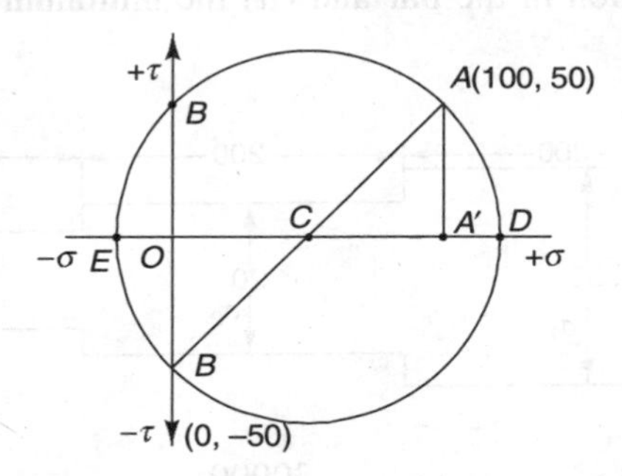

Join A and B which cuts x-axis at point C. With centre C and radius = BC, draw the circle. Principal stresses are:

(a) Major = OD = 6.1 cm ≈ 120.2 N/mm^2 (Tensile)
(b) Minor = OE = 1.05 cm = − 21 N/mm^2 (Compressive)

20. A copper rod of 15 mm diameter and 0.8 m long is heated through 50°C. What is its expansion when it is free to expand? Suppose the expansion is prevented by gripping it at both ends, find the stress, its nature and the force applied by the grips when

(a) the grips do not yield
(b) one grip yield back by 0.5 mm

Take $\alpha_c = 18.5 \times 10^{-6}$ per °C and $E_c = 1.25 \times 10^5$ N/mm^2

$$\text{Free expansion } \delta L = \alpha \Delta T L$$
$$= 18.5 \times 10^{-6} \times 50 \times 800$$
$$= 0.74 \text{ mm}$$

Case 1: *Grips do not yield*

$$\sigma = \frac{\delta L}{L} \cdot E = \frac{0.74}{800} \times 1.25 \times 10^5$$

$$= 115.63 \text{ N/mm}^2 \text{ (compressive)}$$

$$F = \sigma \times A = 20.43 \text{ kN (compression)}$$

Case 2: *Grip yield* = 0.5.

Net contraction = 0.74 – 5 = 0.24 mm

$$\sigma = \frac{0.24}{800} \times E = \frac{0.24}{800} \times 1.25 \times 10^5$$

$$= 37.5 \text{ N/mm}^2$$

$$F = \sigma \times A = 6.63 \text{ kN (compressive)}$$

21. A 550 mm long bar of variable cross section is subjected to an axial pull of 30000 N. The diameter of the bar is 30 mm over a length of 200 mm, 20 mm over another length of 200 mm and 10 mm over the remaining length of 150 mm. If E = 100 kN/mm^2, determine (i) the elongation in the bar and (ii) the minimum and maximum stresses set up in the bar. (UPTU: Dec. 2005)

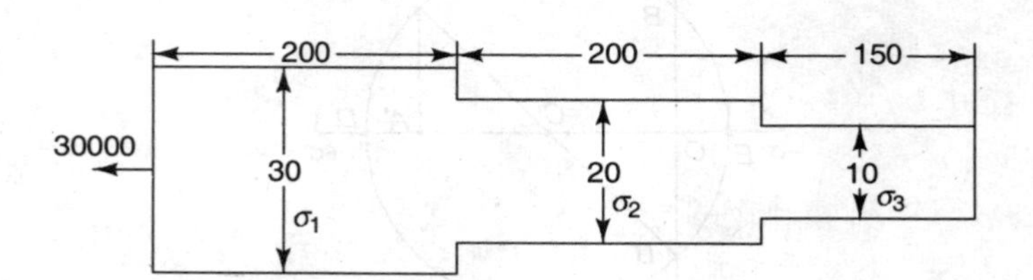

$$\sigma_1 = \frac{30000}{\frac{\pi \times 30^2}{4}} \qquad \sigma_2 = \frac{30000}{\frac{\pi \times 20^2}{4}} \qquad \sigma_3 = \frac{30000}{\frac{\pi \times 10^2}{4}}$$

$$\sigma_1 = \frac{30000}{706.5} \qquad \sigma_2 = \frac{30000}{314.15} \qquad \sigma_3 = \frac{30000}{78.5}$$

$$\sigma_1 = 42.43 \frac{\text{N}}{\text{mm}^2} \qquad \sigma_2 = 95.5 \frac{\text{N}}{\text{mm}^2} \qquad \sigma_3 = 382.16 \frac{\text{N}}{\text{mm}^2}$$

Therefore,

$$\sigma_{max} = 382.16 \text{ N/mm}^2$$

and

$$\sigma_{min} = 42.45 \text{ N/mm}^2$$

$$\delta l_1 = \frac{200 \times 42.43}{100 \times 10^{+3}} = 0.085 \text{ mm}, \ \delta l_2 = 0.191 \text{ mm}, \ \delta l_3 = 0.766 \text{ mm}$$

Therefore

$$\delta l = \delta l_1 + \delta l_2 + \delta l_3 = 1.04 \text{ mm}$$

22. The normal stress at a point are $\sigma_x = 10$ MPa and $\sigma_y = 2$ MPa, the shear stress at this point is 4 MPa. The maximum principal stress at this point is

(a) 16 MPa (b) 14 MPa
(c) 11 MPa (d) 10 MPa (GATE: 1998)

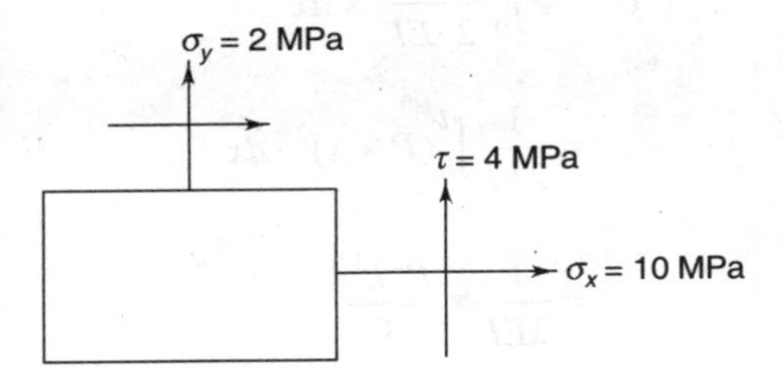

$$\sigma_1 = \frac{\sigma_x + \sigma_y}{2} + \sqrt{\left(\frac{\sigma_x - \sigma_y}{2}\right)^2 + \tau^2}$$

$$= \frac{10+2}{2} + \sqrt{\left(\frac{10-2}{2}\right)^2 + 4^2}$$

$$= 11.66 \text{ MPa}$$

Hence the option (a) is correct.

23. A square bar of side 4 cm and length 100 cm is subjected to an axial load P. The same bar is then used as a cantilever beam and subjected to an end load P. The ratio of the strain energies stored in the bar in the second case to that stored in the first case is

(a) 16 (b) 400
(c) 1000 (d) 2500 (GATE: 1998)

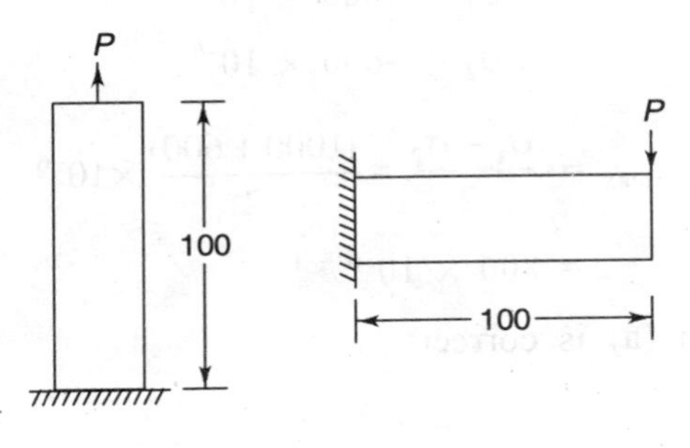

$$U_1 = \frac{1}{2} \text{ force} \times \text{elongation}$$

$$= \frac{1}{2} \times P \times \frac{PL}{AE}$$

$$= \frac{P^2 \times 100}{2 \times 4 \times 4} = \frac{25\,P^2}{8E}$$

$$U_2 = \int_0^l \frac{1}{2} \frac{M^2}{EI} \times dx$$

$$= \frac{1}{2EI} \int_0^l (P \times x)^2 \cdot dx$$

$$= \frac{1}{2EI} \times \frac{P^2 L^3}{3}$$

$$= \frac{P^2 \times 100^3}{2 \times 3 \times E \times \frac{4 \times 4^3}{12}}$$

$$= \frac{15625 P^2}{2E}$$

$$\frac{U_2}{U_1} = \frac{15625}{2} \times \frac{8}{25} = 2500$$

Hence the option (d) is correct.

24. If two principal strains at a point are 1000×10^{-6} and -600×10^{-6}, then the maximum shear strain is

(a) 800×10^{-6} (b) 500×10^{-6}

(c) 1600×10^{-6} (d) 200×10^{-6}

(GATE: 1996)

$$\sigma_1 = 1000 \times 10^{-6}$$

$$\sigma_2 = -600 \times 10^{-6}$$

$$\tau_{max} = \frac{\sigma_1 - \sigma_2}{2} = \frac{(1000 + 600)}{2} \times 10^{-6}$$

$$= 800 \times 10^{-6}$$

Therefore, the option (a) is correct.

25. The stress and strain behaviour of a material is shown in the figure. Its resilence and toughness in N m/m^3 are respectively:

(a) 28×10^4, 76×10^4 (b) 28×10^4, 48×10^4
(c) 14×10^4, 90×10^4 (d) 76×10^4, 104×10^4 (GATE: 2000)

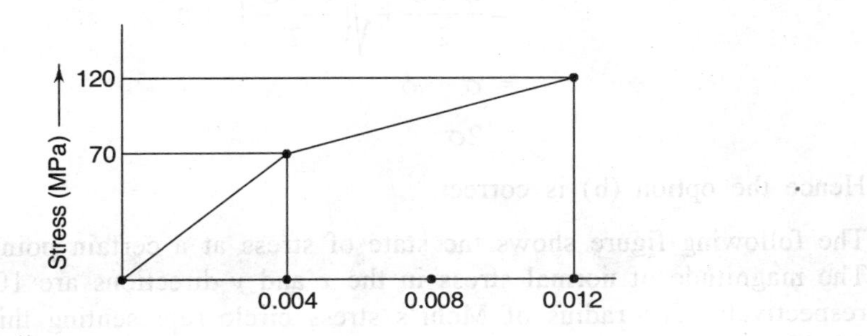

$$\text{Resilence} = \frac{1}{2} \times \text{stress} \times \text{strain}$$

$$= \frac{1}{2} \times 0.004 \times 70 \times 10^6$$

$$= 14 \times 10^4 \text{ N m/m}^3$$

$$\text{Toughness} = \text{Total area of the diagram}$$

$$= 14 \times 10^4 + \frac{1}{2}(120 + 70)(0.012 - 0.004) \times 10^6$$

$$= 14 \times 10^4 + 95 \times 0.008 \times 10^6$$

$$= 14 \times 10^4 + 76 \times 10^4$$

$$= 90 \times 10^4$$

Hence the option (c) is correct.

26. The maximum principal stress for the stress state as shown in the figure

(a) σ (b) 2σ
(c) 3σ (d) $1.5\ \sigma$ (GATE: 2001)

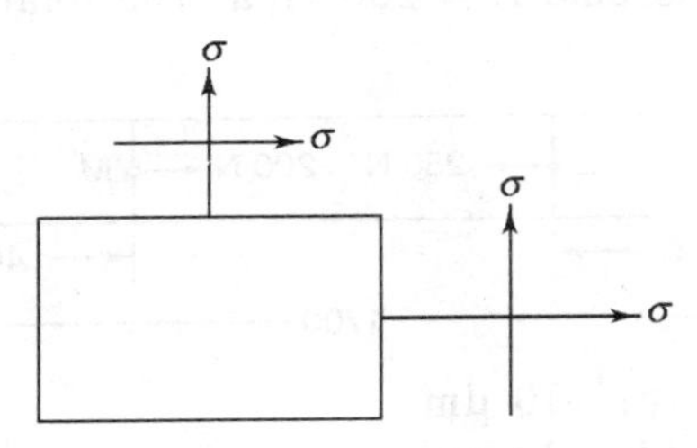

$$\sigma_1 = \frac{\sigma_x + \sigma_y}{2} + \sqrt{\left(\frac{\sigma_x - \sigma_y}{2}\right)^2 + \tau^2}$$

$$= \frac{\sigma + \sigma}{2} + \sqrt{\left(\frac{\sigma - \sigma}{2}\right)^2 + \sigma^2}$$

$$= \sigma + \sigma$$

$$= 2\sigma$$

Hence the option (b) is correct.

27. The following figure shows the state of stress at a certain point in a stressed body. The magnitude of normal stress in the x and y directions are 100 MPa and 20 MPa respectively. The radius of Mohr's stress circle representing this state of stress is

(a) 120 (b) 80
(c) 60 (d) 40 (GATE: 2004)

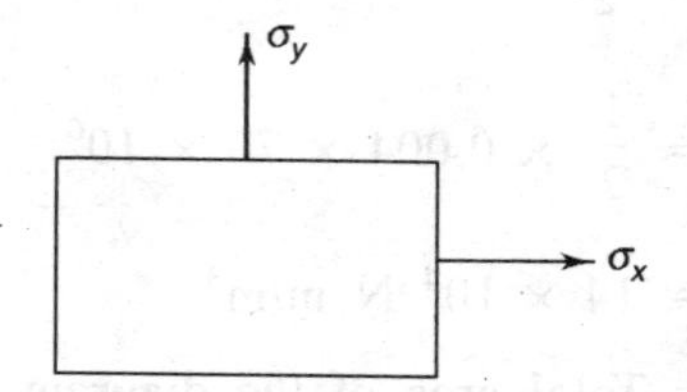

$$\text{Radius} = \sqrt{\left(\frac{\sigma_x - \sigma_y}{2}\right)^2 + \tau^2}$$

$$= \sqrt{\left(\frac{100 - 20}{2}\right)^2 + 0}$$

$$= 40$$

Therefore, the option (d) is correct.

28. The figure below shows a steel rod of 25 mm^2 cross-sectional area. It is loaded at four points K, L, M and N. Assume E = 200 GPa. The total change in length of the rod due to loading is

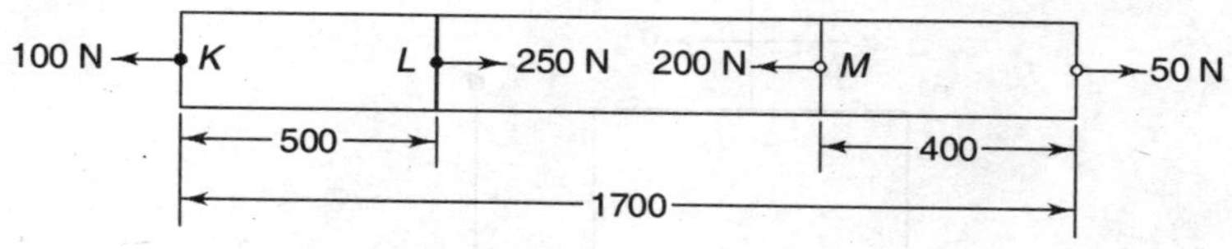

(a) 1 μm (b) −10 μm
(c) 16 μm (d) −20 μm (GATE: 2004)

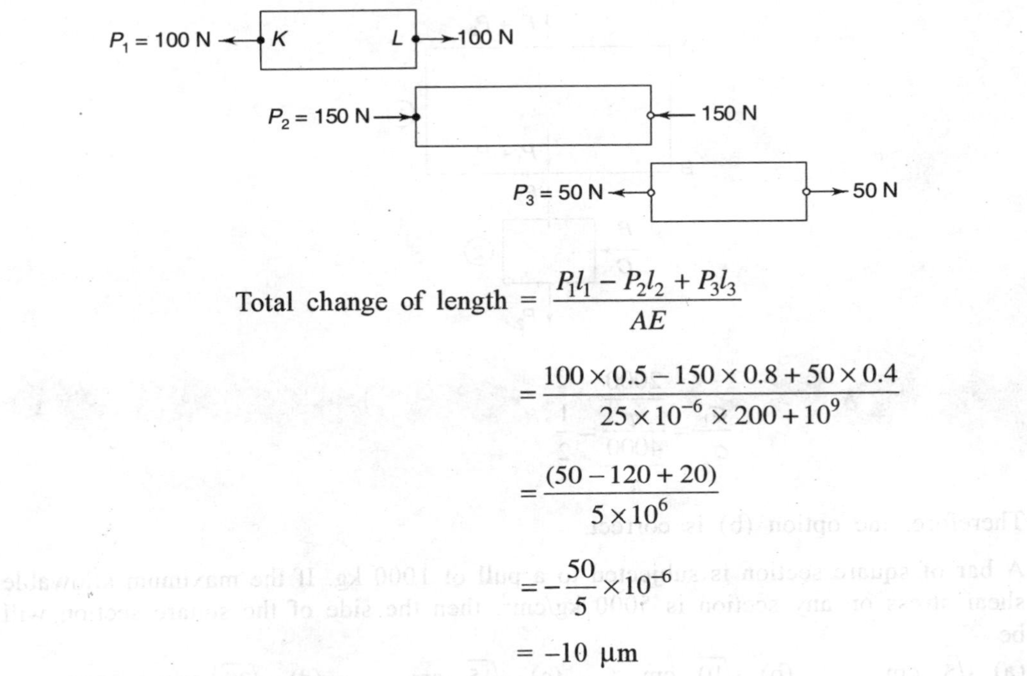

$$\text{Total change of length} = \frac{P_1 l_1 - P_2 l_2 + P_3 l_3}{AE}$$

$$= \frac{100 \times 0.5 - 150 \times 0.8 + 50 \times 0.4}{25 \times 10^{-6} \times 200 + 10^9}$$

$$= \frac{(50 - 120 + 20)}{5 \times 10^6}$$

$$= -\frac{50}{5} \times 10^{-6}$$

$$= -10 \ \mu\text{m}$$

Hence the option (b) is correct.

29. A mild steel bar is 40 cm long. The lengths of part AB and BC of the bar are 20 cm each. It is loaded as shown in the figure. The ratio of the stresses σ_1 in part AB to σ_2 in part BC is P_1 = 1000 kg, P_2 = 1000 kg.

(a) 2 (b) $\frac{1}{2}$ (c) 4 (d) $\frac{1}{4}$

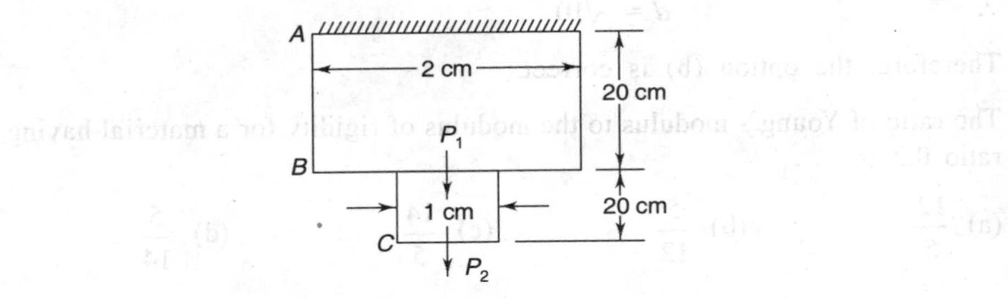

$$\text{Length } BC: \quad \sigma_2 = \frac{P_2}{A_2} = \frac{P_2}{\frac{\pi \times 1^2}{4}} = \frac{4P_2}{\pi} = \frac{4000}{\pi}$$

$$\text{Length } AB: \quad \sigma_1 = \frac{P_1 + P_2}{A_1} = \frac{2000}{\frac{\pi \times 2^2}{4}} = \frac{2000}{\pi}$$

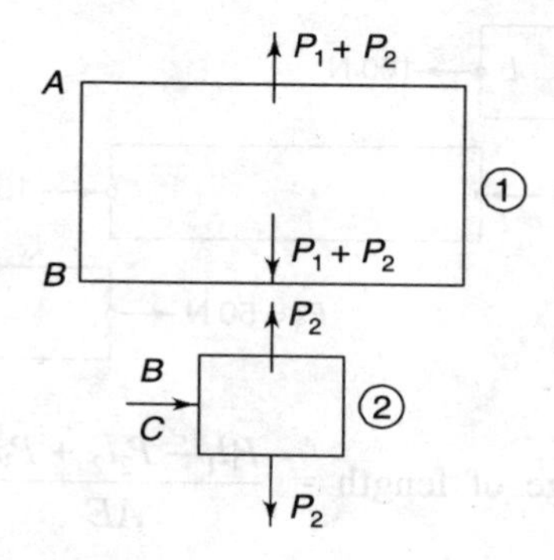

$$\frac{\sigma_1}{\sigma_2} = \frac{\frac{2000}{\pi}}{\frac{4000}{\pi}} = \frac{1}{2}$$

Therefore, the option (b) is correct.

30. A bar of square section is subjected to a pull of 1000 kg. If the maximum allowable shear stress on any section is 5000 kg/cm^2, then the side of the square section will be

(a) $\sqrt{5}$ cm (b) $\sqrt{10}$ cm (c) $\sqrt{15}$ cm (d) $\sqrt{20}$ cm

$$\text{Force} = 1000$$

$$\text{Area} = d \times d = d^2$$

$$\text{Stress} = \frac{1}{2} \times \frac{1000}{d^2} = 5{,}000 \text{ kg/cm}^2$$

$$\therefore \quad d = \sqrt{10}$$

Therefore, the option (b) is correct.

31. The ratio of Young's modulus to the modulus of rigidity for a material having Poisson's ratio 0.2 is

(a) $\frac{12}{5}$ (b) $\frac{5}{12}$ (c) $\frac{14}{5}$ (d) $\frac{5}{14}$

$$E = 2G(1 + v)$$

$$\frac{E}{G} = 2(1 + v)$$

$$= 2(1 + 0.2)$$

$$= 2.4 = \frac{24}{10} = \frac{12}{5}$$

Therefore, the option (a) is correct.

32. A member is formed by connecting a steel bar to an aluminium bar. Find the magnitude of P that will cause contraction of 0.50 mm.

Assume $E_s = 2 \times 10^{11}$ N/m^2 and $E_{A1} = 0.7 \times 10^{11}$ N/m^2.

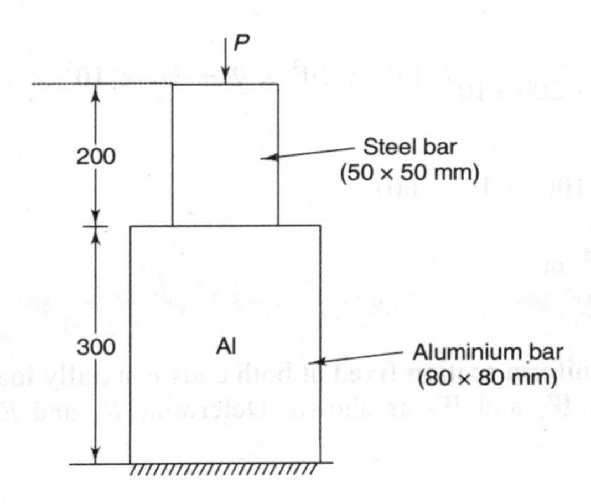

Using the principle of superposition,

$$\delta l = P\left(\frac{l_s}{A_s E_s} + \frac{l_{AL}}{A_{Al} E_{Al}}\right)$$

$$0.50 = P\left(\frac{200}{2500 \times 2 \times 10^5} + \frac{300}{6400 \times 0.7 \times 10^5}\right)$$

$$P = 467.29 \text{ kN}$$

33. Determine the stress in all the three sections and total deformation of the steel rod shown in the figure. Cross-sectional area = 10 cm^2, E = 200 GN/m^2.

(UPTU: 2006–2007)

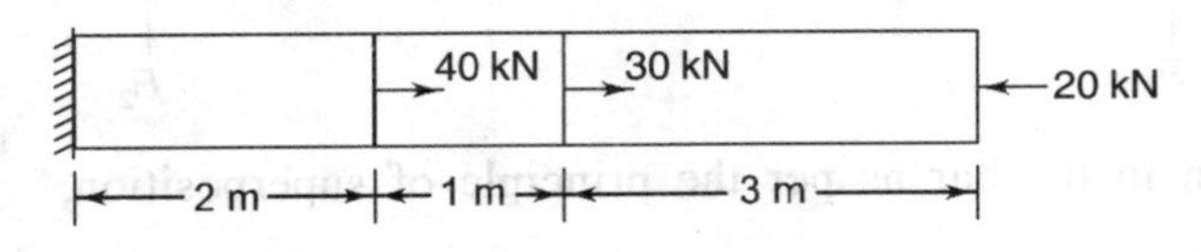

Draw the free body diagram of three sections:

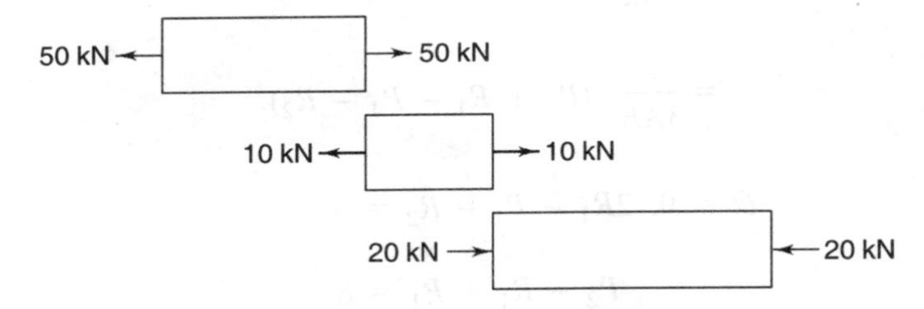

Using the principal of superposition,

$$\delta l = \frac{1}{AE}(P_1l_1 + P_2l_2 - P_3l_3)$$

$$= \frac{1}{10 \times 10^{-4} \times 200 \times 10^9}(50 \times 10^3 \times 2 + 10 \times 10^3 \times 1 - 20 \times 10^3 \times 3)$$

$$= \frac{10^3}{2 \times 10^8}(100 + 10 - 60)$$

$$= 25 \times 10^{-5} \text{ m}$$

$$= 0.25 \text{ mm}$$

34. A vertical bar of uniform section fixed at both ends is axially loaded at two intermediate sections by forces W_1 and W_2 as shown. Determine R_1 and R_2 when $P_1 = 4500$ and $P_2 = 5000$.

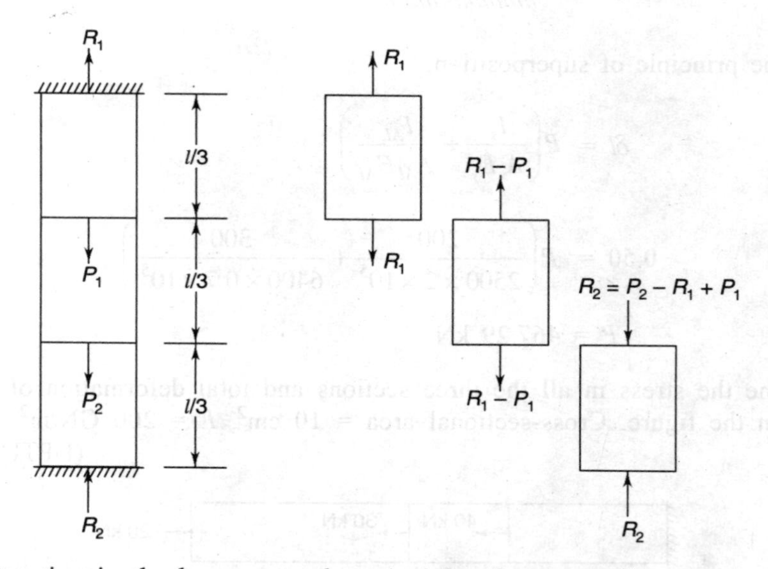

Contraction in the bar as per the principle of superposition,

$$\delta l = \frac{1}{AE}[R_1l_1 + (R_1 - P_1)l_2 - R_2l_3]$$

$$= \frac{l}{3AE}(R_1 + R_1 - P_1 - R_2)$$

But

$$\delta l = 0,\ 2R_1 - P_1 - R_2 = 0 \quad \text{(i)}$$

Also

$$P_2 - R_1 + P_1 = R_2 \quad \text{(ii)}$$

From Eqs. (i) and (ii),

$$R_1 = \frac{2P_1 + P_2}{3} = \frac{9000 + 3000}{3} = 4000 \text{ N}$$

$$R_2 = 4000 + 3500 - 4000$$

$$= 3500 \text{ N}$$

35. A bar ABC is fixed and loaded as shown. Determine reactions at ends and stresses produced in two sections. Assume $E = 2 \times 10^{17}$ N/cm^2.

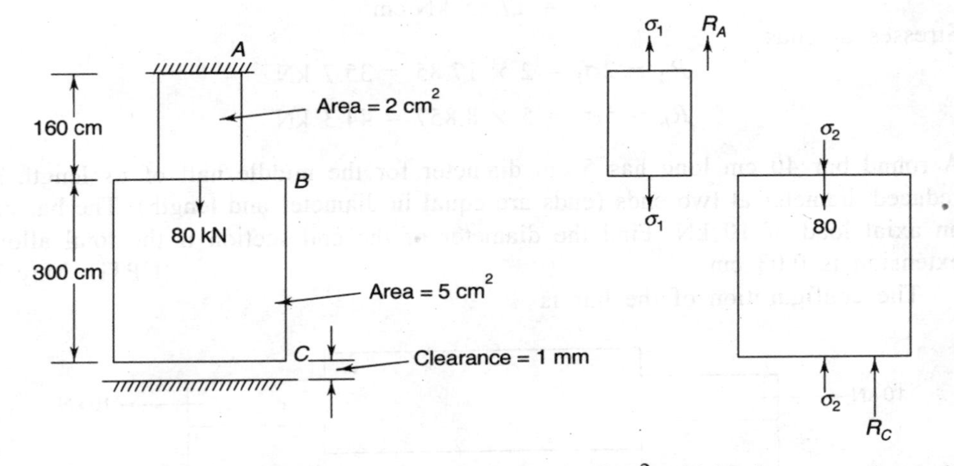

Let stresses in AB and BC are σ_1 and σ_2 in kN/cm^2 which gives net elongation of 1 mm. Using the principle of superposition

$$\delta l = 1 \times 10^{-2} = \delta_1 - \delta_2$$

$$1 \times 10^{-2} = \frac{\sigma_1}{E} \times l_1 - \frac{\sigma_2 \times l_2}{E}$$

$$= \frac{1}{E}(160\ \sigma_1 - 300\ \sigma_2)$$

or

$$\sigma_1 - 1.875\ \sigma_2 = 0.125 \times 10$$

$$= 1.25 \qquad \text{(i)}$$

Let reaction R_A and R_C at end A and end C. Then,

$$\text{reaction } R_A = \sigma_1 \times \text{area} = \sigma_1 \times 2.0 = 2\sigma_1$$

$$\text{reaction } R_C = \sigma_2 \times \text{area} = \sigma_2 \times 5.0 = 5\sigma_2$$

Now

$$R_A + R_C = 80$$

$$2\sigma_1 + 5\sigma_2 = 80$$

$$\sigma_1 + 2.5\sigma_2 = 40$$

Putting the value of σ_1 from Eq. (i),

$$1.875\sigma_2 + 2.5\sigma_2 + 1.25 = 40$$

or

$$\sigma_2 = \frac{38.75}{4.375} = 8.857 \text{ kN/cm}^2$$

and

$$\sigma_1 = 17.85 \text{ kN/cm}^2$$

Stresses at ends

$$R_A = 2\sigma_1 = 2 \times 17.85 = 35.7 \text{ kN}$$

$$R_B = 5\sigma_2 = 5 \times 8.857 = 44.3 \text{ kN}$$

36. A round bar 40 cm long has 5 cm diameter for the middle half of its length and a reduced diameter at two ends (ends are equal in diameter and length). The bar carries an axial load of 10 kN. Find the diameter of the end section if the total allowable extension is 0.03 cm. (UPTU: May 2008)

The configuration of the bar is

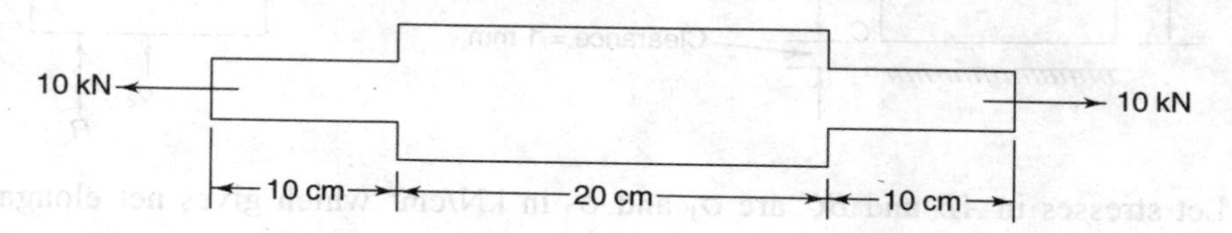

Using the principal of superimposition

$$\Delta L = \frac{F}{E}\left(\frac{l_1}{A_1} + \frac{l_2}{A_2} + \frac{l_3}{A_3}\right)$$

$$0.03 \times 10^{-2} = \frac{10 \times 10^3}{200 \times 10^9}\left(\frac{10 \times 10^{-2}}{A} + \frac{20 \times 10^{-2}}{\frac{\pi \times (0.05)^2}{4}} + \frac{10 \times 10^{-2}}{A}\right)$$

$$= 5 \times 10^{-8} \times 10^{-2} \times 20\left(\frac{1}{A} + \frac{4}{\pi \times 25 \times 10^{-4}}\right)$$

$$3 \times 10^4 = \frac{1}{A} + 509.55$$

$$\frac{1}{A} = 29.49 \times 10^{-3}$$

or $$A = \frac{1}{29.49 \times 10^{-3}}$$

$$\frac{\pi d^2}{4} = \frac{1 \times 10^{-3}}{29.49}$$

or $$d^2 = \frac{4 \times 1 \times 10^{-3}}{29.43 \times \pi}$$

$$= 43.28 \times 10^{-6}$$

or $$d = 6.58 \times 10^{-3} \text{ m}$$

$$= 6.58 \text{ mm}$$

OBJECTIVE TYPE QUESTIONS

Don't jump to conclusions. There are usually not only two sides to every story, but three, four, or more. Give others the benefit of the doubt.

State True or False

1. For elastoplastic materials, deformation disappears fully on removal of the load. (*True/False*)

2. Elastic material comes back to the same length on removal of the load within the elastic limit. (*True/False*)

3. Lack of ductility is brittleness. (*True/False*)

4. Ductile materials can snap off without sufficient warning by elongation. (*True/False*)

5. Toughness is the strength with which the material opposes rupture. (*True/False*)

6. Molecular attraction gives the material the toughness. (*True/False*)

7. The converse of hardness is softness. (*True/False*)

8. The strength of the material does not change under fatigue (repeated loads). (*True/False*)

9. The material does not fail at prolonged lower load at higher temperature (creep's failure). (*True/False*)

10. Mohr's circle is a normal stress and shear stress diagram. (*True/False*)

11. The angle of planes is the same as on Mohr's circle. (*True/False*)

12. Clockwise shear stress is taken negative on Mohr's circle. (*True/False*)

13. Tensile stress is taken negative on Mohr's circle. (*True/False*)

14. Principal planes are the planes having zero shear stress. (*True/False*)

15. The modulus of resilence for shear stress $= \dfrac{\tau_{max}^2}{2G}$ (*True/False*)

16. The modulus of resilence for normal stress $= \dfrac{\sigma_{max}^2}{2E}$ (*True/False*)

17. The maximum stress is also called a proof stress. (*True/False*)

18. Poisson's ratio is the ratio of longitudinal strain to lateral strain. (*True/False*)

19. For axial loading, the maximum shear stress is half the normal stress if the plane is at 45° to the axis. (*True/False*)

20. A pair of mutually perpendicular planes each under a shear stress (τ) produce normal stresses of opposite nature equal in magnitude to shear stress on two mutually perpendicular planes at an angle of 45°. (*True/False*)

21. Complementary shear stresses are balancing stresses. (*True/False*)

22. If an applied shear stress is 25 kN/mm^2 (clockwise), then the complementary shear stress is also 25 kN/mm^2 (clockwise). (*True/False*)

23. The modulus of rigidity is the ratio of shear stress to shear strain. (*True/False*)

24. The volumetric strain is change in volume to the original volume. (*True/False*)

25. $\varepsilon_v = \varepsilon(1 + 2\gamma)$ when ε_v = volumetric strain, ε = longitudinal strain and γ = Poisson's ratio. (*True/False*)

26. The bulk modulus is the ratio of volumetric strain to normal stress. (*True/False*)

27. $E = 3k(1 - 2\gamma)$ where E = Young's modulus, k = bulk modulus and γ = Poisson's ratio. (*True/False*)

28. $E = 2G(1 - \gamma)$ where G = Young's modulus, G = modulus of rigidity and γ = Poisson's ratio. (*True/False*)

29. If a punch has a diameter of 50 mm and the maximum shear stress of steel is 100 kN/mm, then the punch must have a force of 31.4 MN to punch a hole in a 2 mm thick plate of steel. (*True/False*)

30. On Mohr's circle, the angle between the resultant and the normal stress axis is called a shear angle. (*True/False*)

31. The normal stress for sudden loading is twice the normal stress for slowly applied loading of the same magnitude. (*True/False*)

32. If a bar is held between two fixed supports and cooled, there will be a tensile stress in the bar. (*True/False*)

33. If a specimen held by two grips is heated, there will be a compressive stress in the specimen. (*True/False*)

34. A gap is left between two rail lengths to counter misalignment. (*True/False*)

35. A truss has one end supported on the roller support to counter the wind load. (*True/False*)

36. If a composite bar has a higher coefficient of linear expansion material at the top than at the bottom, then the composite bar will bend downwards. (*True/False*)

37. In a reinforced concrete, steel bars would take a more tensile or bending load as compared to the concrete. (*True/False*)

38. In a composite bar, the strain in both materials will be the same. (*True/False*)

39. A strained hardened specimen will have a lower elastic limit. (*True/False*)

40. Compression tests are performed on brittle materials. (*True/False*)

41. Shear failure in brittle materials occurs along the 45°-shear plane. Wood is a brittle material and hence it would have shear failure along the 45°-shear plane. (*True/False*)

42. Bimodulus materials have different values of Young's modulus in tension and compression. (*True/False*)

43. Mild steel has carbon content in the range of 0.1 to 0.3%. *(True/False)*
44. A brittle material can sustain more impact. *(True/False)*
45. Wrought iron can absorb maximum strain energy and cast iron can absorb minimum strain energy. *(True/False)*
46. Surface scratches are the sources of stress concentration. A specimen having scratches develops more stresses due to stress concentration resulting into poor impact energy. *(True/False)*
47. Rupture energy and notch impact strength increase with a rise in temperature. *(True/False)*
48. In the Charpy test for finding impact strength, the specimen is placed as a cantilever beam. *(True/False)*
49. In the Izod test for finding impact strength, the specimen is placed as simply supported beam. *(True/False)*
50. The specimen has a V-shaped notch of 45° and located at the tension side during impact testing. *(True/False)*
51. Diamond is the hardest known material. *(True/False)*
52. Brinell hardness can be accurately carried out for the materials above 400 BHN. *(True/False)*
53. Elongation in a tapered bar = $\frac{1}{6} \times \frac{\rho g}{E} \times l^2$ due to its weight. *(True/False)*
54. Elongation in a uniform bar = $\frac{\rho g}{2E} \times l^2$ due to its weight. *(True/False)*
55. A body can elongate due to its own weight. *(True/False)*
56. Compressive stress and strain diagrams will be in the second quadrant. *(True/False)*
57. For ductile material, the yield shear stress will be half the normal tensile stress yield. *(True/False)*
58. The tensile elastic limit will be more than the compressive elastic limit. *(True/False)*
59. Brittle materials generally rupture after elongation. *(True/False)*
60. Brittle materials do not have a yield point. *(True/False)*
61. If principal stresses are 80 and 20 kN/mm^2, then the maximum shear stress is 30 kN/mm^2 *(True/False)*
62. If a body is subjected to only a shear stress of 40 kN/mm^2, then the principal stresses are 40 and – 40 kN/mm^2. *(True/False)*
63. If a body has a normal stress of 60 kN/mm^2 besides some shear stress, then the principal stresses will be lesser than 60 kN/mm^2. *(True/False)*
64. If principal stresses are 80 and 40 kN/mm^2, then the maximum shear stress can be 30 kN/mm^2. *(True/False)*
65. If a body is subjected to only a shear stress, then both the principal stresses are equal but opposite in direction. *(True/False)*

Multiple Choice Questions

1. The unit of Young's modulus is the same as
   (a) shear (b) strain (c) force
2. The unit of strain is
   (a) that of shear (b) that of force (c) dimensionless
3. The ratio of Young's modulus to the modulus of rigidity for Poission's ratio = 0.2 is
   (a) 12/5 (b) 5/12 (c) 5/14
4. Which of the following materials is highly elastic?
   (a) brass (b) steel (c) rubber
5. A composite section made of two materials has Young's modulus 1 : 2 and length 2 : 1. The ratio of stresses is
   (a) 4 : 1 (b) 1 : 2 (c) 1 : 4
6. Which of the following favours brittle fracture in a ductile material?
   (a) elevated temperature (b) slow rate of straining (c) presence of notch
7. Young's modulus and Poisson's ratio are 1.25×10^5 MPa and 0.34 respectively. The modulus of rigidity is
   (a) 0.4025×10^5 MPa (b) 0.4664 MPa (c) 0.8375×10^5 MPa
8. For a linearly elastic, isentropic and homogeneous material, the number of elastic constants required to relate stress and strain is
   (a) two (b) three (c) four
9. A simply-supported beam is made of two wooden planks of the same width resting one upon the other. The upper plank is half the thickness as compared to the lower plank. The assembly is loaded by a uniformly distributed load on the entire span. The ratio of the maximum stresses developed between top and bottom planks will be
   (a) 1 : 16 (b) 1 : 8 (c) 1 : 4
10. If principal stresses are 2 and 8 kN/mm^2, then a plane may have a normal stress of
    (a) 9 kN (b) 1 kN (c) 6 kN
11. If a body is subjected to only a shear stress of 5 kN/mm^2, then a plane can have a normal stress of
    (a) –3 kN (b) –6 kN (c) 6 kN
12. The material of a rubber balloon has Poisson's ratio = 0.5. If uniform pressure is applied to blow the balloon, the volumetric strain of the material is
    (a) 0.5 (b) 0.2 (c) zero
13. When the strain in a material increases with time under sustained constant stress, the phenomenon is known as
    (a) system hardening (b) hysteresis (c) creep
14. The relation between Young's modulus (E), modulus of rigidity (G) and Poisson's ratio (γ) is given by
    (a) $E = 2G(1 + \gamma)$ (b) $E = 2G(1 - \gamma)$ (c) $E = G(1 + \gamma)$

15. If a test specimen is stressed slightly beyond the yield point and then unloaded, the yield strength will
(a) decrease (b) increase (c) remain the same

16. The material that exhibits the same elastic properties in all the directions at a point is said to be
(a) homogeneous (b) isotropic (c) orthotropic

17. The bolts in a flanged coupling connecting two shafts transmitting power are subjected to
(a) shear force and bending moment
(b) torsion
(c) torsion and bending moment

18. The value of Poisson's ratio for any material cannot exceed
(a) 2 (b) 1 (c) 0.5

19. The rigidity modulus of a material where $E = 2 \times 10^6$ kg/cm^2 and Poisson's ratio = 0.25 will be
(a) 0.8×10^5 kg/cm^2 (b) 0.8×10^6 kg/cm^2 (c) 0.5×10^5 kg/cm^2

20. In engineering material, the rigidity modulus
(a) is higher than Young's modulus
(b) is equal to Young's modulus
(c) is less than the half the value of Young's modulus

21. The relationship between E, G and k is given by

(a) $E = \dfrac{9kG}{3k+G}$ (b) $E = \dfrac{9kG}{G+k}$ (c) $E = \dfrac{9kG}{k+3G}$

22. A steel rod of 1 cm, cross sectional area is 100 cm long and has $E = 2 \times 10^6$ kgf/cm^2. It is subjected to an axial pull of 2000 N. The elongation of the rod will be
(a) 1.0 cm (b) 0.1 cm (c) 0.2 cm

23. If a material had a modulus of elasticity = 2.1×10^6 kgf/cm^2 and $G = 0.8 \times 10^6$ kgf/cm^2, then Poisson's ratio is
(a) 0.26 (b) 0.31 (c) 0.47

24. An elastic bar of length L, cross sectional area A, Young's modulus E and self weight W is hanging vertically. It is subjected to load P axially at the bottom end. The total elongation of the bar is

(a) $\dfrac{WL}{AE} + \dfrac{PL}{AE}$ (b) $\dfrac{WL}{AE} + \dfrac{PL}{2AE}$ (c) $\dfrac{WL}{2AE} + \dfrac{PL}{AE}$

25. A solid metal bar of uniform diameter D and length L is hung vertically. If ρ = density, E = Young's modulus, then total elongation due to its own weight is

(a) $\dfrac{\rho L}{2E}\,g$ (b) $\dfrac{\rho L^2}{2E}\,g$ (c) $\dfrac{\rho E}{2L^2}\,g$

26. The stress-strain curve for an ideally plastic material is

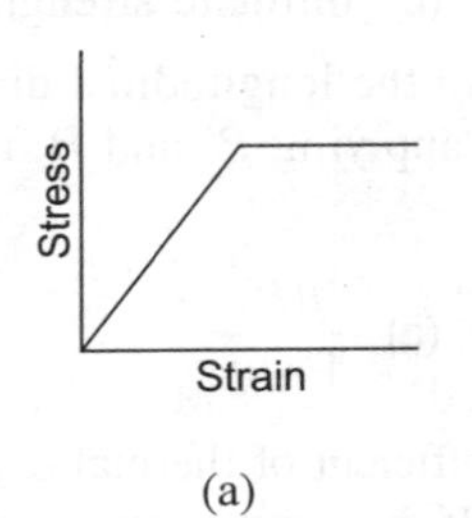

(a)

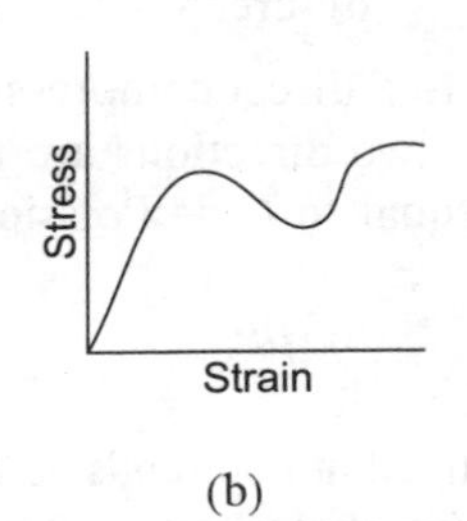

(b)

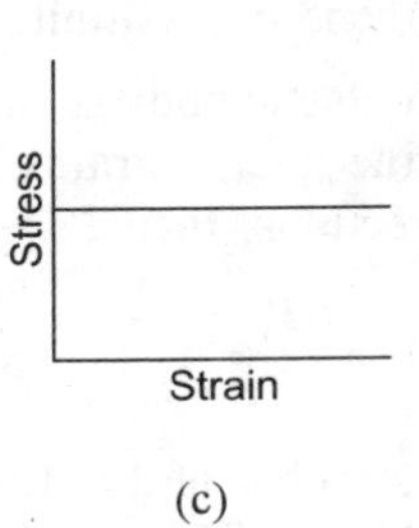

(c)

27. In the following figure the resultant force at point A is
(a) 60 (b) 50 (c) 55

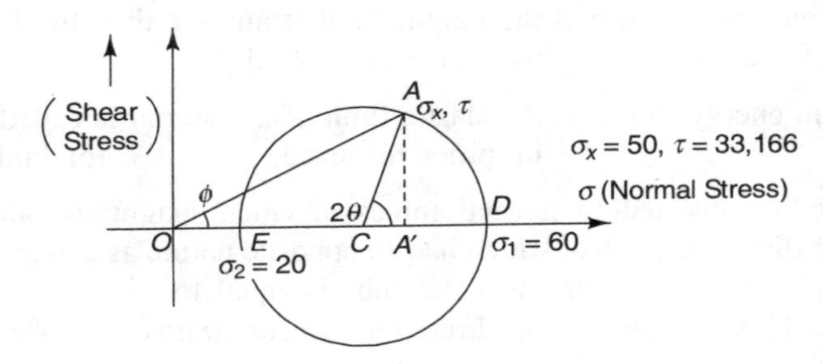

28. In the figure of Q.27, the angle of obliquity is
(a) 30° (b) 38.85 (c) 33.55

29. In the figure of Q.27, the major principal stress is
(a) 50 (b) 60 (c) 55

30. In the figure of Q.27, the minor principal stress is
(a) 40 (b) 20 (c) 60

31. In the figure of Q.27, the angle of plane (θ_1) from the principal plane is
(a) 36.6 (b) 40.6 (c) 35.6

32. A bar with length = 200 mm and diameter = 30 mm on the axial loading has elongation = 0.09 mm and decrease of diameter = 0.0045 mm. Poisson's ratio is

(a) $\frac{1}{4}$ (b) $\frac{1}{3}$ (c) $\frac{1}{2}$

33. If Poison's ratio is 0.5, the elastic modulus of the material is
(a) three times its shear modulus
(b) four times its shear modulus
(c) equal to its shear modulus

34. During the tensile testing of a specimen using a universal testing machine, the parameters actually measured include
(a) load and elongation (b) stress and strain (c) load and stress

35. The stress at which a material fractures under a large number of reversed stress is called
(a) endurance limit (b) creep (c) ultimate strength

36. An elastic body is subjected to a direct compressive stress P_x in the longitudinal direction. If the lateral strains in other two directions are prevented by applying P_y and P_z in these directions, then $P_y = P_z$ is equal to (μ = Poission's ratio)

(a) $\dfrac{\mu P_x}{1-\mu}$ (b) μP_x (c) $\dfrac{\mu P_x}{1+\mu}$

37. A steel bar of 2 m length is fixed at both ends at 20°C. The coefficient of thermal expansion is 11×10^{-6}/°C and the modulus of elasticity is 2×10^6 kg/cm². If the temperature is changed to 18°C, then the bar will experience a stress of
(a) 44 kg/cm² (compressive) (b) 44 kg (tensile) (c) 22 kg (tensile)

38. A prismatic bar of volume V is subjected to a compressive force in the longitudinal direction. If Poisson's ratio = μ and the longitudinal strain = ε then the final volume will be
(a) $V[1 - \varepsilon(1 - 2\mu)]$ (b) $V[1 + \varepsilon(1 - 2\mu)]$ (c) $V[1 - \varepsilon(1 + 2\mu)]$

39. The maximum energy stored at the elastic limit of a material is called the
(a) resilence (b) proof resilence (c) bulk resilence

40. A solid cube is subjected to normal forces of equal magnitude along three mutually perpendicular directions, one of which has an opposite nature as compared to the remaining two forces. The volumetric strain of the cube is equal to
(a) two times linear strain (b) three times linear strain (c) six times linear strain

41. Given that for an element in a body of homogeneous and isotropic material is subjected to a plane stress; ε_x, ε_y and ε_z are normal strains in x, y and z directions and μ is Poisson's ratio. The magnitude of volumetric strain of the element is given by
(a) $\varepsilon_x + \varepsilon_y + \varepsilon_z$ (b) $\varepsilon_x - \mu(\varepsilon_y + \varepsilon_z)$ (c) $\varepsilon_x + \mu(\varepsilon_y + \varepsilon_z)$

42. A steel cube of volume 8000 cc is subjected to an all round stress of 1330 kgf/cm². The bulk modulus of the material is 1.33×10^6 kg/cm². The volumetric change is
(a) 8 cc (b) 6 cc (c) 9 cc

43. The cross section of a bar is subjected to a uniaxial tensile stress P. The tangential stress on a plane included θ to the cross section of the bar would be

(a) $\dfrac{P \sin 2\theta}{2}$ (b) $\dfrac{P}{2 \sin 2\theta}$ (c) $P \times 2 \sin \theta$

44. Two bars of materials A and B of the same length are tightly secured between two unyielding walls and $\alpha_A > \alpha_B$. When temperature rises, the stresses induced are
(a) tension in both materials
(b) compression in both materials
(c) compression in material A and tension in material B

45. A ductile material is one for which the plastic deformation before fracture is
(a) smaller than the elastic deformation
(b) equal to elastic deformation
(c) much larger than the elastic deformation

46. A mild steel bar is in two parts having equal lengths. The area of cross section of part-1 is double that of part-2. If the bar carries an axial load P, then the ratio of elongation in part-1 to that in part-2 will be

(a) 2 (b) 4 (c) 1/2

47. If the bulk modulus is equal to the shear modulus, then Poisson's ratio is

(a) 0.125 (b) 0.250 (c) 0.500

48. A large uniform plate containing a rivet hole is subjected to a uniform uniaxial tension of 95 MPa. The maximum stress in the plate is

(a) 100 MPa (b) 190 MPa (c) 90 MPa

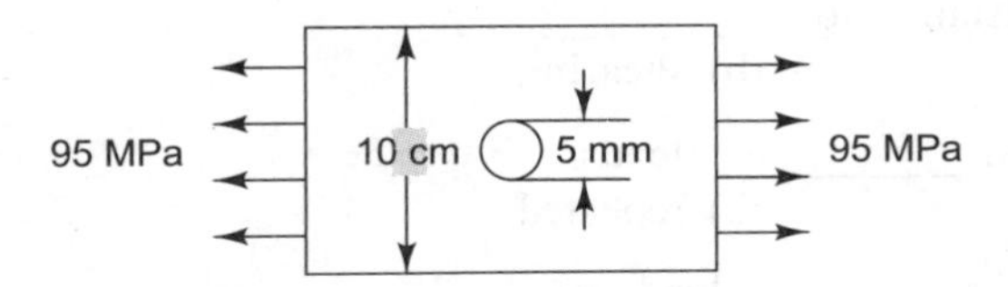

49. A block of steel is loaded by a tangential force on its top surface and the bottom surface is held rigidly. The deformation of the block is due to

(a) shear only (b) bending only (c) shear and bending

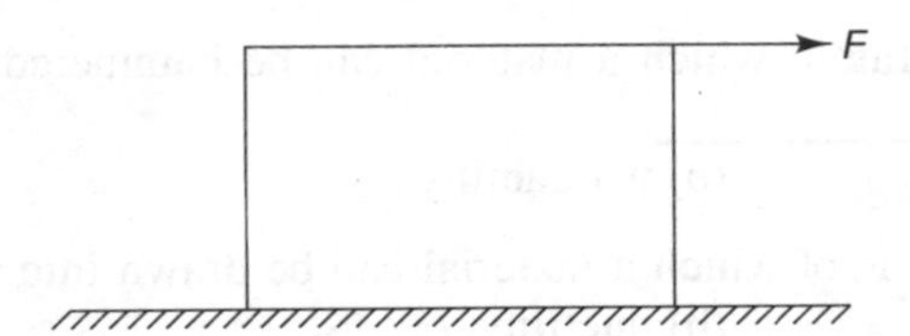

50. A load w is suspended by wire ropes AB and CD as shown. The ropes are of the same material and cross section. Elongation of the ropes is the same under loading. If the stresses are σ_1 and σ_2, then the ratio of σ_1 to σ_2 will be

(a) 3/2 (b) 2/3 (c) 4/3

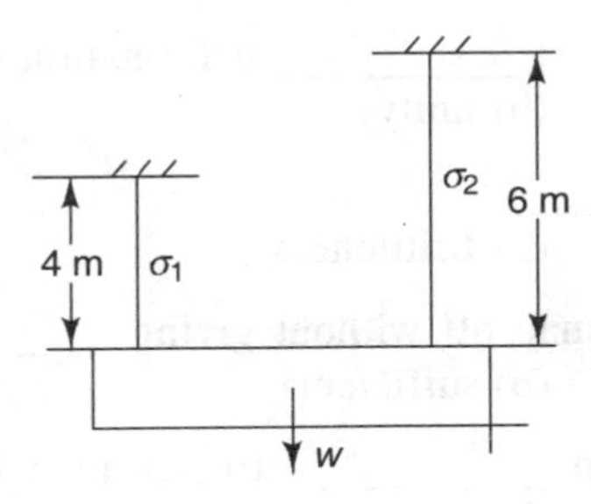

51. A bolt is threaded through a tubular sleeve and the nut is turned up just tight. The nut is turned further, the bolt being put in tension and the sleeve in compression. The distance by which the nut is turned is equal to

(a) the summation of deformation in the bolt and the sleeve
(b) the difference of deformation in the bolt and the sleeve
(c) the deformation in the bolt

52. A 15 m long straight wire is subjected to a tensile stress of 2000 kgf/cm^2. $E = 1.5 \times 10^4$ kgf/cm^2, $\alpha = 16.66 \times 10^{-4}$/°F. The temperature change in °F to produce the same elongation due to the tensile stress is
(a) 40 (b) 80 (c) 120

53. If the dimensions of a bar with square cross section are doubled, then the elongation by its own weight will be
(a) four times (b) three times (c) two times

Fill in the Blanks

1. Hardness is resistance to __________.
(a) penetration (b) shearing

2. Fatigue is due to __________ loads.
(a) sustained (b) repeated

3. Creep is due to __________ loading.
(a) repeated (b) prolonged

4. The area under a stress and strain diagram up to ultimate strength indicates __________.
(a) toughness (b) strain energy

5. The property by virtue of which a material can be hammered or rolled into a thin sheet without rupture is __________.
(a) malleability (b) workability

6. The property by virtue of which a material can be drawn into wire is called __________.
(a) wireability (b) ductility

7. The angle of obliquity is the angle between the resultant and __________ axis.
(a) shear stress (b) normal stress

8. Ductile materials have a __________ elastic limit.
(a) higher (b) lower

9. Brittle materials have almost __________ deformation before rupture.
(a) zero (b) unity

10. Lack of ductility is __________.
(a) softness (b) brittleness

11. Ductile materials do not snap off without giving __________ warning by elongation.
(a) short (b) sufficient

12. Pulling-type loads develop __________ stresses in a body.
(a) tensile (b) compressive

13. Pushing-type loads develop __________ stresses in a body.
(a) tensile (b) compressive

14. Tensile stresses are taken as __________.
(a) positive (b) negative

15. Compressive stresses are taken as __________.
(a) positive (b) negative

16. The ratio of lateral strain to longitudinal strain is called __________.
(a) Poisson's ratio (b) modulus of strain

17. The internal resistance which the body offers to oppose the load is __________.
(a) stress (b) resilence

18. The ratio of shear stress to shear strain is called the modulus of __________.
(a) rigidity (b) elasticity

19. The ratio of normal stress to volumetric strain is called __________ modulus.
(a) elastic (b) bulk

20. The deformation per unit length is called __________.
(a) stress (b) strain

21. The ratio of tensile stress to strain is called __________.
(a) modulus of rigidity (b) Young's modulus

22. Principal planes have __________ shear stress.
(a) maximum (b) zero

23. Brittle materials fail on compression loading due to __________.
(a) shear stress (b) compressive stress

24. The yield point in compression is __________ than in tension.
(a) lower (b) higher

25. Stress in a specimen will be __________ where its cross section is smallest.
(a) maximum (b) least

26. Before rupture, the strain increases at the same load due to __________.
(a) flowing (b) necking

27. After the low yield point __________ hardening takes place.
(a) stress (b) strain

28. Ultimate strength is __________ than yield point.
(a) lower (b) higher

29. Brittle materials almost do not have a __________.
(a) rupture point (b) yield point

30. Yield shear stress is __________ normal yield stress.
(a) equal (b) half

31. The radius of Mohr's circle given in the figure is __________.
(a) 40 (b) 60

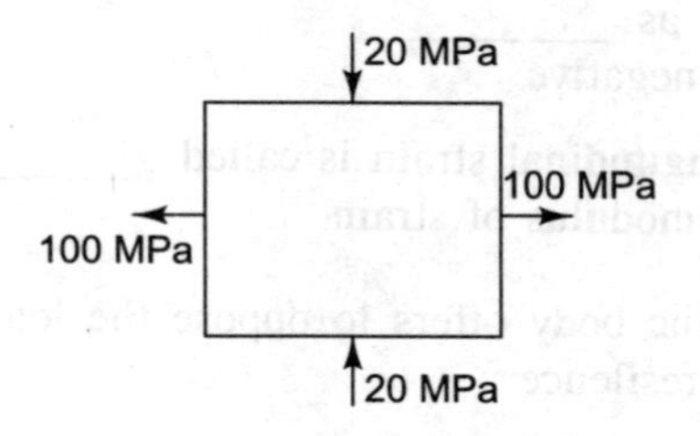

32. For the given stress and strain, resilence and toughness are ________.
(a) 14×10^4, 90×10^4 (b) 14×10^4, 14×10^4

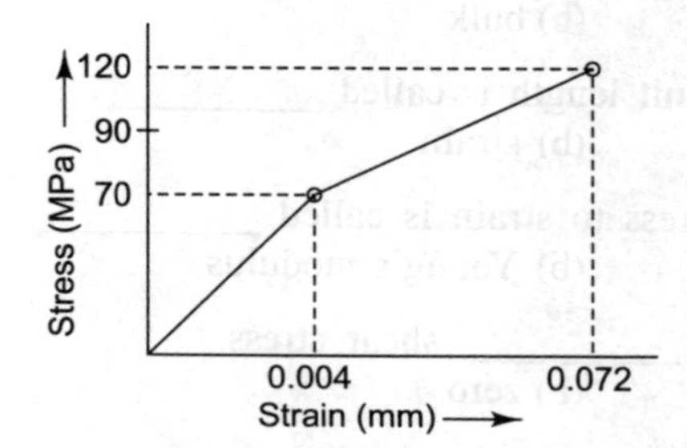

ANSWERS

I'd rather confess I'm wrong and be right than claim I'm right and be wrong.

State True or False

1. False
2. True
3. True
4. False
5. True
6. True
7. True
8. False (The strength changes and the material fails below its strength under repeated loading.)
9. False (The material fails due to prolonged loading which is known as creep's failure.)
10. True
11. False (The angle is double on Mohr's circle.)
12. False (Clockwise shear stress is positive)
13. False (Tensile stress is taken as positive.)
14. True
15. True
16. True
17. True
18. False $\left(\text{Poisson's ratio} = \dfrac{\text{lateral strain}}{\text{longitudinal strain}}\right)$
19. True
20. True
21. True
22. False (Complementary shear stresses have different rotation to applied shear stresses.)
23. True
24. True
25. False ($\varepsilon_v = \varepsilon(1 - 2v)$)
26. False $\left(\text{Bulk modulus } k = \dfrac{\sigma}{\varepsilon_v}\right)$
27. True
28. False ($E = 2G\ (1 + \gamma)$)
29. True ($F = \tau \times \text{Area} = \tau \times \pi D t = 100 \times 3.14 \times 50 \times 2 = 31.4$ MN)
30. False (It is called an oblique angle.)
31. True
32. True
33. True
34. False (The gap is to compensate elongation and contraction on heating and cooling.)
35. False (The end on the roller support permits extention and contraction of the truss on heating and cooling.)
36. True
37. True (Load = normal stress in concrete × area + normal stress in steel bars × area. Since steel can take more normal stress, the combined load permissible increases.)

38. True

39. False (The elastic limit increases due to strain hardening.)

40. True (The brittle materials fail under compressive loads due to shear force developed at 45° to the inclined plane.)

41. True 42. True 43. True 44. False

45. True 46. True 47. True

48. False (The specimen is kept as a simply-supported beam)

49. False (The specimen is kept as a cantilever beam.)

50. True

51. True

52. False (Brinell hardness test is accurate up to 400 BHN.)

53. True 54. True 55. True

56. False (It will be in the third quadrant as both stress and strain are negative.)

57. True

58. False (The compressive elastic limit is more than the tensile elastic limit.)

59. False (Brittle materials have almost zero elongation.)

60. True

61. True ($\tau_{max} = \dfrac{\sigma_1 - \sigma_2}{2} = \dfrac{80 - 20}{2} = 30$ kN/mm^2)

62. True

τ
(0, 40)
(– 40, 0)
(0, 0)
(40, 0)
σ
(0, – 40)

63. False ($\sigma_1 = \dfrac{\sigma_x - \sigma_y}{2} + \sqrt{\left(\dfrac{\sigma_x - \sigma_y}{2}\right)^2 + \tau^2}$.

Here $\sigma_y = 0$. Therefore, $\sigma_1 = \dfrac{\sigma x}{2} + \sqrt{\left(\dfrac{\sigma x}{2}\right)^2 + \tau^2} > \sigma_x > 60$)

64. False $\left(\tau_{max} = \dfrac{\sigma_1 - \sigma_2}{2} = \dfrac{60 - 40}{2} = 20\right)$

65. True ((0, τ) σ = τ (−σ, 0) (0, 0) (σ, 0) (0, −τ))

Multiple Choice Questions

1. (a)

2. (c)

3. $\left(E = 2G(1+\gamma) \text{ or } \dfrac{E}{G} = 2(1+\gamma) = 2(1+0.2) = 2.4 = \dfrac{12}{5}\right)$

4. (b) (Steel has a higher elastic limit and less elongation in the elastic limit.)

5. (c) $\left(\delta l_1 = \delta l_2 \text{ or } \dfrac{\sigma_1 \times l_1}{E_1} = \dfrac{\sigma_2 \times l_2}{E_2} = \text{or } \dfrac{\sigma_1}{\sigma_2} = \dfrac{l_2}{l_1} \times \dfrac{E_1}{E_2} = \dfrac{E_1/E_2}{l_1/l_2} = \dfrac{\frac{1}{2}}{2/1} = \dfrac{1}{4}\right)$

6. (c)

7. (b) $\left(E = 2G(1+\gamma) \text{ or } G = \dfrac{E}{2(1+2\gamma)} = \dfrac{1.25 \times 10^5}{2(1.34)} = 0.4664 \text{ MPa}\right)$

8. (c) (E, G, K and γ)

9. (b) [moment of inertia × (depth)^3]

10. (c) (The normal stress will have any value between 8 and 2 kN/mm^2.)

11. (a) (The principal stress is 5 and −5 kN/mm^2 and normal stress on a plane can have any value in between the principal stresses.)

12. (c) ($\upsilon = 0.5$, $\varepsilon_v = \varepsilon(1 - 2\gamma)$ where ε_v = volumetric strain and ε = linear strain and γ = Poisson's ratio. $\varepsilon_v = \varepsilon(1 - 0.5 \times 2)$. Therefore, $\varepsilon_v = 0$)

13. (c)

14. (a)

15. (b) (

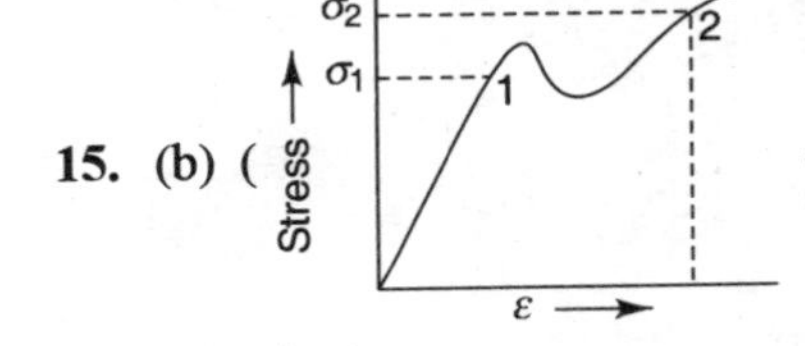

$\sigma_2 > \sigma_1$ and process is known as strain hardening)

16. (b)

17. (a)

18. (c)

19. (b) $\left[G = \dfrac{E}{2(1+\gamma)} = \dfrac{2\times10^6}{2\times1.25} = 0.8\times10^6\right]$ **20.** (c) **21.** (a)

22. (b) $\left(\delta l = \dfrac{\tau l}{AE} = \dfrac{2000 \times 100}{1 \times 2 \times 10^6} = 0.1\text{cm}\right)$

23. (b) $\left(1+\gamma = \dfrac{E}{2G} = \dfrac{2.1\times10^6}{2\times0.8\times10^6} = 1.31 \quad \text{or } \gamma = 0.31\right)$

24. (c) **25.** (b) **26.** (c)

27. (a) $(R^2 = OA^2 = (OA')^2 + (AA')^2 = (50)^2 + (33.163)^2 = 36 \times 10^2, \therefore R = 60)$

28. (c) $\left[\tan\phi = \dfrac{AA'}{OA'} = \dfrac{33.166}{50} = 0.66, \phi = 33.55°\right]$

29. (b) (OD = 60 where shear stress is zero)

30. (b) (OE = 20 where shear stress is zero)

31. (a) $\left[\tan 2\theta_1 = \dfrac{AA'}{CA'} = \dfrac{33.166}{10} = 3.3166. \text{ Therefore, } 2\theta_1 = 73.22° \text{ or } \theta_1 = 36.61°\right]$

32. (b) $\left[\varepsilon_{long} = \dfrac{0.09}{200} = 4.5\times10^{-4}, \varepsilon_{lateral} \dfrac{.0045}{30} = 1.5 \times 10^{-4}. \text{ Therefore, } \gamma = \dfrac{\varepsilon_{lateral}}{\varepsilon_{long}} = \dfrac{1.5\times10^{-4}}{4.5 \times 10^4} = 1/3\right]$

33. (a) $(E = 2G(1 + \gamma) = 2 \times 1.5G = 3G)$

34. (a)

35. (a)

36. (a)
$$\varepsilon_x = \frac{1}{E}[P_x - \mu(P_y + P_z)]$$
$$\varepsilon_y = \frac{1}{E}[P_y - \mu(P_x + P_z)]$$
$$\varepsilon_z = \frac{1}{E}[P_z - \mu(P_x + p_y)]$$

Now $\varepsilon_y = \varepsilon_z$

$$P_y - \mu(P_x + P_z) = P_z - \mu(P_x + P_y) = 0$$
$$P_y = \mu P_x + \mu P_y \qquad \text{(i)}$$
$$P_z = \mu P_x + \mu P_y \qquad \text{(ii)}$$

Eqs. (i) + (ii)

$$P_y + P_z = 2\mu P_x + \mu(P_y + P_z)$$

or $(P_y + P_z)(1 - \mu) = 2\mu P_x \ (P_y = P_z)$

or $P_y = \dfrac{\mu P_x}{1-\mu}$ and $P_2 = \dfrac{\mu P_x}{1-\mu}$

37. (b) ($\delta l = l\alpha\Delta T = 2 \times 100 \times 2 \times 11 \times 10^{-6}$ cm $= 44 \times 10^{-4}$ cm, $\varepsilon = \dfrac{\delta l}{l} = \dfrac{44\times10^{-4}}{2\times100} = 22 \times 10^{-6}$

$\sigma = E \times \varepsilon = 2 \times 10^6 \times 22 \times 10^{-6} = 44$ kgf. It is tensile as bar contracts on cooling.)

38. (a) $\quad (\varepsilon_V = \varepsilon(1 - 2\mu) = \dfrac{\delta V}{V}$

$\therefore \quad \delta V = \varepsilon(1 - 2\mu)V.$

Final volume $= V + \delta V = V - \varepsilon_V(1 - 2\mu)$

$= V[1 - \varepsilon(1 - 2\mu)])$

39. (b)

40. (b) ($V = 3$, $dV = 3l^2 dl$

Volumetric strain $\varepsilon_v = \dfrac{dV}{V} = \dfrac{3l^2}{l^3}dl$

$= 3 \times \dfrac{dl}{l} = 3 \times$ linear strain)

41. (a) ($V = l \times b \times t$, $dV = lb \times dt + bt\,dl + tl db$. Now $\dfrac{dV}{V} = \dfrac{dt}{t} + \dfrac{db}{b} + \dfrac{dl}{l}$, $\varepsilon_v = \varepsilon_l + \varepsilon_b + \varepsilon_t$)

42. (a) ($k = \dfrac{\sigma}{\varepsilon_y}$, $\varepsilon_V = \dfrac{\sigma}{k} = \dfrac{1330}{1.33 \times 10^6}$ $\varepsilon_V = 1 \times 10^{-3}$.

Now $\dfrac{dV}{V} = \varepsilon_V$ or $dV = V \times \varepsilon_V = 8000 \times 1 \times 10^{-3} = 8$ cc)

43. (a)

44. (b) (Both bars expands to create compressive thermal load.)

45. (c)

46. (c) ($\delta l = \dfrac{Pl}{AE}$. Now l = constant, E = constant, P = constant. Therefore, $\dfrac{\delta l_1}{\delta l_2} = \dfrac{A_2}{A_1} = \dfrac{1}{2}$)

47. (a) ($E = 3k(1 - 2\gamma) = 2G(1 + \gamma)$. Now $k = G$, hence $3 - 6\gamma = 2 + 2\gamma$ or $8\gamma = 1$ or $\gamma = \dfrac{1}{8} =$ 0.125)

48. (a) (Force is constant. Hence $\sigma_1 A_1 = \sigma_2 A_2$ or $95 \times 10 \times t = \sigma_2 \times (10 - 0.5)t$ or $\sigma_2 = \frac{95 \times 10}{9.5} = 100$ MPa)

49. (c)

50. (a) $\left[\delta l \text{ is same. Hence } \frac{\sigma_1}{E} \times l_1 = \frac{\sigma_2}{E} \times l_2. \text{ Therefore, } \frac{\sigma_1}{\sigma_2} = \frac{l_2}{l_1} = \frac{6}{4} = \frac{3}{2}\right]$

51. (a) (The bolt will expand due to tensile load and the sleeve will contract due to compressive load. The nut will move the distance equal to the sum of contraction and expansion.)

52. (b) $\left[\delta l = \frac{\sigma}{E} \times l = l \times \Delta T \times 16.66 \times 10^{-4} \; \Delta T = \frac{\sigma}{E} \times \frac{1}{16.66 \times 10^{-4}} = \frac{2000}{1.5 \times 10^4 \times 16.66 \times 10^{-4}} = 80\text{°F}\right]$

53. (a) $\left[\delta l = \frac{\rho g l^2}{2E}, \; \frac{\delta l_1}{\delta l_2} = \frac{l_1^2}{l_2^2} = \frac{1}{4}\right]$

Fill in the Blanks

1. (a) **2.** (b) **3.** (b) **4.** (a)
5. (a) **6.** (b) **7.** (b) **8.** (b)
9. (a) **10.** (b) **11.** (b) **12.** (a)
13. (b) **14.** (a) **15.** (b) **16.** (a)
17. (a) **18.** (a) **19.** (b) **20.** (b)
21. (b) **22.** (b) **23.** (a) **24.** (b)
25. (a) **26.** (b) **27.** (b) **28.** (b)
29. (b) **30.** (b)

31. (b) $\left[\text{radius} = \frac{\sigma_1 - \sigma_2}{2} = \frac{100 + 20}{2} = 60\right]$

32. (a) $\left(\begin{array}{l}\text{resilence} = \frac{1}{2} \times 70 \times 4 \times 10^{-3} \times 10^6 = 14 \times 10^4 \text{ toughness} = \frac{1}{2} \times 70 \times 4 \times 10^3 \\ + \frac{1}{2}(70 + 120) \times 8 \times 10^{+3} = 14 \times 10^4 + 76 \times 10^4 = 90 \times 10^4\end{array}\right)$

CHAPTER 16

Bending Stresses in Beams

Troubles, like a washing machine, twist us and knock us around, but in the end we come out brighter and better than before.

INTRODUCTION

A beam is subjected to bending moment and shear force when an external load is applied to it. Bending moment and shear force vary from section to section depending upon loading. Deformation takes place in the beam due to bending moment and shear. Stresses are generated in the beam as resistance to these deformations. Stress induced due to bending moment is called *bending stress*, while stress induced due to shear force is called *shear stress*.

To understand pure bending, take a beam *CD* supported at *A* and *B* and load *F* at *C* and *D* (Figure 16.1). The shear force diagram and bending moment diagram are as shown in the figure. The portion *AB* of the beam is subjected to pure bending. The bending moment in this portion is $F \times a$. The longitudinal stress is developed due to bending moment. The magnitude of stress depends upon the radius of curvature, and cross section of the beam. The equation which connect these quantities is known as the *bending equation*.

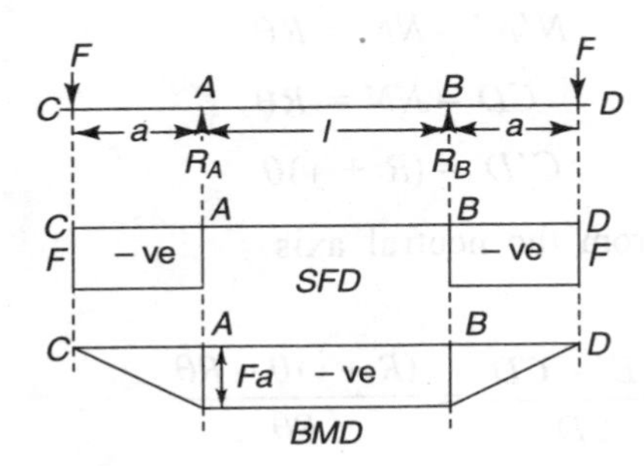

FIGURE 16.1 Shear force diagram and bending moment diagram.

THEORY OF BENDING

The following assumptions are made:

1. The material of the beam is homogeneous and isotropic.
2. Stresses are within the elastic limit. Young's modulus is the same in tension and compression.
3. Bernoulli's assumption: The transverse section remains in the plane even after bending.
4. All longitudinal elements of the beam bend into a circular arc having a common centre of curvature.
5. The radius of curvature is large in comparison to the cross section of the beam.
6. Each layer of the beam is free to expand or contract independently.

Bending can be sagging or hogging (Figure 16.2). In sagging, the beam curves upwards. The layers above the neutral axis (length remains the same) contract, while the layers below the neutral axis expand. In hogging, the layers on top of the neutral axis expand and the layers below the neutral axis contract. Sagging bending moment (BM) is taken positive while hogging bending moment (BM) is taken negative.

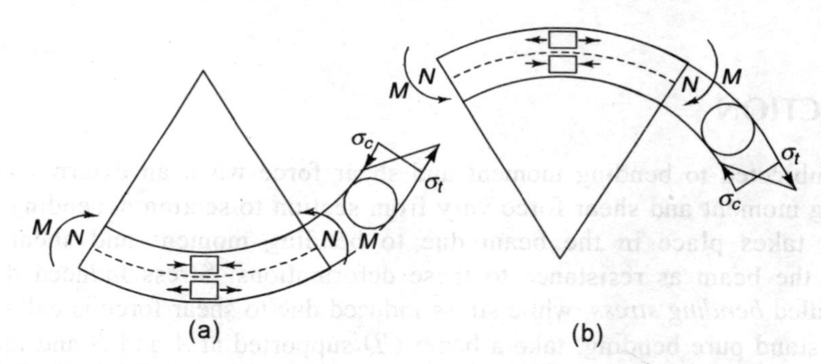

FIGURE 16.2 (a) Sagging bending moment and (b) hogging bending moment.

Bending equation: A bean *ABFE* is subjected to sagging moment (Figure 16.3). Let *R* is the radius of curvature and θ is the angle formed at the centre of curvature by the beam. Select *CD* layer.

$$N'N' = NN. = R\theta$$

$$CD = NN = R\theta$$

$$C'D' = (R + y)\theta$$

where y = distance of *CD* from the neutral axis

Strain in *CD* is

$$\frac{C'D' - CD}{CD} = \frac{(R+y)\theta - R\theta}{R\theta}$$

$$= \frac{y}{R}$$

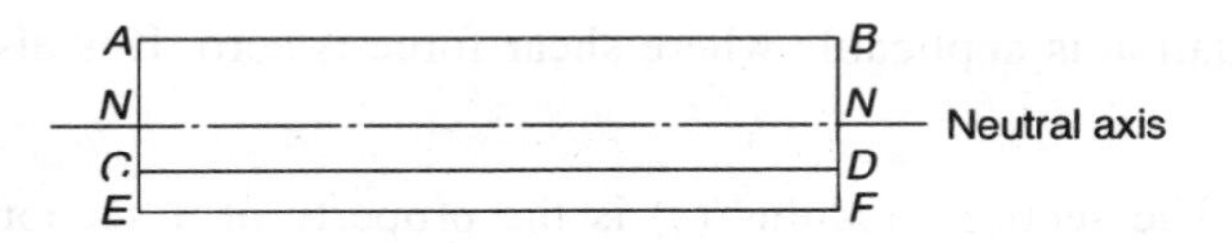

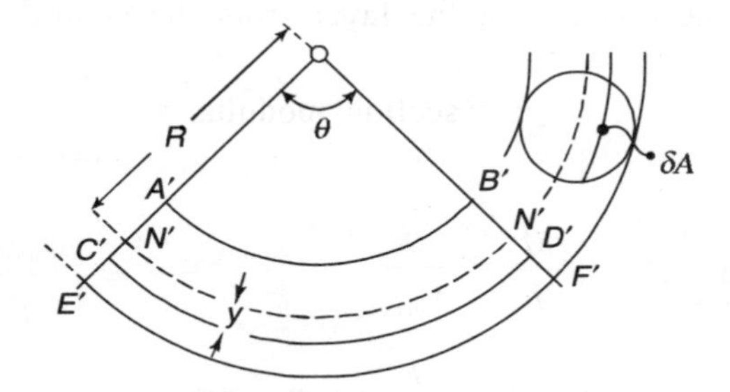

FIGURE 16.3 Bending: Sagging moment.

$$\text{Stress} = E \times \text{strain}$$

$$\sigma = E \times \frac{y}{R}$$

$$\frac{\sigma}{y} = \frac{E}{R}$$

As $\sigma = \dfrac{Ey}{R}$, we select area δA.

$$\text{Force on area } \delta A = \sigma \times \delta A = \frac{E}{R} \times y\delta A$$

$$\text{Moment of force} = F \times y = \frac{E}{R} \times y^2\delta A$$

$$\text{Total moment } M = \Sigma \frac{E}{R} \times y^2\delta A$$

$$= \frac{E}{R} \Sigma y^2\delta A$$

$$M = \frac{EI}{R}$$

where I = moment of inertia of cross section from the neutral axis. Therefore,

$$\frac{M}{I} = \frac{E}{R} = \frac{\sigma}{y}$$

where z is section modulus.

The bending equation is applicable where shear force is zero. It is also called the *flexure formula*.

Section modulus: The section modulus (z) is the property of a section and it is the ratio of moment of inertia to the distance of the layer from the neutral axis.

$$z = \text{section modulus} = \frac{I}{y_{\max}}$$

Since

$$\frac{M}{I} = \frac{\sigma}{y_{\max}}$$

$$M = \frac{I}{y_{\max}} \times \sigma = z\sigma$$

or

$$z = \frac{BM}{\sigma_{\max}}$$

This is vastly used in finding the section modulus for designing a beam for given BM and the maximum stress.

BEAMS OF HETEROGENEOUS MATERIALS (FLITCHED BEAMS)

It is common to reinforce beams to withstand loads greater than those possible without any reinforcement. An important example of a structural member made of two different materials is furnished by a reinforced concrete beam. The concrete beam is reinforced by steel rods placed a short height above its lower face. Since concrete is weak in tension, the entire tensile load coming at the lower part of the beam is taken by steel rods and the upper part carries the compressive load. Reinforcement can be done with the same material or different materials. Reinforcement is usually symmetrically provided about the neutral axis. The following expressions are applicable for a flitched beam:

1. Total resistance moment is the sum of resisting moments of individual materials, i.e.

   $$M = M_1 + M_2$$

2. The radius of curvature is the same. For reinforced wood with steel,

   $$M = z_w\sigma_w + z_s\sigma_s$$

   For concrete reinforced with steel,

   $$M = z_{\text{conc}} \times \sigma_{\text{conc}} + z_s\sigma_s$$

3. The steel strip having width x can be transformed into the wooden strip of $\frac{E_s}{E_w} \times x$ width to simplify the analysis of a flitched beam where E_s is Young's modulus of steel and E_w is Young's modulus of wood. The transformation from steel to an equivalent wooden strip is as shown in Figure 16.4.

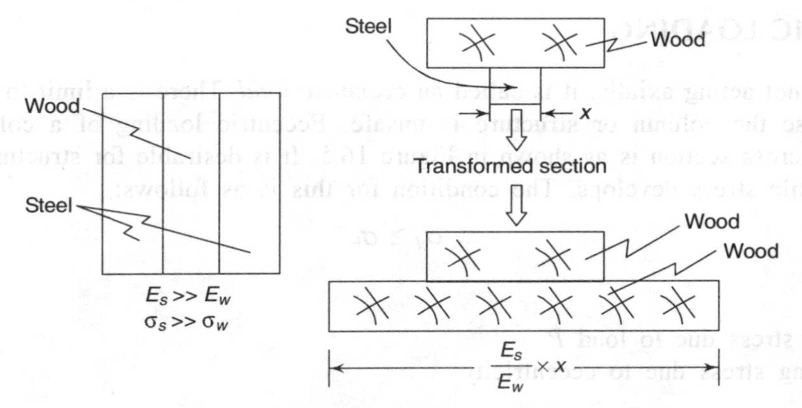

FIGURE 16.4 Transformation of a steel strip to an equivalent wooden strip.

FLEXURAL RIGIDITY AND UNIFORM STRENGTH

EI is termed flexural rigidity. As $EI = M \times R$ and if the radius of curvature $R = 1$, then flexural rigidity is bending moment to produce the unit radius of curvature.

Beams of uniform strength: The bending moment varies along the length of a beam depending upon loading, but the beam is designed to withstand the maximum bending moment which acts at some portion of the beam. Though this leads to a uniform cross section of the beam, additional material is required at some parts of the beam where the bending moment is small thus leads to wastage of material. If a beam is designed so that every extreme fibre along its length is loaded to maximum permissible stress by varying the section, it is known as the beam of uniform strength. If at every section, the extreme fibre stress is equal to σ_{max}, then the section modulus (z) of the beam at any section should be proportional to the bending moment at that section. In order to get a beam of uniform strength, the sections of the beam may be varied by

- Varying the depth and keeping the width constant
- Varying the width and keeping the depth
- Varying both the depth and width
- Varying the diameter of a circular beam

Unit radius of inertia: It is a dimensionless quantity which gives moment of inertia when it is multiplied by the square of area.

$$\text{Unit radius of inertia} = \frac{I}{A^2}$$

$$= \frac{\frac{\pi}{64}D^4}{\left(\frac{\pi D^2}{4}\right)^2} = \frac{1}{4\pi}$$

ECCENTRIC LOADING

If a load is not acting axially, it is called an *eccentric load.* There is a limit to eccentricity, (e) otherwise the column or structure is unsafe. Eccentric loading of a column with a rectangular cross section is as shown in Figure 16.5. It is desirable for structural safety so that no tensile stress develops. The condition for this is as follows:

$$\sigma_d \geq \sigma_b$$

where,

σ_d = direct stress due to load P
σ_b = bonding stress due to eccentricity

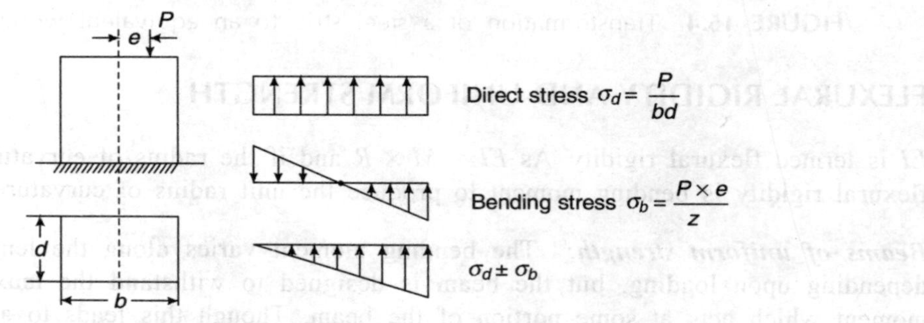

FIGURE 16.5 Eccentric loading.

$$\frac{P}{A} \geq \frac{M}{z}$$

$$\geq \frac{Pe}{z}$$

or

$$e \leq \frac{z}{A}$$

$$\leq \frac{\frac{db^2}{6}}{d \times b} \leq \frac{b}{6}$$

Therefore, eccentricity (e) must be less or equal to $b/6$ from either side of vertical axis of the column so that no tensile stress is developed. Hence the load must act within the middle third (Figure 16.6). This is called the *middle third rule* for a rectangular beam.

For a circular section, let us find eccentricity which is permissible. Refer to Figure 16.7.

$$\sigma_d \geq \sigma_b$$

$$\frac{P}{A} \geq \frac{M}{z} \geq \frac{Pe}{z}$$

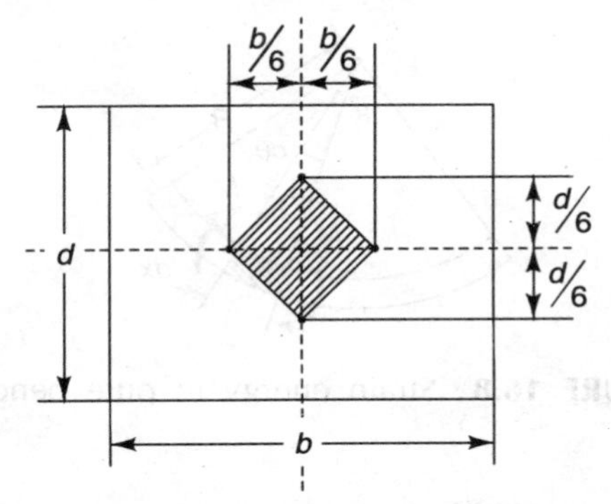

FIGURE 16.6 Load acting within the middle third of a rectangular beam.

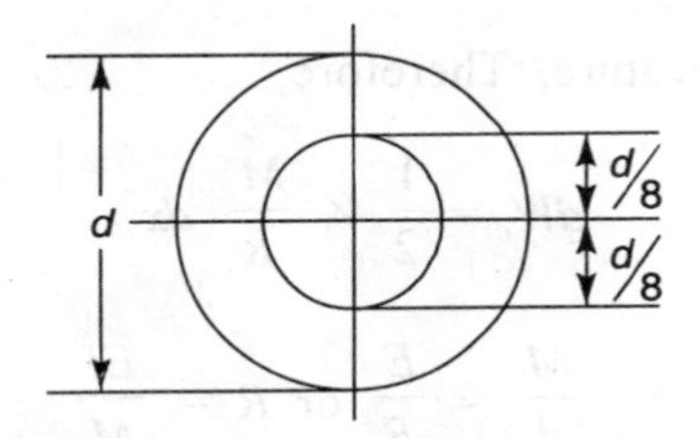

FIGURE 16.7 Load acting within the middle quarter of a circular section.

$$\therefore \qquad e \le \frac{Z}{A}$$

or

$$e \le \frac{\dfrac{\pi d^3}{32}}{\dfrac{\pi d^2}{4}}$$

$$e \le \frac{d}{8}$$

The load must act within the middle quarter of the circle to avoid any tensile stress. This is called the *middle quarter rule* for a circular section.

STRAIN ENERGY IN PURE BENDING

Consider an element of a beam as dx in length (Figure 16.8). The element on application of BM rotates by angle $d\theta$. Since the value of moment is attained gradually, the average value of moment acting through angle $d\theta$ is 1/2 M.

$$\text{External work } dW_e = \frac{1}{2} M d\theta$$

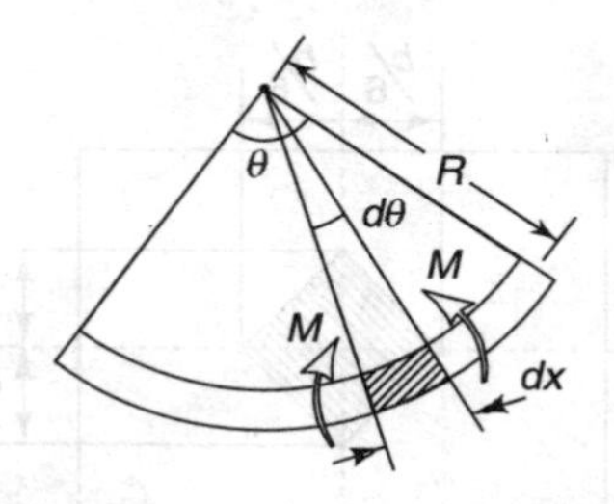

FIGURE 16.8 Strain energy in pure bending.

But $$\frac{dx}{R} = d\theta$$

where R is the radius of curvature. Therefore,

$$dW_e = \frac{1}{2} \times \frac{M}{R}\, dx$$

But $$\frac{M}{I} = \frac{E}{R} \text{ or } R = \frac{EI}{M}$$

∴ $$dW_e = \frac{1}{2} \times \frac{M^2}{EI}\, dx$$

$$W_e = \int_0^l \frac{M^2}{2EI}\, dx = \frac{M^2 l}{2EI}$$

where l = length of beam.

Therefore, $$\text{strain energy} = \frac{M^2 l}{2EI}$$

SOLVED PROBLEMS

1. A flitched timber beam made up of steel and timber has a section as shown in the figure. Determine the resistance of the beam. Take σ_s = 100 MN/m^2 and σ_w = 5 MN/m^2.

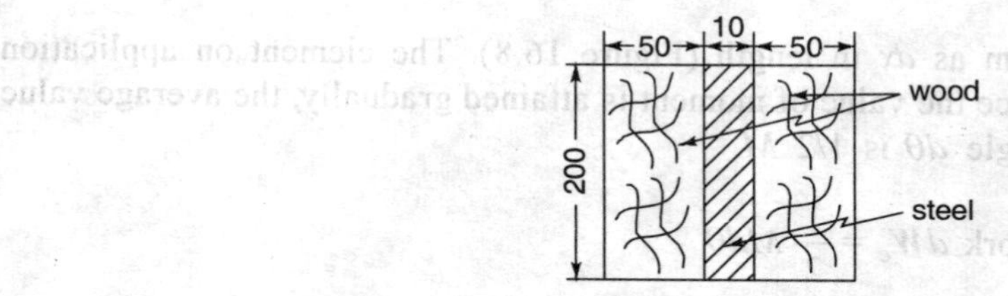

$$\text{Section modulus of wood } z_w = \frac{bd^2}{6} = \frac{100 \times 200^2}{6} \times 10^{-9}$$

$$= 66.6 \times 10^{-5} \text{ m}^3$$

$$\text{Section modulus of steel } z_s = \frac{bd^2}{6} = \frac{10 \times 200^2}{6} \times 10^{-9}$$

$$= 6.66 \times 10^{-5} \text{ m}^3$$

$$\text{Moment of resistance of wood} = \sigma_w z_w$$

$$= 5 \times 66.6 \times 10^{-5} \times 10^6$$

$$= 3330 \text{ N m}$$

$$\text{Moment of resistance of steel} = \sigma_z z_s$$

$$= 100 \times 10^6 \times 6.66 \times 10^{-5}$$

$$= 6660 \text{ N m}$$

$$\text{Moment of resistance} = M_w + M_s$$

$$= 3330 + 6660$$

$$= 9990 \text{ N m.}$$

2. A rectangular beams of cross section (300 × 200) mm² is simply supported in a span of 5 m. What uniformly distributed loads/metre the beam may carry in its two positions (i) when height is 300 mm and (ii) when height is 200 mm if the bending stress is not to exceed 130 N/mm²? (UPTU: Dec. 2005)

(i) $$\text{Section modulus } (z) = \frac{bh^2}{6} = \frac{200 \times 300^2}{6} = \frac{2 \times 9}{6} \times 10^6$$

$$= 3 \times 10^6 \text{ mm}^3$$

$$M = z\sigma_{\max} = 3 \times 10^6 \times 130$$

$$= 39 \times 10^7 \text{ N mm}$$

$$= 39 \times 10^4 \text{ N m}$$

$$M = \frac{wl^2}{8} = 39 \times 10^4$$

or $$\frac{w \times 5^2}{8} = 39 \times 10^4$$

or $$w = \frac{39 \times 8}{25} \times 10^4$$

$$= 12.48 \times 10^4 \text{ N/m}$$

$$= 124.8 \text{ kN/m}$$

(ii) Section modulus $(z) = \frac{bh^2}{6} = \frac{300 \times 200^2}{6} = \frac{3 \times 4}{6} \times 10^6$

$$= 2 \times 10^6 \text{ mm}^3$$

$$M = z\sigma_{max} = 2 \times 10^6 \times 130$$

$$= 26 \times 10^7 \text{ N mm}$$

$$= 26 \times 10^4 \text{ N m}$$

$$\frac{wl^2}{8} = 26 \times 10^4$$

$$\therefore \quad w = \frac{26 \times 8 \times 10^4}{25}$$

$$= 8.32 \times 10^4 \text{ N/m}$$

$$= 83.2 \text{ kN/m}$$

3. A rectangular wooden beam is subjected to a bending moment of 5 kN m. If the depth of the section is to be twice the breadth and stress in wood is not to exceed to 60 N/cm^2. Find the dimension of the cross section of the beam. (UPTU: Dec. 2005)

$$z = \frac{bd^2}{6}$$

Given $d = 2b$. Therefore,

$$z = \frac{b(2b)^2}{6} = \frac{4b^3}{6} = \frac{2b^3}{3}$$

$$M = z\sigma$$

$$5 \times 10^3 \times 100 = \frac{2}{3} b^3 \times 60$$

or $$b^3 = \frac{5 \times 10^5 \times 3}{2 \times 60} = 12500$$

or $$b = 23.2 \text{ cm}$$

$\therefore$ $$d = 46.4 \text{ cm}$$

4. A beam of CI having a section of 50 mm external diameter and 25 mm internal diameter is supported at two points 4 m apart. The beam carries a concentrated load of 100 N at its centre. Find the maximum bending stress induced in the beam. (UPTU: 2002)

$$\text{Maximum BM} = \frac{WL}{4} = \frac{100 \times 4000}{4} = 10^5 \text{ N mm}$$

$$\text{Section modulus} = \frac{\pi}{32}\left(\frac{D^4 - d^4}{D}\right) = \frac{\pi}{32}\left(\frac{50^4 - 25^4}{50}\right)$$

$$= 11.5 \times 10^3 \text{ mm}^3$$

$$\sigma_{max} = \frac{M_{max}}{z} = \frac{10^5}{11.5 \times 10^3}$$

$$= 8.69 \text{ N/mm}^2$$

5. A steel bar 10 mm wide and 8 mm thick is subjected to bending moment. The radius of neutral surface is 100 cm. Determine the maximum and minimum bending stress in the beam. (UPTU: Feb. 2002)

For pure bending,

$$\frac{\sigma}{y} = \frac{E}{R}$$

Given $R = 1000$ mm.

$$y = \frac{h}{2} = \frac{8}{2} = 4 \text{ mm}$$

$$E = 210 \text{ kN/mm}^2 \text{ (For steel)}$$

$$\sigma_{max} = \frac{E}{R} \times y = \frac{210 \times 4}{1000} \times 10^3$$

$$= 840 \text{ N/mm}^2$$

$$\sigma_{min} = 0 \text{ at the neutral axis}$$

6. Three beams have the same length, the same allowable stress and the same bending moment. The cross section of the beams are a square, a rectangle with the depth twice the breadth and a circle. Determine the ratio of weights of the circular and rectangular beams with respect to the square beam. (UPTU: Feb. 2001)

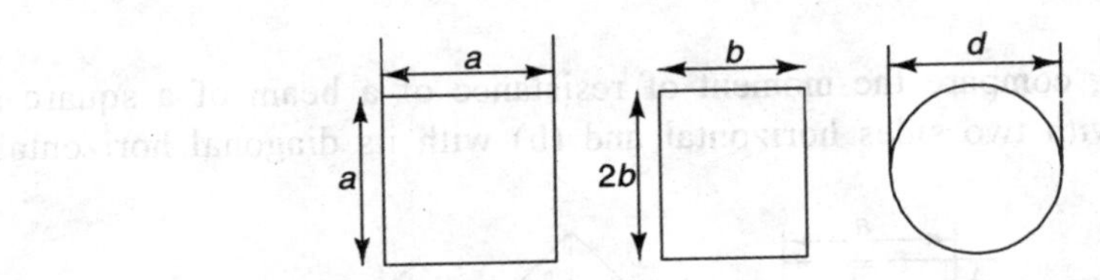

σ_{max} and BM are the same for all three beams. Hence the section modulus will also be the same for all three beams.

$$z_1 = \frac{aa^2}{6} = \frac{a^3}{6}$$

$$z_2 = \frac{b(2b)^2}{6} = \frac{4b^3}{6} = \frac{2b^3}{3}$$

$$z_3 = \frac{\pi}{32} d^3$$

$$z_1 = z_2$$

Therefore, $$\frac{a^3}{6} = \frac{2b^3}{3}$$

$$a^3 = 4b^3$$

$$b = 0.63a$$

$$z_3 = z_1$$

Therefore, $$\frac{\pi \times d^3}{32} = \frac{a^3}{6}$$

$$d^3 = \frac{32}{\pi} \times \frac{a^3}{6}$$

$$d = 1.19a$$

$$\frac{\text{Weight of the rectangular beam}}{\text{Weight of the square beam}} = \frac{2b \times b \times \rho \times l}{a^2 \times \rho} = \frac{2 \times (0.63)^2 a^2}{a^2}$$

$$= 0.79$$

$$\frac{\text{Weight of the circular beam}}{\text{Weight of the square beam}} = \frac{\frac{\pi}{4} d^2 \times l}{a^2 \times l} = \frac{\frac{\pi}{4} \times (1.190)^2 a^2}{a^2}$$

$$= \frac{\pi}{4} \times 1.416 = 1.11.$$

7. For a given stress, compare the moment of resistance of a beam of a square section when placed (a) with two sides horizontal and (b) with its diagonal horizontal.

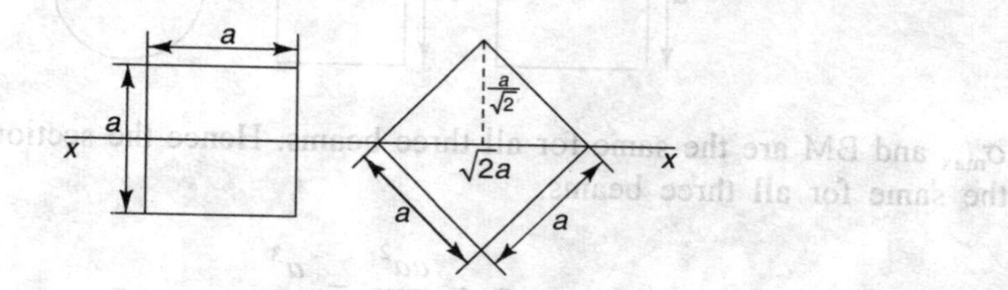

(a) $$I_1 = \frac{a \cdot a^3}{12} = \frac{a^4}{12}$$

$$y_{max} = \frac{a}{2}$$

$$\therefore \quad z_1 = \frac{a^4 \times 2}{12 \times a} = \frac{a^3}{6}$$

(b) $$\text{Height} = \frac{a}{\sqrt{2}}$$

$$\text{Base} = \sqrt{2}\, a$$

$$I_2 = 2 \times \frac{1}{12} \times (\text{base})\ (\text{height})^3$$

$$= 2 \times \frac{1}{12} \times \sqrt{2} \times a \times \frac{a^3}{2 \times \sqrt{2}}$$

$$= \frac{a^4}{12}$$

$$y_{max} = \frac{a}{\sqrt{2}}$$

$$\therefore \quad z_2 = \frac{I}{y_{max}} = \frac{a^4 \times \sqrt{2}}{12 \times a} = \frac{a^3}{6\sqrt{2}}$$

Now $$\frac{M_1}{M_2} = \frac{z_1}{z_2} = \frac{a^3/6}{a^3/6\sqrt{2}} = \sqrt{2} = 1.414$$

8. A rectangular beam of 200 mm in width and 400 mm in depth is simply supported over a span of 4 m and carries a distributed load of 10 kN/m. Determine the maximum bending stress in the beam. (UPTU: 2003–2004)

$$M = \frac{wl^2}{8} = \frac{10 \times 4^2}{8} = 20 \text{ kN-m}$$

$$z = \frac{bd^2}{6} = \frac{0.2 \times (0.4)^2}{6} = 5.33 \times 10^{-3} \text{ m}$$

$$\sigma = \frac{M}{z} = \frac{20}{5.33 \times 10^{-3}} = 3752 \text{ N/m}^2$$

9. A 300 mm deep rectangular beam is simply supported over a span of 5 m wide. What uniformly distributed load per metre the beam may carry if bending stress is not to exceed 110 N/mm^2. Take I = 8.5 × 10^6 mm^2 (UPTU: Dec. 2003)

$$M = \frac{wl^2}{8} = \frac{w \times 25}{8}$$

$$z = \frac{I}{y} = \frac{8.5 \times 10^6}{150} \qquad \left(\because y = \frac{d}{2}\right)$$

$$= 5.67 \times 10^4$$

$$M = z\sigma$$

$$10^3 \times w \times \frac{25}{8} = 5.67 \times 10^4 \times 110$$

$$w = \frac{8 \times 5.67 \times 10^4 \times 110}{25 \times 10^3} = 1996 \text{ N/mm}^2$$

$$= 1.996 \text{ kN/mm}^2$$

10. Calculate the maximum tensile and maximum compressive stress developed in the cross section of a beam subjected to a moment of 30 kN m.

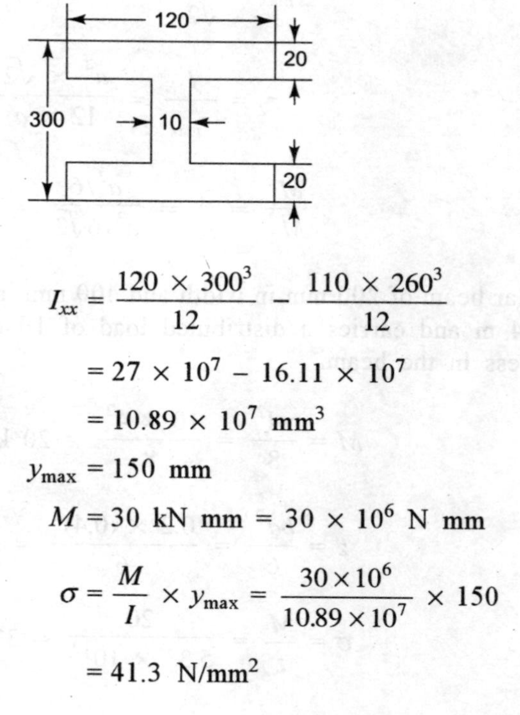

$$I_{xx} = \frac{120 \times 300^3}{12} - \frac{110 \times 260^3}{12}$$

$$= 27 \times 10^7 - 16.11 \times 10^7$$

$$= 10.89 \times 10^7 \text{ mm}^3$$

$$y_{max} = 150 \text{ mm}$$

$$M = 30 \text{ kN mm} = 30 \times 10^6 \text{ N mm}$$

$$\sigma = \frac{M}{I} \times y_{max} = \frac{30 \times 10^6}{10.89 \times 10^7} \times 150$$

$$= 41.3 \text{ N/mm}^2$$

$$\sigma_{max} \text{ (tensile)} = 41.3 \text{ N/mm}^2$$

$$\sigma_{max} \text{ (compressive)} = -41.3 \text{ N/mm}^2$$

11. A short hollow column having 200 mm external diameter and 120 mm internal diameter is subjected to vertical compressive load acting at an eccentricity of 60 mm from the axis of the column. Find the greatest allowable load if the permissible stresses are 80 N/mm² in compression and 20 N/mm² in tension. (UPTU: 2002–2003)

$$D_0 = 200 \text{ mm}, \; D_1 = 120 \text{ mm}, \; e = 60 \text{ mm}$$

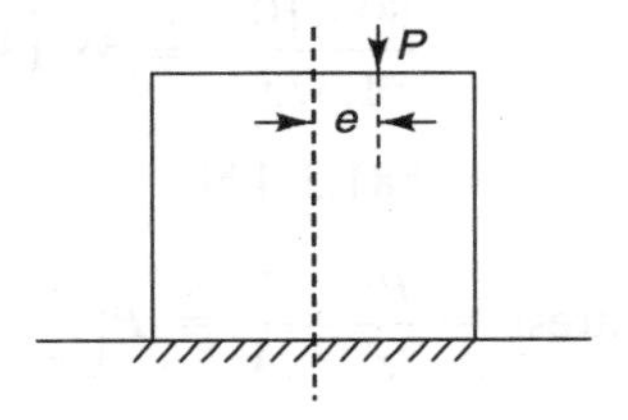

$$\text{Compressive stress } \sigma_c = \frac{P}{A} = \frac{P}{\frac{\pi}{4}(D_0^2 - D_1^2)} = \frac{P}{\frac{\pi}{4}(200^2 - 120^2)}$$

$$= \frac{P}{2 \times 10^4}$$

$$\text{Bending stress } \sigma_b = \frac{Pe}{z}$$

$$z = \text{Section modulus}$$

$$= \frac{\pi}{32} \frac{(D_0^4 - D_1^4)}{D_0}$$

$$= \frac{\pi}{32} \times \frac{(200^4 - 120^4)}{200}$$

$$= \frac{\pi}{32} \times \frac{(16 - 2.07)\,10^6}{2}$$

$$= \frac{\pi}{32} \times \frac{13.93}{2} \times 10^6 = 6.83 \times 10^5 \text{ mm}^3$$

$$\sigma_b = \frac{P \times 60}{6.83 \times 10^5}$$

Total compressive stress $\sigma_1 = \sigma_c + \sigma_b$

$$80 = \frac{P}{2 \times 10^4} + \frac{P \times 60}{6.85 \times 10^5}$$

$$P = \frac{80}{\frac{1}{2 \times 10^4} + \frac{60}{6.83 \times 10^5}} = \frac{80 \times 10^4}{0.5 + 0.875}$$

$$= \frac{80 \times 10^4}{1.375} = 58.13 \times 10^4 \text{ N}$$

$$= 581.3 \text{ kN}$$

$$\text{Total tensile stress} = \frac{Pe}{z} - \frac{P}{A} = P\left(\frac{e}{z} - \frac{1}{A}\right)$$

$$P = \frac{20 \times 10^4}{0.875 - 0.5} = \frac{20}{0.375} \times 10^4 = 53.2 \times 10^4 \text{ N}$$

$$= 532 \text{ kN}$$

Hence maximum permissible load = 532 kN

12. A short column of 20 cm external diameter and 15 cm internal diameter is subjected to a load. The stress measurements indicate that the stress varies from 150 MN/m^2 compressive at one end to 25 MN/m^2 tensile on the other end. Estimate the load and distance of the line of action from the axis of the column.

$$z = \text{Section modulus} = \frac{\pi}{32}\left(\frac{D_0^4 - D_1^4}{D_0}\right) = \frac{\pi}{32}\left(\frac{20^4 - 15^4}{20}\right)$$

$$= 537 \times 10^{-6} \text{ m}^3$$

$$A = \frac{\pi}{4}(20^2 - 15^2)$$

$$= 137.5 \times 10^{-4} \text{ m}^2$$

$$\text{Compressive } \sigma_{max} = 150 = \frac{w}{A} + \frac{we}{z} \quad \text{(i)}$$

$$\text{Tensile } \sigma_{max} = 25 = \frac{we}{z} - \frac{w}{A} \quad \text{(ii)}$$

Subtracting Eq. (ii) from Eq. (i),

$$\frac{2w}{A} = 125$$

$$\therefore \quad w = \frac{125 \times 137.5 \times 10^{-4}}{2} = 859 \text{ kN}$$

Adding Eqs. (i) and (ii)

$$\frac{2 \times we}{z} = 175$$

or

$$e = \frac{175 \times z}{2w} = \frac{175 \times 537 \times 10^{-6}}{2 \times 859 \times 10^{3}} \times 10^{3}$$

$$= 54.7 \text{ mm}$$

13. The moment of inertia of a beam section of 50 cm in depth is 69490 cm^2. Find the longest span over which a beam of the section, when simply supported, could carry a uniformly distributed load of 50 kN per metre run. The maximum flange stress in the material is not to exceed 110 N/mm^2. (UPTU: Sample question)

$$M = \frac{wl^2}{8} = \frac{50l^2}{8} \text{ kN mm}$$

$$I = 69490$$

$$z = \frac{69490}{25} = 2779.6 \text{ cm}^2$$

$$M = \sigma_{max}\, z$$

$$\frac{50}{8} \times l^2 = 110 \times 27796 \times 10^{+2}$$

$$l^2 = \frac{8 \times 110 \times 2779.6 \times 10^2}{50}$$

$$= 48.92 \times 10^2 \text{ cm}$$

$$l = 699.4 \text{ cm}$$

$$\approx 7 \text{ m}$$

14. The beam of symmetrical I section is simply supported over a span of 9 m. If the maximum permissible stress is 75 N/mm^2, what concentrated load can be carried at a distance of 3 m from one support. Take $I = 31 \times 10^6$ mm^4.

(UPTU: Sample question)

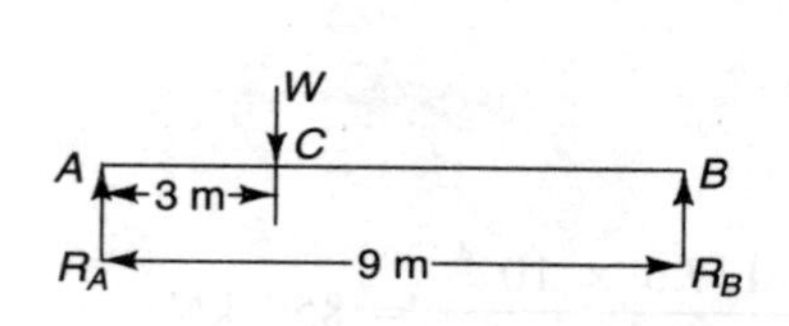

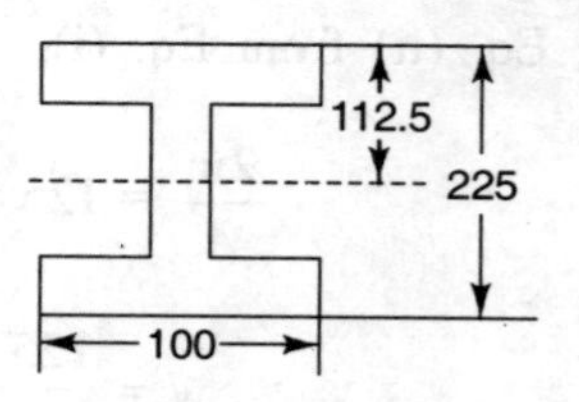

Find R_A and R_B.

$$R_A + R_B = W$$

$$\Sigma M_A = 0$$

$$R_B \times 9 = W \times 3$$

$$\therefore \quad R_B = \frac{W}{3}$$

$$\text{Maximum BM} = R_B \times 6 = \frac{W}{3} \times 6 = 2W \text{ Nm}$$

$$z = \frac{I}{y}$$

$$y = \frac{d}{2} = 112.5$$

Therefore,
$$z = \frac{31 \times 10^6}{112.5} = 27.56 \times 10^4 \text{ mm}^3$$

$$M = z\sigma$$

$$10^3 \times 2W = 27.56 \times 10^4 \times 75$$

or
$$W = \frac{27.56 \times 75}{2} \times 10$$

$$= 10335 \text{ N}$$

$$= 10.335 \text{ kN}$$

15. Find the dimensions of the strongest rectangular beam that can be cut out of a log of 200 mm diameter. (UPTU: Sample question)

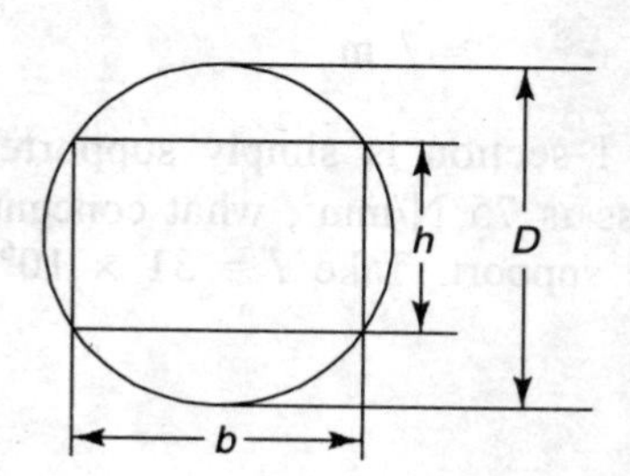

Let a beam of b width and h height be cut from the log of diameter D. Therefore,

$$D^2 = b^2 + h^2$$

Also for the beam, section modulus $z = \frac{bh^2}{\sigma}$

$$= \frac{b(D^2 - b^2)}{6}$$

For maximum z, $\frac{dz}{db} = 0$, i.e.,

$$\frac{1}{6}(D^2 - 3b^2) = 0$$

or $$3b^2 = D^2$$

or $$b = \frac{D}{\sqrt{3}} = 0.577\ D$$

$\therefore$ $$h^2 = D^2 - b^2 = D^2 - \frac{D^2}{3} = \frac{2}{3}D^2$$

or $$h = 0.816\ D$$

Hence the strongest beam has:

$$\text{width} = 0.577D = 0.577 \times 200 = 115.4 \text{ mm}$$
$$\text{height} = 0.8160 = 0.816 \times 200 = 163.2 \text{ mm}$$

16. Consider a composite beam of cross sectional dimensions as shown below. $E_w = 10$ GPa and $E_s = 200$ GPa. If the beam is subjected to a BM of 30 kN m around the horizontal axis, what is the maximum stress in the steel and wood?

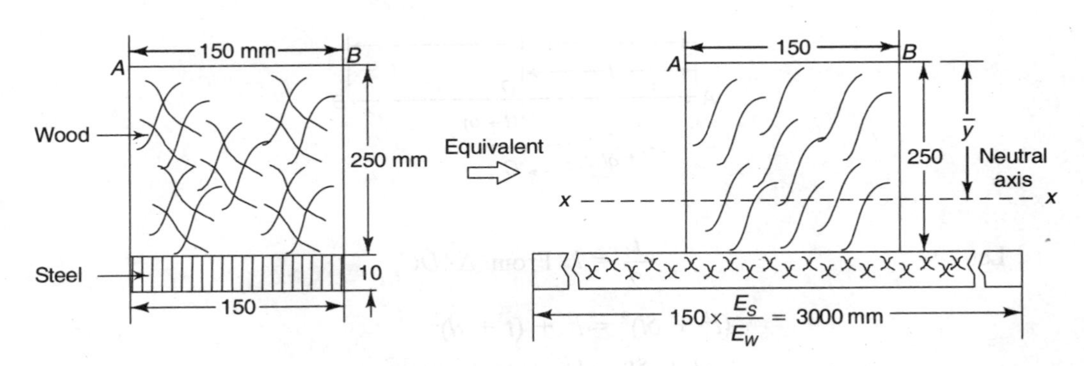

$$\frac{E_s}{E_w} = \frac{200}{10} = 20$$

Hence transformed length of the steel section in length is 150 × 20 = 3000 mm
Now to find the CG from top (AB)

$$\bar{y} = \frac{150 \times 250 \times 125 + 10 \times 3000 \times 255}{150 \times 250 + 10 \times 3000} = 183 \text{ mm}$$

$$I_{xx} = \frac{150 \times 250^3}{12} + (150 \times 250) \times (183 - 125)^2$$

$$+ \frac{3000 \times 10^3}{12} + (10 \times 3000) \times (255 - 183)^2$$

$$= 478 \times 10^6 \text{ mm}^4$$

$$(\sigma_w)_{\max} = \frac{M}{I} \times y_{\text{top}} = \frac{30 \times 10^9 \times 183}{478 \times 10^6} = 11.5 \text{ MPa}$$

$$(\sigma_s)_{\max} = n\sigma_w$$

$$= 20 \times \frac{30 \times 10^9 \times 77}{478 \times 10^6} = 96.7 \text{ MPa}$$

17. A long rod of a uniform rectangular section and thickness t, originally straight is bent into the form of a circular arc and the displacement d of the midpoint of a length L, is measured by means of a dial gauge. If d is regarded as small as compared to L, show that the longitudinal strain (ε) in the rod is given by $\varepsilon = \frac{4td}{L^2}$.

(UPTU: Aug. 2001)

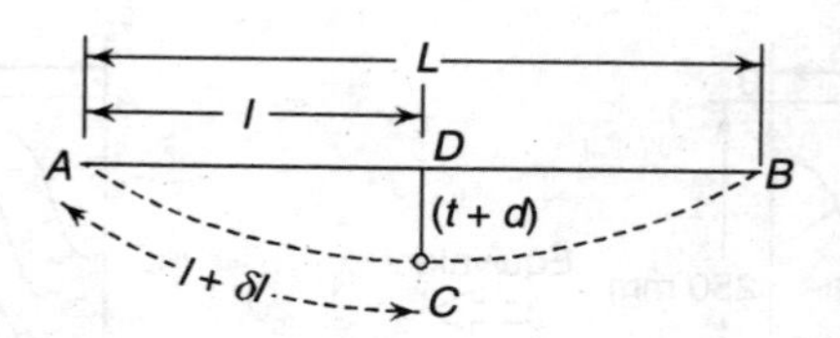

Let $\frac{L}{2} = l$. From ΔADC,

$$(l^2 + \delta l)^2 = l^2 + (t + d)^2$$

$$(l + \delta l) = [l^2 + (t + d)^2]^{1/2}$$

$$1+\frac{\delta l}{l} = \left[1+\left(\frac{t+d}{1}\right)^2\right]^{1/2}$$

$$\frac{\delta l}{l} = \left[1+\left(\frac{t+d}{1}\right)^2\right]^{1/2} - 1$$

$$= 1+\frac{1}{2}\left(\frac{t+d}{l}\right)^2 - 1$$

$$= \frac{1}{2}\left(\frac{t^2+d^2+2dt}{l^2}\right)$$

Neglecting t^2 and d^2 being small,

$$\frac{\delta l}{l} = \frac{dt}{l^2}$$

$$\varepsilon = \frac{dt}{l^2} = \frac{4dt}{L^2}$$

18. Determine the moment of inertia of I section of the following dimensions about an axis passing through the centroid and parallel to the flange.

Top flange: 100 × 10 mm

Web: 10 × 100 mm

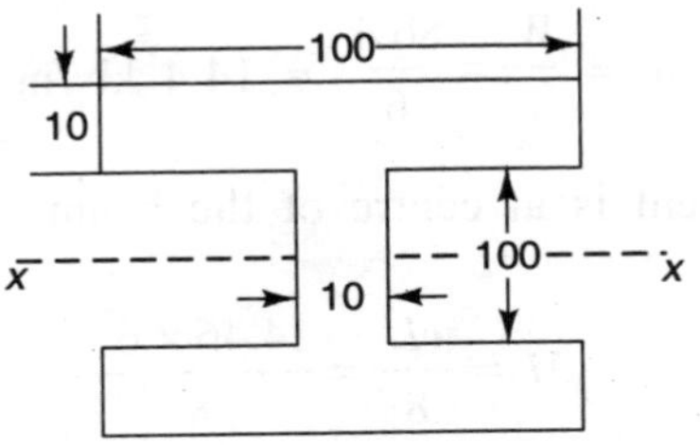

First method:

$$I_{xx} = 2\left[\frac{100 \times 10^3}{12} + (10 \times 100)(55)^2\right] + \frac{10 \times 100^3}{12}$$

$$= 2(8.35 \times 10^3 + 3025 \times 10^3) + 8.33 \times 10^5$$

$$= 6066.70 \times 10^3 + 833 \times 10^3$$

$$= 6899.70 \times 10^3 \text{ mm}^4 \approx 69.0 \times 10^5 \text{ mm}^4$$

Second method:

$$I_{xx} = \frac{100 \times 120^3}{12} - \frac{90 \times 100^3}{12}$$

$$= 144 \times 10^5 - 7.5 \times 10^6$$

$$= 69 \times 10^5 \text{ mm}^4$$

19. Determine the dimensions of a simply supported rectangular steel beam 6 m long to carry a brick wall 250 mm thick and 3 m high, if brick work weighs 19.2 kN/m^2 and maximum permissible bending stress is 800 N/cm^3. The depth of beam is 3/2 times its width.

The free body diagram of beam is

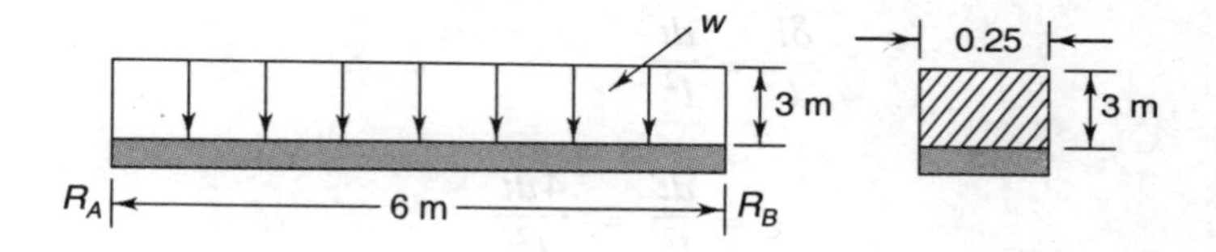

$$\text{Volume of brick wall} = 6 \times 3 \times 0.25$$

$$= 4.5 \text{ m}^3$$

$$\text{Total weight } W = \text{volume} \times \text{density}$$

$$= 4.5 \times 19.2 \text{ kN}$$

$$= 86.4 \text{ kN}$$

$$\therefore \qquad w = \frac{W}{L} = \frac{86.4}{6} = 14.4 \text{ kN/m}$$

Maximum bending moment is at centre of the beam

$$M = \frac{wL^2}{8} = \frac{14.46 \times 6^2}{8}$$

$$= 65.10 \text{ kN m}$$

$$I = \frac{b \cdot d^3}{12} = \left(\frac{2d}{3}\right) \cdot \frac{d^3}{12} = \frac{d^4}{18}$$

Now

$$\frac{M}{I} = \frac{\sigma_b}{d/2}$$

$$\sigma_b = 800 \text{ N/cm}^3$$

$$= 800 \times 10^6 \text{ N/m}^2$$

$$\frac{d^4 \times 2}{18 \times d} = \frac{65.1 \times 10^3}{800 \times 10^6}$$

$$d^3 = \frac{9 \times 65.1 \times 10^{-3}}{800}$$

$$= 0.732 \times 10^{-3}$$

$$d = 0.09 \text{ m}$$

$$= 90 \text{ mm}$$

$$b = \frac{2}{3} \times d = \frac{2}{3} \times 90$$

$$= 60 \text{ mm}$$

20. A cantilever beam of rectangular cross-section is 1.0 m deep and 0.6 m thick. If the beam was to be 0.6 m deep and 1 m thick, then the beam would be weakend by (a) 0.5 times, (b) 0.6 times, (c) 0.7 times, (d) 0.8 times.

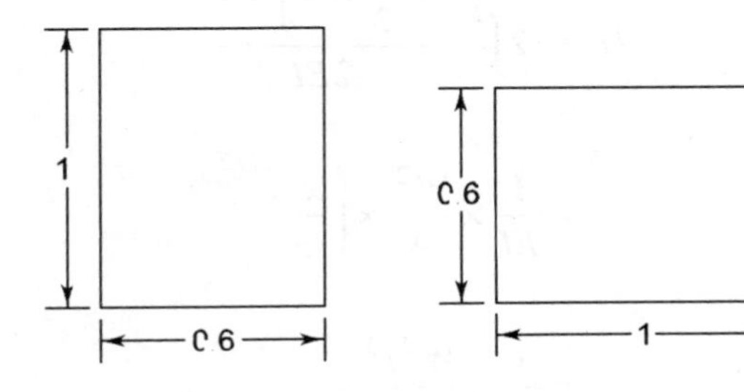

$$M = Z \times \sigma_b$$

$$M \propto Z \text{ if } \sigma_b \text{ is constant}$$

The beam having bigger sectional modulus (Z) will be stronger

$$Z_1 = \frac{bh^2}{6} = \frac{0.6 \times 1}{6} = 0.1$$

$$Z_2 = \frac{1 \times 0.6^2}{6} = 0.06$$

$$\frac{Z_2}{Z_1} = \frac{0.06}{0.1} = 0.6$$

Therefore, option (b) is correct.

21. Total strain energy stored in a simply supported beam of span L and flexural rigidity EI subjected to concentrated load 'W' at the centre is equal to

(a) $\dfrac{W^3L^3}{40\,EI}$ (b) $\dfrac{W^2L^3}{60\,EI}$ (c) $\dfrac{W^2L^3}{96\,EI}$ (d) $\dfrac{W^2L^3}{240\,EI}$

Strain energy

$$U = \int_0^L \frac{M^2 dx}{2EI}$$

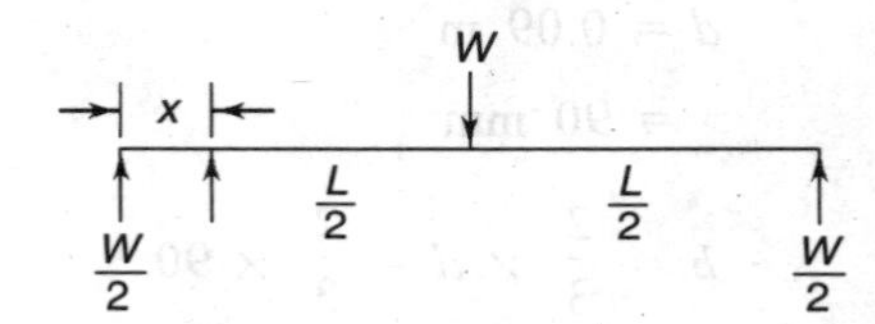

Now

$$M = \frac{W}{2} \times x$$

$\therefore$

$$U = 2\int_0^{L/2} \frac{\left(\frac{W}{2} \times x\right)^2 dx}{2EI}$$

$$= \frac{1}{EI} \times \frac{W^2}{4} \times \left(\frac{x^3}{3}\right)_0^{L/2}$$

$$= \frac{1}{96} \times \frac{W^2L^3}{EI}$$

Option (c) is correct.

22. A CI pipe of wall thickness 10 mm and outside diameter 120 mm carries water and is supported at a distance of 9 m. Calculate the value of maximum bending stress and its nature when water is running full. Take the density of water as 1 g/cc and that for CI as 7 g/cc. (UPTU: 2007–2008)

Weight of water per unit length is

$$\frac{\pi}{4} \times (0.1)^2 \times 1 \times 10^3 \times 9.81 = 308 \text{ N/m}$$

Weight of CI pipe per unit length is

$$\frac{\pi}{4}(0.12^2 - 0.1^2) \times 1 \times 7 \times 10^3 \times 9.81 = 237 \text{ N/m}$$

$$w = \text{uniformly distributed load}$$

$$= 308 + 273 = 545 \text{ N/m}$$

w = 545 N/m

9 m

$$M_{max} = \frac{wl^2}{8} = \frac{545 \times 9 \times 9}{8}$$

$$= 5520 \text{ N m}$$

$$I = \frac{\pi(d_0^4 - d_1^4)}{64} = \frac{\pi(0.12^4 - 0.10^4)}{64}$$

$$= 5.27 \times 10^{-6} \text{ m}^4$$

$$y_{max} = \frac{d_0}{2} = \frac{0.12}{2} = 0.06$$

$$\frac{M}{I} = \frac{\sigma_b}{y_{max}}$$

or
$$\sigma_b = \frac{M}{I} \times y_{max}$$

$$\sigma_b = \frac{5520 \times 0.06}{5.27 \times 10^{-6}} = 62.85 \times 10^6 \text{ Pa}$$

23. Determine the dimensions of a rectangular simply supported steel beam 5 m long to carry an UDL of 10 kN/m, if the maximum permissible bending stress is 1000 N/cm The depth of the beam is 1.5 times its width. (UPTU: May 2008)

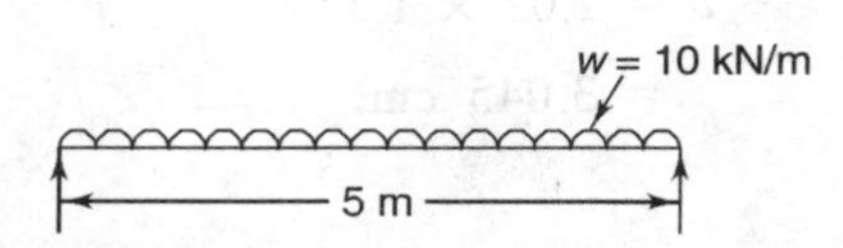

$$\text{Maximum BM} = M = \frac{wl^2}{8}$$

$$= \frac{10 \times 5^2}{8} = 31.25 \text{ kN m}$$

$$I = \frac{bd^3}{12}$$

But $d = 1.5\ b$. Therefore

$$I = \frac{b \times (1.5b)^3}{12} = 0.28\ b^4$$

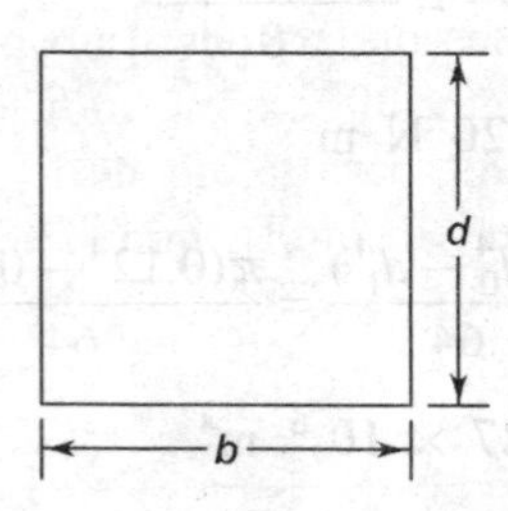

Now

$$\frac{M}{I} = \frac{\sigma}{y} = \frac{\sigma}{d/2} = \frac{\sigma}{1.5b/2}$$

or

$$\frac{31.32}{0.28b^4} = \frac{10^3 \times 10^4}{0.75\,b}$$

or

$$b^3 = \frac{31.32 \times 0.75}{0.28 \times 10^7}$$

$$= 8.38 \times 10^{-6}$$

$$b = 2.03 \times 10^{-2} \text{ m}$$

$$= 2.03 \text{ cm}$$

and

$$d = 2.02 \times 1.5$$

$$= 3.045 \text{ cm}$$

OBJECTIVE TYPE QUESTIONS

The world is full of problems. Wisdom is full of solutions.

State True or False

1. Homogeneous material is the same kind throughout. (*True/False*)

2. Isotropic material has elastic properties which are equal in all directions. (*True/False*)

3. Sagging bending moment is negative. (*True/False*)

4. Hogging bending moment is positive. (*True/False*)

5. The layer at the neutral axis remains unchanged after bending. (*True/False*)

6. The layer farthest from the neutral axis will have maximum stress. (*True/False*)

7. Top and bottom layers of a beam will have the same stress when the beam bends. (*True/False*)

8. The moment of resistance depends upon the section modulus of the beam. (*True/False*)

9. The moment of resistance is proportional to the width of the beam. (*True/False*)

10. The moment of resistance is proportional to the depth of the beam. (*True/False*)

11. The first moment of area of a section from the axis of centroid is unity. (*True/False*)

12. The second moment of area of a section from the neutral axis is moment of inertia. (*True/False*)

13. The centroid of a section lies on the neutral axis. (*True/False*)

14. A section of steel of size 10×1 cm can be transformed into 150×1 cm of wood if $E_s/E_w = 15$. (*True/False*)

15. The radius of gyration for a section having moment of inertia = 100 cm^4 and area = 4 cm^2 will be 10 cm. (*True/False*)

16. The unit radius of inertia of a beam with circular section is $1/4\ \pi$. (*True/False*)

17. The unit radius of inertia of a section with I = 100 cm^2 and area = 5 is 4 π. (*True/False*)

18. Polar moment of inertia is with respect to the axis of rotation of a shaft. (*True/False*)

19. Polar moment of inertia is the difference between moments of inertia about *x-x* axis and *y-y* axis. (*True/False*)

20. In a flitched beam, the total resisting moment is the sum of the resisting moments of the individual materials. (*True/False*)

21. A beam having extreme fibre along its length is loaded to a maximum permissible stress by varying the section is called the beam of uniform strength. (*True/False*)

22. A beam of uniform strength can be designed by varying the depth of the beam. (*True/False*)

23. A beam of uniform strength can be designed by varying the width of the beam. *(True/False)*

24. Beam of uniform strength cannot be designed by varying both the width and depth of the beam. *(True/False)*

25. The load must act in the middle third of the rectangular column for eccentric loading. *(True/False)*

26. The load must act in the middle quarter for a circular cross-sectional column for eccentric loading. *(True/False)*

27. Eccentricity $\geq \frac{b}{6}$ for a column where b is the width of the column for safety. *(True/False)*

28. Eccentricity $\geq \frac{D}{8}$ where D is the diameter of the cross-section of a column for the column to remain safe. *(True/False)*

29. A column has a rectangular section with width = 60 cm. A load is put at 12 cm from the centre. The column is safe. *(True/False)*

30. The strain energy of pure bending is $\frac{Ml}{2EI}$ where M = BM, l = length, E = Young's modulus and I = moment of inertia. *(True/False)*

31. Flitched beams have greater moment of resistance. *(True/False)*

32. A column has a diameter of 80 cm. A load is put 12 cm away from the centre. The column is safe. *(True/False)*

33. $I_{AA} = I_{yy} + Ah^2$ where A = area and h = distance between AA and yy axis. This is known as the theorem of perpendicular axis. *(True/False)*

34. $I_{zz} = I_{xx} + I_{yy}$ is based on the theorem of parallel axis. *(True/False)*

35. Flexural formula $\frac{M}{I} = \frac{E}{R} = \frac{\sigma}{y}$ is applicable even where shear stress is acting. *(True/False)*

Multiple Choice Questions

1. The moment of inertia of a rectangular section with breadth = b and depth = d is

(a) $\frac{bd^2}{12}$ (b) $\frac{b^3d}{12}$ (c) $\frac{bd^3}{12}$

2. The polar moment of inertia of a circular section with diameter = D is

(a) $\frac{\pi D^4}{64}$ (b) $\frac{\pi D^4}{16}$ (c) $\frac{\pi D^4}{32}$

3. The section modulus of a rectangular section is

(a) $\frac{bd^2}{12}$ (b) $\frac{bd^2}{6}$ (c) $\frac{b^2d}{6}$

4. The section modulus is equal to

(a) $M\sigma$ (b) $\frac{M}{\sigma}$ (c) $\frac{\sigma}{M}$

5. The section modulus of a circular section is

(a) $\frac{\pi D^4}{16}$ (b) $\frac{\pi D^3}{32}$ (c) $\frac{\pi D^3}{16}$

6. The radius of gyration for a circular section is

(a) $D/6$ (b) $D/4$ (c) $D/2$

7. The moment of inertia of a hollow rectangular section is

(a) $\frac{b_o d_o^3}{12} + \frac{b_i d_i^3}{12}$ (b) $\frac{b_o d_o^2}{12} - \frac{b_i d_i^3}{12}$ (c) $\frac{b_o d_o^3}{12} - \frac{b_i d_i^3}{12}$

8. The polar moment of inertia of a hollow circular section is

(a) $\frac{\pi}{64}(D_0^4 - D_1^4)$ (b) $\frac{\pi}{32}(D_0^4 + D_1^4)$ (c) $\frac{\pi}{32}(D_0^4 - D_1^4)$

9. The flexural formula is given by

(a) $\frac{M}{I} = \frac{\sigma}{R} = \frac{E}{y}$ (b) $\frac{M}{I} = \frac{\sigma}{y} = \frac{E}{R}$ (c) $\frac{M}{y} = \frac{\sigma}{I} = \frac{E}{R}$

10. The flexural formula is applicable where we have
(a) bending and shear stress
(b) shear stress is zero
(c) bending stress but shear stress is zero

11. Section modulus is given by the ratio of
(a) bending stress to bending moment
(b) shear stress to bending moment
(c) bending moment to maximum bending stress

12. The moment of inertia of a triangular section with base = b and height = h with respect to the base is

(a) $\frac{bh^3}{12}$ (b) $\frac{bh^3}{36}$ (c) $\frac{bh^3}{24}$

13. Strain energy in bending is

(a) $\frac{Ml}{2EI}$ (b) $\frac{M^2 l}{2EI}$ (c) $\frac{M^2 l}{EI}$

14. Eccentricity (e) of a column with respect to the width of section (b) is

(a) $e \le \frac{b}{6}$ (b) $e \ge \frac{b}{6}$ (c) $e \le \frac{b}{4}$

15. Eccentricity (e) of a column with respect to the diameter (D) of the section is

(a) $e \le \dfrac{D}{6}$ (b) $e \le \dfrac{D}{8}$ (c) $e \le \dfrac{D}{10}$

16. $\dfrac{PL^3}{3EI}$ is the deflection under the load P of a cantilever beam (length = L, E = Young's modulus, I = moment of inertia). The strain energy due to bending is

(a) $\dfrac{P^2L^2}{6EI}$ (b) $\dfrac{P^2L^3}{3EI}$ (c) $\dfrac{PL^4}{3EI}$

17. A steel wire of 20 mm diameter is bent into a circular shape of 10 m radius. If $E = 2 \times 10^6$ kg/cm², then the maximum stress induced in the wire is

(a) 2×10^3 kg/cm² (b) 4×10^3 kg/cm² (c) 6×10^3 kg/cm

18. Magnitude of the bending moment at the fixed support of the beam is equal to

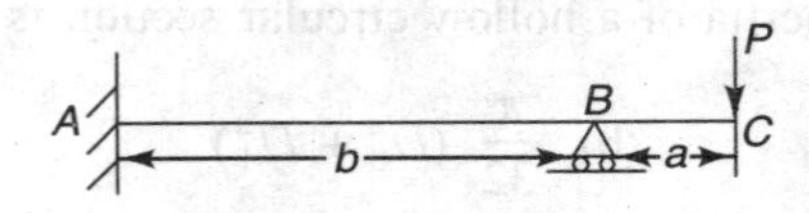

(a) Pa (b) $\dfrac{Pa}{2}$ (c) Pb

19. The deflection of a cantilever beam at free end B applied with a moment M at the same point is

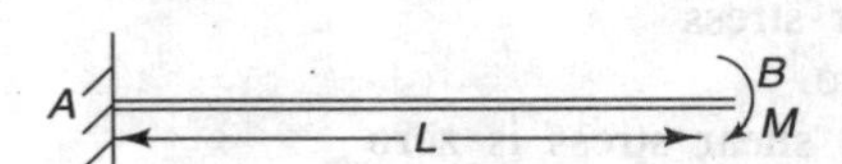

(a) $\dfrac{ML^2}{EI}$ (b) $\dfrac{ML^2}{2EI}$ (c) $\dfrac{ML^2}{3EI}$

20. The reaction at support B of the structure is

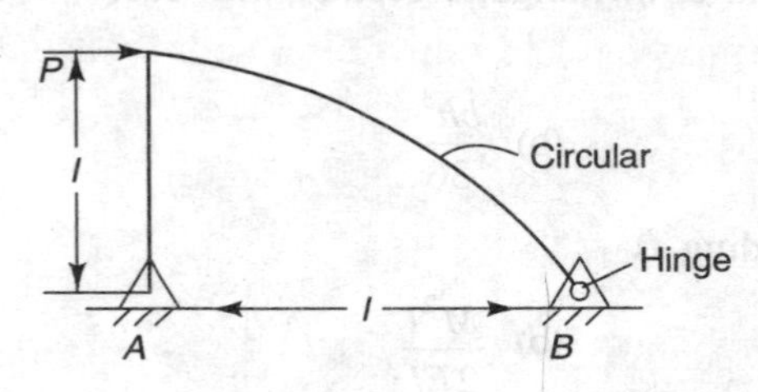

(a) P (b) $\dfrac{P}{2}$ (c) $\sqrt{2}\,P$

21. A cantilever beam is as shown below. The moment to be applied at the free end for zero vertical deflection at that point is

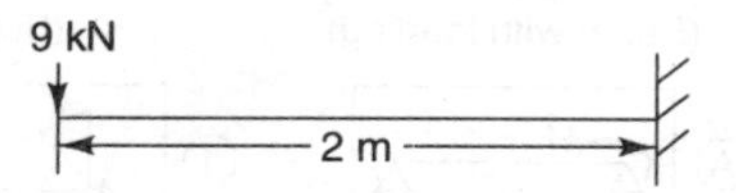

(a) 9 kN m clockwise
(b) 12 kN m clockwise
(c) 12 kN m anticlockwise

22. Match list I and list II and select the correct answer using the codes given below the lists.

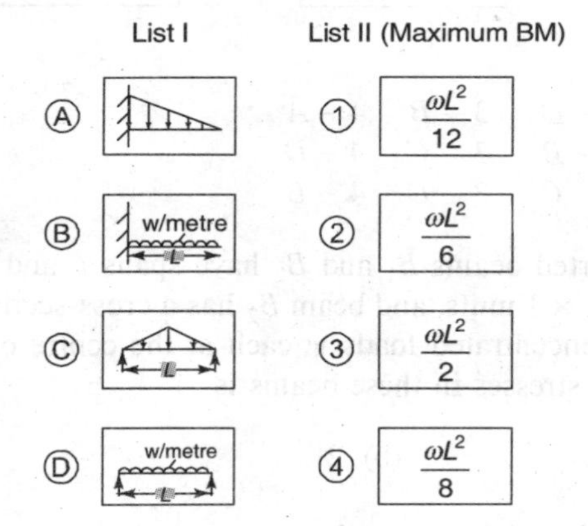

Codes

	A	B	C	D
(a)	2	3	1	4
(b)	1	2	3	4
(c)	4	3	1	2

23. If the area under the shear curve for a beam between two points X_1 and X_2 is k, then the difference between the moments of the two points x_1 and x_2 will be
(a) k (b) $2k$ (c) k^2

24. The ratio of the flexural strength of two beams of a square cross section, the first beam being placed with its top and bottom sides horizontally and the second beam being placed with one diagonal horizontal, is

(a) $\sqrt{2}$ (b) $\frac{1}{\sqrt{2}}$ (c) 2

25. Match list I with list II and select the correct answer using the codes given below the list.

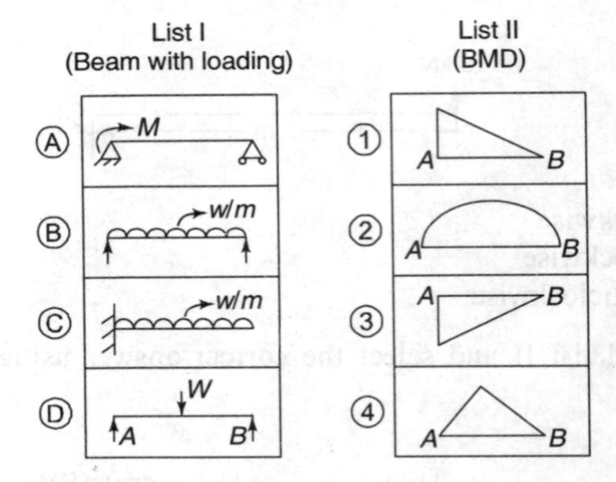

Codes

(a) $1-C$ $2-D$ $3-B$ $4-A$
(b) $1-A$ $2-B$ $3-C$ $4-D$
(c) $1-A$ $2-C$ $3-D$ $4-B$

26. Two simply supported beams B_1 and B_2 have spans L and $2L$ respectively. Beam B_1 has a cross section of 1×1 units, and beam B_2 has a cross-section of 2×2 units. These beams are subjected to concentrated loads w each at the centre of their spans. The ratio of the maximum flexural stresses in these beams is

(a) 4 (b) 2 (c) $\frac{1}{4}$

27. A circular beam of uniform strength can be made by varying the diameter in such a way that

(a) $\frac{M}{z}$ is constant (b) $\frac{\sigma}{y}$ is constant (c) $\frac{E}{R}$ is constant

28. An eccentric load W with eccentricity e is equivalent to
(a) an axial load W (b) a moment equal to $W \times e$ (c) both (a) and (b)

29. For no tension in a section (d = depth of section, k = radius of gyration), the eccentricity must not exceed

(a) $\frac{k^2}{d}$ (b) $\frac{2k^2}{d}$ (c) $\frac{4k^2}{d}$

30. The diameter of kernel of circle section is

(a) $\frac{d}{4}$ (b) $\frac{d}{8}$ (c) $\frac{d}{2}$

31. The diameter of kernel of hollow circular section is

(a) $\frac{D^2+d^2}{4D}$ (b) $\frac{D^2+d^2}{2D}$ (c) $\frac{D^2+d^2}{8D}$

32. In a rectangular section, the stress will be of the same sign throughout if the load acts in the

(a) middle third of the section of the column
(b) first third of the section of the column
(c) last third of the section of the column

33. The brick chimney of a round cross section is stable if the load lies in the

(a) first quarter of the section
(b) middle quarter of the section
(c) last quarter of the section

34. The second moment of a circular area about the diameter is given by

(a) $\frac{\pi D^4}{64}$ (b) $\frac{\pi D^4}{32}$ (c) $\frac{\pi D^4}{16}$

35. A concentrated load of P acts on a simply supported beam of span L at a distance $\frac{L}{3}$ from the left support. The bending moment is given by

(a) $\frac{2}{9}$ PL (b) $\frac{PL}{3}$ (c) $\frac{2PL}{3}$

36. Bars AB and BC, each of negligible mass support load, are shown below. All joints are hinged.

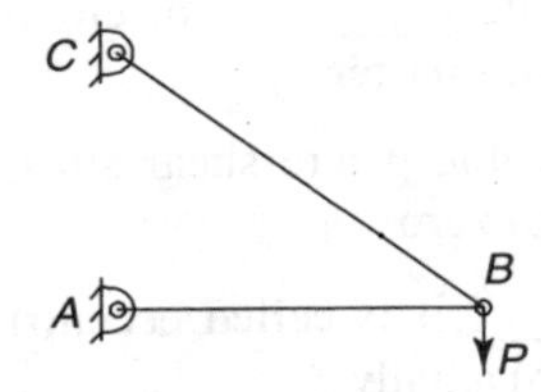

(a) Neither bar is subjected to bending.
(b) AB is in bending but BC is not in bending.
(c) AB is not in bending but BC is in bending.

37. The area moment of inertia of a square of size 1 unit about its diagonal is

(a) $\frac{1}{3}$ (b) $\frac{1}{4}$ (c) $\frac{1}{12}$

Fill in the Blanks

1. Material which has equal elastic properties in all directions is called ________.
(a) homogeneous (b) isotropic

2. Sagging bending moment is ________
(a) positive (b) negative

3. Hogging bending moment is ________.
(a) positive (b) negative

4. The layer at the ________ axis remains unchanged in length during bending.
(a) transverse (b) neutral

5. The top most layer will have ________ stress during sagging moment.
(a) tensile (b) compressive

6. The bottom most layer will have ________ stress during hogging moment.
(a) tensile (b) compressive

7. The section modulus of a rectangular section will be ________

(a) $\frac{bd^2}{6}$ (b) $\frac{bd^2}{12}$

8. The section modulus of a circular section will be ________.

(a) $\frac{\pi D^3}{64}$ (b) $\frac{\pi D^3}{32}$

9. A beam of uniform strength has a value of $\frac{M}{\sigma}$ where M = moment and σ = stress equal to ________.
(a) linear (b) constant

10. The load must lie in the ________ third of the section in a rectangular section column.
(a) first (b) middle

11. The load must lie in the middle ________ of circular section column.
(a) third (b) quarter

12. The flexural formula is applicable where shear stress is ________.
(a) maximum (b) zero

13. If load is not acting ________, it is called eccentric.
(a) horizontally (b) axially

14. Bending moment divided by the section modulus is ________.
(a) shear force (b) permissible stress

15. The moment of inertia of a rectangular section varies ________ with the depth of the section.
(a) parabolically (b) cubically

16. The moment of inertia of a rectangular section varies ________ with the width of the section.
(a) linearly (b) parabolically

17. ________ loading induces both compressive and bending stresses.
(a) Axial (b) Eccentric

18. ________ moment of area about the axis of rotation is called moment of inertia.
(a) First (b) Second

19. Eccentricity must be ________ than one sixth of the width of a column.
(a) greater (b) lesser

20. Eccentricity must be _________ than one eighth of the diameter of a column.
(a) greater (b) lesser

21. Bending moment to produce the unit radius of curvature is called _________.
(a) flexural rigidity (b) bending constant

22. The neutral axis of a section is an axis at which bending moment is _________.
(a) maximum (b) zero

23. In the theory of simple bending, the bending stress in the beam section varies _________.
(a) linearly (b) parabolically

24. When a cantilever is loaded at the free end, maximum compressive stress shall develop at the _________.
(a) bottom (b) top

25. The moment of inertia of a rectangular section is _________.

(a) $\frac{bd^3}{12}$ (b) $\frac{bd^2}{6}$

26. The moment of inertia of a circular section is _________.

(a) $\frac{\pi d^3}{32}$ (b) $\frac{\pi d^4}{64}$

27. If a square sectional beam is kept diagonal-wise, the moment of resistance is reduced by a factor of _________.

(a) 2 (b) $\sqrt{2}$

28. If two beams of the same cross sectional area and one is circular and the other is square, then the ratio of their moment resistance is _________.
(a) 1.44 (b) 0.844

29. When shear force is zero, the bending moment at that point is _________.
(a) minimum (b) maximum

30. The point of contraflexure is the point where bending moment _________.
(a) changes sign (b) maximum

31. The radius of gyration of a rectangular section about the neutral axis is _________.

$\left((a)\ \frac{d}{2\sqrt{3}} \quad (b)\ \frac{d}{2} \right)$

32. If a section has an area of 4 cm^2 and the radius of gyration is 3 cm, then moment of inertia is _________.
(a) 48 cm^4 (b) 36 cm^4

33. A section with the radius of gyration (k) = 4 cm has moment of inertia = 48 cm^4. Its moment of inertia will become 75 cm^4 in case the radius of gyration is changed to _________.
(a) 5 cm (b) 6 cm

ANSWERS

Challenges in life come in three broad categories: easy, difficult, and impossible. Those who take on only the easy have a safe and boring life. Those who take on the difficult have a tough but satisfying life. Those who take on the impossible are remembered.

State True or False

1. True
2. True
3. False (Sagging is positive bending moment)
4. False (Hogging is negative bending moment)
5. True
6. True
7. False (The magnitude of stress at top and bottom layers in the rectangular and circular sections remains the same, but one will be in tension and other in compression depending upon the direction of bending.)
8. True $\left(z = \frac{I}{y} = \frac{M}{\sigma}\right)$
9. True $\left(z = \frac{bd^2}{6}\right)$
10. False ($z \propto d^2$)
11. False (It is zero)
12. True ($I = \Sigma ay^2 = \Sigma ax^2$)
13. True
14. True (Steel: 1 × 10 = Wood: 1 × 150)
15. False $\left(\text{Radius of gyration} = \sqrt{\frac{I}{A}} = \sqrt{\frac{100}{4}} = 5\right)$

16. True $\left(\text{Unit radius of inertia} = \dfrac{I}{A^2} = \dfrac{\dfrac{\pi D^2}{64}}{\left(\dfrac{\pi}{4}D\right)^2} = \dfrac{1}{4\pi}\right)$

17. False $\left(\text{Unit radius of inertia} = \dfrac{I}{A^2} = \dfrac{100}{(5)^2} = 4\right)$

18. True

19. False (It is the sum of moments of inertia about x-x axis and y-y axis.)

20. True

21. True $\left(\text{Section modulus of the beam varies as per } \dfrac{M}{\sigma}.\right)$

22. True

23. True

24. False

25. True

26. True

27. False $\left(e = \leq \dfrac{b}{6}\right)$

28. False $\left(e = \leq \dfrac{D}{8}\right)$

29. False (For safety $e \leq \dfrac{60}{6} \leq 10$, hence the column with eccentricity = 12 is unsafe.)

30. False $\left(\text{Strain energy in pure bending is } \dfrac{M^2 l}{2EI}.\right)$

31. True ($M = M_1 + M_2$ where $M_1 >>> M_2$. For example, M of steel is very high as compared to wood.)

32. False ($e \leq \dfrac{D}{8} \leq \dfrac{80}{8} \leq 10$. Hence loading at an eccentricity of 12 cm is unsafe.)

33. False (It is as per the theorem of parallel axis.)

34. False (It is as per the theorem of perpendicular axis.)

35. False (It is applicable when a beam has pure bending or where shear force is zero.)

Multiple Choice Questions

1. (c) **2.** (c) **3.** (b) **4.** (b)

5. (b) $\left[z = \dfrac{\pi D^4/64}{D/2} = \dfrac{\pi D^3}{32}\right]$ **6.** (b) $\left[k = \sqrt{\dfrac{I}{A}} = \sqrt{\dfrac{\pi D^4/64}{\pi D^2/4}} = \sqrt{\dfrac{D^2}{16}} = \dfrac{D}{4}\right]$

7. (c) **8.** (c) **9.** (b)

10. (c) **11.** (c) **12.** (a)

13. (b) **14.** (a) **15.** (b)

16. (a) $\left[\text{Strain energy} = \dfrac{1}{2}\ P \times \delta = \dfrac{1}{2} \times P \times \dfrac{PL^3}{3EI} = \dfrac{P^2L^3}{6EI}\right]$

17. (a) $\left[\dfrac{\sigma}{y} = \dfrac{E}{R} \text{ or } \sigma = \dfrac{E}{R} \times y,\ R = 1000 \text{ cm},\ y = 1 \text{ cm},\ \sigma = \dfrac{2 \times 10^6 \times 1}{1000} = 2 \times 10^3 \text{ kg/cm}^2\right]$

18. (b)

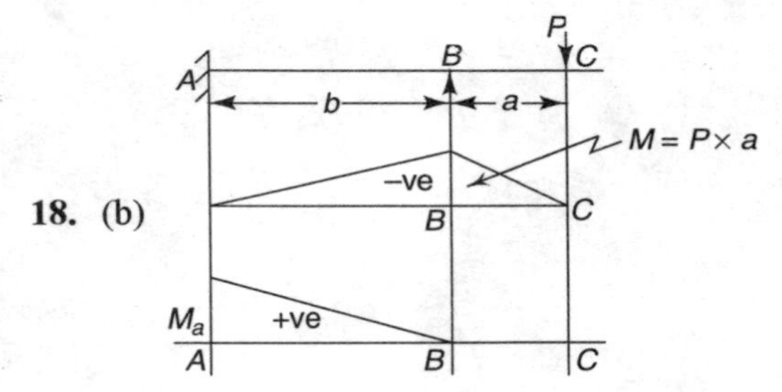

(Deflection at point B is zero, Σ BMD area moment = 0

$$\left(M_a \times \frac{b}{2}\right) \times \frac{2}{3} b - \left(\frac{P \times a + b}{2}\right) \times \frac{b}{3} = 0, \quad M_a = \frac{P_a}{2}$$

19. (b) $\left[EI \dfrac{d^2y}{dx^2} = M\right.$

$$EI \frac{dy}{dx} = Mx + C_1$$

$$C_1 = 0 \text{ as } \quad x = 0, \quad \frac{dy}{dx} = 0$$

$$EI \frac{dy}{dx} = M \times x$$

$$EIy = \frac{Mx^2}{2} + C_2$$

$C_2 = 0$, as $x = 0, y = 0$

$$EIy = \frac{Mx^2}{2}\Bigg]$$

At $x = L$, $y = \dfrac{ML^2}{2EI}$

20. (a)

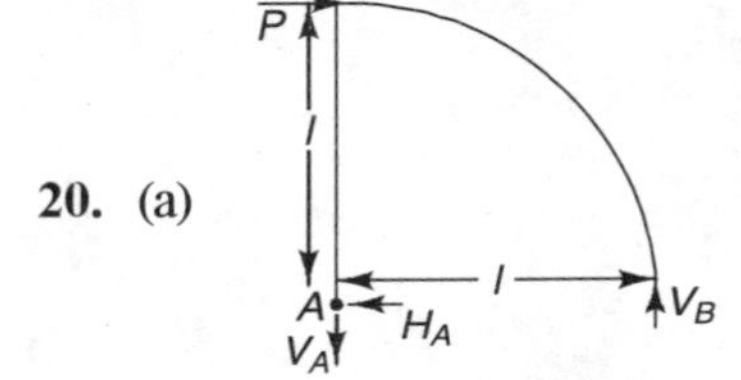

($\Sigma M_A = 0$,

$V_B \times l = P \times l$ or $V_B = P$)

21. (b) (Deflection at free end $y = \dfrac{WL^3}{3EI} = \dfrac{9 \times 2^3}{3EI} = \dfrac{24}{EI}$

Let M be the clockwise moment applied at the free end so that deflection is zero. The deflection of free end due to M is

$$\frac{ML^2}{EI} = \frac{M \times 4}{2EI} = \frac{2M}{EI}$$

Comparing: $\dfrac{24}{EI} = \dfrac{2M}{EI}$

$M = 12$ kN m (clockwise)

22. (a)

23. (a) (The change of bending moment is proportional to the area of the shear force diagram. Since the area of SFD is K, the difference between moments at two points will also equal to k.)

24. (a)

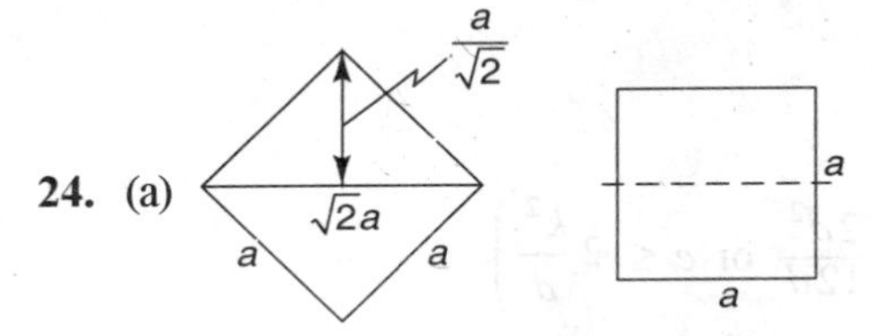

$$\left[z_1 = \frac{a \times a^3 \times 2}{12 \times a} = \frac{a^3}{6} \right.$$

$$z_2 = \frac{2 \times \sqrt{2}\,a}{12} \times \left(\frac{a}{\sqrt{2}}\right)^3 \times \sqrt{\frac{2}{a}}$$

$$= \frac{2 \times \sqrt{2}a \times a^2}{12 \times 2} = \frac{a^3}{6\sqrt{2}}$$

$$\left. \therefore \quad \frac{z_1}{z_2} = \frac{a^3}{6} \times \frac{6\sqrt{2}}{a^3} = \sqrt{2} \right]$$

25. (b)

26. (a)

1 × 1 square, 2 × 2 square

$$z_1 = \frac{1}{6} \qquad z_2 = \frac{2 \times 2^2}{6} = \frac{4}{3}$$

$$M_1 = \frac{WL}{4} \qquad M_2 = \frac{W \times 2L}{4} = \frac{WL}{2}$$

$$\sigma_1 = \frac{M_1}{z_1} = \frac{WL \times 6}{4 \times 1} \qquad \sigma_2 = \frac{WL \times 3}{2 \times 4}$$

$$= \frac{3}{2} WL \qquad = \frac{3}{8} WL$$

$$\frac{\sigma_1}{\sigma_2} = \frac{\frac{3}{2} WL}{\frac{3}{8} WL} = 4$$

27. (a) (Stress (σ) should be uniform throughout the length, $\sigma = \frac{M}{z}$)

28. (c)

29. (b) $\left(k = \sqrt{\frac{I}{A}} = \sqrt{\frac{bd^3}{12 \times bd}} = \frac{d}{\sqrt{12}} \right.$

$$\left. k^2 = \frac{d^2}{12}, \text{ As } e \le \frac{d}{6} \text{ for safety } e \le \frac{d^2}{6d} \text{ or } e \le \frac{2d^2}{12d} \text{ or } e \le 2\frac{k^2}{d} \right)$$

30. (a)

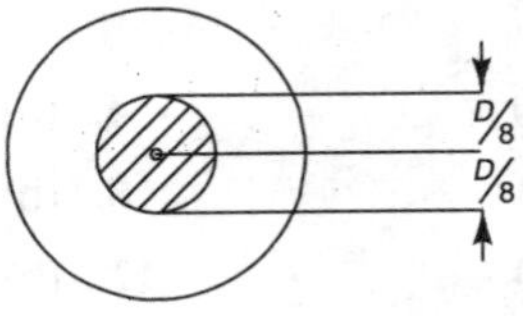

D/8
D/8

($e \leq \frac{D}{8}$, Kernel circle diameter $(2e) = \frac{D}{4}$)

31. (a) (No tensile stress if load is within kernel, i.e. $e \leq \frac{b}{6}$ where b = width of the section.)

32. (a) $\left(e \leq \frac{b}{6}\right)$

33. (b) (No tensile stress if $e \leq \frac{D}{8}$ where D = diameter of section.)

34. (a)

35. (a)

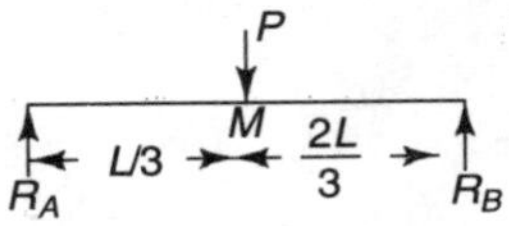

P
M
L/3
2L/3
R_A
R_B

$$\Sigma M_B = 0,\ R_A \times l = P \times \frac{2}{3} l$$

$$R_A = \frac{2}{3} P$$

$$\therefore \quad \Sigma M_C = \frac{2}{3} P \times \frac{1}{3} l = \frac{2\,Pl}{9}$$

36. (a) (All joints are hinged and no movement: equilibrium)

37. (c)

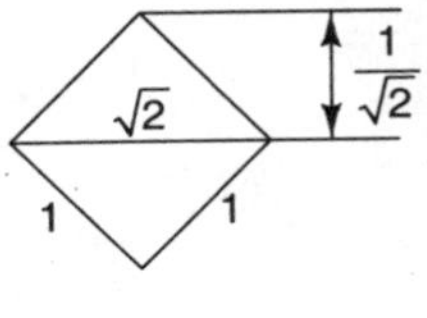

$\sqrt{2}$
$\frac{1}{\sqrt{2}}$
1
1

$$\left[I = 2 \times \frac{b\,d^3}{12} = 2 \times \frac{\sqrt{2} \times \left(\frac{1}{\sqrt{2}}\right)^3}{12} = \frac{1}{12} \right]$$

Fill in the Blanks

1. (b)
2. (a)
3. (b)
4. (b)
5. (b)
6. (b)
7. (a)
8. (b)
9. (b)
10. (b)
11. (b)
12. (b)
13. (b)
14. (b)
15. (b)
16. (a)
17. (b)
18. (b)
19. (b)
20. (b)
21. (a)
22. (b)
23. (a)
24. (a)
25. (a)
26. (b)
27. (b)

28. ($a^2 = \frac{\pi D^2}{4}$ or $a = \sqrt{\frac{\pi}{4}}\ D$

$$z_1 = \frac{a^3}{6} \text{ and } z_2 = \frac{\pi D^3}{32} = \pi \times \left(\sqrt{\frac{4}{\pi}}\ a\right)^3$$

$$z_2 = \frac{8}{32} \times \frac{1}{\sqrt{\pi}}\ a^3 = \frac{1}{4\sqrt{\pi}} \times a^3$$

$$\frac{z_2}{z_1} = \frac{6}{4\sqrt{\pi}} = 0.844$$

29. (b)

30. (a)

31. (a) ($k^2 = \frac{I}{A} = \frac{bd^3}{12 \times bd} = \frac{d^2}{12}$ and hence $k = \frac{d}{2\sqrt{3}}$)

32. (b) ($I = k^2 A = 9 \times 4 = 36$)

33. (a) ($A = \frac{I}{k^2} = \frac{48}{16} = 3$

$$I_{\text{new}} = 75,\ k_{\text{new}}^2 = \frac{75}{3} = 25$$

$k_{\text{new}} = 5$ cm)

CHAPTER 17
Torsion

If every tool in your bag is a hammer, then every problem in the world appears to be a nail.

INTRODUCTION

Circular shafts are widely used in various engineering applications to transmit power. The shafts have to bear torsion, bending and axial forces. If bending moment and axial force do not act, then the shaft is under pure torsion. Under pure torsion, the cross section of the shaft is under pure shear stress only. The product of turning force and its distance from the axis of the shaft is called *torque*. Due to this torque, every cross section of the shaft is subjected to some shear stresses. The theory of pure torsion is used to find out the value of shear stresses at various distances from the centre of the shaft.

THEORY OF PURE TORSION

Certain assumptions are made for working out pure torsion by employing the theory of pure torsion:

1. Material of the shaft is homogeneous and isotropic.
2. Twist is uniform along the length of the shaft.
3. Material is perfectly elastic and obeys Hooke's law. Shear strain is within the elastic limit due to torsion.
4. The shaft has a uniform cross section throughout its length.
5. The cross section remains plane even after its twist.
6. All diameters of the cross section of the shaft remain straight before and after its twist.

A solid shaft of diameter D and length L is subjected to a couple T and its other end is fixed (Figure 17.1). AB is a layer of the shaft which is parallel to the shaft axis before couple application and it is twisted to AC. Let ϕ be the shear angle.

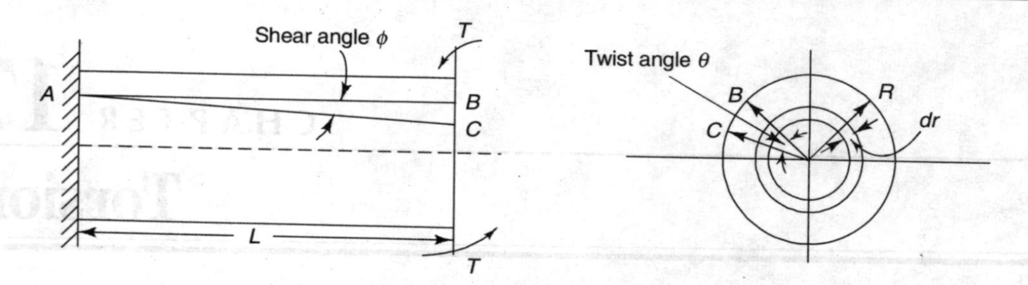

FIGURE 17.1 A shaft subjected to a couple.

$$BC = L\phi$$

$$\phi = \frac{BC}{L}$$

$$\text{Shear strain} = \frac{\text{shear stress at surface}}{\text{modulus of rigidity}}$$

$$\varepsilon_{\text{shear}} = \frac{\tau_s}{G}$$

But

$$\varepsilon_{\text{shear}} = \phi, \text{ so}$$

$$\phi = \frac{\tau_s}{G}$$

In the cross section, the angle of twist = θ

$$BC = \frac{D}{2} \times \theta$$

$$L\phi = \frac{D}{2} \times \theta$$

But $\phi = \dfrac{\tau_s}{G}$. Therefore,

$$L \times \frac{\tau_s}{G} = \frac{D}{2} \times \theta = R\theta \quad \left(\because R = \frac{D}{2}\right)$$

$$\frac{\tau_s}{R} = \frac{G\theta}{L}$$

The maximum shear stress is at the surface and it reduces towards the centre where it becomes zero. Hence

$$\frac{\tau_s}{R} = \frac{\tau}{r}$$

where r varies from zero to R.

Shear stress increases as the angle of twist increases. Hence the shaft can fail due to excessive twist (θ) in the shaft. This is the stiffness criterion for the failure of the shaft.

Consider an elementary ring of the shaft of radius r and thickness dr. Let shear stress in the ring be τ.

$$\frac{\tau_s}{R} = \frac{\tau}{r}$$

or

$$\tau = r\left(\frac{\tau_s}{R}\right)$$

$$\text{Area of the ring} = 2\pi r\, dr$$

$$\therefore \quad \text{Shear resistance in this ring} = \tau \times 2\pi r\, dr$$

$$\text{Torsional moment of resistance} = 2\pi r^2 \tau\, dr$$

$$dT = 2\pi r^3 \left(\frac{\tau_s}{R}\right) dr$$

Integrating the above equation,

$$T = \frac{\tau_s}{R} \times 2\pi \int_0^R r^3 dr$$

$$= \frac{\tau_s}{R} \times 2\pi \left[\frac{r^4}{4}\right]_0^R$$

$$= \tau_s \times \frac{\pi R^3}{2}$$

$$= \tau_s \times \frac{\pi D^3}{16}$$

$$= \frac{\tau_s}{D/2} \times \frac{\pi D^4}{32}$$

$$\therefore \quad T = \frac{\tau_s}{R} \times I_p \quad \left(\because I_p = \text{polar moment of inertia} = \frac{\pi D^4}{32}\right)$$

$$\frac{T}{I_p} = \frac{\tau_s}{R}$$

As torque increases, shear stress increases. The shaft will fail if shear stress exceeds the permissible range of τ when applied torque increases. This is the strength criterion for the failure of the shaft. The torsion formula is as follows:

$$\frac{T}{I_p} = \frac{\tau_s}{R} = \frac{G\theta}{L}$$

or

$$\frac{T}{I_p} = \frac{\tau}{r} = \frac{G\theta}{L}$$

POLAR MODULUS OF SECTION

The polar modulus of a section (z_p) is given as follows:

$$z_p = \frac{I_p}{R}$$

We know

$$\frac{T}{I_p} = \frac{\tau_s}{R}$$

or

$$T = \frac{I_p}{R}\ (\tau_s)$$

$$T = z_p \times (\tau_s) \qquad \left(\because\ z_p = \frac{I_p}{R}\right)$$

As the torque to be transmitted and maximum permissible shear stress of the material of the shaft are known, the polar section modulus can be found out and the shaft can be designed.

TORSIONAL RIGIDITY

Torsional rigidity (GI_p) is derived as follows:

$$\frac{T}{I_p} = \frac{G\theta}{L}$$

or

$$GI_p = \frac{TL}{\theta}$$

If θ = 1 radian and length (L) = 1,

$$GI_p = T$$

Hence torsional rigidity is the torque (T) required to produce a twist of one radian over a unit length of the shaft.

POWER TRANSMITTED BY A SHAFT

Power transmitted by a shaft is the product of average torque and angular displacement per unit time.

$$\text{Work done} = T\theta$$

$$\text{Power, } P = \frac{T\theta}{t} = T\omega$$

where $\omega = \dfrac{\theta}{t}$ = angular velocity in radians/second.

$$P = \frac{2\pi N}{60} \times T$$

as $\omega = \dfrac{2\pi N}{60}$ where N = rpm.

ARRANGEMENT OF SHAFTS

Shafts in Series

When shafts are in series (Figure 17.2), they transmit the same torque. Hence

$$\frac{T}{I_{p1}} = \frac{\tau_{s1}}{r_1} = \frac{G_1\theta_1}{L_1}$$

$$\frac{T}{I_{p2}} = \frac{\tau_{s2}}{r_2} = \frac{G_2\theta_2}{L_2}$$

Also the total angular rotation, $\theta = \theta_1 + \theta_2$

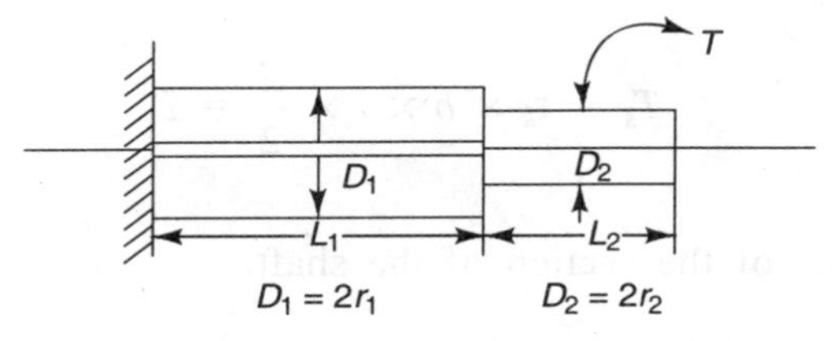

FIGURE 17.2 Shafts in series.

Shafts in Parallel

When shafts are in parallel, torque is distributed between the shafts. For example, in Figure 17.3 torque is distributed in T_1 and T_2 on two shafts, i.e.

$$T = T_1 + T_2$$

But

$$\theta = \theta_1 = \theta_2$$

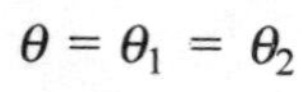

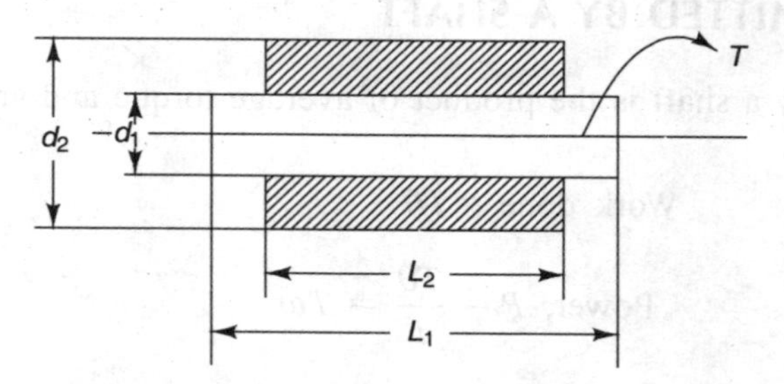

FIGURE 17.3 Shafts in parallel.

Shear failure of key: A key is commonly used for connecting a gear or coupling to a shaft in engineering applications (Figure 17.4). It is a wedge-like part to prevent relative motion between two parts of transmitting torque (T).

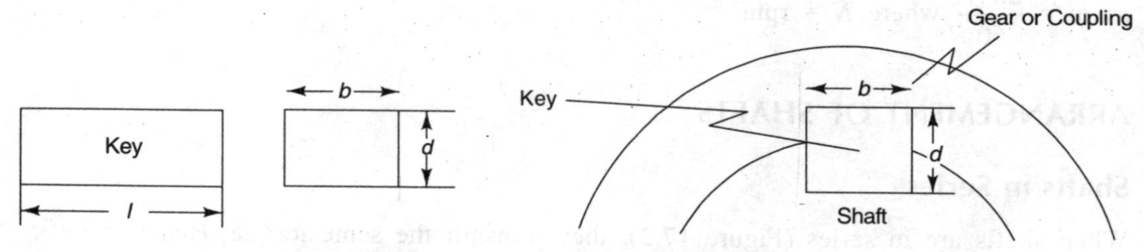

FIGURE 17.4 A gear connected to a key.

If τ_k is safe shear stress of the key, then

$$\text{shear resistance} = \tau_k \times \text{area}$$

$$= \tau_k \times b \times l$$

The maximum torque which can be transmitted by the shaft through the key without shear failure is

$$T_k = \tau_k \times b \times l \times \frac{d}{2} = T$$

where $T = \tau_s \times z_p$

z_p = polar modulus of the section of the shaft.

Shaft coupling: A coupling is used to connect two shafts transmitting torque. It has two flanges, which are connected by a number of bolts (Figure 17.5). Let

n = number of bolts,
D_p = pitch circle,
d_b = bolt diameter,
τ_b = safe stress,
d_s = diameter of shaft and
τ_s = shear stress of shaft.

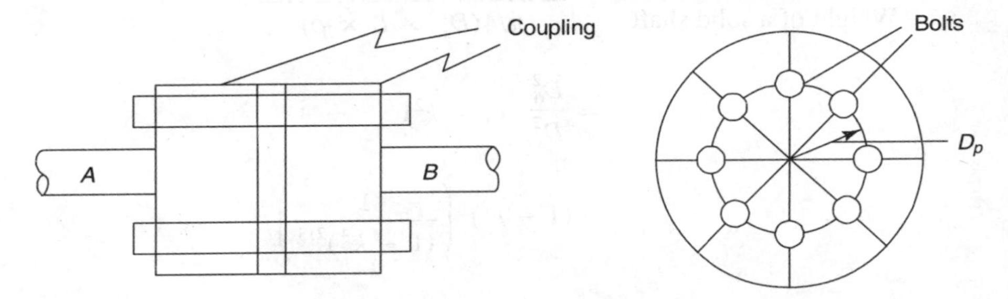

FIGURE 17.5 Shaft coupling.

Torsional moment transmitted by all bolts is

$$T_b = n \times \left(\tau_b \times \frac{\pi}{4} d_b^2 \right) \times \frac{D_p}{2}$$

Torque transmitted by the shaft is

$$T_s = \tau_s \times \frac{\pi d_s^3}{16}$$

$$T_b = T_s$$

$$\therefore \qquad n \times \tau_b \times \frac{\pi}{4} d_b^2 \times \frac{D_p}{2} = \tau_s \times \frac{\pi d_s^2}{16}$$

COMPARISON BETWEEN HOLLOW AND SOLID SHAFTS

If hollow and solid shafts have the same torsional strength, then the polar modulus of sections of both shafts is the same, i.e.

$$(z_p)_{\text{hollow}} = (z_p)_{\text{solid}}$$

$$\frac{\pi}{16} \left(\frac{D_0^4 - D_1^4}{D_0} \right) = \frac{\pi}{16} D_s^3$$

$$D_0^3 (1 - k^4) = D_s^3$$

where $k = \dfrac{D_1}{D_0}$.

$$\frac{D_0}{D_s} = \frac{1}{(1 - k^4)^{1/3}}$$

$$\frac{\text{Weight of a hollow shaft}}{\text{Weight of a solid shaft}} = \frac{\pi/4(D_0^2 - D_1^2) \times L \times \rho}{\pi/4(D_s^2 \times L \times \rho)}$$

$$= \frac{D_0^2}{D_s^2} (1 - k^2)$$

$$= (1 - k^2) \left(\frac{1}{(1 - k^4)^{2/3}}\right)$$

$$\simeq (1 - k^2) \left(1 + \frac{2}{3} k^4\right)$$

$$\simeq 1 - k^2$$

$$< 1$$

If hollow and solid shafts have equal weight, then the torsional strength or polar modulus of section are to be compared.

$$\frac{\pi}{4} (D_0^2 - D_1^2) \times L \times \rho = \pi/4 \times D_s^2 \times L \times R$$

$$D_0^2 - D_1^2 = D_s^2$$

$$\frac{D_0}{D_s} = \left(\frac{1}{1 - k^2}\right)^{1/2}$$

where $k = \dfrac{D_1}{D_0}$

$$\frac{(z_p)_{\text{hollow}}}{(z_p)_{\text{solid}}} = \frac{D_0^3 (1 - k^4)}{D_s^3} = \frac{1}{(1 - k^2)^{3/2}} (1 - k^4)$$

$$\approx (1 - k^4) \left(1 + \frac{3}{2} k^2\right)$$

$$\approx 1 + \frac{3}{2} k^2$$

$$> 1$$

STRAIN ENERGY

Strain energy in torsion in a solid shaft depends upon work done by the applied torque (T). As applied torque (T) increases, the first angle (θ) increases (Figure 17.6). The work done by the torque is equal to the product of torque (since torque is applied gradually from zero to T its value is taken as $T/2$) and the twist angle (θ) which is stored in the shaft as strain energy.

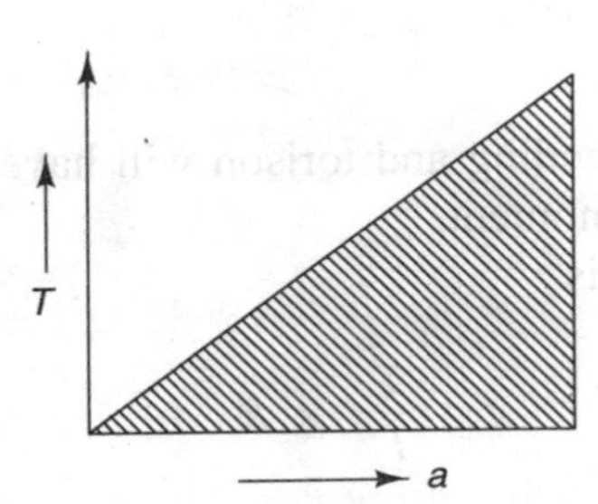

FIGURE 17.6 Torque (T) vs twist angle (θ).

$$\text{Strain energy} = U = \frac{1}{2} T\theta$$

$$\frac{T}{I_p} = \frac{\tau}{R} = \frac{G\theta}{L}$$

$\therefore$
$$T = \frac{\tau \times I_p}{R}$$

and
$$\theta = \frac{\tau}{R} \times \frac{L}{G}$$

Now
$$U = \frac{1}{2} \times \frac{\tau \times I_p}{R} \times \frac{\tau}{R} \times \frac{L}{G}$$

$$= \frac{1}{2} \times \frac{\tau^2}{G} \times \frac{I_p}{R^2} \times L$$

$$= \frac{1}{2} \times \frac{\tau^2}{G} \times \frac{\pi R^4}{2 \times R^2} \times L$$

$\therefore$
$$U = \frac{\tau^2}{4G} (\pi R^2 L) = \frac{\tau^2}{4G} \times V$$

where V = volume.

Hence strain energy in torsion is half of what is in constant pure shear stress. Strain energy for a hollow shaft can be worked on the similar line:

$$U_{\text{hollow}} = \frac{\tau^2}{4G} \times \frac{(D_0^2 + D_1^2)}{D_0^2} \times \text{volume}$$

BENDING AND TORSION

A shaft subjected to combined bending and torison will have shear stress due to torsion and bending stress due to bending moment.

Shear stress due to torsion is

$$\tau = \frac{T}{I_p} \times R$$

$$= \frac{16T}{\pi D^3}$$

Bending stress (σ_x) due to bending moment is

$$\sigma_x = \frac{M}{I} \times \frac{D}{2}$$

$$= \frac{M}{\frac{\pi D^4}{64}} \times \frac{D}{2} = \frac{M \times 32}{\pi D^3}$$

$$\text{Principal stress } \sigma_1 = \frac{\sigma_x}{2} + \sqrt{\left(\frac{\sigma_x}{2}\right)^2 + (\tau)^2}$$

$$= \frac{1}{2} \times \frac{M \times 32}{\pi D^3} + \sqrt{\left(\frac{M \times 32}{2 \times \pi D^3}\right)^2 + \left(\frac{16T}{\pi D^3}\right)^2}$$

$$= \frac{16}{\pi D^3}(M + \sqrt{M^2 + T^2})$$

$$= \frac{1}{z_p}(M + \sqrt{M^2 + T^2})$$

Similarly, $$\sigma_2 = \frac{16}{\pi D^3}(M - \sqrt{M^2 + T^2})$$

$$= \frac{1}{z_p}(M - \sqrt{M^2 + T^2})$$

Also

$$\tan 2\theta = \frac{\tau}{\frac{\sigma_x}{2}} = \frac{T}{M}$$

$$\text{Maximum shear stress} = \frac{\sigma_1 - \sigma_2}{2} = \frac{1}{z_p}\sqrt{M^2 + T^2}$$

If bending moment and torque are acting on a shaft, then the equivalent torque is the torque that produces the same maximum shear stress as produced by the combined bending moment and torque, i.e.

$$\tau_{\max} = \frac{T_e}{z_p}$$

where $T_e = \sqrt{M^2 + T^2}$.

Similarly, the equivalent bending moment can be defined as

$$\sigma_{b(\max)} = \pm\frac{M_e}{z_p}$$

where $M_e = \left(M + \sqrt{M^2 + T^2}\right)$.

If a shaft is subjected to various torques T_1, T_2, T_3, and T_4 as shown in Figure 17.7, then the torque in various sections can be found out as follows.

- Torque in $AB = -T_1$ (negative as anticlockwise)
- Torque in $BC = -T_1 + T_2$
- Torque in $CD = -T_1 + T_2 - T_3$
- Torque in $DE = -T_1 + T_2 - T_3 + T_4 = T_5$

Pure shear formula for torsion $\left(\frac{T}{I_p} = \frac{\tau}{R}\right)$ is not applicable beyond the elastic limit. However, we can find out the maximum fictitious shear stress by using the experimentally found maximum torque at which the shaft fails. This is called the *modulus of rupture*.

$$\text{Modulus of rapture } (\tau_r) = \frac{T_u R}{I_p}$$

T_u = Torque ultimate at failure of the shaft

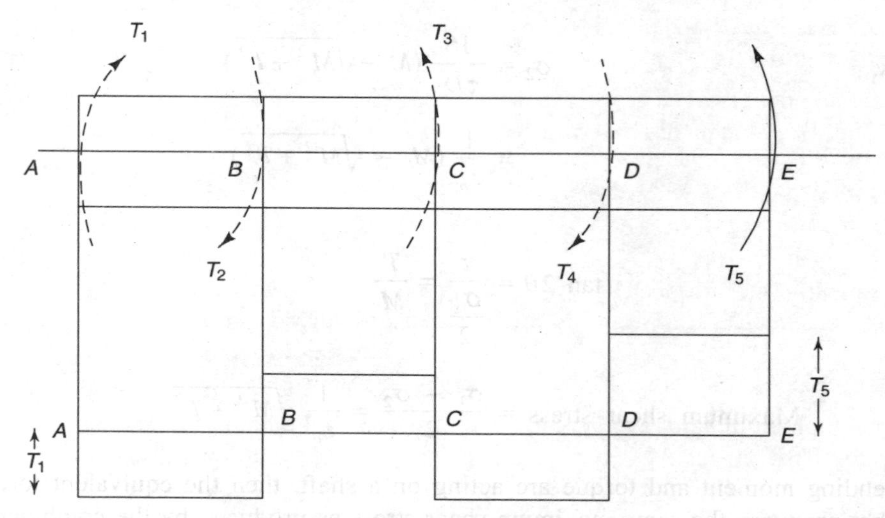

FIGURE 17.7 Torque diagram.

SOLVED PROBLEMS

1. The diameter of a shaft is 20 cm. Find the safe maximum torque which can be transmitted by the shaft if the permissible shear stress in the shaft material is 4000 N/cm^2 and the permissible angle of twist is 0.2° per metre length. Take G = 8×10^6 N/cm^2. If the shaft rotates at 320 rpm, what maximum power can be transmitted by the shaft? (UPTU: Dec. 2005)

The shaft is to be checked from the strength criteria, i.e.

$$\tau_{\text{permissible}} = 4000 \text{ N/cm}^2$$

$$\frac{T_{\max}}{I_p} = \frac{\tau_{\text{permissible}}}{R}$$

or

$$T_{\max} = z_p \times \tau_{\text{per}}$$

$$z_p = \frac{\pi D^3}{16} = \frac{\pi \times 20^3}{16} = \frac{\pi \times 8}{16} \times 10^3 = 1.58 \times 10^3 \text{ cm}^3$$

$$\therefore \quad T_{\max} = 1.58 \times 10^3 \times 4000 = 6.32 \times 10^6 \text{ N cm}$$

Now the shaft is to be checked from the stiffness criteria, i.e.

$$\theta = \frac{0.2}{180}\pi \text{ radian/m}$$

$$\frac{T_{\max}}{I_p} = \frac{G\theta}{L}$$

and

$$I_p = \frac{\pi D^4}{32} = 1.58 \times 10^4$$

Therefore, $$T_{max} = I_p \times \frac{G\theta}{L} = 1.58 \times 10^4 \times 8 \times 10^6 \times \frac{0.2\,\pi}{180 \times 10^2}$$

$$= 4.42 \times 10^6 \text{ N cm}$$

The lowest of two, i.e., $T_{max} = 4.42 \times 10^6$ N cm can be transmitted (P).

$$P = T \times \frac{2\pi N}{60} = \frac{4.42 \times 10^6}{10^2} \times \frac{2\pi \times 320}{60}$$

$$= 1483 \text{ kW}$$

2. Find the power transmitted by a circular solid shaft of steel of 50 mm diameter at 120 rpm, if permissible shear stress is 62.5 N/mm^2

$$T_{max} = z_p \tau_{permissible}$$

$$= \frac{\pi}{16} D^3 \tau_{permissible}$$

$$= \frac{\pi}{16} \times (50)^3 \times \frac{62.5}{1000} = 1534 \text{ N-m}$$

$$P = \frac{2\pi N T_{max}}{60} = \frac{2\pi \times 120 \times 1534}{60}$$

$$= 192.8 \times 10^2 \text{ W} = 19.28 \text{ kW}$$

3. Compare the weights of a solid shaft and a hollow shaft of the same material, same length, same torque and the same allowable shear stores. The internal diameter of the hollow shaft is two-thirds of the outer diameter.

$$T_s = \frac{\pi}{16} D^3 \tau$$

$$T_h = \frac{\pi}{16} \times \frac{D_0^4 - D_1^4}{D_0} \tau$$

$$T_s = T_h \text{ (Given)}$$

$$\therefore \quad \frac{\pi}{16} D^3 \tau = \frac{\pi}{16} \times \frac{D_0^4 - D_1^4}{D_0}$$

or $$D^3 = \frac{D_0^4 - \frac{16}{81} D_0^4}{D_0} = \frac{65}{81} D_0^3$$

or $$D^3 = 0.8\ D_0^3$$

or $$D = 0.93\ D_0$$

$$\frac{\text{Weight of the solid shaft}}{\text{Weight of the hollow shaft}} = \frac{\frac{\pi}{4} D^2 L\rho g}{\frac{\pi}{4}(D_0^2 - D_1^2)\, L\rho g}$$

$$= \frac{D^2}{D_0^2 - \frac{4}{9} D_0^2} = \frac{9}{5} \times \frac{D^2}{D_0^2}$$

$$= \frac{9}{5} \times (0.93)^2 = 1.56$$

4. A solid round shaft is replaced by a hollow shaft, the external diameter of which is $1\frac{1}{4}$ times the internal diameter. Allowing the same intensity of torsional stress in each, compare the weight and stiffness of the solid shaft with that of the hollow shaft.
(UPTU: Aug. 2001)

$$T = z_s \tau = z_h \times \tau$$

or $$z_s = z_h$$

$$\therefore \quad \frac{\pi D^3}{16} = \frac{\pi}{16} \times \frac{D_0^4 - D_1^4}{D_0}$$

$$D^3 = \frac{(1.25)^4 - 1}{1.25} \times D_1^3$$

$$= \frac{2.44 - 1}{1.25} \times D_1^3$$

$$= \frac{1.44}{1.25} \times D_1^3$$

or $$D = 1.05\ D_1$$

$$\frac{\text{Weight of the hollow shaft}}{\text{Weight of the solid shaft}} = \frac{\frac{\pi}{4}(D_0^2 - D_1^2)\rho g L}{\frac{\pi}{4} \times D^2 \rho g L}$$

$$= \frac{D_0^2 - D_1^2}{D^2} = \frac{(1.56 - 1)}{D^2} \times D_1^2$$

$$= \frac{0.56}{(1.05)^2} = \frac{0.56}{1.1} = 0.51$$

$$k = \text{stiffness} = G\theta = \frac{T}{z_p}$$

Here T and L are the same for both shafts.

$$\therefore \quad \frac{k_h}{k_s} = \frac{z_s}{z_h} = \frac{D^3}{\dfrac{D_0^4 - D_1^4}{D_0}}$$

$$= \frac{D^3}{1.15 D_1^3} = \frac{(1.05)^3}{1.15}$$

$$= 1$$

5. A solid circular shaft transmits 75 kW power at 200 rpm. Calculate the shaft diameter, if the twist in the shaft is not to exceed 1° in 2 metre length and the shear strength is limited to 50 MN/m². Take G = 100 GN/m². (UPTU: Dec. 2003)

Let us find the diameter from strength criteria:

$$\text{Power } P = \frac{2\pi NT}{60}$$

$$75 \times 10^3 = \frac{2\pi \times 200}{60} \times T$$

$$T = \frac{75 \times 10^3 \times 60}{2\pi \times 200} = 3581 \text{ N-m}$$

$$T = z_p \tau = \frac{\pi}{16} \times D^3 \tau$$

$$D^3 = \frac{16 \times 3581}{\pi \times 50 \times 10^6} = 364.24 \times 10^{-6}$$

$$D = 71.4 \text{ mm}$$

Now let us we find the diameter from stiffness criteria:

$$\frac{T}{I_p} = \frac{G\theta}{L}$$

$$I_p = \frac{T}{G\theta/L} = \frac{3581}{100 \times 10^9 \times \frac{1}{2} \times \frac{\pi}{180}}$$

$$D^3 = \frac{32}{\pi} \times \frac{3581 \times 2 \times 180}{100 \times 10^9 \times \pi}$$

$$D = 34.7 \text{ mm}$$

Hence we select the larger diameter, i.e., $D = 71.4$ mm.

6. A circular shaft of 10 cm diameter is subjected to a torque of 8×10^3 N m. Determine the maximum shear stress and the consequent principal stresses induced in the shaft.

$$z_p = \frac{\pi D^3}{16} = \frac{\pi \times 10^3}{16} = 1.97 \times 10^2 \text{ cm}^2$$

$$\tau = \frac{T}{z_p} = \frac{8 \times 10^3 \times 100}{1.97 \times 10^2}$$

$$= 4.07 \times 10^3 \text{ N/cm}^2$$

$$= 4.07 \text{ kN/cm}^2$$

For pure shear stress, $\sigma_1 = \tau$ and $\sigma_2 = -\tau$

$\therefore$ $\sigma_1 = 4.07 \text{ kN/cm}^2$

and $\sigma_2 = -4.07 \text{ kN/cm}^2$

7. For one propeller drive shaft, compute the torsional shear stress when it is transmitting a torque of 1.76 kN m. The shaft is a hollow tube having an outside diameter of 60 mm and an inside diameter of 40 mm. Find the stress at both the outer and inner surfaces. (UPTU: 2001–2002)

$$z_p = \frac{\pi}{16} \times \frac{D_0^4 - D_1^4}{D_0}$$

$$= \frac{\pi}{16}\left[\frac{6.0^4 - 4.0^4}{6.0}\right]$$

$$= \frac{\pi}{16}\left[\frac{1296 - 256}{6}\right] = \frac{\pi \times 1040}{16 \times 6}$$

$$= 34.03 \text{ cm}^3$$

$$\tau_s = \frac{T}{z_p} = \frac{1.76 \times 100}{34.03} = 5.17 \text{ kN/cm}^2$$

The stress varies linearly along the diameter of the shaft. The stress at the inner diameter of the shaft will be

$$\tau_i = \tau_s \times \frac{D_1}{D_0} = 5.17 \times \frac{40}{60}$$

$$= 3.45 \text{ kN/cm}^2$$

8. A propeller shaft, 100 mm in diameter, and 45 m in length, transmits 10 mW at 80 rpm. Determine the maximum shearing stress in the shaft. Also calculate the stress at 20 mm, 40 mm, 60 mm and 80 mm in diameter. Show the stress variation.
(UPTU: Special carry over 2005–2006)

$$P = \frac{2\pi N}{60} T \quad (P = \text{power}, T = \text{torque})$$

$$T = \frac{10 \times 10^6 \times 60}{2\pi \times 80} = 1.19 \times 10^6 \text{ N m}$$

$$z_p = \frac{\pi D^3}{16} = \frac{\pi \times (0.1)^3}{16} = 0.197 \times 10^{-3} \text{ m}^3$$

$$T = z_p \, \tau_{max}$$

$$\tau_{max} = \frac{1.19 \times 10^6}{0.197 \times 10^{-3}} = 6.04 \times 10^9 \text{ N/m}^2$$

$$= 6.04 \text{ GN/m}^2$$

$$\frac{\tau_{max}}{\tau} = \frac{D_0}{D}$$

$$\tau = \tau_{max} \times \frac{D}{D_0}$$

when $D_0 = 100$ mm,

$$\tau_{20} = 6.04 \times \frac{20}{100} = 1.21 \text{ GN/m}^2$$

$$\tau_{40} = \frac{6.04 \times 40}{100} = 2.42 \text{ GN/m}^2$$

$$\tau_{60} = \frac{6.04 \times 60}{100} = 3.63 \text{ GN/m}^2$$

$$\tau_{80} = \frac{6.04 \times 80}{100} = 4.84 \text{ GN/m}^2$$

The stress variation is shown as follows.

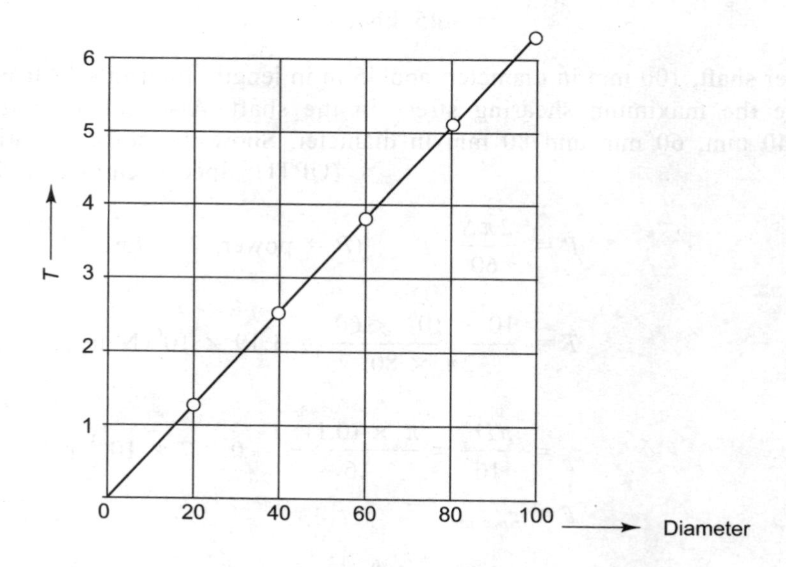

9. A solid shaft of mild steel 200 mm in diameter is to be replaced by a hollow shaft of alloy steel for which the allowable shear stress is 22% greater. If power to be transmitted is to be increased by 20% and the speed of rotation is increased by 6%, determine the maximum internal diameter of the hollow shaft. The external diameter of the hollow shaft is to be 200 mm. (UPTU: Feb. 2001)

Let P_1, τ_1 and N_1 be for the solid shaft and P_2, τ_2 and N_2 be for the hollow shaft. For the solid shaft

$$P_1 = \frac{2\pi N_1 T_1}{60}$$

$$\therefore \qquad T_1 = \frac{60 P_1}{2\pi N_1} = \frac{\pi}{16} \times D^3 \tau_1$$

or $$\tau_1 = \frac{16T_1}{\pi D^3}$$

For the hollow shaft

$$P_2 = \frac{2\pi N_2 T_2}{60}$$

$$T_2 = \frac{60 \times 1.2\, P_1}{2\pi(1.06\, N_1)}$$

$$\therefore \quad \frac{T_2}{T_1} = \frac{60 \times 1.2 P_1}{2\pi(1.06\, N_1)} \times \frac{2\pi N_1}{60\, P_1}$$

$$= 1.132$$

$$\tau_2 = 22\% \text{ more of } \tau_1 = 1.22\ \tau_1$$

$$T_1 = \frac{\pi D^3}{16}\ \tau_1$$

$$T_2 = \frac{\pi(D^4 - d^4)}{16D}\ \tau_2$$

$$\frac{T_2}{T_1} = \frac{D^2 - d^4}{D^4} \times \frac{\tau_2}{\tau_1}$$

$$1.132 = 1.22\left[1 - \left(\frac{D}{d}\right)^4\right]$$

$$1 - \left(\frac{d}{200}\right)^4 = 0.928$$

$$\left(\frac{d}{200}\right)^4 = 0.072$$

or $$\frac{d}{200} = 0.518$$

or $$d = 103.6 \text{ mm}$$

10. A shaft is to be designed for transmitting 100 kW power at 150 rpm. The shaft is supported in bearings 3 m apart and at 1 m from one bearing a pulley exerting a transverse load of 30 kN on the shaft is mounted. Obtain the diameter of the shaft if the maximum direct stress is not to exceed 100 N/m^2. (UPTU: July 2001)

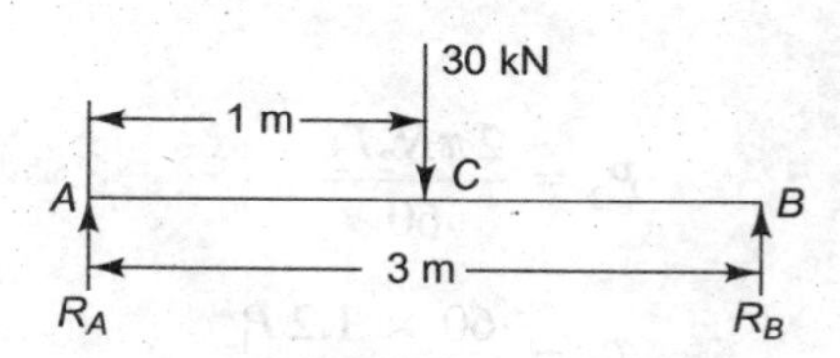

$$R_A + R_B = 30$$

$$\Sigma M_B = 0$$

$$R_A \times 3 = 30 \times 2$$

or $$R_A = 20$$

Maximum bending moment at $C = 20 \times 1$

$$= 20 \text{ kN m}$$

$$\sigma_b = \text{bending stress} = \frac{20}{z_p}$$

As $$z_p = \frac{\pi D^3}{32}$$

$$\sigma_b = \frac{20 \times 32}{\pi D^3} \times 10^6 \text{ N mm}$$

Power transmitted $$P = \frac{2\pi NT}{60}$$

$$100 \times 10^6 = \frac{2\pi \times 150 \times T}{60}$$

$$T = 6.37 \times 10^6 \text{ N mm}$$

$$T = z_p \times \tau$$

$\therefore$ $$\tau = \frac{6.37 \times 10^6 \times 16}{\pi D^3} = \frac{32.42 \times 10^6}{D^3}$$

Direct stress $$\sigma_1 = \frac{\sigma_b}{2} + \sqrt{\left(\frac{\sigma_b}{2}\right)^2 + \tau^2}$$

$$100 = \frac{203.7 \times 10^6}{2D^3} + \sqrt{\left(\frac{203.7 \times 10^6}{2D^3}\right)^2 + \left(\frac{32.42 \times 10^6}{D^3}\right)^2}$$

$$D = 117.5 \text{ mm}$$

11. A compound shaft consists of a 1-m long aluminium bar secured to a 1-m length of a brass bar, the diameter of each being 50 mm. Calculate the maximum torque that can be applied at the lower end if the allowable angle of twist is 1° and allowable shear stress in aluminium is 75 N/mm² and that in brass is 50 N/mm². Take G for brass = 0.34×10^5 N/mm² and that for aluminium as 0.27×10^5 N/mm².

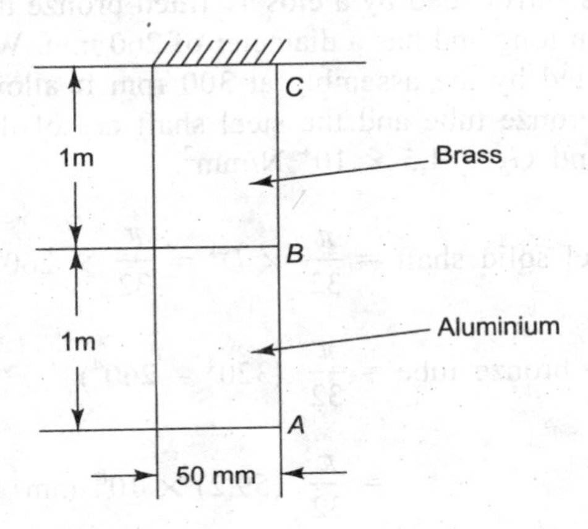

$$\frac{T}{I_p} = \frac{G\theta}{L}$$

For aluminium portion (AB):

$$\theta_1 = \frac{TL}{GI_p} = \frac{T \times 1000}{0.27 \times 10^5 \times \left(\frac{\pi \times 50^4}{32}\right)}$$

$$\theta_1 = 6.03 \times 10^{-8}\ T$$

For brass portion (BC):

$$\theta_2 = \frac{T \times 1000}{0.34 \times 10^5 \times \frac{\pi \times 50^4}{32}}$$

$$= 4.79 \times 10^{-8}\ T$$

Now

$$\theta = \theta_1 + \theta_2$$

$$= (6.03 + 4.79) \times 10^{-8} \times T$$

$$\theta_{\text{allowable}} = 1^\circ = \frac{1 \times \pi}{180}$$

$$\frac{\pi}{180} = 10.82 \times T \times 10^{-8}$$

or
$$T = \frac{\pi}{180} \times \frac{1}{10.82 \times 10^{-8}} = 161 \times 10^3 \text{ N mm}$$

$$= 161 \text{ N m}$$

12. A solid steel shaft is surrounded by a closely fitted bronze tube, which is 30 mm thick. The steel shaft is 8 m long and has a diameter of 260 mm. What is the maximum power that can be transmitted by the assembly at 300 rpm if allowable stress for steel shaft is 16 N/mm^2. The bronze tube and the steel shaft are of the same length. Take G_s = 8.5×10^4 N/mm^2 and $G_b = 4.5 \times 10^4$ N/mm^2.

$$I_{p_s} \text{ for the steel solid shaft} = \frac{\pi}{32} \times D^4 = \frac{\pi}{32} \times 260^4 \text{ mm}^4$$

$$I_{p_b} \text{ for the bronze tube} = \frac{\pi}{32}(320^4 - 260^4)$$

$$= \frac{\pi}{32}(59.2) \times 10^8 \text{ mm}^4$$

The angle of twist of both the shafts will be the same as they are in parallel.

$$\theta = \frac{T_s}{I_{p_s}} \times \frac{L}{G_s} = \frac{T_b}{I_{p_b}} \times \frac{L}{G_b}$$

$$\therefore \quad \frac{T_s}{T_b} = \frac{I_{p_s}}{I_{p_b}} \times \frac{G_s}{G_b}$$

$$= \frac{\frac{\pi}{32} \times 260^4}{\frac{\pi}{32} \times 59.2 \times 10^8} \times \frac{8.5 \times 10^4}{4.5 \times 10^6}$$

$$= 1.46$$

$$\tau_{s\,\max} \text{ for steel} = 16 \text{ N/mm}^2$$

$$\therefore \quad T_s = \frac{\pi}{16} D^3 \times \tau_{s\,\max} = \frac{\pi}{16} \times (260)^3 \times 16$$

$$= 55.22 \times 10^6 \text{ N mm}$$

So, $$T_b = \frac{T_s}{1.46} = \frac{55.22 \times 10^6}{1.46}$$

$$= 37.85 \times 10^6$$

Total torque $T = T_s + T_b$

$$= (55.22 + 37.85) \times 10^6$$

$$= 93.07 \times 10^3 \text{ N m}$$

$$P = \frac{2\pi NT}{60} \times \frac{2\pi \times 300}{60} \times 93.07 \times 10^3$$

$$= 2924 \text{ kW}$$

13. A steel wire of 100 m length and 10 mm diameter is twisted 5 rounds. Find the torque required to do that. Take $G = 80 \text{ GN/m}^2$

$$I_p = \frac{\pi}{32} \times D^4 = \frac{\pi}{32} \times (10)^4 = 9.83 \times 10^2 \text{ mm}^4$$

$$= 9.83 \times 10^{-10} \text{ m}^4$$

$$\theta = 2\pi \times 5 = 10\pi = 31.42 \text{ radians}$$

$$\frac{T}{I_p} = \frac{G\theta}{L}$$

$$T = \frac{80 \times 10^9 \times 31.42}{100} \times 9.83 \times 10^{-10}$$

$$= 24.71 \text{ N m}$$

14. A solid shaft is connected to a coupling by a key which transmits 100 kW power. The key is 20 mm long and 150 mm wide. Find the shear stress developed in the key and the shaft if the diameter of the shaft is 60 mm and N is 120 rpm.

$$P = 100 \text{ kW}$$

$$P = \frac{2\pi NT}{60}$$

$$\therefore \quad T = \frac{60P}{2\pi N} = \frac{60 \times 100 \times 10^3}{2\pi \times 120} = 7958 \text{ N m}$$

$$T = z_p \tau_{\text{shaft}}$$

$$\tau_{\text{shaft}} = \frac{7958}{\pi d^3/16} = \frac{7958 \times 16}{\pi \times (60)^3 \times 10^{-9}}$$

$$= 0.188 \times 10^9 \text{ N/m}^2 = 188 \text{ MN/m}^2$$

$$T = \tau_{\text{key}} \times \frac{\text{key area} \times \text{shaft diameter}}{2}$$

$$7958 = \tau_{\text{key}} \times \frac{(300) \times 60}{2} \times 10^{-9}$$

or

$$\tau_{\text{key}} = \frac{7958 \times 2}{300 \times 60 \times 10^{-9}}$$

$$= 884.2 \text{ MN/mm}^2$$

15. A flange coupling has 10 bolts on a pitch circle of 200 mm and it is fitted on a shaft which carries either a tensile load of 500 kN or a torque of 25 kN m. If the maximum allowable stress for the bolt material is 100 N/mm² for the tensile load and 50 N/mm² for the shear load, find the suitable diameter of a coupling bolt. The load on bolts is equal.

A tensile load of 500 kN is shared by 10 bolts equally. Let d_b = bolt diameter.

$$10 \times \text{area of bolt} \times \sigma_{\text{permissible}} = \text{Tensile load}$$

$$10 \times \frac{\pi}{4} \times d_b^2 \times 100 = 500 \times 10^3$$

or

$$d_b^2 = \frac{500}{100} \times \frac{4}{10\pi} \times 10^3$$

$$= 636.6 \text{ mm}^2$$

or

$$d_b = 25.23 \text{ mm}$$

$$\text{Now torque} = 25 \times 10^3 \text{ N m}$$

$$10 \times \text{area} \times \tau \times \text{pitch diameter}/2 = \text{torque}$$

$\therefore$

$$10 \times 50 \times \frac{\pi}{4} \times d_b^2 \times \frac{200}{2} = 25 \times 10^3 \times 10^3$$

or

$$d_b^2 = \frac{25 \times 10^6 \times 4}{10 \times 50 \times \pi \times 100}$$

$$= 636.6$$

or

$$d_b = 25.213 \text{ mm}$$

Hence the bolt diameter in tension and torsion is the same, i.e. 25.21 mm.

16. A solid shaft of 200 mm diameter has the same cross-sectional area as that of a hollow shaft of the same material with an inside diameter of 150 mm. Determine the ratio of the power transmitted by the two shafts at the same speed. (UPTU: 2007–2008)

As speed is the same, we have

$$\frac{P_{\text{solid}}}{P_{\text{hollow}}} = \frac{\dfrac{2\pi N}{60} \times T_s}{\dfrac{2\pi N}{60} \times T_h} = \frac{T_s}{T_h}$$

$$(I_p)_s = \frac{\pi d^3}{16} = \frac{\pi \times 0.2^3}{16}$$

$$(I_p)_{\text{hollow}} = \frac{\pi}{16} \frac{(d_o^4 - d_i^4)}{d_o}$$

As the cross section areas are equal,

$$\frac{\pi d^2}{4} = \pi \times \frac{(d_o^2 - d_i^2)}{4}$$

$$(0.2)^2 = d_o^2 - 0.15^2$$

or $$d_o^2 = 0.04 + 0.0225$$

$$= 0.0625$$

or $$d_o = 0.25$$

Since $T = \tau \times I_p$

$$\therefore \quad \frac{T_s}{T_h} = \frac{(I_p)_s}{(I_p)_h} = \frac{\dfrac{\pi \times 0.2^3}{16}}{\dfrac{\pi \times (0.25^4 - 0.15^4)}{16 \times 0.25}}$$

$$= \frac{0.2^3 \times 0.25}{0.25^4 - 0.15^4} = \frac{8 \times 10^{-3} \times 0.25}{3.9 \times 10^{-3} - 0.5 \times 10^{-3}}$$

$$= 0.588$$

$$\therefore \quad \frac{P_s}{P_h} = 0.588$$

Note: A solid shaft transmits less power as compared to a hollow shaft for the same speed and cross-sectional area.

17. A solid circular shaft transmits 75 kW power at 180 rpm. Calculate the shaft diameter if the twist in the shaft is not to exceed 1 degree in 2 m length and shear stress is limited to 50 MN/m^2. Take the modulus of rigidity G = 100 GN/m^2.

(UPTU: 2006–2007)

$$\text{Power } P = \frac{2\pi NT}{60}$$

or $$75 \times 10^3 = \frac{2\pi \times 180}{60} \times T$$

or $$T = \frac{75\times 10^3}{6\pi}$$

$$= 3980.9 \text{ N m}$$

1. *Applying strength criterion*

$$\frac{T}{I_p} = \frac{\tau}{R}$$

Given τ = 50 × 10^6 N/m^2. Therefore,

$$\frac{3980.9}{I_p} = \frac{50\times 10^6}{R}$$

or $$\frac{I_p}{R} = z_p = \frac{3980.9}{50\times 10^6} = 79.62 \times 10^{-6}$$

$$z_p = \frac{\pi d^3}{16} = 79.62 \times 10^{-6}$$

or $$d^3 = \frac{16\times 79.62}{\pi} = 405.7 \times 10^{-6}$$

$$d = 7.4 \times 10^{-2} \text{ m}$$

$$= 74 \text{ mm}$$

2. *Applying rigidity criterion*

$$\frac{T}{I_p} = \frac{G\theta}{L}$$

$$I_p = \frac{T\times L}{G\times \theta}$$

$$I_p = \frac{3980.9\times 2\times 180}{100\times 10^9 \times \pi} = \frac{\pi d^4}{32}$$

$$d^4 = \frac{32 \times 3980.9 \times 2 \times 180}{10^{11} \times \pi^2}$$

$$= 4.65 \times 10^{-5}$$

$$\therefore \quad d = 0.826 \text{ m}$$

$$= 82.6 \text{ mm}$$

Selecting the bigger diameter of the pipe out of two criteria, d = 82.6 mm.

18. A shaft was initially subjected to a bending moment and then was subjected to torsion. If the magnitude of bending moment is found to be the same as that of torque, then the ratio of maximum bending stress to shear would be

(a) 0.25 (b) 0.50 (c) 2.0 (d) 4.0

$$\sigma_b = \frac{M}{z} = \frac{M}{\frac{\pi d^3}{32}}$$

$$\tau = \frac{T}{z_p} = \frac{T}{\frac{\pi d^3}{16}}$$

Now

$$T = M$$

$$\therefore \quad \tau = \frac{M}{\frac{\pi d^3}{16}}$$

Now

$$\frac{\sigma_b}{\tau} = \frac{\frac{M}{\frac{\pi d^3}{32}}}{\frac{M}{\frac{\pi d^3}{16}}} = \frac{32}{16} = 2$$

Option (c) is correct.

19. A circular solid shaft is subjected to a bending moment of 400 kN m and a twisting moment of 300 kN m. The ratio of maximum principal stress to maximum shear stress is

(a) $\frac{1}{5}$ (b) $\frac{3}{9}$ (c) $\frac{9}{5}$ (d) $\frac{11}{6}$

Principal stresses are

$$\sigma_{1\text{ and }2} = \frac{1}{z_p}[M \pm \sqrt{M^2 + T^2}]$$

$$= \frac{1}{z_p}[400 \pm \sqrt{(400)^2 + (300)^2}]$$

$$= \frac{1}{z_p}[400 \pm 500]$$

$$= \frac{900}{z_p} \text{ and } -\frac{100}{z_p}$$

$$\tau_{max} = \frac{\sigma_1 - \sigma_2}{2} = \frac{900 + 100}{2 \times z_p}$$

$$= \frac{500}{z_p}$$

Now

$$\frac{\sigma_1}{\tau_{max}} = \frac{\frac{900}{z_p}}{\frac{500}{z_p}} = \frac{9}{5}$$

Option (c) is correct.

20. Two shafts of same length and material are joined in series. If the ratio of their diameters is 2, then the ratio of their angle of twist will be
(a) 2 (b) 4 (c) 8 (d) 16

Shafts in series will have the same torque (T).

$$\frac{T}{(I_p)_1} = \frac{G\theta_1}{l} \quad \text{(i)}$$

$$\frac{T}{(I_p)_2} = \frac{G\theta_2}{l} \quad \text{(ii)}$$

$$(I_p)_1 = \frac{\pi d_1^4}{32}$$

$$(I_p)_2 = \frac{\pi\left(\frac{d_1}{2}\right)^4}{32} = \frac{\pi d_1^4}{16 \times 32}$$

Dividing Eq. (i) by Eq. (ii)

$$\frac{\theta_1}{\theta_2} = \frac{(I_p)_2}{(I_p)_1} = \frac{\dfrac{\pi d_1^4}{16 \times 32}}{\dfrac{\pi d_1^4}{32}} = \frac{1}{16}$$

Option (d) is correct.

21. The shafts A and B are made of the same material. The diameter of shaft B is twice that of shaft A. The ratio of power which can be transmitted by shaft A to that of shaft B is

(a) $\frac{1}{2}$ (b) $\frac{1}{4}$ (c) $\frac{1}{8}$ (d) $\frac{1}{16}$ (GATE: 1994)

$$d_B = 2d_A$$

$$z_p = \frac{\pi d^3}{16}$$

$\therefore$ $$(z_p)_B = 2^3(z_p)_A = 8(z_p)_A$$

Now

$$T = z_p \times \tau$$

or $T \propto z_p$ (Torque is proportional to section modulus)

$\therefore$ $$T_B = 8T_A$$

Now

$P \propto T$ (Power is proportional to torque)

$\therefore$ $$P_B = 8P_A$$

or $$\frac{P_A}{P_B} = \frac{1}{8}$$

Hence option (c) is correct.

22. The outside diameter of a hollow shaft is twice its inside diameter. The ratio of its torque carrying capacity to that of a solid shaft of the same material and the same outside diameter is

(a) $\frac{15}{16}$ (b) $\frac{3}{4}$ (c) $\frac{1}{2}$ (d) $\frac{1}{16}$ (GATE: 1993)

Given $d_o = 2d_i$

$$(Z_p)_{\text{hollow}} = \frac{\pi(d_o^4 - d_i^4)}{32 \times \frac{d_o}{2}}$$

$$= \frac{\pi}{16} \times d_o^3 \left(1 - \left(\frac{d_1}{d_0}\right)^4\right)$$

$$= \frac{\pi d_o^3}{16}\left[1 - \left(\frac{1}{2}\right)^4\right]$$

$$= \frac{15}{16} \times \frac{\pi d_o^3}{16}$$

$$(z_p)_{\text{solid}} = \frac{\pi d_o^3}{16}$$

$$T_h \propto (z_p)_{\text{hollow}}$$

$$T_s \propto (z_p)_{\text{solid}}$$

$$\frac{T_h}{T_s} = \frac{(z_p)_{\text{hollow}}}{(z_p)_{\text{solid}}}$$

$$= \frac{\frac{15}{16} \times \frac{\pi}{16} d_o^3}{\frac{\pi}{16} \cdot d_o^3}$$

$$= \frac{15}{16}$$

Option (a) is correct.

23. A solid circular shaft of 60 mm diameter transmits a torque of 1600 N m. The value of maximum shear stress developed is

(a) 37.72 MPa (b) 47.72 MPa (c) 57.72 MPa (d) 67.72 MPa (GATE: 2004)

$$z_p = \frac{\pi d^3}{16} = \frac{\pi \times (0.06)^3}{16}$$

$$= 4.24 \times 10^{-5}$$

$$T = z_p \times \tau$$

$$\tau = \frac{1600}{4.24 \times 10^{-5}}$$

$$= 37.72 \text{ MPa}$$

Option (a) is correct.

24. A torque of 10 N m is transmited through a stepped shaft as shown in the figure. The torsional stiffness of individual sections of lengths MN, NO and OP are 20 N m/rad, 30 N m/rad and 60 N m/rad respectively. The angular deflection between ends M & P of the shaft is

(a) 0.5 rad (b) 1.0 rad (c) 5.0 rad (d) 10.0 rad

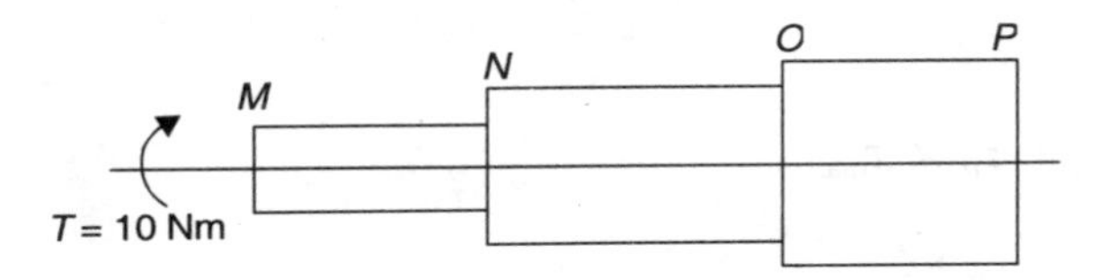

$$\frac{T}{I_p} = \frac{G\theta}{l}$$

or
$$T = \frac{G\theta \times I_p}{l}$$

$$= \frac{\theta}{l} \times (G \times I_p) = \text{constant}$$

Since the stepped shaft can be considered three shafts in series with $T_1 = T_2 = T_3$ and $G \times I_p$ = torsional stiffnesses which are given for shafts in series in the problem

$$\therefore \quad \theta = \frac{\text{Torque}}{\text{Torsional stiffness}}$$

$$\therefore \quad \theta_1 = \frac{10}{20} = \frac{1}{2} \text{ rad}$$

$$\theta_2 = \frac{10}{30} = \frac{1}{3} \text{ rad}$$

$$\theta_3 = \frac{10}{60} = \frac{1}{6} \text{ rad}$$

Angular deflection $= \theta_1 + \theta_2 + \theta_3$

$$= \frac{1}{2} + \frac{1}{3} + \frac{1}{6} = \frac{3+2+1}{6} = 1.0 \text{ rad}$$

Option (b) is correct.

25. A solid shaft can resist a bending moment of 3.0 kN m and a twisting moment of 4.0 kN m together, the maximum torque that can be applied is
(a) 7 kN m (b) 3.5 kN m (c) 4.5 kN m (d) 5 kN m (GATE: 1996)

$$\tau_{max} = \frac{1}{z_p}\sqrt{m^2 + T^2}$$

$$= \frac{1}{z_p}\sqrt{3^2 + 4^2}$$

$$= \frac{5}{z_p}$$

But torque $T_{max} = z_p \times \tau_{max}$

$$= \frac{z_p \times 5}{z_p} = 5 \text{ kN m}$$

Option (d) is correct.

26. Design a circular solid shaft to transmit 80 kW power at 200 rpm, if the twist in the shaft is not to exceed 2° in 3 m length of shaft and maximum shear stress is limited to 70 MN/m^2. Take the modulus of rigidity G = 90 GN/m^2. (UPTU: May 2008)

$$\text{Power } P = \frac{2\pi N \times T}{60}$$

or

$$T = \frac{60 \times 80}{2 \times \pi \times 200}$$

$$= 3.82 \text{ kN m}$$

1. *Strength criteria*

$$T = z_p \ \tau$$

or

$$z_p = \frac{3.82 \times 10^3}{70 \times 10^{16}}$$

$$= 0.0545 \times 10^{-3}$$

But

$$z_p = \frac{\pi d^3}{16}$$

$\therefore$

$$\frac{\pi d^3}{16} = 545 \times 10^{-6}$$

or $$d^3 = \frac{16 \times 545 \times 10^{-6}}{\pi}$$

$$= 0.14 \text{ m}$$

2. *Rigidity criteria*

$$\frac{T}{I_p} = \frac{G\theta}{l}$$

or $$I_p = \frac{T \times l}{G \times \theta} = \frac{3.82 \times 3}{90 \times 10^9 \times \frac{2 \times \pi}{180}}$$

$$= 3.65 \times 10^{-6}$$

$$\frac{\pi d^4}{32} = 3.65 \times 10^{-6}$$

or $$d^4 = \frac{32}{\pi} \times 3.65 \times 10^{-6}$$

$$= 37.2 \times 10^{-6} = 3720 \times 10^{-8}$$

or $$d = 0.078 \text{ m}$$

We have to take bigger of two sizes as arrived by different criteria.

$\therefore$ $$d = 0.14 \text{ m}$$

$$= 14 \text{ cm}$$

OBJECTIVE TYPE QUESTIONS

Future dividends are created in the factories of present investments.

State True or False

1. The shaft is under pure torsion if bending moment and axial force are not acting. (*True/False*)
2. Shear stress due to torsion is maximum at the surface and zero at the centre when a torque acts at a shaft. (*True/False*)
3. Torsion formula $\frac{T}{I_p} = \frac{\tau}{R}$ is applicable for all values of shear stress. (*True/False*)
4. Shear stress due to torsion varies parabolically from the centre to the surface. (*True/False*)
5. If torque is increased by two times, the twist angle increases by four times. (*True/False*)
6. If a shaft under torsion has a shear stress of 100 N/m^2 at the surface, then shear stress at half the radius from the centre is 75 N/m^2. (*True/False*)
7. $\frac{T}{I_p} = \frac{\tau}{R}$ gives the value of maximum torque from stiffness criteria of failure of a shaft. (*True/False*)
8. $\frac{T}{I_p} = \frac{G\theta}{L}$ gives the value of maximum torque from strength criteria of failure of a shaft. (*True/False*)
9. GI_p (where G = modulus of rigidity and I_p is polar moment of inertia) is called torsional rigidity. (*True/False*)
10. Torsional rigidity is torque required to produce a twist of one degree over a unit length of shaft. (*True/False*)
11. If a torque of 20 kN m is applied on a shaft, which produces a twist of 4 radians in a 2-metre shaft, then torsional rigidity is 10 kN m^2/radian. (*True/False*)
12. If a shaft has a torsional rigidity of 5 kN m^2/radian and it has a twist of 2 radian/m, then the value of the torque is 10 kN m. (*True/False*)
13. The value of polar modulus of the section of a solid shaft is $\frac{\pi d^4}{32}$ where d = diameter. (*True/False*)
14. The value of polar modulus of the section for a hollow shaft is $\frac{\pi}{16}(D_0^3 - D_1^3)$, where D_0 = external diameter and D_1 = internal diameter. (*True/False*)
15. Power transmitted by a shaft is $\frac{2\pi NT}{60}$ where N = rpm and T = torque. (*True/False*)
16. If two shafts are in series, both the shafts will transmit the same torque. (*True/False*)

17. If two shafts are in series, the twist angle will be the same and equal to the total twist. *(True/False)*

18. If two shafts are connected in parallel, the total torque transmitted is the sum of torques in each shaft. *(True/False)*

19. If two shafts are connected in parallel, the twist in each is equal and equal to the total twist. *(True/False)*

20. The strain energy in torsion is equal to pure shear strain. *(True/False)*

21. If a hollow shaft and a solid shaft have the same torsional strength, the weight of the hollow shaft is lesser than the solid shaft. *(True/False)*

22. If a hollow shaft and a solid shaft have the same weight, the strength of the hollow shaft is more than the solid shaft. *(True/False)*

23. If a shaft is subjected to torque = T and bending moment = M then equivalent torsion is $\frac{1}{2}\sqrt{M^2 + T^2}$. *(True/False)*

24. If a torque acting on a shaft produces shear stress (= τ), then the principal stress is half the shear stress. *(True/False)*

25. The modulus of rupture is the maximum fictitious shear stress found out from the torsion formula for the maximum torque at failure of the shaft. *(True/False)*

26. A coupling is used to connect two shafts transmitting torque. *(True/False)*

27. A key is a wedge-like part to prevent relative motion between two parts transmitting torque. *(True/False)*

28. In Figure (a), the torque in AB portion of the shaft is 10 kN m (anticlockwise) *(True/False)*

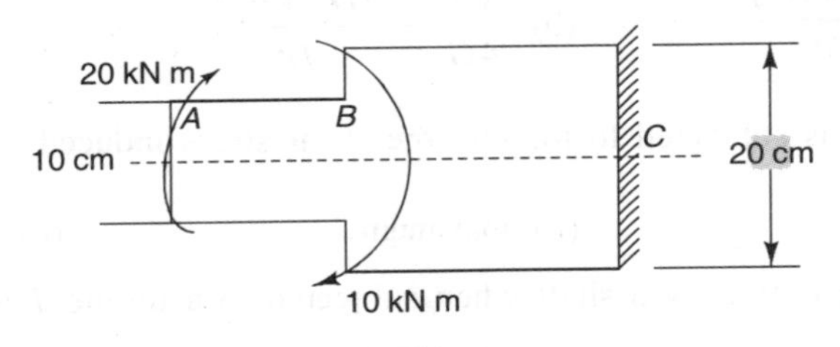

FIGURE (a)

29. In Figure (a), the torque in BC portion of the shaft is 30 kN m. *(True/False)*

30. In Figure (a), the shear stress on central axis is 200 N/cm^2. *(True/False)*

Multiple Choice Questions

1. The outside diameter of a hollow shaft is twice its inside diameter. The ratio of its torque carrying capacity to that of a solid shaft of the same material and the same outside diameter is

(a) $\frac{15}{16}$ (b) $\frac{3}{4}$ (c) $\frac{1}{2}$

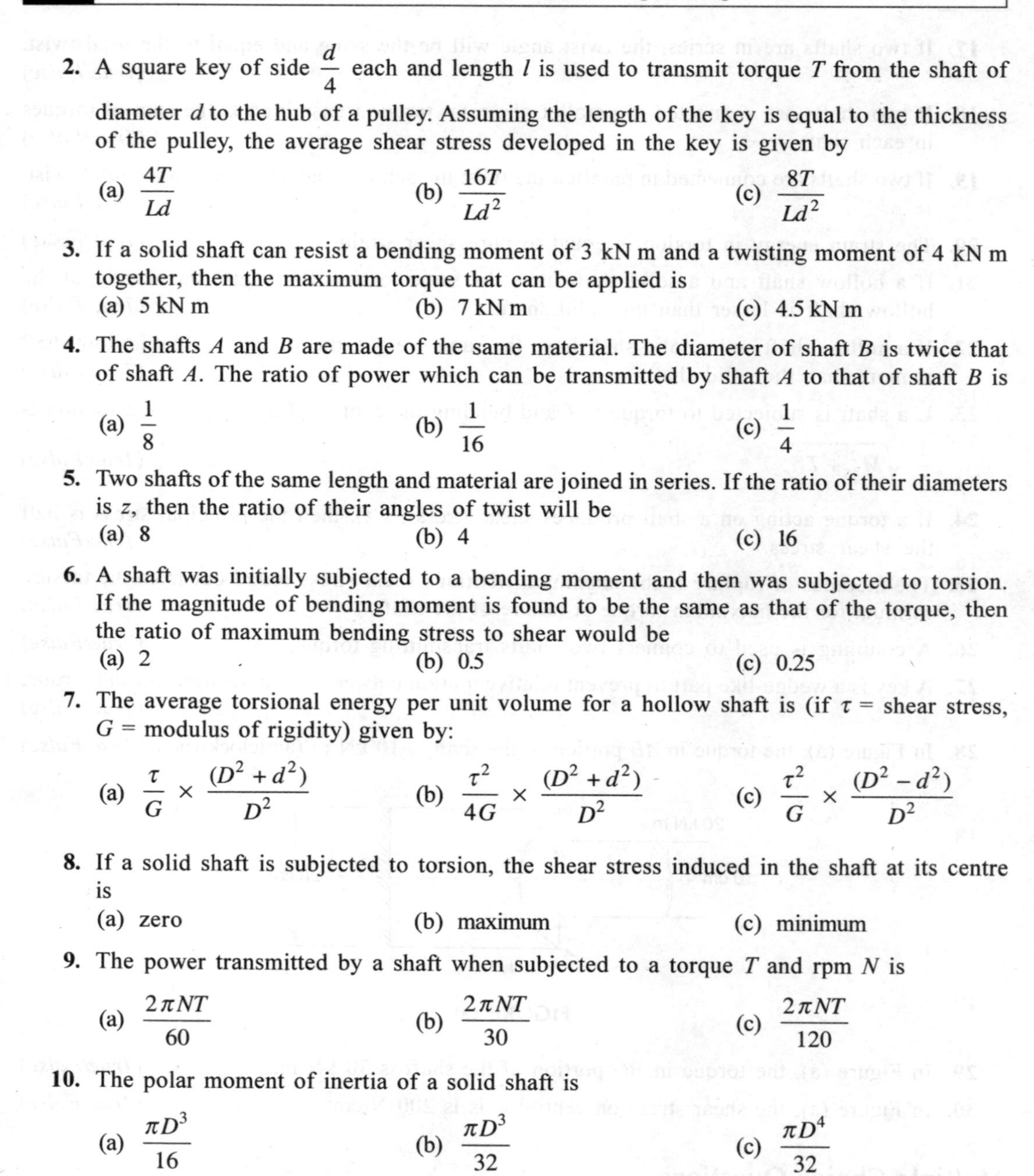

2. A square key of side $\frac{d}{4}$ each and length l is used to transmit torque T from the shaft of diameter d to the hub of a pulley. Assuming the length of the key is equal to the thickness of the pulley, the average shear stress developed in the key is given by

(a) $\frac{4T}{Ld}$ (b) $\frac{16T}{Ld^2}$ (c) $\frac{8T}{Ld^2}$

3. If a solid shaft can resist a bending moment of 3 kN m and a twisting moment of 4 kN m together, then the maximum torque that can be applied is

(a) 5 kN m (b) 7 kN m (c) 4.5 kN m

4. The shafts A and B are made of the same material. The diameter of shaft B is twice that of shaft A. The ratio of power which can be transmitted by shaft A to that of shaft B is

(a) $\frac{1}{8}$ (b) $\frac{1}{16}$ (c) $\frac{1}{4}$

5. Two shafts of the same length and material are joined in series. If the ratio of their diameters is z, then the ratio of their angles of twist will be

(a) 8 (b) 4 (c) 16

6. A shaft was initially subjected to a bending moment and then was subjected to torsion. If the magnitude of bending moment is found to be the same as that of the torque, then the ratio of maximum bending stress to shear would be

(a) 2 (b) 0.5 (c) 0.25

7. The average torsional energy per unit volume for a hollow shaft is (if τ = shear stress, G = modulus of rigidity) given by:

(a) $\frac{\tau}{G} \times \frac{(D^2 + d^2)}{D^2}$ (b) $\frac{\tau^2}{4G} \times \frac{(D^2 + d^2)}{D^2}$ (c) $\frac{\tau^2}{G} \times \frac{(D^2 - d^2)}{D^2}$

8. If a solid shaft is subjected to torsion, the shear stress induced in the shaft at its centre is

(a) zero (b) maximum (c) minimum

9. The power transmitted by a shaft when subjected to a torque T and rpm N is

(a) $\frac{2\pi NT}{60}$ (b) $\frac{2\pi NT}{30}$ (c) $\frac{2\pi NT}{120}$

10. The polar moment of inertia of a solid shaft is

(a) $\frac{\pi D^3}{16}$ (b) $\frac{\pi D^3}{32}$ (c) $\frac{\pi D^4}{32}$

11. The polar moment of inertia of a hollow shaft is

(a) $\frac{\pi}{32}(D_0^3 - D_1^3)$ (b) $\frac{\pi}{16} \times \frac{(D_0^4 - D_1^4)}{D_0}$ (c) $\frac{\pi}{32}(D_0^4 - D_1^4)$

12. The polar modulus of the section of a hollow shaft is

(a) $\frac{\pi}{16} \times \frac{(D_0^4 - D_1^4)}{D_1}$ (b) $\frac{\pi}{16} (D_0^3 - D_1^3)$ (c) $\frac{\pi}{16} \times \frac{(D_0^4 - D_1^4)}{D_0}$

13. The shafts are designed on the basis of
(a) strength only (b) stiffness only (c) both (a) and (b)

14. Torque transmitted by a solid shaft of diameter (D) when subjected to a shear stress (τ) is

(a) $\frac{\pi}{16} \times \tau \times D^2$ (b) $\frac{\pi}{16} \times \tau \times D^3$ (c) $\frac{\pi}{32} \times \tau \times D^3$

15. Shafts A and B of the same material and length with polar moments of inertia of 60 cm^4 and 20 cm^4 respectively are subjected to the same torque. If the angle of twist in shaft A is 2 radians, then the angle of twist in shaft B will be

(a) $\frac{2}{3}$ (b) 4 (c) 6

16. If a torque produces a twist of 4 and 2 radians in shaft A and shaft B having the same diameter and material, then the ratio of their length is

(a) $\frac{1}{2}$ (b) 2 (c) 8

17. If a torque produces a twist of 6 radians and 3 radians in shafts A and B having the same diameter and length, then the ratio of their modulus of rigidity is

(a) $\frac{1}{2}$ (b) 2 (c) $\frac{1}{4}$

18. The modulus of rupture is
(a) force at which rupture takes place
(b) friction stress calculated from the torsional formula from the torque at which rupture takes place
(c) large modulus of the section which a torque can have at rupture

19. A torque of 5 N cm is required to produce a twist of 180°. The torque to produce a twist of 5 turns in the same shaft is
(a) 25 N cm (b) 50 N cm (c) 100 N cm

20. Torsion formula $\left(\frac{T}{I_p} = \frac{\tau}{R} = \frac{G\theta}{L}\right)$ is only applicable when a shaft is subjected to

(a) axial load and torsion load
(b) bending load and torsion load
(c) torsion load only

21. Torsion formula is only applicable up to
(a) ultimate strength (b) yield point (c) elastic limit

22. Shear strain energy due to torsion is
(a) equal to pure shear strain energy
(b) twice to pure shear strain energy
(c) half of pure shear strain energy

23. Two shafts are connected in parallel with the diameter of the inner shaft is half of the outer diameter of the hollow shaft. If the twist in the hollow shaft is 4 radians, then the twist in the inner shaft is
(a) 8 radians (b) 2 radians (c) 4 radians

24. If stress in shafts *A* and *B* connected in parallel is 50 N/cm^2 and 25 N/cm^2, and the polar section moduli of shafts are 100 cm^3 and 50 cm^3, then the total torque transmitted by them is
(a) 5.75 kN cm (b) 6 kN cm (c) 5 kN cm

25. If two shafts *A* and *B* are connected in series and have twist angles of 3 and 2 radians/length, then the total twist is
(a) 3 radians (b) 5 radians (c) 1 radian

Fill in the Blanks

1. Angle of twist is ________ proportional to torque.
(a) directly (b) inversely
2. For the same material, length and torque, a hollow shaft has a ________ weight than a solid shaft.
(a) less (b) more
3. For the same material, length and weight, a hollow shaft has ________ strength than solid stress.
(a) less (b) more
4. A shaft is designed for ________ criteria.
(a) strength (b) strength and stiffness
5. The torsion formula is applicable up to ________.
(a) ultimate strength (b) limit of elasticity
6. Two shafts are joined in series with a ________.
(a) key (b) coupling
7. Two shafts are joined in parallel with a ________.
(a) key (b) coupling
8. The torque which produces the same maximum shear stress as produced by the combined torsion and bending moment is called ________ torque.
(a) effective (b) equivalent

9. If M = bending moment and T = torque, then the equivalent torque is given by ________.

(a) $(M + T)$ (b) $\sqrt{M^2 + T^2}$

10. If bending moment is 4 kN m and torque is 3 kN m, then equivalent torque is ________.

(a) 7 kN m (b) 5 kN m

11. If a shaft ruptures at a torque of 100 kN m and the polar modulus of section is 25 cm^3, then the modulus of rupture is ________.

(a) 400 kN/cm^2 (b) 250 kN/cm^2

12. If torque T is applied on two identical shafts in series, the torque in each shaft is ________.

(a) T (b) $T/2$

13. If shafts A and B are in torque and each has torque T, then the total torque transmitted is ________.

(a) T (b) $2T$

14. If an angle of twist in two shafts (connected together) remains the same for all torques, then the shafts are connected in ________.

(a) series (b) parallel

15. Torsion rigidity is the torque required to produce a twist of ________ radian over a unit length of the shaft.

(a) π (b) one

16. If a coupling has 8 bolts each having area = 10 cm^2 at distance of 50 cm from the centre of a shaft, the torque transmitted by the coupling is ________ (permissible shear stress = 5 kN/cm^2).

(a) 20 MN cm (b) 100 MN cm

17. Shear stress in a shaft due to torsion varies ________ from centre to top.

(a) linearly (b) uniformly

ANSWERS

There's a light at the end of every tunnel; the sun returns after every storm.

State True or False

1. True
2. True
3. False (Torsion formula is valid upto limit of elasticity)
4. False (Shear stress varies linearly from centre ($\tau = 0$) to surface (τ = maximum))
5. False $\left(T \text{ is directly proportional to twist angle, i.e. } \frac{T}{I_p} = \frac{G\theta}{L} \right)$
6. False $\left(\frac{\tau}{r} = \frac{\tau_{max}}{R},\ \tau = \frac{R}{2} \times \frac{\tau_{max}}{R} = \frac{100}{2} = 50 \right)$
7. False (The formula gives maximum torque from strength criteria.)
8. False (The formula gives maximum torque from stiffness criteria.)
9. True
10. False (Twist angle is to be one radian instead of one degree.)
11. True (Twist = 4/2 = 2 radian/unit length. Torsion rigidity $= \frac{T}{\theta} = \frac{20}{2} = 10 \text{ kN m}^2$)
12. True ($T = (GI_p)\frac{\theta}{L} = 5 \times 2 = 10$ kN m)
13. False $\left(z_p = \frac{\pi d^3}{16} \right)$
14. False $\left(z_p = \frac{\pi}{16} \times \frac{D_0^4 - D_1^4}{D_0} \right)$
15. True
16. True
17. False ($\theta = \theta_1 + \theta_2$)
18. True (Total $T = T_1 + T_2$)
19. True ($\theta = \theta_1 = \theta_2$)
20. False $\left(U_{\text{Torsion}} = \frac{\tau^2}{4G} \text{ while } U_{\text{shear}} = \frac{\tau^2}{2G} \right)$

21. True

22. True

23. False ($T_{eqv} = \sqrt{M^2 + T^2}$)

24. False (Major principal stress = τ, minor principal stress = $-\tau$)

25. True

26. True

27. True

28. False (Torque in AB part is –20 kN m.)

29. False [Torque in BC part is –10 kN m ($T = -20 + 10$)]

30. False (Shear stress at the centre is zero.)

Multiple Choice Questions

1. (a) $$\left(z_{\text{hollow}} = \frac{\pi}{16}\left(\frac{D_0^4 - \left(\frac{D_0}{2}\right)^4}{D_0}\right) = \frac{\pi}{16} \times \frac{15}{16} D_0^3\right.$$

$$\left. z_{\text{solid}} = \frac{\pi}{16} D_0^3 \quad \therefore \frac{z_{\text{hollow}}}{z_{\text{solid}}} = \frac{15}{16} \right)$$

2. (c) [Shear force in the key = $\tau \times$ area

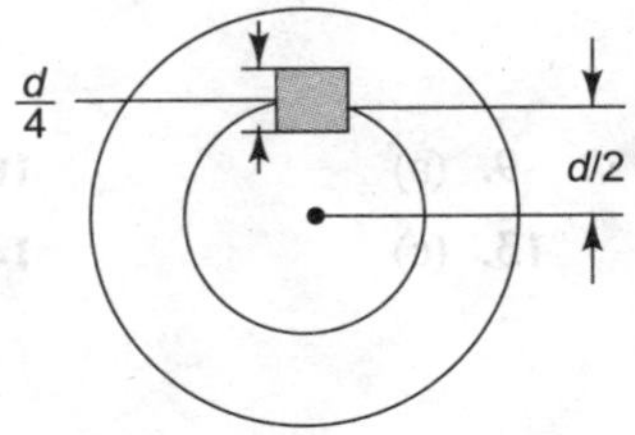

$$\text{SF} = \tau \times \frac{d}{4} \times l$$

$$\text{Torque} = \text{SF} \times \frac{d}{2}$$

$$T = \tau \times \frac{d}{4} \times l \times \frac{d}{2}$$

$$\therefore \tau = \frac{8T}{d^2 l}$$]

3. (a) $(T_{\text{equivalent}} = \sqrt{T^2 + M^2} = \sqrt{9+16} = \sqrt{25} = 5 \text{ kN m})$

4. (a) $\left(z_{p_1} = \dfrac{\pi d^3}{16} \text{ and } z_{p_2} = \dfrac{\pi(2d^3)}{16} = 8 \times \dfrac{\pi d^3}{16} = 8\, z_{p_1} \right.$

$\therefore T_2 = 8T_1$

$P_2 = 8P_1$ as $P = \dfrac{2\pi NT}{60}$

$\left. \therefore \dfrac{P_1}{P_2} = \dfrac{1}{8} \right)$

5. (c) $[T = I_p \dfrac{G}{L} \theta = \dfrac{\pi D^4}{32} \times \dfrac{G}{L} \theta$

$\dfrac{T_1}{T_2} = \left(\dfrac{D_1}{D_2}\right)^4 = (2)^4 = 16]$

6. (a) $(M = z\sigma_b = \dfrac{\pi D^3}{32} \sigma_b$

$T = z_p \tau = \dfrac{\pi D^3}{16} \pi$

$M = T$, then $\dfrac{\pi D^3}{32} \sigma_b = \dfrac{\pi D^3}{16} \tau$

$\sigma_b = 2\tau)$

7. (b) **8.** (a) **9.** (a) **10.** (c)

11. (c) **12.** (c) **13.** (c) **14.** (b)

15. (c) $(T = I_p G \times \dfrac{\theta}{L}$

$I_{p1}\theta_1 = I_{p2}\theta_2$

$\dfrac{\theta_1}{\theta_2} = \dfrac{I_{p_2}}{I_{p_2}}$

$\dfrac{\theta_1}{\theta_2} = \dfrac{20}{60} = \dfrac{1}{3}$ or $\dfrac{2}{\theta_2} = \dfrac{1}{3}$

$\theta_2 = 6)$

16. (b) $(T = I_p G \times \frac{\theta}{L} \therefore \frac{\theta_1}{L_1} = \frac{\theta_2}{L_2}$

$\frac{L_1}{L_2} = \frac{\theta_1}{\theta_2} = \frac{4}{2} = 2)$

17. (a) $(T = I_p G \times \frac{\theta}{L}$

$\therefore G_1\theta_1 = G_2\theta_2$

$\therefore \frac{G_1}{G_2} = \frac{\theta_2}{\theta_1} = \frac{3}{6} = \frac{1}{2})$

18. (b)

19. (b) $(T \propto \theta$ or $\frac{T_1}{T_2} = \frac{\theta_1}{\theta_2}$

$T_2 = T_1 \times \frac{\theta_2}{\theta_1}$

$= 5 \times \frac{5}{1/2} = 5 \times 5 \times 2 = 50$ N cm)

20. (c) **21.** (c) **22.** (c)

23. (c) (Twist in parallel connection is constant.)

24. (a) $(T_A = z_A \tau_A = 100 \times 50 = 5000$ N cm

$T_B = z_B \tau_B = 50 \times 25 = 750$ N cm

$T = T_A + T_B = 5750$ N cm)

25. (b) $(\theta = \theta_1 + \theta_2 = 3 + 2 = 5)$

Fill in the Blanks

1. (a) **2.** (a) **3.** (b) **4.** (b)

5. (b) **6.** (b) **7.** (a) **8.** (b)

9. (b)

10. (b) $(T_{\text{equivalent}} = \sqrt{T^2 + M^2} = \sqrt{3^2 + 4^2} = 5)$

11. (a) $(\tau_{\text{rupture}} = \frac{T}{z_p} = \frac{100 \times 100}{25} = 400 \frac{kN}{\text{cm}^2})$

12. (a) ($T = T_1 = T_2$)

13. (b) ($T = T_1 + T_2 = 2T_1$)

14. (b)

15. (b)

16. (a) (Force/bolt = Area × stress

$= 10 \times 5 = 50$ kN

Total Force = $8 \times 50 = 400$ kN

Torque = Force × $\frac{d}{2}$ = 400×50 kN

= 20×10^3 kN cm)

17. (a)

CHAPTER 18

Fluid Dynamics

Every day is a new beginning. Take a deep breath and start again.

INTRODUCTION

There are three states of matter, i.e. (i) solid (ii) liquid and (iii) gas. The liquid and gas are both fluids as compared to solid as they lack the ability to resist deformation. Fluid cannot resist deformation force and it starts to flow under the action of shear force. Its shape will change continuously as long as the shear force is applied. When a fluid is in motion, shear stresses are developed and fluid particles move relative to one another. As fluid particles move with different velocities, there is shear force in the moving fluid. In practice, we are concerned with fluid flow past solid boundaries such as (i) aeroplanes (ii) cars (iii) pipe walls and (iv) river channels, etc. and shear force is present in these situations. Fluid dynamics is a branch of fluid mechanics which deals with fluid flow. It has several subdisciplines including (i) hydrodynamics (the study of liquids in motion which are incompressible) and (ii) aerodynamics (the study of air and other gases in motion which are compressible). Fluid dynamics has a wide range of applications, including calculating forces (drag and lift) on aircraft; streamlining of vehicles, ships and missiles; measuring of flow from pipes, vessels and rivers; and predicting weather patterns. Fluids are considered to obey the continuum assumption which considers fluids to be continuous. Properties such as density, temperature and velocity are taken to be well-defined at infinitesimal points and these are assumed to change continuously from one point to another. Three conservation laws (mass, momentum and energy) are used to solve fluid dynamics problems using the concept of a control volume (a specified volume in space in which fluid can flow in and out). The Bernoulli's theorem is a fundamental theorem on conservation of total energy and interchangeability of types of energy during fluid flow, i.e. kinetic energy to potential or pressure energy and vice-versa in fluid flow. The term open channel flow represents flows through channels (rivers and canals) that are open to the atmosphere, i.e. liquid flows with a free surface as compared to flow through a pipe in constant contact with pipe wall.

Mechanical energy can be either potential energy on account of height from the ground or kinetic energy on account of velocity. The potential energy or kinetic energy of water can be converted into mechanical work with the use of hydraulic turbines. These machines operate on the principle of rate change of linear and angular momentum. As per Newton's second law, the time rate of change of linear momentum of the fluid will exert a force on the vanes of the rotor of the turbine. It is similar to a large force which acts on the ball when a cricket ball strikes the bat momentarily or passengers in moving bus who fall forward when the bus suddenly stops as a large force acts due to the change of momentum. The vanes on a waterwheel or rotor of hydraulic turbine convert kinetic energy of water into mechanical works as shown in Figure 18.1. A vane is a flat or curved plate. A large number of vanes are fixed on the rim of a wheel to form a waterwheel or turbine rotor. A pump on other hand is used to impart energy to fluid in a fluid system. It is run by using an external source such as electric motor so that it can impart pressure energy or kinetic energy or both energies to the fluid. Centrifugal pumps operate by rotating rotor blades called *impeller* inside a casing filled with water, thereby imparting energy to the water moving in them. A reciprocating pump is a positive displacement pump. It has a piston that executes reciprocating motion in a closely fitting cylinder. The liquid is drawn in and raised by the actual displacement of the piston inside the cylinder. Pumps are used for transporting liquids through pipes.

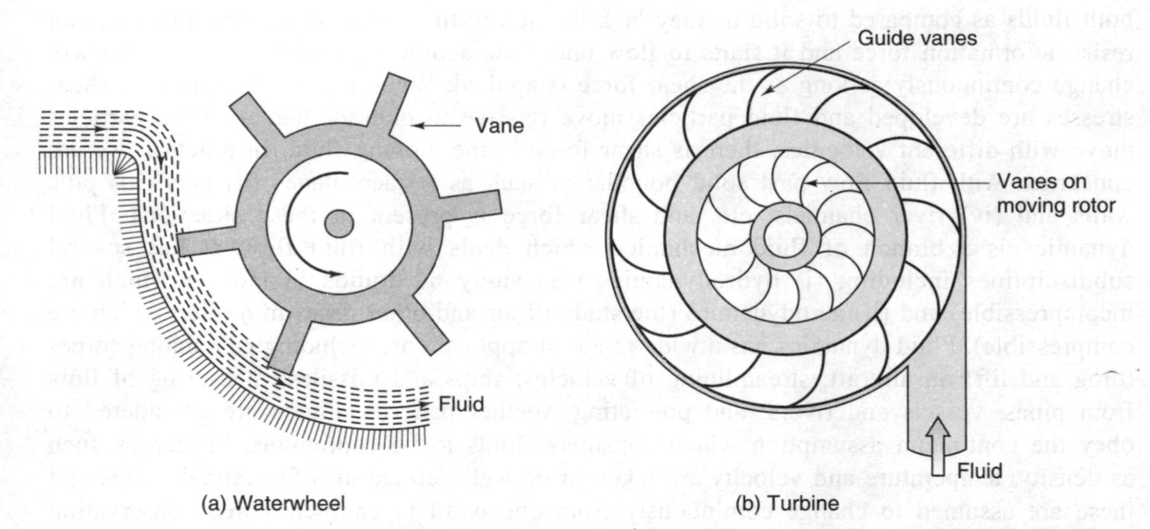

FIGURE 18.1 The vanes on waterwheel and turbine perform mechanical work when water moves over them and change its direction of flow.

COMPRESSIBLE AND INCOMPRESSIBLE FLUIDS

Fluid dynamics deals with both incompressible and compressible fluids. Liquid such as water is considered as incompressible while air and gases are considered compressible.

The incompressible fluids have constant density while compressible fluids have variable density. Although, there is no such thing in reality as an incompressible fluid, we use this term where the change in density with pressure is so small as to be negligible as in the case with most of liquids. We can also consider gases to be incompressible, when the pressure variation is small, resulting change in density as negligible. For an airplane flying at speed below 100 m/s, we may consider the air to be of constant density. Any object moving through the air with velocity of sound (1200 km/h), we treat the air as a compressible fluid.

IDEAL FLUID

An ideal fluid is usually defined as a fluid in which there is no friction. Ideal fluid is inviscid, i.e. its viscosity is zero. The internal forces at any section of ideal fluid flow are always normal to section and purely pressure forces exist. Although such a fluid does not exist in reality, many fluids are considered to have frictionless flow at sufficient distances from solid boundaries. In a real fluid, either liquid or gas, tangential or shearing forces always develop whenever there is a motion relative to a boundary, thus creating fluid friction as these forces oppose the motion of one particle past another. These frication forces give rise to a fluid property called *viscosity*.

VISCOSITY AND NEWTON'S LAW OF VISCOSITY

The *viscosity* of a fluid is a measure of its resistance to shear or angular deformation. For example, lubricating oil has high viscosity and resistance to shear which makes it cohesive and sticky whereas patrol has low viscosity. The friction forces in flowing fluid result from cohesion and momentum interchange between molecules. The viscosity of liquids decrease while it increases for gases as temperature increases. Water has about 50 times more viscosity than air. The most common unit is poise (10^{-3} pascal-second) while one-hundredth of a poise is a centipoise. The viscosity of some common fluids are as per Table 18.1.

Table 18.1 Viscosity of fluids

S.No.	*Fluids*	*Viscosity (centipoise)*
1.	Water	1
2.	Air	0.018
3.	Petrol	0.29
4.	Mercury	1.5
5.	Corn oil	72
6.	SAE 30 oil	290
7.	Tomato sauce	50,000

When fluid is applied shear force, it is deformed and the rate of shear $\left(\frac{d\phi}{dt}\right)$ depends upon shear stress (τ). As shear stress $\left(\tau = \frac{\text{Force}}{\text{Area}}\right)$ increases, the rate of shear or deformation increases. Hence

$$\tau \propto \frac{d\phi}{dt}$$

or

$$\tau = \mu \times \frac{d\phi}{dt}$$

where μ = viscosity

In Figure 18.2, a depth of fluid is trapped between two horizontal plates, one fixed and other moving horizontally at constant velocity (u) has been shown. If the velocity of the top plate is small enough, the fluid particles will move parallel to it and their speed will vary linearly from zero at the bottom stationary plate to speed (u) at the top plate. Each layer of fluid moves slower to its top layer as friction between layers will give rise to a force resisting its relative motion. Hence, fluid will apply on the top plate a force [shear stress (τ) $\times$ area of plate] in the direction opposite to its motion, thereby an external force is required to keep the top plate moving at constant speed (u). The rate of shear deformation $\left(\frac{d\phi}{dt}\right)$ can be seen equal to the $\frac{du}{dy}$. Hence, Newton expressed the viscous stress τ as differential equation as

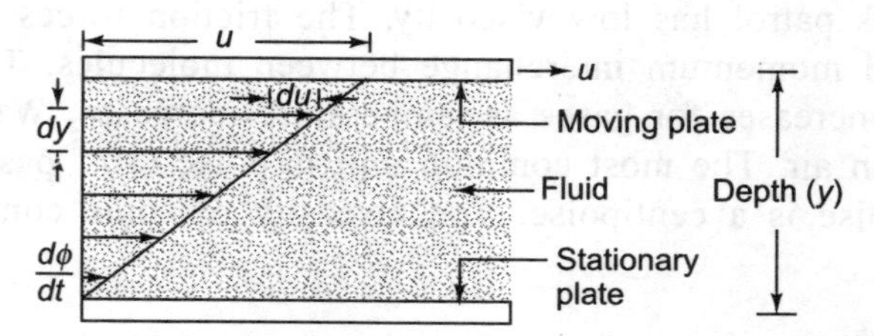

FIGURE 18.2 Laminar shear of fluid between two plates.

$$\tau = \mu . \frac{d\phi}{dt} = \mu . \frac{du}{dy}$$

where

$$\tau = \frac{F}{A} = \frac{\text{Viscous force}}{\text{Area of plate}}$$

$$\frac{du}{dy} = \text{Velocity gradient}$$

$$\mu = \text{Dynamic viscosity.}$$

All fluids do not obey Newton's law of viscosity. The fluids obeying Newton's law of viscosity are called *Newtonian fluids*. Newtonian fluids include (i) water (ii) alcohols and

(iii) organic solvents. Non-Newtonian fluids do not obey Newton's law of viscosity and they are (i) honey (ii) toothpaste (iii) paint (iv) blood (v) ketchup (vi) syrup and (vii) many polymers. The behaviours of Newtonian and non-Newtonian fluids are depicted in Figure 18.3.

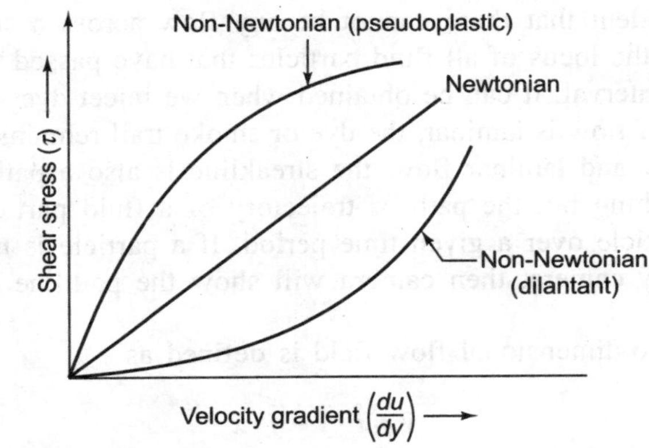

FIGURE 18.3 Newtonian and non-Newtonian fluids.

Example 18.1 An air track supports a cart that rides on a thin cushion of air of 1 mm thick and 0.04 m^2 in area. The viscosity of air is 18×10^{-6} Pa.s. Find the force required to move the cart at constant speed of 0.2 m/s.

$$\tau = \mu.\frac{du}{dy}$$

$$\tau = \frac{\text{Force}}{\text{Area}} = \frac{F}{A}$$

$$\therefore \quad F = A.\mu.\frac{du}{dy}$$

$$F = (0.04) \times (18 \times 10^{-6}) \times \frac{0.2}{1 \times 10^{-3}}$$

$$F = 144 \times 10^{-6} \text{ N}$$

STREAMLINES, STREAKLINE AND PATHLINES

A streamline is a line drawn in the flow at a particular instant of time such that the velocity vector of a fluid particle at any point on the streamline is tangent to the streamline at that point. If v_x, v_y and v_z are the components of velocity vector at a given point and two points on streamline are separated by dx, dy and dz components of a distance, then we have

$$\frac{dx}{v_x} = \frac{dy}{v_y} = \frac{dz}{v_z} \tag{18.1}$$

The integration of Eq. (18.1) at given instant of time will give the equation of streamline. If a flow is steady, the streamline and pathlines coincide.

A stream surface is a surface in space at a given time constructed with streamlines so that velocity vector is tangent to that surface at given point. It is made of many streamlines adjacent to one another. A stream surface that closes to form a tube or conduit is called a *stream tube*. It is evident that there cannot be any flow across a stream tube.

The streakline is the locus of all fluid particles that have passed through given point in space during a time interval. It can be obtained when we inject dye into a transparent fluid or smoke into a gas. If flow is laminar, the dye or smoke trail remains coherent and it forms a streakline. In steady and laminar flow, the streakline is also a pathline and streamline.

A pathline is nothing but the path or trajectory of a fluid particle. It gives the travel history of a fluid particle over a given time period. If a particle is made luminous and its motion is recorded by camera, then camera will show the pathline of the particle.

Example 18.2 A two-dimensional flow field is defined as

$$\vec{\mathbf{v}} = \hat{i} \cdot y - \hat{j} \cdot x$$

Define the equation of streamline passing through the point (1, 0).

The streamline has equation

$$\frac{dx}{v_x} = \frac{dy}{v_y}$$

or

$$\frac{dx}{dy} = \frac{v_x}{v_y} \tag{i}$$

Given $v_x = y$ and $v_y = -x$

Hence, Eq. (i) can be written

$$\frac{dx}{dy} = \frac{y}{-x}$$

or

$$y\ dy = -\ x.dx \tag{ii}$$

Integrating the Eq. (ii), we get

$$x^2 + y^2 = \text{constant} = k \tag{iii}$$

As streamline pass through point (1, 0), we have

$$1 + 0 = k$$

Putting the value of k in Eq. (iii), we have

$$x^2 + y^2 = 1$$

The streamline is nothing but a circle with radius of 1 as shown in Figure 18.4.

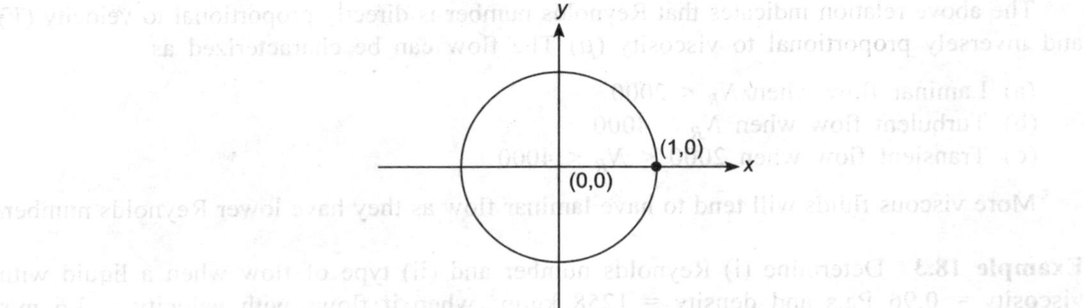

FIGURE 18.4 A circular streamline.

LAMINAR AND TURBULENT FLOW

Laminar or streamline flow occurs when a fluid flows in parallel layers with no disruption between the layers. At low velocities, a fluid tends to flow without lateral mixing and adjacent layers slide past one another. There are no cross currents perpendicular to the direction of flow. There are no eddies and swirls in fluid. In laminar flow, the motion of the fluid particles takes place in very orderly manner with all particles moving in straight lines parallel to the pipe wall as shown in Figure 18.5(a).

Turbulent flow is a type of fluid flow in which the fluid undergoes irregular fluctuations, i.e. fluid particles do not move in laminar layers and there is intermixing. In turbulent flow, the speed and direction of the fluid at a point is continuously undergoing changes in both magnitude and direction. The flow of wind and rivers is generally turbulent and it has air or water swirls and eddies as shown in Figure 18.5(b).

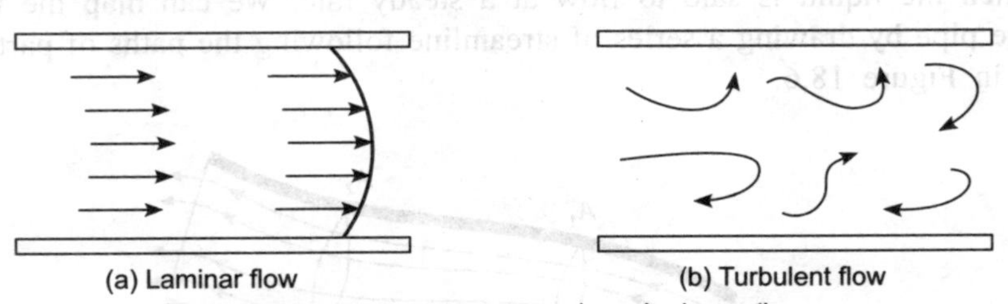

FIGURE 18.5 Laminar and turbulent flows.

Laminar and turbulent flows can be characterized and quantified using Reynolds number. The Reynolds number is given by

$$N_R = \frac{\text{Inertial force}}{\text{Viscous force}}$$

$$= \frac{V \cdot D \cdot \rho}{\mu}$$

where V = velocity of flow
D = diameter of pipe
ρ = density of fluid
μ = viscosity of fluid

The above relation indicates that Reynolds number is directly proportional to velocity (V) and inversely proportional to viscosity (μ) The flow can be characterized as

(a) Laminar flow when $N_R < 2000$
(b) Turbulent flow when $N_R > 4000$
(c) Transient flow when $2000 < N_R < 4000$

More viscous fluids will tend to have laminar flow as they have lower Reynolds number.

Example 18.3 Determine (i) Reynolds number and (ii) type of flow when a liquid with viscosity = 0.96 Pa.s and density = 1258 kg/m^3 when it flows with velocity = 3.6 m/s through a pipe of diameter of 0.15 m.

$$N_R = \frac{V.D.\rho}{\mu}$$

$$N_R = \frac{3.6 \times 0.15 \times 1258}{0.96}$$

$$N_R = 708$$

As $N_R < 2000$, the flow is laminar.

STEADY FLOW AND CONTINUITY EQUATION

When a liquid flows through a pipe in such a way that it completely fills the pipe and as much liquid enters one end of the pipe as it leaves the other end of the pipe in same period of time, then the liquid is said to flow at a steady rate. We can map the flow of liquid through the pipe by drawing a series of streamline following the paths of particles of liquid as shown in Figure 18.6.

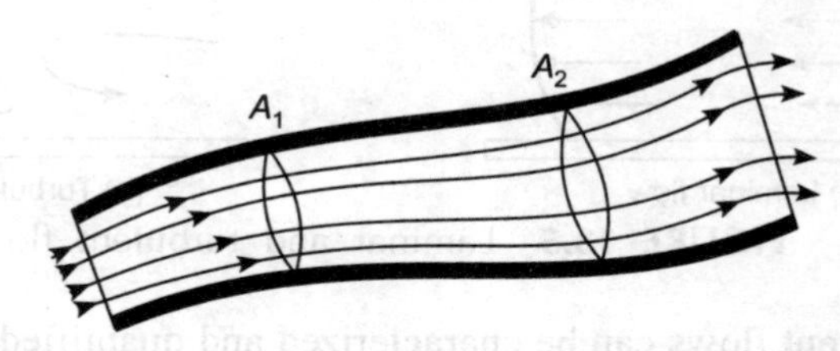

FIGURE 18.6 Streamline of a liquid flowing through a pipe at steady rate.

The instantaneous velocity of a liquid particle can be determined by drawing tangent to the streamlines. The rate of flow may be represented by the density of streamline, i.e. the number of streamlines passing through a surface of unit area perpendicular to the flow direction. Streamlines are closer when liquid is moving rapidly and they are father apart whereever liquid is moving slowly. As the liquid is incompressible, the volume of liquid flowing through any section of pipe in any interval of time remains constant. Consider two planes as shown in the figure where pipe has areas as A_1 and A_2 perpendicular to the streamlines. The volume of liquid Q_1 passing through area A_1 in unit time is

$$Q_1 = A_1 \times V_1 \qquad V_1 = \text{velocity of liquid}$$

Similarly, the volume of liquid Q_2 passing through A_2 in unit time is

$$Q_2 = A_2 \times V_2$$

Since $Q_1 = Q_2$ for steady flow, we have

$$A_1V_1 = A_2V_2$$

In the steady or streamline flow, the total quantity of liquid flowing into imaginary volume element of pipe must be equal to the quantity of liquid leaving the volume element. As shown in Figure 18.7(a), when flow is shown by streamlines then the same number of streamlines enter the imaginary volume selected in space as the number of streamlines leave the volume. Streamline flow is laminar and there is no circulation of the liquid about any point in the pipe. When the flow is laminar and we keep a small peddle wheel in the flow, no rotation of peddle wheel is produced by the liquid flow as shown in Figure 18.7(b).

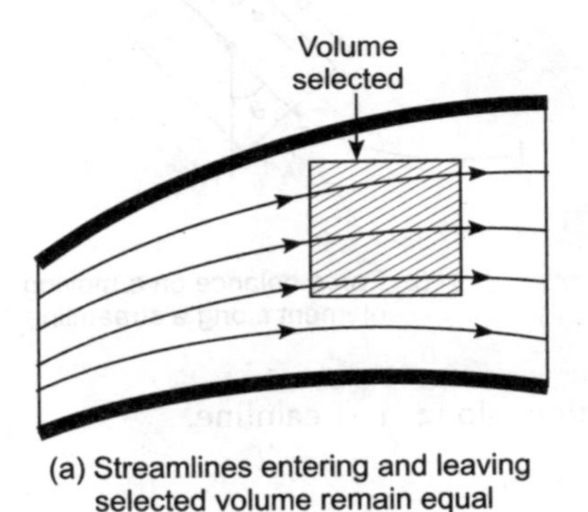

(a) Streamlines entering and leaving selected volume remain equal

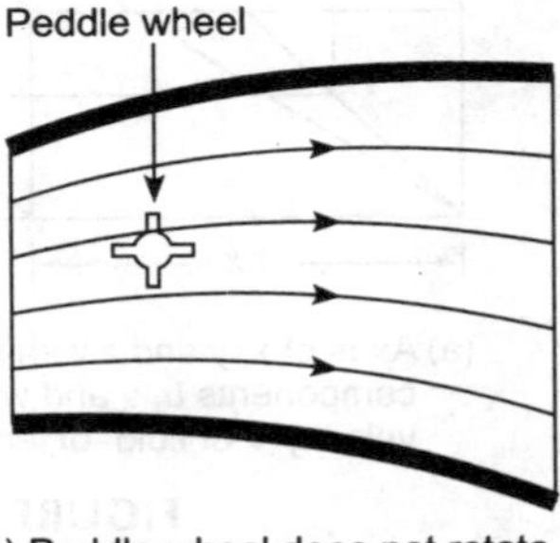

(b) Peddle wheel does not rotate in laminar flow

FIGURE 18.7 Steady or laminar flow.

Example 18.4 Water flows out of horizontal pipe at the rate of 2 m^3/s. Determine the velocity of the water of a point where cross-sectional area is 0.1 m^2.

$$Q = A.V$$

$$Q = 0.1 \times V$$

or

$$V = \frac{2}{0.1} = 20 \text{ m / s.}$$

EULER'S EQUATION ALONG A STREAMLINE

Euler's equation is applicable to inviscid (zero viscosity) or ideal liquid flow and equation represents conservation of (i) mass (ii) momentum and (iii) energy. It is a relation between

(i) velocity (ii) pressure and (iii) density of moving fluid. Euler's equation along a streamline is derived by applying Newton's second law of motion to a fluid element (length *ds*) moving along a streamline. The element accelerates due to net force, i.e. $\Sigma F = m \times \frac{dV}{dt}$. It is based on assumptions:

(a) The fluid is non-viscous.
(b) The fluid is homogeneous and incompressible, i.e. density is constant.
(c) Velocity of flow is uniform over section.
(d) Flow is continuous, steady and along the streamline.
(e) No energy or force (except gravity or pressure force) is involved in the flow.

Let us consider a steady flow of an ideal fluid along a streamline and a small element *ds* of the flowing fluid as shown in Figure 18.8.

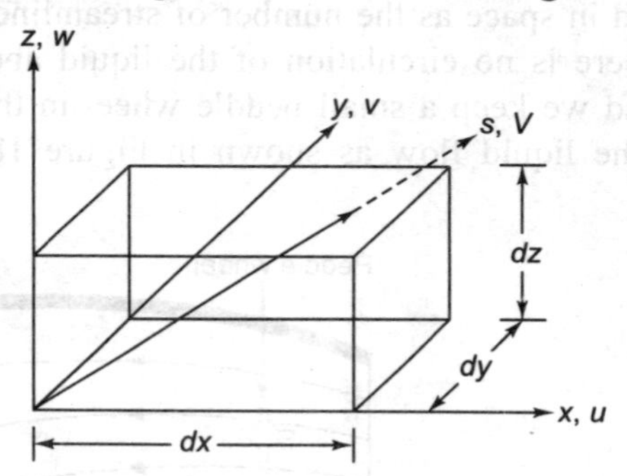

(a) Axes of *x*, *y* and *z* with velocity components *u*, *v* and *w* of velocity *V* of fluid of length *ds*

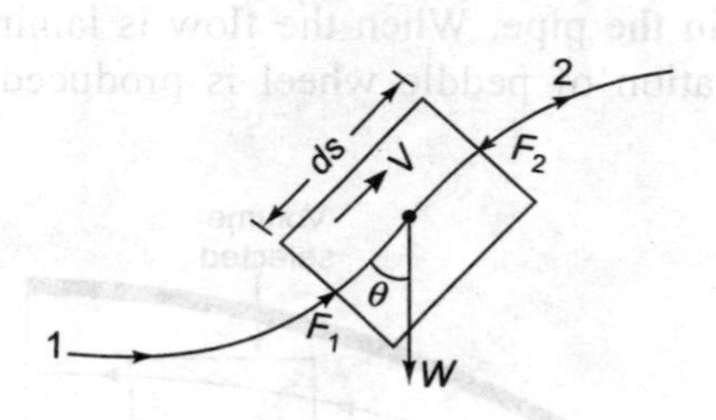

(b) Force balance on a moving element along a streamline 1–2

FIGURE 18.8 Euler's equation along a streamline.

The forces acting on fluid element *ds* are:

(a) Pressure force acting upward on *ds*

$$F_1 = \text{Pressure} \times \text{cross-section}$$
$$= P \times \Delta A$$

(b) Pressure force acting downward on *ds*

$$F_2 = -\left(P + \frac{dP}{ds} \times ds\right) \times \Delta A$$

(c) Weight acting downward on *ds*

$$= -W \cos\theta = -\rho(ds \times \Delta A) \cdot g \cdot \cos\theta$$

$$= -\rho \cdot ds \cdot \Delta A \cdot g \cdot \frac{dz}{ds}$$

The fluid element has net force along ds is:

$$\begin{aligned}\Sigma F &= F_1 + F_2 + W\cos\theta \\ &= P.\Delta A - \left(P + \frac{dP}{ds}\cdot ds\right)\cdot \Delta A - \rho\cdot ds\cdot \Delta A\cdot g\cdot \frac{dz}{ds} \\ &= -\frac{dP}{ds}\cdot ds\cdot \Delta A - \rho\cdot ds\cdot \Delta A\cdot g\cdot \frac{dz}{ds} \\ &= -dP\cdot \Delta A - \rho\cdot g\cdot \Delta A\cdot dz\end{aligned}$$

Now, according to Newton's second law, the net force ΣF will accelerate the fluid element.

$$\begin{aligned}\Sigma F &= \text{Mass} \times \text{Acceleration} \\ &= \rho(ds\cdot \Delta A)\frac{dV}{dt}\end{aligned}$$

or

$$-dP\cdot \Delta A - \rho\cdot g\cdot \Delta A\cdot dz = \rho\cdot ds\cdot \Delta A\cdot \frac{dV}{dt}$$

or

$$\rho\cdot ds\cdot \frac{dV}{dt} = -dP - \rho\cdot g\cdot dz \qquad (18.2)$$

Now we can write

$$\begin{aligned}\frac{dV}{dt} &= \frac{dV}{ds}\cdot \frac{ds}{dt} \\ &= V\cdot \frac{dV}{ds}\end{aligned}$$

Putting the value of $\frac{dV}{dt}$ in Eq. (18.2), we have

$$\rho\cdot V\cdot dV = -dP - \rho\cdot g\cdot dz$$

or

$$\frac{dP}{\rho} + g\cdot dz + V\cdot dV = 0 \qquad (18.3)$$

Equation (18.3) is the Euler's equation for motion in the form of a differential equation.

BERNOULLI'S PRINCIPLE

Bernoulli's principle states that for an inviscid flow, an increase of the speed of fluid occurs simultaneously with a decrease in pressure or a decrease in the fluid's potential energy. It is based on conservation of energy in fluid flow. For a non-viscous and incompressible fluid in steady flow, the sum of pressure, potential and kinetic energy per unit volume is constant at any point. It can be specified as

$$\frac{P}{\rho} + \frac{V^2}{2} + gz = \text{constant} \tag{18.4}$$

Equation (18.4) is a special form of the Euler's equation derived along a fluid flow streamline. The Euler's equation is once integrated, we will get Bernoulli's equation. The Euler's equation is

$$\frac{dP}{\rho} + V \cdot dV + g \cdot dz = 0$$

or

$$\frac{d}{ds}\left(\frac{P}{\rho} + \frac{V^2}{z} + g \cdot z\right) = 0 \tag{18.5}$$

On integration of Eq. (18.5), we get

$$\frac{P}{\rho} + \frac{V^2}{z} + gz = 0 \tag{18.6}$$

From Eq. (18.6), it can be seen that

(a) $\frac{P}{\rho}$ is flow or pressure energy

(b) $\frac{V^2}{2}$ is kinetic energy

(c) gz is potential energy.

Therefore, the Bernoulli's equation can be viewed as conservation of mechanical energy principle, i.e. the sum of kinetic, potential and flow energies of a fluid particle is constant along a streamline during steady flow when the compressibility and frictional effects are negligible. However, Bernoulli's equation is commonly used in practice for a variety of real fluid problems (steady and incompressible with negligible friction force) for analysis with reasonable accuracy. It is often convenient to represent the level of mechanical energy graphically using heights to facilitate visualization of various energy terms of Bernoulli's equation by dividing each term by gravity g as

$$\frac{P}{\rho \cdot g} + \frac{V^2}{2g} + z = \text{Constant}$$

$$= \text{Head, } H$$

Figure 18.9 shows a typical fluid flow which has

(a) Total head $= H$

(b) Pressure head $= \frac{P}{\rho \cdot g}$

(c) Velocity head = $\dfrac{V^2}{2g}$

(d) Potential or elevation head = z.

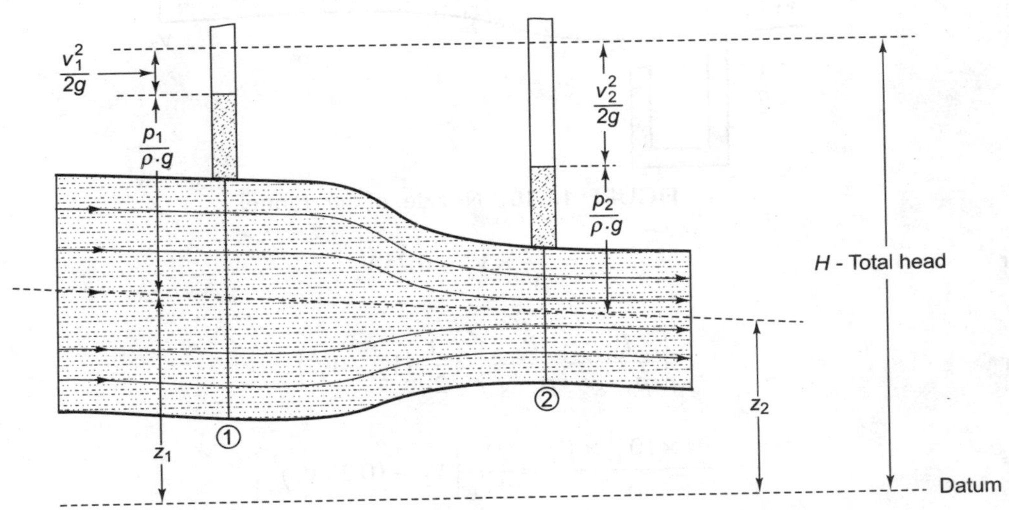

FIGURE 18.9 Total head including various heads.

The Bernoulli's equation can be written between two points 1 and 2 for the above flow as

$$\frac{P_1}{\rho g} + \frac{V_1^2}{2g} + z_1 = \frac{P_2}{\rho g} + \frac{V_2^2}{2g} + z_2 = H$$

The applications of Bernoulli's equation are

(a) Velocity measurement by a pitot tube
(b) Flow rate measurement by a venturi tube
(c) Flow from a small hole (orifice) on a side of water tank
(d) Flow in a syphon.

Example 18.5 A '*U*' tube manometer containing water is connected to a nozzle of an air tunnel that discharges to the atmosphere as shown in Figure 18.10. The area ratio is $\dfrac{A_2}{A_1} = 0.25$. For given operation condition, the level difference of manometer is $h = 94$ mm. Take water density = 1000 kg/m^3 and air density = 1.23 kg/m^3.

Apply Bernoulli's equation between section 1 and section 2 of the figure.

$$\frac{P_1}{\rho_a \cdot g} + \frac{V_1^2}{2g} + z_1 = \frac{P_2}{\rho_a \cdot g} + \frac{V_2^2}{2g} + z_2$$

and $$A_1 V_1 = A_2 V_2$$

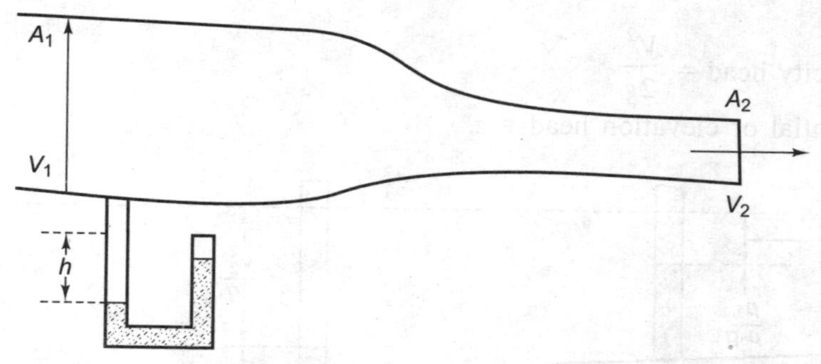

FIGURE 18.10 Nozzle of air tunnel.

or
$$V_1 = \frac{A_2}{A_1} \times V_2 = 0.25\, V_2$$

or
$$\left(\frac{P_1}{\rho_a \cdot g} - \frac{P_2}{\rho_a \cdot g}\right) = \frac{1}{2g}\left(V_2^2 - V_1^2\right)$$

or
$$\frac{94 \times 10^{-3} \times \rho_g}{\rho_a} = \frac{1}{2 \cdot g}\left[V_2^2 - (0.25\, V_2)^2\right]$$

or
$$V_2^2 = \frac{2g \times 94 \times 10^{-3}}{\rho_a} \rho_g \times \frac{1}{\left(1 - 0.25^2\right)}$$

or
$$V_2 = \sqrt{\frac{2 \times 9.81 \times 94 \times 10^{-3} \times 10^3}{1.23} \times \frac{1}{0.9375}}$$

or
$$V_2 = \sqrt{1599.38} \simeq 40 \text{ m/s.}$$

Example 18.6 A person holds his hand out of car window while driving through still air at speed of V_{car}. What is the maximum pressure on his hand? Does it remain same in water?

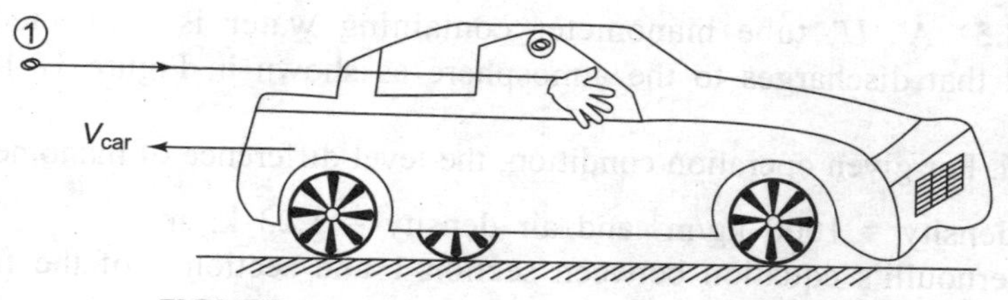

FIGURE 18.11 A car moving on a road.

Apply Bernoulli's equation on air stream where ① is moving air while ⓪ is stationary air

$$\frac{P_1^2}{\rho g} + \frac{P_2^2}{2g} + z_1 = \frac{P_0}{\rho g} + \frac{V_0^2}{2g} + z_2$$

Here

$$z_1 = z_2$$
$$P_1 = P_{atm}$$
$$V_1 = V_{car}$$
$$V_0 = 0$$

$$\therefore \quad \frac{P_{atm}}{\rho_a g} + \frac{V_{car}}{2g} = \frac{P_0}{\rho_a g}$$

or

$$P_0 = P_{atm} + \frac{1}{2} \cdot \rho_a \cdot V_{car}^2$$

As pressure P_0 depends upon density of fluid. The pressure will be more when hand is put in water moving with same speed.

Example 18.7 Water (density = 1000 kg/m^3) flows through a hose with velocity of 1 m/s (see Figure 18.12). As it leaves the constricted area, the velocity increases to 20 m/s. What is the pressure in the hose?

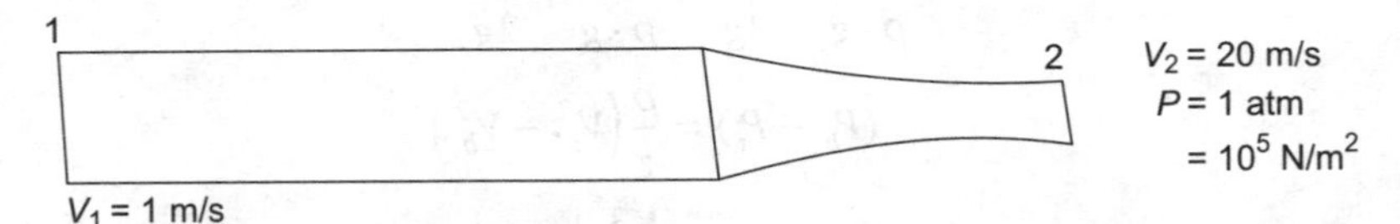

FIGURE 18.12 A hose with nozzle.

$$\frac{P_1}{\rho \cdot g} + \frac{V_1^2}{2g} + z_1 = \frac{P_2}{\rho \cdot g} + \frac{V_2^2}{2g} + z_2 \qquad \text{but } z_1 = z_2$$

or

$$\frac{P_1}{\rho \cdot g} + \frac{1^2}{2 \cdot g} = \frac{10^5}{\rho \cdot g} + \frac{20^2}{2 \cdot g}$$

or

$$P_1 = \frac{\rho}{2} \times \left(20^2 - 1^2\right) + 10^5$$
$$= \frac{10^3}{2} \times 399 + 10^5$$
$$= 1.995 \times 10^5 + 10^5$$
$$= (2.995) \times 10^5 \text{ N/m}$$
$$= 2.995 \text{ atm}$$

Example 18.8 The flow of air around an airplane wing is as shown in Figure 18.13.

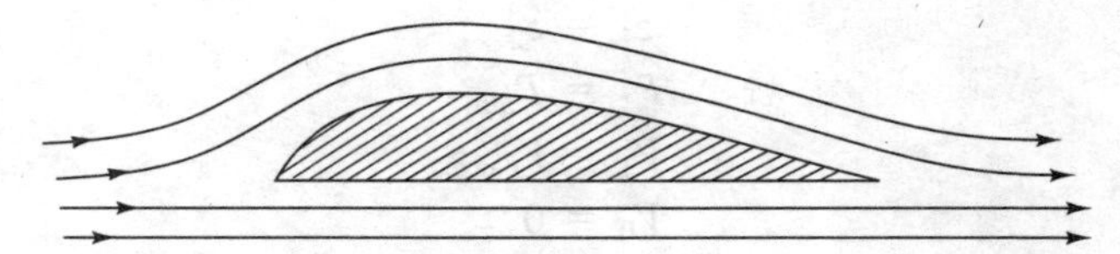

FIGURE 18.13 Air flow around the wing of the airplane.

Consider flow of air (density = 1.3 kg/m^3) with velocity 250 m/s on the top of the wing and 220 m/s over the bottom. If the area of wings = 10 m^2, find (i) pressure difference between the top and bottom wing, and (ii) upward force for lifting on wings.

Apply Bernoulli's equation at bottom and top of wing.

$$\frac{P_b}{\rho \cdot g} + \frac{V_b^2}{2g} + z_1 = \frac{P_t}{\rho \cdot g} + \frac{V_t^2}{2g} + z_2$$

If

$$z_1 = z_2$$

$$\frac{P_b}{\rho \cdot g} + \frac{V_b^2}{2g} = \frac{P_t}{\rho \cdot g} + \frac{V_t^2}{2g}$$

$$(P_b - P_t) = \frac{\rho}{z}\left(V_t^2 - V_b^2\right)$$

$$= \frac{1.3}{2}(250^2 - 220^2)$$

$$= 9165 \text{ N/m}^2$$

Lift force

$$F = \left(P_b - P_t\right) \times \text{Area}$$

$$= 9165 \times 10$$

$$= 91650 \text{ N}$$

TORRICELLI'S THEOREM

Torricelli's theorem is used to determine the flow of a liquid out of an orifice at the base of a tank. It is based on Bernoulli's equation. Consider an orifice (c) at the base of the tank as shown in Figure 18.14.

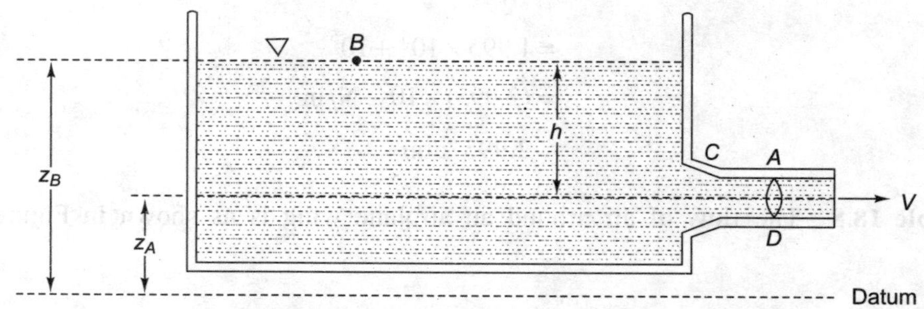

FIGURE 18.14 Torricelli's theorem.

The area of orifice is A and it is at depth of 'h' from point 'B' at top of tank and water surface is open to atmosphere. At both positions (i) 'B' at top of the tank and (ii) 'D' at the emerging stream, the pressure existing is atmospheric pressure. It may be observed that stream emerging from orifice (c) narrows as it emerges from the tank at C. The portion of the stream with parallel sides is called the *vena contracta*. The vena contracta or narrowing of stream occurs as the liquid is being accelerated and it has not yet reached its final highest velocity, i.e. higher velocity needs lesser area for same flow. If tank is sufficiently large, we can assume that the flow does not change the level of liquid in the tank with the flow and liquid at 'B' is almost at rest, i.e. velocity of liquid of 'B' is zero. On applying Bernoulli's equation between points 'A' and 'B', we have

$$\frac{P_A}{\rho \cdot g} + \frac{V_A^2}{2g} + z_A = \frac{P_B}{\rho \cdot g} + \frac{V_B^2}{2g} + z_B$$

or

$$0 + \frac{V_A^2}{2g} + z_A = 0 + 0 + z_B$$

or

$$V_A = \sqrt{2g(z_B - z_A)}$$
$$= \sqrt{2gh}$$

The above relation shows that the liquid emerging from the orifice has same velocity which it gains when it falls freely from a height. Hence, a unit volume of liquid would lose potential energy and gain equal amount of kinetic energy.

Potential energy = kinetic energy

$$\rho \cdot g \cdot h = \frac{1}{2}\rho \cdot V^2$$

or

$$V = \sqrt{2g \cdot h}$$

The above relation gives us the theoretical value of velocity. The real velocity of liquid with friction losses can be given by multiplying the coefficient of velocity as:

$$V_{\text{actual}} = C_V \times \sqrt{2gh}$$

Where

$$C_V = \text{Coefficient of velocity}$$
$$= \leq 1$$

The discharge (Q) through the orifice can be found out by multiplying the actual velocity (V_{actual}) with the area of jet emerging from the orifice. This is the area at vena contracta and it is determined by multiplying coefficient of contraction (C_c) to the area of orifice (A_{orifice}).

$$A_{\text{jet}} = C_c \times A_{\text{orifice}}.$$

The actual discharge from the orifice is given by

$$Q_{\text{actual}} = C_V \times C_c \times A_{\text{orifice}} \times \sqrt{2gh}$$
$$= C_d \times A_{\text{orifice}} \times \sqrt{2gh}$$

where C_d = Discharge coefficient

Vena Contracta and Contraction Coefficient

Depending on the geometry of the orifice, the flow fields can differ as shown in Figure 18.15.

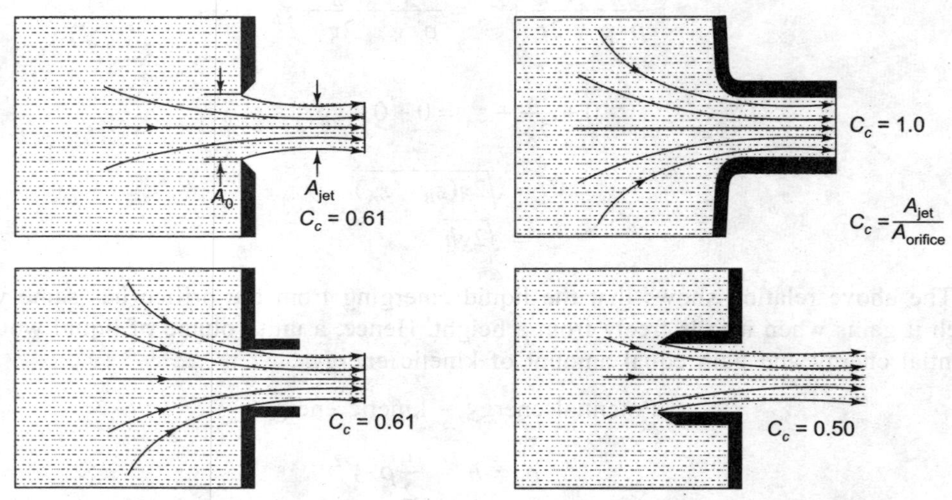

FIGURE 18.15 Flow through different geometry of orifice.

The vena contracta is the cross-section of the jet where the streamlines are straight and parallel. This is the section at which pressure is equal to atmospheric pressure. The contraction coefficient is defined as

$$C_c = \frac{A_{\text{jet}}}{A_{\text{orifice}}}$$

Discharge Coefficient

The discharge coefficient (C_d) combines contraction and velocity coefficients as follows:

$$C_d = C_c \times C_V$$

The discharge through the orifice can be given as:

$$Q = C_V \times C_V \times A_{\text{orifice}} \times \sqrt{2 \cdot g \cdot h}$$
$$= C_d \times A_{\text{orifice}} \times \sqrt{2gh}$$

Example 18.9 A tank containing water has an orifice to one vertical side. If the centre of the orifice is 4.9 m below the surface level in the tank, find the velocity of discharge.

$$\rho \cdot g \cdot h = \frac{1}{2} \rho \cdot V^2$$

or

$$V = \sqrt{2g \cdot h}$$

$$= \sqrt{2 \times 9.81 \times 4.9}$$

$$= 9.8 \text{ m/s}$$

Example 18.10 Water stands at a depth 'H' of a tank having vertical walls. A hole is made on one of the walls at a depth of 'h' below the water surface. Find (i) at what distance 's' from the foot of the wall does the emerging stream of water strike the floor as shown in Figure 18.16, and (ii) for what value of 'h' is the range maximum.

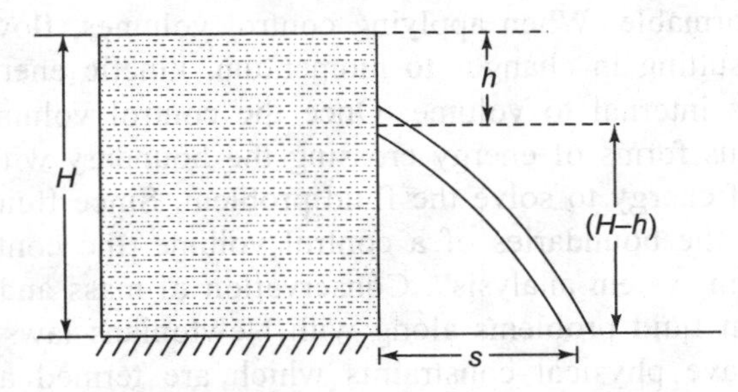

FIGURE 18.16 Torricelli's theorem.

$$\rho g h = \frac{1}{2} \rho V^2$$

or

$$V = \sqrt{2gh}$$

Let 't' is time for jet of water emerging from hole to reach the ground

$$H - h = \frac{1}{2} \cdot g \cdot t^2$$

or

$$t = \sqrt{\frac{2(H-h)}{g}}$$

Now, we have horizontal velocity V and time 't' for water to travel. Hence,

$$s = V \times t$$

$$= \sqrt{2 \cdot g \cdot h} \sqrt{\frac{2(H-h)}{g}}$$

$$= \sqrt{4h(H-h)}$$

$$= 2\sqrt{h \cdot (H-h)}$$

For maximum s, we must have $\frac{ds}{dh} = 0$

$$\frac{ds}{dh} = 2 \times \frac{1}{2} \times [h(H-h)]^{-1/2} \times [H - 2h] = 0$$

or
$$h = \frac{H}{2}$$

The hole should be in the middle for maximum range.

CONTROL VOLUME ANALYSIS

A fluid dynamic system can be analysed using a control volume which is an imaginary surface enclosing a volume of interest. The control volume can be fixed or moving and it can be rigid and deformable. When applying control volumes, flow is allowed to enter or leave the volume, resulting in changes to momentum, kinetic energy and other physically measurable properties internal to volume. Once the control volume and its boundary are established, the various forms of energy crossing the boundary with the fluid can be dealt with equation form of energy to solve the fluid problem. Since fluid flow problems usually treat a fluid crossing the boundaries of a control volume, the control volume approach is referred to as an "open system analysis". Conservation of mass and conservation of energy are always satisfied in fluid problems along with Newtonians laws of motion. In addition, each problem will have physical constraints which are termed as boundary conditions. These must be satisfied before a solution to the problem will be consistent with the physical results. The difference between a control mass system (closed system) and a control volume system (open system) is as shown in Figure 18.17.

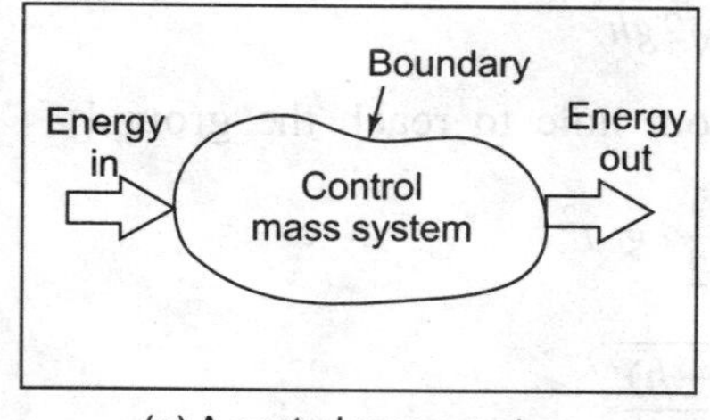

(a) A control mass system

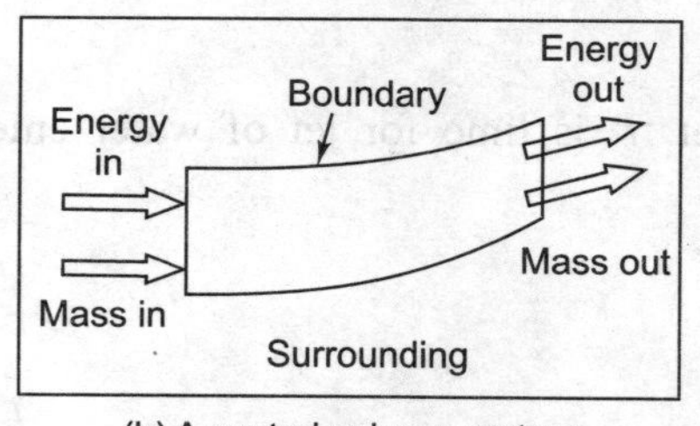

(b) A control volume system

FIGURE 18.17 Difference between a control mass and control volume system.

(a) ***Volume and mass flow rate:*** An arbitrary surface(s) is as shown in Figure 18.18. Let **n** be defined as the unit vector normal to elemental area dA. The amount of fluid swept through dA in time dt is

$$dV = V \cdot dt \cdot dA \cos\theta = (V \cdot \mathbf{n}) dA \cdot dt$$

The integral $\frac{dV}{dt}$ is the total volume rate of flow through surface S.

$$Q = \int (V \cdot \mathbf{n}) dA = \int_s V_n \cdot dA$$

where V_n = normal component of velocity. The mass flow is given by

$$\dot{m} = \int_s \rho (V \cdot \mathbf{n}) dA = \int_s \rho V_n \, dA$$

If density ρ and velocity V_n are constant, then

$$\dot{m} = \rho \cdot V \int_s dA = \rho \cdot V \cdot A$$

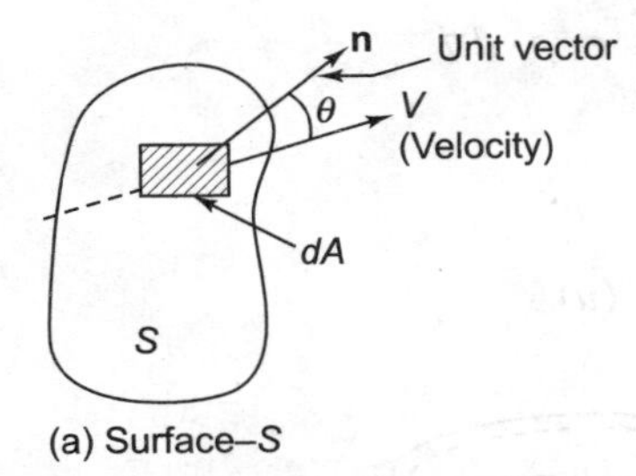

(a) Surface–S

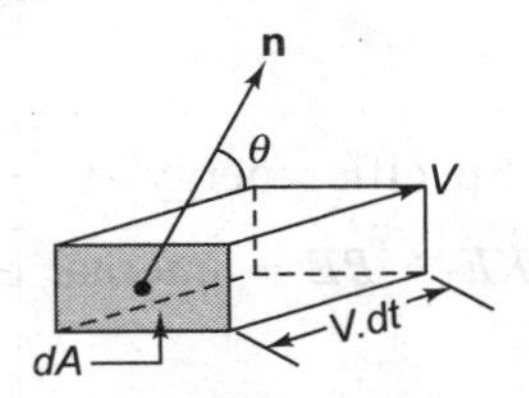

(b) Flow through small area dA

FIGURE 18.18 Volume and mass flow rate.

(b) ***Conservation of mass:***

$$\dot{m} = \int_{CS} \rho \cdot (V \cdot \mathbf{n}) dA + \frac{d}{dt} \int_{CV} \rho \cdot d\forall$$

where $dm = \rho \cdot d\forall$, and $\forall$ = volume, CS = control surface and CV = control volume

In case mass is constant, then $\dot{m} = 0$

$$\int_{CS} \rho \cdot (V \cdot \mathbf{n}) dA = -\frac{d}{dt} \int_{CV} \rho \cdot d\forall$$

The above relation shows that mass of fluid flowing out of control surface (CS) is equal to decrease of mass in control volume (CV) For steady flow, the mass in control volume remains unchanged. Hence, we have

$$\int_{CS} \rho \cdot (V \cdot \mathbf{n}) dA = 0$$

For incompressible flow, we have

$$\int_{CS} (V \cdot \mathbf{n}) dA = 0$$

If we consider flow through a tube, we have for steady flow

$$\rho_1 A_1 V_1 = \rho_2 A_2 V_2$$

or

$$A_1 V_1 = A_2 V_2 \text{ for incompressible flow.}$$

(c) ***Conservation of momentum:*** Law of conservation of momentum as applied to control volume states that

$$\Sigma F = \frac{d}{dt}\left(\int_{CV} V \cdot \rho \cdot dV\right) + \int_{CS} V \cdot \rho (V \cdot n) dA$$

(d) ***Conservation of energy:*** The law of conservation of energy:

Heat added (dQ) – Work done (dW) = change energy (dE)

$$E_{sys} = \int_{CV} e \cdot \rho \cdot d\forall$$

where e = Specific energy $= \frac{V^2}{2} + gz + u$

$e = KE + PE +$ internal energy (u)

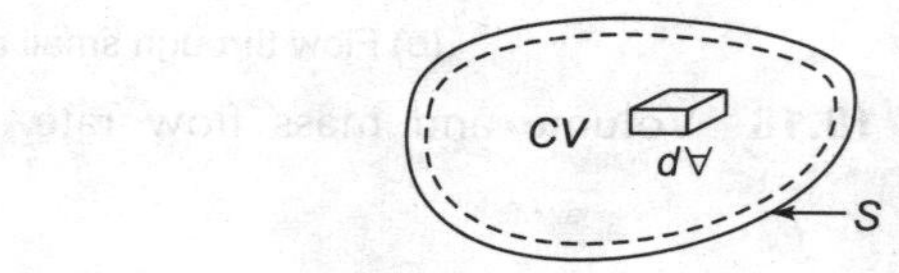

FIGURE 18.19 Conservation of energy in control volume system.

FLOW MEASUREMENT BY VENTURI METER, ORIFICE METER, FLOW NOZZLE AND PITOT TUBE

Venturi meter and orifice meter are commonly used as flow meters for measuring flow rate or velocity of the flowing fluid. These measuring devices work on the principle of Bernoulli's equation in which a restriction to flow is put to cause a decrease in flow area and to increase the velocity of stream using pressure energy. The difference in pressure is measured to determine velocity of flow.

Venturi Meter

The venturi meter has a converging conical inlet, a cylindrical throat and a diverging recovery cone as shown in Figure 18.20. It has no projection in the fluid, no sharp corners and no sudden changes in contour. The venturi meter has a uniform cylindrical section before converging entrance, a throat and divergent outlet. The converging inlet section decreases the area of the fluid flow, causing the velocity to increase and pressure to decrease. The low pressure is measured at the centre of the cylindrical throat as the pressure is lowest at this place. As the fluid enters the diverging section of the meter, the pressure is largely recovered due to the lowering of flow velocity in enlarging area.

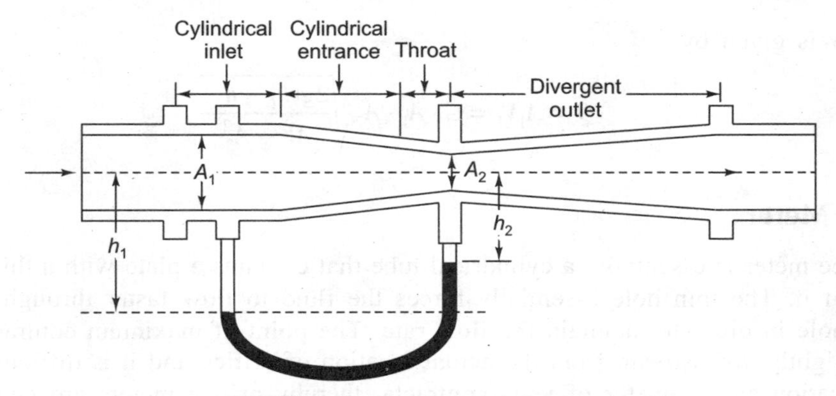

FIGURE 18.20 A venturi meter.

The venturi effect is the reduction in fluid pressure that results when fluid flows through the constricted section in pipe. The fluid velocity has to increase through the constriction so as to satisfy the equation of continuity, thereby its pressure has to decrease for maintaining conservation of fluid energy. The gain in kinetic energy is balanced by a drop of pressure energy. An equation for the drop in pressure due to venturi effect can be derived from a combination of Bernoulli's principle and equation of continuity.

The equation for venturi meter is obtained by applying Bernoulli's equation and equation of continuity, assuming an incompressible flow of fluid. The pressures at section 1 and section 2 are measured by manometer tubes as shown in the figure. As per Bernoulli's equation and equation of continuity, we have

$$\frac{P_1}{\rho g} + \frac{V_1^2}{2g} = \frac{P_2}{\rho \cdot g} + \frac{V_2^2}{2g}$$

and

$$A_1 V_1 = A_2 V_2$$

or

$$\frac{P_1}{\rho} + \frac{V_1^2}{2} = \frac{P_2}{\rho} + \left(\frac{A_1}{A_2}\right)^2 \cdot \frac{V_1^2}{2}$$

or

$$\frac{V_1^2}{2}\frac{\left(A_1^2 - A_2^2\right)}{A_2^2} = \frac{1}{\rho}\left(P_2 - P_1\right) = \frac{g}{\rho}(h_1 - h_2) \times P$$

or

$$V_1 = A_2\sqrt{\frac{2g(h_1 - h_2)}{A_1^2 - A_2^2}}$$

In case, we apply correction coefficient C, we have

$$V_1 = CA_2\sqrt{\frac{2g(h_1 - h_2)}{A_1^2 - A_2^2}}$$

The flow is given by

$$Q = A_1V_1 = C \cdot A_1 \cdot A_2\sqrt{\frac{2g(h_1 - h_2)}{A_1^2 - A_2^2}}$$

Orifice Meter

An orifice meter is essentially a cylindrical tube that contains a plate with a thin hole in the middle of it. The thin hole essentially forces the fluid to flow faster through the reduced area in hole in order to maintain the flow rate. The point of maximum contraction usually occurs slightly downstream from the actual location of orifice and it is difficult to find the exact location and diameter of vena contracta, thereby orifice meters are generally lesser accurate as compared to venturi meter. Beyond the vena contracta, the fluid expands again which ensures that pressure regains at the expense of velocity. Orifice meter is as shown in Figure 18.21 with variable position of vena contracta.

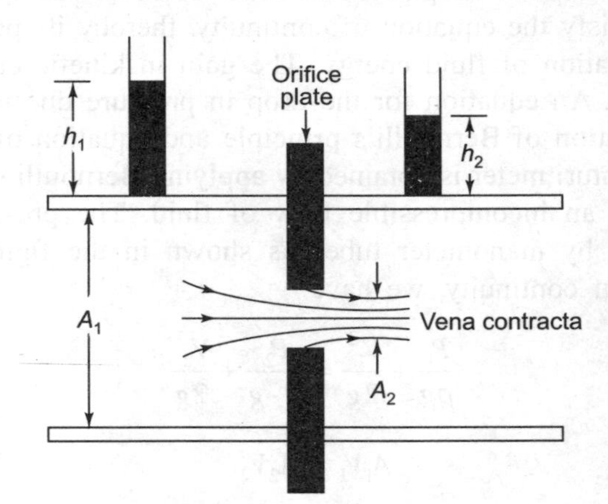

FIGURE 18.21 Orifice meter.

Orifice meter uses the same principle of Bernoulli's equation and equation of continuity to determine the velocity and discharge of the fluid flow

$$\frac{P_1}{\rho} + \frac{V_2^2}{2} = \frac{P_2}{\rho} + \frac{V_2^2}{2} \tag{18.7}$$

and

$$A_1V_1 = A_2V_2 \tag{18.8}$$

From Eqs. (18.7) and (18.8), we have

$$V_1 = C_2 \cdot A_2 \sqrt{\frac{2g(h_1 - h_2)}{A_1^2 - A_2^2}}$$

and

$$Q = C \cdot A_1 A_2 \sqrt{\frac{2g(h_1 - h_2)}{A_1^2 - A_2^2}}$$

Flow Nozzle

The principle of working is same as that of orifice meter and venturi tube. The reduction in flow area is caused by using a nozzle as shown in Figure 18.22. The flow nozzle is relatively simple and cheap. It uses the principle of Bernoulli's equation and equation of continuity. The velocity and flow can be measured as

$$V_1 = C \times A_2 \sqrt{\frac{2g(h_1 - h_2)}{A_1^2 - A_2^2}}$$

$$Q = C \times A_1 \times A_2 \sqrt{\frac{2g(h_1 - h_2)}{A_1^2 - A_2^2}}$$

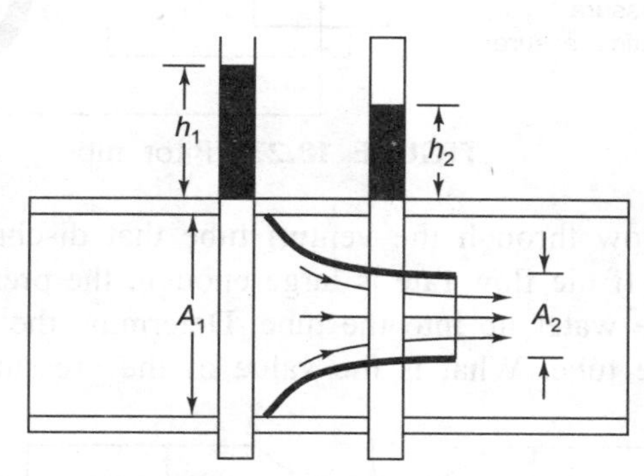

FIGURE 18.22 Flow nozzle.

Pitot-static Tube

A pitot-static tube can measure the fluid flow velocity by converting the kinetic energy in the fluid flow into potential energy. The principle is based on the Bernoulli's equation. The basic pitot tube consists of a tube pointing directly into the fluid flow as shown in Figure 18.23. As the tube contains fluid, a pressure can be measured when the moving fluid is brought to rest or stagnation as there is no outlet for the flow to continue. The tube pressure is the stagnation pressure of the fluid which is also called total pressure or the *pitot pressure*. We have

Stagnation pressure = Static pressure + Dynamic pressure

$$P_t = P_s + \rho\frac{V^2}{2}$$

or
$$V^2 = \frac{2(P_t - P_s)}{\rho}$$

or
$$V = \sqrt{\frac{2(P_t - P_s)}{\rho}} = \sqrt{\frac{2(\text{stagnation pressure} - \text{static pressure})}{\rho}}$$

where $P_t - P_s = \rho gh$ and it can be measured by differential monometer. Hence, we have

$$V = \sqrt{\frac{2\rho_f gh}{\rho}}$$

The pitot tube is also used for measuring velocity of aircraft.

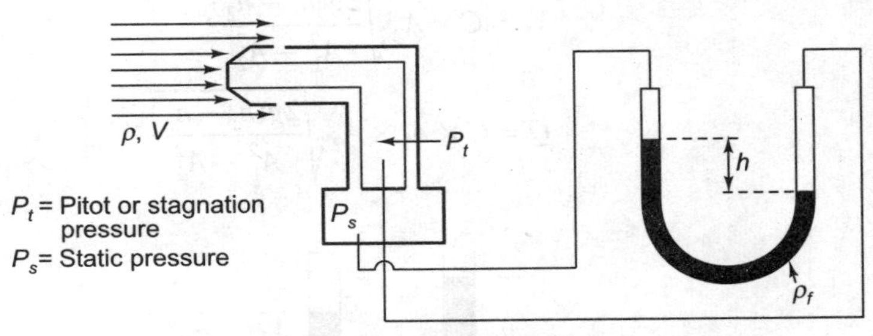

FIGURE 18.23 Pitot tube.

Example 18.11 Air flow through the venturi tube that discharges to the atmosphere as shown in Figure 18.24. If the flow rate is large enough, the pressure in contraction will be low enough to draw the water up into the tube. Determine the flow rate Q needed to just draw the water into the tube. What is the value of the pressure P_1?

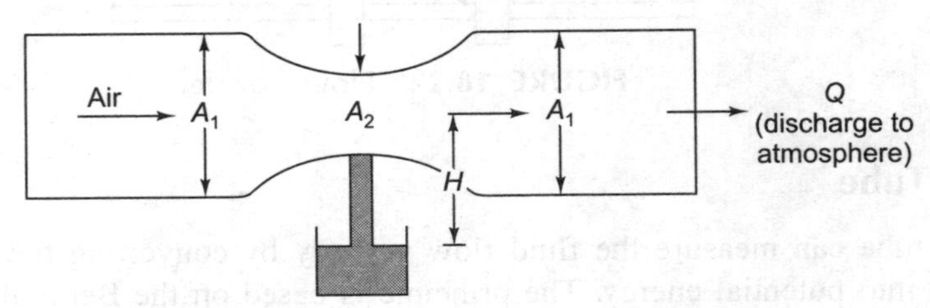

FIGURE 18.24 Flow through a venturi tube.

We know that flow Q in venturi meter is

$$Q = A_1 A_2 \sqrt{\frac{2(P_1 - P_2)}{\rho(A_1^2 - A_2^2)}}$$

But
$$P_2 = P_{atm} - \rho_{H_2O} \times g \times H$$
$$P_1 = P_{atm} \text{ as air being discharged to atmosphere}$$

$\therefore$
$$Q = A_1 \cdot A_2 \sqrt{2 \times g \times H \left(\frac{\rho_{H_2O}}{\rho}\right) \times \frac{1}{\left(A_1^2 - A_2^2\right)}}$$

Example 18.12 Draw the flows through (i) orifice (ii) nozzle and (iii) venturi meter.

The flows through (i) orifice (ii) nozzle and (iii) venturi meter are as shown in Figure 18.25.

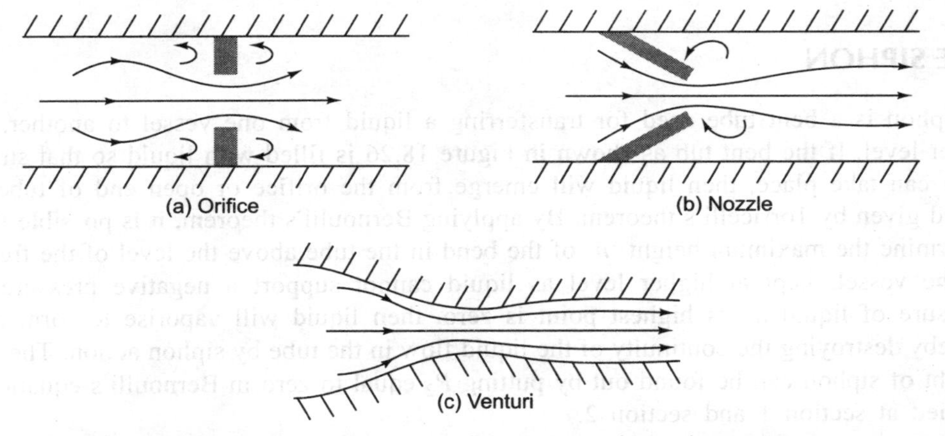

FIGURE 18.25 Flows through (i) orifice (ii) nozzle and (iii) venturi meter.

Example 18.13 A venturi tube is used to measure the flow of water. It has a main diameter of 3 cm tapering down to a throat diameter of 10 cm. The pressure difference "$P_1 - P_2$" is measured to be 18 mm Hg. Calculate the velocity V_1 of the fluid input and flow rate Q.

$$V_1 = A_2 \sqrt{\frac{2(P_1 - P_2)}{\rho(A_1^2 - A_2^2)}}$$

$$P_1 - P_2 = 18 \text{ mm of Hg} = 18 \times 10^{-3} \times \left(1.8 \times 10^5 \frac{P_a}{m}\right)$$

$$= 24 \times 10^2 \text{ Pa}$$

$$A_1 = \frac{\pi \times (3 \times 10^{-2})^2}{4} = \frac{9\pi}{4} \times 10^{-4} \text{ and}$$

$$A_2 = \frac{\pi \times (10^{-2})^2}{4} = \frac{\pi}{4} \times 10^{-4}$$

$$V_1 = \frac{\pi}{4} \times 10^{-4} \sqrt{\frac{2 \times 24 \times 10^2}{10^3 \left(\frac{\pi}{4}\right)^2 \times 10^{-8} (81 - 1)}}$$

$\therefore$

$$V_1 = 24.6 \text{ cm/s}$$

$$Q = A_1 \times V_1$$

$$Q = \frac{9\pi}{4} \times 10^{-4} \times 24.6 \times 10^{-2}$$

$$Q = 174 \times 10^{-6} \text{ m}^3/\text{s}$$

$$Q = 174 \text{ cm}^3/\text{s}$$

THE SIPHON

A siphon is a bent tube used for transferring a liquid from one vessel to another, kept at lower level. If the bent tub as shown in Figure 18.26 is filled with liquid so that streamline flow can take place, then liquid will emerge from the orifice or open end of tube with a speed given by Torricelli's theorem. By applying Bernoulli's theorem, it is possible firstly to determine the maximum height '*h*' of the bend in the tube above the level of the free liquid in the vessel, kept at higher level as liquid cannot support a negative pressure. If the pressure of liquid at its highest point is zero, then liquid will vaporise to form bubbles, thereby destroying the continuity of the liquid flow in the tube by siphon action. The limiting height of siphon can be found out by putting P_2 equal to zero in Bernoulli's equation when applied at section 1 and section 2.

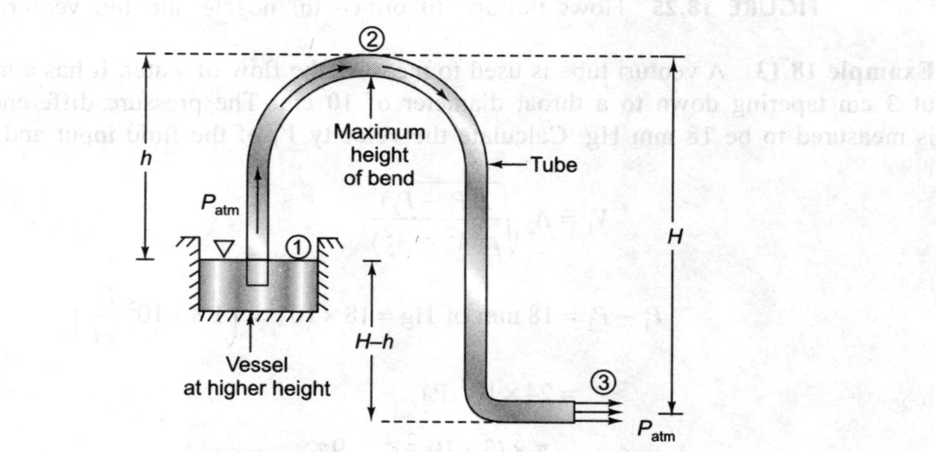

FIGURE 18.26 Working principle of siphon.

$$\frac{P_1}{\rho g} + \frac{V_1^2}{2g} + z_1 = \frac{P_2}{\rho g} + \frac{V_2^2}{2g} + z_2$$

or
$$\frac{P_{atm}}{\rho \cdot g} + 0 + 0 = 0 + \frac{V^2}{2g} + h$$

or
$$h = \frac{P_{atm} - \frac{1}{2}\rho V^2}{g}$$

To get the faster fluid flow, we have to keep lower value of height "h" of the bend.

To get the velocity of water emerging from pipe at section 3, we use Torricelli's theorem. The velocity of water is

$$V_3 = \sqrt{2 \cdot g \cdot (H - h)}$$

Example 18.14 A siphon is used to drain water from a tank as shown in Figure 18.27. If h = 1.0 m, find the speed of outflow at the end of siphon. What is the limitation of the height of the top of siphon above the water surface?

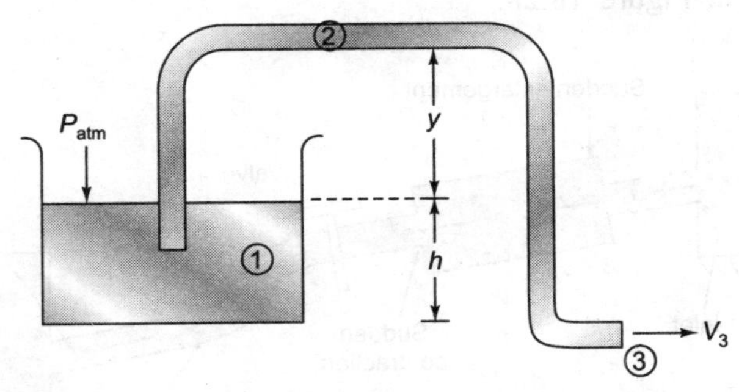

FIGURE 18.27 Find speed of flow and height 'y' of siphon.

In order to avoid cavitation and interruption of siphon action, the pressure at point 2 must be above zero.

$$P_2 > 0$$

But
$$P_2 = P_1 + y \times \rho \times g$$
$$= P_{atm} + y \times \rho \times g$$

∴
$$P_{atm} + y \times \rho \times g > 0$$

where
$$P_{atm} = 1.013 \times 10^5 \text{ Pa}$$

or
$$y < \frac{P_{atm}}{\rho \times g}$$
$$y < \frac{1.013 \times 10^5}{10^3 \times 9.81}$$
$$y < 10.3 \text{ m}$$

Using Torricelli's theorem for point 1 and point 3, we have

$$V_3 = \sqrt{2gh}$$

$$V_3 = \sqrt{2 \times 9.81 \times 1}$$

$$V_3 = 4.43 \text{ m/s}$$

FLOW IN PIPES AND DUCTS

Pipe flow is a type of full flow within a closed conduit. A loss of energy takes place when fluid flows through pipes and ducts. Hence, pumping power is required to maintain the flow. The energy losses can be categorised as major losses and minor losses. The major loss is due to viscous resistance and it is termed loss due to friction. Minor losses are due to change of velocity of fluid flow in pipe. The minor losses take place at (i) inlet to pipe (ii) sudden change in cross-section of the pipe (iii) contraction in pipe (iv) enlargement in pipe (v) obstruction in flow such as valve (vi) outlet and (vii) change at flow direction at elbow and bend as shown in Figure 18.28.

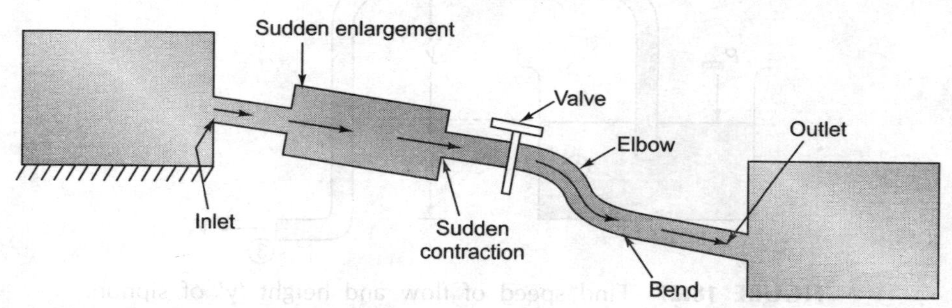

FIGURE 18.28 Changes in velocity and direction lead to minor losses.

Energy Grade Line and Hydraulic Grade Line

In pipe flow without head losses, the total energy at point 1 is equal the total energy at point 2 as per the Bernoulli's equation and as shown in Figure 18.29. However, hydraulic grade line varies depending upon velocity of flow. Wherever velocity is large (in smaller sections of pipe), the hydraulic grade line lowers down as velocity head increases.

$$E_1 = E_2$$

$$\frac{P_1}{\gamma} + \frac{V_1^2}{2g} + z_1 = \frac{P_2}{\gamma} + \frac{V_2^2}{2g} + z_2$$

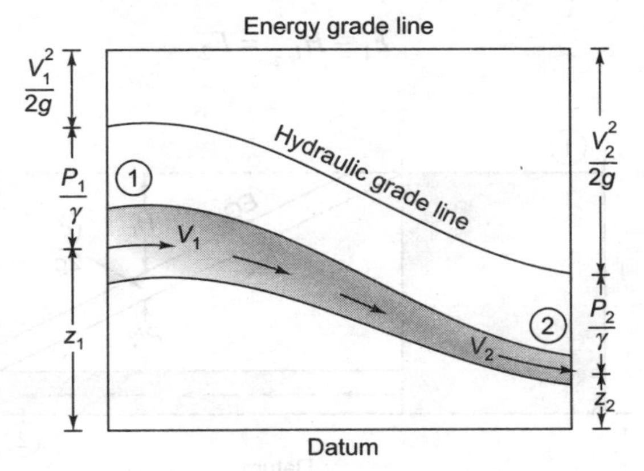

FIGURE 18.29 Energy grade line remains constant and hydraulic grade slopes down with increased velocity in friction less flow (ideal fluid).

In pipe flow with friction, there is head loss (H_l) along the length of pipe. Energy grade line and hydraulic grade line fall along the length of pipe as shown in Figure 18.30.

$$E_1 - H_{l_{1-2}} = E_2$$

$$\frac{P_1}{\gamma} + \frac{V_1^2}{2g} + z_1 - H_{l_{1-2}} = \frac{P_2}{\gamma} + \frac{V_2^2}{2g} + z_2$$

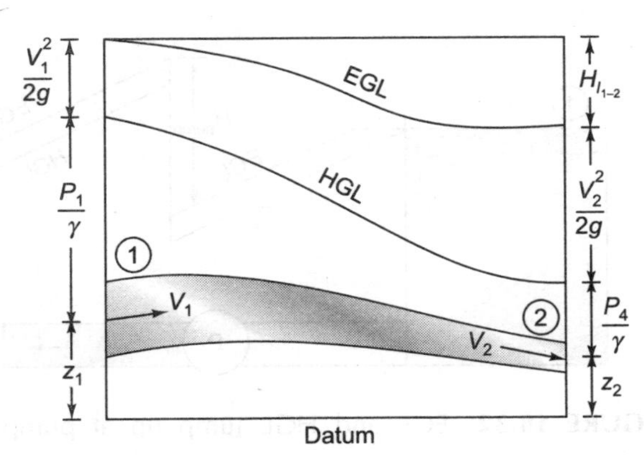

FIGURE 18.30 EGL and HGL slope down along the length of pipe in viscous flow with friction head loss.

In uniform sized pipe, the difference between EGL and HGL remains constant as $\dfrac{V^2}{2g}$.

However, EGL slopes towards flow due to losses as shown in Figure 18.31.

$$E_1 - H_{l_{1-2}} = E_2$$

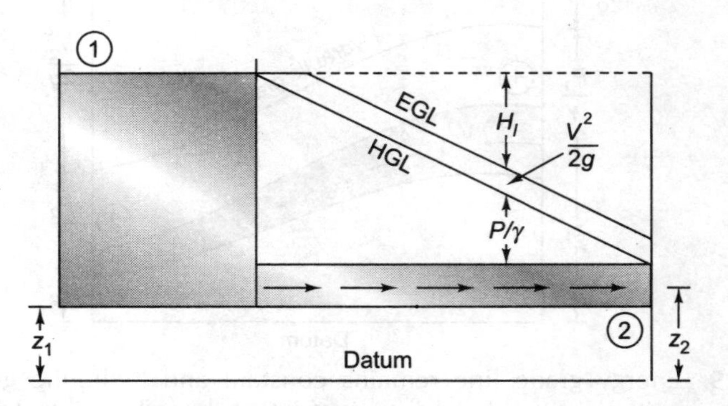

FIGURE 18.31 EGL and HGL slope towards flow due to losses.

If a pump is located in the pipeline, EGL and HGL rise vertically at pump. HGL remains parallel to EGL. EGL and HGL slope downwards and towards flow direction as shown in Figure 18.32.

$$E_1 + H_{\text{pump}} - H_{l_{1-2}} = E_2$$

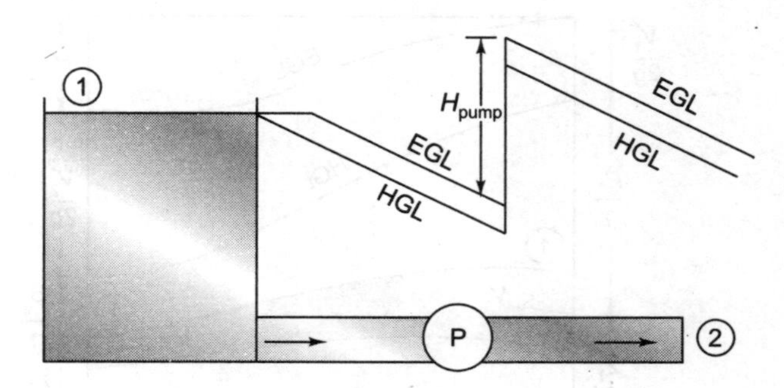

FIGURE 18.32 EGL and HGL jump up at pump location.

If a turbine is located in the pipeline, EGL and HGL drop. After turbine, as liquid is gradually expanding, hence velocity head is constantly decreasing and being converted into pressure head. Therefore, HGL approaches EGL as shown in Figure 18.33.

$$E_1 - H_{\text{turbine}} - H_{l_{1-2}} = E_2$$

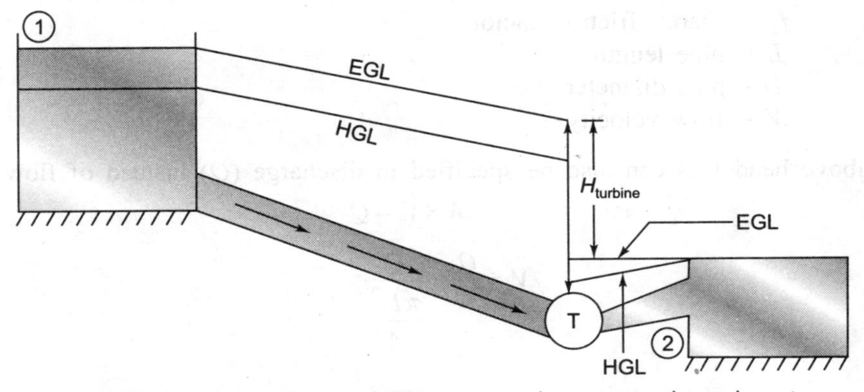

FIGURE 18.33 EGL and HGL jump down at turbine location.

Friction Loss in Pipe Flow

Friction loss is the major loss during fluid flow in pipe and it is the loss of energy or head that occurs due to viscous effect generated by the surface of the pipe as shown in Figure 18.34. The energy or head loss is dependent on the wall shear stress (τ) between water and pipe surface. The shear stress developed depends on whether the flow is turbulent or laminar. In turbulent flow (Reynolds number > 4000), the pressure drop is dependent on the roughness of surface while the roughness effect on laminar flow is negligible. The reason is that a viscous layer is formed near the pipe surface in turbulent flow but not in laminar flow which causes a loss of energy. There are many methods to calculate frication losses but most accepted method is by using the Darcy–Weisbach equation. For circular pipe, head loss due to friction (h_l) is

$$h_l = f_D\left(\frac{L}{D}\right)\left(\frac{V^2}{2g}\right)$$

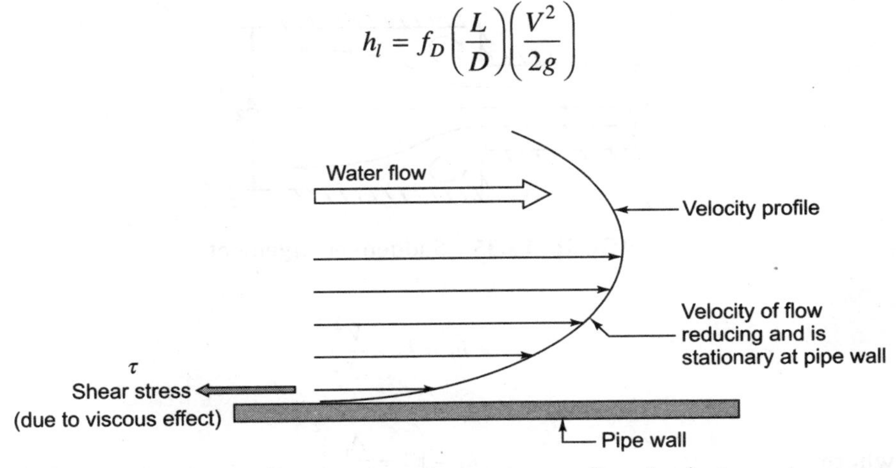

FIGURE 18.34 Viscous effect generated on pipe surface leads to major energy or head loss.

where f_D = Darcy friction factor
L = pipe length
D = pipe diameter
V = flow velocity.

The above head loss can also be specified in discharge (Q) instead of flow velocity

$$A \times V = Q$$

or
$$V = \frac{Q}{A} = \frac{Q}{\frac{\pi D^2}{4}}$$

$$\therefore \quad h_l = f_D \cdot \frac{LV^2}{D \cdot 2g} = f_D \cdot \frac{L}{D^5} \cdot \frac{16Q^2}{2g\pi^2}$$

Minor Head Losses

Minor head losses occur in pipeline due to sudden enlargement, contraction, junction and bend. For fittings, head loss is given by

$$h_l = k_l \cdot \frac{V^2}{2g}$$

where k_l = a constant for fitting

(a) ***Sudden enlargement***: It is as shown in Figure 18.35. The velocity is reduced and pressure increases. Eddies formation gives rise to head loss.

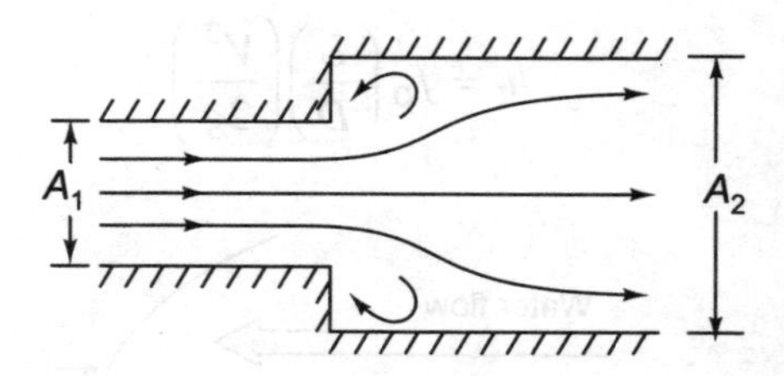

FIGURE 18.35 Sudden enlargement.

$$h_l = k_l \times \frac{V^2}{2g}$$

where
$$k_l = \left(1 - \frac{A_1}{A_2}\right)^2$$

(b) ***Sudden contraction:*** In sudden contraction, flow contracts and forms a vena contracta as shown in Figure 18.36. From experiment, it is seen that the contraction due vena contracta is about 40% for contraction area of pipe (A_2)

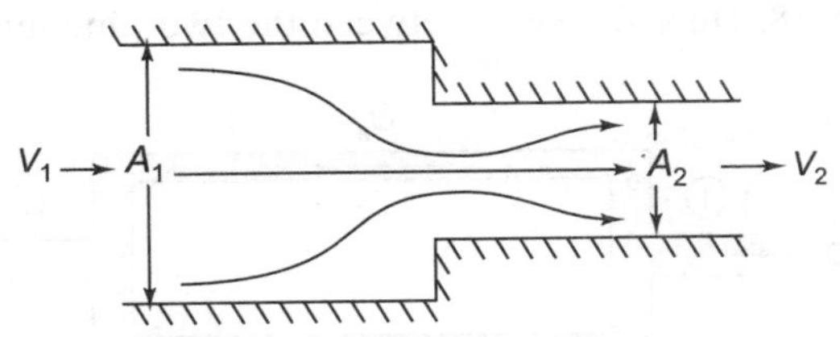

FIGURE 18.36 Head loss due to vena contracta.

$$h_l = \left(1 - \frac{0.6A_2}{A_2}\right)\left[\frac{(V_2/0.6)^2}{2g}\right]$$

$$h_l = 0.44\frac{V_2^2}{2g}$$

Pipes in Series

When two or more pipes of different diameters or roughness are connected in such a way that fluid follows a single flow path throughout the system, then the system represents a series pipeline. In a series pipeline, the total energy loss is the sum of individual minor losses and all pipe friction losses. Figure 18.37 shows a series pipe system and Bernoulli equation can be written between points 1 and 2 as follows:

$$\frac{P_1}{\rho g} + \frac{V_1^2}{2g} + z_1 = \frac{P_2}{\rho g} + \frac{V_2^2}{2g} + z_2 + H_{l_{1-2}}$$

where $\quad H_{l_{1-2}} = (h_f)_A + (h_f)_B + h_{\text{entrance}} + h_1 + h_{\text{expansion}}$

In case $\quad P_1 = P_2 = P_{\text{atm}} \quad$ and $\quad V_1 = V_2$

$$H_{l_{1-2}} = z_1 - z_2$$

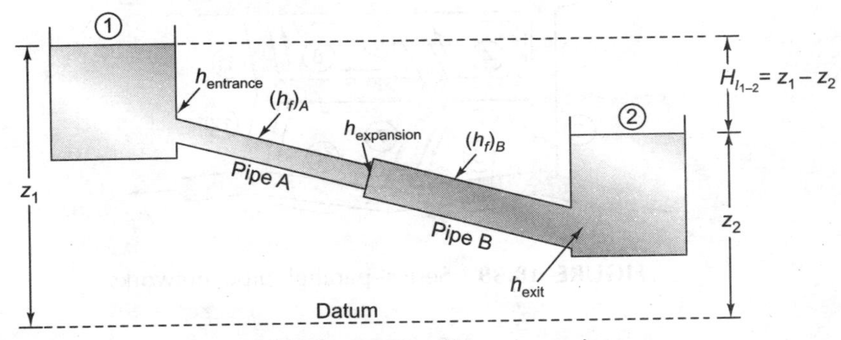

FIGURE 18.37 Pipes in series.

Pipes in Parallel

A combination of two or more pipes connected between two points so that the discharge divides at the first junction and rejoins at the next junction is known as *pipes in parallel* as shown in Figure 18.38. Here the head between the two junctions is the same for all pipes.

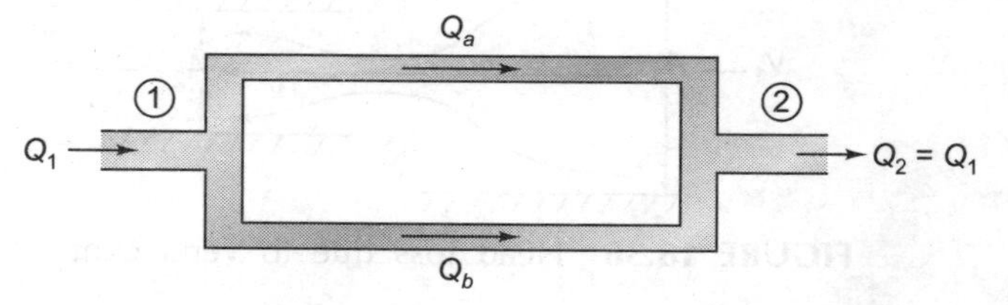

FIGURE 18.38 Pipes in parallel.

Pipes in parallel must have

$$Q_1 = Q_a + Q_b = Q_2$$

The energy equation between point 1 and point 2 can be written as

$$\frac{P_1}{\rho g} + \frac{V_1^2}{2g} + z_1 = \frac{P_2}{\rho g} + \frac{V_2^2}{2g} + z_2 + H_{l_{1-2}}$$

and $$H_{l_{1-2}} = (h_l)_a + (h_l)_b$$

Series–Parallel Pipe Networks

In a series–parallel pipe network as shown in Figure 18.39, the rules similar to analysis of an electric circuit are applied to solve it. The following rules are applicable for solving:

(a) The net flow into any junction is zero, i.e. $\Sigma\ Q_i = 0$
(b) The net head loss around any loop must be equal to zero, i.e. $\Sigma h_l = 0$

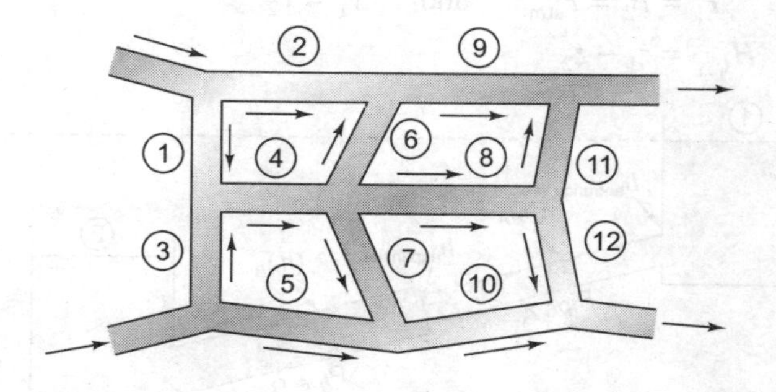

FIGURE 18.39 Series–parallel pipe networks.

The application of these rules leads to a complex set of equations which have to be solved numerically.

Example 18.15 A horizontal pipe carries water from a reservoir as shown in Figure 18.40. Pipe diameter = 20 cm, rate of flow = 0.6 m^3/s and length L = 2000 m. Take friction coefficient = 0.02. Find pressure at point 2.

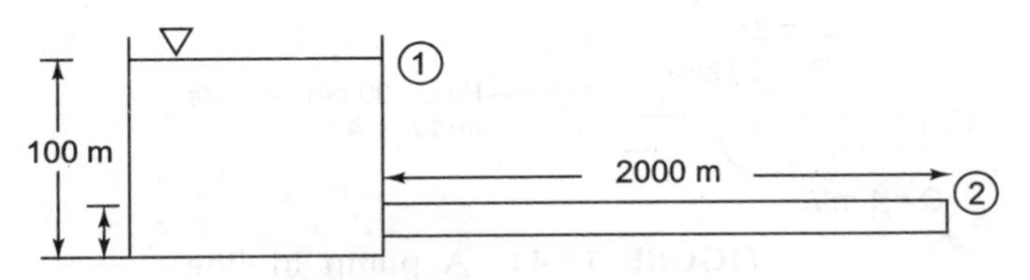

FIGURE 18.40 Pipe flow.

The velocity of water through pipe is

$$V = \frac{Q}{A} = \frac{0.06}{\frac{\pi}{4} \times (0.2)^2} = 1.91 \text{ m/s}$$

The head loss in pipe by Darcy equation is

$$h_l = f_D \frac{L \cdot V^2}{D \cdot 2g}$$

$$h_l = \frac{0.02 \times 2000 \times (1.91)^2}{0.2 \times 2 \times 9.81}$$

$$h_l = 37.2 \text{ m}$$

Now applying Bernoulli's equation between point 1 and point 2.

$$\frac{P_1}{\rho g} + \frac{V_1^2}{2g} + z_1 = \frac{P_2}{\rho g} + \frac{V_2^2}{2g} + z_2 + h_l$$

$$P_1 = P_{\text{atm}} = 0 \text{ (gauge pressure)}$$

$$V_1 = V_2$$

$$z_1 = \frac{P_2}{\rho g} + z_2 + h_l$$

or
$$\frac{P_2}{\rho g} = 100 - 20 - 37.2 = 42.8 \text{ m}$$

or
$$P_2 = \rho \cdot g \times 42.8$$

$$P_2 = 10^3 \times 9.81 \times 42.8$$

$$P_2 = 419.87 \text{ kPa}$$

Example 18.16 A pipe 100 cm in diameter carries water of 1 m^3/s. A pump is used to move the water from elevation of 25 m to 40 m. The pressure at section 1 is 80 kPa gauge and pressure at section 2 is 380 kPa gauge, what is power of pump in kilowatt? Assume $h_l = 4$ m.

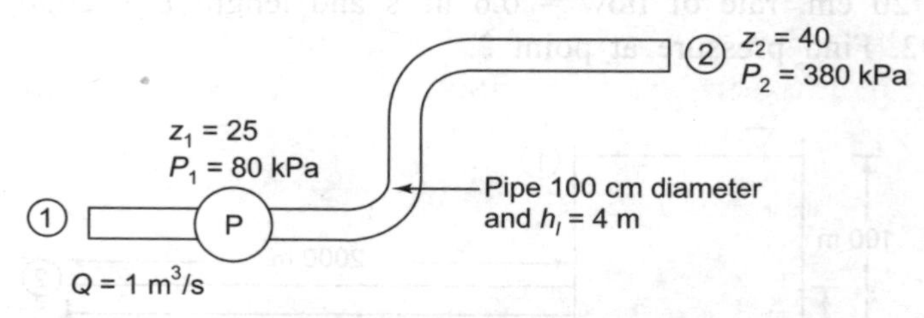

FIGURE 18.41 A pump in line.

$$\frac{P_1}{\rho \cdot g} + \frac{V_1^2}{2g} + z_1 + h_{\text{pump}} = \frac{P_2}{\rho \cdot g} + \frac{V_2^2}{2g} + z_2 + h_l$$

Now
$$V_1 = V_2$$

or
$$\frac{(P_2 - P_1)}{\rho g} + (z_2 - z_1) + h_l = h_{\text{pump}}$$

or
$$h_{\text{pump}} = \frac{10^3 \times (380 - 80)}{10^3 \times 9.81} + (40 - 25) + 4$$

$$h_{\text{pump}} = \frac{300 \times 10^3}{10^3 \times 9.81} + 15 + 4$$

$$h_{\text{pump}} = 30.58 + 19 = 49.58 \text{ m}$$

Power is given by
$$p = \rho \cdot g \cdot h_{\text{pump}} \times Q$$
$$= 10^3 \times 9.81 \times 49.58 \times 1$$
$$= 485.9 \text{ kW}$$

Example 18.17 A small hydroelectric power plant takes a discharge of 15 m^3/s through an elevation drop of 65 m. The head loss through the intake, penstock and outlet has been found out as 3 m. The efficiency of turbine is 90%. What is the rate of power generation?

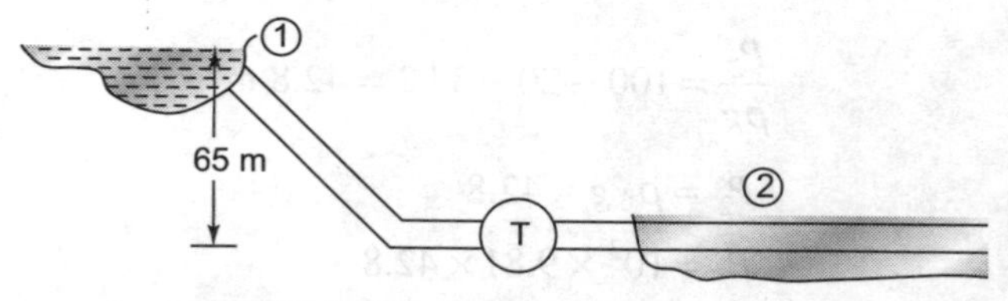

FIGURE 18.42 Turbine in pipeline.

$$\frac{P_1}{\rho g} + \frac{V_1^2}{2g} + z_1 = \frac{P_2}{\rho g} + \frac{V_2^2}{2g} + z_2 + h_T - h_l$$

Here
$$V_1 = V_2 = 0$$
$$P_1 = P_2 = P_{atm}$$
$$h_T = (z_1 - z_2) - h_l$$
$\therefore$
$$= 65 - 3 = 62 \text{ m}$$

Power output from turbine is

$$p = \eta_{turbine} \times (h_T \times \rho \times g) \times Q$$
$$= \frac{90}{100} \times (62 \times 10^3 \times 9.81) \times 15$$
$$= 8.21 \text{ MW}$$

Example 18.18 Water flows from the reservoir on the left to the reservoir on the right at a rate of 0.45 m^3/s as shown in Figure 18.43. What elevation in the left reservoir is required to produce this flow? Sketch the HGL and EGL for the system. Take f_D = 0.02.

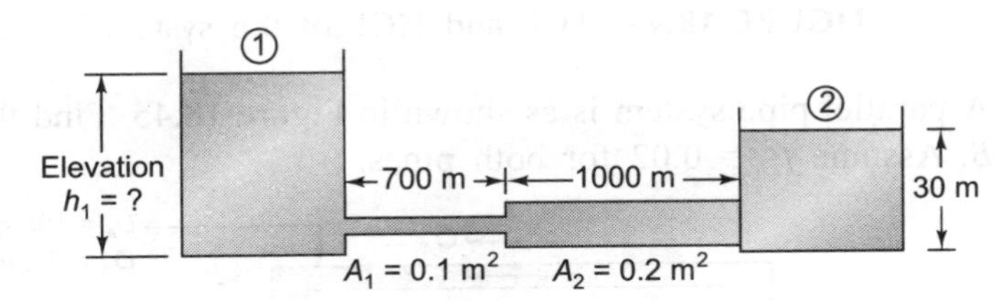

FIGURE 18.43 Two pipes in series.

$$V_{A_1} = \frac{Q}{A_1} = \frac{0.45}{0.1}$$
$$= 4.5 \text{ m/s}$$
$$V_{A_2} = \frac{Q}{A_2} = \frac{0.45}{0.2}$$
$$= 2.25 \text{ m/s}$$

Apply Bernoulli's equation between points 1 and 2.

$$\frac{P_1}{\rho g} + \frac{V_1^2}{2g} + z_1 = \frac{P_2}{\rho g} + \frac{V_2^2}{2g} + z_2 + h_{l_{1-2}}$$

$$0 + 0 + h_1 = 0 + 0 + 30 + h_{l_{1-2}}$$

$$h_{l_{1-2}} = f_D \times \frac{L_{A_1} \times V_{A_1}^2}{D_{A_1} \times 2g} + f_D \times \frac{L_{A_2} \times V_{A_2}^2}{D_{A_1} \times 2g}$$

$$= 0.02 \times \frac{700 \times (4.5)^2}{0.356 \times 2 \times 9.81} + \frac{0.02 \times 1000 \times (2.25)^2}{0.505 \times 2 \times 9.81}$$

$$= 4.06 + 10.22$$

$$= 14.28$$

$$h_1 = 30 + 14.28$$

$$\therefore \quad = 44.28 \text{ m}$$

The EGL and HGL are as shown in Figure 18.44.

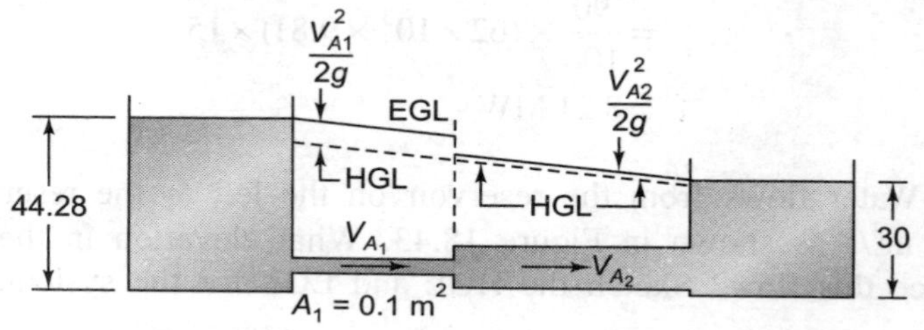

FIGURE 18.44 EGL and HGL of the system.

Example 18.19 A parallel pipe system is as shown in Figure 18.45. Find the flow through pipe A and pipe B. Assume $f_D = 0.02$ for both pipes.

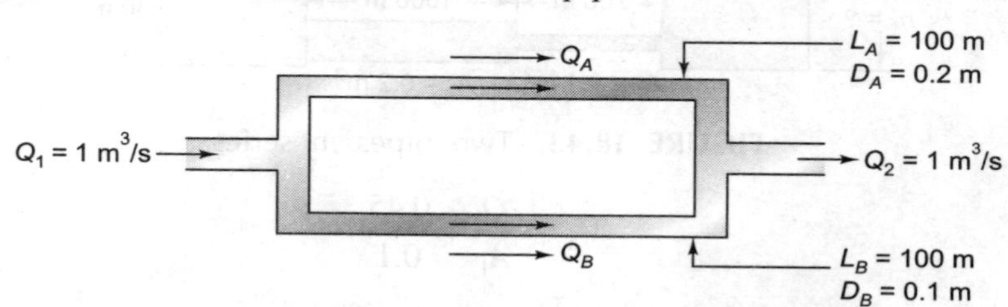

FIGURE 18.45 Parallel pipe system.

$$(h_l)_A = h_1 - h_2 = f_D \times \frac{L_A \times V_A^2}{D_A \times 2g}$$

$$= f_D \times \frac{L_A \times 16\, Q_A^2}{D_A \times 2g \times \pi^2}$$

$$= \frac{0.02 \times 100 \times 16 \times Q_A^2}{0.2 \times 2 \times 9.81 \times \pi^2} = 0.827 \times Q_A^2$$

$$(h_l)_B = h_1 - h_2 = \frac{f_D \times 100 \times 16 \times Q_B^2}{0.1 \times 2 \times 9.81 \times \pi^2}$$

$$= 1.654 \times Q_B^2$$

Now $(h_l)_A = (h_l)_B$

$$0.827\, Q_A^2 = 1.654\, Q_B^2$$

or $$\frac{Q_A}{Q_B} = \sqrt{2} = 1.4$$

Now, we have

$$Q = Q_A + Q_B$$
$$1 = 1.4\, Q_B + Q_B$$
$$Q_B = \frac{1}{1.5} = 0.67 \text{ m}^3/\text{s}$$

$\therefore$ $$Q_A = 1 - 0.67 = 0.33 \text{ m}^3/\text{s}$$

OPEN CHANNEL FLOW

The flow of water in open channel is a familiar sight, whether in natural channels like that of a river or in artificial channels like that of an irrigation ditch. The flow in open channel has a free surface which is the interface between the moving water and an overlying air. The interface has constant atmospheric pressure on it as shown in Figure 18.46.

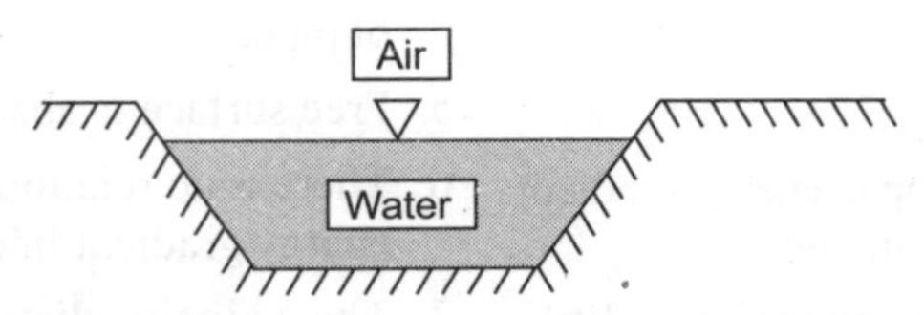

FIGURE 18.46 Open channel flow.

Open channel flow is a complex flow when everything is considered, especially with the variability of natural channels but in many cases, the major features can be expressed in terms of only a few variables, whose behaviour can be described adequately by a simple theory. The principal forces working during open channel flow include (i) inertial force (ii) gravity force and (iii) viscous and those play important role.

Applications of Open Channel Flow

The applications of open channel flow are:

(a) In designing a canal to carry a desired amount of water
(b) Estimation of area likely to be submerged on construction of a dam on a river
(c) Estimation of discharge with the change of depth of flow
(d) Measurement of discharge in river and canal
(e) Ensuring high velocity flow not to take place, thereby preventing any damage to canals

(f) Estimation of change in flow with a change in bed width or bed elevation
(g) Estimation of change in the flow depth due to over-land run off
(h) Estimation of time needed by any flood wave to pass through a given length of a river
(i) Estimation of the amount of sediment carried by a channel
(j) Estimation of the spread of pollutants in a river.

Difference between Open Channel Flow and a Pipe Flow

The differences between open channel flow and a pipe flow are as per Table 18.2.

Table 18.2 Comparison of open channel flow and a pipe flow

Open channel flow	*Pipe flow*
1. Open channel flow occurs due to the slope of the channel.	1. Pipe flow occurs when inlet pressure is higher than outlet pressure in the pipe.
2. Open channel may have regular or irregular shapes. Regular shapes include (i) rectangular (ii) triangular (iii) trapezoidal and (iv) circular.	2. Pipes are generally having regular shape of circular.
3. It can be (i) natural or (ii) artificial.	3. It is always artificial.
4. The hydraulic roughness may vary with the depth of flow.	4. The surface roughness remains constant on outer periphery depending upon the material of pipe.
5. Free surface is present.	5. Free surface is absent.
6. For uniform flow, the drop in energy gradient line is equal to the drop in bed.	6. There is no relation between the drop of the energy gradient line and slope of pipe axis.
7. The maximum velocity occurs at a little distance below the water surface. The shape of velocity profile is dependent on channel roughness.	7. The velocity distribution is symmetrical about the pipe axis with maximum velocity occurring at pipe centre and velocity reduces to zero at pipe wall.

Type of Channel

Open channels can be classified as:

(a) ***Natural and artificial:*** Natural channels have irregular sections of varying shapes, alignment and surface roughness. Examples are (i) rivers (ii) small rivulets and (iii) sheets of water across fields. Artificial open channels are built for some specific purposes such as (i) irrigation (ii) drain ditches and (iii) water gutters. Artificial channels are regular in shape and alignment and they have uniform surface roughness as shown in Figure 18.47.

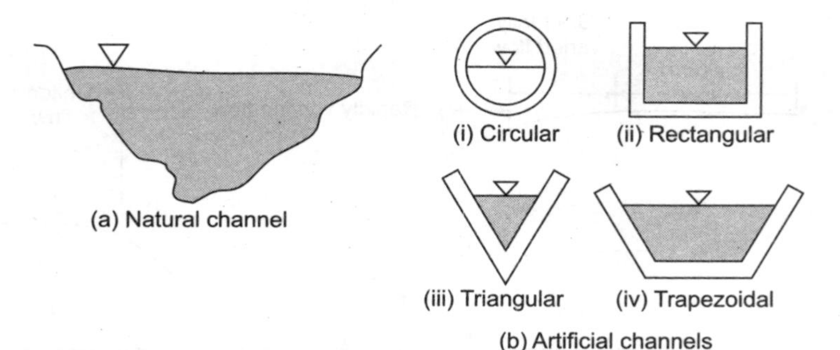

FIGURE 18.47 Natural and artificial open channels.

(b) ***Prismatic and non-prismatic:*** A channel where cross-sectional shape, site and bottom slope are unvarying is termed as prismatic channel. Most of artificial channels are prismatic. All natural channels are non-prismatic.

(c) ***Rigid and mobile boundary channel:*** Rigid channels are non-deformable and their shape remains unchangeable. These are (i) lined canals and (ii) sewers. Deformable channels are termed as mobile boundary channels.

Classification of Flows

(a) ***Steady and unsteady flow:*** A steady flow occurs when flow properties such as discharge (Q), depth of water from surface (y) and velocity (V) at a section remain unchanged with time.

$$\frac{dQ}{dt} = 0, \quad \frac{dV}{dt} = 0 \quad \text{and} \quad \frac{dy}{dt} = 0$$

For unsteady flow, we have

$$\frac{dQ}{dt} \neq 0, \quad \frac{dV}{dt} \neq 0 \quad \text{and} \quad \frac{dy}{dt} \neq 0$$

(b) ***Uniform and non-uniform flow:*** Uniform flow has depth (y), velocity (V) and discharge (Q) unchanged in its longitudinal or flow direction.

$$\frac{dy}{dl} = 0, \quad \frac{dV}{dl} = 0 \quad \text{and} \quad \frac{dQ}{dl} = 0$$

For non-uniform flow, we have

$$\frac{dy}{dl} \neq 0, \quad \frac{dV}{dl} \neq 0 \quad \text{and} \quad \frac{dQ}{dl} \neq 0$$

(c) ***Classification of non-uniform flow:*** Non-uniform flow can be (i) gradually varied flow (ii) rapidly varied flow and (iii) spatially varied flow as shown in Figure 18.48.

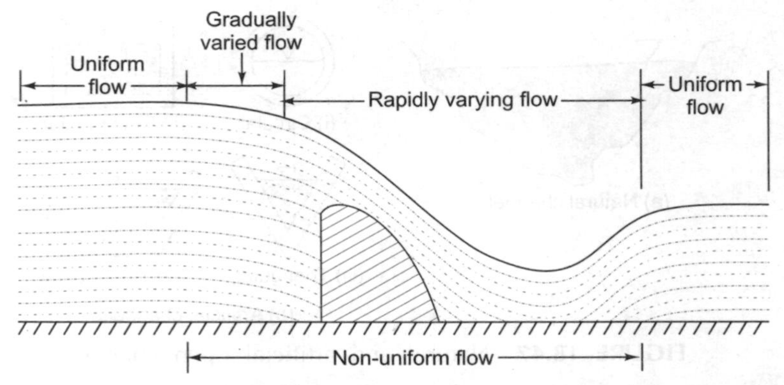

FIGURE 18.48 Types of non-uniform flow.

(i) *Gradually varied flow:* In this type of flow, the change of depth takes place gradually along the length of channel. The flow takes place when backing up of water occurs in the presence of dam or dropping of water surface occurs due to sudden drop in channel bed. The passage of a flood wave in a river is also an example of gradually varied non-uniform flow.

(ii) *Rapidly varied flow:* In rapidly varied non-uniform flow, there is appreciable depth (y) changes occurring over a short length of flow. Example is a hydraulic jump occurring below a spillway or sluice gate.

(iii) *Spatially varied flow:* In this flow, water is continuously added or removed along the channel length. Example is flow in roof gutter and irrigation channel.

Laminar and Turbulent Flow

The laminar flow is smooth and it takes place in different layers with velocity of top layer more than its bottom layer, i.e. flow is dominated by viscosity. In the turbulent flow, the flow takes place in random fashion and viscous forces have little effect. The criterion for determining the type of flow is the Reynolds number (R_e).

$$R_e = \frac{\rho \cdot V \cdot R}{\mu}$$

where V = average velocity of flow

R = hydraulic radius

ρ = density

μ = viscosity

Depending upon Reynolds number, flow is classified as:

(a) Laminar when $R_e < 500$
(b) Transitional when $500 < R_e < 2000$
(c) Turbulent when $R_e > 2000$

The viscous or friction losses can be given by Darcy formula as in pipe flow by

$$h_f = \frac{4 \cdot f \cdot L \cdot V^2}{2 \cdot g \cdot D}$$

However, channel flow does not fill completely the pipe and we take hydraulic radius $R = \frac{D}{4}$. The channel flow has

$$\frac{h_f}{L} = S_0 = \frac{f \cdot V^2}{2 \cdot g \cdot R}$$

where S_0 = Slope of bed.

Geometric Properties for Analysis

For analysis, lemma various geometric properties of channel cross-sections are required. The commonly needed geometric properties are shown in Figure 18.49. These are

(a) Depth of flow (y) from the lowest point of channel section to the free surface
(b) Surface width (B) is the width of channel section at the free surface
(c) Wetted perimeter (p) is the length of wetted surface measured normal to the direction of flow
(d) Area (A) is the cross-sectional area of flow, normal to the direction of flow
(e) Hydraulic radius (R) is the ratio of area (A) to wetted perimeter (p)
(f) Hydraulic mean depth (D_m) is the ratio of area (A) to surface width (B)

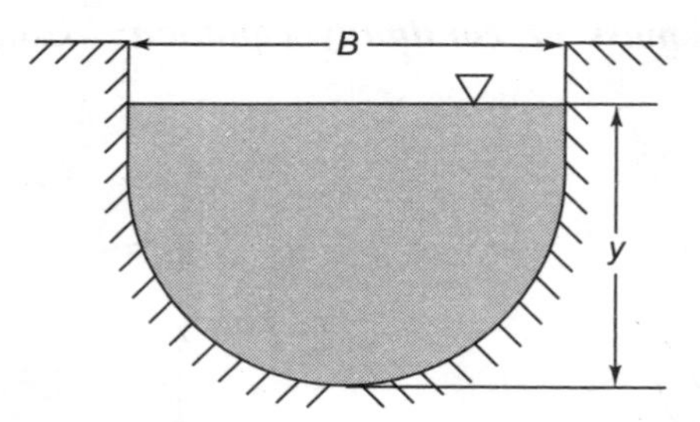

FIGURE 18.49 Geometric properties of a channel.

The geometric properties of various cross-sectional shapes are as per Table 18.3.

Table 18.3 The geometric properties of different shapes of cross-section

Area (A)	$b \times y$	$(b + x)y$	$\frac{1}{3}(\phi - \sin\phi)D^2$
Wetted perimeter (p)	$b + 2y$	$b + 2y\sqrt{1+x^2}$	$\frac{1}{2}\phi D$
Top width (B)	b	$b + 2xy$	$(\sin\phi / 2) \times D$
Hydraulic radius (R)	$by/b + 2y$	$\frac{(b + xy)y}{b + 2y\sqrt{1+x^2}}$	$\frac{1}{4}\left(1 - \frac{\sin\phi}{\phi}\right)D$
Hydraulic mean depth (D_m)	y	$\frac{(b + xy)y}{b + 2xy}$	$\frac{1}{8}\left(\frac{\phi - \sin\phi}{\sin\phi / 2}\right)D$

Fundamental Equation of Flow

The fundamental equations of flow are (i) conservation of mass (ii) conservation of energy and (iii) conservation of momentum.

(a) ***Conservation of mass or continuity equation:*** Consider the flow as shown in Figure 18.50.

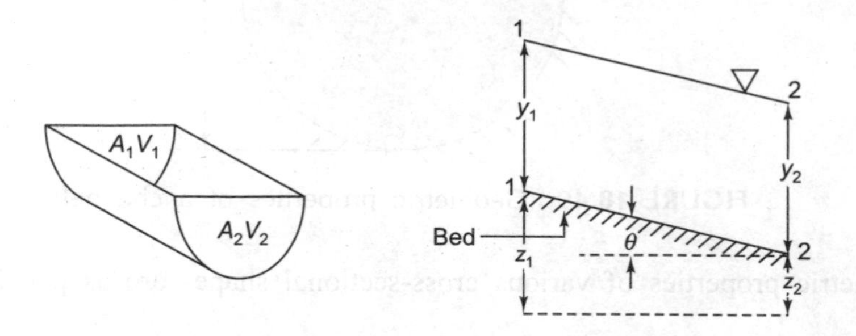

FIGURE 18.50 Small length of channel as control volume.

$$Q_{\text{entering}} = Q_{\text{leaving}}$$

$$A_1V_1 = A_2V_2$$

(b) ***Energy equation or conservation of energy:*** Total energy for liquid moving length (L) is

$$\text{Work done} + KE + PE = \text{constant}$$

or $$P(A_1 L) + \frac{1}{2}(\rho A_1 L) \times V_1^2 + (\rho A_1 L) z_1 = \text{constant}$$

or $$\frac{P_1}{\rho g} + \frac{V_1^2}{2g} + z_1 = \frac{P_2}{\rho g} + \frac{V_2^2}{2g} + z_2$$

The above is the Bernoulli's equation.

(c) ***Momentum equation:*** The change of momentum in a direction is equal to total force

$$F_x = \rho \cdot Q(V_2 - V_1)$$

Energy and Momentum Coefficients

In deriving energy or Bernoulli's equation, the velocity across whole cross-section has been taken as constant which is incorrect as velocity is non-uniform over the cross-section. In order to correct the energy and momentum equations with constant velocity term, we use energy coefficient α and momentum coefficient β as explained below:

$$\text{Energy coefficient} = \alpha = \frac{\int \rho \cdot u^3 dA}{\rho \cdot V^3 A}$$

$$\text{Momentum coefficient} = \beta = \frac{\int \rho \cdot u^2 dA}{\rho \cdot V^2 A}$$

The Bernoulli's equation can be written as

$$\frac{P}{\rho g} + \frac{a \cdot V^2}{2g} + z = \text{constant}$$

The change of momentum equation can be written in term of force in x-direction as

$$F_x = \rho \cdot Q \cdot \beta \cdot (V_{2x} - V_{1x})$$

Velocity Distribution in Open Channel

The velocity in open channel varies across the cross-section because friction acts along the boundary on fluid flow. The velocity distribution is different from pipe flow which is symmetrical with respect to the pipe axis. The typical velocity distributions across certain channel cross-sections are as shown in Figure 18.51.

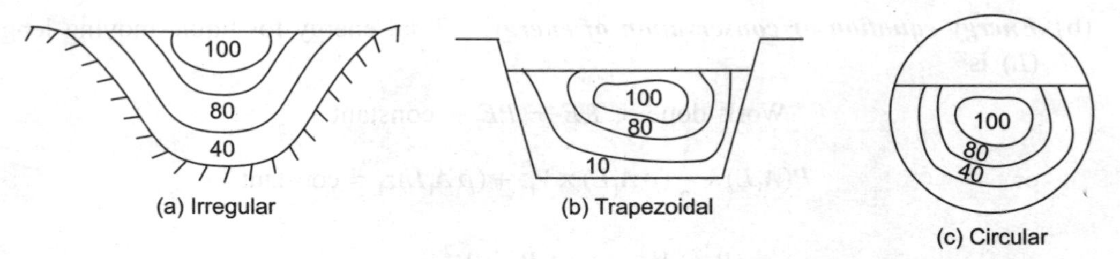

FIGURE 18.51 Velocity distribution in various shape channels.

Example 18.20 A typical cross-section of a channel is as shown in Figure 18.52, find energy and momentum coefficients.

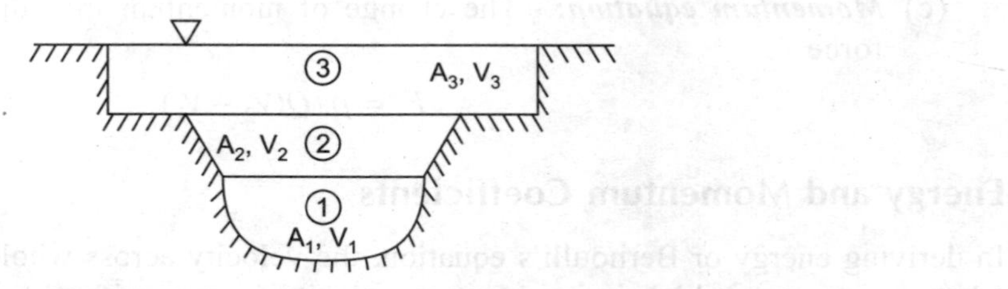

FIGURE 18.52 Energy and momentum coefficients.

$$\alpha = \frac{\int u^3 \cdot dA}{\overline{V}^3 A} = \frac{V_1^3 A_1 + V_2^3 \cdot A_2 + V^3 \cdot A_3}{\overline{V}^3 (A_1 + A_2 + A_3)}$$

where
$$\overline{V} = \frac{Q}{A} = \frac{V_1 A_1 + V_2 A_2 + V_3 A_3}{A_1 + A_2 + A_3}$$

and
$$\beta = \frac{u^2 \cdot dA}{\overline{V}^2 \cdot A} = \frac{V_1^2 \cdot A_1 + V_2^2 \cdot A_2 + V_3^2 \cdot A_2}{\overline{V}^2 (A_1 + A_2 + A_3)}$$

The Chezy Equation

When uniform flow takes place as shown in Figure 18.53, the gravitational forces exactly balance the viscous or resistance forces acting along the boundary of the channel. The forces acting along the direction of flow are:

(a) gravity flow = $\rho \cdot g \cdot A \cdot L \sin \theta$ where A = area
(b) shear force = $\tau_0 \cdot p \cdot L$ where p = perimeter

Equating these forces, we have

$$\rho \cdot g \cdot A \cdot L \, \sin \theta = \tau \cdot p \cdot L$$

or
$$\tau_0 = \frac{\rho \cdot g \cdot A \sin\theta}{p}$$

But
$$\sin\theta = \tan\theta = S_0 = \text{slope of bend}$$

or
$$\tau_0 = \frac{\rho \cdot g \cdot A \cdot S_0}{p} \tag{18.9}$$

Now, the shear stress is proportional to the flow velocity. Hence,

$$\tau_0 \propto V^2$$

or
$$\tau_0 = k \cdot V^2 \tag{18.10}$$

From Eqs. (18.9) and (18.10), we have

$$kV^2 = \frac{\rho \cdot g \cdot A \cdot S_0}{p}$$

or
$$V^2 = \frac{\rho \cdot g}{k} \cdot \left(\frac{A}{p}\right) \cdot S_0$$

or
$$V^2 = \frac{\rho \cdot g}{k} \cdot R \cdot S_0$$

where
$$R = \text{hydraulic radius}$$

or
$$V = \sqrt{\frac{\rho \cdot g}{k} \cdot R \cdot S_0}$$

or
$$V = C\sqrt{R \cdot S_0} \qquad \text{where } C = \sqrt{\frac{\rho \cdot g}{k}}$$

The above is called Chezy equation. C depends on Reynolds number and boundary roughness.

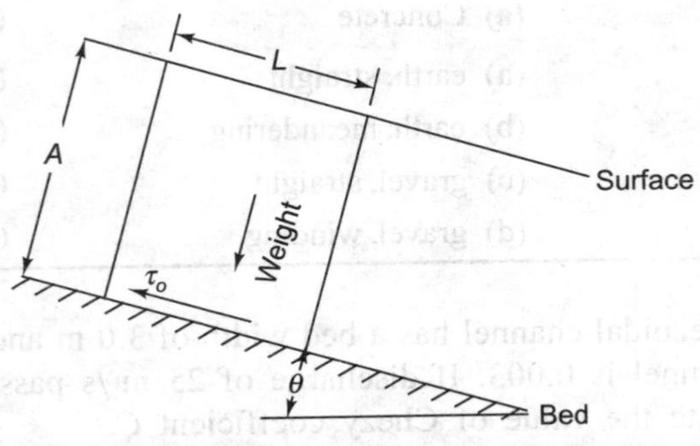

FIGURE 18.53 Forces on a channel length in uniform flow.

The Manning Equation

Many studies have been made to evaluate the value of Chezy coefficient C for different natural and artificial channels and on the basis of these, it is found that Chezy coefficient C can be given as:

$$C = \frac{R^{1/6}}{n}$$

where R = hydraulic radius and
n = Manning's n which depends mainly on surface roughness.

The above relation is called as Manning's formula and we get Manning's equation when value of Chezy coefficient C is put

$$V = C\sqrt{R \cdot S_0}$$

$$V = \frac{R^{1/6}}{n}\sqrt{R \cdot S_0}$$

$$V = \frac{R^{2/3} S_0^{\ 1/2}}{n}$$

In case we put $Q = A \times V$ and $R = A/p$, then Manning's equation becomes

$$Q = \frac{1}{n} \times \frac{A^{5/3}}{p^{2/3}} \times S_0^{\ 1/2}$$

The Manning's n for various surfaces is as per Table 18.4.

Table 18.4 Value of Manning's n

Channel type	*Surface material and form*	*Manning's n*
1. Unlined canal	(a) earth, straight	0.018 to 0.025
	(b) rock, straight	0.025 to 0.045
2. Lined canal	(a) Concrete	0.012 to 0.017
3. River	(a) earth, straight	0.02 to 0.025
	(b) earth, meandering	0.03 to 0.05
	(c) gravel, straight	0.03 to 0.04
	(d) gravel, winding	0.04 to 0.08

Example 18.21 A trapezoidal channel has a bed width of 3.0 m and side slope of 1:1. The bottom slope of the channel is 0.003. If discharge of 25 m³/s passes in this channel at a depth of 1.25 m, estimate the value of Chezy coefficient C.

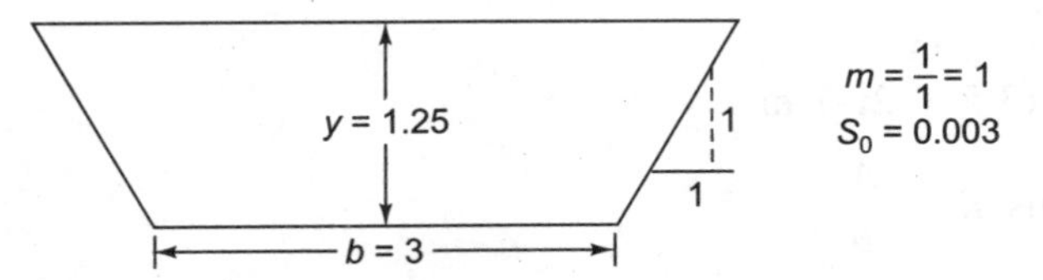

FIGURE 18.54 Trapezoidal channel.

Area $A = [b + my]y$

$$= \left(3 + \frac{1}{1} \times 1.25\right) \times 1.25$$

$$= 4.25 \times 1.25 = 5.31 \text{ m}^2$$

Velocity $V = \dfrac{Q}{A} = \dfrac{25}{5.31}$

$= 4.71$ m/s

Wetted perimeter $p = b + 2y\sqrt{m^2 + 1}$

$$= 3 + 2 \times 1.25\sqrt{1+1}$$

$= 6.54$ m

Hydraulic radius $R = \dfrac{A}{p} = \dfrac{5.31}{6.54}$

$= 0.81$

As per Chezy's equation, we have

$$V = C\sqrt{R \cdot S_0}$$

$$4.71 = C \cdot \sqrt{0.81 \times 0.003}$$

$$C = 101.85$$

Example 18.22 A rectangular channel 3.5 m wide is laid on a slope of 0.0003. Calculate the normal depth of flow for discharge of 5 m^3/s in this channel. The Manning's coefficient 'n' can be taken as 0.01.

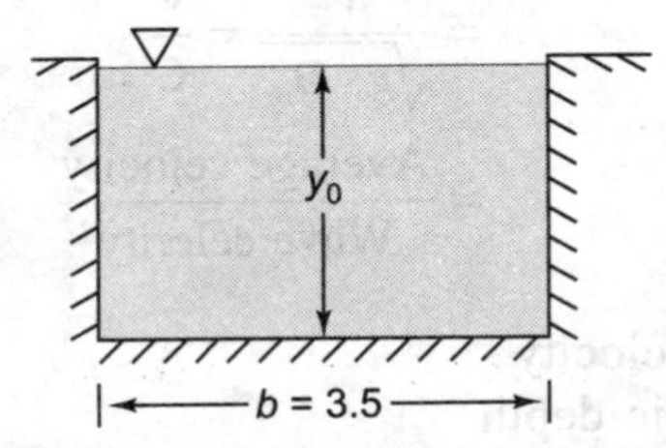

FIGURE 18.55 A rectangular channel.

Area $A = 3.5 \times y_0$ m^2
Perimeter $p = (3.5 + 2y_0)$ m

Hydraulic radius $R = \dfrac{A}{p}$

$$= \frac{3.5y_0}{3.5 + 2y_0}$$

As per Manning's equation, we have

$$Q = \frac{1}{n} \cdot A \cdot R^{2/3} \times S_0^{1/2}$$

$$5 = \frac{1}{0.01} \times (3.5y_0) \times \left(\frac{3.5y_0}{3.5 + 2y_0}\right)^{2/3} \sqrt{(0.0003)}$$

$$\frac{(3.5y_0)^{5/3}}{(3.5 + y_0)^{2/3}} = 2.88$$

or $y_0 = 1.08$ m by trial and error method.

Froude Number

In open channel flow, the gravitation force is the predominant force. On the basis of which the flow can be (i) subcritical (ii) critical and (iii) supercritical. The Froude number is a dimensionless number which is the ratio of inertial and gravitational forces. Depending upon the value of Froude number, the open channel flow can have different flow regimes such as (i) more depth of fluid (y) but less velocity (V) (ii) less depth but more velocity or (iii) constant depth and constant velocity. The gravity force tends to move water downhill while inertial force reflects the willingness of water to do so. The Froude number is

$$F_r = \frac{\text{Inertial force}}{\text{Gravity force}}$$

$$= \frac{V}{\sqrt{g \cdot D_m}} = \frac{V}{C}$$

$$= \frac{\text{Average velocity}}{\text{Wave celerity}}$$

where V = water velocity
D_m = Hydraulic depth
C = Wave celerity.

The Froude number is a measurement of bulk flow characteristics occurring in open channel flow (i) when flow undergoes variation in flow velocity and depth of a cross-section or (ii) when geometry of flow varies due to the presence of boulders in bed slope. The flow can be classified as:

(a) Critical flow when $F_r = 1$, i.e. celerity (surface wave) equals water velocity and surface disturbance remains stationary.
(b) Supercritical flow when $F_r > 1$, i.e. flow is fast or rapid with lesser flow depth. As $V > C$, all surface disturbances will be swept downstream.
(c) Subcritical flow when $F_r < 1$, i.e. flow is slow and tranquil with more flow depth. As $V < C$, the waves created by any surface disturbance at downstream can travel upstream.

In subcritical flow, the flow is controlled from a downstream point and the information is transmitted upstream.

This condition leads to backwater effects. Supercritical flow is controlled upstream and disturbances are transmitted downstream. When tranquil flow gives way to rapid flow, it occurs in smooth transition. However, when rapid flow suddenly decreases to a tranquil flow, there is an abrupt change known as *hydraulic jump* (sudden increase in depth accompanied by much turbulence). The transitions from subcritical to supercritical and from supercritical to subcritical are as shown in Figures 18.56(a) and (b).

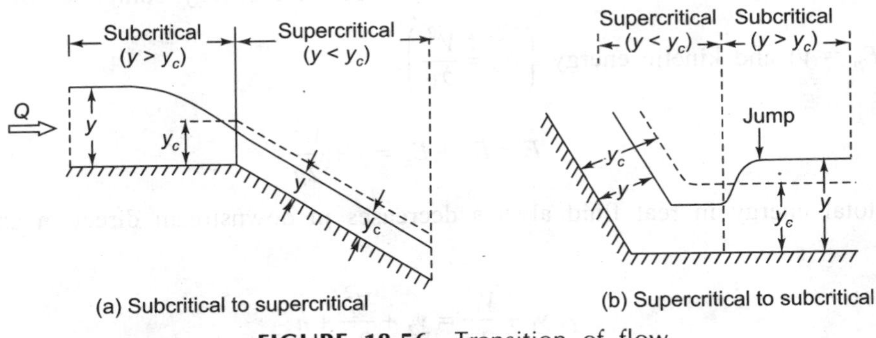

FIGURE 18.56 Transition of flow.

Specific Energy

The energy of flow has (i) potential energy component calculated from the depth of water (y) in the flow, (ii) a pressure $\left(\frac{P}{\gamma}\right)$ component and (iii) a kinetic energy component $\left(\frac{V^2}{2g}\right)$ calculated from the velocity of the flow. The energy of flow per unit weight of water is given by the Bernoulli's equation as

$$E = \frac{P}{\gamma} + \frac{V^2}{zg} + y \qquad \text{(when channel bottom coincide with datum)}$$

For open channel flow, water is open to atmosphere so that pressure (P) can be considered equal at all points in the length of flow. The energy equation can be given as

$$E_1 = \frac{V_1^2}{2g} + y_1 = E_2 = \frac{V_2^2}{2g} + y_2$$

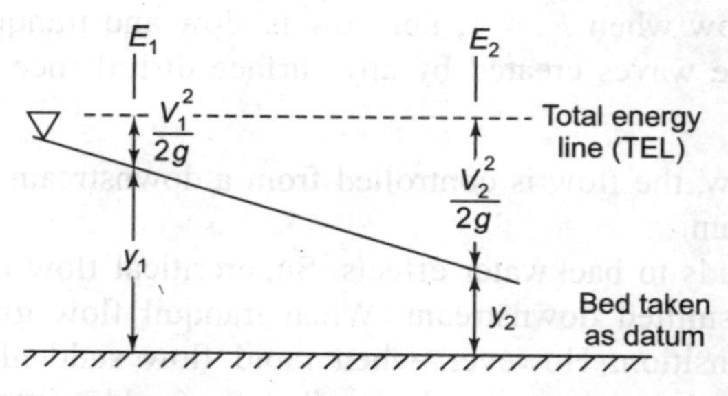

FIGURE 18.57 Concept of specific energy.

The specific energy can be defined as energy per unit weight of flowing liquid above the channel bed as shown in Figure 18.57. The specific energy comprises of potential energy ($E_p = y$) and kinetic energy $\left(E_k = \dfrac{V^2}{2g}\right)$.

$$E = E_p + E_k = y + \frac{V^2}{2g}$$

The total energy in real fluid always decreases in downstream direction as per the equation

$$y_1 + \frac{V_1^2}{2g} = y_2 + \frac{V_2^2}{2g} + h_f$$

where h_f = friction loss head.

Specific energy in uniform flow is constant while in non-uniform flow, it is constant only in horizontal bed channel.

Specific Energy Curve

The specific energy of flow is

$$E = y + \frac{V^2}{2g}$$

But discharge $Q = A.V$

$$\therefore \quad E = y + \frac{\left(\frac{Q}{A}\right)^2}{2g}$$

In case of rectangular channel, the area A is

$$A = y \times b$$

$$\therefore \qquad E = y + \frac{\left(\frac{Q}{y \cdot b}\right)^2}{2g}$$

If q is discharge per unit width, i.e. $q = \frac{Q}{b}$, then we have

$$E = y + \frac{q^2}{2gy^2}$$

In case discharge (q) is constant, then variation of specific energy (E) depends upon depth (y). Specific energy is a cubic parabola with depth (y) and its variation is a shown in Figure 18.58.

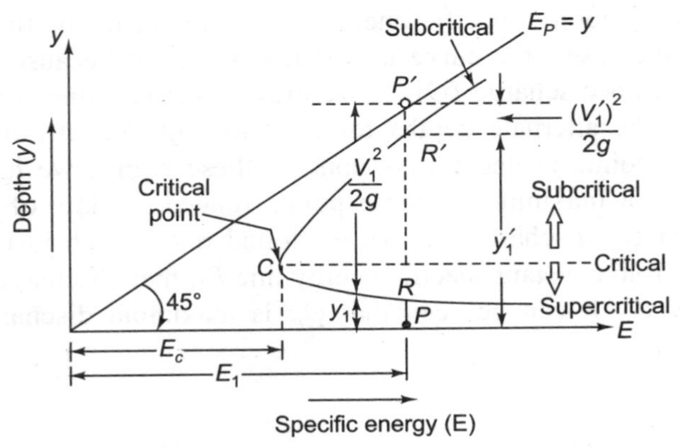

FIGURE 18.58 Specific energy curve.

The specific energy curve contains following informations:

(a) When $q = 0$, then $E_p = y$, i.e. the curve of potential energy is a straight line which passes through the origin and it makes an angle of 45° with each of two axes.

(b) There are two positive roots of specific energy E'. It means at any particular discharge q in a channel, there can be two depths y_1 and y_1', which have same value of specific energy E_1'. These two depths y_1 and y_1' are called alternate depths or *conjugate depths*.

(c) The specific energy curve is asymptotic to horizontal axis for small value of depth (y) and asymptotic to $E_p = y$ line (45° line) for higher value of depth (y').

(d) At certain depth (y_c) of flow called critical depth, the specific energy curve has a point of minimum specific energy. The corresponding flow velocity at minimum specific energy point is called *critical velocity* (V_c).

In specific energy curve as shown in Figure 18.58. The ordinate PP' represents the condition of specific energy E_1. The alternate depths of flow are $PR = y_1$ and $PR' = y_1'$ which have same value 'E_1', of specific energy with same flow 'q_1'. The intercepts $P'R = \frac{V_1^2}{2g}$ and $P'R' = \frac{(V_1')^2}{2g}$ are velocity heads corresponding to depth of 'y_1' and 'y_1'' of the flow. The smaller water depth 'y_1' has higher velocity head while larger water depth 'y_1'' has lower velocity head in order to maintain the same specific energy. It can also be seen that difference $(y_1' - y_1)$ is decreasing with lowering of specific energy value and at a certain value of specific energy $E = E_c$ the two depths (y and y') merge with each other as shown by point C on the curve. Left of point C ($E < E_c$), the flow under given condition is impossible. The condition of minimum specific energy is called *critical flow condition* and corresponding depth (y_c) is called as *critical depth*.

Variable Discharge Condition

Figure 18.59 shows different specific energy curves for different flows in which $Q_1 < Q_2 < Q_3$ …. As flow increases, the curve moves towards right because the specific energy of the flow increases with discharge (Q). If we draw a vertical line for a particular value of specific energy E_1, the alternate depths for discharge Q_1 are at points y_1 and y_1' and for discharge Q_2 are at points y_2 and y_2'. As none of these points, we have critical depth (y_c), hence discharge is not maximum at these points related to flow of Q_1 and Q_2. Now, we draw flow curve for Q_m discharge (shown by dotted line) which just touches at point C to vertical line drawn from constant specific energy line E_1, then discharge Q_m will be maximum for specific energy E_1. It can be seen that Q_m is maximum discharge for a flow having specific energy as E_1.

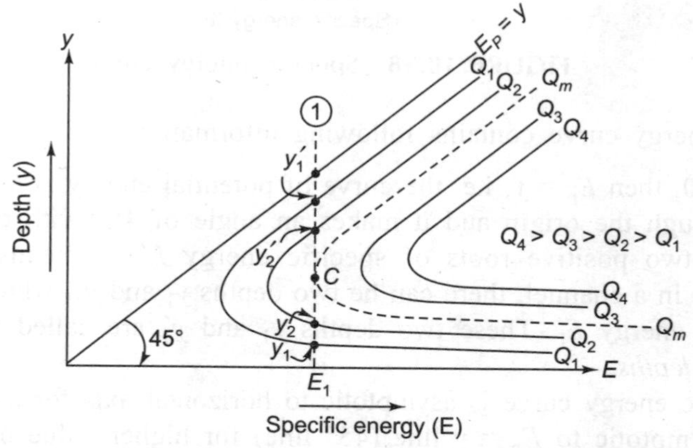

FIGURE 18.59 Specific energy curves for different flows.

$$E_1 = E_c = y_c + \frac{V_c^2}{2g}$$

$$= y_c + \frac{Q^2}{2 \cdot g \cdot A^2}$$

(a) ***Condition for minimum specific energy:*** For given Q, minimum specific energy (E) with respect to depth of flow (y) is when $\frac{dE}{dy} = 0$

$$\frac{dE}{dy} = \frac{d}{dy}\left(y + \frac{Q^2}{2gA^2}\right) = 0$$

or
$$1 + \frac{Q^2}{2g} \times \left(-2A^{-3}\right)\frac{dA}{dy} = 0$$

For rectangular channel, we have

$$A = y \times T \qquad \text{(where } T = \text{top width)}$$

or
$$\frac{dA}{dy} = T$$

∴
$$1 - \frac{Q^2 \times T_c}{g \times A_c^3} = 0$$

or
$$\frac{Q^2}{g} = \frac{A_c^3}{T_c}$$

(b) ***Critical depth:*** The specific energy curve has one point 'C' at which specific energy is minimum. The depth (y_c) at which specific energy is minimum or at which discharge is maximum is called critical depth.

$$\frac{Q^2}{g} = \frac{A_c^3}{T_c}$$

$$\frac{\left(\dfrac{q}{T_c}\right)^2}{g} = \frac{(y_c \times T_c)^3}{T_c}$$

or
$$y_c = \left(\frac{q^2}{g}\right)^{1/3}$$

(c) ***Critical velocity:*** The velocity of flow at critical depth is called critical velocity (V_c).

$$V_c = \frac{Q}{y_c \times T_c} = \frac{q}{\left(\frac{q^2}{g}\right)^{1/3}}$$

$$= q^{1/3} \times g^{1/3}$$

but $$q = V_c \times y_c$$

∴ $$V_c = (V_c \times y_c)^{1/3} \times g^{1/3}$$

or $$V_c^3 = V_c \times y_c \times g$$

or $$V_c^2 = y_c \times q$$

or $$V_c = \sqrt{g \cdot y_c}$$

(d) ***Specific energy at critical depth:*** The specific energy is

$$E_c = y_c + \frac{V_c^2}{2g}$$

$$E_c = y_c + \frac{\left(\frac{q}{y_c}\right)^2}{2g}$$

$$E_c = y_c + \frac{q^2}{g} \cdot \frac{1}{2y_c^2}$$

But $$\frac{q^2}{g} = (y_c)^3$$

∴ $$E_c = y_c + y_c^3 \times \frac{1}{2y_c^2}$$

$$E_c = 1.5 y_c$$

Flow over a Raised Hump

The specific energy equation can be used to solve the raised hump problem. A hump of height 'Δz' and corresponding changes possible on depth of flow at downstream have been shown in Figure 18.60. The Bernoulli's equation can be applied between section 1 and section 2.

$$E_1 = E_2 + \Delta z$$

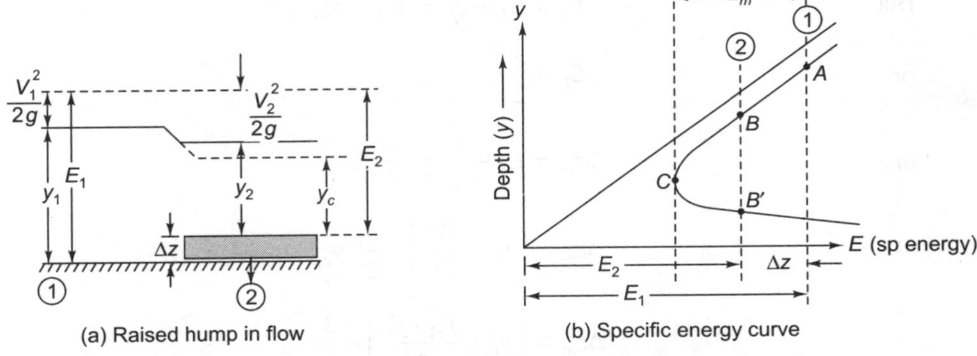

FIGURE 18.60 Raised bed hump and change in depth of flow.

The point 'A' shows specific energy 'E_1' on specific curve in Figure 18.60(b) for flow in section 1 while any point from point B to B' shows the specific energy 'E_2' at section 2 of the channel. All points in the channel between points 1 and 2 must lie on the specific energy curve between point 'A' and 'B' or 'B''.

Following flow conditions at section 2 are possible:

(a) Point from B to C—subcritical flow
(b) Point at C—critical flow
(c) Point from C to B'—supercritical flow

As the value of raised hump (Δz) increases, the depth of y_2 at section 2 will decrease. The minimum depth of flow will reach when point B concoids with critical point C and the height of raised hump cannot be raised as then the flow downstream is impossible. The value of hump height is maximum, i.e. Δz_m, when depth $y_z = y_c$ and specific energy $E_2 = E_c$

$$E_2 = E_1 - \Delta z_m = E_c = y_c + \frac{Q^2}{2g \cdot T^2 \cdot y_c^2}$$

It is clear from Figure 18.60 that if hump height Δz increases more than Δz_m, then flow is not possible in channel with given specific energy. In order to ensure flow at section 2, the only way is to firstly increase the upstream depth of water (section 1). Hence, we can say

(i) When $0 < \Delta z < \Delta z_m$, the upstream water level will remain constant as y_1 while y_2 will decrease with increase of hump height (Δz) till Δz becomes Δz_m
(ii) When $\Delta z > \Delta z_m$, the flow at section 2 will stop and depth of upstream y_1 at section 1 will start increasing while depth y_2 at downstream will continue to remain y_c.

(a) ***Minimum size of hump for critical flow:***

$$\Delta z = \Delta z_m \quad \text{(for critical flow)}$$

$$E_2 = E_c = \frac{3}{2} E_c$$

But $$E_1 = E_2 + \Delta z = E_c + \Delta z_m$$

or $$E_1 = \frac{3}{2} y_c$$

or $$\Delta z_m = E_1 - \frac{3}{2} \cdot y_c$$

or $$\Delta z_m = \left(y_1 + \frac{V_1^2}{2g} \right) - \frac{3}{2} \cdot y_c$$

$$\Delta z_m = \left(y_1 + \frac{F_{r1}^2 \cdot y_1}{2} \right) - \frac{3}{2} \cdot y_c$$

$$\Delta z_m = y_1 \left(1 + \frac{F_{r1}^2}{2} \right) - \frac{3}{2} \cdot y_c$$

or $$\frac{\Delta z_m}{y_1} = 1 + \frac{F_{r1}^2}{2} - \frac{3}{2} \left(\frac{q^2}{g} \right)^{1/3} \times \frac{1}{y_1} \quad \text{as} \quad y_c = \left(\frac{q}{g} \right)^{1/3}$$

$$\frac{\Delta z_m}{y_1} = 1 + \frac{F_{r1}^2}{2} - \frac{3}{2} \left(\frac{q^2}{g y_1^3} \right)^{1/3} = 1 + \frac{F_{r1}^2}{2} - \frac{3}{2} \cdot F_{r_1}^{2/3}$$

(b) ***Energy loss due to hump:***

$$E_1 = E_2 + \Delta z + H_l$$

$$\frac{\Delta z_m}{y_1} + \frac{h_l}{y_1} = 1 + \frac{F_{r1}^2}{2} - \frac{3}{2} \cdot F_{r_1}^{3/2}$$

The energy loss due to shape and friction is equivalent to that of hump placed in the downstream section of the channel flow.

Weirs

Weirs are sharp-crested overflow structures that are built across open channels. These are easy to construct and can measure the discharge accurately. The water level of downstream of weir is always below the weir crest. A typical weir is as shown in Figure 18.61.

$$V_2 = V_c = \sqrt{g y_c}$$

$$Q = A \times V_c = (b \times y_c)\sqrt{g \cdot y_c} \tag{18.11}$$

But $$E_2 = \frac{3}{2} \cdot y_c = h + \frac{V_1^2}{2g} = H$$

or $$y_c = \frac{2}{3} \cdot H \tag{18.12}$$

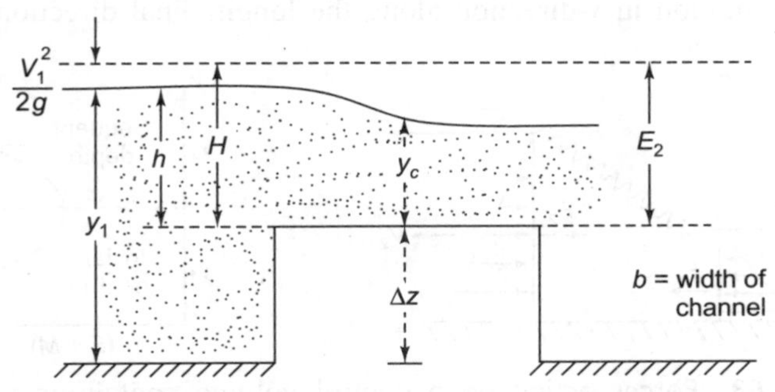

FIGURE 18.61 Flow over a broad crested weir.

From Eqs. (18.11) and (18.12), we have

$$Q = \left(b \times \frac{2}{3} H\right)\sqrt{g \cdot \frac{2}{3} H}$$

$$Q = \left[\frac{2}{3} \cdot \sqrt{\frac{2}{3} g}\right] \cdot b \cdot H^{3/2}$$

$$Q = 1.705 \times b \cdot H^{3/2} \approx 1.705 \cdot b \cdot h^{3/2}$$

In practice, there are energy losses upstream to the weir. To incorporate these losses, a coefficient of discharge (C_d) and velocity (C_v) are introduced. Hence, flow is given by

$$Q = C_d \cdot C_v \times 1.705 \times b \times h^{3/2}$$

Hydraulic Jump

A hydraulic jump flow occurs when a supercritical (high velocity and small depth) meets subcritical flow (low velocity and large depth) causing a jump in flow depth. The occurrence of hydraulic jump takes place (i) when bed slope changes from steep to mild, (ii) at the downstream of a sluice gate and (iii) at the toe of the spillway or weir as shown in Figure 18.62.

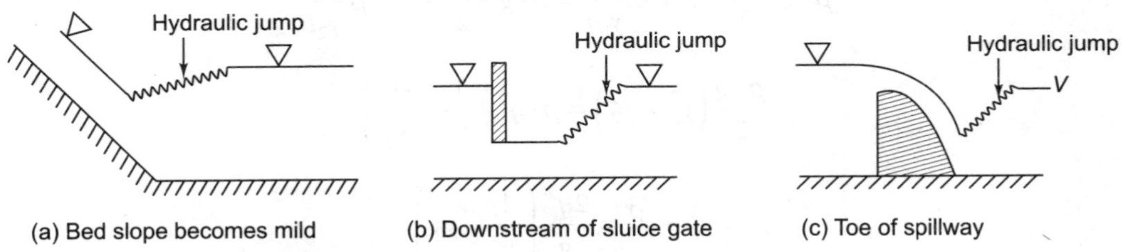

FIGURE 18.62 Occurrence of hydraulic jump.

The jump is the mechanism for the water surfaces to join. They join in extremely turbulent manner which causes a large energy losses. The momentum equation is used for analysis of the flow. Figure 18.63 shows forces applied to the control volume containing the hydraulic jump. Considering motion in x-direction along the longitudinal direction, we have

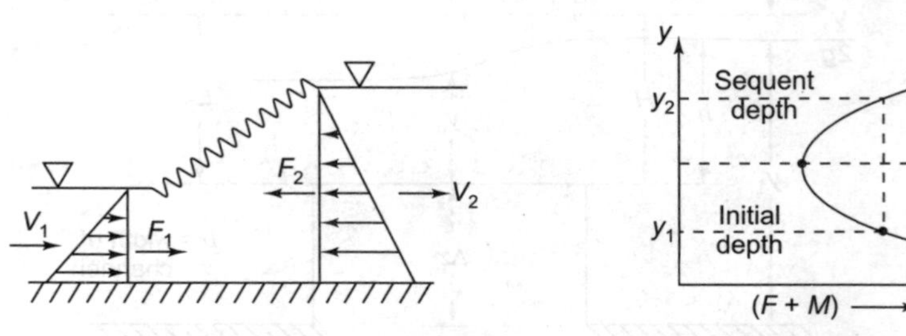

FIGURE 18.63 Forces acting on a control volume containing the hydraulic jump.

$$F_x = \text{resultant force} = F_1 - F_2$$

$$\text{Momentum change} = M_2 - M_1$$

As per Newton's law, the change of momentum is equal to resultant force (F_x).

$$F_x = M_2 - M_1$$

or
$$F_1 - F_2 = M_2 - M_1$$

or
$$F_1 + M_1 = F_2 + M_2$$

For rectangular channel, we have

$$F_1 = \rho \cdot g \cdot \frac{y_1}{2} \times y_1 \times b$$

$$F_2 = \rho \cdot g \cdot \frac{y_2}{2} \times y_2 \times b$$

$$M_1 = \rho \cdot Q \cdot V_1 = \rho \cdot Q \cdot \frac{Q}{y_1 \cdot b}$$

and
$$M_2 = \rho \cdot Q \cdot V_2 = \rho \cdot Q \cdot \frac{Q}{y_2 \cdot b}$$

or
$$\rho \cdot g \cdot \frac{y_1^2}{2} \times b + \rho \frac{Q^2}{y_1 b} = \rho \cdot g \cdot \frac{y_2^2}{2} \times b + \rho \frac{Q^2}{y_2 b}$$

$\therefore$
$$\frac{\rho \cdot g}{2}\left(y_1^2 - y_2^2\right) = \rho \cdot q^2 \left(\frac{1}{y_1} - \frac{1}{y_2}\right)$$

or
$$y_1^2 - y_2^2 = \frac{2q^2}{g}\left(\frac{1}{y_1} - y_2\right)$$

or
$$y_1 y_2 (y_1 + y_2) = \frac{2q^2}{g}$$

or
$$\frac{1}{2} \times \frac{y_2}{y_1}\left(1 + \frac{y_2}{y_1}\right) = \frac{q^2}{g \cdot y_1^3} = F_{r1}^2$$

or
$$\frac{y_2}{y_1} = \frac{1}{2}\left[-1 + \sqrt{1 + 8 \cdot F_{r1}^{2/3}}\right] \qquad (18.13)$$

Equation (18.13) is known as Belanger momentum equation. It can also be written in F_{r2} instead of F_{r1} as

$$\frac{y_1}{y_2} = \frac{1}{2}\left[-1 + \sqrt{1 + 8 \cdot F_{r2}^2}\right]$$

So knowing the discharge and either one of depths on the upstream or downstream side of the jump, the other depth can be easily determined. The energy loss in the jump is

$$\Delta E = \frac{(y_2 - y_1)^3}{4.y_1 y_2}$$

Sluice Gate

Submerged gates have been used for flow control in open channels. A typical flow through a sluice gate is as shown in Figure 18.64. As shown in the figure, we have

y_1 = flow depth upstream of sluice gate
y_2 = gate opening
y_3 = tail water depth

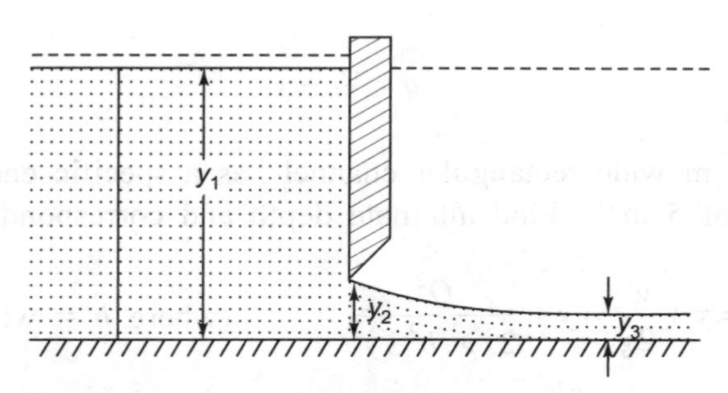

FIGURE 18.64 Working principle of sluice gate.

Applying the conservation of energy between head water and tail water depths, we have

$$y_1 + \frac{V_1^2}{2g} = y_3 + \frac{V_3^2}{2g}$$

But

$$q = V_1 \cdot y_1 = V_3 \cdot y_3$$

$\therefore$

$$y_1 + \frac{q^2}{2gy_1^2} = y_3 + \frac{q^2}{2 \cdot g \cdot y_3^2}$$

or

$$q = y_1 \cdot y_3 \left[\frac{2g}{\sqrt{y_1 + y_3}} \right]$$

To express this equation in terms of the gate opening y_2, a contraction coefficient is defined as:

$$C_c = \frac{y_3}{y_2}$$

$\therefore$

$$q = \frac{C_c}{\sqrt{1 + C_c\left(\frac{y_2}{y_1}\right)}} \times \sqrt{2gy_1} \cdot y_2$$

Now if we take discharge coefficient as:

$$C_d = \frac{C_c}{\sqrt{1 + C_c(y_2 / y_1)}}$$

Then, we have

$$q = \left[C_d \cdot \sqrt{2gy_1} \right] \cdot y_2$$

In case we define reference velocity as:

$$V_2 = C_d \sqrt{2gy_1}$$

Then, we have

$$q = V_2 - y_2$$

Example 18.23 A 2 m wide rectangular channel has a specific energy of 1.5 m, when carrying a discharge of 5 m^3/s. Find alternate depth and corresponding Froude numbers.

$$E = y + \frac{v^2}{2g} = y + \frac{Q^2}{2 \cdot g \cdot b^2 \cdot y^2} \qquad \text{(where } b = \text{width)}$$

$$1.5 = y + \frac{5^2}{2 \cdot g \cdot 2^2 \cdot y^2}$$

or

$$1.5 = y + \frac{0.319}{y^2}$$

$$\therefore \quad y_1 = 1.31 \quad \text{and} \quad y_2 = 0.592$$

$$F_r = \frac{V}{\sqrt{g \cdot y}} = \frac{Q}{2 \cdot y\sqrt{9.81 \cdot y}} = \frac{5}{2 \times \sqrt{9.81} \times y^{3/2}}$$

$$= \frac{0.798}{y^{3/2}}$$

Now, keeping $y = y_1$ and $y = y_2$, we get

$$F_{r_1} = 0.53 \quad \text{and} \quad F_{r_2} = 1.75$$

Example 18.24 Show that at the critical rate of flow, the specific energy in a rectangular channel is equal to 1.5 times the depth of flow.

$$\frac{Q^2}{g} = \frac{A^3}{T} \quad \text{(at critical flow)}$$

$$\frac{V^2}{g} = \frac{A}{T}$$

or

$$\frac{V^2}{g} = \frac{b \cdot y}{b} = y$$

Now

$$E = y + \frac{V^2}{2g} = y + \frac{y}{2}$$

$$E = \frac{3}{2} \cdot y$$

Example 18.25 Calculate critical depth and corresponding specific energy for discharge of 10^2 m^3/s in a rectangular channel with width 5 m.

$$q = \frac{Q}{b} = \frac{10}{5} = 2 \text{ m}^2/\text{s}$$

$$y_c = \left(\frac{q^2}{g}\right)^{1/3}$$

$$y_c = \left(\frac{2}{9.81}\right)^{1/3} = 0.742 \text{ m}$$

$$E_c = 1.5 \times y_c$$

$$E_c = 1.5 \times 0.742 = 1.112 \text{ m}$$

Example 18.26 A rectangular channel has width of 2 m and carries a discharge of 5 m^3/s with depth of 1.5 m. At a certain a small, smooth hump with flat top and of height 0.10 m is proposed to be built. Calculate the likely change in the water surface. Neglect energy loss.

$$q = \frac{5}{2} = 2.5 \text{ m}^2/\text{s}$$

$$V_1 = \frac{Q}{A} = \frac{5}{2 \times 1.5} = 1.667 \text{ m/s}$$

$$\frac{V_1^2}{2g} = \frac{1.667^2}{2 \times 9.81} = 0.142 \text{ m}$$

$$F_{r_1} = \frac{V_1}{\sqrt{g \cdot y_1}} = \frac{1.667}{\sqrt{9.81 \times 1.5}} = 0.435$$

$$E_1 = y_1 + \frac{V_1^2}{2g} = 1.5 + 0.142 = 1.642 \text{ m}$$

At section 2, the specific energy is

$$E_2 = E_1 - \Delta z$$
$$= 1.642 - 0.10$$
$$= 1.542 \text{ m}$$

$$y_c = \left(\frac{q^2}{g}\right)^{1/3} = \left(\frac{2.5^2}{9.81}\right)^{1/3} = 0.861 \text{ m}$$

$\therefore$
$$E_c = 1.5 \cdot y_c = 1.5 \times 0.861 = 1.291 \text{ m}$$

As minimum specific energy at section 2, i.e. E_{c_2} is less than E_2, hence $y_2 > y_c$ and the upstream depth y, will remain unchanged.

$$E_2 = y_2 + \frac{V_2^2}{2g}$$

$$1.542 = y_2 + \frac{(2.5)^2}{2 \times 9.81 \times y_2^2}$$

Solving by trial and error method, we find out

$$y_2 = 1.37 \text{ m}$$

Example 18.27 A 2.5 m wide rectangular channel carries 6 m^3/s flow at a depth of 0.60 m. Find the height of hump to be placed at a section to cause critical flow over the hump. The energy loss over the hump can be taken as 0.05% of the upstream velocity head.

$$q = \frac{Q}{b} = \frac{6}{2.5} = 2.4 \text{ m}^2/\text{s}$$

$$\cdot\ V_1 = \frac{2.4}{y_1} = \frac{2.4}{0.6} = 4 \text{ m/s}$$

$$\frac{V_1^2}{2g} = \frac{4^2}{2 \times 9.81} = 0.816 \text{ m}$$

$$h_l = 0.05\% \times \frac{V_1^2}{2g} = 0.05 \times 0.816$$

$$h_l = 0.041 \text{ m}$$

$$\Delta z_m = E_1 - E_2 - h_l = E_1 - E_c - h_l$$

$$E_c = 1.5 \cdot y_c = 1.5 \times \left(\frac{q^2}{g}\right)^{1/3}$$

$$E_c = 1.5 \times \left(\frac{2.4}{9.81}\right)^{1/3} = 1.256 \text{ m}$$

$$E_1 = y_1 + \frac{v_1^2}{2g} = 0.6 + 0.816$$

$$E_1 = 1.416 \text{ m}$$

$$\Delta z_m = E_1 - E_c - h_l$$

$$\Delta z_m = 1.416 - 1.256 - 0.041 = 0.119 \text{ m}$$

Example 18.28 A stationary hydraulic jump occurs in a rectangular channel with initial and sequent depth being equal to 0.3 m and 1.2 m respectively. Find (i) discharge per unit width and (ii) loss of energy.

$$\frac{y_2}{y_1} = \frac{1}{2}\left[-1 + \sqrt{1 + 8 \cdot F_{r1}^2}\right]$$

$$\frac{1.20}{0.30} = \frac{1}{2}\left[-1 + \sqrt{1 + 8 \cdot F_{r1}^2}\right]$$

$\therefore$
$$F_{r_1} = 3.162$$

$$F_{r_1} = \frac{V_1}{\sqrt{g y_1}} = \frac{V_1}{\sqrt{9.81 \times 0.30}} = 3.162$$

$\therefore$
$$V_1 = 5.424 \text{ m/s}$$

Now discharge $q = V_1 \cdot y_1$

$$q = 5.424 \times 0.3 = 1.627 \text{ m}^2/\text{s}$$

$$\text{Energy loss} = \frac{(y_2 - y_1)^3}{4.y_1 y_2}$$

$$= \frac{(1.20 - 0.30)^3}{4 \times 1.2 \times 0.3} = 0.506 \text{ m}$$

Example 18.29 In a hydraulic jump, Froude number before the jump is 10 and energy loss is 3.10 m. Find sequent depths.

$$\frac{y_2}{y_1} = \frac{1}{2}\left(-1 + \sqrt{1 + 8F_{r_1}^2}\right)$$

$$\frac{y_2}{y_1} = \frac{1}{2}\left(-1 + \sqrt{1 + 10^2 \times 8}\right) = 13.651$$

$$\frac{E_l}{y_1} = \frac{(y_2 - y_1)^2}{4 \cdot y_1 \cdot y_2} = \frac{\left(\frac{y_2}{y_1} - 1\right)^2}{4 \times \frac{y_2}{y_1}}$$

$$\frac{3.10}{y_1} = \frac{(13.651 - 1)^2}{4 \times 13.651}$$

$\therefore$ $$y_1 = 0.0836$$

and $$y_2 = 1.41 \text{ m}$$

Example 18.30 An overflow spillway is 40 m high. At the design energy head of 2.85 m over the spillway, find the sequent depth. Assume $C_d = 0.98$.

$$q = C_d \times 1.705 \times h^{3/2}$$

$$q = 0.98 \times 1.705 \times 2.85^{3/2}$$

$$q = 8.04 \text{ m}^3/\text{s}$$

$$E_1 = 40 + 2.85 = 42.85$$

$$E_1 = y_1 + \frac{V_1^2}{2g} = y_1 + \frac{q^2}{2g.y_1^2}$$

$$42.85 = y_1 + \frac{(8.04)^2}{2g \cdot y_1^2}$$

$$y_1 = 0.282 \text{ m}$$

$$V_1 = \frac{q_1}{y_1} = \frac{8.04}{0.282} = 28.51 \text{ m/s}$$

$$F_{r_1} = \frac{V_1}{\sqrt{g \cdot y_1}} = \frac{28.51}{\sqrt{9.81 \times 0.282}} = 17.1$$

$$\frac{y_2}{y_1} = \frac{1}{2}\left[-1 + \sqrt{1 + 8 \times F_{r1}^2}\right]$$

$$\frac{y_2}{y_1} = \frac{1}{2}\left[-1 + \sqrt{1 + 8 \times 17.1^2}\right]$$

$$\frac{y_2}{0.282} = \frac{1}{2}\left[-1 + \sqrt{1 + 8 \times 17.1^2}\right]$$

$$y_2 = 6.67 \text{ m}$$

Example 18.31 Water flows at a depth of 10 cm with velocity of 6 m/s. Is the flow subcritical or supercritical? What is the alternate depth?

$$F_r = \frac{V}{\sqrt{gy}} = \frac{6}{\sqrt{9.81 \times 0.1}}$$

$$F_r = 6.06 > 1$$

The flow is supercritical

$$E = y + \frac{V^2}{2g} = 0.1 + \frac{6^2}{2 \times 9.81}$$

$$E = 1.935$$

Now

$$E = 1.935 = y' + \frac{q^2}{2 \cdot gy'}$$

$$q = y \times V = 0.1 \times 6 = 0.6 \text{ m}^2\text{/s}$$

$$1.935 = y' + \frac{0.6^2}{2 \cdot g \cdot (y')^2}$$

Using trial and error method, we get

$$y' = 1.93 \text{ m}$$

Example 18.32 The spillway (y_0 = 5 m) has discharge of 1.2 m^3/s per metre of width. What depth y_2 will exist at downstream of the spillway? Assume negligible energy loss over the spillway.

$$y_0 + \frac{q^2}{2g \cdot y_0^2} = y_1 + \frac{q^2}{2 \cdot g \cdot y_1^2}$$

$$5 + \frac{1.2^2}{2 \times 9.81 \times 5^2} = y_1 + \frac{1.2^2}{2 \cdot g \cdot y_1^2}$$

On solving, we get $y_1 = 0.123$ m

$$F_{r_1} = \frac{q}{\sqrt{gy_1}} = \frac{1.2}{\sqrt{9.81 \times (0.123)}} = 8.88$$

$$\frac{y_2}{y_1} = \frac{1}{2}\left[-1 + \sqrt{1 + 8 \times F_{r1}^2}\right]$$

$$\frac{y_2}{y_1} = \frac{1}{2}\left[-1 + \sqrt{1 + 8{,}88^2}\right]$$

or $$y_2 = 1.48 \text{ m}$$

Example 18.33 Water is flowing under a sluice gate (a horizontal rectangular) has depths y_0 and y_1 as 20 m and 1.0 cm respectively. Find the hydraulic loss.

$$y_0 + \frac{V_0^2}{2g} = y_1 + \frac{V_1^2}{2g}$$

$$20 + 0 = 0.1 + \frac{V_1^2}{2 \times 9.81}$$

$$V_1 = \sqrt{19.9 \times 2 \times 9.81}$$

$$V_1 = \sqrt{390.44} = 19.8$$

$$F_{r_1} = \frac{V_1}{\sqrt{g \cdot y_1}} = \frac{19.8}{\sqrt{9.81 \times 0.1}}$$

$$F_{r_1} = \frac{19.8}{0.99} = 20$$

$$\frac{y_2}{y_1} = \frac{1}{2}\left(\sqrt{1 + F_{r1}^2} - 1\right) = \frac{1}{2}\left(\sqrt{1 + 20^2} - 1\right)$$

or $$y_2 = \frac{0.1}{2}(20.2 - 1) = 0.96 \text{ m}$$

$$h_l = \frac{(y_2 - y_1)^2}{4 \cdot y_1 \cdot y_2} = \frac{(0.96 - 0.1)^2}{4 \times 0.96 \times 0.1}$$

$$h_l = 1.926 \text{ m}$$

COMPRESSIBLE FLOW

When a fluid moves at a speed comparable to the speed of sound, density changes become significant and the flow is termed compressible. Such flows are difficult to obtain in liquids as higher pressure ratios of order of 1000 atm are needed to generate sonic velocities. However, we require a pressure ratio of only 2:1 in gases to cause sonic flow. The compressible gas flow, therefore, is quite common and its study is termed as gas dynamics.

Probably, two most important and distinctive effects of compressibility are (i) choking in which the duct flow rate is sharply limited by the sonic condition and (ii) shock waves which are nearly discontinuous property changes in a supersonic flow.

Mach-Number

The Mach-number is the ratio of flow velocity (V) and the speed of sound in the fluid (a)

$$M_a = \frac{V}{a}$$

For small Mach-numbers flows, changes in fluid density are small and the flows can be considered incompressible flows, requiring only a momentum and continuity analysis. However, the density changes are significant when flows have Mach-numbers greater than 0.3. Such flows can be solved with four equations which are (i) momentum equation (ii) continuity equation (iii) energy equation and (iv) equation of state. These equations can be solved with (i) pressure (ii) density (iii) temperature and (iv) flow velocity (P, ρ, T, V). The Mach-number is the dominant parameter in compressible flow analysis. On the basis of Mach-numbers, flows can be classified as

(a) $M_a < 0.3$ – Incompressible flow
(b) $0.3 < M_a < 0.8$ – Subsonic flow where density effects are relevant but no shock waves appear
(c) $0.8 < M_a < 1.2$ – Transonic flow where shock waves first appear
(d) $1.2 < M_a < 3.0$ – Supersonic flow where shock waves are present but subsonic region is not existing
(e) $3.0 < \text{Ma}$ – Hypersonic flow where shock waves and other flow changes are specially strong

Isentropic Process in Compressible Flow

The isentropic approximation is common in compressible flow analysis

$$T \cdot ds = dh - \frac{dP}{\rho}$$

Taking

$$dh = C_P \cdot dT$$

$$T \cdot ds = C_P \cdot dT - \frac{dP}{\rho}$$

or

$$ds = C_p \frac{dT}{T} - \frac{dP}{\rho \times T}$$

$$= C_p \cdot \frac{dT}{T} - R \cdot \frac{dP}{\rho}$$

Integrating, we get

$$\int_1^2 ds = \int_1^2 C_P \frac{dT}{T} - R\int_1^2 \frac{dP}{P}$$

$$s_2 - s_1 = C_P \log\frac{T_2}{T_1} - R\log\frac{P_2}{P_1}$$

For isentropic process, $s_2 - s_1 = 0$

$$\frac{P_2}{P_1} = \left(\frac{T_2}{T_1}\right)^{k/k-1} = \left(\frac{\rho_2}{\rho_1}\right)^k \qquad \text{where } k = \frac{C_P}{C_V}$$

Isentropic Flow

Consider an isentropic flow in a duct which varies in size as shown in Figure 18.65.

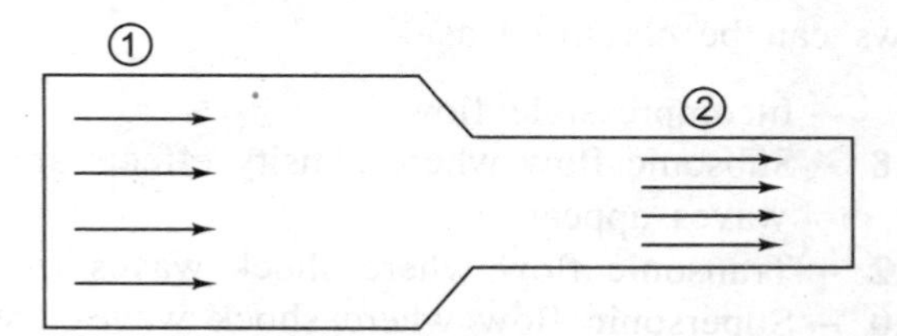

FIGURE 18.65 Applying steady flow energy equation on isentropic flow in a duct.

Applying steady flow energy equation between section 1 and section 2, we have

$$Q - W = \Delta u + \Delta FE + \Delta KE + \Delta PE \tag{18.14}$$

For adiabatic flow $Q = 0$ and for no work is done $W = 0$.

Also $\Delta u + \Delta FE = \Delta H$

where

FE = flow energy

H = enthalpy

u = internal energy

Hence Eq. (18.14) is reduced to

$$0 = \Delta H + \Delta KE + \Delta PE$$

In specific terms, the equation is

$$\Delta h + \Delta ke + \Delta Pe = 0$$

Applying this in section 1 and section 2

$$h_1 + \Delta ke_1 + \Delta Pe_1 = h_2 + \Delta ke_2 + \Delta Pe_2$$

or $$h_1 + \frac{u_1^2}{2} + gz_1 = h_2 + \frac{u_2^2}{2} + gz_2$$

For a gas, we have $h = C_P.T$ where T is in Kelvin (°C + 273)

$$C_P \cdot T_1 + \frac{u_1^2}{2} + gz_1 = C_P \cdot T_2 + \frac{u_2^2}{2} + gz_2$$

The above is Bernoulli's equation for gas.

Stagnation Condition

(a) ***Stagnation temperature:*** If a stream of gas is brought to rest, it is said to stagnate. This occurs on leading edges of any obstacle placed in the flow. Consider a horizontal flow as shown in Figure 18.66 which is brought to stagnation at point 2.

FIGURE 18.66 Flow is brought to stagnation.

In this case we have

$$u_2 = 0, \; z_1 = z_2$$

Applying Bernoulli's equation, we have

$$C_P \cdot T_1 + u_1^2 / 2 = C_P T_2 + 0$$

$$T_2 = T_1 + \frac{u_1^2}{2 \cdot C_P}$$

or $$\Delta T = T_2 - T_1 = \frac{u_1^2}{2 \cdot C_P}$$

But $$\frac{C_P}{C_V} = \gamma \quad \text{and} \quad C_P = \frac{R}{\gamma - 1}$$

$\therefore$ $$\frac{\Delta T}{T_1} = \frac{u_1^2(\gamma - 1)}{(2\gamma \cdot R \cdot T_1)} = \frac{u_1^2(\gamma - 1)}{2 \cdot a^2} \quad \text{as} \quad a^2 = \gamma \cdot R \cdot T$$

or $$\frac{\Delta T}{T_1} = \left(\frac{u_1}{a}\right)^2 \cdot \left(\frac{\gamma - 1}{2}\right)$$

$$\frac{\Delta T}{T_1} = M_a^2 \left(\frac{\gamma - 1}{2}\right) \tag{18.15}$$

If M_a is less than 0.2, then M_a^2 is less than 0.04 which makes $\frac{\Delta T}{T_1}$ less than 0.008. Therefore, rise in temperature is negligible under low Mach-number condition. However, when $M_a = 3$ as it is in fast aircraft and $M_a = 25$ for spacecraft's entry into atmosphere, then ΔT is quite large. In fact, spacecraft can start burning due to high temperature rise in case proper insulation is not provided.

(b) ***Stagnation pressure:***
Equation (18.15) is

$$\frac{\Delta T}{T_1} = M_a^2\left(\frac{\gamma - 1}{2}\right)$$

or $$\frac{T_2 - T_1}{T_1} = M_a^2\left(\frac{\gamma - 1}{2}\right)$$

or $$\frac{T_2}{T_1} = M_a^2\left(\frac{\gamma - 1}{2}\right) + 1$$

or $$\left(\frac{P_2}{P_1}\right)^{\frac{\gamma-1}{\gamma}} = M_a^2\left(\frac{\gamma - 1}{2}\right) + 1$$

or $$\frac{P_2}{P_1} = \left[\frac{M_a^2(\gamma - 1)}{2} + 1\right]^{\frac{\gamma}{\gamma-1}}$$

Here P_2 is stagnation pressure. If we expand right hand terms of the above equation using the binomial theorem, we have

$$\frac{P_2}{P_1} = 1 + \frac{\gamma \cdot M_a^2}{2}\left[1 + \frac{M_a^2}{4} + \frac{M_a^4}{8} + \ ...\right]$$

If $M_a < 0.4$, then we have

$$\frac{P_2}{P_1} = 1 + \frac{\gamma \cdot M_a^2}{2} \tag{18.16}$$

Now $$\Delta P = P_2 - P_1$$

$$\frac{\Delta P}{P_1} = \gamma\,\frac{M_a^2}{2}$$

or $$\Delta P = \frac{\gamma}{2} \times \left(\frac{u_1}{a}\right)^2 \cdot P_1 = \frac{\gamma}{2} \times \left(\frac{u_1^2}{\gamma \cdot R \cdot T}\right) \times P_1$$

$$\Delta P = \frac{\rho_1 u_1^2}{2} \quad \text{as} \quad \rho_1 = \frac{P_1}{RT}$$

or $$u = \left(\frac{2\Delta P}{\rho_1}\right)^{0.5} \tag{18.17}$$

Equation (18.16) for gas and Eq. (18.17) are same.

Example 18.34 A pitot tube reads (i) airstream pressure as 105 kPa and (ii) differential pressure as 20 kPa. The air temperature is 20°C. Find air speed.

$$P_1 = 105 \text{ kPa}$$

$$P_2 = \text{stagnation pressure} = P_1 + \Delta P = 105 + 20 = 125 \text{ kPa}$$

or $$\frac{P_2}{P_1} = \left(M_a^2 \times \frac{\gamma}{2} + 1\right)$$

$$\frac{125}{105} = 1 + M_a^2 \times 1.4$$

or $$M_a = 0.634$$

$$a^2 = \gamma \cdot R \cdot T = 1.4 \times 287 \times 293$$

or $$a = 343 \text{ m/s}$$

$$M_a = \frac{u}{a} = 0.634$$

or $$u = 0.634 \times a = 0.634 \times 343$$

$$u = 217.7 \text{ m/s}$$

Equation of Motion for a Compressible Flow

Consider a one-dimensional control volume as shown in Figure 18.67. Assume (i) flow is one-dimensional (ii) viscosity and heat transfer are neglected (iii) behaviour of flow as a consequence of area changes is only considered and (iv) flow is steady.

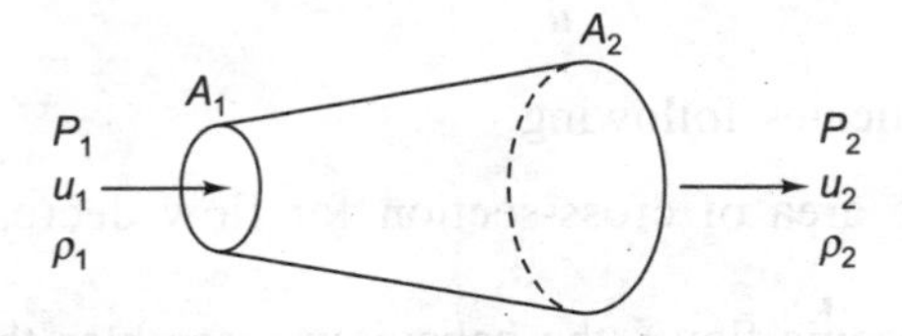

FIGURE 18.67 Control volume in compressible flow.

Applying continuity equation at sections 1 and 2.

$$\rho_1 \cdot u_1 \cdot A_1 = \rho_2 \cdot u_2 \cdot A_2$$

or $$\frac{d}{dx}(\rho \cdot u \cdot A) = 0$$

or
$$\frac{d\rho}{\rho}+\frac{du}{u}+\frac{dA}{A}=0$$

Area Velocity Relation

As per continuity equation, we have

$$\frac{d\rho}{\rho}+\frac{du}{u}+\frac{dA}{A}=0 \tag{18.18}$$

As per Euler equation, we have

$$u\cdot du+\frac{dP}{\rho}=0$$

or
$$u\cdot du=-\frac{dP}{\rho}=-\frac{dP}{d\rho}\cdot\frac{d\rho}{\rho}$$

$$u\cdot du=-a^2\cdot\frac{d\rho}{\rho} \quad \text{as} \quad a^2=\frac{dP}{d\rho}$$

or
$$\frac{d\rho}{\rho}=-M_a^2\cdot\frac{du}{u} \tag{18.19}$$

From Eqs. (18.18) and (18.19), we have

$$\frac{du}{u}+\frac{dA}{A}=M_a^2\cdot\frac{du}{u}$$

or
$$\frac{du}{u}\left(1-M_a^2\right)=-\frac{dA}{A}$$

or
$$\frac{du}{u}=\frac{-\frac{dA}{A}}{1-M_a^2}$$

The above relation indicates following:

(a) When $M_a = 0$, the area of cross-section for flow decreases, the velocity increases and vice-versa.
(b) When $M_a < 1$ (subsonic flow), the behaviour resembles that for incompressible flows (Figure 18.68).
(c) When $M_a > 1$ (supersonic flow), the area decreases, velocity also decreases and vice-versa (Figure 18.68).

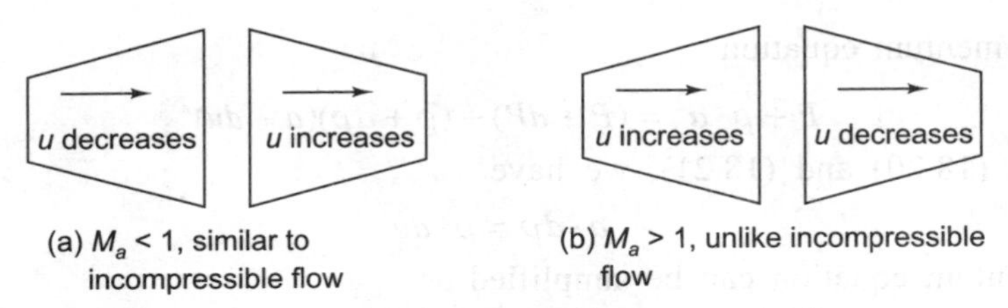

(a) $M_a < 1$, similar to incompressible flow (b) $M_a > 1$, unlike incompressible flow

FIGURE 18.68 Response of subsonic and supersonic flows to area changes.

Wave Propagation

Waves carry information in a flow. These waves travel at the local speed of sound. This brings a sharp contrast between incompressible and compressible flows. In case of incompressible fluid, the speed of sound is infinite, i.e. Mach-number is zero. Due to this, information of flow in the form of pressure, density and velocity changes is conveyed to all parts of the flow instantaneously. The flow is capable to change instantaneously. The incompressible fluid flow changes smoothly to accommodate the presence of a body placed in the flow both at upstream and downstream. This does not happen in compressible flows as in compressible fluids, any disturbance travels only at finite speeds. Signals generated at a point in the flow take a finite time to reach other points of the flow. Any changes in flow condition is transmitted to other parts of the flow by means of waves travelling at the speed of sound. The ratio of flow speed to the speed of sound is called *Mach-number*. It indicates the relative importance of compressibility effect for a given flow. We derive now the speed of sound. Consider a sound wave propagating to the right as shown in Figure 18.69. The fluid at the right is at rest with pressure = P, density = ρ and temperature = T. Due to propagation of sound wave, the fluid is compressed at left and it has pressure = $P + dP$, density = $\rho + d\rho$ and temperature = $T + dT$. To simplify the analysis, it is better to make the wave stationary as shown in Figure 18.69(b) by superposing an equal and opposite speed everywhere in the flow.

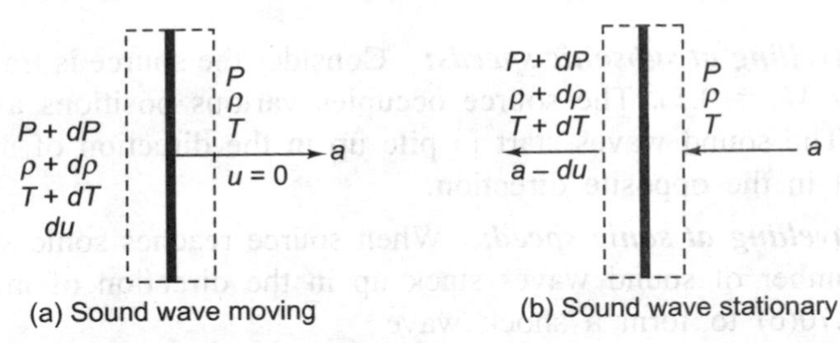

(a) Sound wave moving (b) Sound wave stationary

FIGURE 18.69 Propagation of sound wave in compressible flow.

Now, we apply the equations of mass and momentum to the control volume. As per equation of mass

$$\rho \cdot a = (\rho + d\rho)(a - du) \tag{18.20}$$

As per momentum equation

$$P + \rho \cdot a^2 = (P + dP) + (\rho + d\rho)(a - du)^2 \quad (18.21)$$

From Eqs. (18.20) and (18.21), we have

$$a \cdot d\rho = \rho \cdot du \quad (18.22)$$

The momentum equation can be simplified as

$$P + \rho \cdot a^2 = (P + dP) + (\rho + d\rho)(a - du)^2$$
$$= (P + dP) + \left[(\rho + d\rho)(a - du)\right]$$
$$= (P + dP) + (\rho \cdot a)(a - du)$$

or

$$dP = \rho \cdot a \cdot du \quad (18.23)$$

From Eqs. (18.22) and (18.23), we have

$$dP = a^2 \cdot dP$$

or

$$a^2 = \frac{dP}{d\rho} \quad (18.24)$$

For a perfect gas, the above relation of sound wave (Eq. 18.24) can be reduced to

$$a^2 = \gamma \cdot R \cdot T$$

Now, we study the propagation of sound wave with respect the velocity of sound source (varying from $M_a < 1$ to $M_a > 1$).

(a) ***Stationary source:*** The source emits a sound wave at every second which travels in the form of a circle with its centre at the location of the source as shown in Figure 18.70(a). After three seconds, there are three concentric circles as shown in the figure. The effect of the sound source is felt to an observer within the largest circle.

(b) ***Source travelling at subsonic speeds:*** Consider the source is travelling at subsonic speed (say M_a = 0.5). The source occupies various positions as shown in Figure 18.70(b). The sound waves start to pile up in the direction of motion and start to stretch out in the opposite direction.

(c) ***Source travelling at sonic speed:*** When source reaches sonic speed (M_a = 1), an infinite number of sound waves stack up in the direction of motion as shown in Figure 18.70(c) to form a shock wave.

(d) ***Source travelling at supersonic speeds:*** When source moves with supersonic speeds, it leaves its pressure wave behind. The pressure waves create an angle known as the *Mach wave angle* (μ) as shown in Figure 18.70(d).

$$\mu = \text{arc sin}\left(\frac{a}{u}\right) = \text{arc sin}\left(\frac{1}{M_a}\right)$$

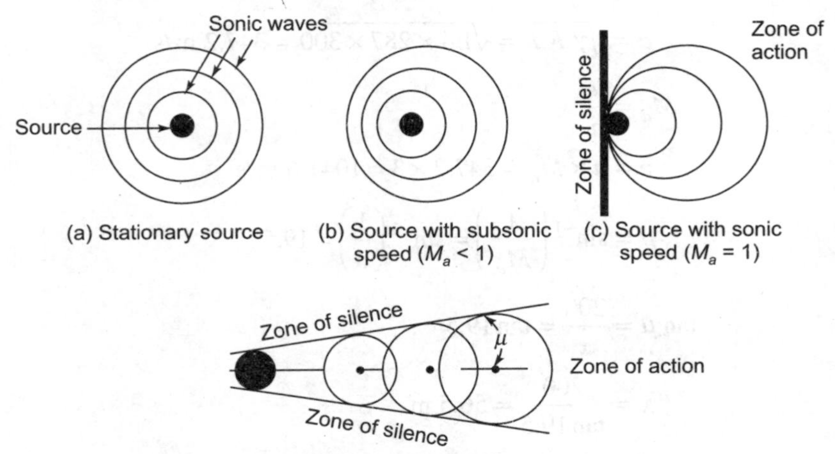

FIGURE 18.70 Propagation of a source of sound at different speeds.

Example 18.35 Determine the speed of sound in Argon (Ar) at 120 °C. Take molecular weight = 40 kg/kmol, R_u = 8314 J/kgmol K and $\gamma = \dfrac{C_P}{C_V} = 1.668$.

The speed of sound $a = \sqrt{\gamma \cdot R \cdot T}$

$$T = 120 + 273 = 393 \text{ K}$$

$$R = \frac{R_u}{M} = \frac{8314}{40}$$

$$R = 207.9 \text{ J/kg K}$$

$$a = \sqrt{1.668 \times 207.9 \times 393}$$

$$a = 318.8 \text{ m/s}$$

Example 18.36 A projectile is travelling at a speed of M = 3 and it passes above an observer at height of 200 m. Find (i) the velocity of projectile and (ii) the distance beyond the observer when the projectile will be firstly heard.

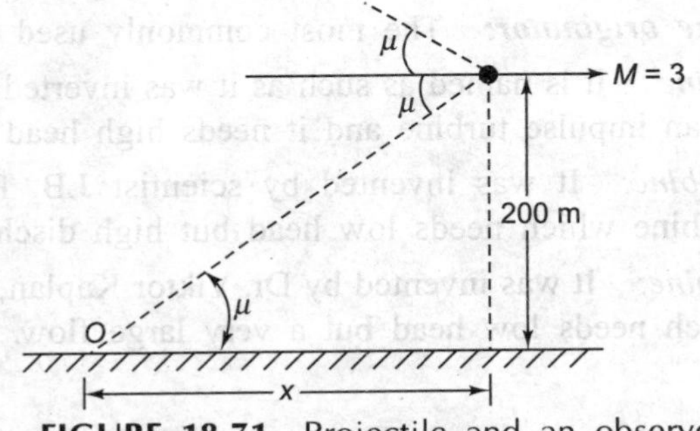

FIGURE 18.71 Projectile and an observer.

$$a = \sqrt{\gamma.R.T} = \sqrt{1.4 \times 287 \times 300} = 347.2 \text{ m/s}$$

$$M_a = \frac{u}{a}$$

$\therefore$
$$u = a \times M_a = 347.2 \times 3 = 1041.6 \text{ m/s}$$

$$\mu = \sin^{-1}\left(\frac{1}{M_a}\right) = \sin^{-1}\left(\frac{1}{3}\right) = 19.5°$$

$$\tan \mu = \frac{200}{x} = \tan 19.5$$

or
$$x = \frac{200}{\tan 19.5} = 56.5 \text{ m}$$

FUNDAMENTAL OF FLUID TURBINES

Fluid or hydraulic turbines are the hydro machines to generate electricity in hydro electrical power plants. Hydro or water power is a conventional renewable source of energy. In hydraulic turbines, the kinetic or potential energy of water is converted into rotary type mechanical energy which is further converted into electric energy by using electric generators. This conversion of hydro energy is clean and free from pollution and it has good environmental effects. Hydraulic turbines are used where appreciable heads of water are available. Water is discharged from high level reservoirs through pipes to hydraulic turbines to harness mechanical work.

A hydraulic turbines is a device that uses kinetic or potential energy of water and converts the same into mechanical energy.

The hydraulic turbines are classified according to (i) head and quantity of water (ii) name of the originator (iii) action of water on moving blades (iv) direction of flow of water in the runner (v) positioning of turbine shaft with respect to ground and (vi) specific speed.

(a) ***Head and quantity of water:*** As per this, turbines can be (i) impulse turbines and (ii) reaction turbines. Impulse turbines require high head and low quantity of water flow. Reaction turbines require low head and high water flow.

(b) ***The name of the originator:*** The most commonly used turbines are:

(i) *Pelton turbine:* It is named as such as it was inverted by Lester Allan Pelton, USA. It is an impulse turbine and it needs high head but low discharge.

(ii) *Francis turbine:* It was invented by scientist J.B. Francis, France. It is a reaction turbine which needs low head but high discharge.

(iii) *Kaplan turbine:* It was invented by Dr. Viktor Kaplan, Austria. It is a reaction turbine which needs low head but a very large flow.

(c) ***Action of water on moving blades:*** On the basis of action of water, the turbines can be (i) impulse and (ii) reaction turbines. In impulse turbines, outside nozzles are used to convert fluid potential energy into the kinetic energy and this kinetic energy is used by moving blades to generate mechanical work. The pressure of water on moving blades remains unchanged. Pelton turbine is an example of impulse turbine. Reaction turbines utilize both velocity and pressure of the fluid on their moving blades. These turbines do not use outside nozzles for conversion of potential energy into kinetic energy. The blades are designed as nozzles to convert the potential energy into kinetic energy when water moves on the blades. Francis and Kaplan turbines are reaction turbines. The principles of impulse and reaction turbines are as explained by diagrams in Figure 18.72.

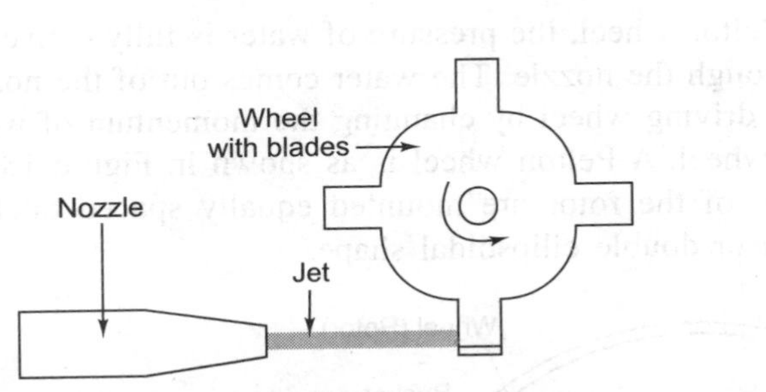

(a) The change of momentum of water jet provides impulse for motion

(b) Reaction of high velocity water exit provides motion

FIGURE 18.72 Principle of impulse and reaction turbines.

(d) ***Direction of flow of water:*** The flow of water can be (i) tangential flow as in Pelton turbine (ii) radial flow as in Francis turbine and (iii) axial flow as in Kaplan turbine.

(e) ***Position of shaft:*** The turbine shaft can either be vertical or horizontal. Pelton turbines have horizontal shafts while other turbines have vertical shafts.

(f) ***Specific speed:*** The turbine can have different head and discharge conditions. In order to compare these different turbines, the concept of specific speed is used. Specific speed is defined as the speed of a geometrical similar turbine which would develop 1.0 kW of power when it is given water having head of 1.0 m. The specific speed (N_s) is:

$$N_s = \frac{N\sqrt{P}}{H^{5/4}}$$

where N = rpm of turbine
P = power of turbine
H = Head

The above relation indicates that turbines with high heads and low discharge will have low specific speeds such as in impulse turbines while turbines with low heads and high discharge will have high specific speeds such as in reaction turbine. The specific speeds of turbines are:

(i) Pelton turbine – 10 to 60
(ii) Francis turbine – 60 to 300
(iii) Kaplan turbine – 300 to 1000

The turbines with high specific speeds work at higher speeds and they are more compact.

Pelton Wheel or Turbine

Pelton wheel is an impulse turbine. In Pelton wheel, the pressure of water is fully converted into kinetic energy by passing water through the nozzle. The water comes out of the nozzle as a high velocity jet which is used for driving wheel by changing the momentum of water while moving on buckets fixed on the wheel. A Pelton wheel is as shown in Figure 18.73. It consists of a rotor. On the periphery of the rotor are mounted equally spaced buckets which have either double hemispherical or double ellipsoidal shape.

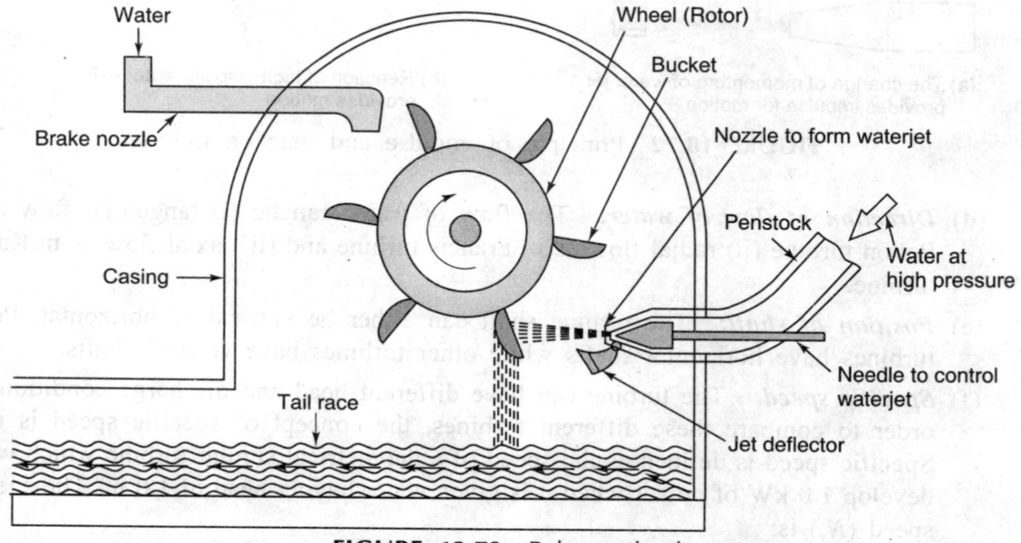

FIGURE 18.73 Pelton wheel.

Water is brought down from a high head source like dam through a pipe system which is called *penstock*. A nozzle is provided at the end of the penstock which converts the potential energy of high head water into a high velocity waterjet. The high velocity jet emerging from the nozzle impinges on the buckets provided on the Pelton wheel and it sets the wheel in motion. The speed of the wheel depends upon the flow rate and velocity of water. The flow rate of water is controlled by means of a needle (spear) in the nozzle as shown in Figure 18.74.

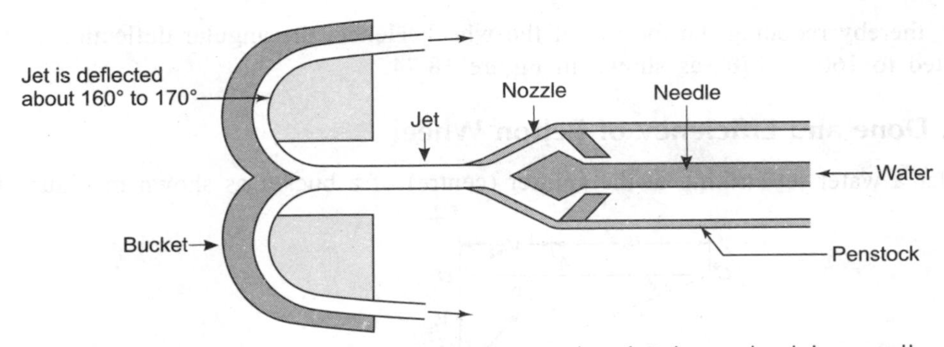

FIGURE 18.74 Control of the jet in nozzle of Pelton wheel by needle.

The turbine operates most efficiently when the wheel rotates at half the velocity of waterjet. When the load on the wheel suddenly reduces, then the jet deflector moves up and partially diverts the jet issuing from the nozzle until the jet needle moves forward and appropriately reduces the jet as shown in Figure 18.75. The arrangement of reducing the size of the jet is necessary because in the event of sudden fall of load requirement, if the jet needle is moved forward in the nozzle to reduce the flow of water, the flow of water may be reduced too quickly which may cause a harmful water hammer phenomenon in the penstock. In practice, the control of jet deflector is linked to the electric generator coupled with the turbine. The complete potential energy available in the water is converted into kinetic energy before the water jet strikes the bucket on the wheel. The pressure on the buckets in the turbine is atmospheric pressure which does not change. Energy transfer occurs purely by impulse action on the bucket with any change of water pressure.

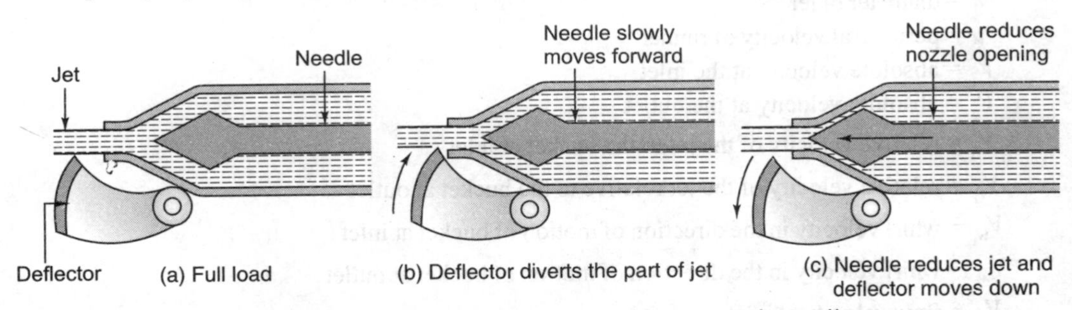

FIGURE 18.75 Working of deflector and needle.

When the water is completely closed with the help of the needle in the nozzle, the runner of turbine tends to keep on rotating due to its inertia for a long time even when jet has stopped. In order to stop the runner in short time, a brake nozzle is provided to stop the rotating wheel by directing a small breaking jet on the buckets in the opposite direction. A casing is provided on the wheel (i) to prevent the splashing of water, (ii) to discharge water to tail race and (iii) to safeguard against accident occurring from the rotating buckets.

The ideal angle of the deflection of the jet on the bucket is 180° for conversion of kinetic energy but this deflection of the jet cannot be achieved in practice. The reason is that 180° deflection will make the jet leaving the bucket strike on the back of succeeding bucket, thereby reducing the motion of the wheel. Hence, the angular deflection of the jet is limited to 160 to 170° as shown in Figure 18.74.

Work Done and Efficiency of Pelton Wheel

Consider a water jet striking at the splitter (centre) of a bucket as shown in Figure 18.76.

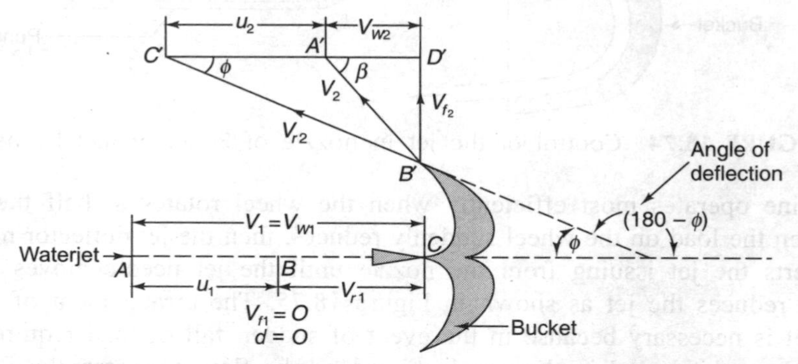

FIGURE 18.76 Velocity triangles at the inlet and outlet.

The velocity triangles of the water jet are drawn at the inlet and outlet of the bucket. The symbols used for determining work done by the jet from velocity triangles are:

N = rpm of the wheel
D = diameter of the Pelton wheel
d = diameter of jet
u = peripheral velocity of runner
V_1 = absolute velocity at the inlet
V_2 = absolute velocity at the outlet
V_{r_1} = relative velocity of the jet to the bucket at inlet
V_{r_2} = relative velocity of the jet relative to the bucket at outlet
V_{w_1} = whirl velocity in the direction of motion of bucket at inlet
V_{w_2} = whirl velocity in the direction of motion of bucket at outlet
V_{f_1} = flow velocity at inlet
V_{f_2} = flow velocity at outlet
α = angle between direction of jet and direction of motion of the bucket at inlet
β = angle made by velocity V_2 with the direction of motion of vane at the outlet
θ = angle made by the relative velocity V_{r_1} in the direction of motion at inlet or vane angle at inlet
ϕ = vane angle at the outlet or angle made by the relative velocity V_{r_2} with the direction of motion at outlet.

The velocity triangle at inlet will be a straight line as shown in the figure. As the jet is entering the bucket in same direction in which bucket is moving, we have

$$V_{f_1} = 0$$

$$V_{w_1} = V_1$$

$$V_{r_1} = V_1 - u_1 = v_1 - u \quad \text{as} \quad u = u_1 = u_2$$

$\therefore$

$$\alpha = 0$$

Vane angle $\theta = 0$

$$V_{r_1} = V_{r_2} \quad \text{(if bucket is smooth)}$$

or

$$V_{r_1} = k \cdot V_{r_2} \quad \text{(where } k < 1 \text{ when bucket is not smooth)}$$

Consider outlet velocity triangle $B'C'D'$, we have

$$C'D' = A'C' + A'D'$$

or

$$B'C' \cos\phi = u_2 + V_{w2}$$

or

$$V_{w2} = V_{r2} \cos\phi - u$$

V_{w2} depends upon angle β. It is negative if $\beta < 90°$ and such angle is used for slow runner. If it is zero ($\beta = 0$), runner will run with medium speed. It is positive when $\beta > 90°$ and such angle is used for fast runner. The force exerted by the jet is determined using impulse momentum equation when applied in the direction of the jet motion

$$F = \rho \cdot a \cdot V_1 (V_{w1} + V_{w2})$$

The work done per second by the jet is

$$W = F \times u$$

$$= \rho \cdot a \cdot V_1 (V_{w1} + V_{w2}) \cdot u$$

The work done per second and per unit weight of water striking the bucket is

$$W = \frac{\text{Work done}}{\text{Weight of water striking}}$$

$$= \frac{\rho \cdot a \cdot V_1 (V_{w1} + V_{w2}) \cdot u}{\rho \cdot a \cdot V_1 \cdot g}$$

$$= \frac{1}{g}(V_{w1} + V_{w2}) \cdot u$$

The hydraulic efficiency (η_h) of turbine is

$$\eta_h = \frac{\text{Work done per second}}{\text{KE of the jet}}$$

$$= \frac{\rho \cdot a \cdot V_1 (V_{w1} + V_{w2}) \cdot u}{\frac{1}{2}(\rho \cdot a \cdot V_1) \cdot V_1^2}$$

$$= \frac{2(V_{w1} + V_{w2}) \cdot u}{V_1^2}$$

But we have

$$V_{w1} = V_1$$

$$V_{w2} = V_{r2} \cos\phi - u$$

$$= k \cdot V_{r1} \cos\phi - u$$

$$= k(V_1 - u)\cos\phi - u$$

$$\eta_h = \frac{2\left[V_1 + k(V_1 - u)\cos\phi - u\right] \cdot u}{V_1^2} \tag{18.25}$$

The efficiency depends upon peripheral velocity u. For maximum efficiency, we have $\frac{d\eta_h}{du} = 0$

$$\therefore \quad \frac{d\eta_h}{du} = \frac{d}{du}\left[\frac{(V_1 + k(V_1 - u)\cos\phi - u) \cdot u}{V_1^2}\right] = 0$$

$$V_1 - 2u = 0$$

or

$$u = \frac{V_1}{2}$$

The above shows that a Pelton wheel will have maximum efficiency when velocity of wheel (u) is half of the velocity of jet (V_1) at the inlet to the wheel. The maximum efficiency is obtained by putting $u = \frac{V_1}{2}$ in Eq. (18.25).

$$(\eta_h)_{\max} = \frac{2\left(V_1 - \frac{V_1}{2}\right)(1 + k\cos\phi) \cdot \frac{V_1}{2}}{V_1^2}$$

$$= \frac{1 + k\cos\phi}{2}$$

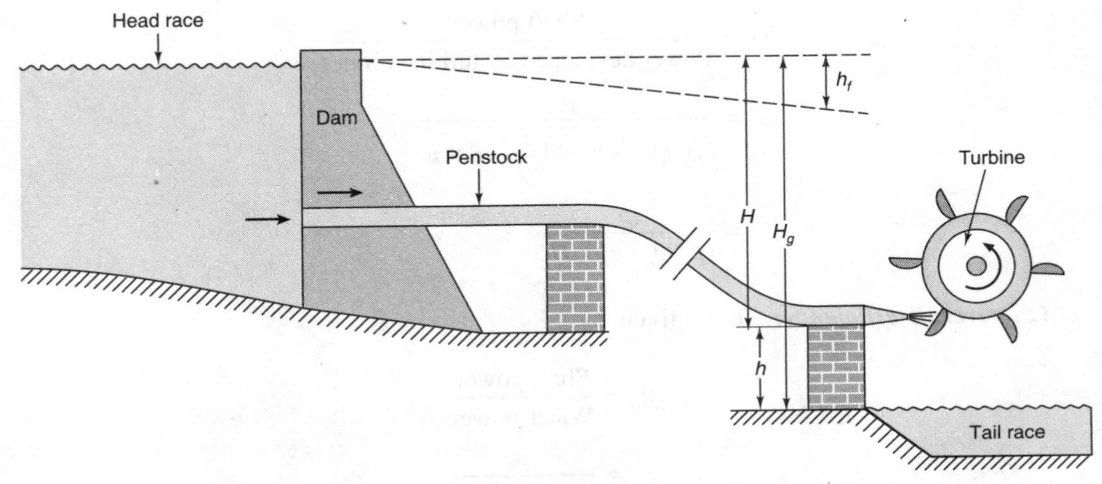

FIGURE 18.77 Layout of impulse turbine and various heads.

Layout of an impulse turbine plant is as shown in Figure 18.77. The various heads are:

(a) ***Gross head:*** The gross or total head is the difference in height of water level between the head race and tail race. It is denoted by H_g.

(b) ***Net head:*** The net head is the head available at the inlet of the turbine. It is denoted by H.

$$H = H_g - h - h_f$$

where h = height of the nozzle above the tail race

h_f = friction head loss in penstock $= \dfrac{u \cdot f \cdot h \cdot V^2}{D \cdot 2g}$ as per Darcy's equation.

The important efficiencies related to the turbine are:

(a) ***Hydraulic efficiency:*** It is the ratio of power developed by the runner to the power available with the water at the inlet of the turbine.

$$\eta_h = \frac{\rho \cdot a \cdot V_1 (V_{w1} \pm V_{w2}) \cdot u}{(\rho \cdot a \cdot V_1) \cdot g \cdot H}$$

$$= \frac{(V_{w1} \pm V_{w2}) \cdot u}{g/H} = \frac{H_r}{H}$$

$$= \frac{\text{Runner head}}{\text{Net head}}$$

(b) ***Mechanical efficiency (η_m):*** Mechanical efficiency is the ratio of the power available at the turbine shaft to the power developed by the turbine runner. It takes into account power loss due to friction in running parts.

$$\eta_m = \frac{\text{Shaft power}}{\text{Power developed by turbine runner}}$$

$$= \frac{p}{(\rho \cdot Q_a \cdot g) \cdot \frac{V_{w1} \pm V_{w2}}{g} \times u}$$

$$= \frac{p}{(\rho \cdot Q_a \cdot g) \times H_r}$$

(c) ***Overall efficiency:*** It is given by

$$\eta_o = \frac{\text{Shaft power}}{\text{Water power}}$$

$$= \frac{p}{(\rho \cdot g \cdot Q) \cdot H}$$

$$= \eta_h \times \eta_m$$

FRANCIS TURBINE

Francis turbine is a reaction turbine in which the runner utilizes both kinetic and potential energies of the water. Both the energies of the water are reduced when it flows through the moving blades of the turbine. The pressure of water varies when it flows over the runner blades and the runner passages are always completely filled up with water.

Layout of Francis Turbine Plant

The general layout of a hydro electric power plant using Francis turbine is as shown in Figure 18.78. The main components of the plant are

(a) ***Penstock:*** It is a large size conduit or pipeline which conveys water from the reservoir to the turbine runner.

(b) ***Turbine:*** It is Francis turbine and it takes water with high pressure from penstock to convert the energy into mechanical work.

(c) ***Draft tube:*** Draft tube is a gradually expanding tube which discharges water from outlet of the turbine to the tail race. The draft tube permits a negative head at the runner of Francis turbine so that power developed by the turbine can be increased. Draft tube converts the kinetic energy of the water emerging from turbine into pressure energy so that the pressure at tail race can be maintained at the atmospheric pressure.

(d) ***Tail race:*** The water emerging out from the draft tube can be removed.

(e) ***Generator:*** The generator is coupled to the turbine shaft so that mechanical energy can be converted into electric energy.

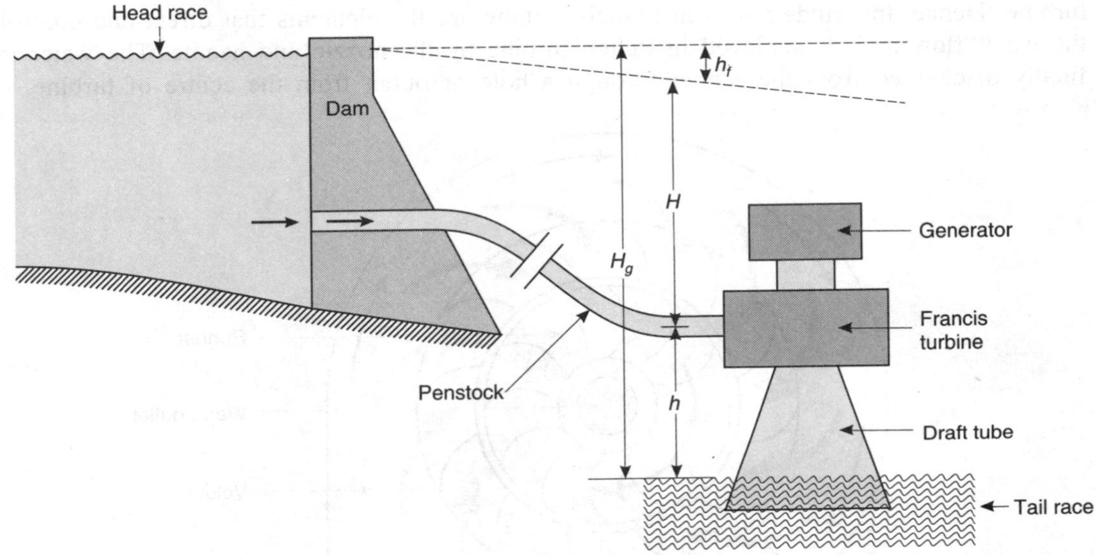

FIGURE 18.78 Layout of Francis turbine and various heads.

Construction and Working of Francis Turbine

The construction of a Francis turbine is as shown in Figure 18.79. The components are:

(a) ***Volute:*** It is a closed passage between the spiral casing and runner whose cross-sectional area gradually decreases so as to increase the velocity of water.

(b) ***Guide vanes:*** The guide vanes guide the water onto the runner at an angle appropriate to the design of the runner blades. The guide vanes can be adjusted from outside.

(c) ***Runner and runner vanes:*** Runner is a rotating disc on which runner vanes are provided. The water moves on the runner vanes and its pressure energy is utilized to impart impulse to the runner using reaction effect.

(d) ***Outlet:*** The water emerges from the outlet which is connected to the draft tube. The water is ultimately discharged to the tail race at atmospheric pressure.

In majority of cases, where we have medium head and discharge of water, Francis turbine is used. Francis is a radial flow turbine. The water entering the volute is mainly diverted inside the turbine by runner vanes. The diversion takes place at right angle to the direction of entry of water. The diversion causes the runner of the turbine to spin around. Water initially enters the volute which is annular channel surrounding the runner and it flows between the guide vanes. The guide vanes are provided to give the flowing water the optimal direction of flow. Water with optimal direction of flow enters the runner and it flows radically along the runner vanes towards the centre of runner. The guide vanes are so arranged on the casing that the pressure energy of the water is largely converted into rotary motion without any part of energy being wasted in eddies and other undesirable flows which may cause energy losses. The guide vanes are adjustable so that the turbine has some degree of adaptability to variations in the flow rate which is necessary to adjust to loading of the turbine. Hence, the guide vanes in Francis turbine are the elements that direct and control the water flow as it is achieved in Pelton turbine by the nozzle and needle. The water is finally discharged from the runner through a hole or outlet from the centre of turbine.

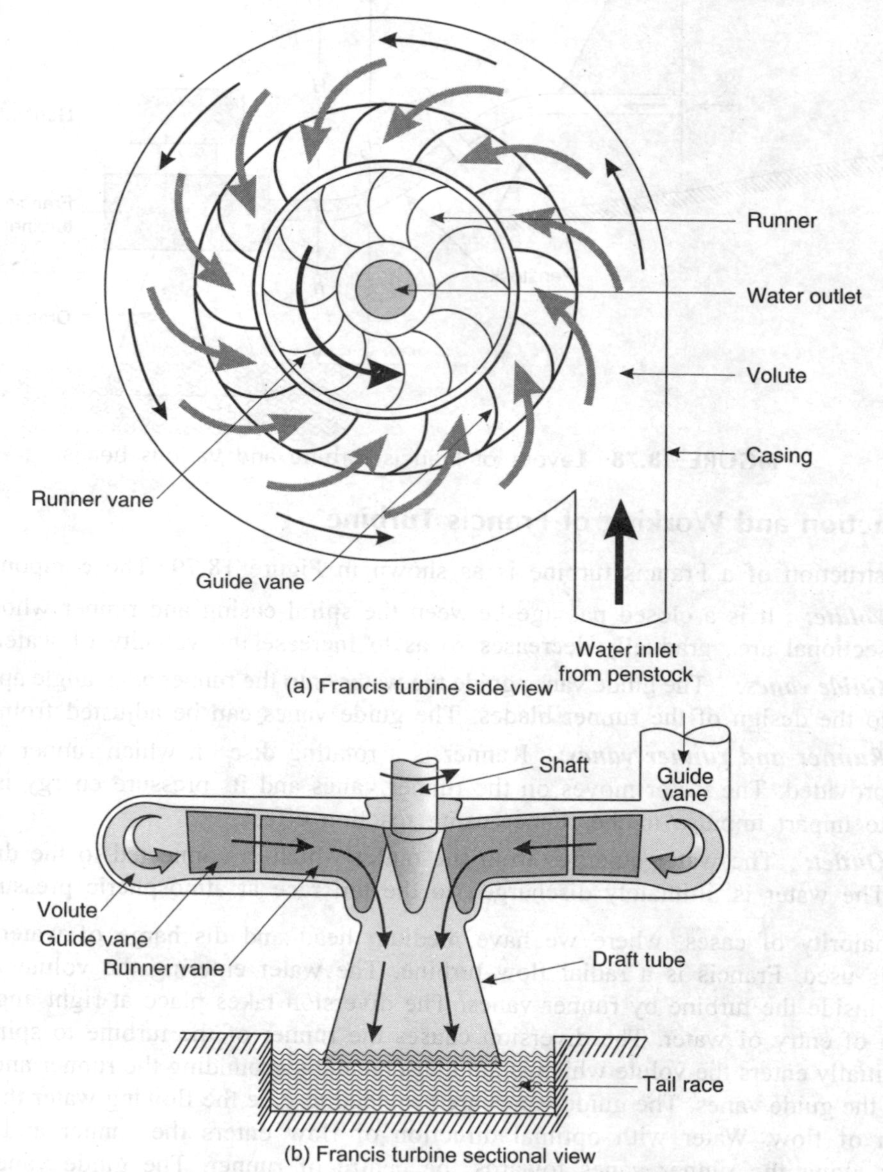

FIGURE 18.79 Construction details of Francis turbine

Francis turbines have purely radial flow runner. The flow can be either inward or outward in the runner as shown in Figure 18.80. In inwardly flow Francis turbine, the water under pressure enters the runner from the guide vanes and flows towards the centre in radial direction before it is finally discharged out of the turbine axially. The water head is partly transformed into kinetic energy and the rest remains as pressure head during the water flow through guide vanes. The difference of pressure on the runner vanes is called the *reaction pressure*. The space or channel between two moving vanes acts as a nozzle to convert pressure energy into kinetic energy during the water flow. The reaction pressure is responsible for the motion of the runner. In Francis turbine, the pressure at inlet is more than at the outlet. The water pressure reduces gradually from inlet to the outlet of the turbine. The runner has to remain always full of water.

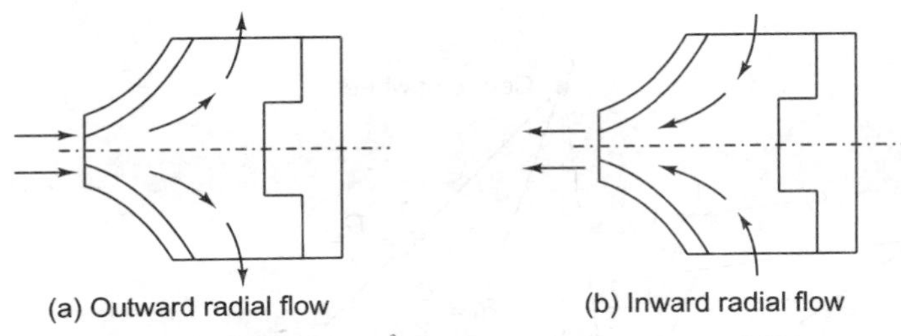

FIGURE 18.80 Types of Francis turbine.

Water after passing through runner vanes is discharged to tail race through a draft tube. The free end of the draft tube is submerged deep in the tail race water so that the entire length of water passage from the head race to tail race is totally enclosed. The draft tube has gradually enlarging section so that water may attain atmospheric pressure by utilizing discharged velocity at the runner outlet which is at negative pressure for obtaining maximum energy from water.

Velocity Diagram and Work Done by Water in Francis Turbine

The Francis turbine has vanes fitted radially to the rim of the wheel. The inlet and outlet velocity diagrams for an inward flow turbine have been drawn as shown in Figure 18.81. The mass of water striking per second on series of vanes is:

$$\dot{m} = \rho \cdot a \cdot V_1$$

Momentum at inlet $= \dot{m} \times V_{w1} = (\rho \cdot a \cdot V_1) \times V_{w1}$

Momentum at outlet $= \dot{m} \times V_{w2} = (\rho \cdot a \cdot V_1) \times V_{w2}$

Angular momentum at inlet $= (\rho \cdot a \cdot V_1) V_{w1} \times R_1$

Angular momentum at outlet $= -(\rho \cdot a \cdot V_1) V_{w2} \times R_2$

Torque exerted by water $= \rho \cdot a \cdot V_1 (V_{w1} \times R_1 + V_{w2} \times R_2)$

$$\begin{aligned}
\text{Work done by water} &= \text{Torque} \times \text{Angular velocity} \\
&= \rho \cdot a \cdot V_1 (V_{w1} \times R_1 + V_{w2} \times R_2) \times w \\
&= \rho \cdot a \cdot V_1 (V_{w1} \times R_1 w + V_{w2} \times R_2 w) \\
&= \rho \cdot a \cdot V_1 (V_{w1} \times u_1 + V_{w2} \times u_2)
\end{aligned}$$

As V_{w2} can be zero for angle $\beta = 0$ and negative when $\beta > 90$, we can write

Work done $= \rho \cdot a \cdot V_1 \left(V_{w1} \times u_1 \pm V_{w2} \times u_2\right)$

$$= \rho \cdot Q\left(V_{w1} \times u_1 \pm V_{w2} \times u_2\right)$$

Work done per unit weight of water $= \dfrac{\rho \cdot Q\left(V_{w1} \times u_1 + V_{w2} \times u_2\right)}{\rho \cdot Q \cdot g}$

$$= \frac{1}{g}\left(V_{w1} \times u_1 + V_{w2} \times u_2\right)$$

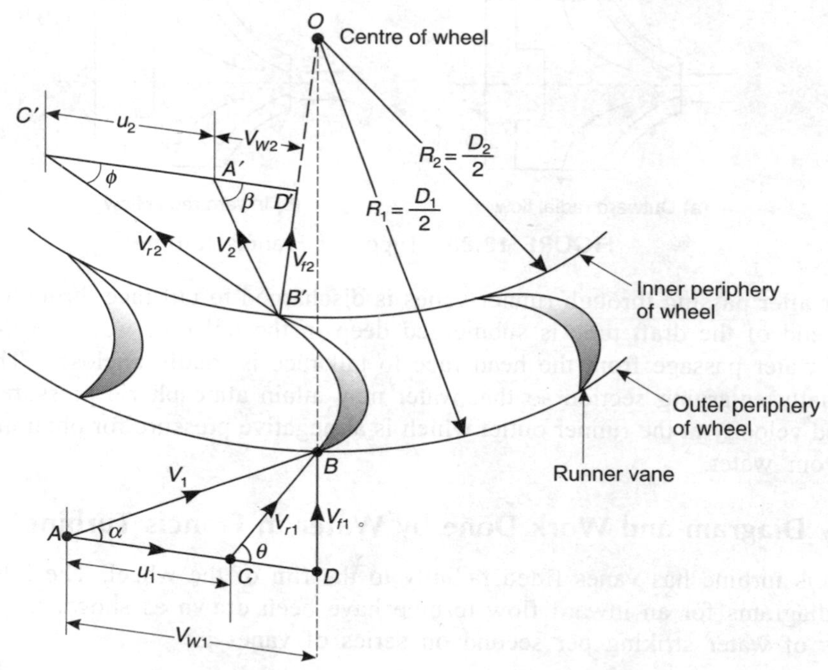

FIGURE 18.81 Velocity diagrams at inlet and outlet of a vane of Francis turbine.

Efficiencies

(a) ***Hydraulic efficiency*** (η_h)

$$\eta_h = \frac{\text{Power developed by the runner}}{\text{Power at the inlet of the runner}}$$

$$= \frac{\rho \cdot Q\left(V_{w1} \times u_1 \pm V_{w2} \times u_2\right)}{\rho \cdot Q \cdot g \cdot H}$$

$$= \frac{V_{w1} \times u_1 \pm V_{w2} \times u_2}{g \cdot H}$$

(b) ***Mechanical efficiency***

$$\eta_m = \frac{\text{Shaft power}}{\text{Power developed by the runner}}$$

$$= \frac{P}{\rho \cdot Q\left(V_{w1} \times u_1 \pm V_{w2} \times u_2\right)}$$

(c) ***Overall efficiency***

$$\eta_o = \eta_h \times \eta_m$$

$$= \frac{\text{Shaft power}}{\text{Water power at the inlet to runner}}$$

Comparison of Francis and Pelton Turbines

The comparison of Francis and Pelton turbines is as given in Table 18.5.

Table 18.5 Comparison of Francis and Pelton turbines

S.No.	*Francis turbine*	*Pelton turbine*
1.	It is a reaction turbine	It is an impulse turbine
2.	Flow inside is radial	Flow is tangential
3.	Lesser head and more flow are suitable for the turbine	More head and lesser flow are suitable for the turbine
4.	Te variation in the operating head can more easily controlled on the turbine	The control of head is difficult in the turbine
5.	The water hammer effect is more frequent	The water hammer effect is less frequent
6.	Cavitation danger is more	No such problem
7.	The overhaul and maintenance work is difficult to perform	Easier to perform overhauling and maintenance work
8.	It requires much cleaner water for operation	Not much cleaner water can be used for operation
9.	The size of turbine is much smaller for a given power and speed	The size is comparatively big for given power and speed
10.	The specific speed is large (51 to 225)	The specific speed is small (85 to 50)
11.	More suitable for variable load	Unsuitable for variable load
12.	Runner, vanes and water passages are completely enclosed in water	The jet strikes one bucket or vane at a time

KAPLAN TURBINE

Kaplan turbine is an axial flow reaction turbine. It is used where comparatively low head and large quantity of water flow is available. The water flow in Kaplan turbine is parallel to the axis of rotation of the shaft. The pressure head of water is converted into kinetic head in the runner of the turbine. The shaft of the Kaplan turbine is always vertical. The lower end of the shaft is made larger and this end is called *boss* or *hub*. The vanes are fixed on to the boss as shown in Figure 18.82. The vanes are fixed on to the boss. Hence, boss is nothing but the runner for Kaplan turbine. The water enters on the runner vanes in axial direction, travels across the vane passage in axial direction and leaves the turbine axially. The pressure at the inlet of the blades is larger than pressure at the exit of the blades. The energy transfer in Kaplan turbine takes place due to reaction effect, i.e. there is change occurring in the magnitude of the relative velocity of water while it flows across the blades.

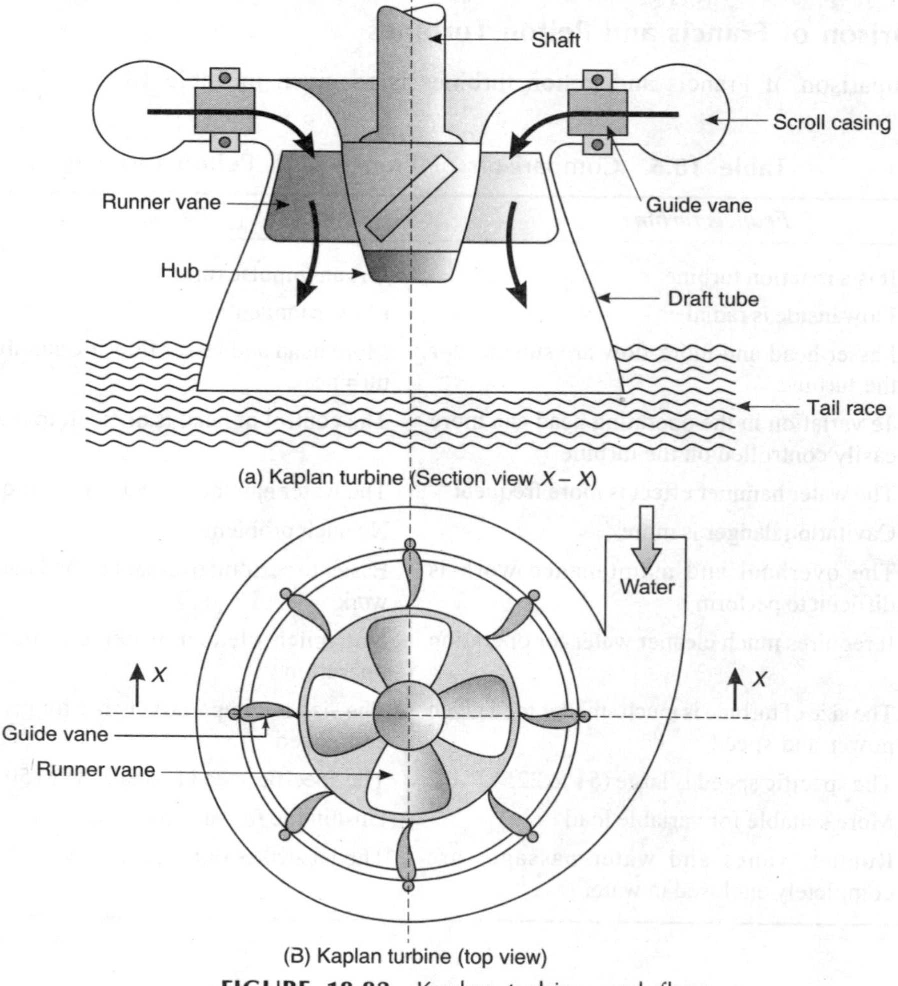

FIGURE 18.82 Kaplan turbine and flow.

In a Kaplan turbine, the number of blades are fewer and the loading on each blade is, therefore, larger. The fewer blades means that there is smaller area of blades which is interacting with water flow for power transfer and this also causes lesser frictional loss. However, the peripheral speed of Kaplan turbine is large as well as its rotor is large. The rotor of Kaplan turbine is enclosed in a cylindrical casing and a small clearance space is maintained between the tip of the blades and the cylindrical casing. The water undergoes a whirl motion while passing through guide vanes ($V_w = \frac{c}{r}$ where c = constant and r = distance of water from the hub). Whirling velocity near the hub is higher and lowest at the end of the blade. This is the reason of twisting the blades so that the flow enters the blades without shock at all loads. The blades can be adjusted for low and high output by a servo mechanism accommodated in the hub. As both guide vanes and runner blades can be adjusted for different angles, the efficiency of Kaplan turbine is maximum and it can operate efficiently at all loads.

Work Done and Efficiencies of Kaplan Turbine

The expression for work done, power and efficiencies in Kaplan turbine are identical as used for Francis turbine

(a) ***Work done***

Work done per second $= \rho \cdot Q\left(V_{w1} \times u_1 - V_{w2} \times u_2\right)$

where Q = area of flow × V_{f1}

$$= \frac{\pi}{4}\left(D_o^2 - D_b^2\right)V_{f1}$$

D_b = diameter of boss

D_o = outside diameter of the runner.

Work done per unit weight of water $= \frac{1}{g}\left(V_{w1} \times u_1 + V_{w2} \times u_2\right)$

(b) ***Hydraulic efficiency***

$$\eta_h = \rho \times Q \frac{\left(V_{w1} \times u_1 \pm V_{w2} \times u_2\right)}{\rho \cdot Q \cdot g \cdot H}$$

$$= \frac{V_{w1} \times u_1 \pm V_{w2} \times u_2}{g \cdot H}$$

(c) ***Overall efficiency***

$$\text{Overall efficiency} = \eta_o = \frac{p}{\rho \cdot g \cdot Q \cdot H}$$

Comparison of Francis and Kaplan Turbines

The comparison of Francis and Kaplan turbines is as per Table 18.6.

Table 18.6 Comparison of Francis and Kaplan turbines

S.No.	*Francis turbine*	*Kaplan turbine*
1.	It is a radially inward flow reaction turbine	It is an axial flow reaction turbine
2.	It has a large number of vanes (about 16 to 24 vanes)	It has a small number of vanes (3 to 8)
3.	It is suitable for medium head ranging from 60 to 250 m.	It is suitable for low head up to 30 m
4.	It is suitable for medium flow rates	It is suitable for large flow rates
5.	The runner vanes cannot be adjusted	The runner vanes can be adjusted
6.	The shaft of turbine can be positioned horizontally or vertically	The shaft can be positioned only in vertical direction
7.	The specific speed varies from 50 to 250	The specific speed varies from 250 to 850
8.	Not as compact as Kaplan turbine for a given power	Most compact as compared to other types of turbine

GOVERNING OF TURBINES

A turbine during operation is always subjected to variation in load which leads to variation to its speed. When load increases, the speed of the turbine decreases and vice-versa. The speed increases or decreases may lead to run away or stalling of turbine. The speed variation also creates surges and water hammer phenomenon in the large column of water in penstock which may cause damage to penstock or turbine. It is, therefore, essential to maintain the turbine speed as close to the mean speed at different loads. The speed of turbine is maintained using governing arrangement.

The governing of a turbine is defined as a way of controlling the speed of the turbine by which the turbine speed is kept almost constant under varying load condition. It is done automatically by means of a governor which regulates the rate of water flow through the turbine so that the turbine can maintain its speed irrespective of load conditions on the turbine. When the load on the turbine decreases, it tends to run fast and it becomes essential to reduce the water flow to maintain constant the turbine speed. Similarly, when the load on turbine increases, the turbine tends to run slow and it becomes necessary to increase the water flow to maintain constant the turbine speed. The process by which the speed of turbine is kept constant under varying load conditions is called *governing*.

Governing Mechanism of Impulse (Pelton) Turbine

The governing of Pelton turbine is done by means of oil pressure governor as shown in Figure 18.82. The governor consists of

(a) Oil sump
(b) Oil pump
(c) Servomotor or relay cylinder
(d) Relay valve or control valve
(e) Centrifugal governor
(f) Spear rod or needle
(g) Pipe system connecting oil sump to control valve and then control valve to servomotor.

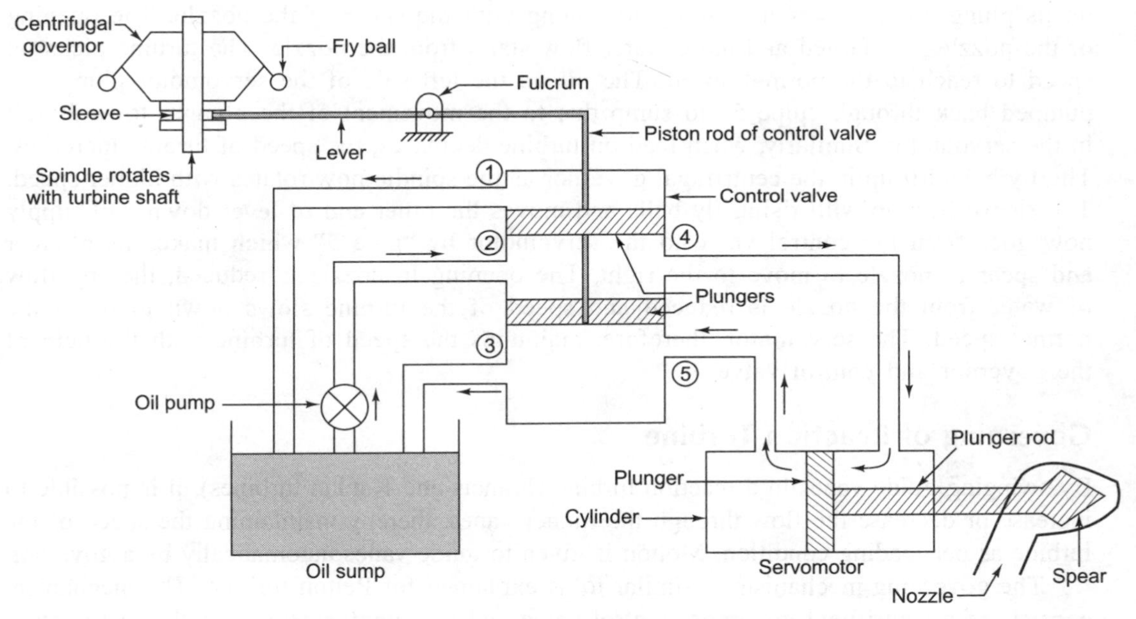

FIGURE 18.83 Governing of Pelton turbine.

A centrifugal governor comprises a spindle which is rotated by the main turbine shaft. The governor has four arms and upper two arms are pivoted with the spindle. The lower ends of the upper arms are provided with fly balls. The upper ends of the lower arms are also connected to the fly balls while their lower ends are connected to a sleeve. The sleeve is free to move up and down on the spindle. The travel of the sleeve acts through a lever and fulcrum arrangement to control the movement of piston rod of control valve. The control valve may be considered as the heart of the government mechanism. It consists of a cylinder and two plungers fixed on the piston rod which is further connected to the lever. The control valve has five pipes connected to it which are shown as numbered 1 to 5 in the figure. Pipes 1 and 3 are meant for bringing back surplus oil to oil sump. Pipe 2 brings oil under pressure to the control valve from the oil sump. Pipes 4 and 5 are meant to connect control valve to the servomotor. The servomotor comprises a cylinder having a plunger. Plunger is connected to spear in the Pelton turbine nozzle with the help of a piston rod. The spear can control the water flow by varying the opening and closing of the nozzle.

When the load on the turbine increases, then the speed of the turbine decreases. As the spindle of centrifugal governor is connected to main shaft of the turbine, the speed of the spindle also decreases, thereby centrifugal forces acting outward on fly balls also decreases. The fly balls move down and the lower arms also move down with the sleeve which is attached to fly balls with lower arms. The lever attached to sleeve also moves down, thereby the portion of lever right of the fulcrum lifts up. The plungers of the control valve move up with the lever, thereby opening "pipe 4" to servomotor. The oil from control valve under pressure rushes to the servomotor. The high pressure oil in servomotor exerts pressure on its plunger and makes it to move left along with the spear of the nozzle. The opening of the nozzle is enlarged and more water flow starts from the nozzle. The turbine picks up speed to reach to the normal speed. The oil on the left side of the servomotor plunger is pumped back through "pipe 5" to sump due to the movement of the plunger towards left in the servomotor. Similarly, when load on turbine decreases, the speed of turbine increases. The fly balls lift up in the centrifugal governor as the spindle now rotates with higher speed. The sleeve lifts up with rising fly balls and moves the other end of lever down. Oil supply now goes from the control valve to the servomotor by "pipe 5" which makes its plunger and spear in nozzle to move to the right. The opening in nozzle is reduced, thereby flow of water from the nozzle is reduced and speed of the turbine slows down to reach the normal speed. The servomotor, therefore, maintains the speed of turbine with the help of the governor and control valve.

Governing of Reaction Turbine

By swinging guide vanes in a reaction turbine (Francis and Kaplan turbines), it is possible to increase or decrease the flow through the runner vanes, thereby maintaining the speed of the turbine as per loading condition. Motion is given to guide vanes automatically by a governor.

The governing mechanism is similar to as explained for Pelton turbine. The mechanism consists of a centrifugal governor, control valve and servomotor. However, the piston shaft of servomotor is connected to a regulating wheel provided in the turbine by means of two rods as shown in Figure 18.84. The regulating wheel can swing the guide vanes about their

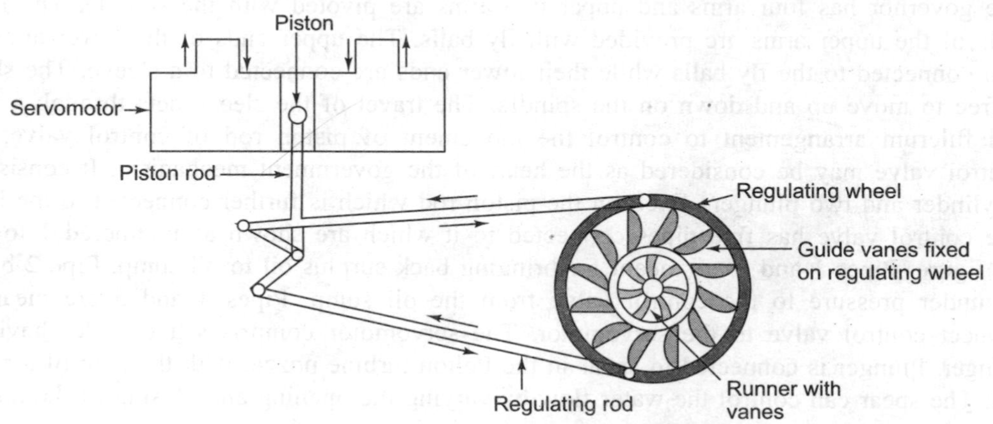

FIGURE 18.84 Guide vanes are adjusted with regulating wheel controlled by servomotor.

fixed axes in the spiral casing. By swinging the regulating wheel under the control of the servomotor, it is possible to reduce or increase the area of passage between two consecutive guides vanes. The variation of the flow area can control the quantity of water which can be supplied to running vanes, thereby ensuring the speed of the runner remains normal.

CENTRIFUGAL PUMPS

A pump is used to impart energy to a fluid system. It is run by using an external source such as electric motor so that it can impart pressure energy or kinetic energy or both energies to a fluid. In pump, the fluid flow takes place from the low pressure side to the high pressure side of the pump.

A centrifugal pump is similar to a Francis turbine but operating is 'in reverse' manner. In a turbine, water drives a runner of the turbine while in a centrifugal pump, a similar runner fitted with vanes known as *impeller* imparts motion to fluid. In centrifugal pump, the pressure for achieving the required "delivery head" is produced by centrifugal acceleration of the fluid in the rotating impeller. On leaving the impeller, the fluid passes through a ring of fixed guide vanes which surround the impeller and known as a *diffuser*. In diffuser, there is a gradually widening passage, the velocity of fluid is reduced and its kinetic energy is converted into pressure energy.

The main parts of a centrifugal pump are:

(a) ***Impeller:*** The rotating part of a centrifugal pump is called *impeller*. It is the main part to raise liquid from a lower to higher level by creating the required pressure with the help of centrifugal action. Whirling motion is imparted to the liquid by means of curved vanes provided on the impeller. The impeller is mounted on a shaft which is coupled to the shaft of an electric motor.

(b) ***Casing:*** The casing is nothing but an airtight chamber surrounding the impeller of the pump with guide vanes. It contains (i) impeller (ii) support for impeller bearings and (iii) suction and discharge arrangements as shown in Figure 18.85. The main purposes of casing are (i) to guide water to and from impeller, and (ii) to partially convert the kinetic energy to pressure energy. The casing has guide vanes mounted on a ring which is called *diffuser*. The diffuser with guide vanes is most efficient arrangement in a casing for the conversion of kinetic energy of water into pressure energy without any eddy losses.

(c) ***Suction pipe:*** The pipe connecting the eye of the impeller to water in sump is called *suction pipe*. The pipe has to be straight so that the pump can continuously lift the water with suction created by the impeller at pump inlet. At lower end, suction pipe is provided with (i) stainer to prevent the entry of any debris along with water into suction pipe and (ii) non-return or foot valve to prevent draining out of water from the suction pipe when the impeller is non-functioning.

(d) ***Delivery pipe:*** The pipe connecting the outlet of the pump to any height where water is discharged or delivered for storage purpose is called *delivery pipe*. The storage reservoir or tank is called *overhead tank*.

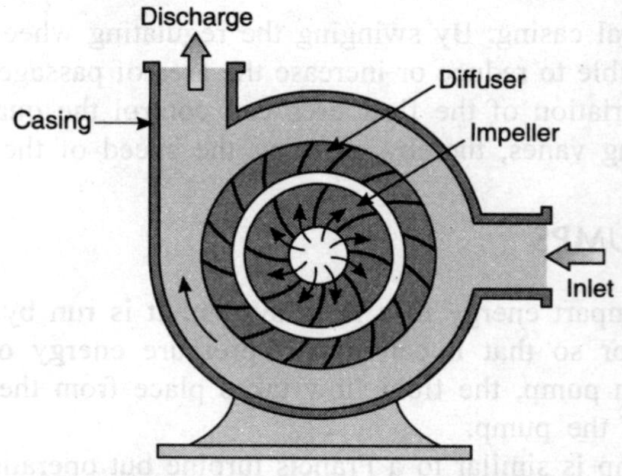

FIGURE 18.85 Centrifugal pump.

Work Done by Centrifugal Pump

The principle of working of a centrifugal pump is totally opposite of a radially flowing reaction turbine

$$\text{Work done per unit weight of water} = -\frac{1}{g}\left(V_{w1} \times u_1 - V_{w2} \times u_2\right)$$

Heads and Efficiencies

A centrifugal pump with various heads is as shown in Figure 18.86.

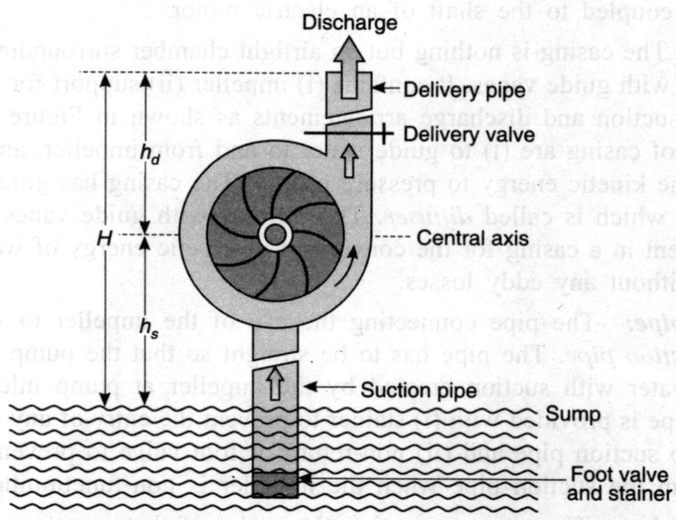

FIGURE 18.86 Centrifugal pump and various heads.

(a) ***Suction head (h_s):*** It is the vertical height between the central axis of the centrifugal pump and water surface in the sump.

(b) ***Delivery head (h_d):*** The vertical distance between the central axis of pump and the water surface in the overhead tank is called *delivery head.*

(c) ***Static head (H_s):*** Sum of the suction and delivery head is called *static head.*

$$H_s = h_s + h_d$$

(d) ***Manometric head (H_m):*** It is the head against which a centrifugal pump has to work to deliver water.

$$H_m = \text{(Theoretical head imparted by impeller)} - \text{(Head loss in the pump)}$$

$$= \frac{V_{w2} \times u_2}{g} - h_l$$

(e) ***Manometric efficiency:***

$$\eta_{\text{man}} = \frac{\text{Manometric head}}{\text{Head imparted by impeller}}$$

$$= \frac{H_m}{V_{w2} \times u_2 / g} = \frac{H_m \times g}{V_{w2} \times u_2}$$

(f) ***Mechanical efficiency:***

$$\eta_m = \frac{\text{Power at the impeller}}{\text{Power at shaft}}$$

$$= \frac{\dot{m} \cdot V_{w2} \times u_2}{1000 \times \text{Shaft power}}$$

(g) ***Volumetric efficiency:***

$$\eta_{\text{vol}} = \frac{Q}{Q + q}$$

where Q = discharge
q = leakage

(h) ***Overall efficiency:***

$$\eta_o = \eta_{\text{man}} \times \eta_m \times \eta_{\text{vol}}$$

RECIPROCATING PUMP

A reciprocating pump is called *positive displacement pump* as it carries out positive displacement of liquid while pushing the liquid from suction to discharge side. Reciprocating pump has low discharge but high head. Reciprocating pump can be (i) single acting when water remains in contact with one side of the piston inside the cylinder and (ii) double acting when water remains in contact with both sides of the piston inside the cylinder.

Construction and Working of Single Acting Reciprocating Pump

The main parts of a single acting reciprocating pump are (i) cylinder (ii) piston (iii) suction valve (iv) delivery valve (v) suction pipe (vi) delivery pipe and (vii) crank and connecting rod mechanism operated by a power source as shown in Figure 18.87. The pump is connected to the sump well with suction pipe and to the delivery point with a delivery pipe. The piston moves to and fro in the cylinder when crack rotates.

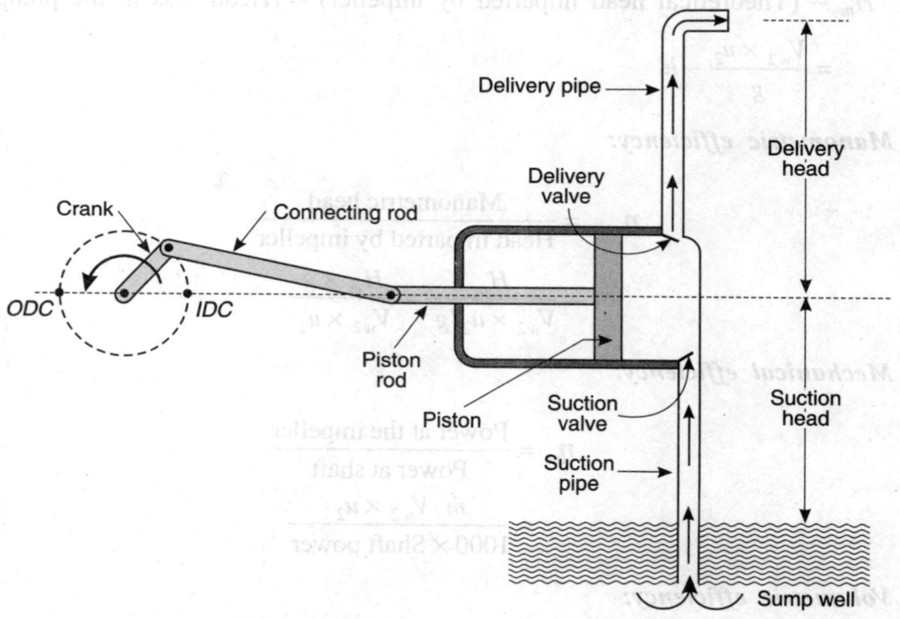

FIGURE 18.87 Single acting reciprocating pump.

The working of the pump is explained as under:

(a) The crank moves from the inner dead centre (IDC) to the outer dead centre (ODC). The piston moves away from the cylinder head in the suction stroke and vacuum is created at the top of the piston. The vacuum causes the suction valve to open and the water is lifted through the suction pipe inside the cylinder due to atmospheric pressure acting on the water in the sump well. The cylinder is completely filled up when the crank reaches ODC.

(b) The crank moves from ODC to IDC. The piston moves towards the cylinder head and compresses the water filled in the cylinder as suction valve is now closed. The delivery valve opens and water is forced into the delivery pipe the water moves in the delivery pipe to the delivery point.

The variation of discharge (Q) through delivery pipe with crank angle (θ) for single acting pump is as shown in Figure 18.88. The discharge is given by the pump in a delivery stroke only. The power input for a given discharge is:

$$\text{Power} = \frac{\rho \cdot g \cdot Q\,(h_s + h_d)}{1000 \times 60}\ \text{kW}$$

where $Q = A.L.N.$
A = Cross-section of cylinder
L = Length of stroke
N = rpm

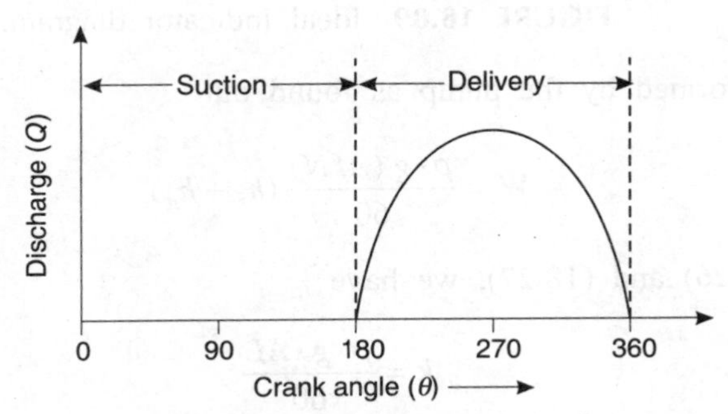

FIGURE 18.88 Variation of discharge.

Indicator Diagram (Ideal)

The indicator diagram for a reciprocating pump is a diagram which is obtained when pressure head of the liquid inside the cylinder is drawn on y-axis and stroke length in crank angle on x-axis. It shows the pressure head of the liquid in the cylinder corresponding to any position during the suction and delivery stroke. It is obtained when the pressure head and stroke length of the piston are drawn for one complete revolution of the crank of two strokes of the piston. If friction and acceleration effects are neglected, we get an ideal indicator diagram as shown in Figure 18.89. The horizontal line '*EF*' represents the atmospheric pressure (H_{atm}). During the suction stroke, the pressure head in the cylinder is constant and is equal to suction head (h_s) below the atmospheric head. It is shown by the horizontal line '*AB*' which is below the line '*EF*' by height given by 'h_s'. During the delivery stroke, the pressure head in the cylinder is constant and is equal to the delivery head (h_d) which is above the atmospheric line by a height of 'h_d'. The horizontal line '*CD*' represents the pressure head during the delivery stroke. The area enclosed by the rectangle *ABCD* on the ideal indicator diagram indicates the work,

$$W = k \times L \times (h_s + h_d) \tag{18.26}$$

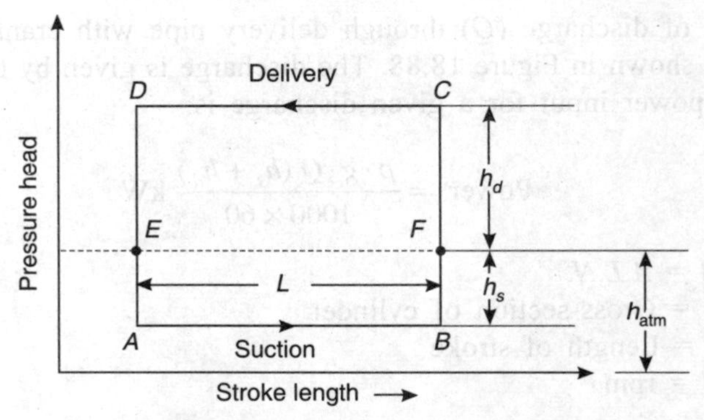

FIGURE 18.89 Ideal indicator diagram.

Now work performed by the pump as found out

$$W = \frac{\rho \cdot g\,(ALN)}{60}(h_s + h_d) \tag{18.27}$$

From Eqs. (18.26) and (18.27), we have

$$k = \frac{\rho \cdot g \cdot AL}{60}$$

Comparison of Centrifugal and Reciprocating Pumps

The comparison of centrifugal and reciprocating pumps is as per Table 18.7.

Table 18.7 Comparison of centrifugal and reciprocating pumps

S.No.	*Centrifugal pump*	*Reciprocating pump*
1.	Smooth and uniform flow	Pulsating flow
2.	Compact	Large size
3.	High speed	Low speed
4.	Simple construction	Complicated construction
5.	No balancing problem	It has balancing problem
6.	More discharge at lower head	Small discharge at higher head
7.	It can handle viscous fluids	It has difficulty in handling viscous fluids
8.	Easy maintenance	Maintenance is difficult
9.	Low efficiency	High efficiency
10.	Initial cost is low	Initial cost is high

OBJECTIVE TYPE QUESTIONS

Share your knowledge, it is a way to achieve immortality.

State True or False

1. Water has viscosity of 1.0 centipoise. *(True/False)*
2. Fluid flow has maximum velocity at boundary surface. *(True/False)*
3. Air can be treated as incompressible fluid when flow has velocity less than 0.3 Mach-number. *(True/False)*
4. Stagnation pressure and temperature are significant when flow is higher than 0.3 Mach-number. *(True/False)*
5. The velocity of flow is higher when it is falling freely as compared to what is determined using Torricelli's theorem. *(True/False)*
6. Energy grade line slopes down in non-viscous flow when bed is sloping. *(True/False)*
7. Hydraulic grade line is at higher level than energy grade line. *(True/False)*
8. The velocity of flow in a pipe is maximum at the centre of pipe. *(True/False)*
9. The pressure head loss in a pipe is determined by Darcy's equation. *(True/False)*
10. An open channel flow has a free surface which is the interface between moving water and an overlying air. *(True/False)*
11. Froude number is a dimensionless number which is the ratio of inertial and gravitational forces. *(True/False)*
12. When Froude number is greater than 1.0, the flow will have more velocity and less depth in open channel. *(True/False)*
13. The flow is critical in open channel when Froude number is 1.0. *(True/False)*
14. Hydro is a renewable source of energy. *(True/False)*
15. Reaction turbines have nozzles to convert fluid potential energy into kinetic energy. *(True/False)*
16. Impulse turbine has low head and high discharge. *(True/False)*
17. The governor reduces the water flow when the lead on the turbine reduces. *(True/False)*
18. In centrifugal pump, the angular momentum of water is increased by the rotating impeller. *(True/False)*
19. The discharge of a reciprocating pump is pulsating. *(True/False)*
20. Kaplan turbine has axial flow while Francis turbine has radial flow. *(True/False)*

21. Kaplan turbine has fewer running vanes. (*True/False*)

22. Pelton turbine has lowest specific speed. (*True/False*)

23. The rotor of a reaction turbine is always submerged in water. (*True/False*)

24. Draft tube is used in impulse turbine. (*True/False*)

Multiple Choice Questions

1. The flow of water in Pelton turbine is ________.

(a) axial (b) radial (c) mixed (d) tangential

2. The range of specific speed of a Pelton turbine is ________.

(a) 12 to 70 (b) 100 to 300 (c) 80 to 220 (d) none of these

3. The number of movable blades is least in ________.

(a) Francis (b) Pelton (c) Kaplan (d) all have equal blades

4. The flow of water in Kaplan turbine is ________.

(a) axial (b) radial (c) mixed (d) tangential

5. The indicator diagram of a reciprocating pump shows for one complete revolution of a crank, the variation in the cylinder of ________.

(a) kinetic head (b) pressure head

(c) kinetic and pressure head (d) none of these

6. Water having a very high head is used for ________.

(a) Kaplan (b) Francis (c) Pelton (d) none of these

7. The discharge (Q) through a single acting reciprocating pump is given by ________.

(a) $\frac{2ALN}{60}$ (b) $\frac{ALN}{60}$ (c) $\frac{AN}{60 \times L}$ (d) none of these

8. The moving runner blades can be adjusted in ________.

(a) Pelton (b) Kaplan (c) Francis (d) none of these

9. In a semicircular shape bucket of Pelton turbine, the water should be ideally deflected by ________.

(a) 90° (b) 120° (c) 160° (d) 180°

10. Euler's equation is ________.

(a) $\frac{dP}{\rho} + \frac{dz}{g} + V \cdot dV = 0$ (b) $\rho \cdot dP + g \cdot dz + V \cdot dV = 0$

(c) $\frac{dP}{\rho} + g \cdot dz + V \cdot dV = 0$ (d) none of these

11. The Chezy's equation for open channel flow is ____________.

(a) $V = \sqrt{R \cdot S_o}$ (b) $V = R\sqrt{C \cdot S_o}$ (c) $V = c \cdot \sqrt{R \cdot S_o}$ (d) none of these

12. Which statement is true for flow in channel?

(a) $F_r < 1$ for subcritical (b) $F_r = 1$ for critical

(c) $F_r > 1$ for subcritical (d) All of these

13. The specific energy of flow in open channel is ____________.

(a) $E = y + \frac{V^2}{2g}$ (b) $E = y + \frac{V^2}{2g} + \frac{P}{\gamma}$ (c) $E = y \cdot g + V^2$ (d) none of these

14. The stagnation temperature is ____________.

(a) $T_s = T_1\left[\frac{M_a^2(\gamma - 1)}{2} + 1\right]$ (b) $T_s = \frac{T_1}{2}\left[M_a^2(\gamma + 1) + 1\right]$

(c) $T_s = T_1\left[M_a^2(\gamma - 1) + 2\right]$ (d) none of these

15. The stagnation pressure is ____________.

(a) $P_s = P_1\left[1 + \frac{\gamma M_a^2}{2}\right]$ (b) $P_s = P_1\left[1 + \frac{\gamma M_a^2}{2}\right]$

(c) $P_s = P_1\left[1 + \gamma M_a^2\right]$ (d) none of these

16. The Bernoulli's equation for gas is ____________.

(a) $C_P \cdot T + \frac{u^2}{2} + gz = C$ (b) $C_V \cdot T + \frac{u^2}{2g} + z = C$

(c) $\frac{C_P \cdot T}{g} + \frac{u^2}{2} + z = C$ (d) none of these

17. The sonic velocity (a) of a fluid is ____________.

(a) $a = \gamma \cdot R \cdot T$ (b) $a = \frac{\gamma \cdot T}{R}$ (c) $a^2 = \gamma \cdot R \cdot T$ (d) none of these

18. The manometric efficiency of a centrifugal pump is ____________.

(a) $\frac{H_m}{g(V_{w2} \times u_2)}$ (b) $\frac{g \cdot H_m}{V_{w2} \times u_2}$ (c) $\frac{g(V_{w2} \times u_2)}{H_m}$ (d) none of these

19. A Kaplan turbine has __________.
    (a) Fewer blades (b) axial flow (c) maximum efficiency (d) all of these
20. A Pelton turbine has __________.
    (a) lowest specific speed (b) high head
    (c) low discharge (d) all of these

Fill in the Blanks

1. The pressure energy in a Pelton turbine is converted into __________ using a __________.
2. Pelton turbine is used where __________ is high and __________ is low.
3. The running wheel is stopped in a Pelton turbine using a __________ nozzle.
4. The kinetic energy and __________ energy change when water flows through the movable vanes of the reaction turbine.
5. All passages in a reaction turbine remain fully __________ with water in reaction turbine.
6. The draft tube in a reaction turbine is to convert outlet velocity of water into __________ energy.
7. Governing is a way of __________ the speed of the turbine.
8. The vertical height between the axis of pump to free surface of water in the sump is called __________ head.
9. The vertical height between the axis of pump to the point of discharge of water is called __________ head.
10. The sum of delivery head and suction head is called __________ head.
11. The graph drawn between pressure head in the cylinder on y-axis and stroke length of the piston on x-axis for one complete revolution of crank under ideal condition is called __________ diagram.
12. When Froude number is less than 1.0, the channel flow is __________.
13. When Froude number is more than 1.0, the channel flow is __________.
14. When Reynolds number > 2000, the flow in pipe is __________.
15. The head losses in pipe is proportion to square of __________ and proportion to __________ of the pipe.
16. The Mach-number is the ratio of __________ of the body to __________ velocity of fluid.

ANSWERS

When you know better, you do better.

State True or False

1. True	**2.** False	**3.** True	**4.** True
5. False	**6.** False	**7.** False	**8.** True
9. True	**10.** True	**11.** True	**12.** False
13. True	**14.** True	**15.** False	**16.** False
17. True	**18.** True	**19.** True	**20.** True
21. True	**22.** True	**23.** True	**24.** False

Multiple Choice Questions

1. (d)	**2.** (a)	**3.** (c)	**4.** (a)
5. (b)	**6.** (c)	**7.** (b)	**8.** (b)
9. (d)	**10.** (c)	**11.** (b)	**12.** (d)
13. (a)	**14.** (a)	**15.** (a)	**16.** (a)
17. (c)	**18.** (b)	**19.** (d)	**20.** (d)

Fill in the Blanks

1. kinetic energy, nozzle	**2.** head, discharge
3. brake	**4.** pressure
5. submerged	**6.** pressure
7. controlling	**8.** suction
9. delivery	**10.** static
11. indicator	**12.** supercritical
13. subcritical	**14.** turbulent
15. velocity, length	**16.** velocity, sonic

Bibliography

Arora, C.P., *Thermodynamics*, Tata McGraw-Hill, New Delhi, 2000.

Bansal, R.K., *Engineering Mechanics*, Laxmi Publications(P) Ltd., New Delhi, 2004.

Bryson, Bill, *A Short History of Nearly Everything*, Black Swan, Great Britain, 2004.

David, Maria, *Mottos for Success,* Aurora Production AG, Switzerland, 2000.

Hawking, Stephen, *A Brief History of Time*, Bantams Books, UK, 1987.

Khurmi, R.S., *Strength of Materials,* S. Chand & Company, New Delhi, 2003.

Khurmi, R.S., *A Textbook of Engineering Mechanics*, S. Chand and Company, New Delhi, 2007.

Kumar, D.S.S., *Mechanical Engineering*, S.K. Kataria & Sons, Delhi, 2005.

Kumar, K.L., *Engineering Mechanics*, Tata McGraw-Hill, New Delhi, 2007.

Mariam, J.L. and L.G. Kraige, *Engineering Mechanics Dynamics*, John Wiley & Sons, New York, 1998.

Nag, P.K., *Engineering Thermodynamics*, Tata McGraw-Hill, New Delhi, 2002.

Rajput, R.P., *Comprehensive Mechanical Engineering*, Laxmi Publication, New Delhi, 2005.

Rao, Y.V.C., *Theory and Problems of Thermodynamics*, New Age International, New Delhi, 2000.

Shames, Irving H., *Engineering Mechanics*, PHI, New Delhi, 2008.

Singh, Onkar, S.S. Bhavikatti and Suresh Chandra, *Introduction to Mechanical Engineering*, New Age International, New Delhi, 2005.

Singhal, R.K., *Mechanical Engineering*, S.K. Kataria and Sons, Delhi, 2004.

Sonntage, R., C. Borgnakke and G., J. Van Wylen, *Fundamental of Thermodynamics*, John Wiley & Sons, Singapore, 2004.

Timoshenko, S. and D.H. Young, *Engineering Mechanics*, McGraw-Hill Book Company, Singapore, 1956.

Timoshenko, S., *Strength of Materials*, CBS Publisher & Distributors, New Delhi, 2002.

Verma, H.C., *Concepts of Physics*, Bharti Bhawan, Patna, 2000.

Yadav, R., *Fundamentals of Engineering Thermodynamics*, Central Publishing House, Allahabad, 2005.

Index

2-stroke diesel engine, 210
2-stroke engine, 210
2-stroke petrol engine, 208
2-stroke spark ignition engine, 210
4-stroke engine, 210

ACME threads, 351
Adiabatic mixing, 54
Air-standard cycle, 198
Anergy, 104
Angle of repose, 344
Angular impulse-momentum, 616
Angular impulse-momentum equation, 584
Angular momentum, 588

Beam, 380
  types, 380
Beam engine mechanism, 260
Beams of uniform strength, 737
Belt-pulley arrangement
  corss belt drive, 348
  power, 349
  rope drive, 348
  slack side, 346
  straight belt drive, 348
  tight side, 346
  V-belt drive, 348
Bending, 784
  equation, 734
  hogging, 734
  sagging, 734
  strain energy, 739
  theory, 734
Bending and torsion, 784
Bending equation, 734
Bending moment, 382
  contraflexure points, 384
  hogging, 382
  inflection, 384
  sagging, 382
  slope, 386
Bending moment diagram, 383, 384, 733
  rules, 383
Bending stress, 733
Bernoulli's principle, 829
Black hole, 304
Boiler, 52
Boiling point, 149
Bottom dead centre, 199
Bow's notation, 418
Boyles' law, 6
Brake power, 211
Brake thermal efficiency, 212
Brinell hardness, 688
Bulk modulus, 678
Bulk modulus of elasticity, 679

Carnot cycle, 94
Carnot refrigeration cycle, 96
Carnot theorem, 96
Carnot vapour cycle, 179

Centre of gravity, 451
Centre of mass, 451, 462, 476
  uniform semicircle wire, 478
  uniform semicircular plate, 479
  uniform straight rod, 477
Centrifugal pumps, 917
Centroid, 452
  axis of symmetry, 455
  composite areas, 456
  composite volumes, 461
  geometrical shapes, 493
  orthogonal axes of symmetry, 456
  plane area, 451
Centroidal axes, 453
Charles' law, 7
Charpy test, 687
Chasles theorem, 539
Chezy equation, 866
Clausius inequality, 98
Clausius statement, 93
Clearance length, 199
Clearance volume, 199
Coefficient of performance, 91
Columns, 664
Combustion chamber, 53
Complementary shear stresses, 675
Composite bar, 673
Compound stresses, 683
Compound stresses (2-D system), 683
Compressible flow, 888
Compressibility factor, 8
Compression ignition, 207, 210
Compression ratio, 199
Compressor, 50
Concurrent force system
  block, 345
  ladder, 345
  wedge, 345
Concurrent forces, 307
Condenser, 52
Cone of friction, 344
Conjugate depths, 873
Constant volume cycle, 199
Constant volume thermometer, 11
Coplanar concurrent force system
  resultant of, 308
Coplanar force system, 306
Coulomb's law of friction, 343
Couple, 307
Critical depth, 874
Critical flow condition, 874
Critical point, 8, 149
Cross belt drive, 348
Cycle, 5

D'Alembert's principle, 620
  rotary motion, 623
Deformation
  under external load, 669, 670
  under own weight, 669
Degrees of freedom, 255
Diesel cycle, 202, 203
Differential pulley block, 288
Diffuser, 53
Discharge coefficient, 836
Double slider-crank chain, 263
Dryness factor, 150
Dryness fraction, 149
Dual cycle, 204
Dynamic friction, 343

Eccentric loading, 738
Efficiency
  variation of, 273
Elastic constants, 678
Energy, 104
  stored, 10
  transit, 10
Enthalpy, 48
Entropy, 100
  generation, 102
Equations of angular motion, 587
Equilibrium
  block, 345
  ladder, 345
  wedge, 345
Equivalent mechanisms, 258
Equivalent torque, 785
Extensive properties, 6
External combustion, 206

Euler's equation, 827

Factor of safety, 676
First law of motion, 303
First law of thermodynamics, 44
  application, 45
  limitations, 55
Flexural rigidity, 737
Flitched beams, 736
Flow nozzle, 843
Flow process, 48
Fluid dynamics, 820
Forces
  applied, 310
  non-applied, 310
Force system
  collinear, 306
  coplanar, 306
Four strokes of a CI engine, 207
  combustion and power stroke, 207
  compression stroke, 207
  exhaust stroke, 207
  suction stroke, 207
Four strokes of SI engines
  combustion and power stroke, 207
  compression stroke, 207
  exhaust stroke, 207
  suction stroke, 206
Four-bar chain, 259
Four-bar mechanism, 554
Francis turbine, 906
Free body diagram, 310
Friction, 342
  static, 342
Friction power, 211
Froude number, 870
Fundamental laws of mechanics, 303

Gas turbine plant, 90
Gauge pressure, 12
Graphical method
  force polygon, 419
  funicular polygon, 419
  space diagram, 419
Grashof's law, 267

Hand pump mechanism, 262
Hardness, 688
Heat energy, 1
Heat engine, 89, 178
Heat pump, 91
Heat pump cycle, 178
Heat reservoir, 89
Hogging bending moment, 383
Hooke's law, 665, 667
Hump, 876, 877
Hydraulic efficiency, 905, 910, 913
Hydraulic jump, 871, 879

Ideal fluid, 821
Impeller, 820
Imperfect truss, 417
Impulse-momentum equation, 584, 616
  applications, 608
Indicated power, 211
Indicated thermal efficiency, 212
Indicator diagram, 921
Intensive properties, 6
Internal combustion, 206
Internal energy, 47
Inversion, 259
Irreversibility, 105
Irreversible process, 6
Isentropic, 889, 890
Isolated system, 45
Izod test, 687

Kaplan turbine, 912
Kelvin–Plank statement, 93
Kinematic chain, 251
Kinematic link, 246
Kinematic pair, 247
  classification of, 247
  closed pair, 247
  cylindrical pair, 247
  higher pair, 247
  lower pair, 247

rolling pair, 247
screw pair, 247
sliding pair, 247
spheric pair, 247
turning pair, 247
unclosed pair, 247
wrapping pair, 247
Kinematics, 584
Kinetic energy-based on centre of mass, 602
Kinetic energy of rotation, 587
Kinetics, 584
rigid body, 584
Kutzbach criterion, 254

Lami's theorem, 310
Laminar, 825
Laminar flow, 862
Lap angle, 348
Latent heating, 149
Law of a machine, 271
Law of conservation of angular momentum, 589
Law of degradation of energy, 105
Law of pressure, 7
Law of transmissibility of force, 305
Laws of mechanics, 303
first law of motion, 303
second law of motion, 303
third law of motion, 303
Linear impulse-momentum principle, 612
Links, 246
classification, 246
types of, 246
Load
bending, 664
centric, 664
combined, 664
concentrated, 664
eccentric, 664
impact, 664
repeated, 664
point, 382
static, 664
sustained, 664
types, 381
uniformly distributed, 382
uniformly varying, 382
Load intensity, 383
Loss of effort in friction, 274

Machine, 252, 253
classification of, 269
compound, 269
simple, 269
Mach-number, 889
Macroscopic approach, 6
Manning equation, 868
Mass centres
bodies, 496
Mass moment of inertia, 480
bodies, 495
rectangular plate, 482
uniform circular plate, 484
uniform circular ring, 483
uniform hollow sphere, 486
uniform rod, 481
uniform solid cone, 488
uniform solid cylinder, 485
uniform solid sphere, 487
Materials
types, 663
Maximum mechanical advantage, 271
Mean effective pressure, 211
Mechanical advantage
variation of, 272
Mechanical efficiency, 212, 905, 911
Mechanical energy conservation, 598
Mechanism, 252, 253
Mechanisms
Ackermann—Steering gear, 259
complex mechanism, 253
compound mechanism, 253
coupling rod of a locomotive, 259
simple mechanism, 253
types of, 253
Melting point, 149
Metal,
properties of, 687
creep, 687
fatigue, 687

hardness, 687
malleability, 687
toughness, 687
Method of joints, 420
Method of section, 420
Microscopic approach, 6
Middle quarter rule, 739
Middle third rule, 738
Minor head losses, 852
Mobility, 254
Modified Rankine cycle, 182
Modulus of resilence, 682, 683
Modulus of rigidity, 675
Modulus of rupture, 785
Mohr's circle, 685, 686
Mohr's theorem, 389
Mollier diagram, 152
Moment, 307
Moment area method, 389
Moment arm, 307
Moment centre, 307
Moment of area
principal axes, 491
Moment of inertia, 464, 586
circular section, 471
hollow circular section, 473
hollow rectangular section, 472
I-section, 476
lamina, 466
parallel axis theorem, 467
product of area, 468
radius of gyration, 468
rectangular section, 468
theorem of the perpendicular axis, 467
triangular section, 474
Moment of volume, 459
Motion
plane, 537
rotational, 536
translation, 535
Moving frame of reference, 529

Newton's law of gravitation, 304
Newtonian fluids, 822
Non-concurrent forces, 307
Nozzle and diffuser, 52
NTP, 12

Oldham's coupling, 264
Open channel flow, 859
Open system, 2
Orifice meter, 842
Oscillating-cylinder engine mechanism, 262
Otto cycle, 199
Overall efficiency, 906, 911, 913

Pantograph, 268
Parallelogram law of forces, 306
Path, 4
Pathline, 824
Pelton, 914
Pelton wheel, 900
Penstock, 900
Perfect gas, 7
Perpetual motion machine, 94
Pipe flow, 848, 851
Pipes in parallel, 854
Pipes in series, 853
Pitot pressure, 843
Pitot-static tube, 843, 844
Plane truss, 416, 418
analysis of, 418
graphical method, 419
imperfect truss, 417
method of joints, 420
method of section, 420
perfect truss, 416
redundant truss, 417
PMM–I, 55
PMM–II, 94
Poisson's ratio, 677
Polar modulus of section, 778
Polygon law of forces, 308, 309
Power cycles, 178, 198
Pressure, 12
Principal planes, 684
Principal stresses, 684

Principle of entropy increase, 102
Principle of moments, 307
Principle of superposition, 670
Process, 4
adiabatic, 15, 47
isentropic, 15
isobaric, 14, 46
isochoric, 14, 46
isothermal, 14, 47
polytropic, 47
quasi static, 6
Proof resilence, 682
Properties of metal, 687
creep, 687
fatigue, 687
hardness, 687
malleability, 687
toughness, 687
Pump, 51
Pure substance, 148
Pure torsion
theory of, 775

Quasi static process, 6
Quick-return mechanism, 260

Rankine cycle, 180, 182
Reaction pressure, 909
Reciprocating piston
acceleration, 551
Reciprocating-engine mechanism, 260
Redundent truss, 417
Refrigeration, 178
Refrigeration cycles, 198
Regenerated Rankine cycle, 182
Reheat Rankine cycle, 183
Resilence, 681, 682
Resistance thermometer, 11
Resolution of force, 308
Reversibility of a machine, 278
Reversible process, 6
Rigid body, 528
kinematics, 528
rotary motion, 585
Rockwell hardness, 688
Rope drive, 348
Rotation of axes, 490

Sagging bending moment, 383
Saturation pressure, 149
Saturation state, 149
Saturation temperature, 149
Scotch yoke, 263
Screw jack, 349
Second law of motion, 303
Second law of thermodynamics, 89, 93
Section modulus, 736
Sensible heating, 148
Separating and throttling calorimeter, 153, 154
Separating calorimeter, 153, 154
Series–parallel pipe networks, 854
Shaft
coupling, 781
hollow, 781
in parallel, 779, 780
in series, 779
power transmitted, 779
solid, 781
Shear failure of key, 780
Shear force, 381, 383
Shear force diagram, 383, 384, 733
rules, 383
Shear strain, 674, 776
Shear stress, 674, 675, 733, 777
Single pulley, 279
Sink, 89
Siphon, 846
Slider-crank chain, 260
Sluice gate, 881
Space truss, 416
Spark ignition, 206
Spark ignition engine, 206, 210
Specific energy curve, 872
Specific heat, 9
at constant pressure, 9
at constant volume, 9
Square thread, 350
Stagnation pressure, 892
Stagnation temperature, 891

State, 3
Steady flow energy equation, 49
Steam tables, 152
Sterling cycle, 205, 206
Stored energy, 49
STP, 12
Straight belt drive, 348
Strain, 665, 674
  diagram, 666
  lateral, 665
  longitudinal, 665
Strain energy, 681, 682
  in pure bending, 739
Strain hardening, 667
Streakline, 824
Streamline, 823
Stream tube, 824
Stress, 665
  diagram, 665, 666
Struts, 664
Sudden contraction, 853
Sudden enlargement, 852
Superheated steam, 149
Supports, 417
  hinged support, 417
  roller support, 417
  types, 381
Swept length, 199
Swept volume, 199
System, 2
  closed, 2
  isolated, 45
  open, 2
System analysis, 6
System of pulleys, 280
  first-order, 280
  second-order, 280, 283
  third-order, 280, 286

Tensile stresses, 664
Tension members, 664
Thermal stresses, 668
Thermocouple, 11
Thermodynamic cycles, 198
Thermodynamic equilibrium, 5
Thermodynamics, 1
Thermodynamic temperature scale, 97
Thermometric property, 11
Third law of motion, 304
Third law of thermodynamics, 104
Throttling, 54
Throttling calorimeter, 153, 154
Ties, 664
Top dead centre, 199
Torricelli's theorem, 834
Torque, 586, 775
Torsion, 775, 784
  strain energy, 783
  strain energy in, 783
  theory, 775
Torsion formula, 778
Torsional rigidity, 778
Triangle law of forces, 309
Triple point, 149
Truss, 416
  double, 421
  imperfect, 417
  perfect, 416
  plane, 416
  redundent, 417
  single, 421
  space, 416
Turbine, 50, 820
Turbulent, 825
  flow, 862
Twin paradox, 305

Uniform strength, 737
Unit radius of inertia, 737

V-belt drive, 348
V-threads, 350
Vacuum pressure, 12
Vapour cycles, 178
Varignon's theorem, 307
Vena contracta, 853
Venturi meter, 840, 841

Viscosity, 821
Volume of revolution, 460
Volumetric efficiency, 212
Volumetric strain, 678

Waterwheel, 820
Wave propagation, 895
Weirs, 878
Wheel and axle, 290
Wheel and differential axle, 291
Whitworth quick-return mechanism, 261
Work-energy equation, 598
  for a rigid body, 604
Work-energy principle, 597
Worm and worm wheel, 293

Young's modulus, 665, 667

Zeroth law of thermodynamics, 1, 10